Parametric Modeling

with Autodesk® Inventor® 2027

Randy H. Shih
Oregon Institute of Technology

PUBLICATIONS

SDC Publications
P.O. Box 1334
Mission, KS 66222
913-262-2664
www.SDCpublications.com
Publisher: Stephen Schroff

ISBN-13: 978-1-63057-800-8
ISBN-10: 1-63057-800-2

Printed and bound in the United States of America.

Preface

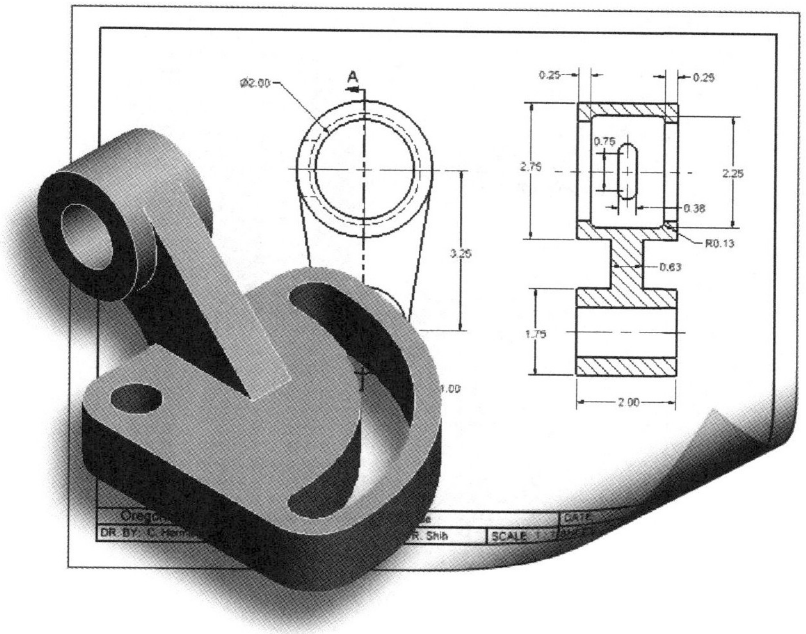

The primary goal of ***Parametric Modeling with Autodesk Inventor 2027*** is to introduce the aspects of designing with **Solid Modeling** and **Parametric Modeling**. This text is intended to be used as a practical training guide for students and professionals. This text uses Autodesk Inventor 2027 as the modeling tool and the chapters proceed in a pedagogical fashion to guide you from constructing basic solid models to building intelligent mechanical designs, creating multi-view drawings and assembly models. This text takes a hands-on, exercise-intensive approach to all the important *Parametric Modeling* techniques and concepts. This textbook contains a series of sixteen tutorial style lessons designed to introduce beginning CAD users to Autodesk Inventor. This text is also helpful to Autodesk Inventor users upgrading from a previous release of the software. The solid modeling techniques and concepts discussed in this text are also applicable to other parametric feature-based CAD packages. The basic premise of this book is that the more designs you create using Autodesk Inventor, the better you learn the software. With this in mind, each lesson introduces a new set of commands and concepts, building on previous lessons. This book does not attempt to cover all of Autodesk Inventor's features, only to provide an introduction to the software. It is intended to help you establish a good basis for exploring and growing in the exciting field of **Computer Aided Engineering**.

Acknowledgments

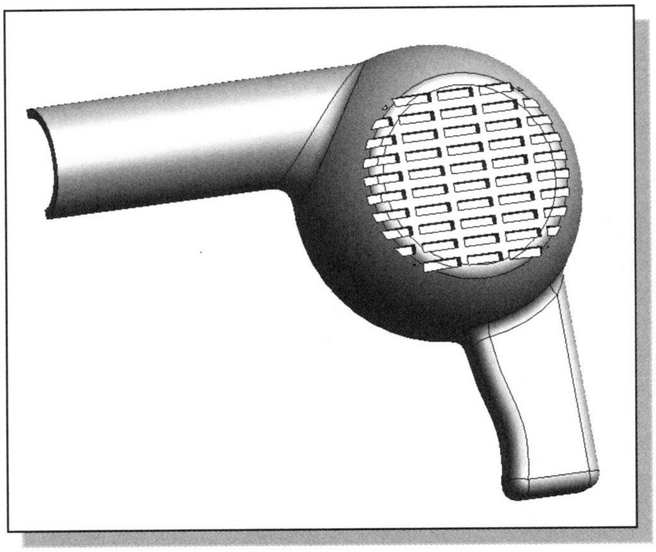

This book would not have been possible without a great deal of support. First, special thanks to two great teachers, Prof. George R. Schade of University of Nebraska-Lincoln and Mr. Denwu Lee from Taiwan, who showed me the fundamentals, the intrigue, and the sheer fun of Computer Aided Engineering.

The effort and support of the editorial and production staff of SDC Publications is gratefully acknowledged.

I am grateful that the Department of Mechanical and Manufacturing Engineering at Oregon Institute of Technology has provided me with an excellent environment in which to pursue my passions in teaching and research.

Finally, truly unbounded thanks are due to my wife Hsiu-Ling and our daughter Casandra for their understanding and encouragement throughout this project.

Randy H. Shih
Klamath Falls, Oregon
Spring, 2026

Table of Contents

Chapter 3
Constructive Solid Geometry Concepts

Chapter 4
Model History Tree

Chapter 5
Parametric Constraints Fundamentals

Chapter 6
Geometric Construction Tools

Chapter 7
Parent/Child Relationships and the BORN Technique

Chapter 8
Part Drawings and 3D Model-Based Definition

Chapter 9
Datum Features and Auxiliary Views

Chapter 10
Introduction to 3D Printing

Chapter 11
Symmetrical Features in Designs

Chapter 12
Advanced 3D Construction Tools

Chapter 13
Sheet Metal Designs

Chapter 14
Assembly Modeling - Putting It All Together

Chapter 15
Content Center and Basic Motion Analysis

Chapter 17
Introduction to Stress Analysis

Appendix

Index

Autodesk Inventor Certified User Examination Overview

The Autodesk Inventor Certified User examination is a performance-based exam. The examination is comprised of approximately 30 questions to be completed in 50 minutes. The test items will require you to use the Autodesk Inventor software to perform specific tasks and then answer questions about the tasks.

Performance-based testing is defined as ***Testing by Doing***. This means you actually perform the given task then answer the questions regarding the task. Performance-based testing is widely accepted as a better way of ensuring the users have the skills needed, rather than just recalling information.

The Autodesk Inventor Certified User examination is designed to test specific performance tasks in the following seven sections:

Section 1: User Interface

Objectives: Primary Environments, UI Navigation/Interaction, Graphics Window Display, Navigation Control.

Certification Examination Performance Task	Covered in this book on Chapter – Page
Screen Layout	1-9, 1-10
Inventor Overview	1-9
New File	1-10
Application Menu	1-12
Mouse Buttons	1-14
Inventor Help System	1-15
Dynamic Viewing – Quick Keys	2-20
Viewing Tools	2-21
2D/3D Navigation Wheels	2-23
Display Modes	2-25
Browser	4-6

Section 2: File Management

Objectives: Project Files.

Certification Examination Performance Task	Covered in this book on Chapter – Page
Data Management	1-16
New Inventor Project	1-17
Content of the Inventor Project File	1-20
Activate an Inventor Project File	2-5

Section 3: Sketches

Objectives: Creating 2D Sketches, Draw Tools, Sketch Constraints, Pattern Sketches, Modify Sketches, Format Sketches, Sketch Doctor, Shared Sketches, Sketch Parameters.

Section 4: Parts

Objectives: Creating parts, Work Features, Pattern Features, Part Properties.

Certification Examination Performance Task	Covered in this book on Chapter – Page

Section 5: Assemblies

Objectives: Creating Assemblies, Viewing Assemblies, Animation Assemblies, Adaptive Features, Parts, and Subassemblies.

Section 6: Drawings

Objectives: Create drawings.

Certification Examination Performance Task	Covered in this book on Chapter – Page

Section 7: Visualization

Objectives: Create Rendered Images, Animate an Assembly.

Certification Examination Performance Task	Covered in this book on Chapter – Page

Autodesk Inventor Certified User Reference Guide

- Every effort has been made to cover the exam objectives included in the Autodesk Inventor Certified User Examination. However, the format and topics covered by the examination are constantly changing. Students planning to take the Certified User Examination are advised to visit the Autodesk website and obtain information regarding the format and details about the Autodesk Inventor Certified User Examination.

Tips about Taking the Autodesk Certified User Examination

1. **Study:** The first step to maximize your potential on an exam is to sufficiently prepare for it. You need to be familiar with the Autodesk Inventor package, and this can only be achieved by doing designs and exploring the different commands available. The Autodesk Inventor Certified User exam is designed to measure your familiarity with the Autodesk Inventor software. You must be able to perform the given task and answer the exam questions correctly and quickly.

2. **Make Notes**: Take notes of what you learn either while attending classroom sessions or going through study material. Use these notes as a review guide before taking the actual test.

3. **Time Management**: Manage the time you spend on each question. Always remember you do not need to score 100% to pass the exam. Also keep in mind that some questions are weighed more heavily and may take more time to answer.

4. **Use Common Sense**: If you are unable to get the correct answer and unable to eliminate all distracters, then you need to select the best answer from the remaining selections. This may be a task of selecting the best answer from amongst several correct answers, or it may be selecting the least incorrect answer from amongst several poor answers.

5. **Use the Autodesk Inventor Help system**: If you get confused and cannot think of the answer, remember the Autodesk Inventor help system is a great tool to confirm your considerations.

6. **Take Your Time:** The examination has a time limit. If you encounter a question you cannot answer in a reasonable amount of time, use the Save As feature to save a copy of the data file, and mark the question for review. When you review the question, open your copy of the data file and complete the performance task. After you verify that you have entered the answer correctly, unmark the question so it no longer appears as marked for review.

7. **Be Cautious and Don't Act in Haste:** Devote some time to ponder and think of the correct answer. Ensure that you interpret all the options correctly before selecting from available choices. Don't go into panic mode while taking a test. Use the ***Review*** screen to ensure you have reviewed all the questions you may have marked for review. When you are confident that you have answered all questions, end the examination to submit your answers for scoring. You will receive a score report once you have submitted your answers.

8. **Relax before the exam:** In order to avoid last minute stress, make sure that you arrive 10 to 15 minutes early and relax before taking the exam.

Chapter 1

Getting Started

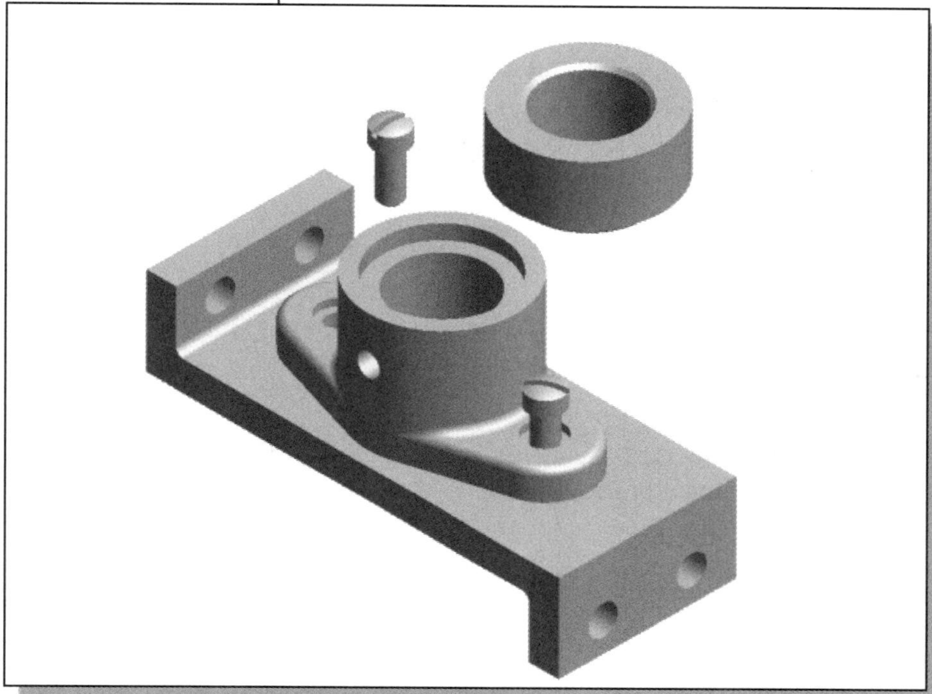

Learning Objectives

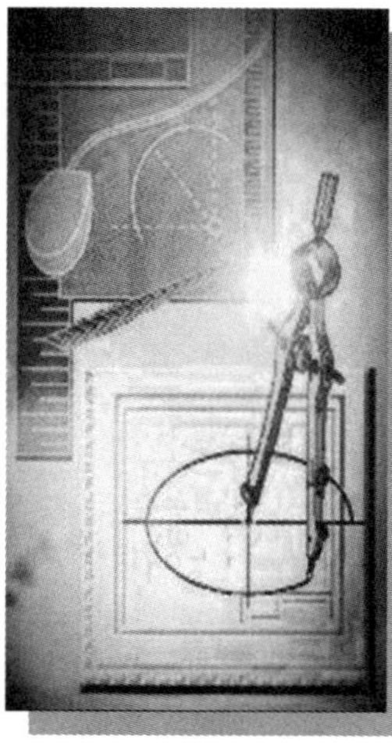

- Development of Computer Geometric Modeling
- Feature-Based Parametric Modeling
- Startup Options and Units Setup
- Autodesk Inventor Screen Layout
- User Interface & Mouse Buttons
- Autodesk Inventor Online Help

Autodesk Inventor Certified User Exam Objectives Coverage

Section 1: User Interface

Objectives: Primary Environments, UI Navigation/Interaction, Graphics Window Display, Navigation Control.

Section 2: File Management

Objectives: Project Files.

Introduction

The rapid changes in the field of **Computer Aided Engineering** (CAE) have brought exciting advances in the engineering community. Recent advances have made the long-sought goal of **concurrent engineering** closer to a reality. CAE has become the core of concurrent engineering and is aimed at reducing design time, producing prototypes faster, and achieving higher product quality. Autodesk Inventor is an integrated package of mechanical computer aided engineering software tools developed by *Autodesk, Inc.* Autodesk Inventor is a tool that facilitates a concurrent engineering approach to the design and stress-analysis of mechanical engineering products. The computer models can also be used by manufacturing equipment such as machining centers, lathes, mills, or rapid prototyping machines to manufacture the product. In this text, we will be dealing only with the solid modeling modules used for part design and part drawings.

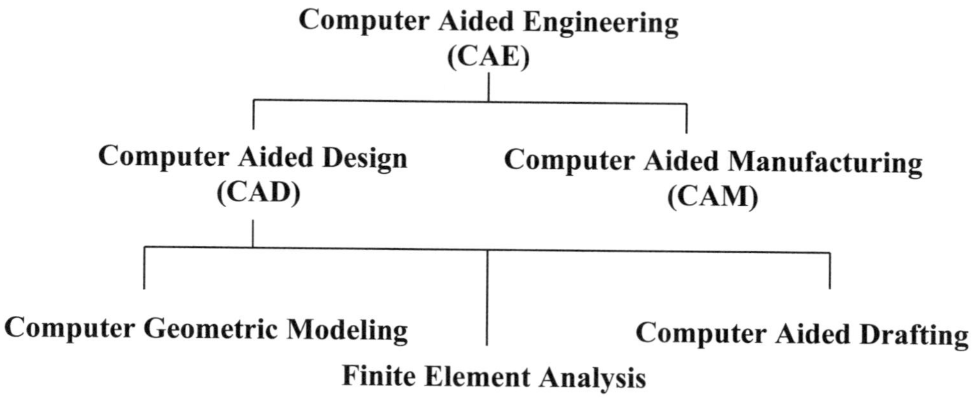

Development of Computer Geometric Modeling

Computer geometric modeling is a relatively new technology, and its rapid expansion in the last fifty years is truly amazing. Computer-modeling technology has advanced along with the development of computer hardware. The first-generation CAD programs, developed in the 1950s, were mostly non-interactive; CAD users were required to create program-codes to generate the desired two-dimensional (2D) geometric shapes. Initially, the development of CAD technology occurred mostly in academic research facilities. The Massachusetts Institute of Technology, Carnegie-Mellon University, and Cambridge University were the leading pioneers at that time. The interest in CAD technology spread quickly and several major industry companies, such as General Motors, Lockheed, McDonnell, IBM, and Ford Motor Co., participated in the development of interactive CAD programs in the 1960s. Usage of CAD systems was primarily in the automotive industry, aerospace industry, and government agencies that developed their own programs for their specific needs. The 1960s also marked the beginning of the development of finite element analysis methods for computer stress analysis and computer aided manufacturing for generating machine tool paths.

The 1970s are generally viewed as the years of the most significant progress in the development of computer hardware, namely the invention and development of **microprocessors**. With the improvement in computing power, new types of 3D CAD programs that were user-friendly and interactive became reality. CAD technology quickly expanded from very simple **computer aided drafting** to very complex **computer aided design**. The use of 2D and 3D wireframe modelers was accepted as the leading-edge technology that could increase productivity in industry. The developments of surface modeling and solid modeling technologies were taking shape by the late 1970s, but the high cost of computer hardware and programming slowed the development of such technology. During this period, the available CAD systems all required room-sized mainframe computers that were extremely expensive.

In the 1980s, improvements in computer hardware brought the power of mainframes to the desktop at less cost and with more accessibility to the general public. By the mid-1980s, CAD technology had become the main focus of a variety of manufacturing industries and was very competitive with traditional design/drafting methods. It was during this period of time that 3D solid modeling technology had major advancements, which boosted the usage of CAE technology in industry.

The introduction of the *feature-based parametric solid modeling* approach, at the end of the 1980s, elevated CAD/CAM/CAE technology to a new level. In the 1990s, CAD programs evolved into powerful design/manufacturing/management tools. CAD technology has come a long way, and during these years of development, modeling schemes progressed from two-dimensional (2D) wireframe to three-dimensional (3D) wireframe, to surface modeling, to solid modeling and, finally, to feature-based parametric solid modeling.

The first-generation CAD packages were simply 2D **computer aided drafting** programs, basically the electronic equivalents of the drafting board. For typical models, the use of this type of program would require that several views of the objects be created individually as they would be on the drafting board. The 3D designs remained in the designer's mind, not in the computer database. Mental translations of 3D objects to 2D views are required throughout the use of these packages. Although such systems have some advantages over traditional board drafting, they are still tedious and labor intensive. The need for the development of 3D modelers came quite naturally, given the limitations of the 2D drafting packages.

The development of three-dimensional modeling schemes started with three-dimensional (3D) wireframes. Wireframe models are models consisting of points and edges, which are straight lines connecting between appropriate points. The edges of wireframe models are used, similar to lines in 2D drawings, to represent transitions of surfaces and features. The use of lines and points is also a very economical way to represent 3D designs.

The development of the 3D wireframe modeler was a major leap in the area of computer geometric modeling. The computer database in the 3D wireframe modeler contains the locations of all the points in space coordinates, and it is typically sufficient to create just one model rather than multiple views of the same model. This single 3D model can then be viewed from any direction as needed. Most 3D wireframe modelers allow the user to create projected lines/edges of 3D wireframe models. In comparison to other types of 3D modelers, the 3D wireframe modelers require very little computing power and generally can be used to achieve reasonably good representations of 3D models. However, because surface definition is not part of a wireframe model, all wireframe images have the inherent problem of ambiguity. Two examples of such ambiguity are illustrated.

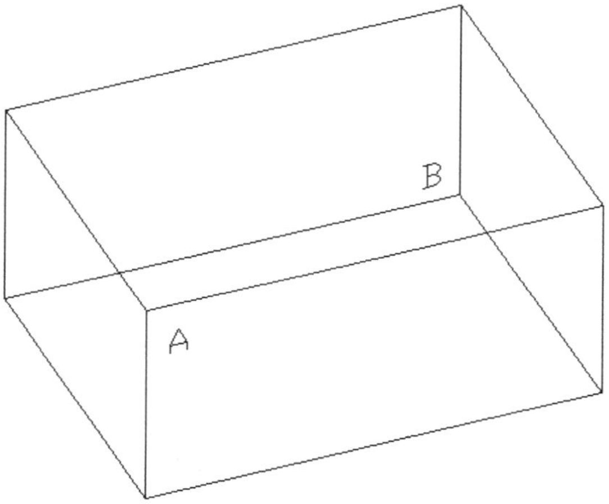

Wireframe Ambiguity: Which corner is in front, A or B?

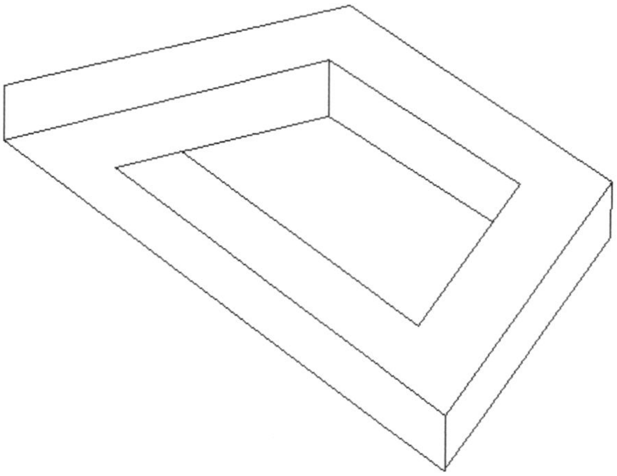

A non-realizable object: Wireframe models contain no surface definitions.

Surface modeling is the logical development in computer geometry modeling to follow the 3D wireframe modeling scheme by organizing and grouping edges that define polygonal surfaces. Surface modeling describes the part's surfaces but not its interiors. Designers are still required to interactively examine surface models to ensure that the various surfaces on a model are contiguous throughout. Many of the concepts used in 3D wireframe and surface modelers are incorporated in the solid modeling scheme, but it is solid modeling that offers the most advantages as a design tool.

In the solid modeling presentation scheme, the solid definitions include nodes, edges, and surfaces, and it is a complete and unambiguous mathematical representation of a precisely enclosed and filled volume. Unlike the surface modeling method, solid modelers start with a solid or use topology rules to guarantee that all of the surfaces are stitched together properly. Two predominant methods for representing solid models are **constructive solid geometry** (CSG) representation and **boundary representation** (B-rep).

The CSG representation method can be defined as the combination of 3D solid primitives. What constitutes a "primitive" varies somewhat with the software but typically includes a rectangular prism, a cylinder, a cone, a wedge, and a sphere. Most solid modelers also allow the user to define additional primitives, which are shapes typically formed by the basic shapes. The underlying concept of the CSG representation method is very straightforward; we simply **add** or **subtract** one primitive from another. The CSG approach is also known as the machinist's approach, as it can be used to simulate the manufacturing procedures for creating the 3D object.

In the B-rep representation method, objects are represented in terms of their spatial boundaries. This method defines the points, edges, and surfaces of a volume, and/or issues commands that sweep or rotate a defined face into a third dimension to form a solid. The object is then made up of the unions of these surfaces that completely and precisely enclose a volume.

By the 1980s, a new paradigm called *concurrent engineering* had emerged. With concurrent engineering, designers, design engineers, analysts, manufacturing engineers, and management engineers all work together closely right from the initial stages of the design. In this way, all aspects of the design can be evaluated, and any potential problems can be identified right from the start and throughout the design process. Using the principles of concurrent engineering, a new type of computer modeling technique appeared. The technique is known as the *feature-based parametric modeling technique*. The key advantage of the *feature-based parametric modeling technique* is its capability to produce very flexible designs. Changes can be made easily, and design alternatives can be evaluated with minimum effort. Various software packages offer different approaches to feature-based parametric modeling, yet the end result is a flexible design defined by its design variables and parametric features.

Feature-Based Parametric Modeling

One of the key elements in the Autodesk Inventor solid modeling software is its use of the **feature-based parametric modeling technique**. The feature-based parametric modeling approach has elevated solid modeling technology to the level of a very powerful design tool. Parametric modeling automates the design and revision procedures by the use of parametric features. Parametric features control the model geometry by the use of design variables. The word ***parametric*** means that the geometric definitions of the design, such as dimensions, can be varied at any time during the design process. Features are predefined parts or construction tools for which users define the key parameters. A part is described as a sequence of engineering features, which can be modified and/or changed at any time. The concept of parametric features makes modeling more closely match the actual design-manufacturing process than the mathematics of a solid modeling program. In parametric modeling, models and drawings are updated automatically when the design is refined.

Parametric modeling offers many benefits:

- **We begin with simple, conceptual models with minimal detail; this approach conforms to the design philosophy of "shape before size."**
- **Geometric constraints, dimensional constraints, and relational parametric equations can be used to capture design intent.**
- **The ability to update an entire system, including parts, assemblies and drawings, after changing one parameter of complex designs.**
- **We can quickly explore and evaluate different design variations and alternatives to determine the best design.**
- **Existing design data can be reused to create new designs.**
- **Quick design turn-around.**

One of the key features of Autodesk Inventor is the use of an assembly-centric paradigm, which enables users to concentrate on the design without depending on the associated parameters or constraints. Users can specify how parts fit together and the Autodesk Inventor *assembly-based fit function* automatically determines the parts' sizes and positions. This unique approach is known as the **Direct Adaptive Assembly approach**, which defines part relationships directly with no order dependency.

The *Adaptive Assembly approach* is a unique design methodology that can only be found in Autodesk Inventor. The goal of this methodology is to improve the design process and allows you, the designer, to **Design the Way You Think**.

Getting Started with Autodesk Inventor

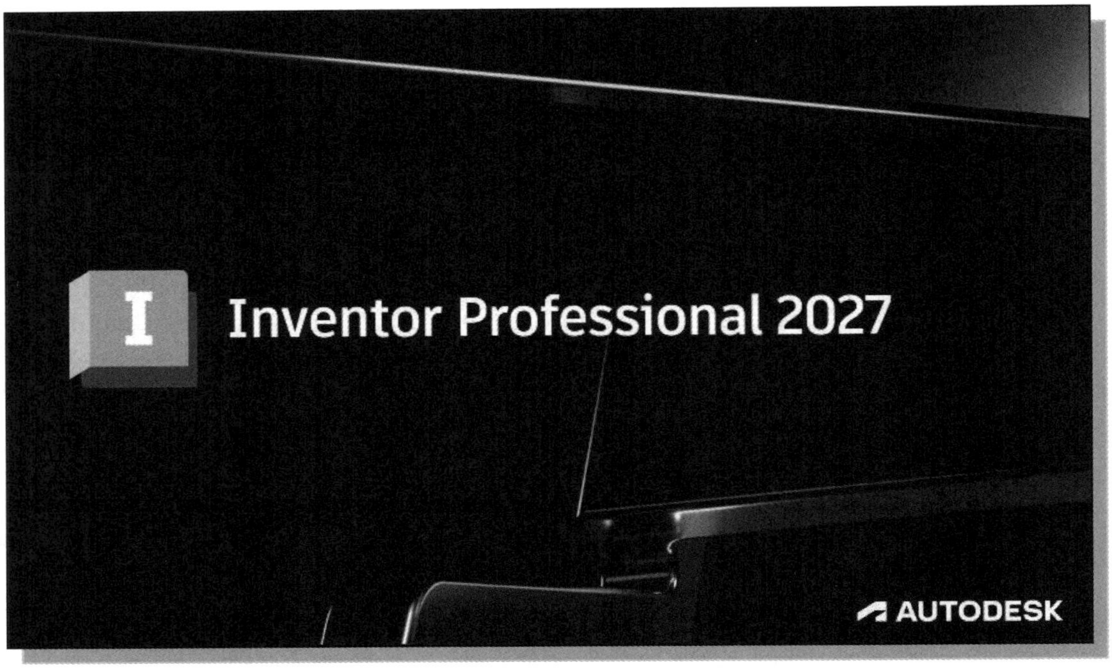

Autodesk Inventor is composed of several application software modules (these modules are called *applications*), all sharing a common database. In this text, the main concentration is placed on the solid modeling modules used for part design. The general procedures required in creating solid models, engineering drawings, and assemblies are illustrated.

How to start Autodesk Inventor depends on the type of workstation and the particular software configuration you are using. With most *Windows* systems, you may select **Autodesk Inventor** on the *Start* menu or select the **Autodesk Inventor** icon on the desktop. Consult your instructor or technical support personnel if you have difficulty starting the software. The program takes a while to load, so be patient.

The tutorials in this text are based on the assumption that you are using Autodesk Inventor's default settings. If your system has been customized for other uses, contact your technical support personnel to restore the default software configuration.

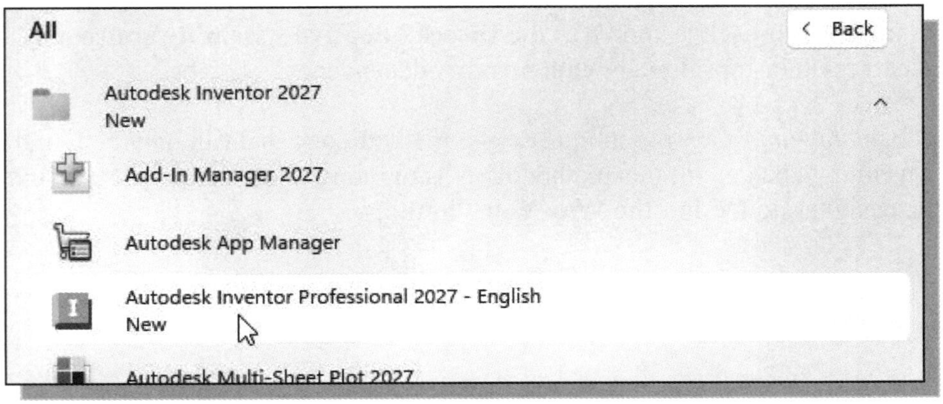

The Screen Layout and Getting Started

Once the program is loaded into the memory, the Inventor window appears on the screen with the *Get Started* options activated.

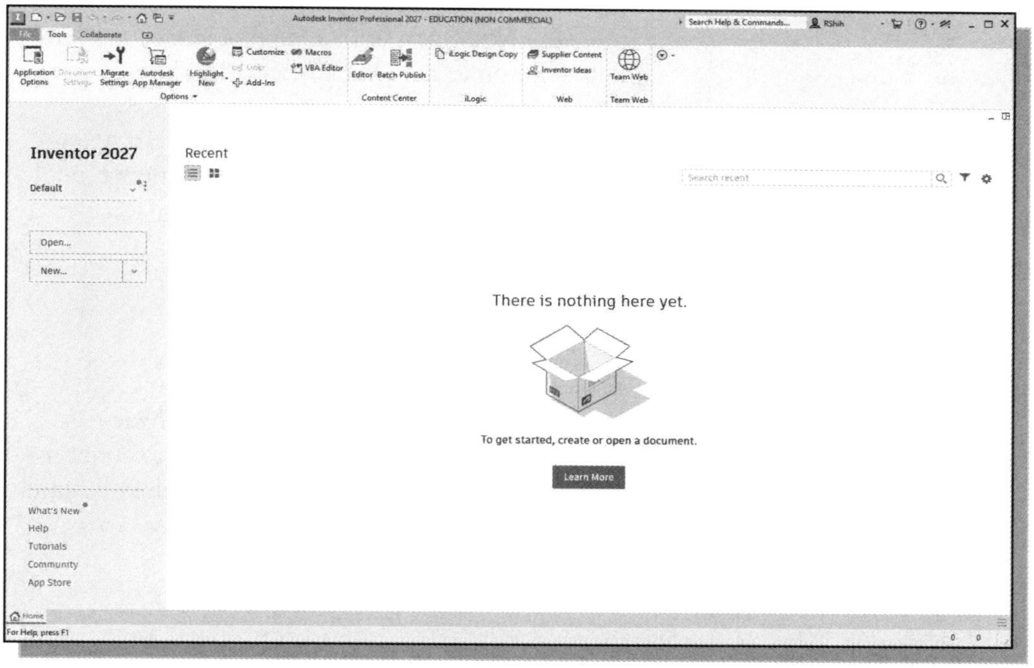

❖ Note that the *Inventor Browser* section contains helpful information in regard to using the Inventor software. For example, clicking the **What's New** option will the list of new features that are included in this release of Autodesk Inventor.

❖ You are encouraged to browse through the different information available in the *Inventor Browser* section.

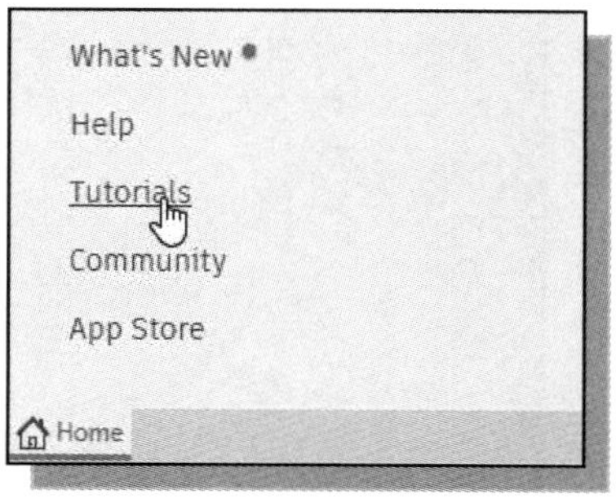

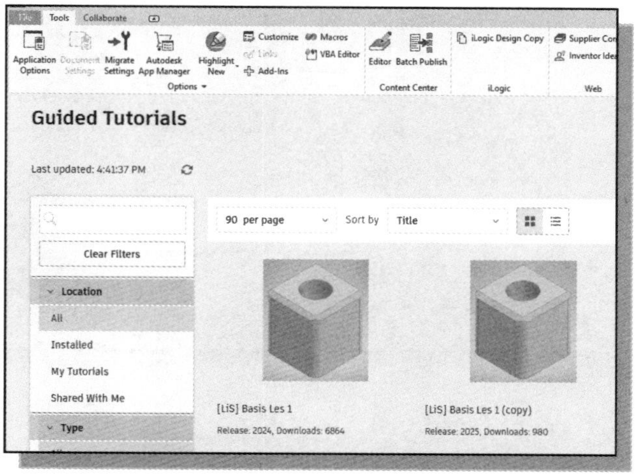

The New File Dialog Box and Units Setup

When starting a new CAD file, the first thing we should do is to choose the units we would like to use. We will use the English (feet and inches) setting for this example.

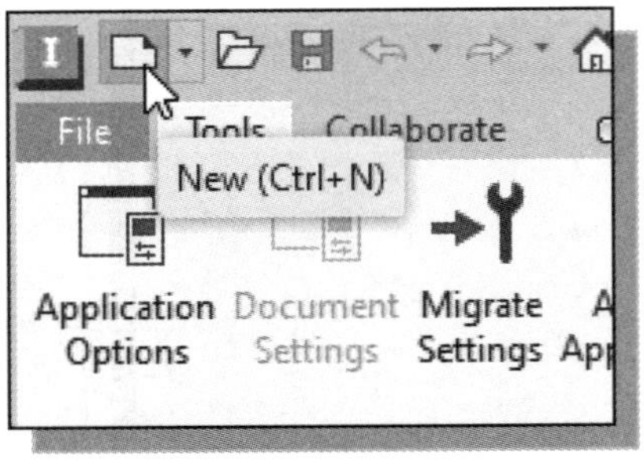

- Select the **New** icon with a single click of the left-mouse-button in the *Quick Access* toolbar.

❖ Note that the **New** option allows us to start a new modeling task, which can be creating a new model or several other modeling tasks.

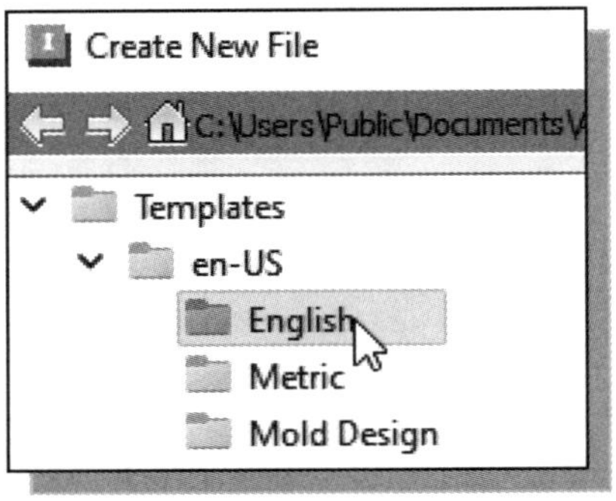

- Select the **en-US->English** tab in the *New File* dialog box as shown. Note the default tab contains the file options which are based on the default units chosen during installation.

- Select the **Standard(in).ipt** icon as shown. The different icons are templates for the different modeling tasks. The **idw** file type stands for drawing file, the **iam** file type stands for assembly file, and the **ipt** file type stands for part file. The **ipn** file type stands for assembly presentation.

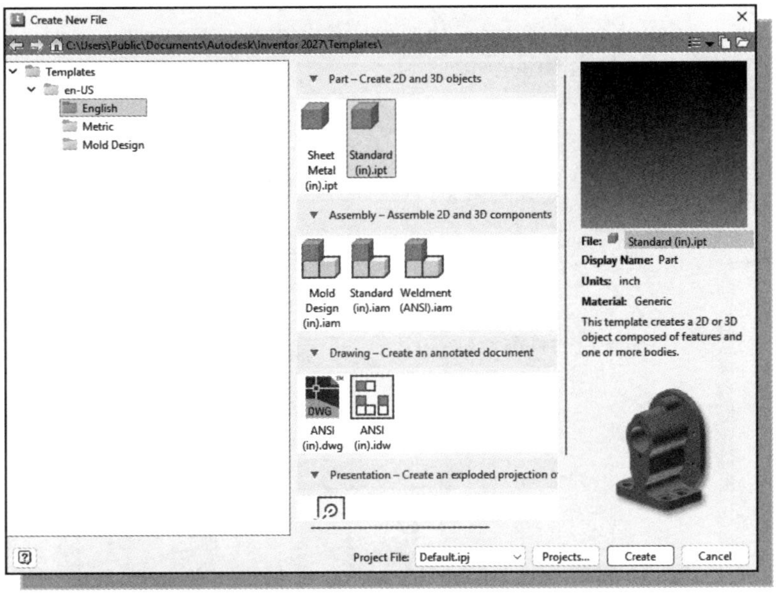

- Click **Create** in the *Create New File* dialog box to accept the selected settings.

The Default Autodesk Inventor Screen Layout

The default Autodesk Inventor drawing screen contains the *pull-down* menus, the *Standard* toolbar, the *Features* toolbar, the *Sketch* toolbar, the *drawing* area, the *browser* area, and the *Status Bar*. A line of quick text appears next to the icon as you move the *mouse cursor* over different icons. You may resize the Autodesk Inventor drawing window by clicking and dragging the edges of the window, or relocate the window by clicking and dragging the window title area.

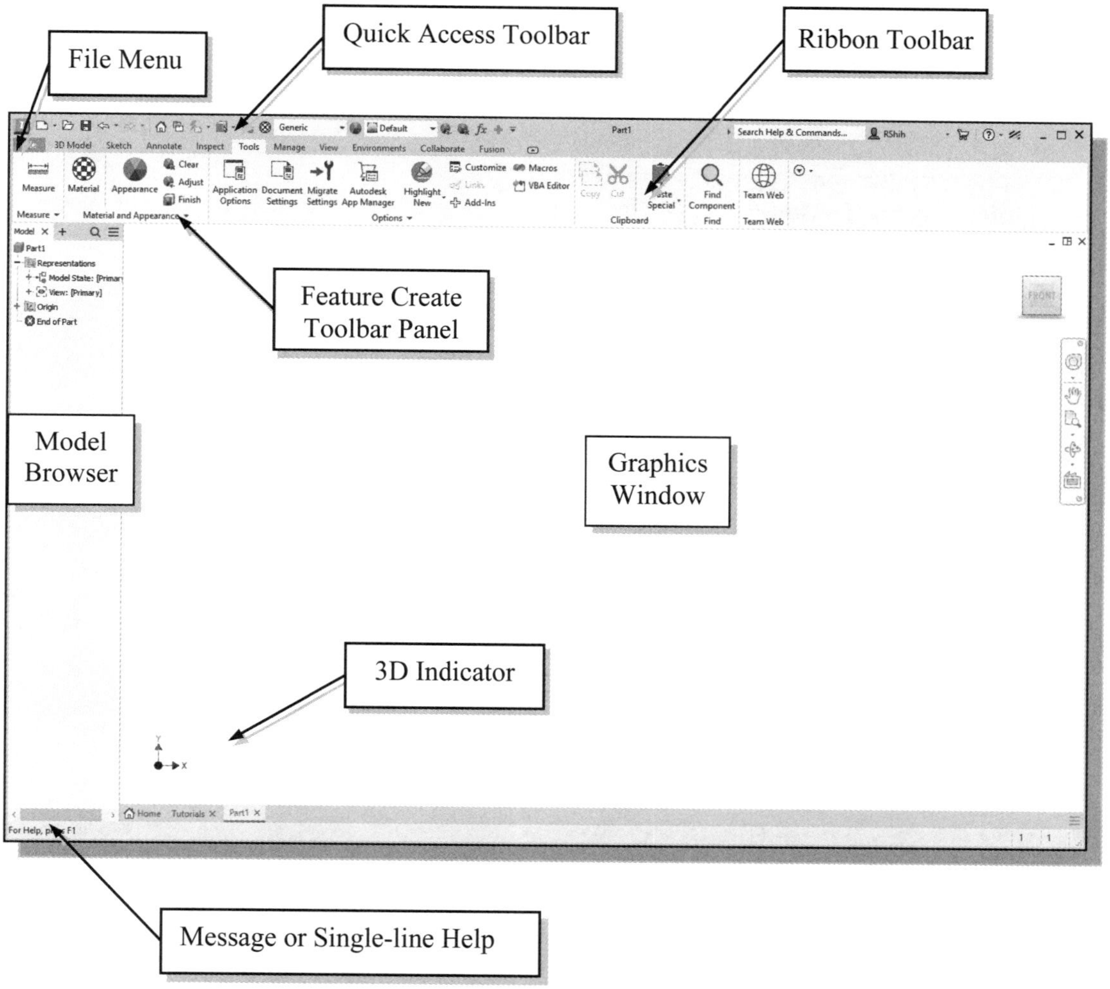

- The ***Ribbon Toolbar*** is a relatively new feature in Autodesk Inventor; the *Ribbon Toolbar* is composed of a series of tool panels, which are organized into tabs labeled by task. The *Ribbon* provides a compact palette of all of the tools necessary to accomplish the different modeling tasks. The drop-down arrow next to any icon indicates additional commands are available on the expanded panel; access the expanded panel by clicking on the drop-down arrow.

- **File Menu**

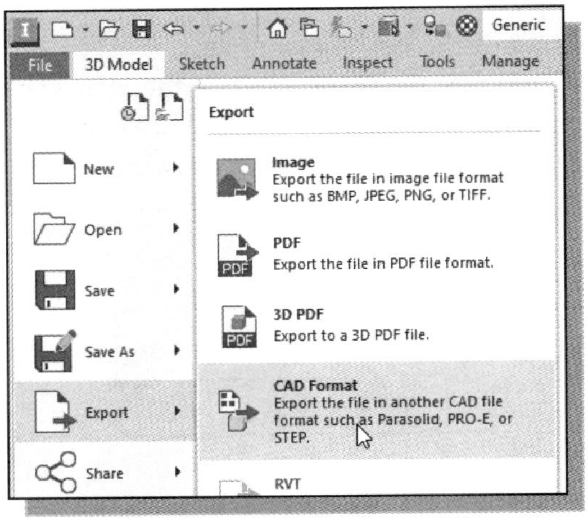

The *File* menu at the upper left corner of the main window contains tools for all file-related operations, such as Open, Save, Export, etc.

- **Quick Access Toolbar**

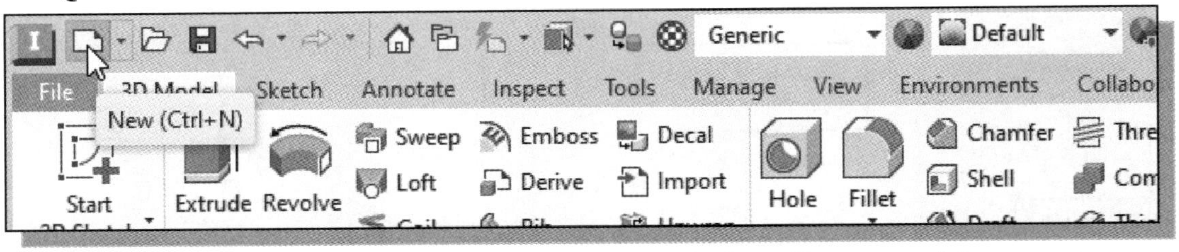

The *Quick Access* toolbar at the top of the *Inventor* window allows us quick access to file-related commands and to Undo/Redo the last operations.

- **Ribbon Tabs and Tool panels**

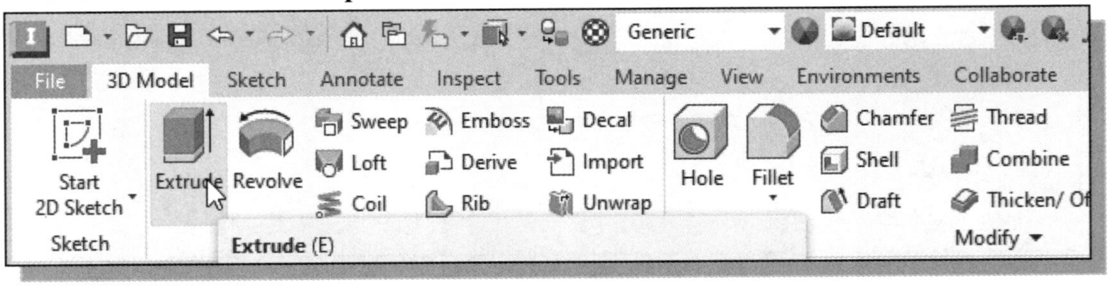

The *Ribbon* is composed of a series of tool panels, which are organized into tabs labeled by task. The assortments of tool panels can be accessed by clicking on the tabs.

- **Online Help Panel**

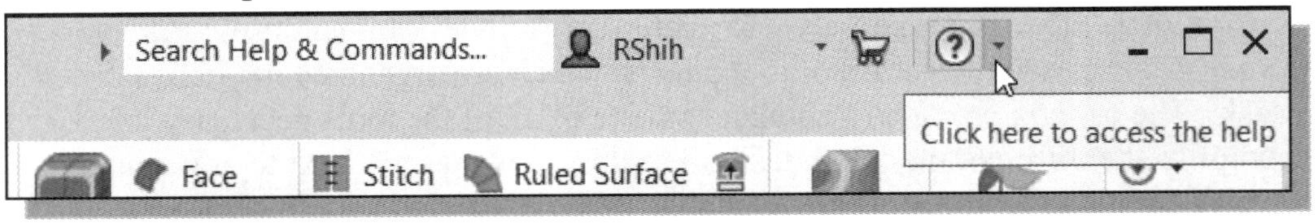

The *Help* options panel provides us with multiple options to access online help for Autodesk Inventor. The ***Online Help system*** provides general help information, such as command options and command references.

- **3D Model Toolbar**

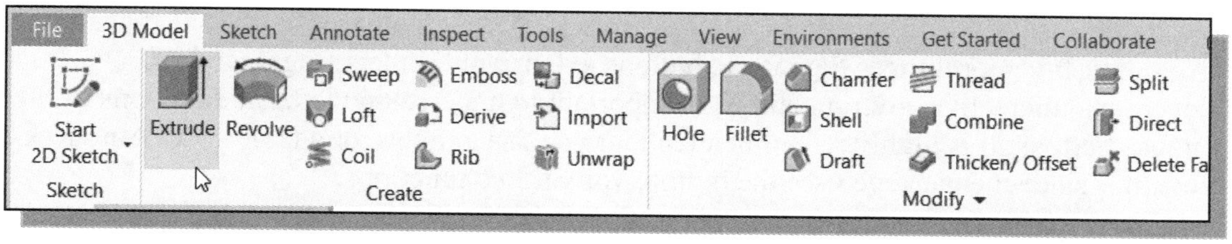

The *3D Model* toolbar provides tools for creating the different types of 3D features, such as Extrude, Revolve, Sweep, etc.

- **Graphics Window**

The *graphics window* is the area where models and drawings are displayed.

- **Message and Status Bar**

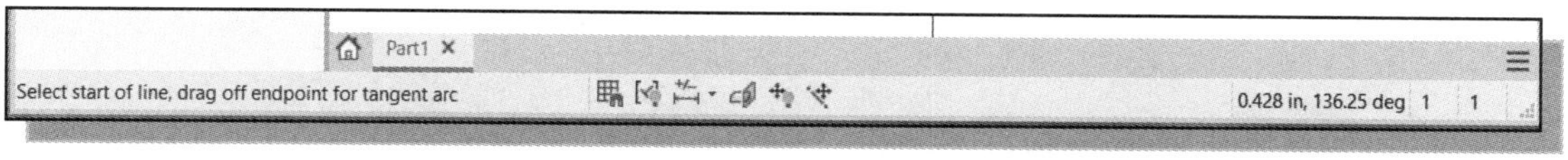

The *Message and Status Bar* area shows a single-line help when the cursor is on top of an icon. This area also displays information pertinent to the active operation. For example, in the figure above, the coordinates and length information of a line are displayed while the *Line* command is activated.

Mouse Buttons

Autodesk Inventor utilizes the mouse buttons extensively. In learning Autodesk Inventor's interactive environment, it is important to understand the basic functions of the mouse buttons. It is highly recommended that you use a mouse or a tablet with Autodesk Inventor since the package uses the buttons for various functions.

- **Left mouse button**
  The **left mouse button** is used for most operations, such as selecting menus and icons, or picking graphic entities. One click of the button is used to select icons, menus and form entries, and to pick graphic items.

- **Right mouse button**
  The **right mouse button** is used to bring up additional available options. The software also utilizes the **right mouse button** the same as the **ENTER** key and is often used to accept the default setting to a prompt or to end a process.

- **Middle mouse button/wheel**
  The middle mouse button/wheel can be used to **Pan** (hold down the wheel button and drag the mouse) or **Zoom** (turn the wheel) realtime.

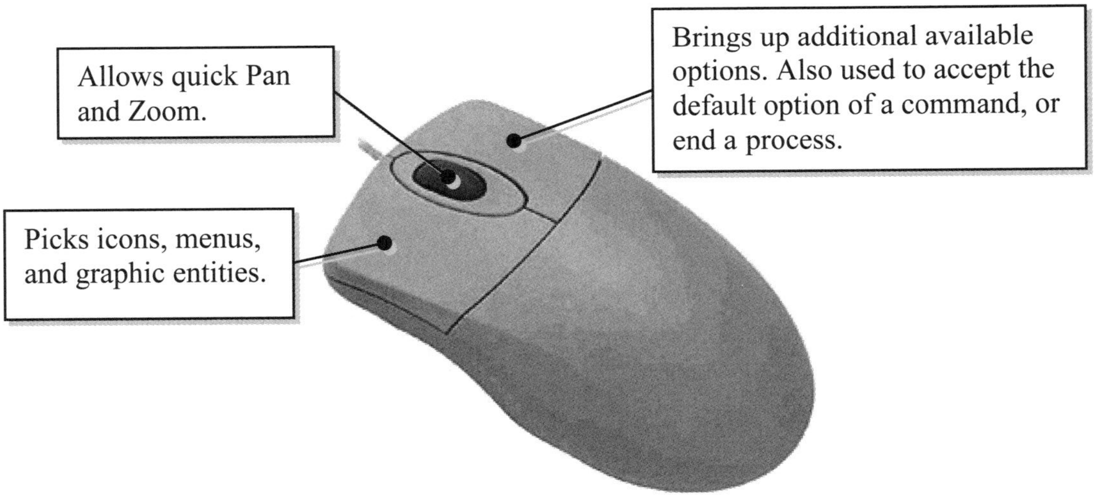
Allows quick Pan and Zoom.
Brings up additional available options. Also used to accept the default option of a command, or end a process.
Picks icons, menus, and graphic entities.

[Esc] – Canceling Commands

The [**Esc**] key is used to cancel a command in Autodesk Inventor. The [**Esc**] key is located near the top-left corner of the keyboard. Sometimes, it may be necessary to press the [**Esc**] key twice to cancel a command; it depends on where we are in the command sequence. For some commands, the [**Esc**] key is used to exit the command.

Autodesk Inventor Help System

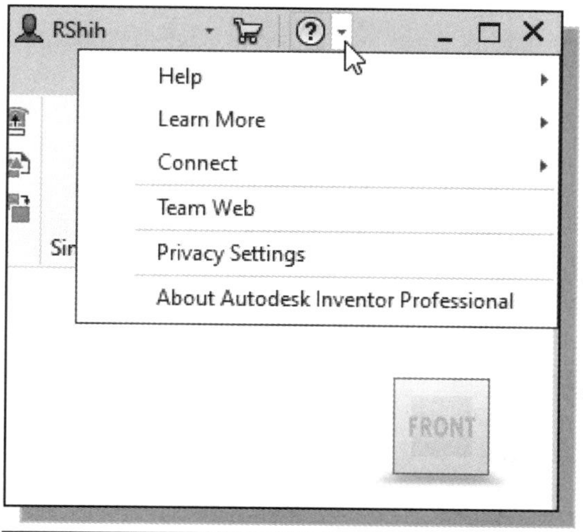

❖ Several types of help are available at any time during an Autodesk Inventor session. Autodesk Inventor provides many help functions, such as:

- Use the ***Help*** button near the upper right corner of the *Inventor* window.
- Help quick-key: Press the [**F1**] key to access the ***Inventor Help* system**.

- Use the ***Info Center*** to get information on a specific topic.

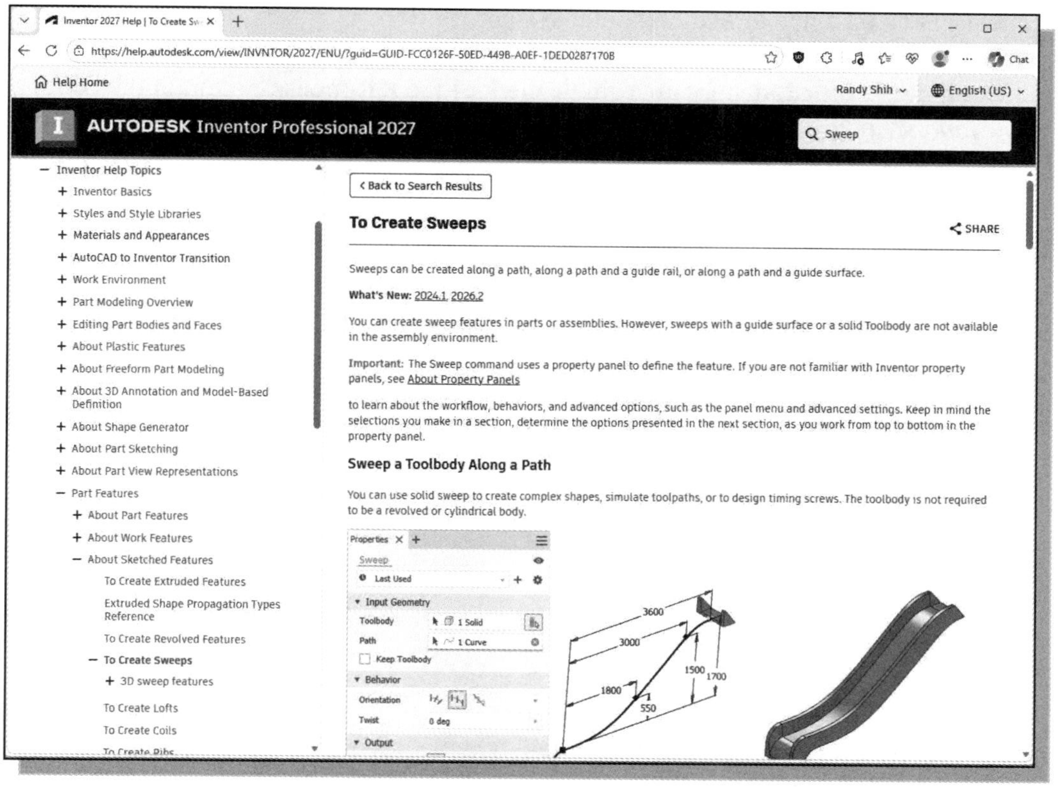

- Use the Internet to access information on the Autodesk community website.

Data Management Using Inventor Project files

With Autodesk Inventor, it is quite feasible to create designs without any regard to using the Autodesk Inventor data management system. Data management becomes critical for projects involving complex designs, especially when multiple team members are involved, or when we are working on integrating multiple design projects, or when it is necessary to share files among the design projects. Autodesk Inventor provides a fairly flexible data management system. It allows one person to use the basic option to help manage the locations of the different design files, or a team of designers can use the data management system to manage their projects stored on a networked computer system.

The Autodesk Inventor data management system organizes files based on **projects**. Each project is identified with a main folder that can contain files and folders associated to the design. In Autodesk Inventor, a *project* file (.ipj) defines the locations of all files associated with the project, including templates and library files.

The Autodesk Inventor data management system uses two types of projects:

- **Single-user Project**
- **Autodesk Vault Project** (installation of ***Autodesk Vault*** software is required)

The ***single-user project*** is for simpler projects where all project files are located on the same computer. The ***Autodesk Vault project*** is more suitable for projects requiring multiple users using a networked computer system.

- Click on the **Projects** icon with a single click of the left-mouse-button in the *[File]* → *[Manage]* pull-down menu.

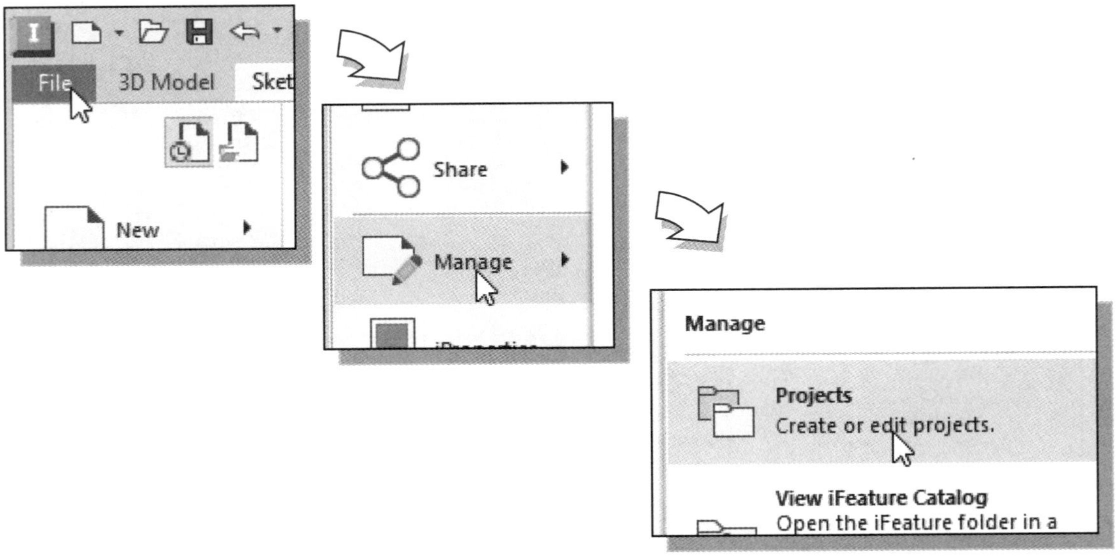

❖ The **Projects Editor** appears on the screen, which allows us to access the settings of new or existing Inventor Single User Project or Autodesk Vault Project.

❖ In the *Projects Editor*: the **Default** project is available.

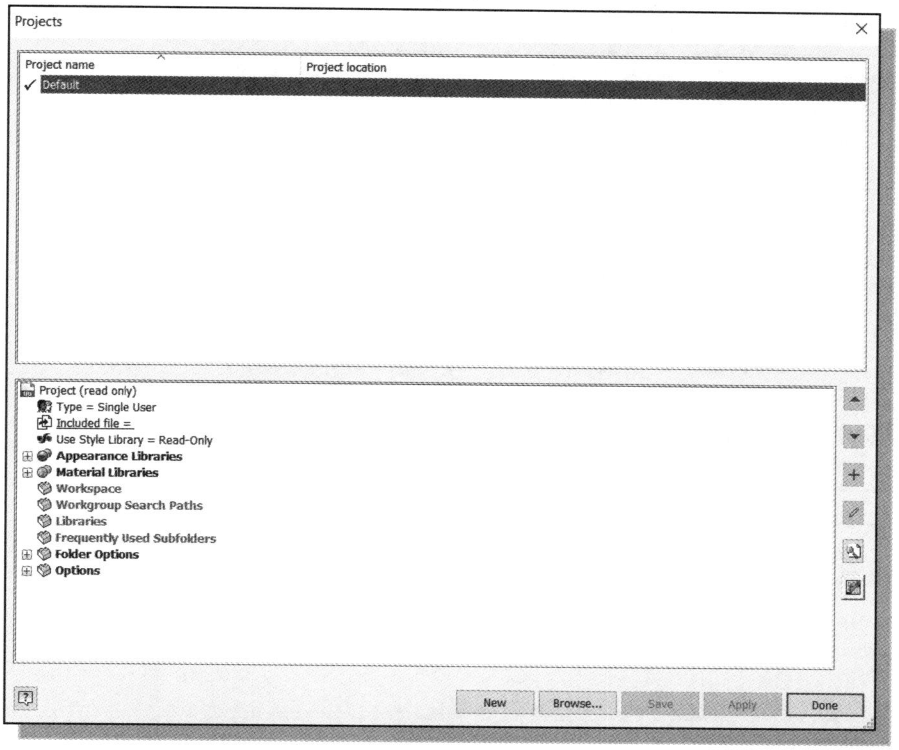

The **Default** project is automatically active by default, and the *default project* does not define any location for files. In other words, the data management system is not used. Using the *default project*, designs can still be created and modified, and any model file can be opened and saved anywhere without regard to project and file management.

Set up of a New Inventor Project

In this section, we will create a new *Inventor* project for the chapters of this book using the *Inventor* built-in **Single User Project** option. Note that it is also feasible to create a separate project for each chapter.

1. Click **New** to begin the setup of a new *project file*.

2. The *Inventor project wizard* appears on the screen; select the **New Single User Project** option as shown.

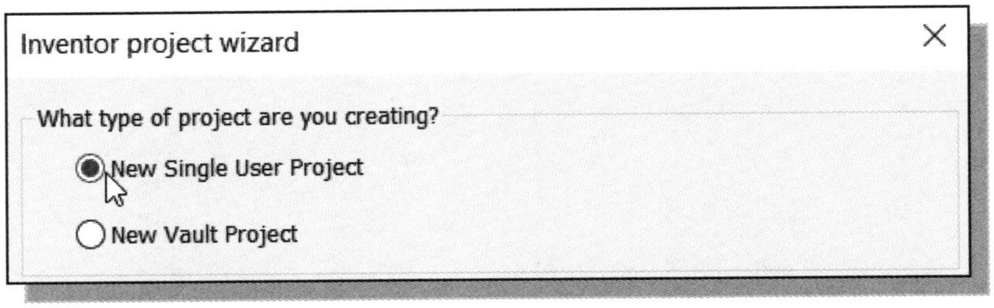

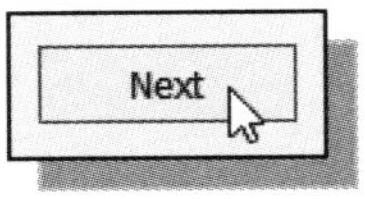

3. Click **Next** to proceed with the next setup option.

4. In the *Project File Name* input box, enter **Parametric-Modeling** as the name of the new project.

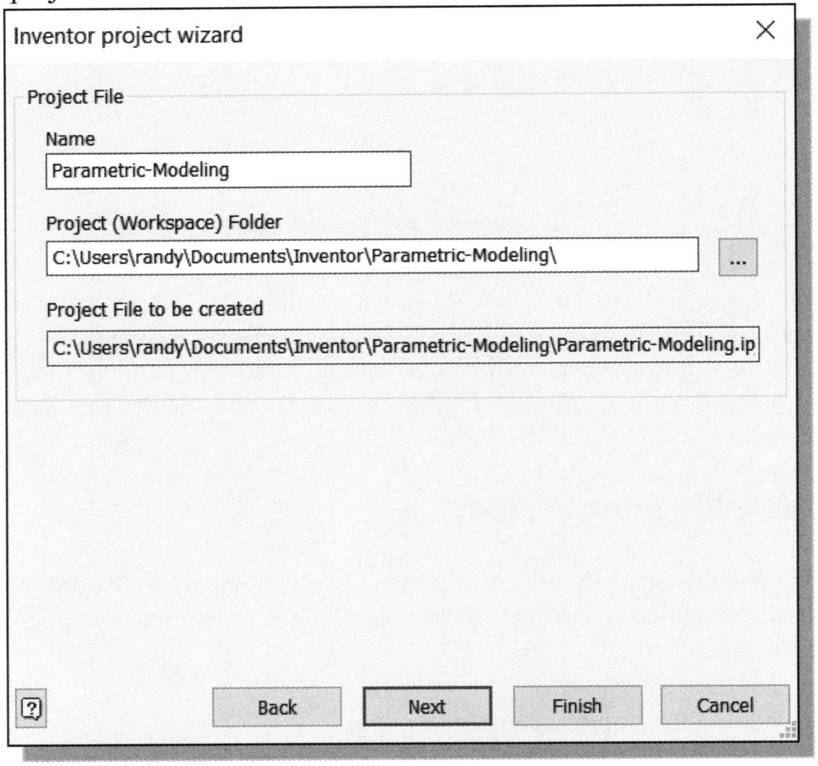

5. In the *Project Folder* input box, note the default folder location, such as **C:\Users\Documents\Inventor \Parametric Modeling\,** and choose a preferred folder name as the folder name of the new project.

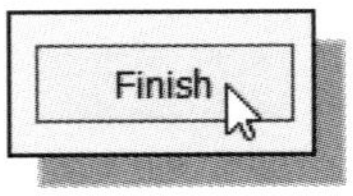

6. Click **Finish** to proceed with the creation of the new project.

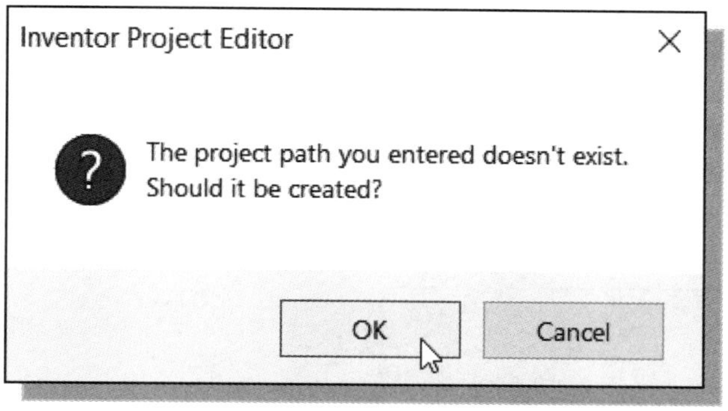

7. A *warning message* appears on the screen, indicating the specified folder does not exist. Click **OK** to create the folder.

8. A second *warning message* appears on the screen, indicating that the newly created project cannot be made active since an *Inventor* file is open. Click **OK** to close the message dialog box.

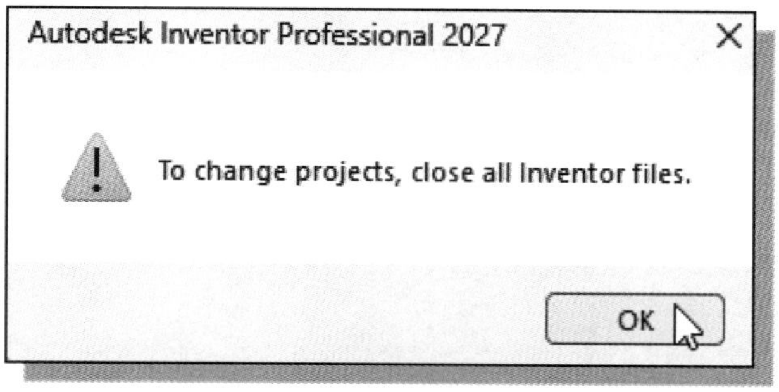

- The new project has been created and its name appears in the project list area as shown in the figure.

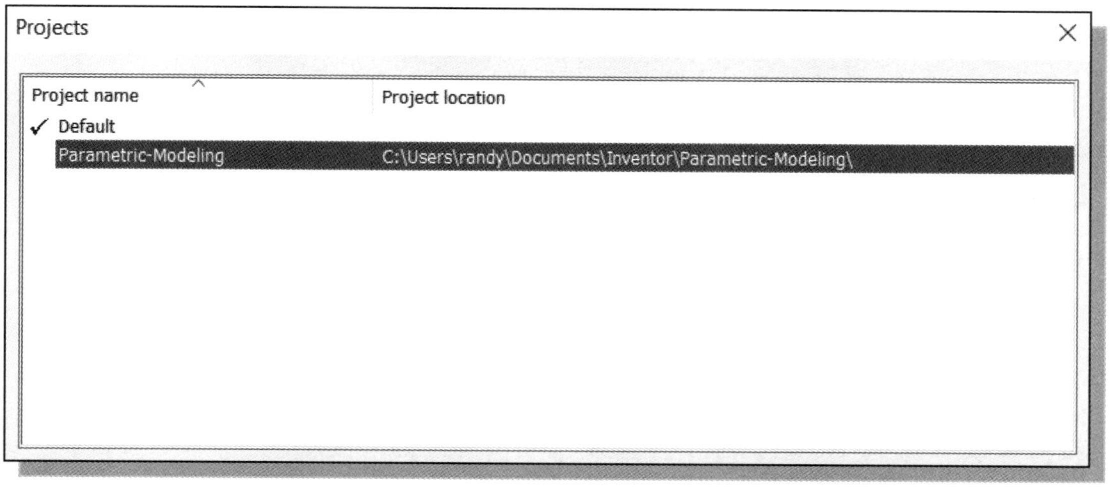

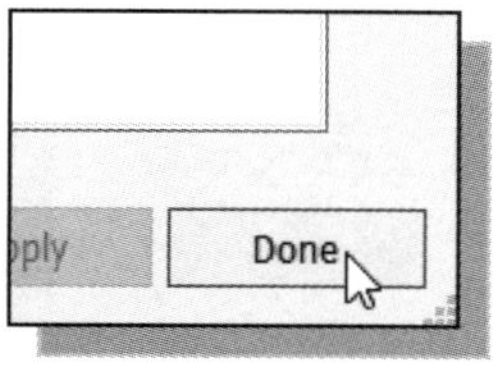

9. Click **Done** to exit the *Inventor Projects Editor* option.

The Content of the Inventor Project File

An *Inventor* project file is actually a text file in .xml format with an **.ipj** extension. The file specifies the paths to the folder containing the files in the project. To assure that links between files work properly, it is advised to add the locations for folders to the project file before working on model files.

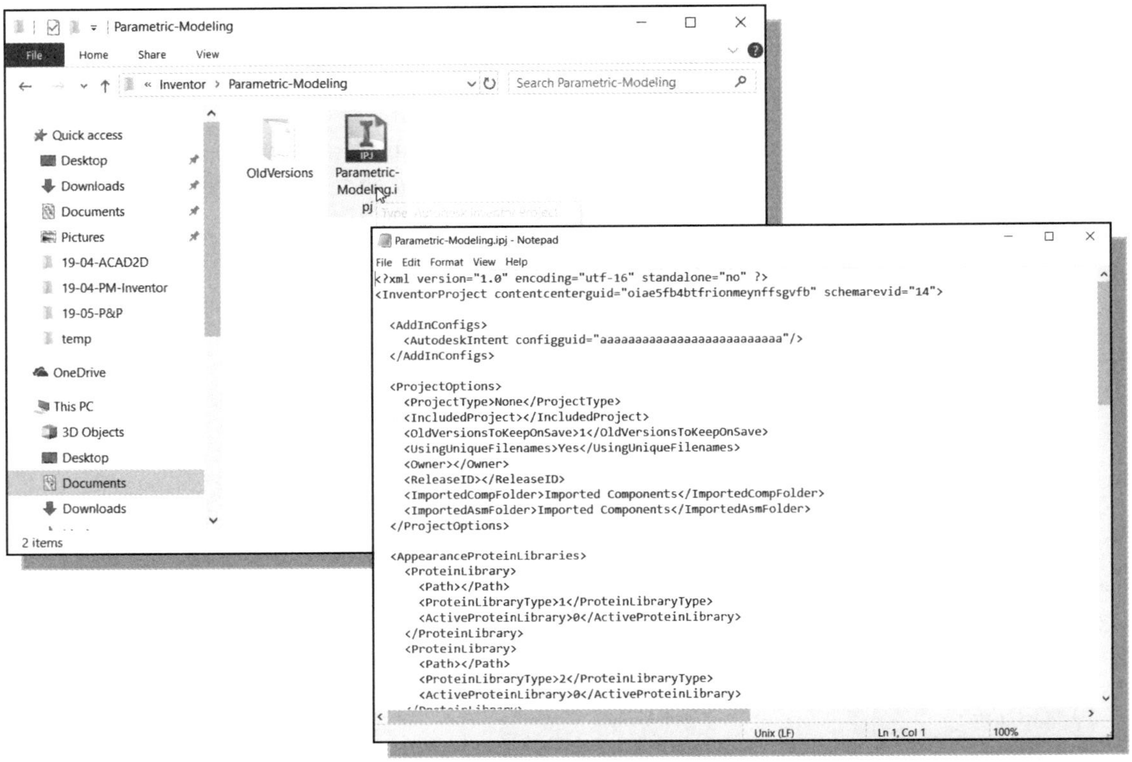

Leaving Autodesk Inventor

To leave the *Application* menu, use the left-mouse-button and click on **Exit Autodesk Inventor** from the pull-down menu.

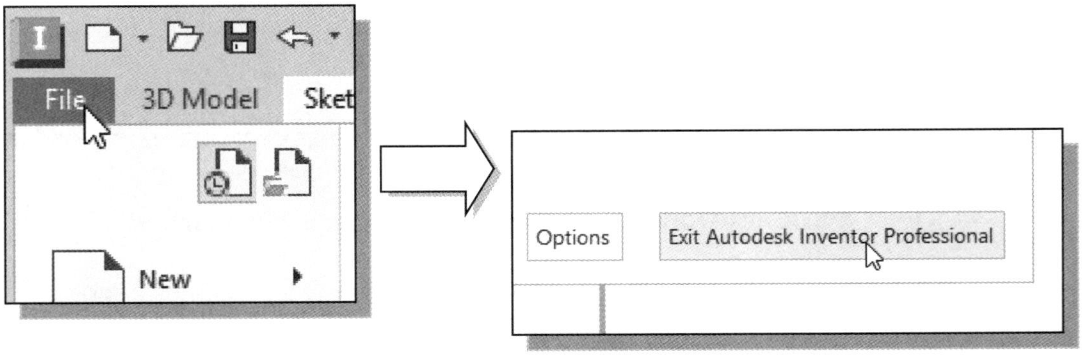

Chapter 2
Parametric Modeling Fundamentals

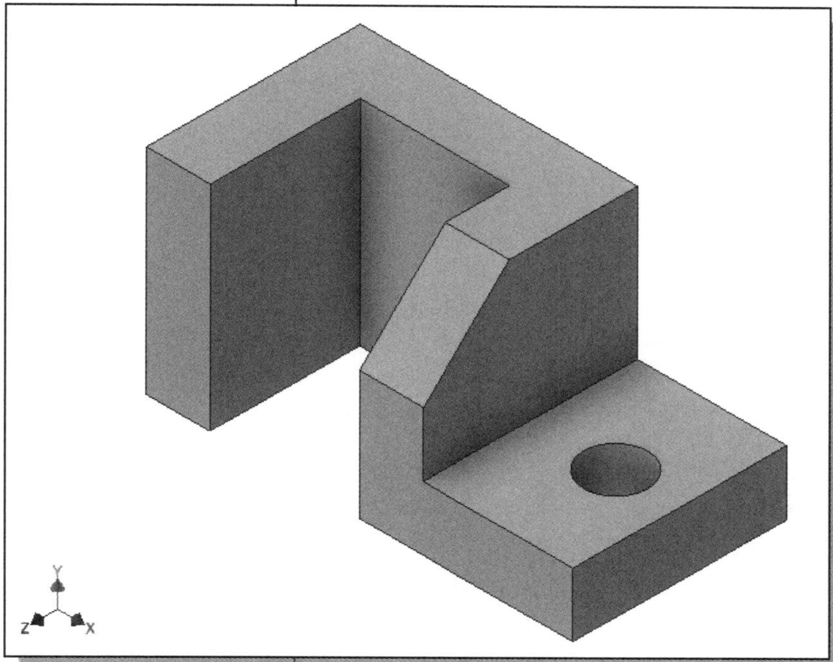

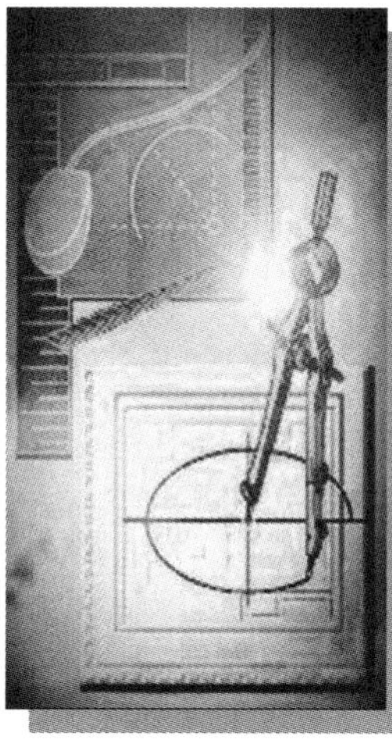

Learning Objectives

- Create Simple Extruded Solid Models
- Understand the Basic Parametric Modeling Procedure
- Create 2D Sketches
- Understand the "Shape before Size" Design Approach
- Use the Dynamic Viewing Commands
- Create and Edit Parametric Dimensions

Autodesk Inventor Certified User Exam Objectives Coverage

Parametric Modeling Basics:

Section 3: Sketches

Objectives: Creating 2D Sketches, Draw Tools, Sketch Constraints, Pattern Sketches, Modify Sketches, Format Sketches, Sketch Doctor, Shared Sketches, Sketch Parameters.

Section 4: Parts

Objectives: Creating parts, Work Features, Pattern Features, and Part Properties.

Introduction

The **feature-based parametric modeling** technique enables the designer to incorporate the original **design intent** into the construction of the model. The word ***parametric*** means the geometric definitions of the design, such as dimensions, can be varied at any time in the design process. Parametric modeling is accomplished by identifying and creating the key features of the design with the aid of computer software. The design variables, described in the sketches as parametric relations, can then be used to quickly modify/update the design.

In Autodesk Inventor, the parametric part modeling process involves the following steps:

1. **Create a rough two-dimensional sketch of the basic shape of the base feature of the design.**

2. **Apply/modify constraints and dimensions to the two-dimensional sketch.**

3. **Extrude, revolve, or sweep the parametric two-dimensional sketch to create the base solid feature of the design.**

4. **Add additional parametric features by identifying feature relations and complete the design.**

5. **Perform analyses on the computer model and refine the design as needed.**

6. **Create the desired drawing views to document the design.**

The approach of creating two-dimensional sketches of the three-dimensional features is an effective way to construct solid models. Many designs are in fact the same shape in one direction. Computer input and output devices we use today are largely two-dimensional in nature, which makes this modeling technique quite practical. This method also conforms to the design process that helps the designer with conceptual design along with the capability to capture the ***design intent***. Most engineers and designers can relate to the experience of making rough sketches on restaurant napkins to convey conceptual design ideas. Autodesk Inventor provides many powerful modeling and design-tools, and there are many different approaches to accomplishing modeling tasks. The basic principle of **feature-based modeling** is to build models by adding simple features one at a time. In this chapter, the general parametric part modeling procedure is illustrated; a very simple solid model with extruded features is used to introduce the Autodesk Inventor user interface. The display viewing functions and the basic two-dimensional sketching tools are also demonstrated.

The Adjuster Design

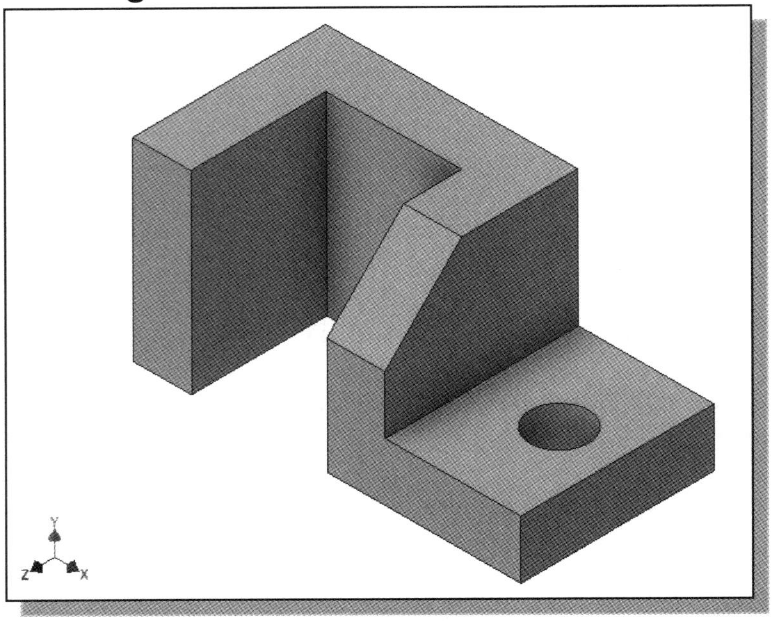

Starting Autodesk Inventor

1. Select the **Autodesk Inventor** option on the *Start* menu or select the **Autodesk Inventor** icon on the desktop to start Autodesk Inventor. The Autodesk Inventor main window will appear on the screen.

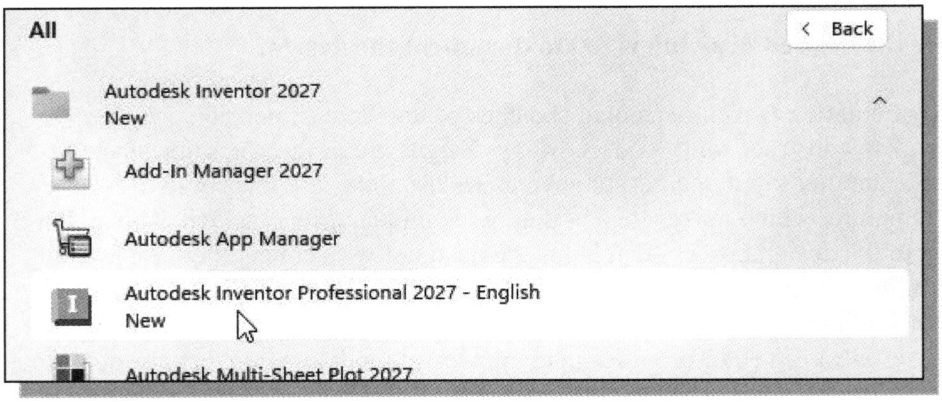

2. Select the **Settings** option next to the *Projects and Settings* icon with a single click of the left-mouse-button.

3. In the *Projects List*, set the ***Parametric-Modeling*** project to be the active project by double-clicking on it as shown.

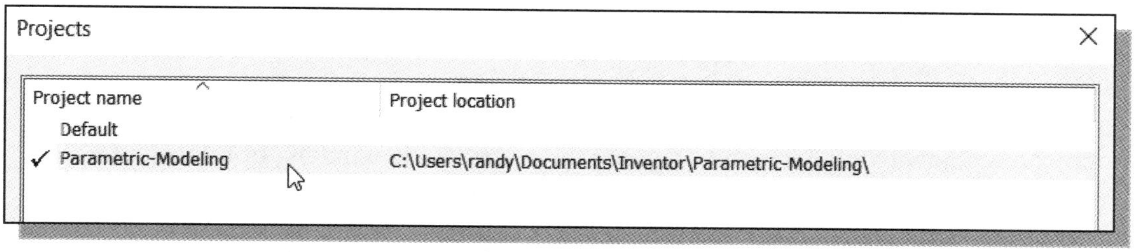

- Note that Autodesk Inventor will keep this activated project as the default project until another project is activated.

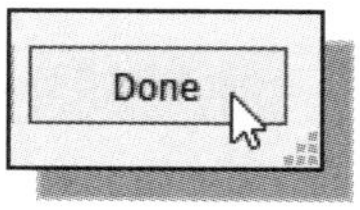

4. Click **Done** to accept the setting and end the *Projects Editor*.

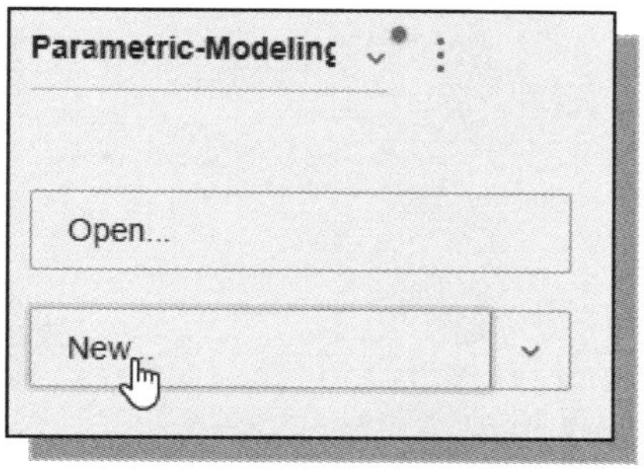

5. Select the **New File** icon with a single click of the left-mouse-button as shown in the figure.

- Notice the ***Parametric-Modeling*** project name is displayed above as the active project.

6. Select the **en-US → English** tab as shown below. When starting a new CAD file, the first thing we should do is choose the units we would like to use. We will use the English setting (inches) for this example.

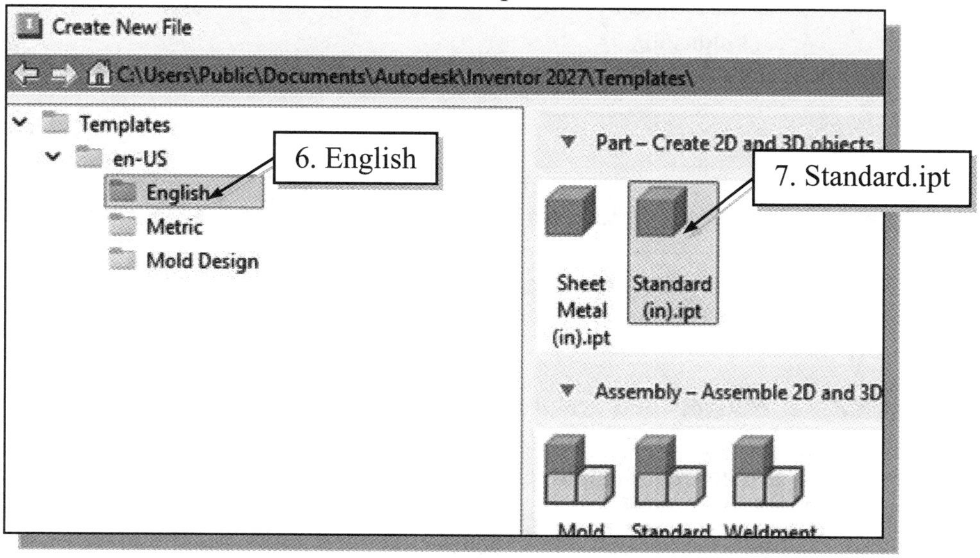

7. Select the **Standard(in).ipt** icon as shown.

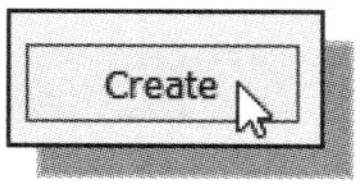

8. Pick **Create** in the *New File* dialog box to accept the selected settings.

The Default Autodesk Inventor Screen Layout

The default Autodesk Inventor drawing screen contains the *pull-down* menus, the *Standard* toolbar, the *3D Model* toolbar, the *Sketch* toolbar, the *Graphics Window*, the *browser* area, and the *Status Bar*. A line of quick text appears next to the icon as you move the *mouse cursor* over different icons. You may resize the *Autodesk Inventor* drawing window by clicking and dragging the edges of the window, or relocate the window by clicking and dragging the window title area.

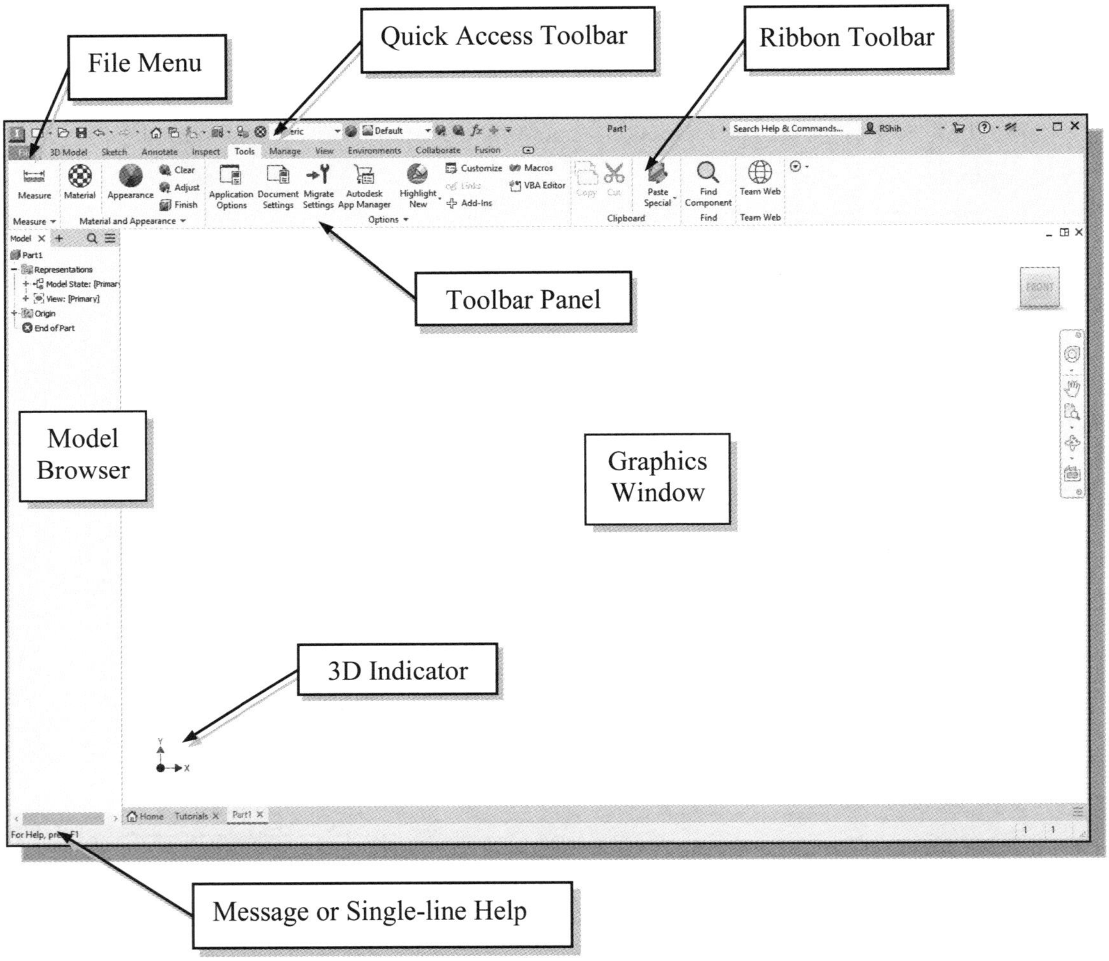

The ***Ribbon Toolbar*** is a relatively new feature in Autodesk Inventor; the *Ribbon Toolbar* is composed of a series of tool panels, which are organized into tabs labeled by task. The *Ribbon* provides a compact palette of all of the tools necessary to accomplish the different modeling tasks. The drop-down arrow next to any icon indicates additional commands are available on the expanded panel; access the expanded panel by clicking on the drop-down arrow.

Sketch Plane – It is an XY monitor, but an XYZ World

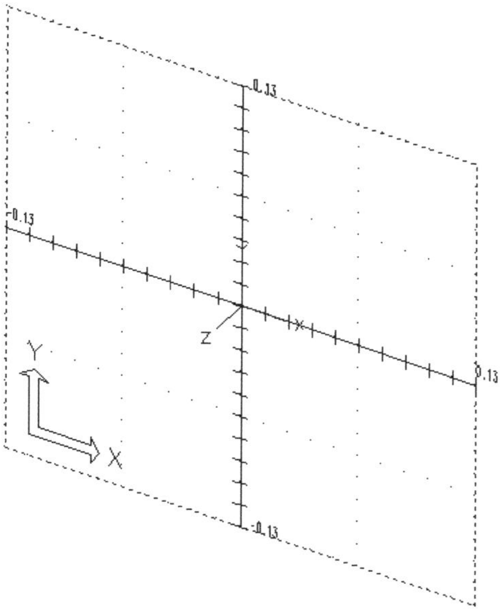

Design modeling software is becoming more powerful and user friendly, yet the system still does only what the user tells it to do. When using a geometric modeler, we therefore need to have a good understanding of what its inherent limitations are.

In most 3D geometric modelers, 3D objects are located and defined in what is usually called **world space** or **global space**. Although a number of different coordinate systems can be used to create and manipulate objects in a 3D modeling system, the objects are typically defined and stored using the world space. The world space is usually a **3D Cartesian coordinate system** that the user cannot change or manipulate.

In engineering designs, models can be very complex, and it would be tedious and confusing if only the world coordinate system were available. Practical 3D modeling systems allow the user to define **Local Coordinate Systems (LCS)** or **User Coordinate Systems (UCS)** relative to the world coordinate system. Once a local coordinate system is defined, we can then create geometry in terms of this more convenient system.

Although objects are created and stored in 3D space coordinates, most of the geometric entities can be referenced using 2D Cartesian coordinate systems. Typical input devices such as a mouse or digitizer are two-dimensional by nature; the movement of the input device is interpreted by the system in a planar sense. The same limitation is true of common output devices, such as displays and plotters. The modeling software performs a series of three-dimensional to two-dimensional transformations to correctly project 3D objects onto the 2D display plane.

The Autodesk Inventor ***sketching plane*** is a special construction approach that enables the planar nature of the 2D input devices to be directly mapped into the 3D coordinate system. The *sketching plane* is a local coordinate system that can be aligned to an existing face of a part, or a reference plane.

Think of the sketching plane as the surface on which we can sketch the 2D sections of the parts. It is similar to a piece of paper, a whiteboard, or a chalkboard that can be attached to any planar surface. The first sketch we create is usually drawn on one of the established datum planes. Subsequent sketches/features can then be created on sketching planes that are aligned to existing **planar faces of the solid part** or **datum planes.**

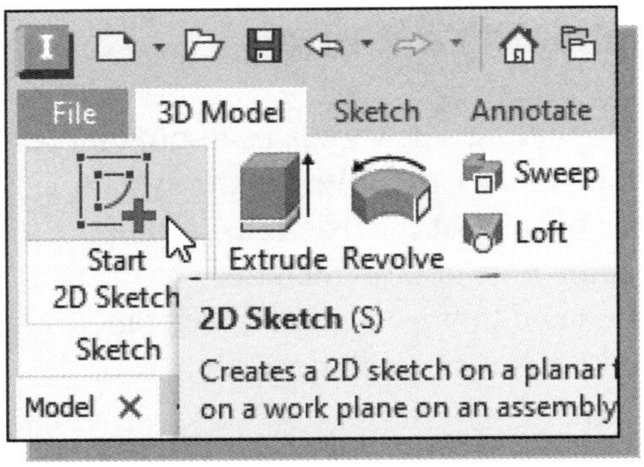

1. Activate the **Start 2D Sketch** icon with a single click of the left-mouse-button.

2. Move the cursor over the edge of the *XY Plane* in the graphics area. When the *XY Plane* is highlighted, click once with the **left-mouse-button** to select the *Plane* as the sketch plane for the new sketch.

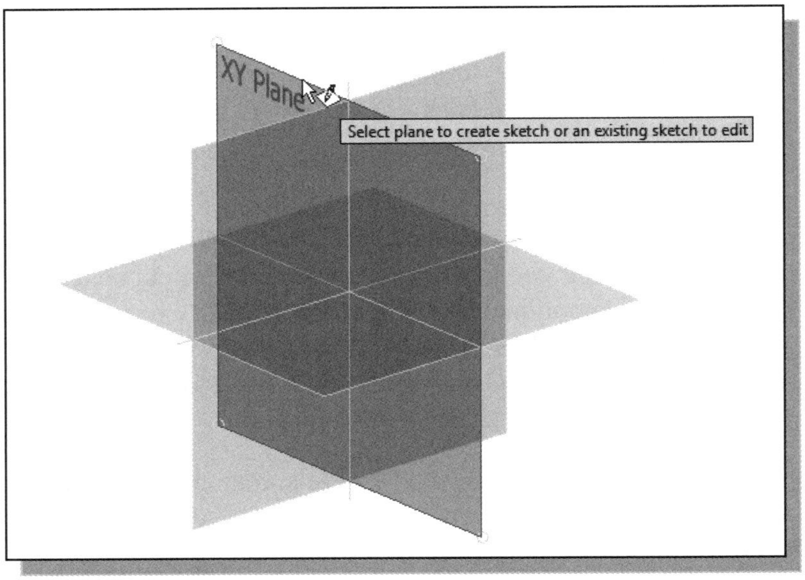

- The *sketching plane* is a reference location where two-dimensional sketches are created. Note that the *sketching plane* can be any planar part surface or datum plane.

3. Confirm the main *Ribbon* area is switched to the **Sketch** toolbars; this indicates we have entered the 2D Sketching mode.

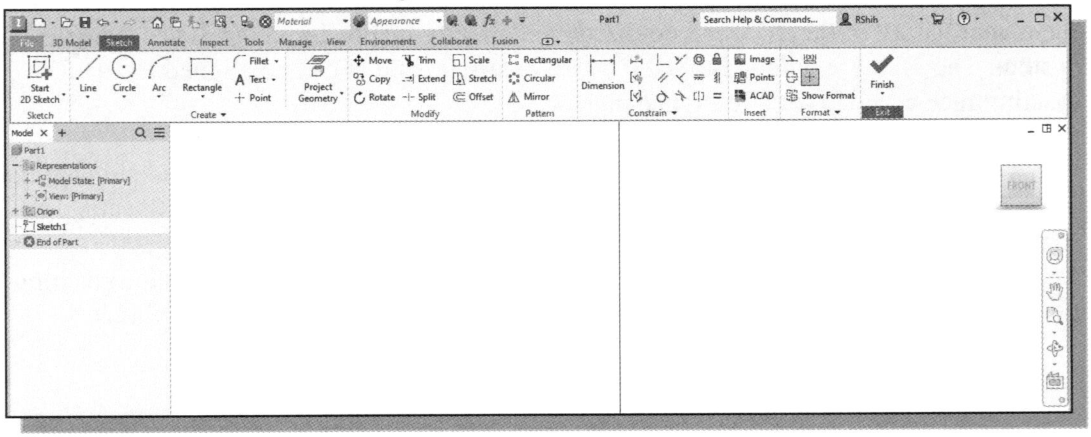

Create A Rough Sketch

Quite often during the early design stage, the shape of a design may not have any precise dimensions. Most conventional CAD systems require the user to input the precise lengths and locations of all geometric entities defining the design, which are not available during the early design stage. With *parametric modeling*, we can use the computer to elaborate and formulate the design idea further during the initial design stage. With Autodesk Inventor, we can use the computer as an electronic sketchpad to help us concentrate on the formulation of forms and shapes for the design. This approach is the main advantage of *parametric modeling* over conventional solid-modeling techniques.

As the name implies, a ***rough sketch*** is not precise at all. When sketching, we simply sketch the geometry so that it closely resembles the desired shape. Precise scale or lengths are not needed. Autodesk Inventor provides many tools to assist us in finalizing sketches. For example, geometric entities such as horizontal and vertical lines are set automatically. However, if the rough sketches are poor, it will require much more work to generate the desired parametric sketches. Here are some general guidelines for creating sketches in Autodesk Inventor:

- **Create a sketch that is proportional to the desired shape.** Concentrate on the shapes and forms of the design.

- **Keep the sketches simple.** Leave out small geometry features such as fillets, rounds and chamfers. They can easily be placed using the Fillet and Chamfer commands after the parametric sketches have been established.

- **Exaggerate the geometric features of the desired shape.** For example, if the desired angle is 85 degrees, create an angle that is 50 or 60 degrees. Otherwise, Autodesk Inventor might assume the intended angle to be a 90-degree angle.

- **Draw the geometry so that it does not overlap.** The geometry should eventually form a closed region. *Self-intersecting* geometry shapes are not allowed.

- **The sketched geometric entities should form a closed region.** To create a solid feature, such as an extruded solid, a closed region is required so that the extruded solid forms a 3D volume.

➢ **Note:** The concepts and principles involved in *parametric modeling* are very different, and sometimes they are totally opposite, to those of conventional computer aided drafting. In order to understand and fully utilize Autodesk Inventor's functionality, it will be helpful to take a *Zen* approach to learning the topics presented in this text: **Have an open mind and temporarily forget your experiences using conventional Computer Aided Drafting systems.**

Step 1: Creating a Rough Sketch

The *Sketch* toolbar provides tools for creating the basic geometry that can be used to create features and parts.

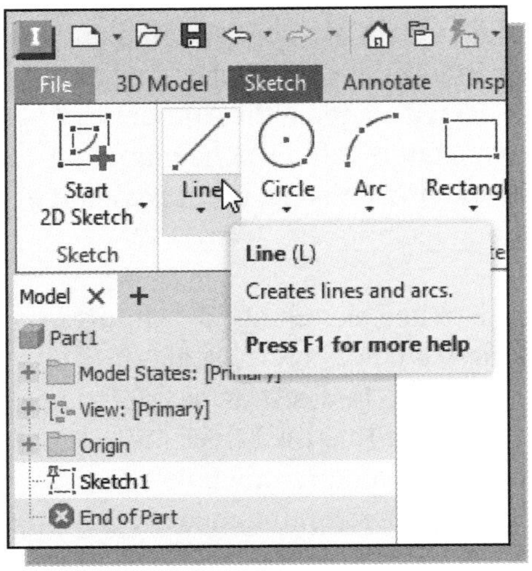

1. Move the graphics cursor to the **Line** icon in the *Sketch* toolbar. A *Help-tip box* appears next to the cursor and a brief description of the command is displayed at the bottom of the drawing screen: "*Creates Straight lines and arcs.*"

2. Select the icon by clicking once with the **left-mouse-button**; this will activate the Line command. Autodesk Inventor expects us to identify the starting location of a straight line.

Graphics Cursors

Notice the cursor changes from an arrow to a crosshair when graphical input is expected.

1. Left-click a starting point for the shape, roughly just below and to the right of the center of the graphics window.

2. As you move the graphics cursor, you will see a digital readout next to the cursor and also in the *Status Bar* area at the bottom of the window. The readout gives you the cursor location, the line length, and the angle of the line measured from horizontal. Move the cursor around and you will notice different symbols appear at different locations.

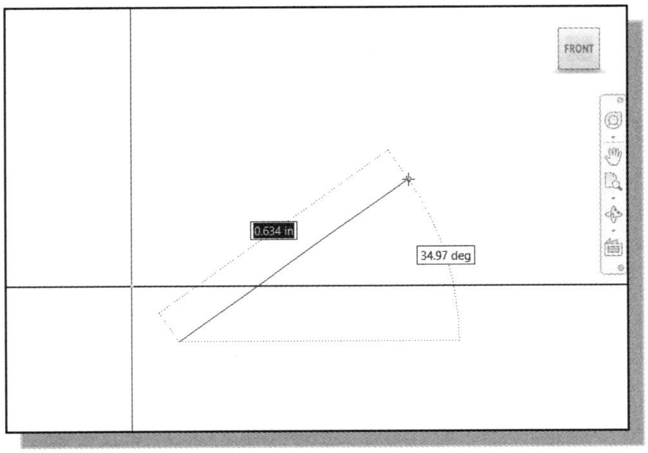

➢ The readout displayed next to the cursor is called the ***Dynamic Input***. This option is part of the **Heads-Up Display** option that is now available in Inventor. Note that *Dynamic Input* can be used for entering precise values, but its usage is somewhat limited in *parametric modeling*.

3. Move the graphics cursor toward the right side of the graphics window and create a horizontal line as shown below (**Point 2**). Notice the geometric constraint symbol, a short horizontal line indicating the geometric property, is displayed.

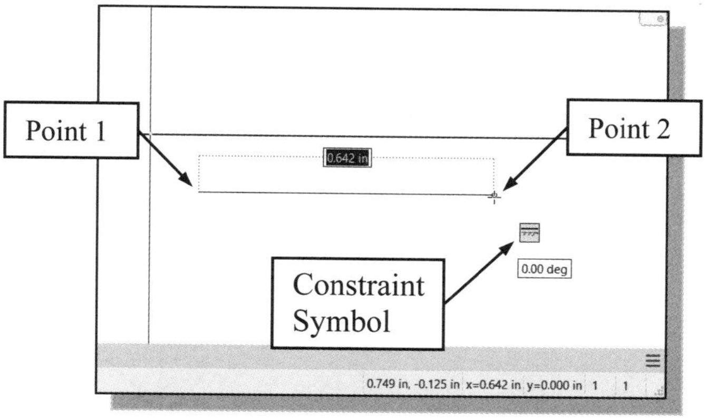

Geometric Constraint Symbols

Autodesk Inventor displays different visual clues, or symbols, to show you alignments, perpendicularities, tangencies, etc. These constraints are used to capture the *design intent* by creating constraints where they are recognized. Autodesk Inventor displays the governing geometric rules as models are built. To prevent constraints from forming, hold down the [**Ctrl**] key while creating an individual sketch curve. For example, while sketching line segments with the Line command, endpoints are joined with a Coincident *constraint*, but when the [**Ctrl**] key is pressed and held, the inferred constraint will not be created.

	**Vertical**	indicates a line is vertical
	**Horizontal**	indicates a line is horizontal
- - -	**Dashed line**	indicates the alignment is to the center point or endpoint of an entity
	**Parallel**	indicates a line is parallel to other entities
	**Perpendicular**	indicates a line is perpendicular to other entities
	**Coincident**	indicates the cursor is at the endpoint of an entity
	**Concentric**	indicates the cursor is at the center of an entity
	**Tangent**	indicates the cursor is at tangency points to curves

1. Complete the sketch as shown below, creating a closed region ending at the starting point (**Point 1**). Do not be overly concerned with the actual size of the sketch. Note that all line segments are sketched horizontally or vertically.

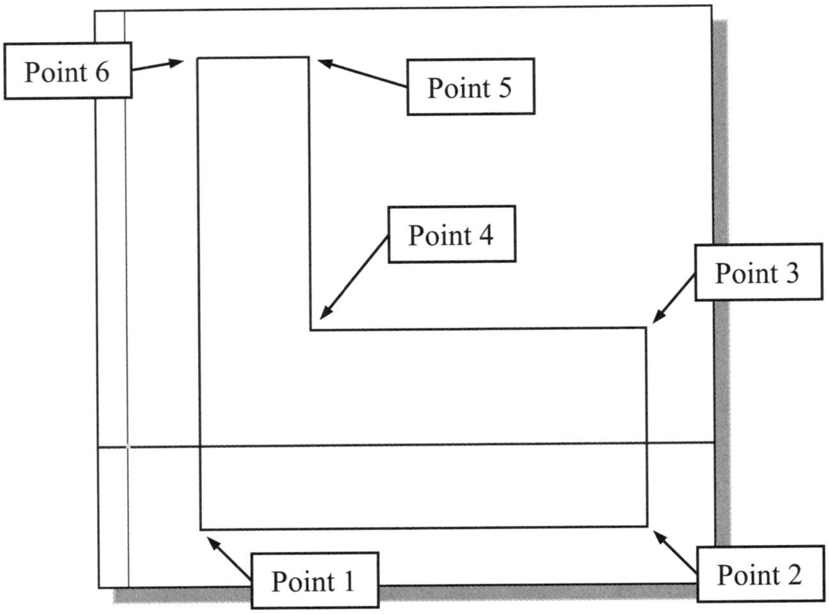

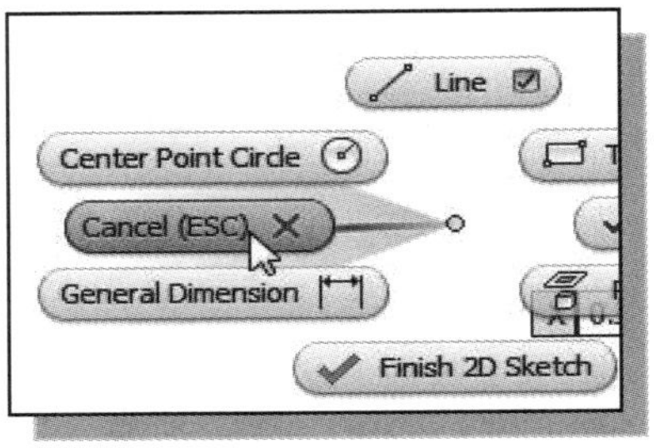

2. Inside the graphics window, click once with the **right-mouse-button** to display the option menu. Select Cancel [Esc] in the pop-up menu, or hit the [**Esc**] key once to end the Sketch Line command.

Step 2: Apply/Modify Constraints and Dimensions

As the sketch is made, Autodesk Inventor automatically applies some of the geometric constraints (such as horizontal, parallel, and perpendicular) to the sketched geometry. We can continue to modify the geometry, apply additional constraints, and/or define the size of the existing geometry. In this example, we will illustrate adding dimensions to describe the sketched entities.

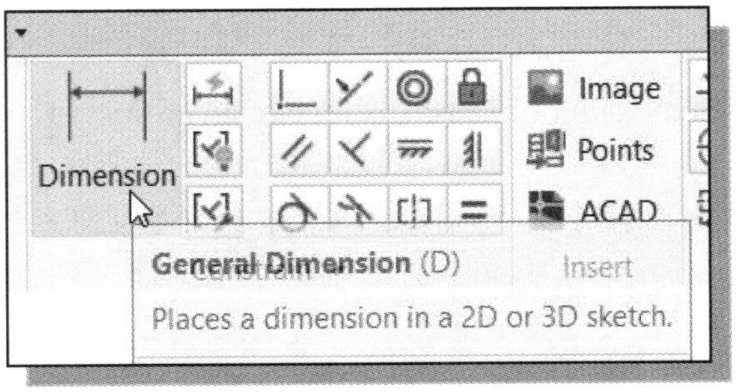

1. Move the cursor to the *Constrain* toolbar area; it is the toolbar next to the *2D Draw* toolbar. Note the first icon in this toolbar is the General Dimension icon. The Dimension command is generally known as **Smart Dimensioning** in parametric modeling.

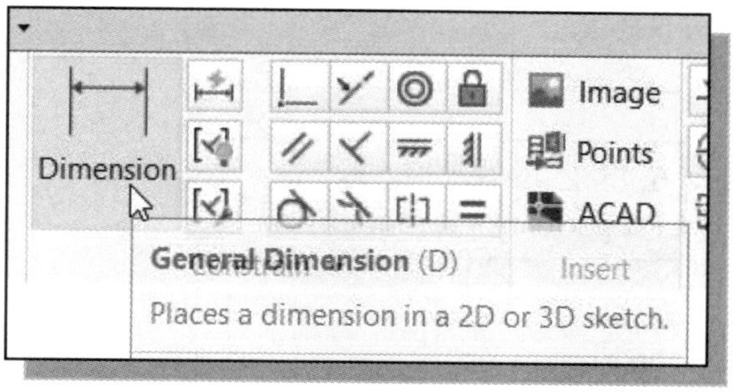

2. Move the cursor on top of the Dimension icon. The Smart Dimensioning command allows us to quickly create and modify dimensions. **Left-click** once on the icon to activate the Dimension command.

3. The message "*Select Geometry to Dimension*" is displayed in the *Status Bar* area at the bottom of the *Inventor* window. Select the bottom horizontal line by left-clicking once on the line.

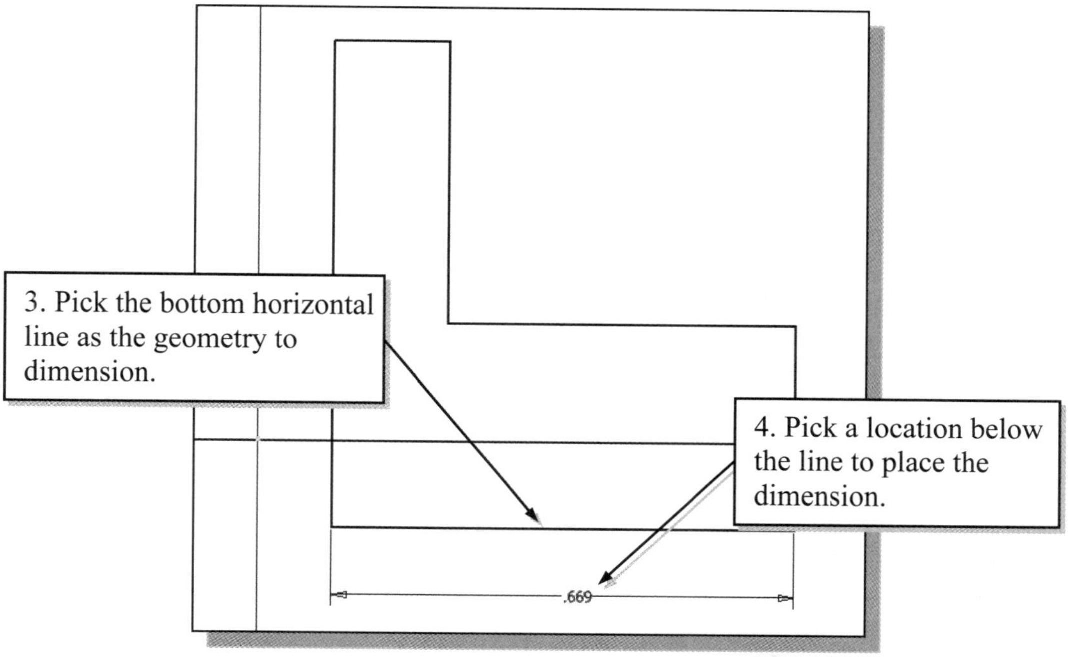

4. Move the graphics cursor below the selected line and **left-click** to place the dimension. (Note that the value displayed on your screen might be different than what is shown in the figure above.)

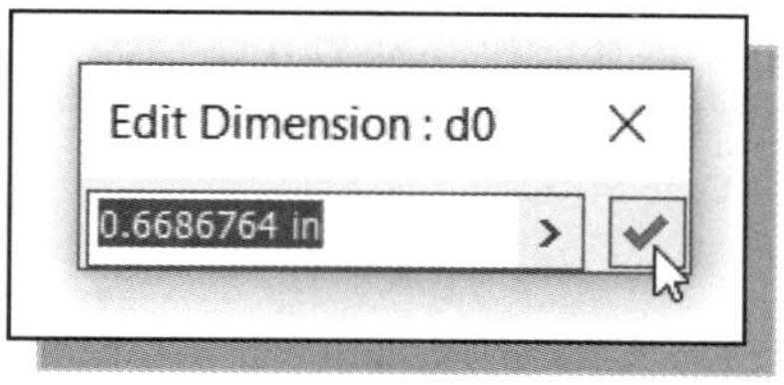

5. Accept the *default value* by clicking on the **Accept** button as shown.

❖ The General Dimension command will create a length dimension if a single line is selected.

6. The message "*Select Geometry to Dimension*" is displayed in the *Status Bar* area, located at the bottom of the *Inventor* window. Select the top-horizontal line as shown below.

7. Select the bottom-horizontal line as shown below.

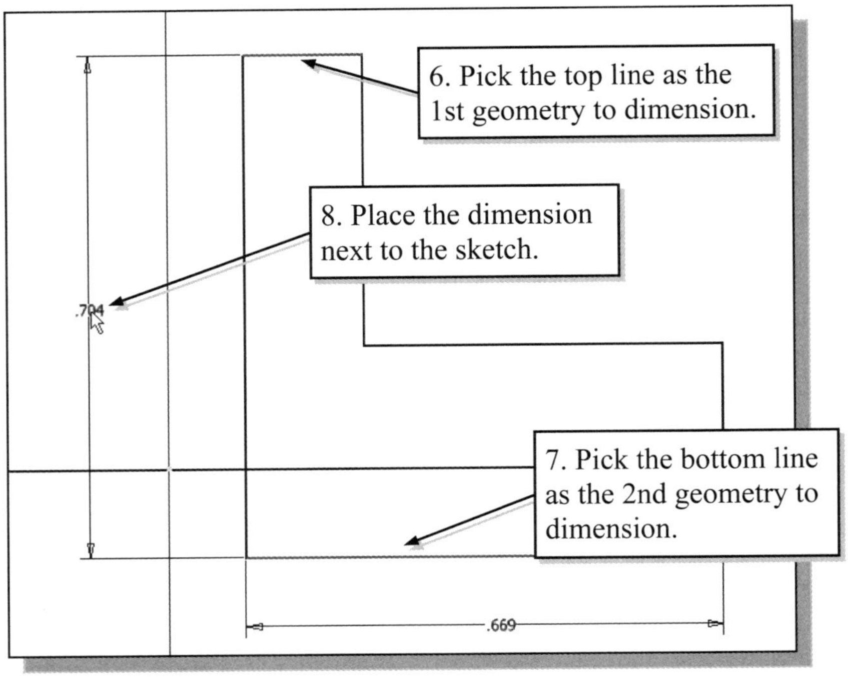

8. Pick a location to the left of the sketch to place the dimension.

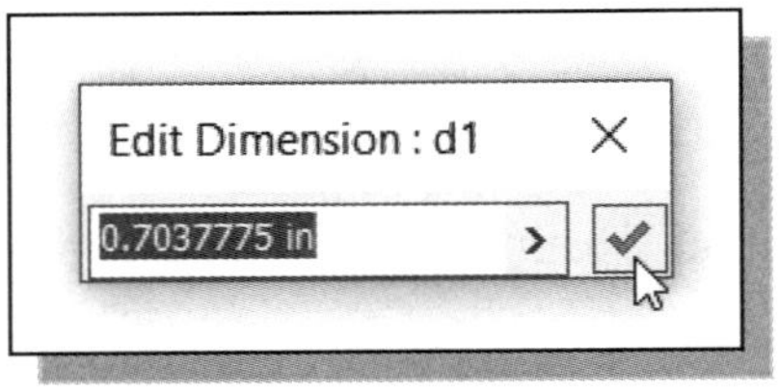

9. Accept the default value by clicking on the **Accept** button.

❖ When two parallel lines are selected, the General Dimension command will create a dimension measuring the distance between them.

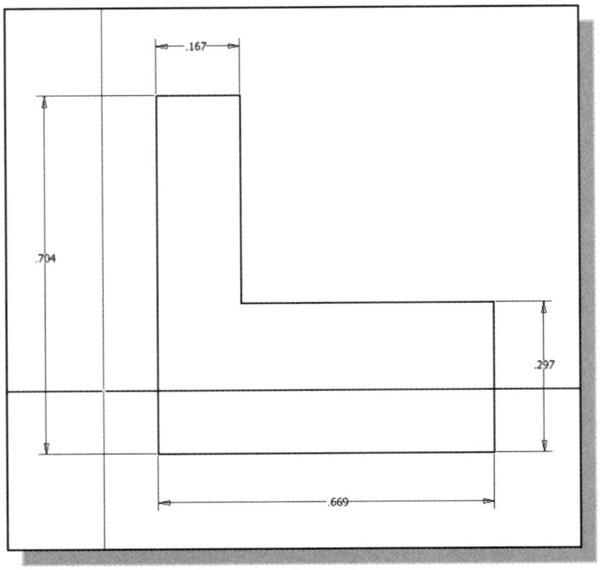

10. On your own, repeat the above steps and create additional dimensions (accepting the default values created by Inventor) so that the sketch appears as shown.

Dynamic Viewing Functions – Zoom and Pan

Autodesk Inventor provides a special user interface called *Dynamic Viewing* that enables convenient viewing of the entities in the graphics window.

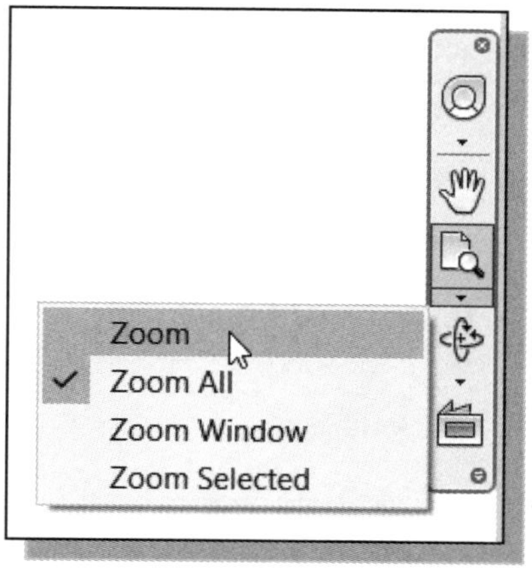

1. Click on the **Zoom** icon located in the *Navigation* bar as shown.

2. Move the cursor near the center of the graphics window.

3. Inside the graphics window, **press and hold down the left-mouse-button**, then move downward to enlarge the current display scale factor.

4. Press the [**Esc**] key once to exit the Zoom command.

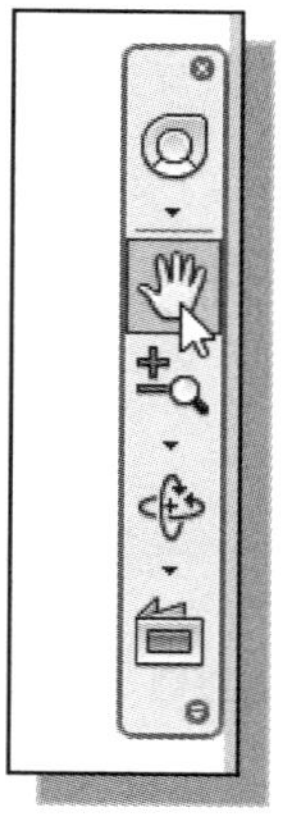

5. Click on the **Pan** icon located above the Zoom command in the *Navigation* bar. The icon is the picture of a hand.

➢ The Pan command enables us to move the view to a different position. This function acts as if you are using a video camera.

6. On your own, use the Zoom and Pan options to reposition the sketch near the center of the screen.

Modifying the Dimensions of the Sketch

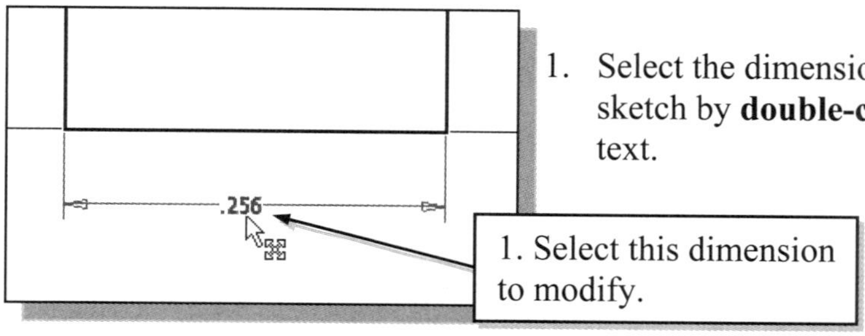

1. Select the dimension that is at the bottom of the sketch by **double-clicking** on the dimension text.

2. In the *Edit Dimension* window, the current length of the line is displayed. Enter **2.5** to set the length of the line.

3. Click on the **Accept** icon to accept the entered value.

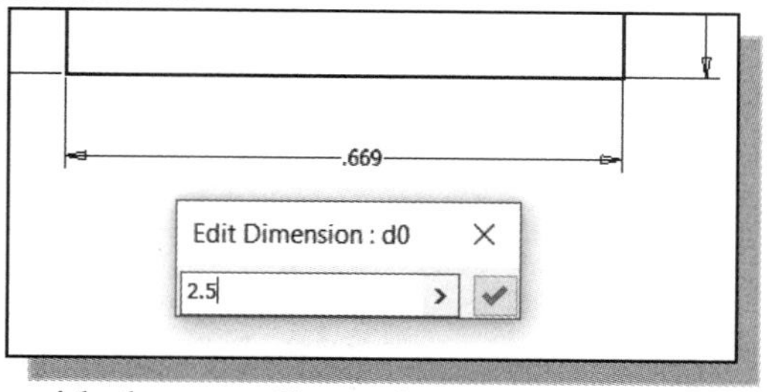

➢ Autodesk Inventor will now update the profile with the new dimension value.

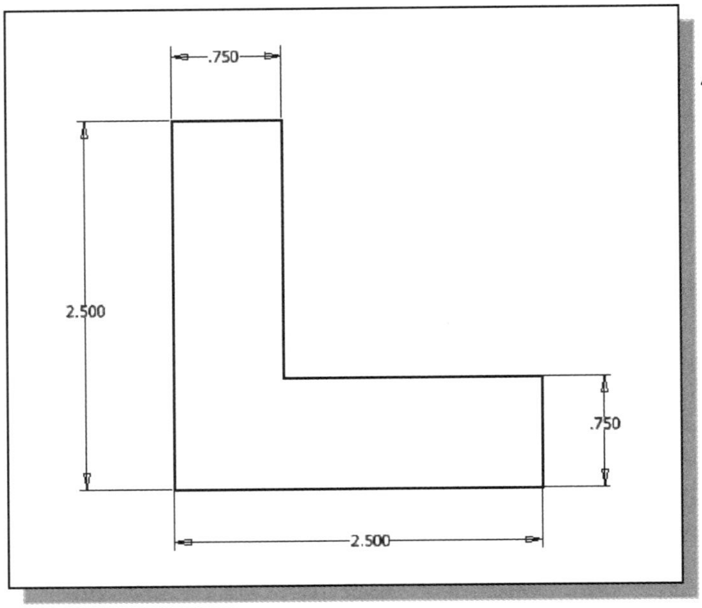

4. On your own, repeat the above steps and adjust the dimensions so that the sketch appears as shown.

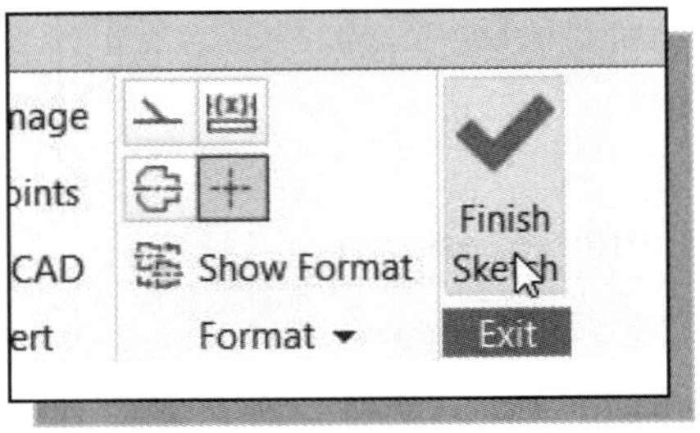

5. In the *Ribbon* toolbar, click once with the **left-mouse-button** to select **Finish Sketch** in the *Ribbon* area to end the Sketch option.

Step 3: Completing the Base Solid Feature

Now that the 2D sketch is completed, we will proceed to the next step: create a 3D part from the 2D profile. Extruding a 2D profile is one of the common methods that can be used to create 3D parts. We can extrude planar faces along a path. We can also specify a height value and a tapered angle. In Autodesk Inventor, each face has a positive side and a negative side; the current face we are working on is set as the default positive side. This positive side identifies the positive extrusion direction, and it is referred to as the face's ***normal***.

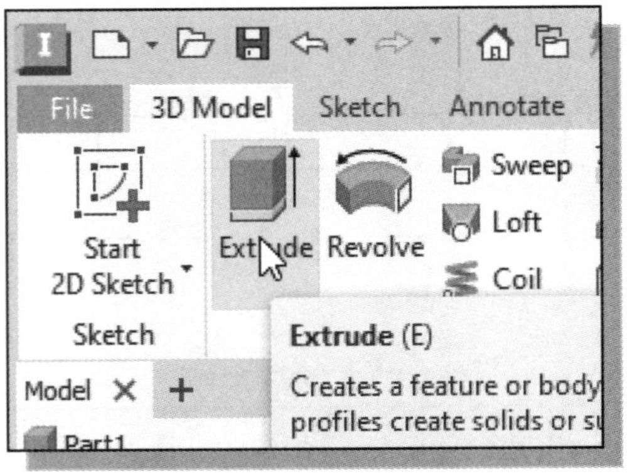

1. In the 3D Model tab, select the **Extrude** command by clicking the left-mouse-button on the icon as shown.

2. In the *Extrude* edit box, enter **2.5** as the extrusion distance. Notice that the sketch region is automatically selected as the extrusion profile.

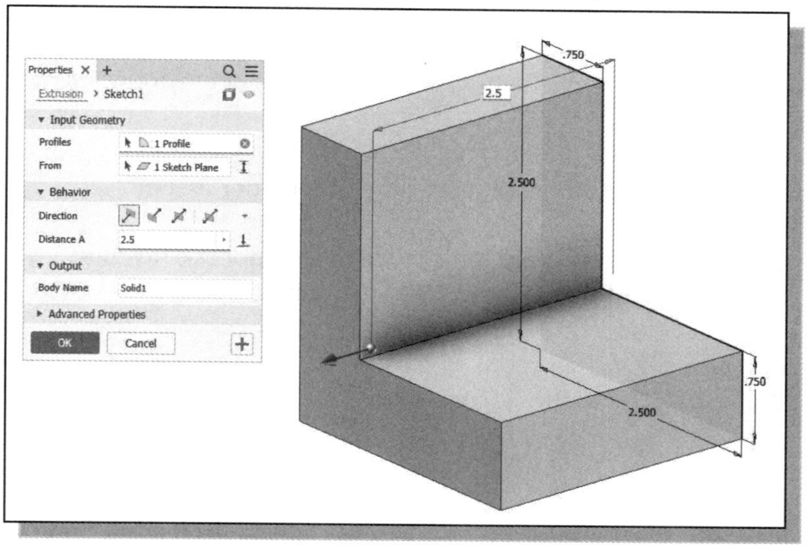

3. Click on the **OK** button to proceed with creating the 3D part.

➢ Note that all dimensions disappeared from the screen. All parametric definitions are stored in the **Autodesk Inventor database** and any of the parametric definitions can be re-displayed and edited at any time.

Isometric View

Autodesk Inventor provides many ways to display views of the three-dimensional design. Several options are available that allow us to quickly view the design to track the overall effect of any changes being made to the model. We will first orient the model to display in the *isometric view* by using the pull-down menu.

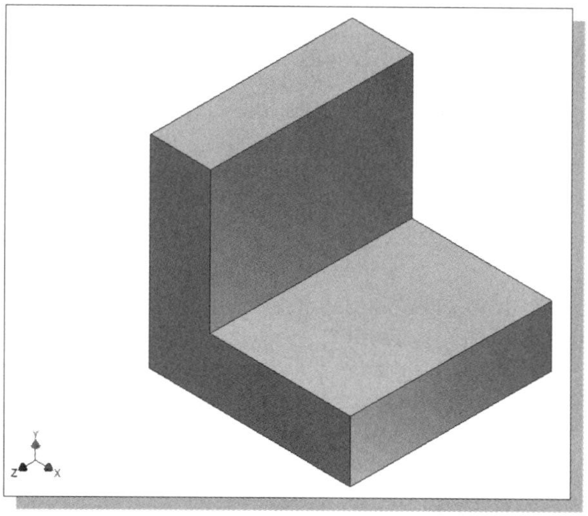

1. Hit the function key **[F6]** once to automatically adjust the display and also reset the display to the *isometric* view.

❖ Note that most of the view-related commands can be accessed in the ViewCube and/or the *Navigation* bar located to the right side of the graphics window.

❖ Dynamic Rotation of the 3D Block – Free Orbit

The Free Orbit command allows us to:

- Orbit a part or assembly in the graphics window. Rotation can be around the center mark, free in all directions, or around the X/Y-axes in the *3D-Orbit* display.
- Reposition the part or assembly in the graphics window.
- Display isometric or standard orthographic views of a part or assembly.

The Free Orbit tool is accessible while other tools are active. Autodesk Inventor remembers the last used mode when you exit the Orbit command.

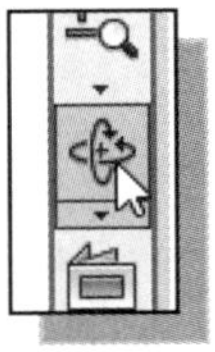

1. Click on the **Orbit** icon in the *Navigation* bar.

➢ The *3D Orbit* display is a circular rim with four handles and a center mark. *3D Orbit* enables us to manipulate the view of 3D objects by clicking and dragging with the left-mouse-button:

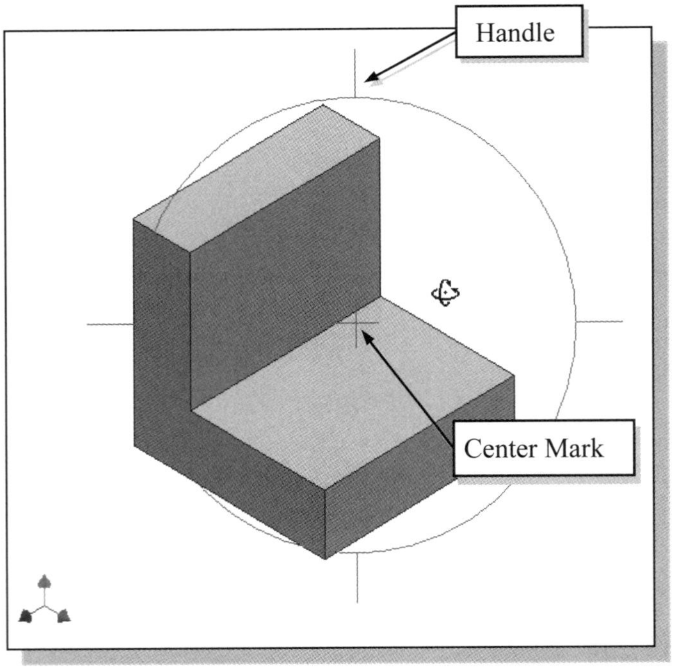

- Drag with the left-mouse-button near the center for free rotation.
- Drag on the handles to orbit around the horizontal or vertical axes.
- Drag on the rim to orbit about an axis that is perpendicular to the displayed view.
- Single left-click to align the center mark of the view.

2. Inside the *circular rim*, press down the left-mouse-button and drag in an arbitrary direction; the 3D Orbit command allows us to freely orbit the solid model.

3. Move the cursor near the circular rim and notice the cursor symbol changes to a single circle. Drag with the left-mouse-button to orbit about an axis that is perpendicular to the displayed view.

4. Single left-click near the top-handle to align the selected location to the center mark in the graphics window.

5. Activate the **Constrained Orbit** option by clicking on the associated icon as shown.

❖ *The* Constrained Orbit *can be used to rotate the model about axes in Model Space, equivalent to moving the eye position about the model in latitude and longitude.*

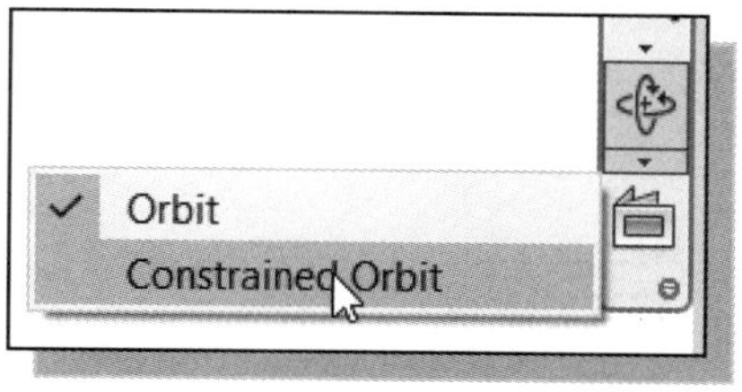

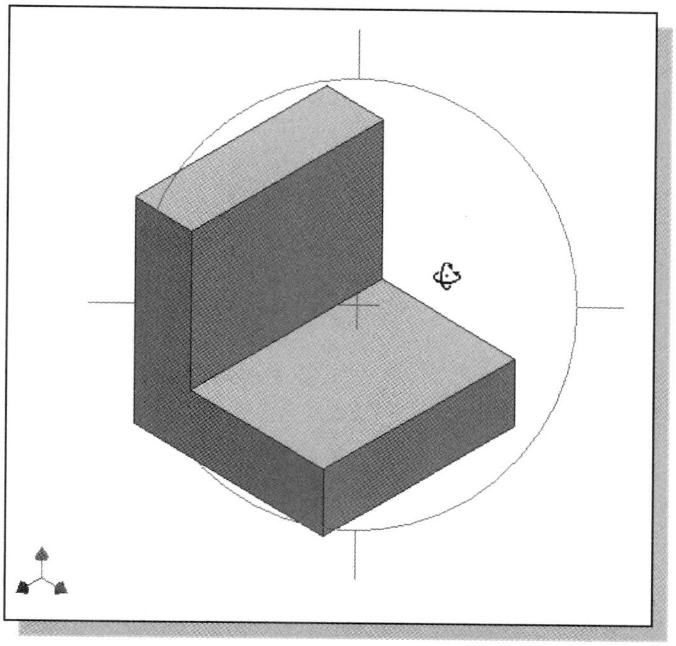

6. On your own, use the different options described in the above steps and familiarize yourself with both of the 3D Orbit commands. Reset the display to the *Isometric* view as shown in the above figure before continuing to the next section.

❖ Note that while in the 3D Orbit mode, a horizontal marker will be displayed next to the cursor if the cursor is away from the circular rim. This is the **exit marker**. Left-clicking once will allow you to exit the 3D Orbit command.

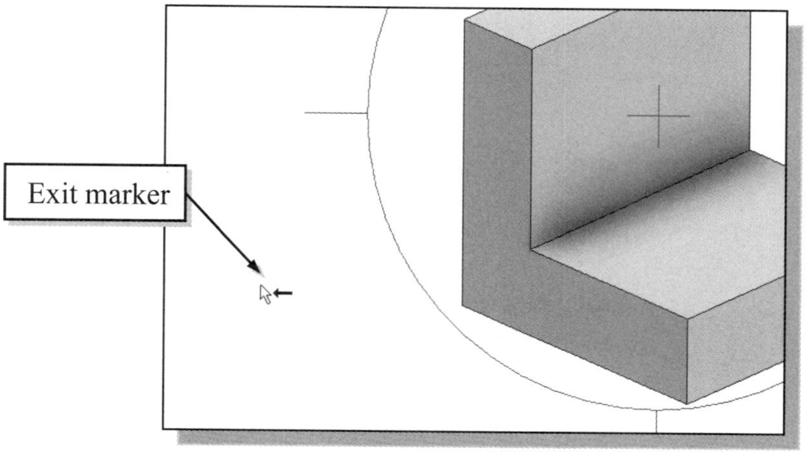

Dynamic Viewing – Quick Keys

We can also use the function keys on the keyboard and the mouse to access the *Dynamic Viewing* functions.

❖ Panning – (1) F2 and drag with the left-mouse-button

Hold the **F2** function key down and drag with the left-mouse-button to pan the display. This allows you to reposition the display while maintaining the same scale factor of the display.

Pan

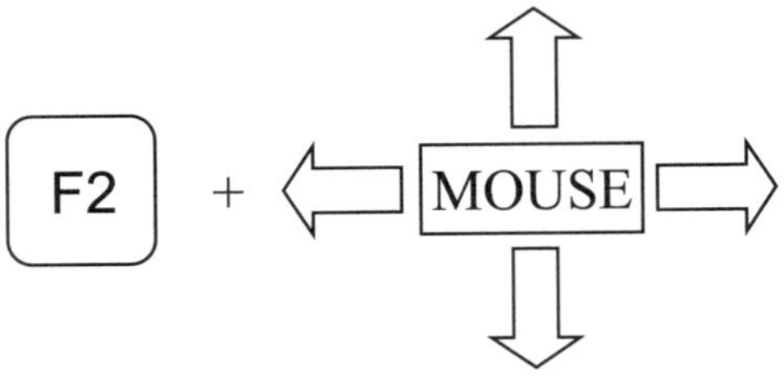

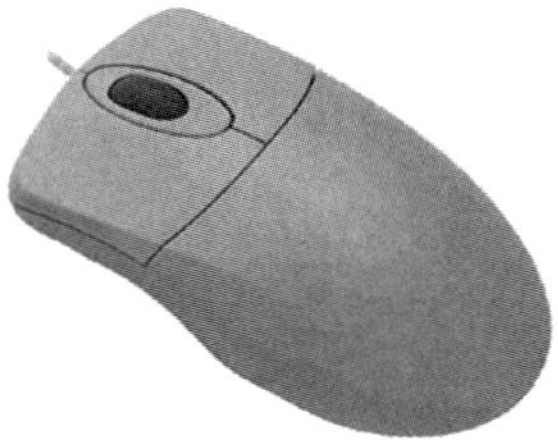

(2) Press and drag with the mouse wheel

Pressing and dragging with the mouse wheel can also reposition the display.

❖ Zooming – (1) F3 and drag with the left-mouse-button

Hold the **F3** function key down and drag with the left-mouse-button vertically on the screen to adjust the scale of the display. Moving upward will reduce the scale of the display, making the entities display smaller on the screen. Moving downward will magnify the scale of the display.

Zoom

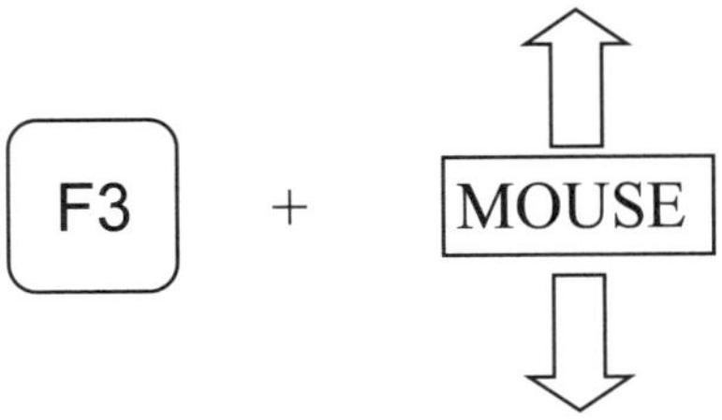

(2) Turning the mouse wheel

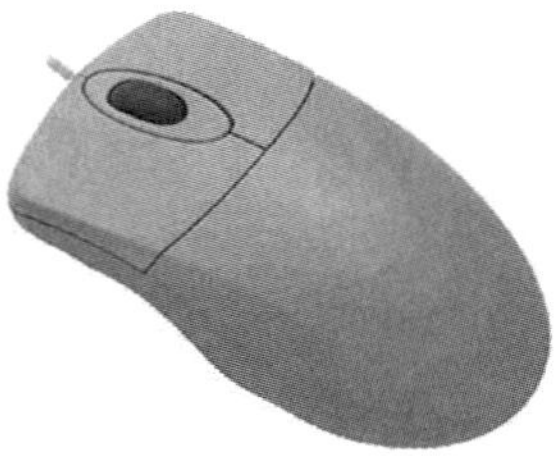

Turning the mouse wheel can also adjust the scale of the display. Turning forward will reduce the scale of the display, making the entities display smaller on the screen. Turning backward will magnify the scale of the display.

❖ 3D Dynamic Rotation – Shift and drag with the middle-mouse-button

Hold the **Shift** key down and drag with the middle-mouse-button to orbit the display. Note that the 3D dynamic rotation can also be activated using the **F4** function key and the left-mouse-button.

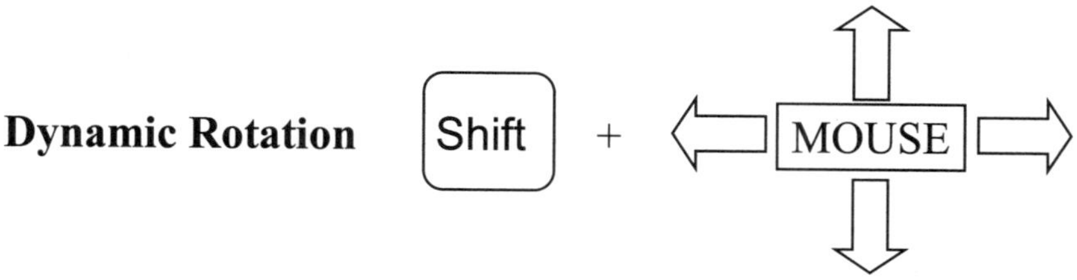

Viewing Tools – Standard Toolbar

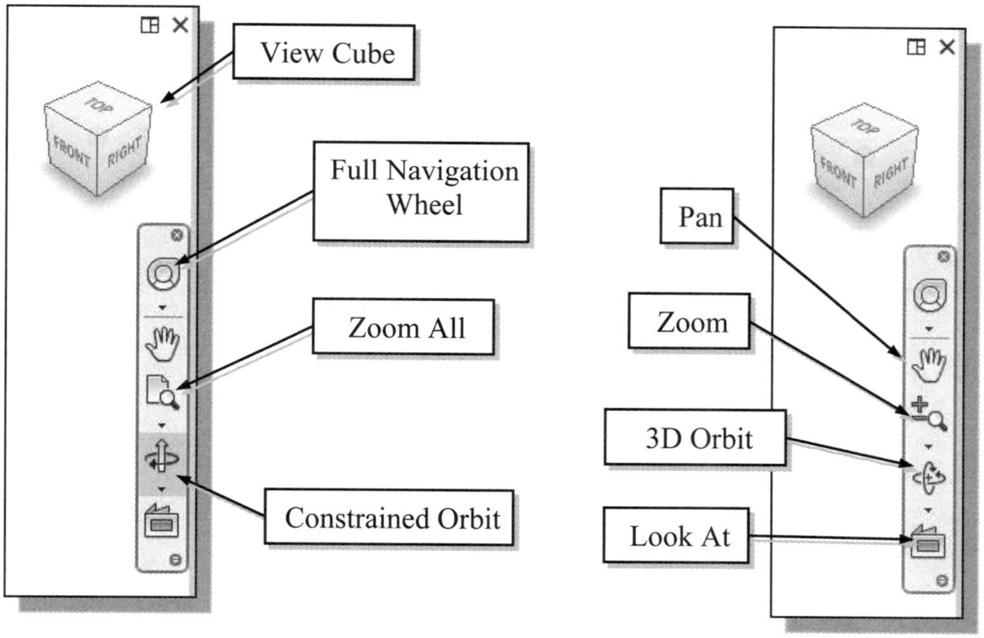

Zoom All – Adjusts the view so that all items on the screen fit inside the graphics window.

Zoom Window – Use the cursor to define a region for the view; the defined region is zoomed to fill the graphics window.

Zoom – Moving upward will reduce the scale of the display, making the entities display smaller on the screen. Moving downward will magnify the scale of the display.

Pan – This allows you to reposition the display while maintaining the same scale factor of the display.

Zoom Selected – In a part or assembly, zooms the selected edge, feature, line, or other element to fill the graphics window. You can select the element either before or after clicking the Zoom button. (Not used in drawings.)

Orbit – In a part or assembly, adds an orbit symbol and cursor to the view. You can orbit the view planar to the screen around the center mark, around a horizontal or vertical axis, or around the X and Y axes. (Not used in drawings.)

Look At – In a part or assembly, zooms and orbits the model to display the selected element planar to the screen or a selected edge or line horizontal to the screen. (Not used in drawings.)

View Cube – The ViewCube is a 3D navigation tool that appears, by default, when you enter Inventor. The ViewCube is a clickable interface which allows you to switch between standard and isometric views.

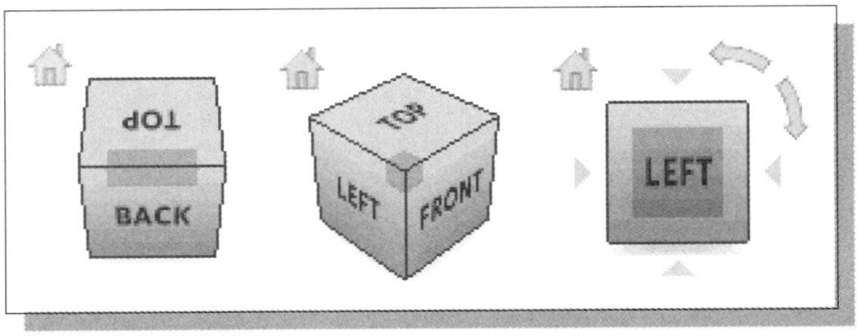

Once the ViewCube is displayed, it is shown in one of the corners of the graphics window over the model in an inactive state. The ViewCube also provides visual feedback about the current viewpoint of the model as view changes occur. When the cursor is positioned over the ViewCube, it becomes active and allows you to switch to one of the available preset views, roll the current view, or change to the Home view of the model.

1. Move the cursor over the ViewCube and notice the different sides of the ViewCube become highlighted and can be activated.

2. Single left-click when the front side is activated as shown. The current view is set to view the front side.

3. Move the cursor over the counterclockwise arrow of the ViewCube and notice the orbit option becomes highlighted.

4. Single left-click to activate the counterclockwise option as shown. The current view is orbited 90 degrees; we are still viewing the front side.

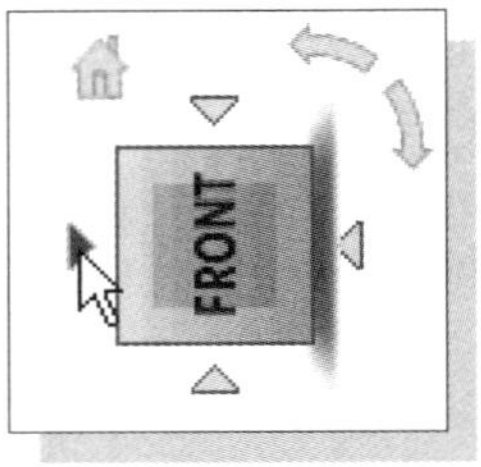

5. Move the cursor over the left arrow of the ViewCube and notice the orbit option becomes highlighted.

6. Single left-click to activate the left arrow option as shown. The current view is now set to view the top side.

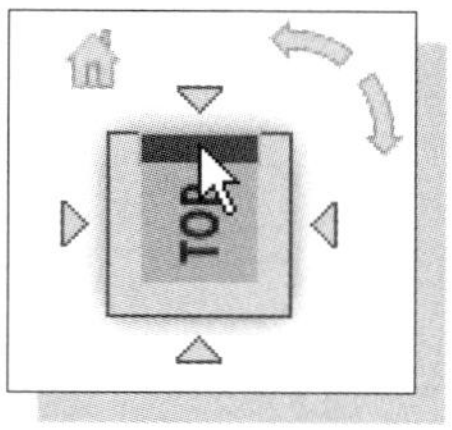

7. Move the cursor over the top edge of the ViewCube and notice the roll option becomes highlighted.

8. Single left-click to activate the roll option as shown. The view will be adjusted to roll 45 degrees.

9. Move the cursor over the ViewCube and drag with the left-mouse-button to activate the **Free Rotation** option.

10. Move the cursor over the home icon of the ViewCube and notice the Home View option becomes highlighted.

11. Single left-click to activate the **Home View** option as shown. The view will be adjusted back to the default ***isometric view***.

Full Navigation Wheel – The Navigation Wheel contains tracking menus that are divided into different sections known as wedges. Each wedge on a wheel represents a single navigation tool. You can pan, zoom, or manipulate the current view of a model in different ways. The 3D Navigation Wheel and 2D Navigation Wheel (mostly used in the 2D drawing mode) have some or all of the following options:

Zoom – Adjusts the magnification of the view.
Center – Centers the view based on the position of the cursor over the wheel.
Rewind – Restores the previous view.
Forward – Increases the magnification of the view.
Orbit – Allows 3D free rotation with the left-mouse-button.
Pan – Allows panning by dragging with the left-mouse-button.
Up/Down – Allows panning with the use of a scroll control.
Walk – Allows *walking*, with linear motion perpendicular to the screen, through the model space.
Look – Allows rotation of the current view vertically and horizontally.

3D Full Navigation Wheel

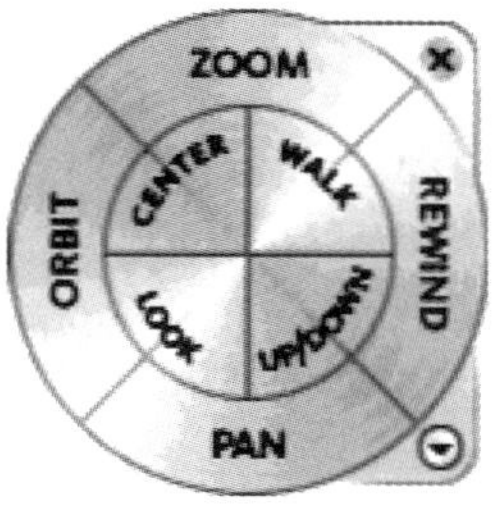

2D Full Navigation Wheel

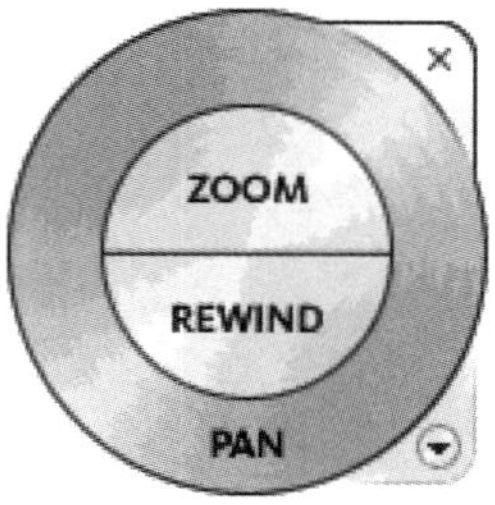

1. Activate the **Full Navigation Wheel** by clicking on the icon as shown.

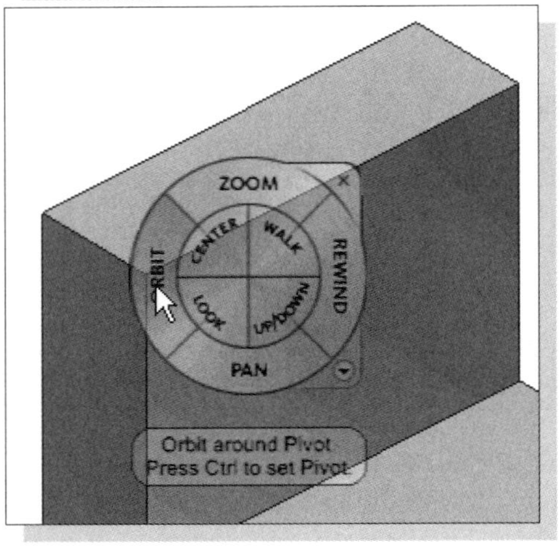

2. Move the cursor in the graphics window and notice the **Full Navigation Wheel** menu follows the cursor on the screen.

3. Move the cursor on the **Orbit** option to highlight the option.

4. Click and drag with the left-mouse-button to activate the **Free Rotation** option.

5. Drag with the left-mouse-button and notice the **ViewCube** also reflects the model orientation.

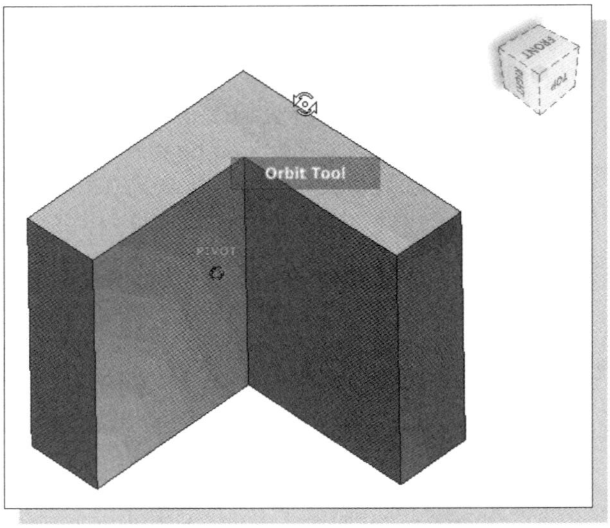

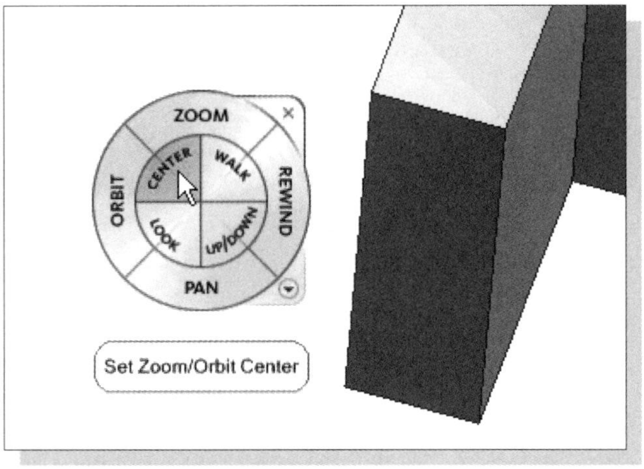

6. Move the cursor to the left side of the model and click the **Center** option as shown. The display is adjusted so the selected point is the new **Zoom/Orbit** center.

7. On your own, experiment with the other available options.

Display Modes

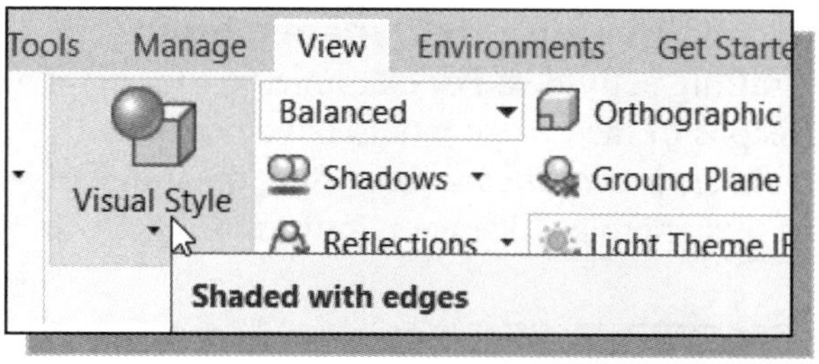

- The **Visual Style** in the *View* tab has eleven display-modes ranging from very realistic renderings of the model to very artistic representations of the model. The more commonly used display modes are as follows:

❖ **Realistic Shaded Solid:**

The *Realistic Shaded Solid* display mode generates a high-quality shaded image of the 3D object.

❖ **Standard Shaded Solid:**

The *Standard Shaded Solid* display option generates a shaded image of the 3D object that requires fewer computer resources compared to the realistic rendering.

❖ **Wireframe Image:**

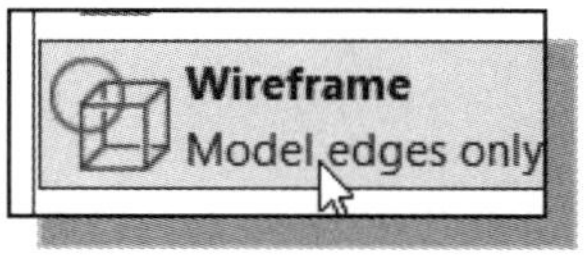

The *Wireframe Image* display option allows the display of the 3D objects using the basic wireframe representation scheme.

❖ **Wireframe with Hidden-Edges:**

The *Wireframe with Hidden-Edges* option can be used to generate an image of the 3D object with all the back lines shown as hidden-lines.

Orthographic vs. Perspective

Besides the above basic display modes, we can also choose orthographic view or perspective view of the display. Click on the triangle icon next to the *display mode button* on the *View* toolbar.

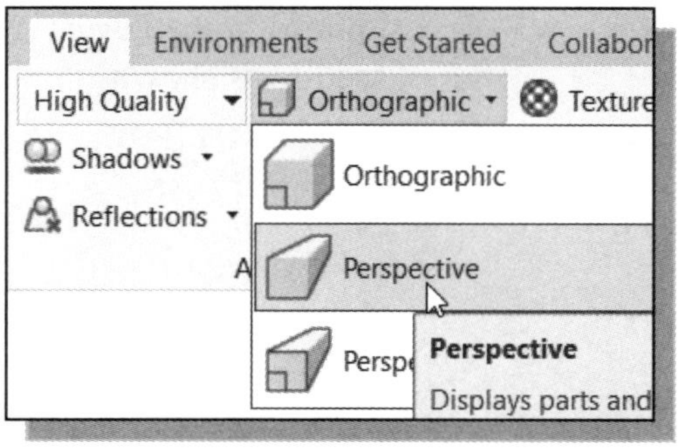

❖ **Orthographic**
The first icon allows the display of the 3D object using the parallel edges representation scheme.

❖ **Perspective**
The second icon allows the display of the 3D object using the perspective, nonparallel edges, and representation scheme.

Disable the Heads-Up Display Option

The **Heads-Up Display** option in Inventor provides mainly the **Dynamic Input** function, which can be quite useful for 2D drafting activities. For example, in the use of a 2D drafting CAD system, most of the dimensions of the design would have been determined by the documentation stage. However, in *parametric modeling*, the usage of the *Dynamic Input* option is quite limited, as this approach does not conform to the "**shape before size**" design philosophy.

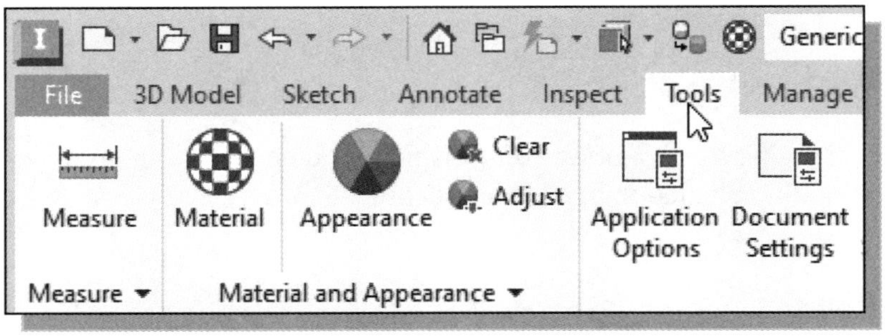

1. Select the **Tools** tab in the *Ribbon* as shown.

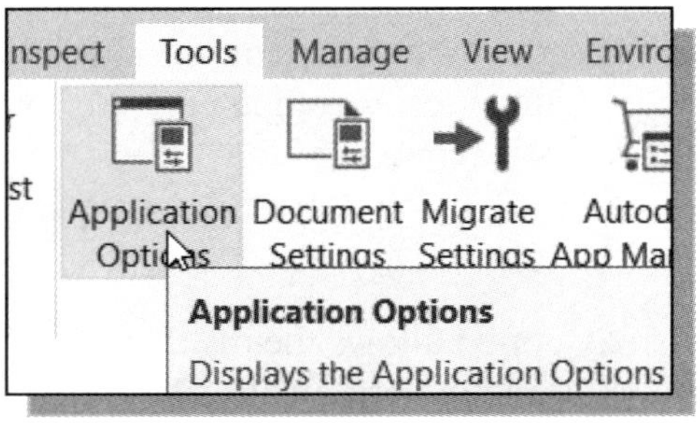

2. Select **Application Options** in the options toolbar as shown.

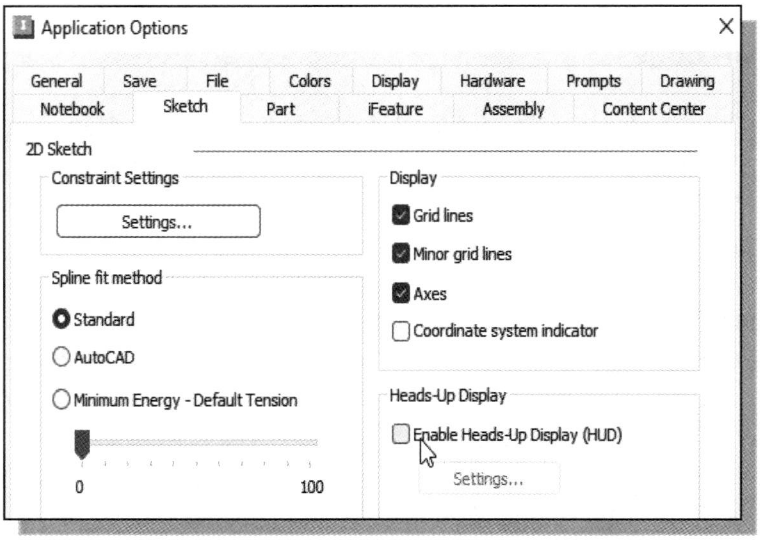

3. Select the **Sketch** tab to display the sketch related settings.

4. In the *Heads-Up Display* section, turn **OFF** the **Enable Heads-Up Display**, and switch **On** the **Grid lines**, **Minor grid lines** and **Axes** options in the *Display* section as shown.

5. On your own, examine the other sketch settings that are available.

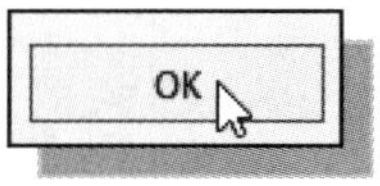

6. Click **OK** to accept the settings.

Step 4-1: Adding an Extruded Feature

1. Activate the *3D Model* tab and select the **Start 2D Sketch** command by left-clicking once on the icon.

2. In the *Status Bar* area, the message "*Select plane to create sketch or an existing sketch to edit*" is displayed. Autodesk Inventor expects us to identify a planar surface where the 2D sketch of the next feature is to be created. Notice that Autodesk Inventor will automatically highlight feasible planes and surfaces as the cursor is on top of the different surfaces. Pick the top horizontal face of the 3D solid object.

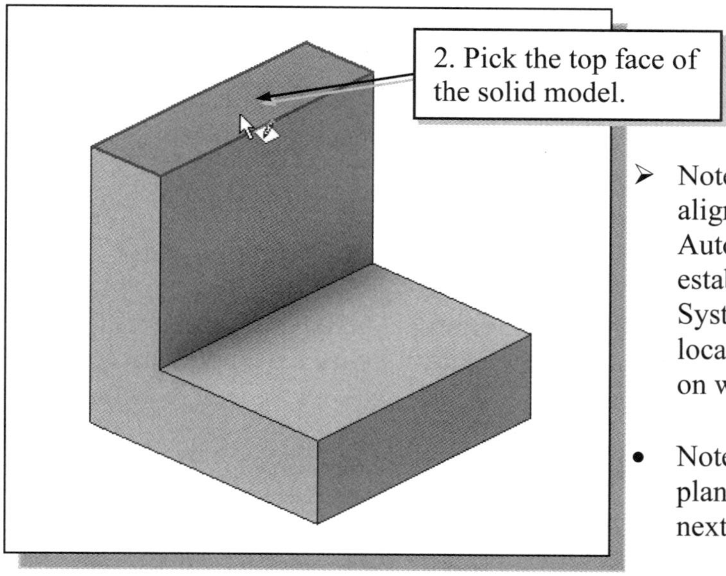

- ➢ Note that the sketch plane is aligned to the selected face. Autodesk Inventor automatically establishes a User-Coordinate-System (UCS) and records its location with respect to the part on which it was created.

- Note the orientation of the sketch plane should be as shown on the next page, adjust if needed.

- Next, we will create and profile another sketch, a rectangle, which will be used to create another extrusion feature that will be added to the existing solid object.

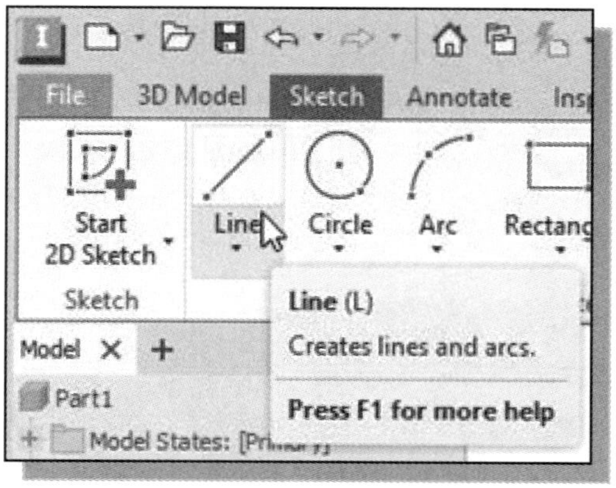

3. Select the **Line** command by clicking once with the **left-mouse-button** on the icon in the *Sketch* tab on the Ribbon.

4. Create a sketch with segments perpendicular/parallel to the existing edges of the solid model as shown below.

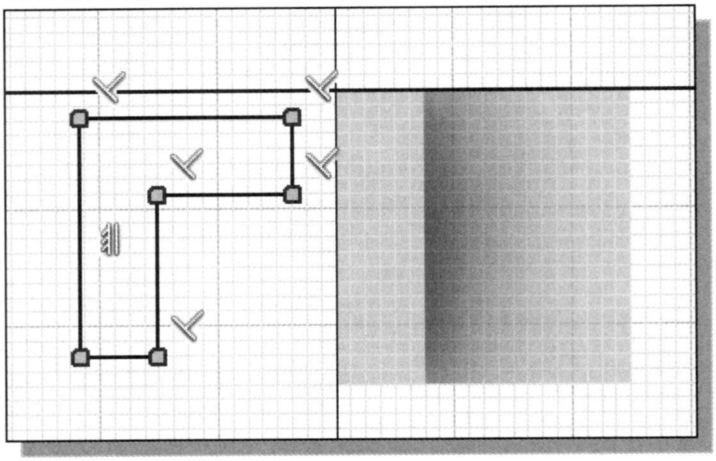

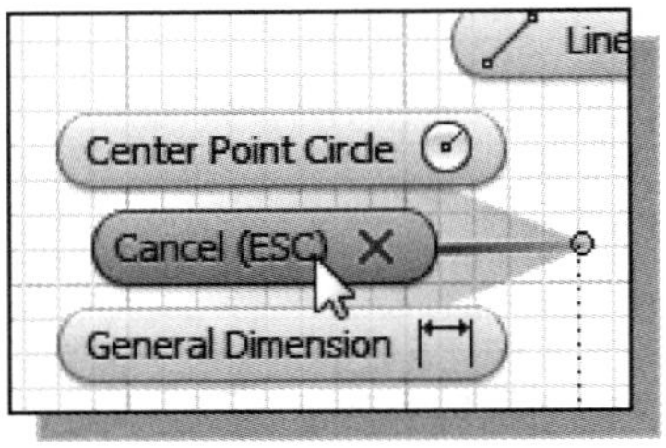

5. Inside the graphics window, click once with the **right-mouse-button** to display the option menu. Select **Cancel (ESC)** in the pop-up menu to end the Line command.

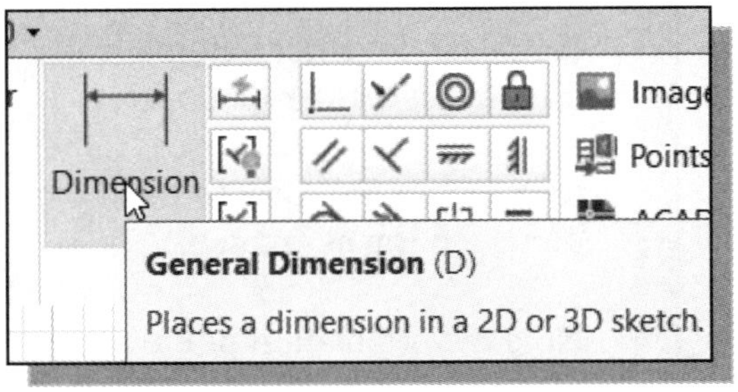

6. Select the **General Dimension** command in the *Sketch* toolbar. The General Dimension command allows us to quickly create and modify dimensions. Left-click once on the icon to activate the General Dimension command.

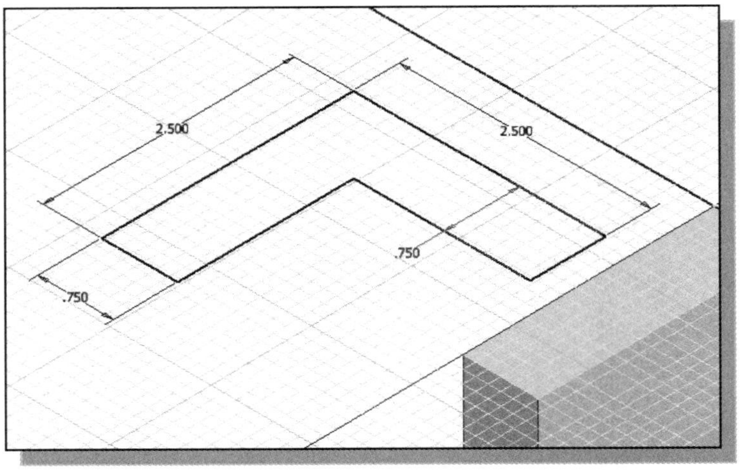

7. Hit the function key **[F6]** once to switch the display to the isometric display as shown.

8. Create the **four dimensions** to describe the size of the sketch as shown in the figure.

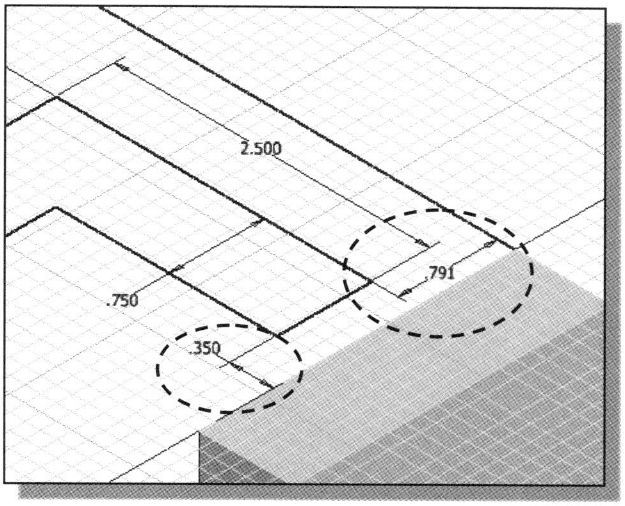

9. Create the two location dimensions to describe the position of the sketch relative to the top corner of the solid model as shown.

10. On your own, modify the two location dimensions to **0.0** and adjust the size dimensions as shown in the figure below.

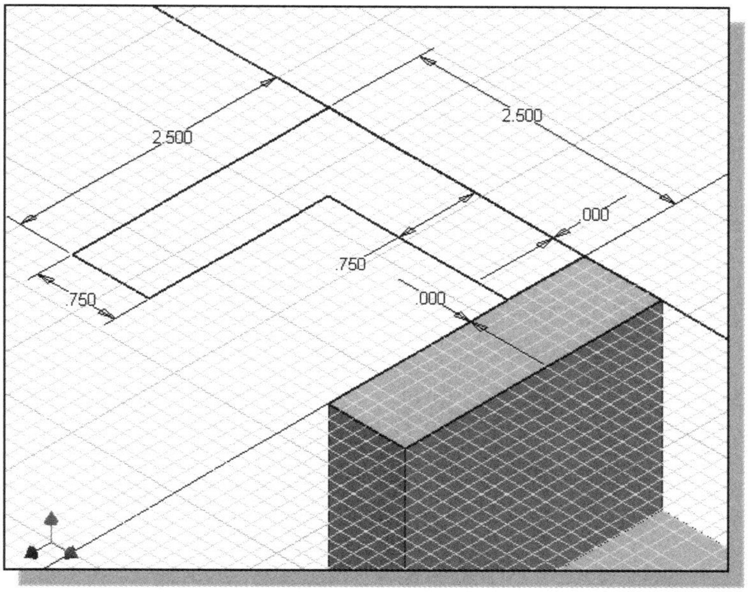

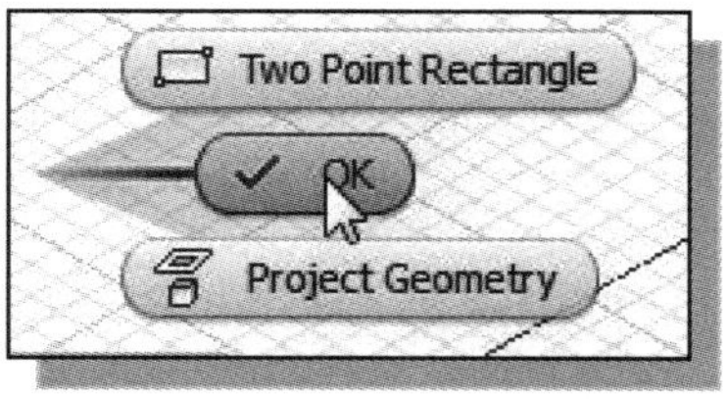

11. Inside the graphics window, click once with the **right-mouse-button** to display the option menu. Select **OK** in the pop-up menu to end the General Dimension command.

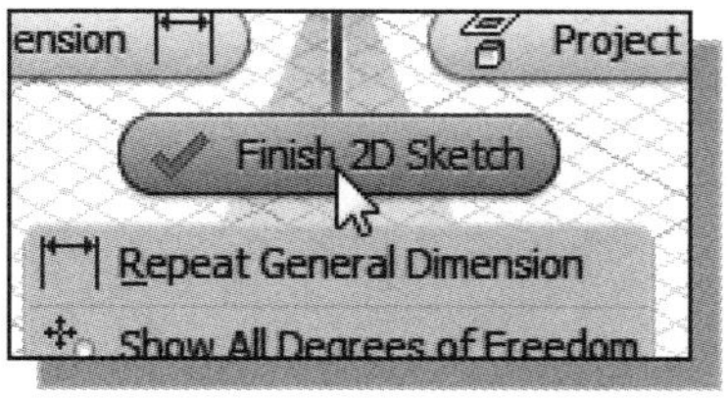

12. Inside the graphics window, click once with the **right-mouse-button** to display the option menu. Select **Finish 2D Sketch** in the pop-up menu to end the Sketch option.

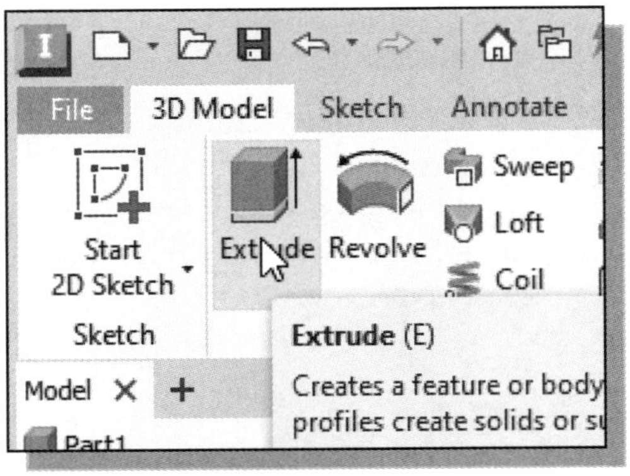

13. In the *3D Model* tab select the **Extrude** command by left-clicking on the icon.

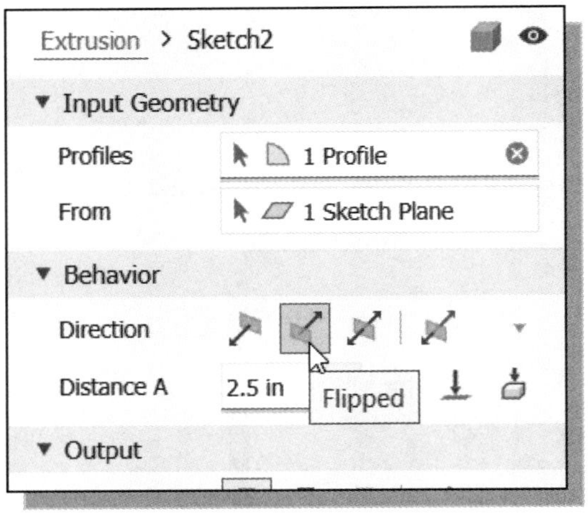

14. In the *Extrude* dialog box, enter **2.5** as the extrude distance as shown.

15. Click on the **Flipped icon** to reverse the *Extrusion Direction* as shown.

16. Confirm the *Boolean* option is set to **Join** and click on the **OK** button to proceed with creating the extruded feature.

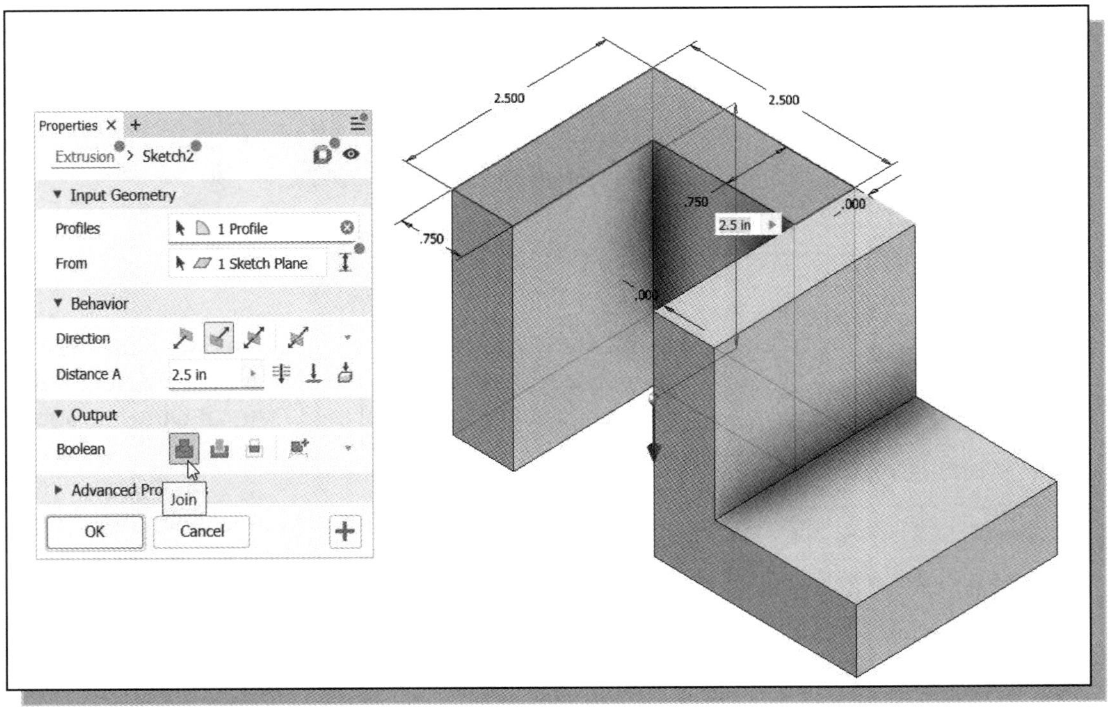

Step 4-2: Adding a Cut Feature

Next, we will create and profile a circle, which will be used to create a cut feature that will be added to the existing solid object.

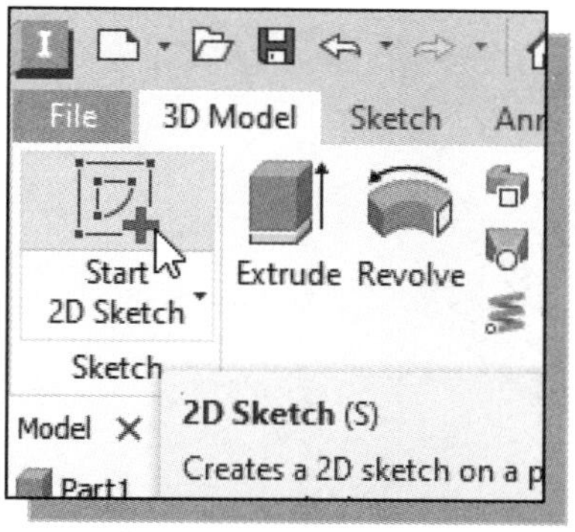

1. In the *3D Model* tab select the **Start 2D Sketch** command by left-clicking once on the icon.

2. In the *Status Bar* area, the message "*Select plane to create sketch or an existing sketch to edit.*" is displayed. Autodesk Inventor expects us to identify a planar surface where the 2D sketch of the next feature is to be created. Pick the top horizontal face of the 3D solid model as shown.

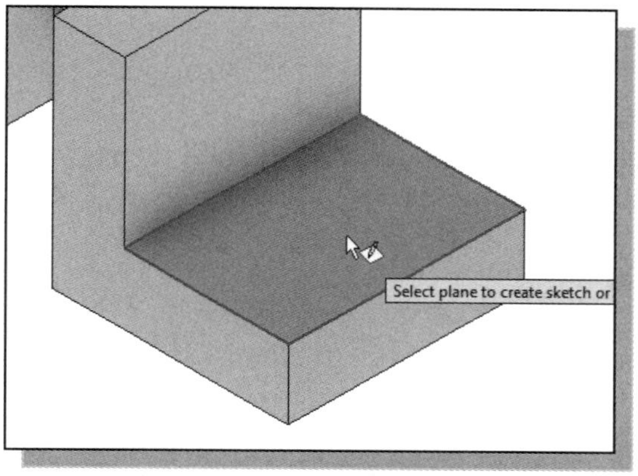

➢ Note that the sketch plane is aligned to the selected face. Autodesk Inventor automatically establishes a User-Coordinate-System (UCS) and records its location with respect to the part on which it was created.

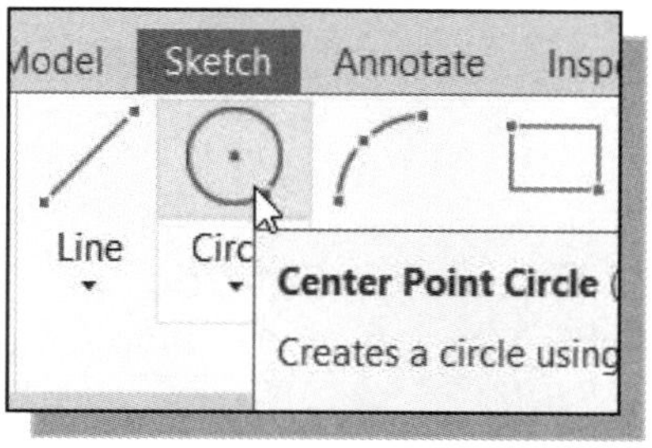

3. Select the **Center point circle** command by clicking once with the **left-mouse-button** on the icon in the *Sketch* tab on the Ribbon.

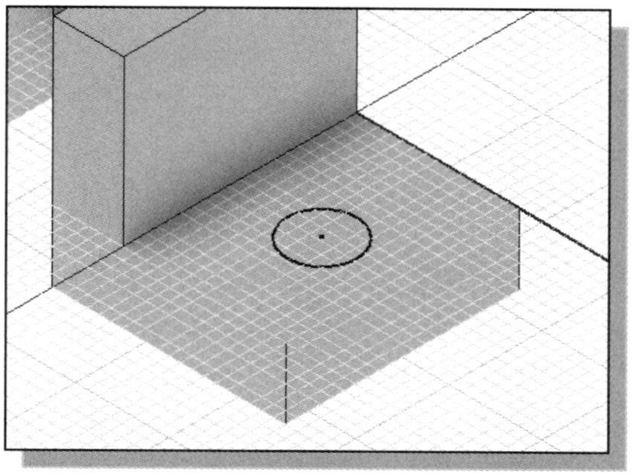

4. Create a circle of arbitrary size on the top face of the solid model as shown.

5. On your own, create and modify the dimensions of the sketch as shown in the figure.

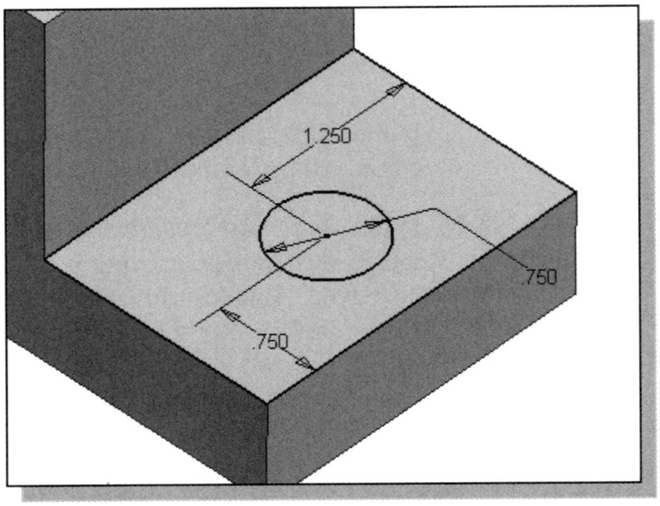

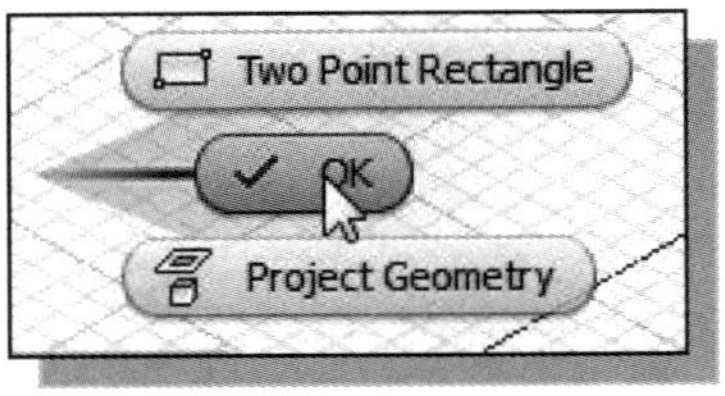

6. Inside the graphics window, click once with the **right-mouse-button** to display the option menu. Select **OK** in the pop-up menu to end the General Dimension command.

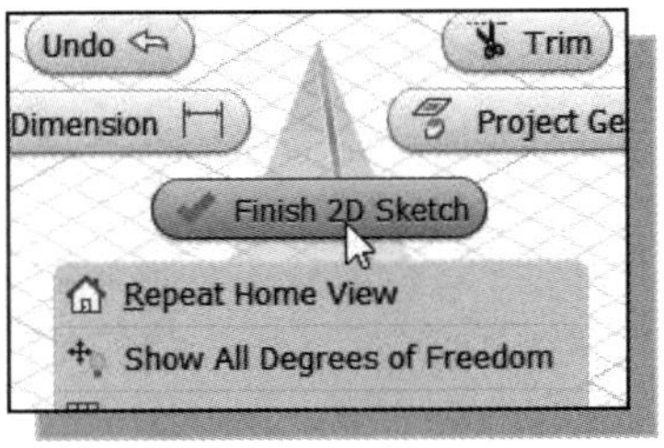

7. Inside the graphics window, click once with the **right-mouse-button** to display the option menu. Select **Finish 2D Sketch** in the pop-up menu to end the Sketch option.

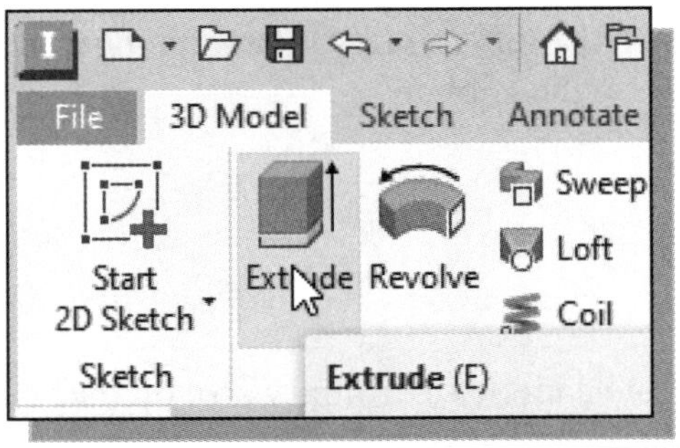

8. In the *3D Model* tab select the **Extrude** command by clicking the left-mouse-button on the icon.

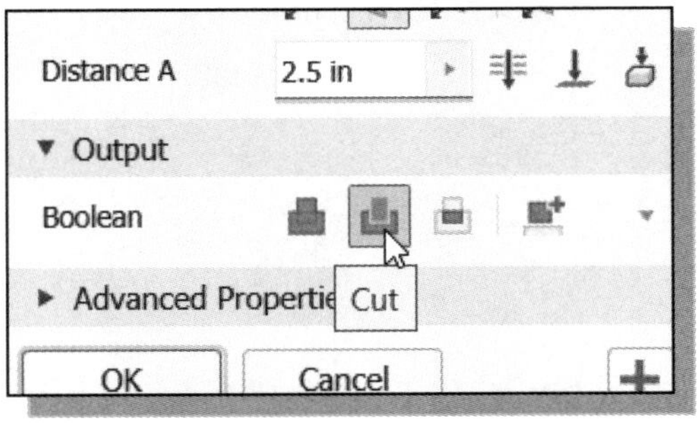

9. Select the **CUT** option in the *Output - Boolean option* list to set the extrusion operation to ***Cut***.

10. Set the *Extents* option to **Through All** as shown. The *All* option instructs the software to calculate the extrusion distance and assures the created feature will always cut through the full length of the model.

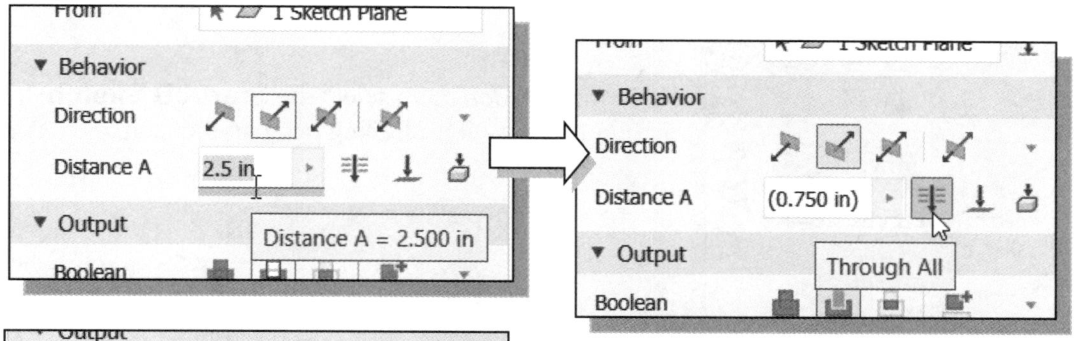

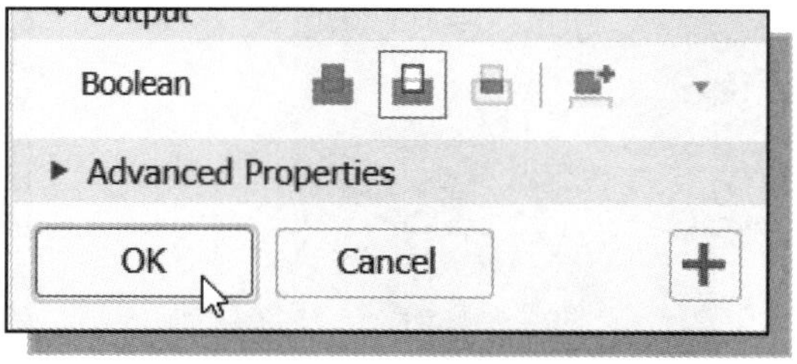

11. Click on the **OK** button to proceed with creating the extruded feature.

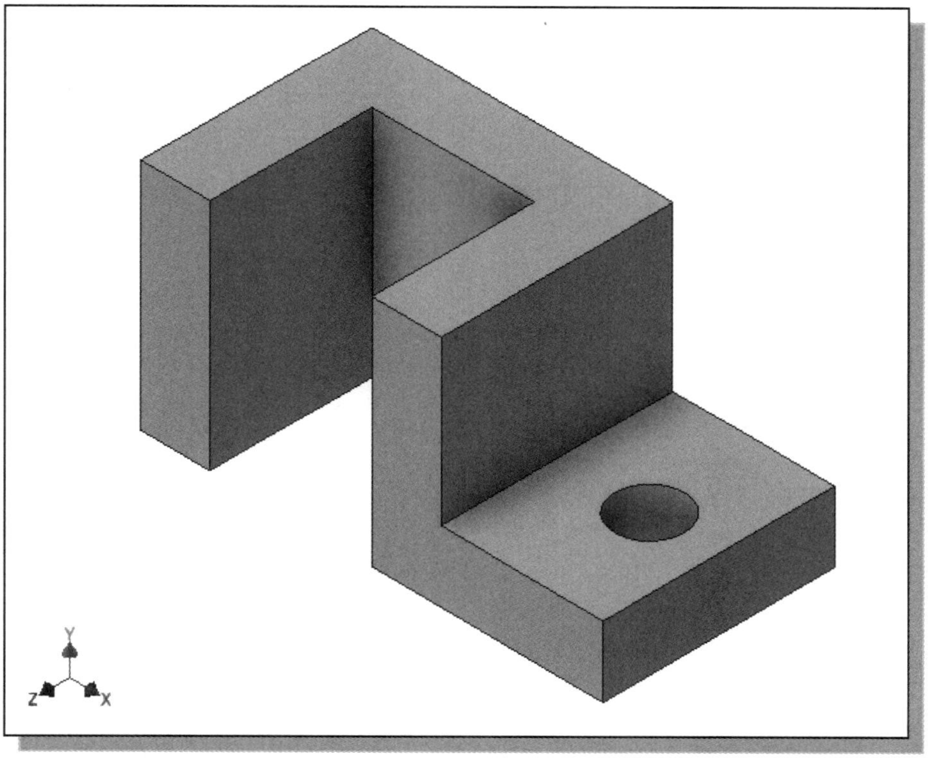

- In *Autodesk Inventor*, the **Extrude** command can be used to create solid features by either adding or subtracting extruded features.

Step 4-3: Adding another Cut Feature

Next, we will create and profile a triangle, which will be used to create a cut feature that will be added to the existing solid object.

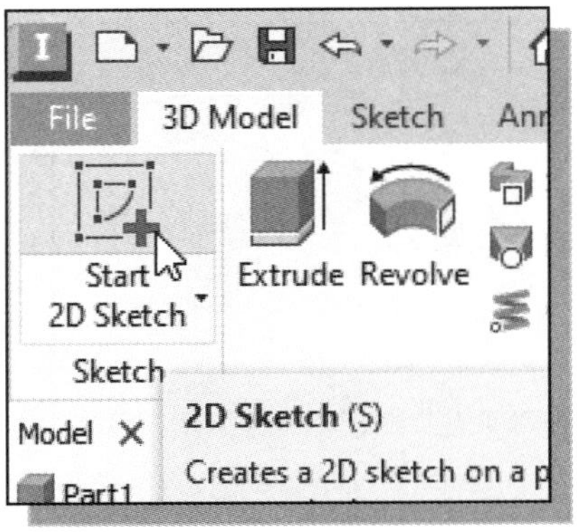

1. In the *3D Model* tab select the **Start 2D Sketch** command by left-clicking once on the icon.

2. In the *Status Bar* area, the message "*Select plane to create sketch or an existing sketch to edit.*" is displayed. Autodesk Inventor expects us to identify a planar surface where the 2D sketch of the next feature is to be created. Pick the vertical face of the 3D solid model next to the horizontal section as shown.

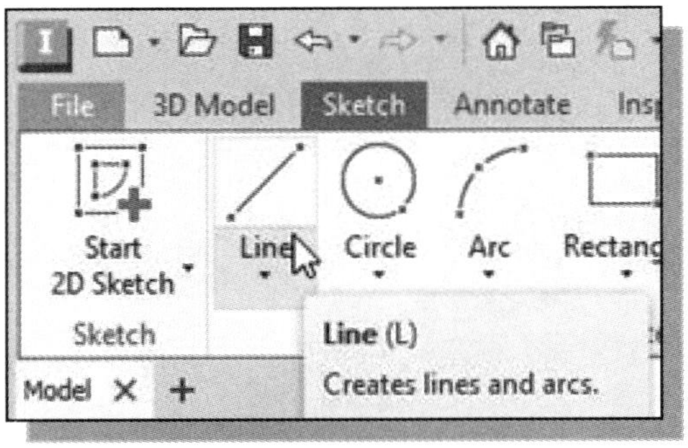

3. Select the **Line** command by clicking once with the **left-mouse-button** on the icon in the *Sketch* ribbon.

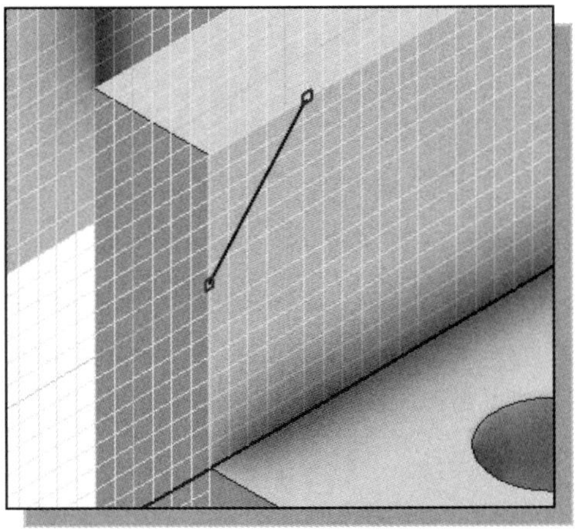

4. Create an **inclined line segment** near the upper left corner of the model as shown. (Do not align the endpoints of the new line to the midpoint of the existing edges.)

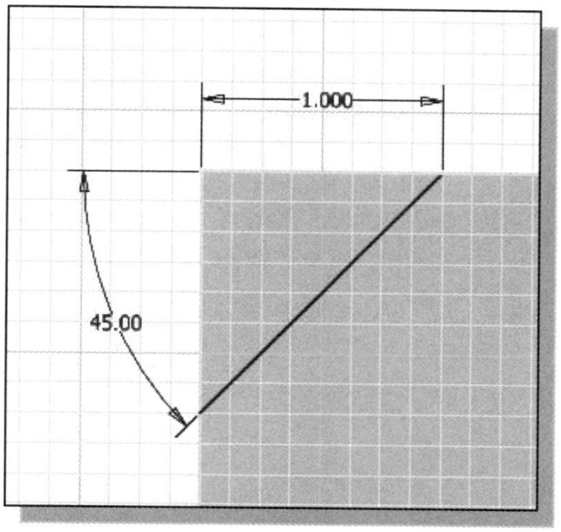

5. On your own, create and modify the two dimensions of the sketch as shown in the figure. (Hint: create the angle dimension by selecting the two adjacent lines and place the angular dimension inside the desired quadrant.)

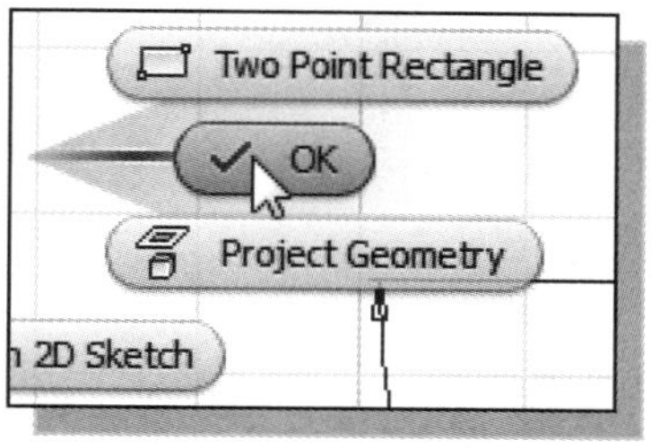

6. Inside the graphics window, click once with the **right-mouse-button** to display the option menu. Select **OK** in the pop-up menu to end the General Dimension command.

7. Select **Finish 2D Sketch** in the *Ribbon toolbar* to end the Sketch option.

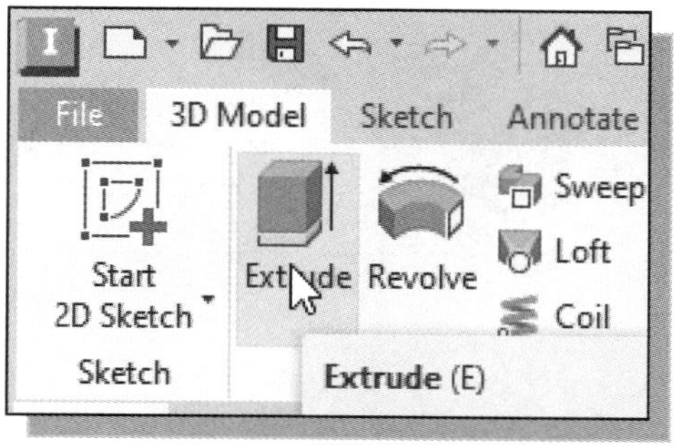

8. In the *3D Model* tab, select the **Extrude** command by left-clicking the icon.

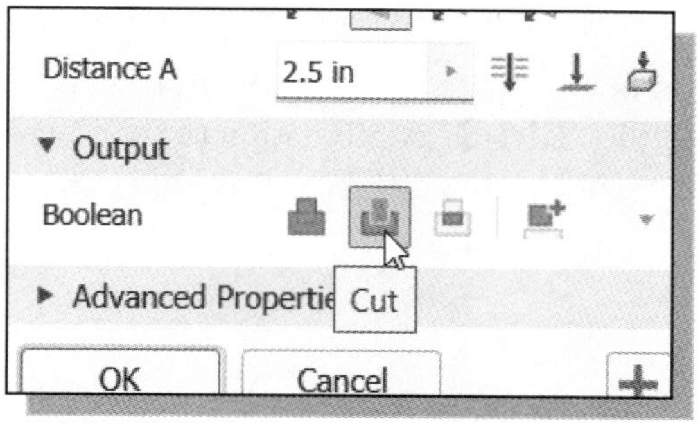

9. Select the **CUT** option in the *Boolean option* list to set the extrusion operation to ***Cut***.

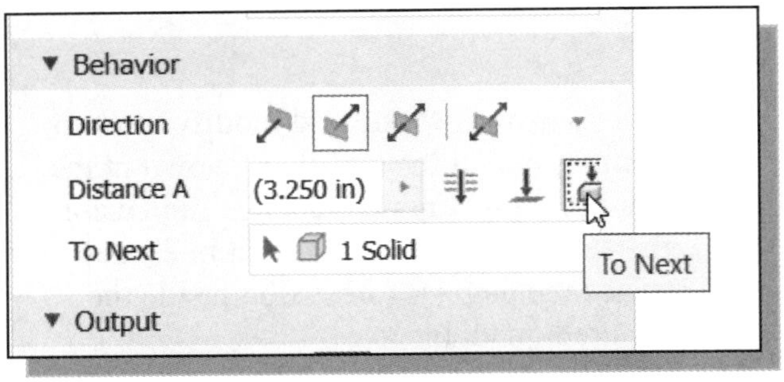

10. Set the *Extents* option to **To Next** as shown. The *To Next* option instructs the software to calculate the extrusion distance and assures the created feature will always cut through the proper length of the model.

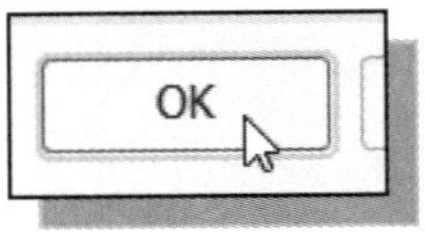

11. Click on the **OK** button to proceed with creating the extruded feature.

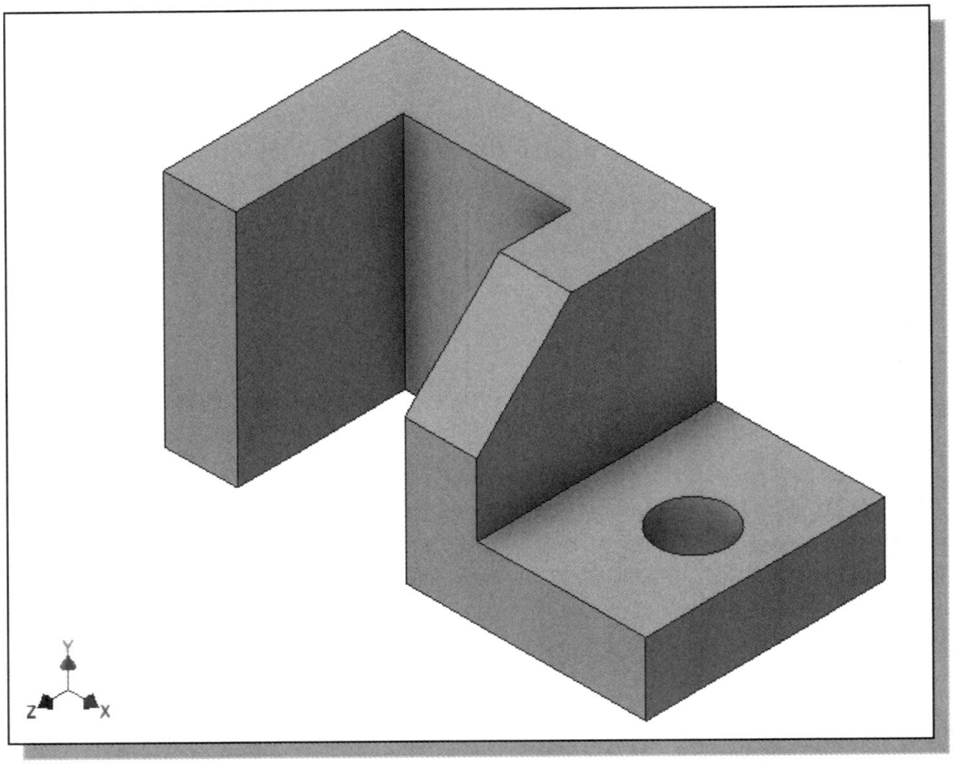

Save the Model

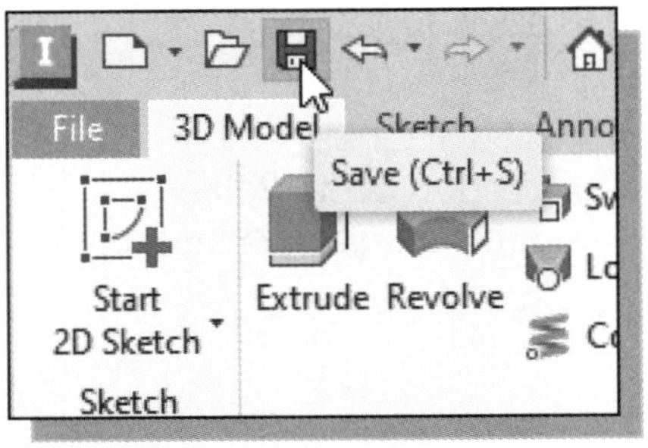

1. Select **Save** in the *Quick Access* toolbar, or you can also use the "**Ctrl-S**" combination (hold down the "Ctrl" key and hit the "S" key once) to save the part.

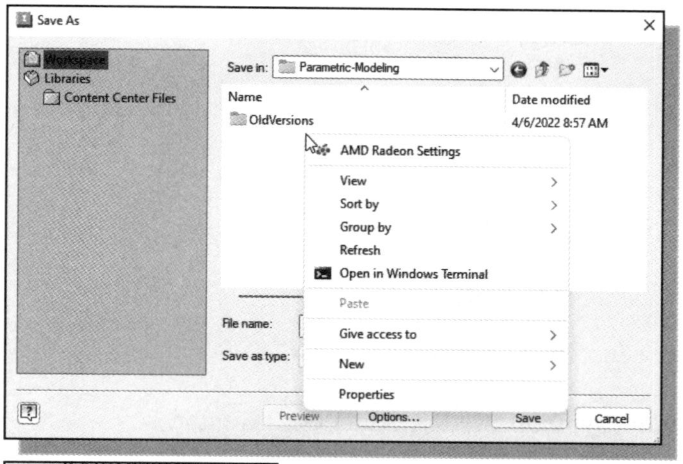

2. In the *Save As* dialog box, **right-click** once in the *list area* to bring up the *option menu*.

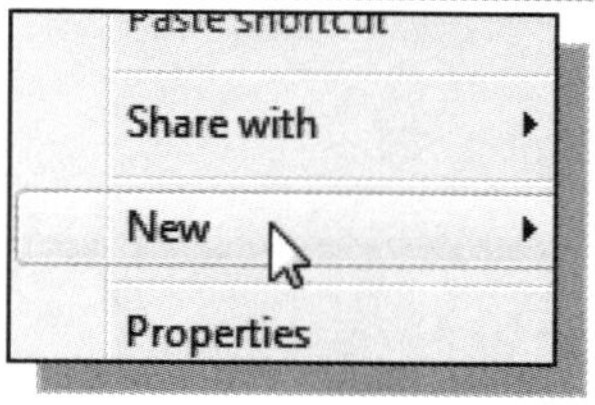

3. In the *option list*, select **New** as shown.

4. In the second *option list*, select **Folder** to create a subfolder.

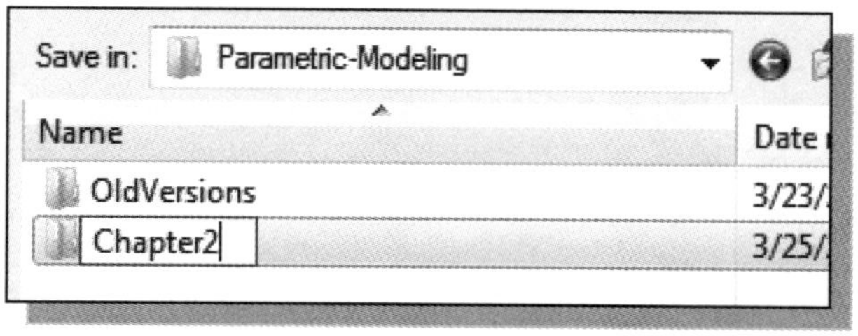

5. Enter **Chapter2** as the new folder name as shown.

6. **Double-click** on the Chapter2 folder to open it.

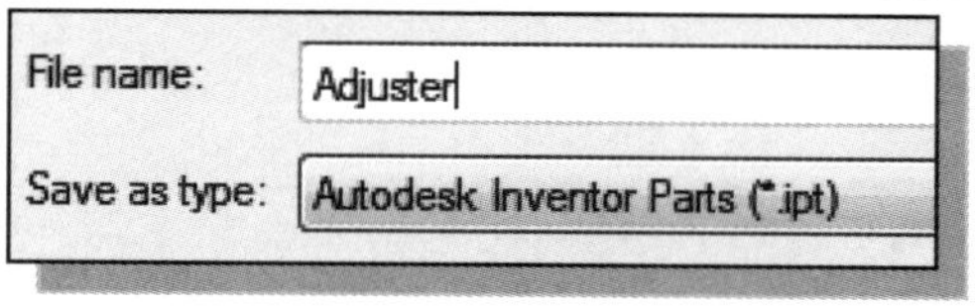

7. In the *file name* editor box, enter **Adjuster** as the file name.

8. Click on the **Save** button to save the file.

❖ You should form a habit of saving your work periodically, just in case something goes wrong while you are working on it. In general, you should save your work at an interval of every 15 to 20 minutes. You should also save before making any major modifications to the model.

Review Questions: (Time: 20 minutes)

1. What is the first thing we should set up in Autodesk Inventor when creating a new model?
2. Describe the general *parametric modeling* procedure.
3. Describe the general guidelines in creating *Rough Sketches*.
4. What is the main difference between a rough sketch and a *profile*?
5. List two of the geometric constraint symbols used by Autodesk Inventor.
6. What was the first feature we created in this lesson?
7. How many solid features were created in the tutorial?
8. How do we control the size of a feature in parametric modeling?
9. Which command was used to create the last cut feature in the tutorial? How many dimensions do we need to fully describe the cut feature?
10. List and describe three differences between parametric modeling and traditional 2D Computer Aided Drafting techniques.

Exercises: Create and save the exercises in the Chapter2 folder.
(Time: 150 minutes. All dimensions are in inches.)

1. **Inclined Support** (Thickness: **.5**)

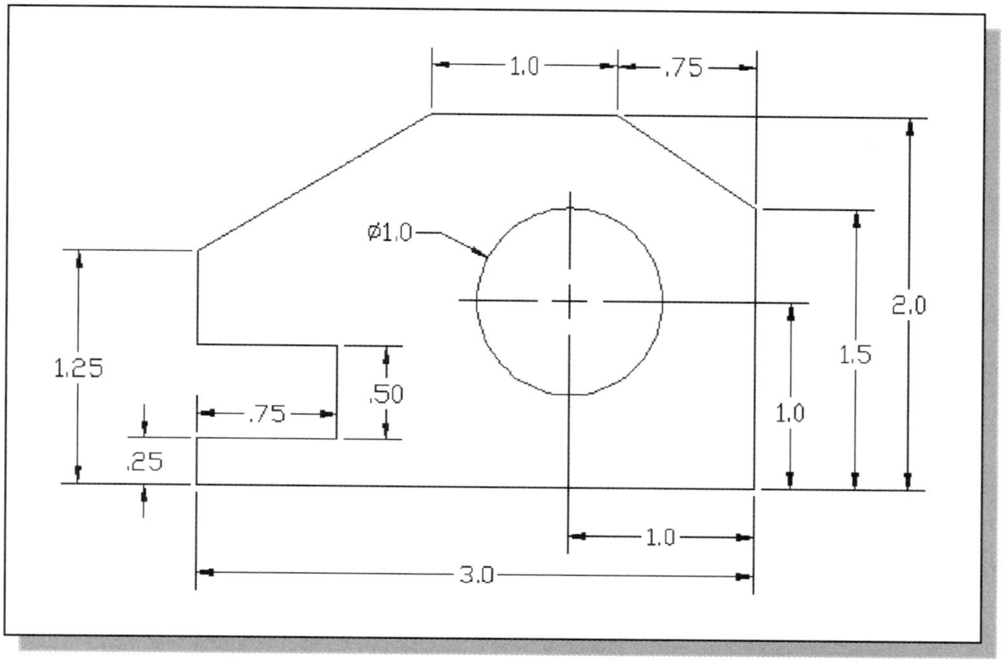

2. **Spacer Plate** (Thickness: **.125**)

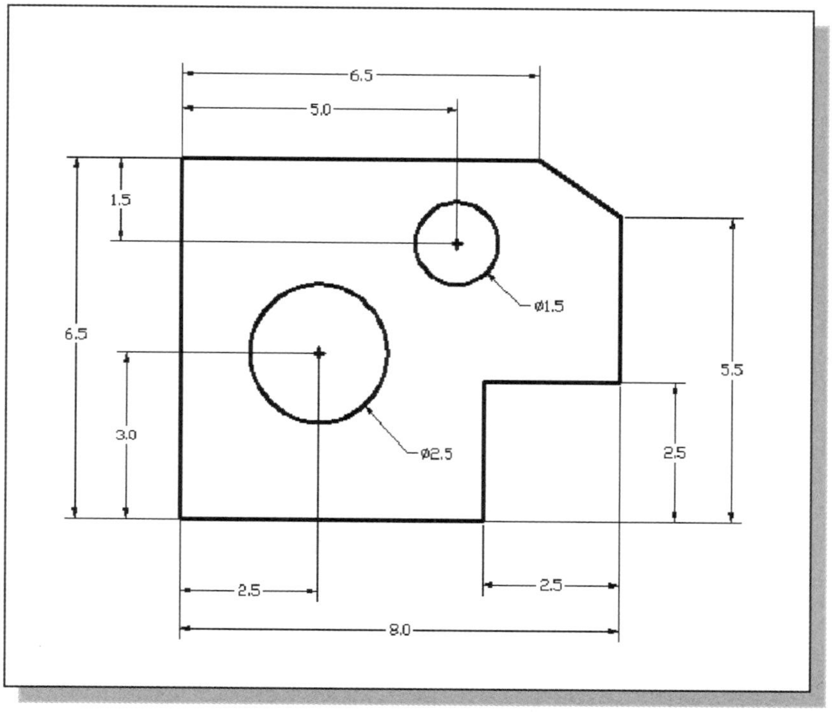

3. **Positioning Stop**

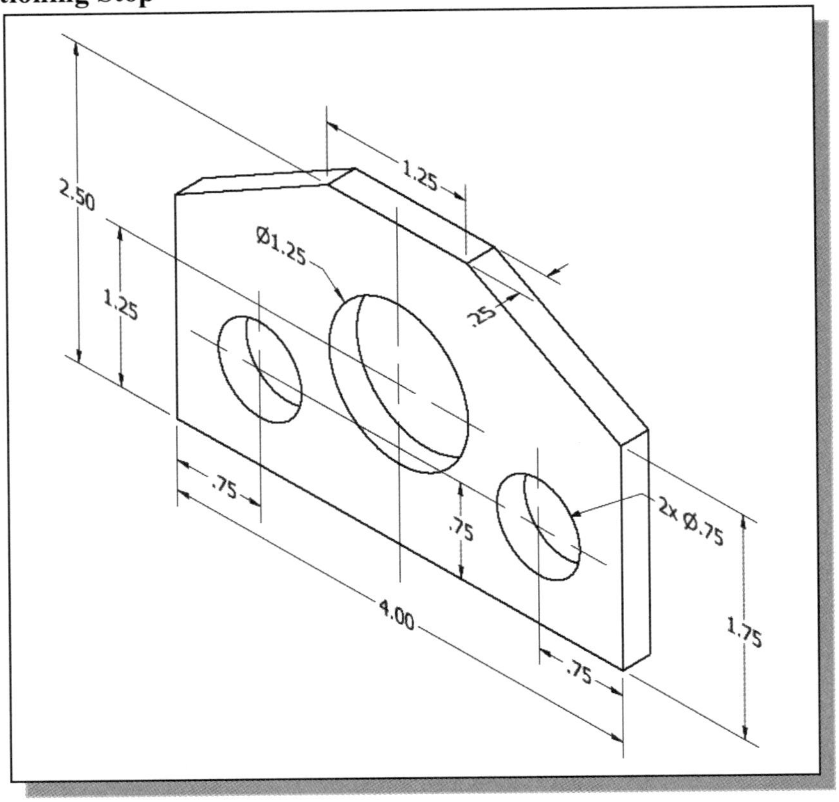

4. **Guide Block**

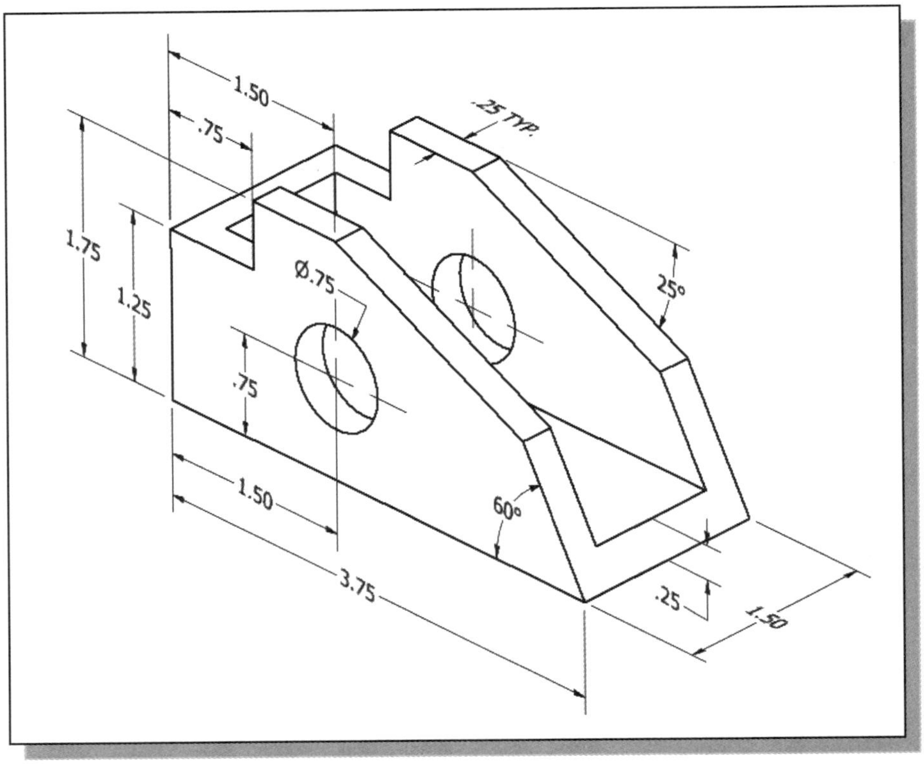

5. **Slider Block**

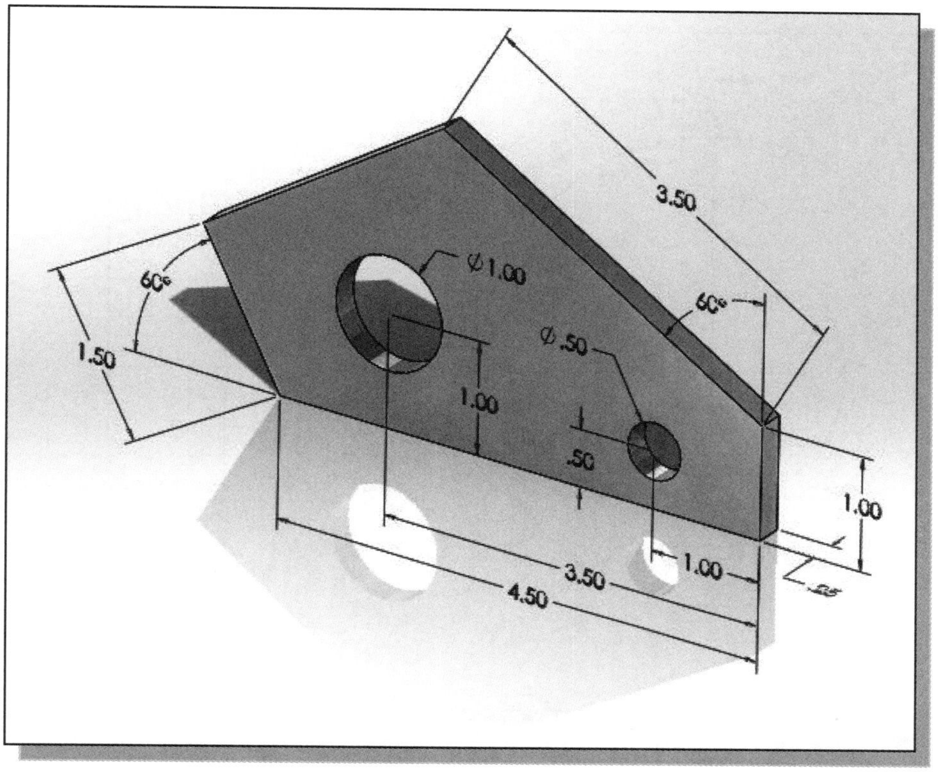

6. **Angle Lock**

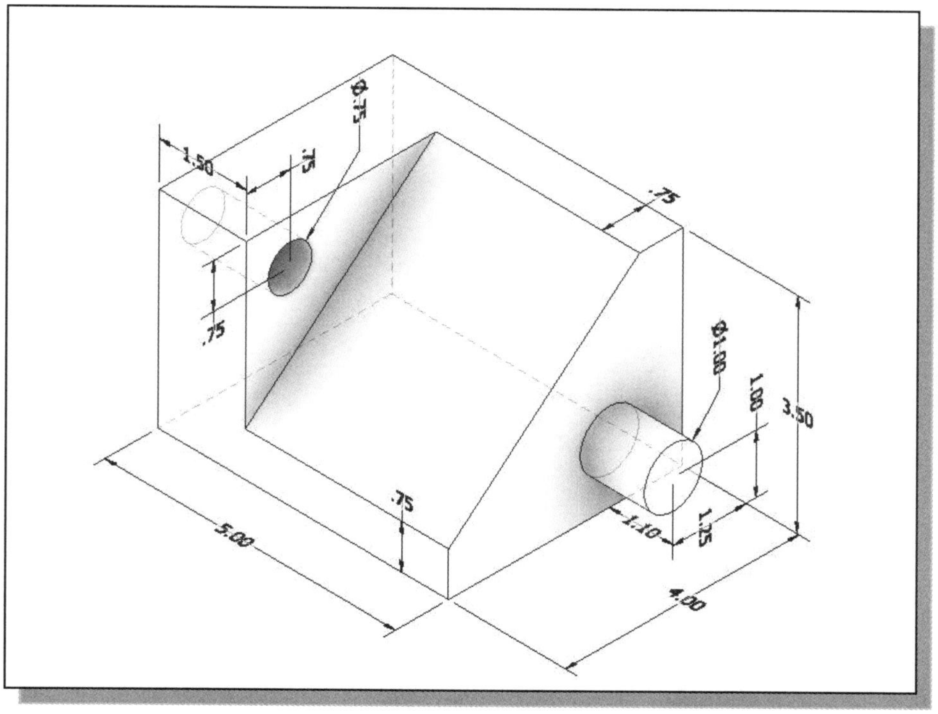

7. **Coupler**

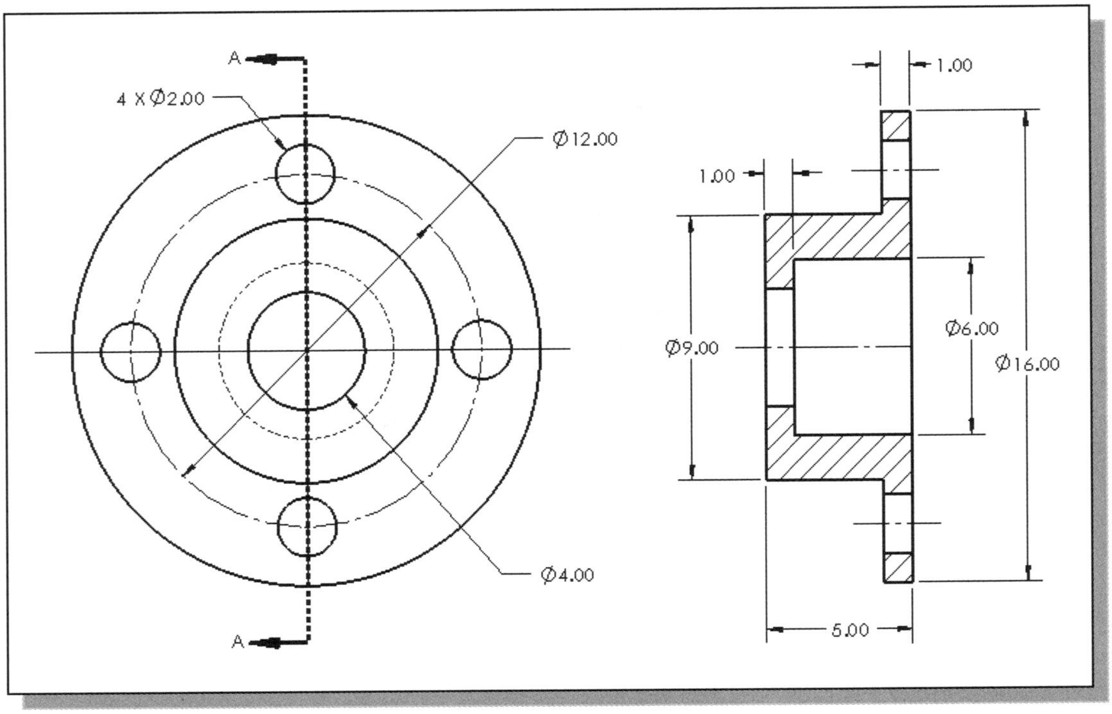

Chapter 3
Constructive Solid Geometry Concepts

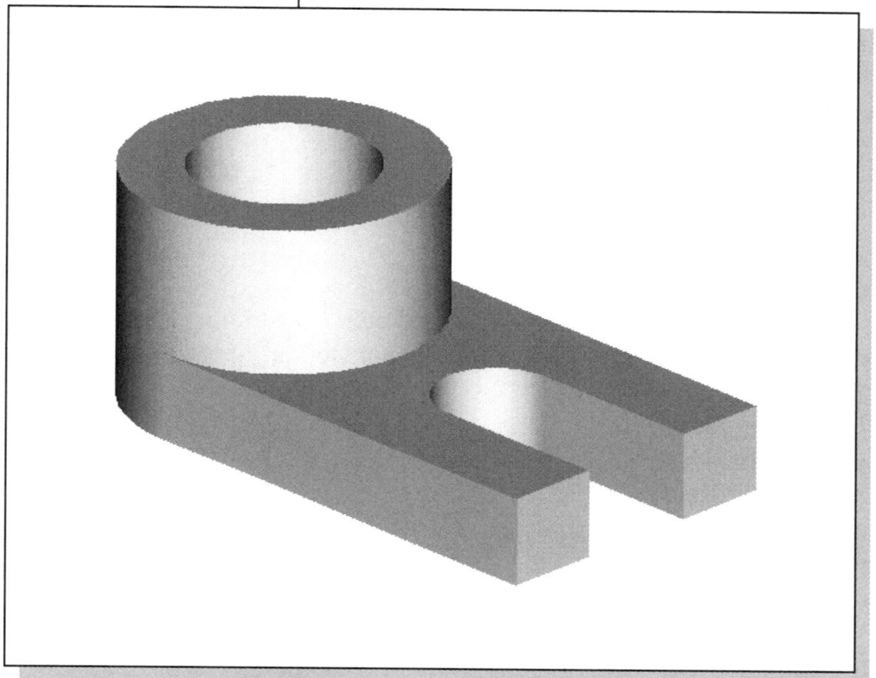

Learning Objectives

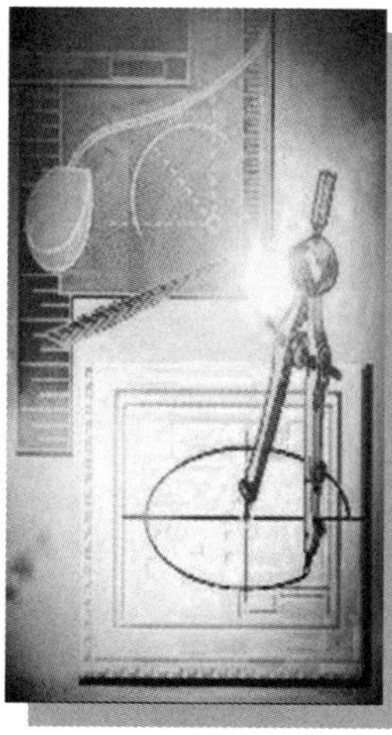

- Understand Constructive Solid Geometry Concepts
- Create a Binary Tree
- Understand the Basic Boolean Operations
- Set up GRID and SNAP Intervals
- Understand the Importance of Order of Features
- Create Placed Features
- Use the Different Extrusion Options

Autodesk Inventor Certified User Exam Objectives Coverage

Parametric Modeling Basics:

Section 3: Sketches

Objectives: Creating 2D Sketches, Draw Tools, Sketch Constraints, Pattern Sketches, Modify Sketches, Format Sketches, Sketch Doctor, Shared Sketches, Sketch Parameters.

Section 4: Parts

Objectives: Creating parts, Work Features, Pattern Features, Part Properties.

Introduction

In the 1980s, one of the main advancements in **solid modeling** was the development of the **Constructive Solid Geometry** (CSG) method. CSG describes the solid model as combinations of basic three-dimensional shapes (**primitive solids**). The basic primitive solid set typically includes Rectangular-prism (Block), Cylinder, Cone, Sphere, and Torus (Tube). Two solid objects can be combined into one object in various ways using operations known as **Boolean operations**. There are three basic Boolean operations: **JOIN (Union)**, **CUT (Difference)**, and **INTERSECT**. The **JOIN** operation combines the two volumes included in the different solids into a single solid. The **CUT** operation subtracts the volume of one solid object from the other solid object. The **INTERSECT** operation keeps only the volume common to both solid objects. The CSG method is also known as the **Machinist's Approach**, as the method is parallel to machine shop practices.

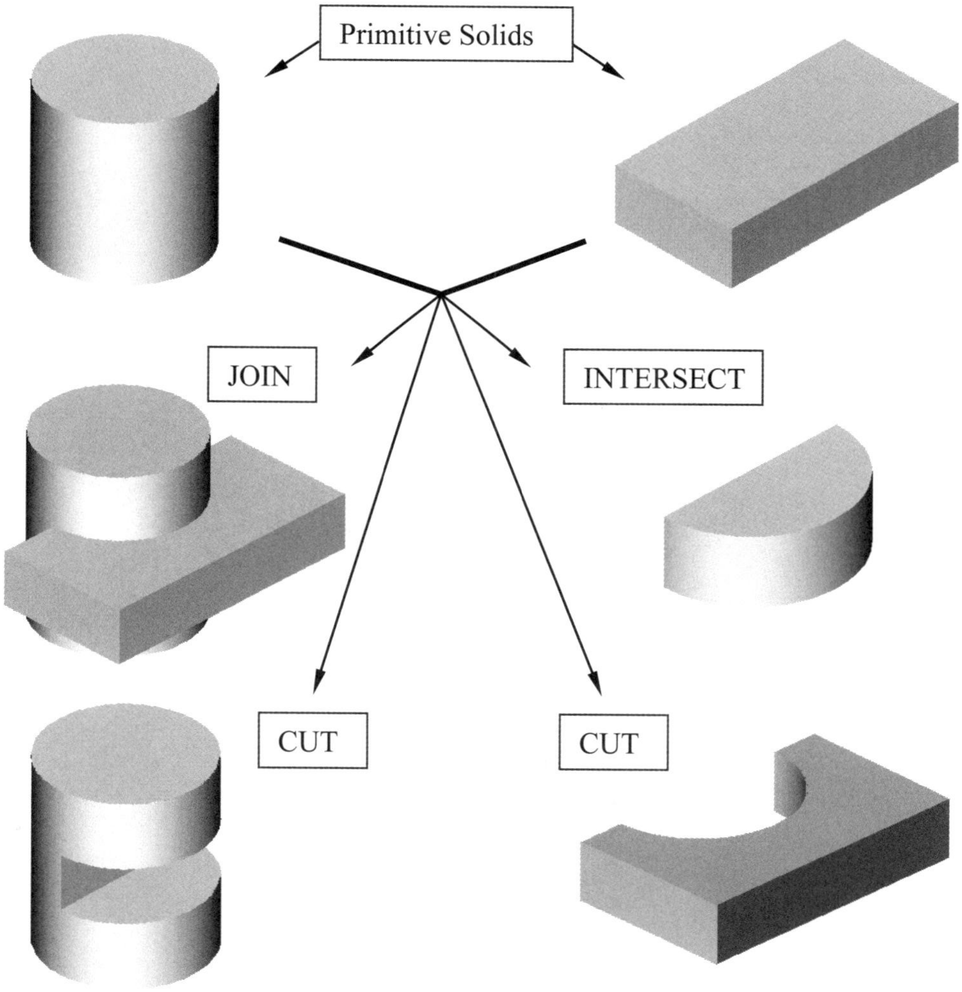

Binary Tree

The CSG is also referred to as the method used to store a solid model in the database. The resulting solid can be easily represented by what is called a **binary tree**. In a binary tree, the terminal branches (leaves) are the various primitives that are linked together to make the final solid object (the root). The binary tree is an effective way to keep track of the *history* of the resulting solid. By keeping track of the history, the solid model can be re-built by re-linking through the binary tree. This provides a convenient way to modify the model. We can make modifications at the appropriate links in the binary tree and re-link the rest of the history tree without building a new model.

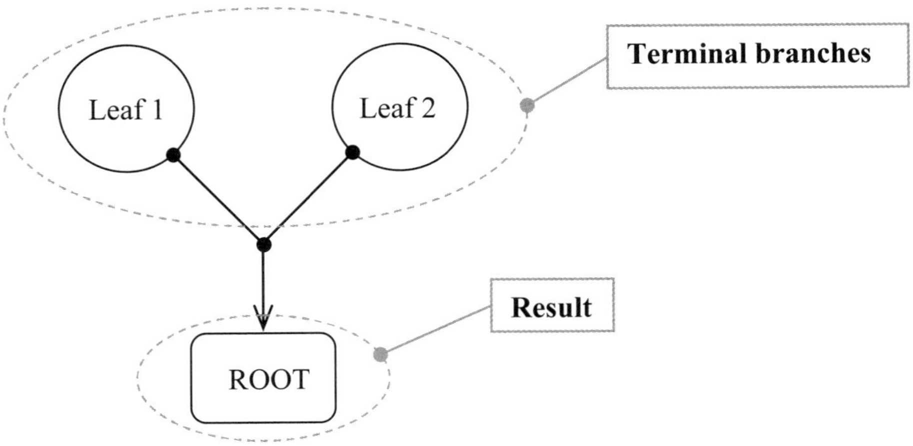

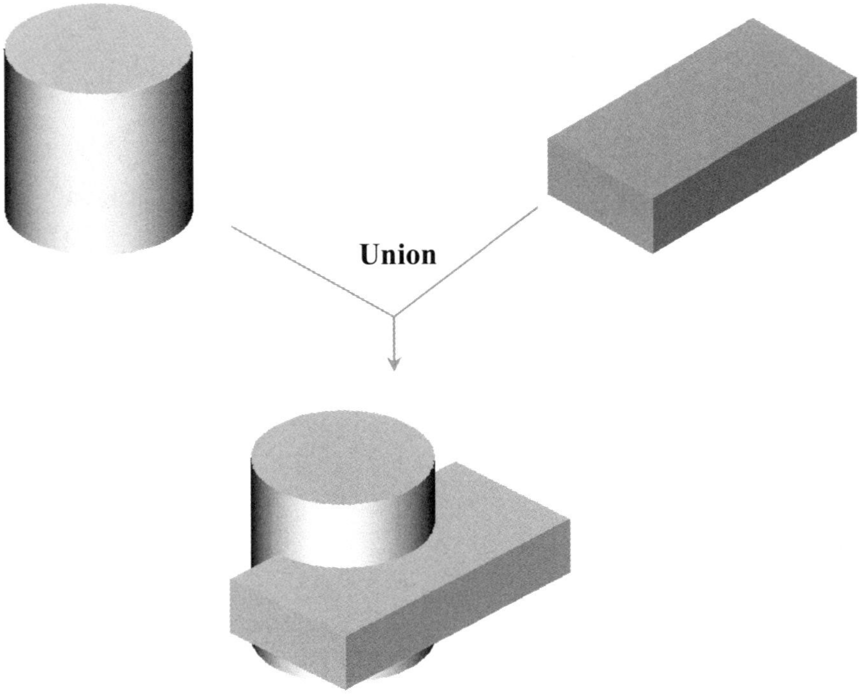

The Locator Design

The CSG concept is one of the important building blocks for feature-based modeling. In Autodesk Inventor, the CSG concept can be used as a planning tool to determine the number of features that are needed to construct the model. It is also a good practice to create features that are parallel to the manufacturing process required for the design. With parametric modeling, we are no longer limited to using only the predefined basic solid shapes. In fact, any solid features we create in Autodesk Inventor are used as primitive solids; parametric modeling allows us to maintain full control of the design variables that are used to describe the features. In this lesson, a more in-depth look at the parametric modeling procedure is presented. The equivalent CSG operation for each feature is also illustrated.

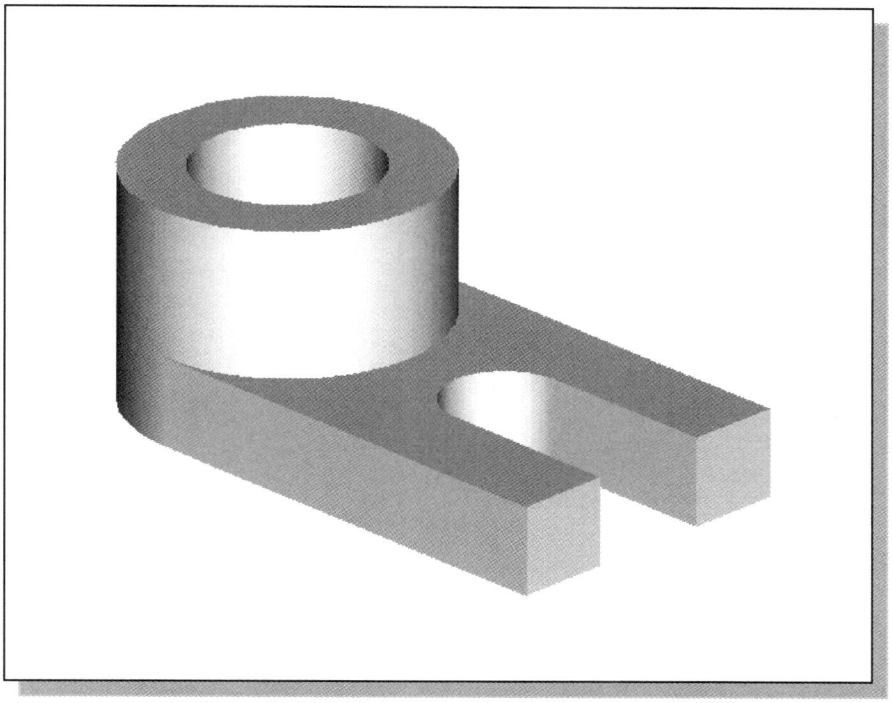

- Before going through the tutorial, on your own make a sketch of a CSG binary tree of the ***Locator*** design using only two basic types of primitive solids: cylinder and rectangular prism. In your sketch, how many *Boolean operations* will be required to create the model? What is your choice of the first primitive solid to use, and why? Take a few minutes to consider these questions and do the preliminary planning by sketching on a piece of paper. Compare the sketch you make to the CSG binary tree steps shown on the next page. Note that there are many different possibilities in combining the basic primitive solids to form the solid model. Even for the simplest design, it is possible to take several different approaches to creating the same solid model.

Modeling Strategy – CSG Binary Tree

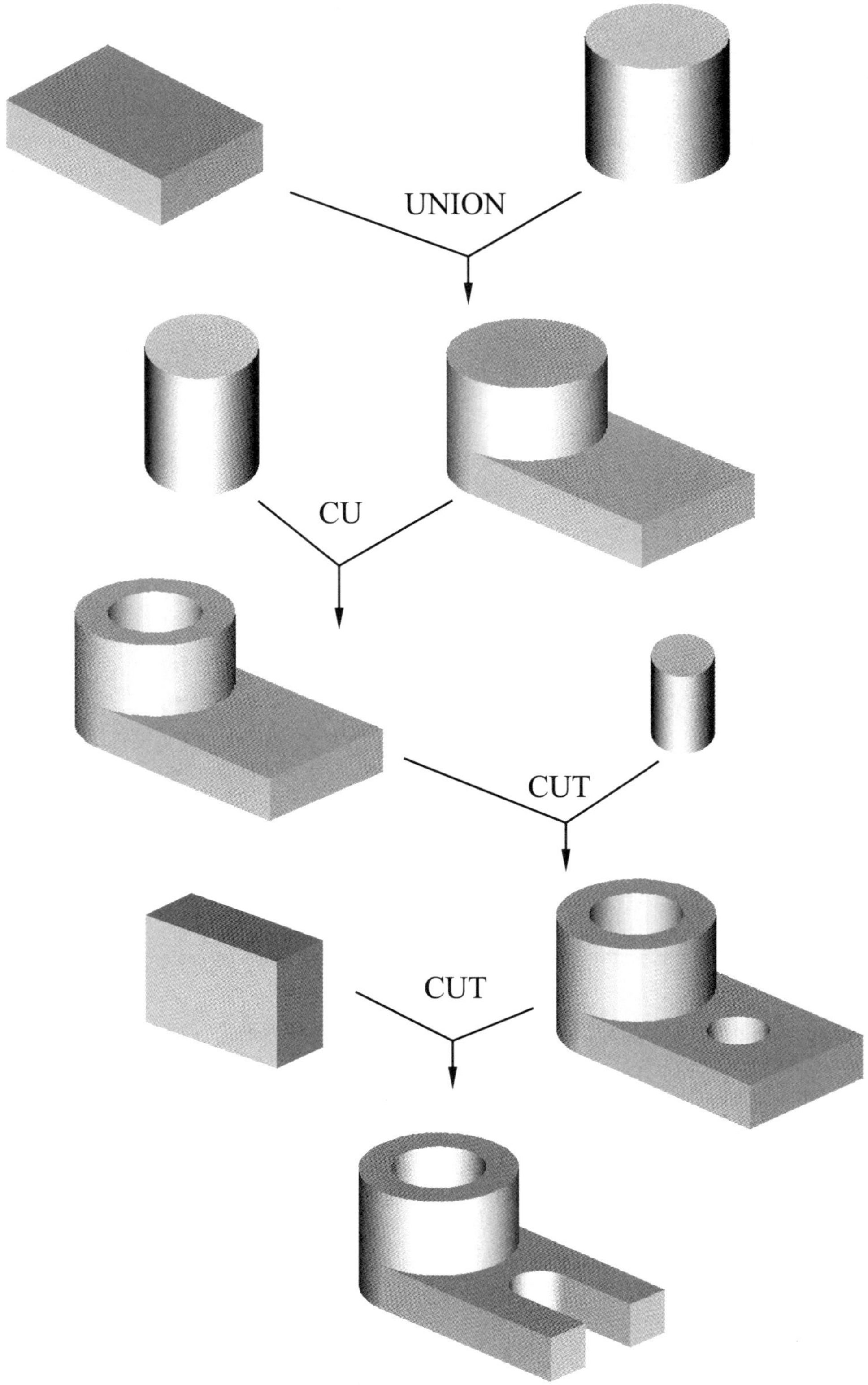

Starting Autodesk Inventor

1. Select the **Autodesk Inventor** option on the *Start* menu or select the **Autodesk Inventor** icon on the desktop to start Autodesk Inventor. The Autodesk Inventor main window will appear on the screen.

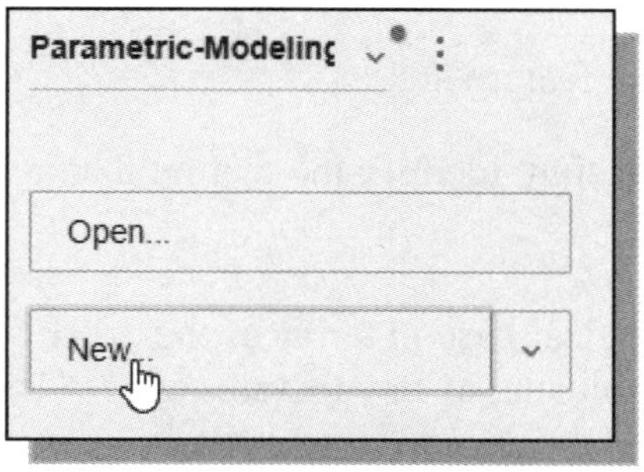

2. Select the **New File** icon with a single click of the left-mouse-button as shown.

❖ Every object we construct in a CAD system is measured in units. We should determine the value of the units within the CAD system before creating the first geometric entities. For example, in one model, a unit might equal one millimeter of the real-world object; in another model, a unit might equal an inch. In Autodesk Inventor, the ***Choose Template*** option is used to control how Autodesk Inventor interprets the coordinate and angle entries.

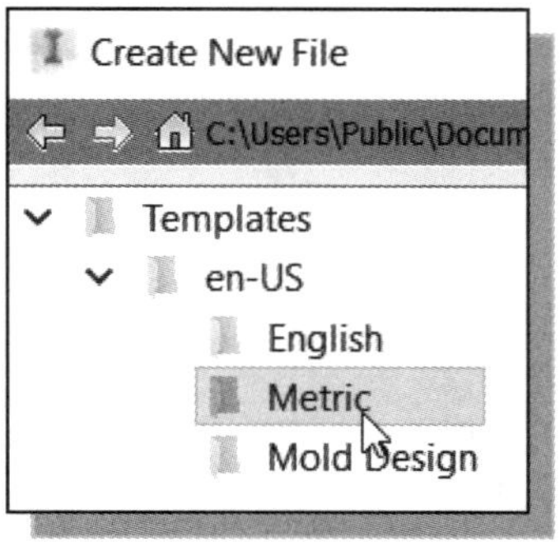

3. Select the **Metric** tab as shown below. We will use the millimeter (mm) setting for this example.

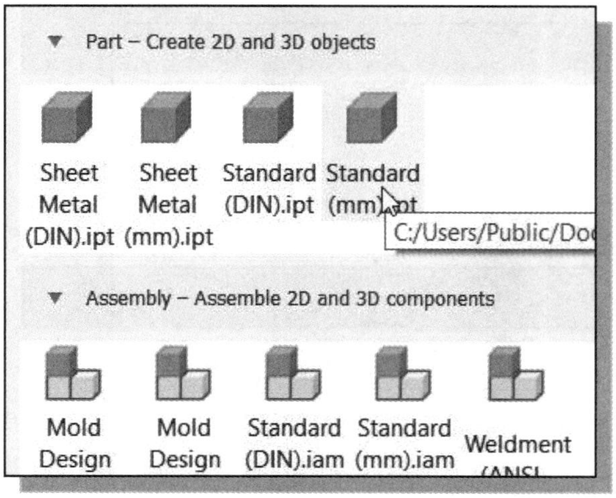

4. In the *New File – Part Template* area, select the **Standard(mm).ipt** icon as shown.

5. Confirm the ***Parametric-Modeling*** project is activated; note the **Projects** button is available to view/modify the active project.

6. Pick **Create** in the *Startup* dialog box to accept the selected settings.

Base Feature

In *parametric modeling*, the first solid feature is called the **base feature**, which usually is the primary shape of the model. Depending upon the design intent, additional features are added to the base feature.

Some of the considerations involved in selecting the base feature are:

- **Design intent** – Determine the functionality of the design; identify the feature that is central to the design.

- **Order of features** – Choose the feature that is the logical base in terms of the order of features in the design. Consider creating the design using as few features as possible.

- **Ease of making modifications** – Select a base feature that is more stable and is less likely to be changed.

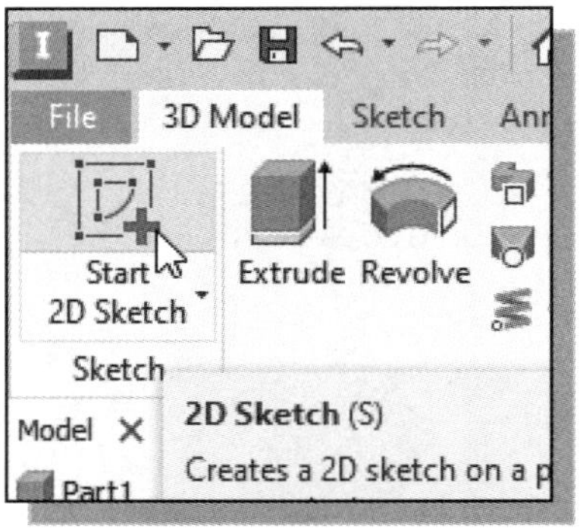

1. Activate the **Start 2D Sketch** icon with a single click of the left-mouse-button.

2. Move the cursor over the edge of the *XZ Plane* in the graphics area. When the *XZ Plane* is highlighted, click once with the **left-mouse-button** to select the *Plane* as the sketch plane for the new sketch.

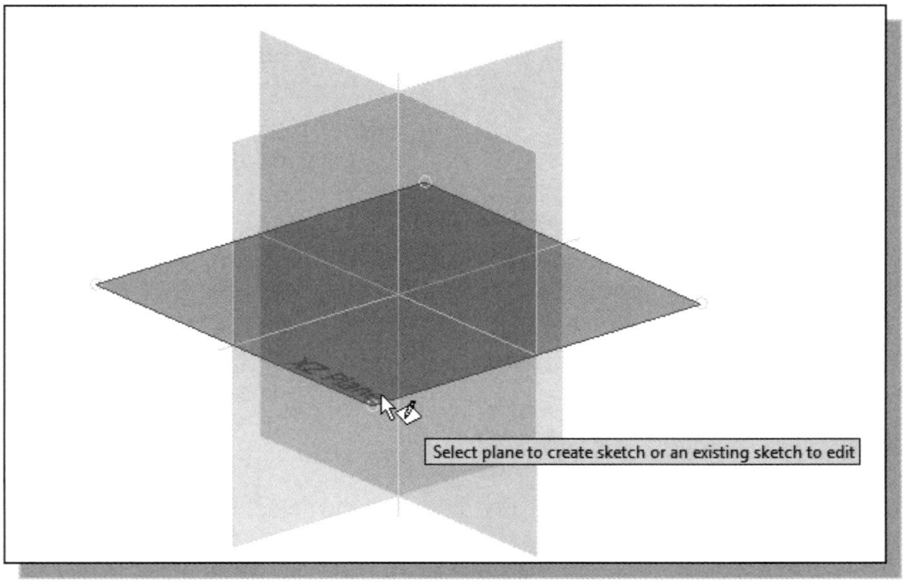

GRID Display Setup

1. In the *Ribbon* toolbar panel, select **[Tools] → [Document Settings]**.

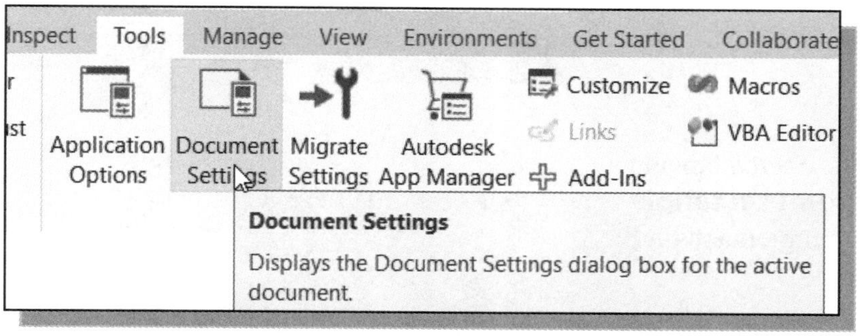

2. In the *Document Settings* dialog box, click on the **Sketch** tab as shown in the below figure.

3. Set the *X* and *Y Snap Spacing* to **5 mm**.

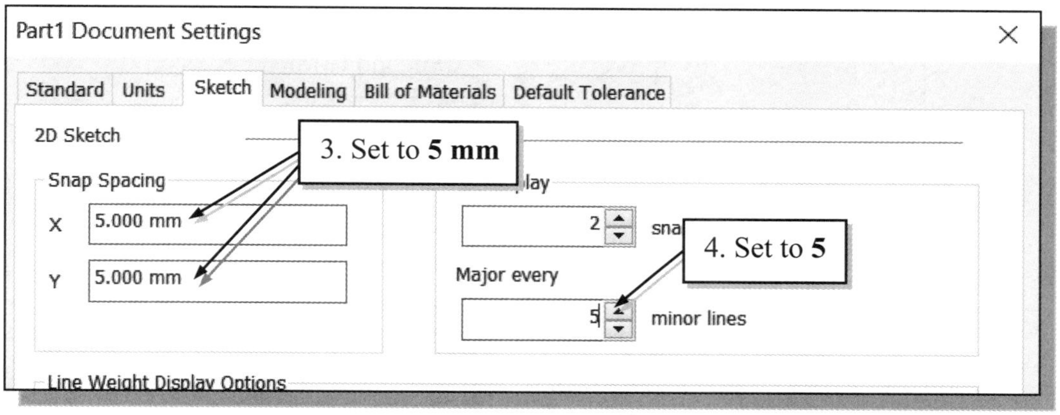

4. Change *Grid Display* to display one *major line* every **5** *minor lines*.

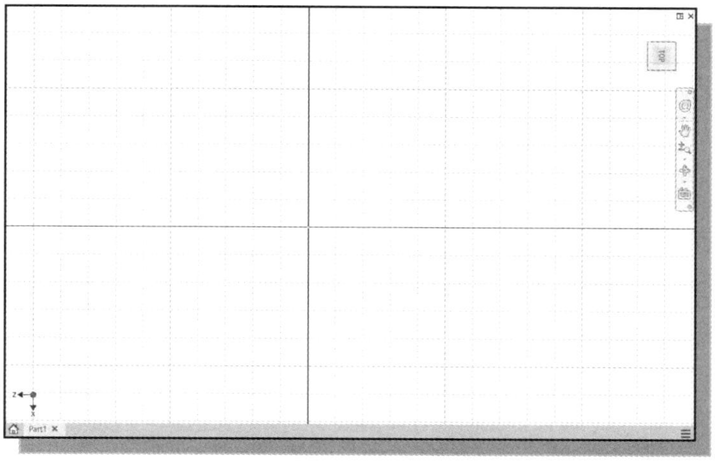

5. Pick **OK** to exit the *Sketch Settings* dialog box.

➢ Note that although the **Snap to grid** option is also available, its usage in parametric modeling is not recommended.

➢ On your own, use the dynamic **Zoom** function to view the grid setup. Refer to Page 2-26 on how to switch on the *grid lines display* options if necessary.

➢ A rectangular block will be first created as the base feature of the ***Locator*** design.

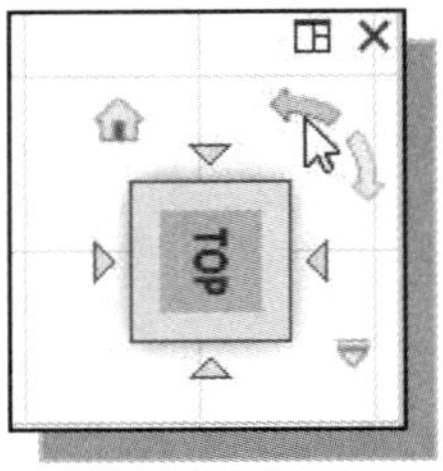

6. Click on the **rotate-left arrow** on the view cube to rotate the display.

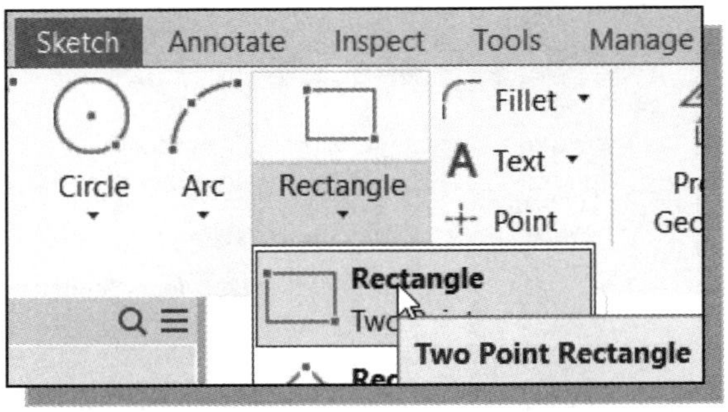

7. Switch back to the *Sketch* tab and select the **Two point rectangle** command by clicking once with the **left-mouse-button**.

8. Create a rectangle of arbitrary size by selecting two locations on the screen as shown below.

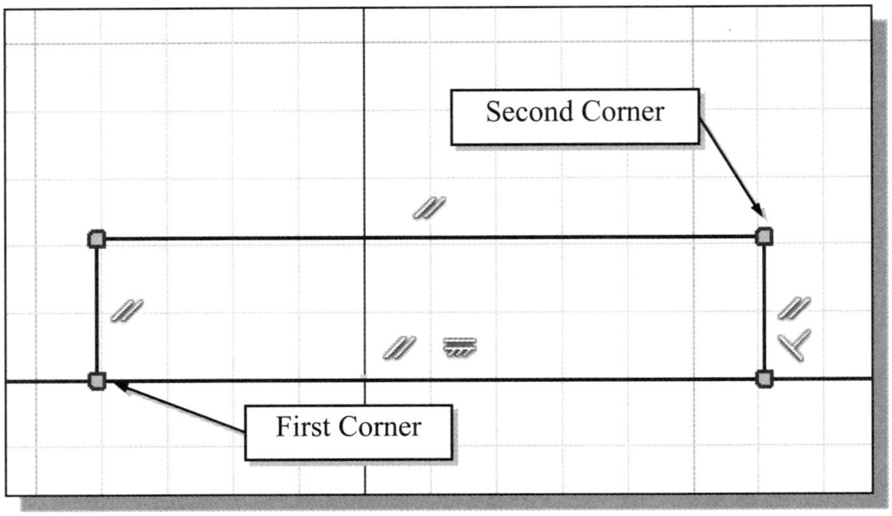

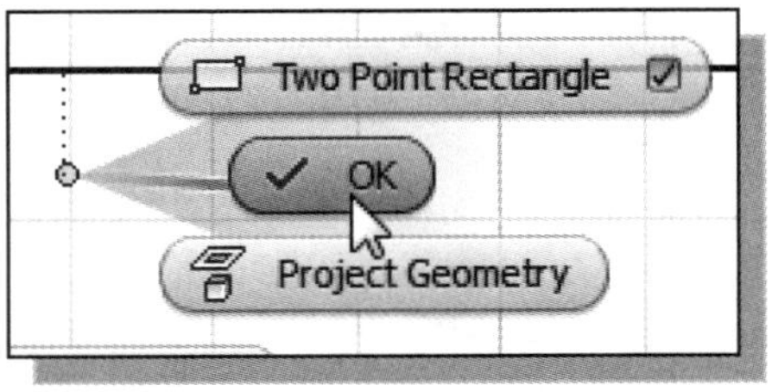

9. Inside the graphics window, click once with the **right-mouse-button** to bring up the option menu.

10. Select **OK** to end the Rectangle command.

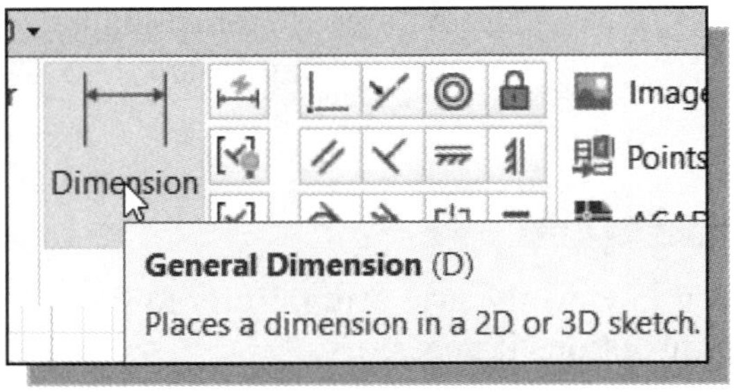

11. Activate the **General Dimension** command by clicking once with the left-mouse-button. The General Dimension command allows us to quickly create and modify dimensions.

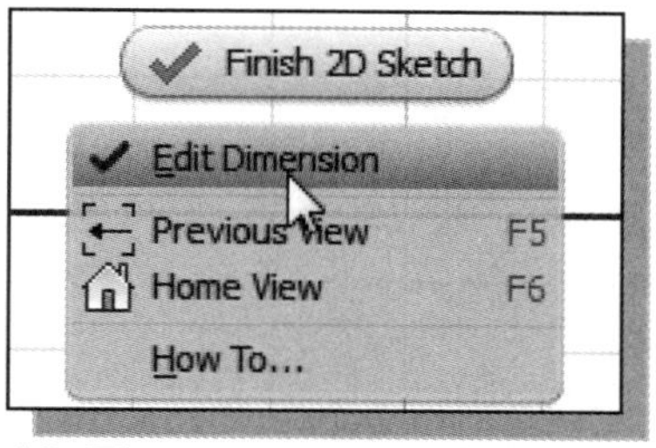

12. Inside the graphics window, click once with the right-mouse-button to bring up the option menu and click **Edit Dimension** to turn **OFF** the editing option while creating dimensions.

13. The message "*Select Geometry to Dimension*" is displayed in the *Status Bar* area at the bottom of the Autodesk Inventor window. Select the bottom horizontal line by left-clicking once on the line.

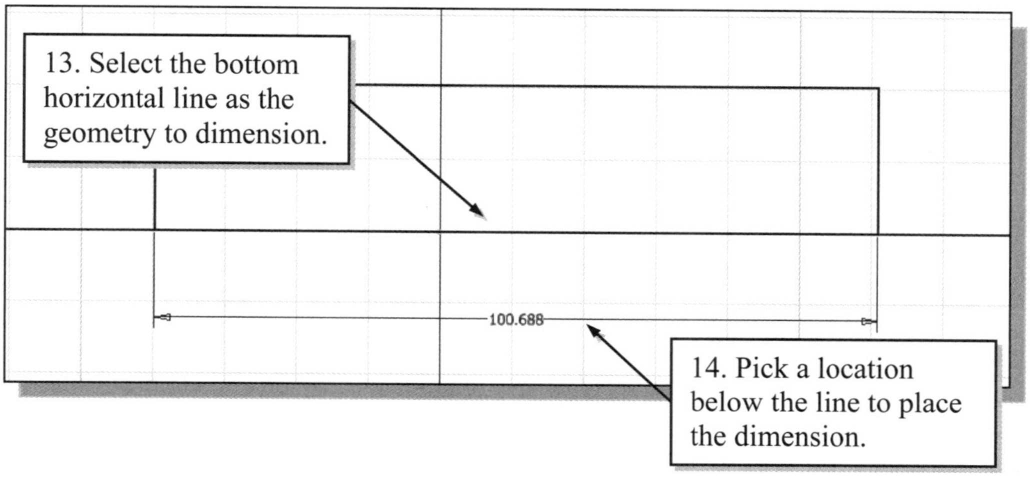

14. Move the graphics cursor below the selected line and left-click to place the dimension. (Note that the value displayed on your screen might be different than what is shown in the above figure.)

15. On your own, create a vertical size dimension of the sketched rectangle as shown.

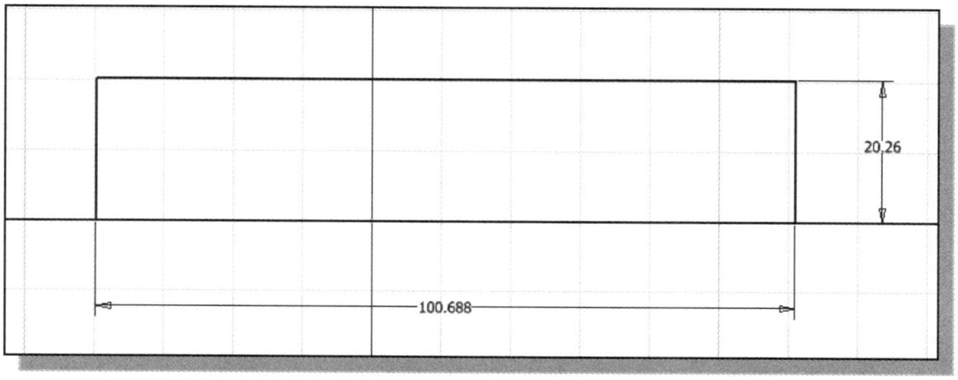

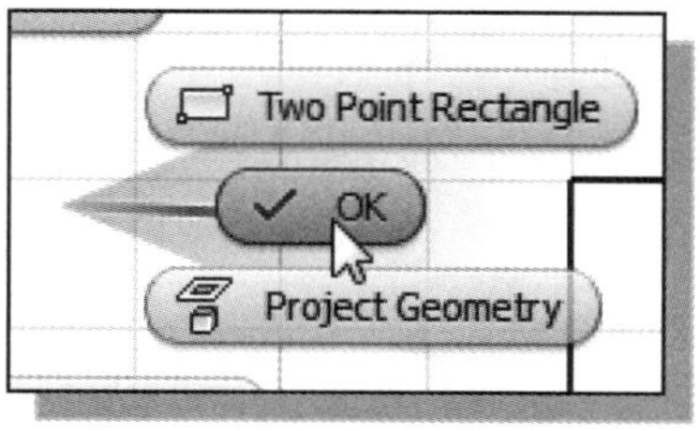

16. Inside the graphics window, click once with the right-mouse-button to bring up the option menu and click **OK** to end the *Dimension* command.

Model Dimensions Format

1. In the *Ribbon* tabs, select
   [Tools] → [Document Settings].

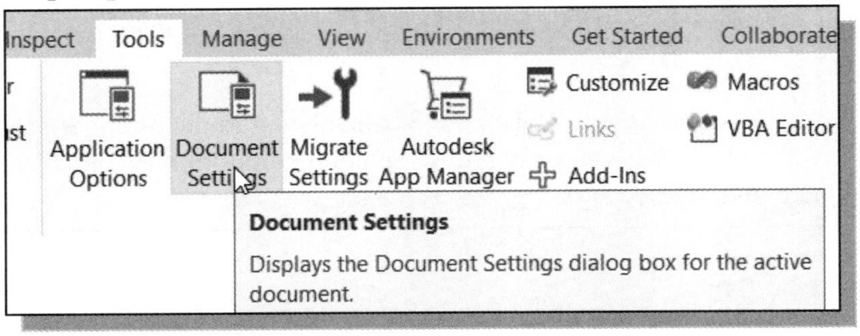

2. In the *Document Settings* dialog box, set the *Modeling Dimension Display* to **Display as value** as shown in the figure.

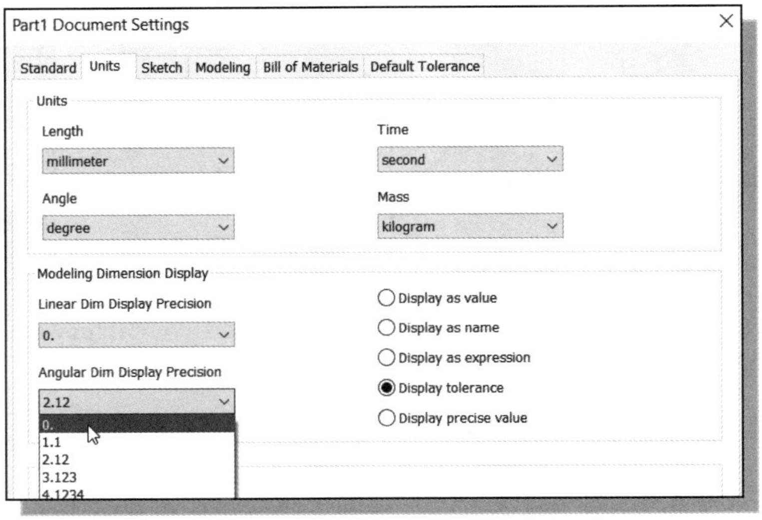

3. Also set the precision to ***no digits*** after the decimal point for both the *linear dimension* and *angular dimension* displays as shown in the above figure.

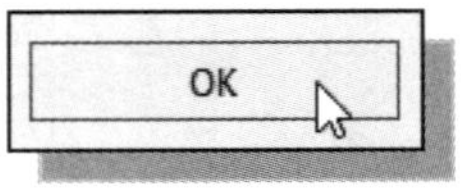

4. Pick **OK** to exit the *Document Settings* dialog box.

Modifying the Dimensions of the Sketch

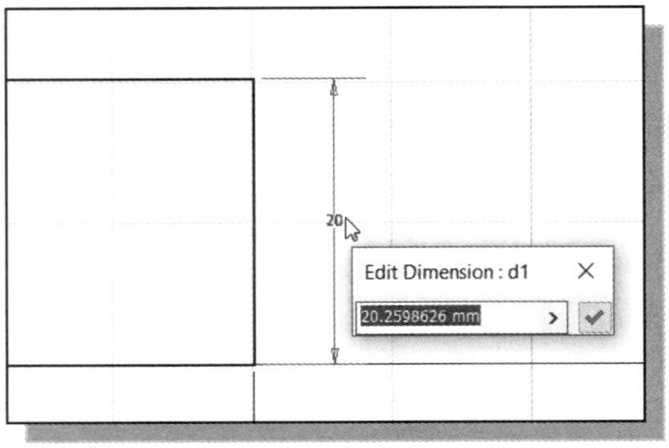

1. Select the height dimension that is to the right side of the sketch by ***double-clicking*** with the left-mouse-button on the dimension text.

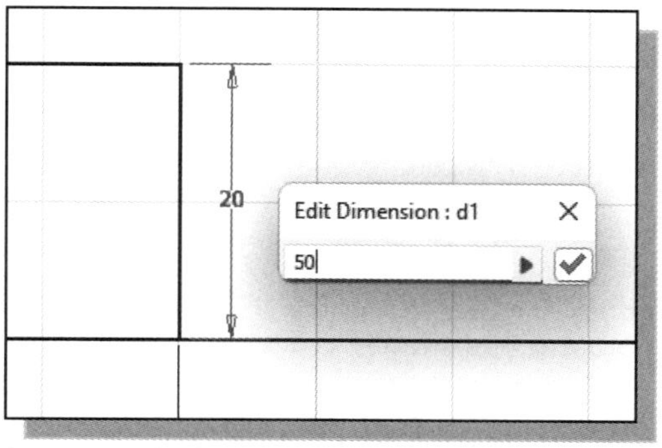

2. In the *Edit Dimension* window, the current length of the line is displayed. Enter **50** to set the selected length of the sketch to 50 millimeters.

3. Click on the **Accept** icon to accept the entered value.

➢ Autodesk Inventor will now update the profile with the new dimension value.

4. On your own, repeat the above steps and adjust the dimensions so that the sketch appears as shown below. Also **exit** the **Dimension** command.

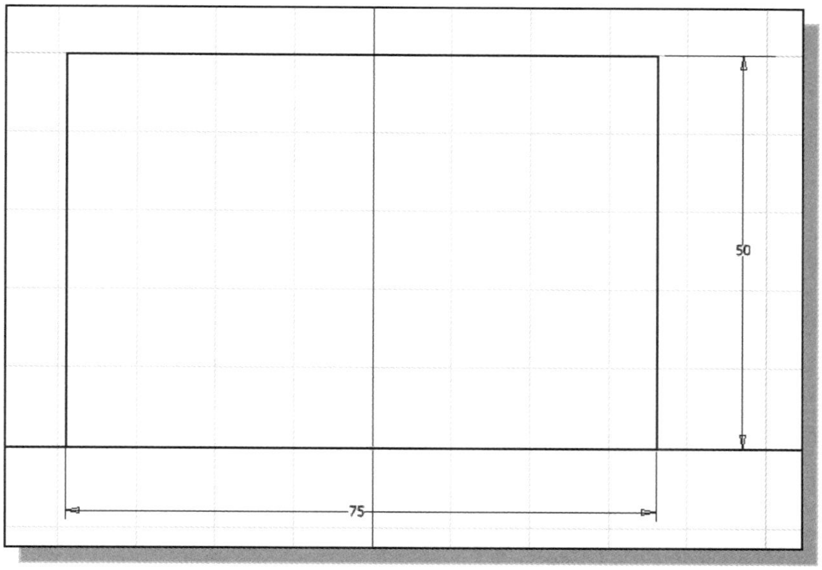

Repositioning Dimensions

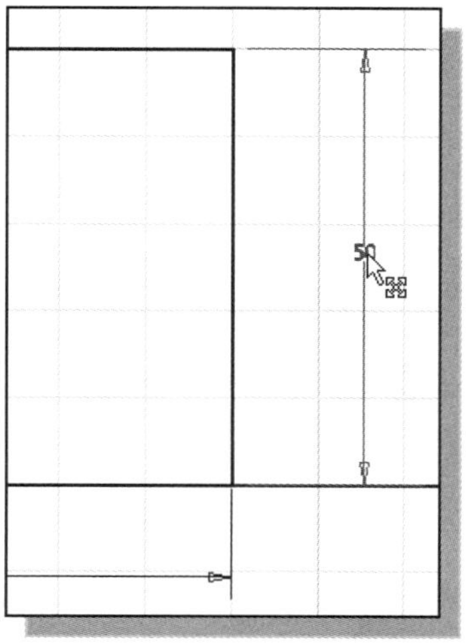

1. Move the cursor near the vertical dimension; note that the dimension is highlighted. Move the cursor slowly until a small arrows marker appears next to the cursor, as shown in the figure.

2. Drag with the **left-mouse-button** to reposition the selected dimension.

3. Repeat the above steps to reposition the horizontal dimension.

Using the Measure Tools

Autodesk Inventor also provides several measuring tools that allow us to measure area, perimeter and additional information of the constructed 2D sketches.

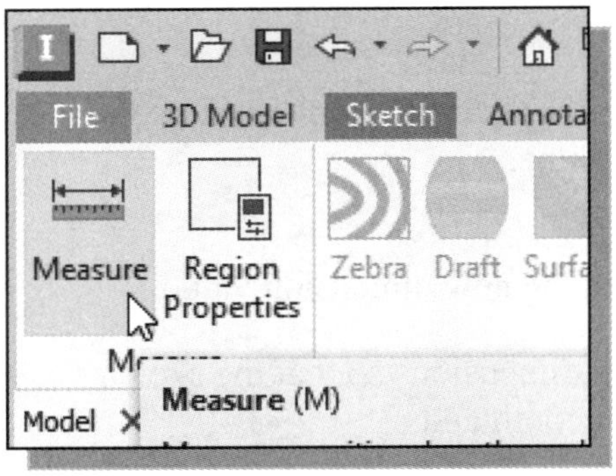

1. In the ***Inspect*** *Ribbon tab*, left-click once on the **Measure** option as shown.

- Note that **other** measurement options are also available in the toolbar.

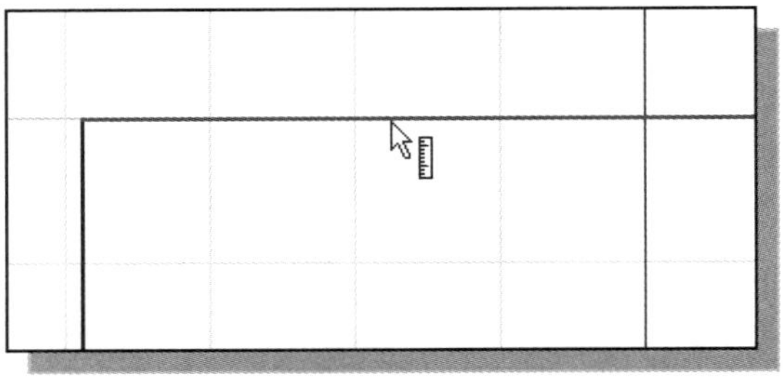

2. Click on the top edge of the rectangle as shown.

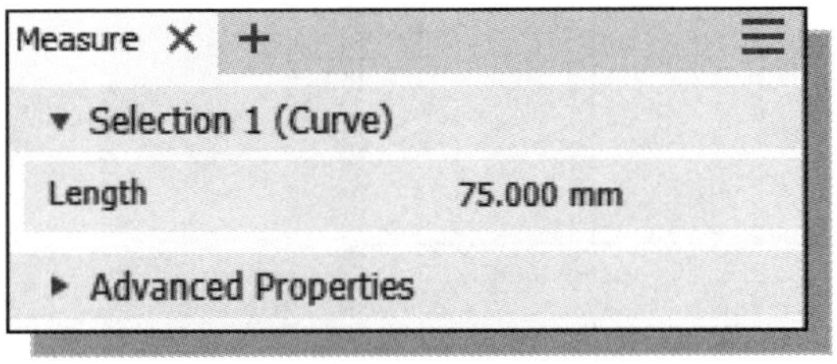

3. The associated length measurement of the selected geometry is displayed in the *Length* dialog box as shown.

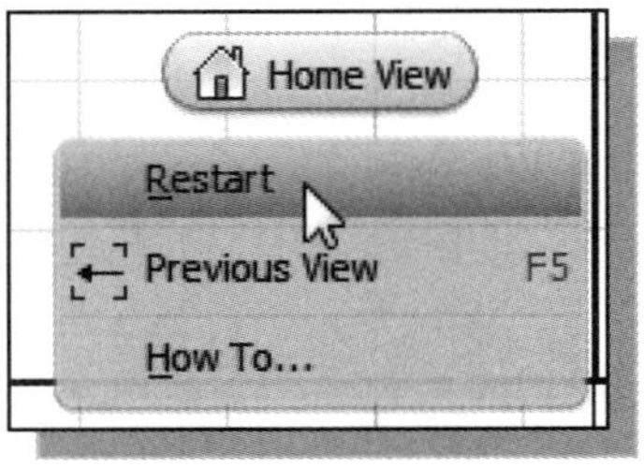

4. Inside the *graphics window*, right-click once to bring up the **Option menu** and select **Restart** as shown.

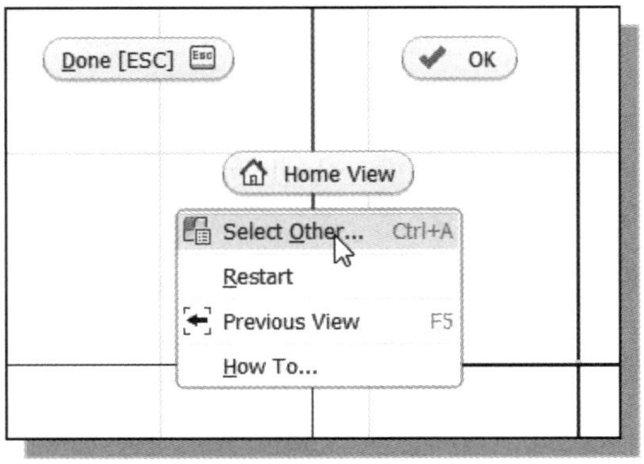

5. Move the cursor on top of any of the edges and click once with the right-mouse-button to bring up the **Option menu** and choose **Select Other...** as shown.

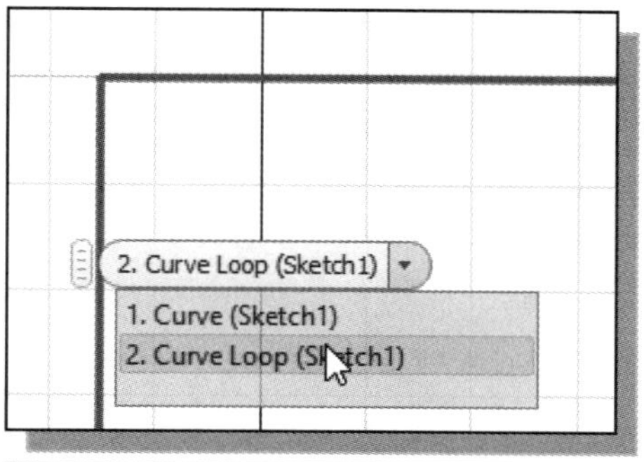

6. In the *Selection list*, left-click once to pick the **Curve Loop** option as shown.

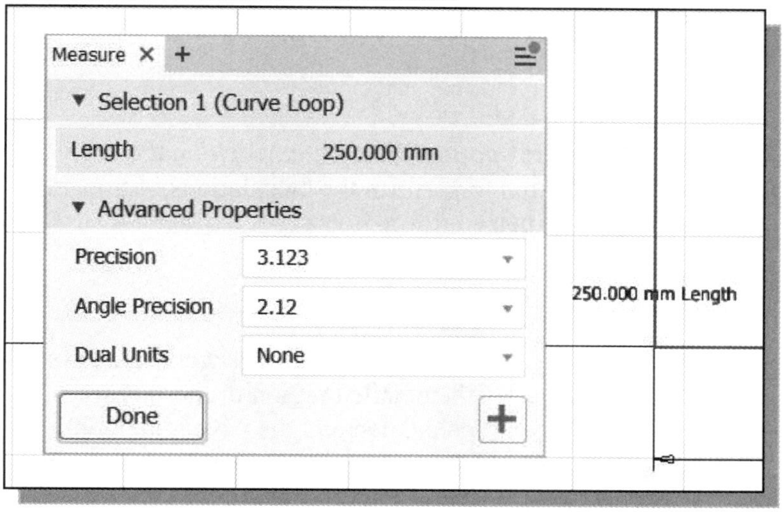

- The *perimeter* of the rectangle is displayed as shown.

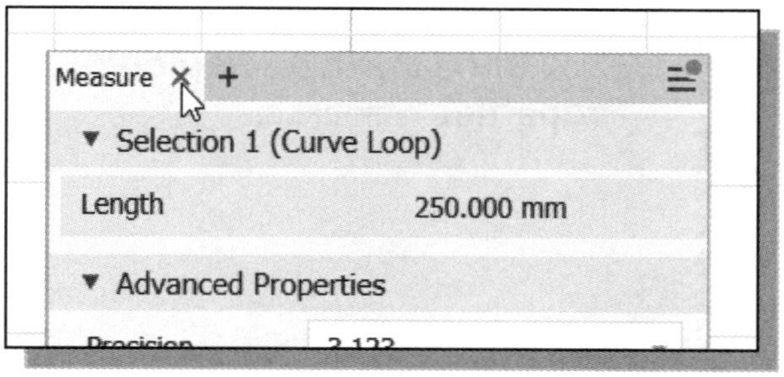

7. Click on the [**X**] icon to end the Measure command as shown.

8. In the *Inspect Ribbon tab*, left-click once on the **Region Properties** option as shown.

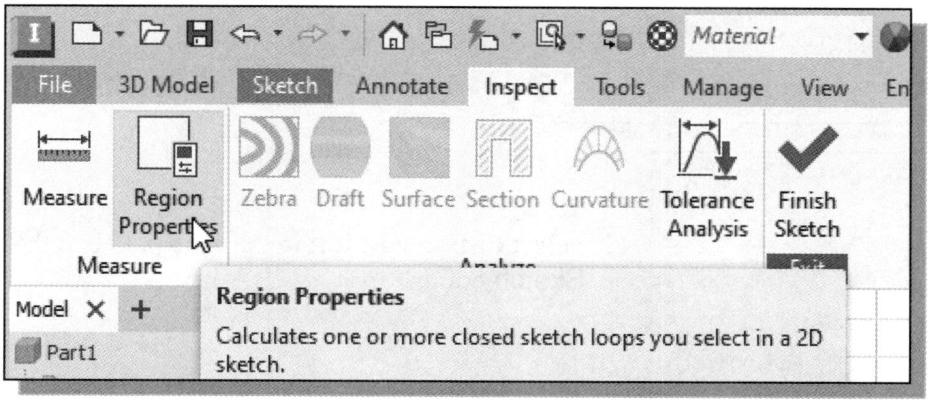

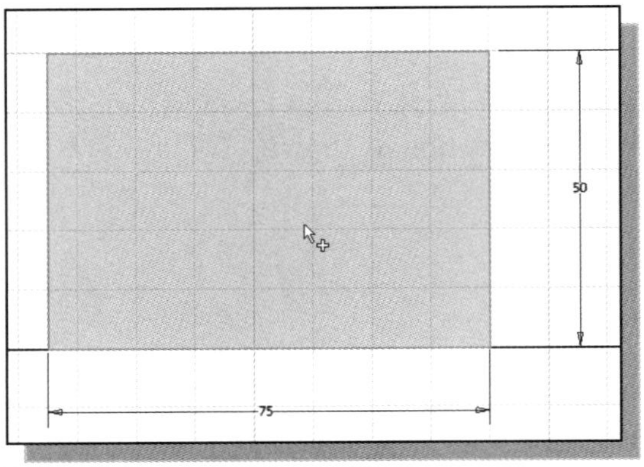

9. Click on the inside of the rectangle; notice the region is highlighted as the cursor is moved inside the rectangle, as shown.

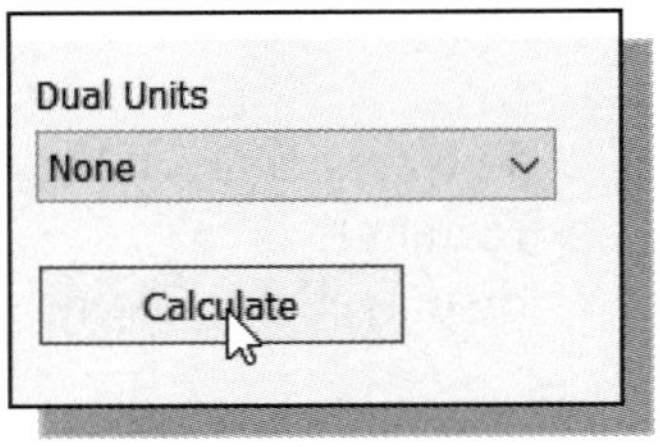

10. In the ***Region Properties*** dialog box, click on the **Calculate** button to perform the calculations of the associated geometry information.

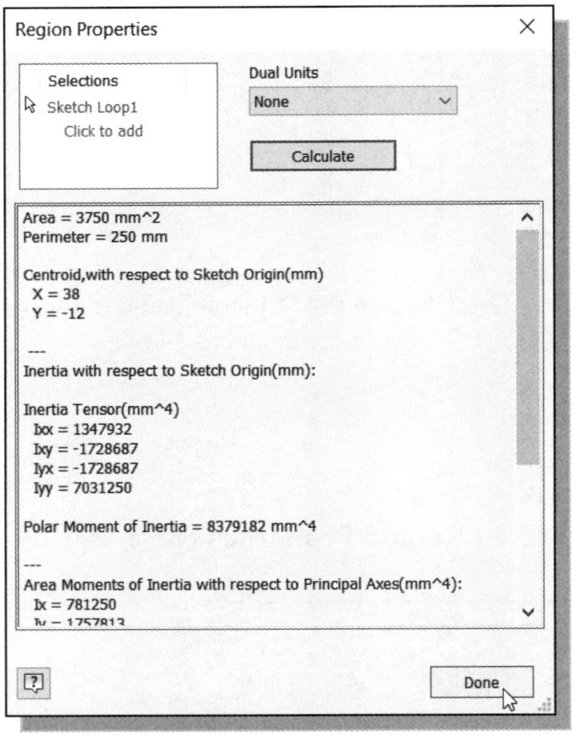

❖ In the ***Region Properties*** dialog box, the detailed region properties are calculated and displayed, including the *Area Moments of Inertia*, *Area* and *Perimeter*.

11. Click **Done** to exit the **Region Properties** command.

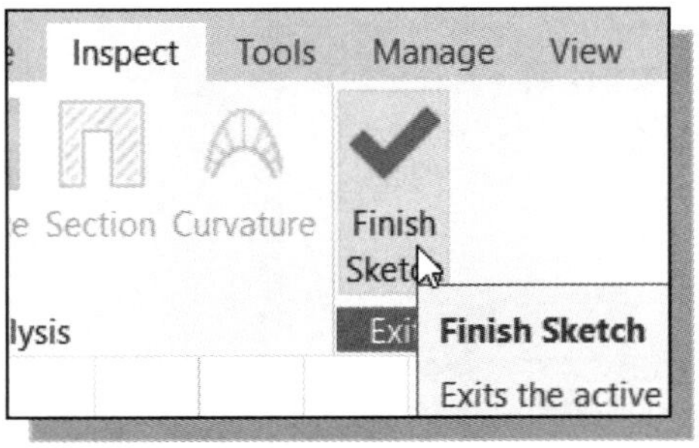

12. Select **Finish Sketch** in the *Ribbon* to end the Sketch option.

Completing the Base Solid Feature

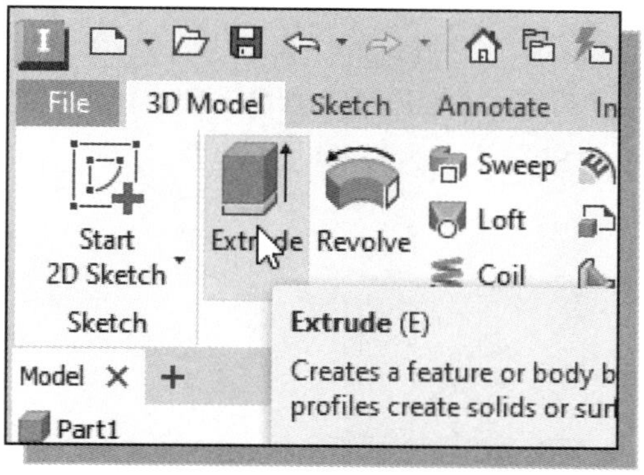

1. In the *3D Model tab* select the **Extrude** command by clicking the left-mouse-button on the icon.

2. In the *Extrude* pop-up window, enter **15** as the extrusion distance. Notice that the sketch region is automatically selected as the extrusion profile.

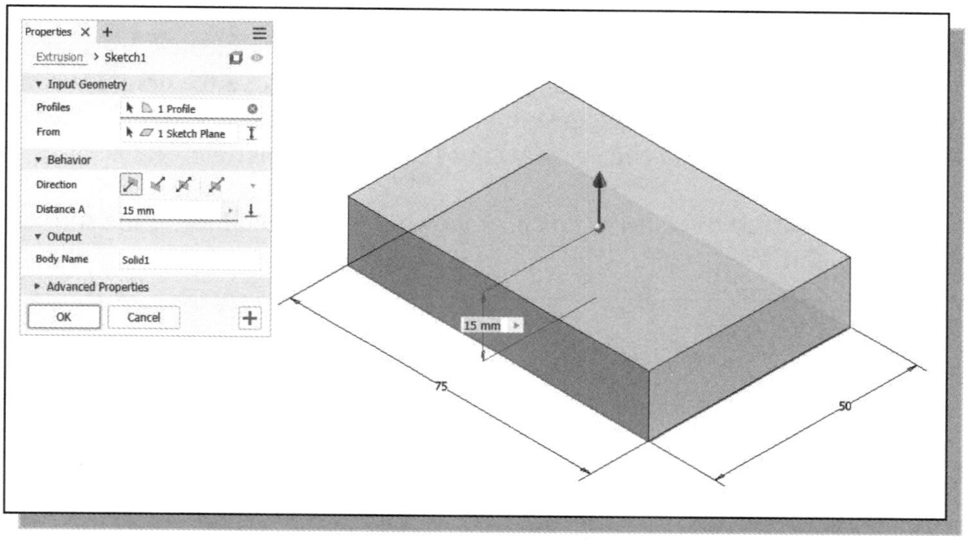

3. Click on the **OK** button to proceed with creating the 3D part. Use the *Dynamic Viewing* options to view the created part. Press **F6** to change the display to the isometric view as shown before going to the next section.

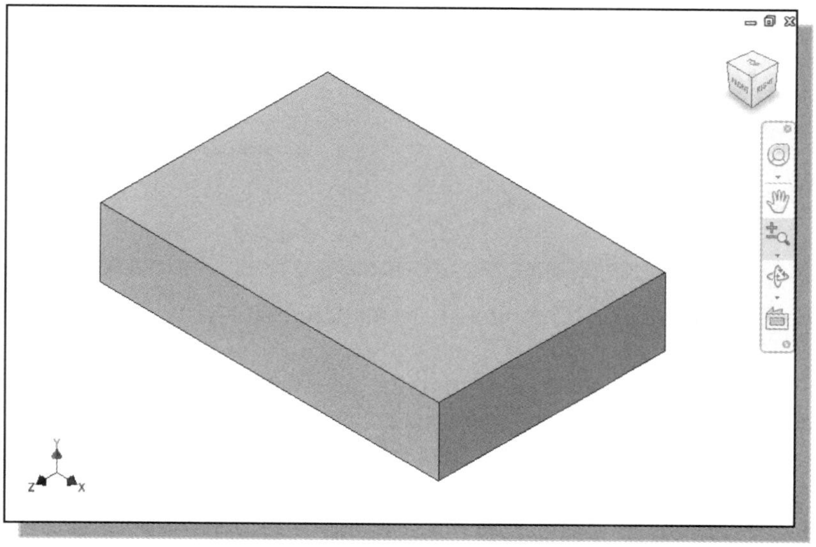

Creating the Next Solid Feature

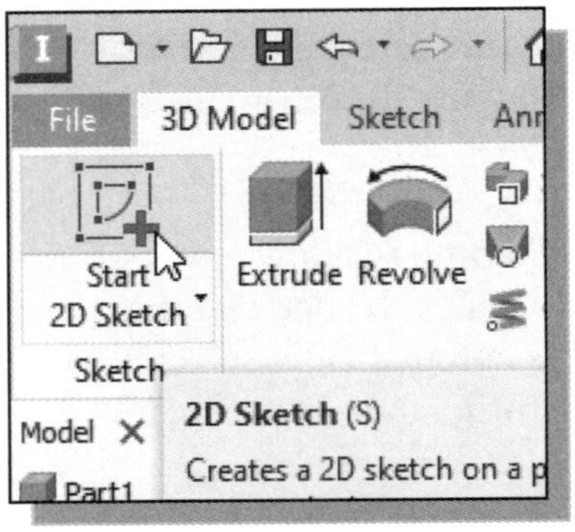

1. In the *3D Model tab* select the **Start 2D Sketch** command by left-clicking once on the icon.

2. In the *Status Bar* area, the message "*Select plane to create sketch or an existing sketch to edit*" is displayed. Autodesk Inventor expects us to identify a planar surface where the 2D sketch of the next feature is to be created. Move the graphics cursor on the 3D part and notice that Autodesk Inventor will automatically highlight feasible planes and surfaces as the cursor is on top of the different surfaces.

3. Use the **ViewCube** to adjust the display viewing the bottom face of the solid model as shown below.

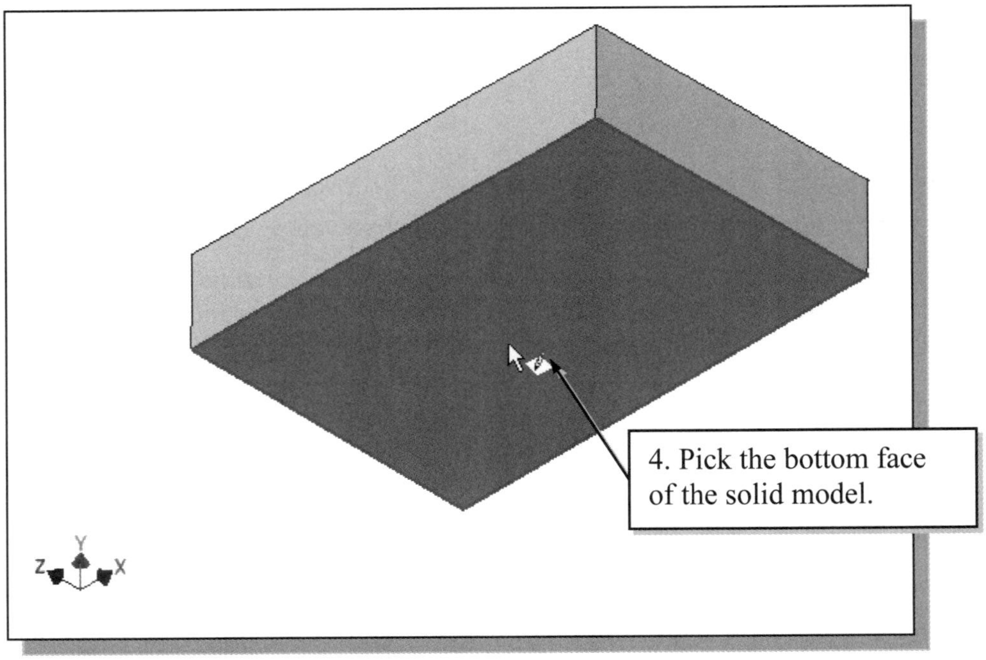

4. Pick the bottom face of the 3D model as the sketching plane.

➢ Note that the sketching plane is aligned to the selected face. Autodesk Inventor automatically establishes a User-Coordinate-System (UCS) and records its location with respect to the part on which it was created.

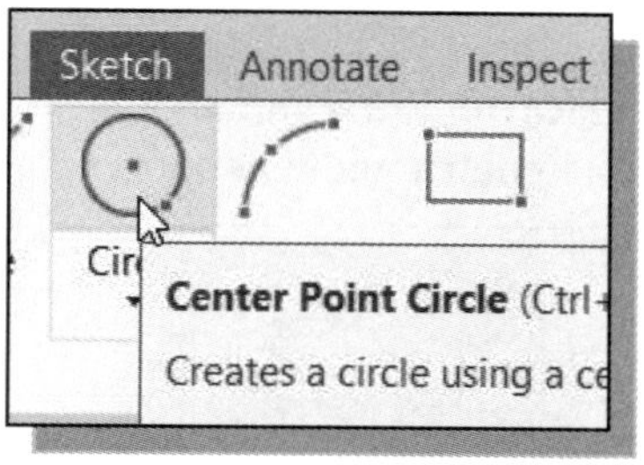

5. Select the **Center Point Circle** command by clicking once with the left-mouse-button on the icon in the *Sketch* tab.

➢ We will align the center of the circle to the midpoint of the base feature.

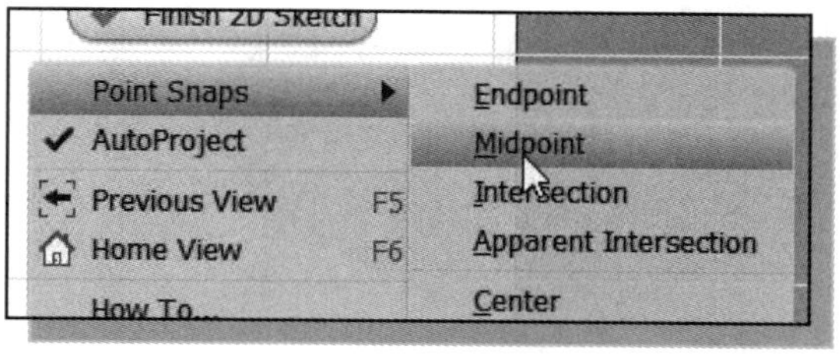

6. Inside the graphics window, click once with the **right-mouse-button** to bring up the option menu and choose the snap to **Midpoint** option.

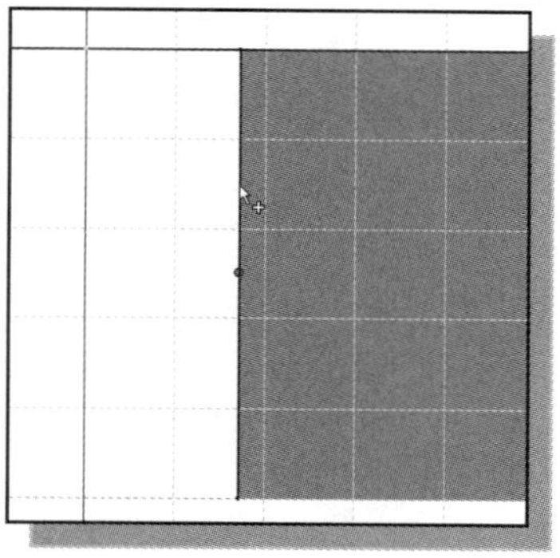

7. Move the cursor along the left edge of the base feature.

8. Note the green dot on the line, which is the **midpoint** of the line. Left click once to select the midpoint.

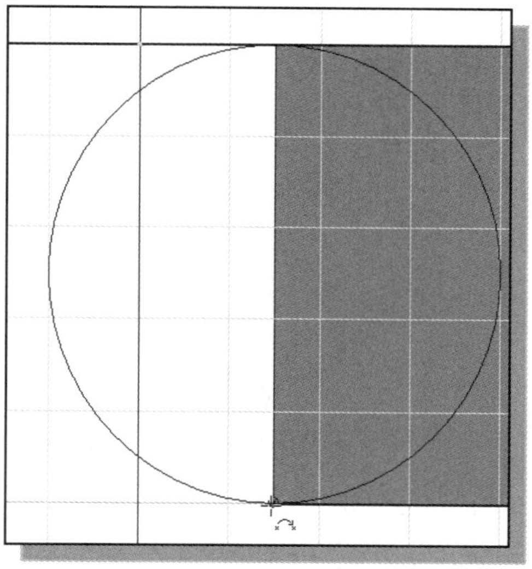

9. Select the bottom corner of the base feature to create a circle as shown in the figure.

10. Inside the *graphics window*, click once with the right-mouse-button to display the option menu. Select **OK** to end the Circle command.

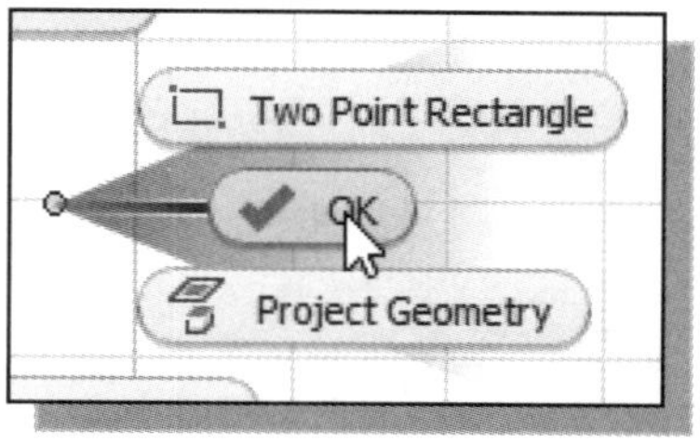

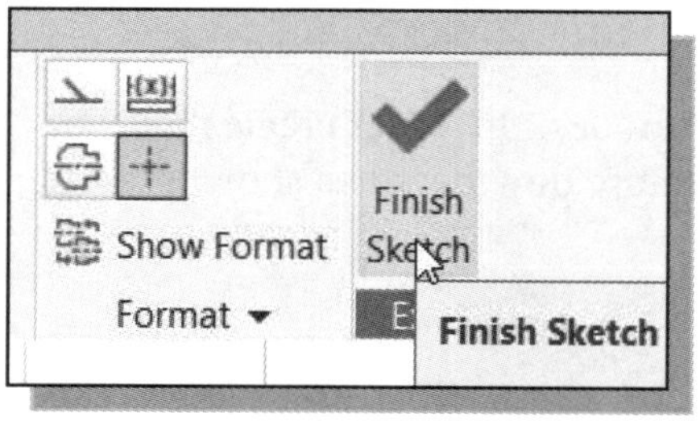

11. In the *Ribbon* toolbar, select **Finish Sketch** to exit the Sketch mode.

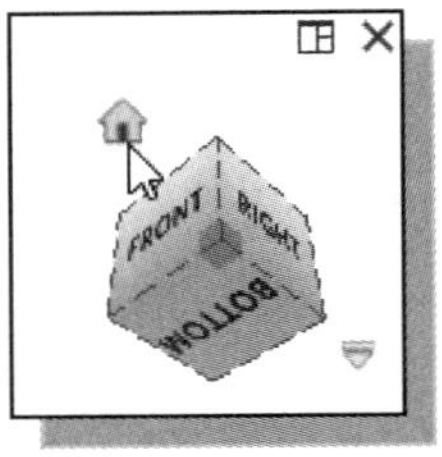

12. Press the function key **F6** once or select **Home View** in the **ViewCube** to change the display to the isometric view as shown.

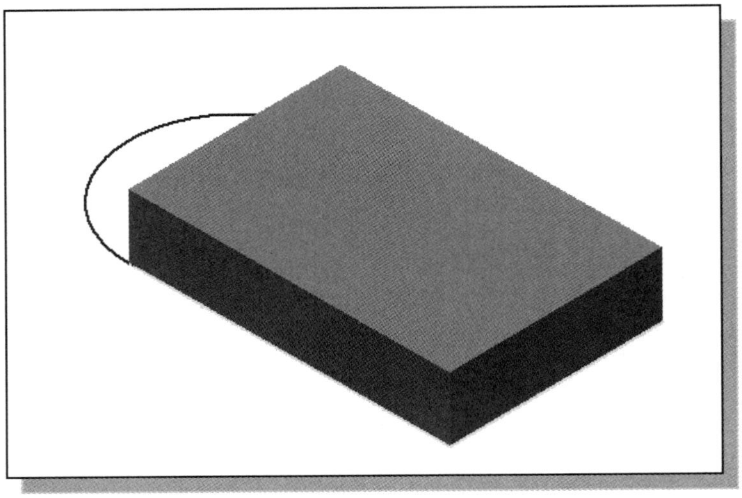

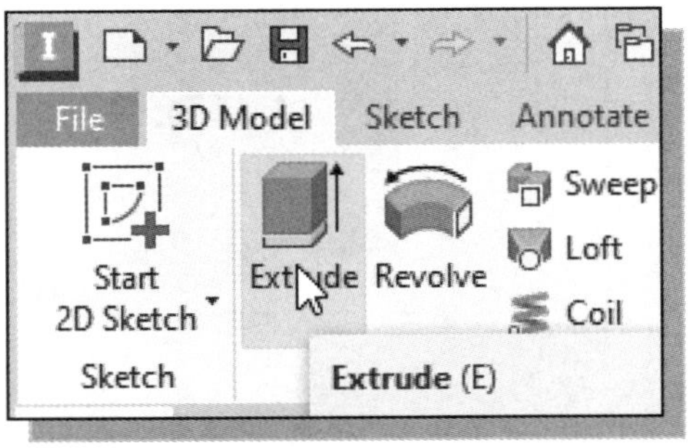

13. In the *3D Model tab*, select the **Extrude** command by left-clicking the icon.

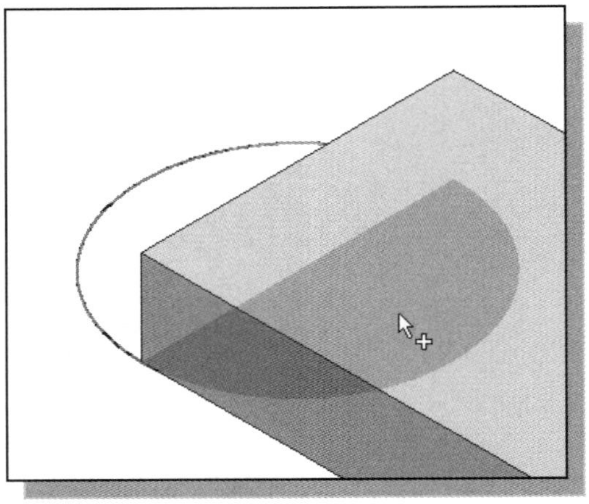

14. Autodesk Inventor next expects us to select the region to be used to create the feature. First select inside the semi-circle region under the solid feature as shown.

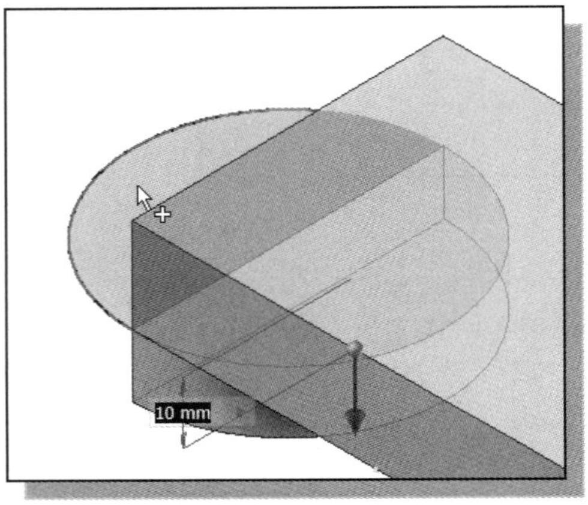

15. Select inside the other semi-circle region outside the solid feature as shown.

- Note that Autodesk Inventor creates the extruded feature downward as shown.

16. In the *Extrude* pop-up control, enter **40** as the blind extrusion distance as shown below. Set the solid operation to **Join** and click on the **Flip direction** button to reverse the direction of extrusion (upward) as shown below.

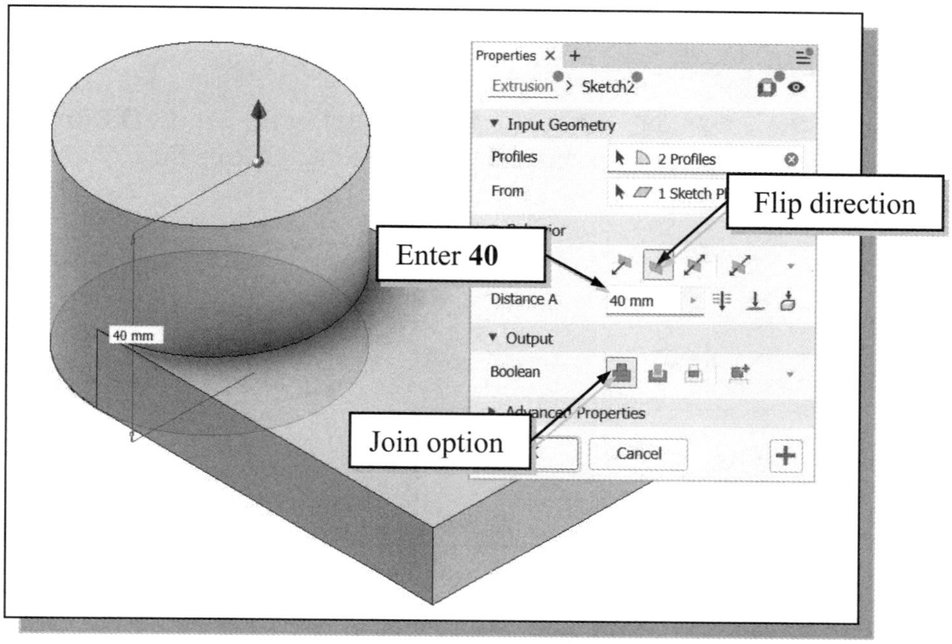

➢ Note that most of the settings can also be set through the icons displayed on the screen.

17. Click on the **OK** button to proceed with the *Join* operation.

- The two features are joined together into one solid part; the *CSG-Union* operation was performed.

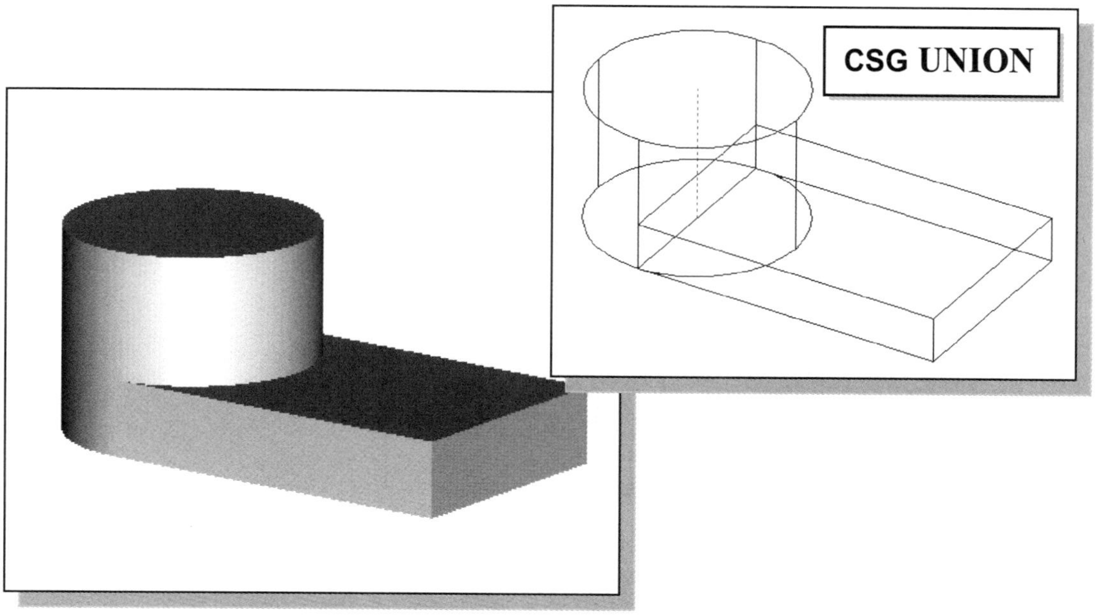

Creating a Cut Feature

We will create a circular cut as the next solid feature of the design. We will align the sketch plane to the top of the last cylinder feature.

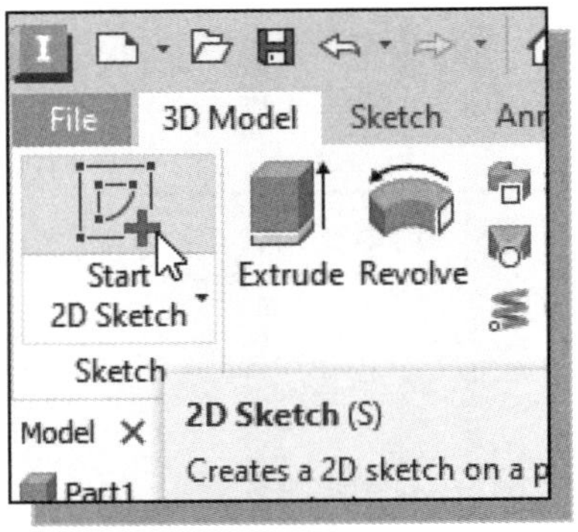

1. In the *3D Model tab* select the **Start 2D Sketch** command by left-clicking once on the icon.

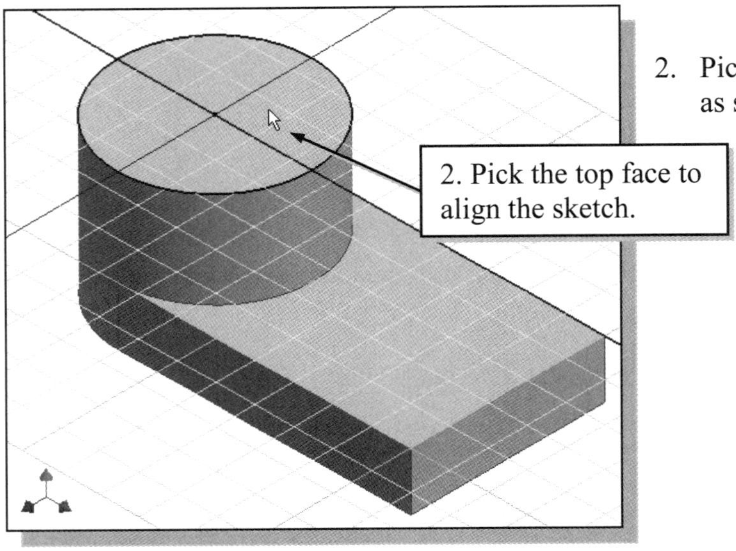

2. Pick the top face of the cylinder as shown.

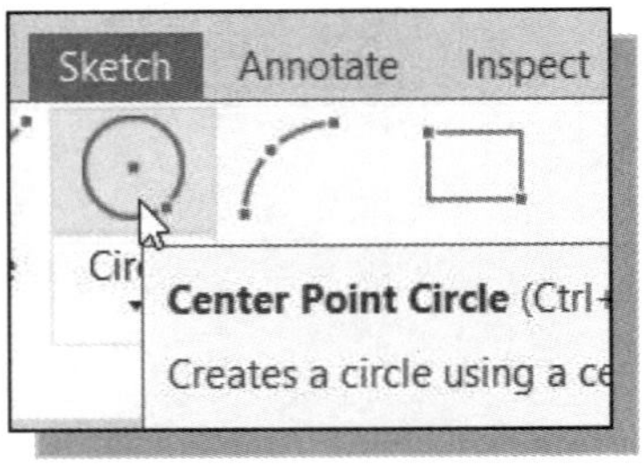

3. Select the **Center Point Circle** command by clicking once with the **left-mouse-button** on the icon in the *Sketch tab* of the Ribbon toolbar.

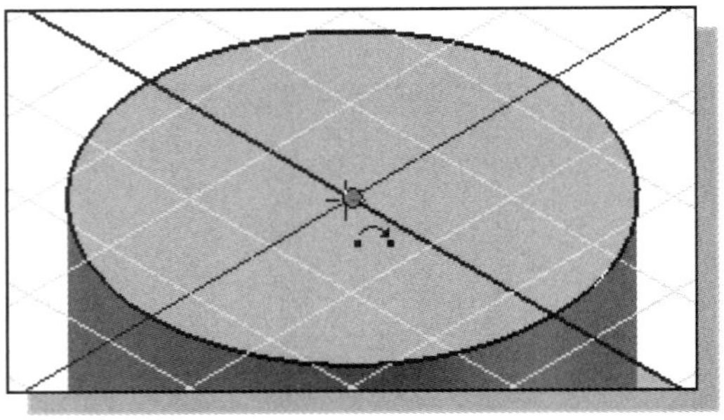

4. Select the **Center** point of the top face of the 3D model as shown.

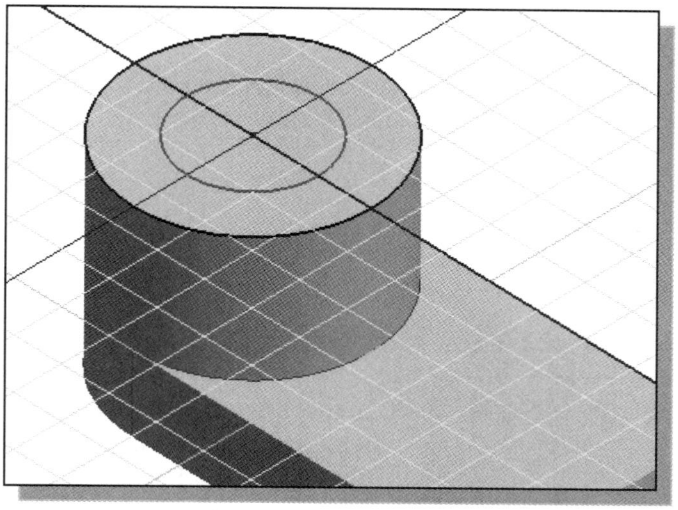

5. Sketch a circle of arbitrary size inside the top face of the cylinder as shown to the left.

6. Use the right-mouse-button to display the option menu and select **OK** in the pop-up menu to end the Circle command.

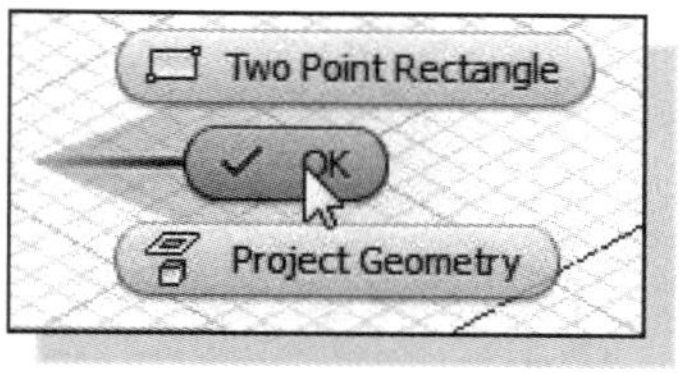

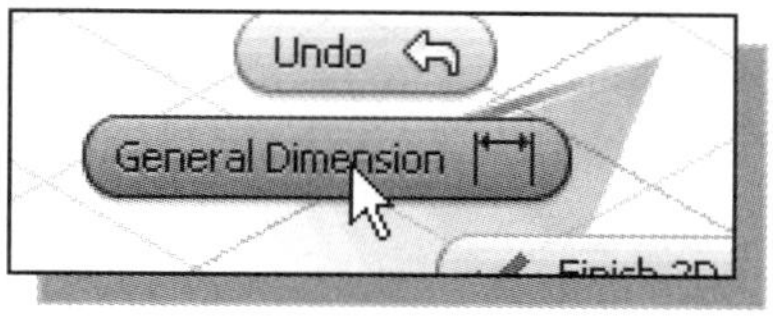

7. Inside the graphics window, click once with the **right-mouse-button** to display the option menu. Select the **General Dimension** option in the pop-up menu.

8. Create a dimension to describe the size of the circle and set it to **30mm**.

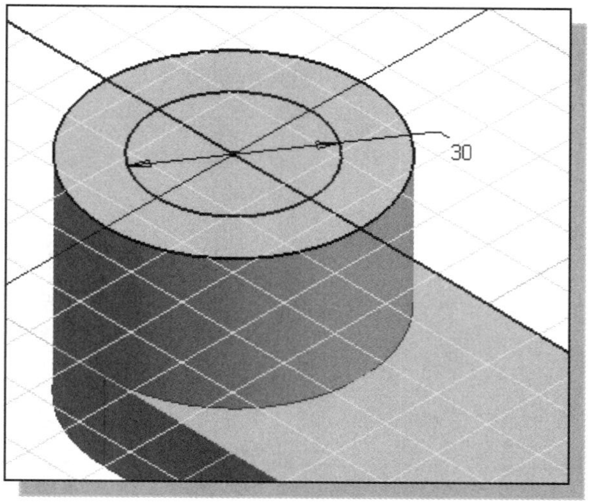

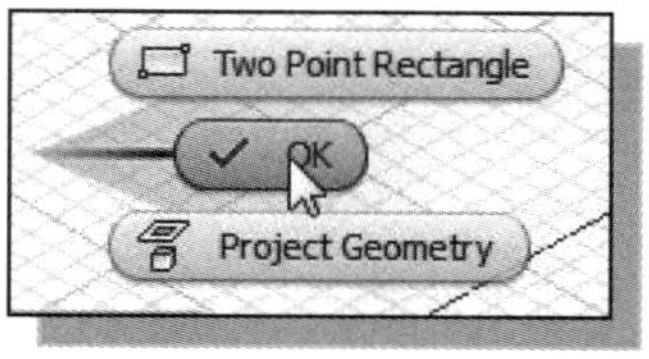

9. Inside the graphics window, click once with the right-mouse-button to display the option menu. Select **OK** in the pop-up menu to end the Dimension command.

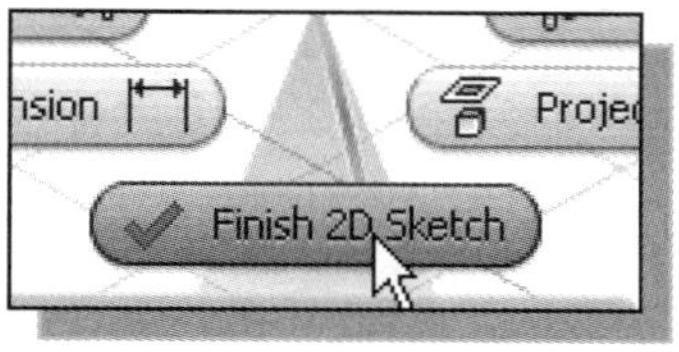

10. Inside the graphics window, click once with the right-mouse-button to display the option menu. Select **Finish 2D Sketch** in the pop-up menu to end the Sketch option.

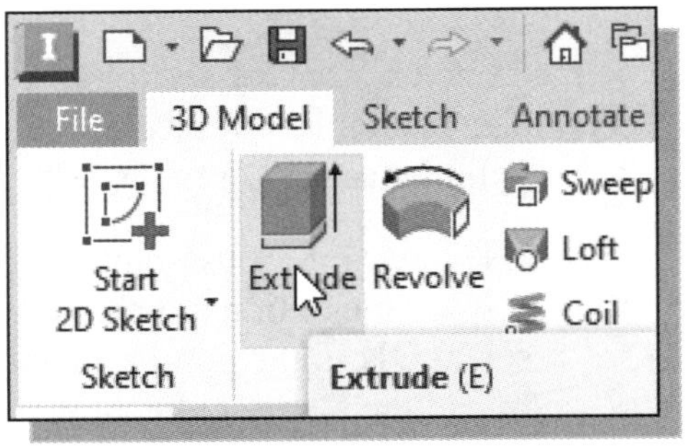

11. In the *3D Model tab*, select the **Extrude** command by left-clicking on the icon.

12. Click on the inside of the sketched circle as the profile to be extruded.

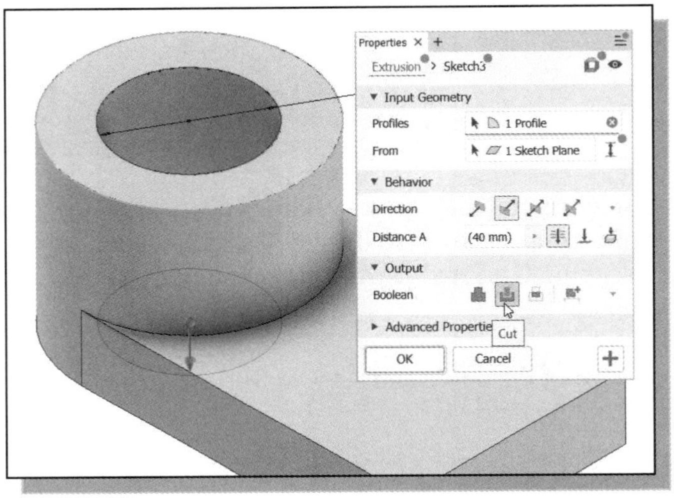

13. In the *Extrusion* pop-up window, set the operation option to **Cut**. Select **Through All** as the *Extents* option, as shown below. Confirm the arrowhead points downward.

14. Click on the **OK** button to proceed with the *Cut* operation.

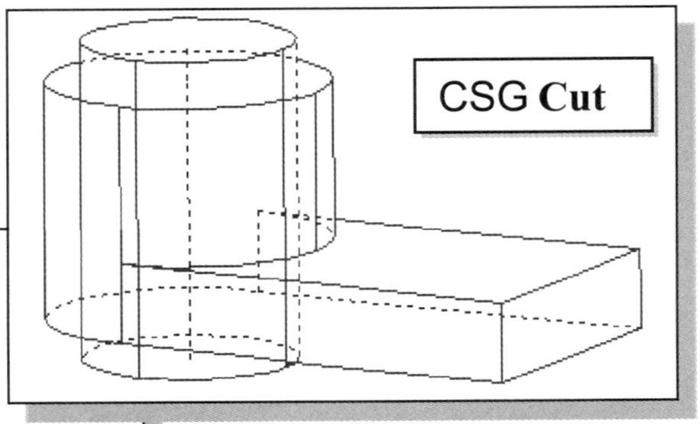

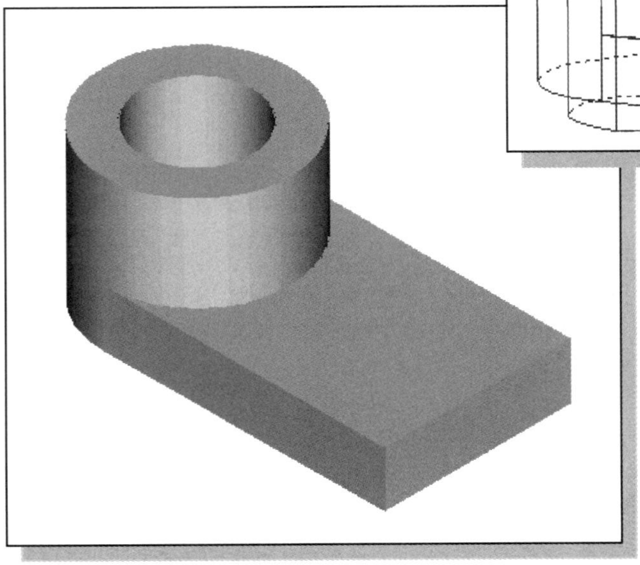

- The circular volume is removed from the solid model; the *CSG-Cut* operation resulted in a single solid.

Sketched Features vs. Placed Features

In Autodesk Inventor, there are two types of geometric features: **placed features** and **sketched features**. The last cut feature we created is a *sketched feature*, where we created a rough sketch and performed an extrusion operation. We can also create a hole feature, which is a placed feature. A *placed feature* is a feature that does not need a sketch and can be created automatically. Holes, slots, fillets, chamfers, and shells are all placed features.

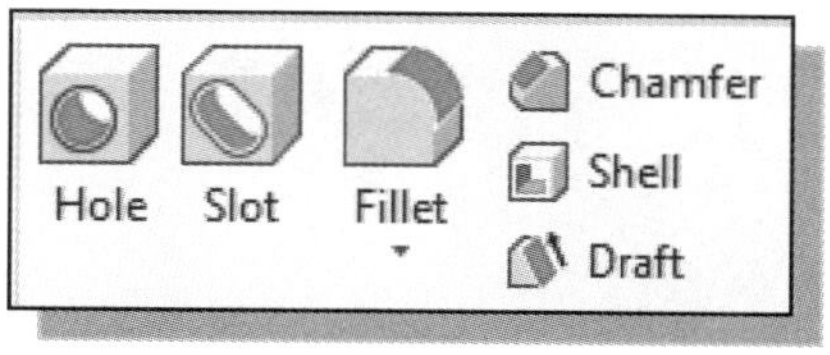

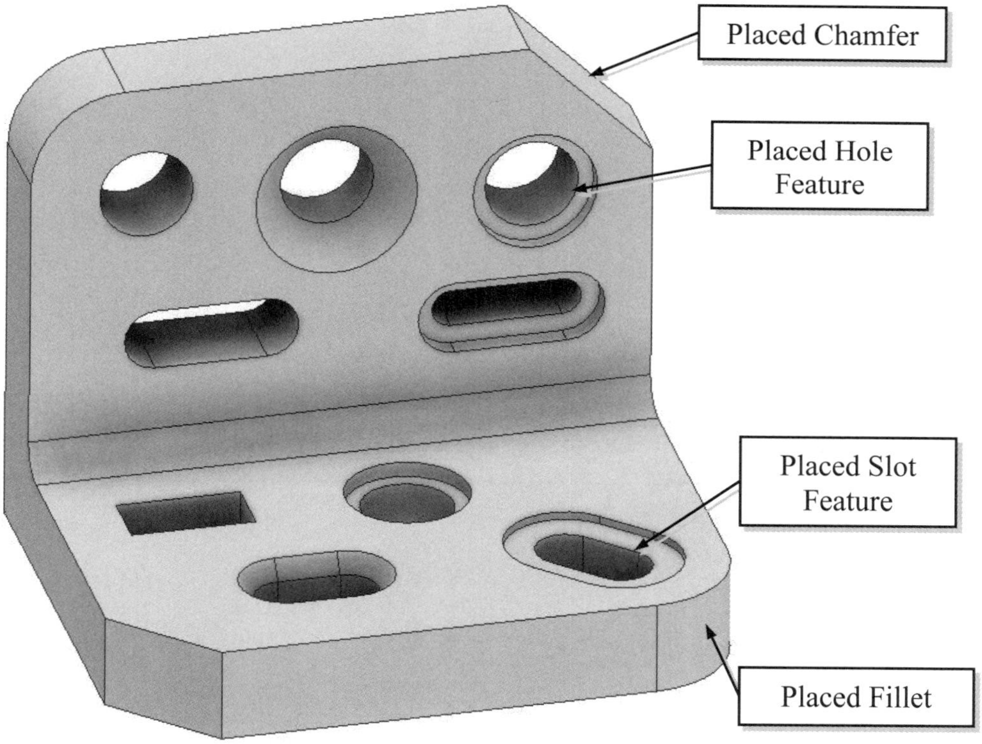

Hole tool: Quickly place holes using presets, or specify the hole dimensions, drill point, termination, and thread type options using the property panel.

Slot tool: Quickly place slots using presets, or specify the slot dimensions. Other options, such as sharp corner, fillets, counter bore, counter sink options.

Fillet tool: Quickly place rounded corners, both curving inward and outward. Basic option is done by selecting edges; more complicated options such as varying radii can also be created.

Creating a Placed Hole Feature

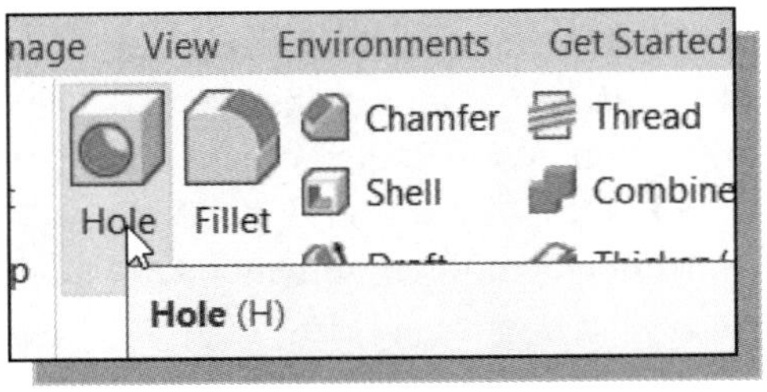

1. In the *Create* toolbar, select the **Hole** command by left-clicking on the icon.

2. Pick a location inside the top horizontal surface of the base feature as shown.

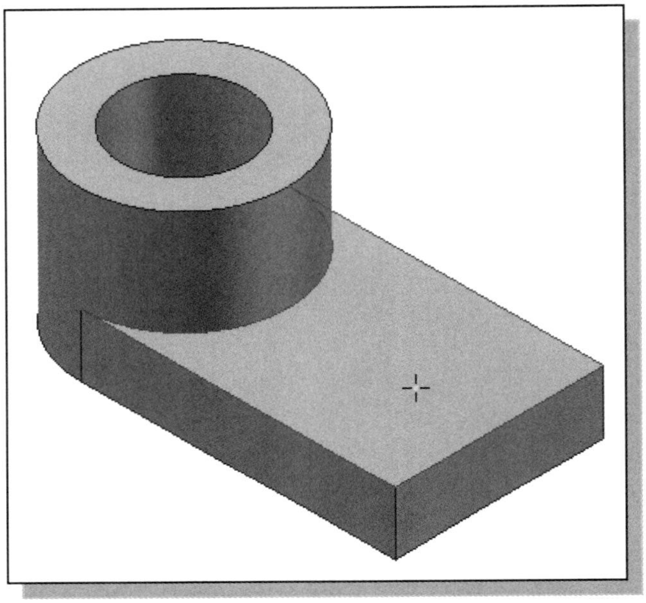

3. Enter **20 mm** as the diameter of the hole as shown. **Do Not** click the OK button yet.

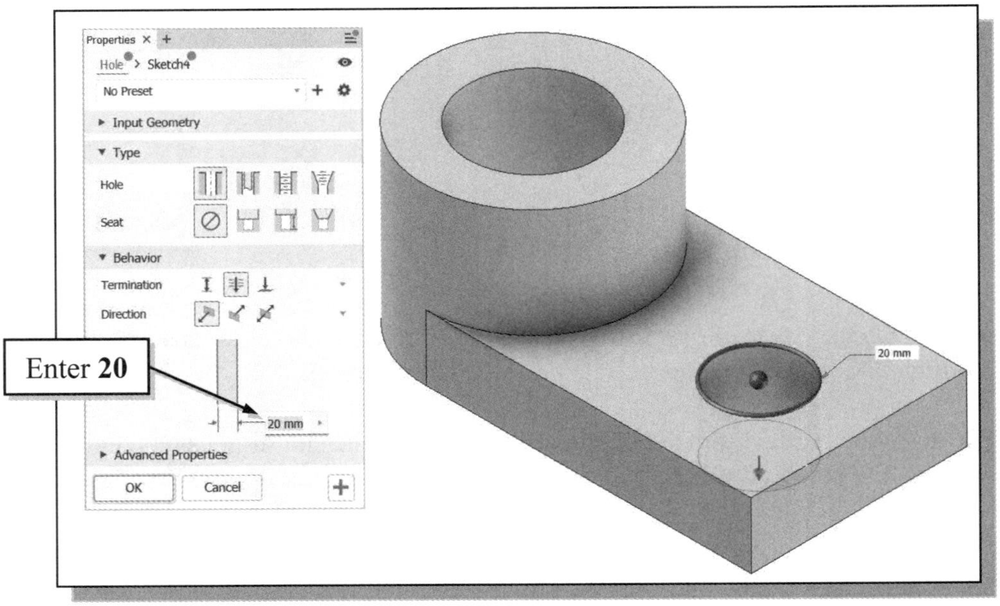

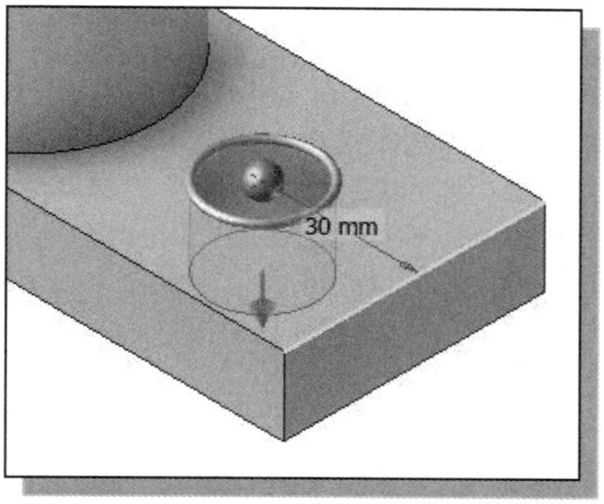

4. Pick the **right-edge** of the top face of the base feature as shown. This will be used as the first reference for placing the hole on the plane.

5. Enter **30** mm as the distance as shown. **Do Not** click the OK button yet.

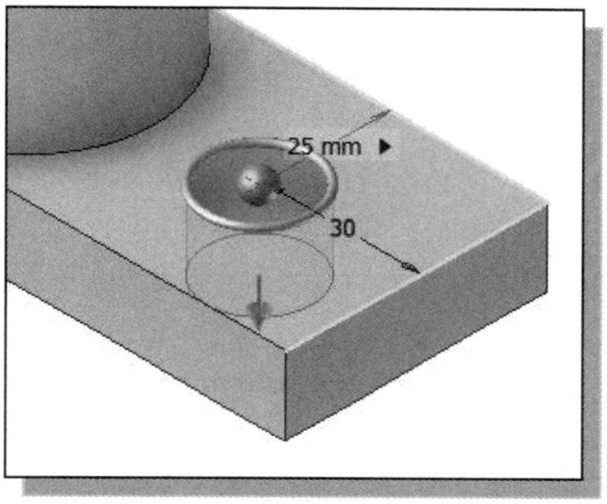

6. Pick the **adjacent edge** of the top face as shown. This will be used as the second reference for placing the hole on the plane.

7. Enter **25** mm as the distance as shown.

8. In *Holes* dialog box, set the *Termination* option to **Through All**.

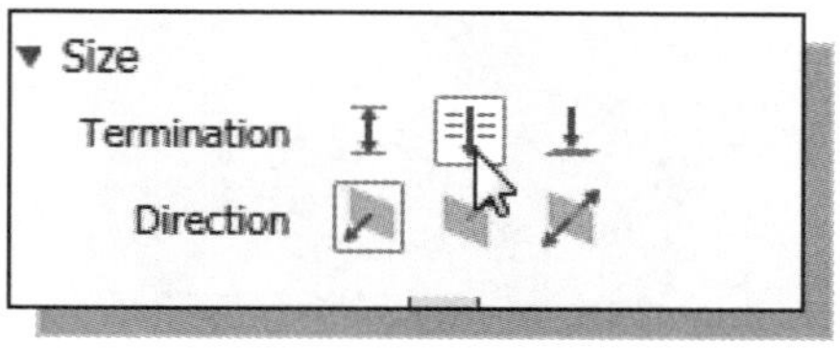

9. Click on the **OK** button to proceed with the *Hole* feature.

- The circular volume is removed from the solid model; the *CSG-Cut* operation resulted in a single solid.

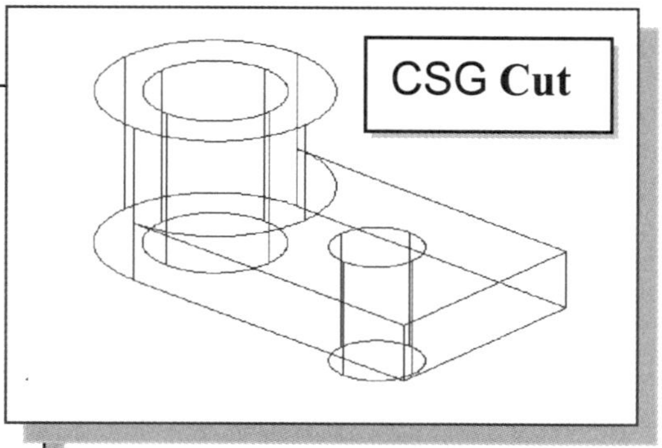

Creating a Rectangular Cut Feature

Next create a rectangular cut as the last solid feature of the ***Locator***.

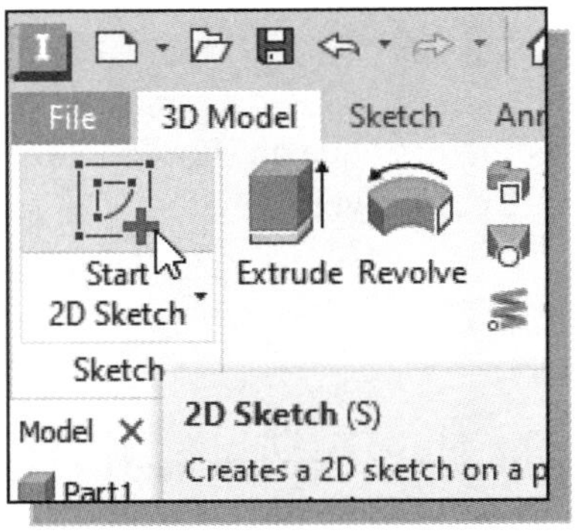

1. In the *3D Model* tab select the **Start 2D Sketch** command by left-clicking once on the icon.

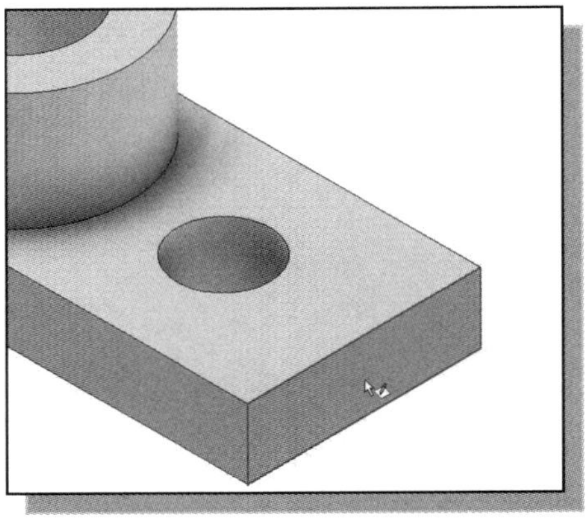

2. Pick the right face of the base feature as shown.

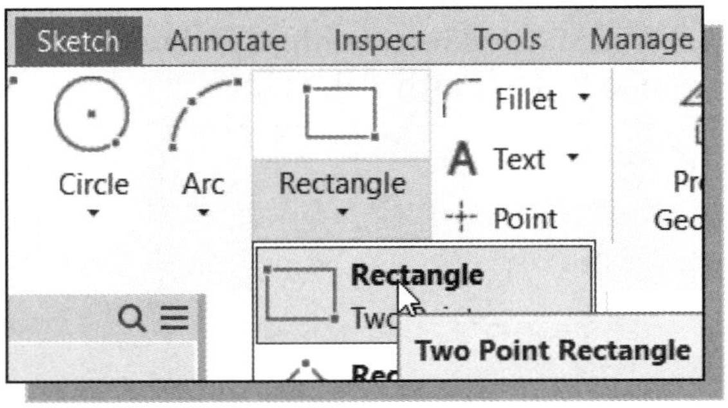

3. Select the **Two point rectangle** command by clicking once with the left-mouse-button on the icon in the *Sketch tab* of the Ribbon toolbar.

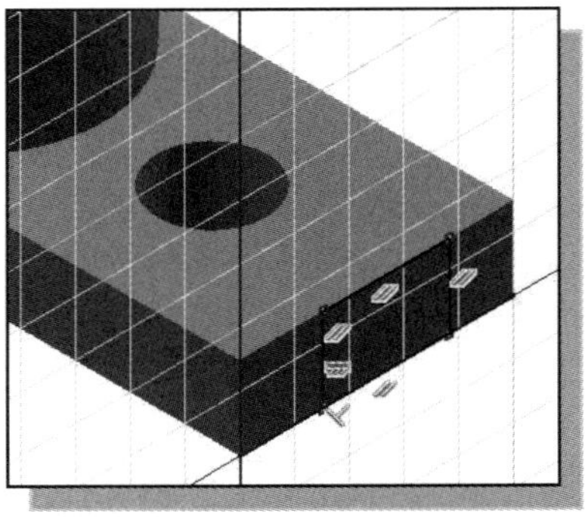

4. Create a rectangle that is aligned to the top and bottom edges of the base feature as shown. (Hit [**F6**] to set the display orientation if necessary.)

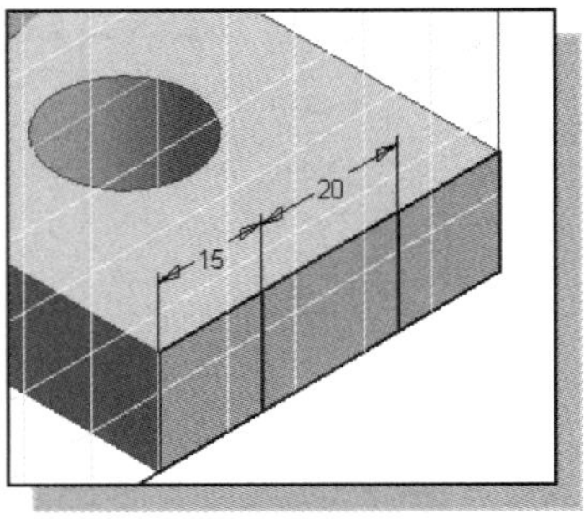

5. On your own, create and modify the two dimensions as shown.

6. Select **Finish Sketch** in the *Ribbon* toolbar to end the Sketch option.

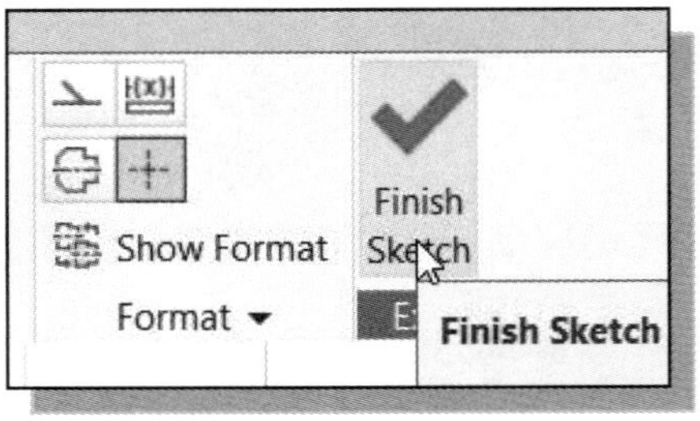

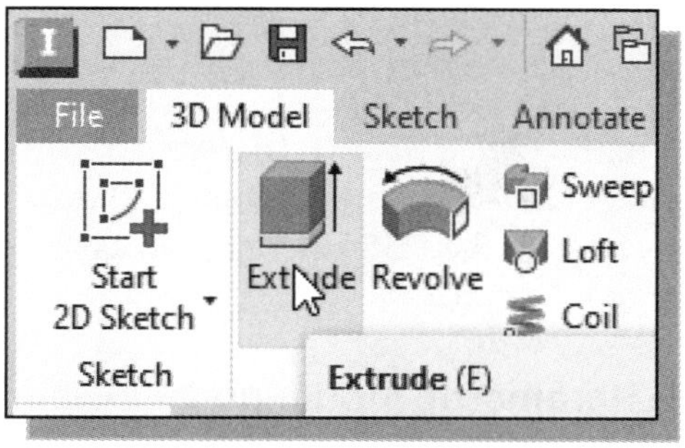

7. In the *3D Model* tab, select the **Extrude** command by left-clicking on the icon.

8. In the *Extrude* pop-up window, the **Profile** button is pressed down; Autodesk Inventor expects us to identify the profile to be extruded. Move the cursor inside the rectangle we just created and left-click once to select the region as the profile to be extruded.

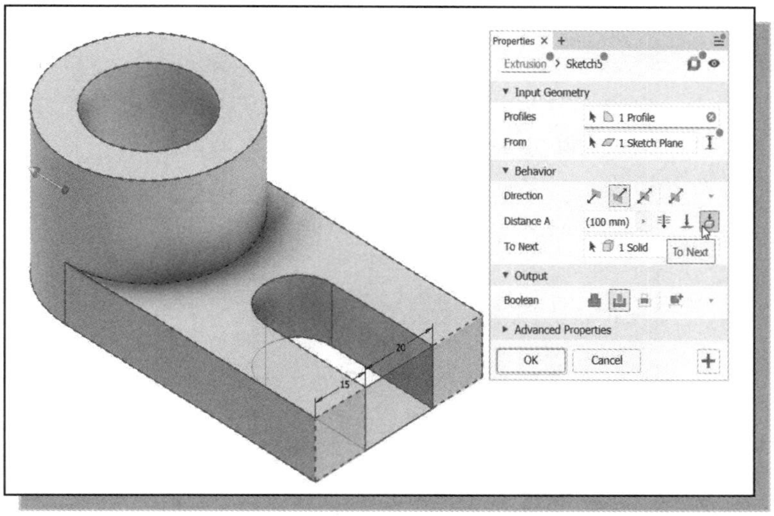

9. In the *Extrude* pop-up window, set the operation option to **Cut**. Select **To Next** as the *Distance* option as shown. Set the arrowhead points toward the center of the solid model.

10. Click on the **OK** button to create the *Cut* feature and complete the design.

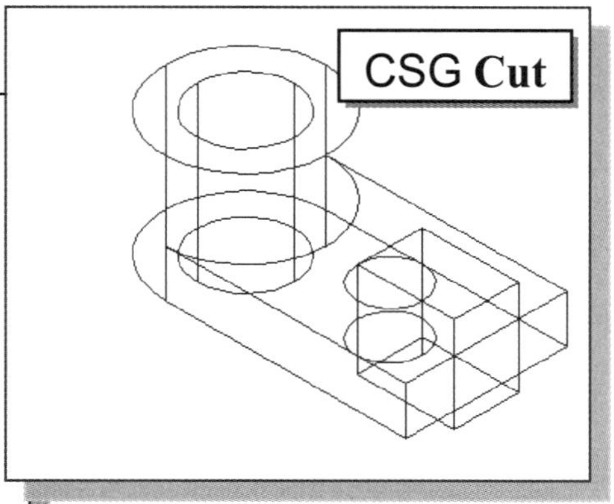

Save the Model

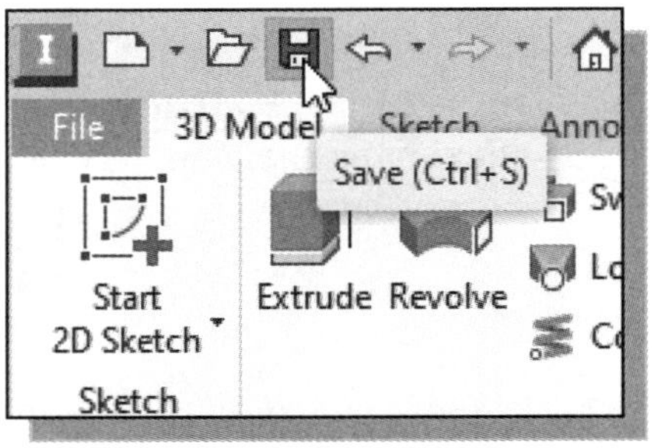

1. Select **Save** in the *Quick Access Toolbar*, or you can also use the "**Ctrl-S**" combination (hold down the "Ctrl" key and hit the "S" key once) to save the part.

2. Switch to the **Parametric Modeling** *folder* if it is not the current folder.

3. In the *Save As* dialog box, **right-click** once in the *list area* to bring up the *option menu*.

4. In the *option list*, select **New** as shown.

5. In the second *option list*, select **Folder** to create a subfolder.

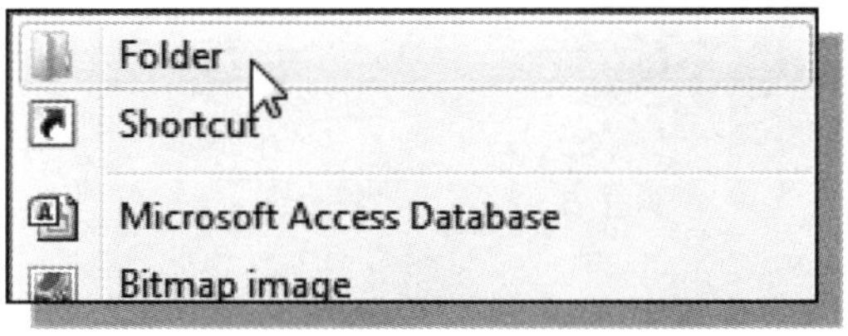

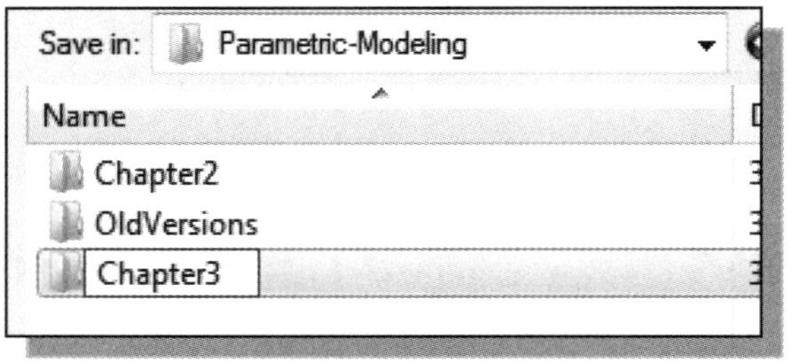

6. Enter **Chapter3** as the new folder name as shown.

7. **Double-click** on the Chapter3 folder to open it.

8. In the *file name* editor box, enter **Locator** as the file name.

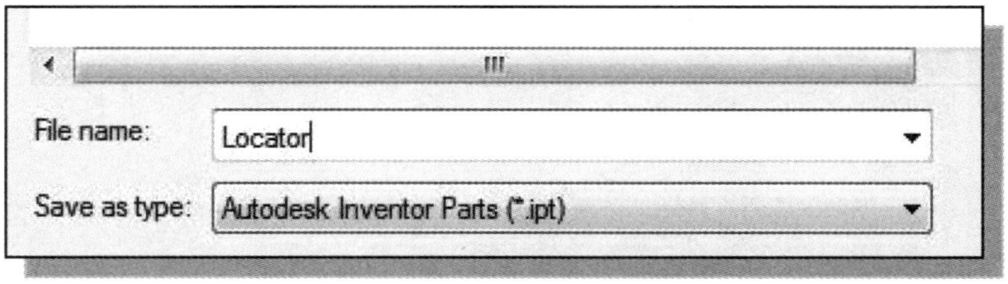

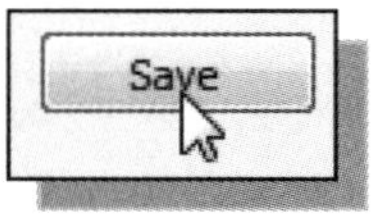

9. Click on the **Save** button to save the file.

Review Questions: (Time: 20 minutes.)

1. List and describe three basic *Boolean operations* commonly used in computer geometric modeling software.

2. What is a *primitive solid*?

3. What does *CSG* stand for?

4. Which *Boolean operation* keeps only the volume common to the two solid objects?

5. What is the main difference between a *CUT feature* and a *HOLE feature* in Autodesk Inventor?

6. Create the following 2D Sketch and measure the associated area and perimeter.

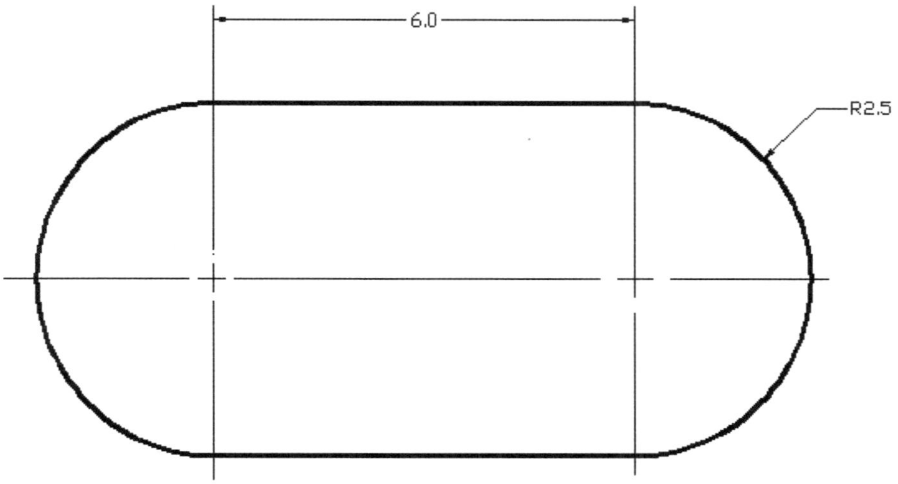

7. Using the CSG concepts, create *Binary Tree* sketches showing the steps you plan to use to create the two models shown on the next page:

Exercises: Create and save the exercises in the Chapter3 folder. (Time: 180 minutes.)

1. **Latch Clip** (Dimensions are in inches. Thickness: **0.25** inches.)

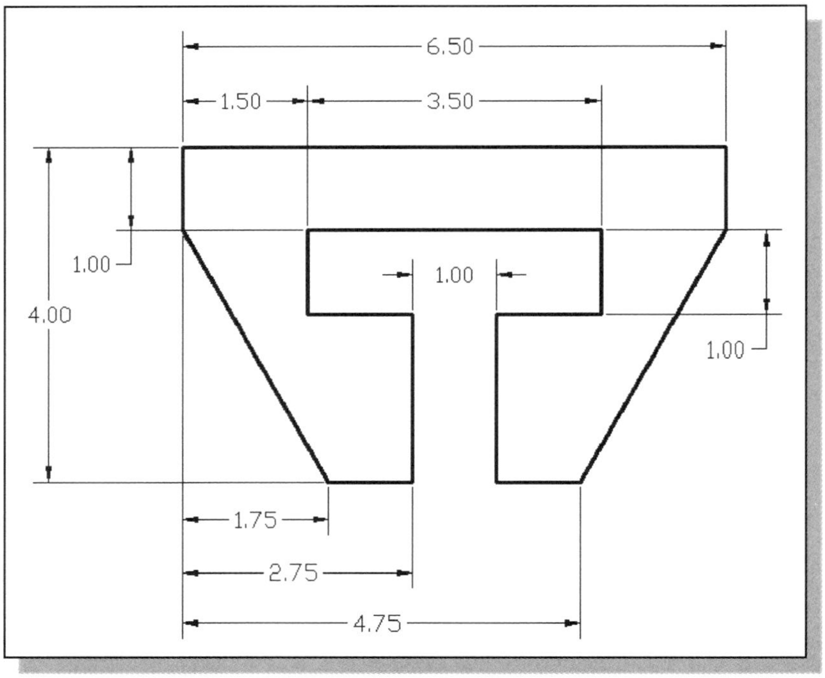

2. **Guide Plate** (Dimensions are in inches. Thickness: **0.25** inches. Boss height **0.125** inches.)

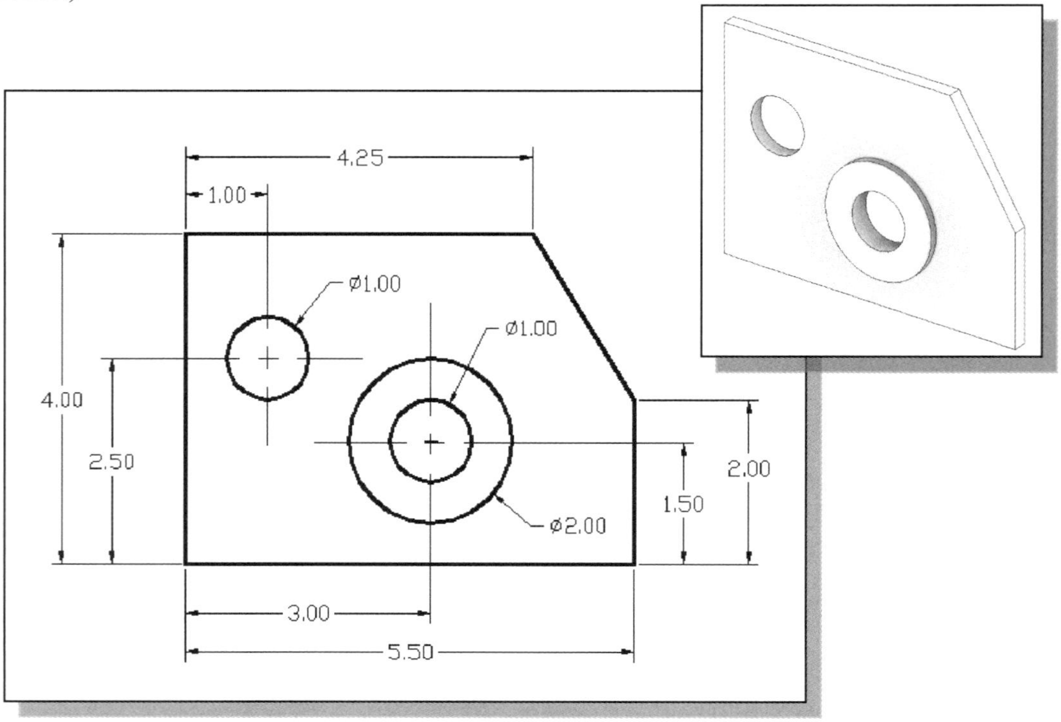

3. **Angle Slider** (Dimensions are in Millimeters.)

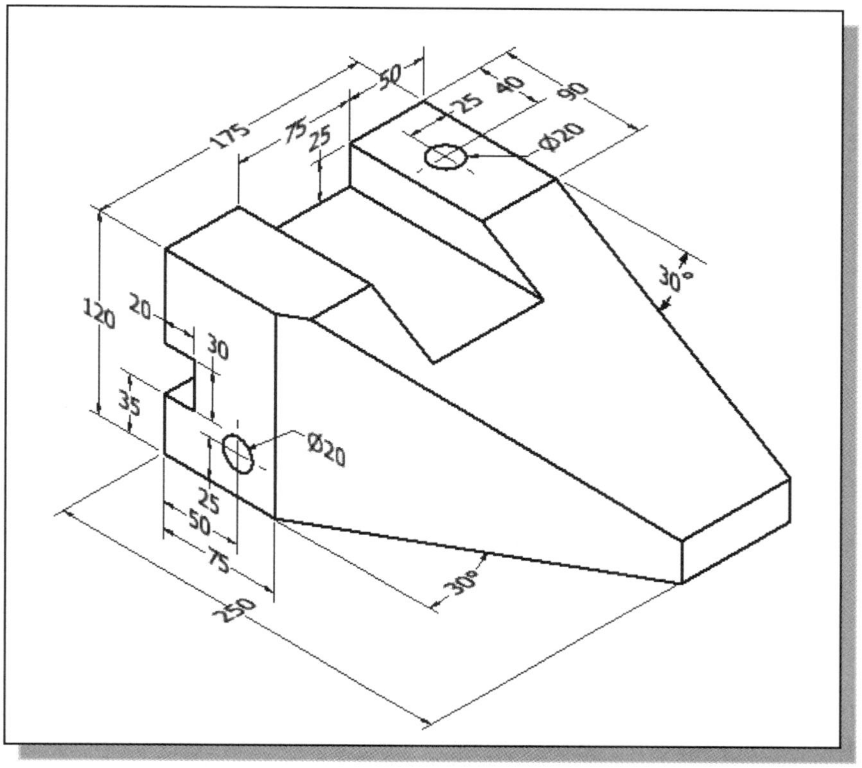

4. **Coupling Base** (Dimensions are in inches.)

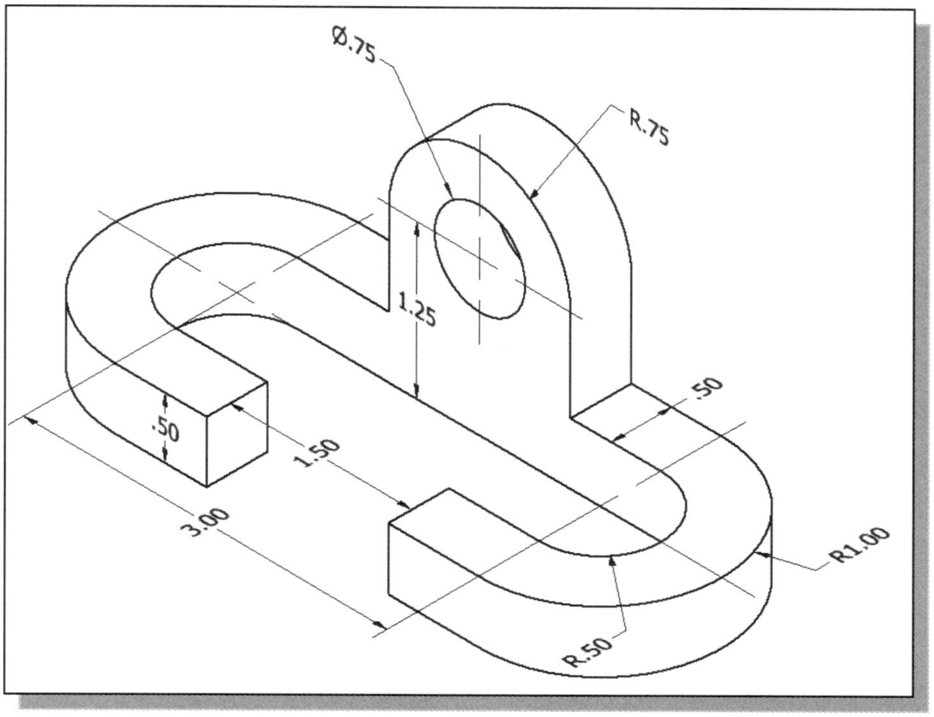

5. **Indexing Guide** (Dimensions are in inches.)

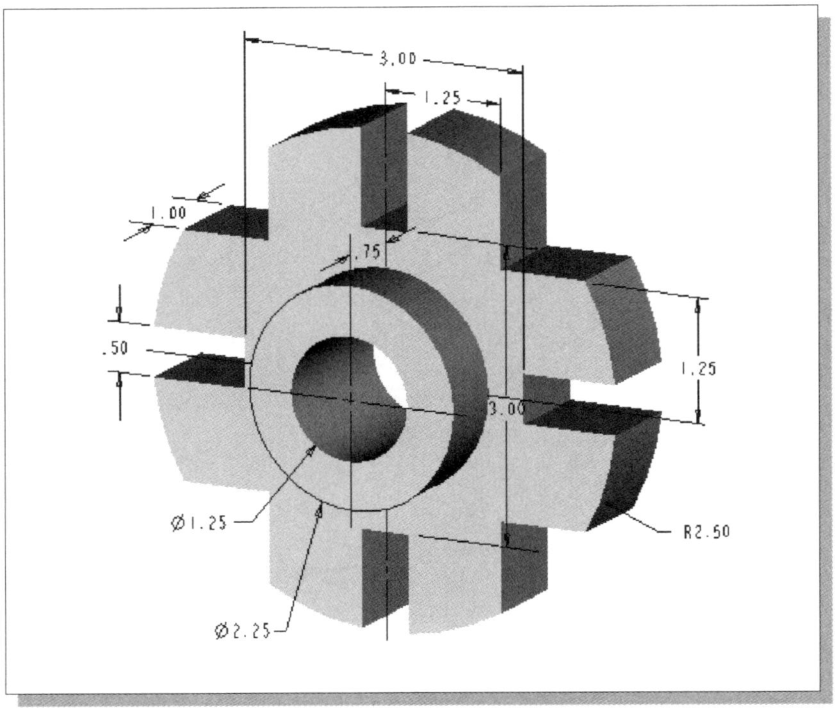

6. **L-Bracket** (Dimensions are in inches.)

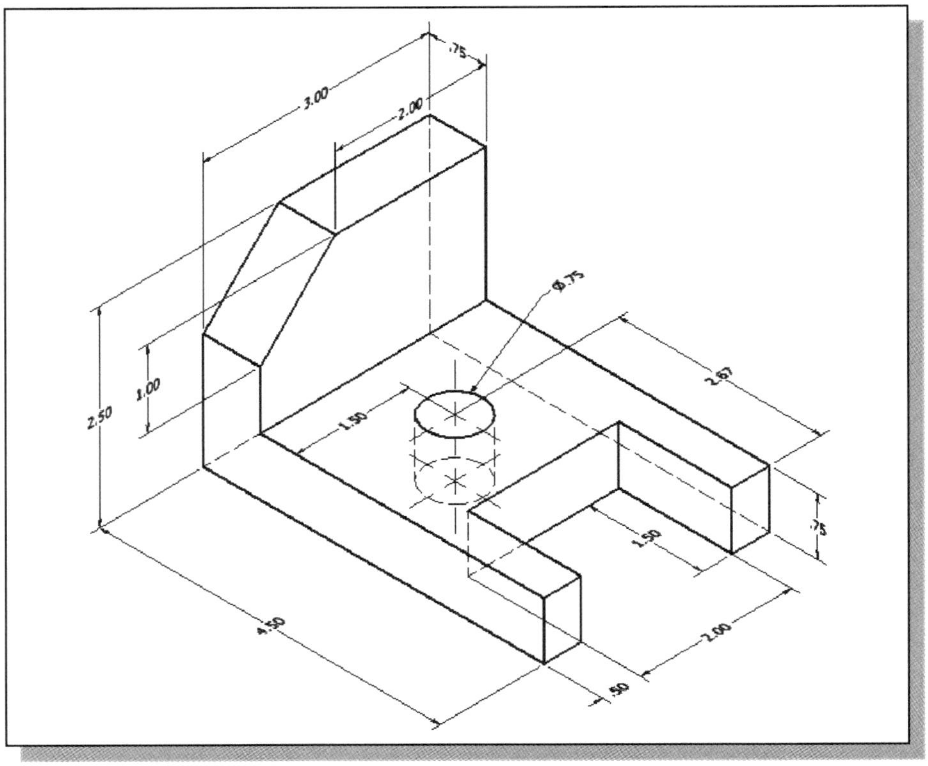

7. **Slide Base** (Dimensions are in inches.)

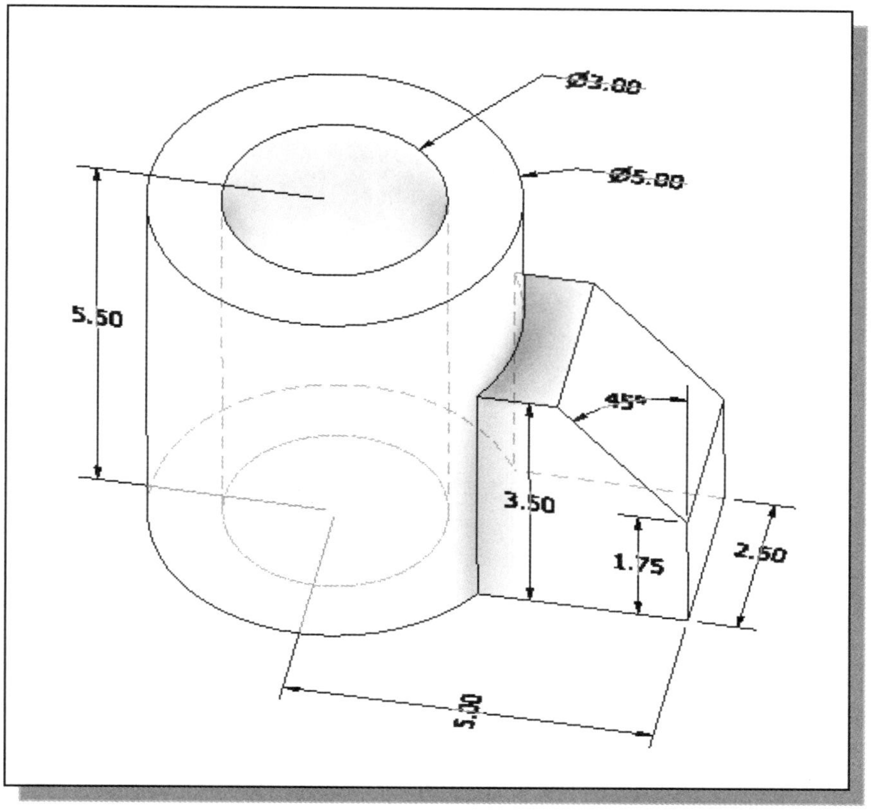

Notes:

Chapter 4
Model History Tree

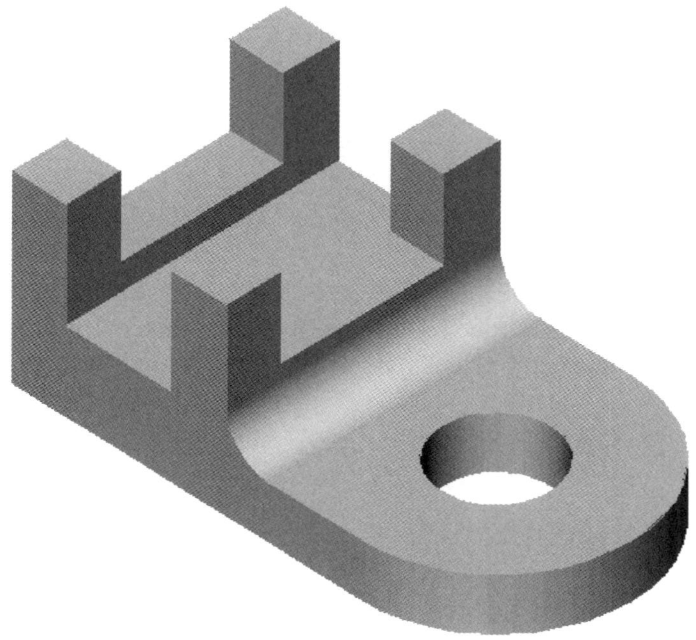

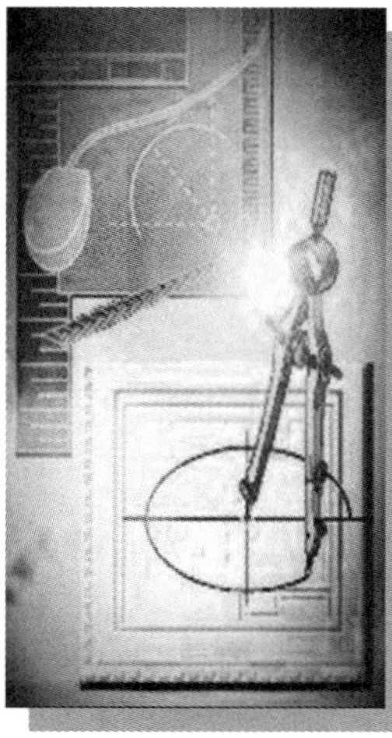

Learning Objectives

- Understand Feature Interactions
- Use the Part Browser
- Modify and Update Feature Dimensions
- Perform History-Based Part Modifications
- Change the Names of Created Features
- Implement Basic Design Changes

Autodesk Inventor Certified User Exam Objectives Coverage

Parametric Modeling Basics:

Section 3: Sketches

Objectives: Creating 2D Sketches, Draw Tools, Sketch Constraints, Pattern Sketches, Modify Sketches, Format Sketches, Sketch Doctor, Shared Sketches, Sketch Parameters.

Section 4: Parts

Objectives: Creating parts, Work Features, Pattern Features, Part Properties.

Introduction

In Autodesk Inventor, the **design intents** are embedded into features in the **history tree**. The structure of the model history tree resembles that of a **CSG binary tree**. A CSG binary tree contains only *Boolean relations*, while the **Autodesk Inventor history tree** contains all features, including *Boolean relations*. A history tree is a sequential record of the features used to create the part. This history tree contains the construction steps, plus the rules defining the design intent of each construction operation. In a history tree, each time a new modeling event is created previously defined features can be used to define information such as size, location, and orientation. It is therefore important to think about your modeling strategy before you start creating anything. It is important, but also difficult, to plan ahead for all possible design changes that might occur. This approach in modeling is a major difference in **FEATURE-BASED CAD SOFTWARE**, such as Autodesk Inventor, from previous generation CAD systems.

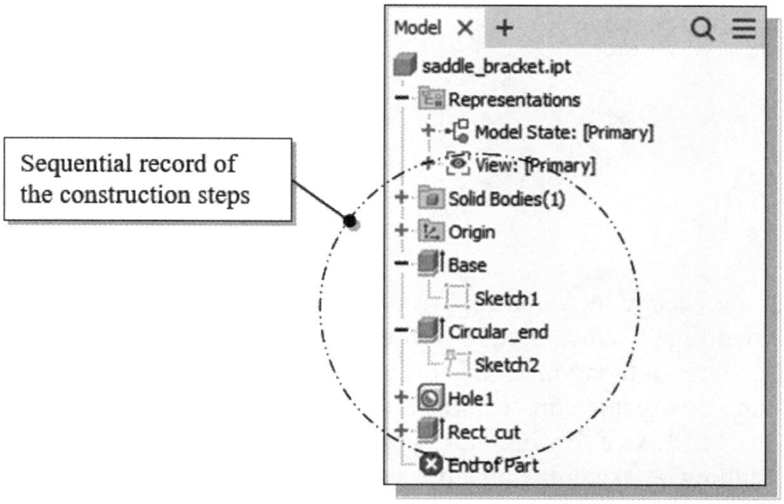

Feature-based parametric modeling is a cumulative process. Every time a new feature is added, a new result is created, and the feature is also added to the history tree. The database also includes parameters of features that were used to define them. All of this happens automatically as features are created and manipulated. At this point, it is important to understand that all of this information is retained, and modifications are done based on the same input information. In parametric modeling, it is important to consider the order of necessary features to create the model; avoid adding extra items in the history tree. Also consider creating as few features as possible to make the design creation more efficient and design editing easier.

In Autodesk Inventor, the history tree gives information about modeling order and other information about the feature. Part modifications can be accomplished by accessing the features in the history tree. It is therefore important to understand and utilize the feature history tree to modify designs. Autodesk Inventor remembers the history of a part, including all the rules that were used to create it, so that changes can be made to any operation that was performed to create the part. In Autodesk Inventor, to modify a feature we access the feature by selecting the feature in the ***browser*** window.

The Saddle Bracket Design

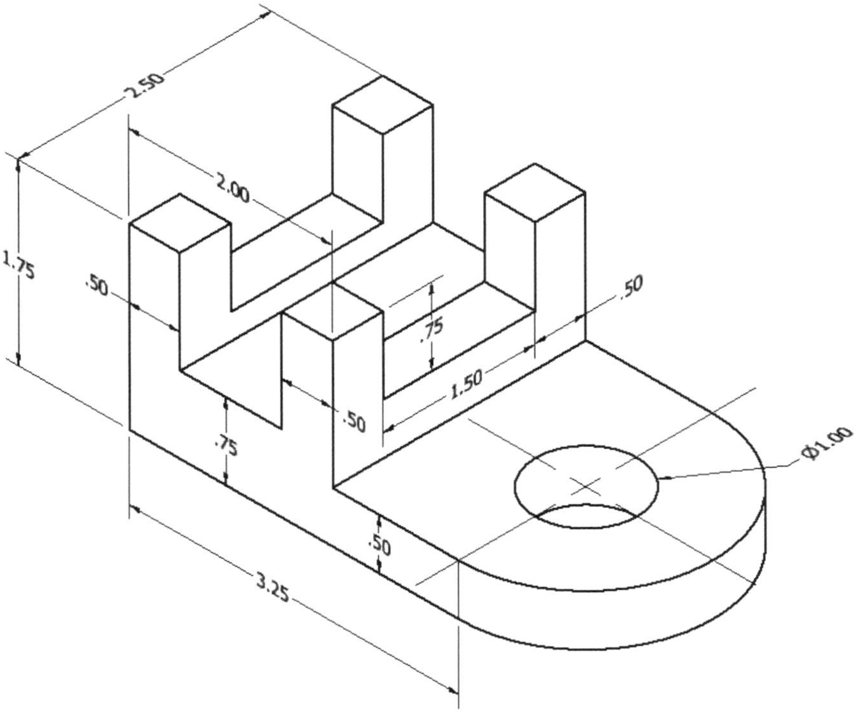

❖ Based on your knowledge of Autodesk Inventor so far, how many features would you use to create the design? Which feature would you choose as the **BASE FEATURE**, the first solid feature, of the model? What is your choice in arranging the order of the features? Would you organize the features differently if additional fillets were to be added in the design? Take a few minutes to consider these questions and do preliminary planning by sketching on a piece of paper. You are also encouraged to create the model on your own prior to following through the tutorial.

Starting Autodesk Inventor

1. Select the **Autodesk Inventor** option on the *Start* menu or select the **Autodesk Inventor** icon on the desktop to start Autodesk Inventor. The Autodesk Inventor main window will appear on the screen.

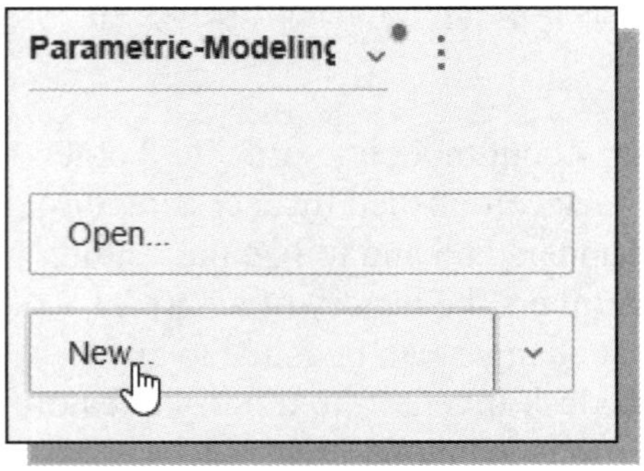

2. Once the program is loaded into memory, select the **New File** icon with a single click of the left-mouse-button as shown.

3. Select the **en-US->English** tab, and in the *Part template* area select **Standard(in).ipt**.

4. Click **Create** in the *New File* dialog box to accept the selected settings to start a new model.

Modeling Strategy

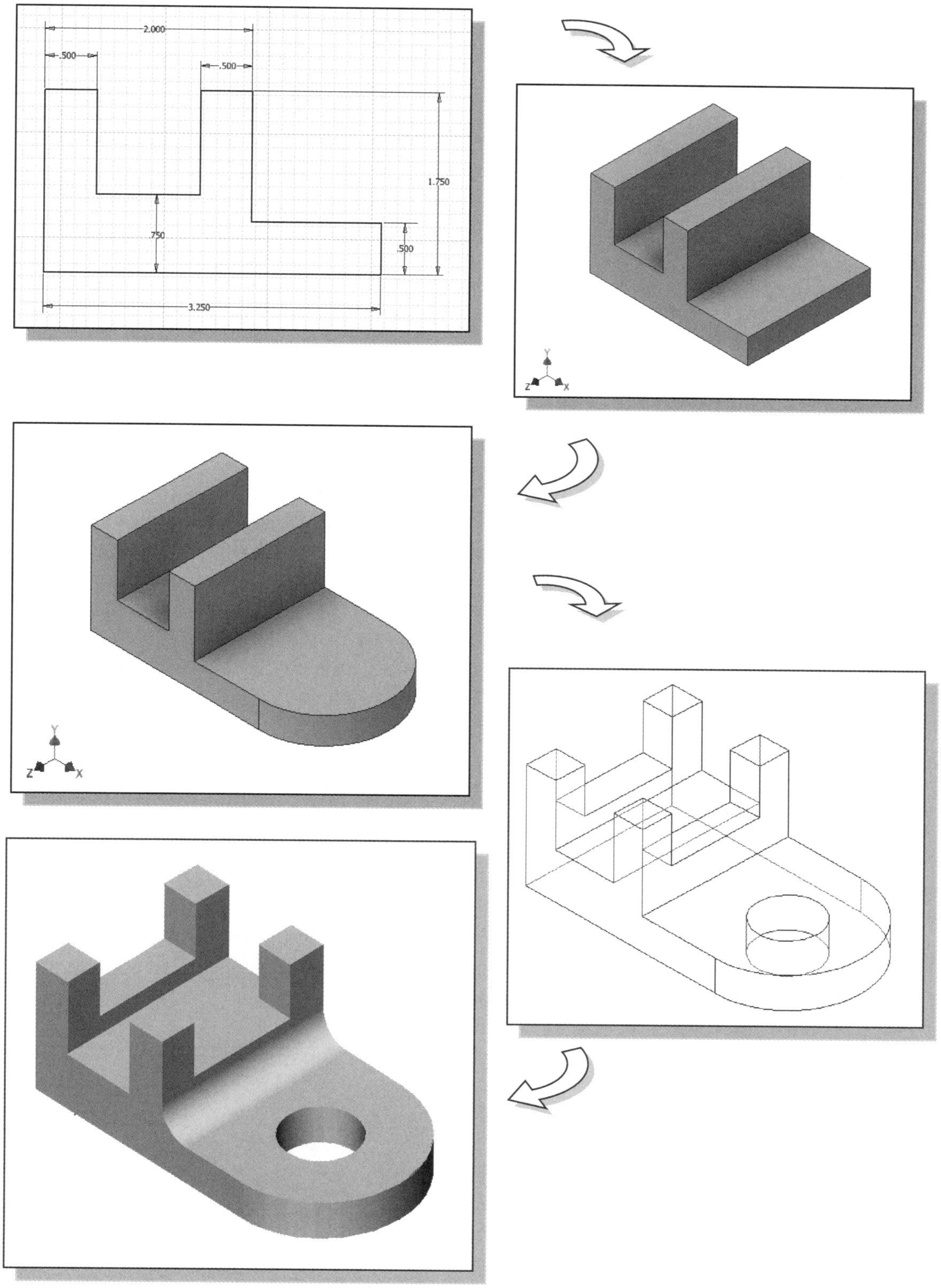

The Autodesk Inventor Browser

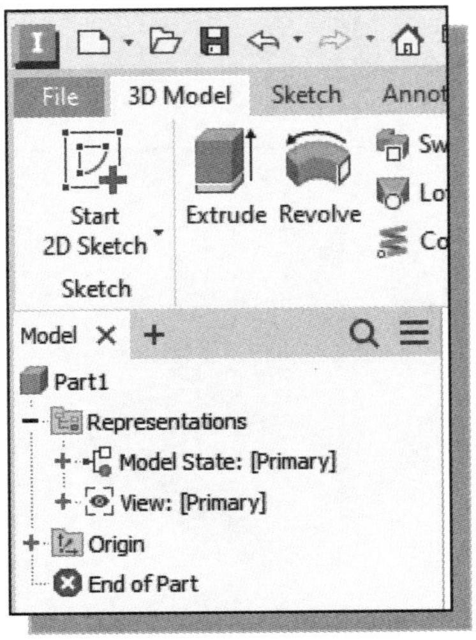

- In the Autodesk Inventor screen layout, the ***browser*** is located to the left of the graphics window. Autodesk Inventor can be used for part modeling, assembly modeling, part drawings, and assembly presentation. The *browser* window provides a visual structure of the features, constraints, and attributes that are used to create the part, assembly, or scene. The *browser* also provides right-click menu access for tasks associated specifically with the part or feature, and it is the primary focus for executing many of the Autodesk Inventor commands.

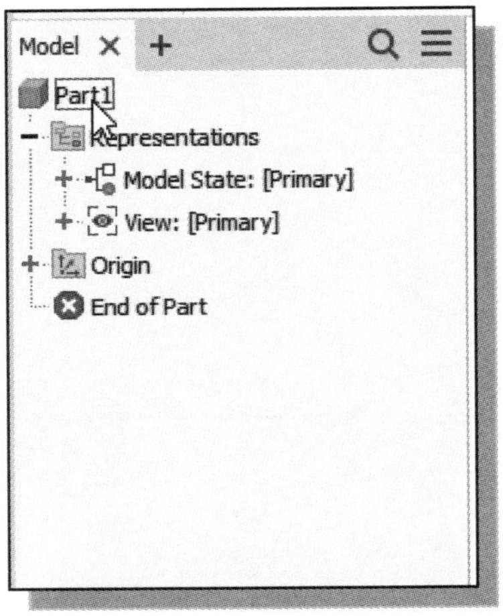

- The first item displayed in the *browser* is the name of the part, which is also the file name. By default, the name "Part1" is used when we first started Autodesk Inventor. The *browser* can also be used to modify parts and assemblies by moving, deleting, or renaming items within the hierarchy. Any changes made in the *browser* directly affect the part or assembly and the results of the modifications are displayed on the screen instantly. The *browser* also reports any problems and conflicts during the modification and updating procedure.

Creating the Base Feature

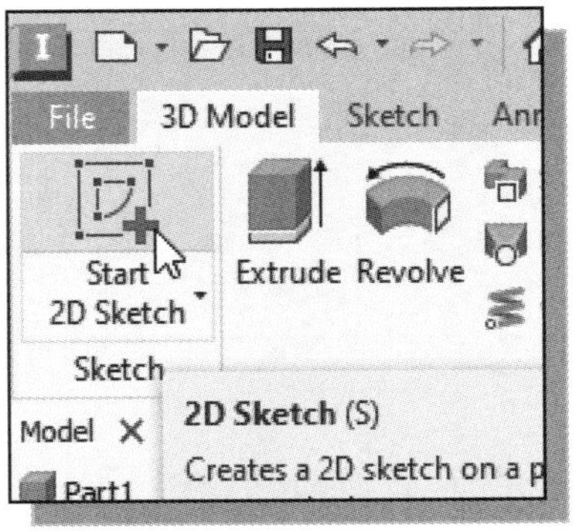

1. Move the graphics cursor to the **Start 2D Sketch** icon in the *Sketch toolbar* under the *3D Model tab*. A *Help-tip box* appears next to the cursor and a brief description of the command is displayed at the bottom of the drawing screen.

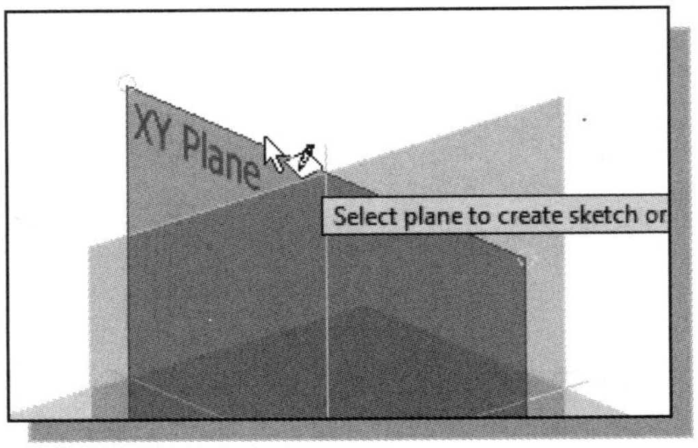

2. Move the cursor over the edge of the *XY Plane* in the graphics area. When the *XY Plane* is highlighted, click once with the **left-mouse-button** to select the *Plane* as the sketch plane for the new sketch.

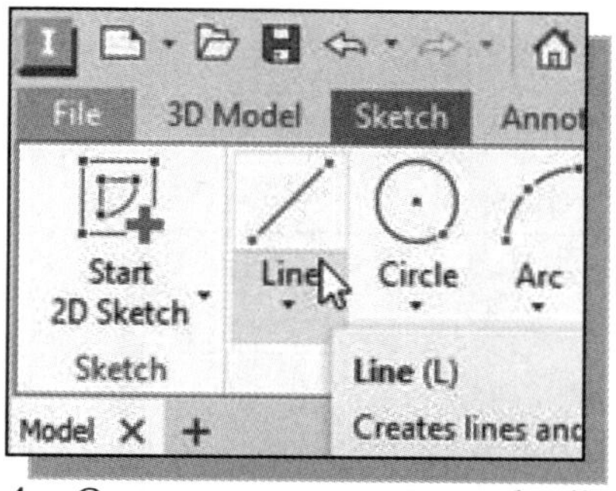

3. Select the **Line** icon by clicking once with the left-mouse-button; this will activate the Line command.

4. On your own, create and adjust the geometry by adding and modifying dimensions as shown below.

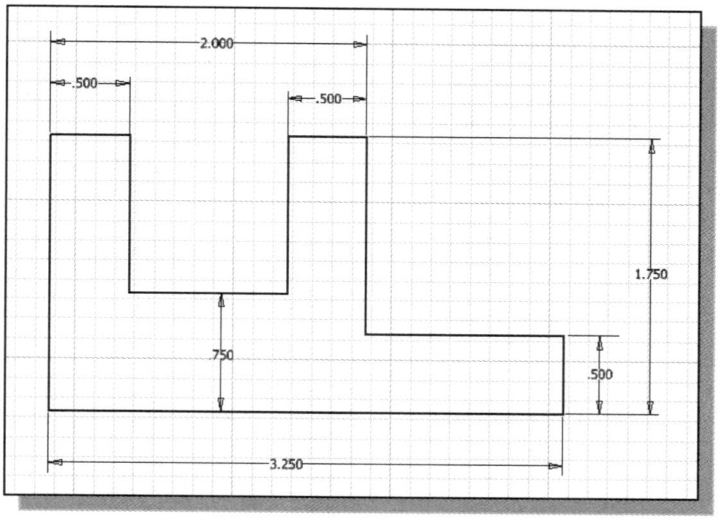

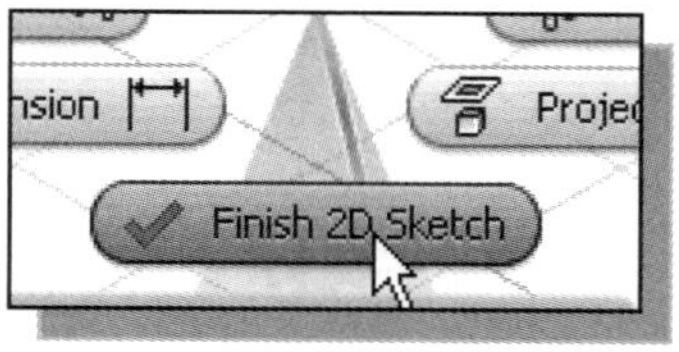

5. Inside the graphics window, click once with the right-mouse-button to display the option menu. Select **Finish 2D Sketch** in the pop-up menu to end the Sketch option.

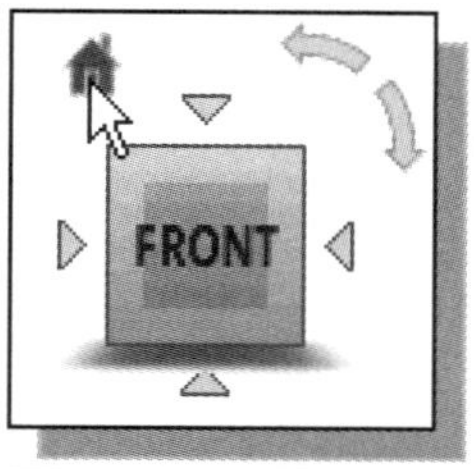

6. On your own, use the dynamic viewing functions to view the sketch. Click the home view icon to change the display to the *isometric* view before proceeding to the next step.

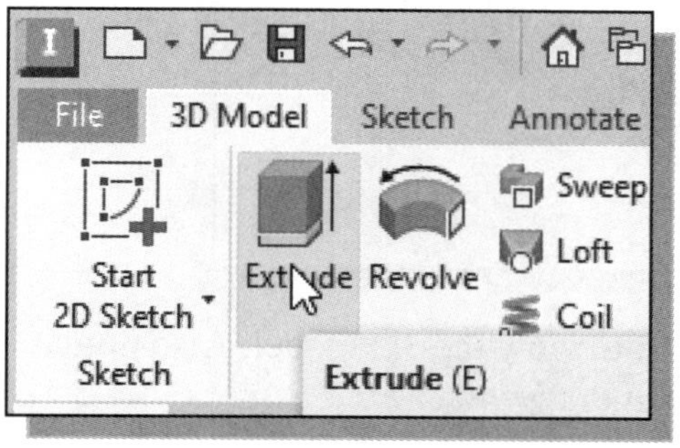

7. In the *Sketch toolbar* under the *3D Model tab*, select the **Extrude** command by left-clicking on the icon.

8. In the *Distance* option box, enter **2.5** as the total extrusion distance.

9. In the *Extrude* pop-up window, left-click once on the **Symmetric** icon. The **Symmetric** option allows us to extrude in both directions of the sketched profile.

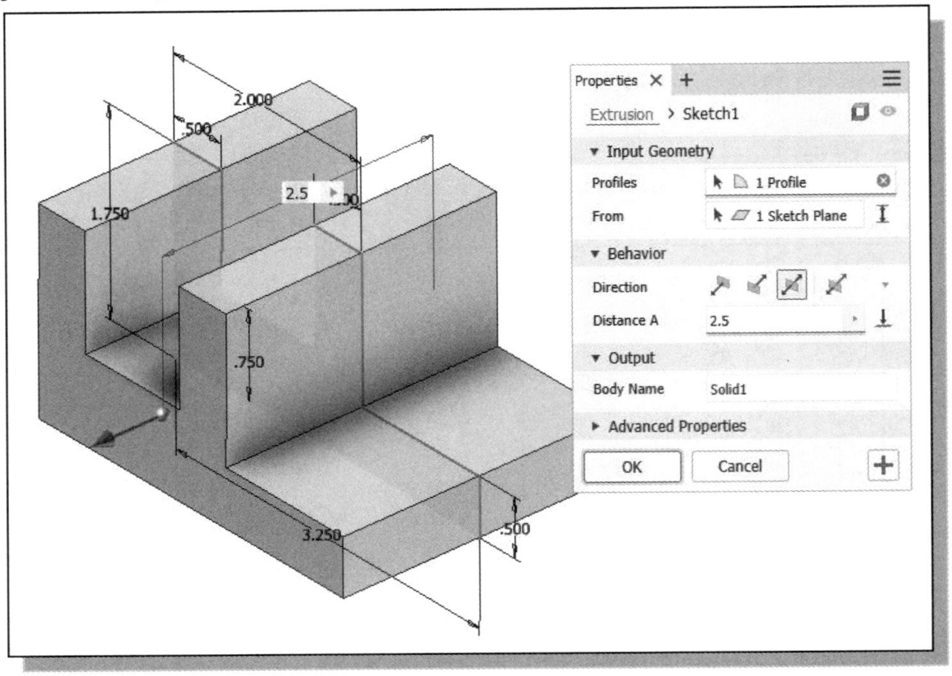

10. Click on the **OK** button to accept the settings and create the base feature.

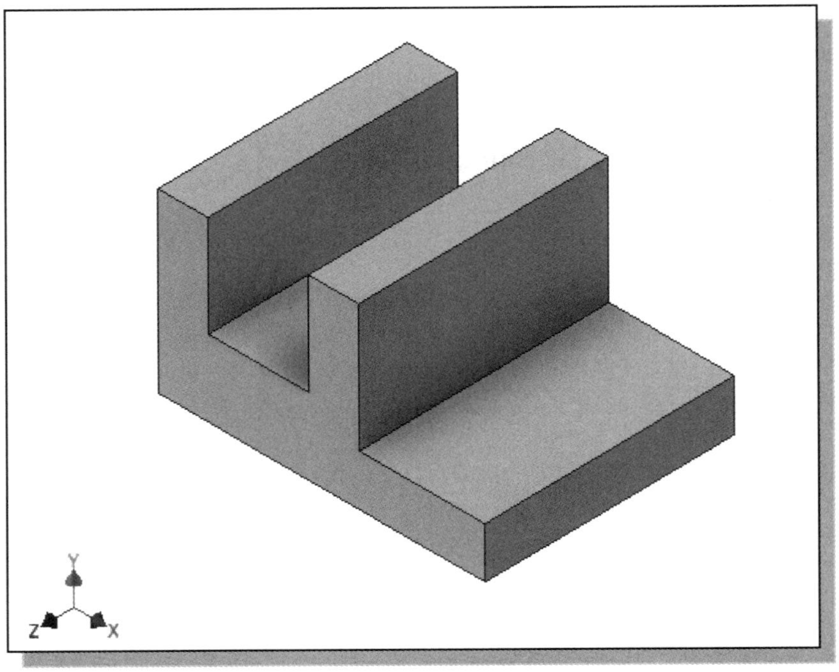

➢ On your own, use the *Dynamic Viewing* functions to view the 3D model. Also notice the extrusion feature is added to the *Model Tree* in the *browser* area.

Adding the Second Solid Feature

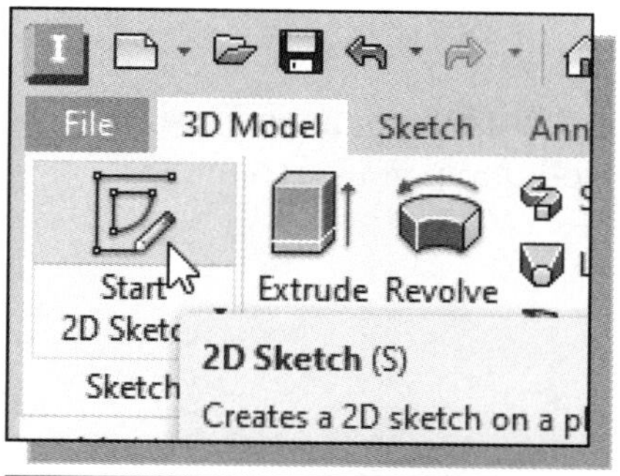

1. In the *Sketch toolbar* under the *3D Model tab* select the **Start 2D Sketch** command by left-clicking once on the icon.

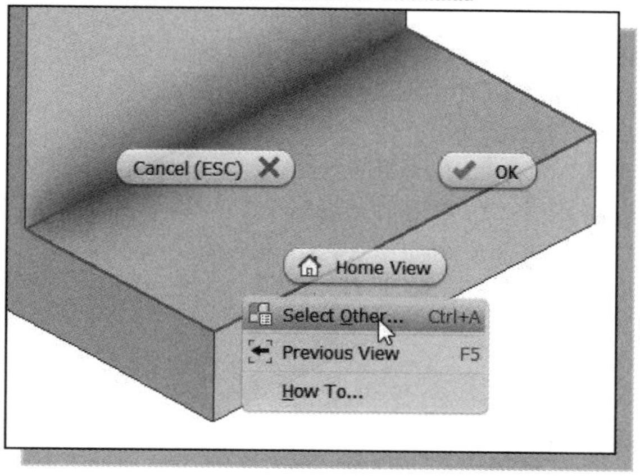

2. In the *Status Bar* area, the message "*Select plane to create sketch or an existing sketch to edit.*" is displayed. Move the cursor inside the upper horizontal face of the 3D object as shown below.

3. Click once with the **right-mouse-button** to bring up the option menu and choose **Select Other** to switch to the next feasible choice.

4. On your own, click on the down arrow to examine all possible surface selections.

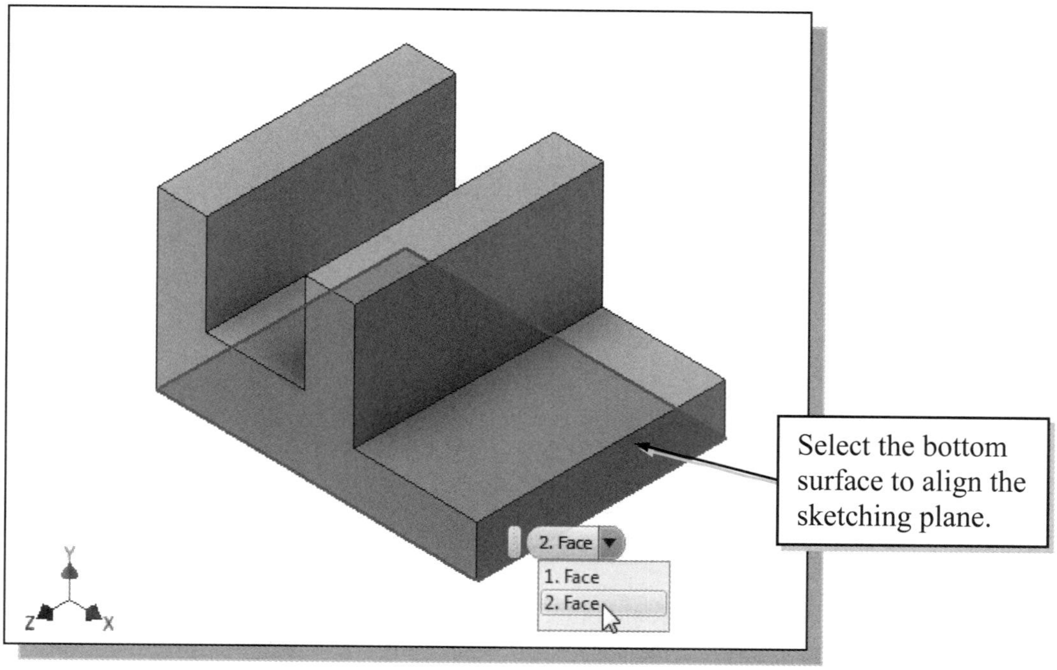

5. Select the **bottom horizontal face** of the solid model when it is highlighted as shown in the above figure.

Creating a 2D Sketch

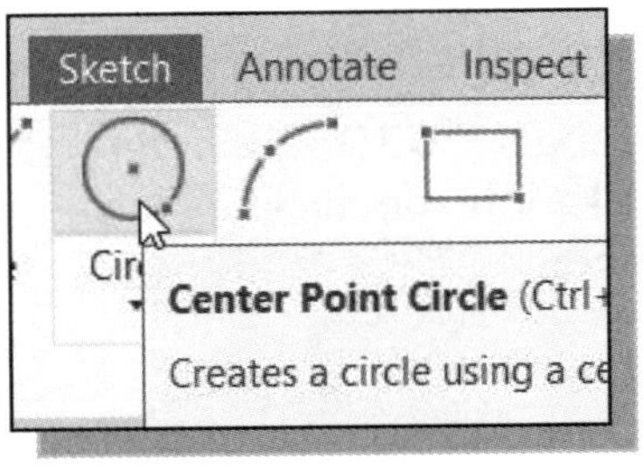

1. Select the **Center Point Circle** command by clicking once with the left-mouse-button on the icon in the *Sketch* tab.

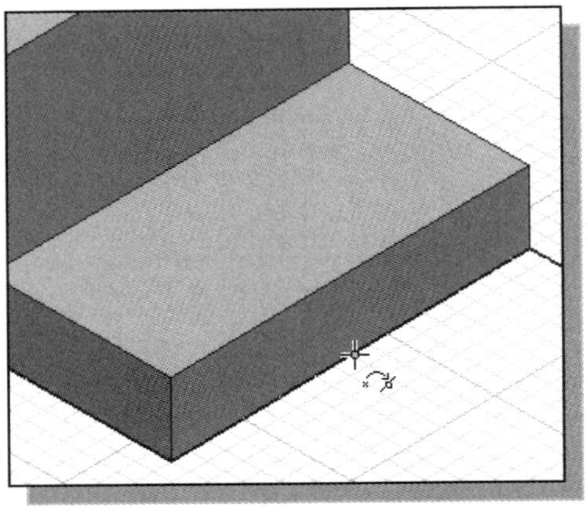

➢ We will align the center of the circle to the midpoint of the base feature.

2. On your own, use the snap to midpoint option to pick the midpoint of the edge when the midpoint is displayed with GREEN color as shown in the figure. (Hit [F6] to set the display orientation if necessary.)

3. Select the front corner of the base feature to create a circle as shown below.

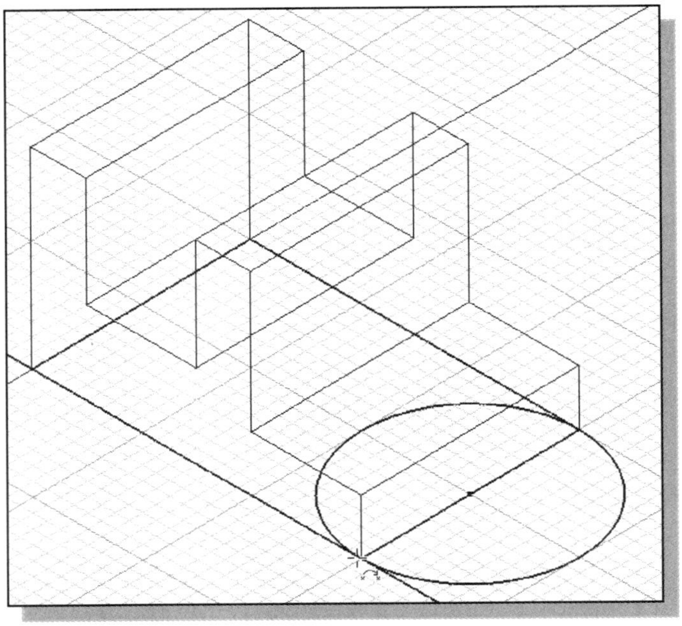

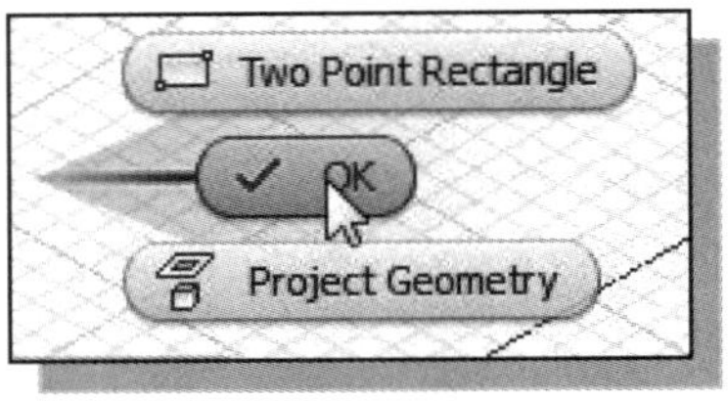

4. Inside the graphics window, click once with the right-mouse-button to display the option menu. Select **OK** in the pop-up menu to end the Circle command.

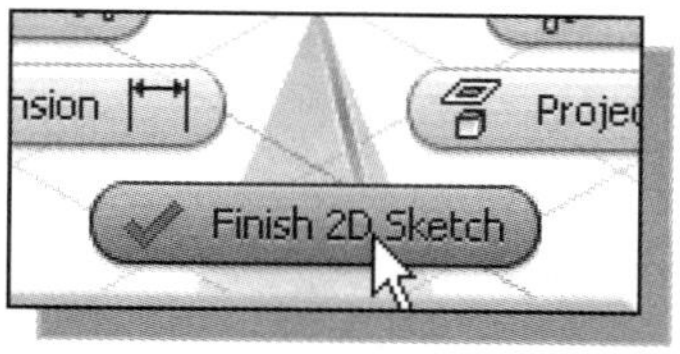

5. Inside the graphics window, click once with the right-mouse-button to display the option menu. Select **Finish 2D Sketch** in the pop-up menu to end the Sketch option.

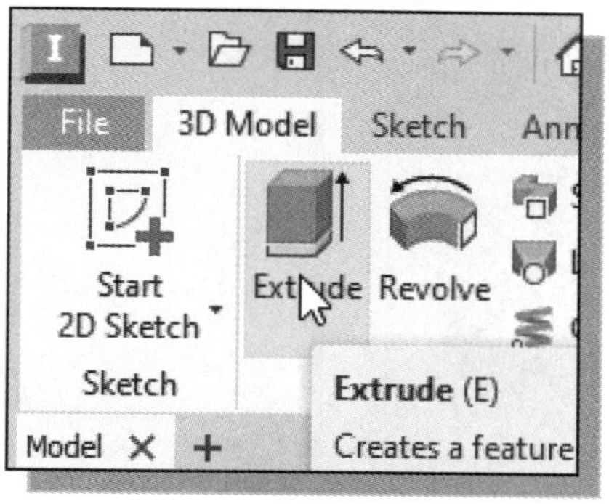

6. In the *Features* toolbar (the toolbar that is located to the left side of the graphics window), select the **Extrude** command by clicking the left-mouse-button on the icon.

7. Move the cursor to the outside **semi-circle** we just created and left-click once to select the region as the **profile** to be extruded.

8. In the *Extrude* pop-up control, set the operation option to **Join.**

9. Also set the *Distance* option to **To Selected Face** as shown below.

10. Select the top face of the base feature as the termination surface for the extrusion.

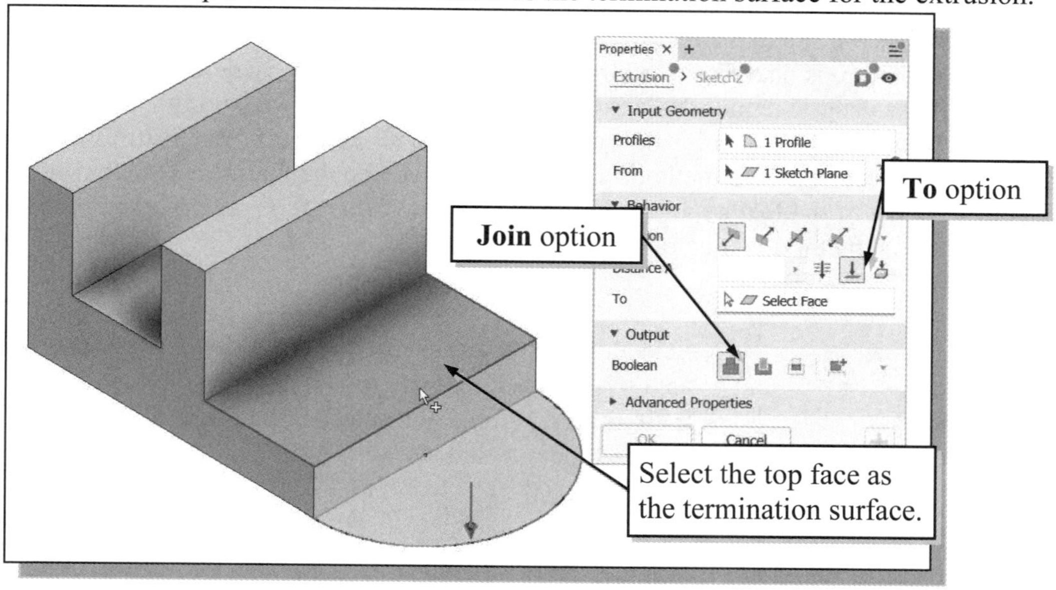

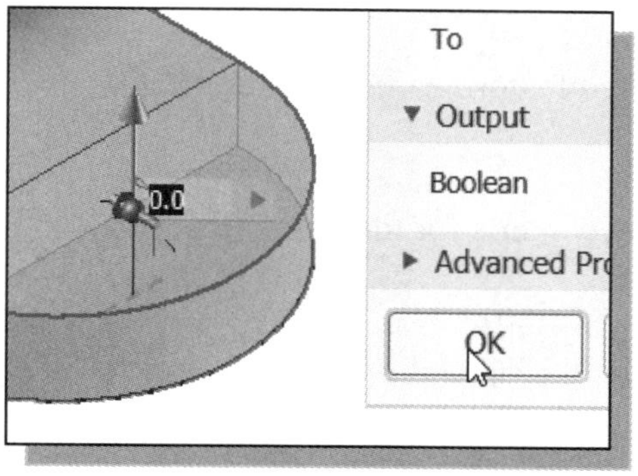

11. Click on the **OK** button to proceed with the *Join* operation.

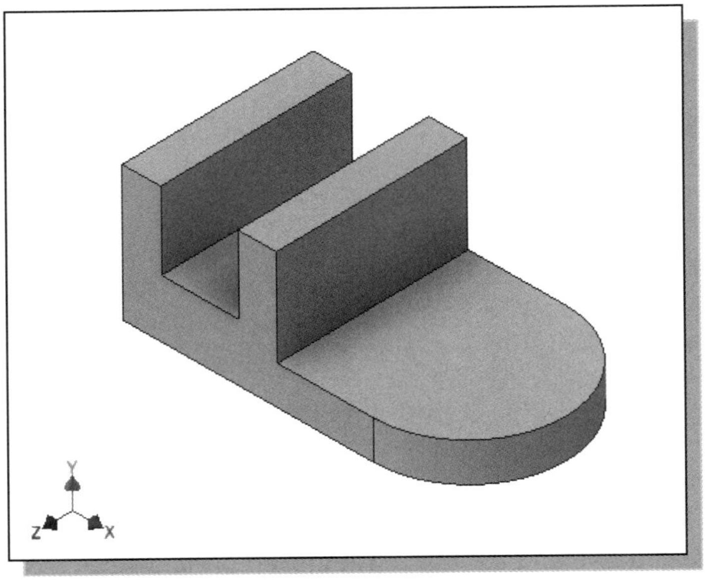

Renaming the Part Features

Currently, our model contains two extruded features. The feature is highlighted in the display area when we select the feature in the *browser* window. Each time a new feature is created, the feature is also displayed in the *Model Tree* window. By default, Autodesk Inventor will use generic names for part features. However, when we begin to deal with parts with a large number of features, it will be much easier to identify the features using more meaningful names. Two methods can be used to rename the features: 1. **Clicking** twice on the name of the feature and 2. Using the **Properties** option. In this example, the use of the first method is illustrated.

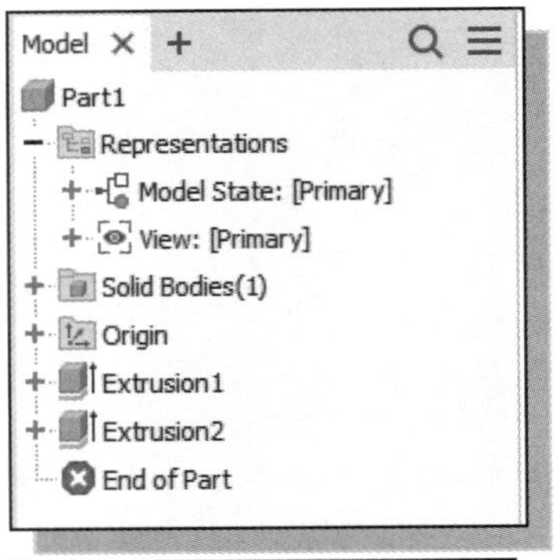

1. Select the first extruded feature in the *model browser* area by left-clicking once on the name of the feature, **Extrusion1**. Notice the selected feature is highlighted in the graphics window.

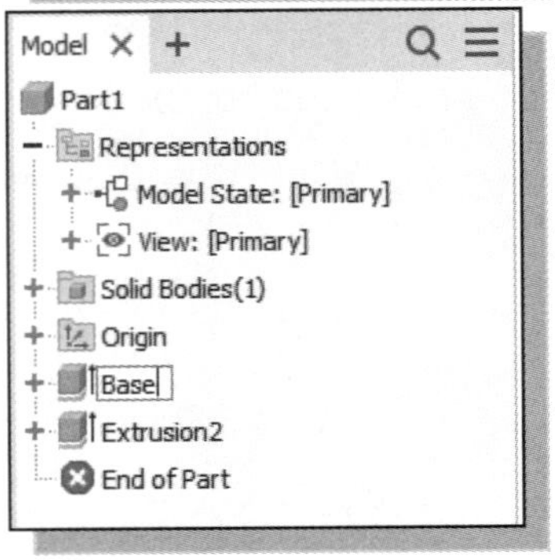

2. Left-click again on the feature name to enter the *Edit* mode as shown.

3. Enter **Base** as the new name for the first extruded feature.

4. On your own, rename the second extruded feature to **Circular_End**.

Adjusting the Width of the Base Feature

One of the main advantages of parametric modeling is the ease of performing part modifications at any time in the design process. Part modifications can be done through accessing the features in the history tree. Autodesk Inventor remembers the history of a part, including all the rules that were used to create it, so that changes can be made to any operation that was performed to create the part. For our *Saddle Bracket* design, we will reduce the size of the base feature from 3.25 inches to 3.0 inches, and the extrusion distance to 2.0 inches.

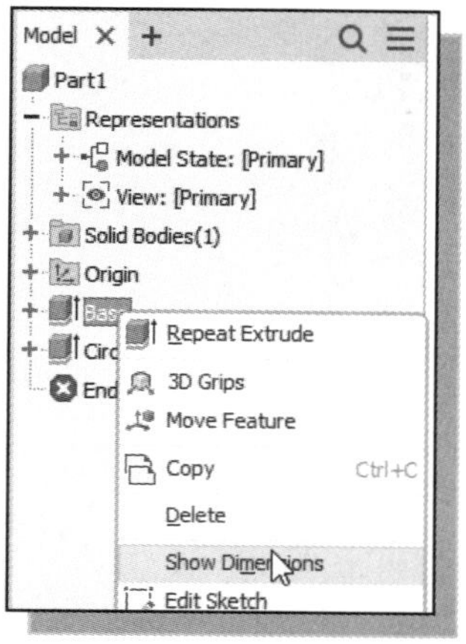

1. Select the first extruded feature, **Base**, in the *browser* area. Notice the selected feature is highlighted in the graphics window.

2. Inside the *browser* area, **right-mouse-click** on the first extruded feature to bring up the option menu and select the **Show Dimensions** option in the pop-up menu.

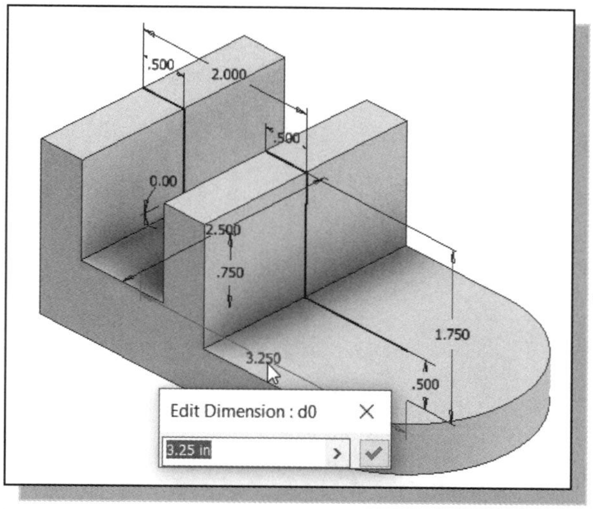

3. All dimensions used to create the **Base** feature are displayed on the screen. Select the overall width of the **Base** feature, the **3.25** dimension value, by double-clicking on the dimension text as shown.

4. Enter **3.0** in the *Edit Dimension* window.

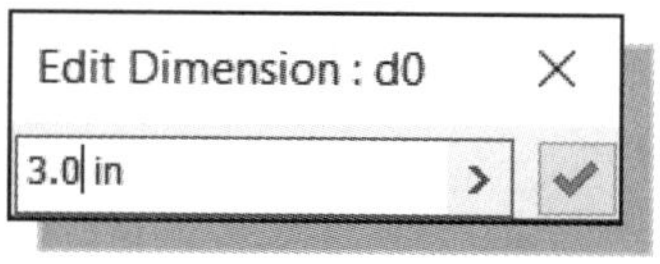

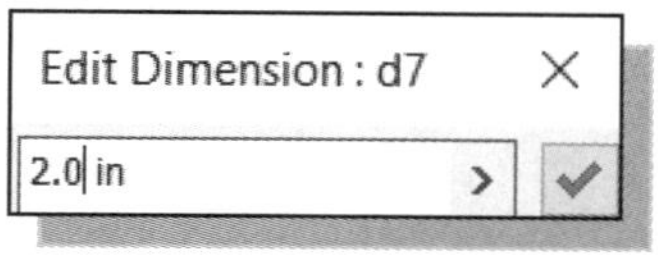

5. On your own, repeat the above steps and modify the extruded distance from **2.5** to **2.0**.

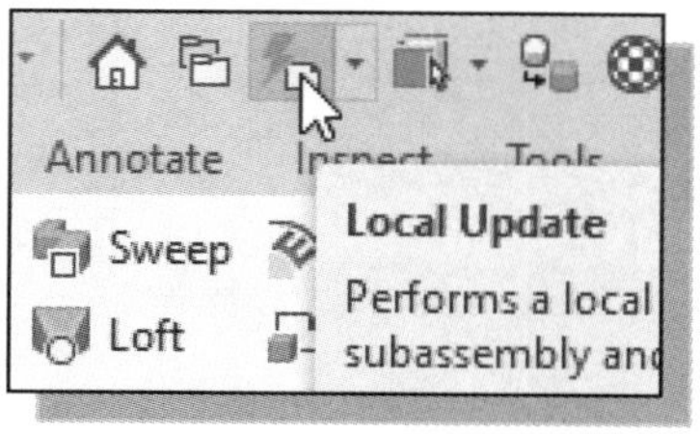

6. Click **Local Update** in the *Quick Access Toolbar*.

➢ Note that Autodesk Inventor updates the model by re-linking all elements used to create the model. Any problems or conflicts that occur will also be displayed during the updating process.

Adding a Placed Feature

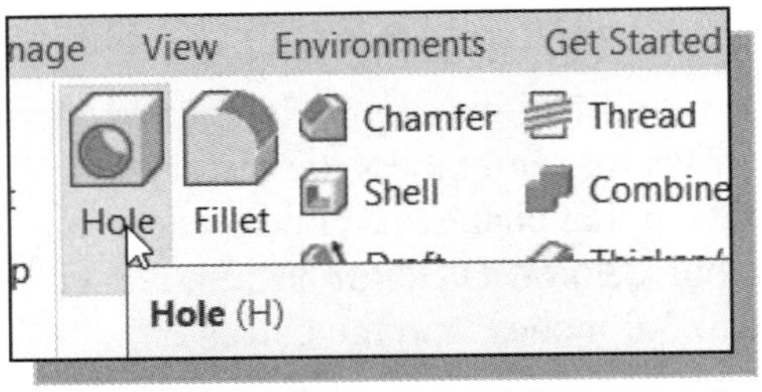

1. In the *Sketch tab*, select the **Hole** command by left-clicking on the icon.

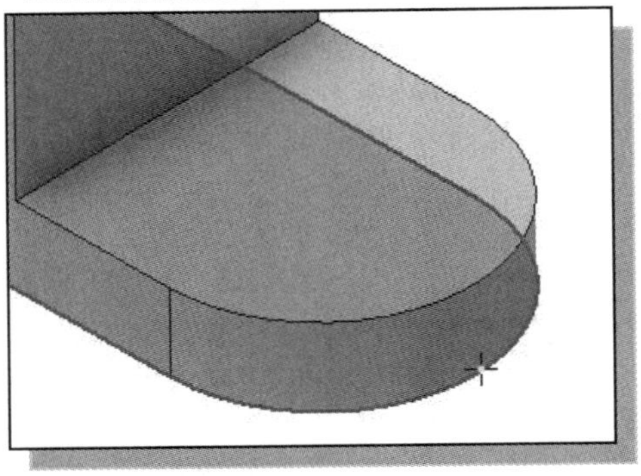

2. Pick the **bottom plane** of the solid model as the placement plane as shown.

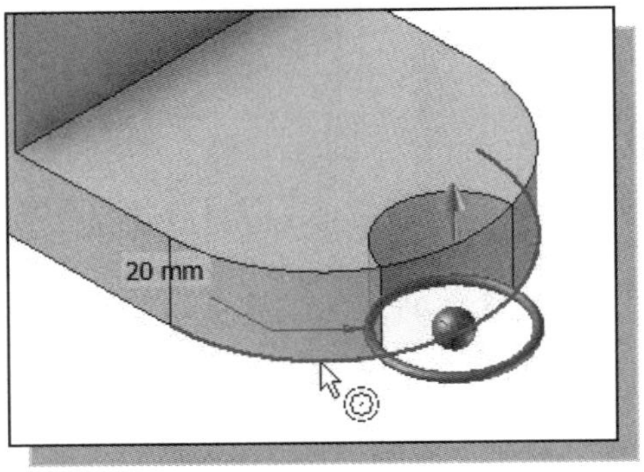

3. Pick the **bottom arc** and notice the *concentric reference symbol* appears indicating the center of the hole will be aligned to the center of the selected arc.

4. Set the hole diameter to **0.75 in** as shown.

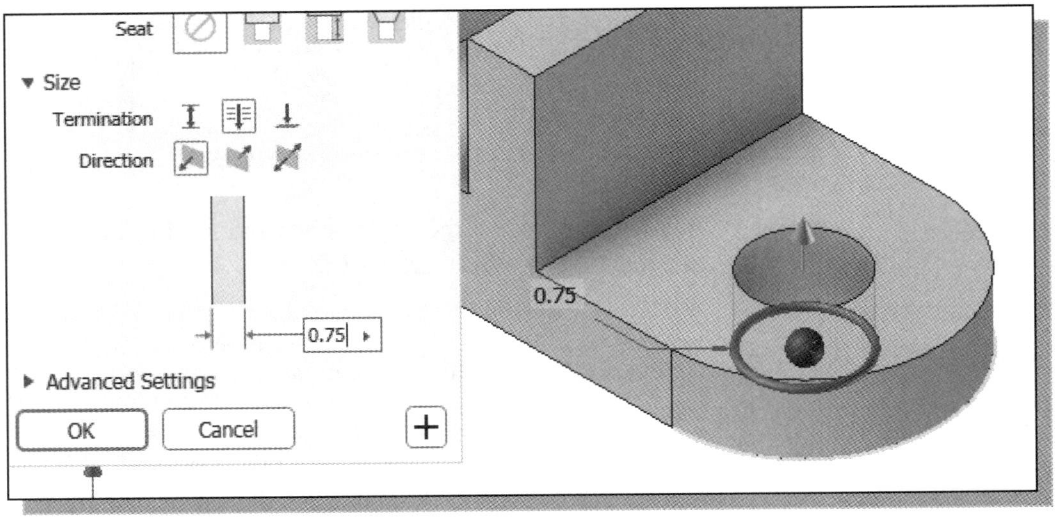

5. Set the termination option to **Through All** as shown.

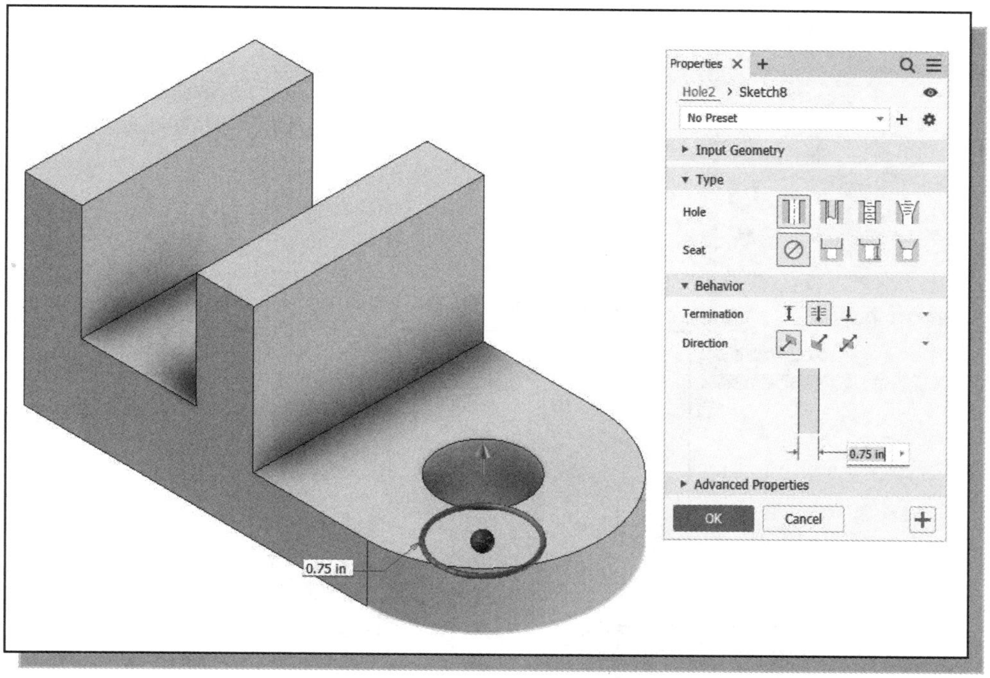

6. Click **OK** to accept the settings and create the *Hole* feature.

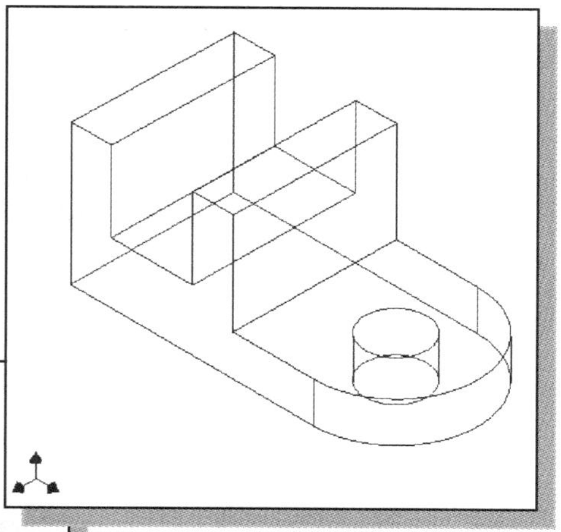

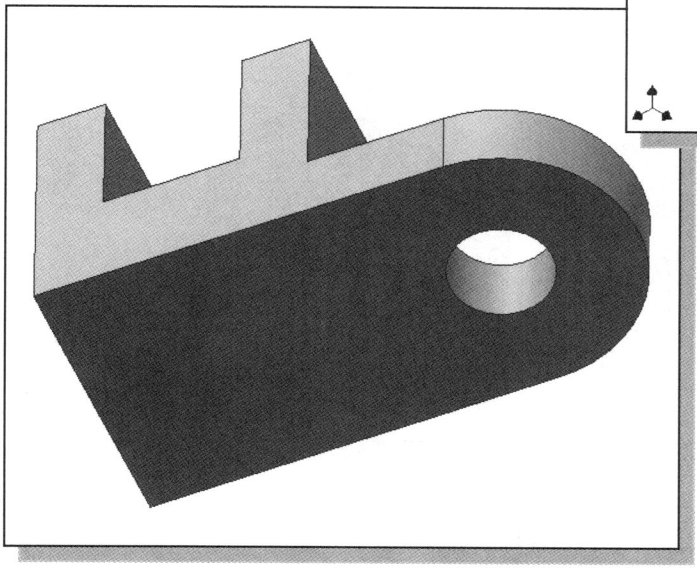

Creating a Rectangular Cut Feature

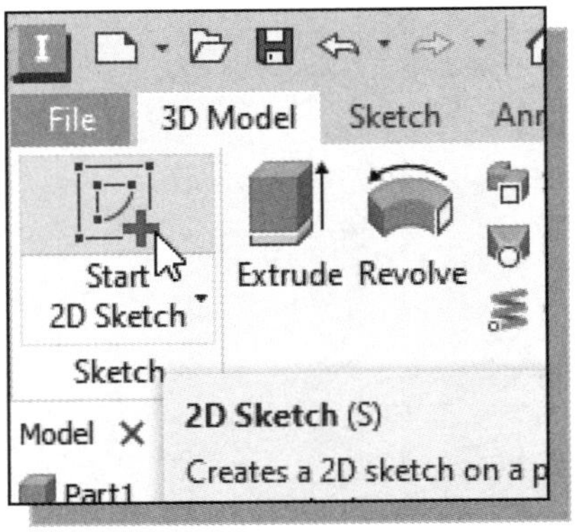

1. In the *Sketch toolbar* under the *3D Model tab* select the **Start 2D Sketch** command by left-clicking once on the icon.

2. Pick the **vertical face** of the solid as shown. (Note the alignment of the origin of the sketch plane.)

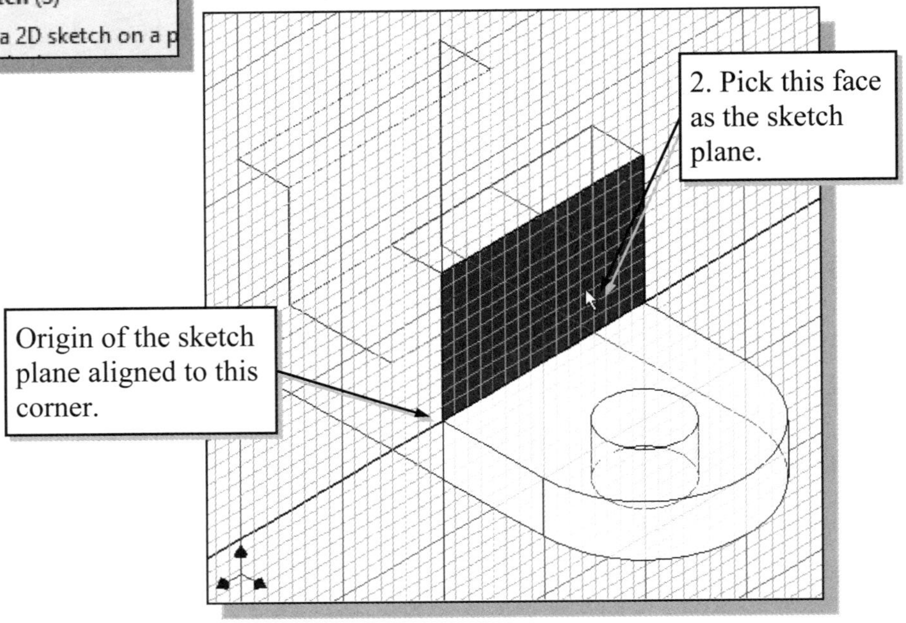

➢ On your own, create a rectangular cut (**1.0** x **0.75**) feature (**To Next** option) as shown and rename the feature to **Rect_Cut**.

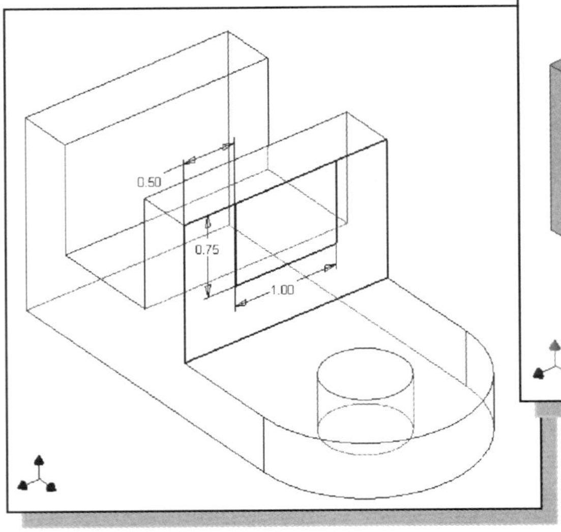

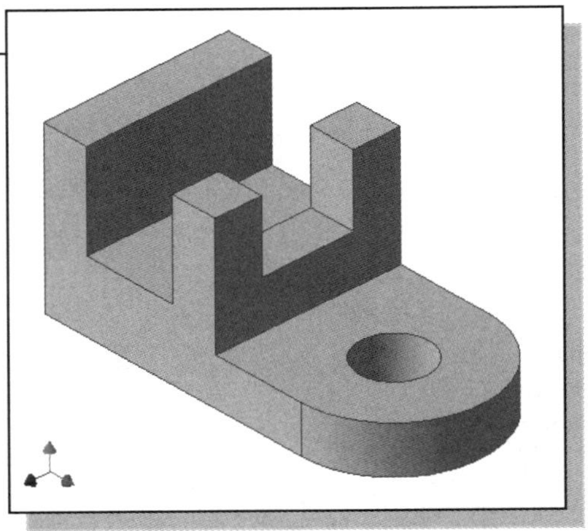

History-Based Part Modifications

Autodesk Inventor uses the *history-based part modification* approach, which enables us to make modifications to the appropriate features and re-link the rest of the history tree without having to reconstruct the model from scratch. We can think of it as going back in time and modifying some aspects of the modeling steps used to create the part. We can modify any feature that we have created. As an example, we will adjust the depth of the rectangular cutout.

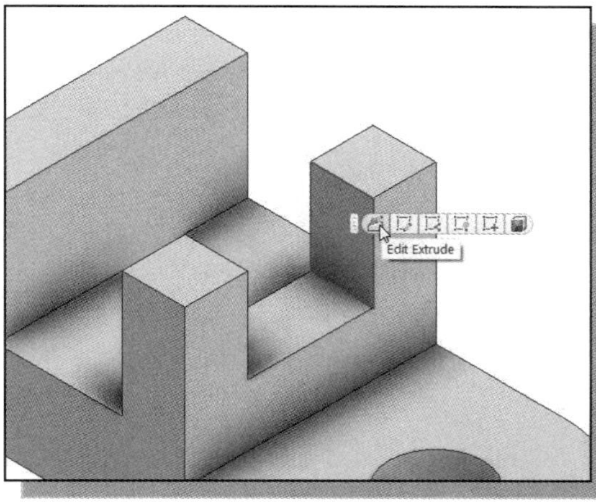

1. In the *Graphics* window, select the last cut feature, **Rect_Cut**, on the inside surface of the feature as shown.

2. Select **Edit Feature** in the pop-up menu. Notice the *Extrude* dialog box appears on the screen.

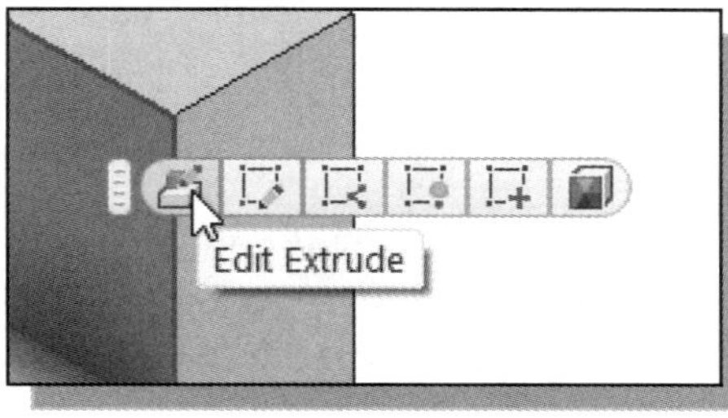

- Note that the feature pop-up options are also available through the Model History Tree, as shown on the next page.

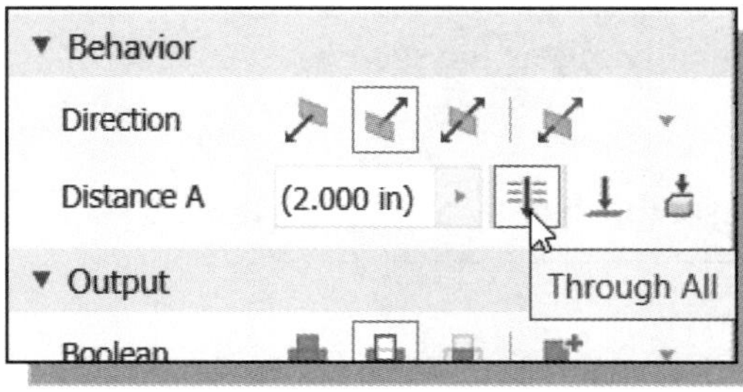

3. In the *Extrude* dialog box, set the termination *Extents* to the **Through All** option.

4. Click on the **OK** button to accept the settings.

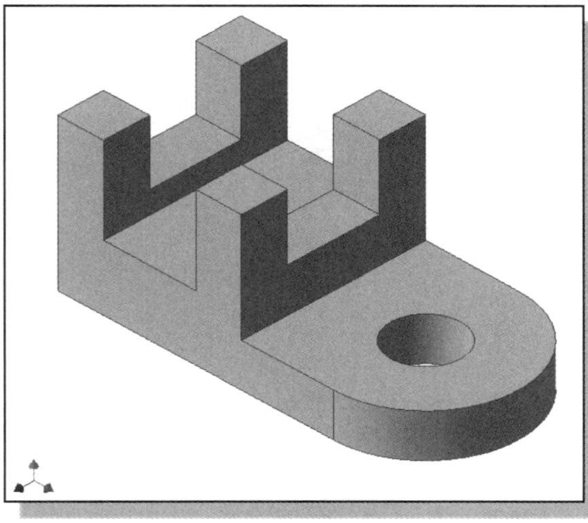

- As can be seen, the history-based modification approach is very straight forward, and it only took a few seconds to adjust the cut feature to the **Through All** option.

A Design Change

Engineering designs usually go through many revisions and changes. Autodesk Inventor provides an assortment of tools to handle design changes quickly and effectively. We will demonstrate some of the tools available by changing the **Base** feature of the design.

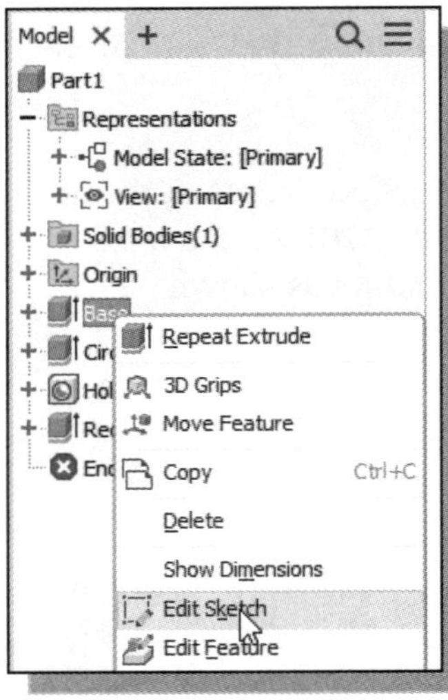

1. In the *browser* window, select the **Base** feature by left-clicking once on the name of the feature.

2. In the *browser*, right-click once on the **Base** feature to bring up the option menu; then pick **Edit Sketch** in the pop-up menu.

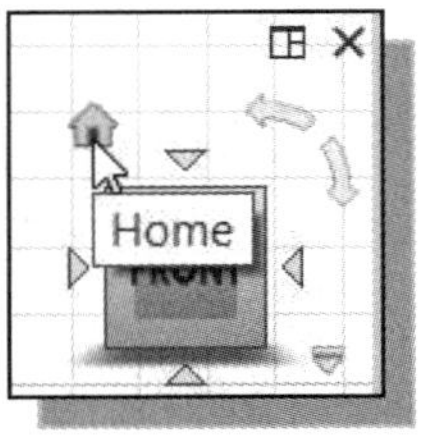

3. Click **Home** to reset the display to ***Isometric***.

❖ Autodesk Inventor will now display the original 2D sketch of the selected feature in the graphics window. We have literally gone back to the point where we first created the 2D sketch. Notice the feature being modified is also highlighted in the desktop *browser*.

4. Click on the **Look At** icon in the *Standard* toolbar area.

- The **Look At** command automatically aligns the *sketch plane* of a selected entity to the screen.

5. Select any line segment of the 2D sketch to reset the display to align to the 2D sketch.

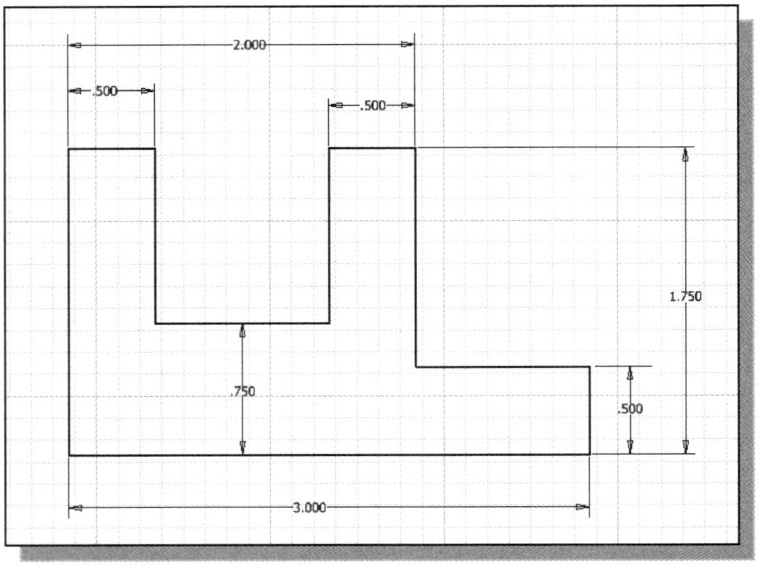

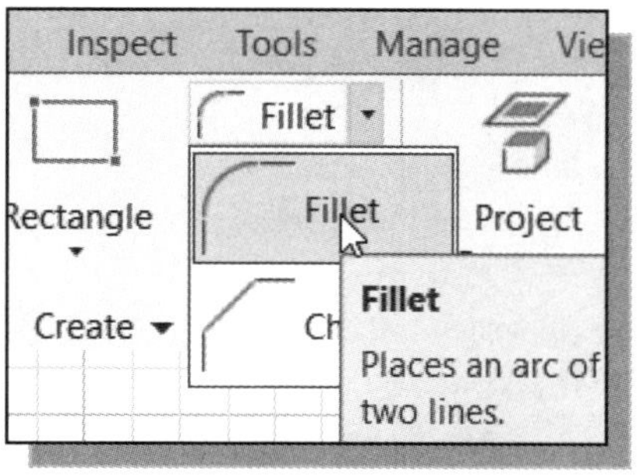

6. Select the **Fillet** command in the *2D Sketch* toolbar.

7. In the graphics window, enter **0.25** as the new radius of the fillet.

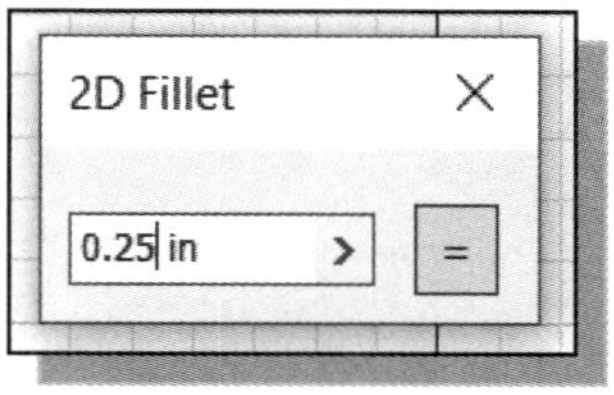

8. Select the two edges as shown to create the fillet.

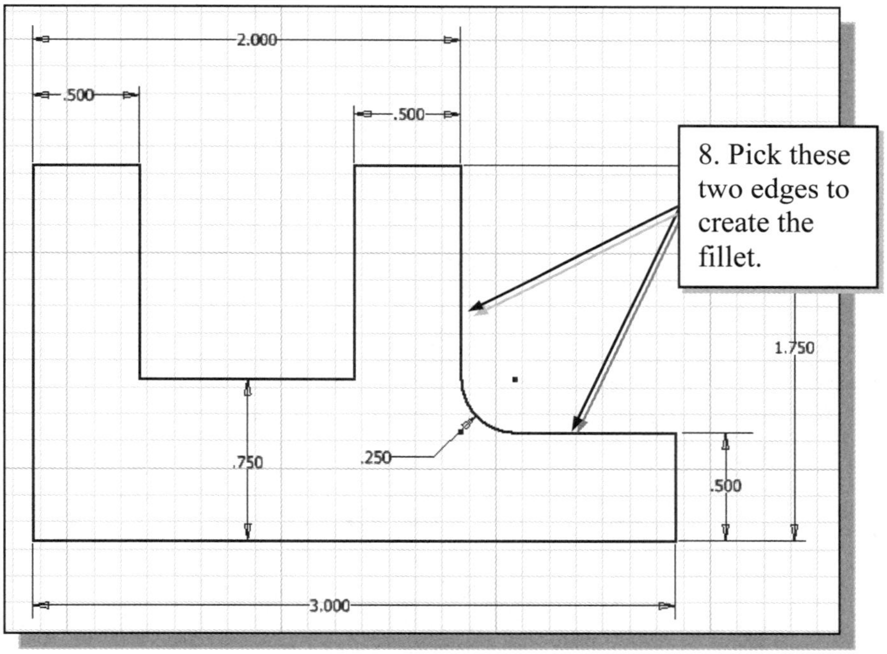

- Note that the fillet is created automatically with the dimension attached. The attached dimension can also be modified through the history tree.

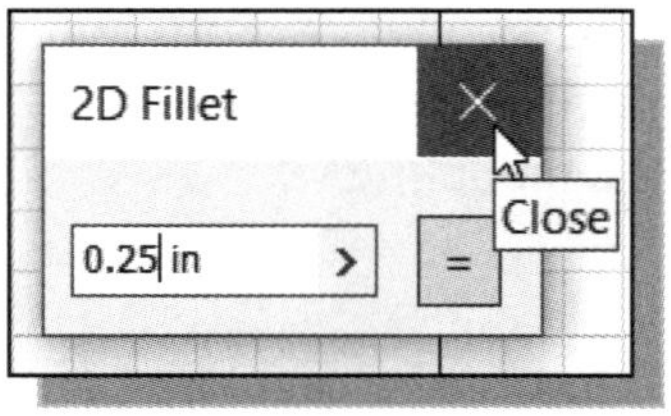

9. Click on the [**X**] icon in the *2D Fillet* window to end the Fillet command.

10. Select **Finish Sketch** in the *Ribbon* toolbar to end the Sketch option.

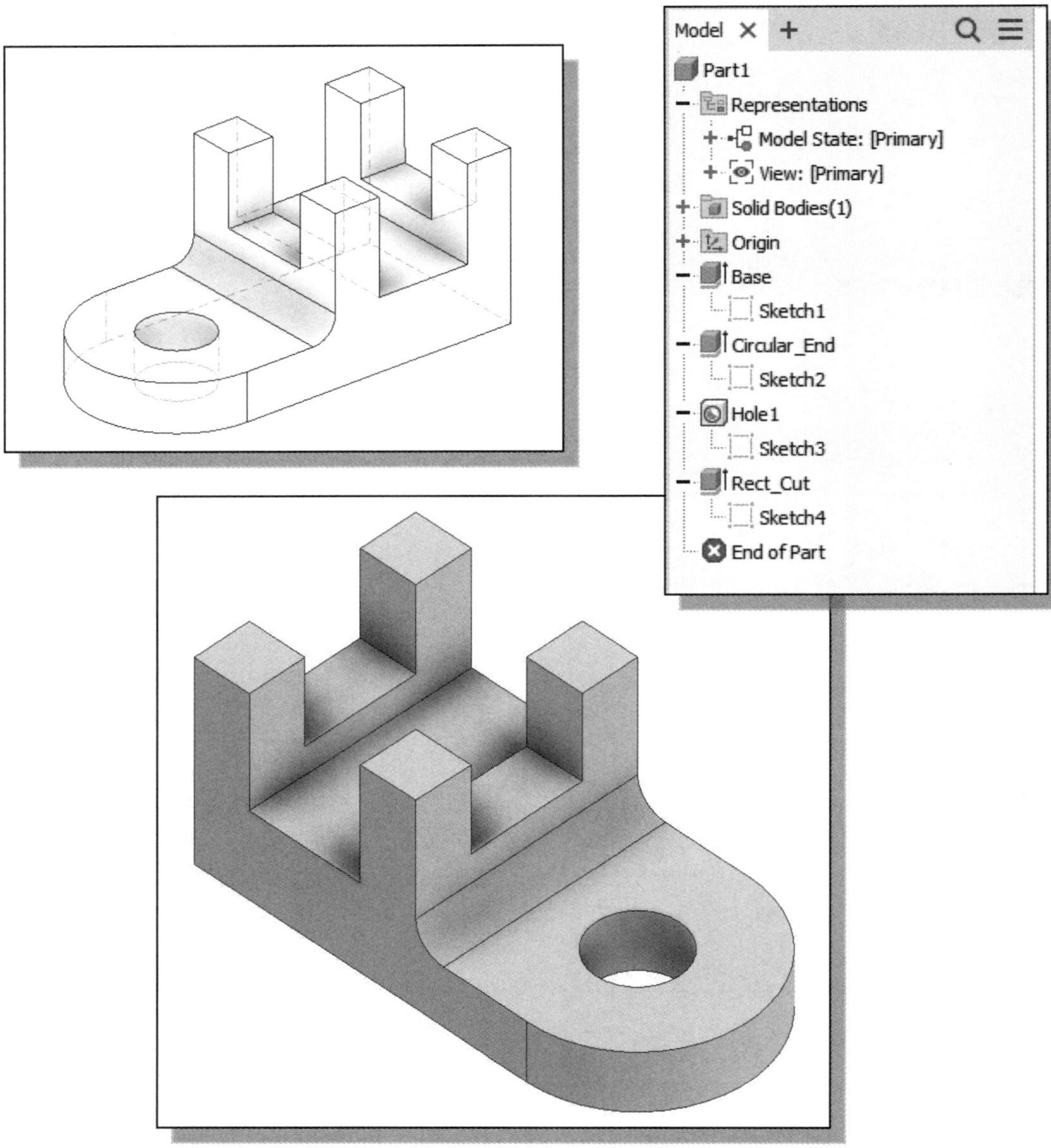

❖ In a typical design process, the initial design will undergo many analyses, testing, and reviews. The *history-based part modification* approach is an extremely powerful tool that enables us to quickly update the design. At the same time, it is quite clear that PLANNING AHEAD is also important in doing feature-based modeling.

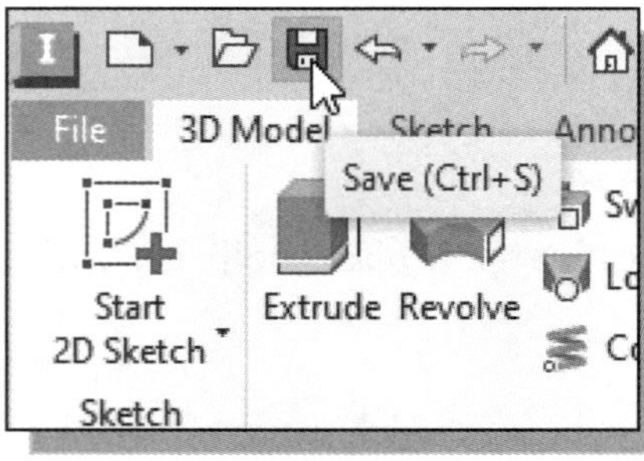

11. On your own, create the Chapter4 folder and save the model as *Saddle_Bracket.ipt*.

Assigning and Calculating the Associated Physical Properties

Autodesk Inventor models have properties called **iProperties**. The *iProperties* can be used to create reports and update assembly bills of materials, drawing parts lists, and other information. With *iProperties*, we can also set and calculate physical properties for a part or assembly using the material library. This allows us to examine the physical properties of the model, such as weight or center of gravity.

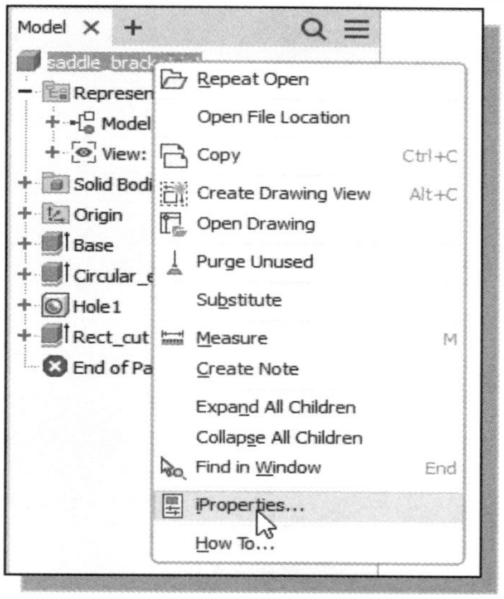

1. In the *browser*, **right-click** once on the ***part name*** to bring up the option menu; then pick **iProperties** in the *pop-up* menu.

2. On your own, look at the different information listed in the *iProperties* dialog box.

3. Click on the **Physical** tab; this is the page that contains the physical properties of the selected model.

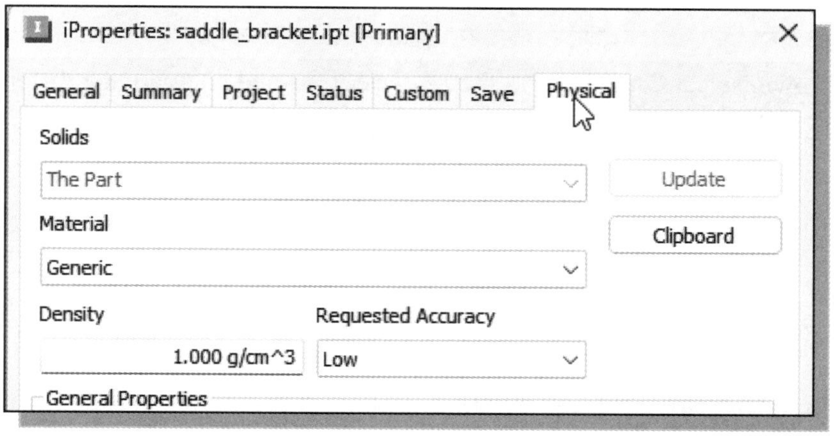

- Note that the *Material* option is not assigned, and none of the physical properties are shown.

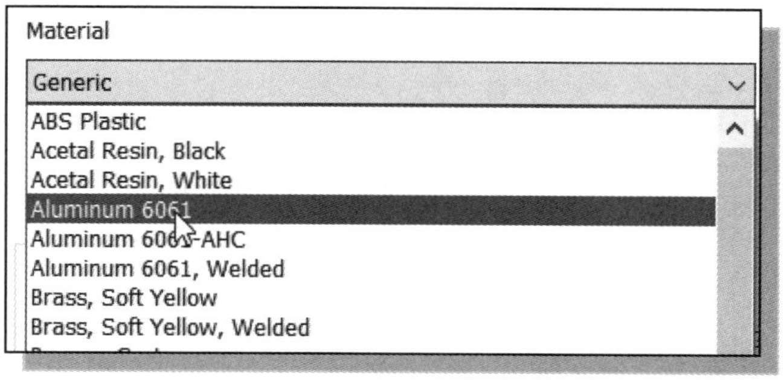

4. Click the down-arrow in the *Material* option to display the material list, and select **Aluminum-6061** as shown.

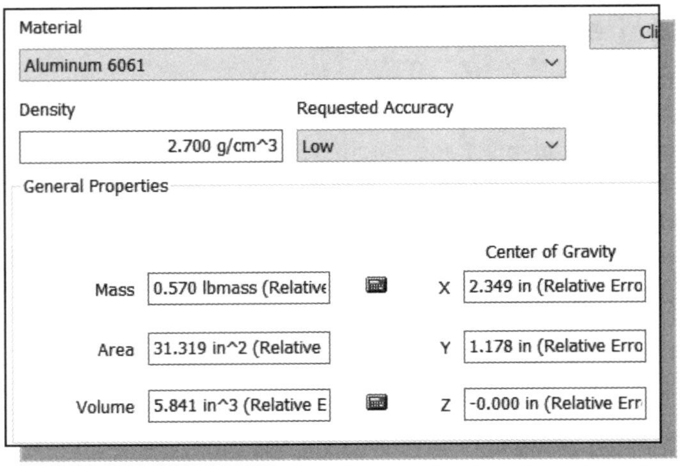

❖ Note the *General Properties* area now has the *Mass*, *Area*, *Volume* and the *Center of Gravity* information of the model based on the density of the selected material.

5. Click on the **Global** button to display the *Mass Moments Inertial* of the design.

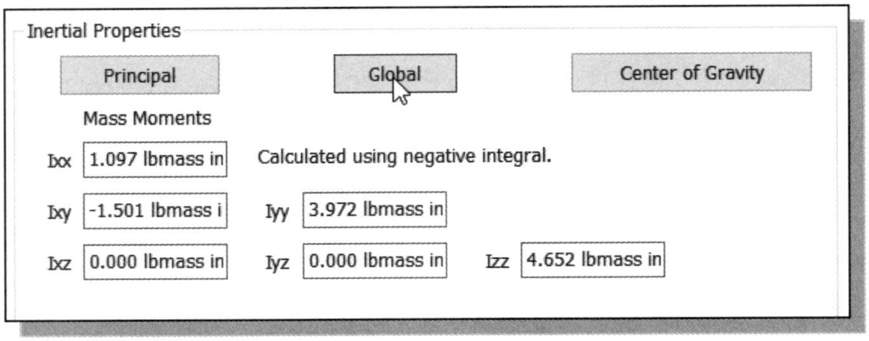

6. On your own, select **Cast Iron** as the *Material* type and compare the differences in using the different materials.

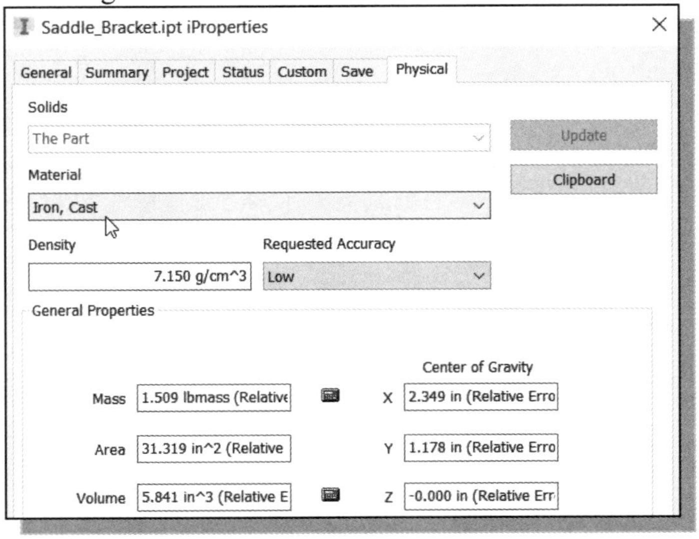

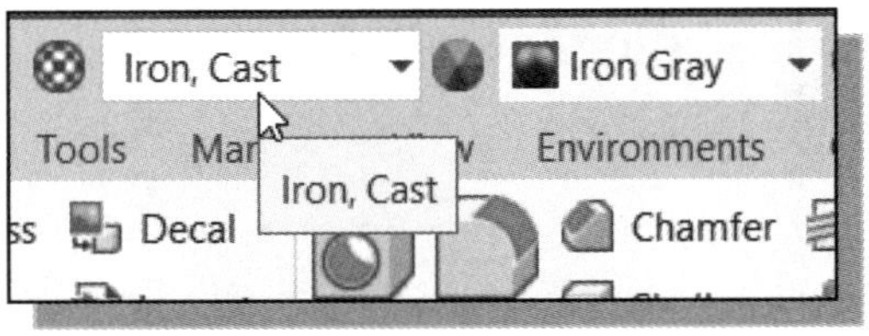

❖ Also note the *Material* can be assigned through the quick access menu as shown.

Review Questions: (Time: 30 minutes.)

1. What are stored in the Autodesk Inventor *History Tree*?

2. When extruding, what is the difference between *Distance* and *Through All*?

3. Describe the *history-based part modification* approach.

4. What determines how a model reacts when other features in the model change?

5. Describe the steps to rename existing features.

6. Describe two methods available in Autodesk Inventor to *modify the dimension values* of parametric sketches.

7. Create *History Tree sketches* showing the steps you plan to use to create the two models shown on the next page:

Ex.1)

Ex.2)

Exercises: Create and save the exercises in the Chapter4 folder. (Time: 180 minutes.)

1. **C-Clip** (Dimensions are in inches. Plate thickness: **0.25 inches**.)

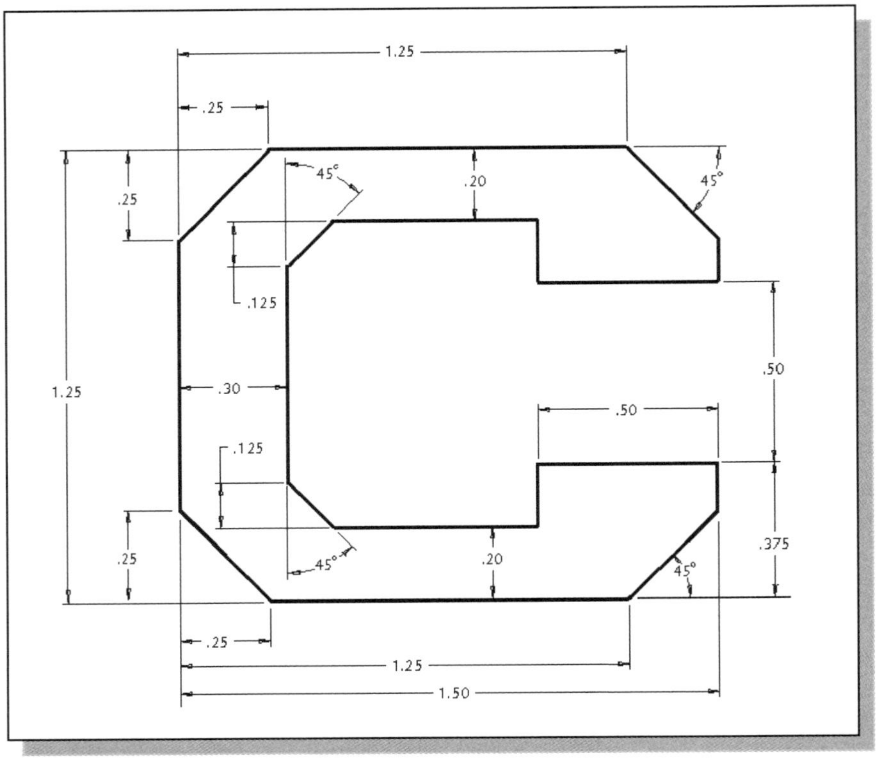

2. **Tube Mount** (Dimensions are in inches.)

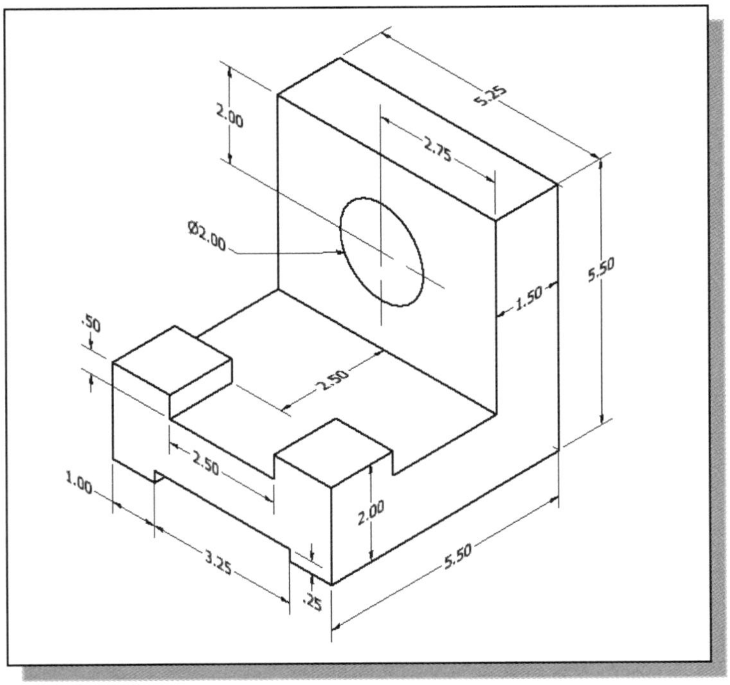

3. **Hanger Jaw** (Dimensions are in inches. Volume =?)

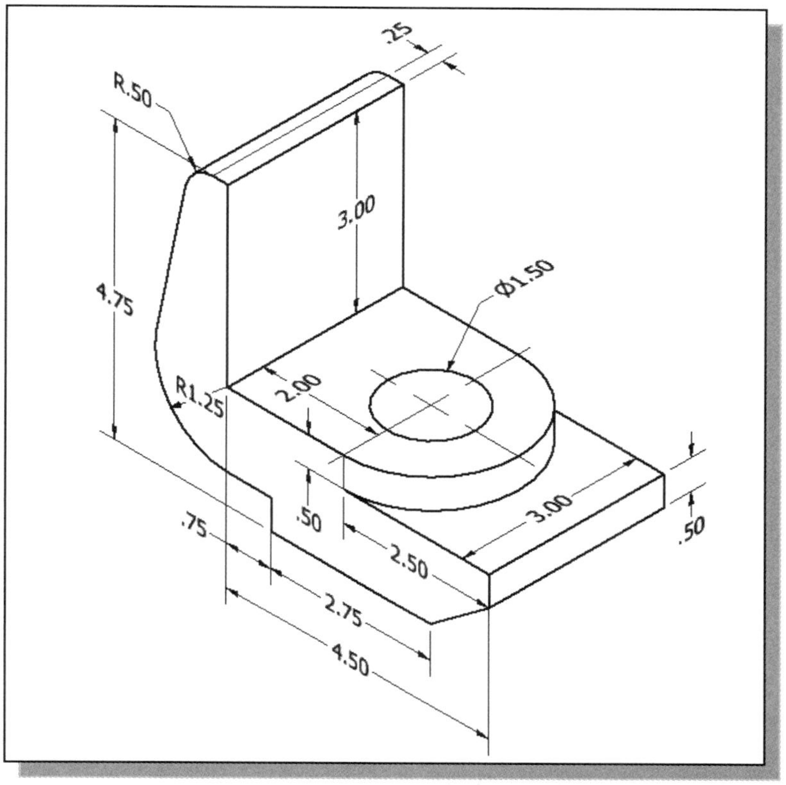

4. **Transfer Fork** (Dimensions are in inches. Material: **Cast Iron.** Volume =?)

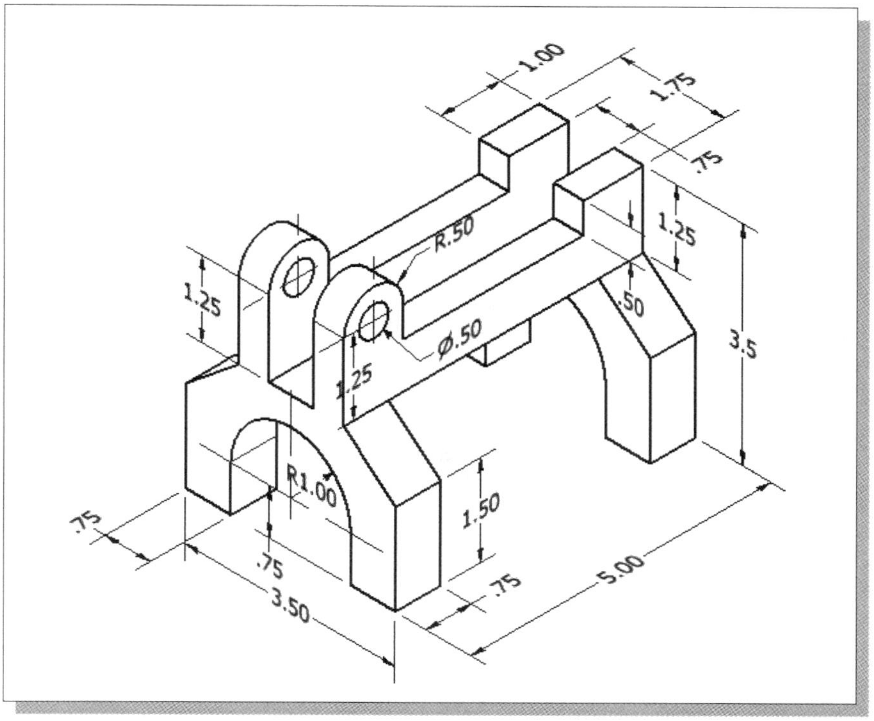

5. **Guide Slider** (Material: **Cast Iron**. Weight and Volume =?)

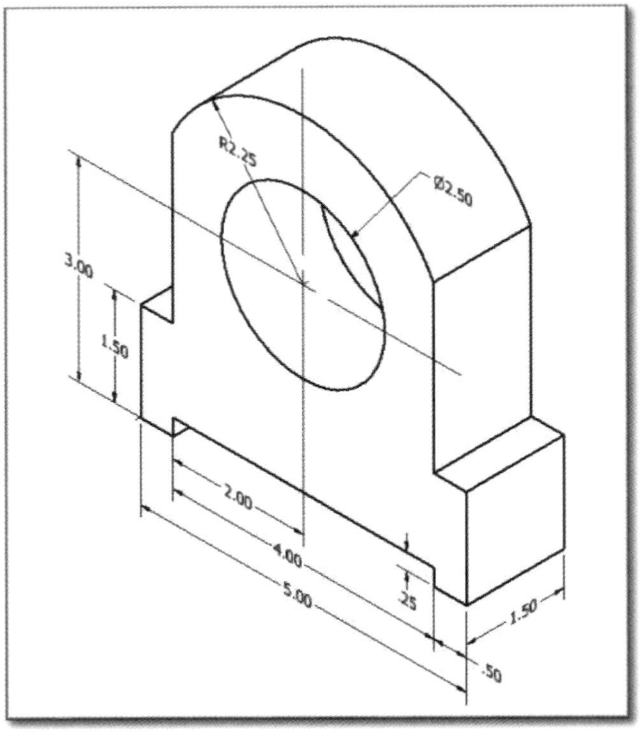

6. **Shaft Guide** (Material: **Aluminum-6061**. Mass and Volume =?)

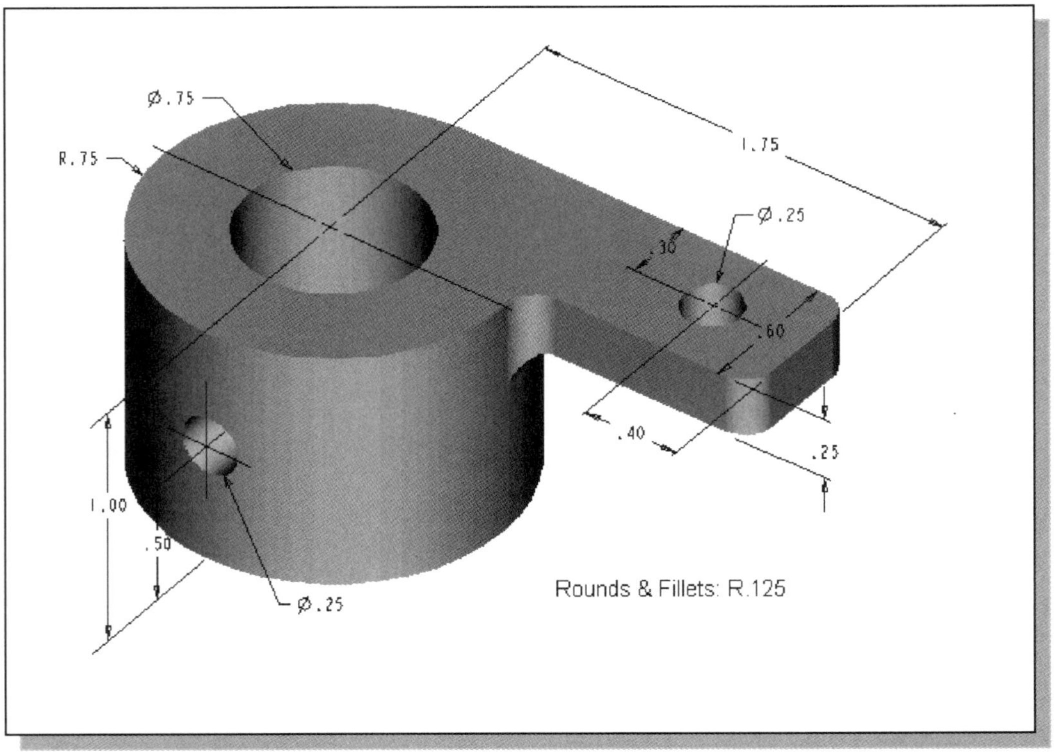

7. **Idler Mount** (Dimensions are in inches. Center to center distance: 6.0)

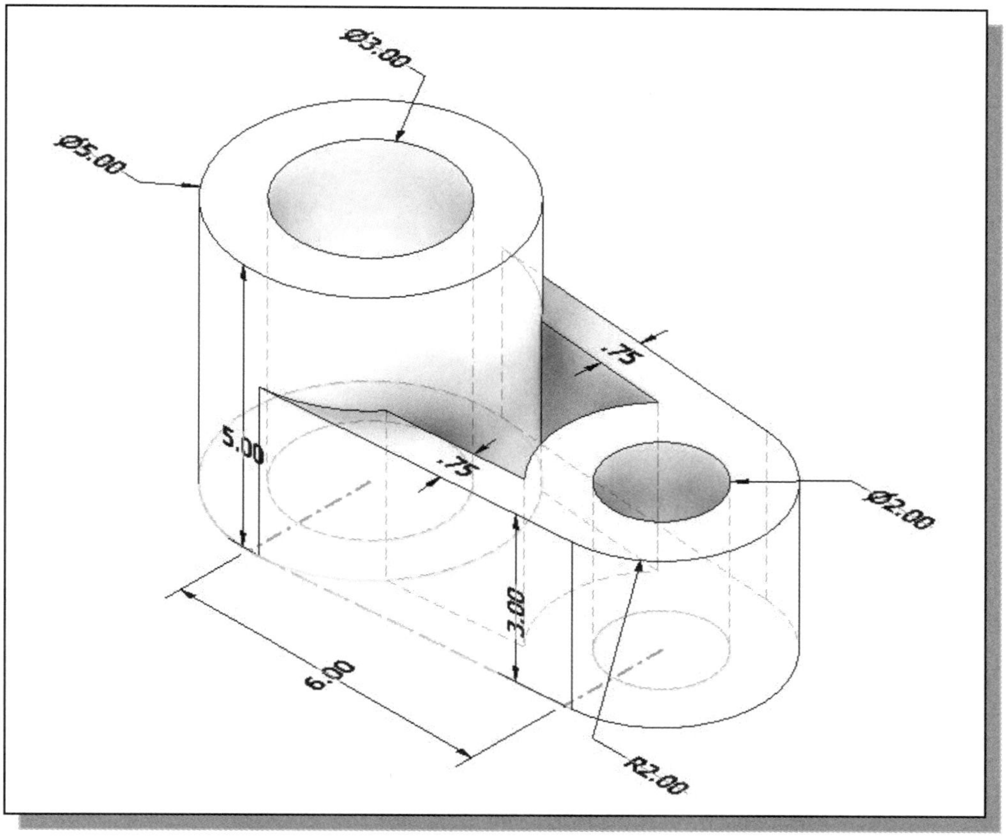

Notes:

Chapter 5
Parametric Constraints Fundamentals

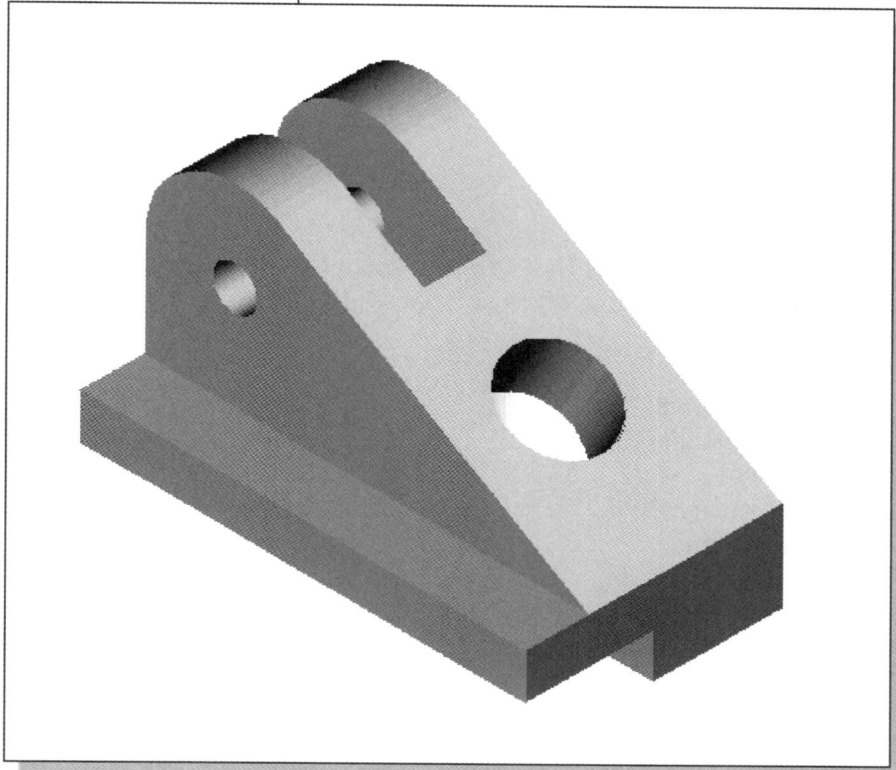

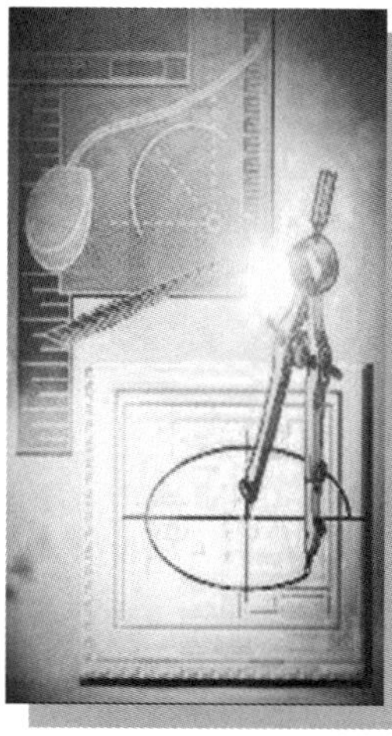

Learning Objectives

- Create Parametric Relations
- Use Dimensional Variables
- Display, Add, and Delete Geometric Constraints
- Understand and Apply Different Geometric Constraints
- Display and Modify Parametric Relations
- Create Fully Constrained Sketches

Autodesk Inventor Certified User Exam Objectives Coverage

Parametric Modeling Basics

Section 3: Sketches

Objectives: Creating 2D Sketches, Draw Tools, Sketch Constraints, Pattern Sketches, Modify Sketches, Format Sketches, Sketch Doctor, Shared Sketches, Sketch Parameters.

Constraints and Relations

A primary and essential difference between parametric modeling and previous generation computer modeling is that parametric modeling captures the *design intent*. In the previous lessons, we have seen that the design philosophy of "*shape before size*" is implemented through the use of Autodesk Inventor's Profile and Dimension commands. In performing geometric constructions, dimensional values are necessary to describe the **SIZE** and **LOCATION** of constructed geometric entities. Besides using dimensions to define the geometry, we can also apply geometric rules to control geometric entities. More importantly, Autodesk Inventor can capture design intent through the use of **geometric constraints**, **dimensional constraints**, and **parametric relations**. In Autodesk Inventor, there are two types of constraints: **geometric constraints** and **dimensional constraints**. For part modeling in Autodesk Inventor, constraints are applied to *2D sketches*. **Geometric constraints** are **geometric restrictions** that can be applied to geometric entities; for example, *horizontal*, *parallel*, *perpendicular*, and *tangent* are commonly used *geometric constraints* in parametric modeling. **Dimensional constraints** are used to describe the SIZE and LOCATION of individual geometric shapes. One should also realize that depending upon the way the constraints are applied, the same results can be accomplished by applying different constraints to the geometric entities. In Autodesk Inventor, **parametric relations** are user-defined mathematical equations composed of dimensional variables and/or *design variables*. In parametric modeling, features are made of geometric entities with both relations and constraints describing individual design intent. In this lesson, we will discuss the fundamentals of parametric relations and geometric constraints.

Create a Simple Triangular Plate Design

In parametric modeling, **geometric properties** such as *horizontal*, *parallel*, *perpendicular*, and *tangent* can be applied to geometric entities automatically or manually. By carefully applying proper **geometric constraints**, very intelligent models can be created. This concept is illustrated by the following example.

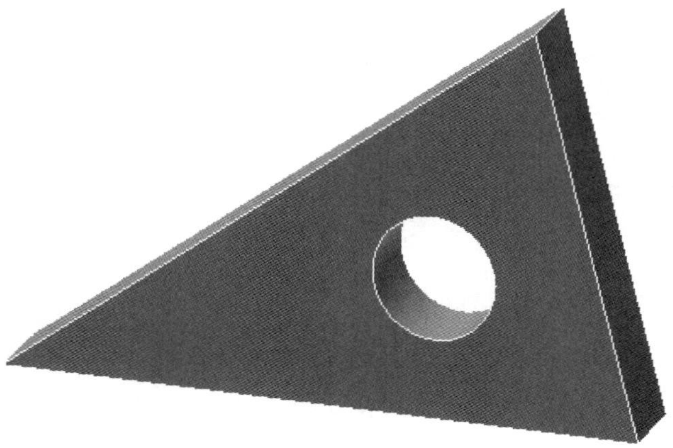

Fully Constrained Geometry

In Autodesk Inventor, as we create 2D sketches, geometric constraints such as *horizontal* and *parallel* are automatically added to the sketched geometry. In most cases, additional constraints and dimensions are needed to fully describe the sketched geometry beyond the geometric constraints added by the system. Although we can use Autodesk Inventor to build partially constrained or totally unconstrained solid models, the models may behave unpredictably as changes are made. In most cases, it is important to consider the design intent and to add proper constraints to geometric entities. In the following sections, a simple triangle is used to illustrate the different tools that are available in Autodesk Inventor to create/modify geometric and dimensional constraints.

Starting Autodesk Inventor

1. Select the **Autodesk Inventor** option on the *Start* menu or select the **Autodesk Inventor** icon on the desktop to start Autodesk Inventor. The Autodesk Inventor main window will appear on the screen. Once the program is loaded into memory, the *Startup* dialog box appears at the center of the screen.

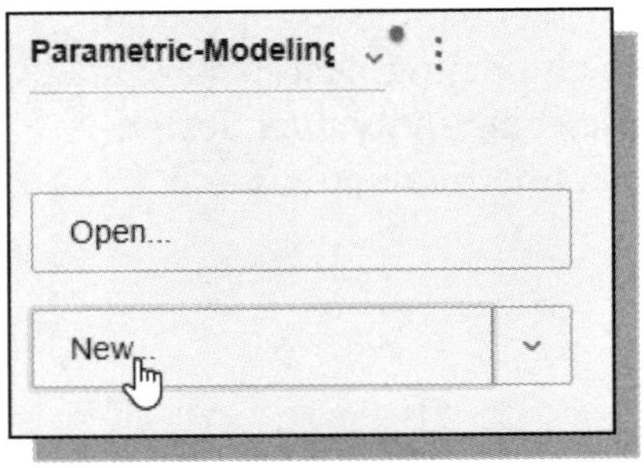

2. Select the **New** icon with a single click of the left-mouse-button in the left panel as shown.

3. Select the **en-US→English** tab and in the *Part template* area, select **Standard(in).ipt**.

4. Click **Create** in the *New File* dialog box to accept the selected settings to start a new model.

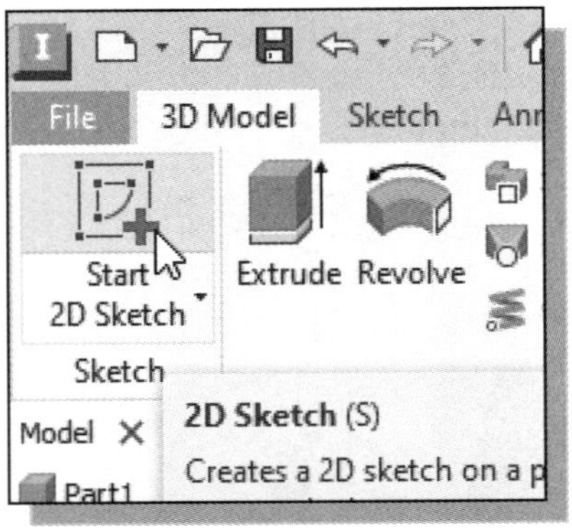

5. Click once with the left-mouse-button to select the **Start 2D Sketch** command.

6. Click once with the **left-mouse-button** to select the *XY Plane* as the sketch plane for the new sketch.

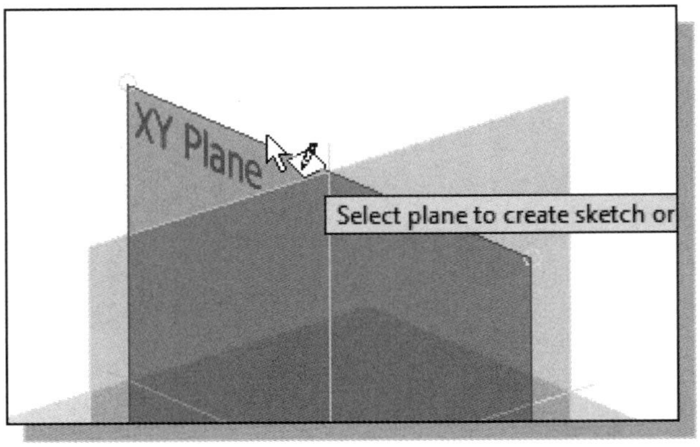

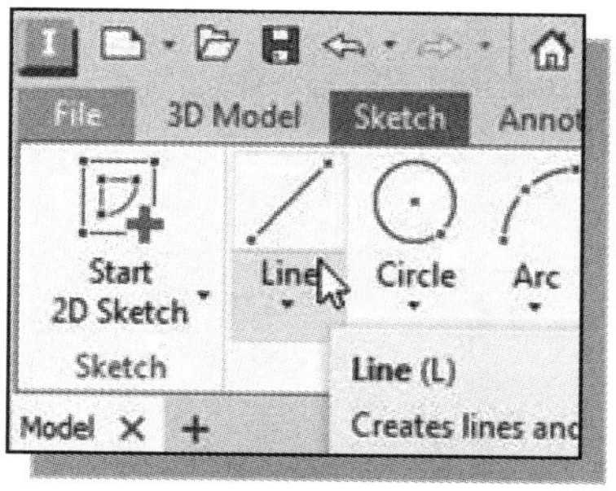

7. Click the **Line** icon in the *Sketch* tab to activate the command. A Help-tip box appears next to the cursor, and a brief description of the command is displayed at the bottom of the drawing screen: "Creates Straight line and arcs."

8. Create a triangle of arbitrary size positioned near the center of the screen as shown below. (Note that the base of the triangle is **horizontal**.) Hit the [**Esc**] key once to end the line command.

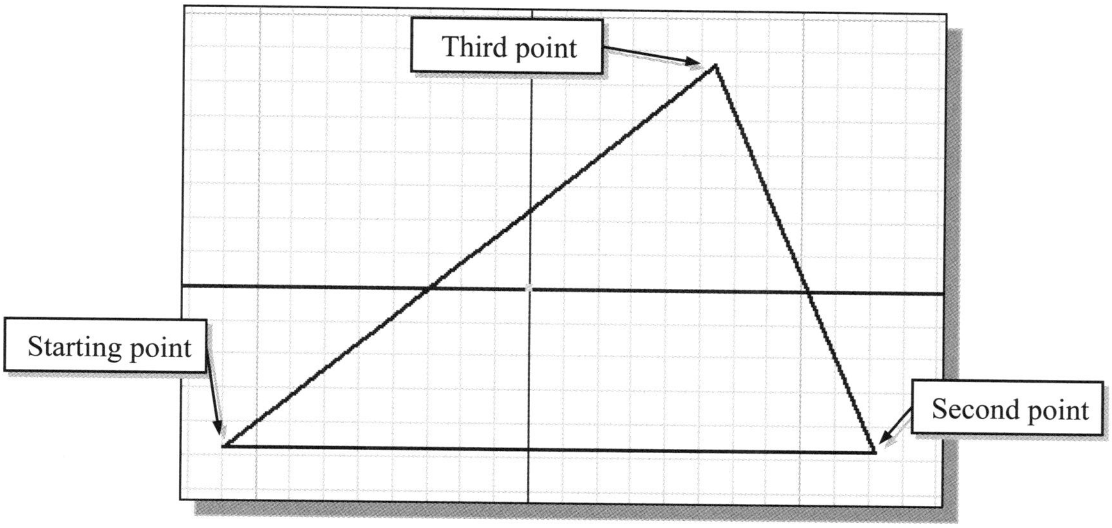

Displaying Existing Constraints

1. Select the **Show Constraints** command in the *Constrain* toolbar. This icon allows us to display constraints that are already applied to the 2D profiles. Left-click once on the icon to activate **Show Constraints**.

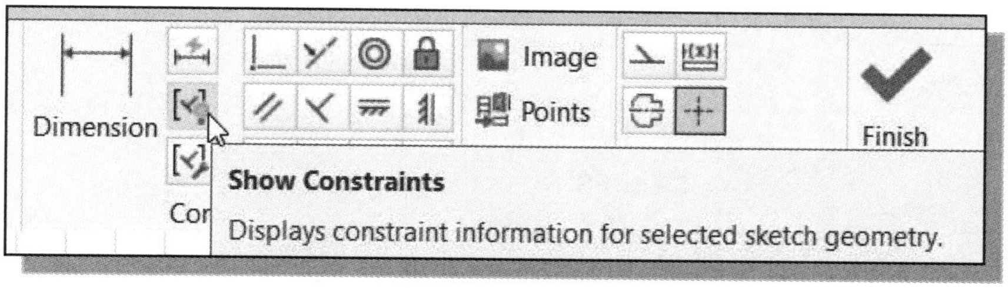

➢ In *parametric modeling*, constraints are typically applied as geometric entities are created. Autodesk Inventor will attempt to add proper constraints to the geometric entities based on the way the entities were created. Constraints are displayed as symbols next to the entities as they are created. The current profile consists of three line entities, three straight lines. The horizontal line has a **Horizontal constraint** applied to it.

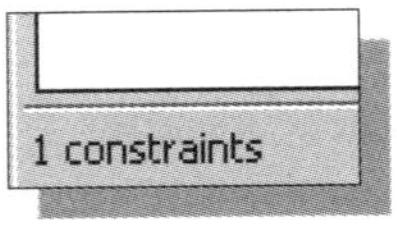

2. Select the horizontal line and notice the number of constraints applied is displayed in the message area.

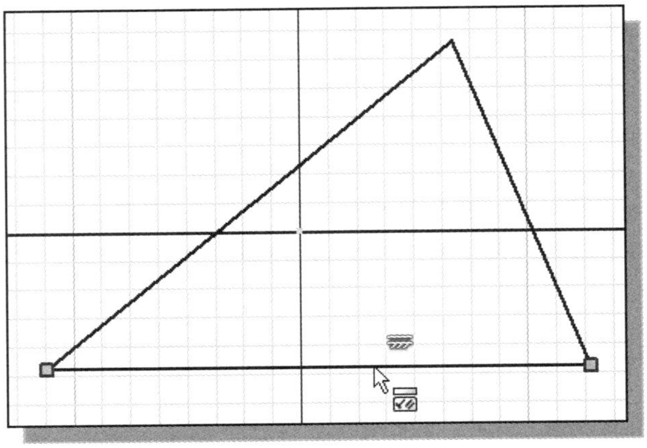

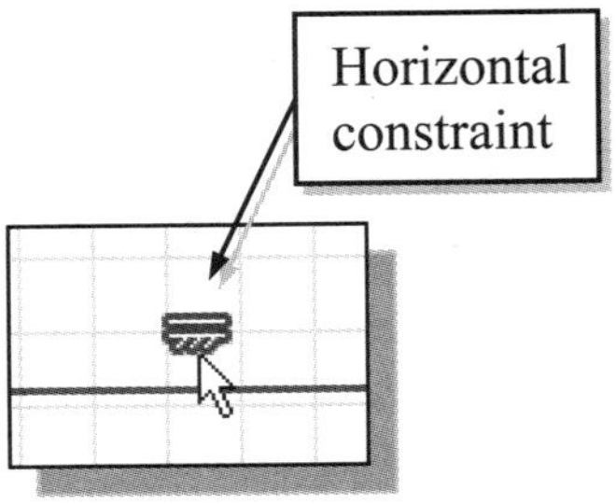

3. On your own, move the cursor on top of the other two lines and notice no additional constraints exist on the other two entities.

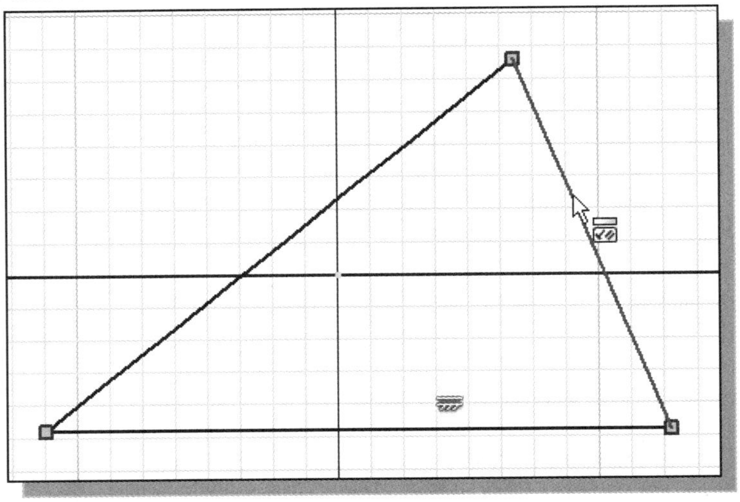

4. Move the cursor on top of the displayed constraint and notice the highlighted endpoint/line indicating the location/entities where the constraints are applied.

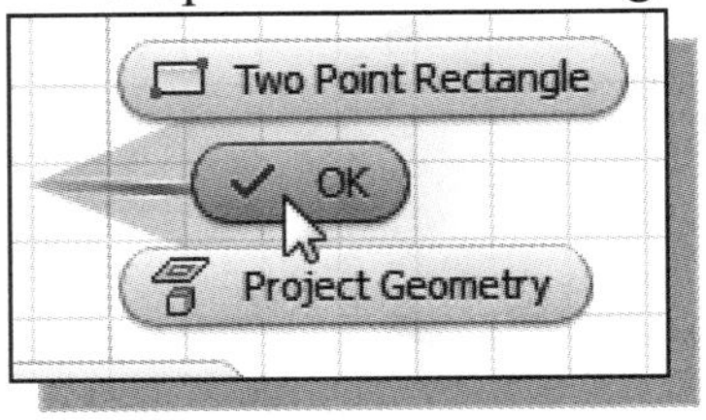

5. Inside the graphics window, right-click to bring up the option menu and select **OK** to end the Show Constraints command.

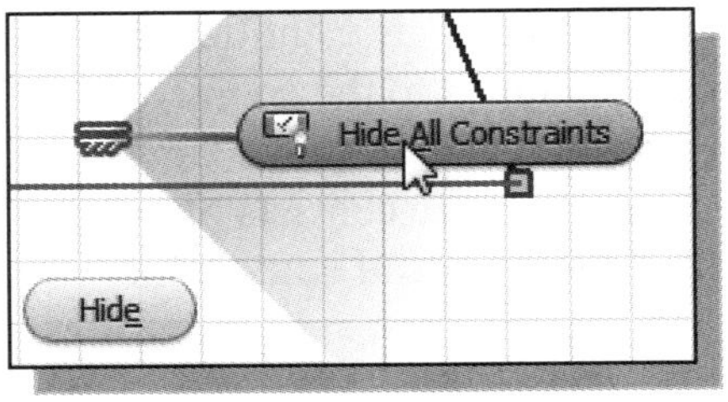

6. Right-click on the **horizontal** constraint icon to bring up the option list and select the **Hide All Constraints** option as shown.

Applying Geometric/Dimensional Constraints

In Autodesk Inventor, twelve types of constraints are available for 2D sketches.

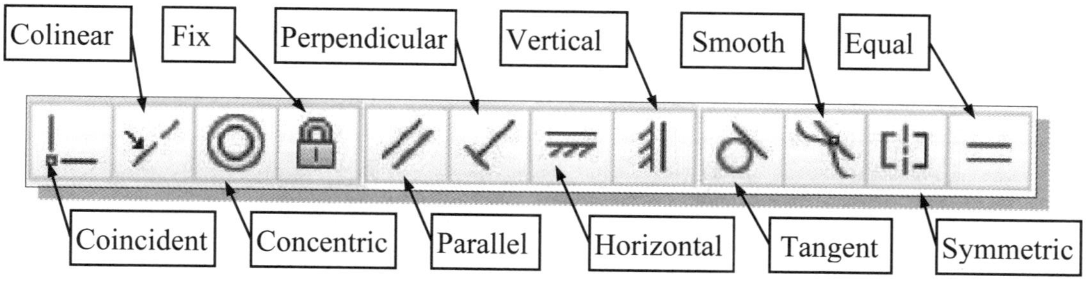

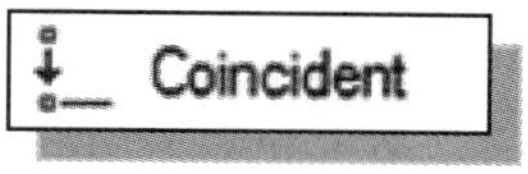

Coincident constraint: Constrains two points together or one point to a curve.

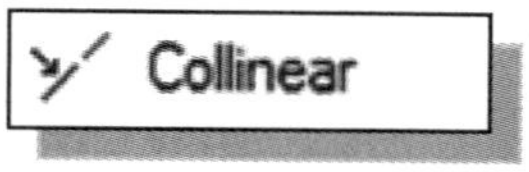

Collinear constraint: Causes two lines or ellipse axes to lie along the same line.

Concentric constraint: Constrains two arcs, circles, or ellipses to the same center point.

Fixed location constraint: Constrains points or curves to a fixed location relative to the sketch coordinate system.

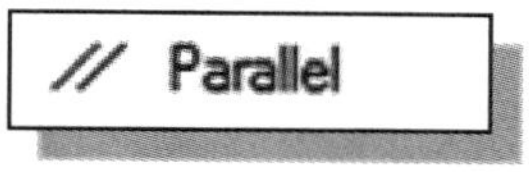

Parallel constraint: Causes selected lines or ellipse axes to lie parallel to one another.

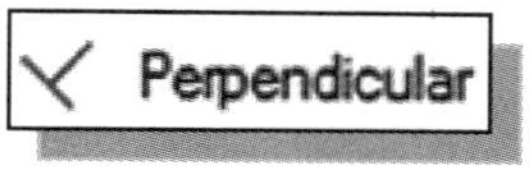

Perpendicular constraint: Causes selected curves or ellipse axes to lie at right angles to one another.

Horizontal constraint: Causes lines, ellipse axes, or pairs of points to lie parallel to the X-axis of the sketch coordinate system.

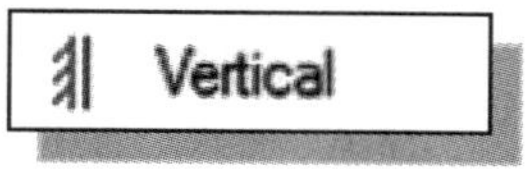

Vertical constraint: Causes lines, ellipse axes, or pairs of points to lie parallel to the Y-axis of the sketch coordinate system.

Tangent constraint: Constrains two curves to be tangent to one another.

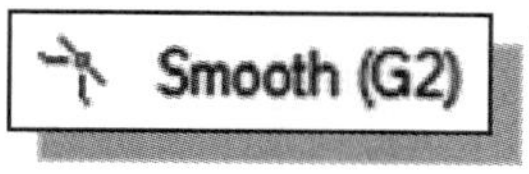

Smooth (G2) constraint: Creates a curvature continuous (G2) condition between a spline and a line, arc, or spline.

Symmetric constraint: Constrains lines or curves to become symmetrically constrained about a selected line.

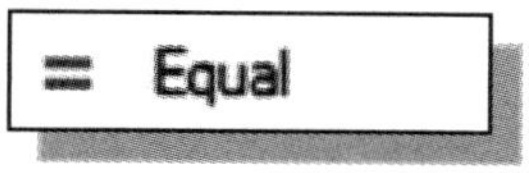

Equal constraint: Selected arcs/circles constrained to the same radius or selected lines to the same length.

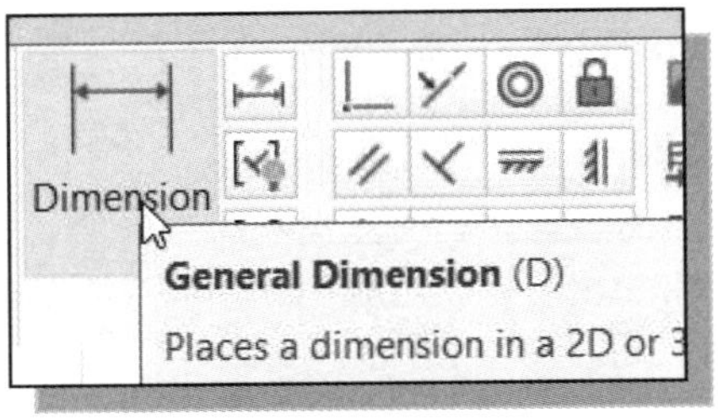

1. Select the **General Dimension** command in the *Sketch* toolbar. The General Dimension command allows us to quickly create and modify dimensions. Left-click once on the icon to activate the General Dimension command.

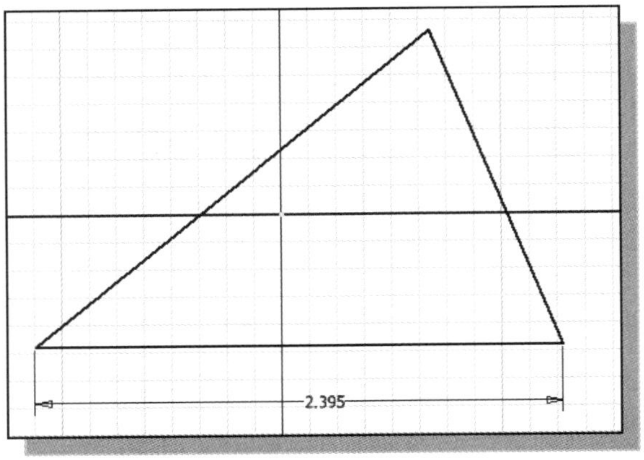

2. On your own, create the dimension as shown in the figure below. (Note that the displayed value might be different on your screen.)

❖ Note, *Inventor* identifies the need for additional dimensions at the bottom of the screen.

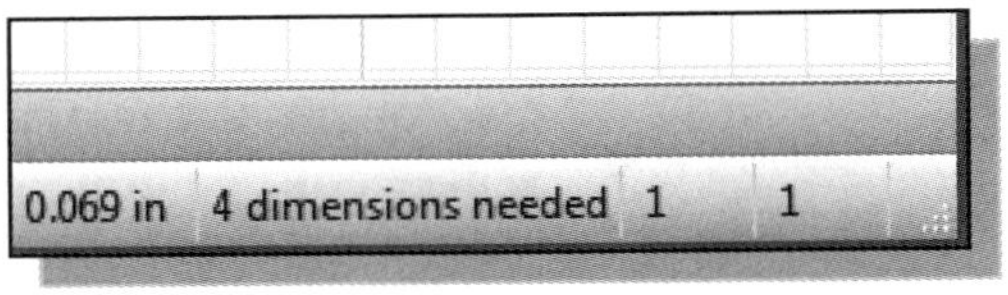

3. Move the cursor on top of the different **Constraint** icons. A *Help-tip box* appears next to the cursor and a brief description of the command is displayed at the bottom of the drawing screen as the cursor is moved over the different icons.

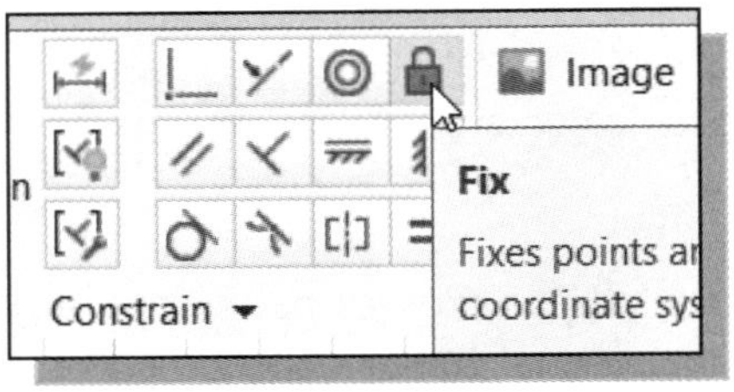

4. Click on the **Fix** constraint icon to activate the command.

5. Pick the **lower right corner** of the triangle to make the corner a fixed point.

6. Inside the graphics window, right-click to bring up the option menu and select **OK** to end the Fix Constraints command.

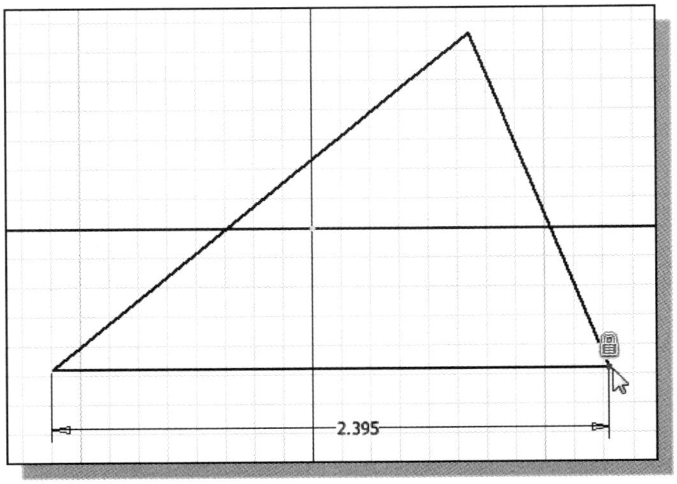

➢ Geometric constraints can be used to control the direction in which changes can occur. For example, in the current design we are adding a horizontal dimension to control the length of the horizontal line. If the length of the line is modified to a greater value, Autodesk Inventor will lengthen the line toward the left side. This is due to the fact that the **Fix** constraint will restrict any horizontal movement of the horizontal line toward the right side.

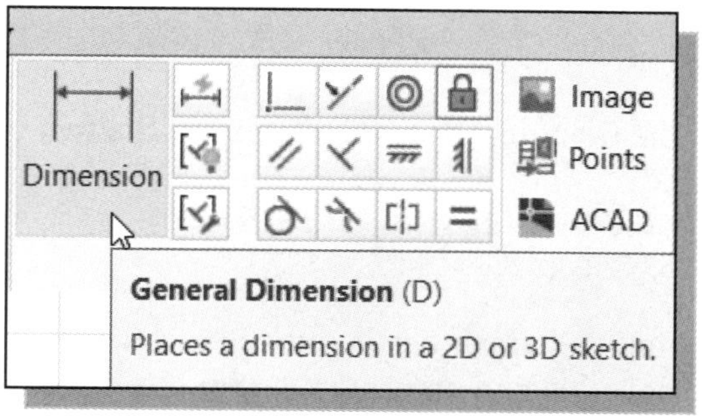

7. Select the **General Dimension** command in the *2D Sketch* toolbar.

8. Click on the dimension text to open the *Edit Dimension* window.

9. Enter a value that is greater than the displayed value to observe the effects of the modification.

10. On your own, use the **Undo** command to reset the dimension value to the previous value.

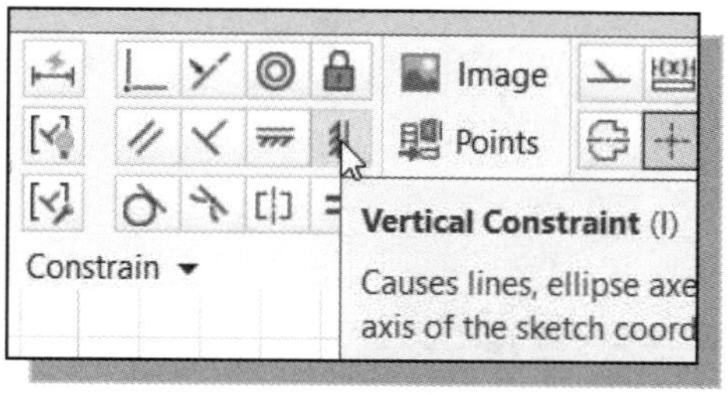

11. Select the **Vertical** constraint icon in the *2D Constraints* toolbar.

12. Pick the inclined line on the right to make the line vertical as shown in the figure below.

13. Hit the [**Esc**] key once to end the Vertical Constraint command.

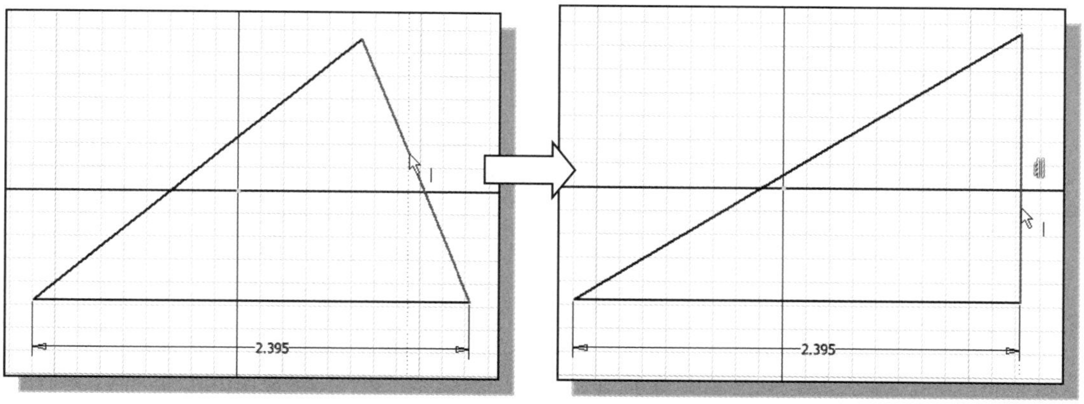

➢ You should think of the constraints and dimensions as defining elements of the geometric entities. How many more constraints or dimensions will be necessary to fully constrain the sketched geometry? Which constraints or dimensions would you use to fully describe the sketched geometry?

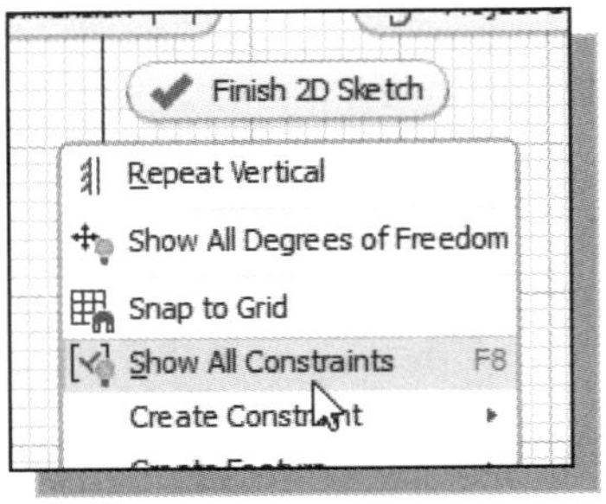

14. Inside the graphics window, click once with the **right-mouse-button** to display the option menu. Select **Show All Constraints** in the pop-up menu to show all the applied constraints. Note that function key **F8** can also be used to activate this command.

15. Move the cursor on top of the top corner of the triangle. (Note the display of the two Coincident constraints at the top corner.)

16. Hit [**F9**] once to hide the constraints.

17. Drag the top corner of the triangle and note that the corner can be moved to a new location. Release the mouse button at a new location and notice the corner is adjusted only in an upward or downward direction. Note that the two adjacent lines are automatically adjusted to the new location.

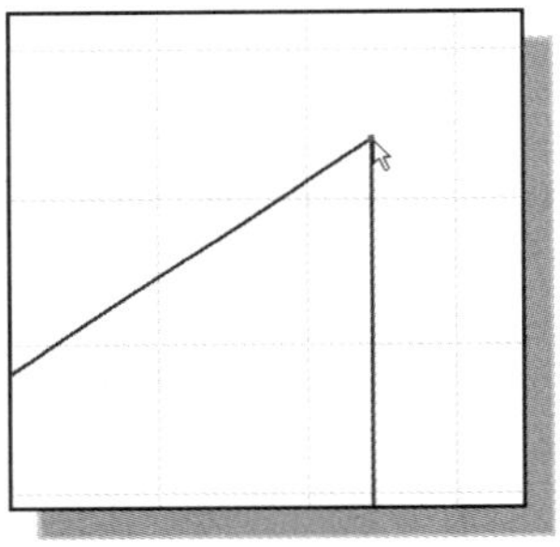

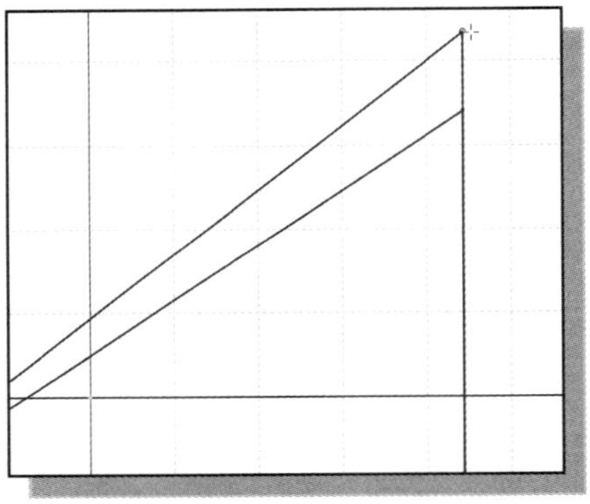

18. On your own, experiment with dragging the other corners to new locations.

- The three constraints that are applied to the geometry provide a full description for the location of the two lower corners of the triangle. The Vertical constraint, along with the Fix constraint at the lower right corner, does not fully describe the location of the top corner of the triangle. We will need to add additional information, such as the length of the vertical line or an angle dimension.

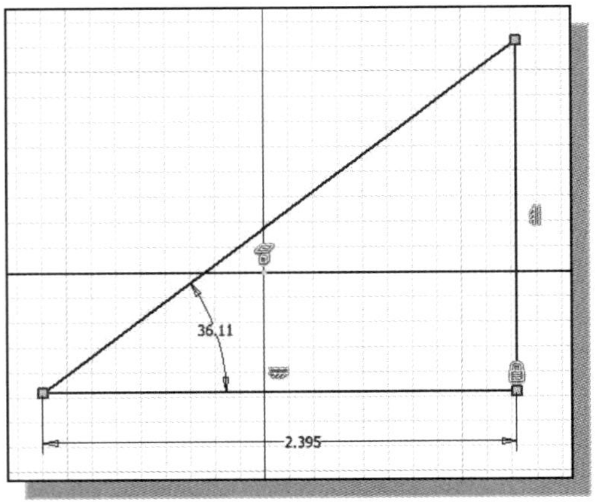

19. On your own, add an angle dimension to the left corner of the triangle.

20. Press the [**Esc**] key once to exit the General Dimension command.

21. On your own, try changing the angle to **45°** and observe the adjustment of the triangle with the adjustment.

- Note the sketched geometry is now fully constrained with the added dimension.

Over-Constraining and Driven Dimensions

We can use Autodesk Inventor to build partially constrained or totally unconstrained solid models. In most cases, these types of models may behave unpredictably as changes are made. However, Autodesk Inventor will not let us over-constrain a sketch; additional dimensions can still be added to the sketch, but they are used as references only. These additional dimensions are called ***driven dimensions***. *Driven dimensions* do not constrain the sketch; they only reflect the values of the dimensioned geometry. They are enclosed in parentheses to distinguish them from normal (parametric) dimensions. A *driven dimension* can be converted to a normal dimension only if another dimension or geometric constraint is removed.

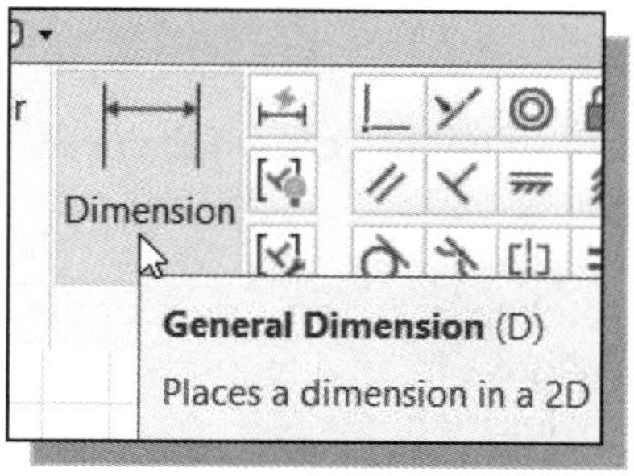

1. Select the **General Dimension** command in the *Sketch* toolbar.

2. Select the **vertical line**.

3. Pick a location that is to the right side of the triangle to place the dimension text.

4. A warning dialog box appears on the screen stating that the dimension we are trying to create will over-constrain the sketch. Click on the **Accept** button to proceed with the creation of a driven dimension.

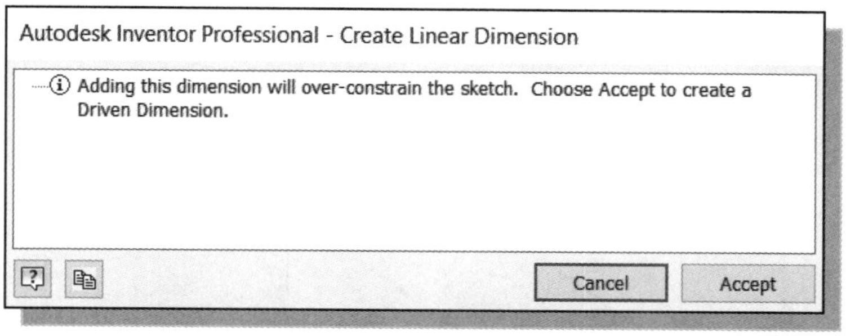

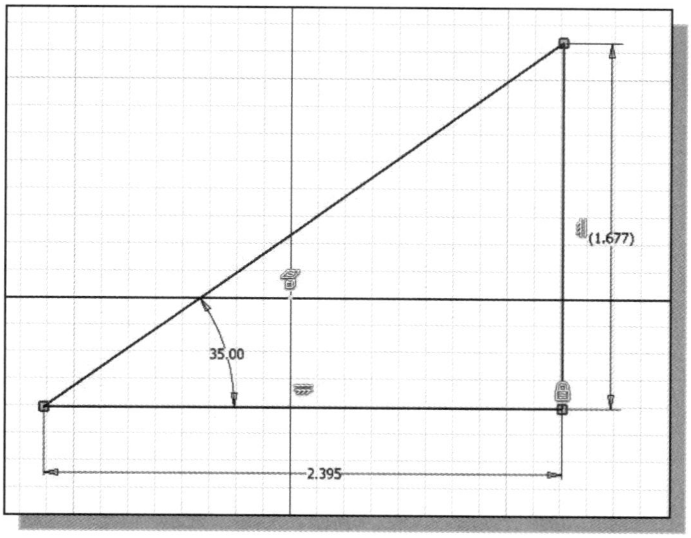

5. On your own, modify the angle dimension to **35°** and observe the changes of the 2D sketch and the driven dimension.

❖ Note that Inventor has indicated the sketch is **Fully Constrained**.

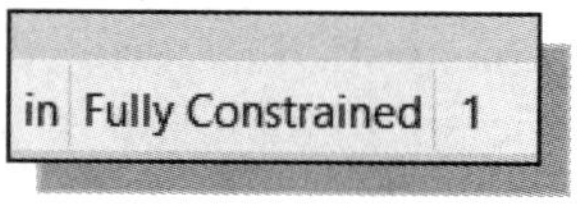

Deleting Existing Constraints

1. On your own, display all the active constraints if they are not already displayed. (Hint: Use the Show Constraints command in the *Sketch* toolbar or Show All Constraints in the option menu.)

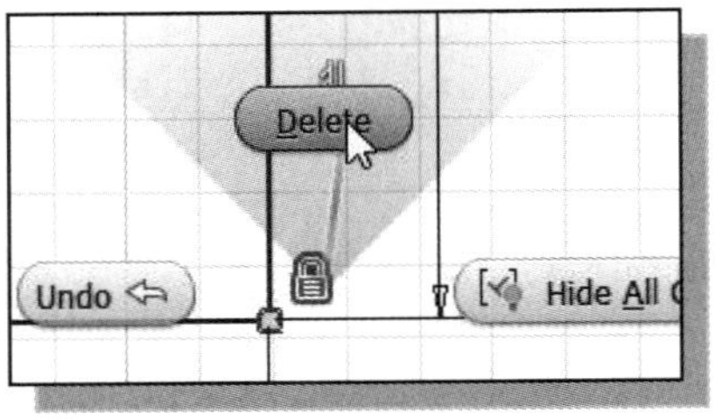

2. Move the cursor on top of the **Fix** constraint icon and **right-click** once to bring up the option menu.

3. Select **Delete** to remove the Fix constraint that is applied to the lower right corner of the triangle.

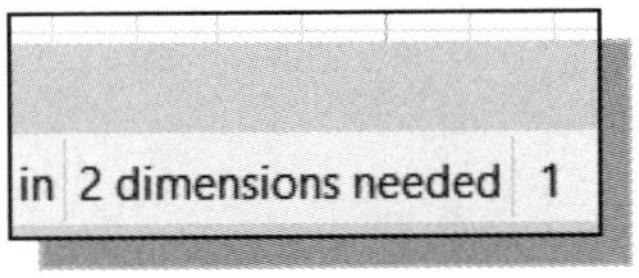

❖ Note the removal of the **Fix** constraint has caused the need for **two additional dimensions**.

4. On your own, hide all the constraints with the [**F9**] key and drag the top corner of the triangle upward and note that the entire triangle is free to move in all directions. Release the mouse button to move the triangle to the new location.

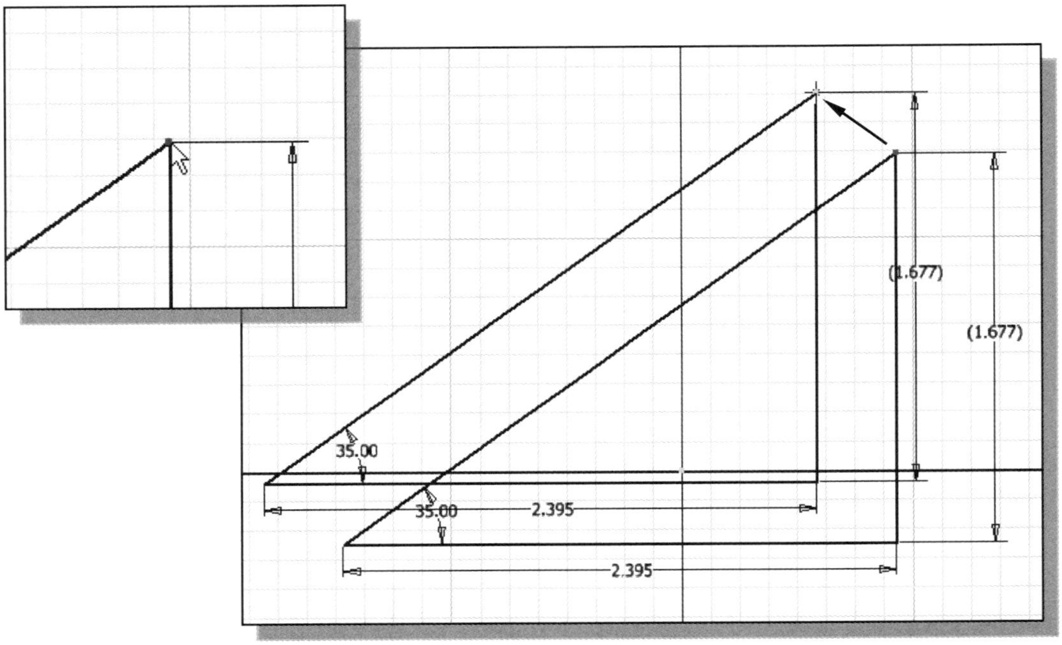

5. On your own, delete the reference dimension on the right and experiment with dragging the corners and the line segments to move the triangle to new locations.

❖ **Dimensional constraints** are used to describe the SIZE and LOCATION of individual geometric shapes. **Geometric constraints** are **geometric restrictions** that can be applied to geometric entities. The constraints applied to the triangle are sufficient to maintain its size and shape, but the geometry can be moved around; its location definition isn't complete.

Using the Auto Dimension Command

In Autodesk Inventor, the Auto Dimension command can be used to assist in creating a fully constrained sketch. **Fully constrained** sketches can be updated more predictably as design changes are implemented. The general procedure for applying dimensions to sketches is to use the General Dimension command to add the more critical dimensions, and then use the Auto Dimension command to add the additional dimensions/constraints to fully constrain the sketch. The Auto Dimension command can also be used to apply the missing dimensions that are needed. It is also important to realize that different sets of dimensions and geometric constraints can be applied to the same sketch to accomplish a fully constrained geometry.

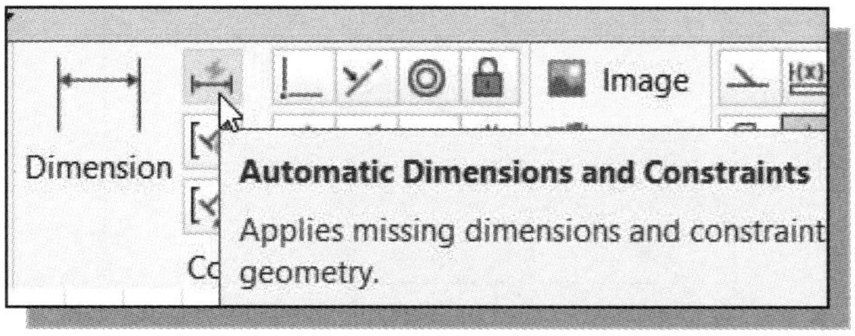

1. Click on the **Auto Dimension** icon in the *Constrain* toolbar.

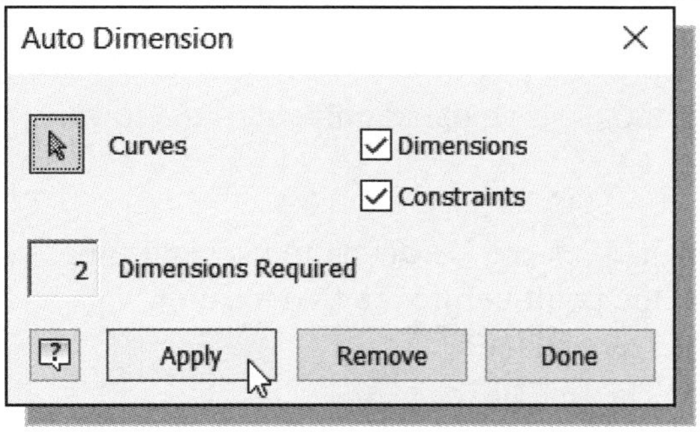

❖ Note that Autodesk Inventor indicates the sketch is missing two locational dimensions.

2. Click **Apply** to the Auto Dimension command.

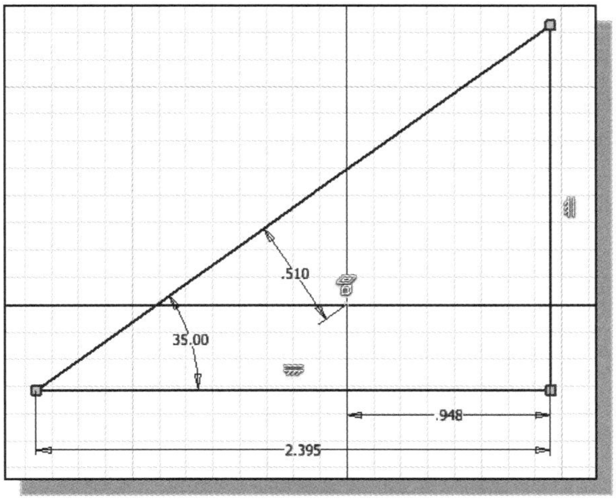

- Note that the Auto Dimension command added two dimensions measuring from the **Origin**, the intersection of the horizontal and vertical axes.

➢ The two newly added dimensions do not change the size of the triangle; they are used to position the triangle.

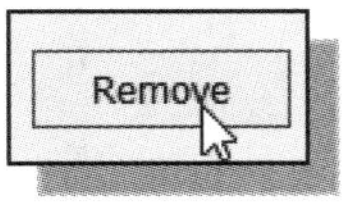

3. Click **Remove** to undo the Auto Dimension command. Note that only the dimensions created by the Auto Dimension command are affected.

4. Click **Done** to close the dialog box and exit the Auto Dimension command.

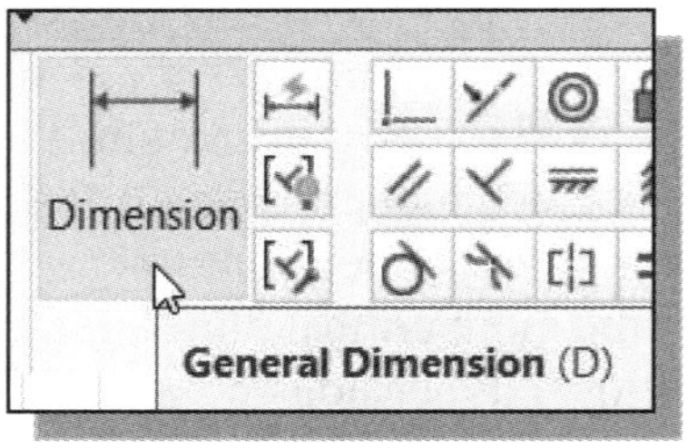

5. Activate the **General Dimension** command in the *Constrain* toolbar.

6. On your own, create the two locational dimensions, measuring the lower right corner to the origin as shown.

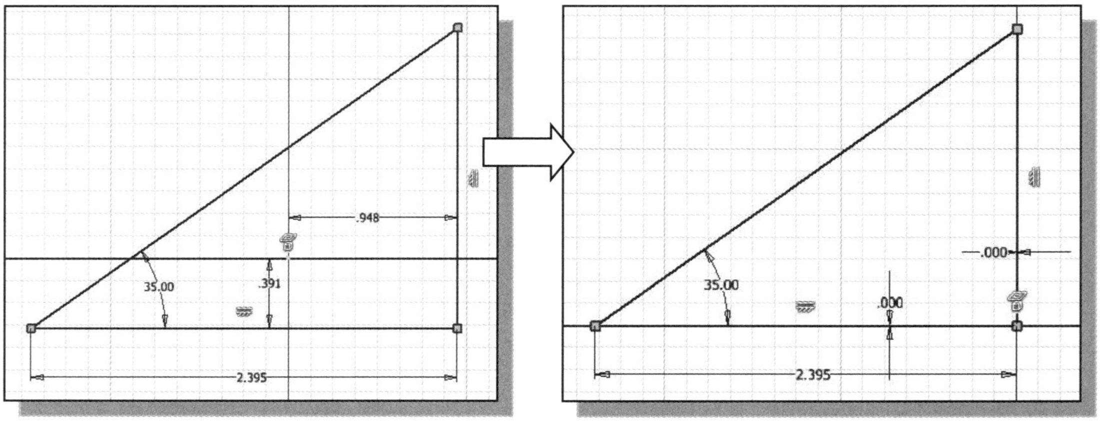

7. On your own, set the two dimensions to **0.0** and observe the alignment of the lower right corner of the triangle to the *Origin*.

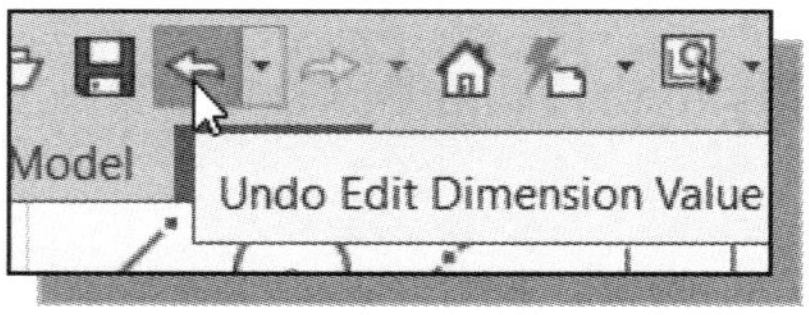

8. On your own, click the **Undo** button several times to return to the point before the two location dimensions were added.

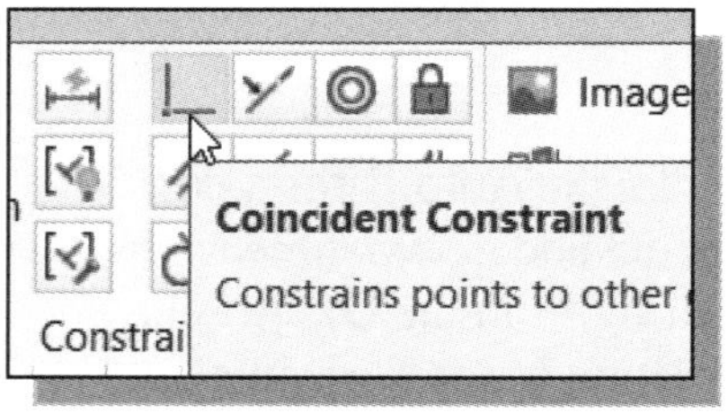

9. Activate the **Coincident Constraint** command in the *Constrain* toolbar.

10. Select the **lower right corner** of the triangle and the **Origin** to fully constrain the sketch as shown.

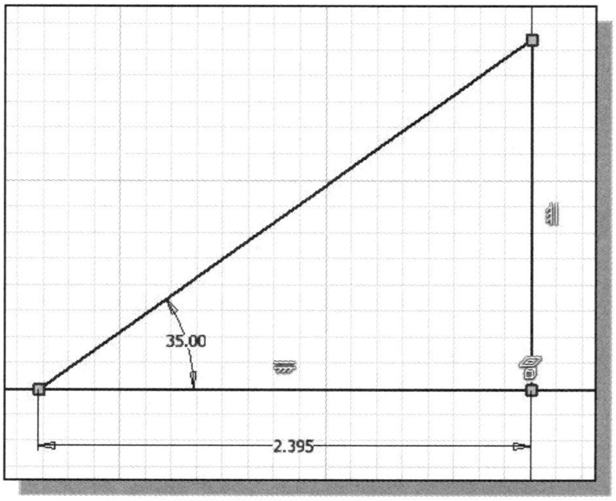

➢ In parametric modeling, it is desired to create fully constrained sketches.

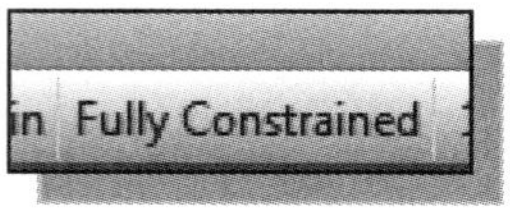

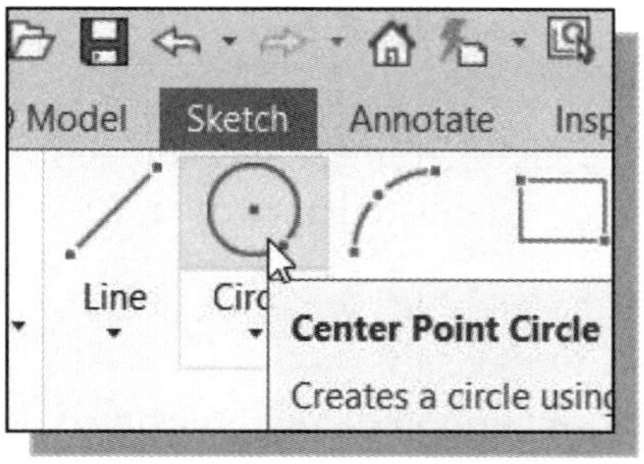

11. Select the **Center Point Circle** command by clicking once with the left-mouse-button on the icon in the *Sketch* toolbar.

12. On your own, create a circle of arbitrary size inside the triangle as shown below.

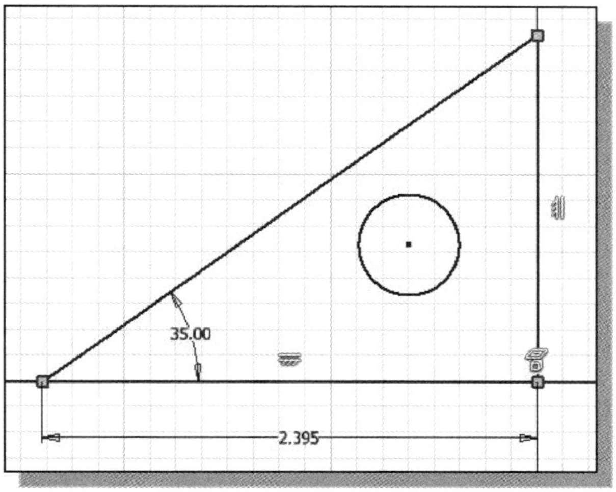

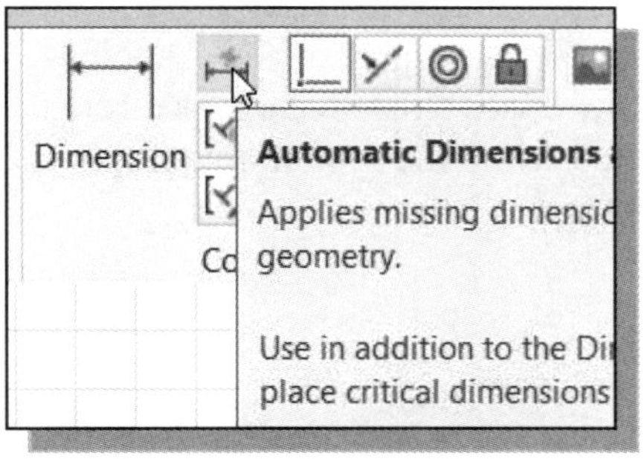

13. Click on the **Auto Dimension** icon in the *Sketch* panel.

❖ Note that *Autodesk Inventor* confirms that the sketch is not fully constrained and "*3 Dimensions Required*" to fully constrain the circle. What are the dimensions and/or constraints that can be applied to fully constrain the circle?

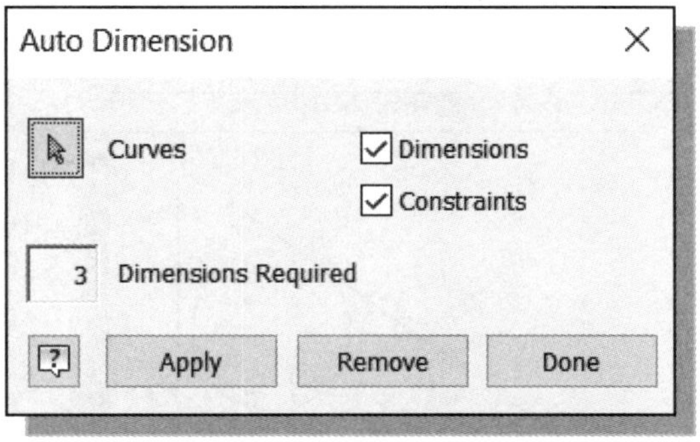

14. Click **Done** to exit the Auto Dimension command.

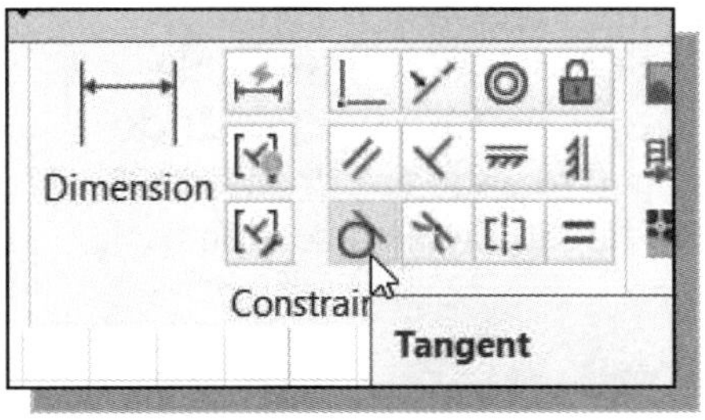

15. Click on the **Tangent** constraint icon in the *Sketch* toolbar.

16. Pick the circle by left-clicking once on the geometry.

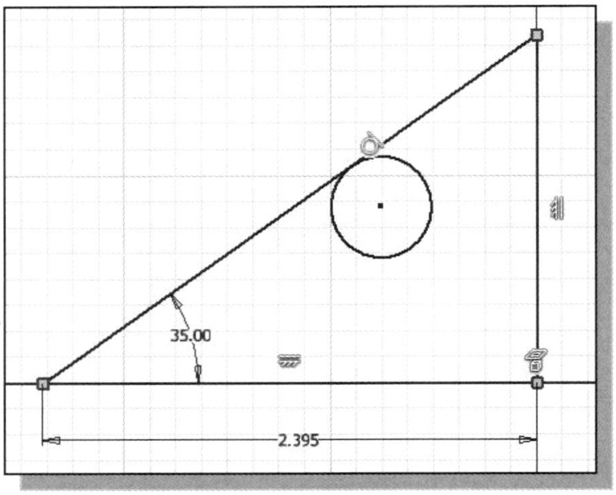

17. Pick the inclined line. The sketched geometry is adjusted as shown.

18. Inside the graphics window, click once with the right-mouse-button to display the option menu. Select **OK** in the pop-up menu to end the **Tangent** command.

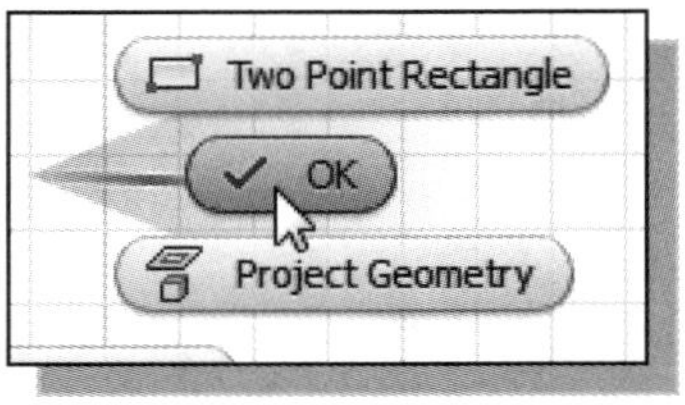

- How many more constraints or dimensions do you think will be necessary to fully constrain the circle? Which constraints or dimensions would you use to fully constrain the geometry?

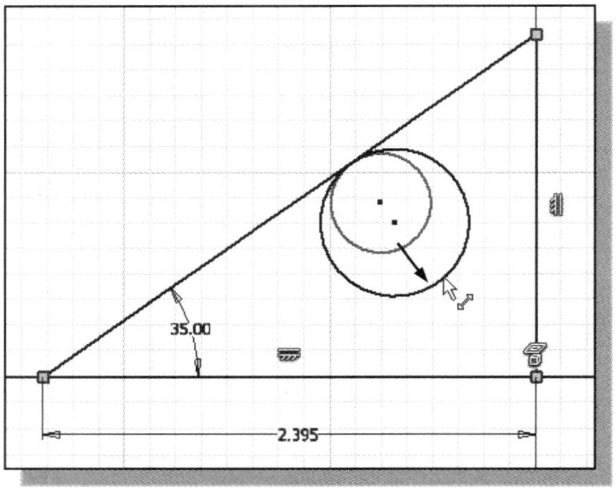

19. Move the cursor on top of the right side of the circle, and then drag the circle toward the right edge of the graphics window. Notice the size of the circle is adjusted while the system maintains the **Tangent** constraint.

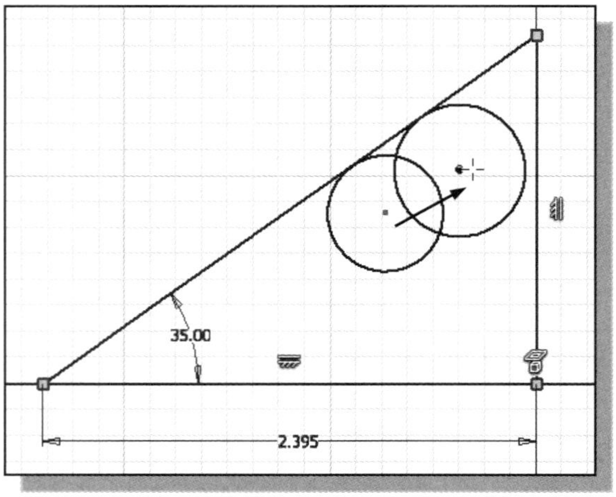

20. Drag the center of the circle toward the upper right direction. Notice the **Tangent** constraint is always maintained by the system.

➢ On your own, experiment with adding additional constraints and/or dimensions to fully constrain the sketched geometry. Use the **Undo** command to undo any changes before proceeding to the next section.

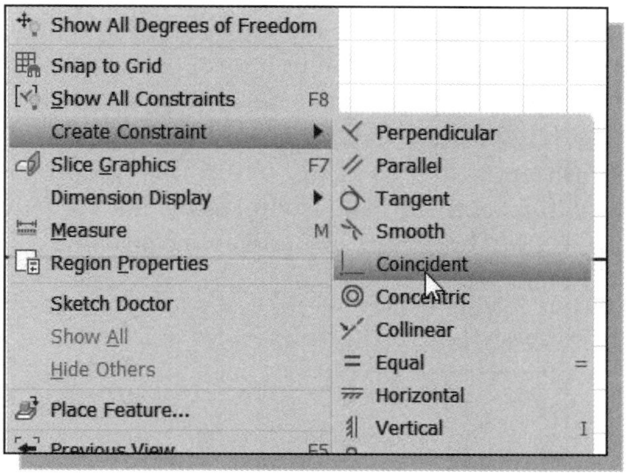

21. Inside the *graphics window*, click once with the **right-mouse-button** to display the option menu. Select **Create Constraint → Coincident** in the pop-up menus.

- The option menu is a quick way to access many of the commonly used commands in Autodesk Inventor.

22. Pick the **vertical line**.

23. Pick the center of the circle to align the **center** of the circle and the vertical line.

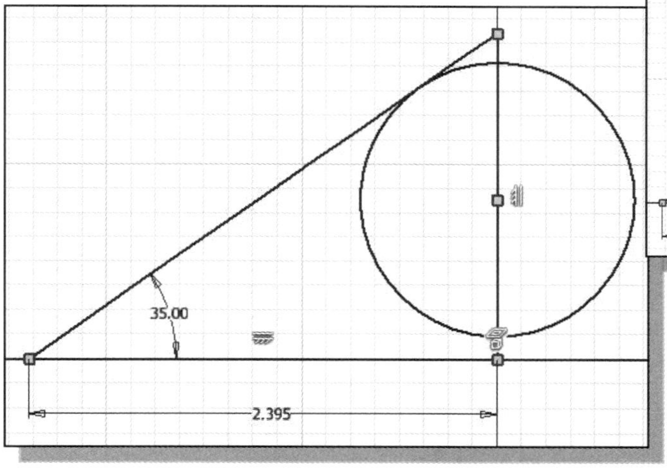

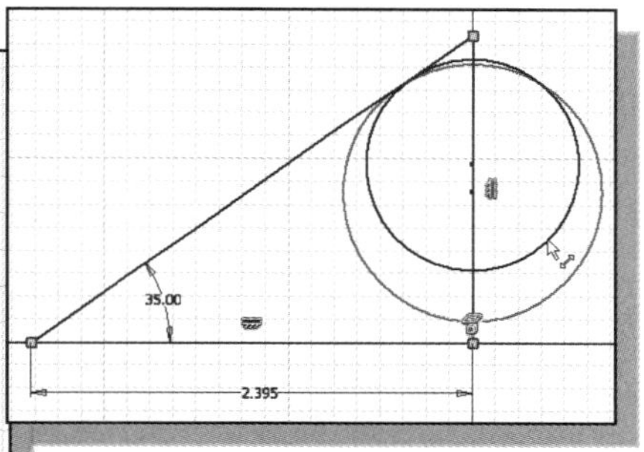

- Is the circle fully constrained?

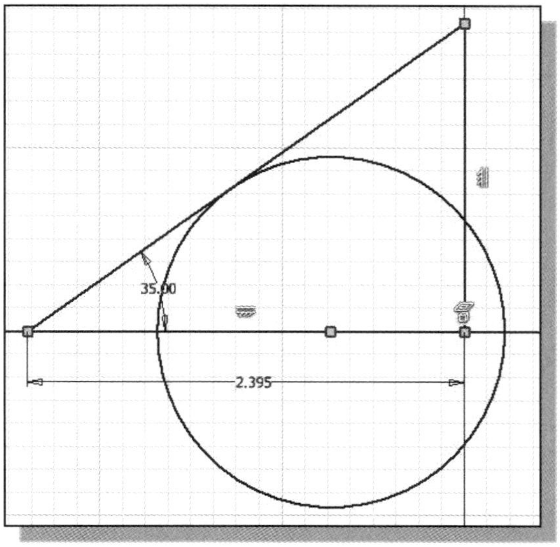

24. On your own, **delete** the Coincident constraint we just applied and add a Coincident constraint between the center of the circle and the horizontal line.

- How many more constraints or dimensions do you think will be necessary to fully constrain the circle? Which constraints or dimensions would you use to fully constrain the geometry?

❖ The application of different constraints affects the geometry differently. The design intent is maintained in the CAD model's database and thus allows us to create very intelligent CAD models that can be modified and revised fairly easily. On your own, experiment and observe the results of applying different constraints to the triangle. For example: (1) adding another **Fix** constraint to the top corner of the triangle; (2) deleting the horizontal dimension and adding another **Fix** constraint to the left corner of the triangle; and (3) adding another **Tangent** constraint and adding the size dimension to the circle.

25. On your own, modify the 2D sketch as shown below.

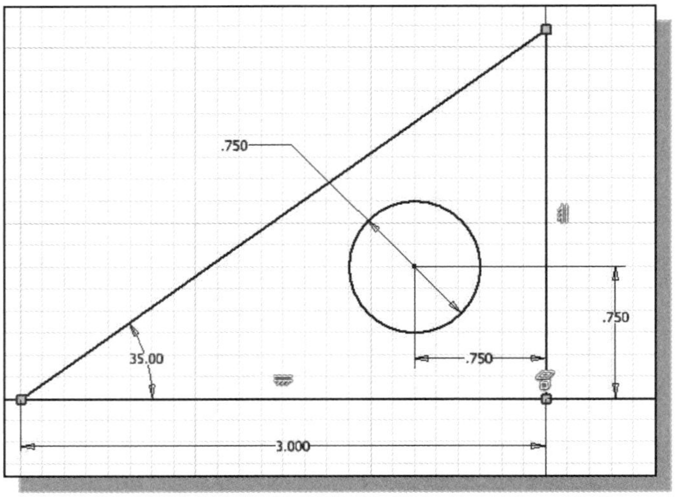

➢ On your own, use the **Extrude** command and create a 3D solid model with a plate thickness of **0.25**. Also experiment with modifying the parametric relations and dimensions through the *part browser*.

Constraint and Sketch Settings

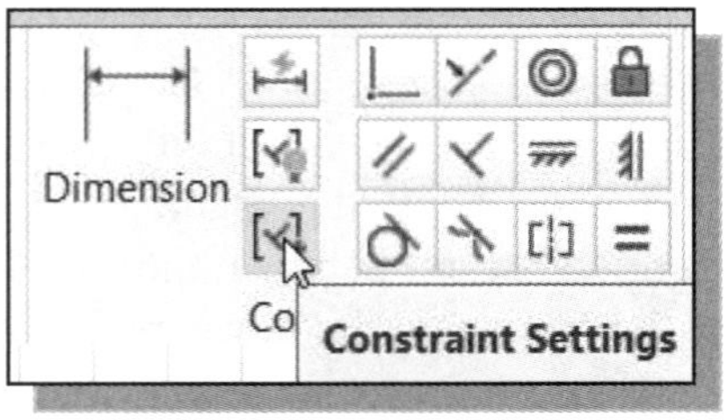

- Click on the **Constraint Settings** icon to access the constraint settings dialog box.

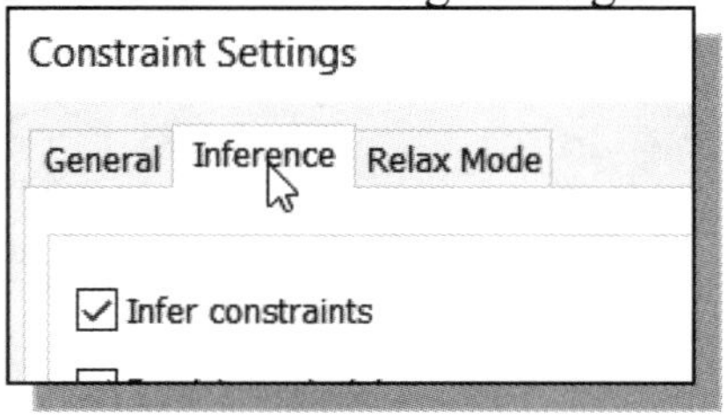

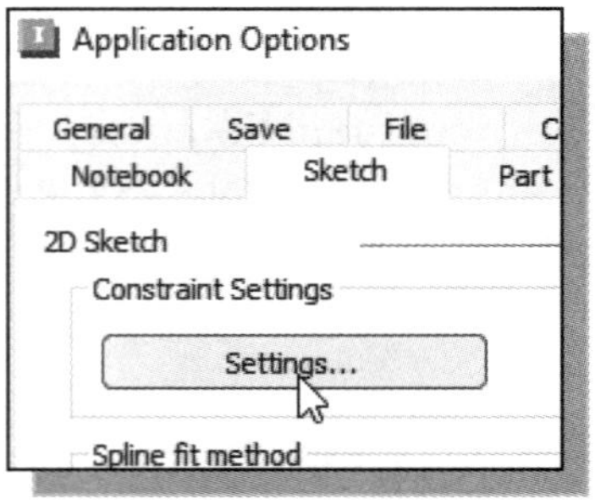

- Note the settings can also be accessed through **Application Options** in the **Tools** pull-down menu. Switch to the **Sketch** tab then click on **Constraint Settings** to display and/or modify the constraint settings.

Parametric Relations

In parametric modeling, dimensions are design parameters that are used to control the sizes and locations of geometric features. Dimensions are more than just values; they can also be used as feature control variables. This concept is illustrated by the following example.

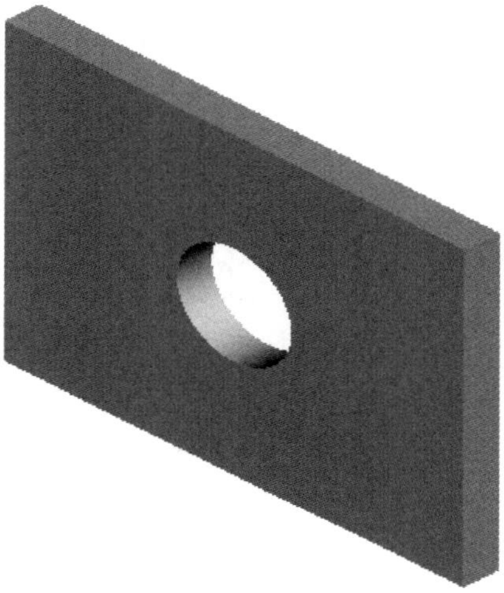

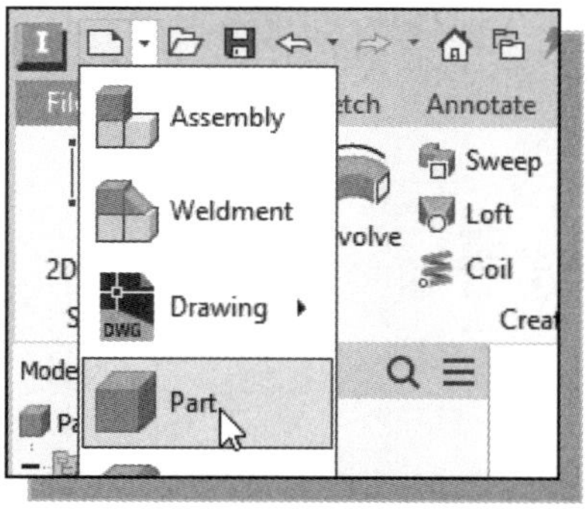

1. Start a new drawing by left-clicking once on the **New → Part** icon in the *Standard* toolbar.

- Another graphics window appears on the screen. We can switch between the two models by clicking on the different graphics windows.

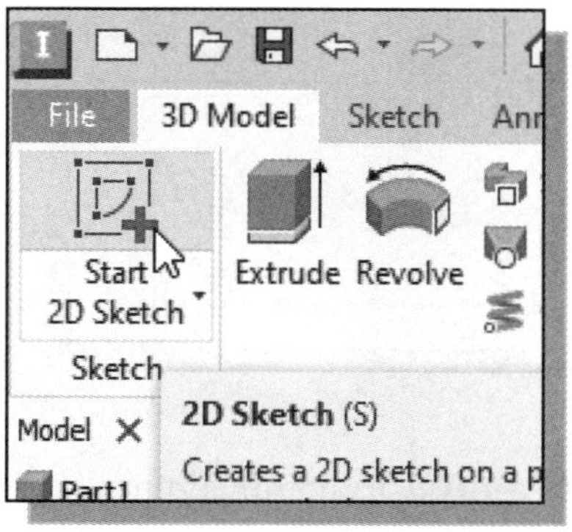

2. Select the **Start 2D Sketch** command with the left-mouse-button.

3. Select the *XY Plane* as the sketch plane for the new sketch with the **left-mouse-button**.

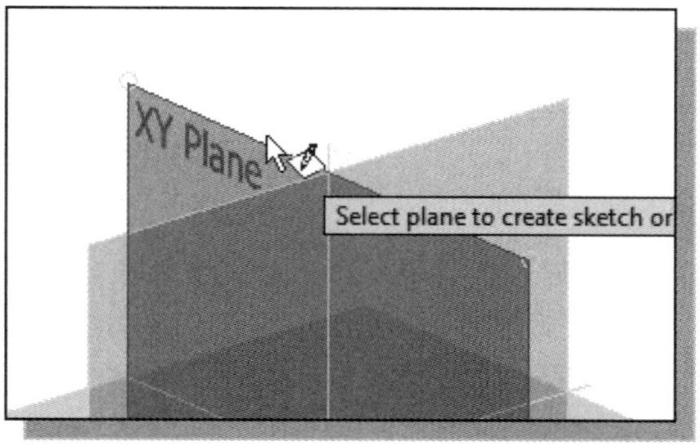

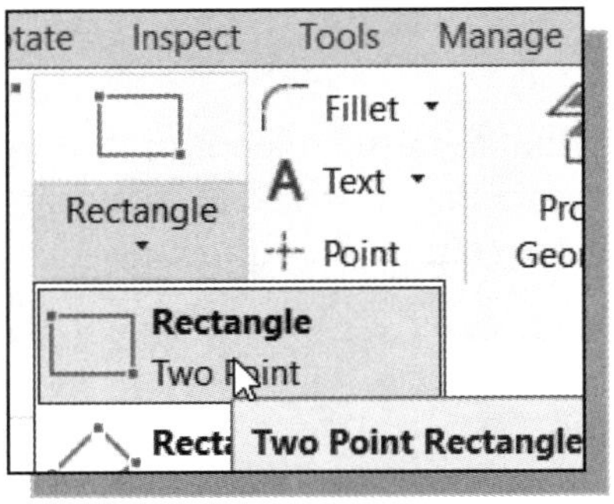

4. Select the **Two point rectangle** command by clicking once with the left-mouse-button on the icon in the *Sketch* toolbar.

5. Create a rectangle of arbitrary size positioned near the center of the screen.

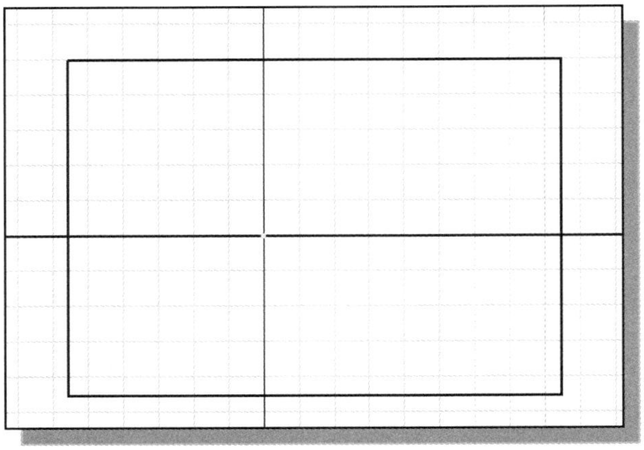

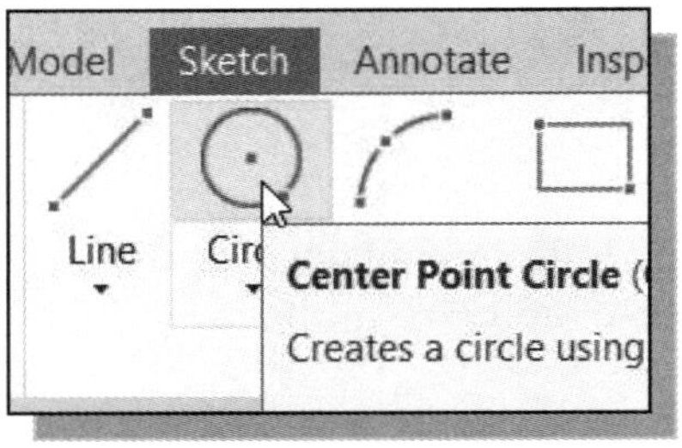

6. Select the **Center Point Circle** command by clicking once with the left-mouse-button on the icon in the *Sketch* toolbar.

7. Create a **circle** of arbitrary size inside the rectangle as shown.

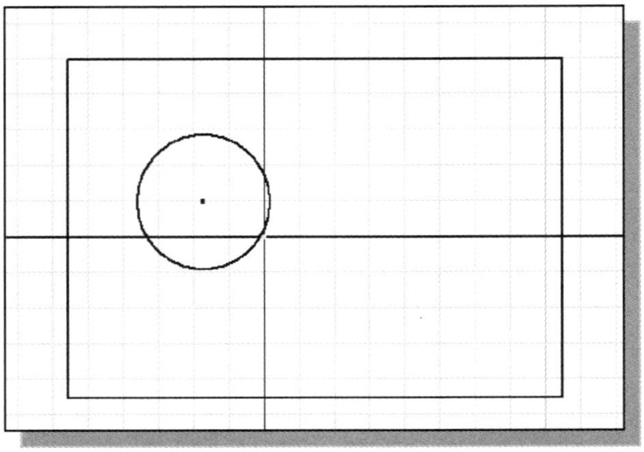

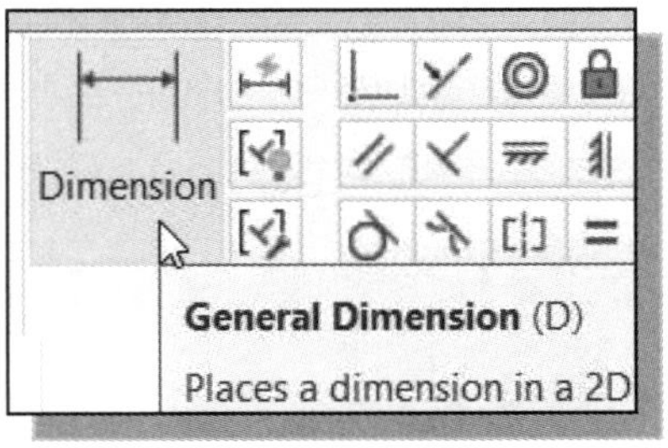

8. Select the **General Dimension** command in the *Sketch* toolbar.

9. On your own, create and adjust the geometry by modifying the dimensions as shown below. (Note: Align the center of the circle to the origin to fully constrain the sketch.)

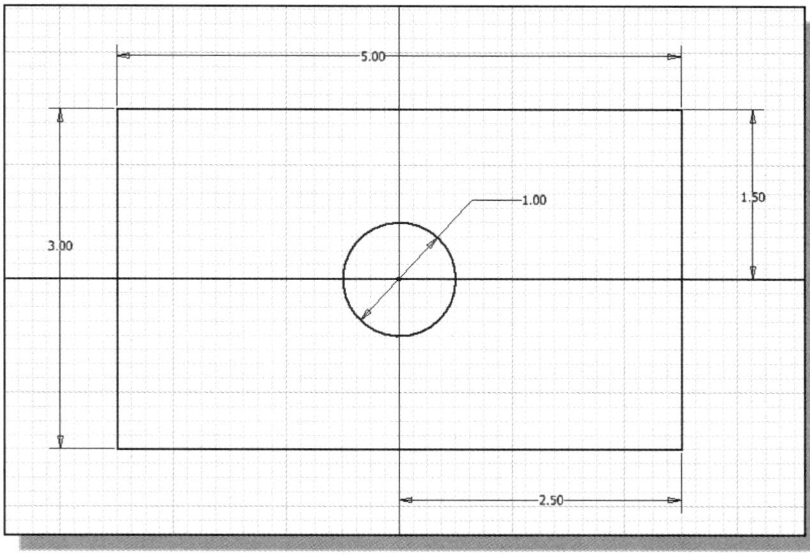

- On your own, change the overall width of the rectangle to **6.0** and the overall height of the rectangle to **3.6** and observe the location of the circle in relation to the edges of the rectangle. Adjust the dimensions back to **5.0** and **3.0** as shown in the above figure before continuing.

Dimensional Values and Dimensional Variables

Initially in Autodesk Inventor, values are used to create different geometric entities. The text created by the Dimension command also reflects the actual location or size of the entity. Each dimension is also assigned a name that allows the dimension to be used as a control variable. The default format is "dxx," where the "xx" is a number that Autodesk Inventor increments automatically each time a new dimension is added.

Let us look at our current design, which represents a plate with a hole at the center. The dimensional values describe the size and/or location of the plate and the hole. If a modification is required to change the width of the plate, the location of the hole will remain the same as described by the two location dimensional values. This is okay if that is the design intent. On the other hand, the *design intent* may require (1) keeping the hole at the center of the plate and (2) maintaining the size of the hole to be one-third of the height of the plate. We will establish a set of parametric relations using the dimensional variables to capture the design intent described in statements (1) and (2) above.

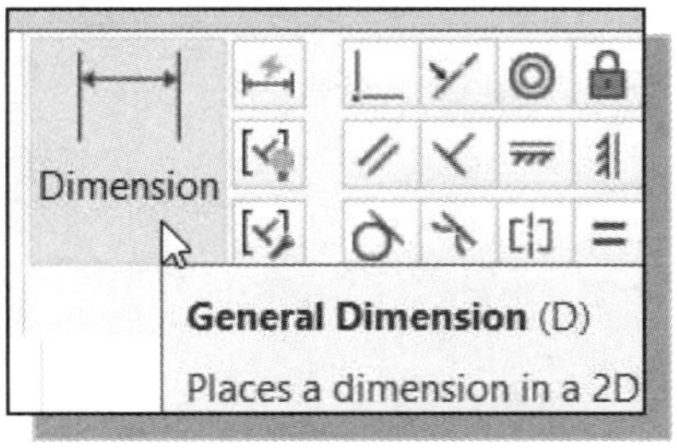

1. Left-click once on the **General Dimension** icon to activate the General Dimension command.

2. Click on the **width dimension** of the rectangle to display the *Edit Dimension* window.

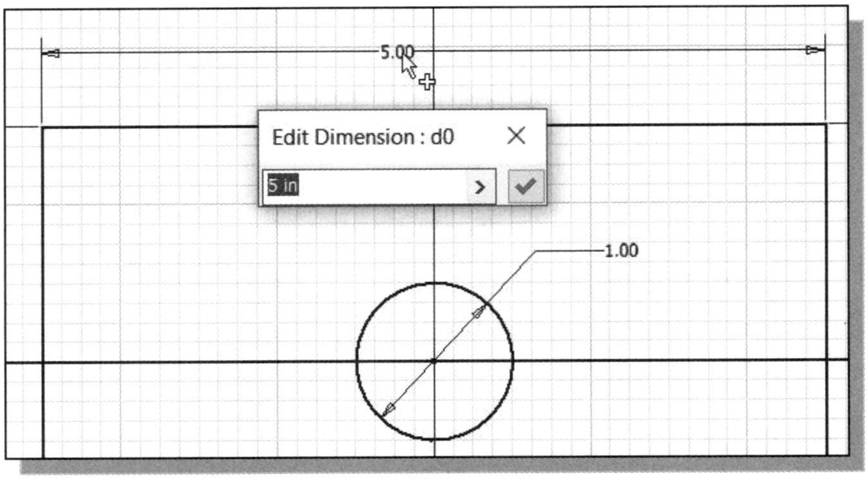

- Notice the ***variable name*** **d0** is displayed in the title area of the *Edit Dimension* window and also in the cursor box when the cursor is moved near the text box.

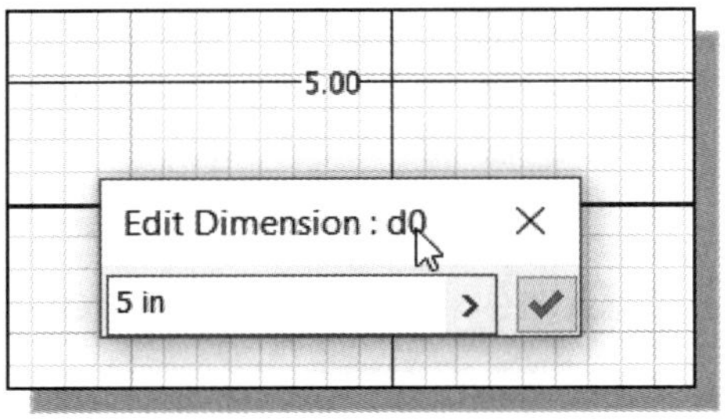

3. Click on the ***check mark*** button to close the *Edit Dimension* window.

Parametric Equations

Each time we add a dimension to a model, that value is established as a parameter for the model. We can use parameters in equations to set the values of other parameters.

1. Click on the **horizontal location dimension** of the circle to display the *Edit Dimension* window.

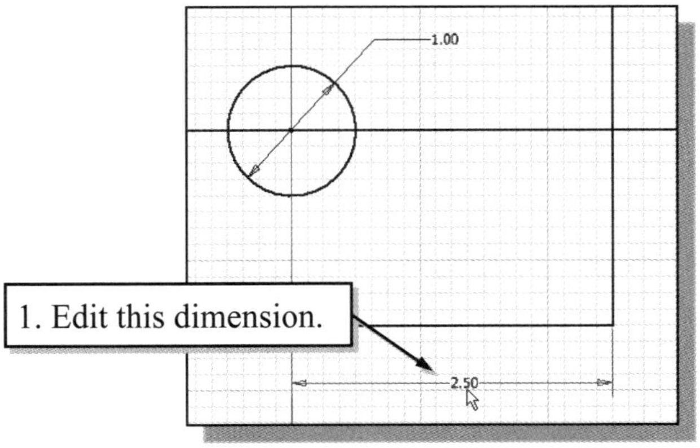

2. Click on the width dimension of the rectangle **d0** (value of **5.0**). Notice the selected variable name is automatically entered in the *Edit Dimension* window.

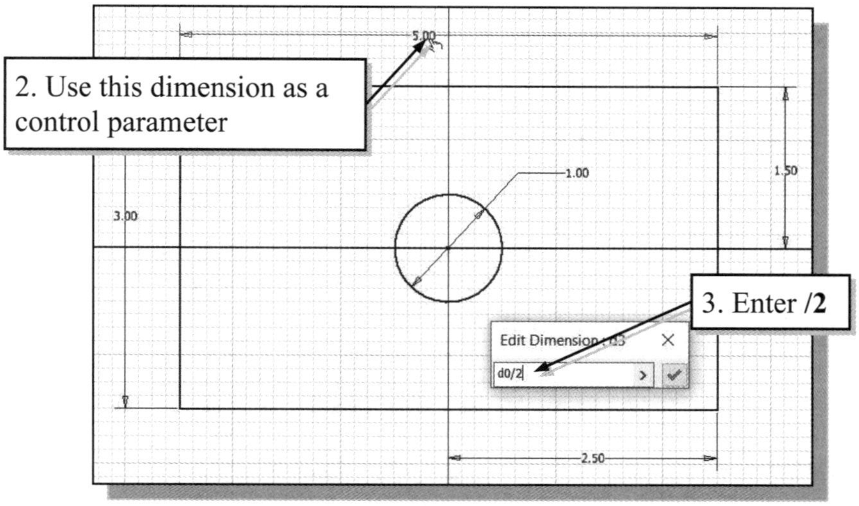

3. In the *Edit Dimension* window, enter **/2** to set the horizontal location dimension of the circle to be one-half of the width of the rectangle.

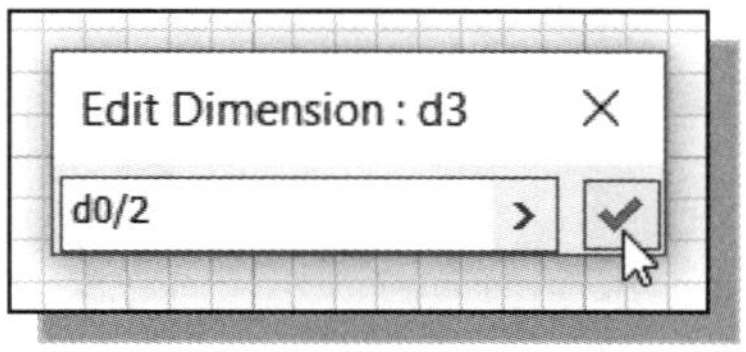

4. Click on the ***check mark*** button to close the *Edit Dimension* window.

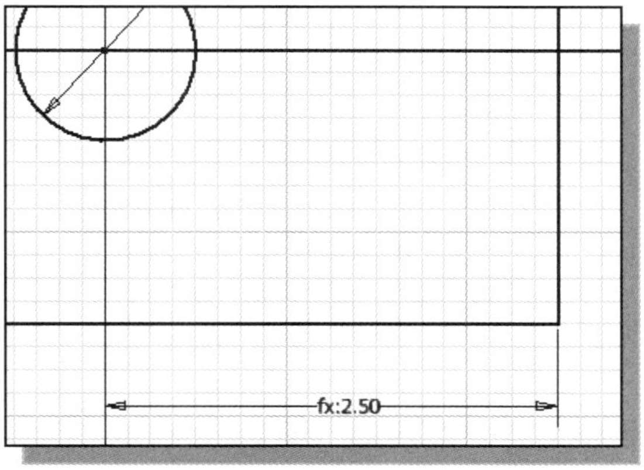

❖ Notice the derived dimension values are displayed with **fx** in front of the numbers. The parametric relations we entered are used to control the location of the circle; the location is based on the height and width of the rectangle.

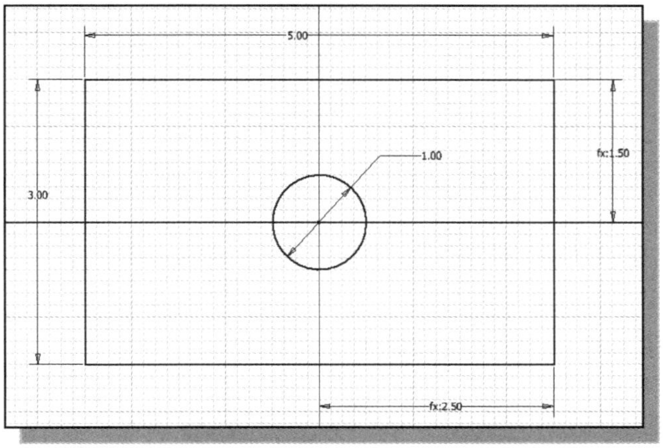

5. On your own, repeat the above steps and set the vertical location dimension to one-half of the height of the rectangle.

Viewing the Established Parameters and Relations

1. In the *Manage* toolbar select the **Parameters** command by left-clicking once on the icon.

- The **Parameters** command can be used to display all dimensions used to define the model. We can also create additional parameters as design variables, which are called ***user parameters***.

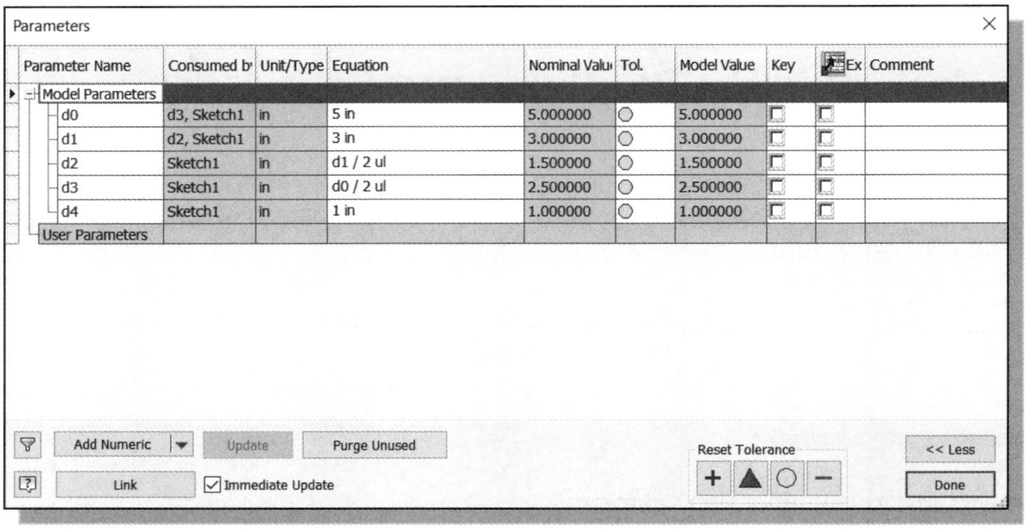

2. Click on the **1.0** value in the equation section of the *Parameters* window. Enter **d1/3** as the parametric relation to set the size of the circle to be one-third of the height of the rectangle as shown below.

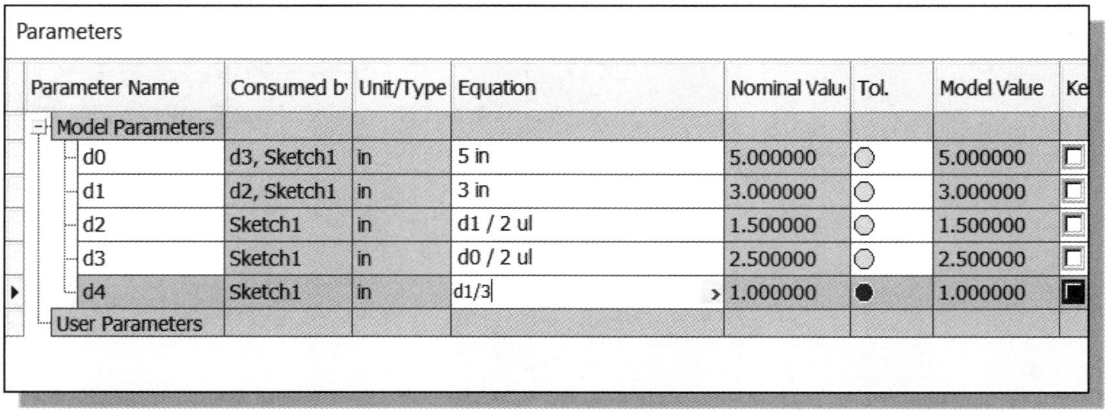

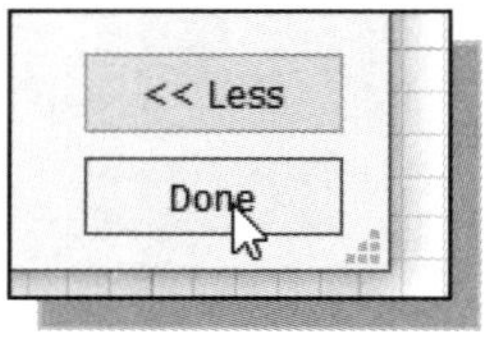

3. Click on the **Done** button to accept the settings.

4. On your own, change the dimensions of the rectangle to **6.0** x **3.6** and observe the changes to the location and size of the circle. (Hint: Double-click the dimension text to bring up the *Edit Dimension* window.)

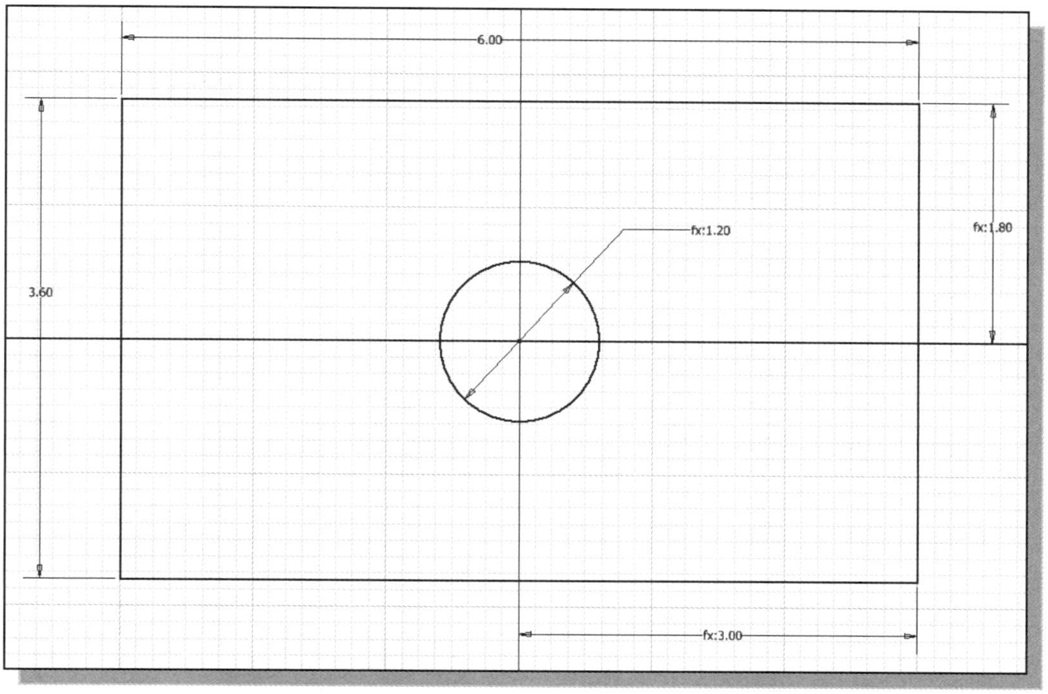

❖ *Autodesk Inventor* automatically adjusts the dimensions of the design, and the parametric relations we entered are also applied and maintained. The dimensional constraints are used to control the size and location of the hole. The design intent, previously expressed by statements (1) and (2) at the beginning of this section, is now embedded into the model.

➢ On your own, use the **Extrude** command and create a 3D solid model with a plate thickness of **0.25**. Also, experiment with modifying the parametric relations and dimensions through the *part browser*.

Saving the Model File

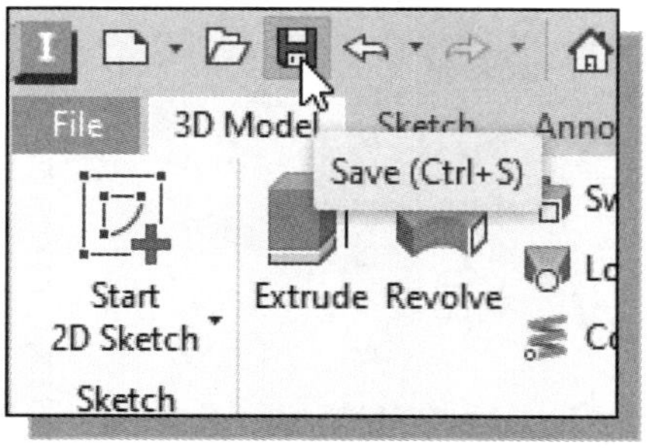

1. Select **Save** in the *Standard* toolbar. We can also use the "**Ctrl-S**" combination (press down the "Ctrl" key and hit the "S" key once) to save the part.

2. In the pop-up window, enter **Plate** as the name of the file.

3. Click on the **SAVE** button to save the file.

Using the Measure Tools

Besides using the measure tools to get geometric information at the 2D level, the measure tools can also be used on 3D models.

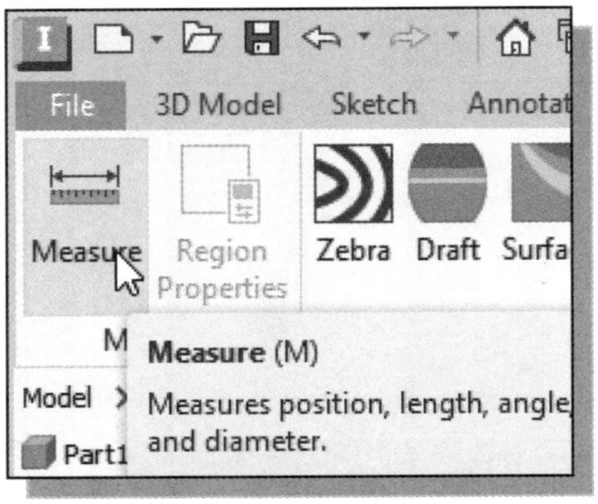

1. In the *Inspect Ribbon tab*, left-click once on the **Measure** option as shown.

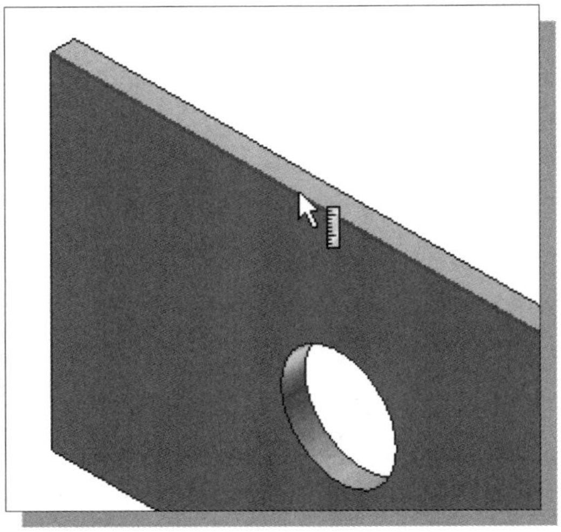

2. Click on the top edge of the rectangular plate as shown.

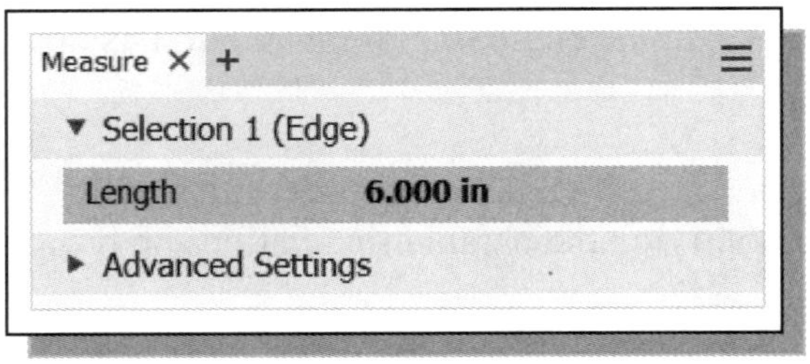

3. The associated length measurement of the selected geometry is displayed in the *Length* dialog box as shown.

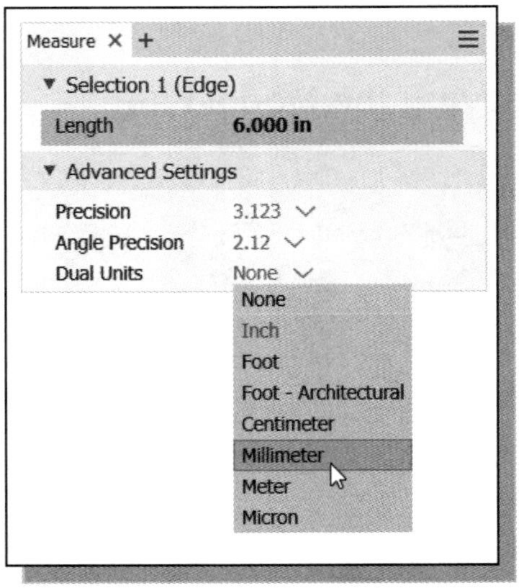

4. Click on the **triangular icon** of the **Advanced Settings** option list to display the available options.

5. Choose [**Dual Units**] → [**Millimeter**] to also display the measurement in mm.

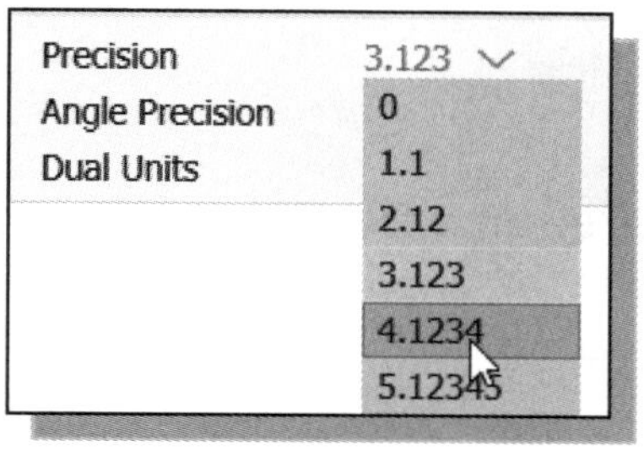

6. On your own, set the precision to show 4 decimal places as shown.

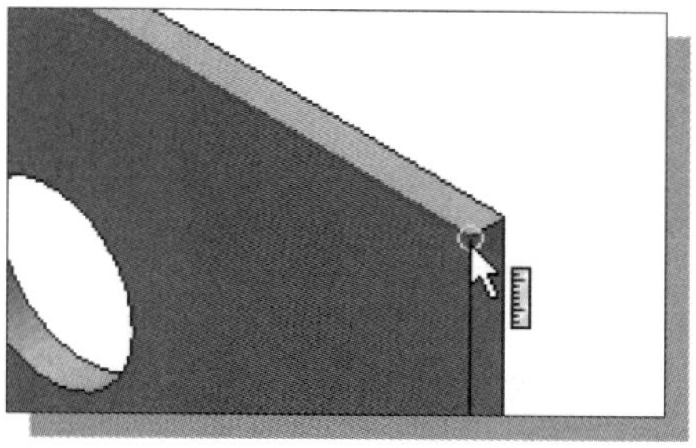

7. Click the **top-right corner** of the 3D model as shown.

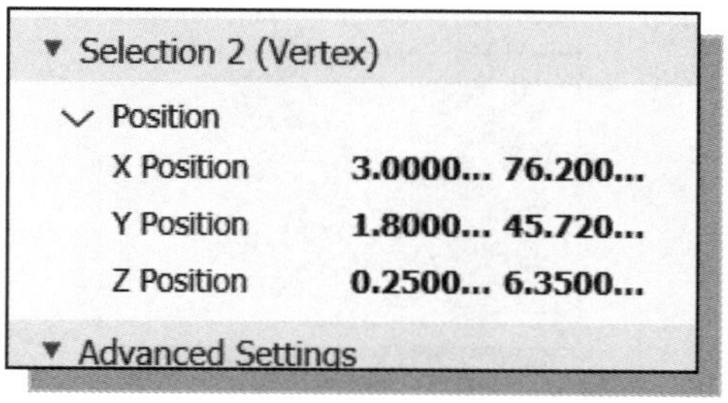

❖ Notice the information regarding the selected point is displayed in the *Measure* dialog box. The absolute position of the point is displayed. (Note the displayed numbers may be different on your screen.)

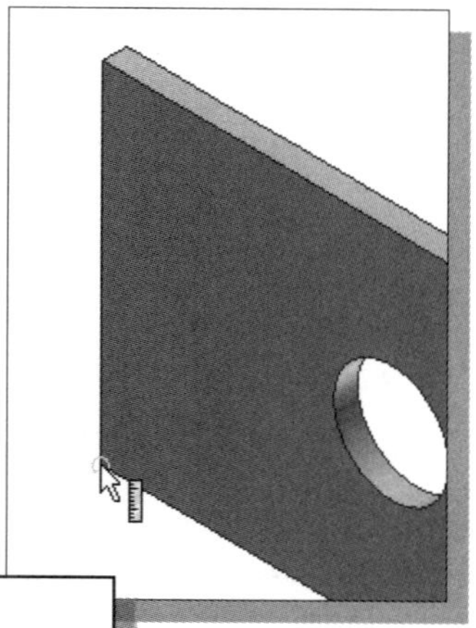

8. Select the **front left bottom corner** as the second location for the Measure Length command. The distance in between the two selected objects is calculated and displayed as shown.

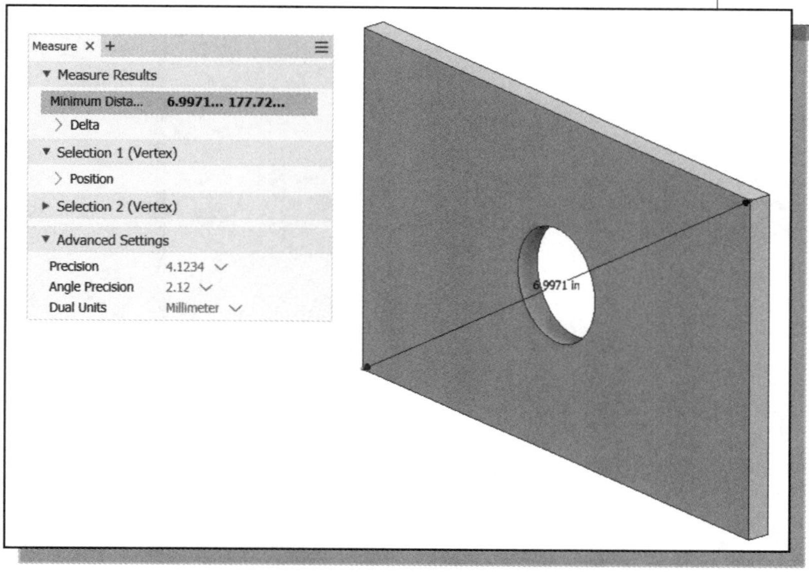

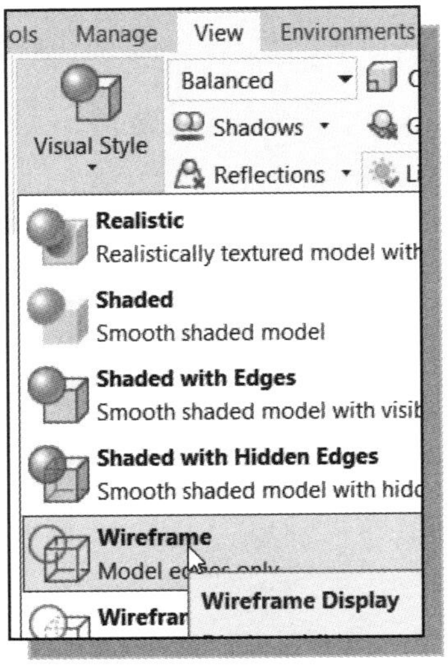

9. Set the display option to **Wireframe Display** by clicking the associated icon in the *Display* toolbar as shown.

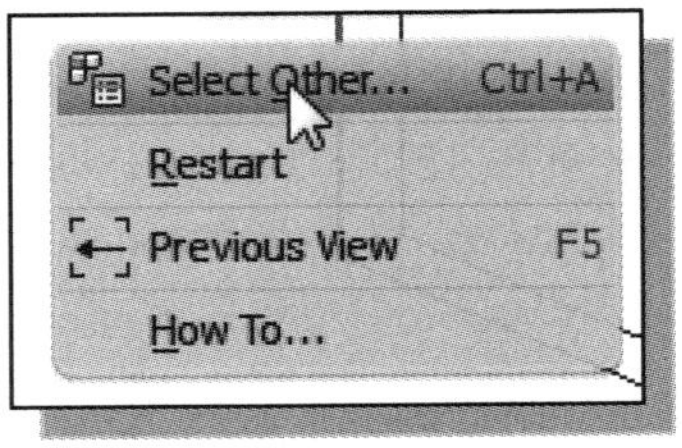

10. Select the left vertical plane by clicking on the selection arrows. (Hint: Use the **Select Other** option if you are having difficulty selecting it.)

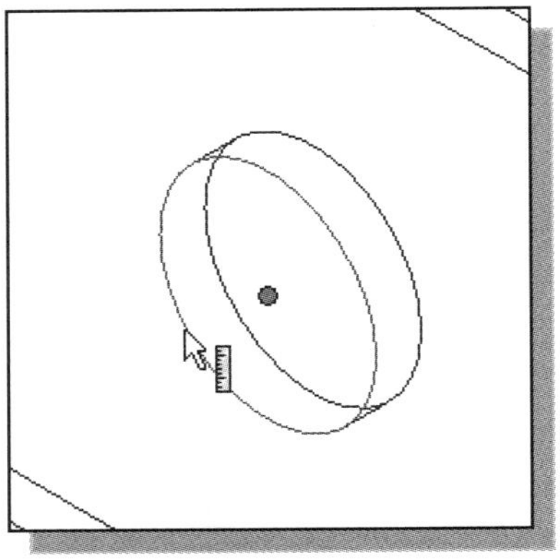

11. Select the circular hole in the front face; notice the center point is highlighted (the actual point used for the calculation) as shown.

❖ Different types of entities can be selected and the measurements are calculated accordingly.

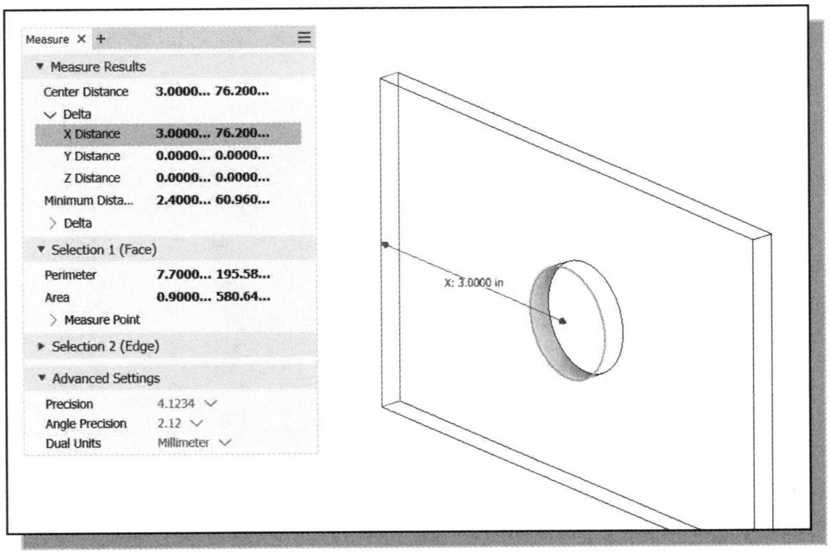

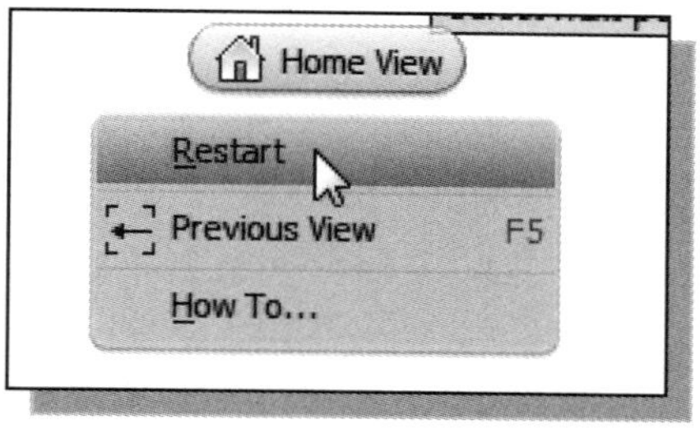

12. On your own, reset the **Measure** option by selecting **Restart** in the option list as shown.

13. Click on the front face of the plate model to display the area of the selected surface.

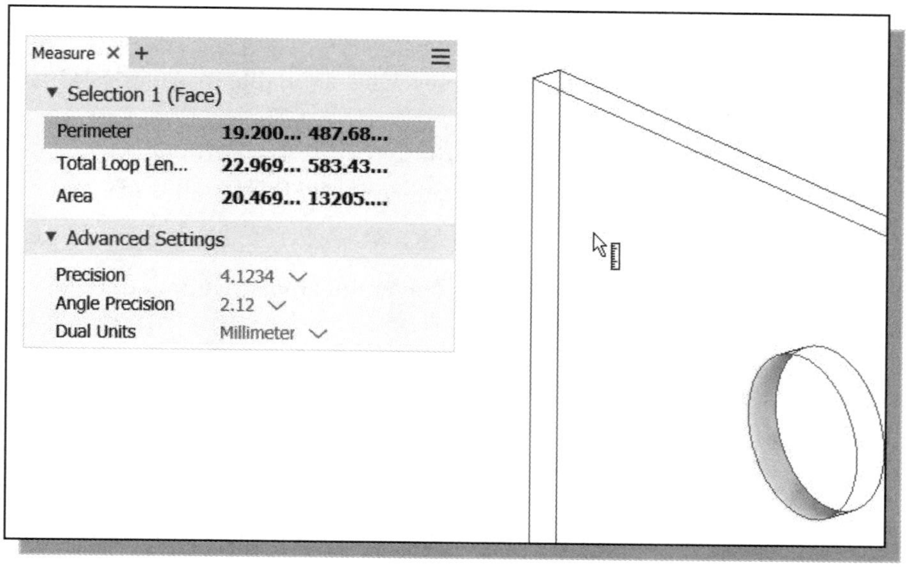

14. Note that we can also measure cylindrical surfaces by selecting the cylindrical surface as shown.

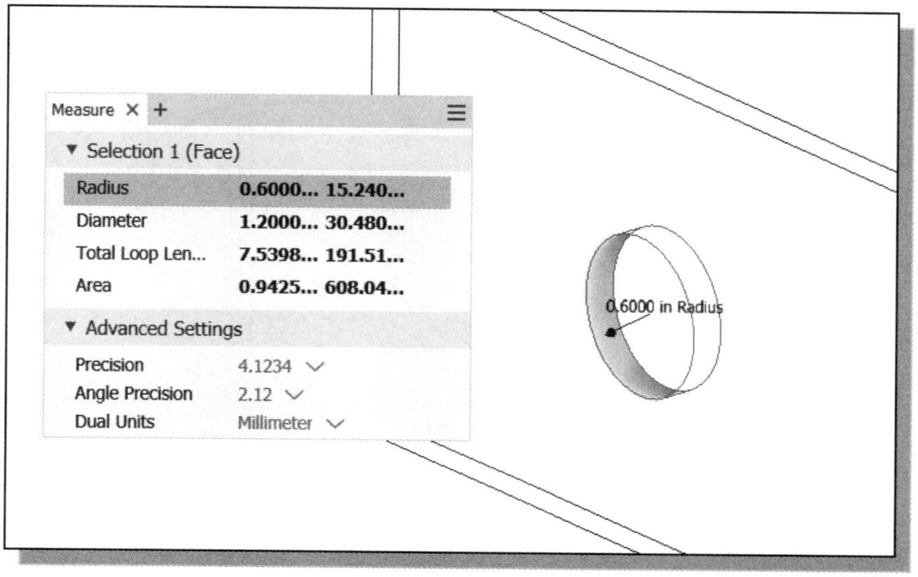

15. On your own, experiment with the different **measure tools** available.

Review Questions: (Time: 30 minutes)

1. What is the difference between *dimensional* constraints and *geometric* constraints?
2. How can we confirm that a sketch is fully constrained?
3. How do we distinguish between derived dimensions and regular dimensions on the screen?
4. Describe the procedure to Display/Edit user-defined equations.
5. List and describe three different geometric constraints available in Autodesk Inventor.
6. Does Autodesk Inventor allow us to build partially constrained or totally unconstrained solid models? What are the advantages and disadvantages of building these types of models?
7. How do we display and examine the existing constraints that are applied to the sketched entities?
8. Describe the advantages of using parametric equations.
9. Can we delete an applied constraint? How?
10. Create the following 2D Sketch and measure the associated area and perimeter.

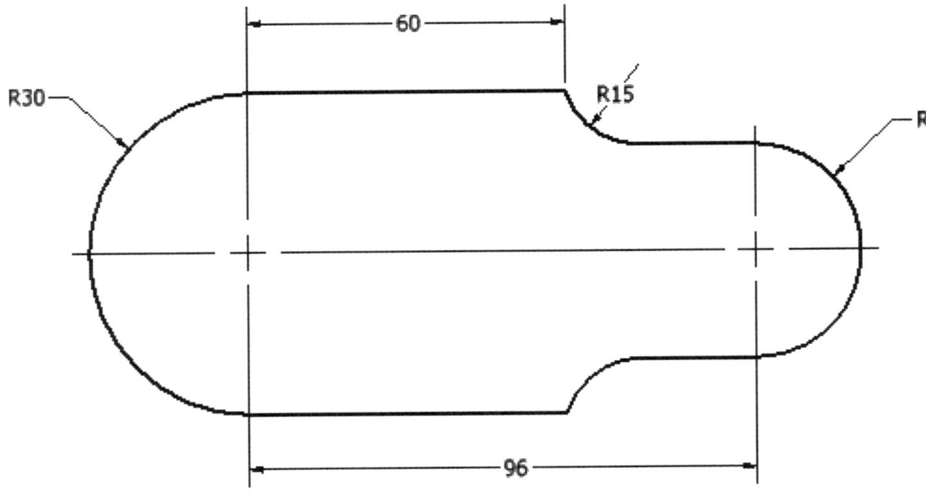

11. Describe the purpose and usage of the Auto Dimension command.

Exercises: Create and save the exercises in the Chapter5 folder.
(Time: 120 minutes.)
(Create and establish three parametric relations for each of the following designs.)

1. **Swivel Base** (Dimensions are in millimeters. Base thickness: **10 mm.** Boss: **5 mm.**)

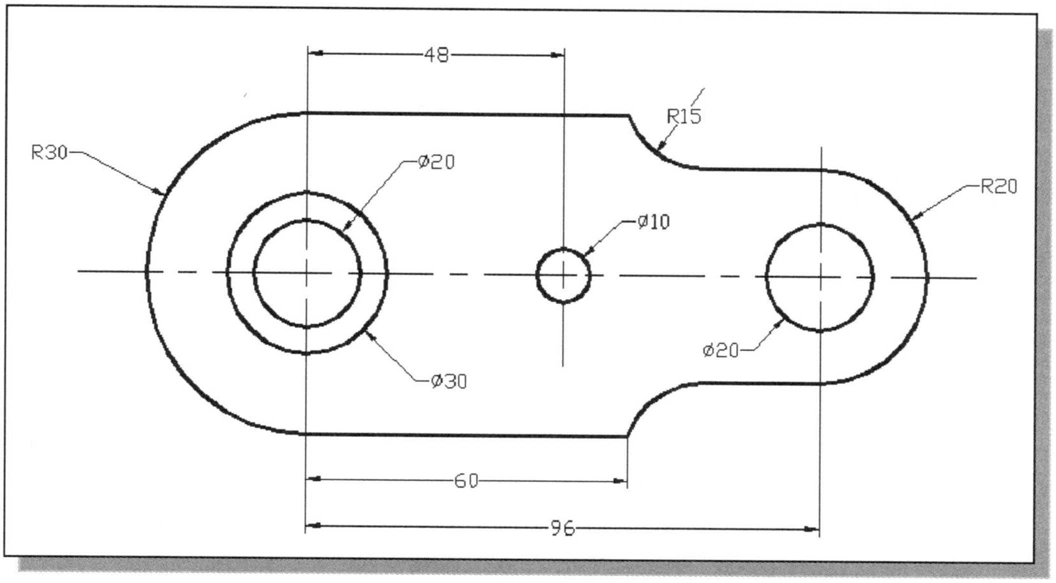

2. **Anchor Base** (Dimensions are in inches.)

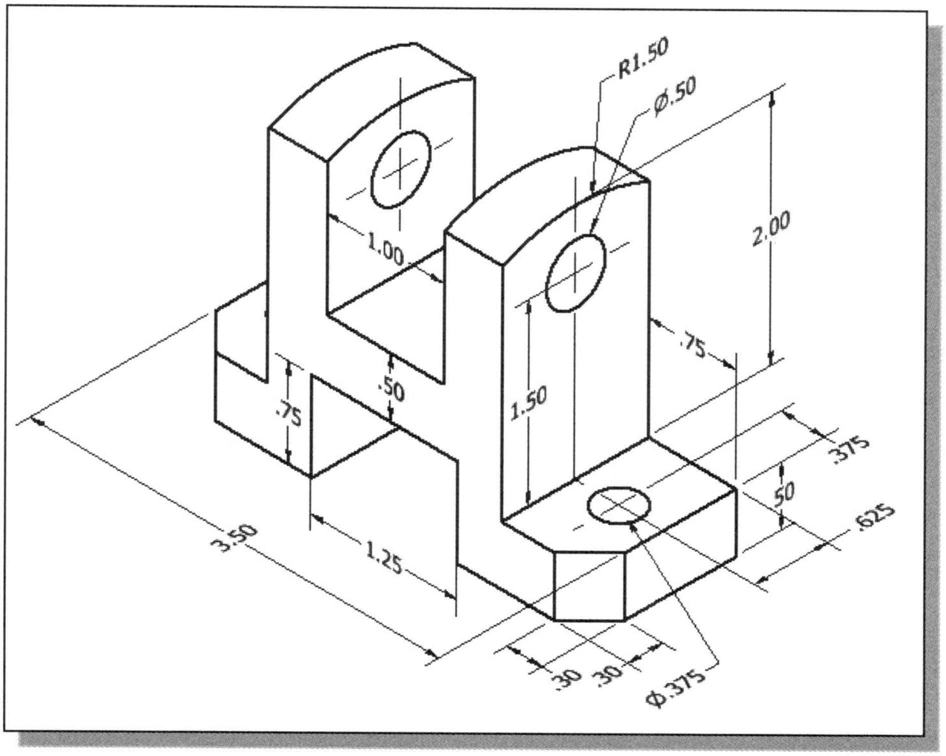

3. **Wedge Block** (Dimensions are in inches.)

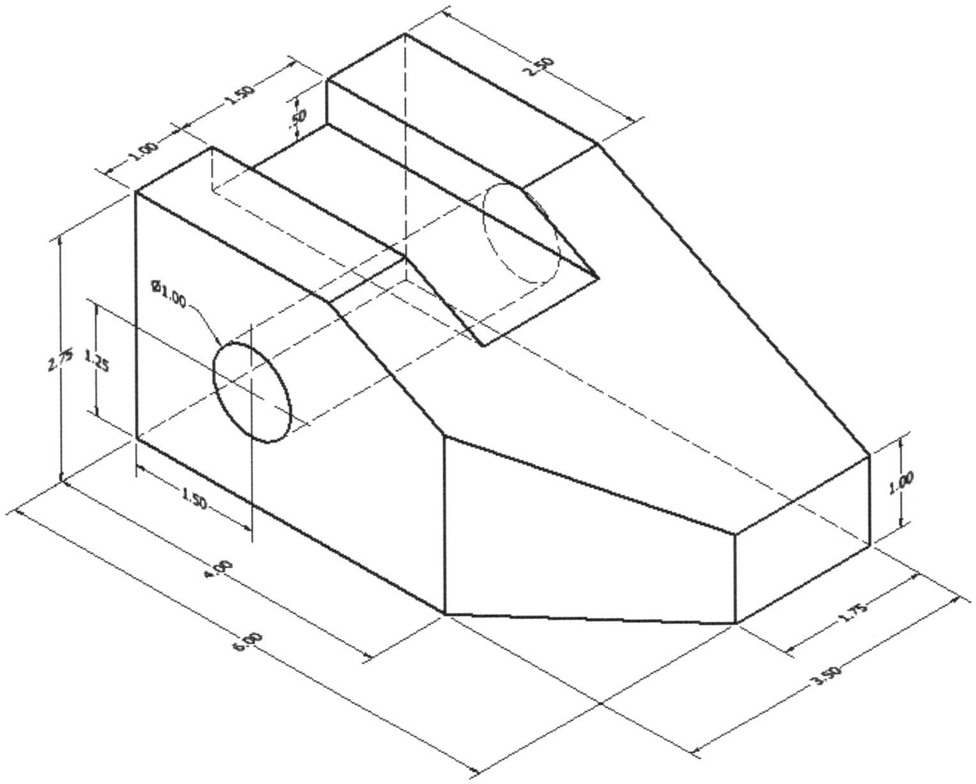

4. **Hinge Guide** (Dimensions are in inches.)

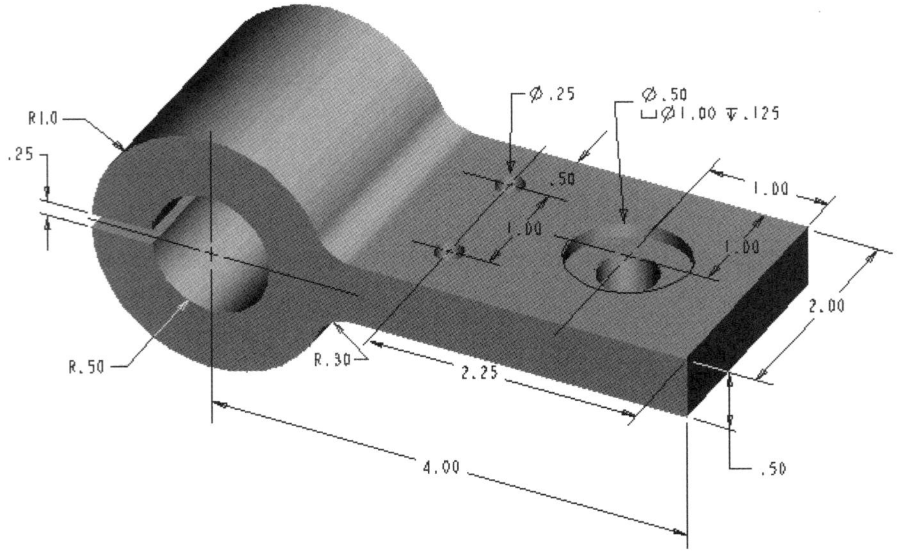

5. **Pivot Holder** (Dimensions are in inches.)

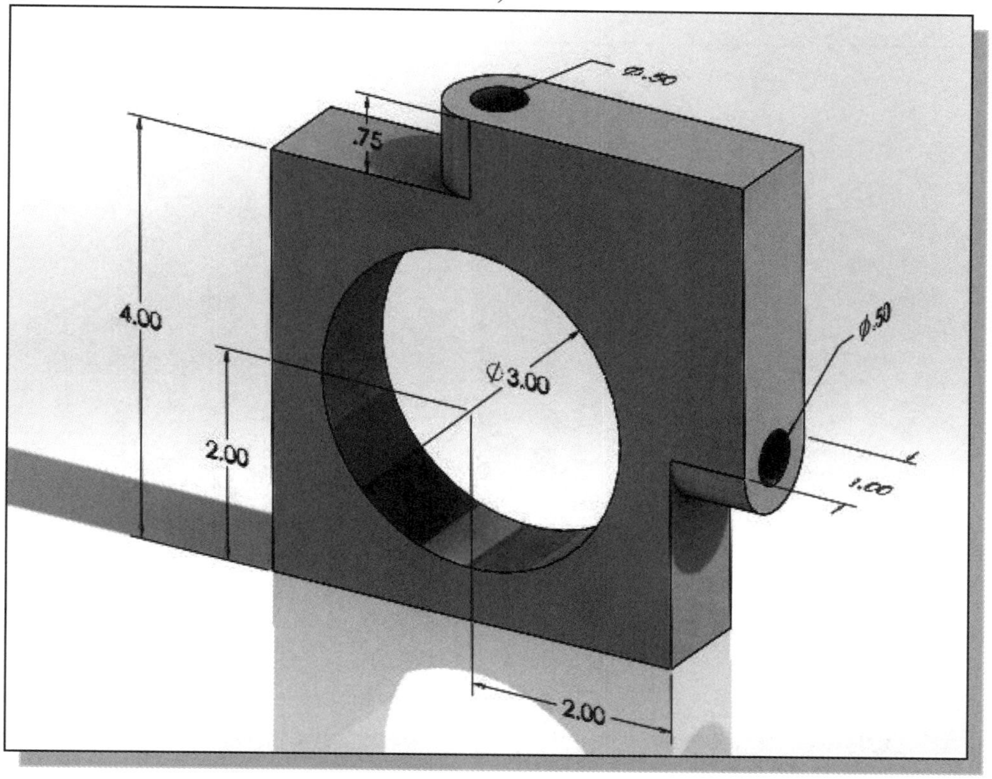

6. **Support Fixture** (Dimensions are in inches.)

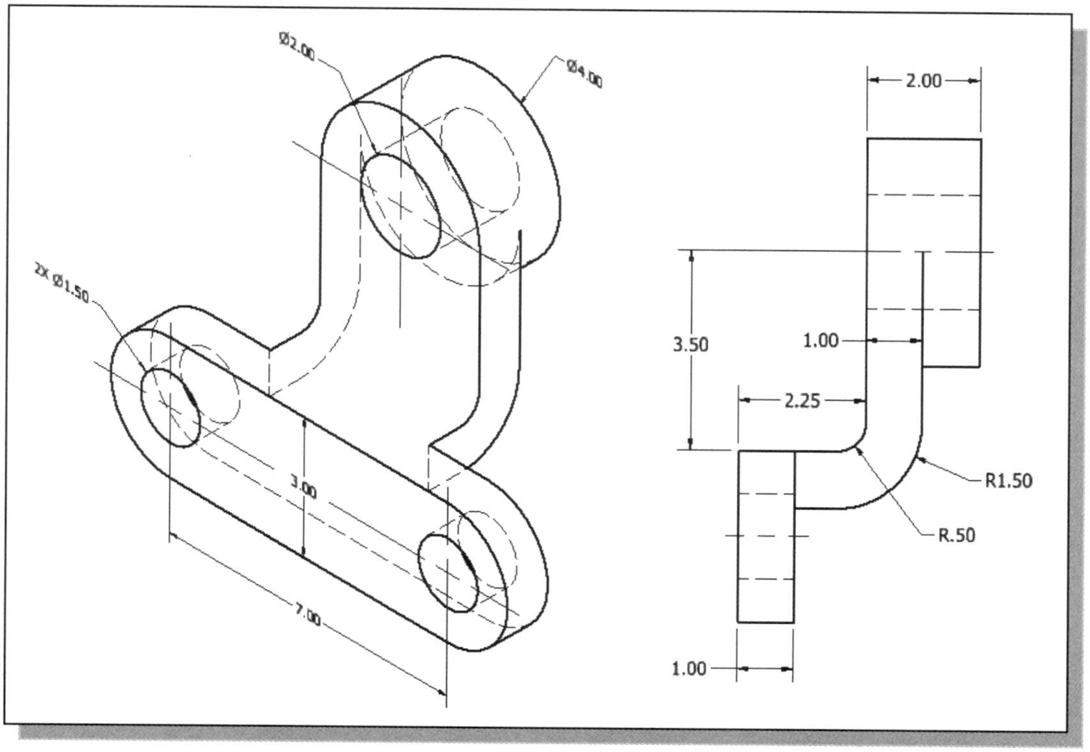

7. **Extruder Base** (Dimensions are in mm.)

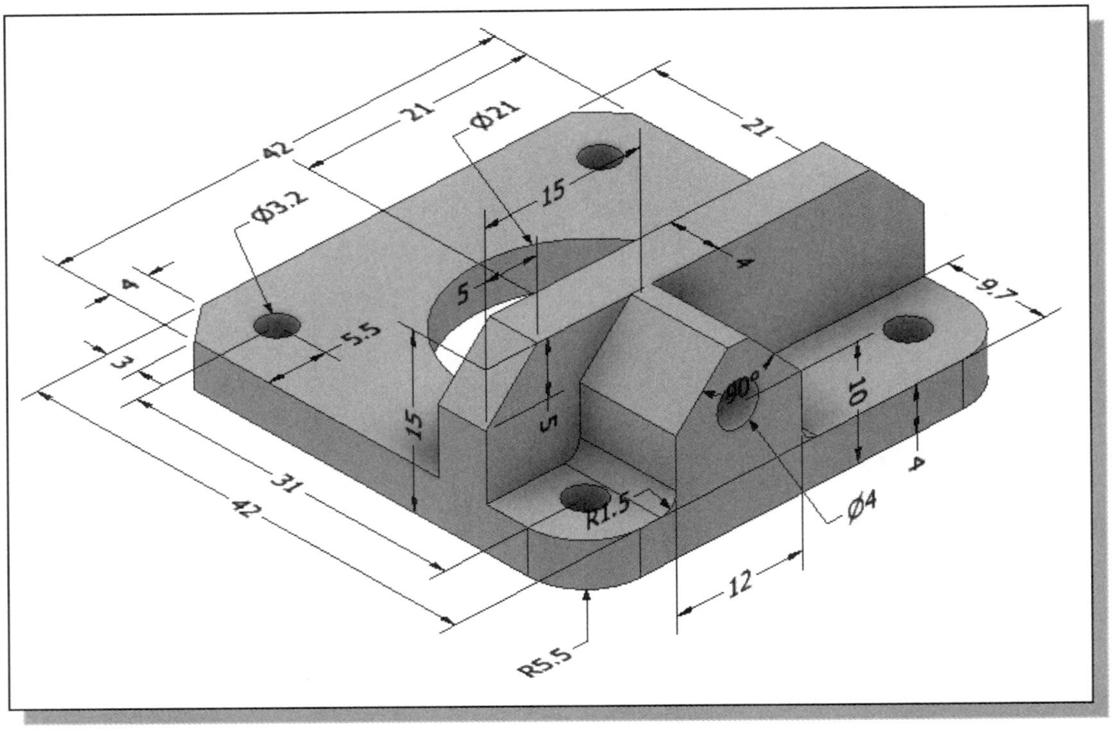

Chapter 6
Geometric Construction Tools

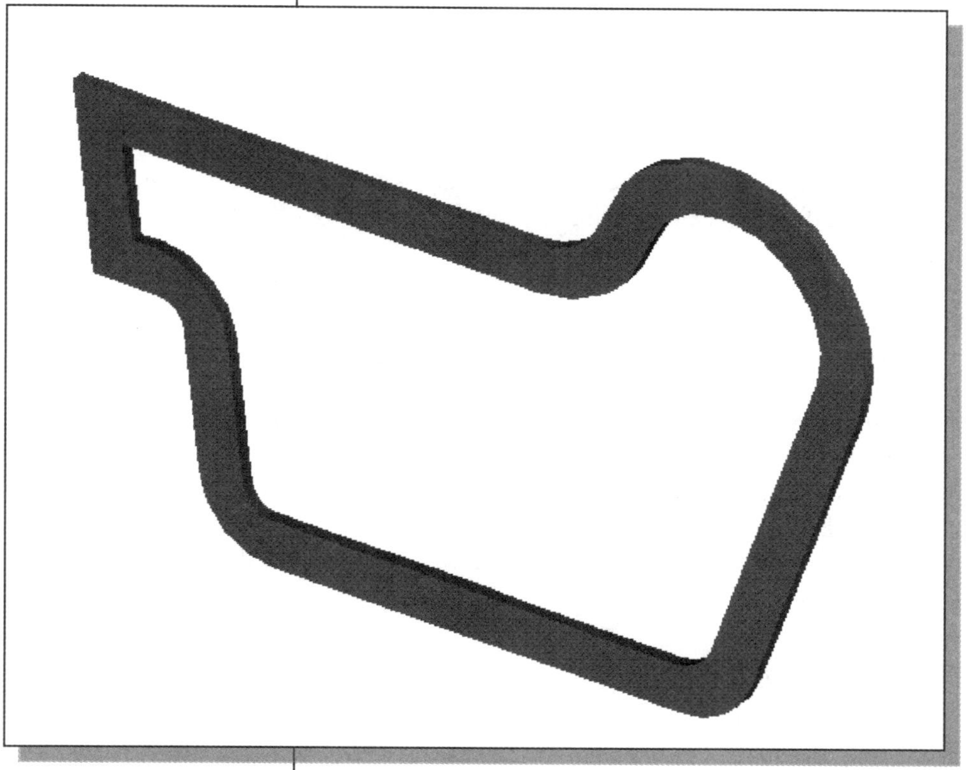

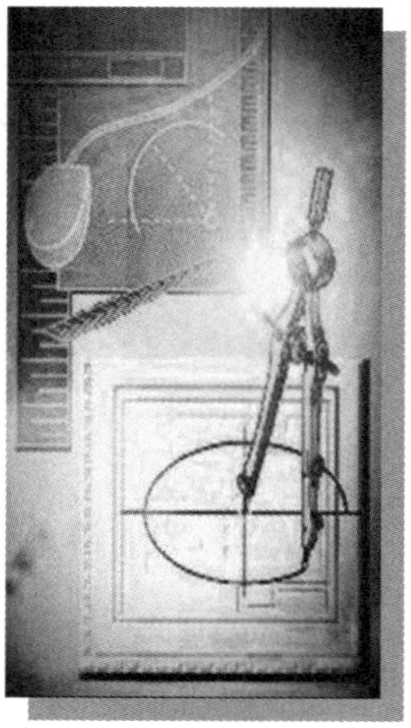

Learning Objectives

- Apply Geometry Constraints
- Use the Trim/Extend Command
- Use the Offset Command
- Understand the Profile Sketch Approach
- Create Projected Geometry
- Understand and Use Reference Geometry
- Edit with Click and Drag
- Use the Auto Dimension Command

Autodesk Inventor Certified User Exam Objectives Coverage

Parametric Modeling Basics

Section 3: Sketches

Objectives: Creating 2D Sketches, Draw Tools, Sketch Constraints, Pattern Sketches, Modify Sketches, Format Sketches, Sketch Doctor, Shared Sketches, Sketch Parameters.

Section 4: Parts

Objectives: Creating parts, Work Features, Pattern Features, Part Properties.

Autodesk Inventor Certified User Reference Guide

Introduction

The main characteristics of solid modeling are the accuracy and completeness of the geometric database of the three-dimensional objects. However, working in three-dimensional space using input and output devices that are largely two-dimensional in nature is potentially tedious and confusing. Autodesk Inventor provides an assortment of two-dimensional construction tools to make the creation of wireframe geometry easier and more efficient. Autodesk Inventor includes two types of wireframe geometry: ***curves*** and ***profiles***. Curves are basic geometric entities such as lines, arcs, etc. Profiles are a group of curves used to define a boundary. A *profile* is a closed region and can contain other closed regions. Profiles are commonly used to create extruded and revolved features. An *invalid profile* consists of self-intersecting curves or open regions. In this lesson, the basic geometric construction tools, such as Trim and Extend, are used to create profiles. The Autodesk Inventor's ***profile sketch*** approach to creating profiles is also introduced. Mastering the geometric construction tools along with the application of proper constraints and parametric relations is the true essence of *parametric modeling*.

The Gasket Design

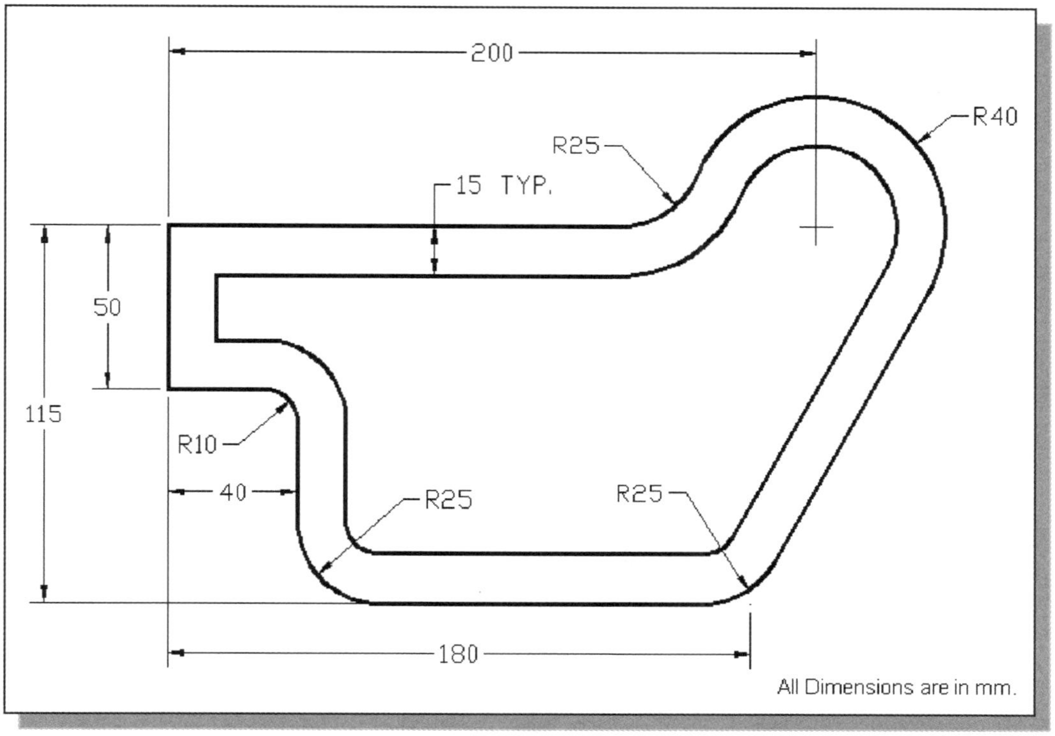

❖ Based on your knowledge of Autodesk Inventor so far, how would you create this design? What is the most difficult geometry involved in the design? Take a few minutes to consider a modeling strategy and do preliminary planning by sketching on a piece of paper. You are also encouraged to create the design on your own prior to following through the tutorial.

Modeling Strategy

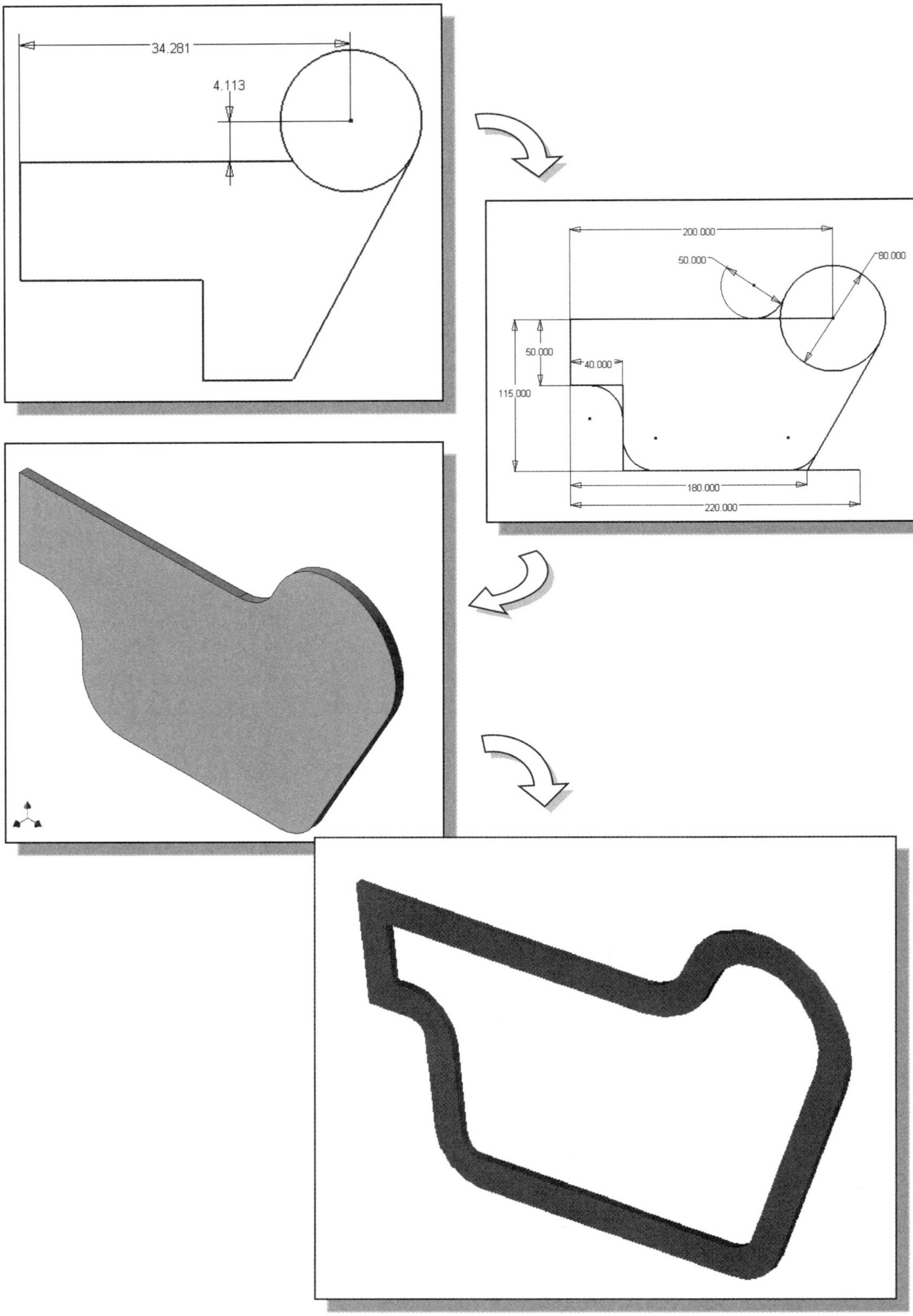

Starting Autodesk Inventor

1. Select the **Autodesk Inventor** option on the *Start* menu or select the **Autodesk Inventor** icon on the desktop to start Autodesk Inventor. The Autodesk Inventor main window will appear on the screen.

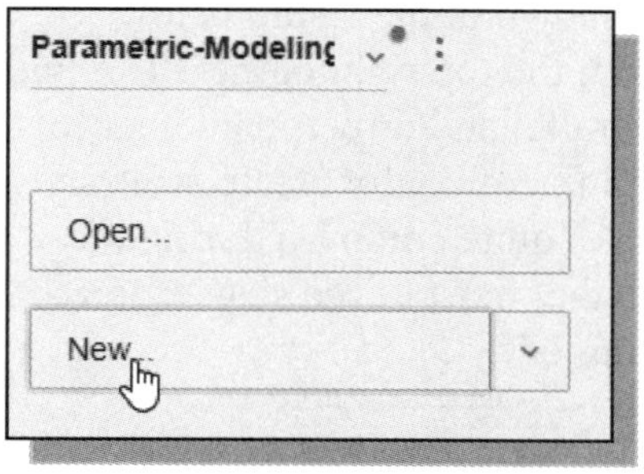

2. Select the **New File** icon with a single click of the left-mouse-button as shown.

3. Select the **Metric** tab as shown. We will use the millimeter (mm) setting for this design.

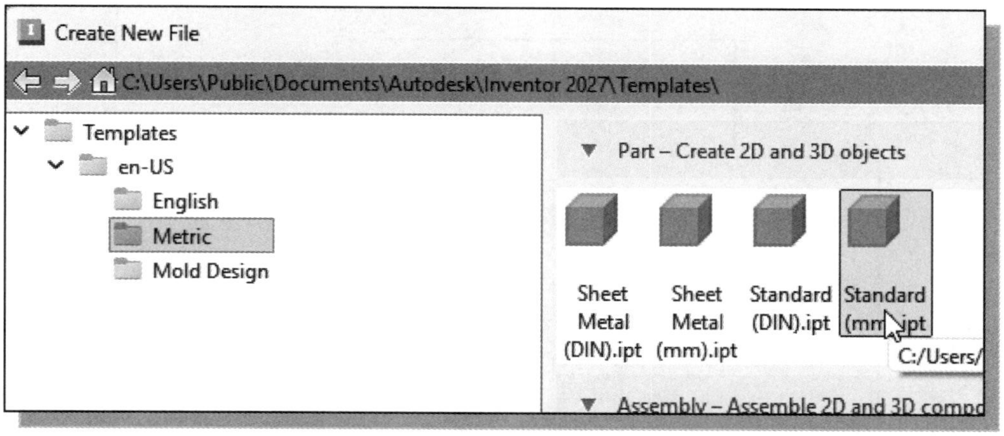

4. In the *New File* dialog box, use the scroll bar on the right and select the **Standard(mm).ipt** icon as shown.

5. Click **Create** in the *New File* dialog box to accept the selected settings to start a new model.

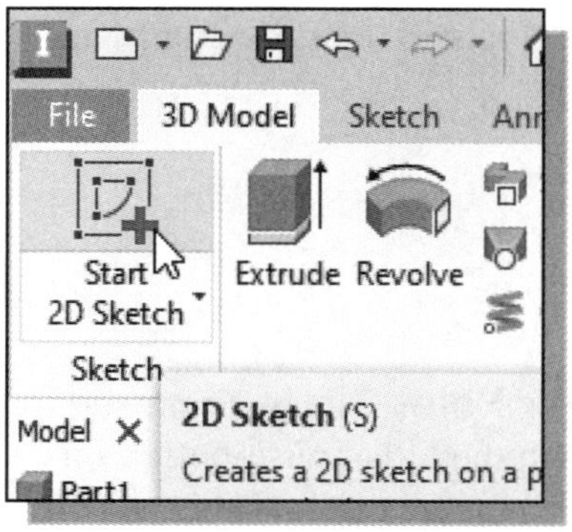

6. Click once with the left-mouse-button to select the **Start 2D Sketch** command. Click once with the **left-mouse-button** to select the *XY Plane* as the sketch plane for the new sketch.

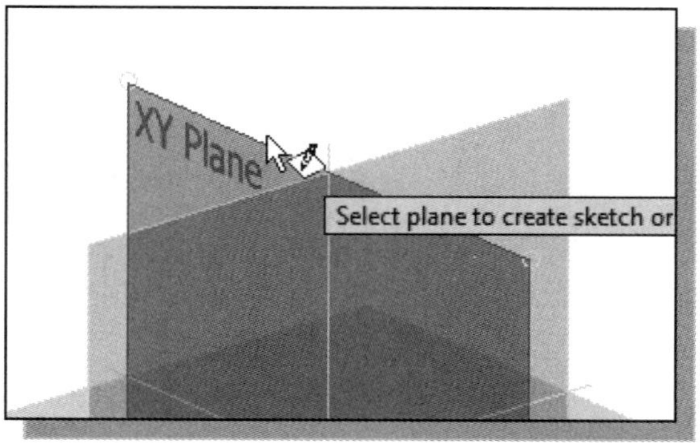

Create a 2D Sketch

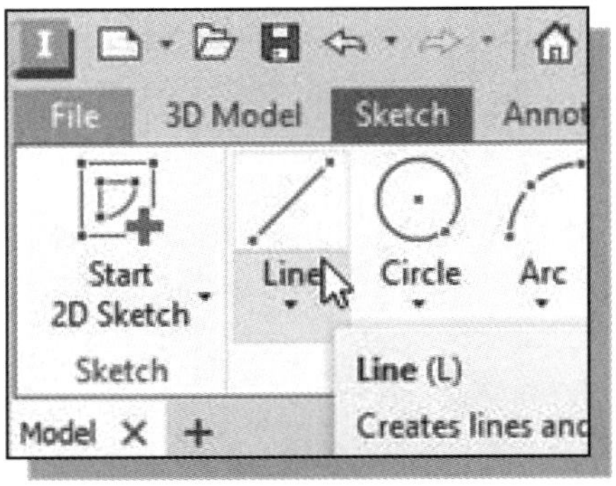

1. Click on the **Line** icon in the *Sketch* tab on the Ribbon.

2. Create a sketch as shown in the figure below. Start the sketch from the top right corner. The line segments are all parallel and/or perpendicular to each other. We will intentionally create arbitrary line segments, as it is quite common during the initial design stage that most of the shapes and forms are undetermined.

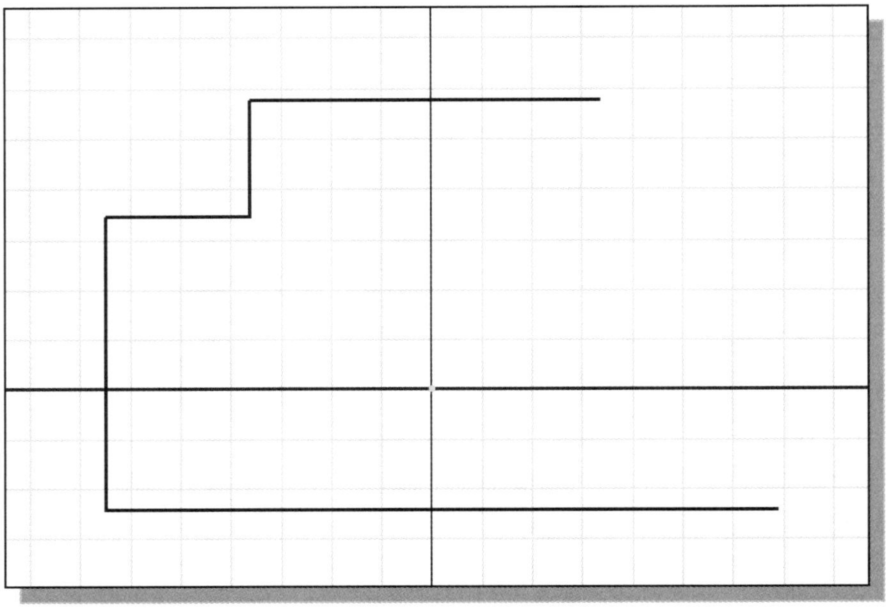

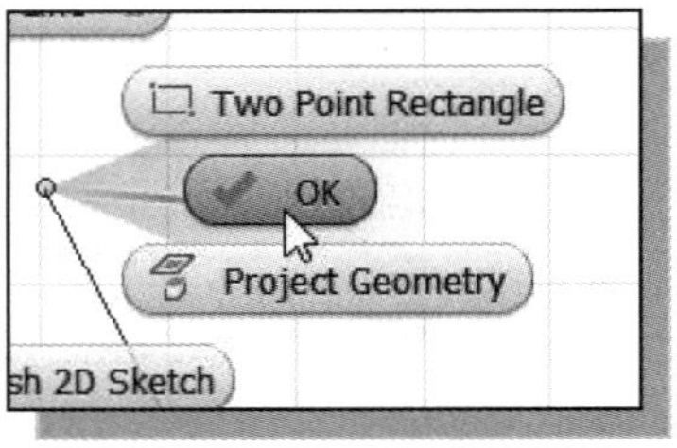

3. Inside the graphics window, right-click to bring up the option menu, and select **OK** to end the **Line** command.

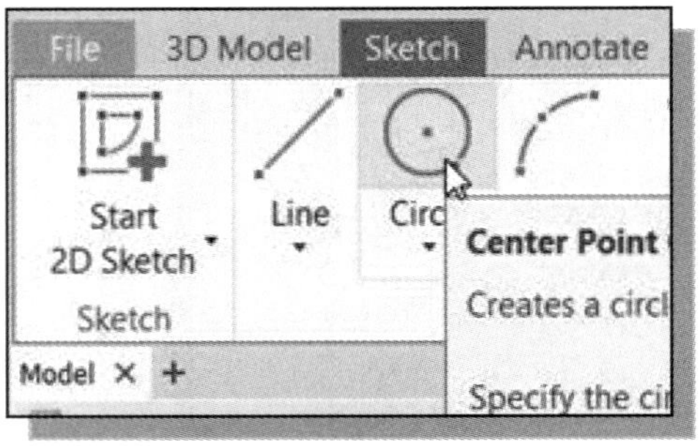

4. Select the **Center Point Circle** command by clicking once with the left-mouse-button on the icon in the *Sketch* tab on the Ribbon.

5. Pick a location that is above the bottom horizontal line as the center location of the circle.

6. Move the cursor toward the right and create a circle of arbitrary size by clicking once with the left-mouse-button.

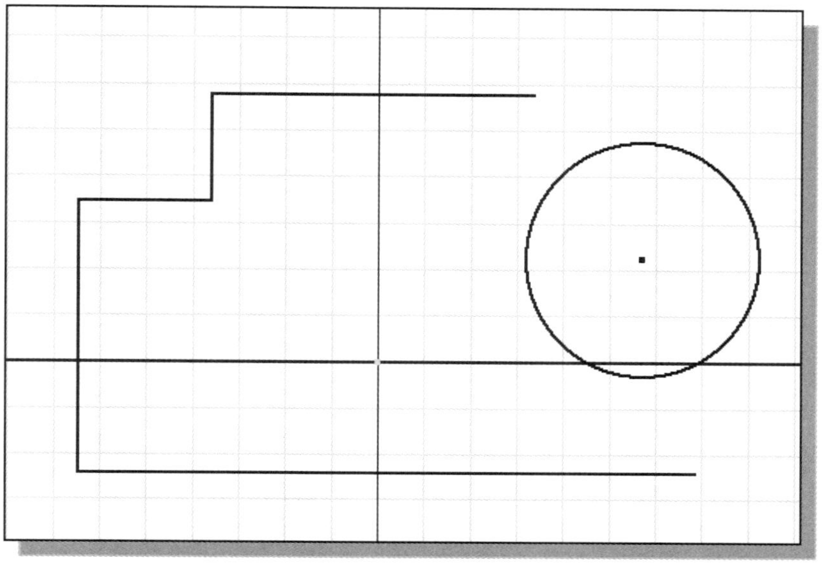

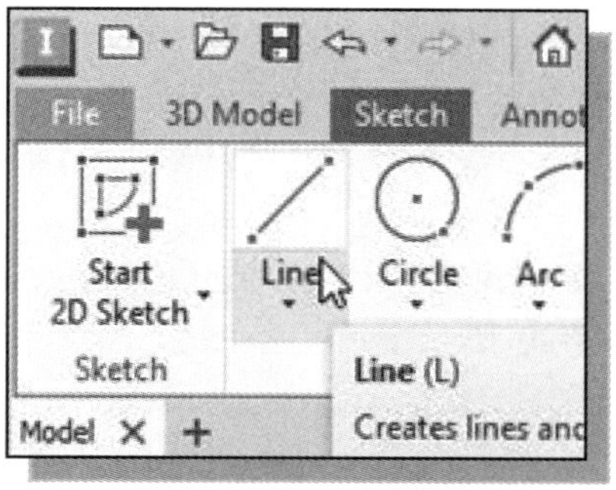

7. Click on the **Line** icon in the *Sketch* tab on the Ribbon.

8. Move the cursor near the upper portion of the circle and pick a location on the circle when the Coincident constraint symbol is displayed.

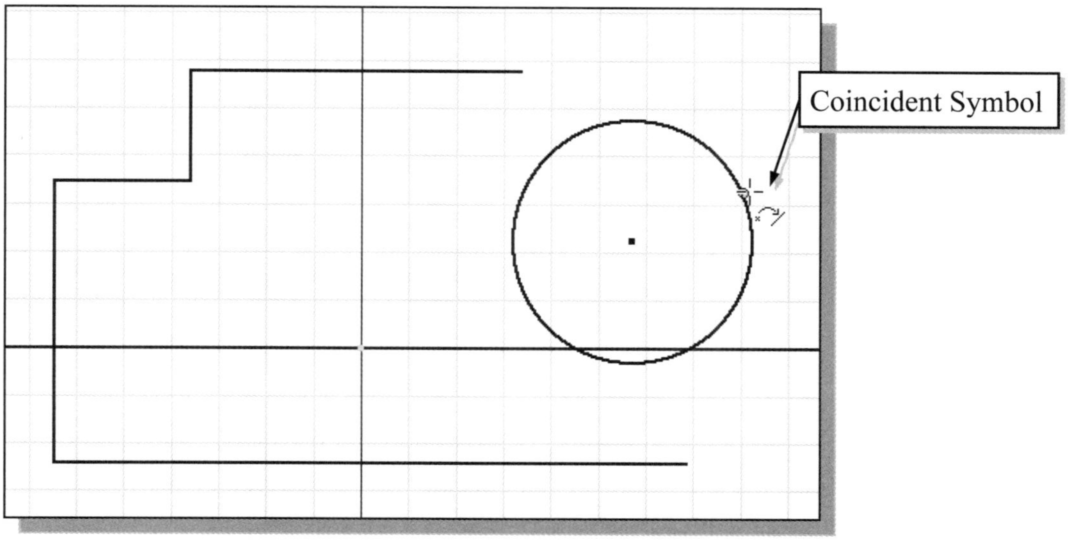

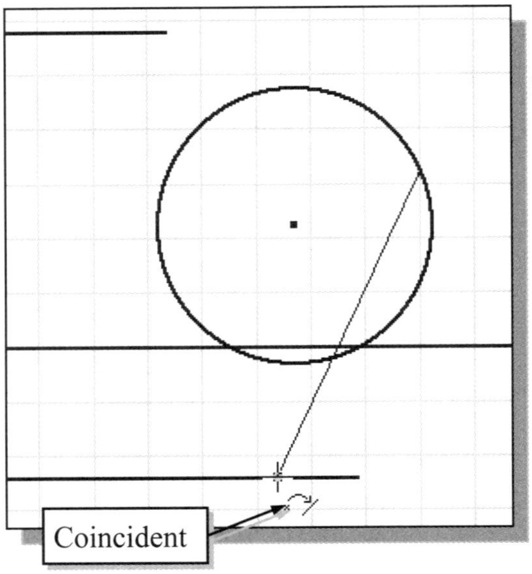

9. For the other end of the line, select a location that is on the lower horizontal line and about one-third from the right endpoint. Notice the **Coincident** constraint symbol is displayed when the cursor is on the horizontal line.

10. Inside the graphics window, right-click to bring up the option menu and select **OK** to end the **Line** command.

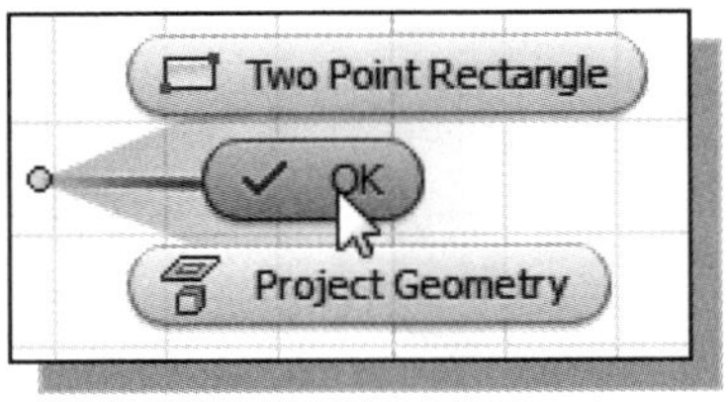

Edit the Sketch by Dragging the Sketched Entities

In Autodesk Inventor, we can click and drag any under-constrained curve or point in the sketch to change the size or shape of the sketched profile. As illustrated in the previous chapter, this option can be used to identify under-constrained entities. This *Editing by dragging* method is also an effective visual approach that allows designers to quickly make changes.

1. Move the cursor on the lower left vertical edge of the sketch. Click and drag the edge to a new location that is toward the right side of the sketch.

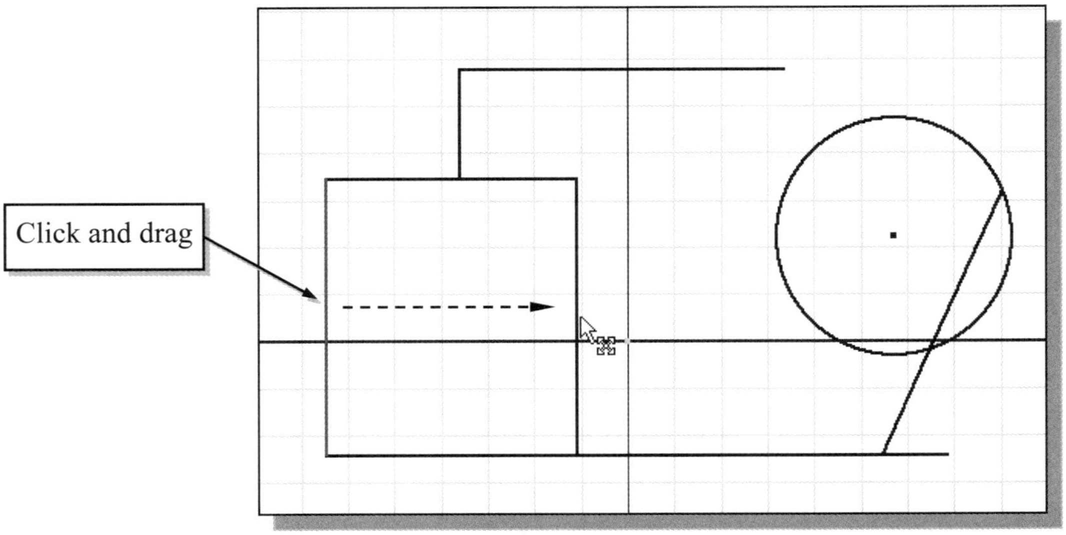

❖ Note that we can only drag the vertical edge horizontally; the connections to the two horizontal lines are maintained while we are moving the geometry.

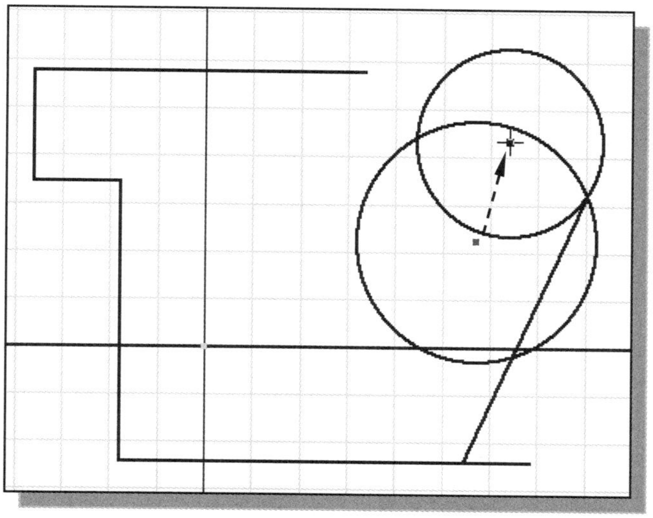

2. Click and drag the center point of the circle to a new location.

❖ Note that as we adjust the size and the location of the circle, the connection to the inclined line is maintained.

3. Click and drag the lower endpoint of the inclined line to a new location.

❖ Note that as we adjust the size and the location of the inclined line the location of the bottom horizontal edge is also adjusted.

❖ Note that several changes occur as we adjust the size and the location of the inclined line. The location of the bottom horizontal line and the length of the vertical line are adjusted accordingly.

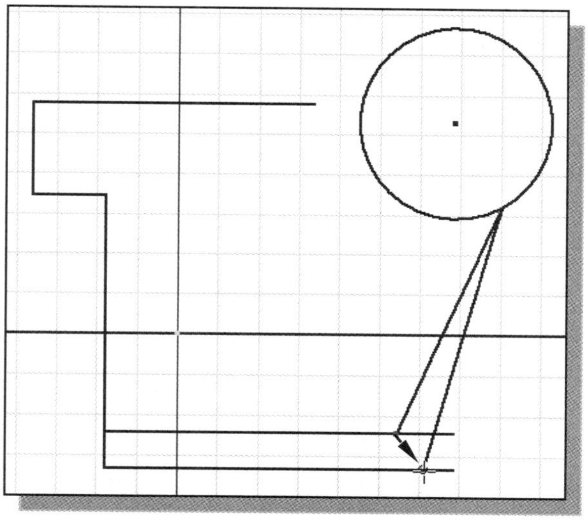

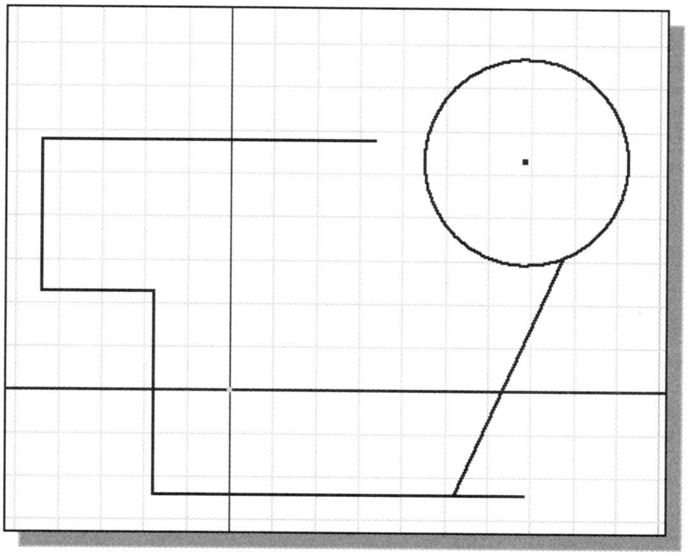

4. On your own, adjust the sketch so that the shape of the sketch appears roughly as shown.

❖ The *Editing by dragging* method is an effective approach that allows designers to explore and experiment with different design concepts.

Add Additional Constraints

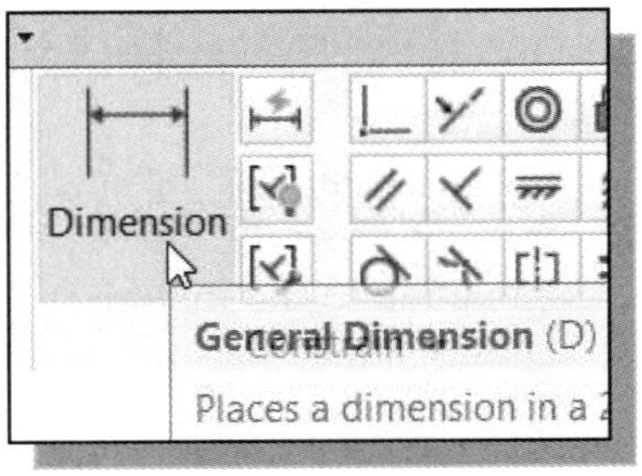

1. Choose **General Dimension** in the *Sketch* panel.

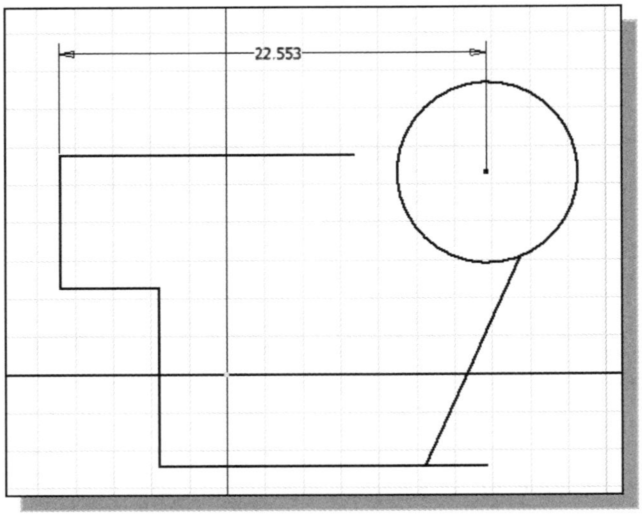

2. Add the horizontal location dimension, from the top left vertical edge to the center of the circle as shown. (Do not be overly concerned with the dimensional value; we are still working on creating a *rough sketch*.)

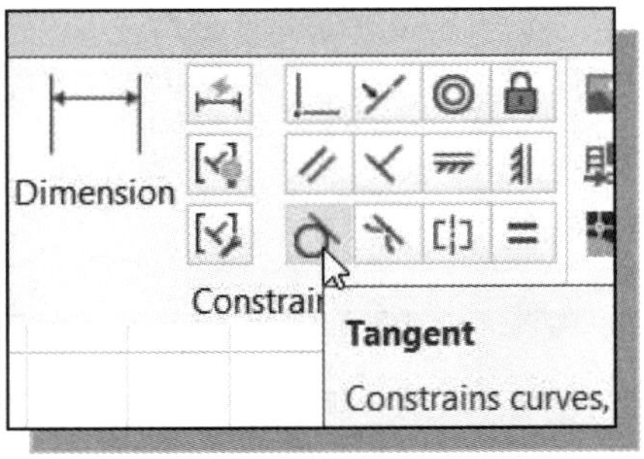

3. Click on the **Tangent** constraint icon in the *Sketch* panel.

4. Pick the inclined line by left-clicking once on the geometry.

5. Pick the circle. The sketched geometry is adjusted as shown below.

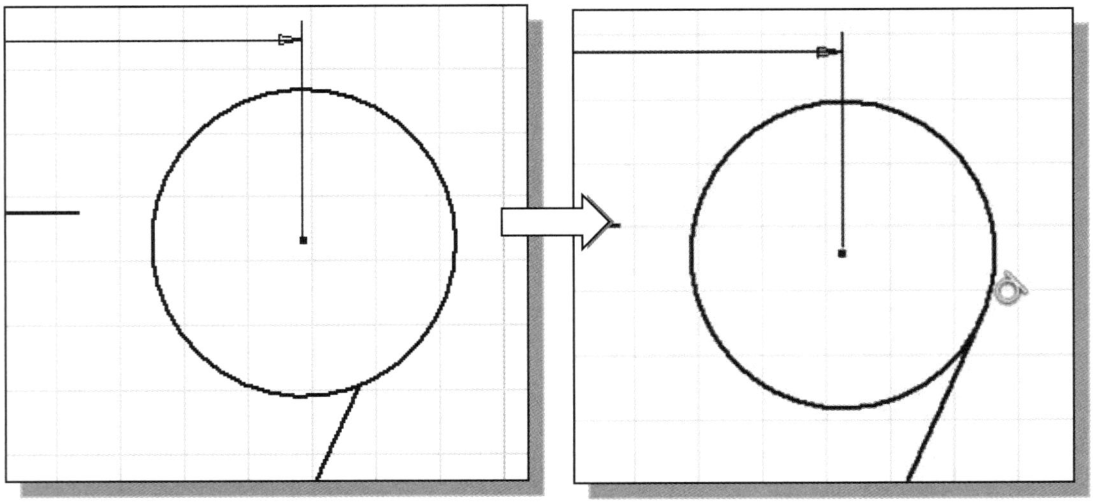

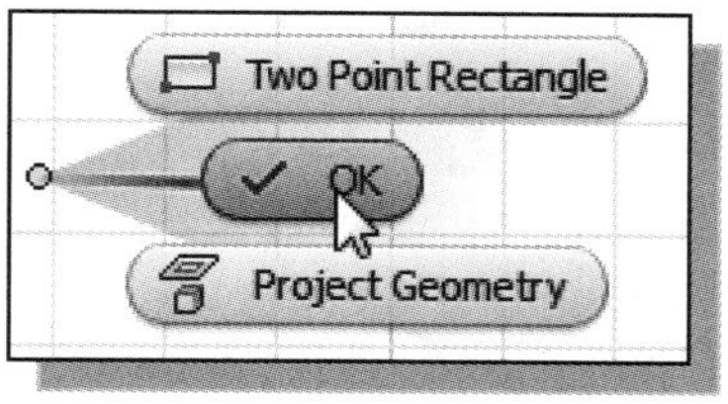

6. Inside the graphics window, right-click to bring up the option menu.

7. In the option menu, select **OK** to end the Tangent command.

8. Use the drag and drop approach, of the circle center, to observe the relations of the sketched geometry.

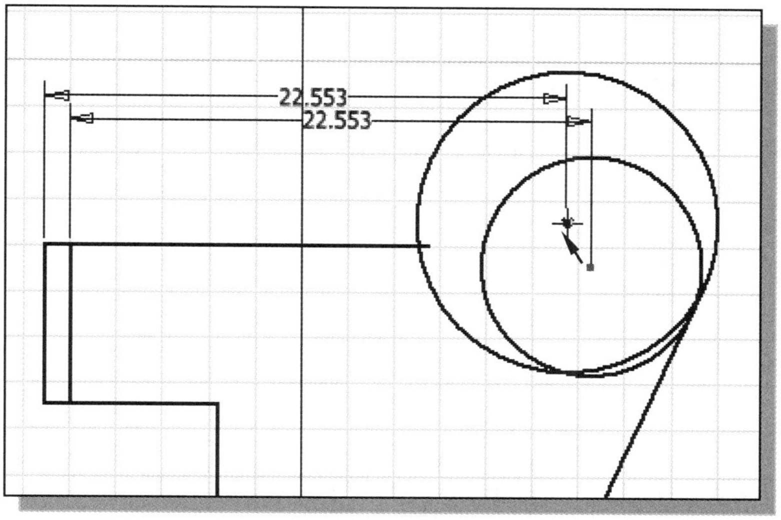

❖ Note that the dimension we added now restricts the horizontal movement of the center of the circle. The tangent relation to the inclined line is maintained.

Use the Trim and Extend Commands

In the following sections, we will illustrate using the Trim and Extend commands to complete the desired 2D profile.

The **Trim** and **Extend** commands can be used to shorten/lengthen an object so that it ends precisely at a boundary. As a general rule, Autodesk Inventor will try to clean up sketches by forming a closed region sketch. Also note that while we are in either **Trim** or **Extend**, we can press the [**Shift**] key to switch to the opposite operation.

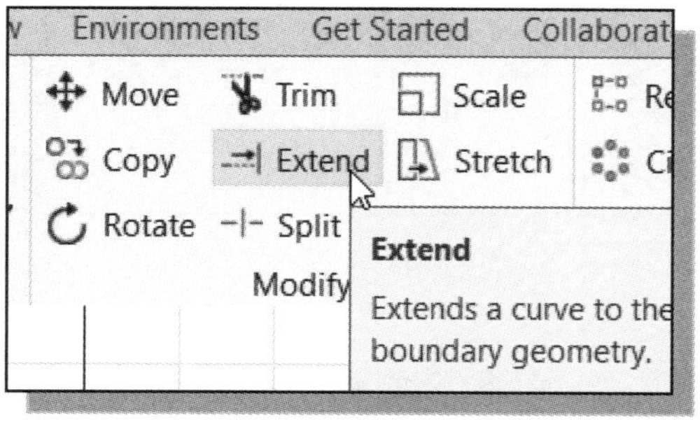

1. Choose **Extend** in the *Modify toolbar* panel. The message "*Extend curves*" is displayed in the prompt area.

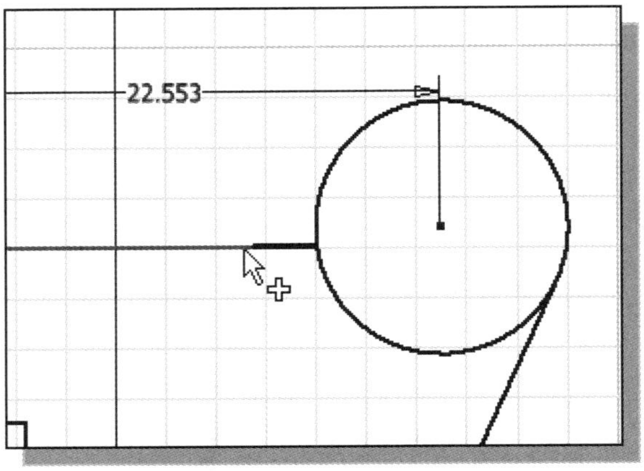

2. We will first extend the top horizontal line to the circle. Move the cursor near the right-hand endpoint of the top horizontal line. Autodesk Inventor will automatically display the possible result of the selection.

3. Next, we will trim the bottom horizontal line to the inclined line. The **Trim** operation can be activated by selecting the **Trim** icon in the *Sketch* panel or by pressing down the [**Shift**] key while we are in the **Extend** command. Move the cursor near the right-hand endpoint of the bottom horizontal line and select the line and press down the [**Shift**] key. Autodesk Inventor will display a dashed line indicating the portion of the line that will be trimmed.

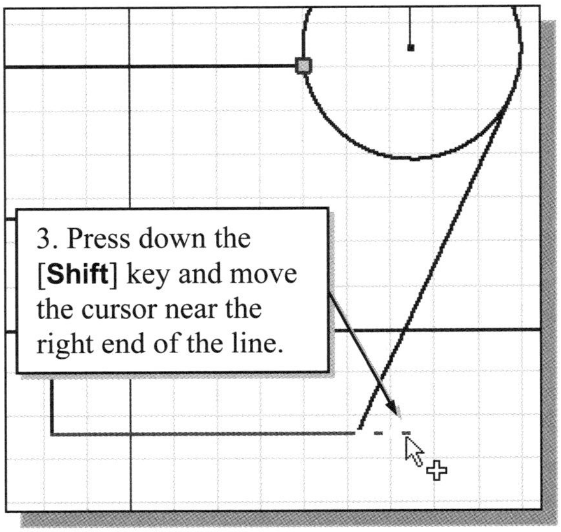

4. Left-click once on the line to perform the **Trim** operation.

5. On your own, create the vertical location dimension as shown below.

6. Adjust the dimension to **0.0** so that the horizontal line and the center of the circle are aligned horizontally.

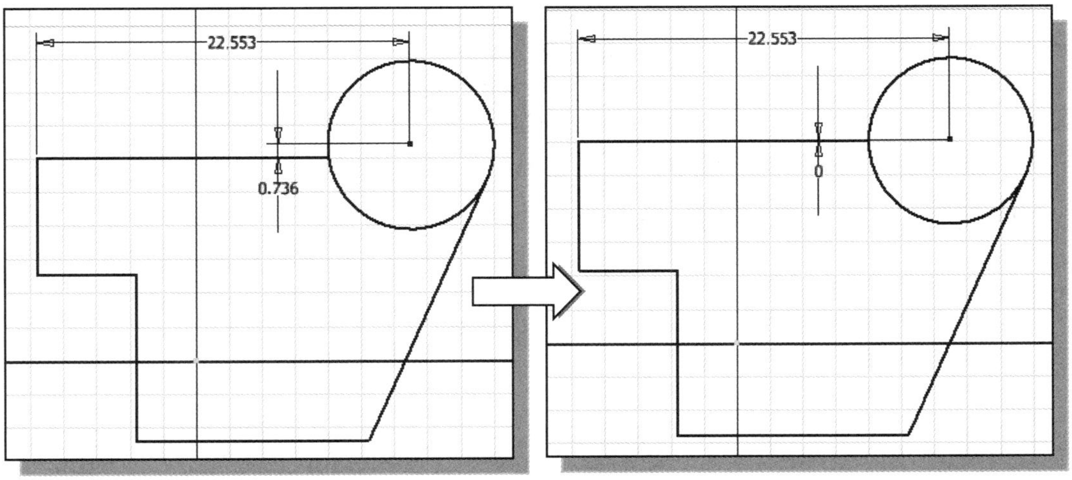

7. On your own, create the **height dimension** to the left and use the **Show Constraints** command to examine the applied constraints. Confirm that a **Perpendicular constraint** is applied to the horizontal and vertical lines as shown.

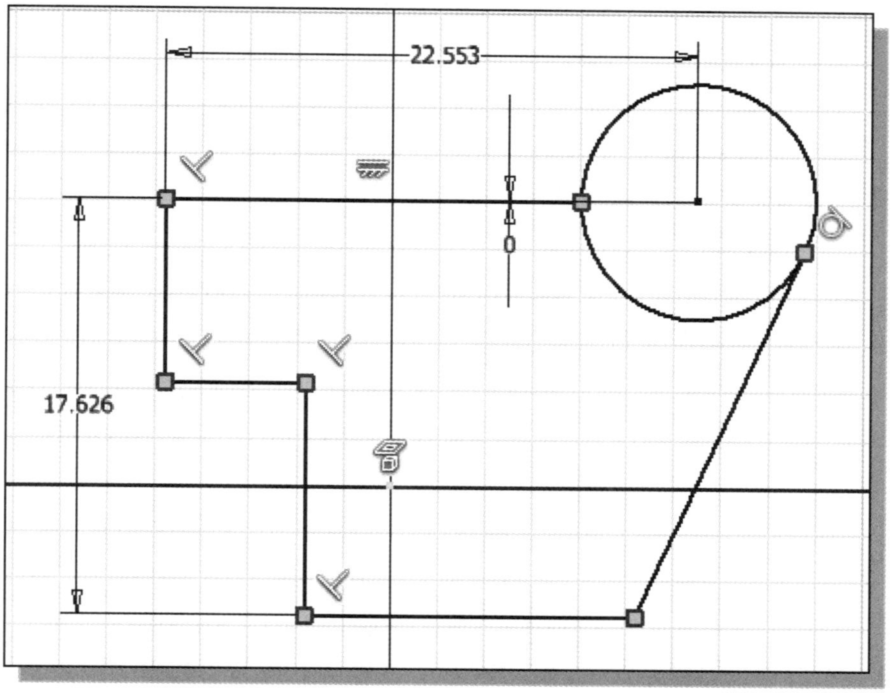

The Auto Dimension Command

In Autodesk Inventor, we can use the **Auto Dimension** command to assist in creating a fully constrained sketch. Fully constrained sketches can be updated more predictably as design changes are implemented. The general procedure for applying dimensions to sketches is to use the **General Dimension** command to add the desired key dimensions, and then use the **Auto Dimension** command as a quick way to calculate all other sketch dimensions necessary. Autodesk Inventor remembers which dimensions were added by the user and which were calculated by the system, so that automatic dimensions do not replace the desired dimensions.

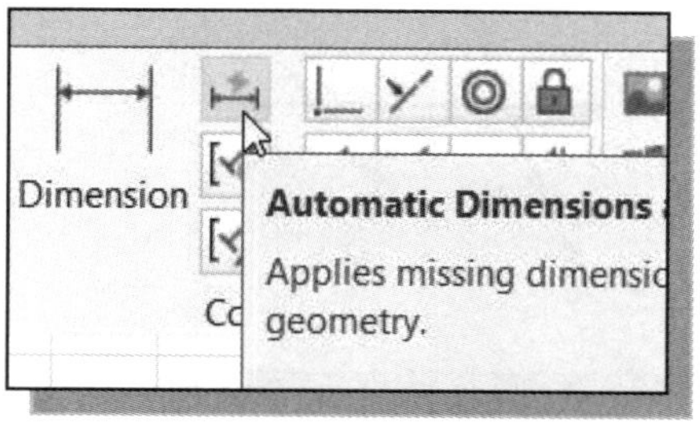

1. Click on the **Auto Dimension** icon in the *Sketch* panel.

- Note that 6 dimensions are needed to fully constrain the sketch, as indicated in the *Status Bar*.

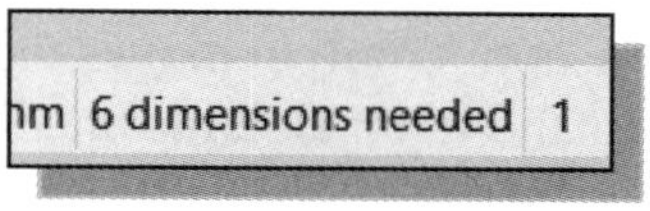

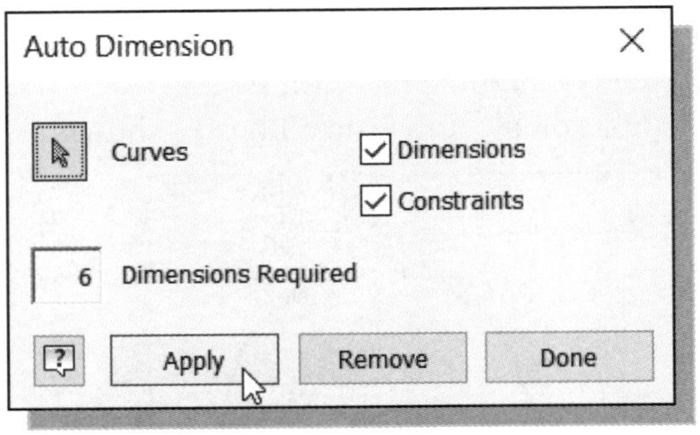

2. The *Auto Dimension* dialog box appears on the screen. Confirm that the **Dimensions** and **Constraints** options are switched *ON* as shown.

3. Click on the **Apply** button to proceed with the Auto Dimension command.

➢ Note that the system automatically calculates additional dimensions for all created geometric entities. Note that at times, the system applied dimensions might not be adequate for the designs. It is best to apply all the key dimensions prior to using the **Auto Dimension** command.

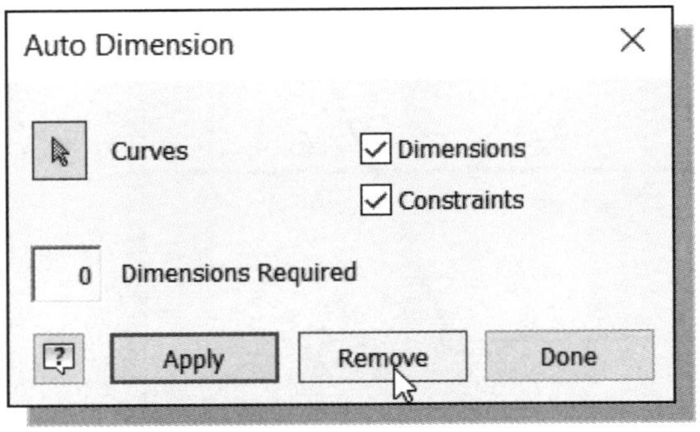

➢ The dimensions created by the **Auto Dimension** command can also be **Removed**.

4. Click on the **Remove** button to undo the dimensions created by the *Auto Dimension* command.

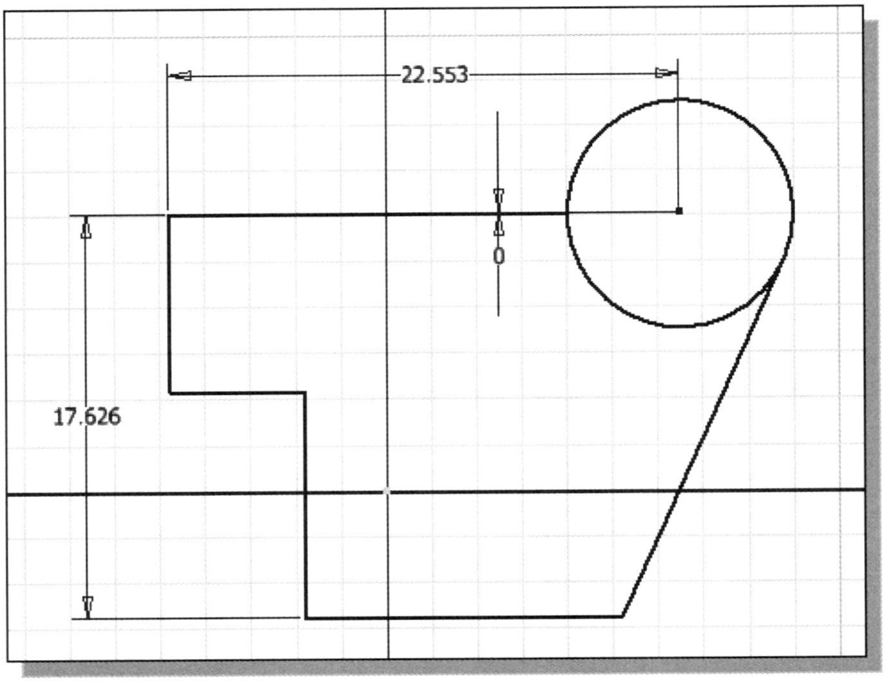

➢ Note the **Remove** option only removes the dimensions created by the **Auto Dimension** command but maintains the dimensions created by the user.

5. On your own, **trim** the circle and add the additional dimensions as shown.

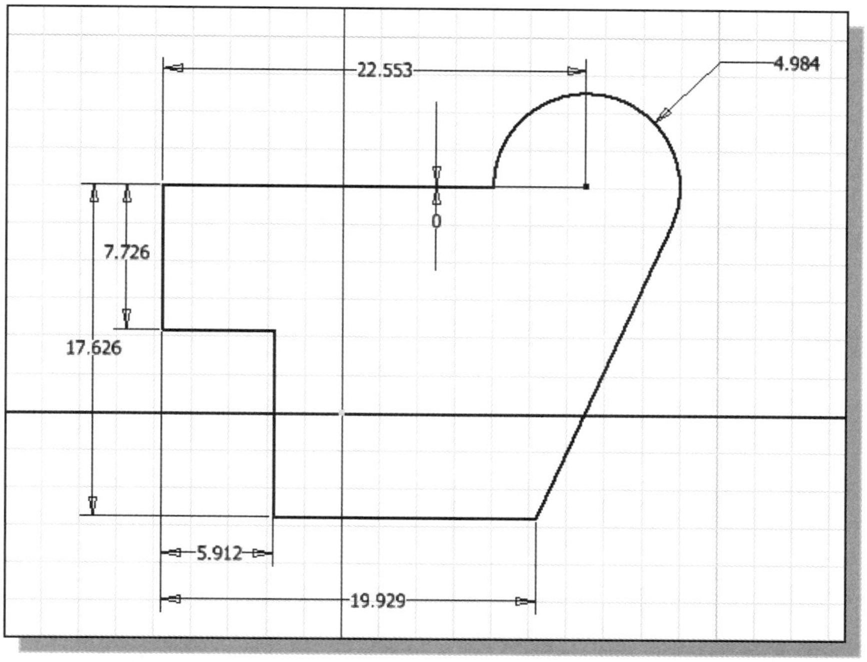

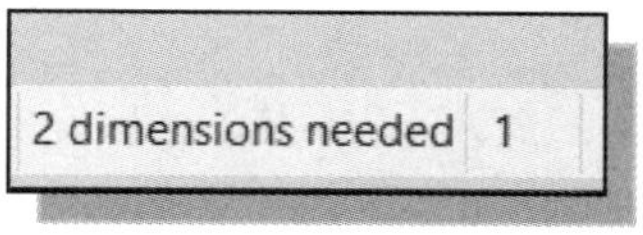

6. Note the display in the *Status Bar* indicates two additional dimensions are still needed to fully constrain the sketch.

Create Fillets and Completing the Sketch

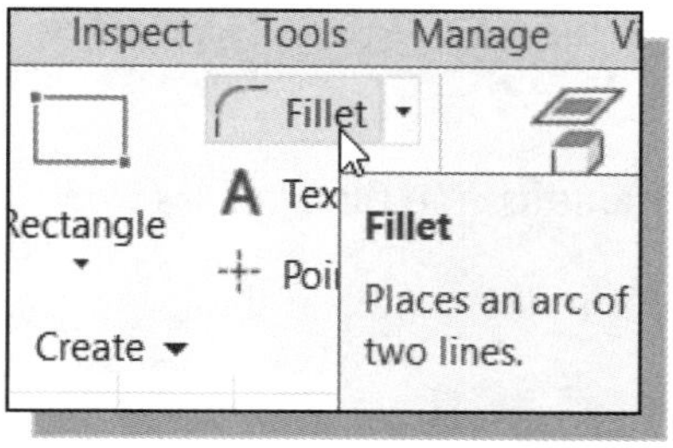

1. Click on the **Fillet** icon in the *Sketch* panel.

2. The *2D Fillet* radius dialog box appears on the screen. Use the **default** radius value and click on the top horizontal line and the arc to create a fillet as shown.

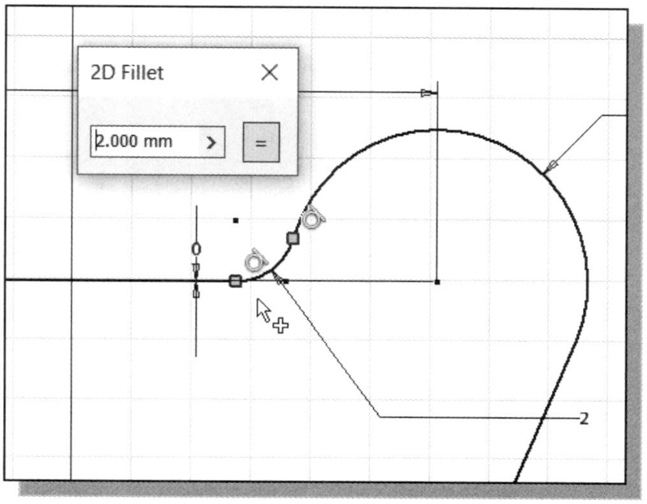

3. On your own, create the three additional fillets as shown in the below figure. Note that the Equal constraint is activated, and all rounds and fillets are created with the constraint.

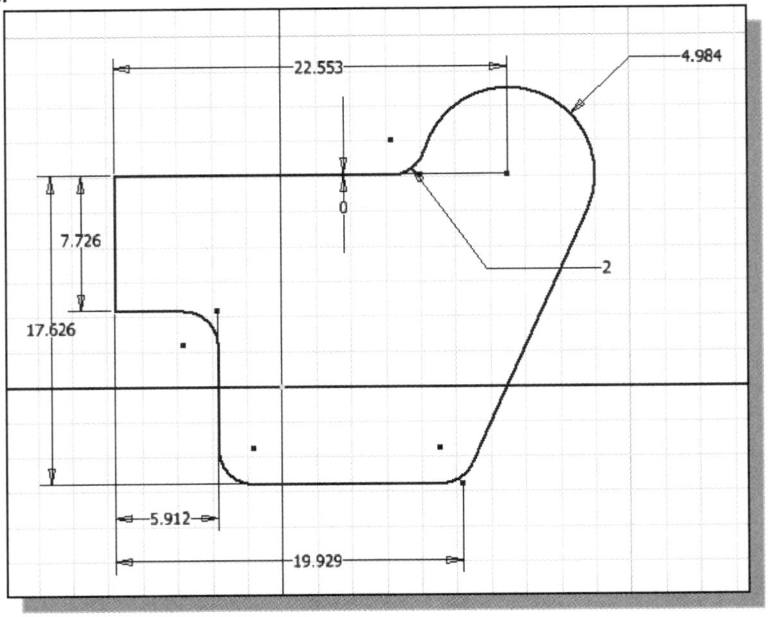

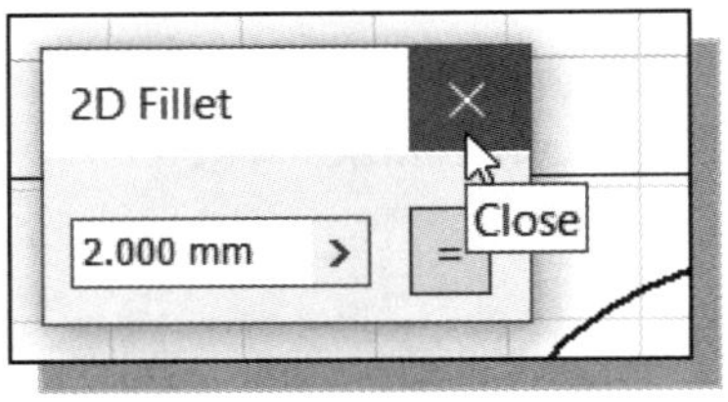

4. Click on the [**X**] icon to close the *2D Fillet* dialog box and end the Fillet command.

Fully Constrained Geometry

1. Click on the **Fix** constraint icon in the *Sketch Panel.*
2. Apply the **Fix** constraint to the center of the large arc as shown below.

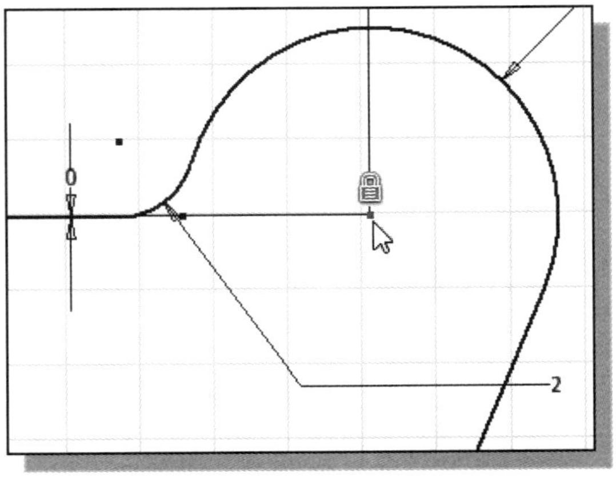

➢ Note that the sketch is now fully constrained.

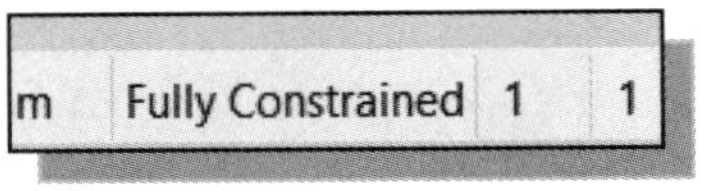

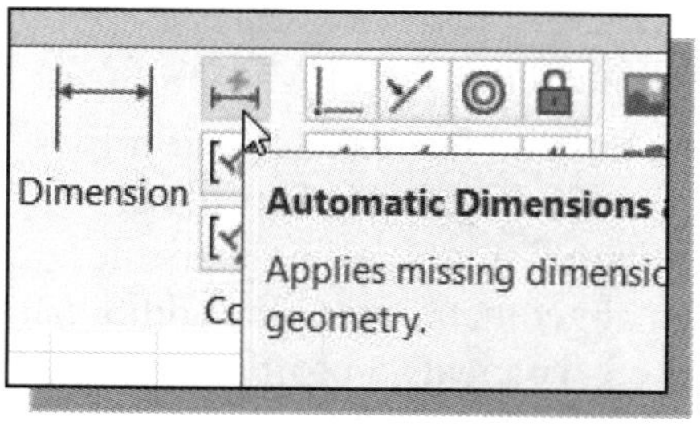

3. Click on the **Auto Dimension** icon in the *Sketch* panel.

➢ Note that the system automatically recalculates if any additional dimension is needed for all created geometric entities. The applied Fix constraint provided the previously missing location dimensions for the created sketches.

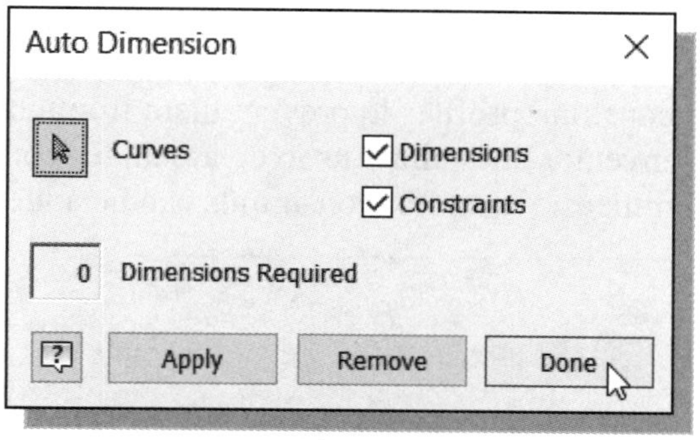

4. Click **Done** to exit the Auto Dimension command.

5. On your own, complete the sketch by adjusting the dimensions to the desired values as shown below. (Hint: Change the larger values first.)

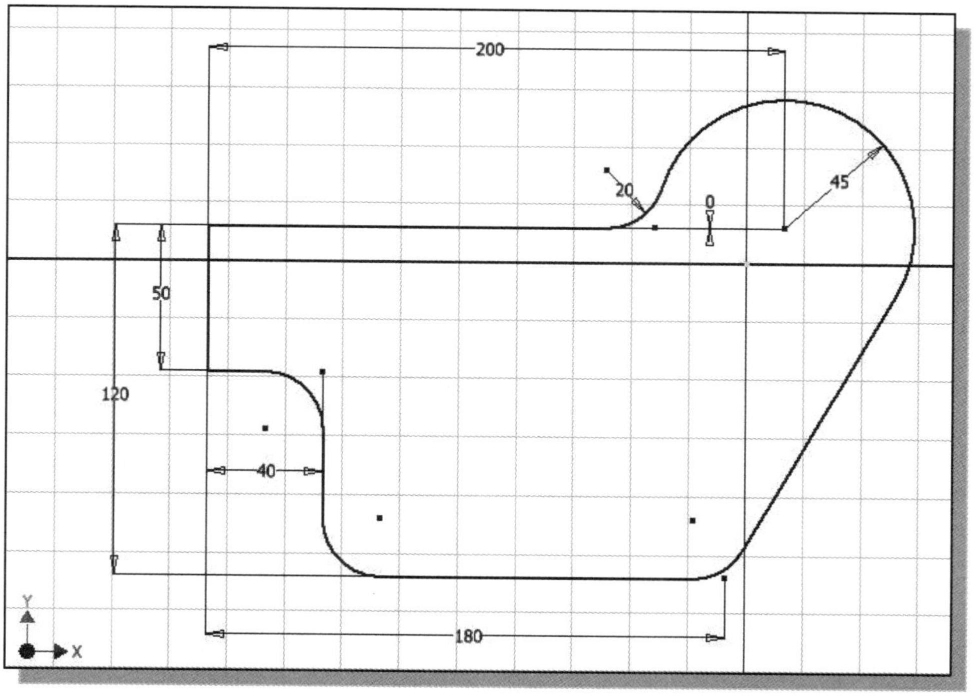

❖ Note that by applying proper geometric and dimensional constraints to the sketched geometry a *constraint network* is created that assures the geometry shape behaves predictably as changes are made.

Profile Sketch

In Autodesk Inventor, ***profiles*** are closed regions that are defined from sketches. Profiles are used as cross sections to create solid features. For example, **Extrude**, **Revolve**, **Sweep**, **Loft**, and **Coil** operations all require the definition of at least a single profile. The sketches used to define a profile can contain additional geometry since the additional geometry entities are consumed when the feature is created. To create a profile we can create single or multiple closed regions, or we can select existing solid edges to form closed regions. A profile cannot contain self-intersecting geometry; regions selected in a single operation form a single profile. As a general rule, we should dimension and constrain profiles to prevent them from unpredictable size and shape changes. Autodesk Inventor does allow us to create under-constrained or non-constrained profiles; the dimensions and/or constraints can be added/edited later.

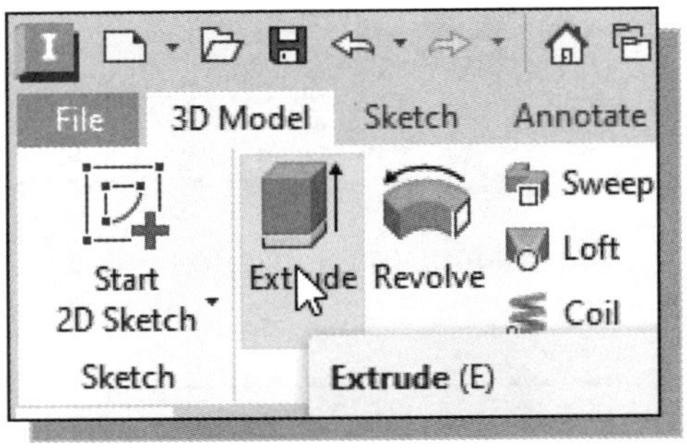

1. In the *3D Model* tab, select the **Extrude** command by left-clicking once on the icon.

❖ Note that Autodesk Inventor automatically highlights the closed region of the sketch. The defining geometry now forms the **profile** required for the *Extrude* operation.

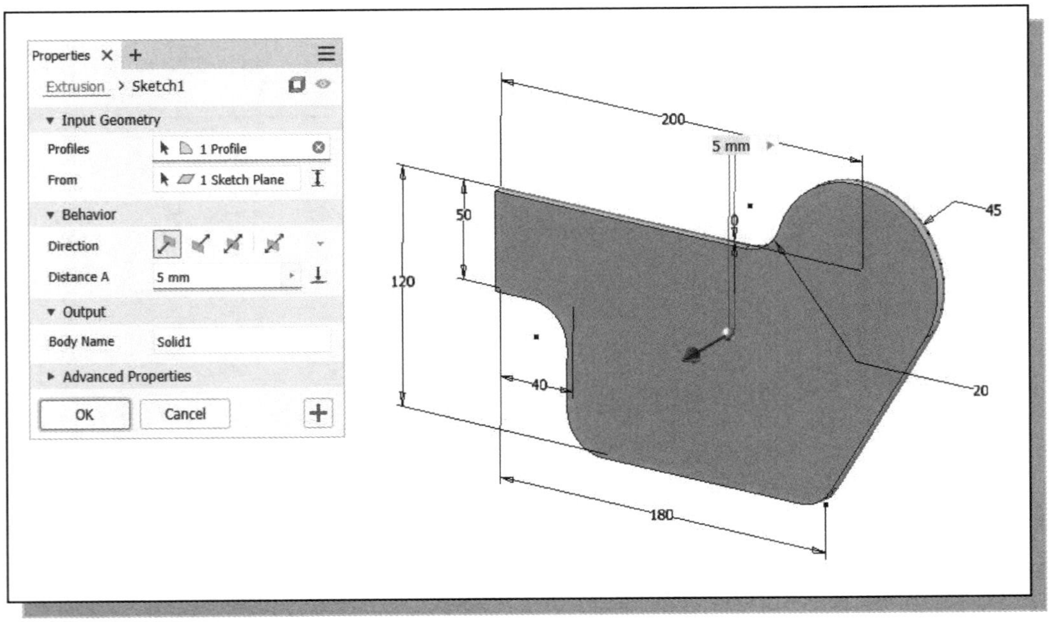

2. In the *Extrude* dialog box, enter **5 mm** as the extrusion distance as shown.

3. Click on the **OK** button to accept the settings and create the solid feature.

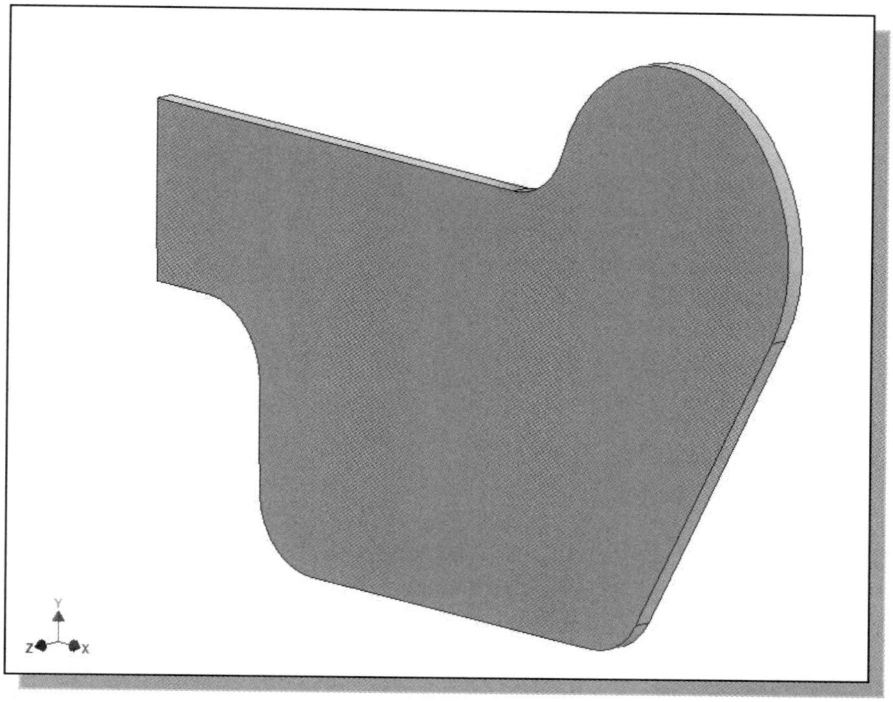

❖ Note that all the sketched geometry entities and dimensions are consumed and have disappeared from the screen when the feature is created.

Redefine the Sketch and Profile

Engineering designs usually go through many revisions and changes. Autodesk Inventor provides an assortment of tools to handle design changes quickly and effectively. We will demonstrate some of the tools available by changing the base feature of the design. The profile used to create the extrusion is selected from the sketched geometry entities. In Autodesk Inventor, any profile can be edited and/or redefined at any time. It is this type of functionality in parametric solid modeling software that provides designers with greater flexibility and the ease to experiment with different design considerations.

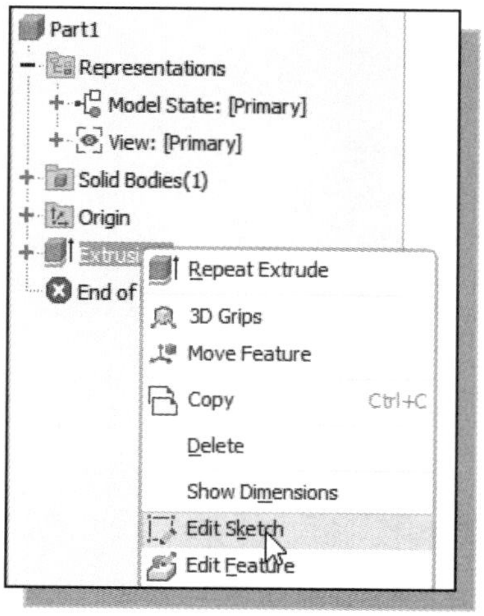

1. In the *browser*, **right-click** once on the **Extrusion1** feature to bring up the option menu, and then pick **Edit Sketch** in the pop-up menu.

2. On your own, create a circle and a rectangle of arbitrary sizes, positioned as shown in the figure below. We will intentionally under-constrain the new sketch to illustrate the flexibility of the system.

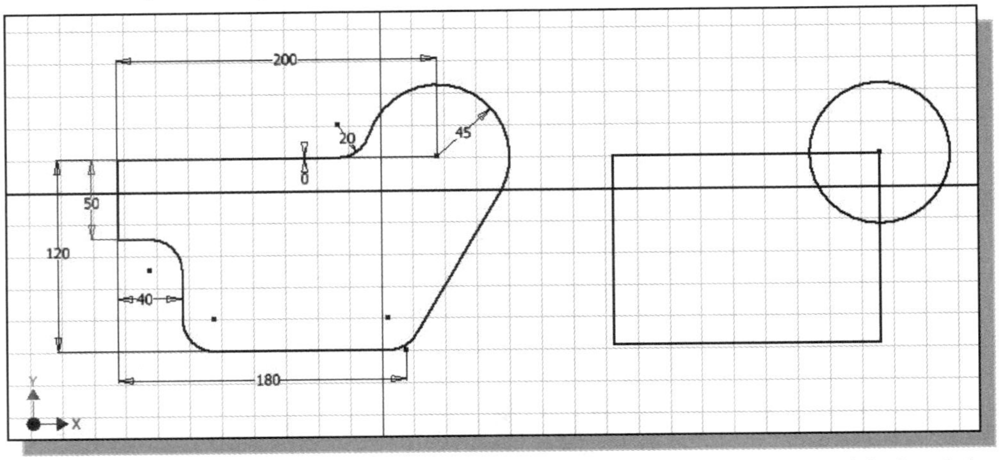

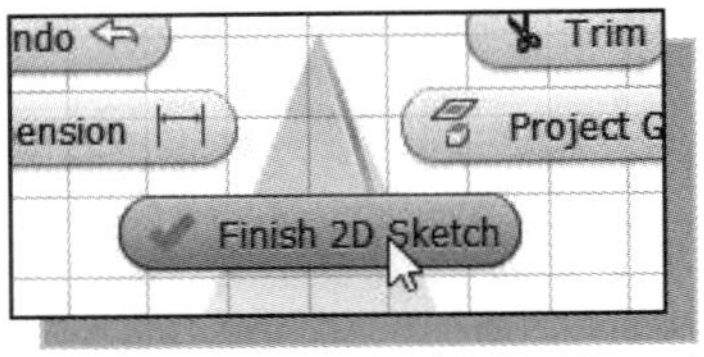

3. Inside the *graphics window*, click once with the right-mouse-button to display the option menu. Select **Finish 2D Sketch** in the pop-up menu to end the Sketch option.

➢ Note that, at this point, the solid model remains the same as before. We will next redefine the profile used for the base feature.

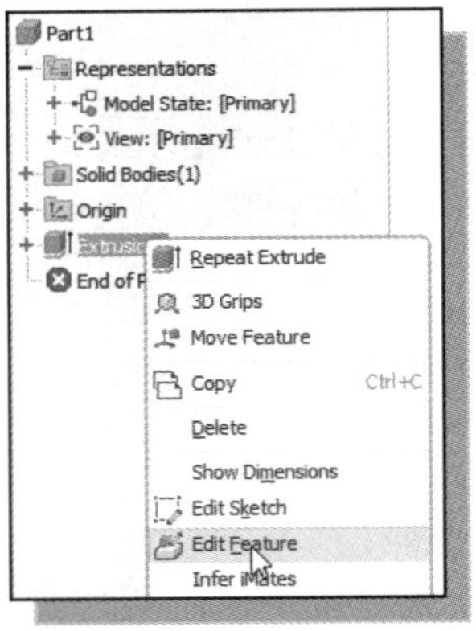

4. In the *browser*, right-click once on the Extrusion1 feature to bring up the option menu, and then pick **Edit Feature** in the pop-up menu.

❖ Autodesk Inventor will now display the 2D sketch of the selected feature in the graphics window. We have literally gone back in time to the point where we defined the extrusion feature.

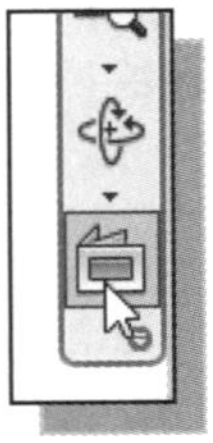

5. Click on the **Look At** icon in the *Standard* toolbar area.

6. Click on the front face of the gasket design.

- The **Look At** command automatically aligns the *sketch plane* of a selected entity to the screen.

7. Select any line segment of the 2D sketch.

8. Click on the **Zoom All** icon in the *Standard* toolbar.

➢ We have literally gone back to the point where we first defined the 2D profile. The original sketch and the new sketch we just created are wireframe entities recorded by the system as belonging to the same **SKETCH**, but only the **PROFILED** entities are used to create the feature.

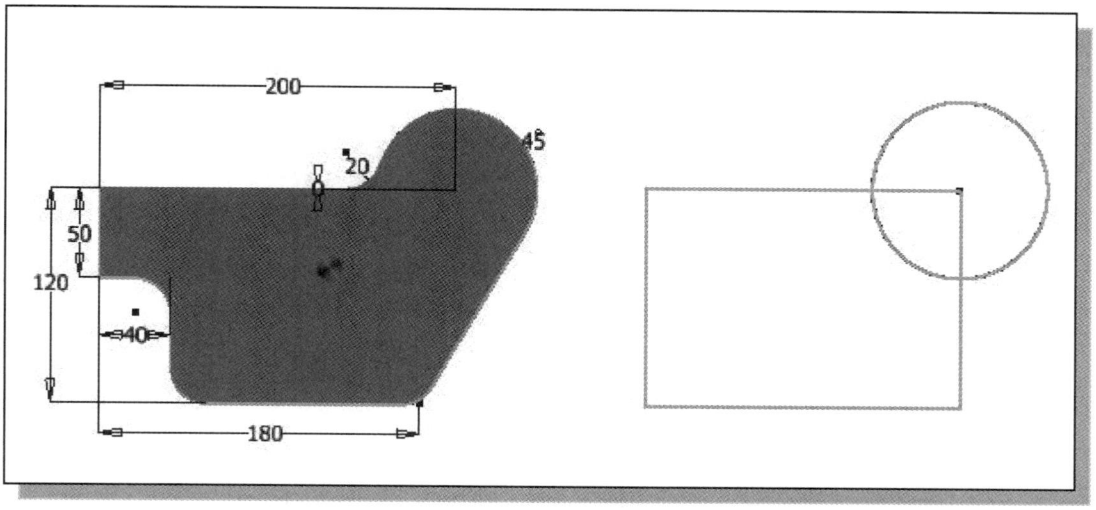

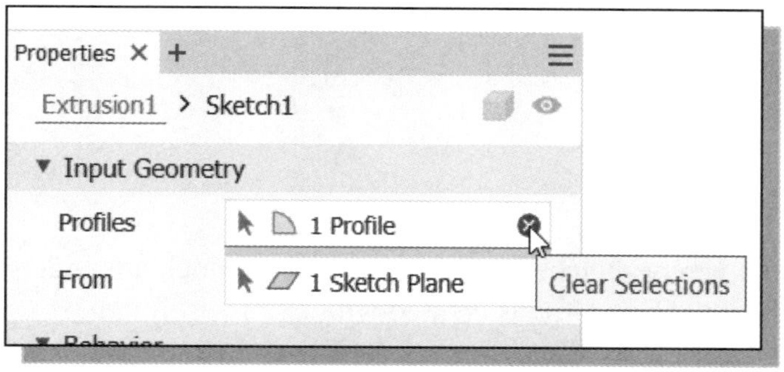

9. In the *Extrude* control panel, click on the **[x]** button to clear the current selected 2D profile.

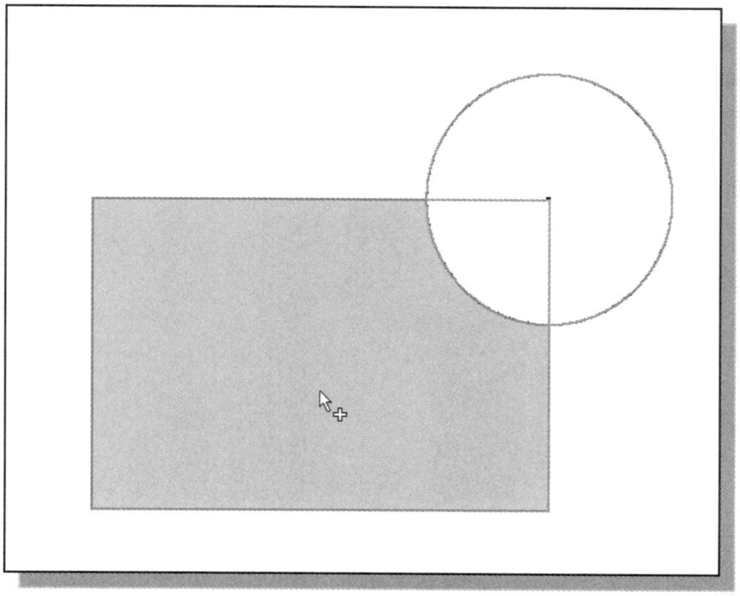

10. Click inside the lower region of the rectangle of the sketch on the right side of the graphics window.

➢ Autodesk Inventor automatically selects the geometry entities that form a closed region to define the profile.

11. Click the inside regions of the circle and complete the profile definition as shown in the figure below.

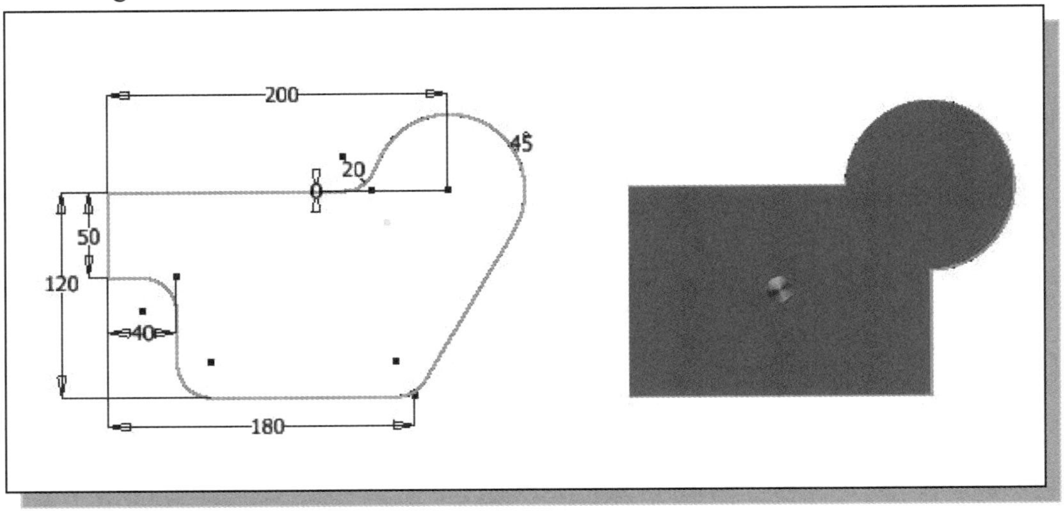

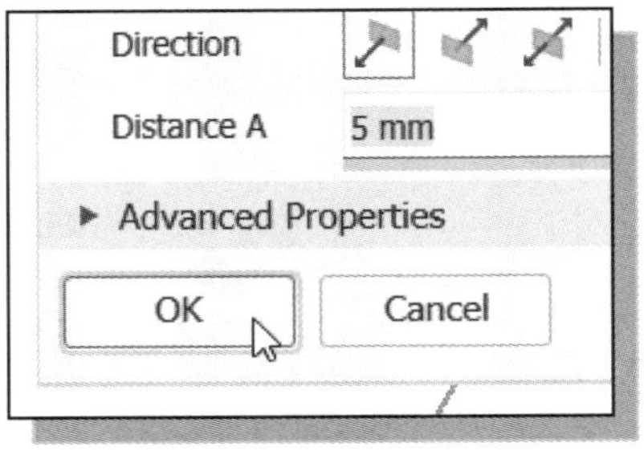

12. In the *Extrude* control panel, click on the **OK** button to accept the settings and update the solid feature.

- The feature is recreated using the newly sketched geometric entities, which are under-constrained and with no dimensions. The profile is created with extra wireframe entities by selecting multiple regions. The extra geometry entities can be used as construction geometry to help in defining the profile. This approach encourages engineering content over drafting technique, which is one of the key features of Autodesk Inventor over other solid modeling software.

13. On your own, repeat the above steps and reset the profile back to the original gasket sketch.

Create an Offset Cut Feature

To complete the design, we will create a cutout feature by using the **Offset** command. First, we will set up the sketching plane to align with the front face of the 3D model.

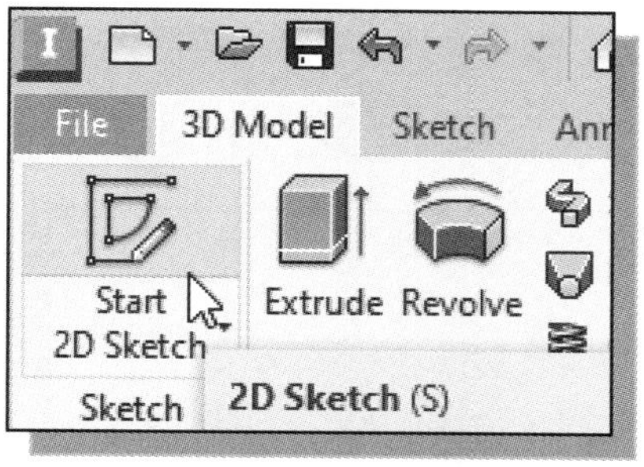

1. In the *3D Model* tab select the **Start 2D Sketch** command by left-clicking once on the icon.

2. In the *Status Bar* area, the message "*Select plane to create sketch or an existing sketch to edit.*" is displayed. Select the **front face** of the 3D model in the graphics window.

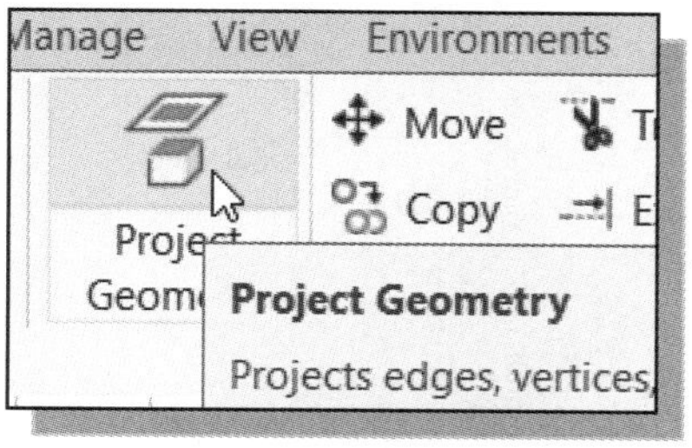

3. Click on the **Project Geometry** icon in the *2D Create* panel.

4. Select the front face of the gasket model, and notice the outline of the design has been projected onto the sketching plane.

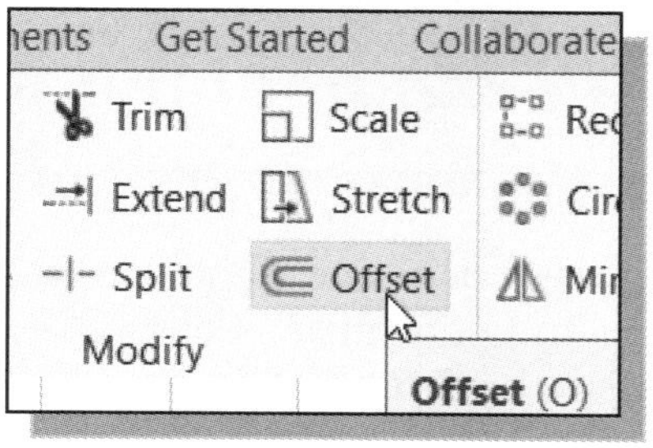

5. Click on the **Offset** icon in the *Sketch* tab on the Ribbon.

6. *Right-click* once to bring up the option menu and activate "**Loop select**" if necessary, as shown.

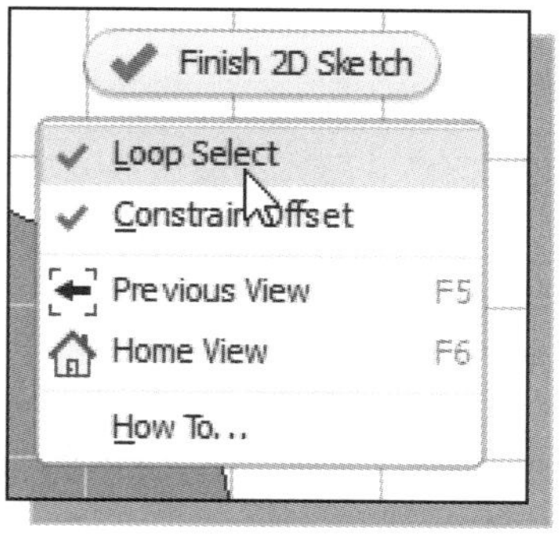

7. Select any edge of the **front face** of the 3D model. Autodesk Inventor will automatically select all of the connecting geometry to form a closed region.

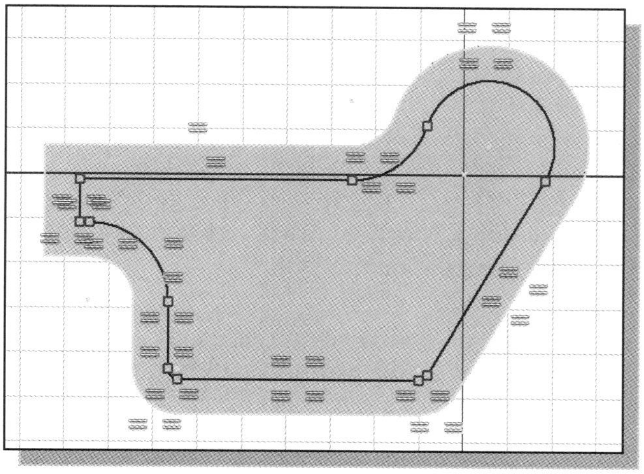

8. Move the cursor toward the inside of the selected region and notice an offset copy of the outline is displayed.

9. Left-click once to create the offset profile as shown.

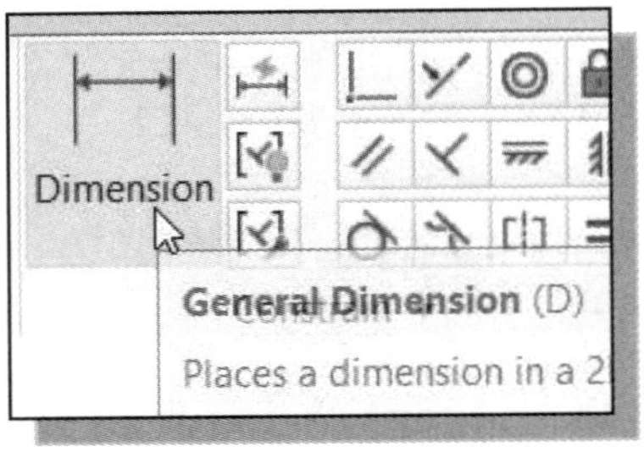

10. On your own, use the **General Dimension** command to create the offset dimension as shown in the figure below.

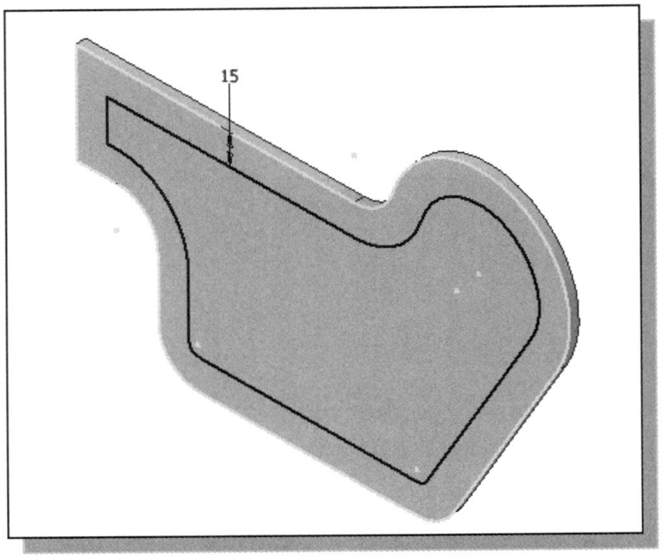

11. Modify the offset dimension to **15 mm** as shown in the figure to the right.

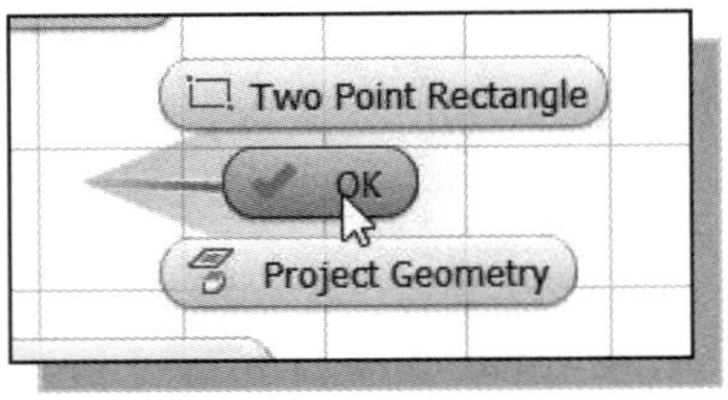

12. Inside the graphics window, click once with the right-mouse-button to display the option menu. Select **OK** in the pop-up menu to end the General Dimension command.

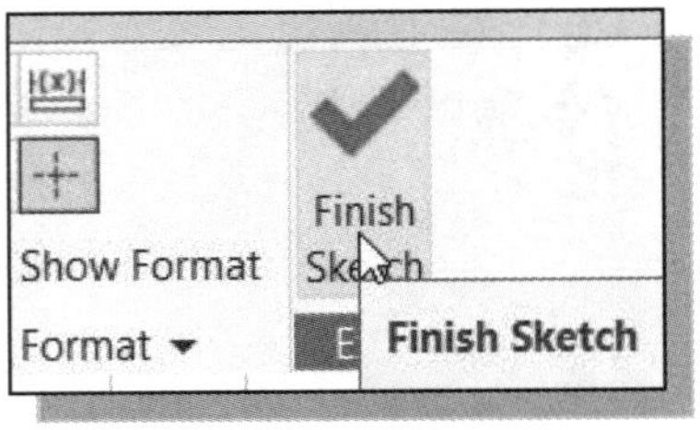

13. Inside the graphics window, click once with the right-mouse-button to display the option menu. Select **Finish Sketch** in the pop-up menu to end the Sketch option.

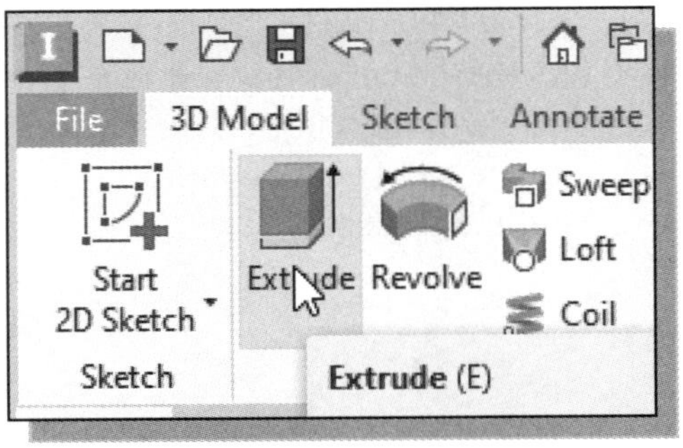

14. In the *3D Model* toolbar (the toolbar that is located to the left side of the graphics window), select the **Extrude** command by left-clicking on the icon.

15. Select the **inside region** of the offset geometry as the profile for the extrusion.

16. Inside the *Extrude* dialog box, select the **Cut** operation, set the *Extents* to **All** and set the direction of the cutout as shown in the figure below.

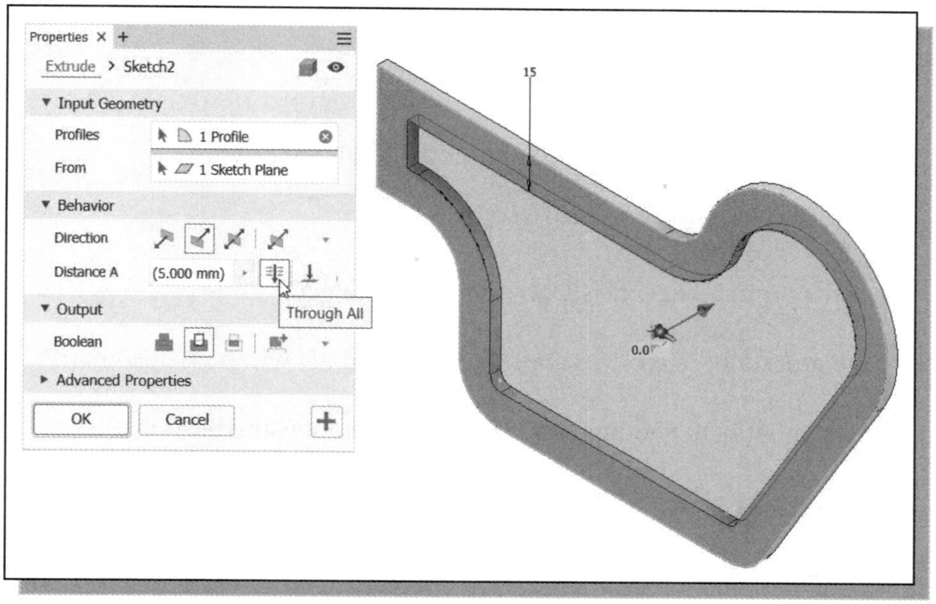

17. In the *Extrude* dialog box, click on the **OK** button to accept the settings and create the solid feature.

➢ The offset geometry is associated with the original geometry. On your own, adjust the overall height of the design to **150 millimeters** and confirm that the offset geometry is adjusted accordingly.

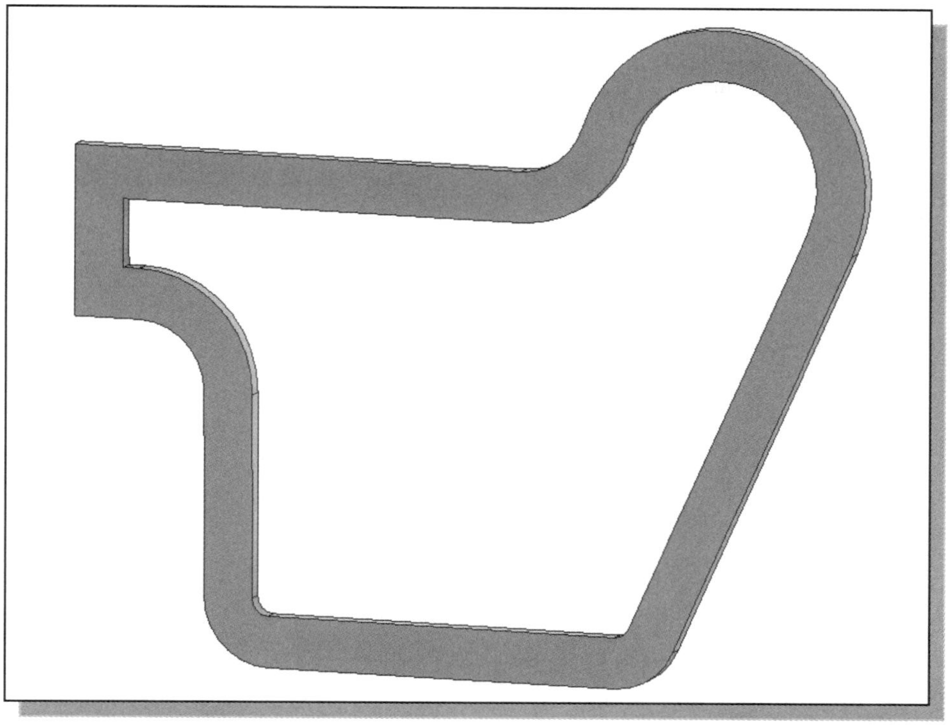

Review Questions: (Time: 25 minutes)

1. What are the two types of wireframe geometry available in Autodesk Inventor?
2. Can we create a profile with extra 2D geometry entities?
3. How do we access the Autodesk Inventor's **Edit Sketch** option?
4. How do we create a *profile* in Autodesk Inventor?
5. Can we build a profile that consists of self-intersecting curves?
6. Describe the procedure to create a copy of a sketched 2D wireframe geometry.
7. Can we create additional entities in a 2D sketch, without using them at all?
8. How do we align the *sketch plane* of a selected entity to the screen?
9. Describe the steps we used to switch existing profiles in the tutorial.
10. Describe the advantages of using the Offset command vs. creating a separate sketch.

Exercises: Create and save the exercises in the Chapter6 folder. (Time: 180 minutes.)

1. **V-slide Plate** (Dimensions are in inches. Plate Thickness: **0.25**)

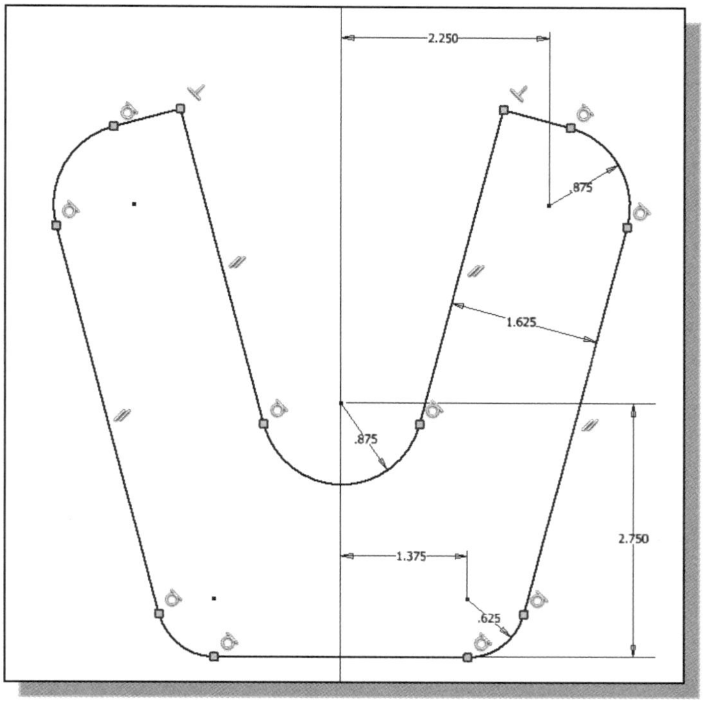

2. **Shaft Support** (Dimensions are in millimeters. Note the two R40 arcs at the base share the same center.)

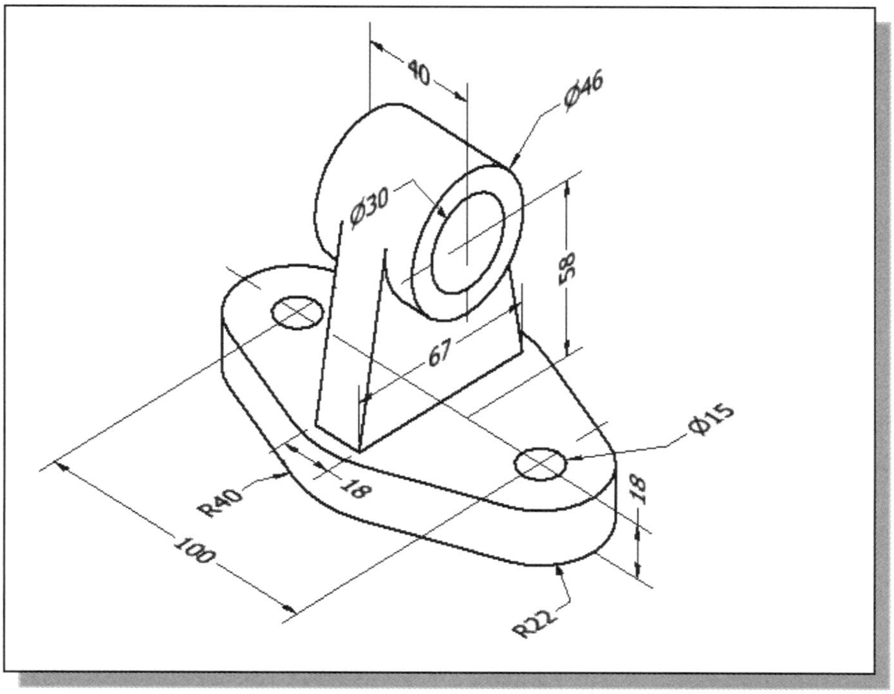

3. **Vent Cover** (Thickness: **0.125** inches. Hint: Use the Ellipse command.)

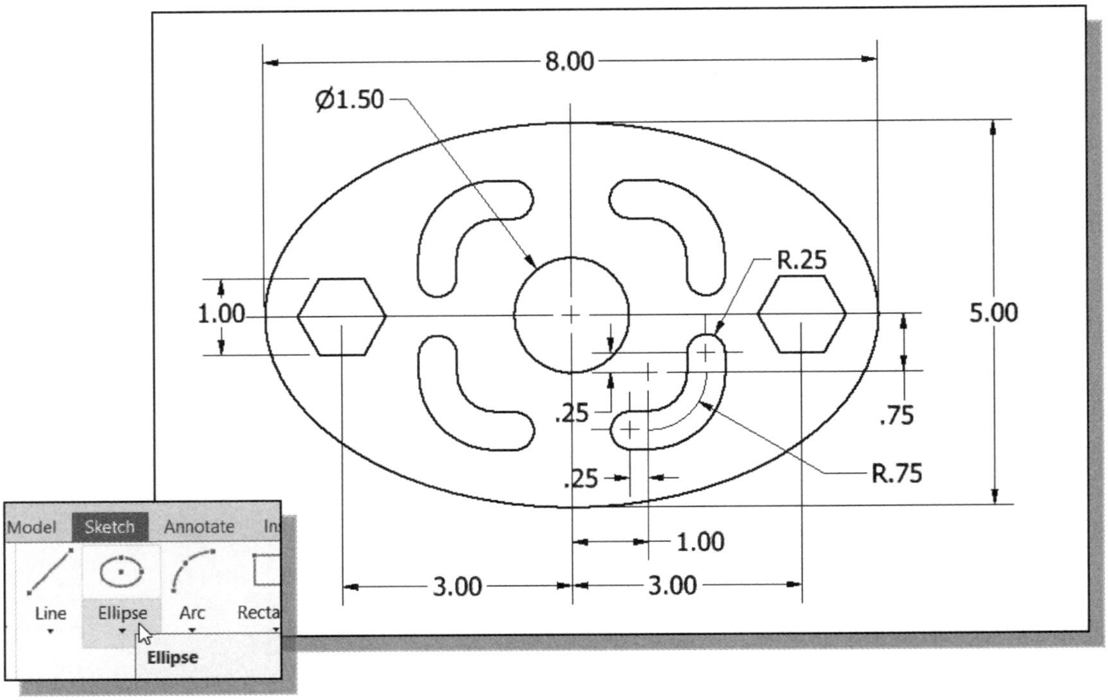

4. **Anchor Base** (Dimensions are in inches.)

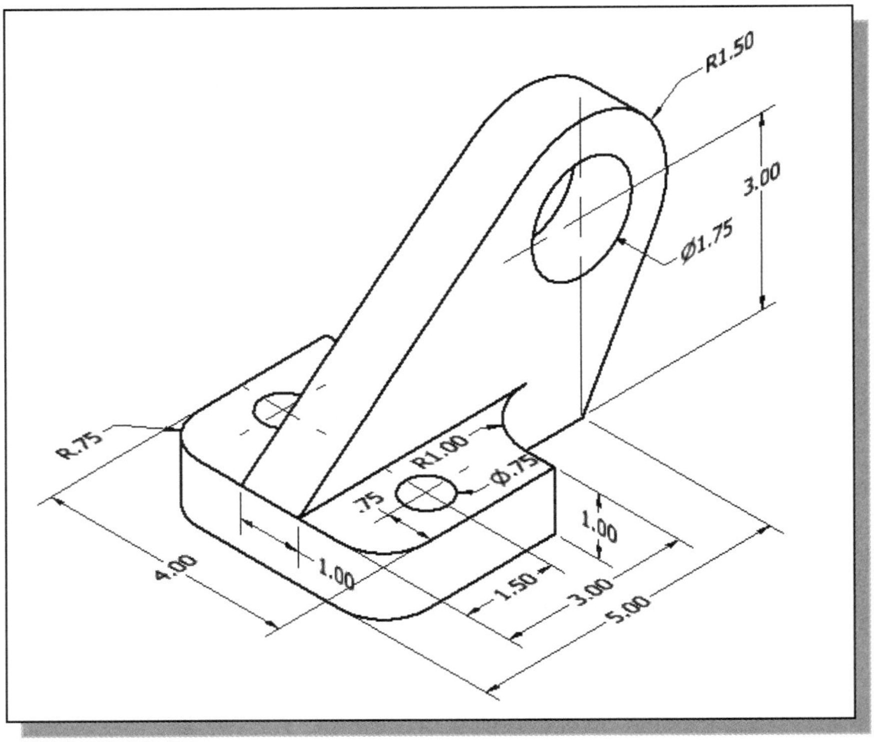

5. **Tube Spacer** (Dimensions are in inches.)

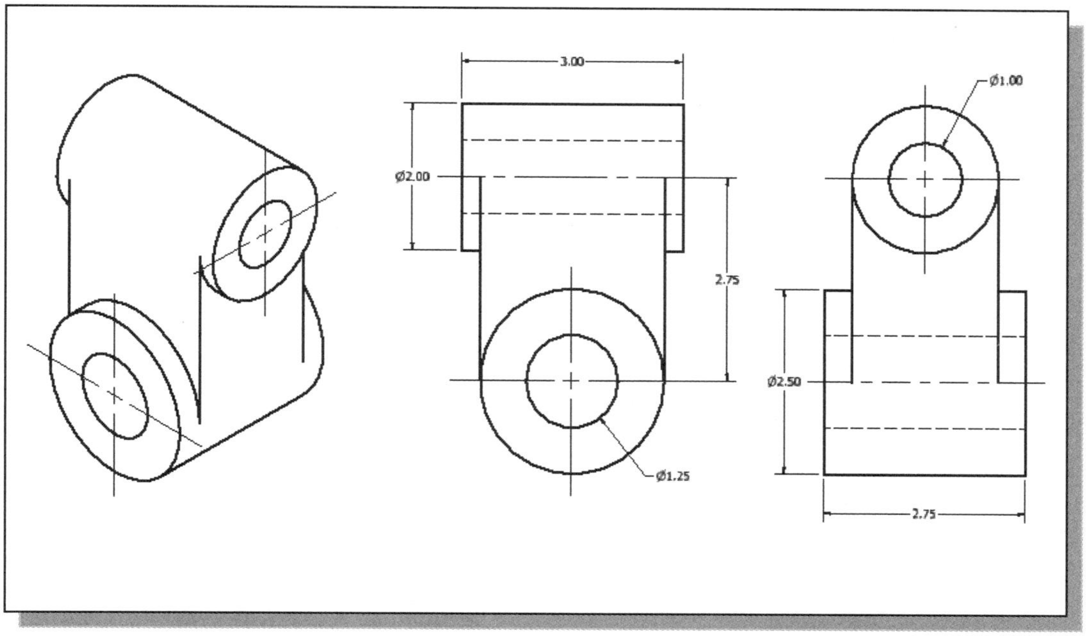

6. **Pivot Lock** (Dimensions are in inches. The circular features in the design are all aligned to the two centers at the base.)

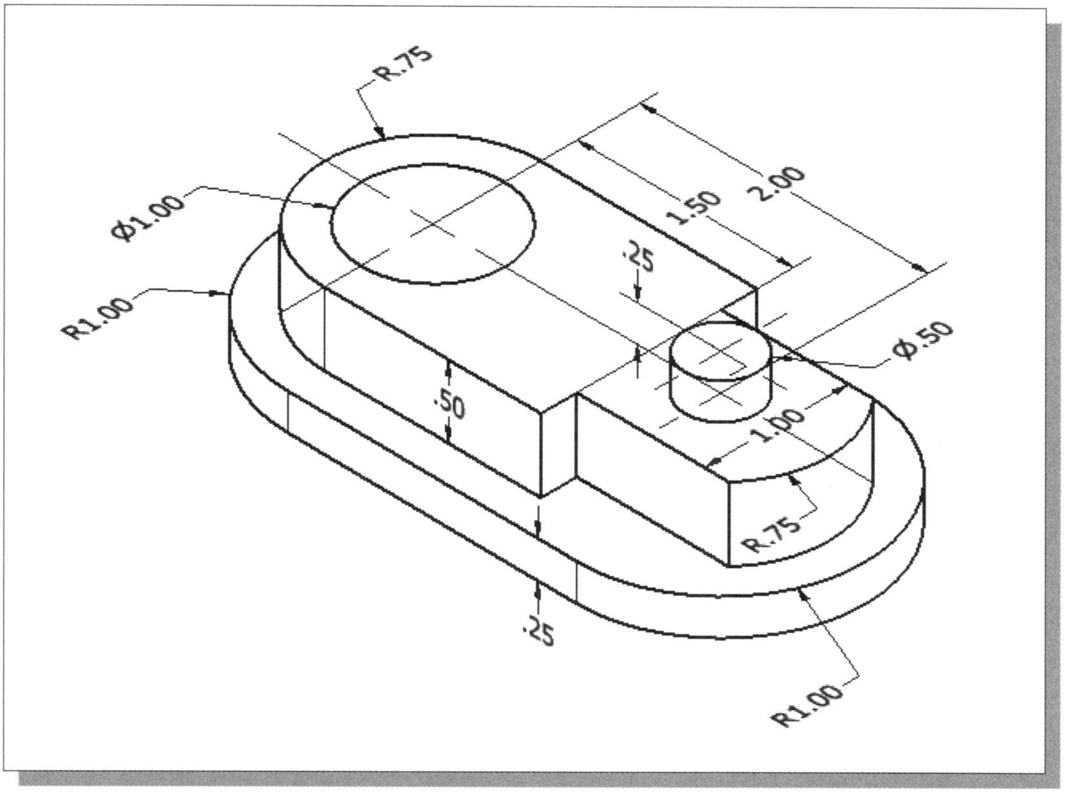

7. **Lock-ring** (Dimensions are in inches.)

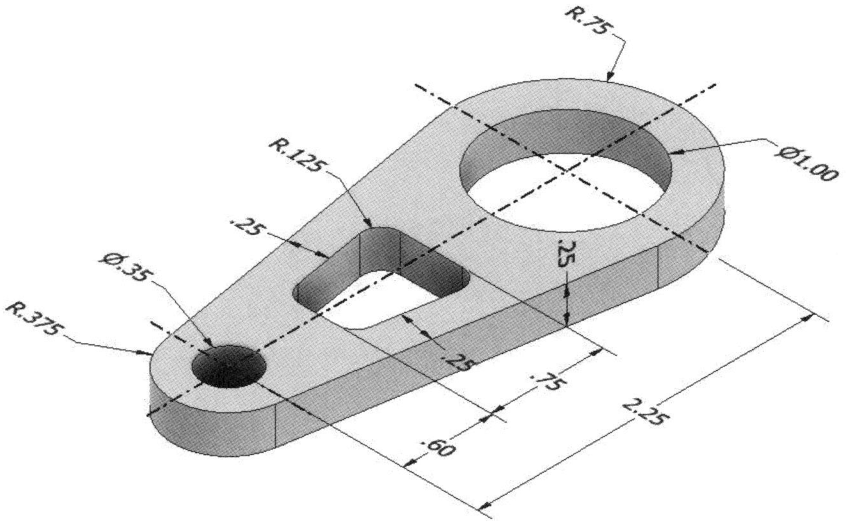

Chapter 7
Parent/Child Relationships and the BORN Technique

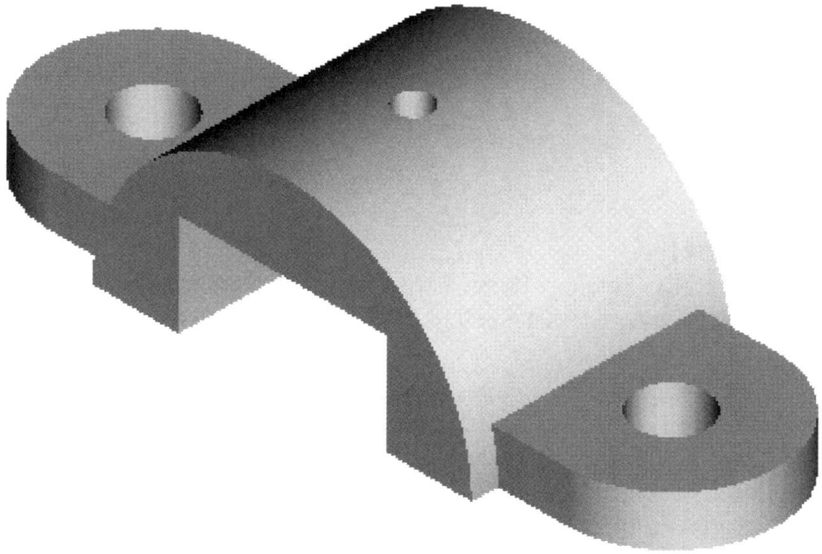

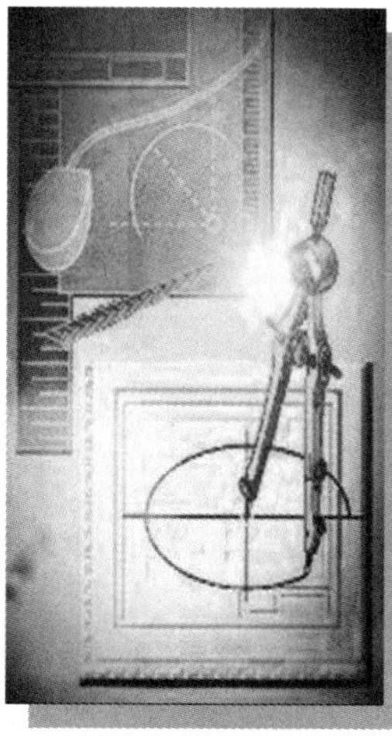

Learning Objectives

- Understand the Concept and Usage of the BORN Technique
- Understand the Importance of Parent/Child Relations in Features
- Use the Suppress Feature Option
- Resolve Undesired Feature Interactions

Autodesk Inventor Certified User Exam Objectives Coverage

Parametric Modeling Basics

Section 3: Sketches

Objectives: Creating 2D Sketches, Draw Tools, Sketch Constraints, Pattern Sketches, Modify Sketches, Format Sketches, Sketch Doctor, Shared Sketches, Sketch Parameters.

Section 4: Parts

Objectives: Creating parts, Work Features, Pattern Features, Part Properties.

Introduction

The parent/child relationship is one of the most powerful aspects of ***parametric modeling***. In Autodesk Inventor, each time a new modeling event is created, previously defined features can be used to define information such as size, location, and orientation. The referenced features become **Parent** features to the new feature, and the new feature is called the **Child** feature. The parent/child relationships determine how a model reacts when other features in the model change, thus capturing design intent. It is crucial to keep track of these parent/child relations. Any modification to a parent feature can change one or more of its children.

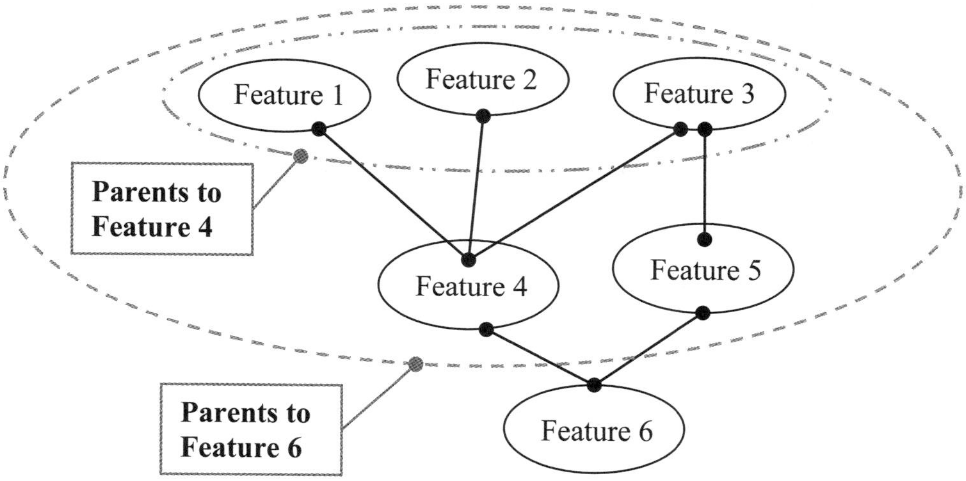

Parent/child relationships can be created *implicitly* or *explicitly*; implicit relationships are implied by the feature creation method, and explicit relationships are entered manually by the user. In the previous chapters, we first select a sketching plane before creating a 2D profile. The selected surface becomes a parent of the new feature. If the sketching plane is moved, the child feature will move with it. As one might expect, parent/child relationships can become quite complicated when the features begin to accumulate. It is therefore important to think about modeling strategy before we start to create anything. The main consideration is to try to plan ahead for possible design changes that might occur which would be affected by the existing parent/child relationships. Parametric modeling software, such as Autodesk Inventor, also allows us to adjust feature properties so that any feature conflicts can be quickly resolved.

The BORN Technique

In *parametric modeling*, the ***base feature*** (the *first solid feature* of the solid model) is the center of all features and is considered the key feature of the design. All subsequent features are built by referencing the first feature. Much emphasis is placed on the selection of the *base feature*.

A more advanced technique of creating solid models is what is known as the "**Base Orphan Reference Node**" (**BORN**) technique. The basic concept of the BORN technique is to use a *Cartesian coordinate system* as the first feature prior to creating any solid features. With the *Cartesian coordinate system* established, we then have three mutually perpendicular datum planes (namely the *XY*, *YZ*, and *ZX planes*), three datum axes and a datum point available to use as sketching planes. The three datum planes can also be used as references for dimensions and geometric constructions. Using this technique, the first node in the history tree is called an "orphan," meaning that it has no history to be replayed. The technique of using the reference geometry in this "base node" is therefore called the "Base Orphan Reference Node" (BORN) technique.

Autodesk Inventor automatically establishes a set of reference geometry when we start a new part, namely a *Cartesian coordinate system* with three work planes, three work axes, and a work point. All subsequent solid features can then use the coordinate system and/or reference geometry as sketching planes. The *base feature* is still important, but the *base feature* is no longer the ONLY choice for creating subsequent solid features. This approach provides us with more options while we are creating parametric solid models. More importantly, this approach provides greater flexibility for part modifications and design changes. This approach is also very useful in creating assembly models, which will be illustrated in the later chapters of this text.

The U-Bracket Design

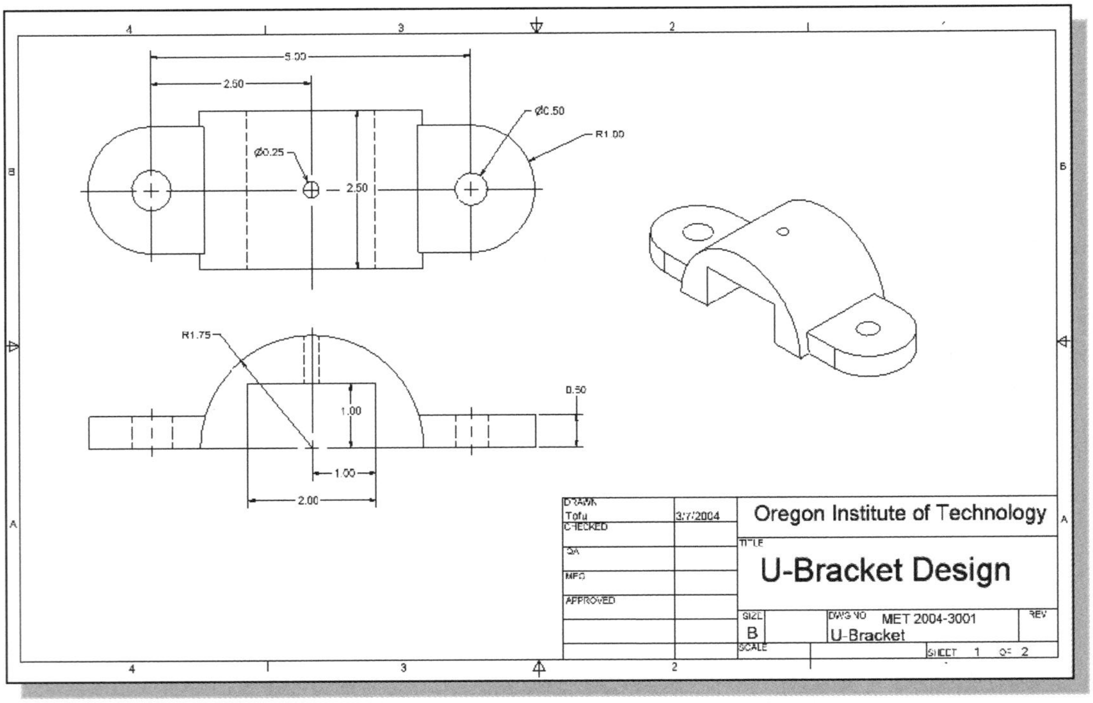

➤ Based on your knowledge of Inventor so far, how many features would you use to create the model? Which feature would you choose as the **base feature**? What is your choice for arranging the order of the features? Would you organize the features differently if the rectangular cut at the center is changed to a circular shape (Radius: 1.25 inch)?

Sketch Plane Settings

1. Select the Autodesk Inventor option on the *Start* menu or select the Autodesk Inventor icon on the desktop to start Autodesk Inventor. The Autodesk Inventor main window will appear on the screen.

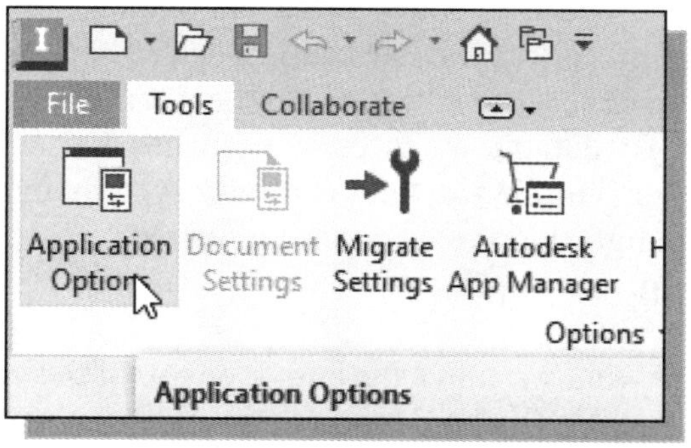

2. Select **Application Options** in the **Tools** pull-down menu.

- The **Application Options** menu allows us to set behavioral options, such as *color*, *file locations*, etc.

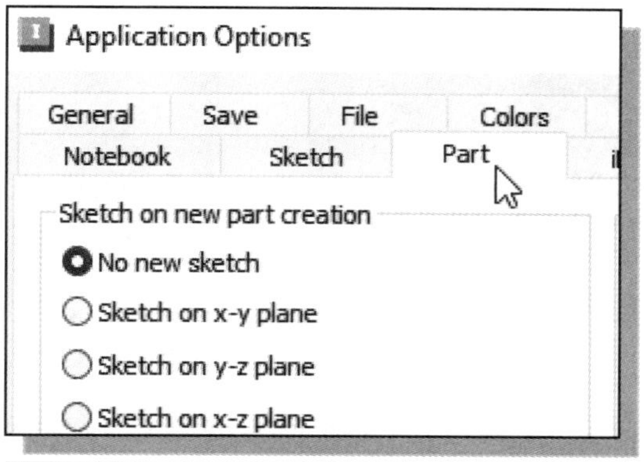

3. Click on the **Part** tab to display and/or modify the default sketch plane settings.

- Note that a sketch plane can be aligned to one of the work planes during new part creation. Confirm the **No new sketch** option is set as shown.

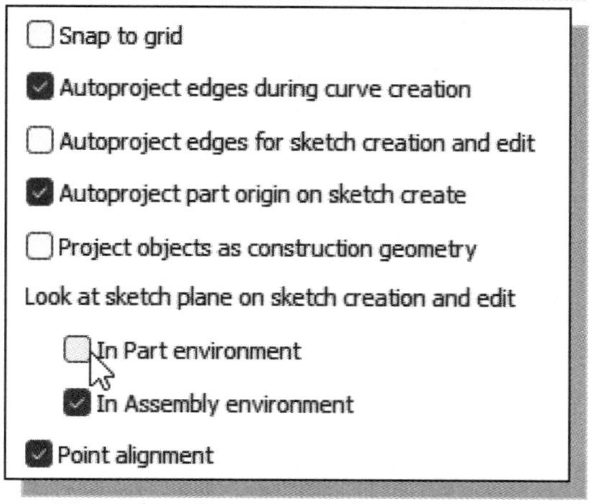

4. Click on the **Sketch** tab to examine/modify the default sketching settings.

5. Turn ***OFF*** the ***Look at sketch plane on sketch creation - In Part environment*** option as shown.

6. Click on the **OK** button to accept the setting.

- Note that the new settings will take effect when a new part file is created.

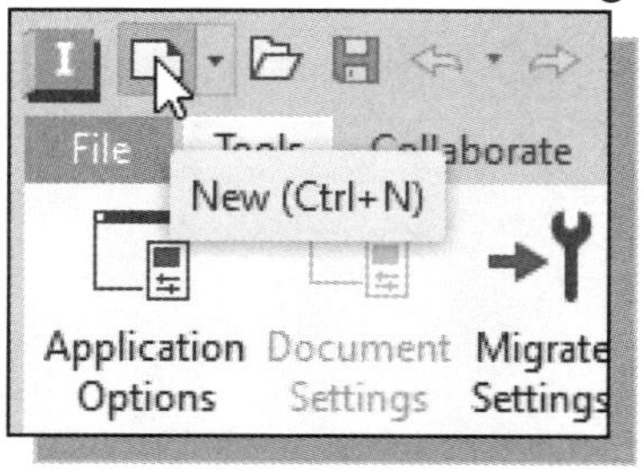

7. Click on the **New** icon in the *Standard* toolbar.

8. On your own, open a new ***English-****units* **standard (in) part** file.

Apply the BORN Technique

1. In the *Part Browser* window, click on the [**+**] symbol in front of the ***Origin*** feature to display more information on the feature.

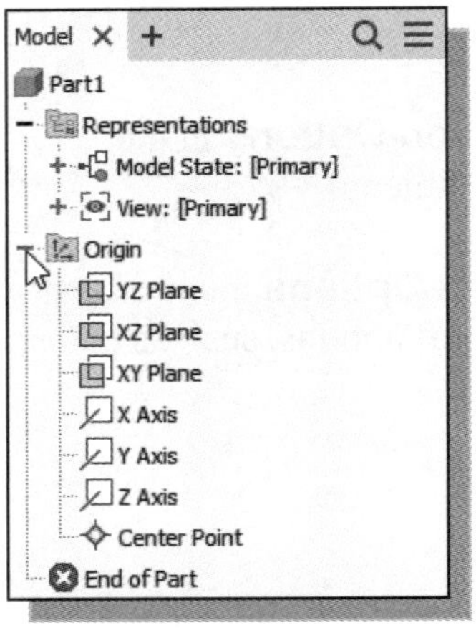

❖ In the *Part Browser* window, notice a new part name appeared with seven work features established. The seven work features include three *work planes*, three *work axes*, and a *work point*. By default, the three work planes and work axes are aligned to the **world coordinate system** and the work point is aligned to the *origin* of the **world coordinate system**.

2. Inside the *browser* window, move the cursor on top of the third work plane, the **XY Plane**. Notice a rectangle, representing the work plane, appears in the graphics window.

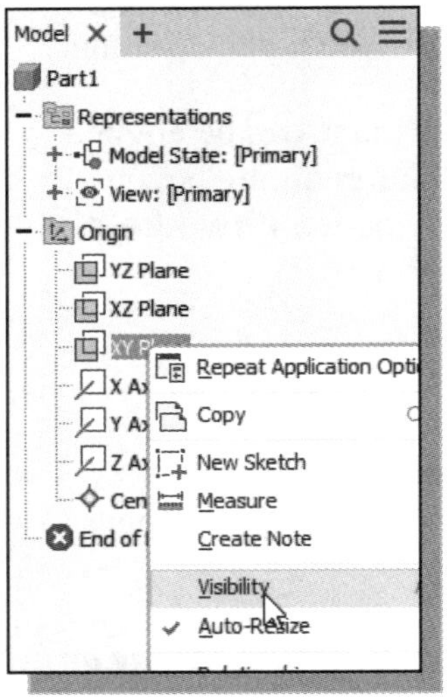

3. Inside the *browser* window, click once with the right-mouse-button on XY Plane to display the option menu. Click on **Visibility** to toggle on the display of the plane.

4. On your own, repeat the above steps and toggle ***ON*** the display of all of the **work planes** and the **center point** on the screen.

5. On your own, use the *Dynamic Viewing* options (ViewCube, 3D Orbit, Zoom and Pan) to view the default work features.

❖ By default, the basic set of work planes is aligned to the world coordinate system; the work planes are the first features of the part. We can now proceed to create solid features referencing the three mutually perpendicular datum planes.

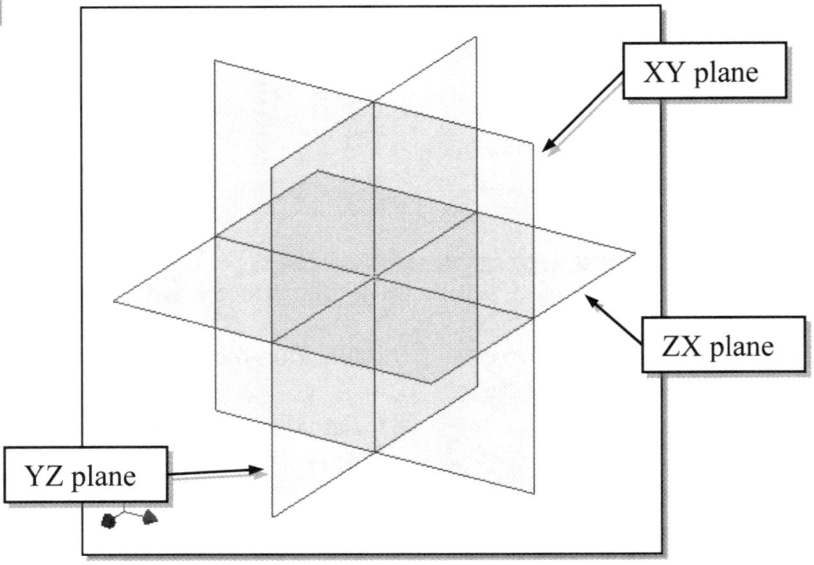

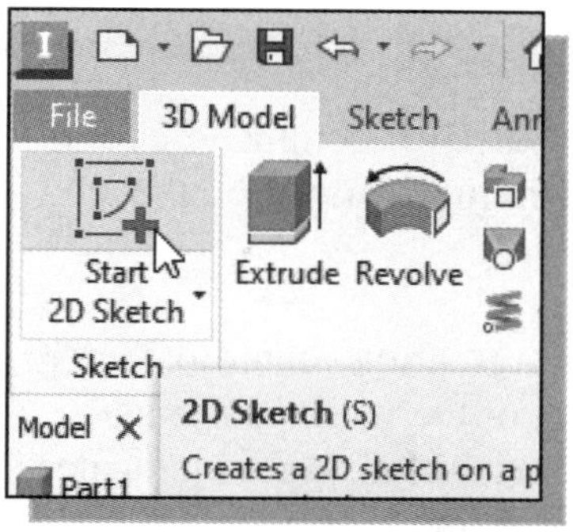

6. In the *Sketch* toolbar select the **Start 2D Sketch** command by left-clicking once on the icon.

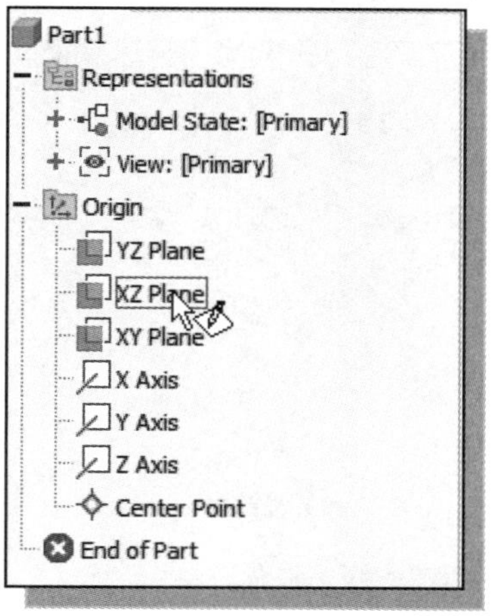

7. In the *Status Bar* area, the message "*Select plane to create sketch or an existing sketch to edit.*" is displayed. Autodesk Inventor expects us to identify a planar surface where the 2D sketch of the next feature is to be created. Move the graphics cursor on top of the **XZ Plane**, inside the *browser* window as shown, and notice that Autodesk Inventor will automatically highlight the corresponding plane in the graphics window. Left-click once to select the XZ Plane as the sketching plane.

❖ Autodesk Inventor allows us to identify and select features in the graphics window as well as in the *browser* window.

- Note that since both of the ***Look at sketch plane on sketch creation*** options are turned ***OFF***, the view display stays at the default ***isometric view*** .

➢ Note the alignment of the sketch plane is set to the XZ plane as shown.

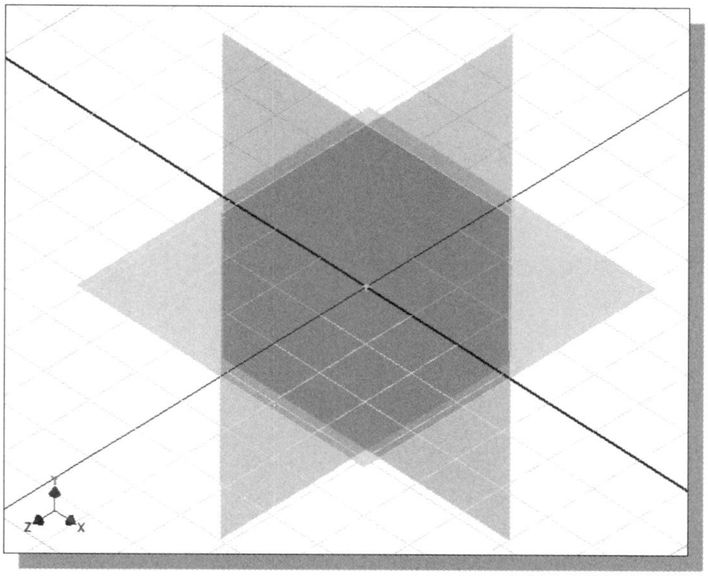

Create the 2D Sketch for the Base Feature

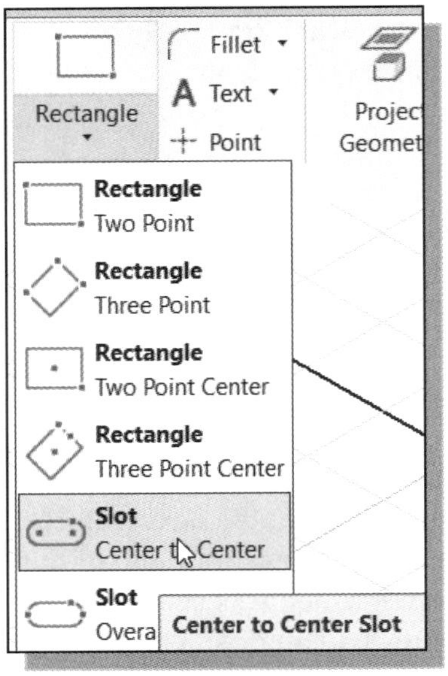

1. Select the **Center to Center Slot** command by clicking once with the **left-mouse-button** on the icon in the *Draw* toolbar.

2. Create a horizontal line, representing center to center distance of the slot, in front of the work planes as shown below. (Note the **Horizontal** symbol indicates the alignment of the sketched line.)

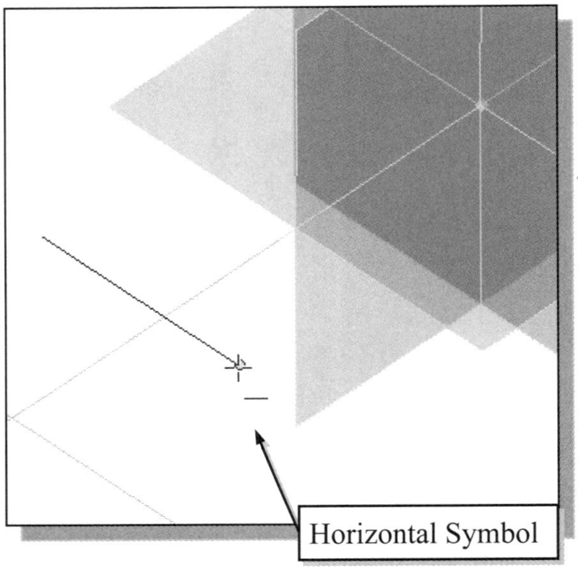

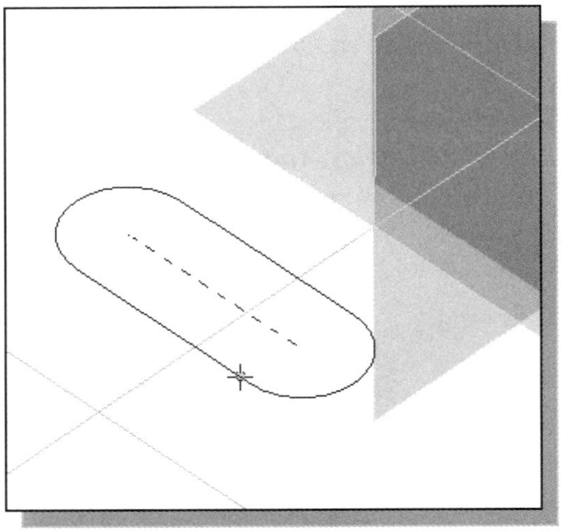

3. Move the cursor outward to define the size of the center to center slot; click once with the left-mouse-button to create the shape.

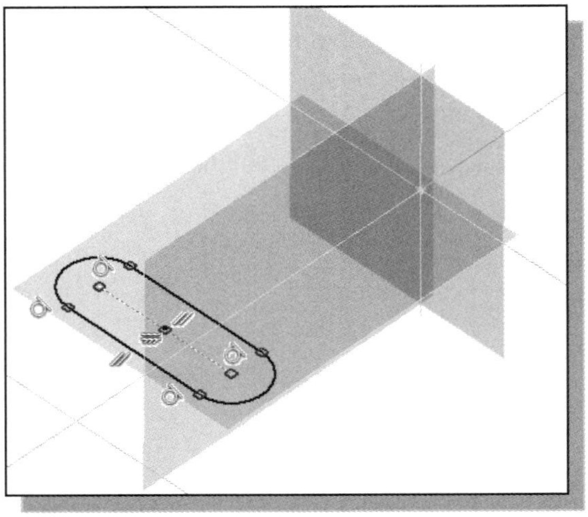

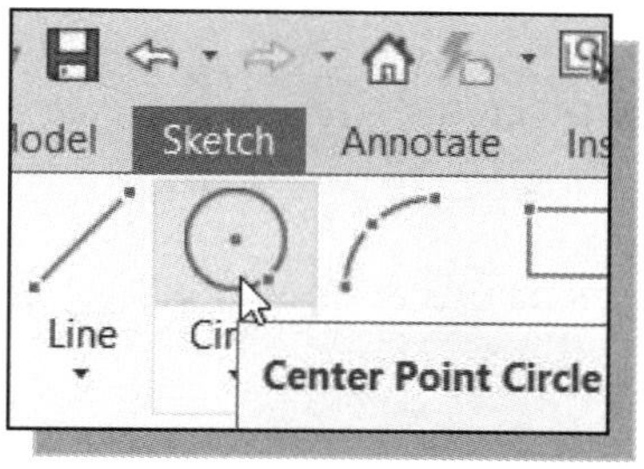

4. Choose **Center Point Circle** in the *Draw* toolbar.

5. On your own, create the two inner circles; the center points of the circles are coincident to the centers of the overall slot. (Attach the center to the centerline.)

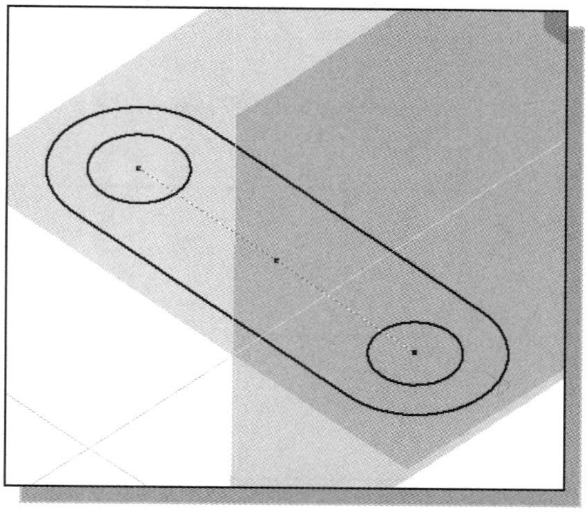

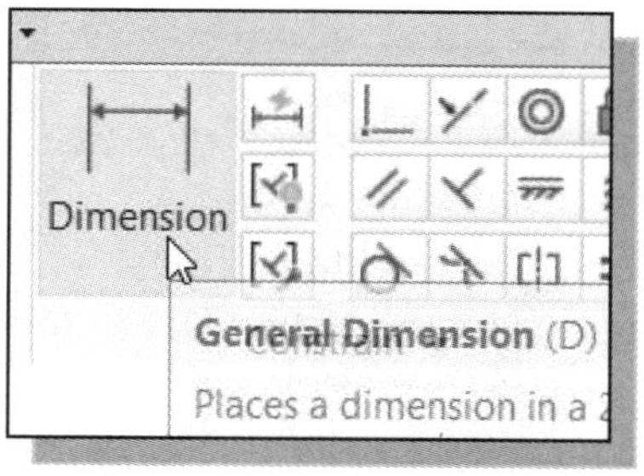

6. Left-click once on the **General Dimension** icon to activate the General Dimension command.

7. On your own, set the display orientation to the Top view position as shown in the below figure.

8. On your own, create the dimensions, referencing the origin point, to fully constrain the sketch as shown. (Do not be overly concerned with the actual numbers displayed; the dimensions will be adjusted in the next steps.)

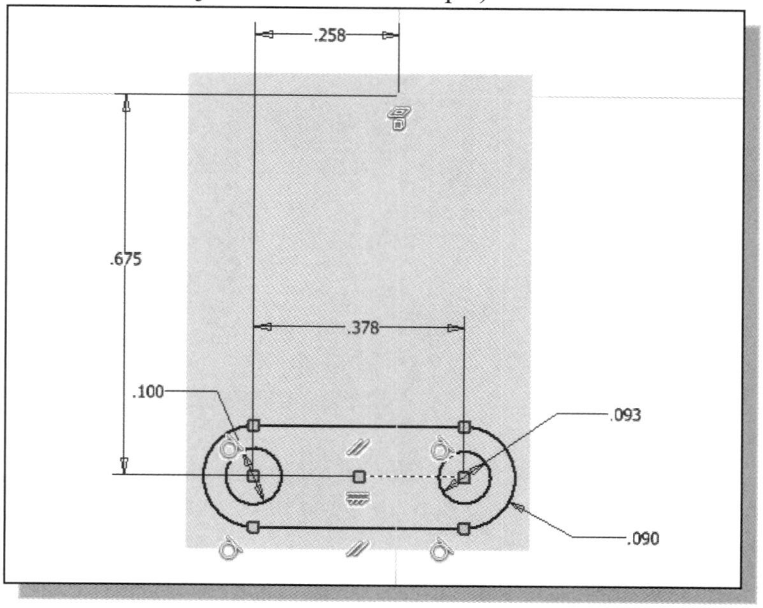

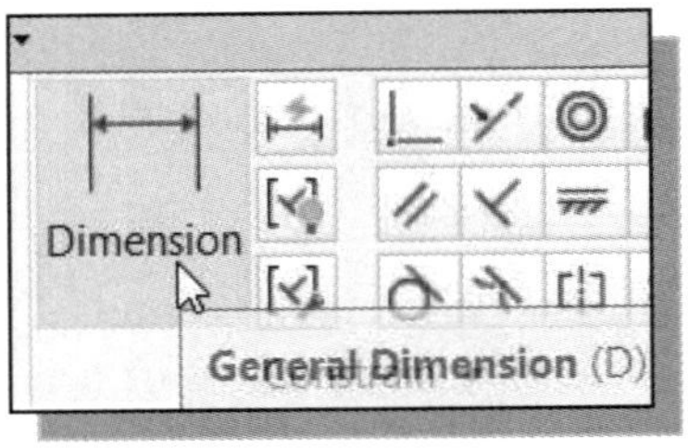

9. Choose **General Dimension** in the *Constrain* panel.

10. On your own, adjust the dimensions as shown in the figure.

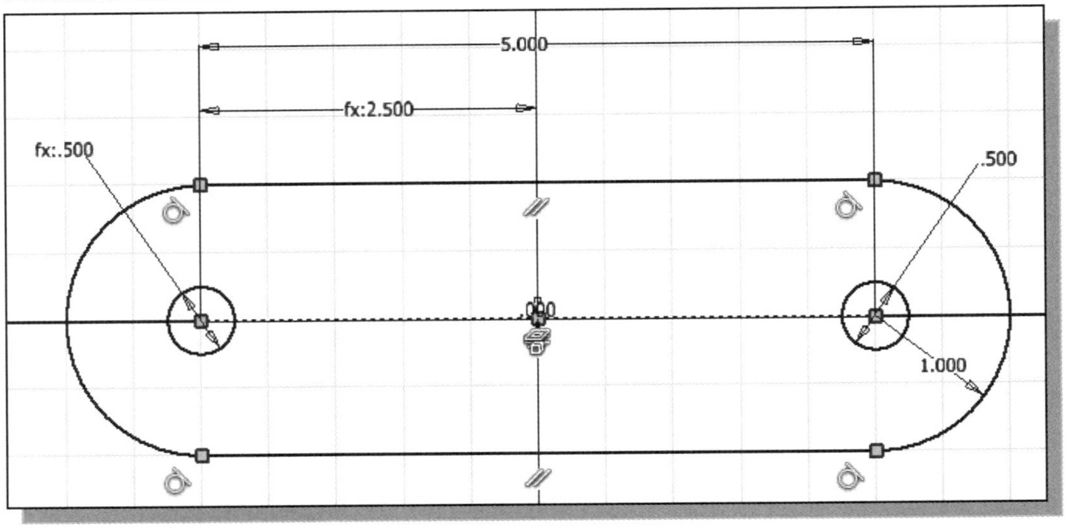

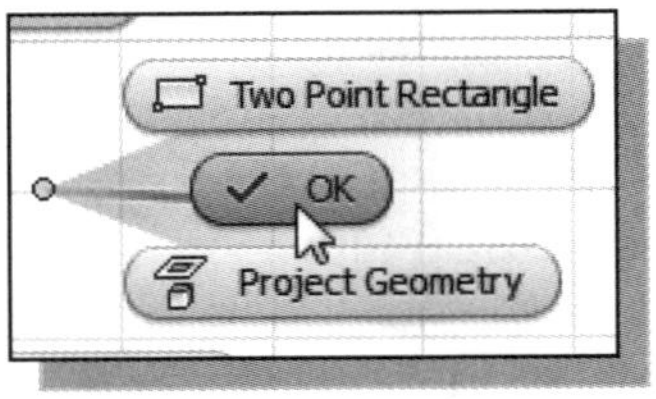

11. Inside the graphics window, click once with the right-mouse-button to display the option menu. Select **OK** in the pop-up menu to end the **General Dimension** command.

12. On your own, use the **Parameters** command to examine the two parametric equations used.

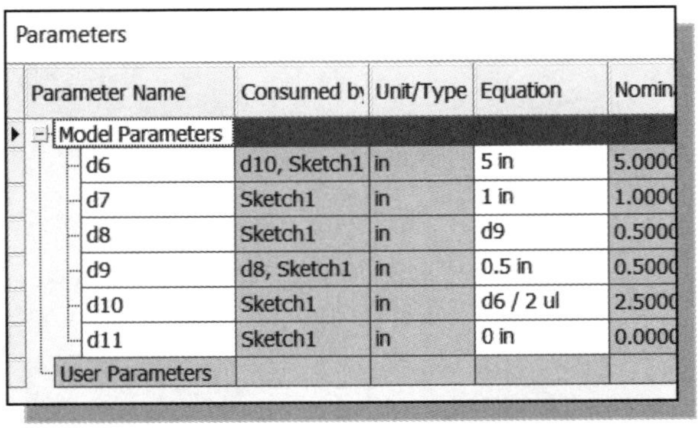
Parameters

Parameter Name	Consumed b	Unit/Type	Equation	Nomin
Model Parameters				
d6	d10, Sketch1	in	5 in	5.000
d7	Sketch1	in	1 in	1.000
d8	Sketch1	in	d9	0.500
d9	d8, Sketch1	in	0.5 in	0.500
d10	Sketch1	in	d6 / 2 ul	2.500
d11	Sketch1	in	0 in	0.000
User Parameters				

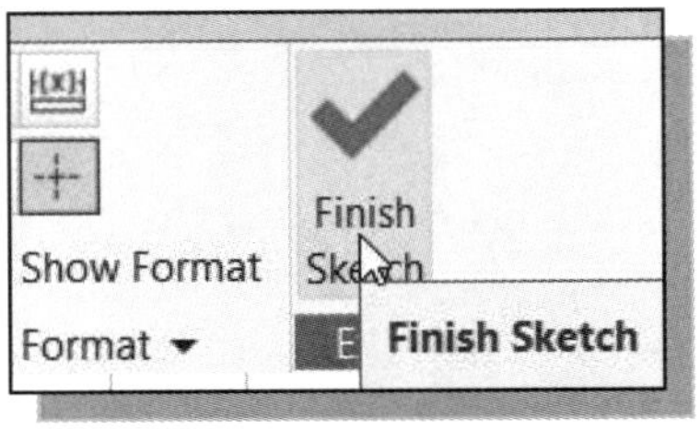

13. In the *Ribbon toolbar* area, select **Finish Sketch** in the pop-up menu to end the **Sketch** option.

Create the First Extrude Feature

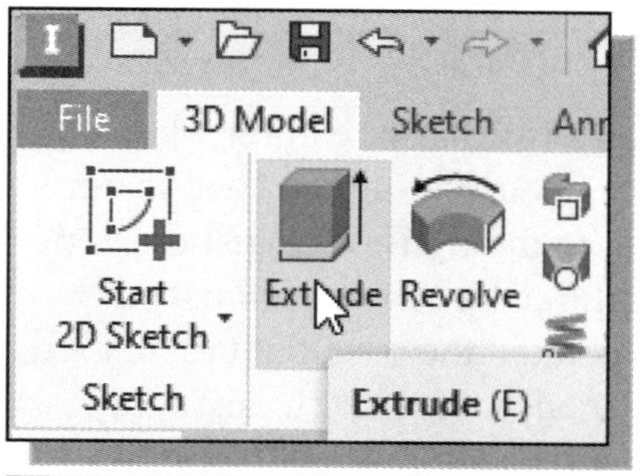

1. In the *Create* toolbar, select the **Extrude** command by clicking once with the left-mouse-button on the icon.

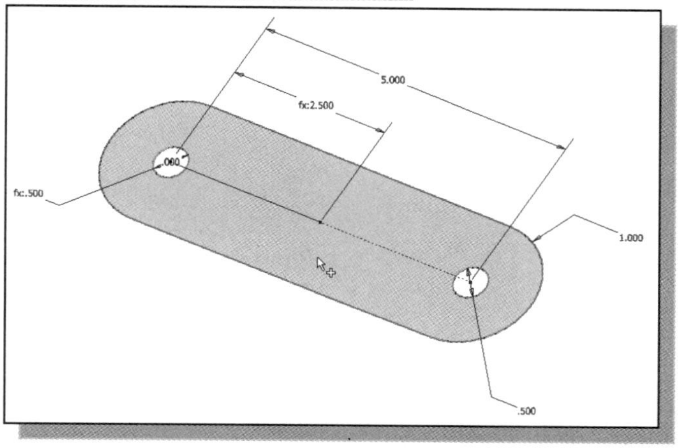

2. Select the inside regions of the sketch to define the profile of the extrusion as shown.

3. In the *Distance* option box, enter **0.5** as the extrusion distance.

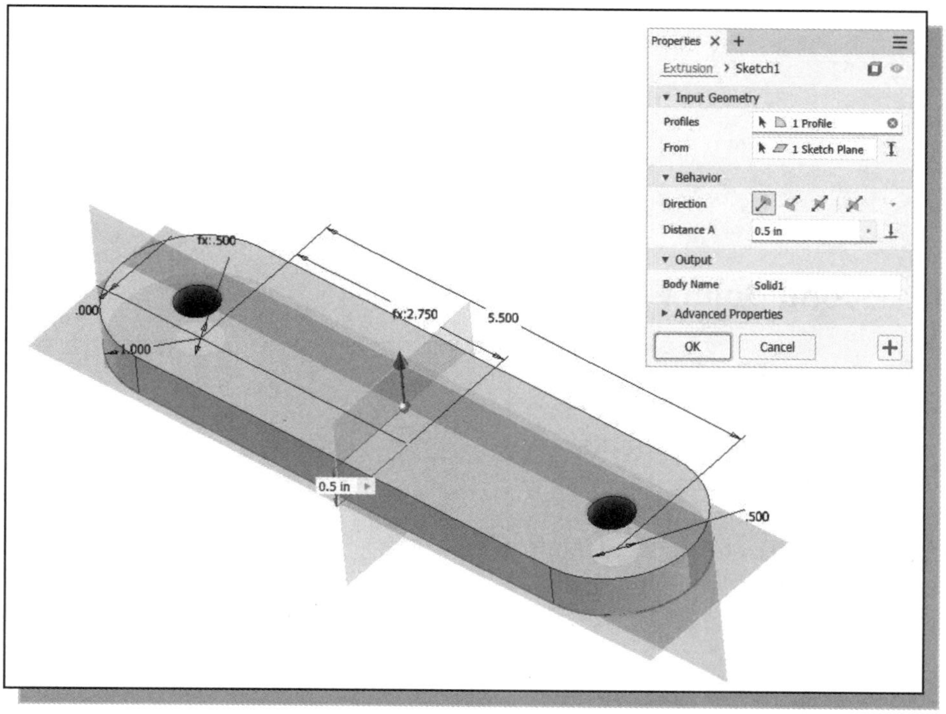

4. In the *Extrude* pop-up control, click on the **OK** button to create the base feature.

The Implied Parent/Child Relationships

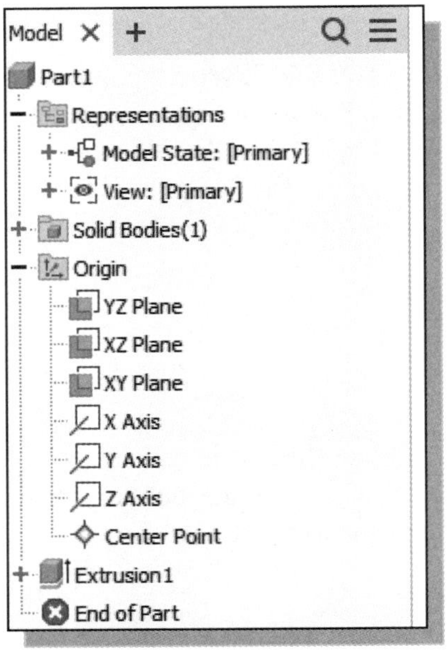

- The *Model Tree* shows two main features: the ***Origin*** (default work features) and the base feature (**Extrusion1** shown in the figure) we just created. The parent/child relationships were established implicitly when we created the base feature: (1) XZ workplane was selected as the sketch plane; (2) Center Point was used as the reference point to align the 2D sketch of the base feature.

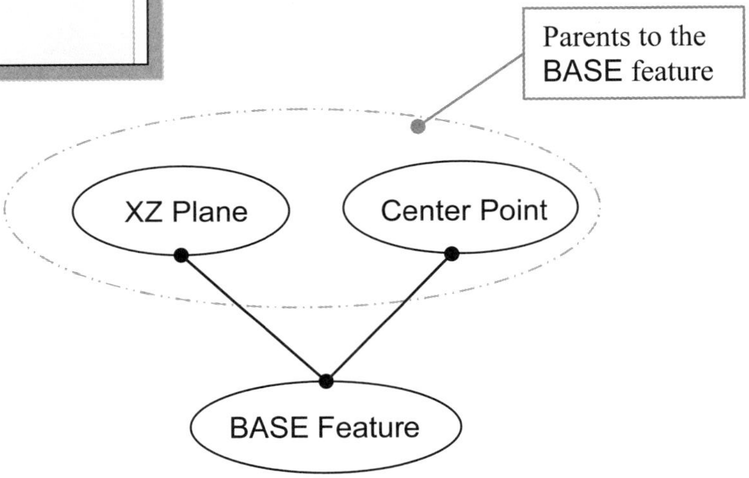

Create the Second Solid Feature

For the next solid feature, we will create the top section of the design. Note that the center of the base feature is aligned to the Center Point of the default work features. This was done intentionally so that additional solid features can be created referencing the default work features. For the second solid feature, the XY workplane will be used as the sketch plane.

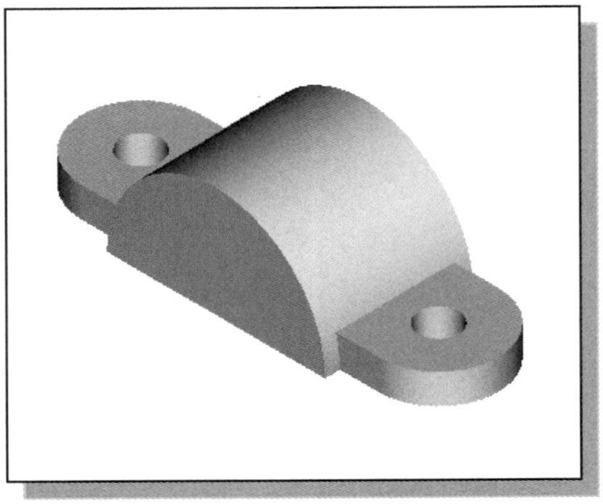

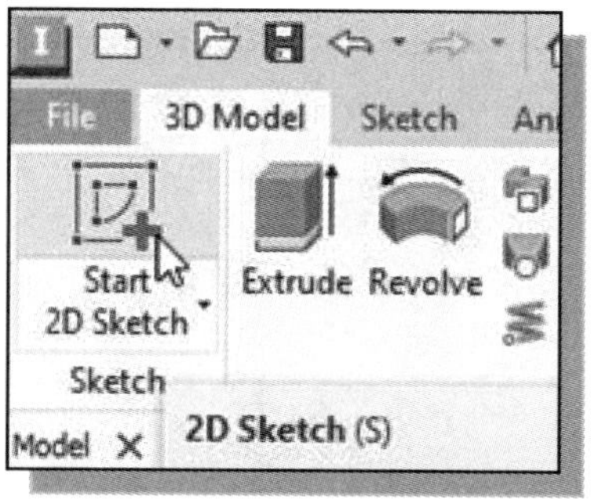

1. In the *Sketch* toolbar select the **Start 2D Sketch** command by left-clicking once on the icon.

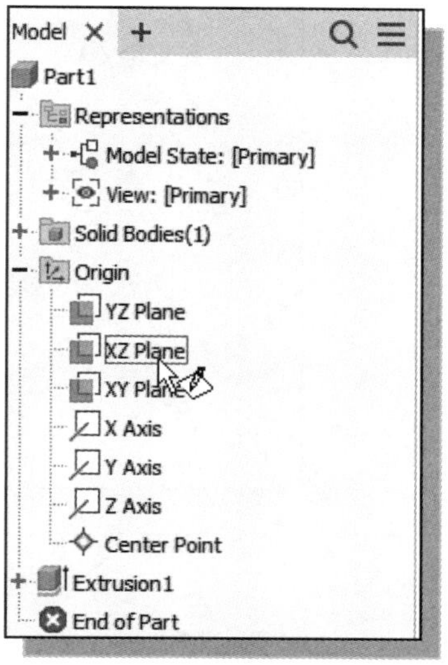

2. In the *Status Bar* area, the message "*Select face, work plane, sketch or sketch geometry.*" is displayed. Pick the **XY Plane** by clicking the work plane name inside the *browser* as shown.

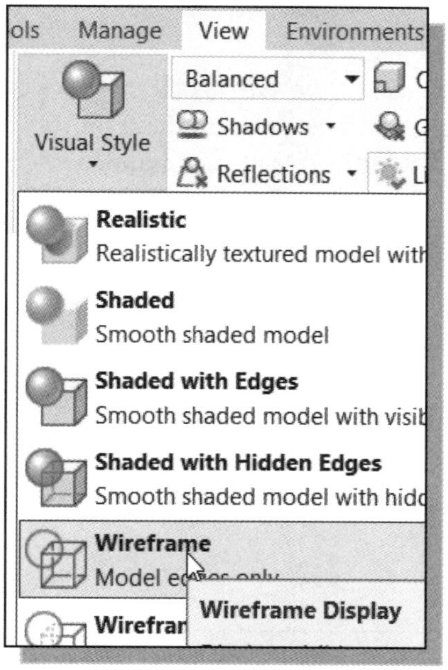

3. Select the **View** tab in the *Ribbon* panel as shown.

4. Select **Wireframe** display mode under the *Visual Style* icon as shown.

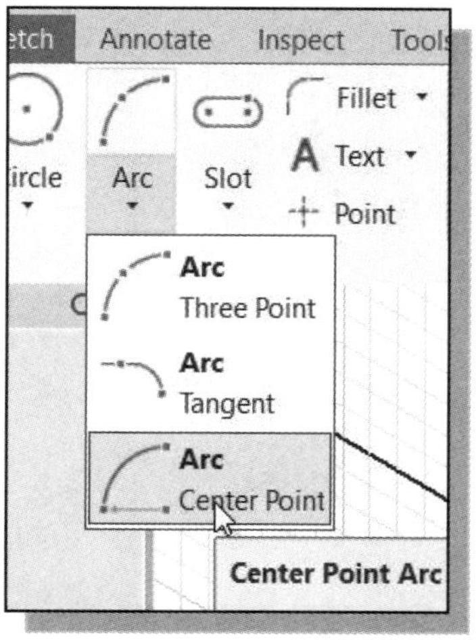

5. Select the **Sketch** tab in the *Ribbon*.

6. Select the **Center point arc** command by clicking once with the left-mouse-button on the icon in the *icon stack* as shown.

7. Pick the **center point** and watch for the Green dot for alignment as the center location of the new arc.

8. On your own, create a semi-circle of arbitrary size, with the center point aligned to the origin and both endpoints aligned to the X-axis, as shown below.

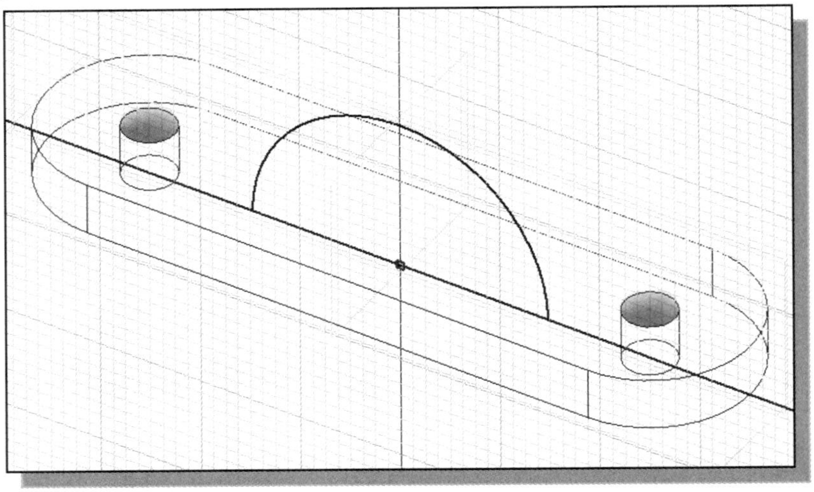

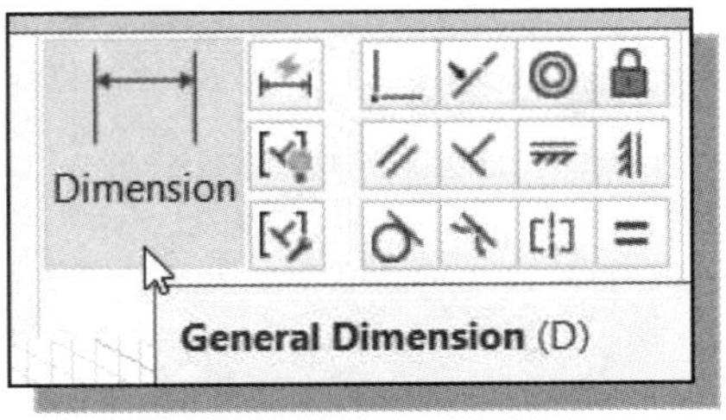

9. Move the cursor on top of the **General Dimension** icon. Left-click once on the icon to activate the General Dimension command.

10. On your own, create and adjust the radius of the arc to **1.75**.

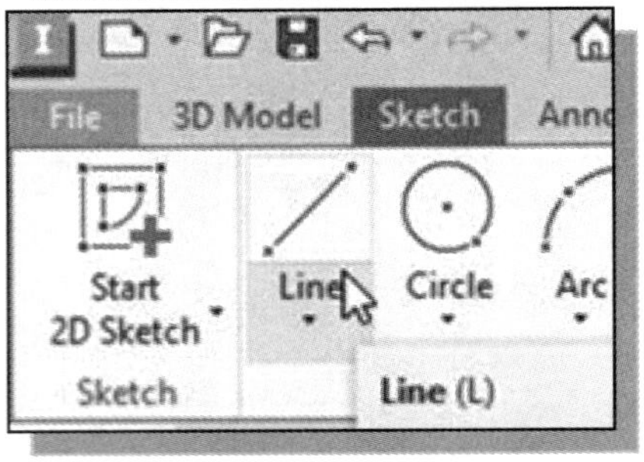

11. Select the **Line** command in the *Draw* toolbar.

12. Create a line connecting the two endpoints of the arc as shown in the figure below. (Hint: Use the **Coincident** constraint to help align the line to the origin if necessary.)

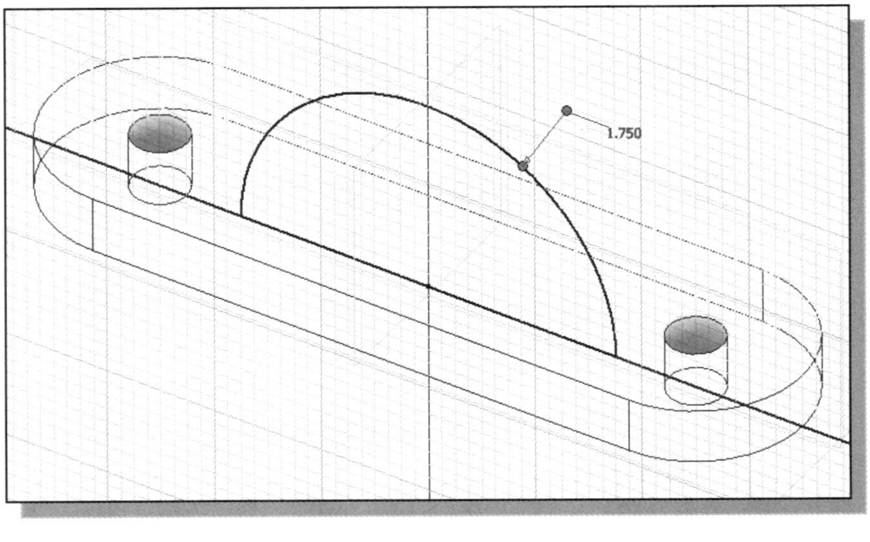

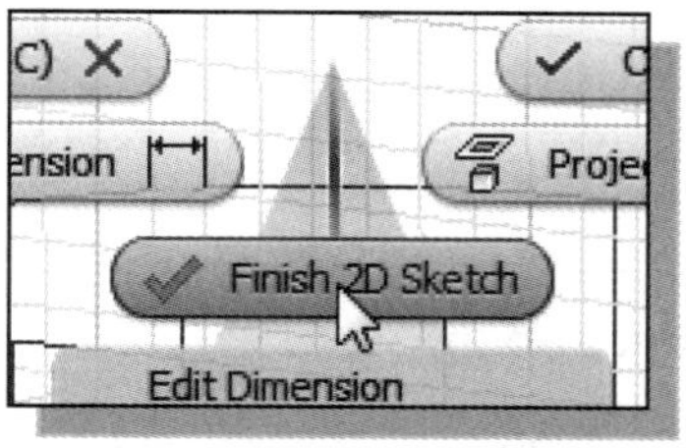

13. Inside the graphics window, click once with the right-mouse-button to display the option menu. Select **Finish 2D Sketch** in the pop-up menu to end the Sketch option.

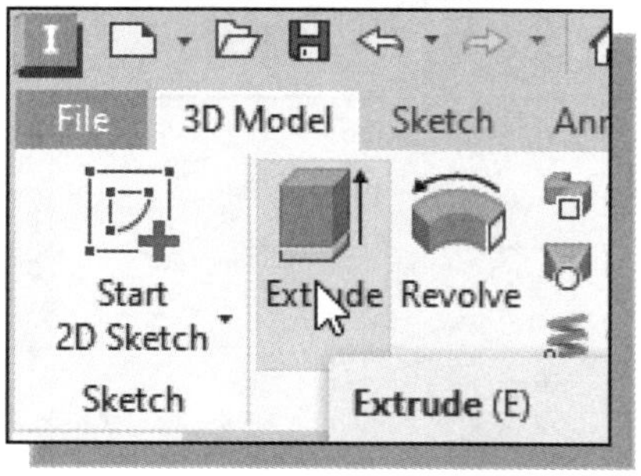

14. In the *Create* toolbar, select the **Extrude** command by clicking once with the left-mouse-button on the icon.

15. Select the inside region of the sketched arc-line curves as the profile to be extruded.

16. In the *Extrude* dialog box, set to the **Symmetric** option.

17. In the *Distance* value box, set the extrusion distance to **2.5**.

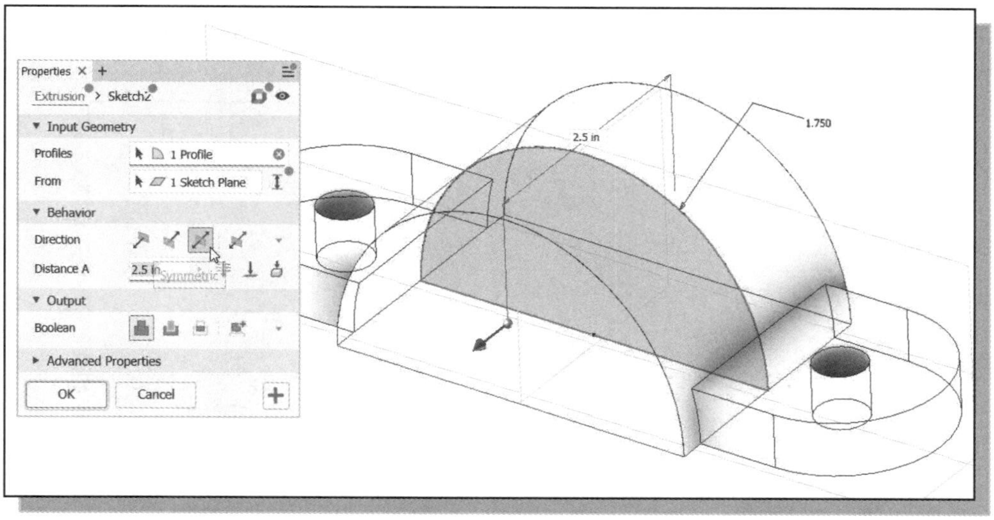

18. Click on the **OK** button to proceed with the Extrude operation.

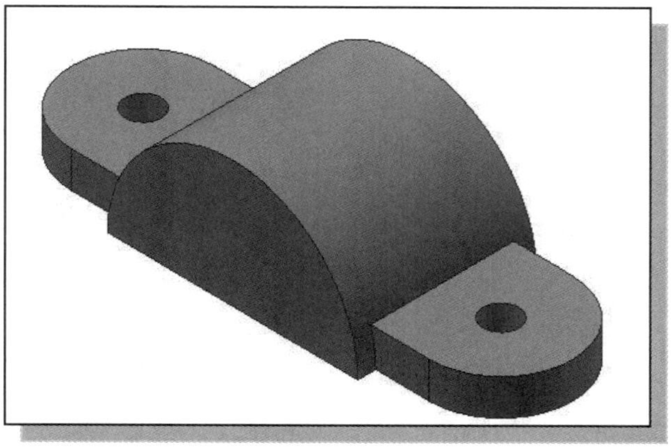

Create a Cut Feature

A rectangular cut will be created as the next solid feature.

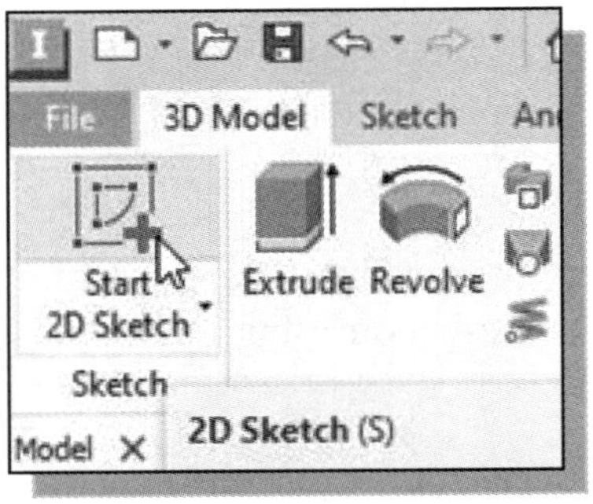

1. In the *Sketch* toolbar select the **Start 2D Sketch** command by left-clicking once on the icon.

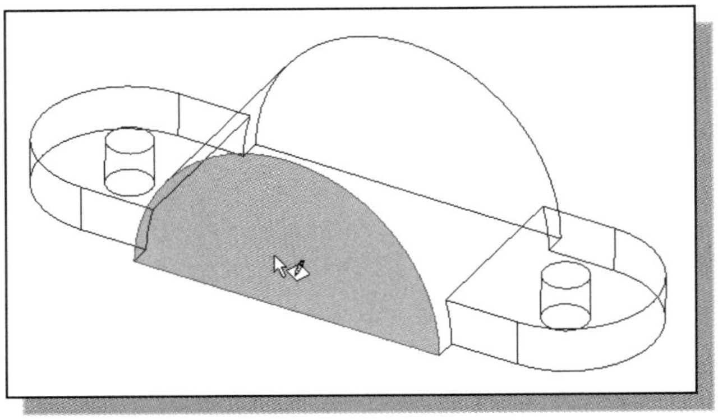

2. In the *Status Bar* area, the message "*Select plane to create sketch or an existing sketch to edit*" is displayed. Pick the front vertical face of the solid model shown.

3. On your own, create a rectangle (2.0 X 1.0) and the associated dimensions as shown below. (Hint: Use the **YZ Plane** to assure the proper alignment of the 2D sketch.)

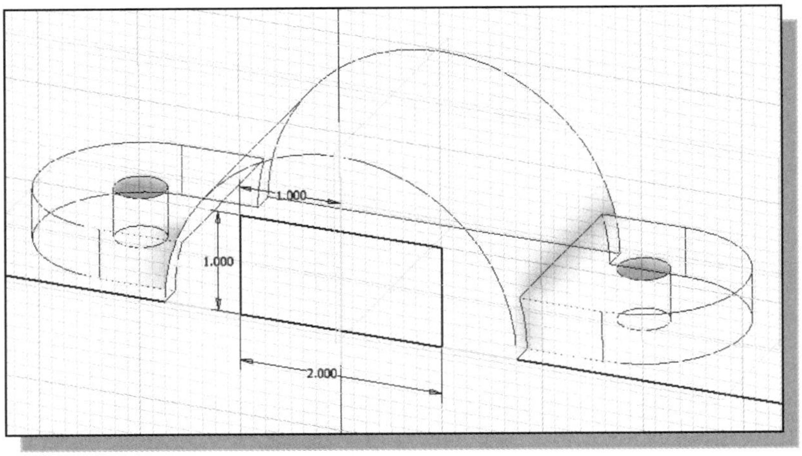

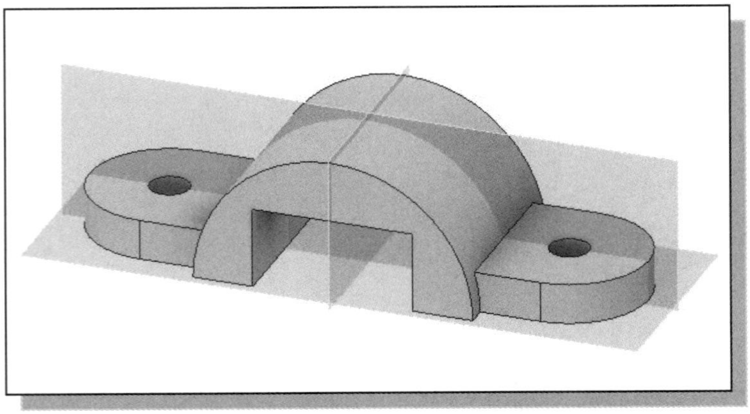

4. On your own, use the **Extrude** command and create a cutout that cuts through the entire 3D solid model as shown.

The Second Cut Feature

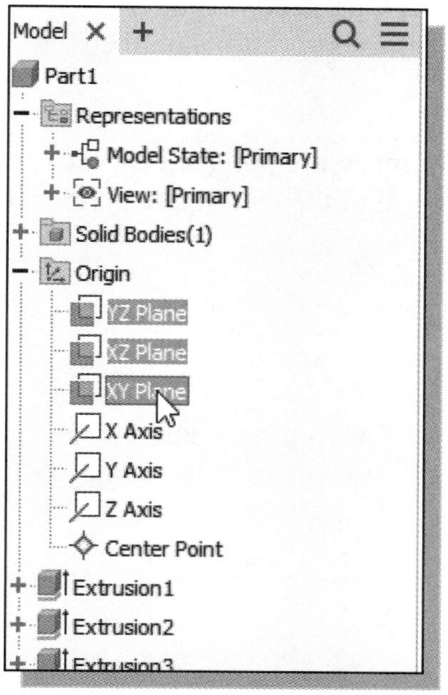

1. In the *Model Tree area*, select all of the **Datum Planes** while holding down the [**SHIFT**] key.

2. *Right-click* on any of the selected items to display the option menu and turn **OFF** the visibility of the selected items as shown.

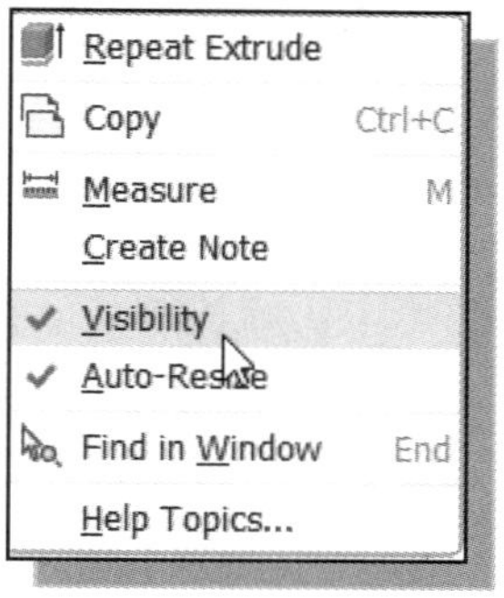

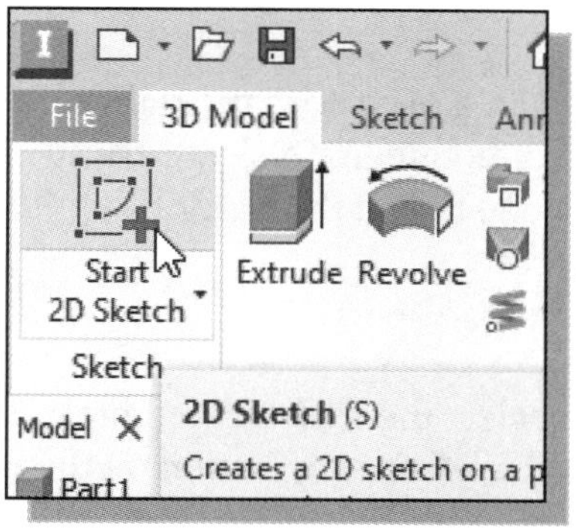

3. In the *Sketch* toolbar select the **Start 2D Sketch** command by left-clicking once on the icon.

4. In the *Status Bar* area, the message "*Select face, work plane, sketch or sketch geometry*" is displayed. Select the horizontal face of the last cut feature as the *sketching plane.*

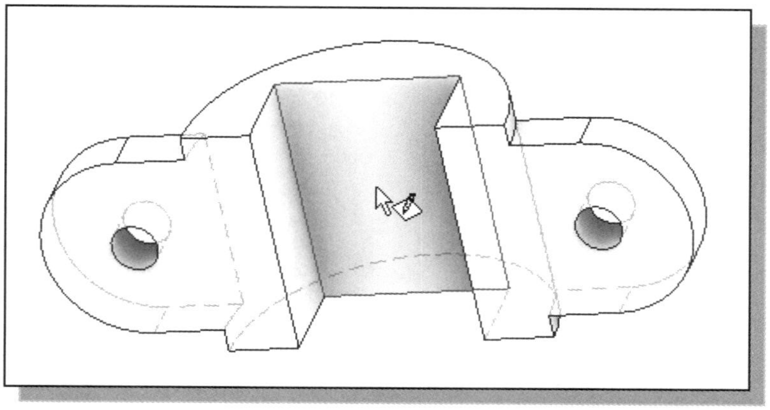

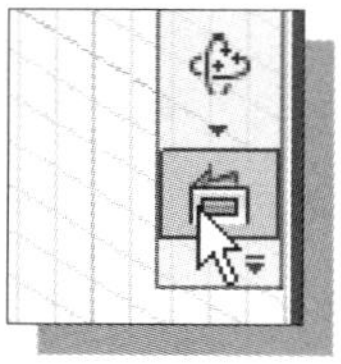

5. In the *Standard* toolbar area, click on the **Look At** button.

6. Select the highlighted sketching plane as shown.

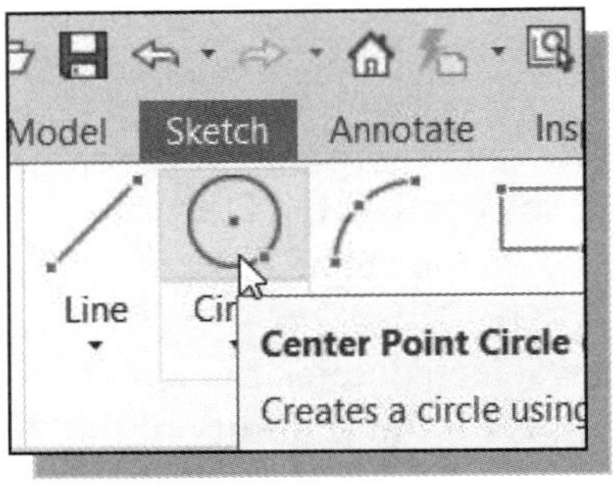

7. Select the **Center Point Circle** command by clicking once with the left-mouse-button on the icon in the *Draw* panel.

8. Pick the **center point**; this will align the center of the new circle to the origin of the coordinate system.

9. On your own, create a circle of arbitrary size.

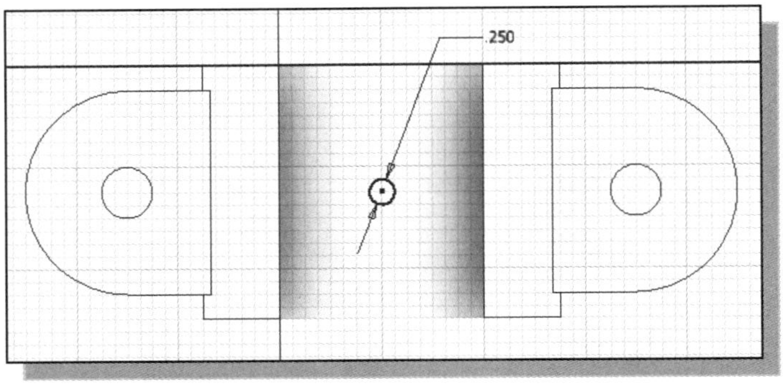

10. On your own, add the size dimension of the circle and set the dimension to **0.25**.

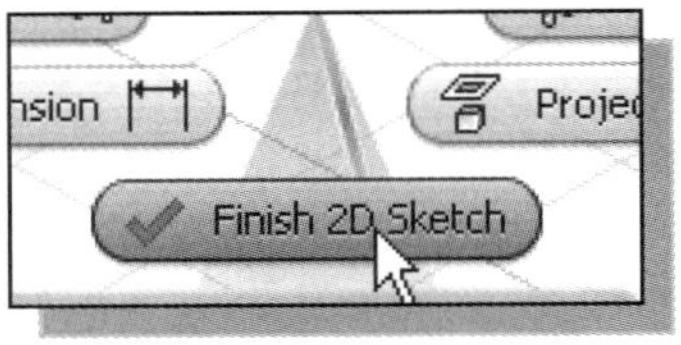

11. Inside the graphics window, click once with the right-mouse-button to display the option menu. Select **Finish 2D Sketch** in the pop-up menu to end the Sketch option.

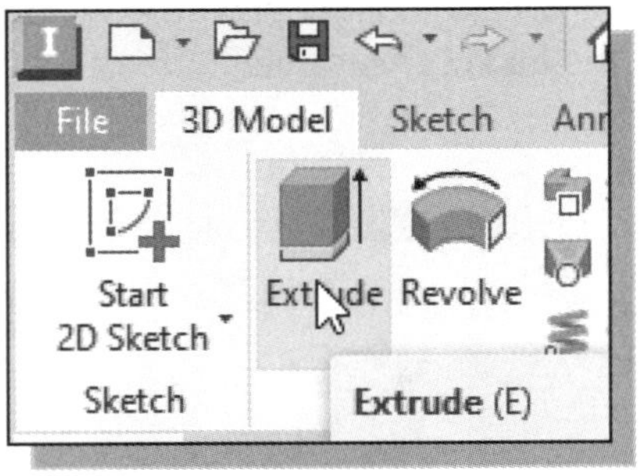

12. In the *Create toolbar*, select the **Extrude** command by clicking once with the left-mouse-button on the icon.

13. Select the inside region of the sketched circle as the profile to be extruded.

14. In the *Extrude* control panel, set to the **Cut – Through All** option.

15. Click on the **OK** button to proceed with creating the cut feature.

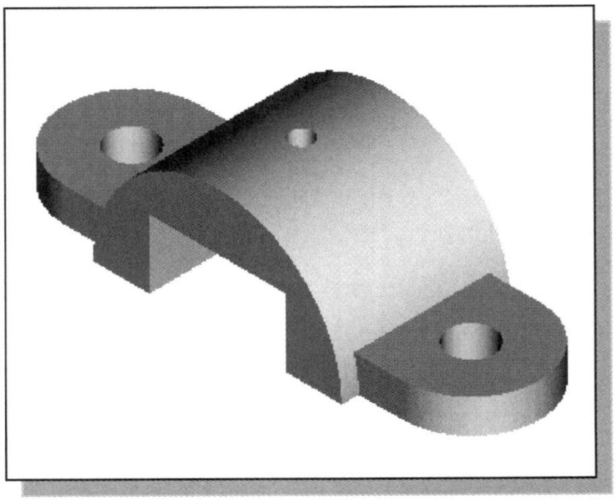

Examine the Parent/Child Relationships

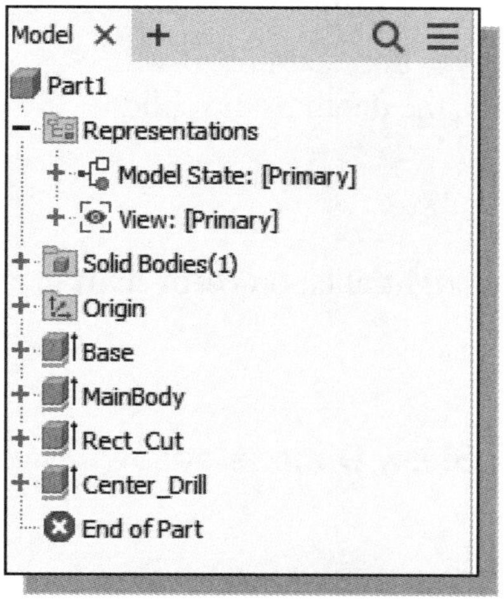

1. On your own, rename the feature names to **Base**, **MainBody**, **Rect_Cut** and **Center_Drill** as shown in the figure.

The *Model Tree* window now contains seven items: the ***Origin*** (default work features) and four solid features. All of the parent/child relationships were established implicitly as we created the solid features. As more features are created, it becomes much more difficult to make a sketch showing all the parent/child relationships involved in the model. On the other hand, it is not really necessary to have a detailed picture showing all the relationships among the features. In using a feature-based modeler, the main emphasis is to consider the interactions that exist between the **immediate features**. Treat each feature as a unit by itself and be clear on the parent/child relationships for each feature. Thinking in terms of *features* is what distinguishes *feature-based modeling* and the previous generation solid modeling techniques. Let us take a look at the last feature we created, the **Center_Drill** feature. What are the parent/child relationships associated with this feature? (1) Since this is the last feature we created, it is not a parent feature to any other features. (2) Since we used one of the surfaces of the rectangular cutout as the sketching plane, the **Rect_Cut** feature is a parent feature to the **Center_Drill** feature. (3) We also used the Origin as a reference point to align the center; therefore, the ***Origin*** is also a parent to the **Center_Drill** feature.

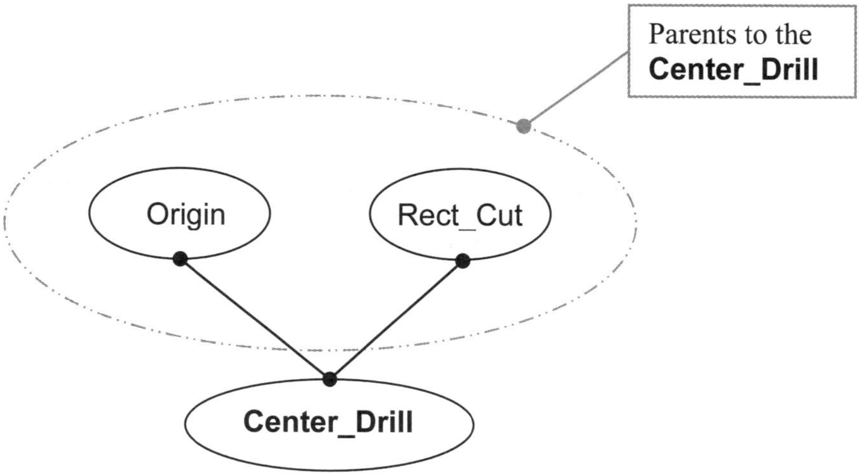

Modify a Parent Dimension

Any changes to the parent features will affect the child feature. For example, if we modify the height of the **Rect_Cut** feature from 1.0 to 0.75, the depth of the child feature (**Center_Drill** feature) will be affected.

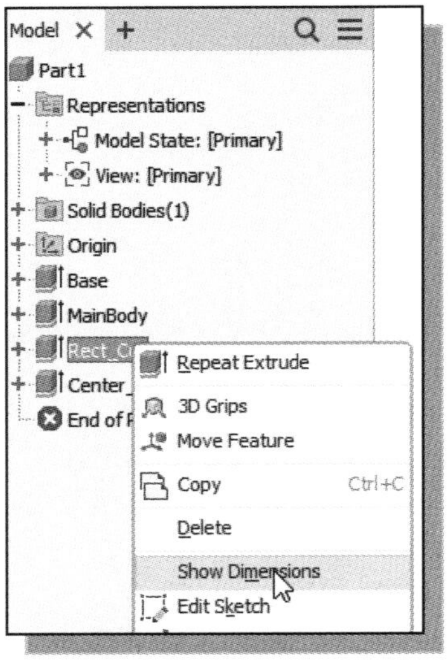

1. In the *Model Tree* window, right-click on **Rect_Cut** to bring up the option menu.

2. In the option menu, select **Show Dimensions**.

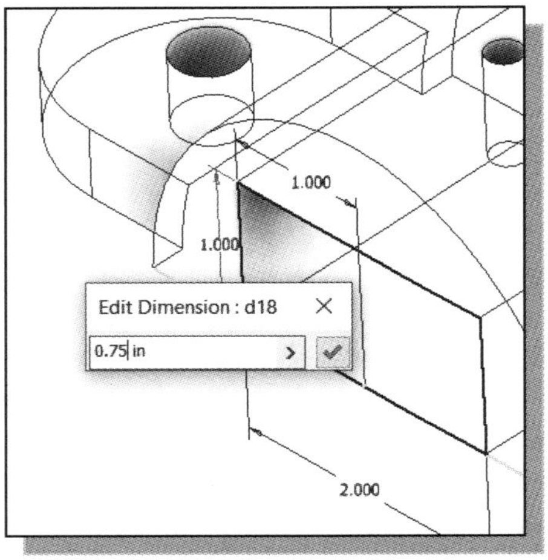

3. Select the height dimension (1.0) by double-clicking on the dimension.

4. Enter **0.75** as the new height dimension as shown.

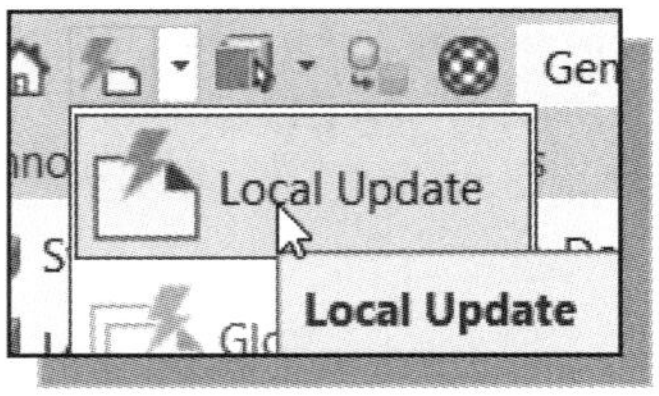

5. Click on the **Update** button in the *Quick Access Toolbar* area to proceed with updating the solid model.

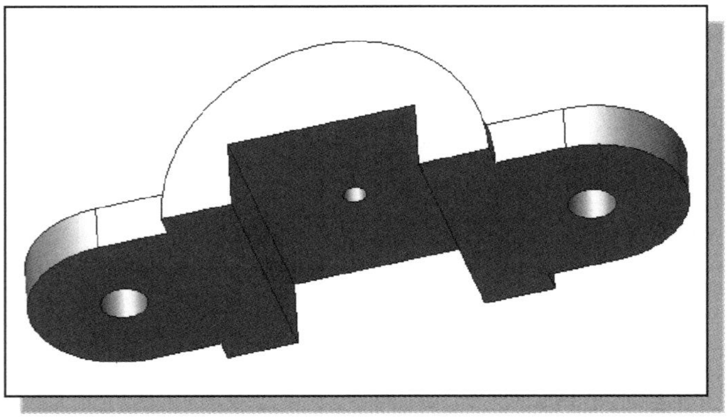

➢ Note that the position of the **Center_Drill** feature is also adjusted as the placement plane is lowered. The drill-hole still goes through the main body of the *U-Bracket* design. The parent/child relationship assures the intent of the design is maintained.

6. On your own, adjust the height of the **Rect_Cut** feature back to **1.0** inch before proceeding to the next section.

A Design Change

Engineering designs usually go through many revisions and changes. For example, a design change may call for a circular cutout instead of the current rectangular cutout feature in our model. Autodesk Inventor provides an assortment of tools to handle design changes quickly and effectively. In the following sections, we will demonstrate some of the more advanced tools available in Autodesk Inventor, which allow us to perform the modification of changing the rectangular cutout (2.0 × 1.0 inch) to a circular cutout (radius: 1.25 inch).

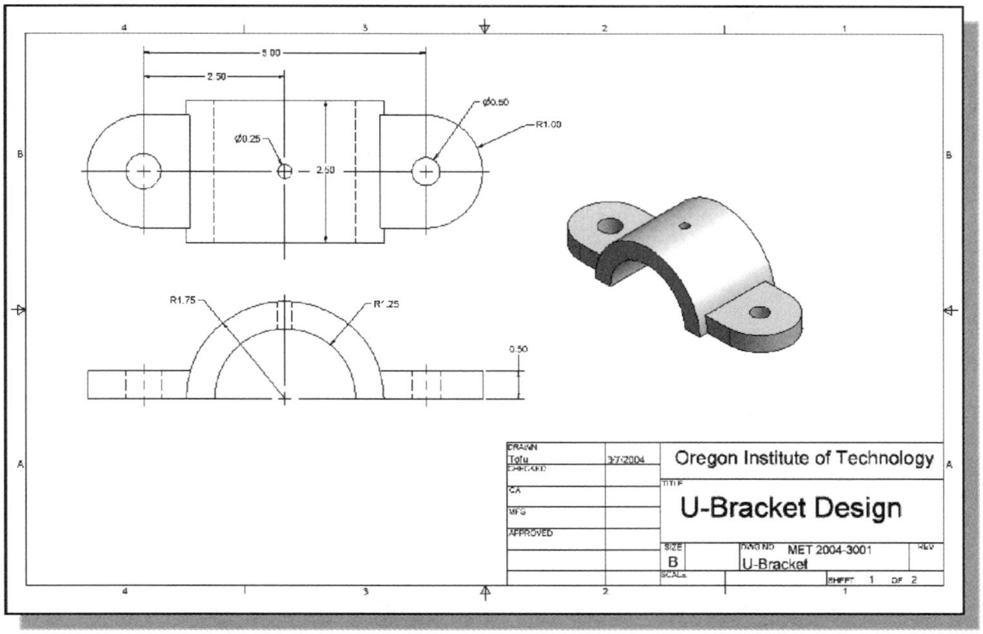

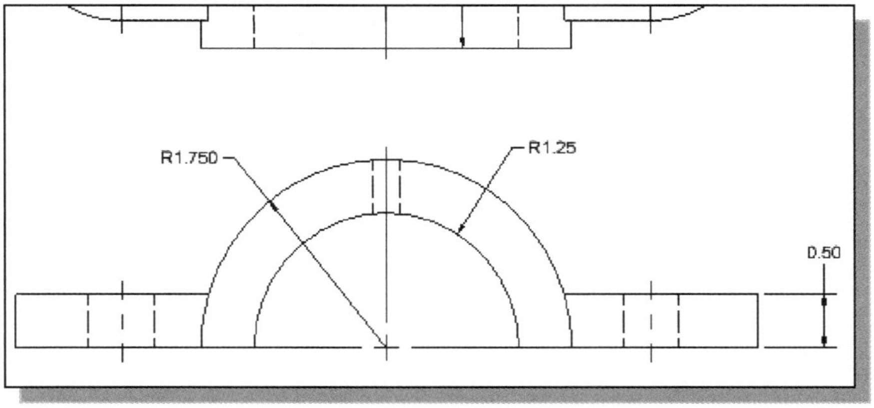

❖ Based on your knowledge of Autodesk Inventor so far, how would you accomplish this modification? What other approaches can you think of that are also feasible? Of the approaches you came up with, which one is the easiest to do and which is the most flexible? If this design change was anticipated right at the beginning of the design process, what would be your choice in arranging the order of the features? You are encouraged to perform the modifications prior to following through the rest of the tutorial.

Feature Suppression

With Autodesk Inventor, we can take several different approaches to accomplish this modification. We could (1) create a new model, or (2) change the shape of the existing cut feature using the **Redefine** command, or (3) perform **feature suppression** on the rectangular cut feature and add a circular cut feature. The third approach offers the most flexibility and requires the least amount of editing to the existing geometry. **Feature suppression** is a method that enables us to disable a feature while retaining the complete feature information; the feature can be reactivated at any time. Prior to adding the new cut feature, we will first suppress the rectangular cut feature.

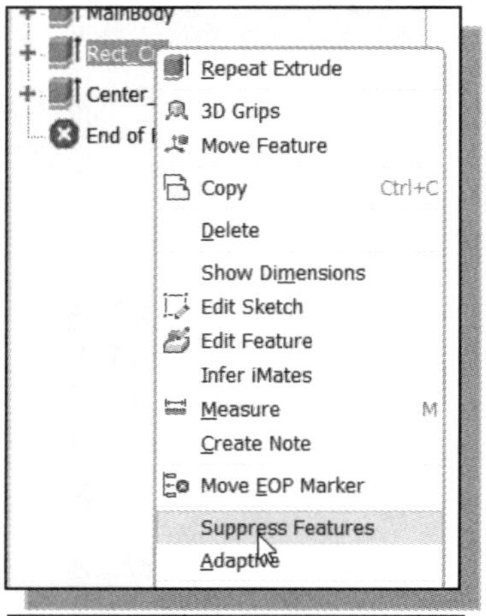

1. Move the cursor inside the *Model Tree* window. Click once with the right-mouse-button on **Rect_Cut** to bring up the option menu.

2. Pick **Suppress** in the pop-up menu.

❖ With the *Suppress* command, the Rect_Cut and Center_Drill features have disappeared in the display area. The child feature cannot exist without its parent(s), and any modification to the parent (Rect_Cut) influences the child (Center_Drill).

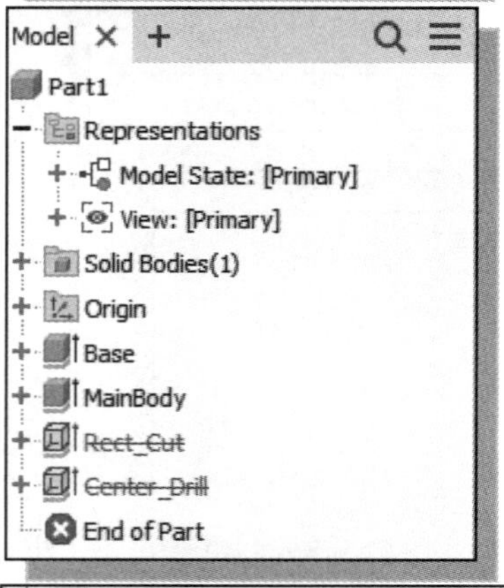

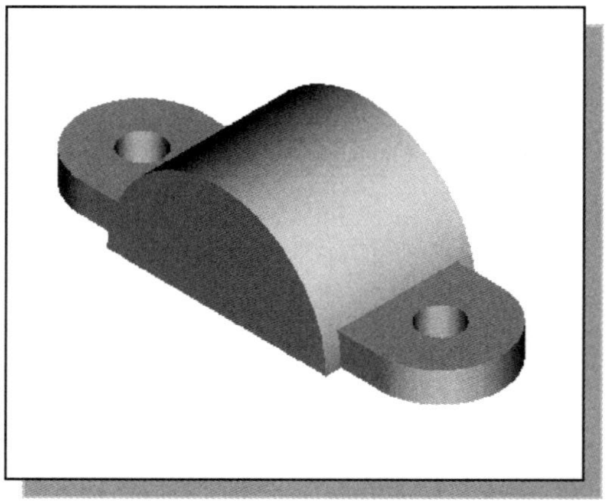

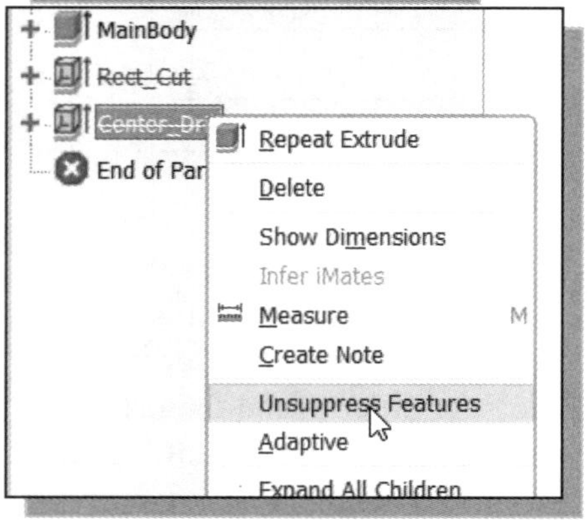

3. Move the cursor inside the *Model Tree* window. Click once with the right-mouse-button on top of **Center_Drill** to bring up the option menu.

4. Pick **Unsuppress Features** in the pop-up menu.

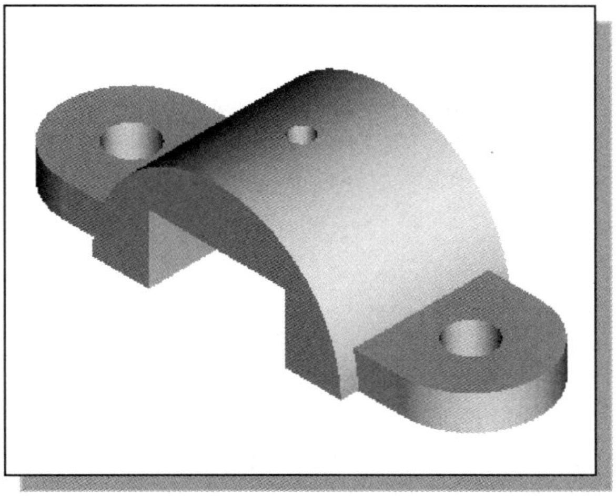

➢ In the display area and the *Model Tree* window, both the **Rect_Cut** feature and the **Center_Drill** feature are re-activated. The child feature cannot exist without its parent(s); the parent (Rect_Cut) must be activated to enable the child (Center_Drill).

A Different Approach to the Center_Drill Feature

The main advantage of using the BORN technique is to provide greater flexibility for part modifications and design changes. In this case, the Center_Drill feature can be placed on the XZ workplane and therefore not be linked to the Rect_Cut feature.

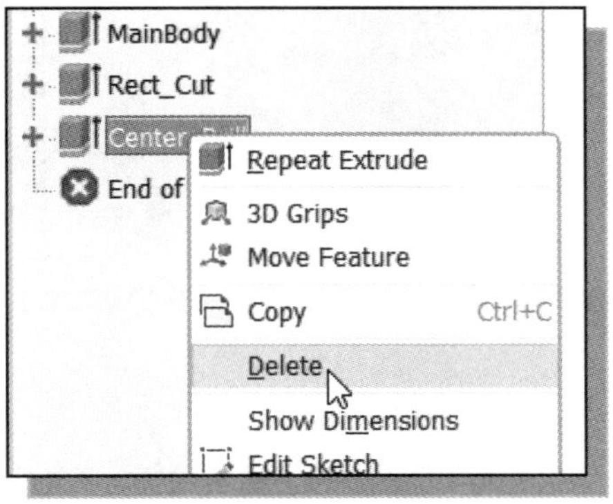

1. Move the cursor inside the *Model Tree* window. Click once with the **right-mouse-button** on top of **Center_Drill** to bring up the option menu.

2. Pick **Delete** in the pop-up menu.

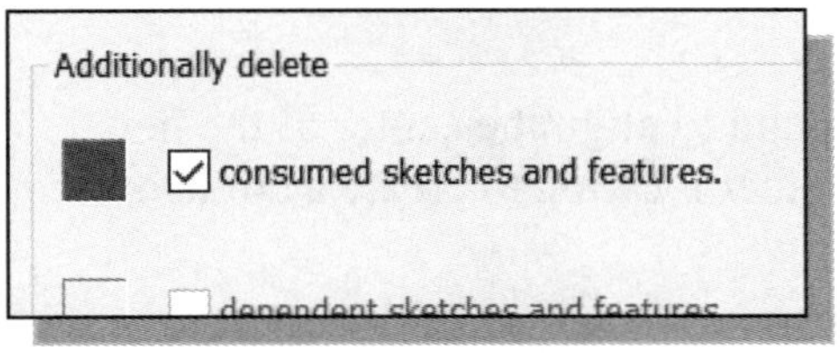

3. In the *Delete Features* window, confirm the **consumed sketches and features** option is switched ***ON***.

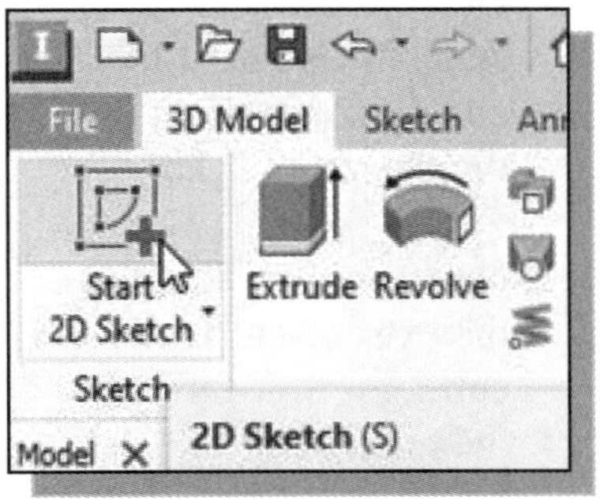

4. Click **OK** to proceed with the Delete command.

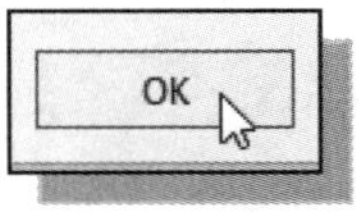

5. In the *Sketch* toolbar select the **Start 2D Sketch** command by left-clicking once on the icon.

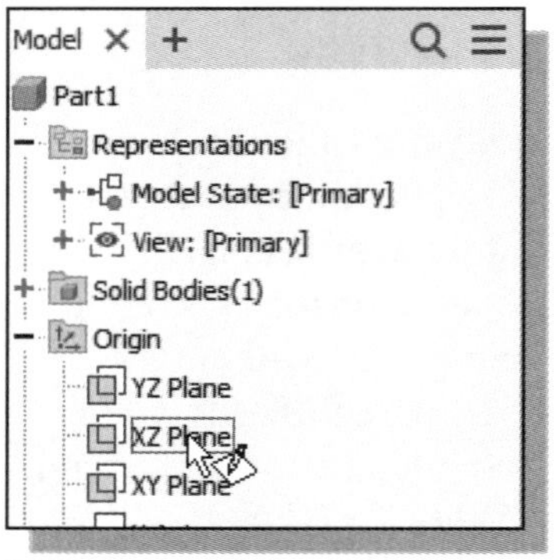

6. In the *Status Bar* area, the message "*Select plane to create sketch or an existing sketch to edit*" is displayed. Pick the **XZ Plane** in the *Model Tree* window as shown.

7. Inside the graphics area, single left-click to activate the **Home View** option as shown. The view will be adjusted back to the default ***isometric view***.

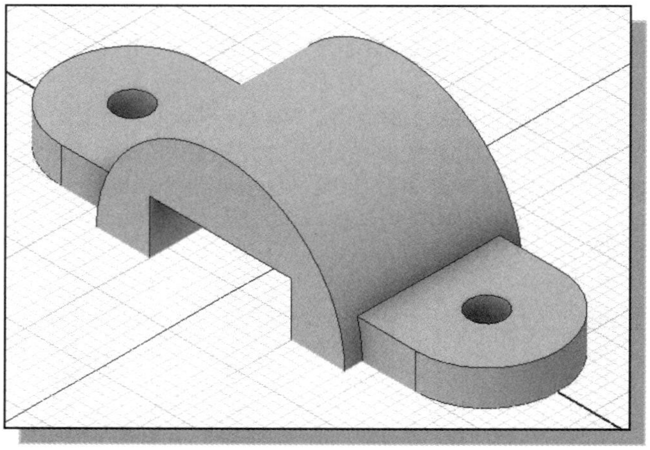

➢ Note the alignment of the sketch plane is set to the XZ plane as shown.

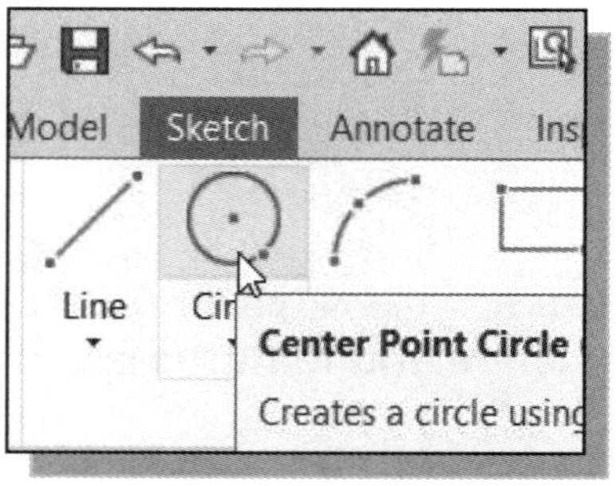

8. Select the **Center Point Circle** command by clicking once with the left-mouse-button on the icon in the *Draw* panel.

9. Pick the center point to align the center of the new circle. Select another location to create a circle of arbitrary size.

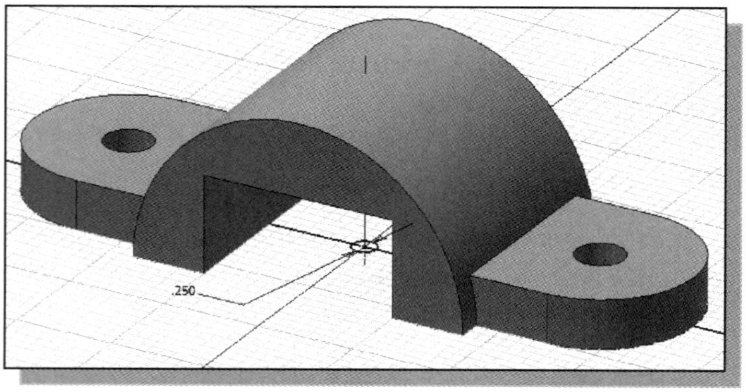

10. On your own, add the size dimension of the circle and set the dimension to **0.25**.

11. On your own, complete the extrude cut feature, cutting upward through the main body of the design.

Suppress the Rect_Cut Feature

Now the new **Center_Drill** feature is no longer a child of the **Rect_Cut** feature, any changes to the **Rect_Cut** feature do not affect the **Center_Drill** feature.

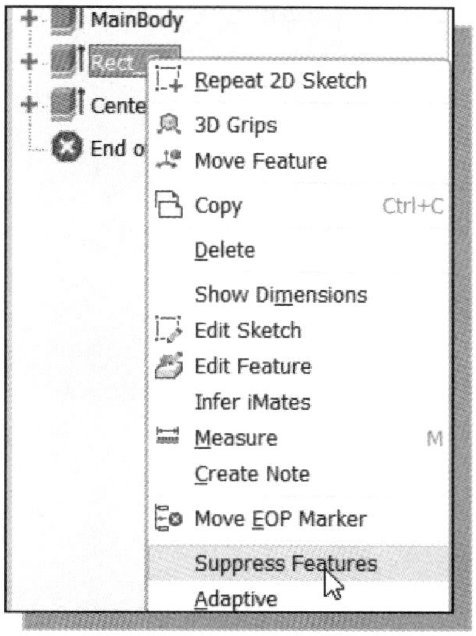

1. Move the cursor inside the *Model Tree* window. Click once with the right-mouse-button on top of **Rect_Cut** to bring up the option menu.

2. Pick **Suppress Features** in the pop-up menu.

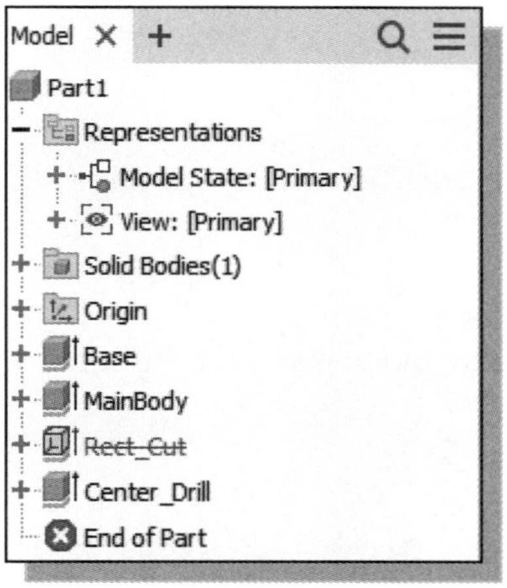

❖ The **Rect_Cut** feature is now disabled without affecting the ***new*** **Center_Drill** feature.

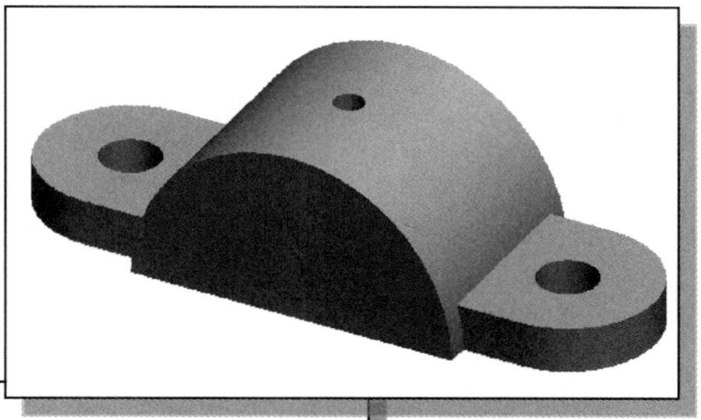

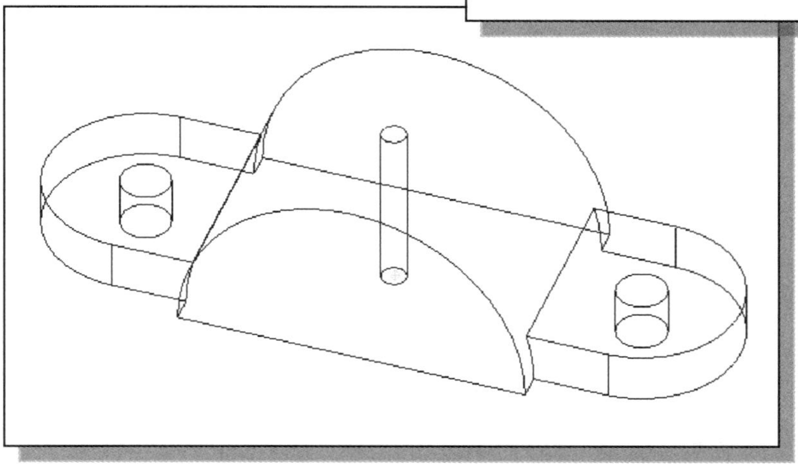

Create a Circular Cut Feature

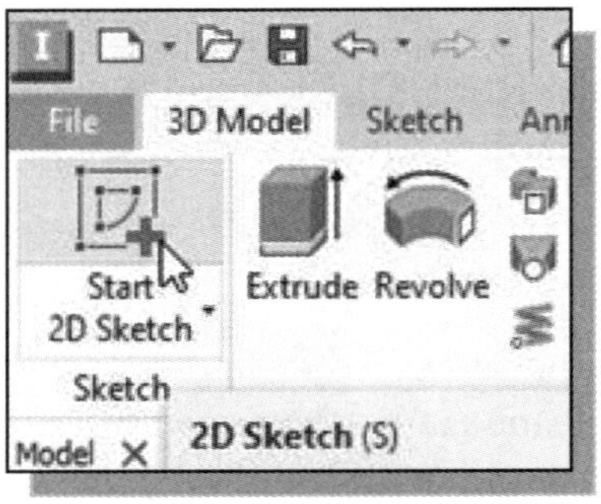

1. In the *Sketch* toolbar select the **Start 2D Sketch** command by left-clicking once on the icon.

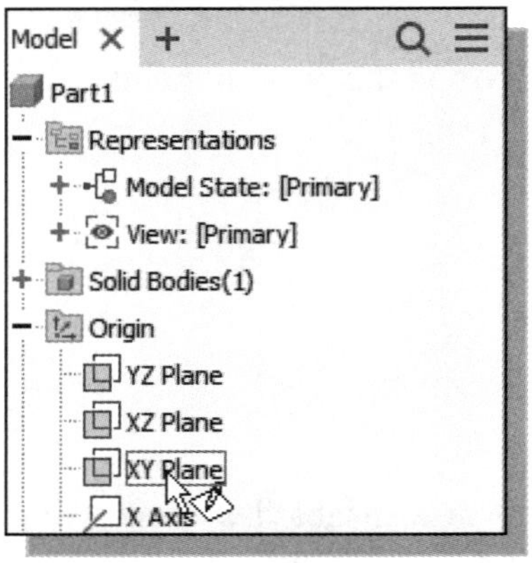

2. In the *Status Bar* area, the message "*Select face, work plane, sketch or sketch geometry*" is displayed. Pick the **XY Plane** in the *Model Tree* window as shown.

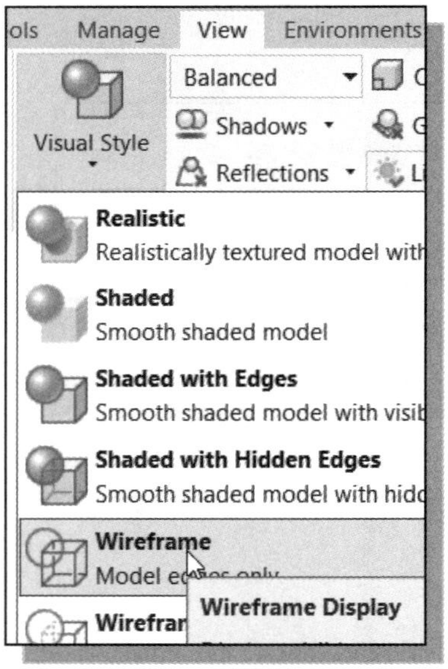

3. Activate the **View** tab in the *Ribbon* panel as shown.

4. Select **Wireframe** display mode under the *Visual Style* icon as shown.

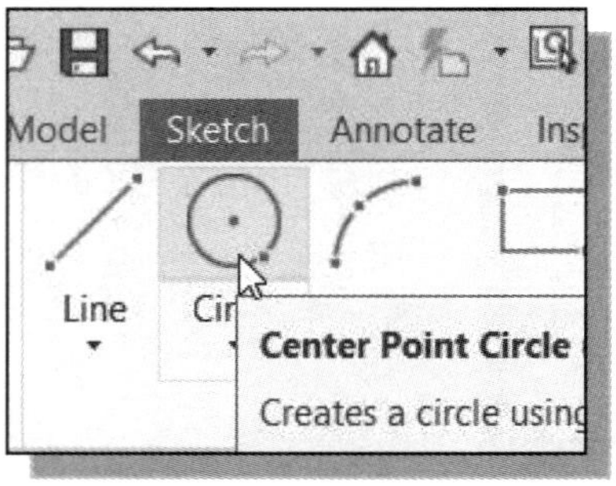

5. Select the **Center Point Circle** command by clicking once with the left-mouse-button on the icon in the *Draw* panel.

6. Pick the **center point** at the origin to align the center of the new circle.

7. On your own, create a circle of arbitrary size.

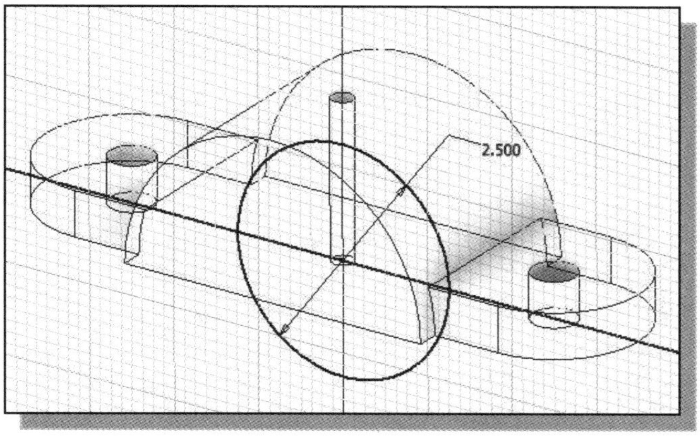

8. On your own, create the size dimension of the circle and set the dimension to **2.5** as shown in the figure.

9. On your own, complete the **cut** feature using the **Symmetric - All** command.

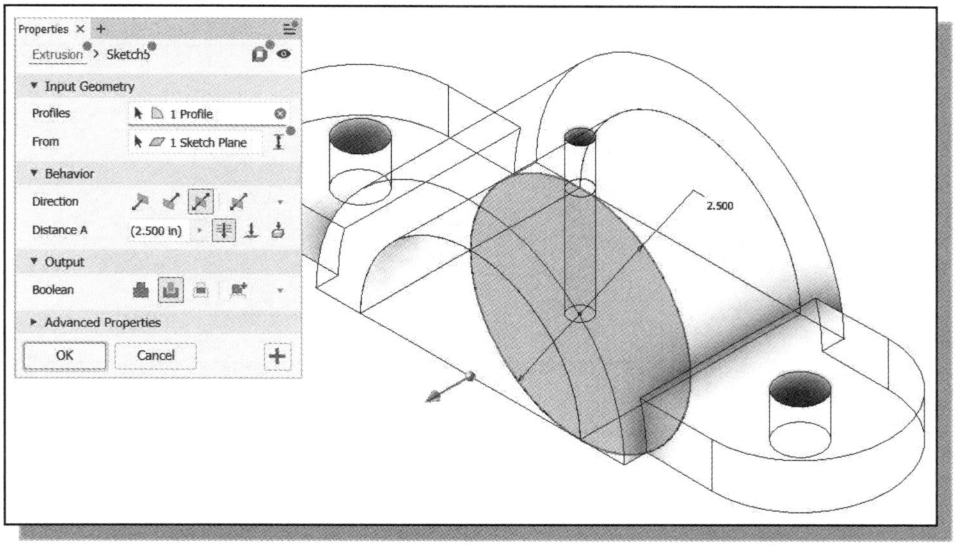

❖ Note that the parents of the Circular_Cut feature are the XY Plane and the Center Point.

➢ On your own, save the model as **U-Bracket**; this model will be used again in the next chapter.

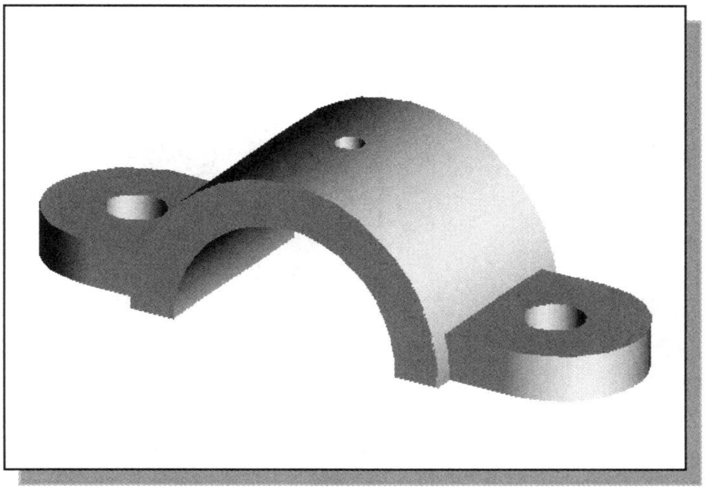

A Flexible Design Approach

In a typical design process, the initial design will undergo many analyses, tests, reviews and revisions. Autodesk Inventor allows the users to quickly make changes and explore different options of the initial design throughout the design process.

The model we constructed in this chapter contains two distinct design options. The *feature-based parametric modeling* approach enables us to quickly explore design alternatives, and we can include different design ideas into the same model. With parametric modeling, designers can concentrate on improving the design, and the design process becomes quicker and requires less effort. The key to successfully using parametric modeling as a design tool lies in understanding and properly controlling the interactions of features, especially the parent/child relations.

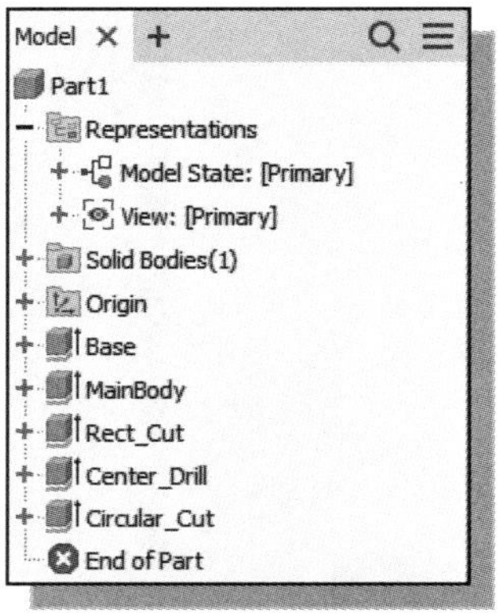

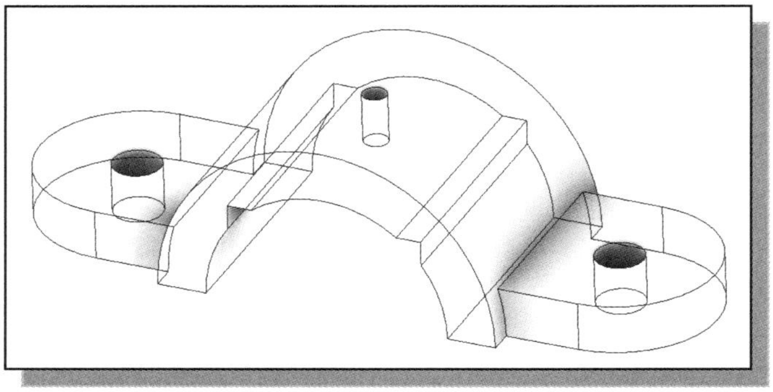

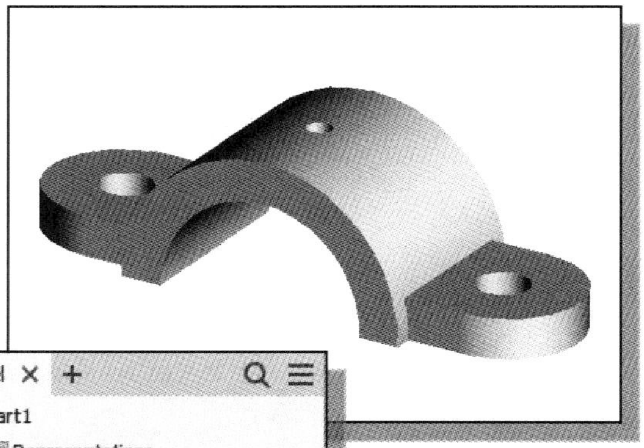

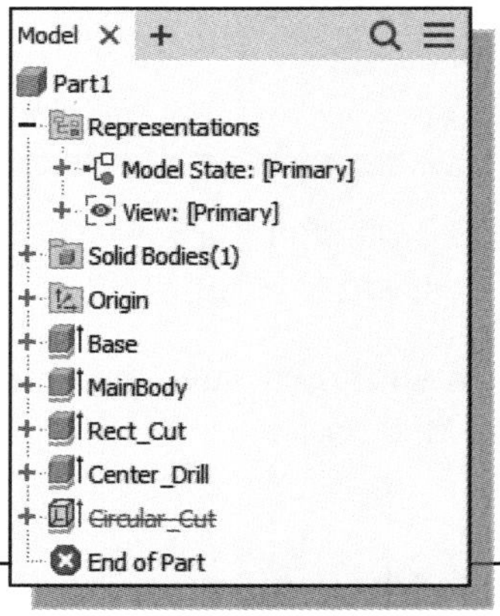

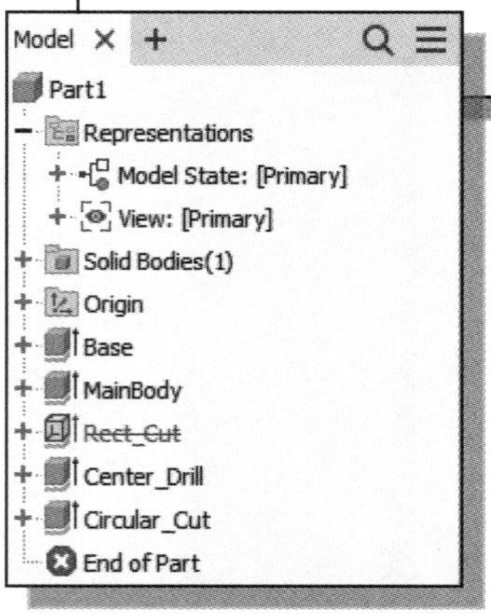

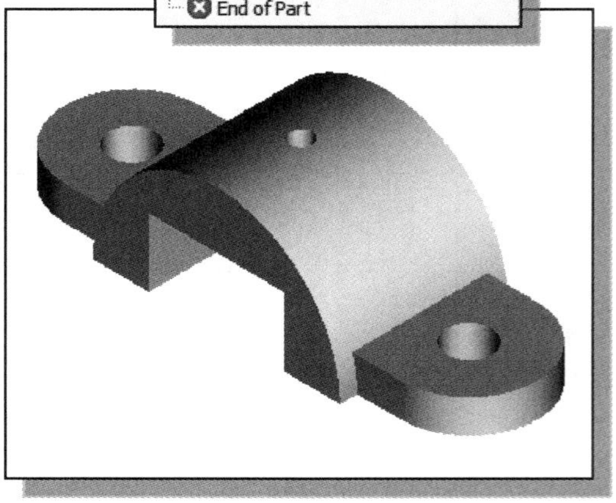

View and Edit Material Properties

The *Inventor Material Library* provides many commonly used materials. The material properties listed in the *Material Library* can be edited and new materials can be added.

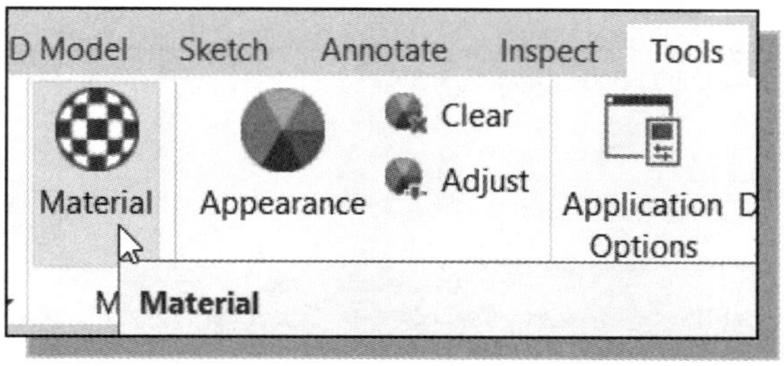

1. In the *Tools* tab, click **Material** as shown.

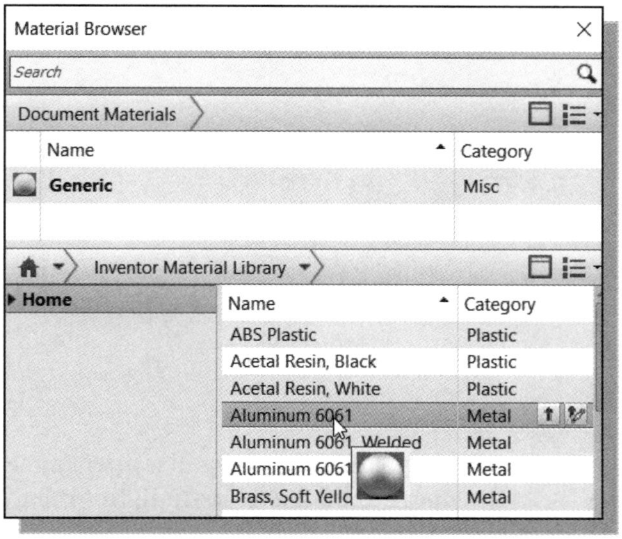

2. In the *Inventor Material* library group, select **Aluminum-6061**.

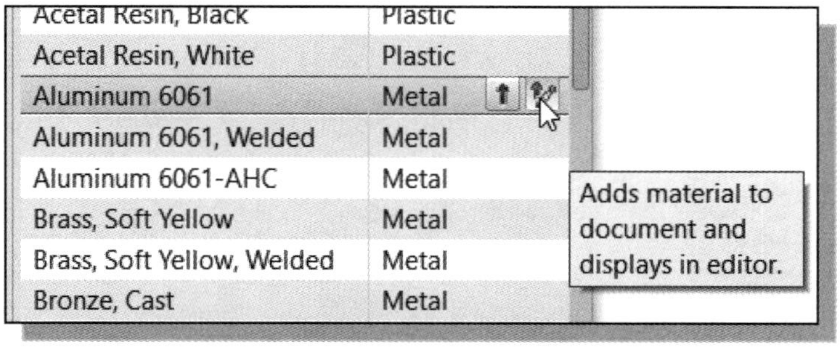

3. Click the **Add Material to document and display in editor** icon as shown.

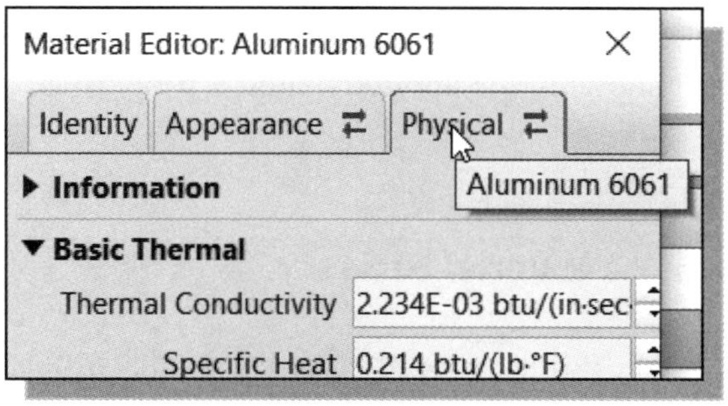

4. Click on the **Physical Aspect** tab to view the associated material properties.

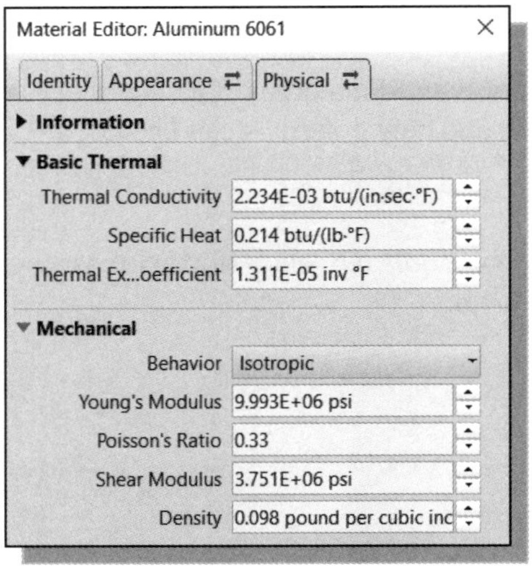

5. Expand the **Mechanical** properties list as shown.

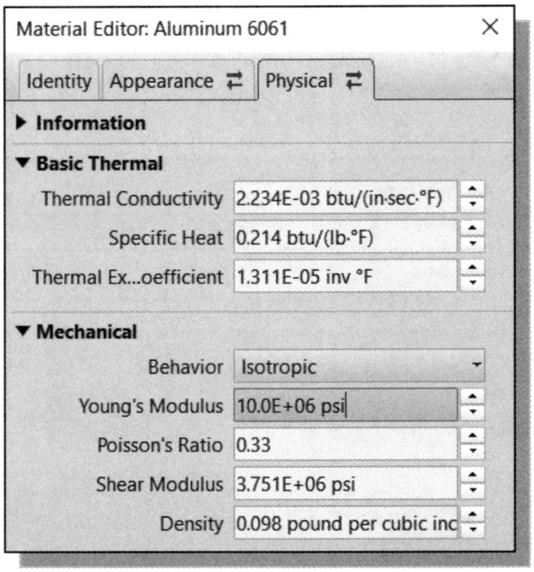

6. Enter **10.0E6 psi** as the new value for ***Young's Modulus***.

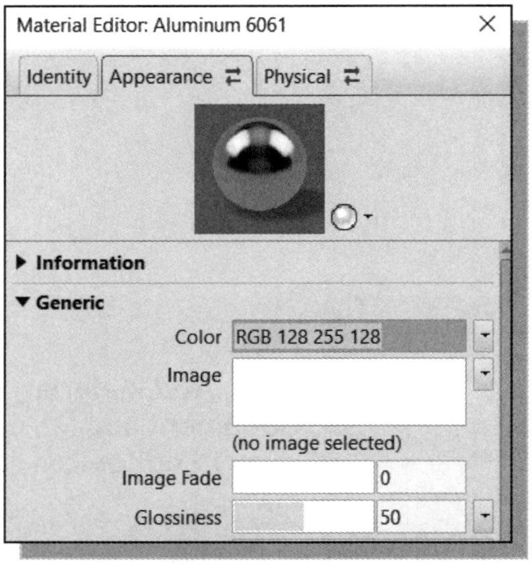

7. On your own, edit the material **Appearance - Glossiness** to **50** and color to **light green** as shown in the figure.

8. Click **OK** to accept the changes and close the editor.

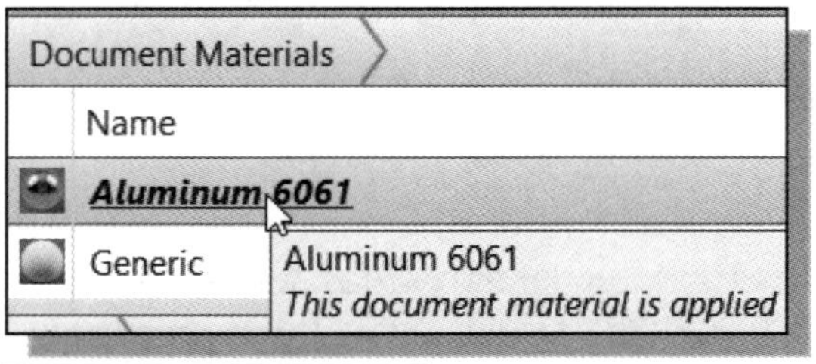

❖ Note the **Aluminum-6061** material in the document material list is highlighted showing the applied material properties.

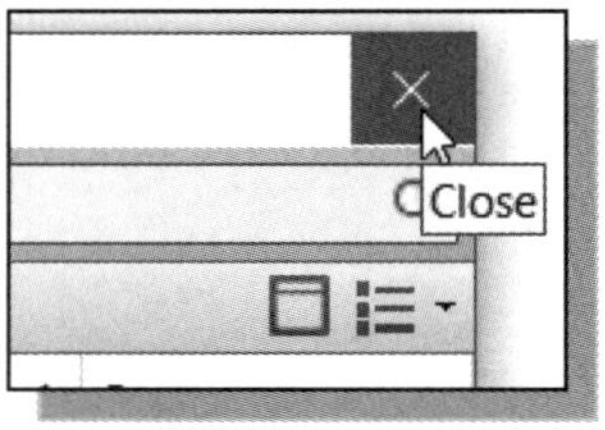

9. Click **Close** to exit the **Material Browser**.

- Note the new material setting is assigned to the model which can be shown under any of the shaded display modes.

Review Questions: (Time: 25 minutes)

1. Why is it important to consider the parent/child relationships between features?
2. Describe the procedure to suppress a feature.
3. What is the basic concept of the BORN technique?
4. What happens to a feature when it is suppressed?
5. How do you identify a suppressed feature in a model?
6. What is the main advantage of using the BORN technique?
7. Create sketches showing the steps you plan to use to create the models shown on the next page:

Exercises: Create and save the exercises in the Chapter7 folder.
(Time: 180 minutes. Dimensions are in inches unless otherwise stated.)

1. **Swivel Yoke** (Material: **Cast Iron**) Hint: Use the **Slot command**.

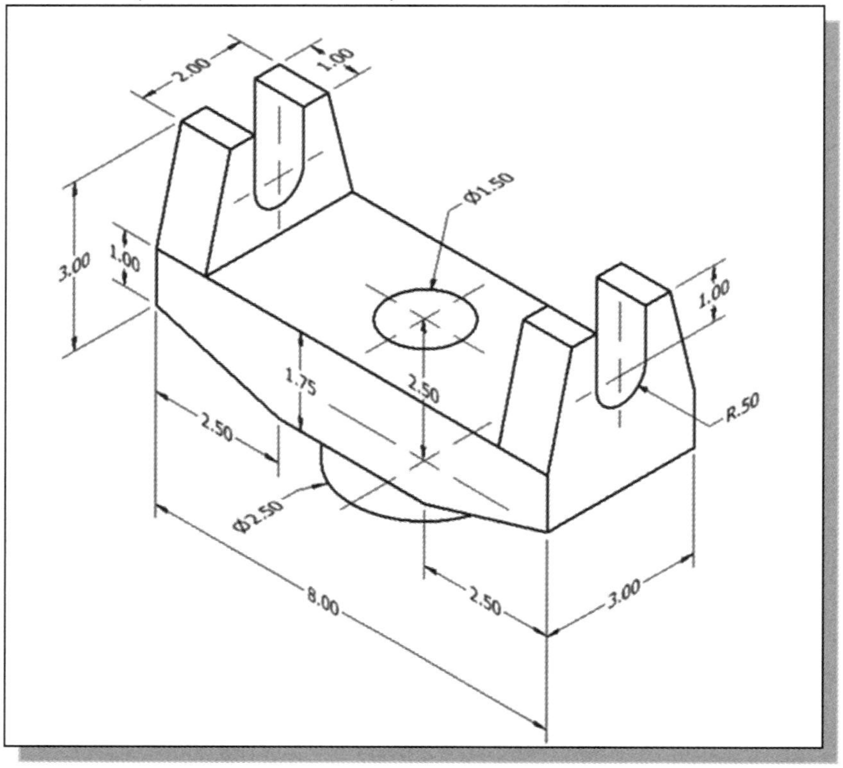

2. **Angle Bracket** (Material: **Carbon Steel**) Hint: Use the **Slot command**.

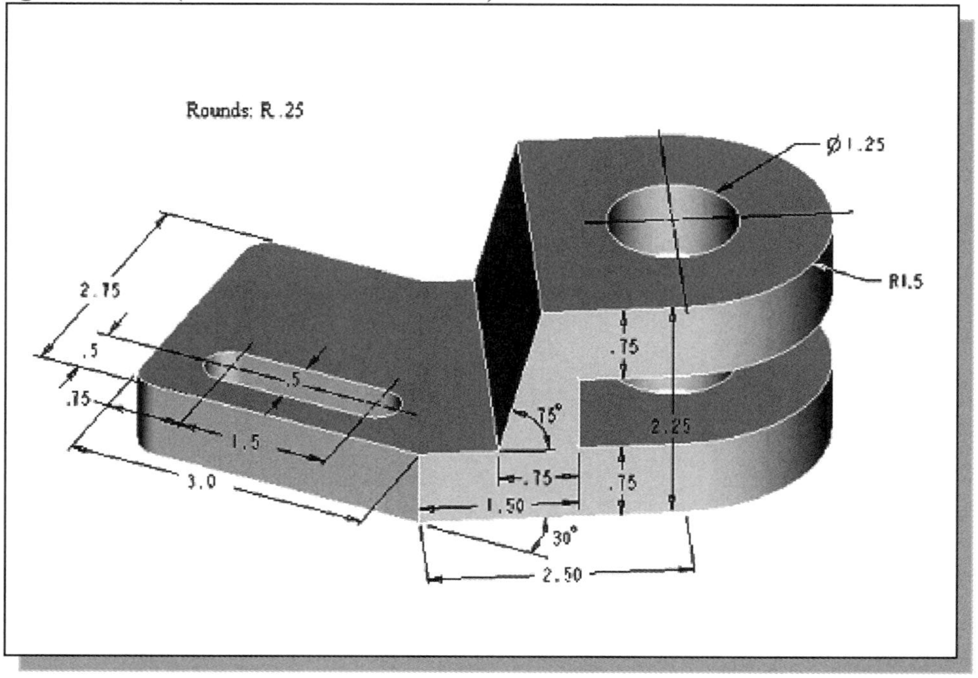

3. **Connecting Rod** (Material: **Carbon Steel**) Hint: Use the **Slot command**.

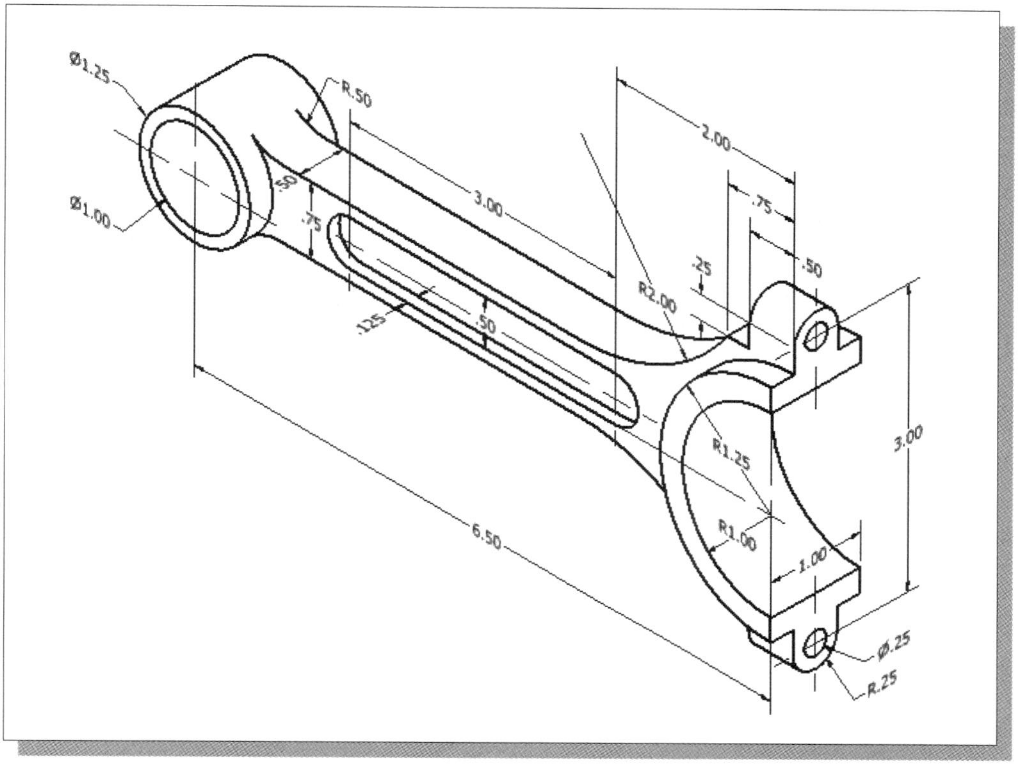

4. **Tube Hanger** (Material: **Aluminum 6061**)

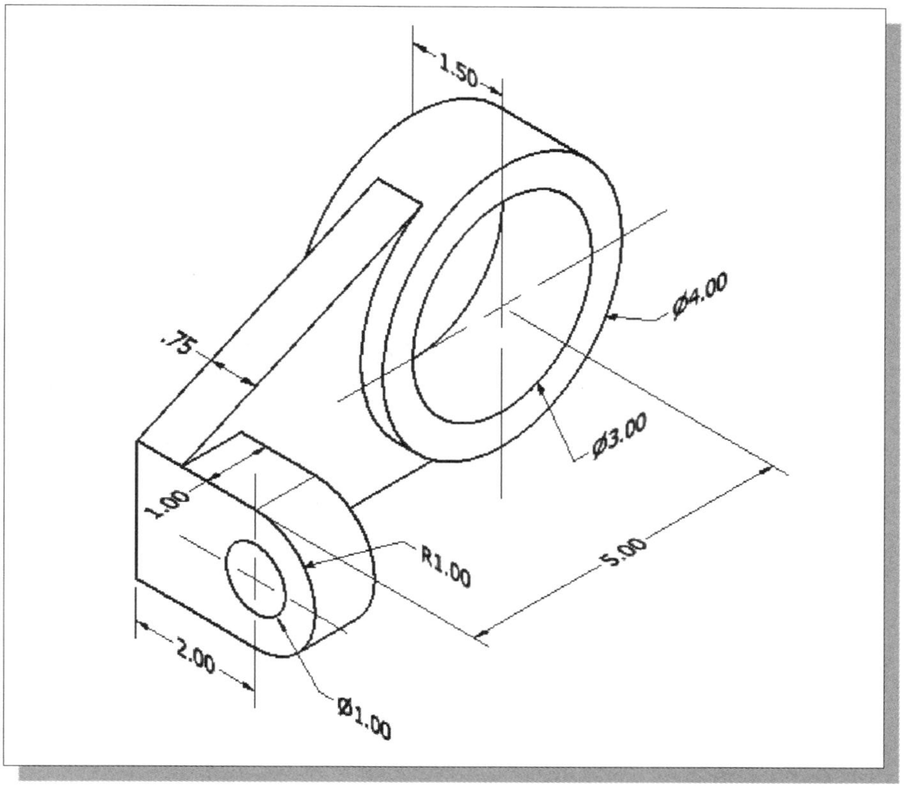

5. **Angle Latch** (Dimensions are in millimeters. Material: **Brass**)

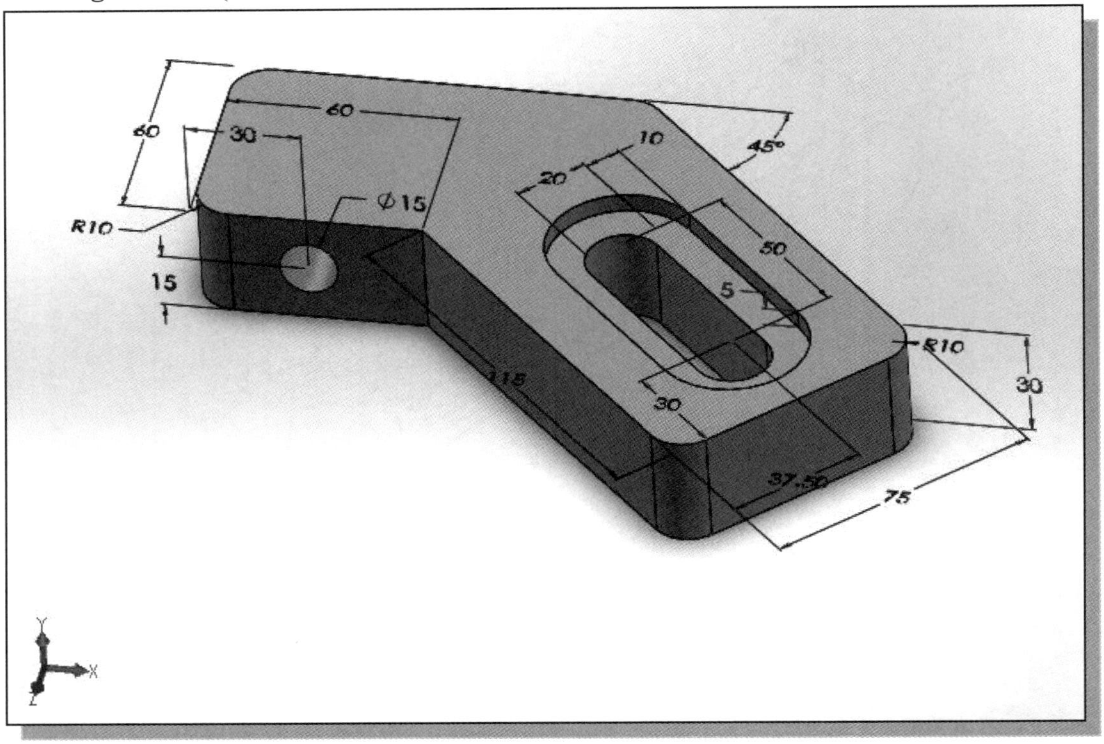

6. **Inclined Lift** (Material: **Mild Steel**)

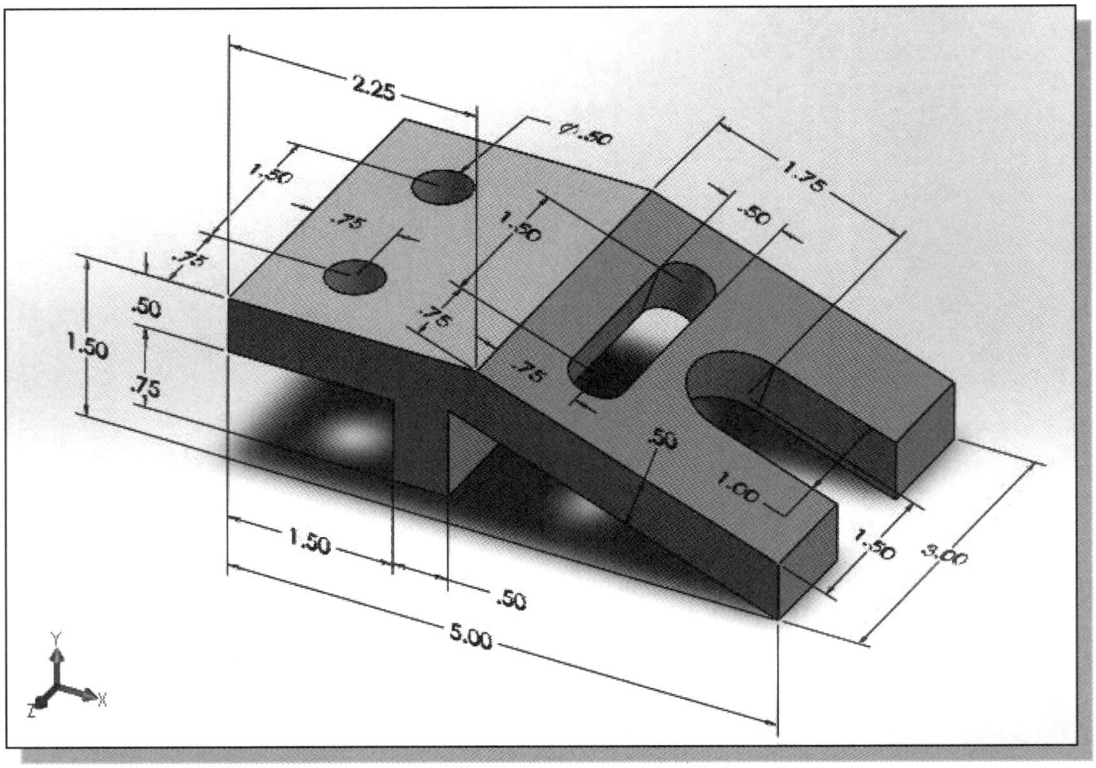

7. **Support Mount** (Material: **Mild Steel**)

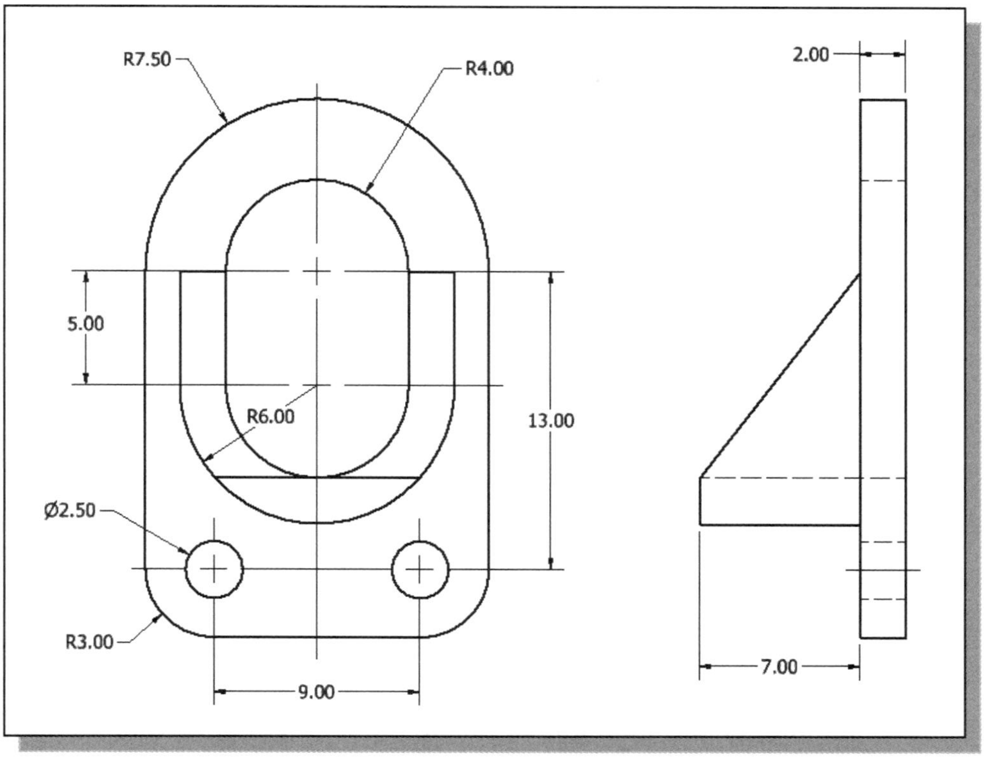

Notes:

Chapter 8
Part Drawings and 3D Model-Based Definition

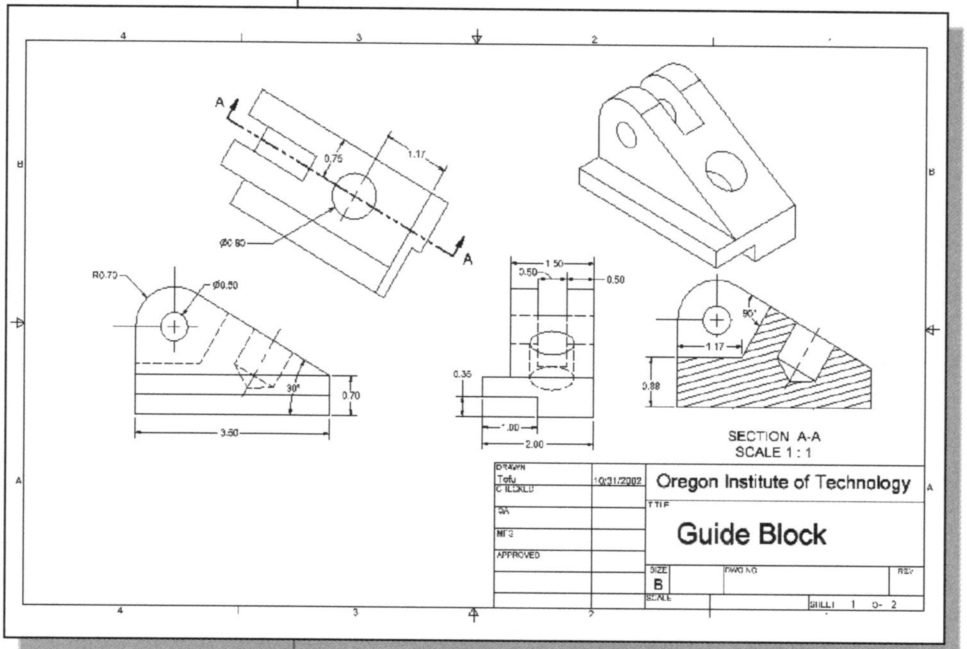

Learning Objectives

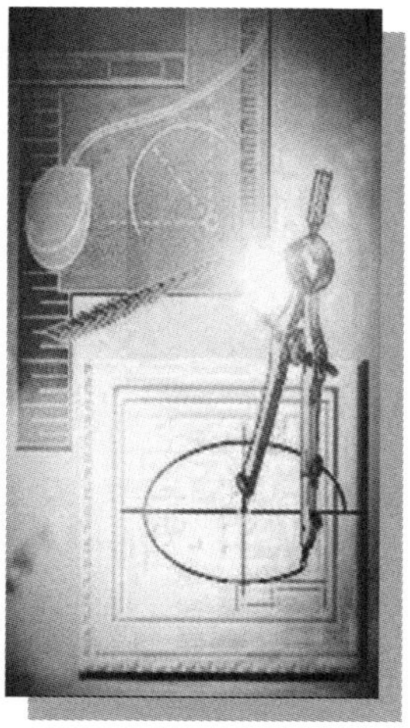

- **Create Drawing Layouts from Solid Models**
- **Understand Associative Functionality**
- **Use the Default Borders and Title Block in the Layout Mode**
- **Arrange and Manage 2D Views in Drawing Mode**
- **Display and Hide Feature Dimensions**
- **Create Reference Dimensions**
- **Create 3D Model-Based Definition on Solid Models**

Autodesk Inventor Certified User Exam Objectives Coverage

Parametric Modeling Basics

Section 6: Drawings

Objectives: Create drawings.

Drawings from Parts and Associative Functionality

With the software/hardware improvements in solid modeling, the importance of two-dimensional drawings is decreasing. Drafting is considered one of the downstream applications of using solid models. In many production facilities, solid models are used to generate machine tool paths for *computer numerical control* (CNC) machines. Solid models are also used in *rapid prototyping* to create 3D physical models out of plastic resins, powdered metal, etc. Ideally, the solid model database should be used directly to generate the final product. However, the majority of applications in most production facilities still require the use of two-dimensional drawings. Using the solid model as the starting point for a design, solid modeling tools can easily create all the necessary two-dimensional views. In this sense, solid modeling tools are making the process of creating two-dimensional drawings more efficient and effective.

Autodesk Inventor provides associative functionality in the different Autodesk Inventor modes. This functionality allows us to change the design at any level, and the system reflects it at all levels automatically. For example, a solid model can be modified in the *Part Modeling Mode* and the system automatically reflects that change in the *Drawing Mode*. We can also modify a feature dimension in the *Drawing Mode*, and the system automatically updates the solid model in all modes.

In this lesson, the general procedure of creating multi-view drawings is illustrated. The *U_Bracket* design from the last chapter is used to demonstrate the associative functionality between the model and drawing views.

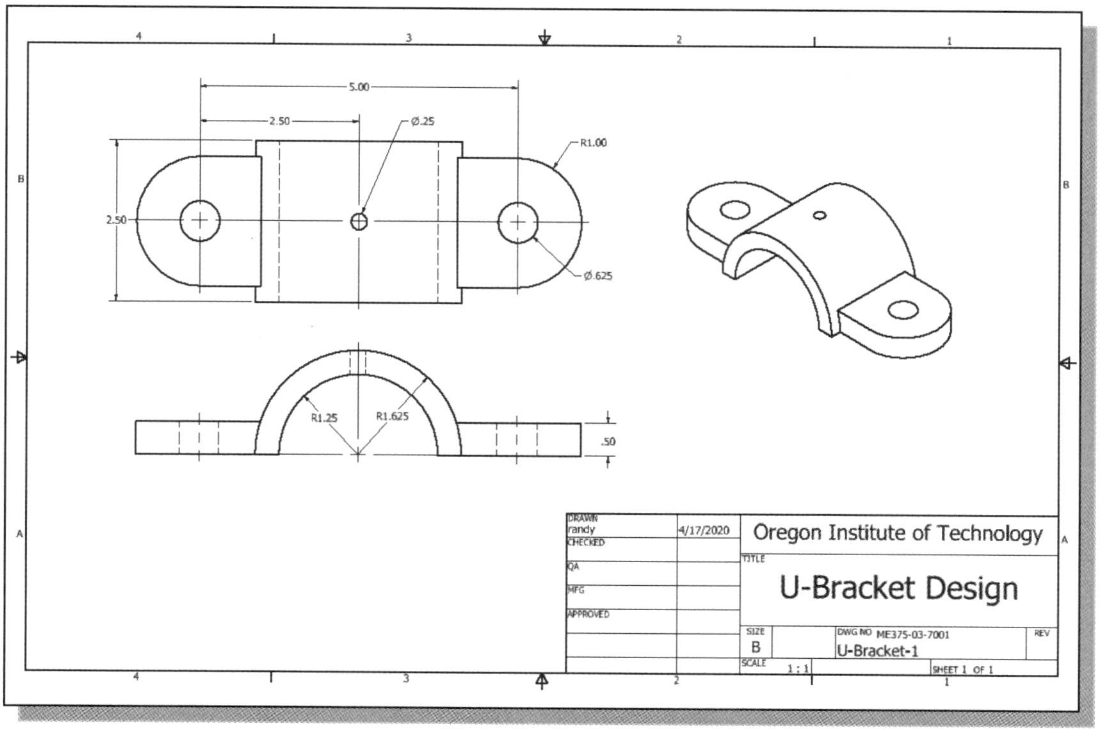

3D Model-Based Definition

Model-based definition (MBD) is the practice of using 3D models within 3D CAD software to provide specifications for individual components and assemblies. The types of information included are geometric dimensioning and tolerancing (GD&T), component level materials, assembly level bills of materials, engineering configurations, design intent, etc. With this new set of tools, the mental translation of 3D models to two-dimensional (2D) drawings for communicating information and then back to 3D models for manufacturing is replaced with a representative 3D prototype that provides all necessary information to communicate and manufacture the product. This approach means faster and more precise product communication, which can greatly improve the product development process.

Autodesk Inventor provides users with the ability to access powerful digital product information for communication and in support of operations such as inspection, manufacturing, or marketing. With software/hardware improvements, it is now feasible to use the solid modeling software to document and communicate all production and manufacturing information in a three-dimensional (3D) environment. In *Autodesk Inventor 2025*, exciting tools are now available for documenting and communicating product designs. We can apply 3D-based annotation, in conjunction with engineering drawing conventions, directly to the solid model or assembly. This new solids-based documentation approach provides the ability to create, manage, and deliver process-specific information without the need for paper-based documentation. In this lesson, the general procedure of creating 3D model-based definition is also illustrated.

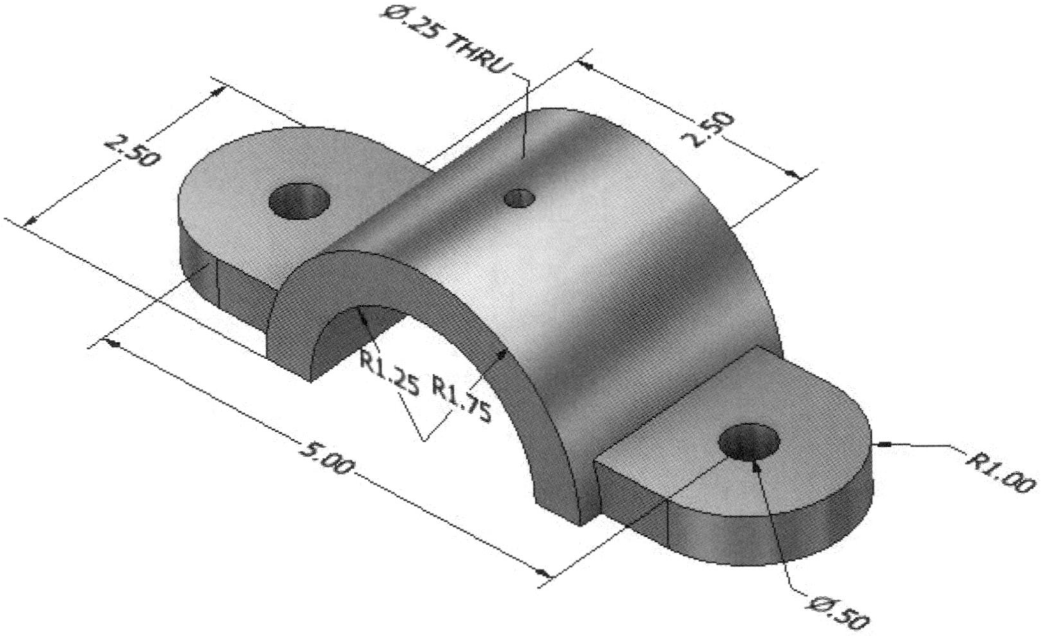

Starting Autodesk Inventor

1. Select the **Autodesk Inventor** option on the *Start* menu or select the **Autodesk Inventor** icon on the desktop to start Autodesk Inventor. The Autodesk Inventor main window will appear on the screen.

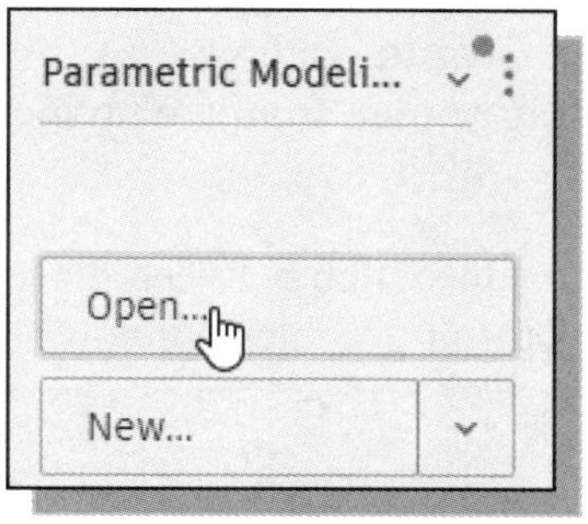

2. Once the program is loaded into memory, select the **Open** option as shown.

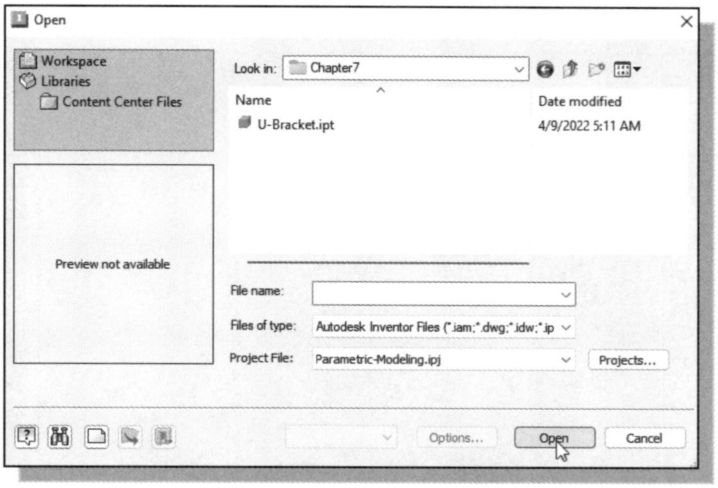

3. In the *File name* list box, select the **U-Bracket.ipt** file. (Use the **Projects** option to locate the file if the wrong project is displayed in the *Project File* box.)

4. Click on the **Open** button to accept the selected settings.

Drawing Mode – 2D Paper Space

Autodesk Inventor allows us to generate 2D engineering drawings from solid models so that we can plot the drawings to any exact scale on paper. An engineering drawing is a tool that can be used to communicate engineering ideas/designs to manufacturing, purchasing, service, and other departments. Until now we have been working in ***model space*** to create our design in ***full size***. We can arrange our design on a two-dimensional sheet of paper so that the plotted hardcopy is exactly what we want. This two-dimensional sheet of paper is known as ***paper space*** in *AutoCAD* and *Autodesk Inventor*. We can place borders and title blocks, objects that are less critical to our design, on *paper space*. In general, each company uses a set of standards for drawing content, based on the type of product and also on established internal processes. The appearance of an engineering drawing varies depending on when, where, and for what purpose it is produced. However, the general procedure for creating an engineering drawing from a solid model is fairly well defined. *In Autodesk Inventor*, creation of 2D engineering drawings from solid models consists of four basic steps: drawing sheet formatting, creating/positioning views, annotations, and printing/plotting.

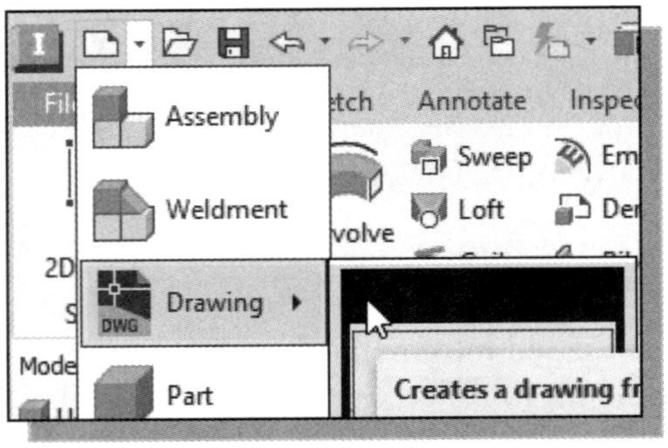

1. Click on the **drop-down arrow** next to the **New File** icon in the *Quick Access* toolbar area to display the available New File options.

2. Select **Drawing → Create**, without choosing any drawing option, from the option list.

➢ Note that a new graphics window appears on the screen. We can switch between the solid model and the drawing by clicking the corresponding tabs or graphics windows.

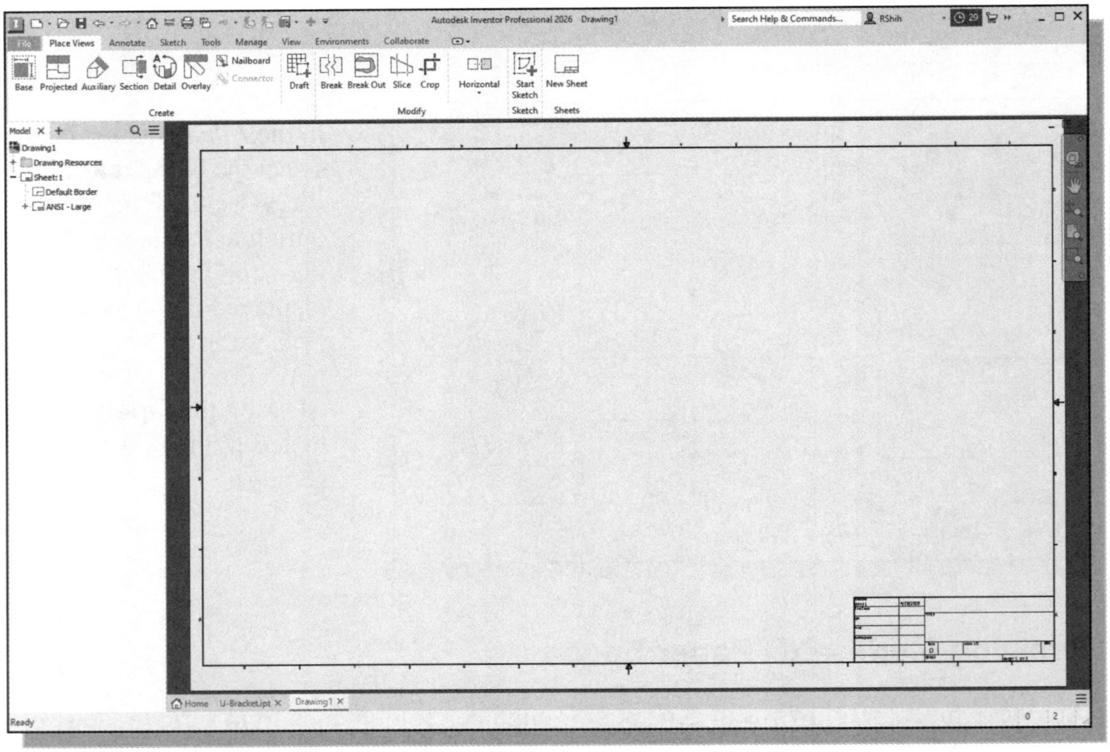

❖ In the graphics window, *Autodesk Inventor* displays a default drawing sheet that includes a title block. The drawing sheet is placed on the 2D paper space, and the title block also indicates the paper size being used.

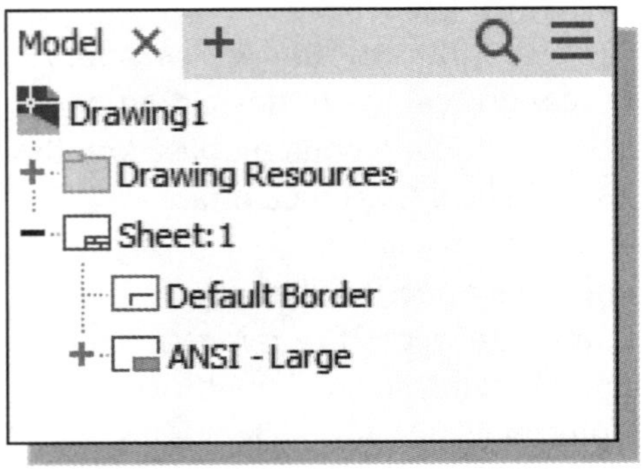

➢ In the *browser* area, the Drawing1 icon is displayed at the top, which indicates that we have switched to ***Drawing Mode***. **Sheet1** is the current drawing sheet that is displayed in the graphics window.

Drawing Sheet Format

1. Choose the **Manage** tab in the *Ribbon* toolbar.

2. Click **Styles Editor** in the *Styles and Standards* toolbar as shown.

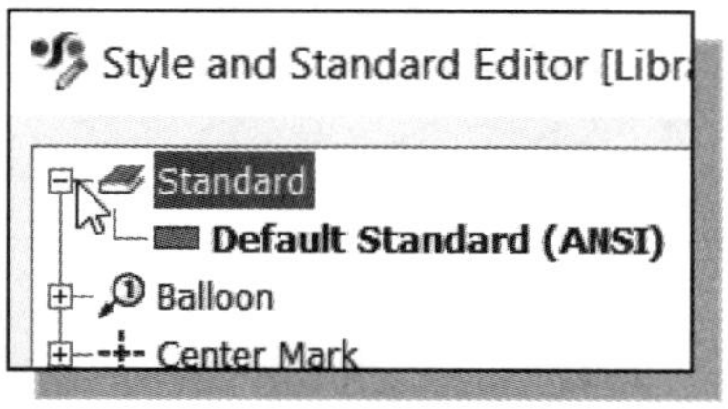

3. Click on the [+] sign in front of **Standard** to display the current active standard. Note that there can only be **one active standard** for each drawing.

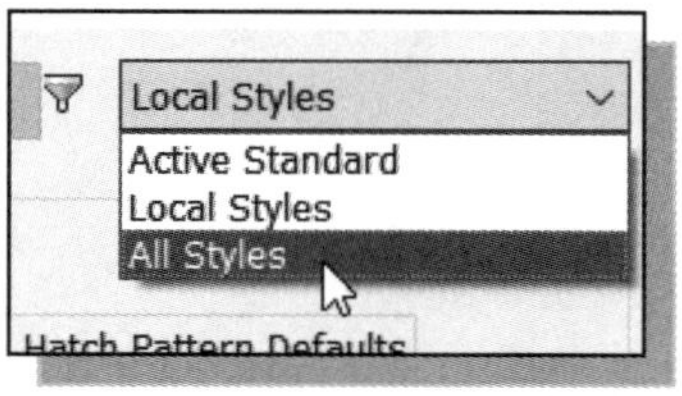

4. Set the *Filter Styles Setting* to display **All Styles**. Note that besides the default ANSI drafting standard, other standards, such as ISO, GB, BSI, DIN, and JIS, are also available.

5. Set the *Filter Styles Setting* to display **Active Standard** and note only the active standard is displayed.

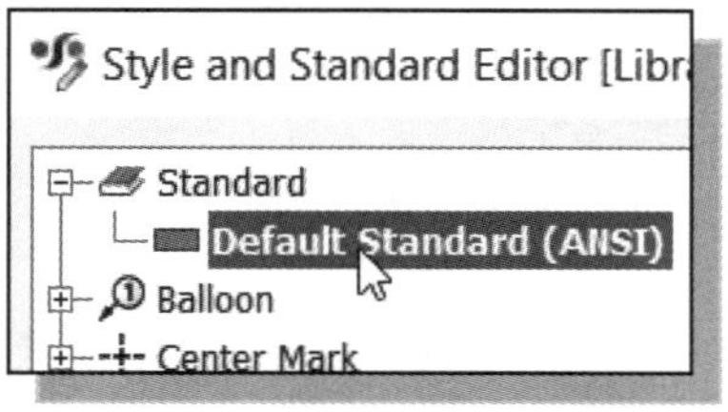

6. Click **Default Standard (ANSI)** to toggle the display of detailed settings for the current standard.

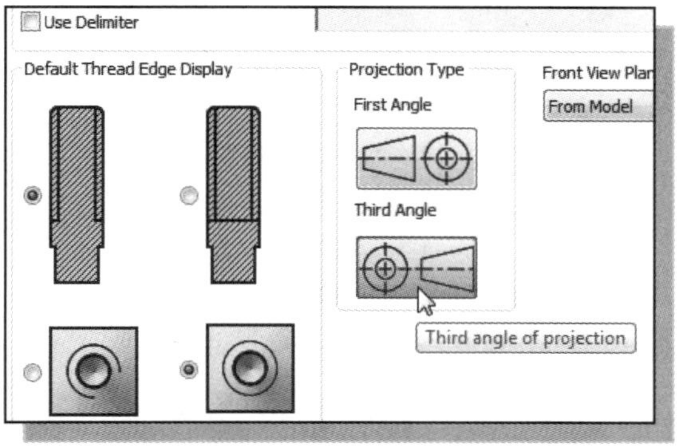

7. In the *View Preferences* page, confirm that the *Projection Type* is set to **Third Angle of projection**.

❖ Notice the different settings available in the *General* option window, such as the *Units* setting and the *Line Weight* setting.

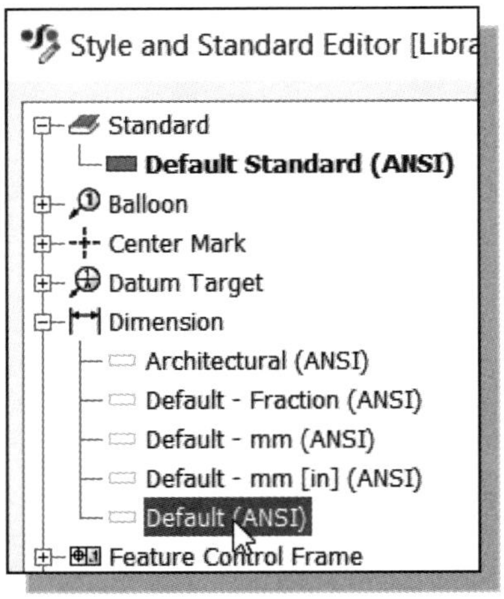

8. Choose **Default (ANSI)** in the **Dimension** list as shown.

➢ Note that the default *Dimension Style* in Inventor is based on the ANSI Y14.5-1994 standard.

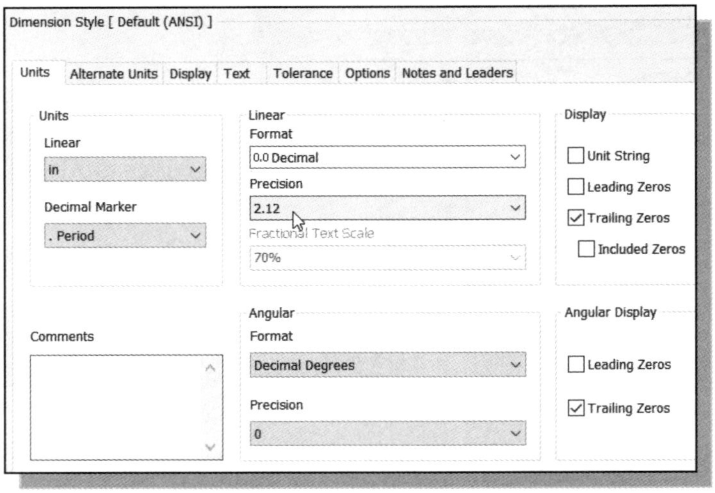

9. The **Units** tab contains the settings for linear/angular units. Note that the *Linear* units are set to Decimal and the *Precision* for linear dimensions is set to two digits after the decimal point.

10. Click on the **Text** tab to display and examine the settings for dimension text. Note that the default *Dimension Style*, DEFAULT-ANSI, cannot be modified. However, new *Dimension Styles* can be created and modified.

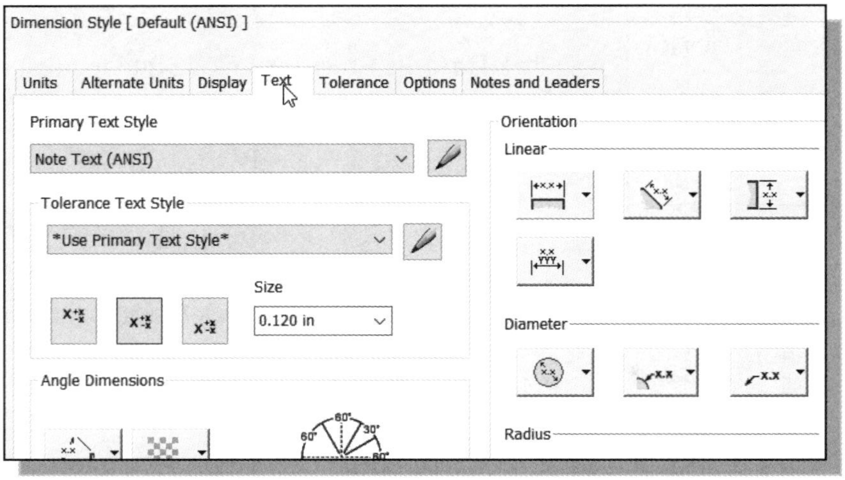

➢ On your own, click on the other tabs and examine the other *Settings* available.

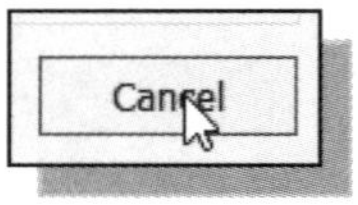

11. Click on the **Cancel** button to exit the *Style and Standard Editor* dialog box.

Using the Pre-defined Drawing Sheet Formats

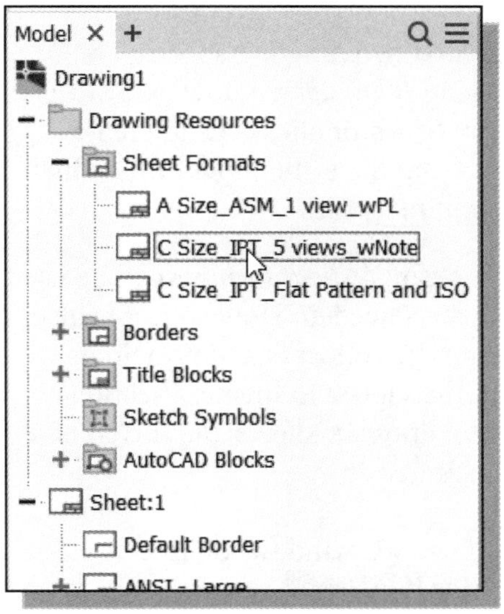

1. Inside the *Drawing Browser* window, click on the **[+]** symbol in front of **Drawing Resources** to display the available options.

2. Click on the **[+]** symbol in front of **Sheet Formats** to display the available pre-defined sheet formats.

❖ Notice several pre-defined *sheet formats*, each with a different view configuration, are available in the *browser* window.

3. Inside the *browser* window, **double-click** on the **C size IPT 5 view** sheet format.

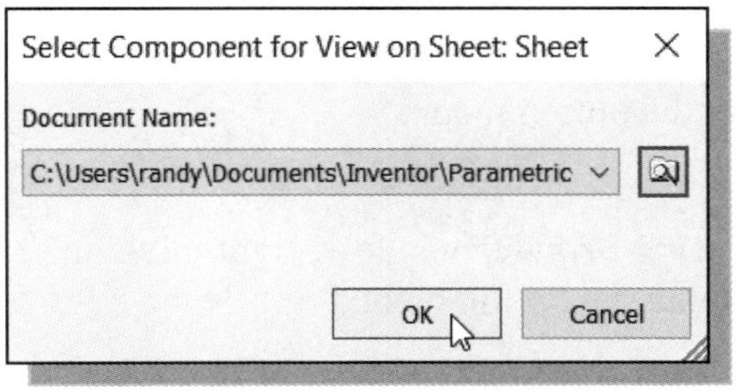

4. Click on the **OK** button to accept the default part file and generate the 2D views.

➢ The ***U-Bracket*** model is the only model opened. By default, all of the 2D drawings will be generated from this model file.

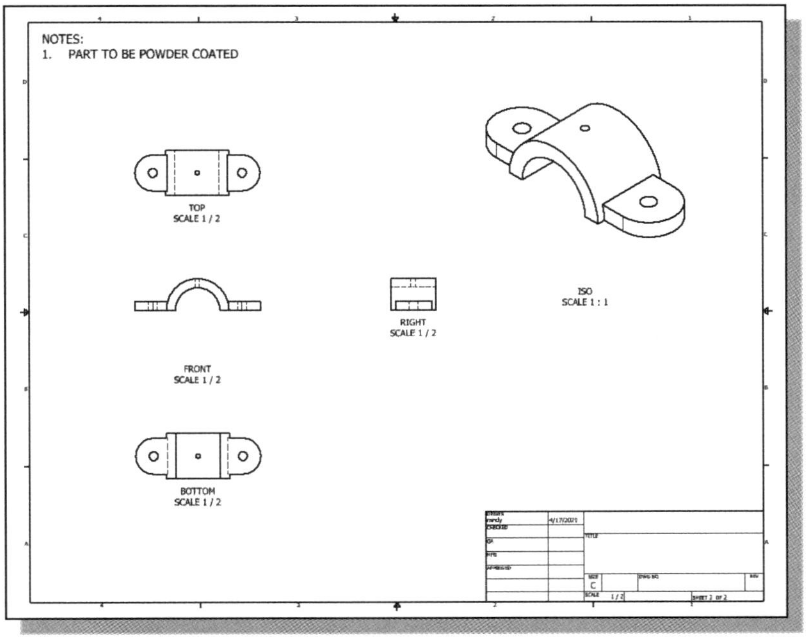

❖ We have created a C-size drawing of the *U-Bracket* model. Autodesk Inventor automatically generates and positions five of the pre-defined views of the model inside the title block.

Activate, Delete and Edit Drawing Sheets

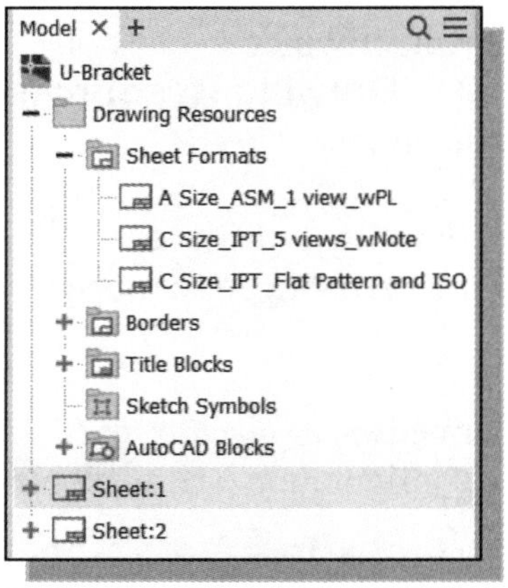

❖ Note that we have created two drawing sheets, displayed in the *Drawing Browser* window as Sheet1, and Sheet2. Autodesk Inventor allows us to create multiple 2D drawings from the same model file, which can be used for different purposes.

➢ In most cases, the pre-defined *sheet formats* can be used to quickly set up the views needed. However, it is also important to understand the concepts and principles involved in setting up the views. In the next sections, the procedures to set up drawing sheets and different types of views are illustrated.

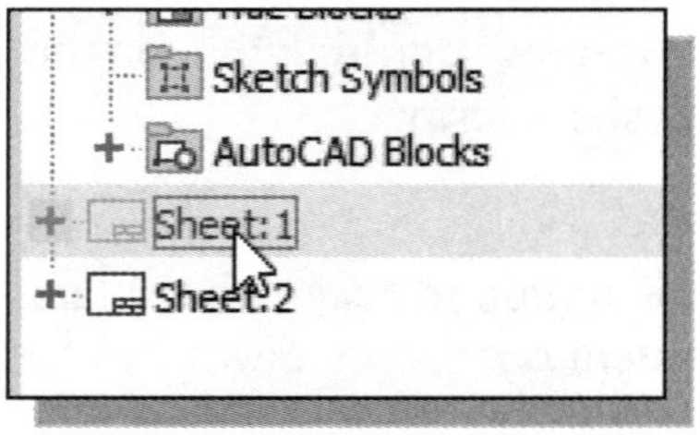

1. Inside the *Drawing Browser* window, double-click on **Sheet1** to activate this drawing sheet.

   • Alternatively, use the right-mouse-click and select **Activate** in the option menu.

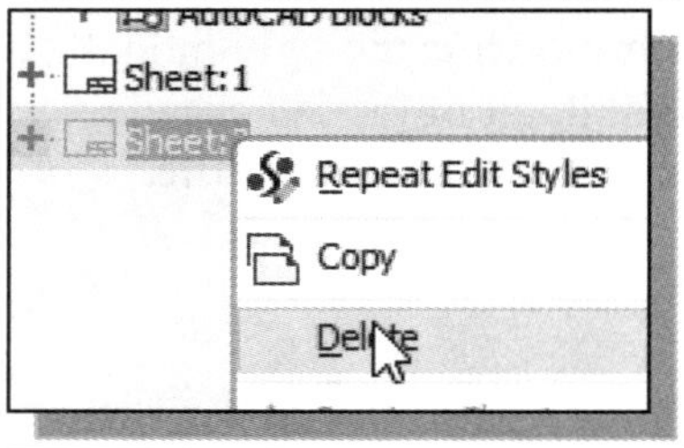

2. Inside the *Drawing Browser* window, right-click on **Sheet2** to display the option menu.

3. Select **Delete Sheet** in the option menu to remove the Sheet2 drawing.

4. In the *warning window*, click on the **OK** button to proceed with deleting the drawing.

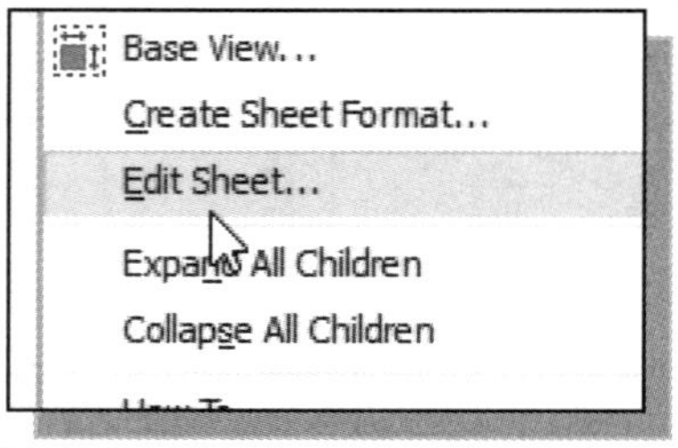

5. Inside the *Drawing Browser* window, **right-click** on **Sheet1** to display the *option menu.*

6. Select **Edit Sheet** in the option menu to display the settings for the *Sheet1* drawing.

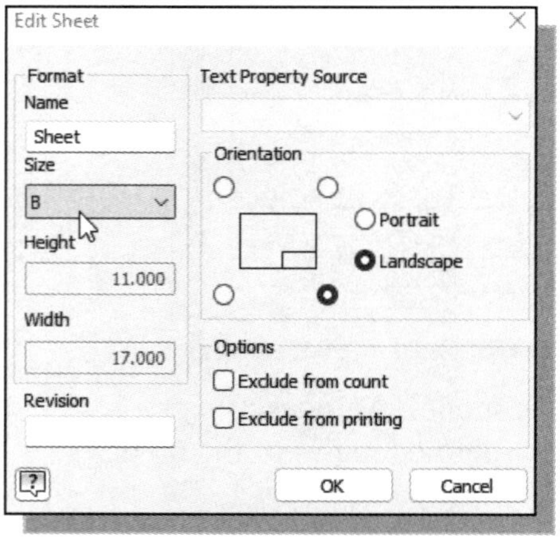

7. Set the sheet size to **B-size** and click on the **OK** button to exit the *Edit Sheet* dialog box.

Add a Base View

In *Autodesk Inventor Drawing Mode*, the first drawing view we create is called a **base view**. A *base view* is the primary view in the drawing; other views can be derived from this view. When creating a *base view*, Autodesk Inventor allows us to specify the view to be shown. By default, Autodesk Inventor will treat the *world XY plane* as the front view of the solid model. Note that there can be more than one *base view* in a drawing.

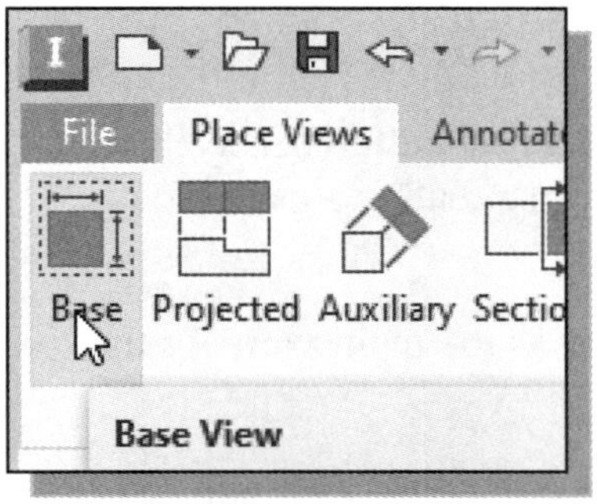

1. Click on the **Base View** in the *Place Views* panel to create a base view.

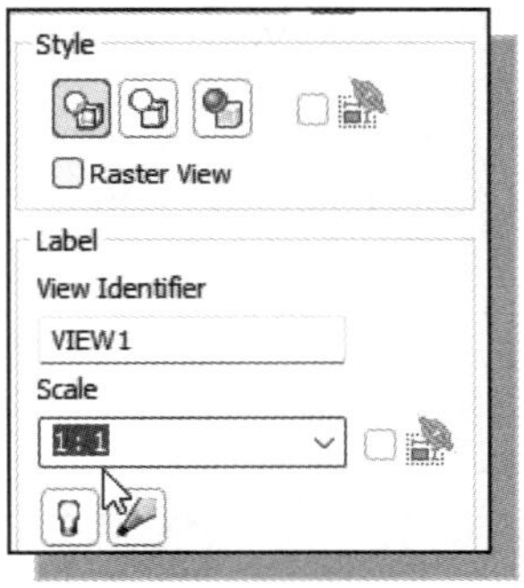

2. In the *Drawing View* dialog box, set the **Scale** to **1 : 1** and Style to **Hidden Line** as shown in the figure below.

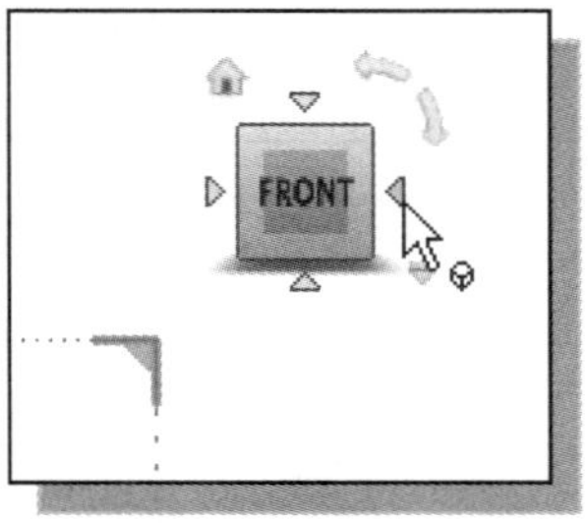

3. In the *graphics area*, click on the arrows to switch to different views; set it to the **Front View** as shown in the below figure.

4. Inside the *graphics window,* drag and place the ***base*** view near the left center of the graphics window as shown below. Click **OK** to place the *base view* and close the *Drawing View* dialog box.

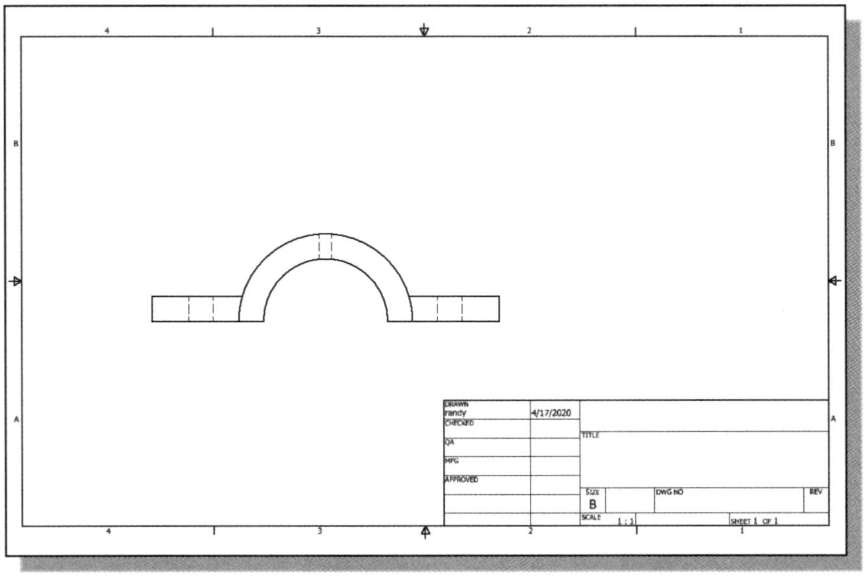

Create Projected Views

In *Autodesk Inventor Drawing Mode*, **projected views** can be created with a first-angle or third-angle projection, depending on the drafting standard used for the drawing. We must have a base view before a projected view can be created. Projected views can be orthographic projections or isometric projections. Orthographic projections are aligned to the base view and inherit the base view's scale and display settings. Isometric projections are not aligned to the base view.

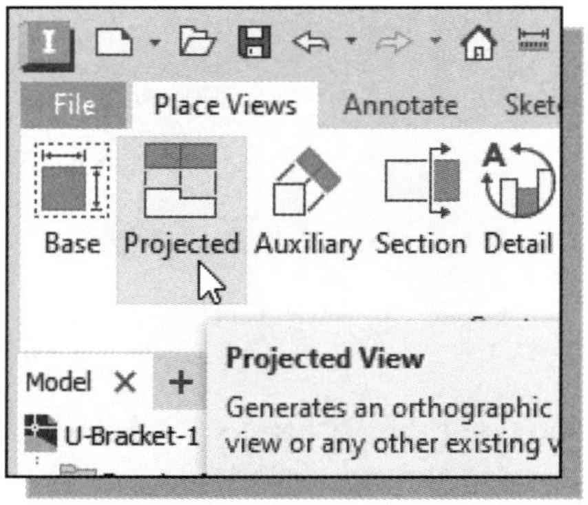

1. Click the **Projected View** button in the *Place Views* panel; this command allows us to create projected views.
2. Select the **base view** as the main view for the projected views.
3. Move the cursor **above** the *base view* and select a location to position the projected side view of the model.
4. Move the cursor toward the upper right corner of the title block and select a location to position the isometric view of the model as shown below.

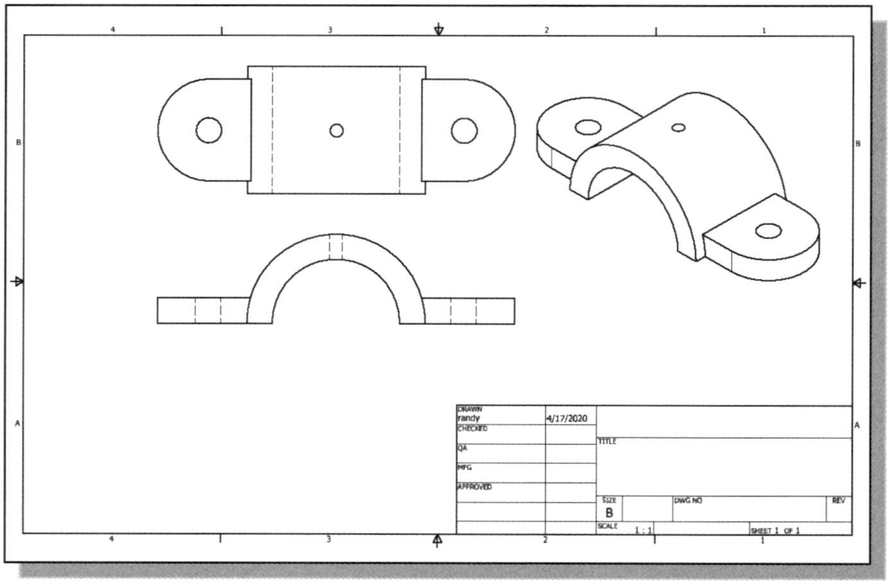

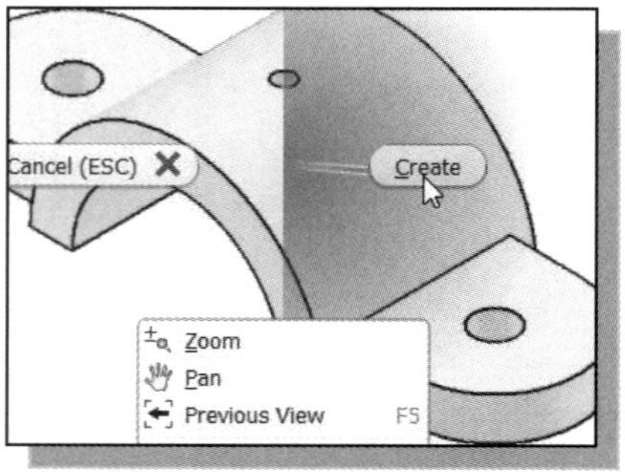

5. Inside the *graphics window*, right-click once to bring up the **option menu**.
6. Select **Create** to proceed with creating the two projected views.

Adjust the View Scale

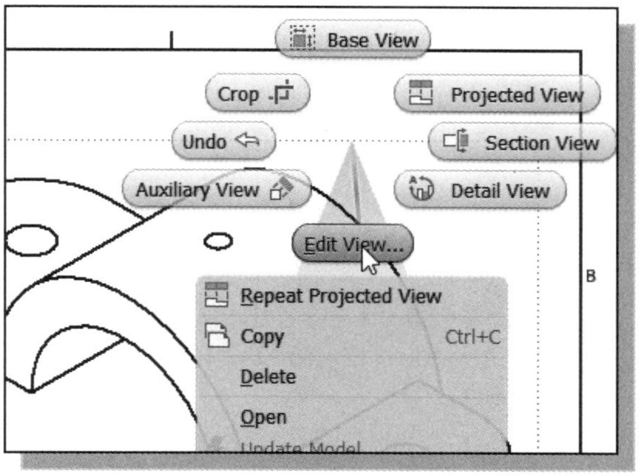

1. Move the cursor on top of the *isometric view* and watch for the box around the entire view indicating the view is selectable as shown in the figure. **Right-click** once to bring up the *option menu.*

2. Select **Edit View** in the option menu.

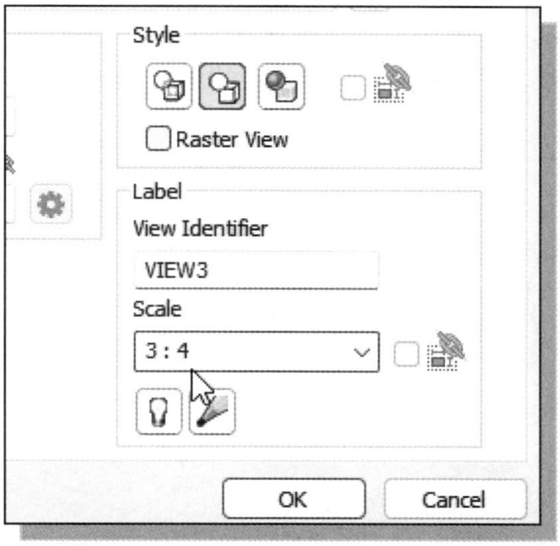

3. Inside the *Drawing View* dialog box set the *Scale* to **3:4** as shown in the figure.

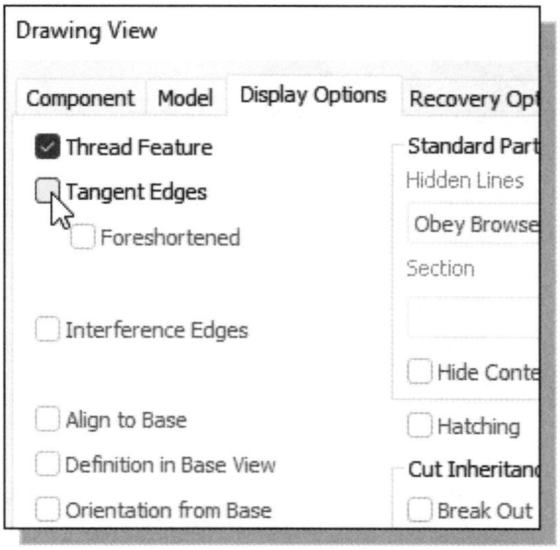

4. Click on the **Display Options** tab and turn ***OFF*** the **Tangent Edges** option as shown.

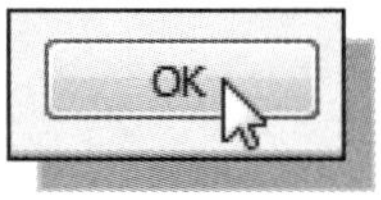

5. Click on the **OK** button to accept the settings and proceed with updating the drawing views.

Repositioning Views

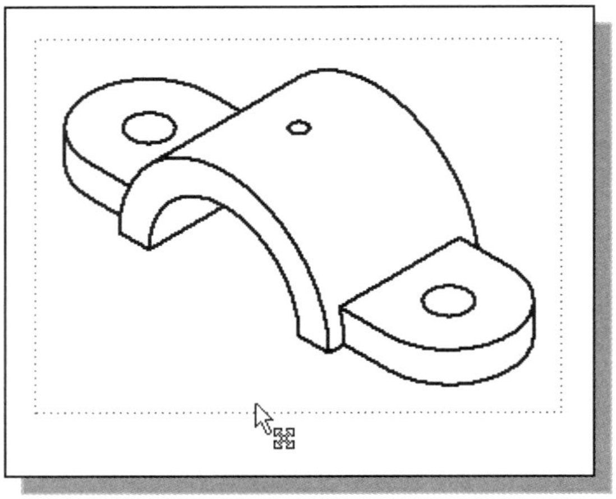

1. Move the cursor on top of the isometric view and watch for the **four-arrow Move symbol** as the cursor is near the border indicating the view can be dragged to a new location as shown in the figure.

2. Press and hold down the left-mouse-button and reposition the view to a new location.

3. On your own, reposition the views we have created so far. Note that the top view can be repositioned only in the vertical direction. The top view remains aligned to the base view, the front view.

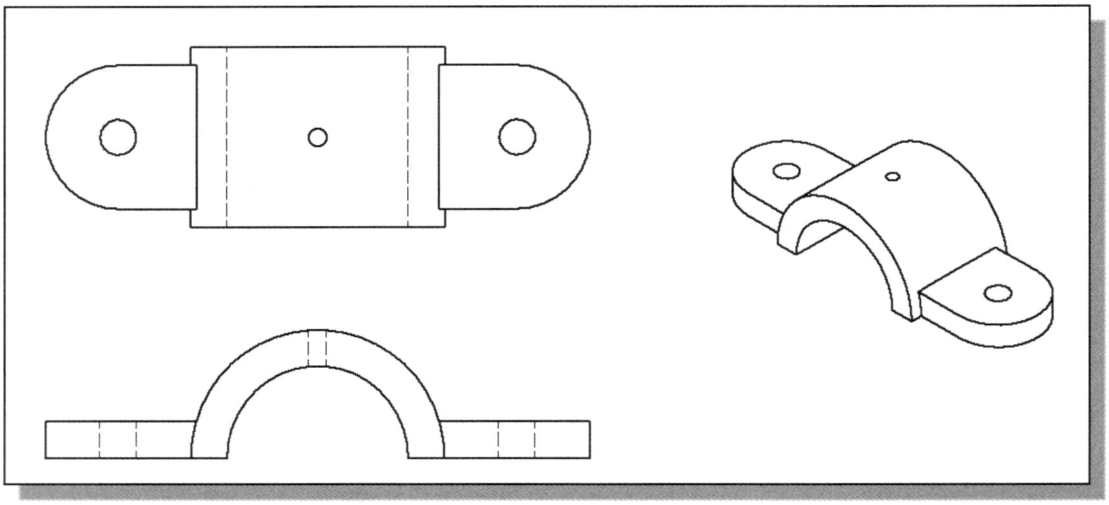

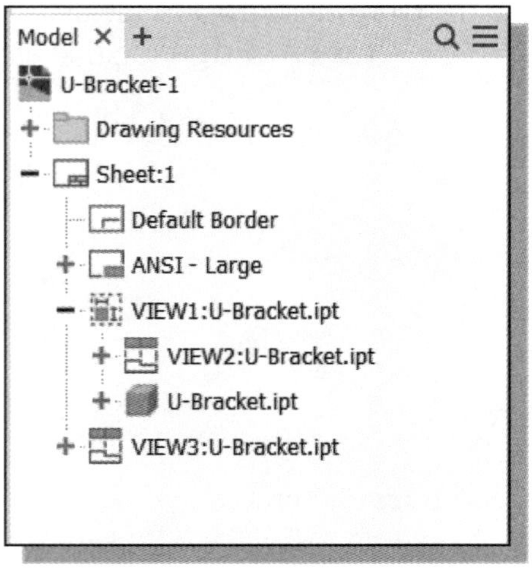

➢ Note that in the *Drawing Browser* area, a hierarchy of the created views is displayed under Sheet1. The base view, View1, is listed as the first view created, with View2 linked to it. The top view, View2, is projected from the base view, View1. The implied parent/child relationship is maintained by the system. Drawing views are associated with the model and the drawing sheets. As we create views from the base view, they are nested beneath the base view in the *browser*.

Display Feature Dimensions

By default, feature dimensions are not displayed in 2D views in Autodesk Inventor. We can change the default settings while creating the views or switch on the display of the parametric dimensions using the option menu.

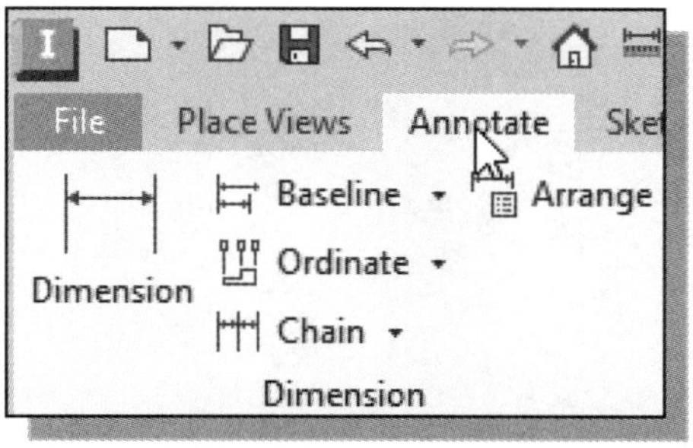

1. Select **Annotate** by left-clicking once in the *Ribbon* toolbar system.

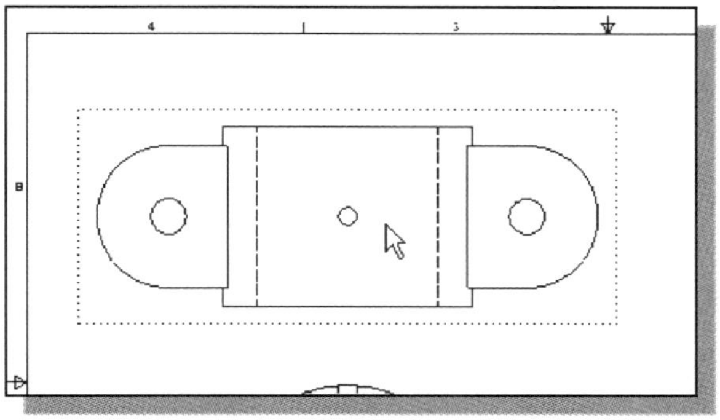

2. Move the cursor on top of the ***top*** view of the model and watch for the box around the entire view indicating the view is selectable as shown in the figure.

3. Inside the graphics window**, right-click** once on the top view to bring up the option menu.

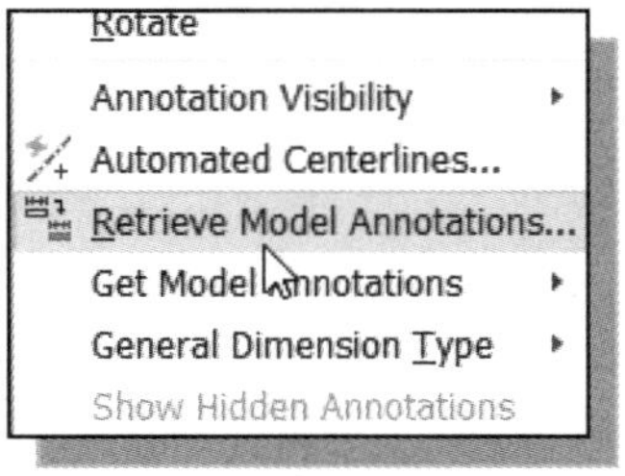

4. Select **Retrieve Model Annotations** to display the parametric dimensions used to create the model. Note the command can also be accessed through the *Ribbon* toolbar.

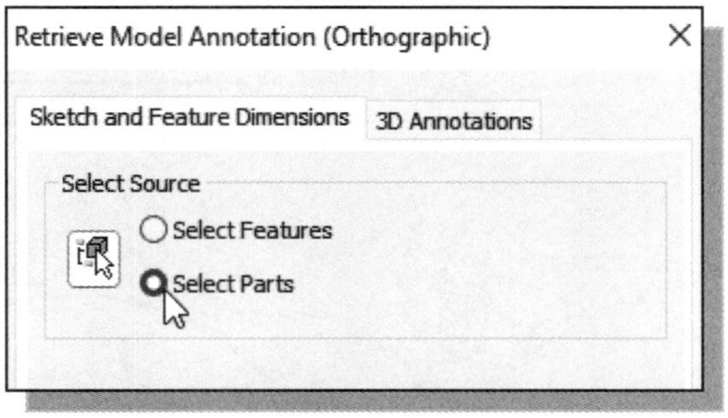

5. In the *Sketch and Features Dimensions* tab, set the *Select Source* option to **Select Parts** as shown.

6. Move the *Retrieve Model annotations* dialog box by pressing the title of the box and dragging with the left-mouse button to the right side of the view.

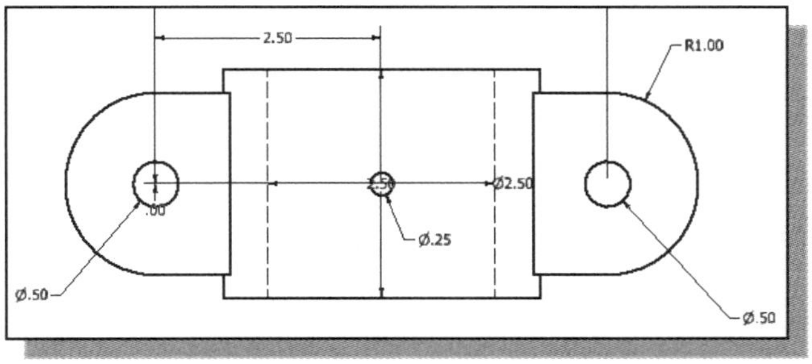

➢ Note that several of the dimensions used to create the part are now displayed in the selected view.

➢ The system now expects us to select the dimensions to retrieve.

7. On your own, select the dimensions to retrieve by left-clicking once on the dimensions as shown. (Note that only selected dimensions will be retrieved.)

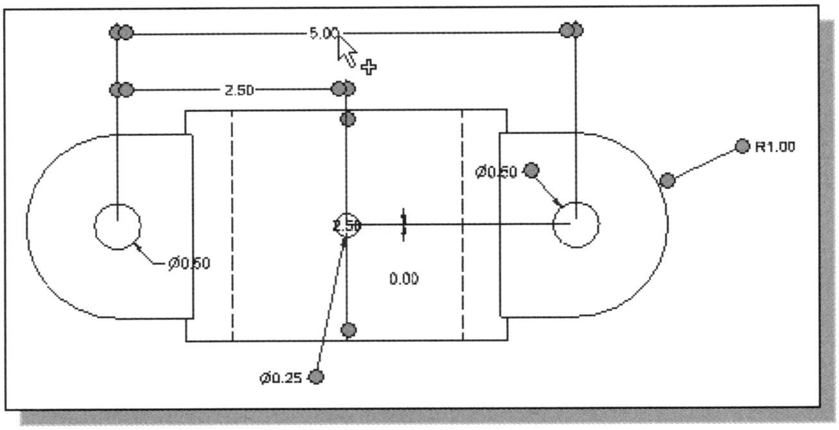

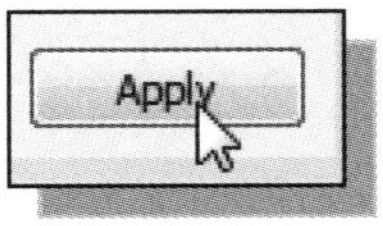

8. Click on the **Apply** button to proceed with retrieving the selected dimensions.

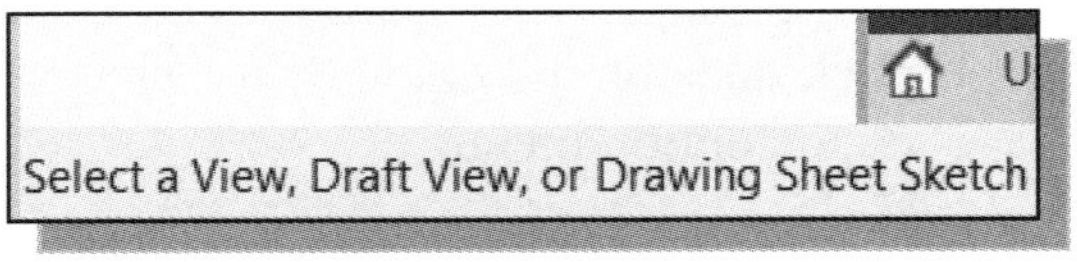

9. In the *message* area, the Inventor system expects us to **Select a View** as shown.

10. Select the ***front*** view.

11. On your own, retrieve the three dimensions as shown.

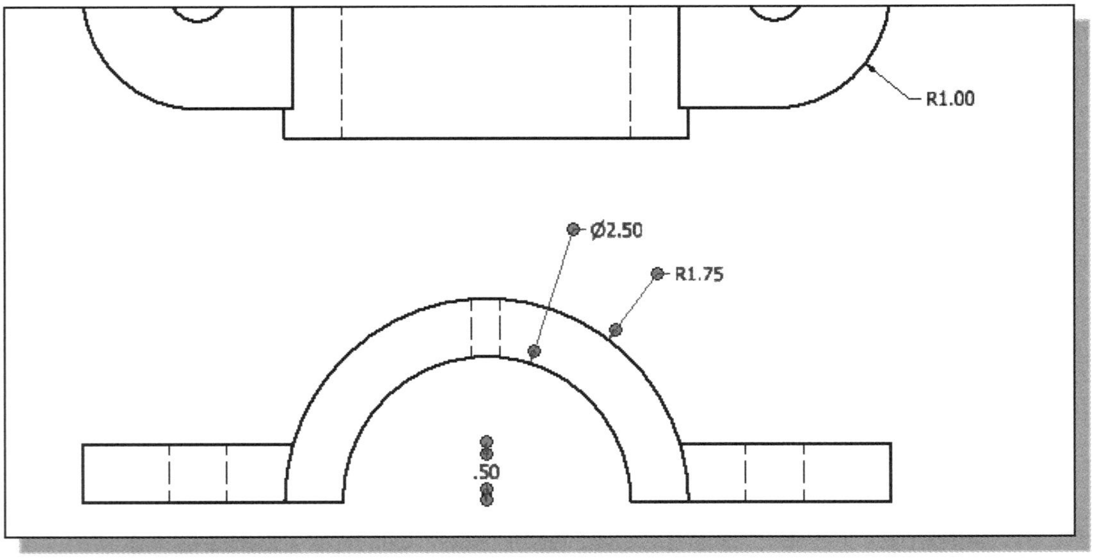

12. Click on the **OK** button to end the Retrieve Model Annotations command.

Repositioning and Hiding Feature Dimensions

1. Move the cursor on top of the width dimension text **5.00**, and watch for when the dimension text becomes highlighted with the four-arrow symbol indicating the dimension is selectable.

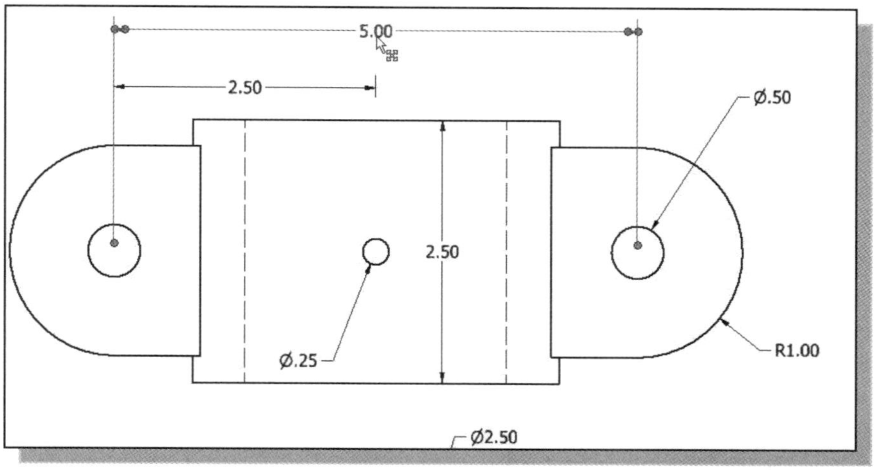

2. Reposition the dimension by using the left-mouse-button and drag the dimension text to a new location.

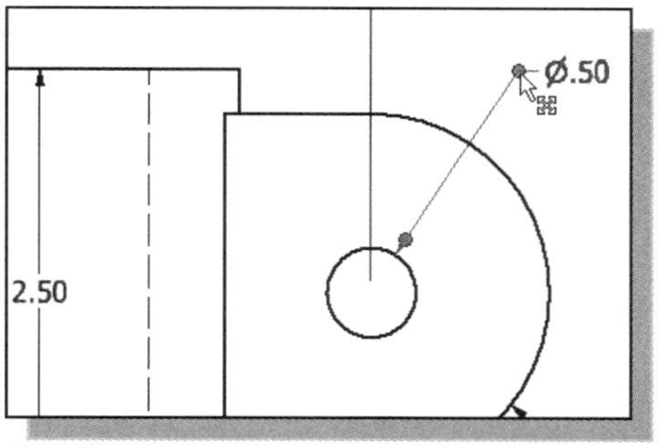

3. Move the cursor on top of the diameter dimension **0.5** and drag the grip point green dot associated with the dimension to reposition the dimension. Note that we can also drag on the dimension text, which only repositions the text.

4. On your own, reposition the dimensions displayed in the ***top*** view as shown in the figure below.

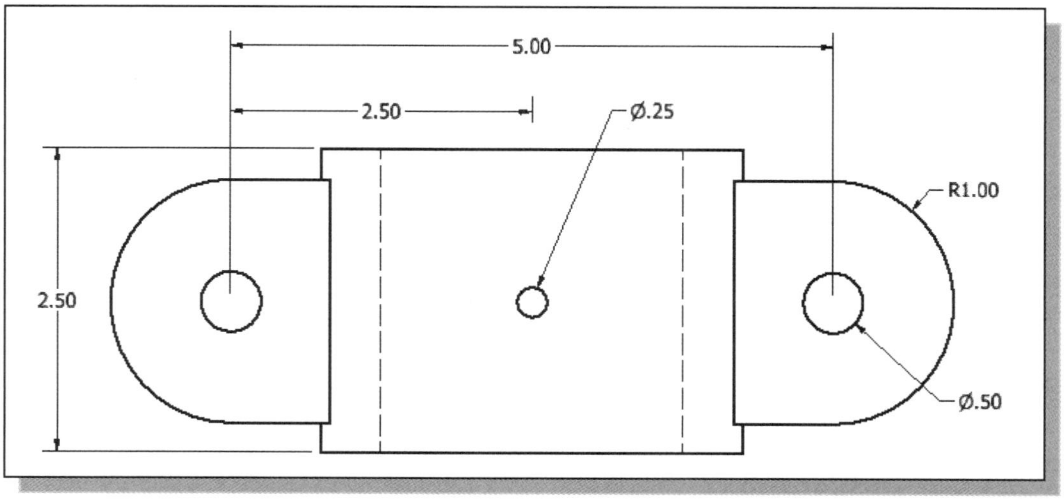

5. Move the cursor on top of the radius dimension **R 1.75** and notice two green grip points appear. The grip points can be used to reposition the dimension.

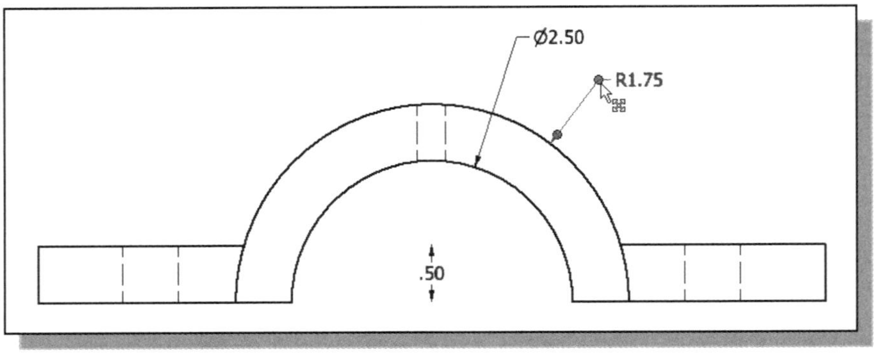

6. Use the left-mouse-button and drag the grip point near the center of the view and notice the arrowhead is automatically adjusted to the inside, as shown in the figures below.

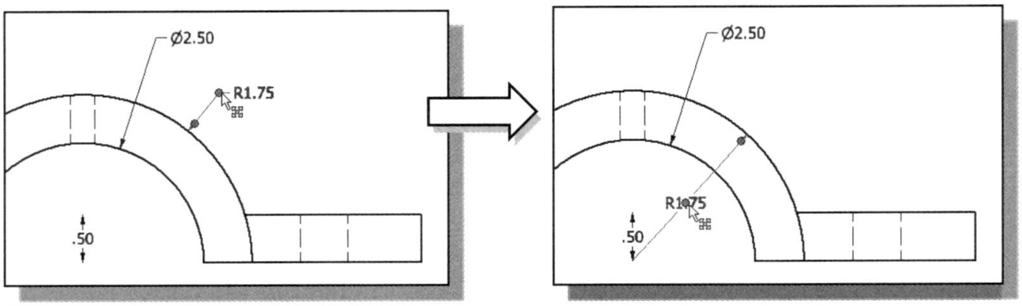

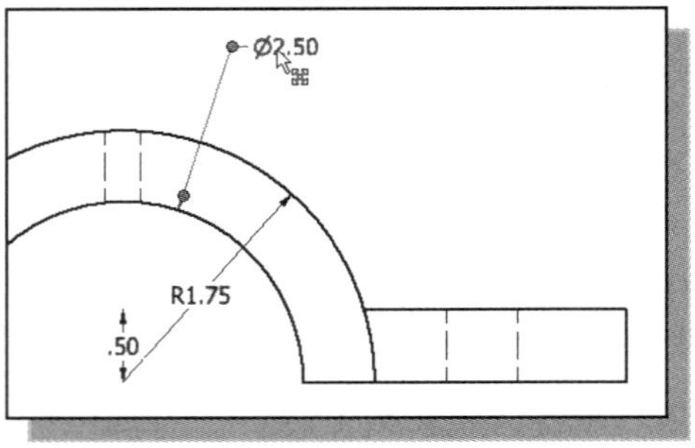

7. Move the cursor on top of the diameter dimension **2.50** and left-mouse-click once to select the dimension.

8. Right-click once on the dimension text to bring up the **option menu**.

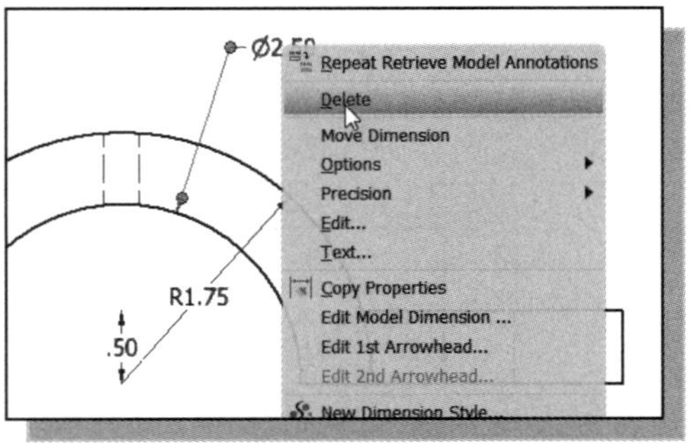

9. Select **Delete** to remove the dimension from the display.

➢ Note that the feature dimension is deleted from the display, but the removed dimension still remains in the database. In other words, the feature dimension is turned off or is hidden. Any feature dimensions can be removed from the display just as they can be displayed.

Add Additional Dimensions – Reference Dimensions

Besides displaying the **feature dimensions**, dimensions used to create the features, we can also add additional **reference dimensions** in the drawing. *Feature dimensions* are used to control the geometry, whereas *reference dimensions* are controlled by the existing geometry. In the drawing layout, therefore, we can ***add*** *or* ***delete*** *reference dimensions*, but we can only *hide* the *feature dimensions*. One should try to use as many *feature dimensions* as possible and add *reference dimensions* only if necessary. It is also more effective to use *feature dimensions* in the drawing layout since they were created when the model was built. Note that additional *Drawing Mode* entities, such as lines and arcs, can be added to drawing views. Before *Drawing Mode* entities can be used in a reference dimension, they must be associated to a *drawing view*.

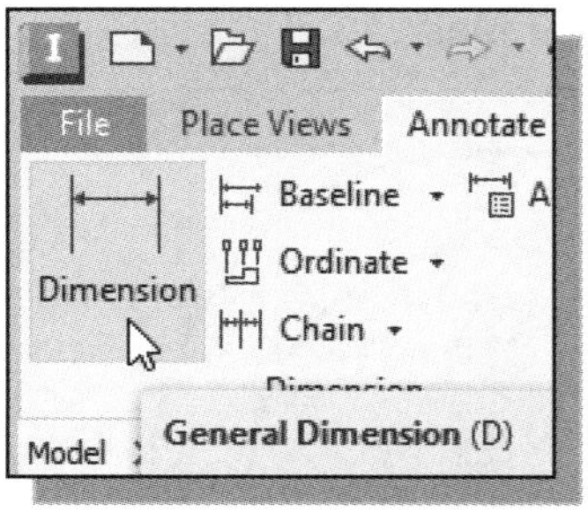

1. Click on the **General Dimension** button.

➢ Note the **General Dimension** command is similar to the **Smart Dimensioning** command in the *3D Modeling Mode.*

2. In the prompt area, the message "*Select first object*:" is displayed. Select the smaller arc of the front view.

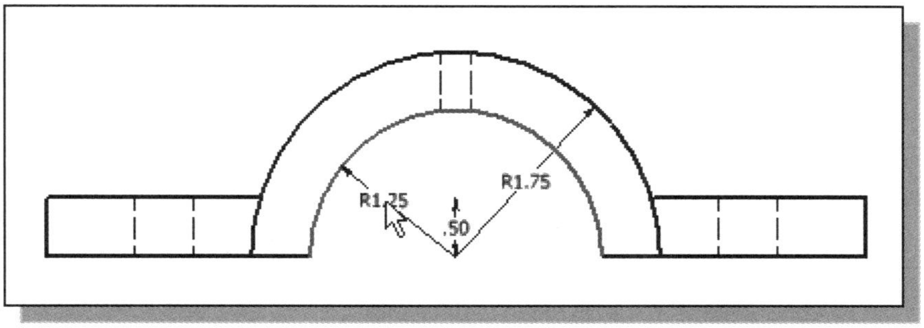

3. Place the dimension text on the inside of the arc as shown in the above figure.

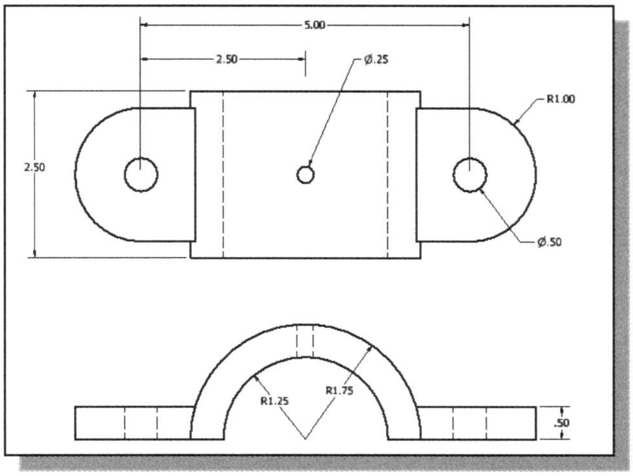

4. On your own, position the necessary dimensions for the design as shown in the figure.

➢ Note the extension lines can also be repositioned by dragging the associated grip points.

Add Center Marks and Center Lines

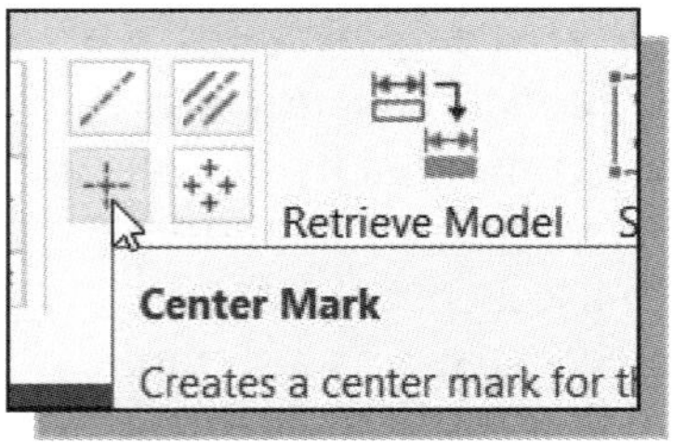

1. Click on the **Center Mark** button in the *Drawing Annotation* window.

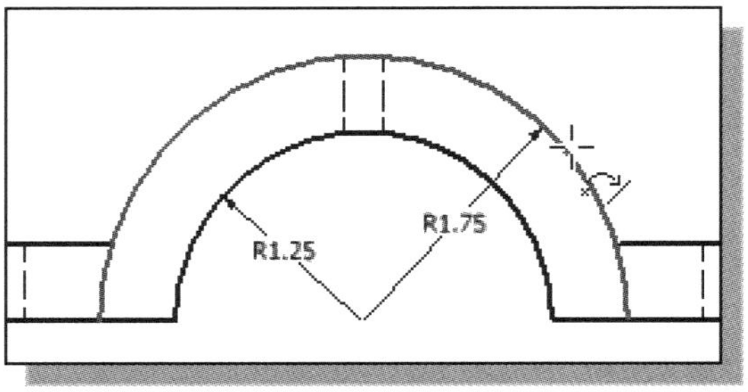

2. Click on the larger arc in the front view to add the center mark.

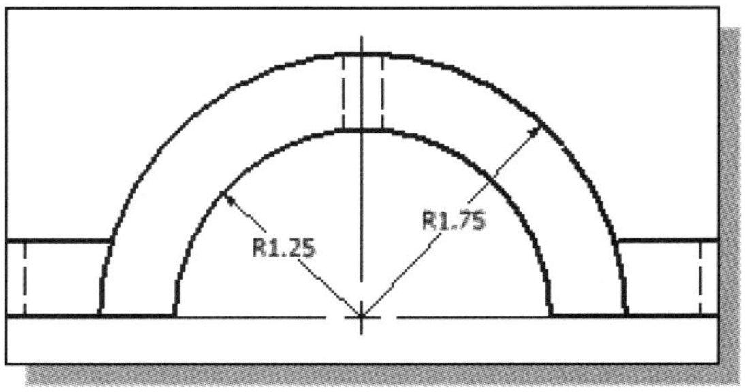

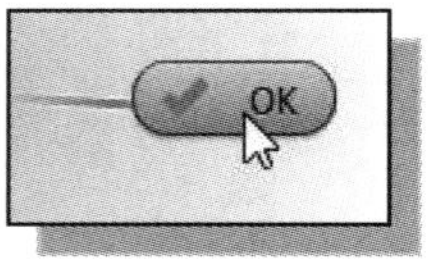

3. Inside the *graphics window*, click once with the **right-mouse-button** to display the option menu. Select **OK** in the pop-up menu to end the Center Mark command.

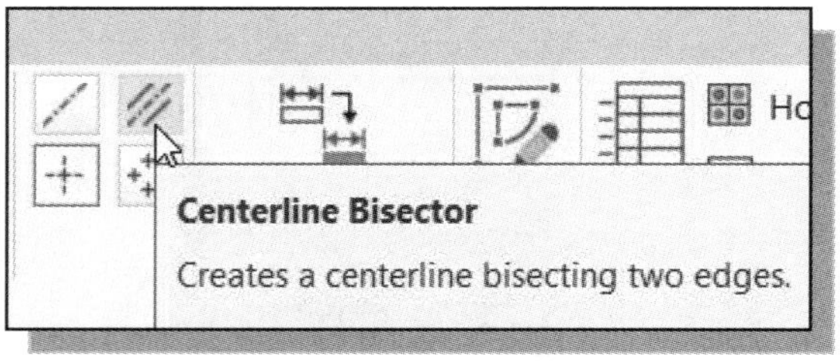

4. Select **Centerline Bisector** from the option list.

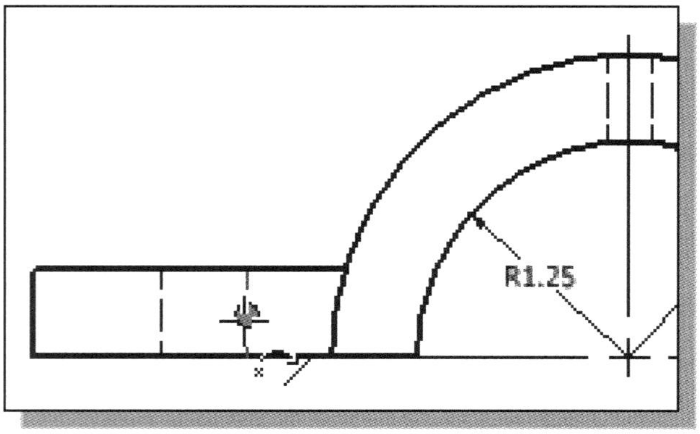

5. Click on the two hidden edges of one of the *drill* features of the front view as shown in the figure.

6. On your own, repeat the above step and create another centerline on the right side of the front view as shown.

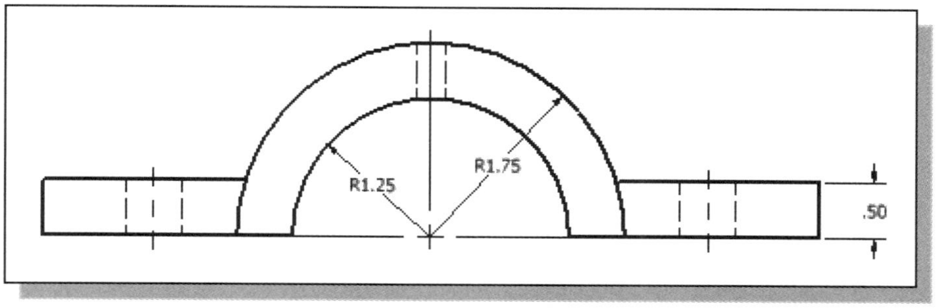

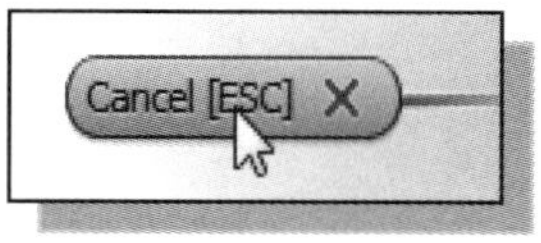

7. Inside the graphics window, click once with the right-mouse-button to display the option menu. Select **Cancel [ESC]** in the pop-up menu to end the Centerline Bisector command.

8. On your own, repeat the above steps and create additional centerlines as shown in the figure below.

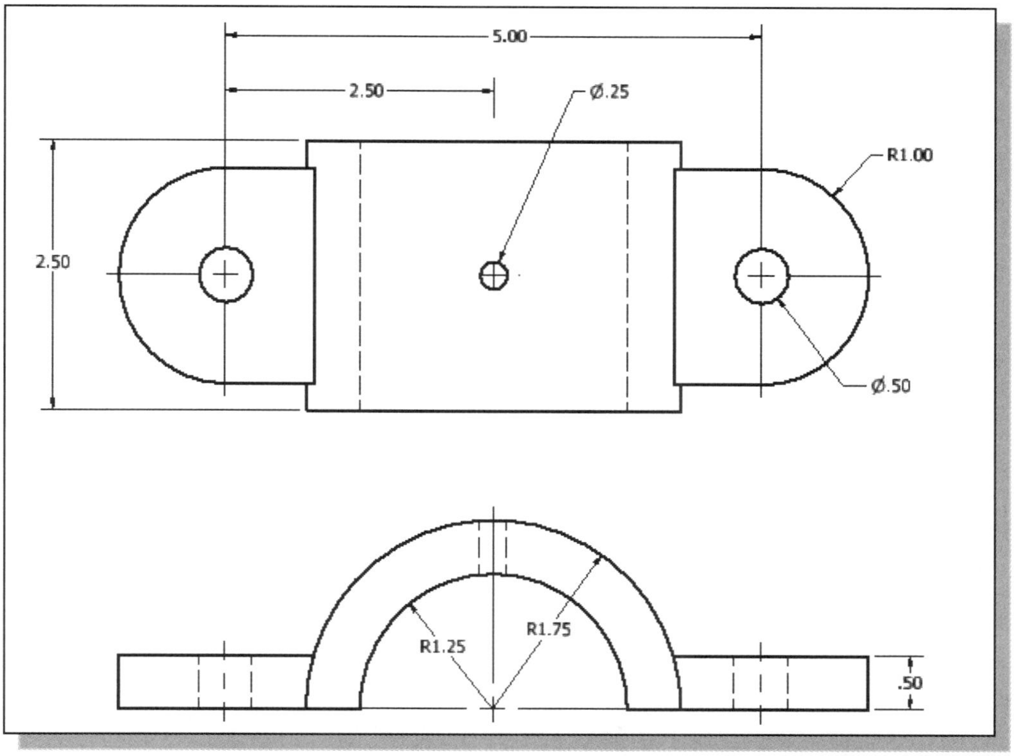

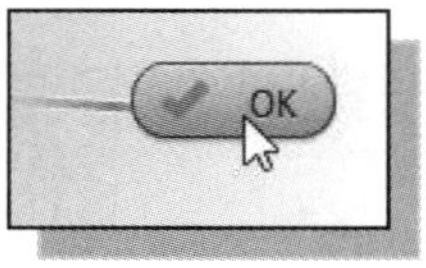

9. Inside the graphics window, click once with the right-mouse-button to display the option menu. Select **OK** in the pop-up menu to end the Centerline command.

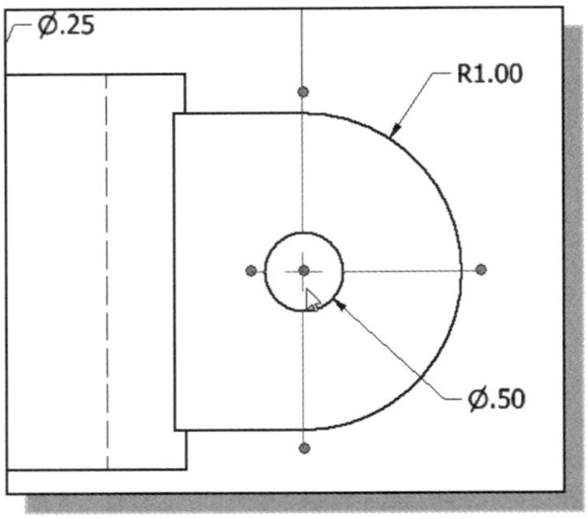

10. Click on the right centerlines in the *top* view as shown.

11. Adjust the length of the horizontal centerline by dragging on one of the grip points as shown.

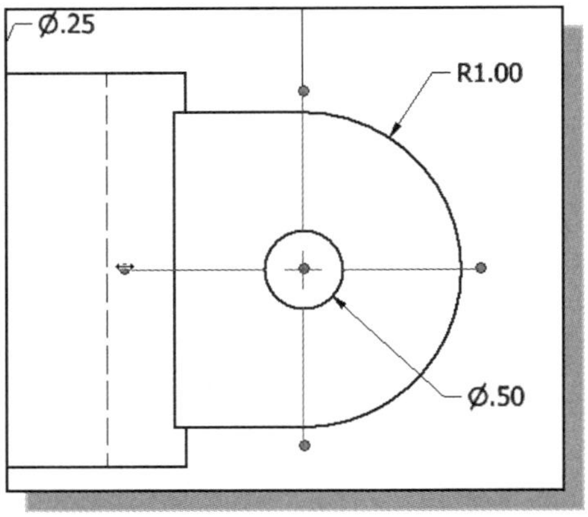

12. On your own, repeat the above steps and adjust the dimensions/centerlines as shown below.

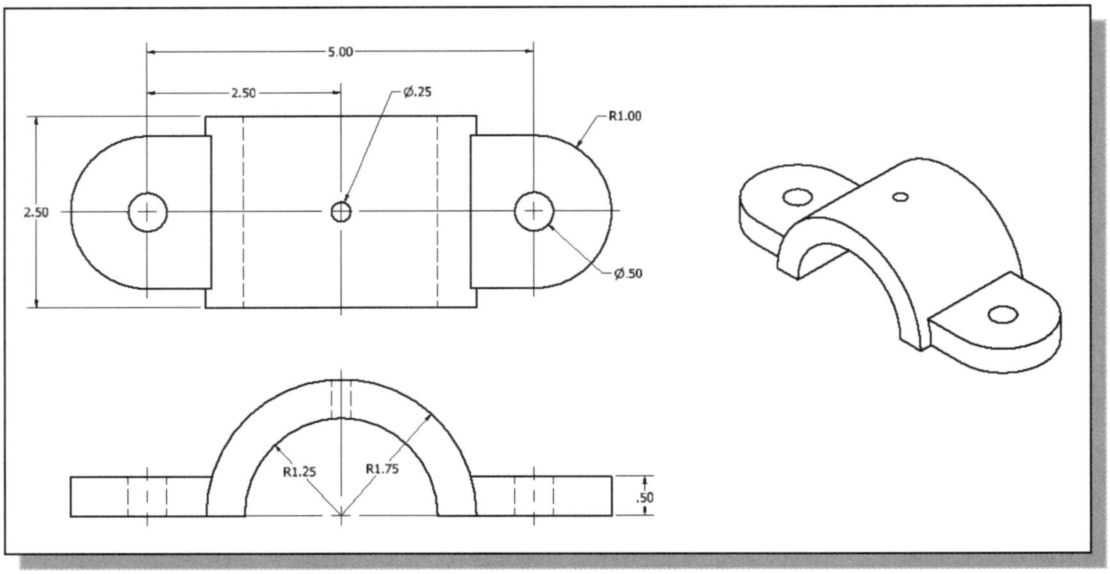

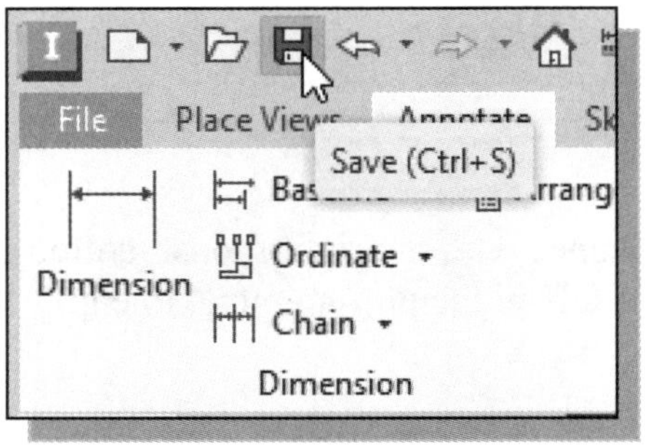

13. Click on the **Save File** icon in the *Standard* toolbar as shown.

14. Click the **Save** button to use the default drawing name.

Complete the Drawing Sheet

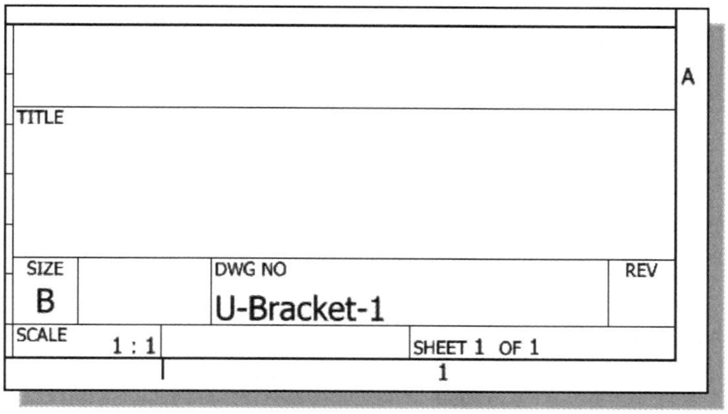

1. On your own, use the **Zoom** and the **Pan** commands to adjust the display as shown; this is so that we can complete the title block.

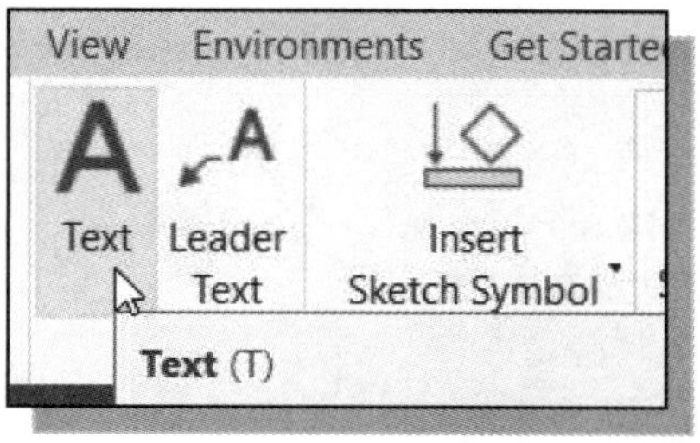

2. In the *Drawing Annotation* window, click on the **Text** button.

3. Pick a location that is inside the top block area as the location for the new text to be entered.

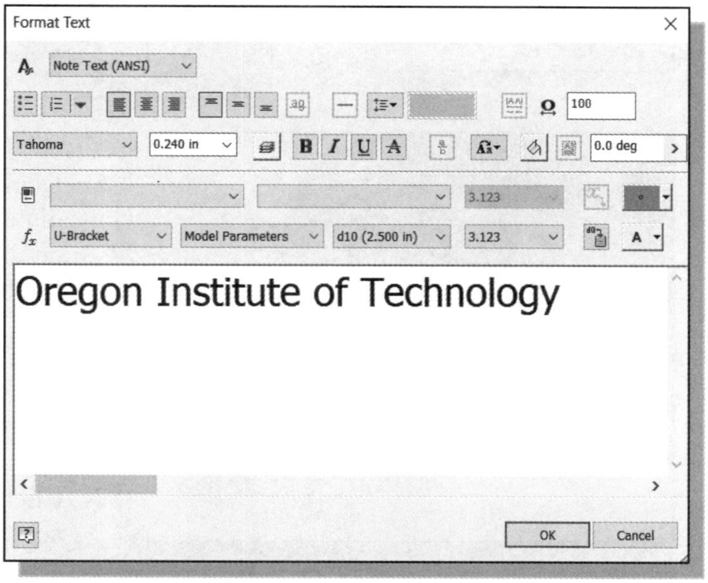

4. In the *Format Text* dialog box, enter the name of your organization. Also note the different settings available.

5. Click **OK** to proceed.

6. On your own, repeat the above steps and complete the title block.

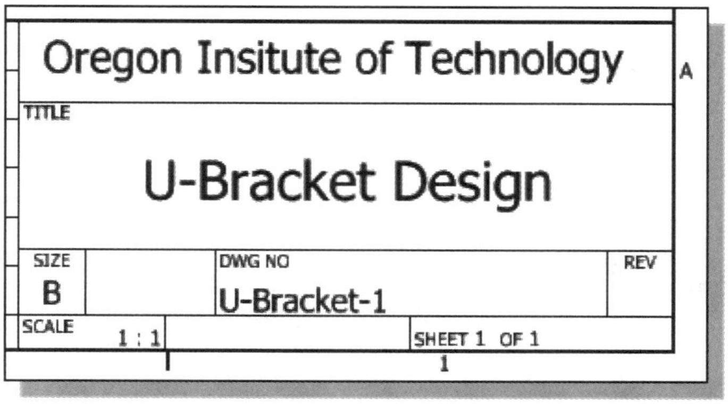

Associative Functionality – Modifying Feature Dimensions

Autodesk Inventor's *associative functionality* allows us to change the design at any level, and the system reflects the changes at all levels automatically.

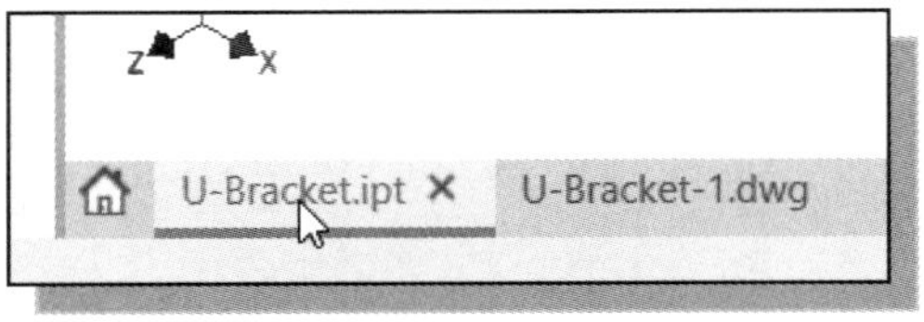

1. Click on the ***U-Bracket*** part window or tab to switch to the *Solid Model.*

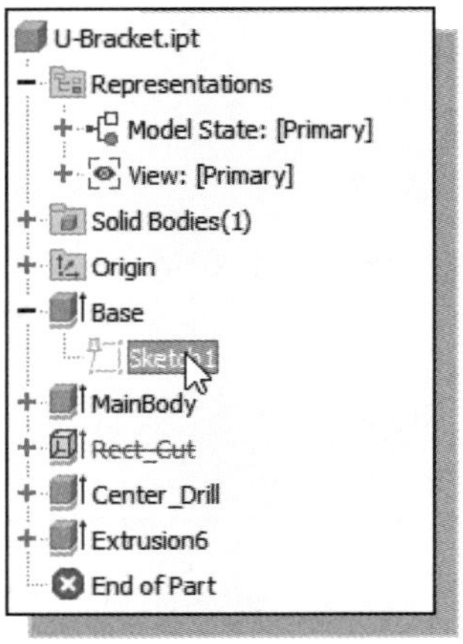

2. In the *browser* window, click once on the [+] sign in front of the **Base** (Extrusion1) to expand the menu list.

3. Select **Sketch1** to display the sketch and dimensions in the graphics area.

4. Choose **Make Sketch visible** as shown.

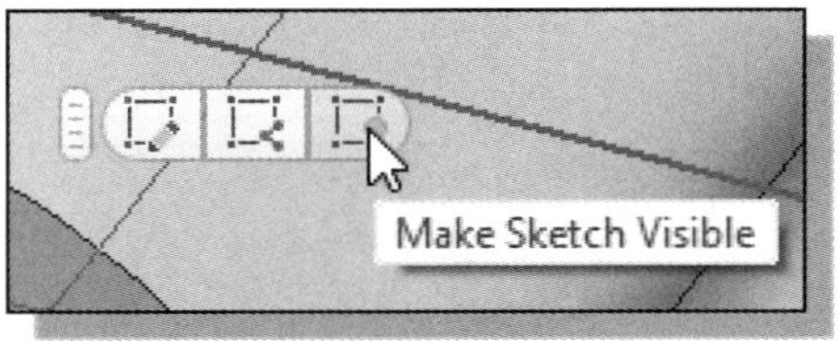

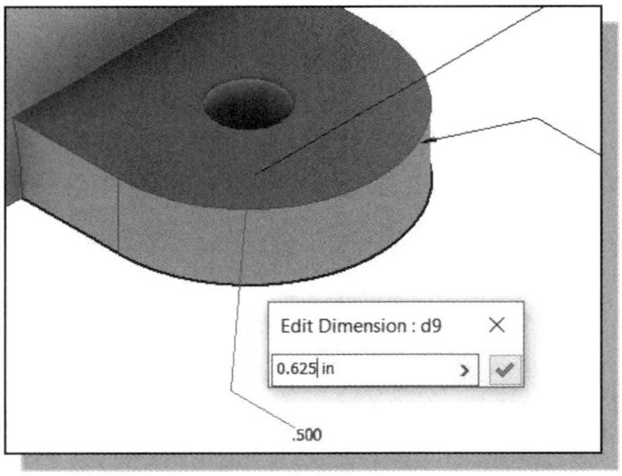

5. Double-click on the diameter dimensions (**0.50**) of the drill feature on the base feature as shown in the figure.

6. In the *Edit Dimension* dialog box, enter **0.625** as the new diameter dimension.

7. Click on the **check mark** button to accept the new setting.

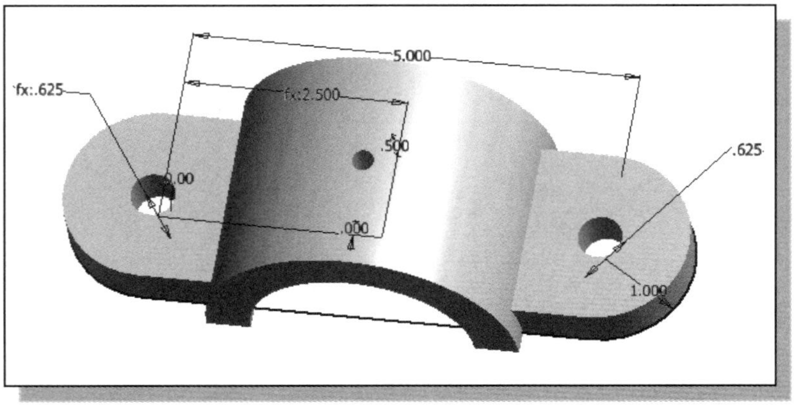

8. On your own, confirm the diameter of the other drill feature is also set to **0.625** as shown.

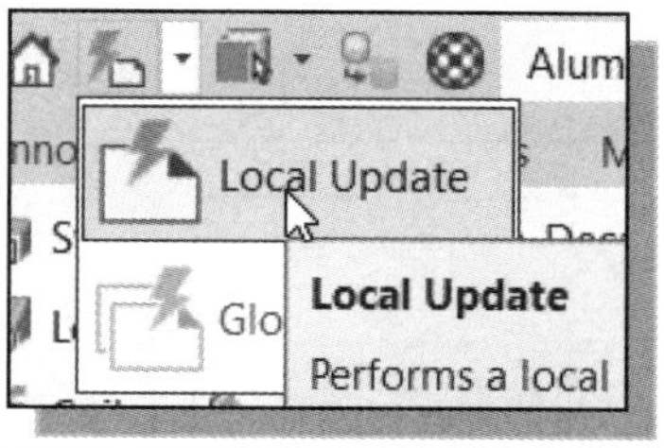

9. Click on the **Update** button in the *Standard* toolbar area to proceed with updating the solid model.

10. Click on the ***U-Bracket-1*** drawing graphics window or tab to switch to the *Multi-View Drawing*.

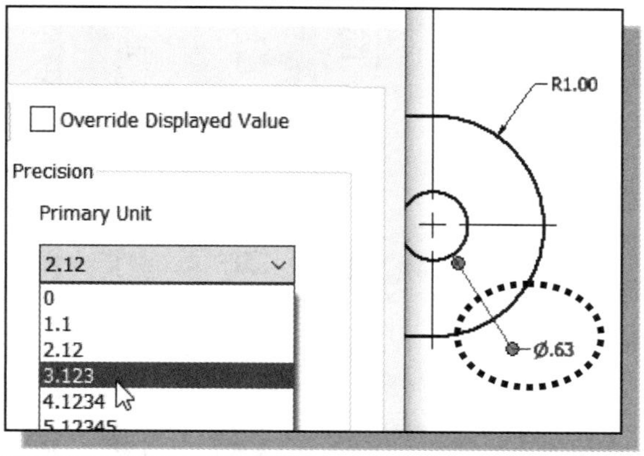

11. Inside the *graphics window*, double-click on the **0.63** dimension in the ***top*** view to bring up the *Precision and Tolerance* dialog box.

12. Set the *Precision* option to **3 digits after the decimal point** as shown.

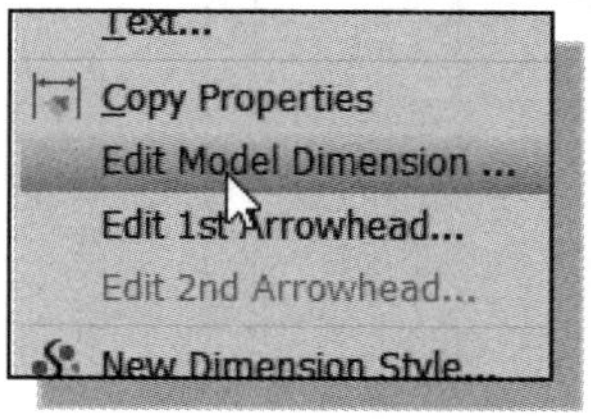

13. Inside the graphics window, right-click once on the **R 1.75** dimension in the *front* view to bring up the option menu.

14. Select **Edit Model Dimension** in the pop-up menu.

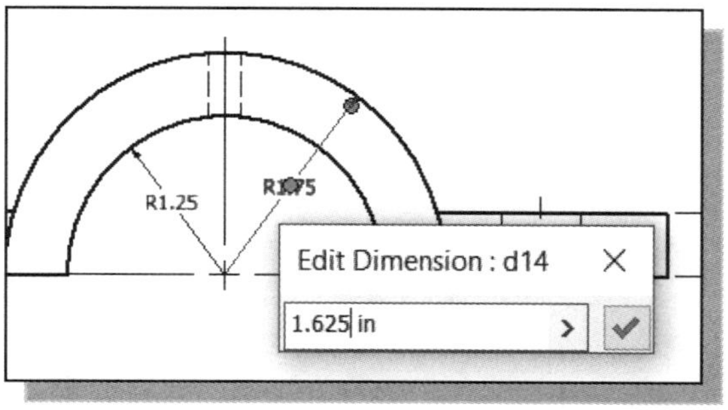

15. Change the dimension to **1.625**.

16. Click on the **check mark** button to accept the setting.

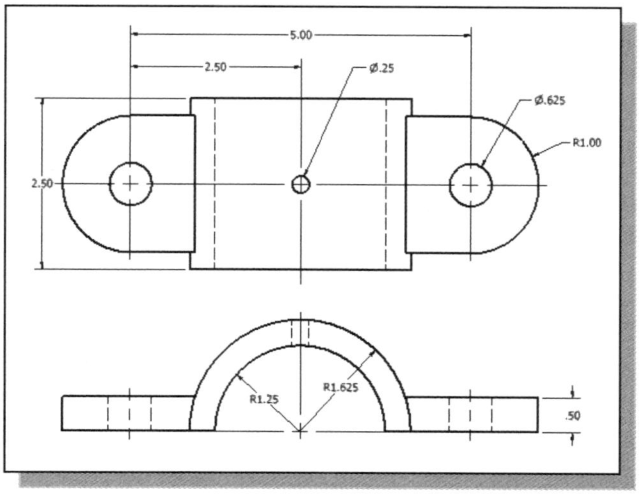

- Note the geometry of the cut feature is updated in all views automatically.

- On your own, switch to the *Part Modeling Mode* and confirm the design is updated as well.

- The completed multi-view drawing should appear as shown on the next page.

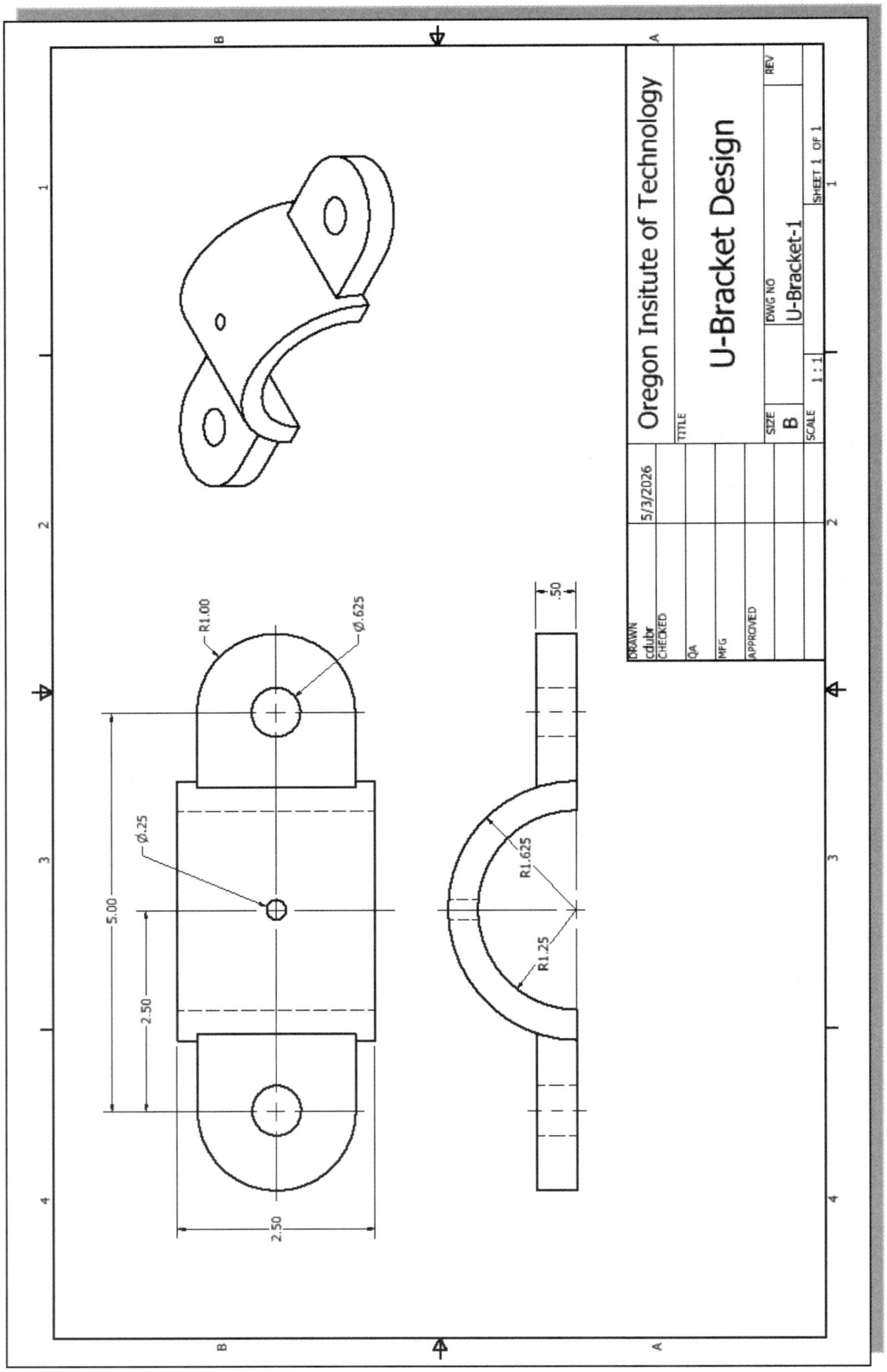
Oregon Insitute of Technology
U-Bracket Design
DRAWN
cdubr
5/3/2026
CHECKED
QA
MFG
APPROVED
TITLE
SIZE
B
DWG NO
U-Bracket-1
REV
SCALE 1 : 1
SHEET 1 OF 1
R1.00
Ø.625
Ø.25
5.00
2.50
2.50
.50
R1.625
R1.25

3D Model-Based Definition

The 3D Model-Based Definition (MBD) functionality is also available in Autodesk Inventor 2025. Modern 3D CAD applications allow for the insertion of engineering information such as dimensions, GD&T, notes and other product details directly within the 3D digital data set for components and assemblies. MBD uses such capabilities to establish the 3D digital data set as the source of these specifications and design authority for the product. The 3D digital data set may contain enough information to manufacture and inspect a product without the need for 2D engineering drawings.

Autodesk has partnered with *Sigmetrix/Advanced Dimensional Management*, leading providers of tolerance analysis and GD&T software and training, to help users apply semantic GD&T properly per the rules of the ASME Y14.5 and applicable ISO GPS standards. Autodesk Inventor 2025 supports both ASME and ISO standards. Inventor 2025 provides API access to all the MBD data, which means that downstream software can directly access the Product Manufacturing Information from the model.

The new Annotate ribbon provides a suite of tools for annotating models and exporting data sets to 3D PDF and other 3D formats. The 3D Annotation tools are divided into three categories: Geometric Annotation, General Annotation, and Notes.

Geometric Annotation tools are related to GD&T and GPS. The **Tolerance Feature** tool is used to apply geometric tolerances to features. The **DRF** tool is used to define Datum Reference Frames and datum systems. The **Tolerance Advisor** tool provides feedback on completeness of the GD&T scheme, error messages and warnings as the GD&T is applied.

General Annotation includes the **Dimension** tool, the **Hole/Thread Note** tool, and the **Surface Texture** tool. The *Dimension* tool is used to apply dimensions and directly-toleranced dimensions. The *Hole/Thread Note* is used to annotate complex holes and threaded holes. The *Surface Texture* tool is used to define surface texture requirements.

Notes are used to apply notes and general profile tolerances. Local notes may be defined using the *Leader Text* tool. General notes may be defined by the *General Note* tool. Default profile tolerances may be defined by the *General Profile Note* tool.

With this set of MBD tools, the mental translation of 3D models to two-dimensional (2D) drawings for communicating information and then back to 3D models for manufacturing is replaced by a simplified approach. The 3D annotation approach means faster and more precise product communication, which can improve the product development process.

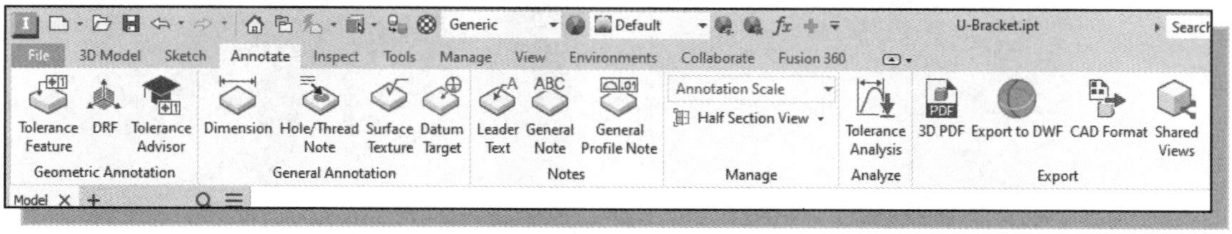

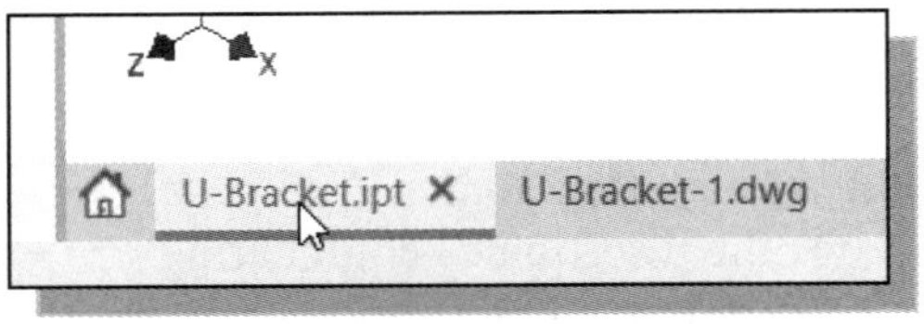

1. Click on the ***U-Bracket*** part window or tab to switch to the *Solid Model.*

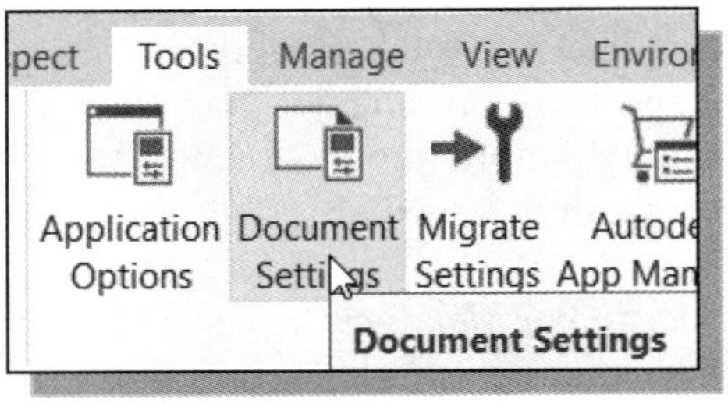

2. Click on the **Tools** tab and pick the *Document Settings* as shown.

3. Confirm the *Annotations Standard* is set to **ASME** as the *annotation* standard as shown.

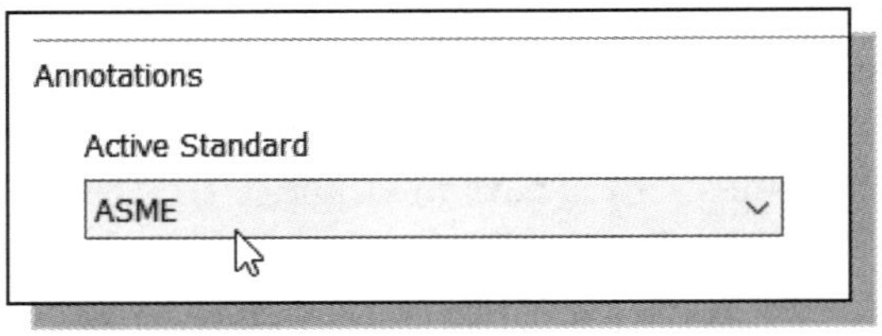

4. Click **Close** to accept the settings.

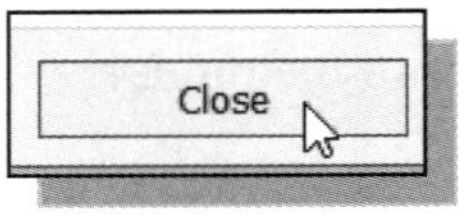

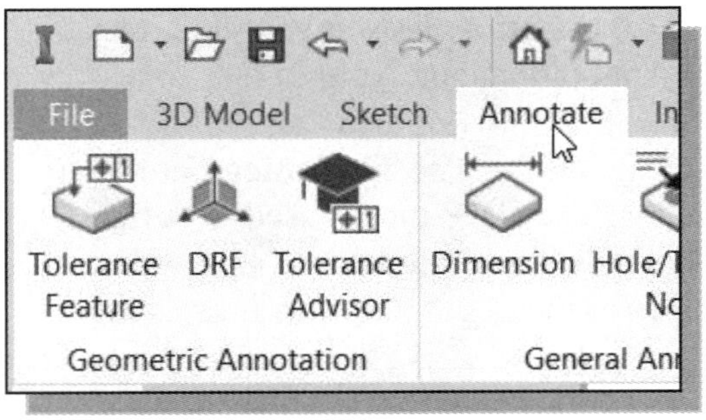

5. Click on the **Annotate** tab to switch to the *MBD toolbar set* as shown.

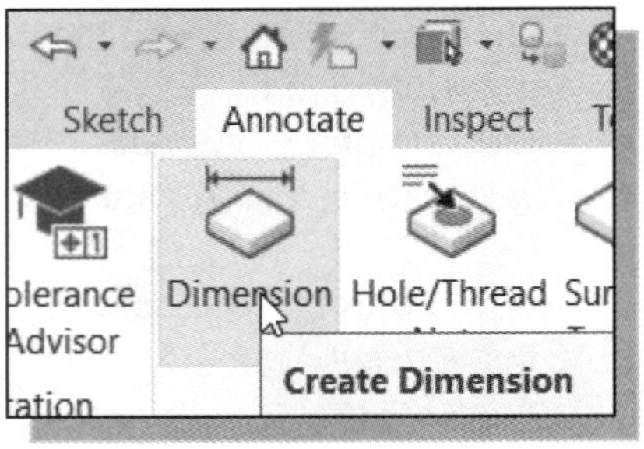

6. Activate the **Create Dimension** command in the *General Annotation* toolbar as shown.

7. Click on the **large arc** on the front section of the solid model as shown.

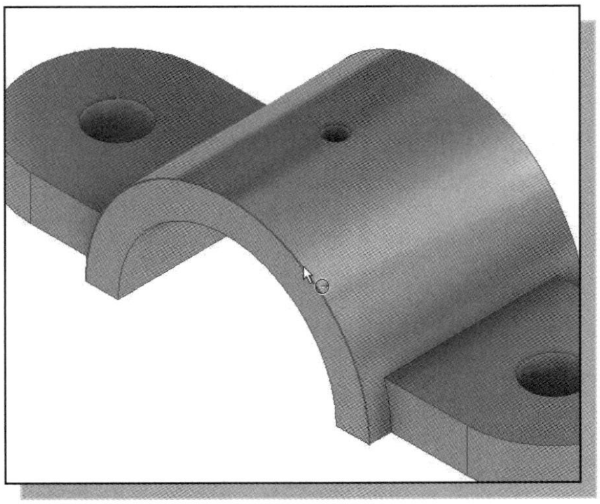

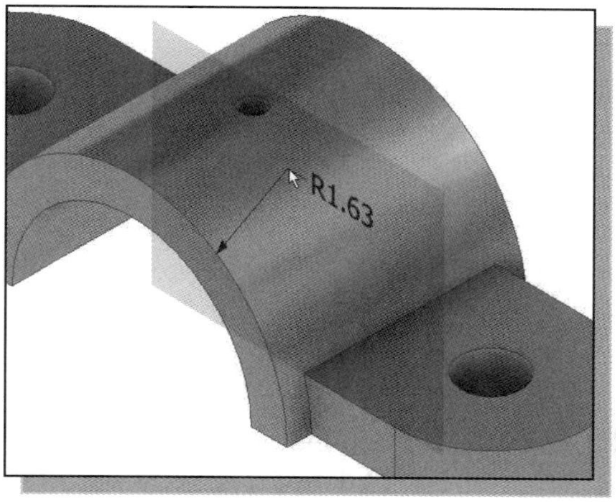

8. On your own, place the dimension above the arc as shown in the figure.

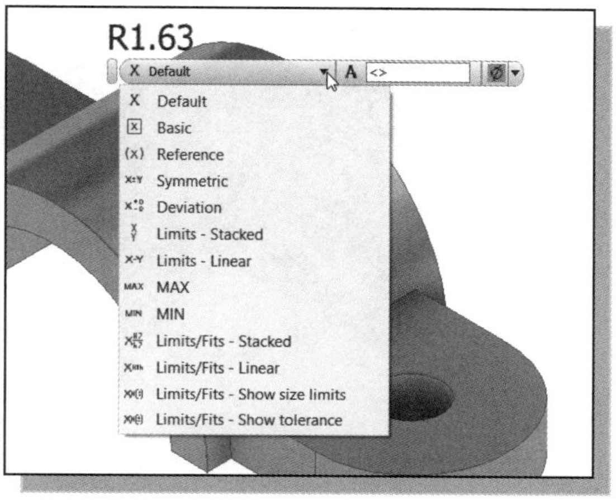

9. Display the first option list and note the different GD&T options available for the associated dimension.

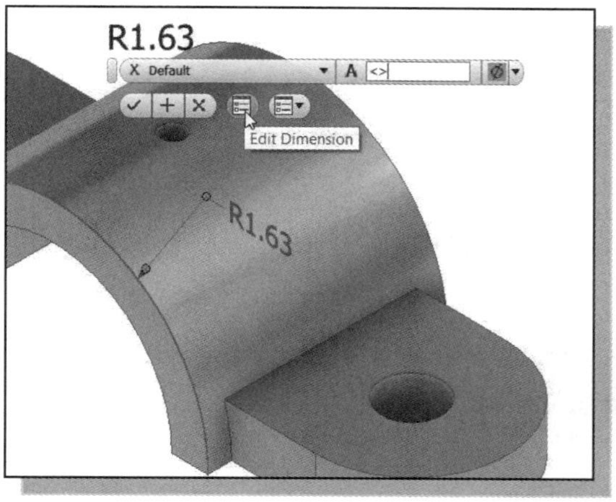

10. On your own, use the **Edit Dimension** icon and change the number of digits displayed as shown.

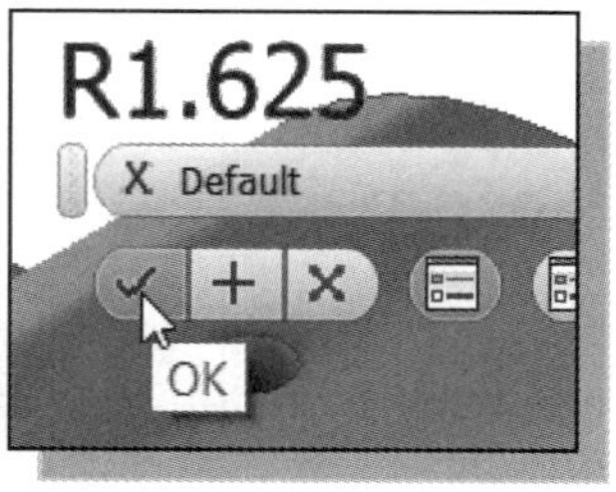

11. Click **OK** to accept the setting and create the dimension as shown.

12. On your own, repeat the above steps and add the other arc dimension as shown.

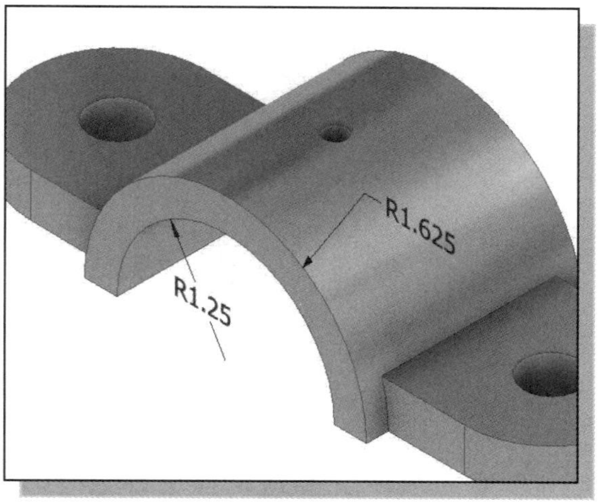

➢ Note that the radius and diameter dimensions are placed on the same plane of the selected arc and circle. The 3D annotation command will automatically select the placement plane if the selected geometry lies on a specific plane.

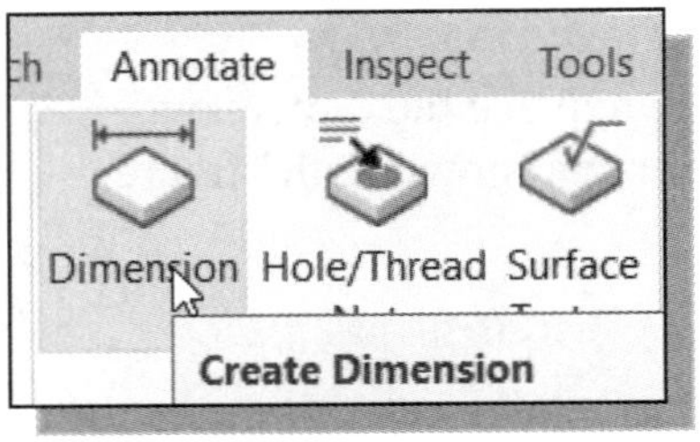

13. Activate the **Create Dimension** command in the *General Annotation* toolbar as shown.

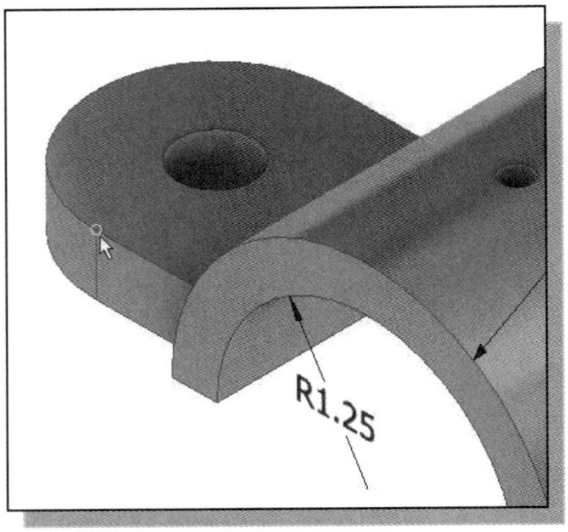

14. Click on the arc endpoint on the left section of the solid model as shown.

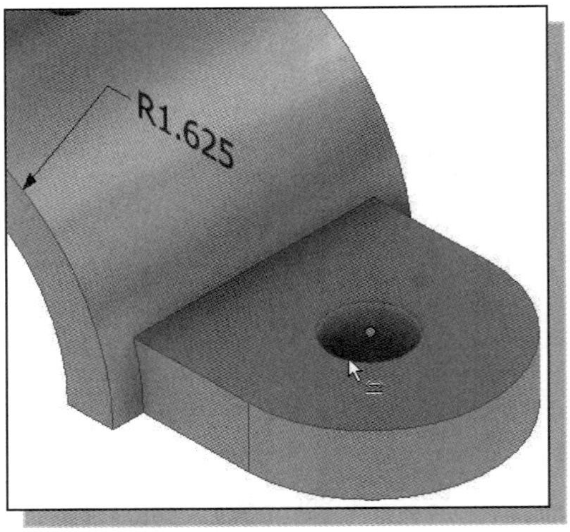

15. Click on the top circle on the right section of the solid model as shown.

➢ Note that by selecting a circle, we also set the placement plane for the dimension.

16. On your own, place the dimension in front of the solid model as shown.

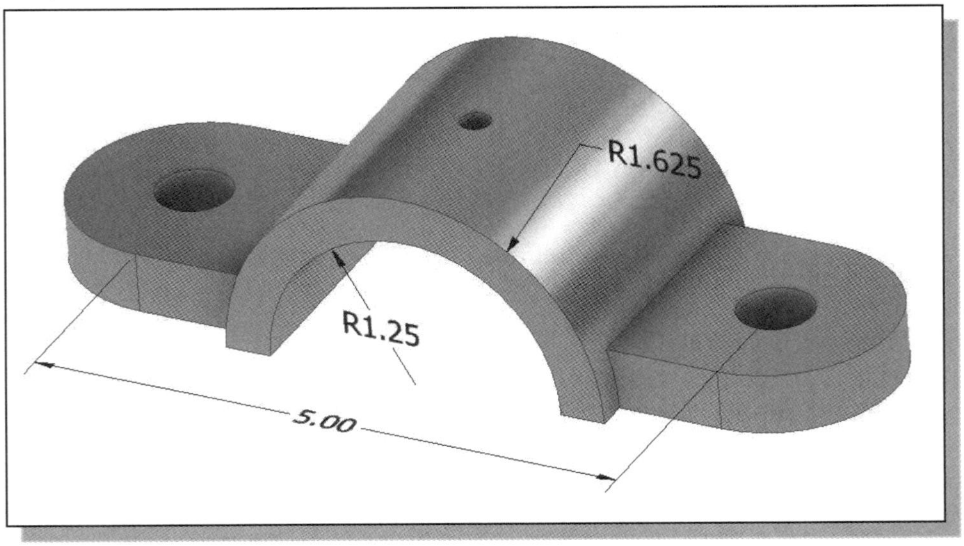

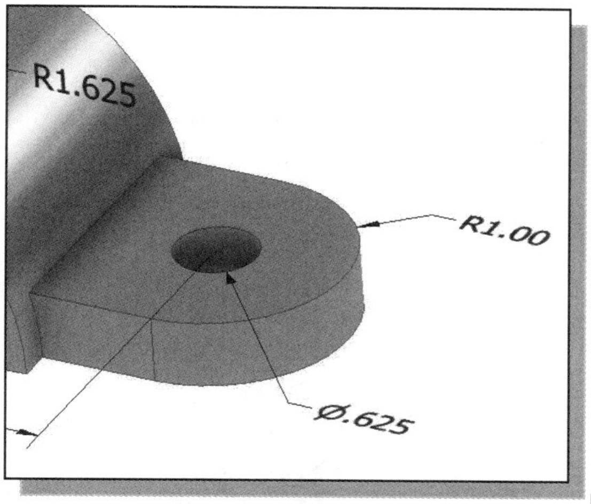

17. On your own, repeat the above steps and create the arc and hole dimensions as shown.

18. On your own, use the dynamic rotation option to view the bottom of the design as shown.

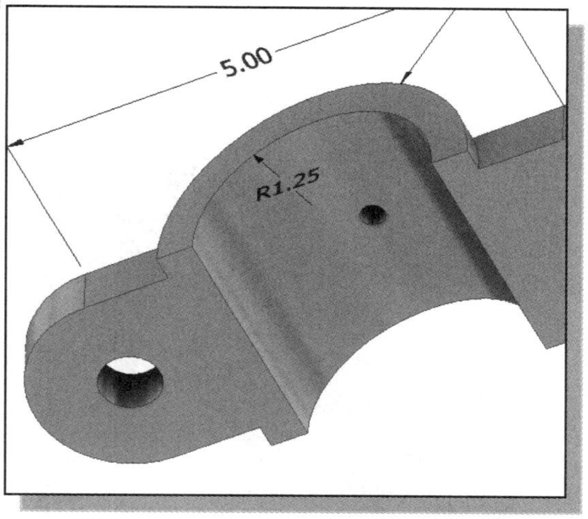

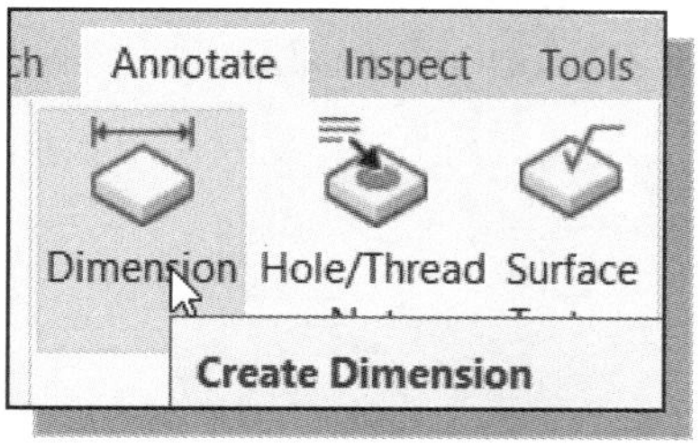

19. Activate the **Create Dimension** command in the *General Annotation* toolbar as shown.

20. On your own, select the straight edge and create the dimension as shown.

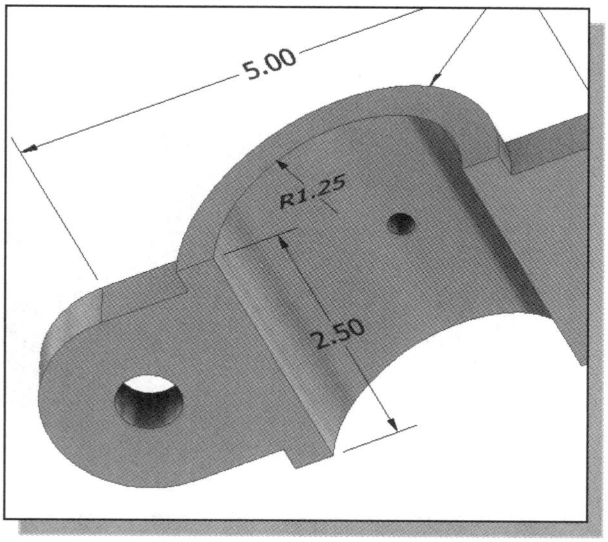

➢ Note the **General Dimension** command in *3D Annotation* behaves similarly to the **Smart Dimensioning** command in the *3D Modeling Mode. Inventor* will automatically create the proper dimensions based on the selected geometry.

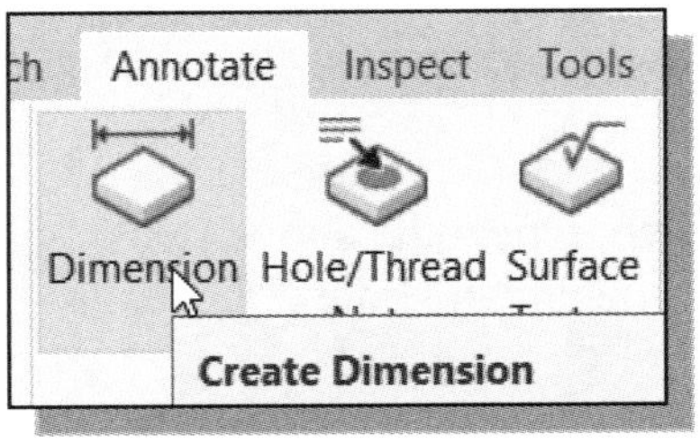

21. Activate the **Create Dimension** command in the General Annotation toolbar as shown.

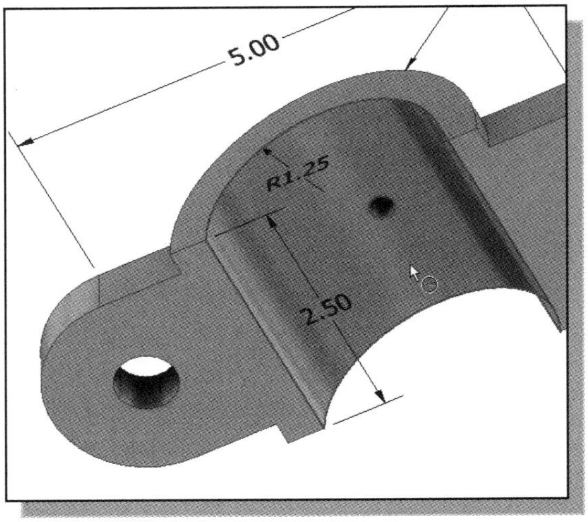

22. Select the inside **cylindrical surface** as shown.

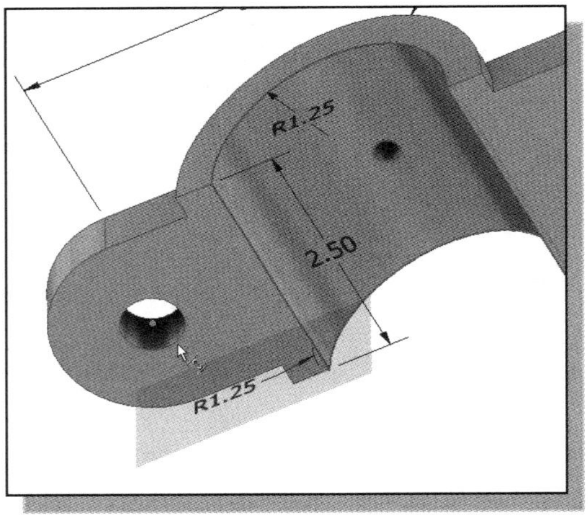

23. Click on the **bottom circle** on the left section of the solid model as shown.

24. On your own, place the dimension as shown.

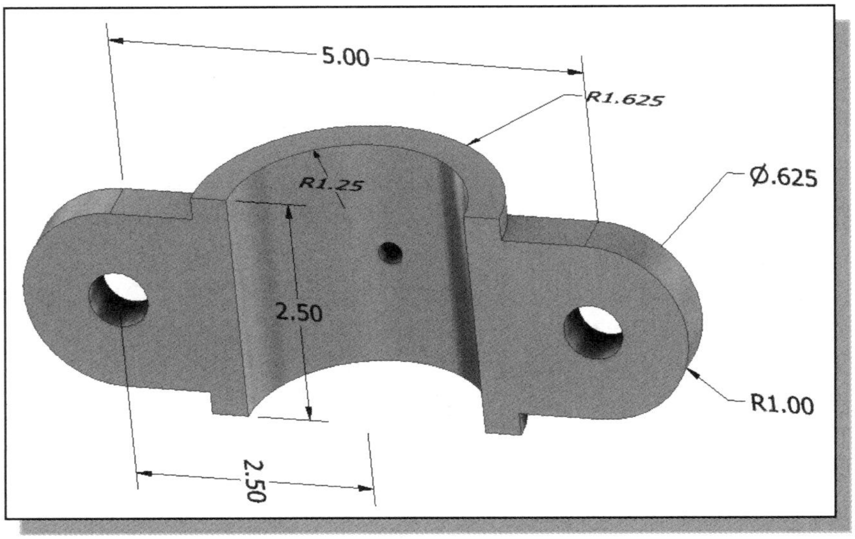

25. Press the function key **F6** once or select **Home View** in the **ViewCube** to change the display back to the default isometric view.

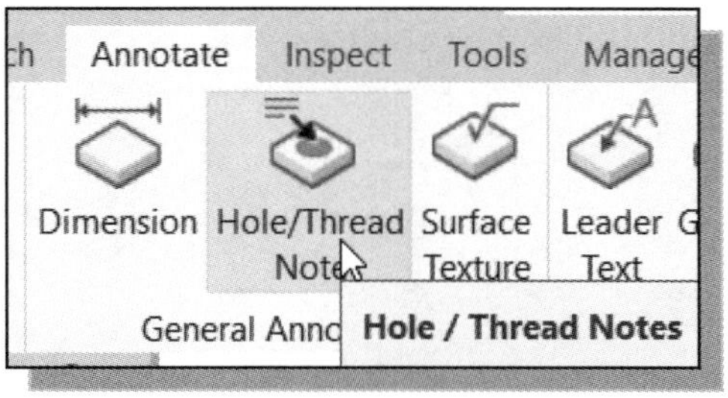

26. Activate the **Hole/Thread Note** command in the *General Annotation* toolbar as shown.

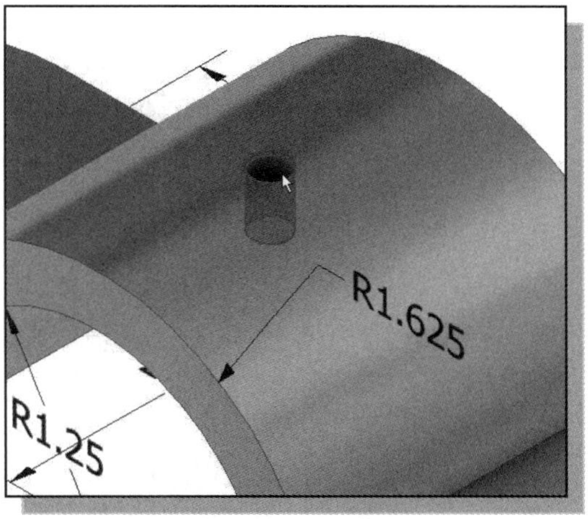

27. Select the **center drill hole** as shown.

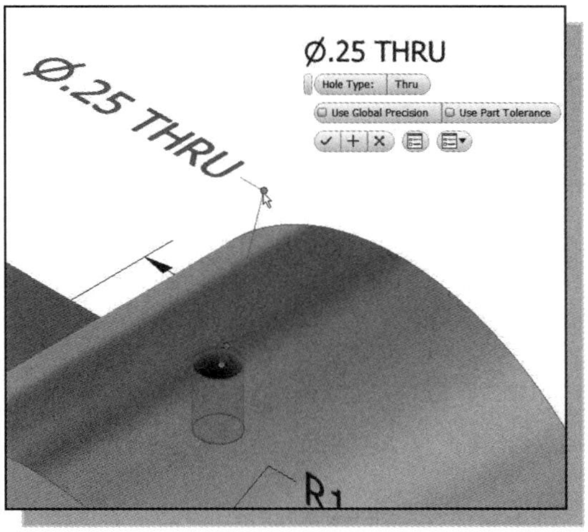

28. Place the dimension toward the left side as shown. Note the different options available of the selected hole feature.

29. Move the cursor on top of the diameter value and **left-click** on the text to display additional dimension options.

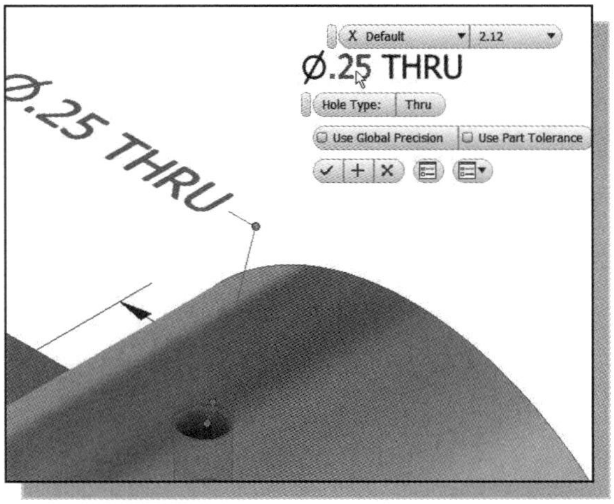

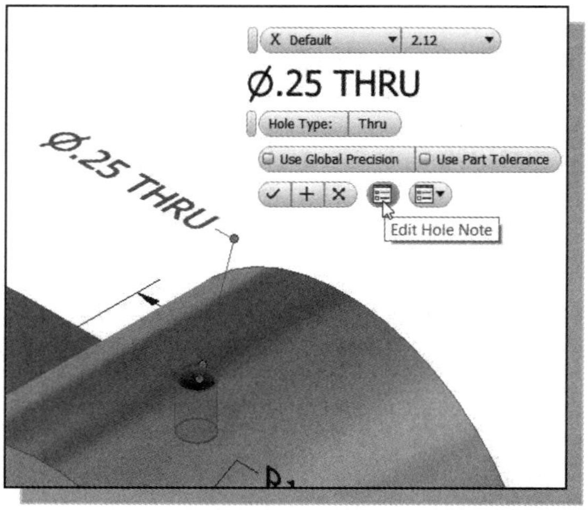

30. Click on the **Edit Hole Note** icon as shown. We will use this option to identify the related features for the small drill hole at the center.

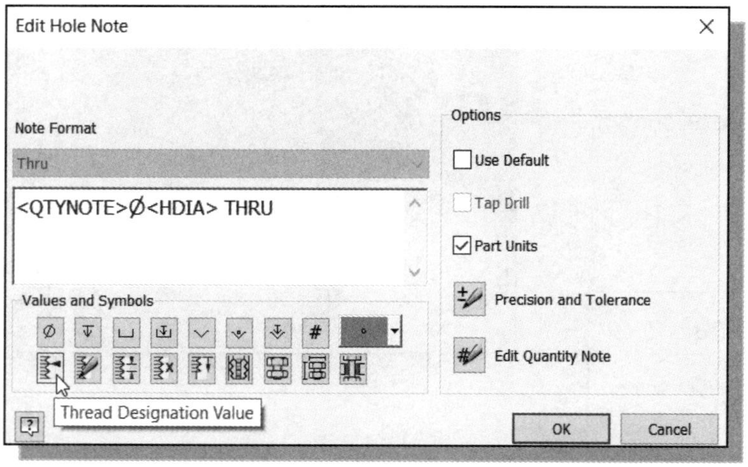

31. Note that we can add additional threads information to the *Hole Note*.

32. Click **OK** to close the *Editor*.

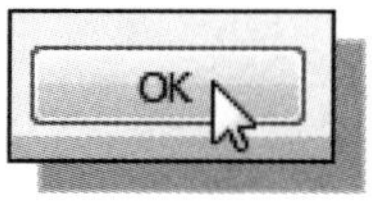

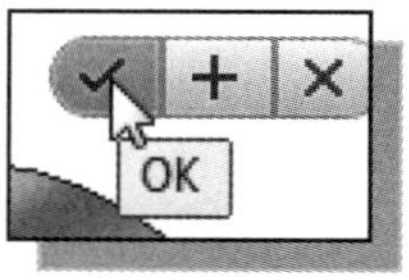

33. Click **OK** to accept the settings and create the dimension and complete the addition of the 3D model-based definition.

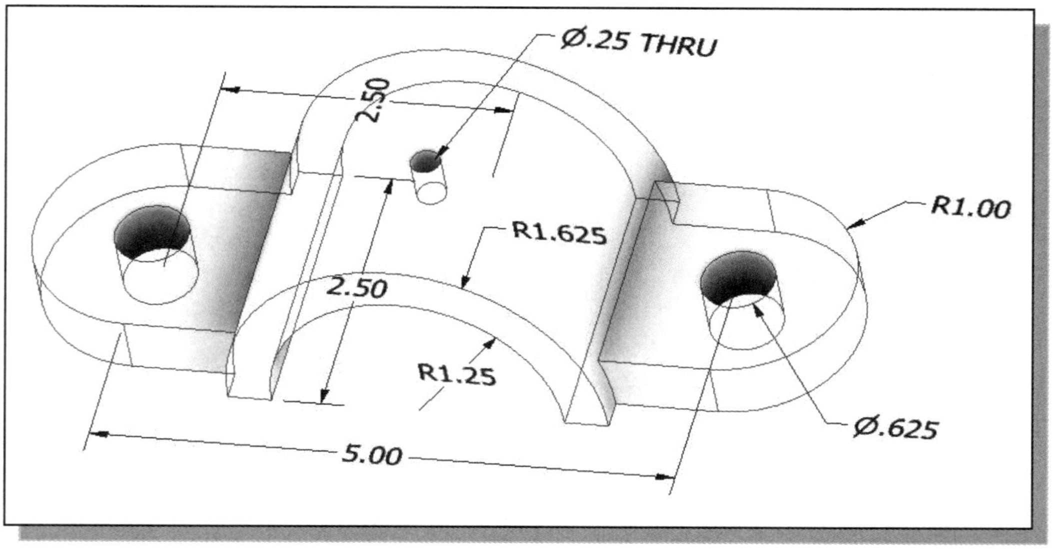

Review Questions: (Time: 25 minutes)

1. What does Autodesk Inventor's *associative functionality* allow us to do?
2. How do we move a view on the *Drawing Sheet*?
3. How do we display feature/model dimensions in the drawing mode?
4. What is the difference between a *feature dimension* and a *reference dimension*?
5. How do we reposition dimensions?
6. What is a *base view*?
7. Can we delete a drawing view? How?
8. Can we adjust the length of centerlines in the drafting mode of Inventor? How?
9. Describe the purpose and usage of the Leader Text command.
10. Describe the advantages of using the **3D Model Based Definition** approach as a documentation tool over the traditional multiview drawing approach.

Exercises: Create the Solid models and the associated 2D drawings and also create the associated MBD of the following exercises. (Time: 180 minutes)

1. **Slide Mount** (Dimensions are in inches.)

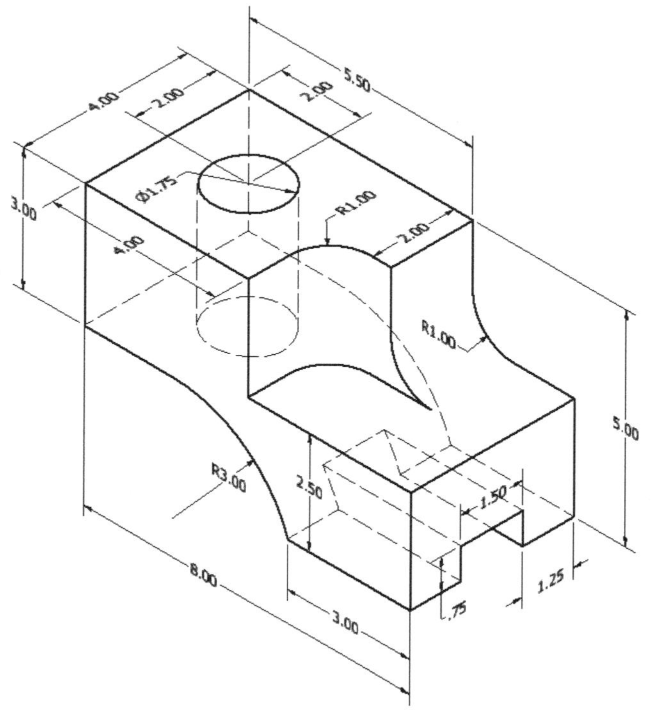

2. **Corner Stop** (Dimensions are in inches.)

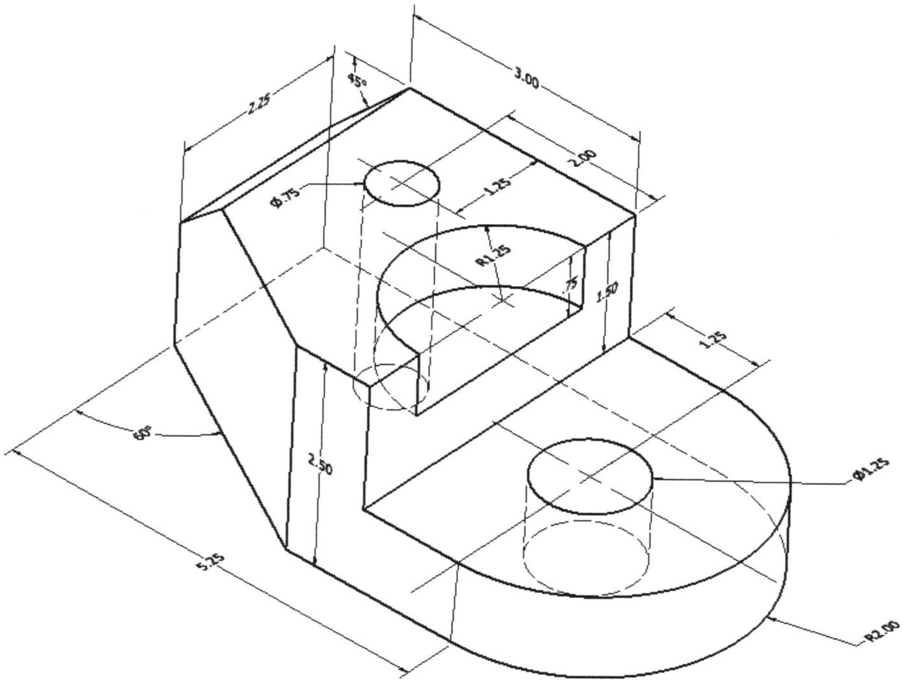

3. **Switch Base** (Dimensions are in inches.)

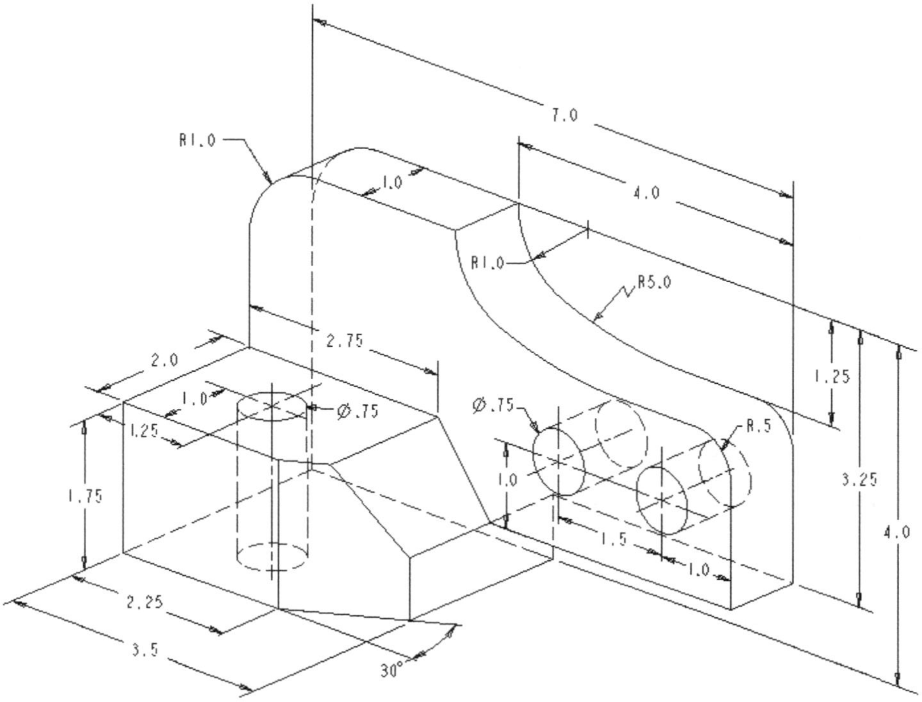

4. **Angle Support** (Dimensions are in inches.)

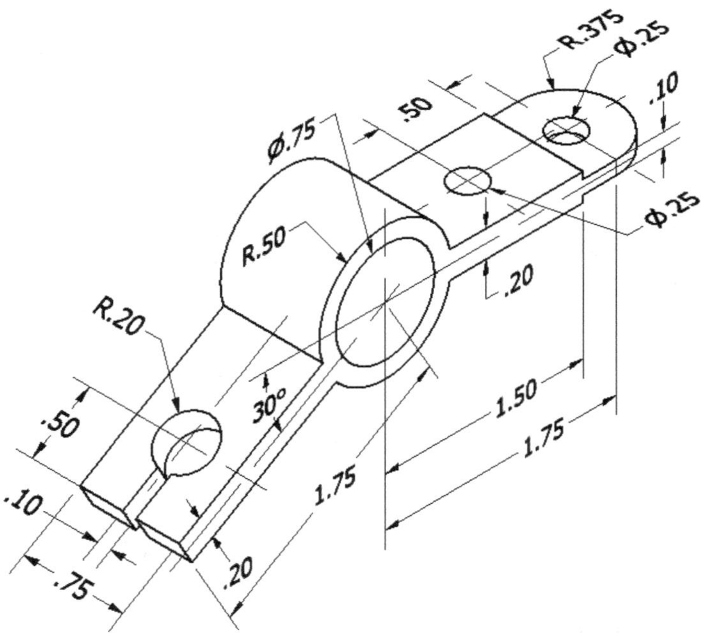

5. **Block Base** (Dimensions are in inches. Plate Thickness: 0.25)

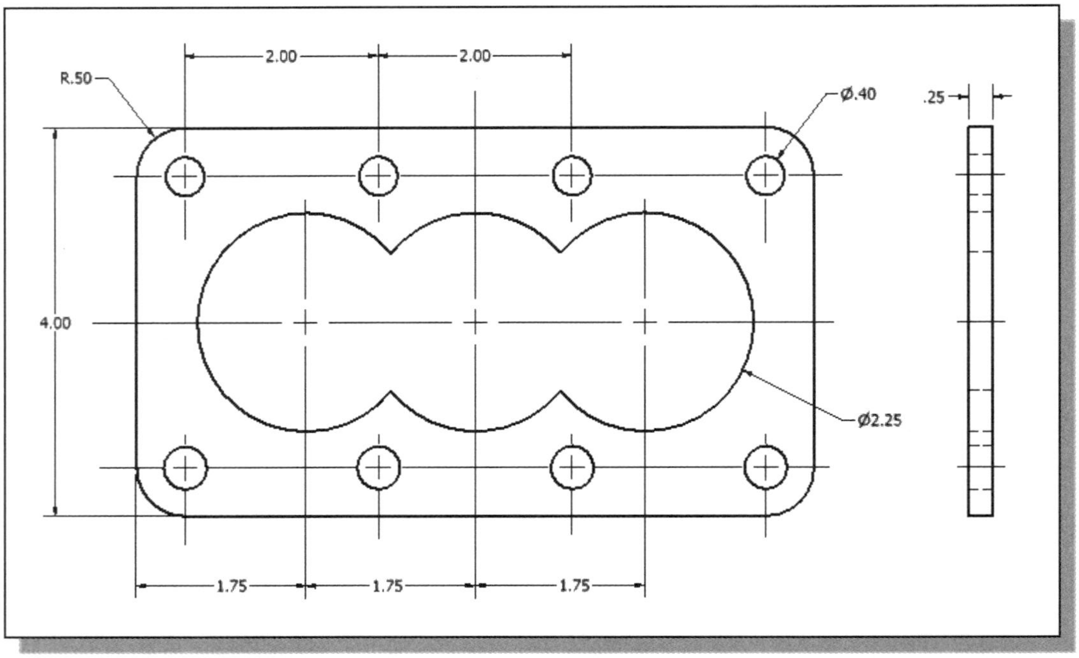

6. **Shaft Guide** (Dimensions are in inches.)

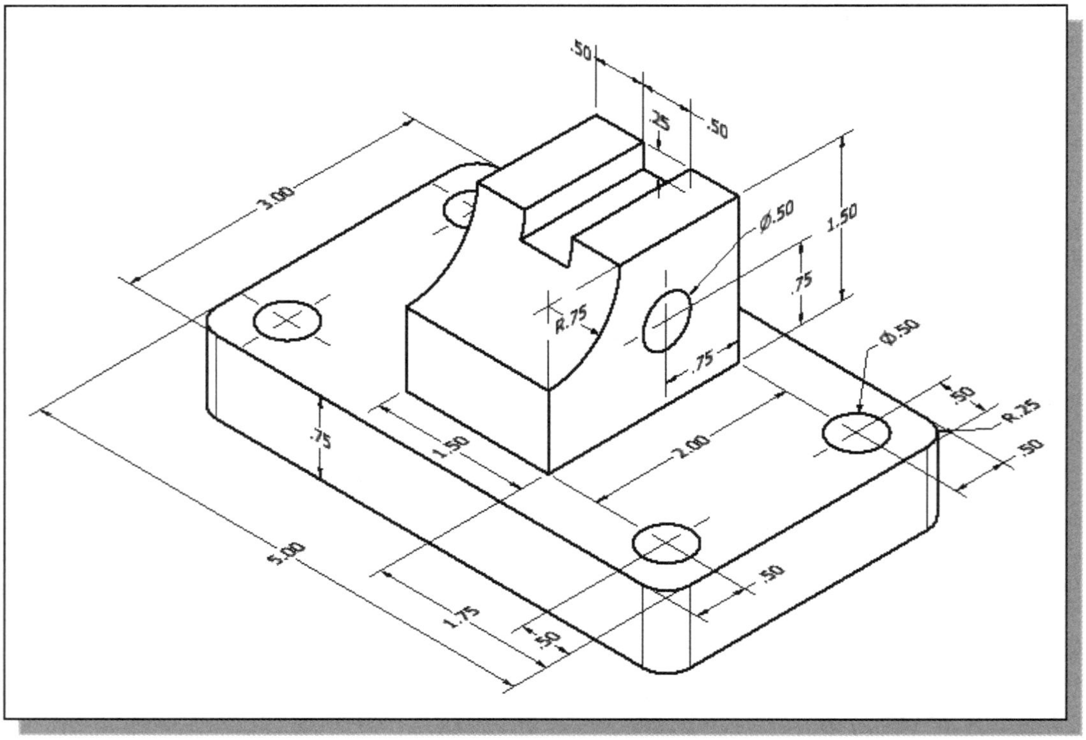

Chapter 9
Datum Features and Auxiliary Views

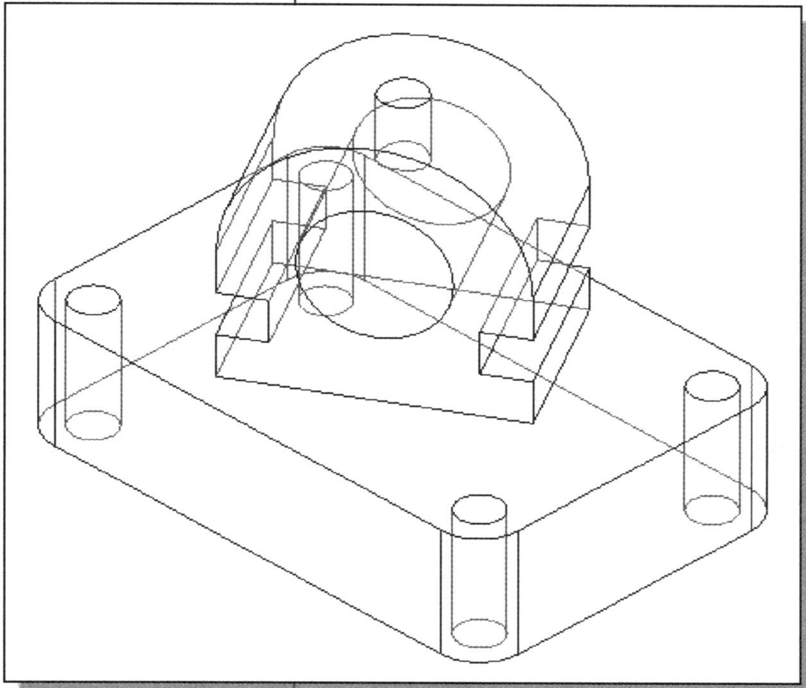

Learning Objectives

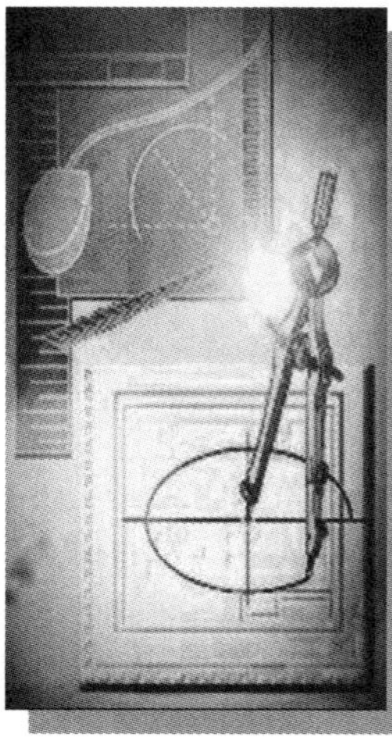

- Understand the Concepts and the Use of Work Features
- Use the Different Options to Create Work Features
- Create Auxiliary Views in 2D Drawing Mode
- Create and Adjust Centerlines
- Create Shaded Images in 2D Drawing Mode

Autodesk Inventor Certified User Exam Objectives Coverage

Parametric Modeling Basics

Section 3: Sketches

Objectives: Creating 2D Sketches, Draw Tools, Sketch Constraints, Pattern Sketches, Modify Sketches, Format Sketches, Sketch Doctor, Shared Sketches, Sketch Parameters.

Section 4: Parts

Objectives: Creating parts, Work Features, Pattern Features, Part Properties.

Section 6: Drawings

Objectives: Create drawings.

Work Features

Feature-based parametric modeling is a cumulative process. The relationships that we define between features determine how a feature reacts when other features are changed. Because of this interaction, certain features must, by necessity, precede others. A new feature can use previously defined features to define information such as size, shape, location and orientation. Autodesk Inventor provides several tools to automate this process. Work features can be thought of as user-definable datum, which are updated with the part geometry. We can create work planes, axes, or points that do not already exist. Work features can also be used to align features or to orient parts in an assembly. In this chapter, the use of the **Offset** option and the **Angled** option to create new work planes, surfaces that do not already exist, is illustrated. By creating parametric work features, the established feature interactions in the CAD database assure the capturing of the design intent. The default work features, which are aligned to the origin of the coordinate system, can be used to assist the construction of the more complex geometric features.

Auxiliary Views in 2D Drawings

An important rule concerning multiview drawings is to draw enough views to accurately describe the design. This usually requires two or three of the regular views, such as a front view, a top view and/or a side view. However, many designs have features located on inclined surfaces that are not parallel to the regular planes of projection. To truly describe the feature, the true shape of the feature must be shown using an **auxiliary view**. An *auxiliary view* has a line of sight that is perpendicular to the inclined surface, as viewed looking directly at the inclined surface. An *auxiliary view* is a supplementary view that can be constructed from any of the regular views. Using the solid model as the starting point for a design, auxiliary views can be easily created in 2D drawings. In this chapter, the general procedure of creating auxiliary views in 2D drawings from solid models is illustrated.

The Rod-Guide Design

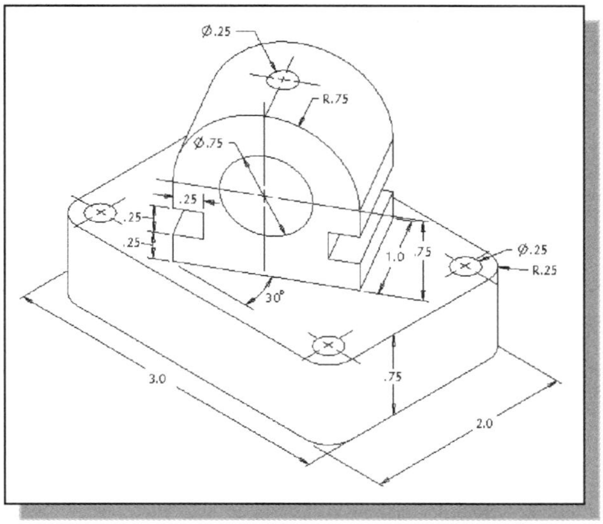

❖ Based on your knowledge of Autodesk Inventor so far, how would you create this design? What are the most difficult features involved in the design? Take a few minutes to consider a modeling strategy and do preliminary planning by sketching on a piece of paper. You are also encouraged to create the design on your own prior to following through the tutorial.

Modeling Strategy

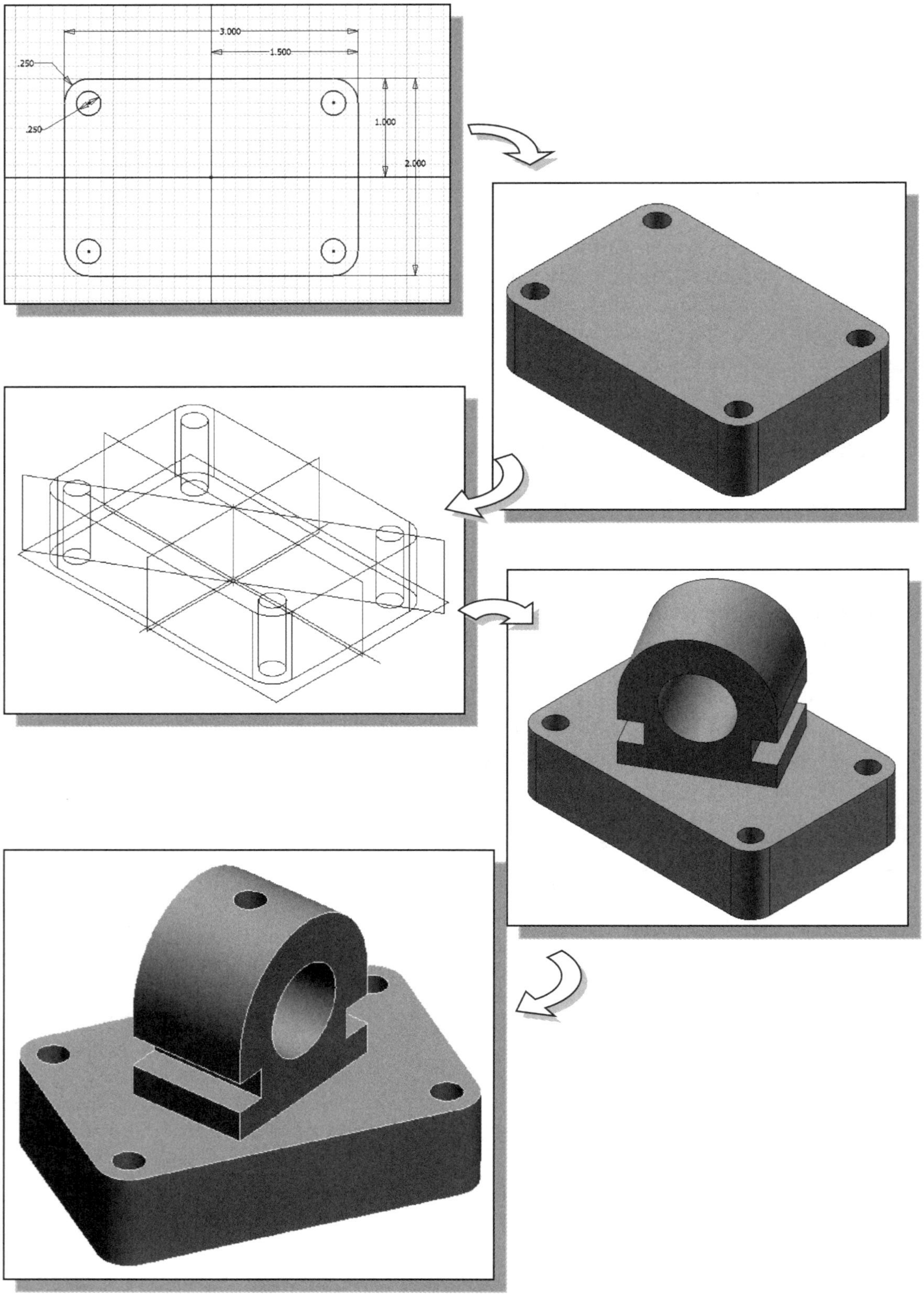

Starting Autodesk Inventor

1. Select the **Autodesk Inventor** option on the *Start* menu or select the **Autodesk Inventor** icon on the desktop to start Autodesk Inventor. The Autodesk Inventor main window will appear on the screen.

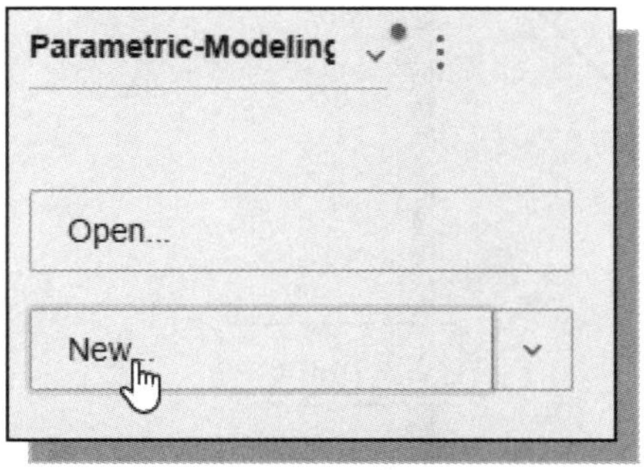

2. Select the **New File** icon with a single click of the left-mouse-button as shown.

3. On your own, confirm the project is set to the ***Parametric-Modeling*** project.

4. In the *New File* dialog box, select the **English** units set and then select **Standard(in).ipt**.

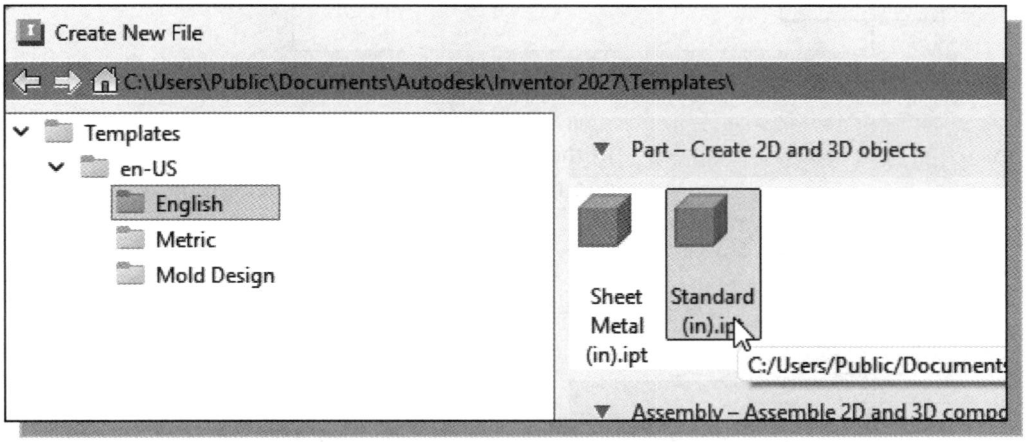

5. Click **Create** in the *New File* dialog box to accept the selected settings.

Apply the BORN Technique

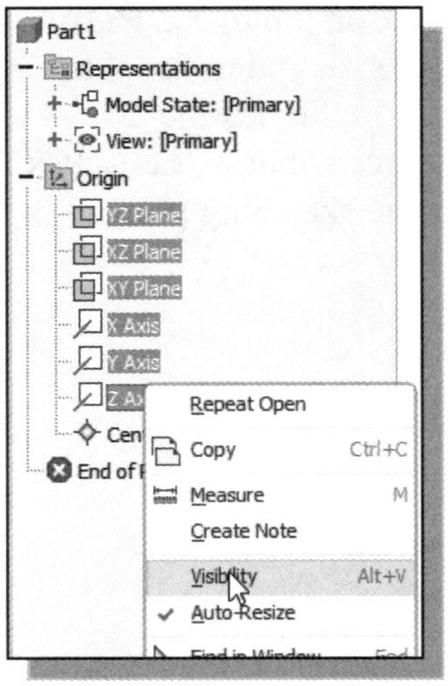

1. In the *Part Browser* window, click on the [**+**] symbol in front of the **Origin** feature to display more information on the feature.

2. Inside the *browser* window, select all of the work features by holding down the **[Control]** or **[Shift]** keys and click with the left-mouse-button.

3. Move the right-mouse-button on any of the work features to display the option menu. Click on **Visibility** to toggle ***ON*** the display of the selected work features.

4. On your own, use the dynamic viewing options (3D Rotate, Zoom and Pan) to view the work features established.

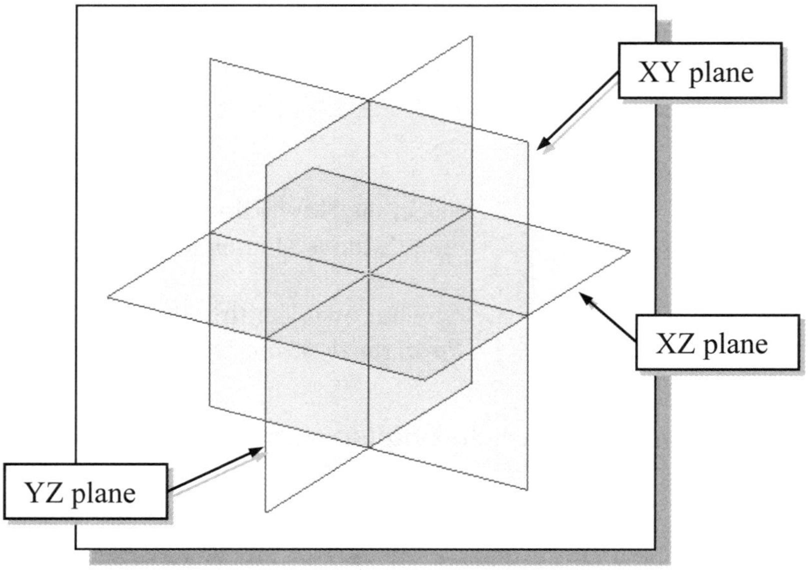

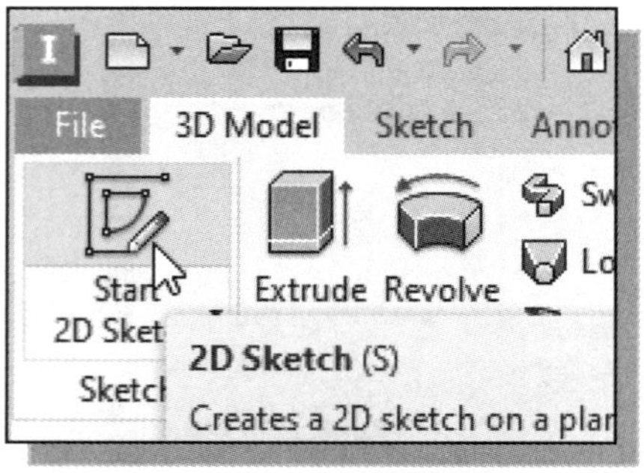

5. In the *Sketch* toolbar select the **Start 2D Sketch** command by left-clicking once on the icon.

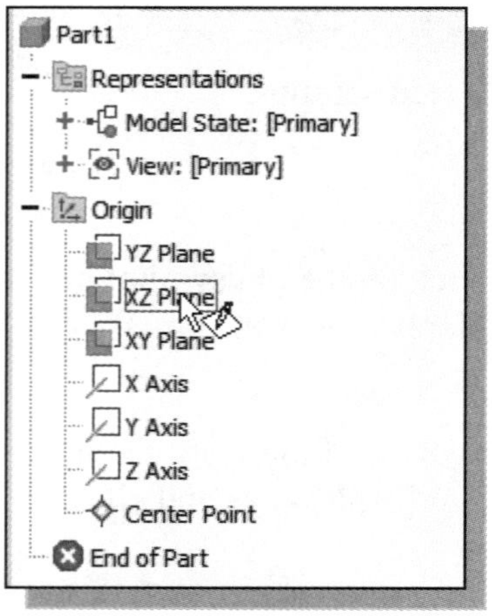

6. In the *Status Bar* area, the message "*Select plane to create sketch or an existing sketch to edit*" is displayed. *Autodesk Inventor* expects us to identify a planar surface where the 2D sketch of the next feature is to be created. Move the graphics cursor on top of **XZ Plane**, inside the *browser* window as shown, and notice that Autodesk Inventor automatically highlights the corresponding plane in the graphics window. Left-click once to select the XZ Plane as the sketching plane.

7. Single left-click to activate the **Home View** option as shown. The view will be adjusted back to the default ***isometric view***.

Creating the Base Feature

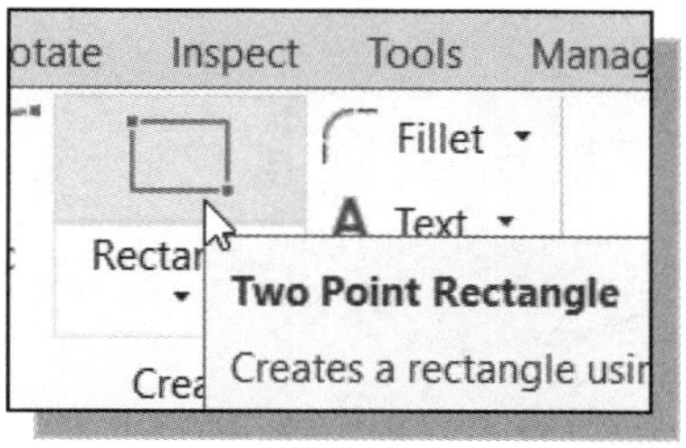

1. Select the **Two point rectangle** command by clicking once with the left-mouse-button on the icon in the *Sketch* toolbar.

2. Create a rectangle of arbitrary size with the center point near the center of the rectangle as shown.

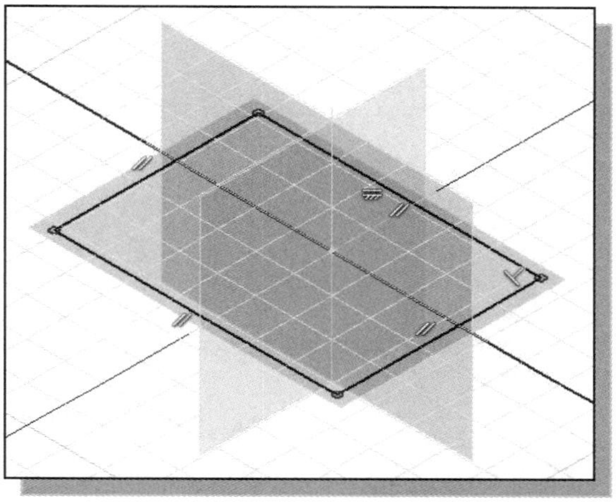

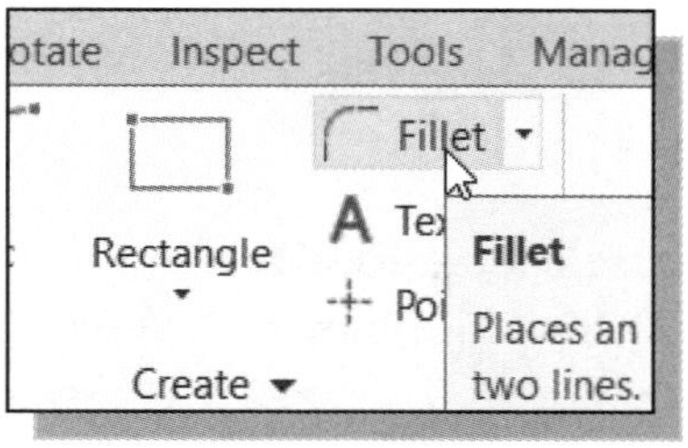

3. Click on the **Fillet** icon in the *Sketch* panel.

4. The *2D Fillet* radius dialog box appears on the screen. Use the default radius value and create four rounded corners of the rectangle.

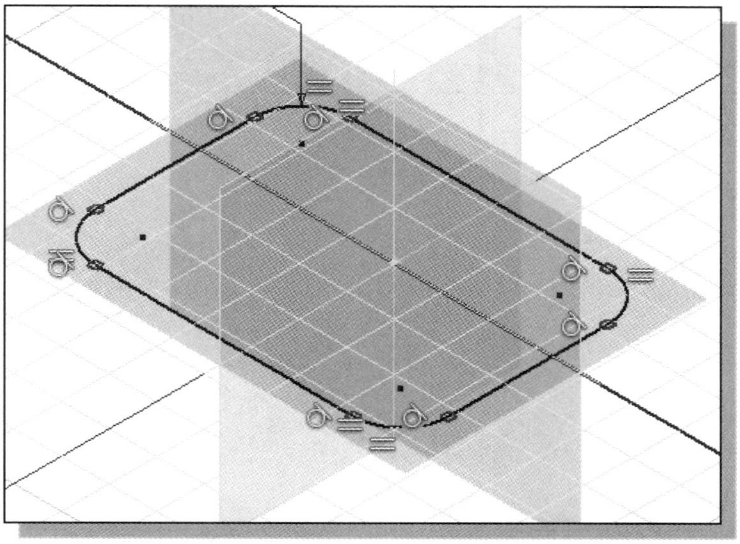

5. On your own, create four circles of the same diameter and with the centers aligned to the centers of the arcs. Also create and modify the six dimensions as shown in the figure.

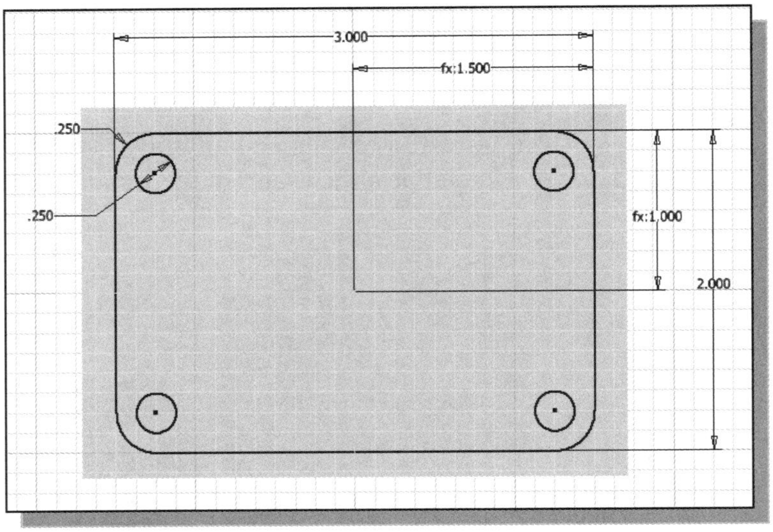

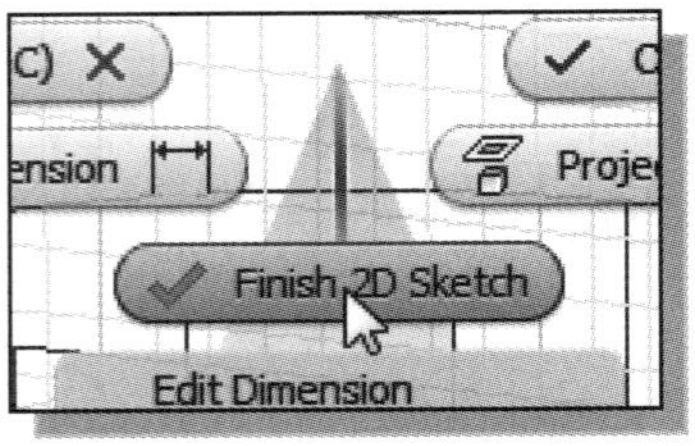

6. Inside the graphics window, click once with the **right-mouse-button** and select **Finish 2D Sketch** in the pop-up menu to end the Sketch option.

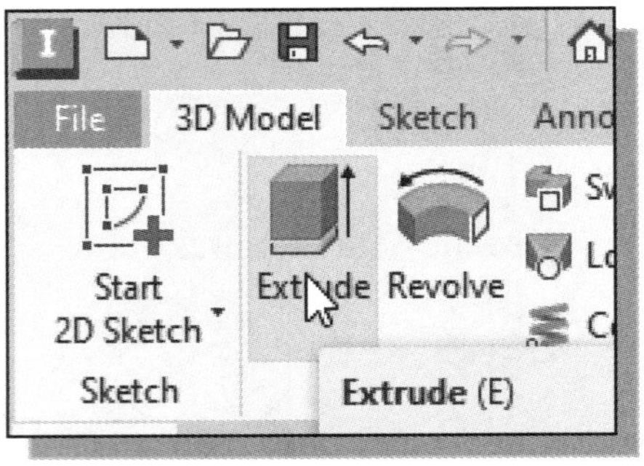

7. In the *3D Model* tab, select the **Extrude** command by clicking the left-mouse-button on the icon.

8. Select the inside regions of the 2D sketch to create a profile as shown.

9. In the *Extrude* pop-up window, enter **0.75** as the extrusion distance and create the feature.

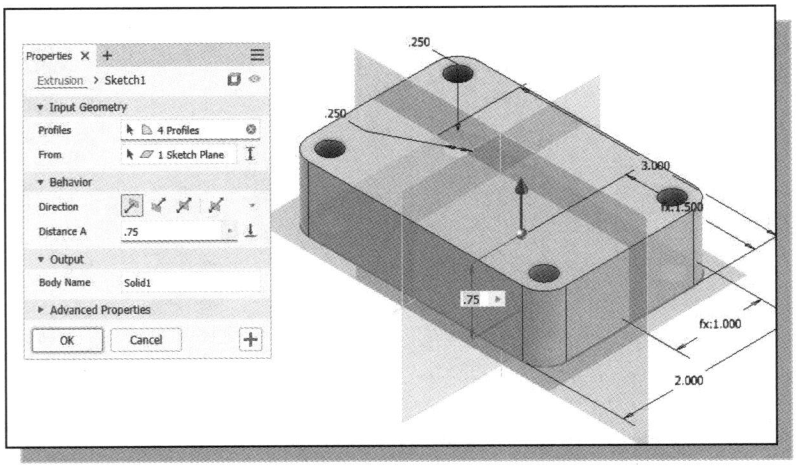

Create an Angled Work Plane

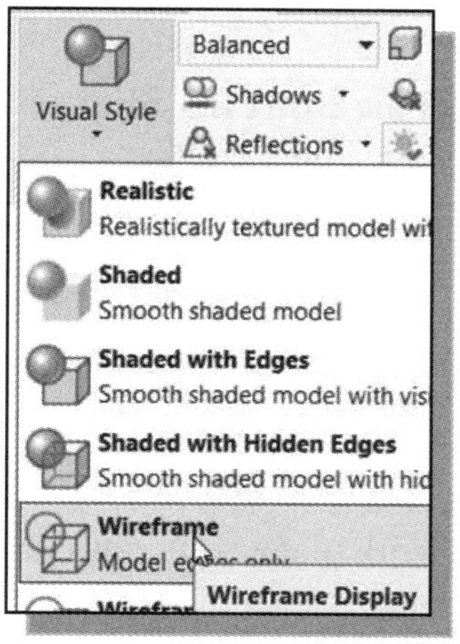

1. Activate the **View tab** in the *Ribbon* panel as shown.

2. Select **Wireframe** display mode under the *Visual Style* icon as shown.

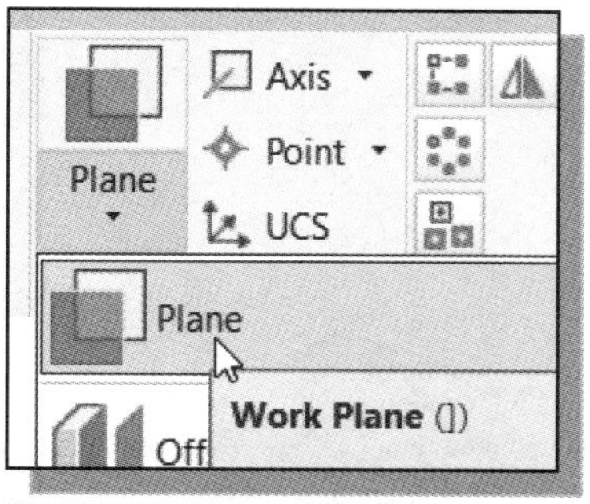

3. Activate the *3D Model* tab and select the **Work Plane** command by left-clicking the icon as shown.

4. In the *Status Bar* area, the message "*Define work plane by highlighting and selecting geometry*" is displayed. Autodesk Inventor expects us to select any existing geometry to be used as a reference to create the new work plane.

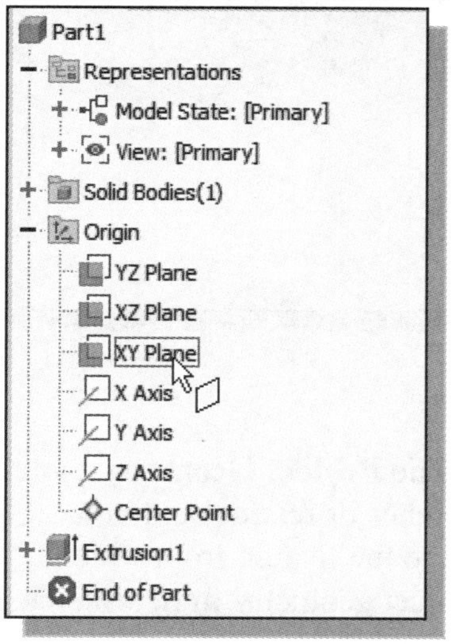

5. Inside the *browser* window, left-click once to select the **XY Plane** as the first reference of the new work plane.

6. Inside the *browser* window, left-click once to select the **Y Axis** as the second reference of the new work plane.

7. In the *Angle* pop-up window, enter **30** as the rotation angle for the new work plane.

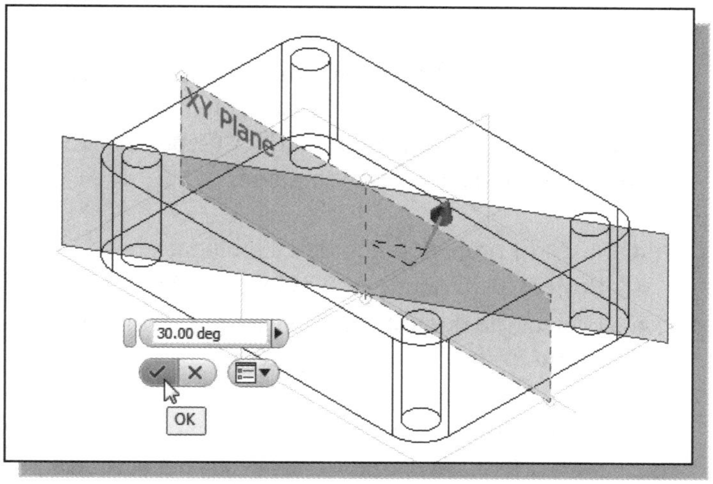

❖ Note that the *angle* is measured relative to the selected reference plane, XY Plane.

8. Click on the **check mark** button to accept the setting.

Create a 2D Sketch on the Work Plane

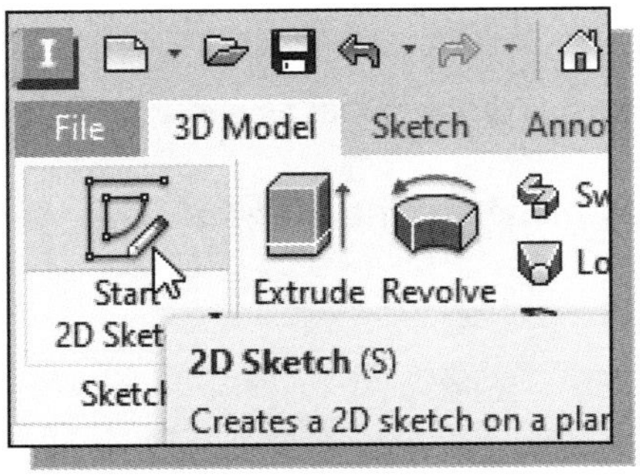

1. In the *Sketch* toolbar select the **Start 2D Sketch** command by left-clicking once on the icon.

2. In the *Status Bar* area, the message "*Select face, work plane, sketch or sketch geometry*" is displayed. Pick the **work plane** by clicking the work plane name inside the *browser* as shown below.

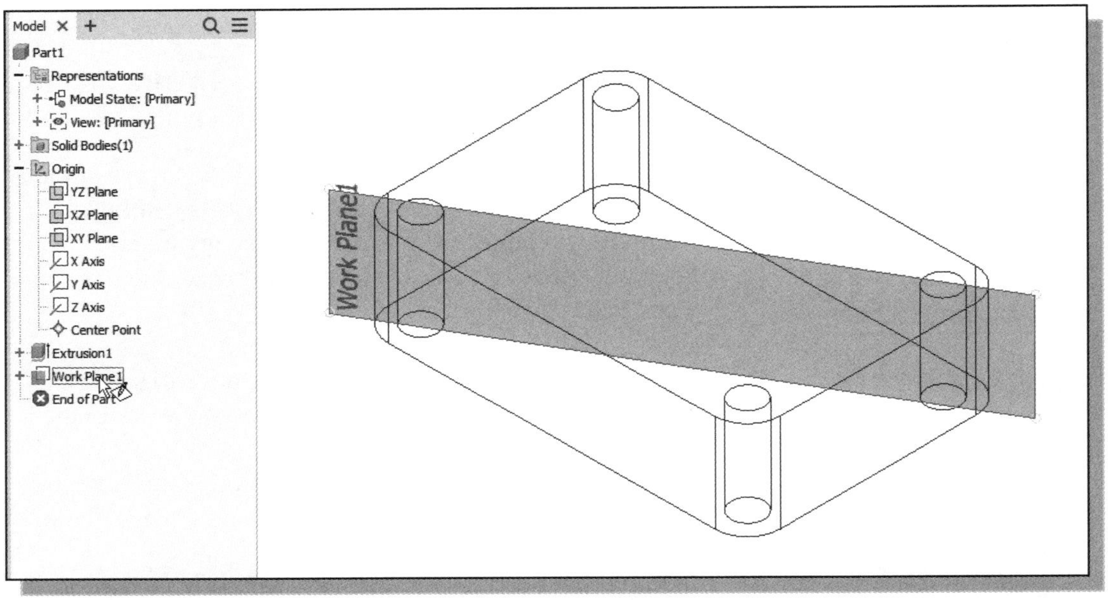

Use the Projected Geometry Option

Projected geometry is another type of *reference geometry*. The **Project Geometry** tool can be used to project geometry from previously defined sketches or features onto the sketch plane. The position of the projected geometry is fixed to the feature from which it was projected. We can use the **Project Geometry** tool to project geometry from a sketch or feature onto the active sketch plane.

Typical uses of projected geometry include:

- Project a silhouette of a 3D feature onto the sketch plane for use in a 2D profile.
- Project the default center point onto the sketch plane to constrain a sketch to the origin of the coordinate system.
- Project a sketch from a feature onto the sketch plane so that the projected sketch can be used to constrain a new sketch.

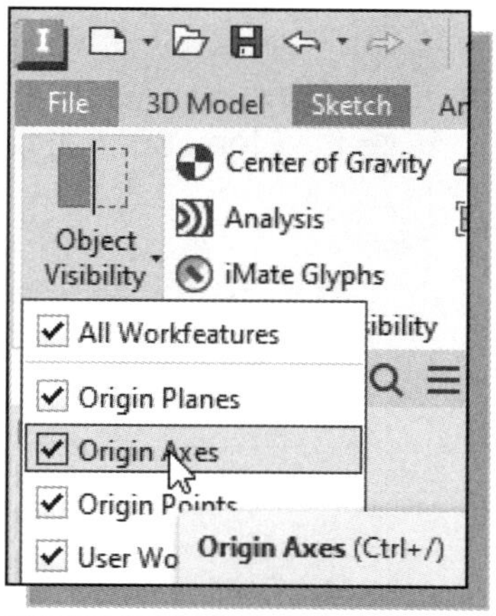

1. Choose **View** in the *Ribbon* tabs.

2. Select **Object Visibility → Origin Axes** in the options list to toggle *OFF* the axes.

3. Note that quick-key option [CTRL]+[/] is also available to toggle *ON/OFF* the different reference geometry.

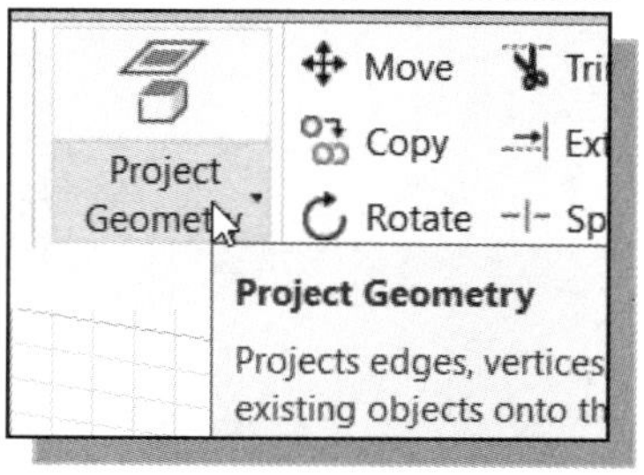

4. Select the **Project Geometry** command in the *Sketch* panel. The *Project Geometry* command allows us to project existing features to the active sketching plane.

5. Select the top back edge of the base feature to create a projected line on the sketching plane.

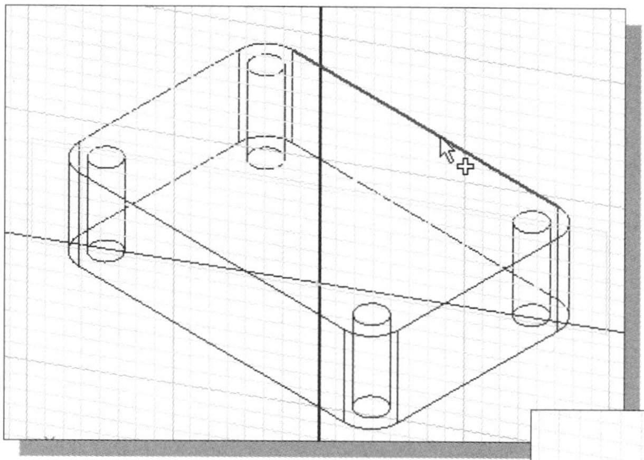

6. Inside the *graphics area*, click once with the right-mouse-button and pick **OK** to end the command.

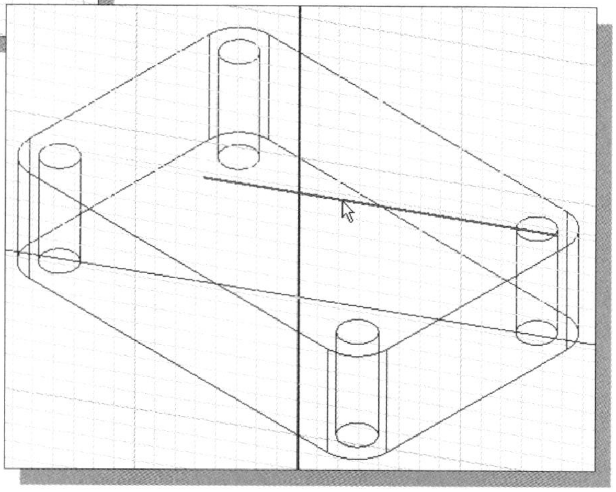

➢ The projected line will be used in the 2D profile of the next solid feature.

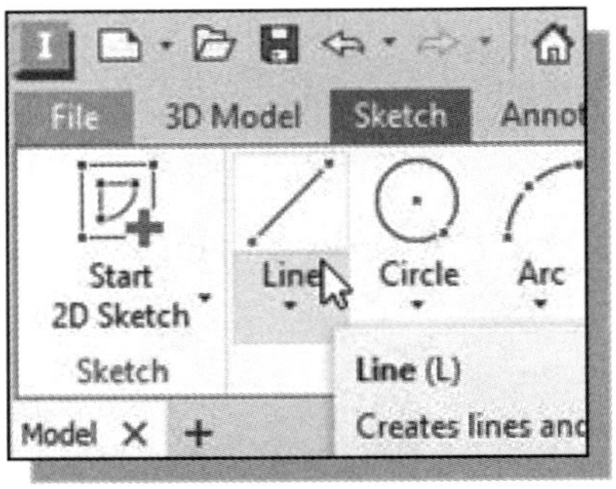

7. In the *Sketch* toolbar, click on the **Line** icon with the left-mouse-button to activate the Line command.

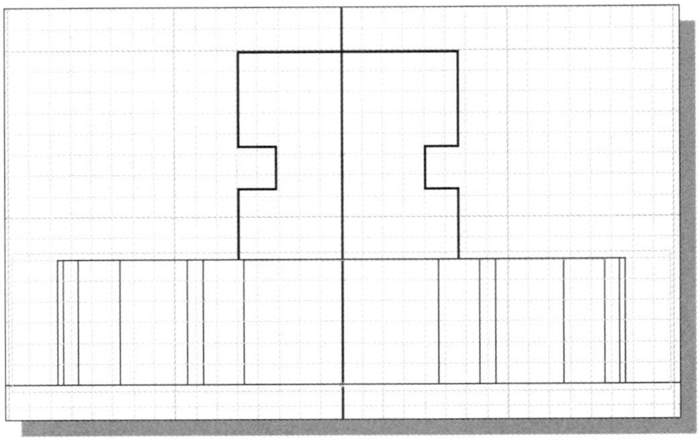

8. Create a rough sketch using the projected edge as the bottom line as shown in the figure. (Note that all edges are either horizontal or vertical.)

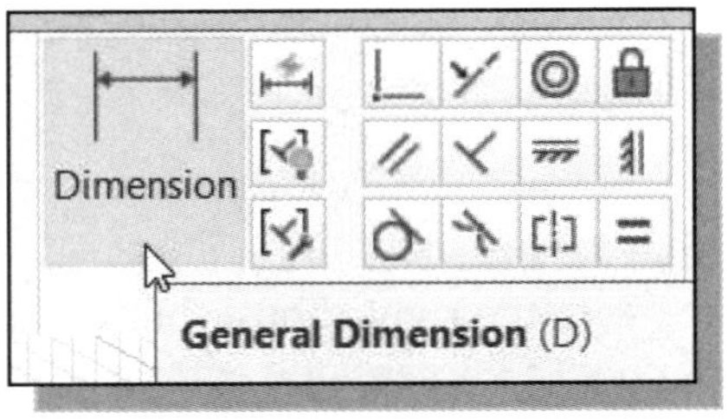

9. Left-click once on the **General Dimension** icon to activate the General Dimension command.

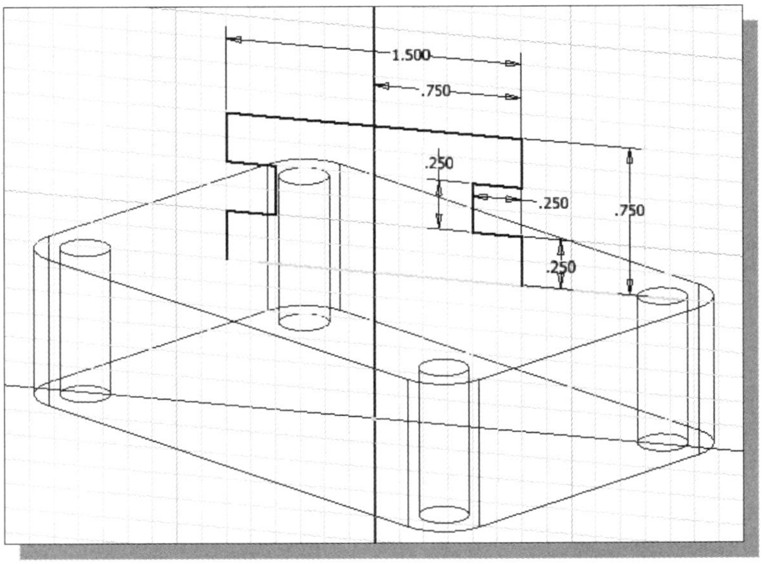

10. On your own, create and modify the dimensions as shown; note that the sketch is symmetrical vertically.

➢ Note that the projected line is used only as a reference; the gap at the bottom of the 2D sketch indicates the 2D sketch does not form a closed region profile.

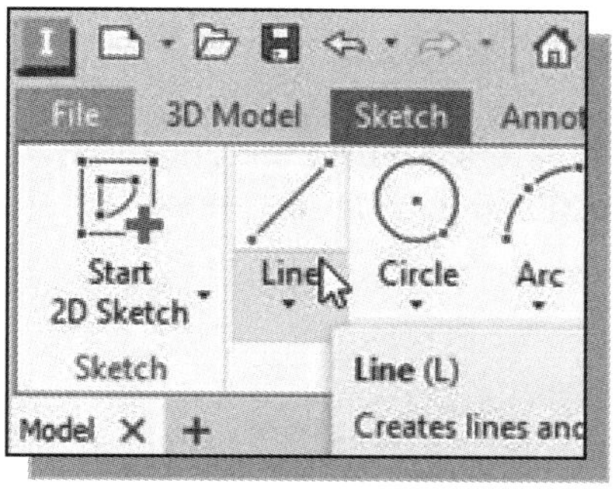

11. Select the **Line** command in the *2D Sketch* panel.

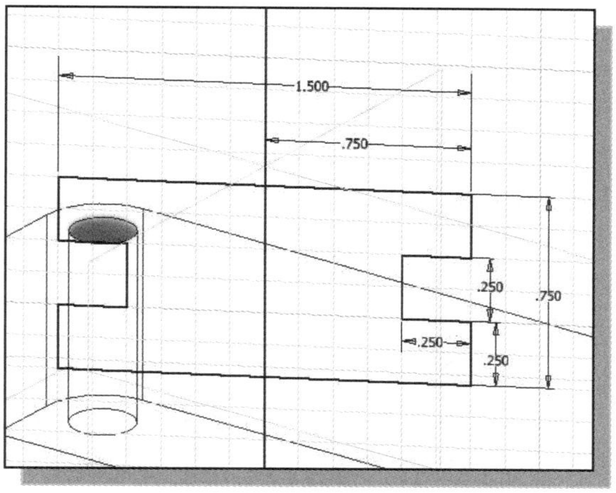

12. On your own, create an additional **bottom line** connecting the bottom of the 2D sketch as shown in the figure.

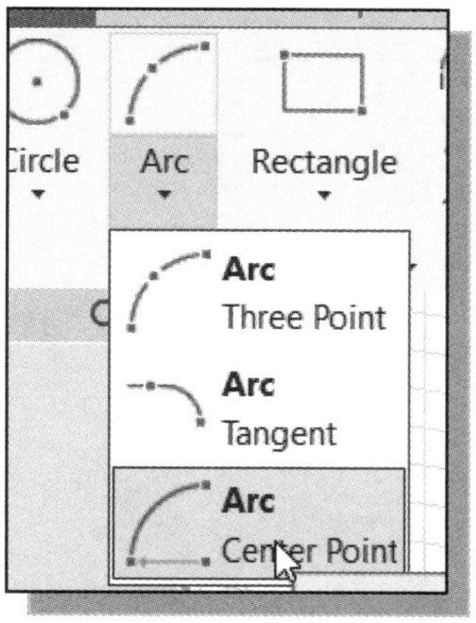

13. Select the **Center point Arc** option in the *2D Sketch* panel as shown.

14. On your own, create the arc so that the center point is aligned to the mid-point of the top edge.

15. On your own, add a **0.75** circle, and complete the 2D sketch as shown in the figure.

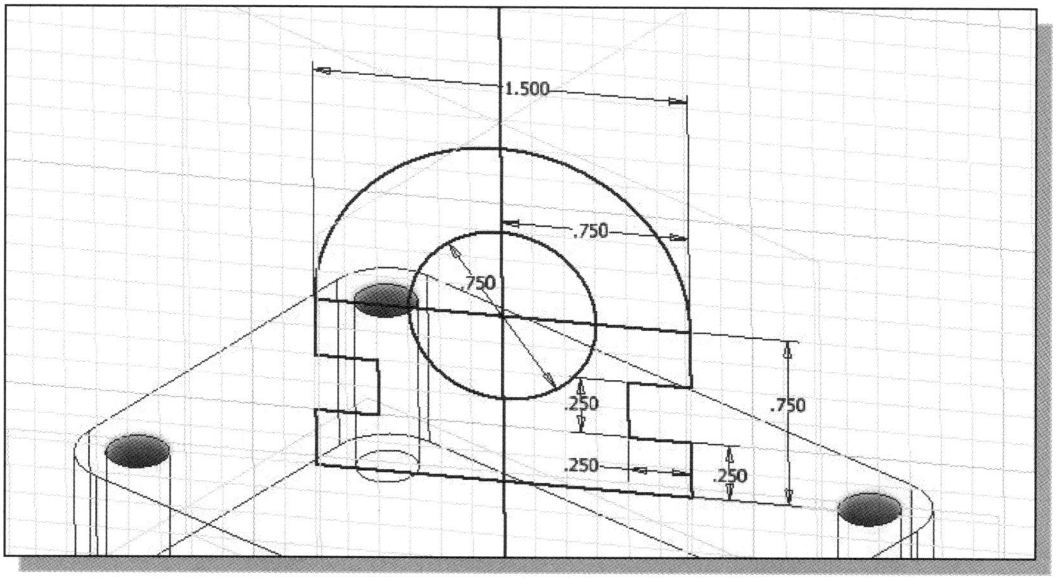

Complete the Solid Feature

1. Inside the graphics window, click once with the right-mouse-button and select **Finish Sketch** in the pop-up menu to end the Sketch option.

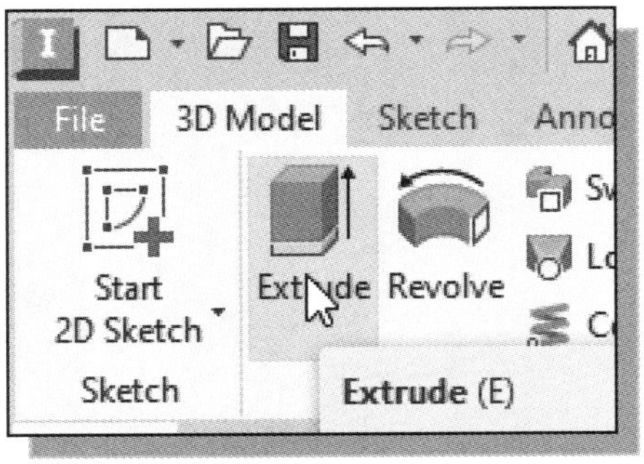

2. In the *3D Model* toolbar, select the **Extrude** command by left clicking on the icon.

3. Select the inside regions of the 2D sketch to create a profile as shown.

4. In the *Extrude* pop-up window, enter **1.0** as the extrusion distance.

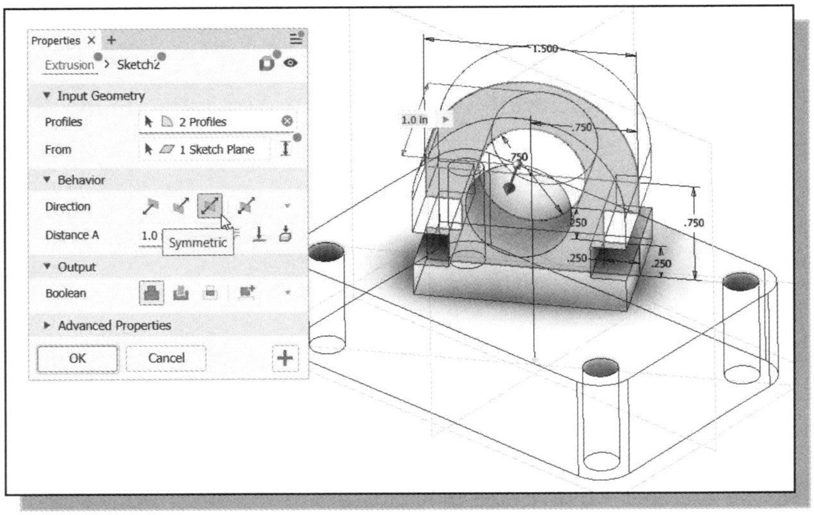

5. Set the extrusion direction to **Symmetric** as shown.

6. Click on the **OK** button to proceed with creating the feature.

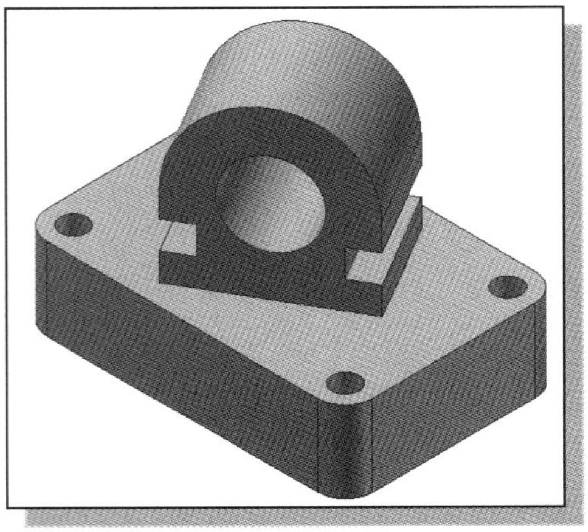

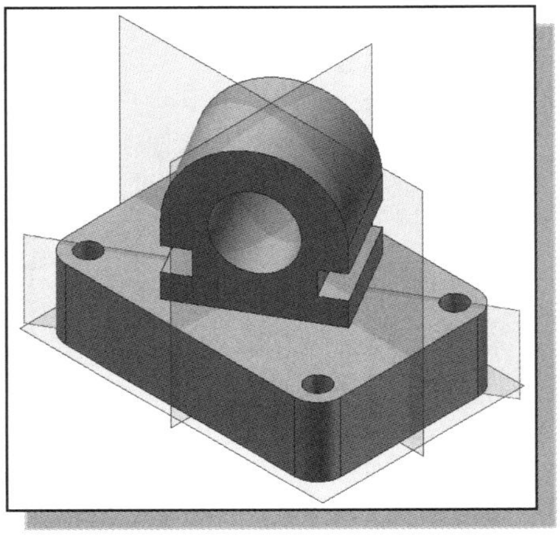

Create an Offset Work Plane

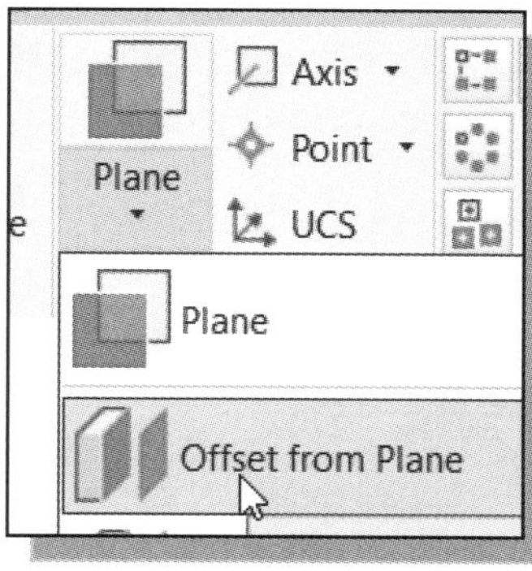

1. In the *3D Model* toolbar, select the **Offset from Plane** command by left-clicking the icon.

❖ In the *Status Bar* area, the message "*Define work plane by highlighting and selecting geometry*" is displayed. Autodesk Inventor expects us to select any existing geometry, which will be used as a reference to create the new work plane.

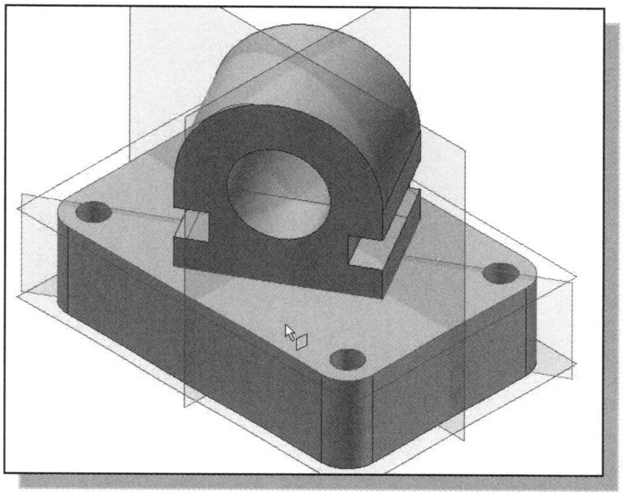

2. Inside the graphics window, select the **top plane** of the base feature as the reference of the new work plane.

3. Set the value in the *Offset* pop-up window to **0.75** as the offset distance for the new work plane.

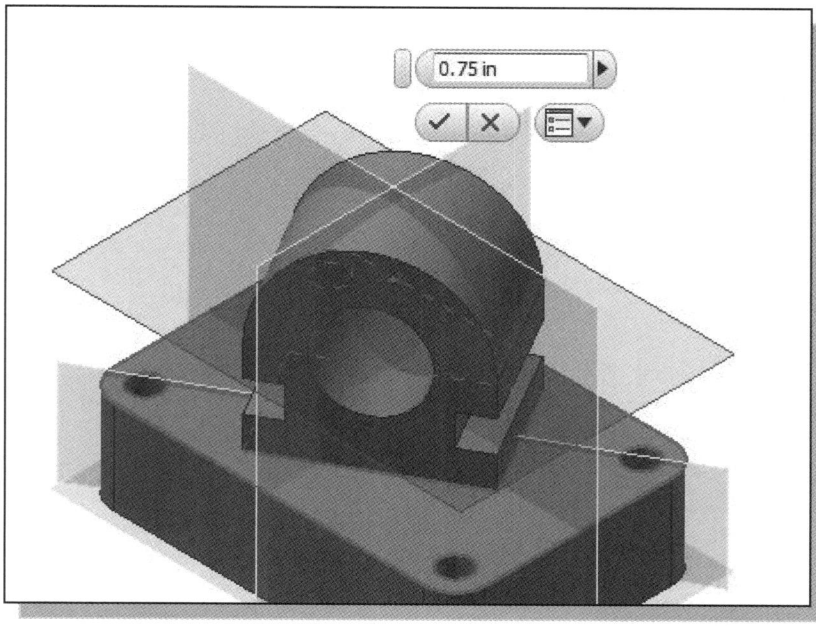

4. Click on the **check mark** button to accept the setting and create the reference plane.

Create another Cut Feature Using the Work Plane

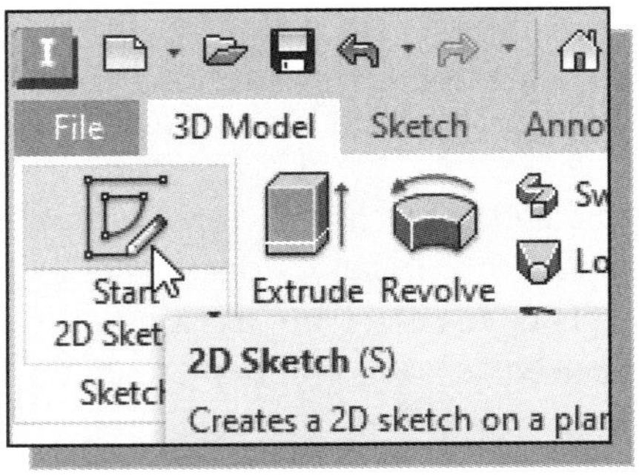

1. In the *Sketch* toolbar select the **Start 2D Sketch** command by left-clicking once on the icon.

2. In the *Status Bar* area, the message "*Select face, work plane, sketch or sketch geometry*" is displayed. Pick the work plane we just created by clicking on one of the edges of the work plane as shown below.

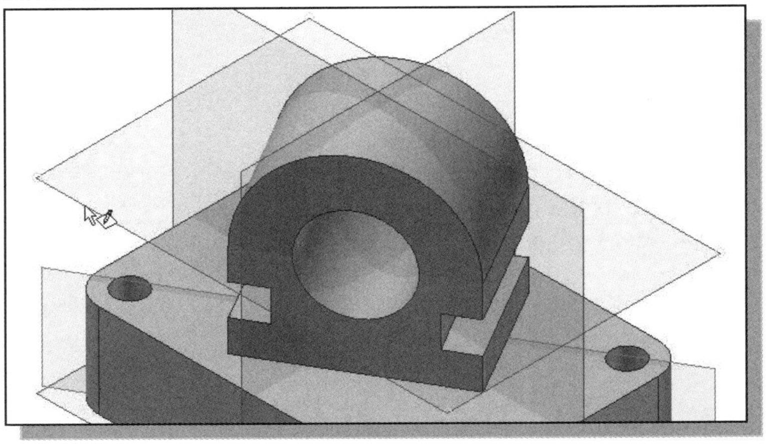

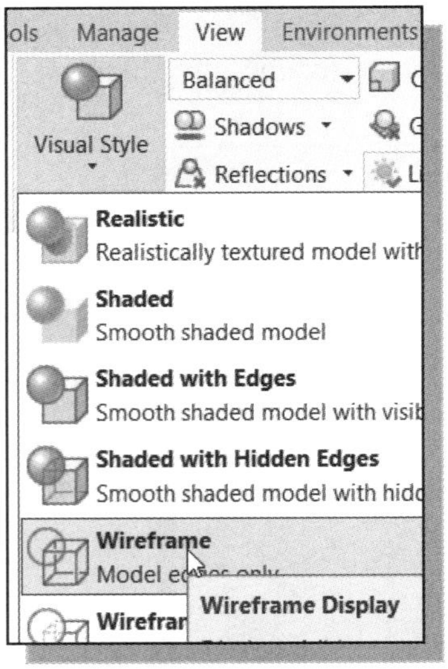

3. Activate the **View tab** in the *Ribbon* panel as shown.

4. Select **Wireframe** display mode under the *Visual Style* icon as shown.

5. Use the **[Ctrl+]]** quick-key combination to toggle off the display of the *Origin Planes*.

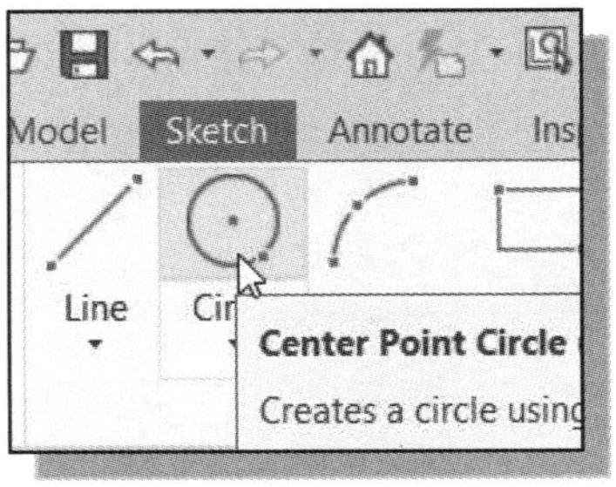

6. Select the **Center Point Circle** command by clicking once with the left-mouse-button on the icon in the *Sketch* tab on the Ribbon.

7. On your own, create a **circle** with the center point aligned to the projected **Center Point** as shown in the figure below.

8. Using the Dimension command, set the circle to **Ø0.25**.

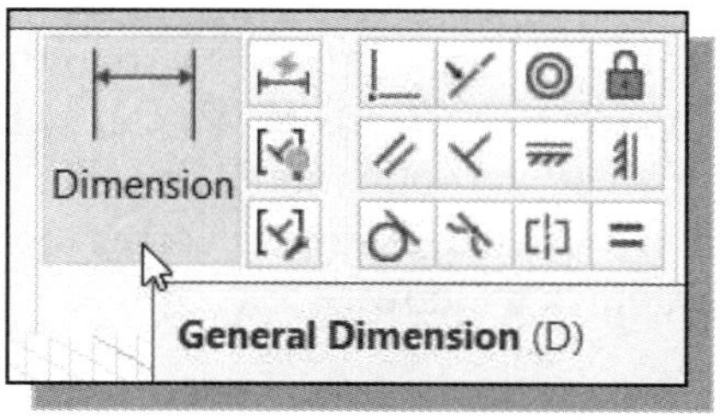

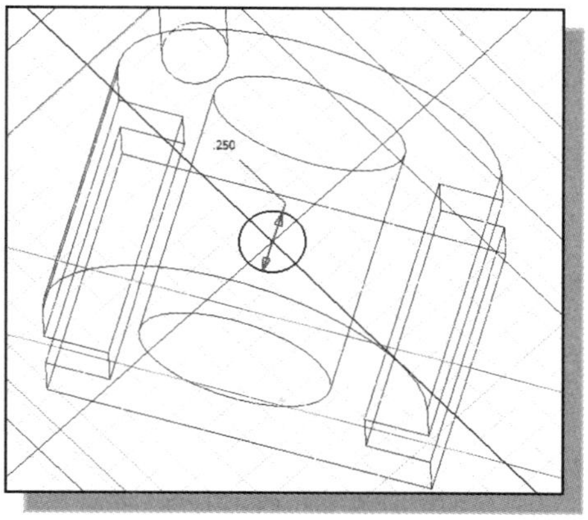

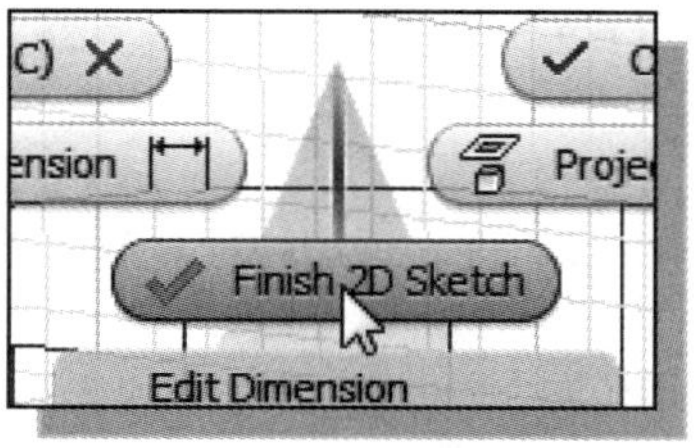

9. Inside the graphics window, click once with the right-mouse-button and select **Finish 2D Sketch** in the pop-up menu to end the **Sketch** option.

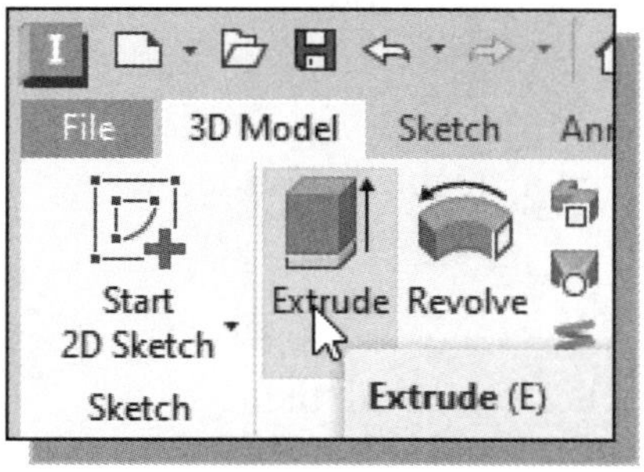

10. In the *3D Model* toolbar, select the **Extrude** command by left clicking on the icon.

11. On your own, complete the **cut** feature as shown.

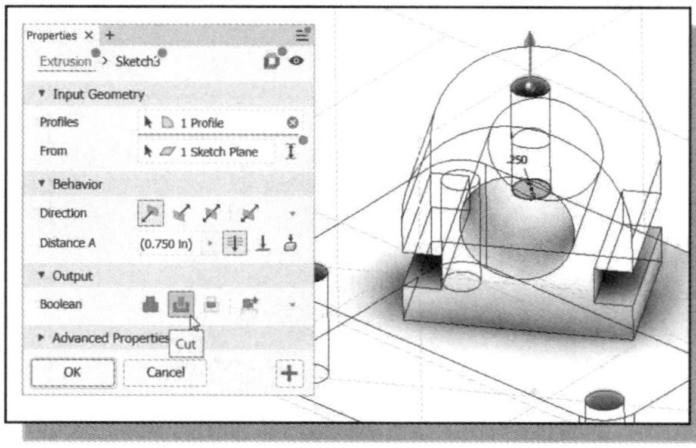

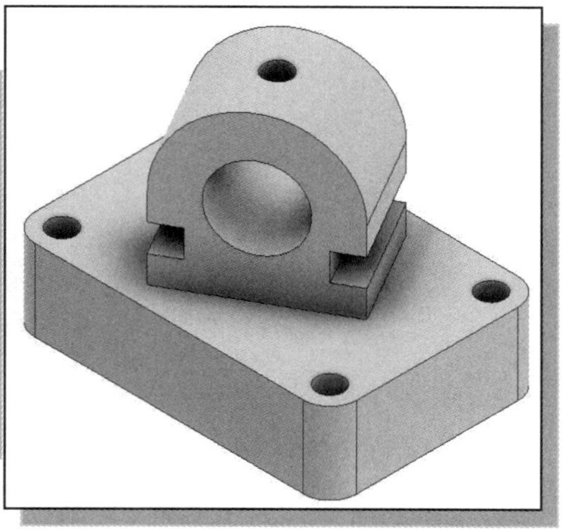

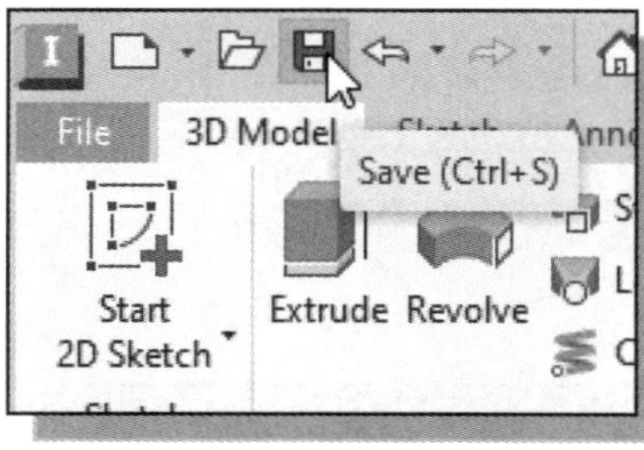

12. Save the design as ***Rod-Guide.ipt*** in the **Chapter 9** folder.

Start a New 2D Drawing

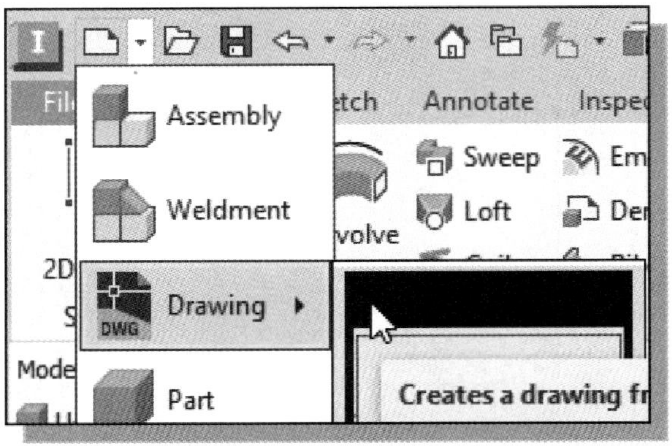

1. Select **New File → Drawing**, without choosing any drawing option, in the *Quick Access* toolbar as shown.

➢ Note that a new graphics window appears on the screen. We can switch between the solid model and the drawing by clicking the corresponding graphics windows.

❖ In the graphics window, Autodesk Inventor displays a default drawing sheet that includes a title block. The drawing sheet is placed on the 2D paper space, and the title block also indicates the paper size being used.

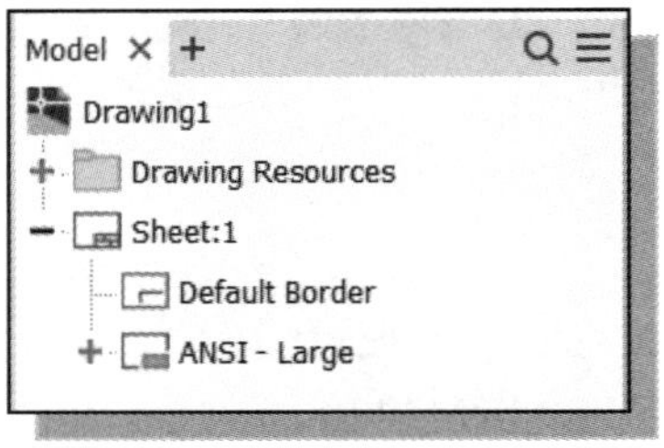

➢ In the *browser* area, the **Drawing1** icon is displayed at the top, which indicates that we have switched to ***Drawing Mode***. **Sheet:1** is the current drawing sheet that is displayed in the graphics window.

❖ Different types of pre-defined borders and title blocks are available in Autodesk Inventor.

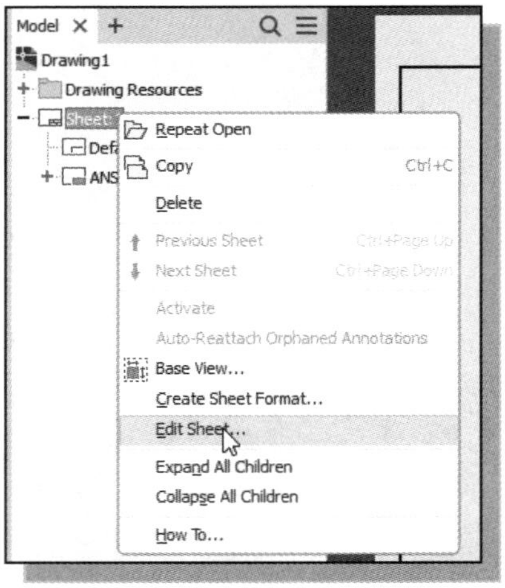

2. In the *browser* area, click once with the right-mouse-button on **Sheet:1** and select **Edit Sheet** in the pop-up menu.

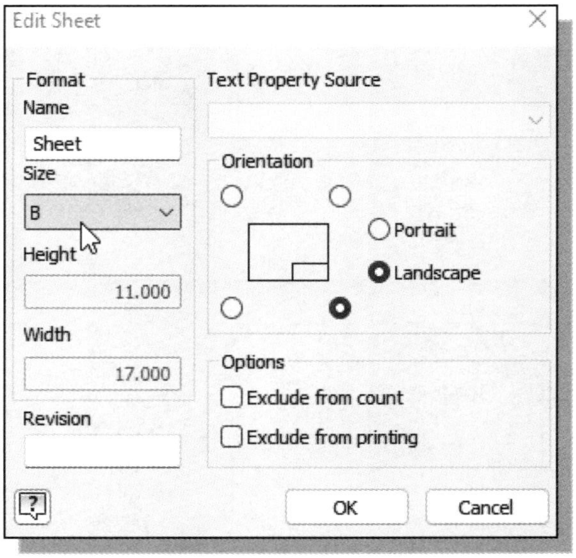

3. In the ***Edit Sheet*** window, set the size option to **B** size as shown.

4. Click on the **OK** button to accept the settings.

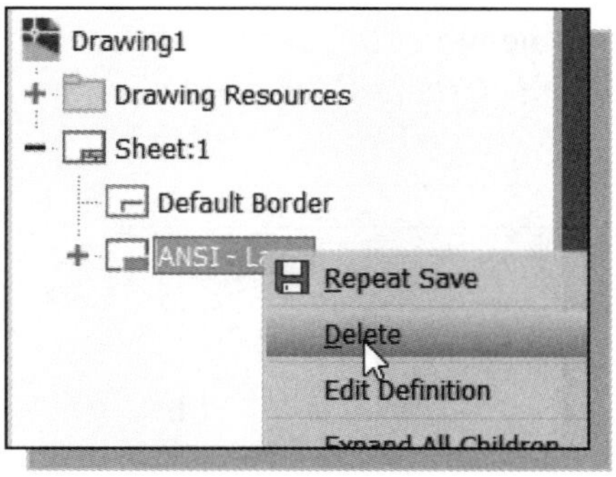

5. In the *browser* area, click once with the right-mouse-button on **ANSI-Large** and select **Delete** in the pop-up menu.

❖ Before applying a different title block, the existing title block must be removed.

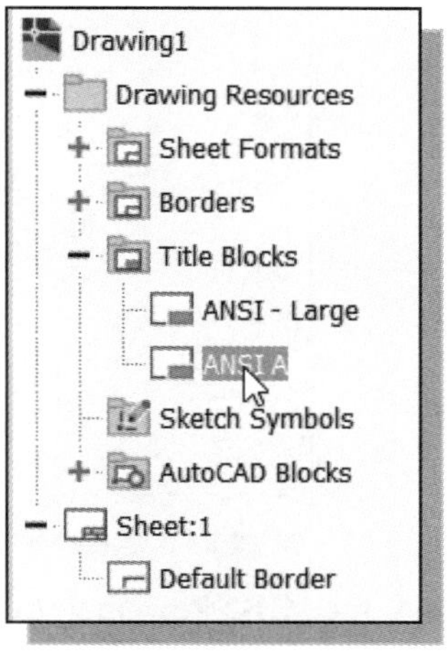

6. In the *browser* area, click on the [>] symbol in front of the **Drawing Resources** and **Title Blocks** options to expand the lists.

7. **Double-click** on the **ANSI-A** title block to place a copy of this title block into the current sheet.

Add a Base View

In Autodesk Inventor *Drawing Mode*, the first drawing view we create is called a **base view**. A *base view* is the primary view in the drawing; other views can be derived from this view. When creating a *base view*, Autodesk Inventor allows us to specify the view to be shown. By default, Autodesk Inventor will treat the *world XY plane* as the front view of the solid model. Note that there can be more than one *base view* in a drawing.

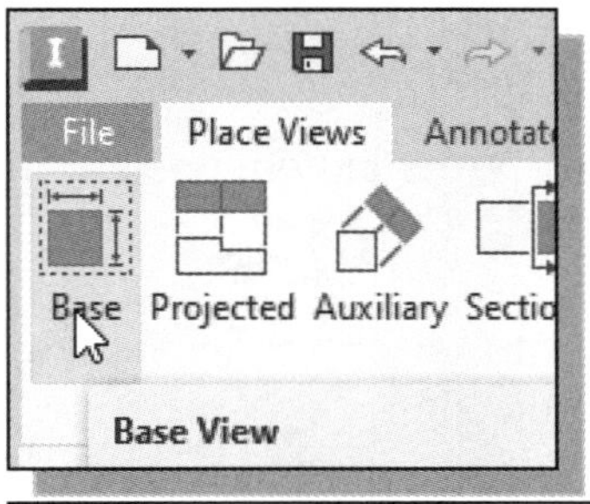

1. Click on the **Base View** in the *Drawing Views* panel to create a base view.

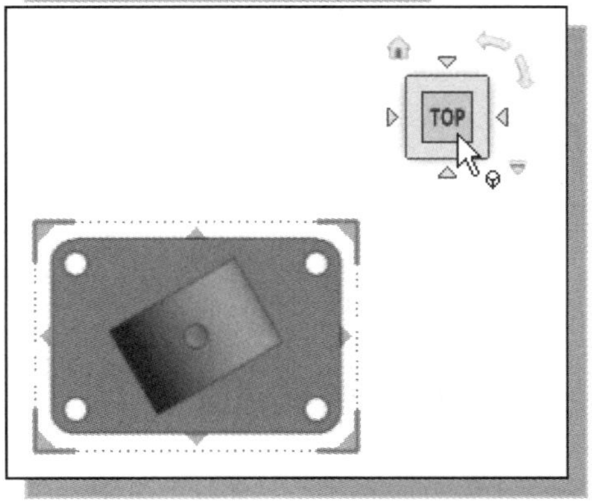

2. In the *Drawing View* dialog box, confirm that the settings are set to **Top** view with **Hidden Line** displayed and **Scale to 1 : 1** as shown.

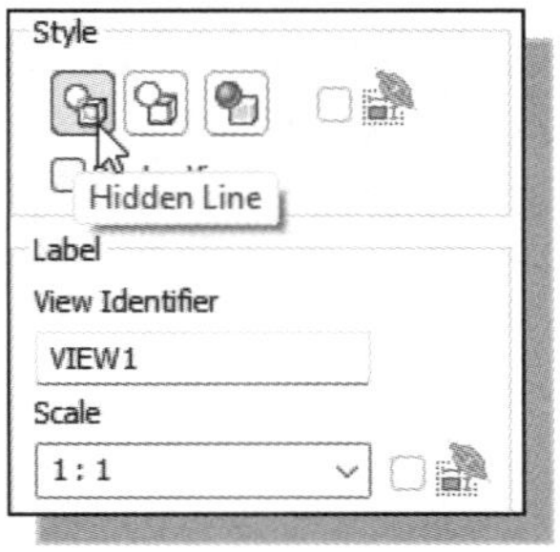

3. Inside the graphics window, place the **base view** near the upper left corner of the graphics window as shown below. (If necessary, drag the *Create View* dialog box to another location on the screen.)

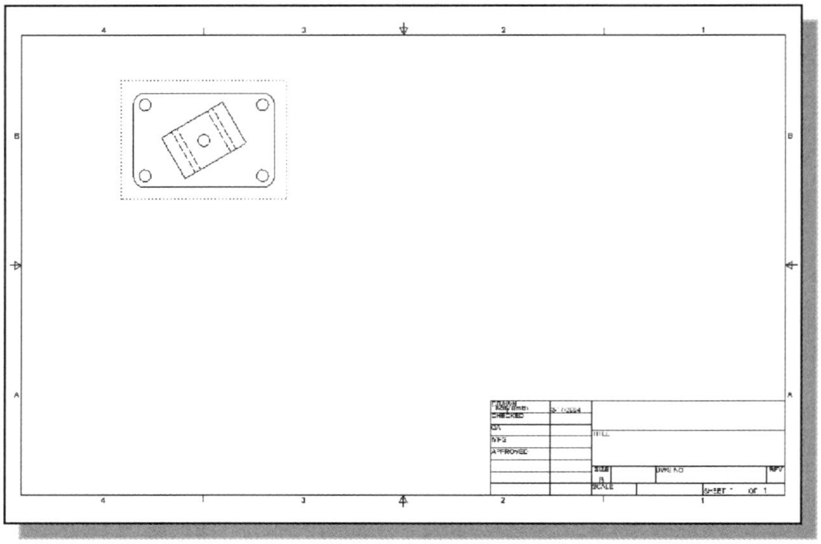

4. Click the [**OK**] key once to end the *Base View* command.

Create an Auxiliary View

In Autodesk Inventor *Drawing Mode*, the **Projected View** command is used to create standard views such as the *top* view, *front* view or *isometric* view. For non-standard views, the **Auxiliary View** command is used. *Auxiliary views* are created using orthographic projections. Generally, orthographic projections are aligned to the base view and inherit the base view's scale and display settings.

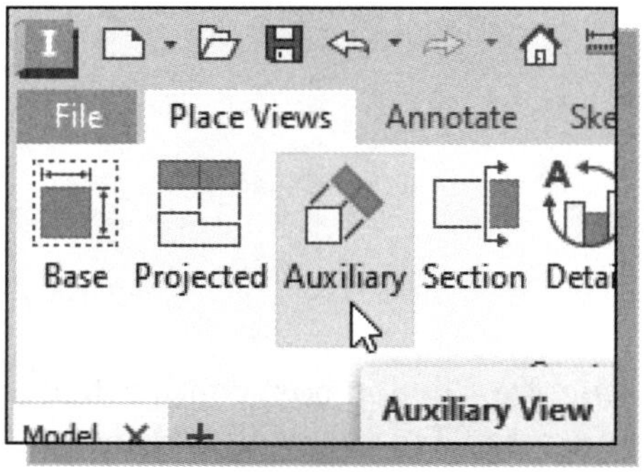

1. Click on the **Auxiliary View** button in the *Drawing Views* panel.

2. Click on the **base view** to select the view as the referenced view for projection.

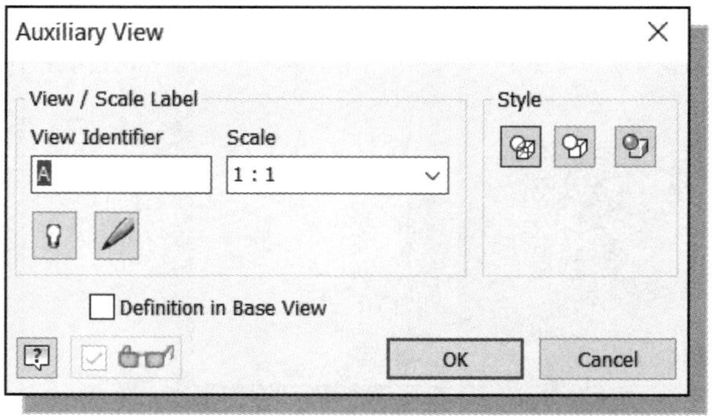

3. Confirm the settings in the *Auxiliary View* window are set as shown. (**DO NOT** click on the **OK** button yet.)

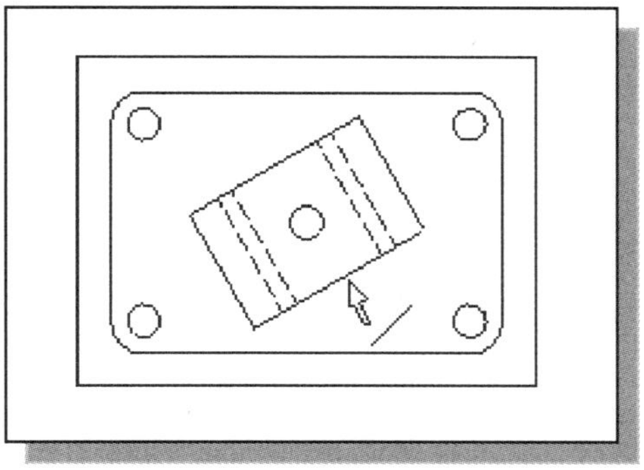

4. Pick the front edge of the upper section of the model as shown.

❖ The orthographic projection direction will be perpendicular to the selected edge.

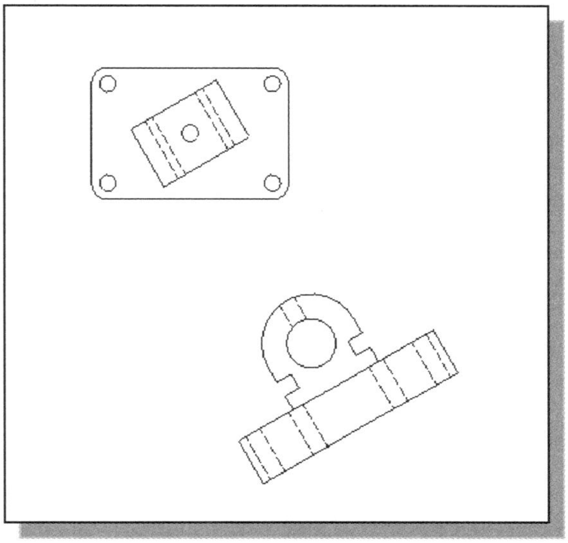

5. Move the cursor below the base view and click once with the **left mouse button** to select a location to position the auxiliary view of the model as shown.

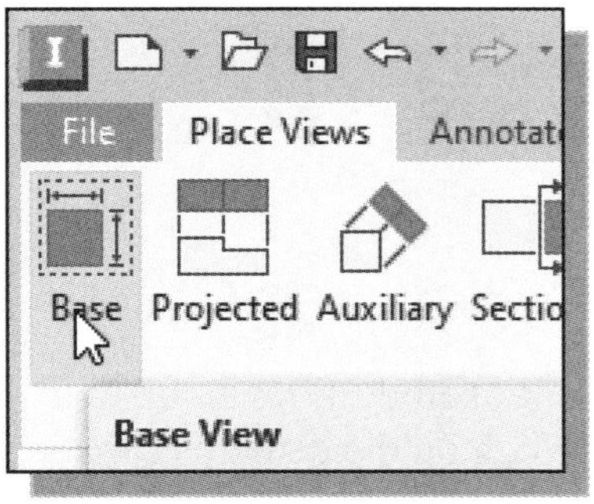

6. Click on **Base View** in the *Drawing Views* panel to create a base view.

7. In the *Drawing View* dialog box, confirm that the settings are set to **Iso Top Right**, **Scale 1 : 1** and **Hidden Line Removed**. (**DO NOT** click on the **OK** button at this point.)

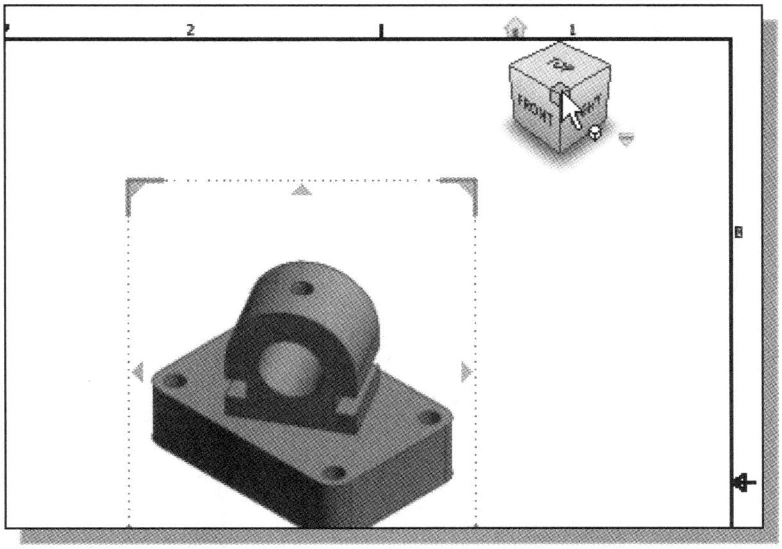

8. Drag and drop with the mouse and position the *isometric* view toward the right side of the title block as shown below.

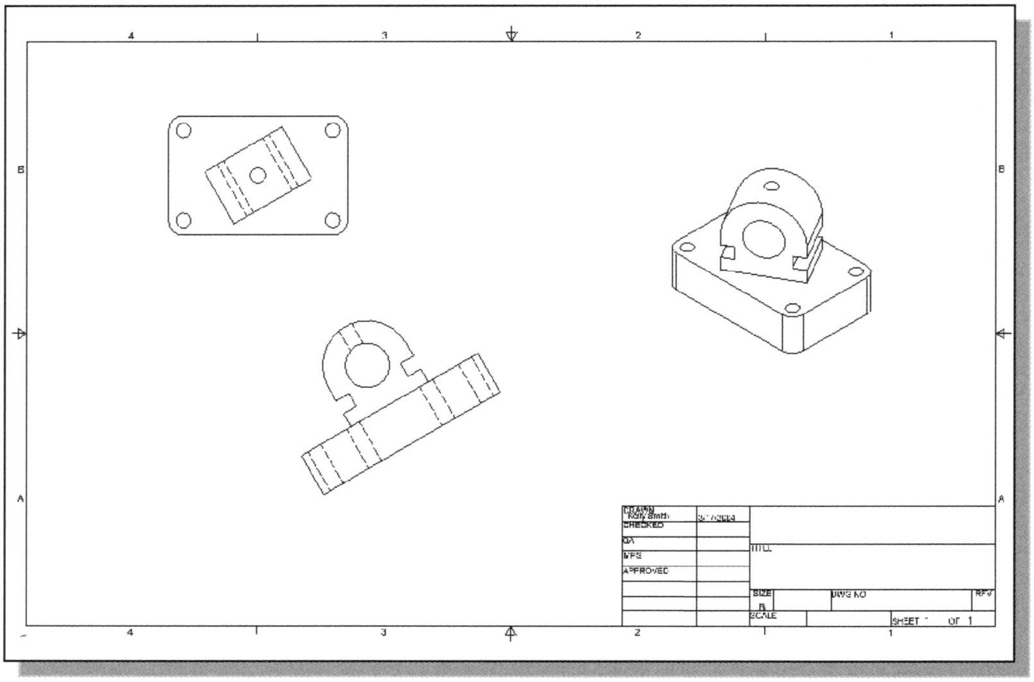

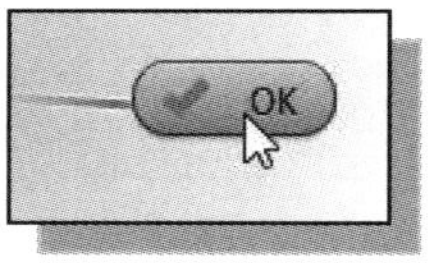

9. Right-click and choose **OK** to end the command.

Display Feature Dimensions

By default, feature dimensions are not displayed in 2D views in Autodesk Inventor. We can change the default settings while creating the views or switch on the display of the parametric dimensions using the option menu.

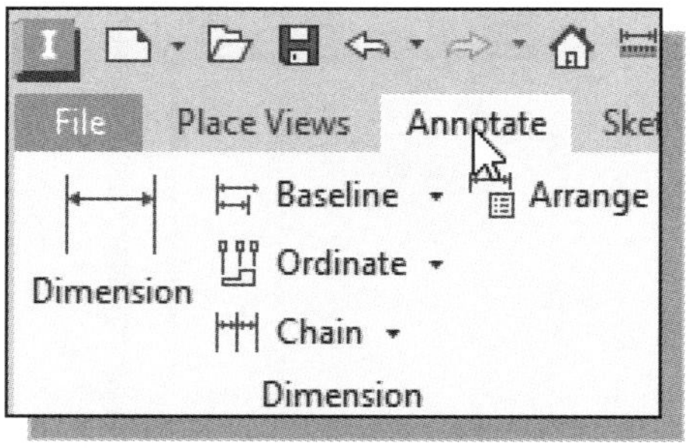

1. Select the **Annotate** tab by left-clicking once in the *Ribbon* toolbar area.

2. Move the cursor on top of the ***top*** view of the model. Watch for the box around the entire view indicating the view is selectable as shown in the figure.

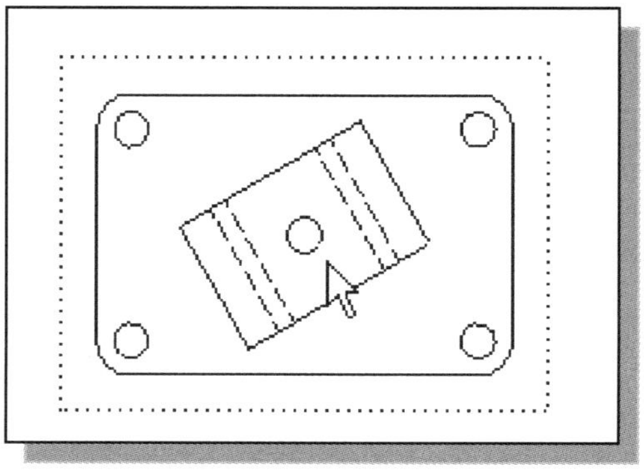

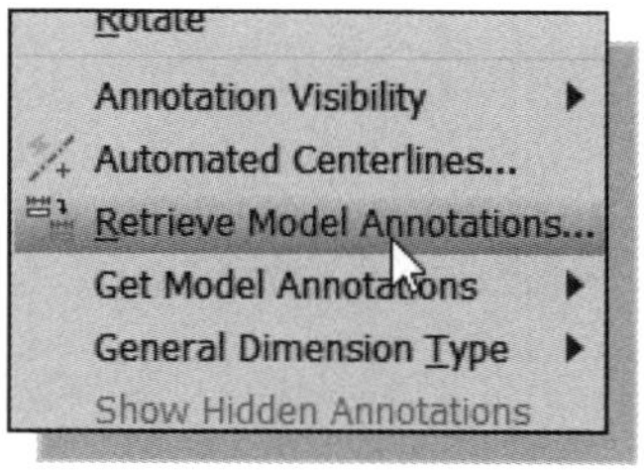

3. Inside the *graphics window*, **right-click** once to bring up the option menu.

4. Select **Retrieve Model Annotations** to display the parametric dimensions used to create the model.

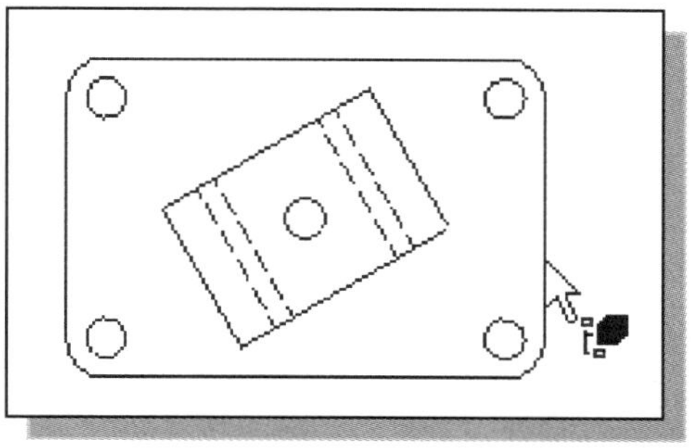

5. Move the cursor to the ***top*** view and select anywhere on the *Rod-Guide* part as shown.

6. In the *Sketch and Feature Dimensions* tab, set the *Select Source* option to **Select Parts** as shown.

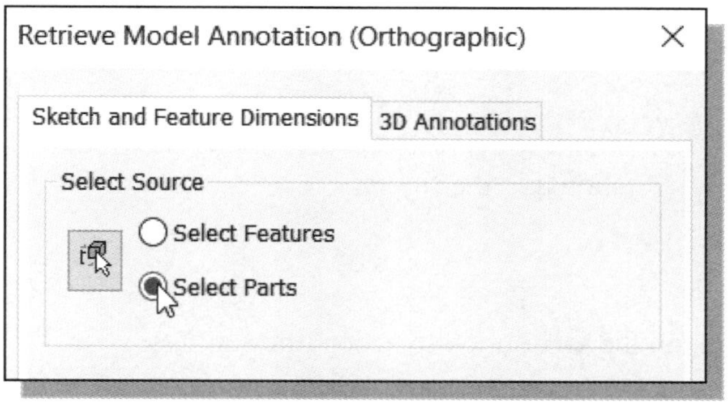

➢ Note that all the dimensions used to create the part are now displayed in the selected view.

➢ The system now expects us to select the dimensions to be retrieved.

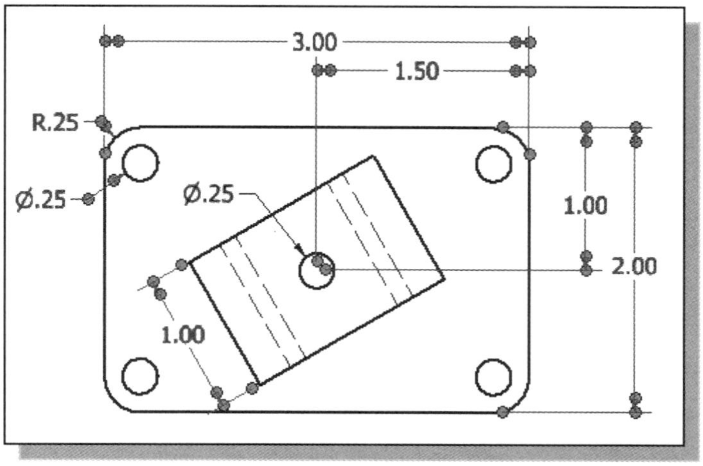

7. On your own, select the dimensions in the top view as shown.

8. Click on the **OK** button to end the Retrieve Dimensions command.

Adjust the View Scale

1. Move the cursor on top of the isometric view and watch for the box around the entire view indicating the view is selectable as shown in the figure. Right-click once to bring up the option menu. Select **Edit View** in the option menu as shown.

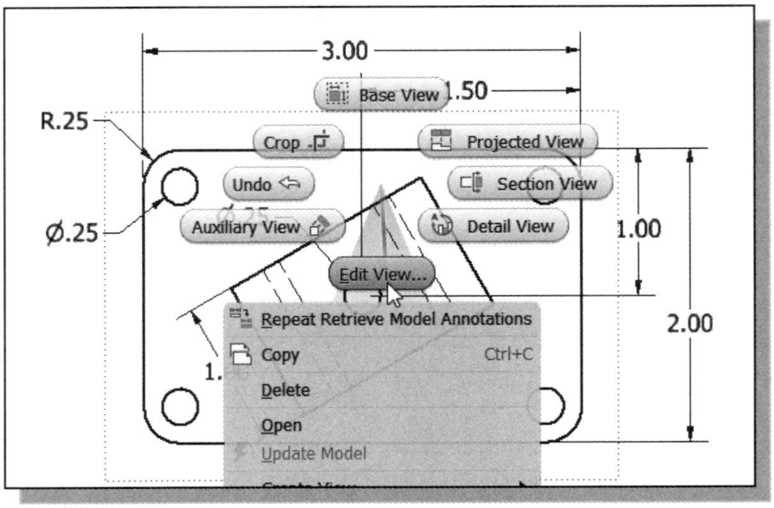

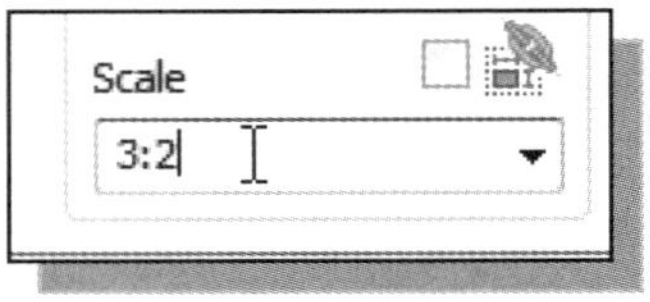

2. Inside the *Drawing View* dialog box set the *Scale* to **3:2** as shown in the figure.

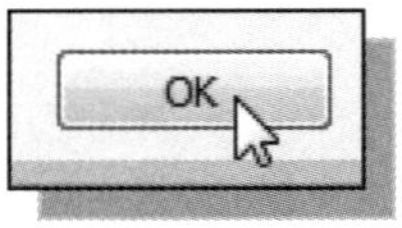

3. Click on the **OK** button to accept the settings.

4. On your own, reposition the views and dimensions by clicking and dragging the individual entities.

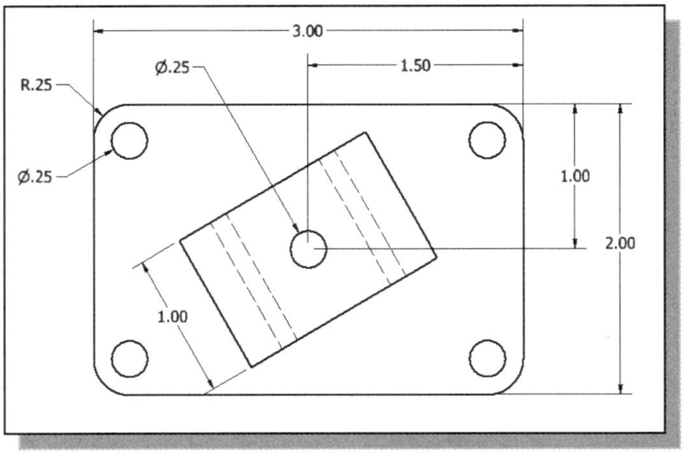

❖ Note that in parametric modeling software, dimensions are always associated with the geometry, even in the *2D Drawing Mode*.

Retrieving Dimensions in the Auxiliary View

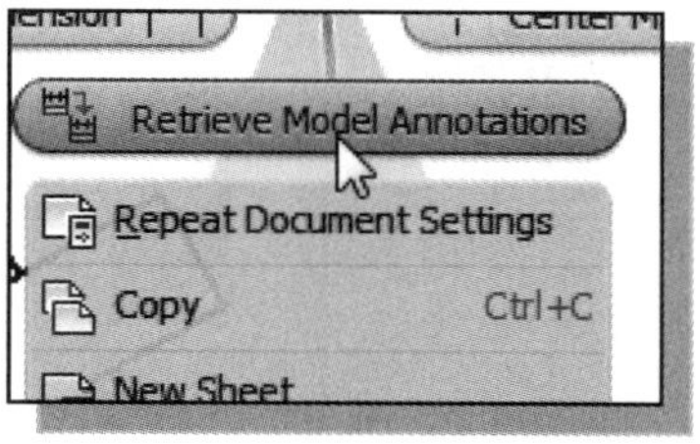

1. Inside the *graphics window*, **right-click** once in a blank area to bring up the option menu.

2. Select **Retrieve Model Annotations** to display the parametric dimensions used to create the model.

3. Select the **auxiliary view** to retrieve the associated dimension.

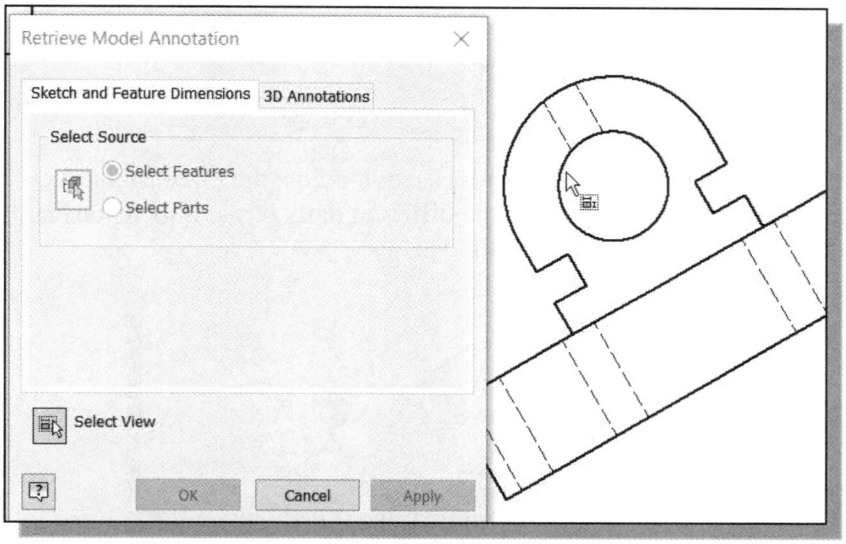

4. Select the six dimensions as shown in the below figure.

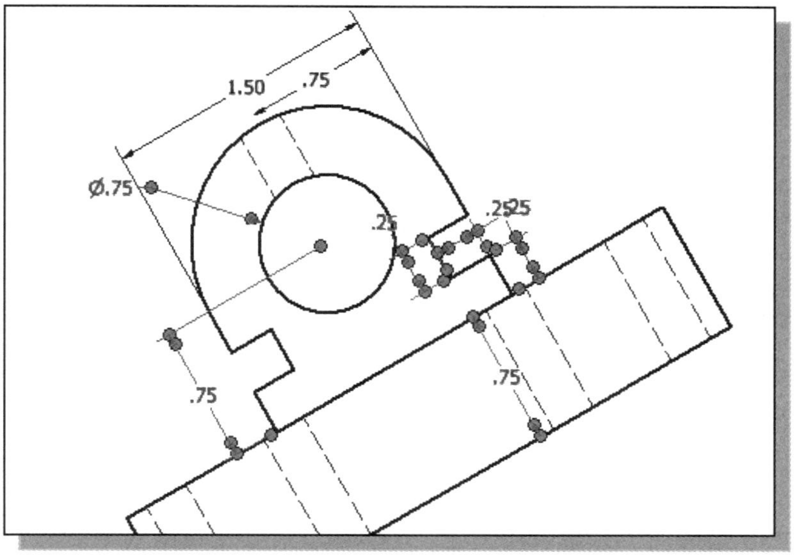

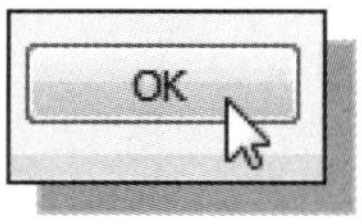

5. Click **OK** to accept the selection.

6. Reposition the dimensions by clicking and dragging the individual entities.

➢ Note that additional options are available through the **Options** menu.

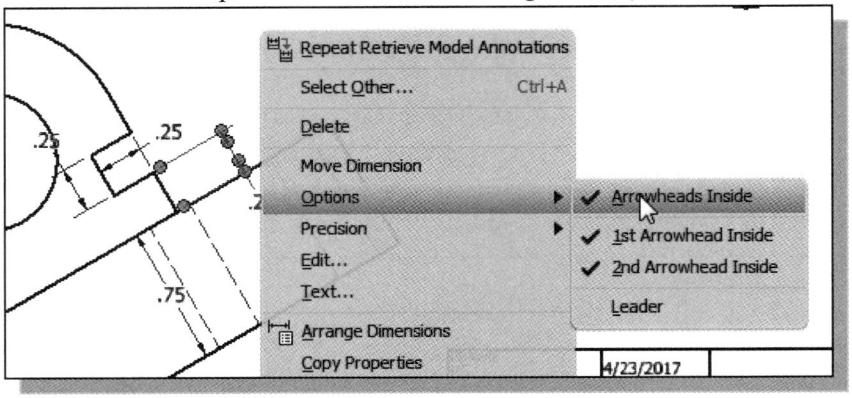

➢ Note that the different grip points can also be used to adjust the dimensions. You are encouraged to experiment with dragging the different parts of the dimensions and understand how to control the displayed dimensions.

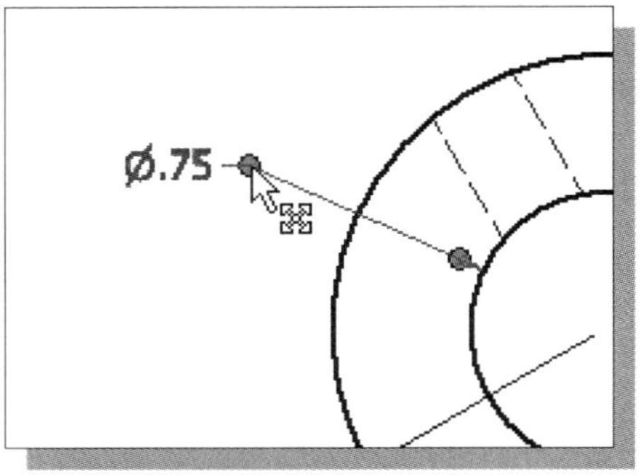

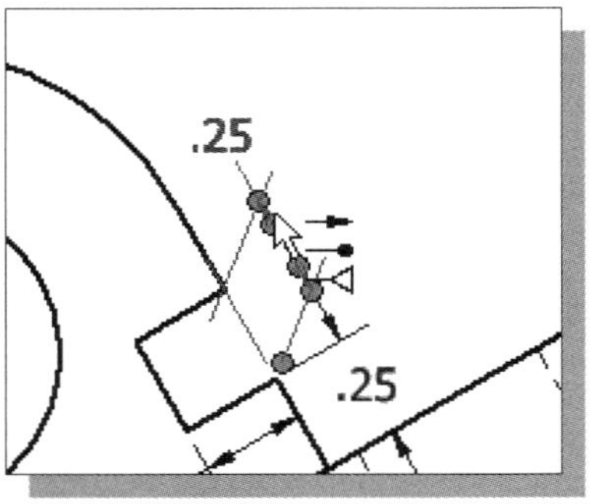

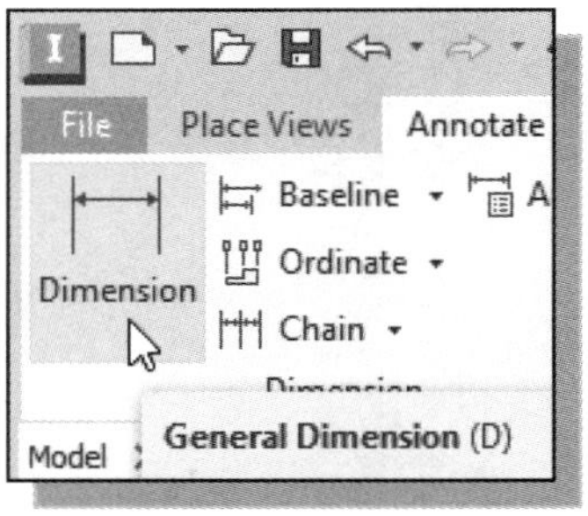

7. On your own, create and position the angle dimension for the design as shown in the figure below.

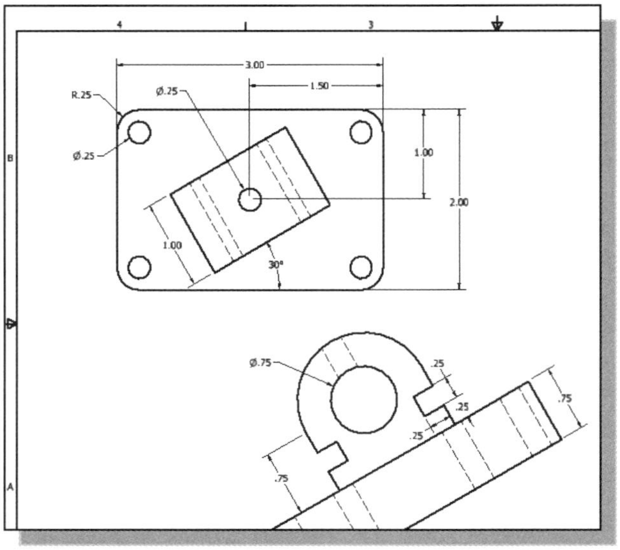

Add Center Marks and Center Lines

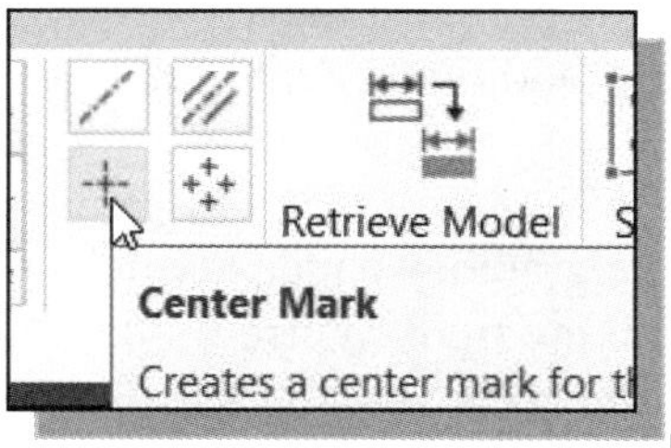

1. Click on the **Center Mark** button in the *Drawing Annotation* window.

2. Click on the five circles in the top view to add the center marks as shown.

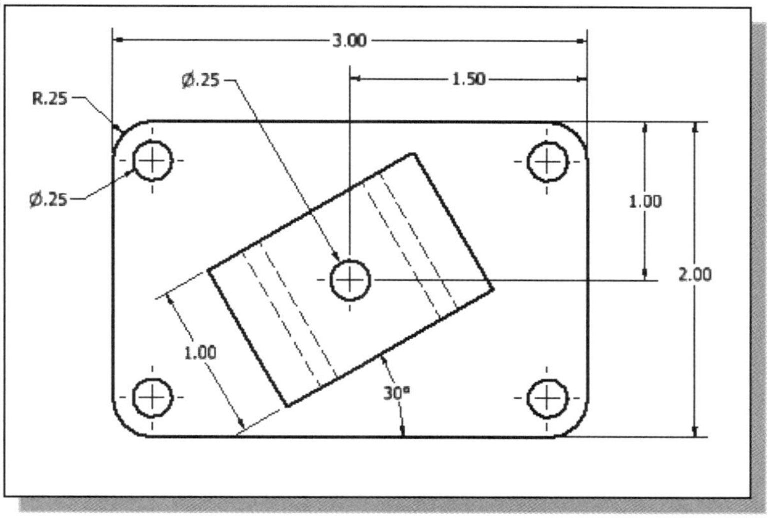

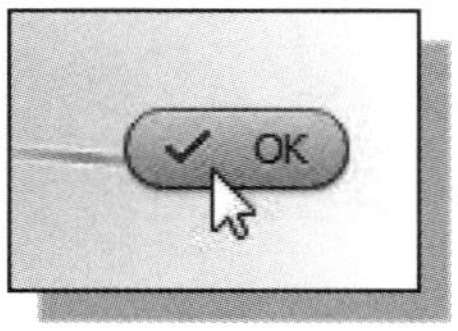

3. Inside the graphics window, click once with the right-mouse-button to display the option menu. Select **OK** in the pop-up menu to end the Center Mark command.

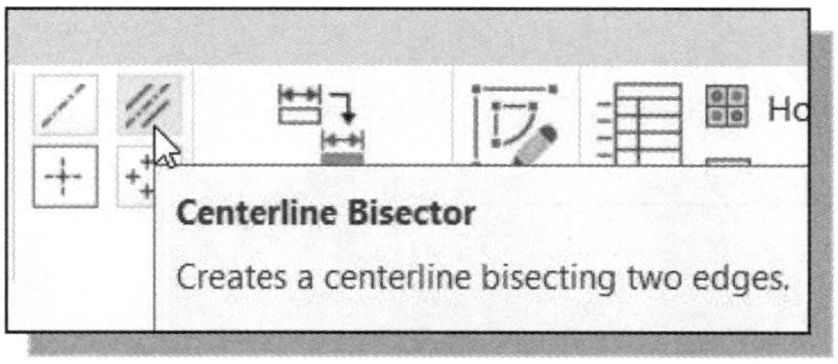

4. Select **Centerline Bisector** in the Ribbon toolbar area.

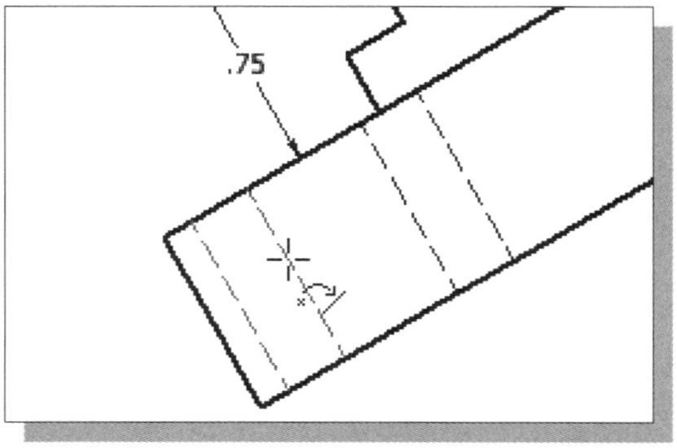

5. Inside the graphics window, click on the two hidden edges of one of the *drill* features and create a center line in the auxiliary view as shown in the figure.

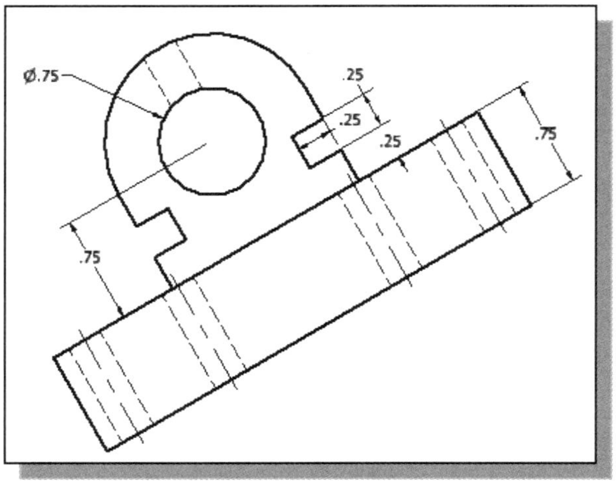

6. On your own, repeat the above step and create additional centerlines as shown.

7. Inside the graphics window, click once with the right-mouse-button to display the option menu. Select **Cancel** in the pop-up menu to end the Centerline Bisector command.

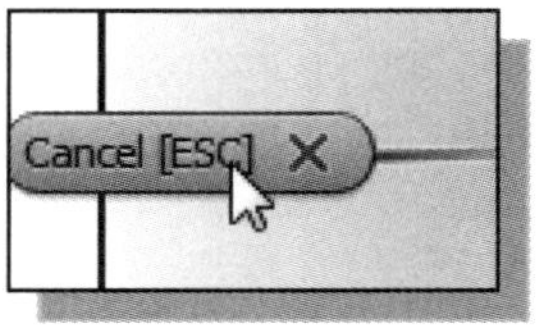

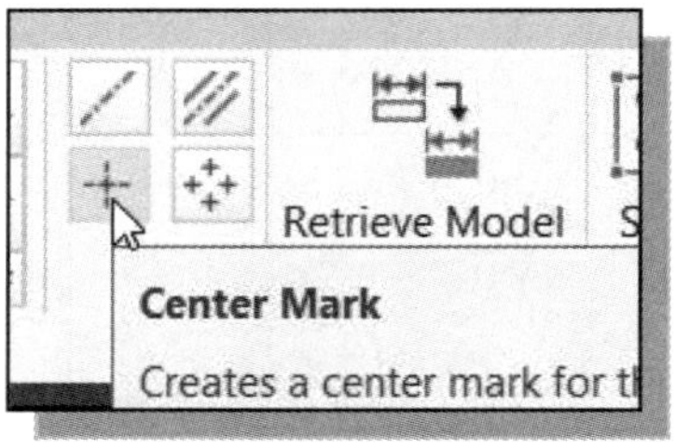

8. Select **Center Mark** in the *Drawing Annotation* window.

9. Click on the arc in the auxiliary view to create the centerlines as shown.

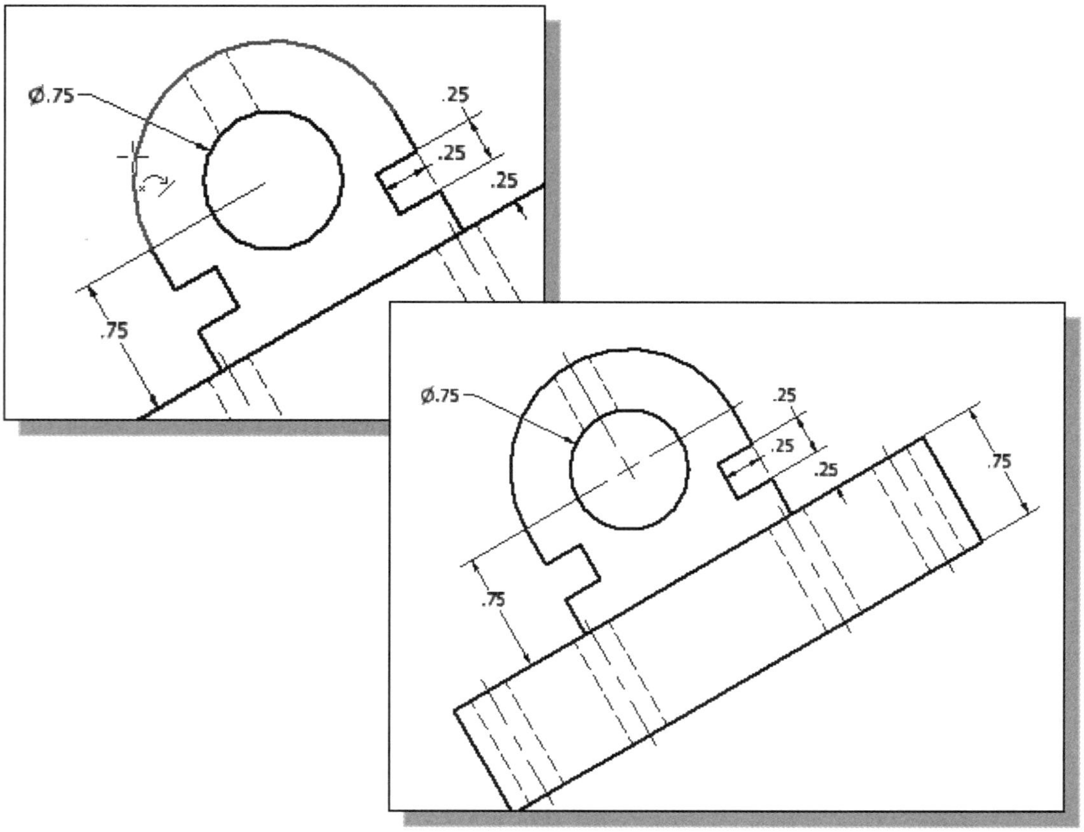

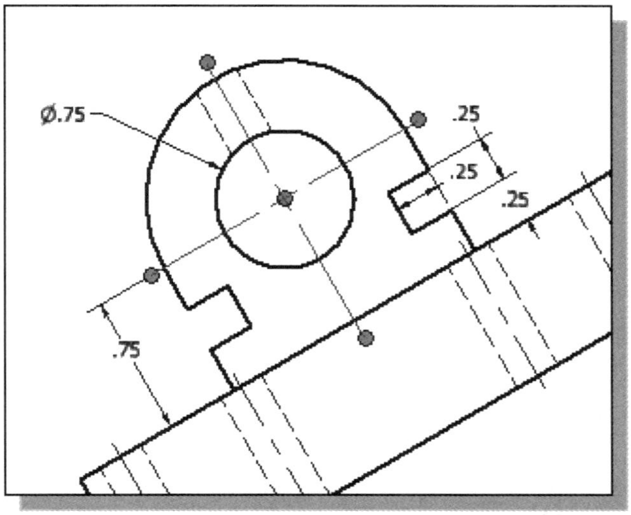

10. Hit the [**ESC**] key once to end the Center Mark command.

11. Click on the **centerlines** in the auxiliary view as shown.

12. Adjust the length of the horizontal centerline by dragging on one of the grip points as shown.

13. On your own, repeat the above steps and adjust the dimensions/centerlines as shown below.

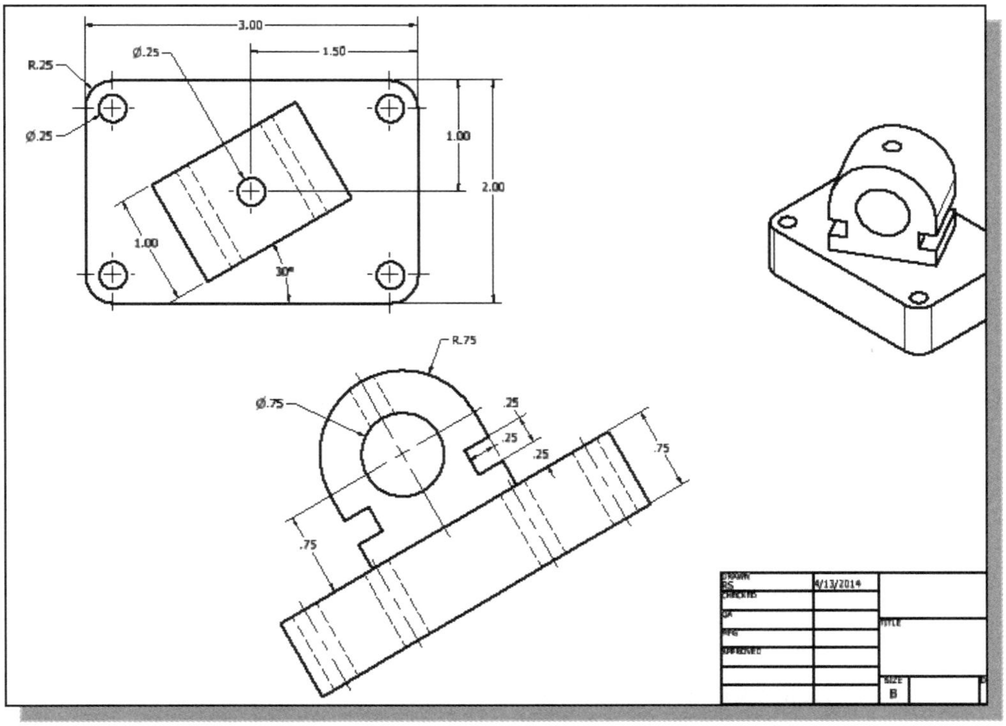

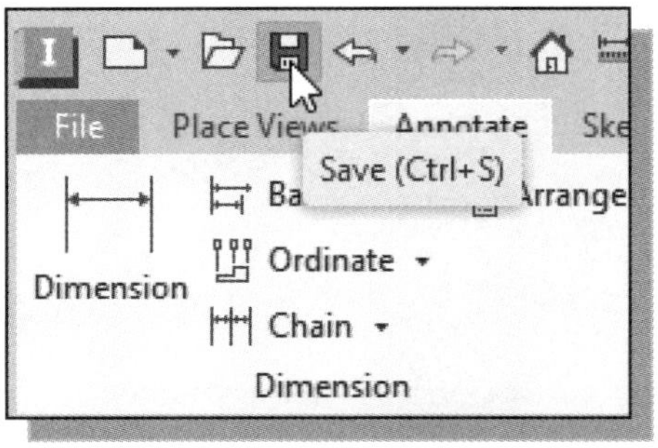

14. Click on the **Save** icon in the *Standard* toolbar and save the drawing as ***Rod-Guide.dwg*** in the **Chapter 9** folder.

Complete the Title Block with iProperties

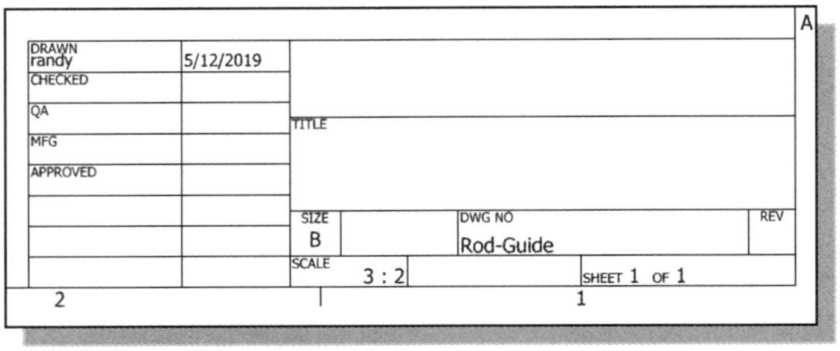

1. On your own, use the **Zoom** and **Pan** commands to adjust the display to work on the title block area.

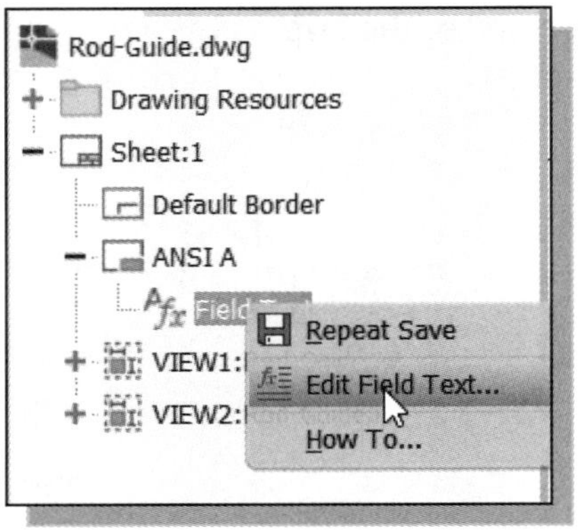

2. In the *Browser* window, right-click on the **Field Text** under the ANSI title block.

3. In the *Edit Property Fields* window, click the **iProperties** icon to bring up the *iProperty dialog box*.

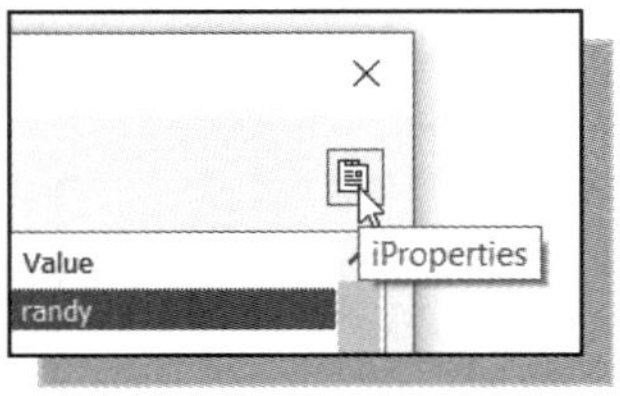

4. In the *Summary* tab, enter **Rod-Guide Design** in the *Title box*, and enter the name of your organization in the *Company box* as shown.

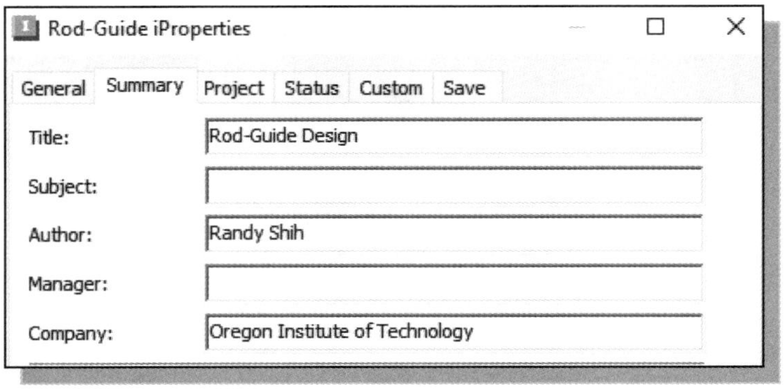

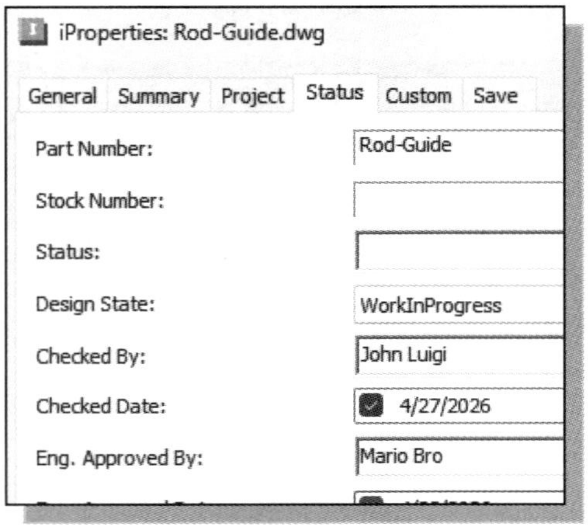

5. In the *Status* tab, enter your instructor's name in the *Checked By box* and set the *Checked Date* as shown.

6. Click **OK** to accept the settings.

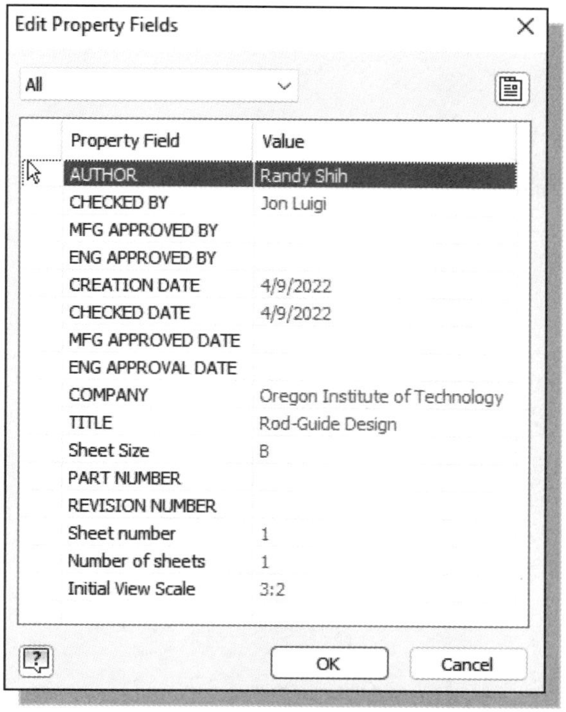

7. Note that the items in the Property fields list have been updated to reflect the changes.

8. On your own, open up the *iProperty dialog box*, and experiment with filling in the other items listed in the Property fields list.

9. Click **OK** to accept the settings.

- Note that any information entered in the iProperty dialog box will be automatically placed in the *Title block* as shown.

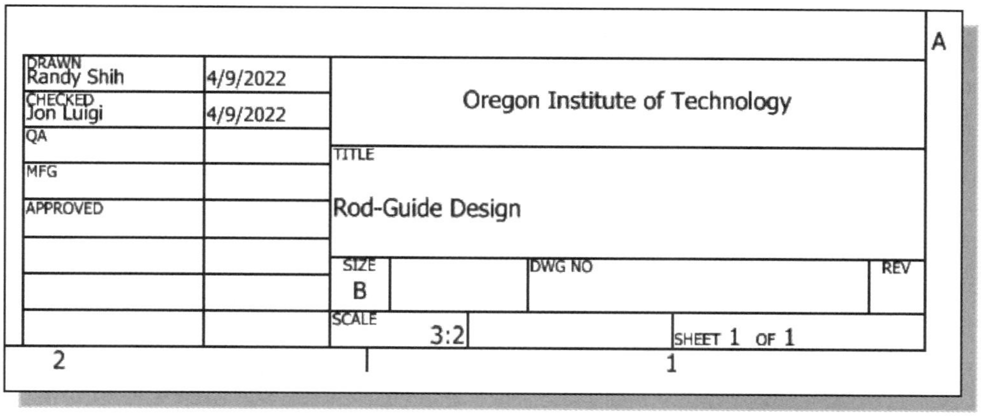

10. On your own, also create a general note at the lower left corner of the border as shown.

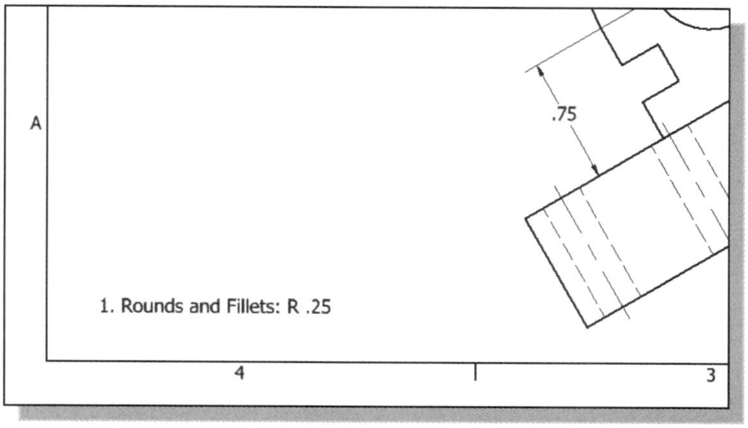

Edit the Isometric View

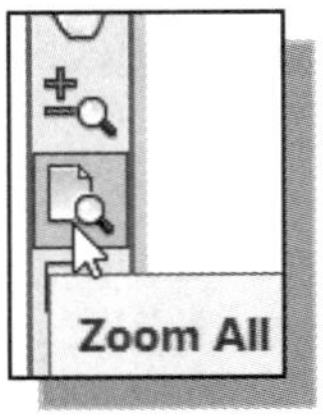

1. Click on the **Zoom All** button in the *Standard* toolbar.

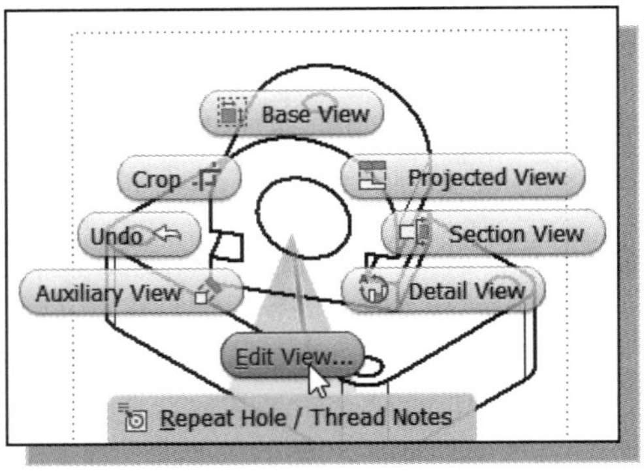

2. Right-click inside the *isometric* view to bring up the option menu as shown.
3. Select **Edit View** in the option menu.

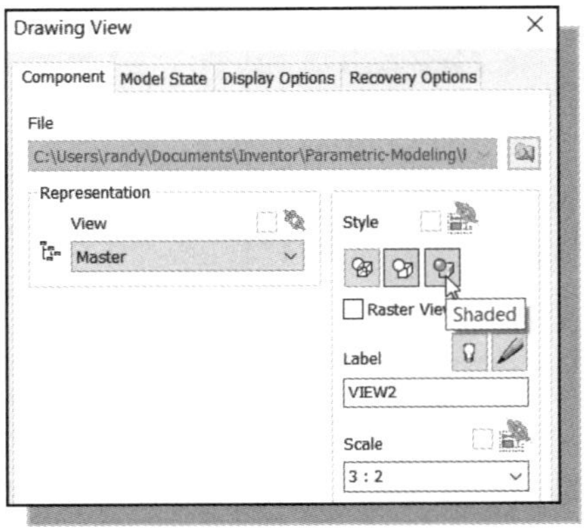

4. In the *Drawing View* dialog box, set the scale to **3:2** and the display *Style* to **Shaded**.
5. Click on the **OK** button to accept the settings.

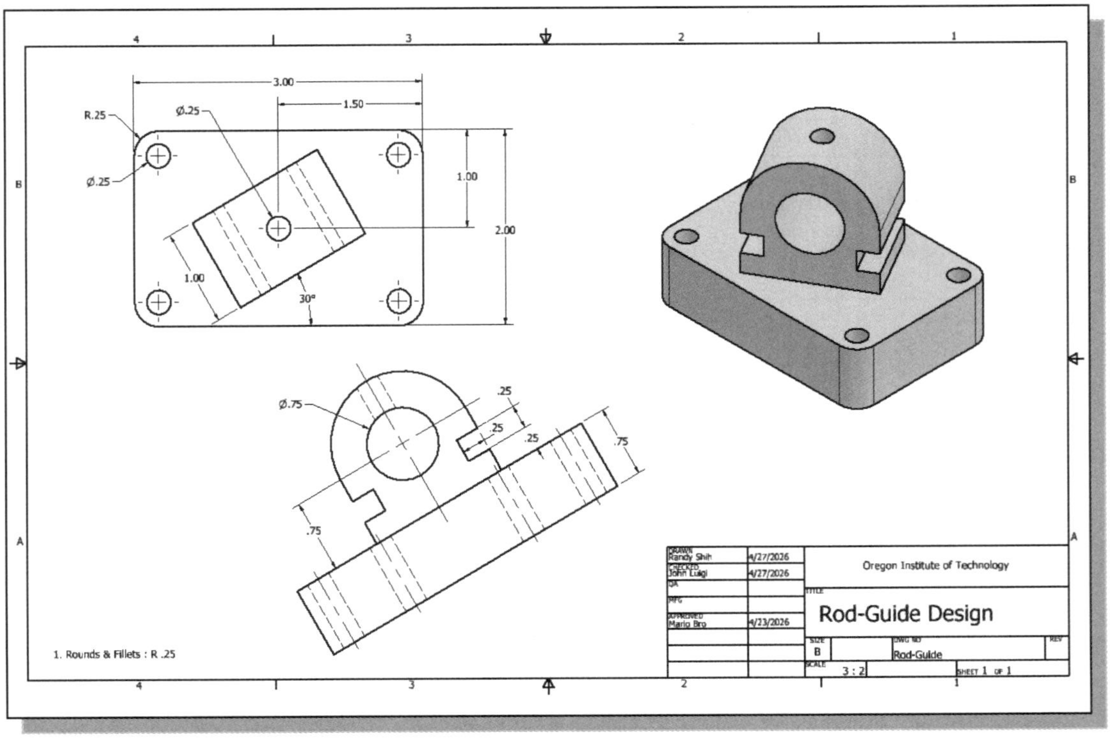

Review Questions: (Time: 25 minutes)

1. What are the different types of work features available in Autodesk Inventor?
2. Why are work features important in parametric modeling?
3. Describe the purpose of auxiliary views in 2D drawings.
4. What are the required elements in order to generate an auxiliary view?
5. Can we use a different Title Block in the Drawing Mode? How?
6. Describe the different methods used to create centerlines in the chapter.
7. Can we change the *View Scale* of existing views? How?
8. What is the main difference between an auxiliary view and a projected view in Autodesk Inventor?
9. Describe the steps to change the *display style* of a drawing view.
10. Describe the difference between the centerlines created with the **Centerline Bisector** and the **Center Mark** commands.

Exercises: Create the Solid models and the associated 2D drawings and save the exercises in the Chapter9 folder. (Time: 250 minutes.)

1. **Rod Slide** (Dimensions are in inches. **2.5** inches is the overall height.)

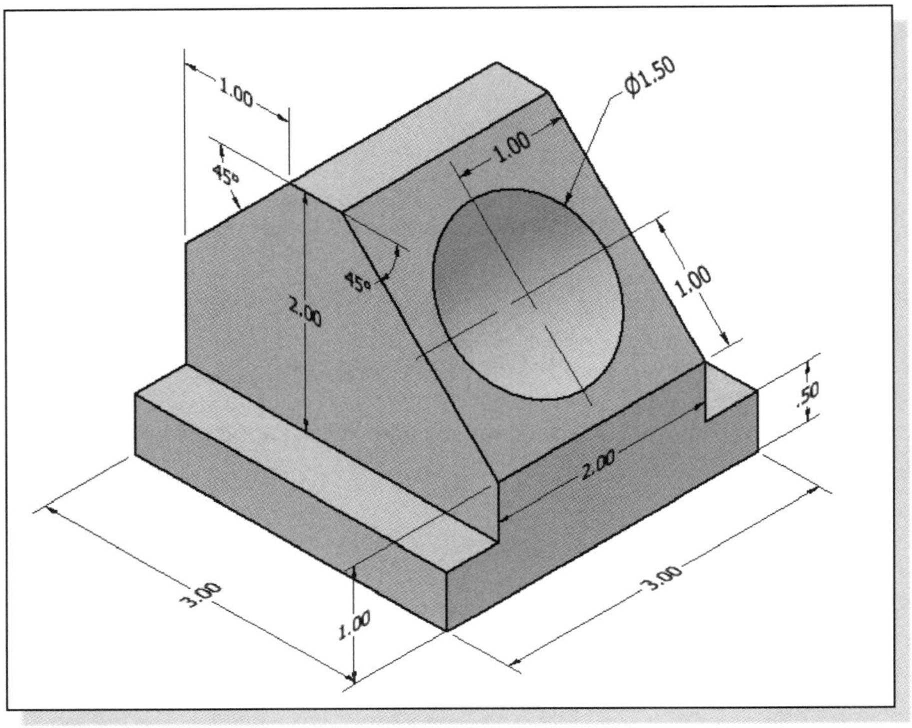

2. **Anchor Base** (Dimensions are in inches.) Hint: Use the **Slot command**.

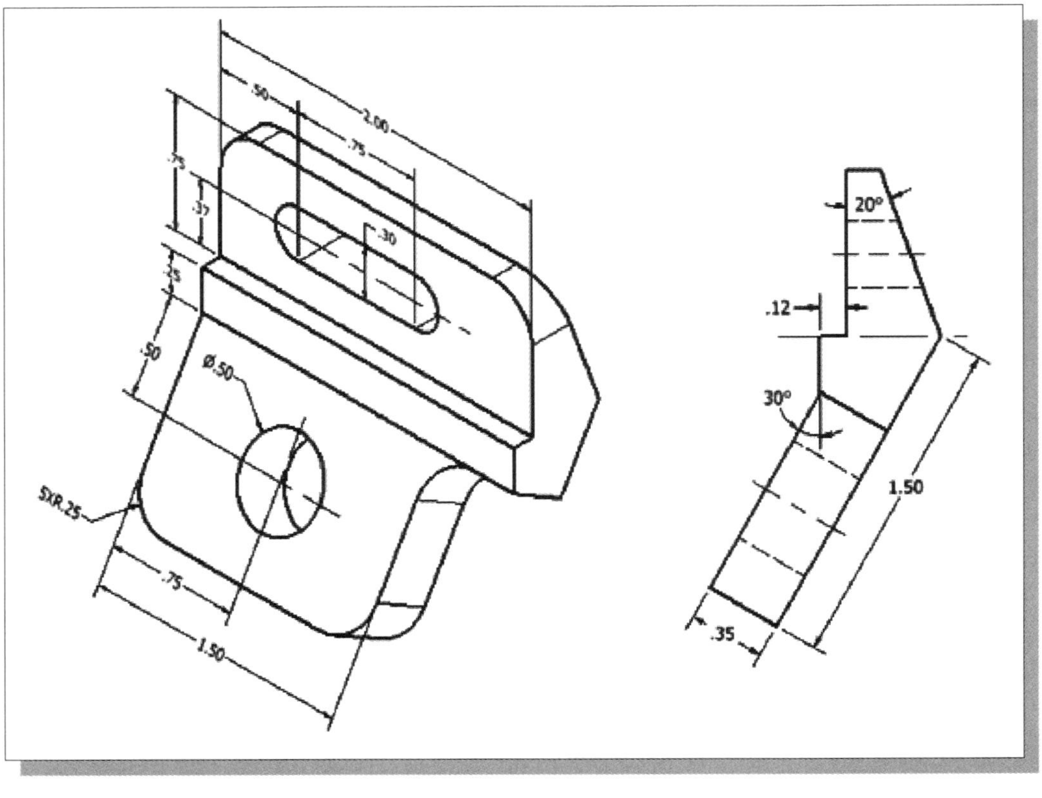

3. **Bevel Washer** (Dimensions are in inches.)

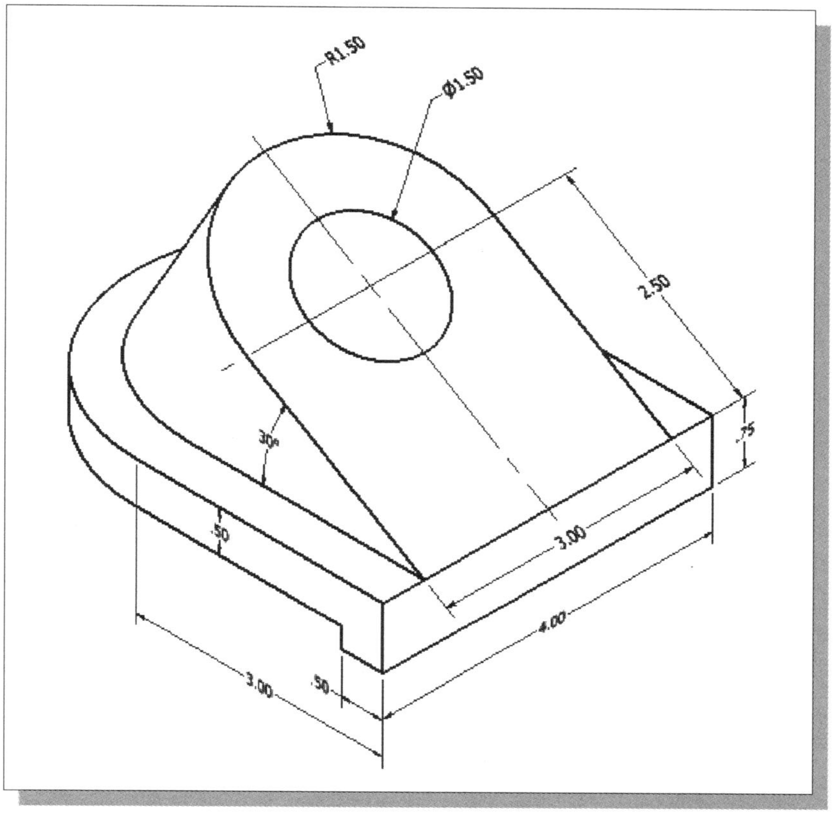

4. **Angle V-Block** (Dimensions are in inches.)

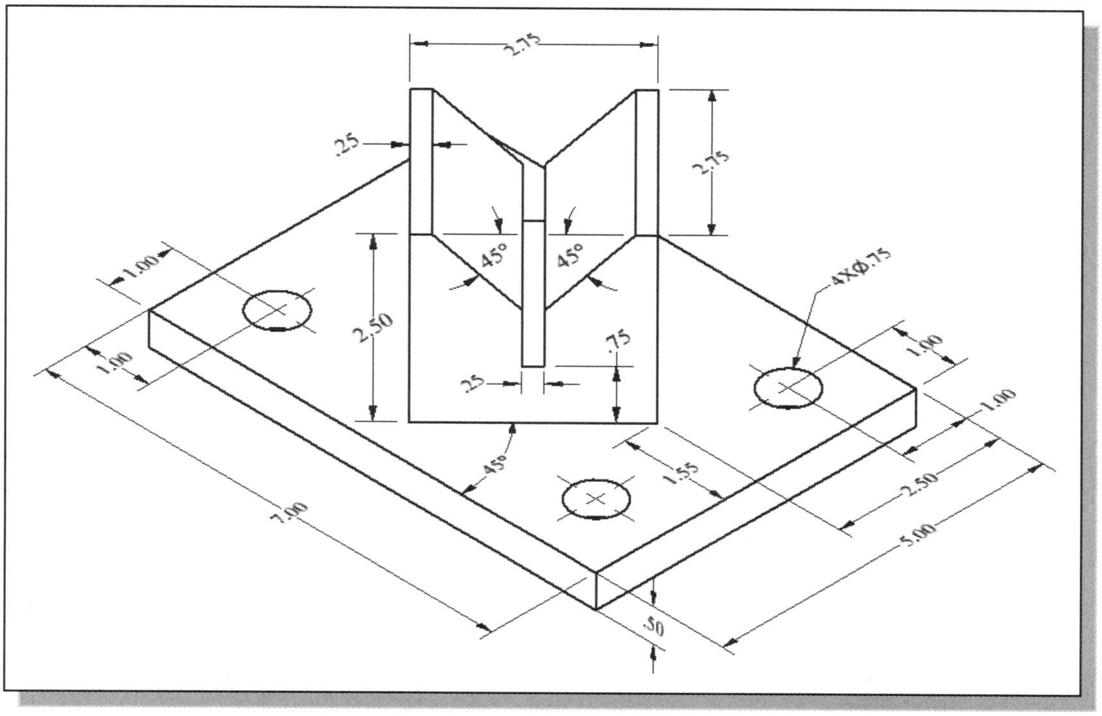

5. **Angle Support** (Dimensions are in millimeters.)

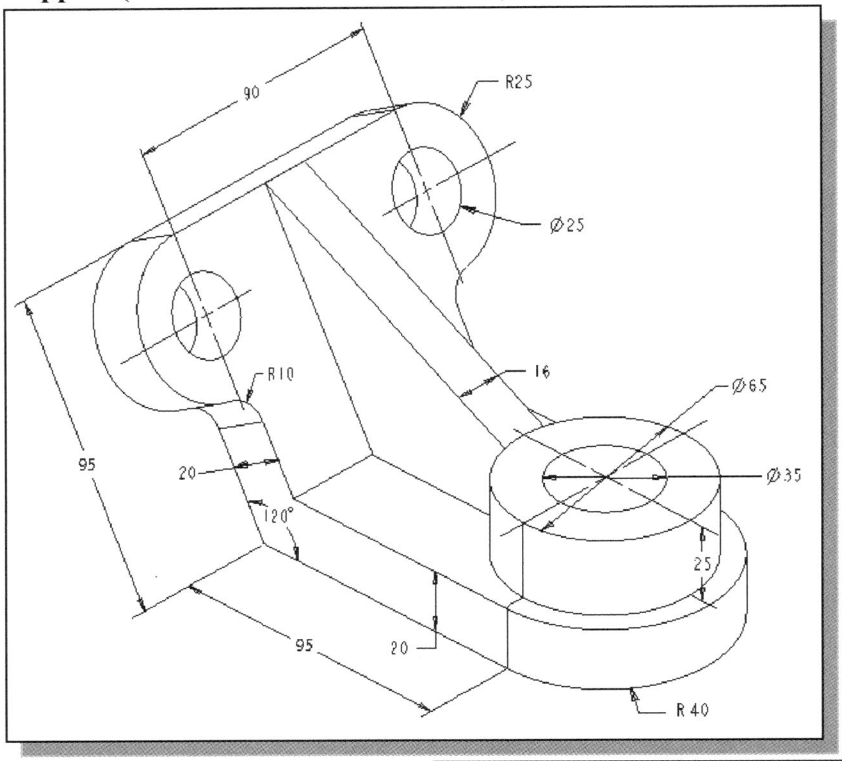

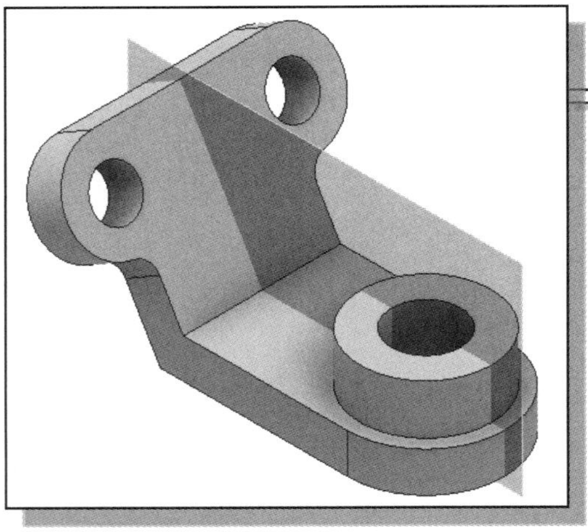

Create a sketch, with two lines, on the datum plane as shown.

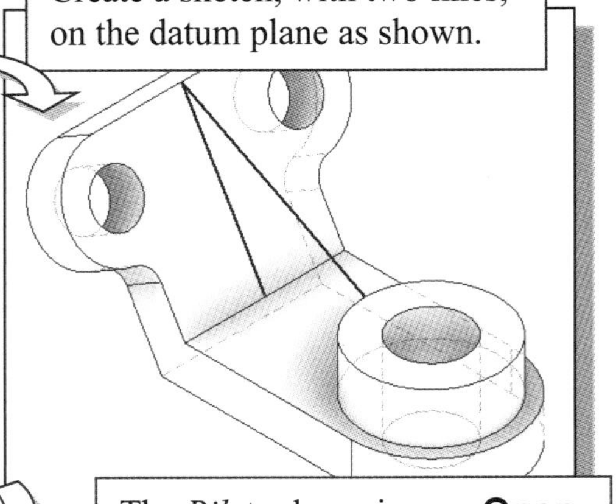

The *Rib* tool requires an **Open Curve** as the sketch profile.

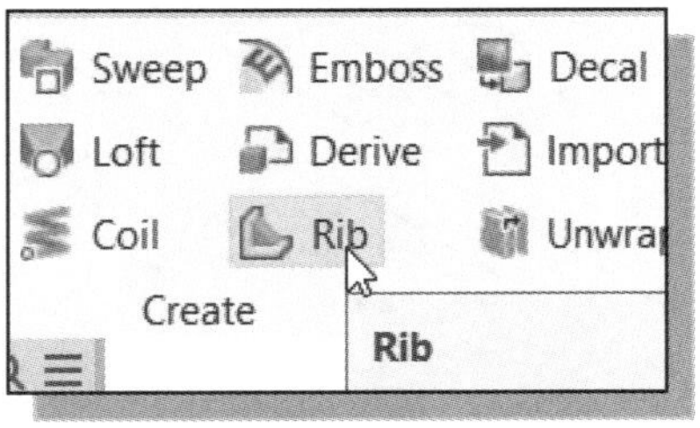

Use the *Rib* tool, *parallel to sketch plane* and extrude in *both directions*.

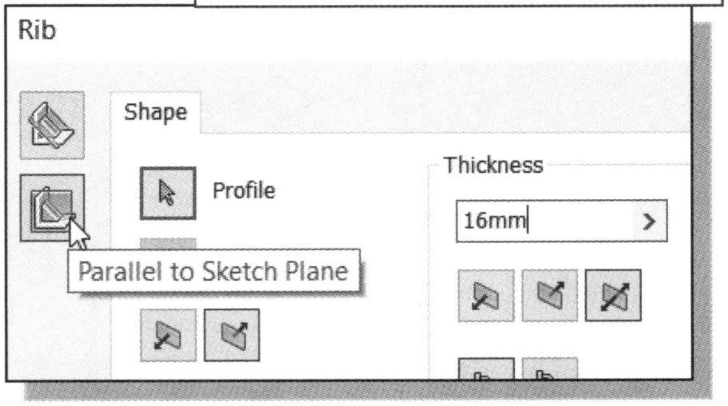

6. **Jig Base** (Dimensions are in millimeters.)

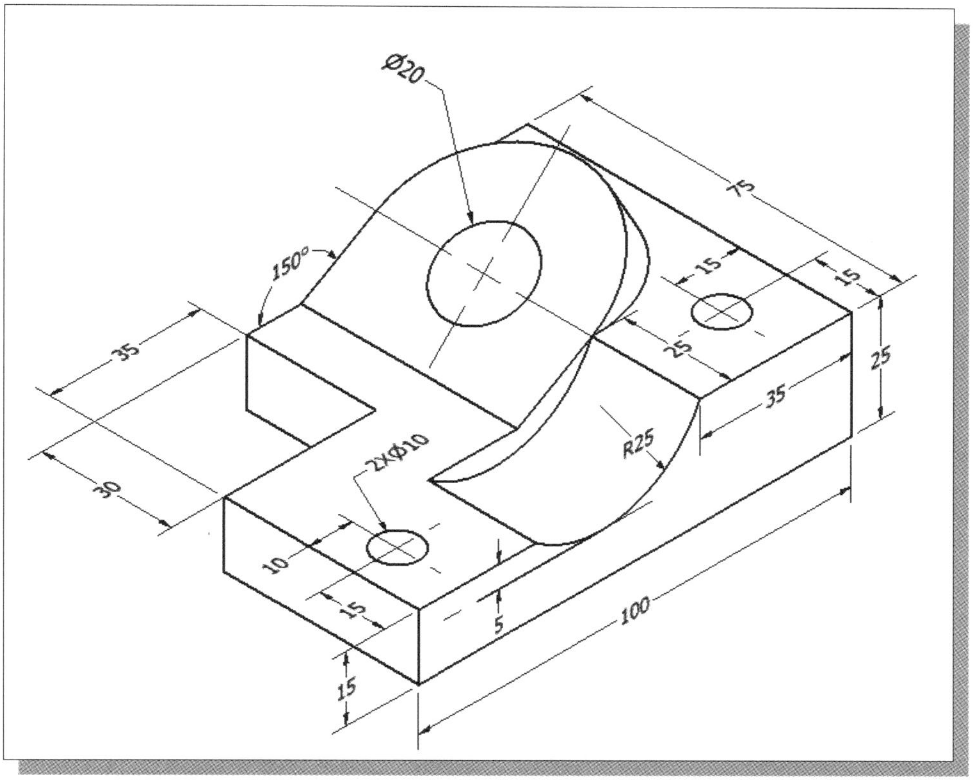

7. **Angle Lock**

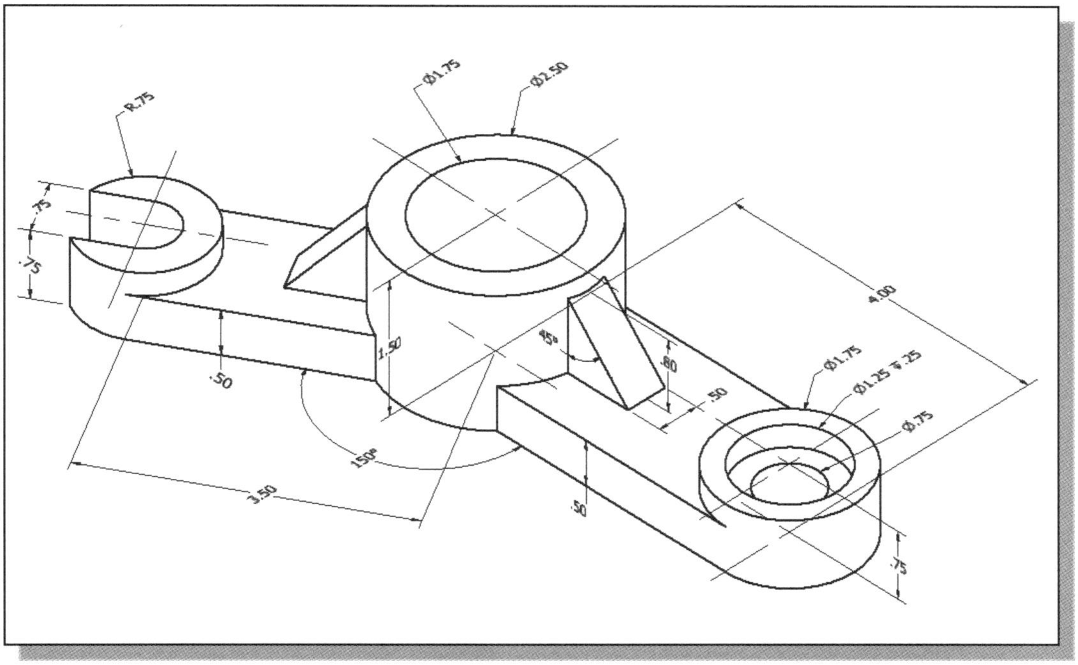

Notes:

Chapter 10
Introduction to 3D Printing

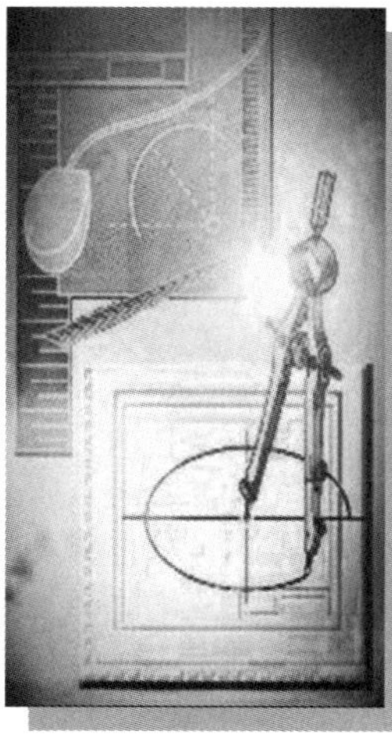

Learning Objectives

- ♦ **Understand the History and Development of 3D Printing**
- ♦ **Be aware of the Primary types of 3D Printing Technologies**
- ♦ **Be able to identify the commonly used Filament types for Fused Filament Fabrication**
- ♦ **Understand the general procedure for 3D Printing**

What is 3D Printing?

3D Printing is a type of *Rapid Prototyping* (RP) method. Rapid prototyping refers to the techniques used to quickly fabricate a design to confirm/validate/improve conceptual design ideas. 3D printing is also known as "**Additive Manufacturing**" and construction of parts or assemblies is usually done by addition of material in thin layers.

Prior to the 1980s, nearly all metalworking was produced by machining, fabrication, forming, and mold casting; the majority of these processes require the removal of material rather than adding it. In contrast to *Additive Manufacturing* technology, traditional manufacturing processes can be described as **Subtractive Manufacturing**. The term *Additive Manufacturing* gained wider acceptance in the 2010s. As the various additive processes continue to advance and become more mature, it is quite clear that they will compete with material removal as the main manufacturing process for many applications in the very near future.

The basic principle behind *3D printing* is that it is an additive process. 3D printing is a radically different manufacturing method based on advanced technology that create parts directly, by adding material layer by layer at the sub millimeter scale. One way to think about 3D Printing is the additive process is really performing "**2D printing over and over again**."

A number of limitations exist in traditional manufacturing processes, which can be labor intensive, requiring expensive tooling, designing of fixtures, and the assembly of parts. The *3D printing* technology provides a way to create parts with complex geometric shapes quite easily using thin layers. The traditional *subtractive manufacturing* can also be quite wasteful as excess materials are cut and removed from large stock blocks, while the *3D printing* process basically uses only the material needed for the parts. *3D printing* is an enabling technology that encourages and drives innovation with unprecedented design freedom while being a tooling-less process that reduces costs and lead times. The relatively fast turnaround time also makes *3D printing* ideal for prototyping. Components with intricate geometry and complex features can also be designed specifically for *3D printing* to avoid complicated assembly requirements. *3D printing* is also an energy efficient technology that can provide better environmental friendliness in terms of the manufacturing process itself and the type of materials used for the product. There are quite a few different techniques to 3D print an object. *3D Printing* brings together two fundamental innovations: the manipulation of objects in the digital format and the manufacturing of objects by addition of material in thin layers.

The term **3D-printing** originally referred only to the smaller 3D printers with moveable print heads similar to an inkjet printer. Today, the term **3D-printing** is used interchangeably with **Additive Manufacturing**, as both refer to the technology of creating parts through the process of adding/forming thin layers of materials.

Development of 3D Printing Technologies

The earliest 3D printing technology was first invented in the 1980s; at that time it was generally called **Rapid Prototyping** (**RP**) technology. This is because the process was originally conceived as a fast and time-effective method for creating prototypes for product development in industry. In 1981, Dr. Hideo Kodama of Nagoya Municipal Industrial Research Institute invented two methods of creating three-dimensional plastic models with photo-hardening polymer through the use of a UV Laser. In 1986, the first US patent for **stereolithography** apparatus (**SLA**) was issued to Charles Hull, who first invented his SLA machine in 1983. Chuck Hull went on to co-found *3D Systems Corporation*, which is one of the largest companies in the 3D printing sector today. Chuck Hull also designed the **STL** (**ST**ereo**L**ithography) file format, which is widely used by 3D printing software performing the digital slicing and infill strategies common to the additive manufacturing processes. The first available commercial RP system, the **SLA-1** by *3D Systems* (as shown in the figure below), was made available in 1987.

The 1980s also mark the birth of many RP technologies worldwide. In 1989, Carl Deckard of University of Texas developed the **Selective Laser Sintering** (**SLS**) process. In 1989, Scott Crump, one of the founders of *Stratasys Inc*., also created **Fused Deposition Modeling** (**FDM**). In Europe, Hans Langer started *EOS GmbH* in Germany; the company focuses on the **Laser Sintering** (**LS**) process. The *EOS Systems Corp.* also developed the **Direct Metal Laser Sintering** (**DMLS**) process. Today, *3D Systems*, *EOS* and *Stratasys* are still the main leaders in the *Additive Manufacturing* industry.

During the 1990s, the *3D printing* sector started to show signs of distinct diversification with two specific areas of emphasis which are much more clearly defined today. First, there was the high end of 3D printing, still very expensive systems, which were geared towards part production for relatively complex designs. For example, in 1995, *Sciaky Inc* developed an additive welding process based on its proprietary **Electron Beam Additive Manufacturing** (**EBAM**) technology. Many *RP* system companies, such as *Solidscape*, *ZCorporation*, *Arcam* and *Objet Geometries* were all launched in the 1990s. At the other end of the spectrum, some of the 3D printing system manufacturers started to develop smaller desktop systems in the 1990s.

The idea of creating low-cost desktop 3D printers also intrigued many technology professionals and hobby enthusiasts during the late 1990s. In 2004, a retired professor, Dr Adrian Bowyer (person on the left in the below photo), started the **RepRap** (*Replication Rapid-Prototyper*) project of an open source, self-replicating 3D printer (**RepRap 1.0 - Darwin**). This set the stage for what was to come in the following years. It was around 2007 that the open-source *3D printing* movement started gaining visibility and momentum. In January of 2009, the first commercially available open-source 3D printer, the **BFB RapMan** 3D printer, became available. The *Makerbot Industries* also came out with their *Makerbot 3D printer* in April of 2009. Since then, a host of low-cost desktop 3D printers have emerged each year.

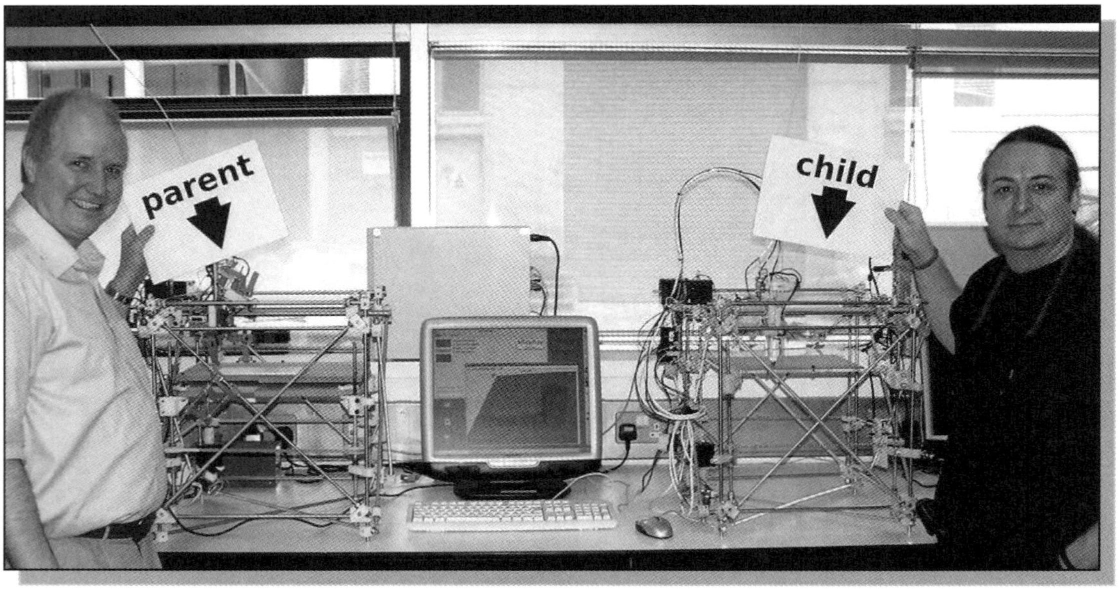

In the beginning of the 2010s, alternative 3D printing processes, such as those using **Polymer Resins**, became available at the desktop level of the market. The **B9 Creator**, using **Digital Light Processing** (**DLP**) technology, by *B9Creations*, came first in June of 2012, followed by the **Form 1** desktop printer by *Formlabs Inc.* Both 3D printers were launched via KickStarter's crowdfunding website and both enjoyed huge success. 2012 was also the year that many different mainstream media took note of the exciting 3D printing technology, which dramatically increased its visibility and awareness to the general public. 2013 was also a year of significant growth and consolidation; one of the most notable moves was the acquisition of **Makerbo**t by *Stratasys*. Currently, the new developments in 3D printing are concentrated more on faster print-speed, multi-color, multi-material using single or multiple extruder printers and new technologies to shorten the 3D printing time.

As a result of the market divergence, the price of desktop 3D printers continues to go down each year. Today, very capable fully assembled desktop 3D printers, such as BamBu X1C, Prusa I3 MK4, can be acquired for around $1000. Fully assembled smaller desktop 3D printers, such as XYZprinting's DA Vinci mini 3D Printer and M3D's Micro 3D, can be acquired for less than $350. Unassembled desktop 3D printer kits can even be acquired for under $250.

Another trend that happened in the 2010s is the availability of **3D Printing Services**. 3D printing services are growing quite rapidly in the US. For example, many state public libraries, such as in California and Oregon, are now providing 3D printing services to the general public, and **UPS** started its worldwide 3D printing services in May of 2016. This trend is spreading throughout the US, with many more companies planning to provide 3D printing services in the very near future. It is now quite feasible, and perhaps more economical, to 3D print designs without owning or ever touching a 3D printer, but understanding of the technology is still needed to increase productivity.

As the exponential adoption rate continues on all fronts, more and more technologies, materials, applications, and online services will continue to emerge. It is predicted that the development of 3D printing will continue in the years to come, and 3D printing will eventually become the mainstream manufacturing method in industries and homes.

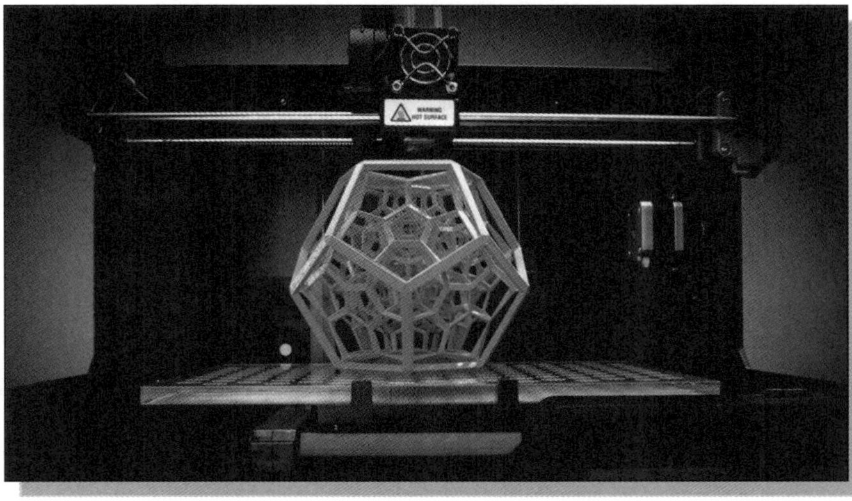

Primary Types of 3D Printing Processes

There are quite a few different techniques to 3D print an object. The different types of 3D printers each employ a different technology that processes different materials in different ways. For example, some 3D printers process powdered materials (nylon, plastic, ceramic and metal), which utilize a light/heat source to sinter/melt/fuse layers of the powder together in the defined shape. Others process polymer resin materials and again utilize a light/laser to solidify the resin in thin layers. **Stereolithography** (**SLA** or **SL**), **Fused Deposition Modeling** (**FDM** or **FFF**) and **Laser Sintering** (**LS** or **SLS**) represent the three primary types of 3D printing processes; the majority of the other 3D printing technologies are variations of the three main types.

Stereolithography

Stereolithography (**SLA** or **SL**) is widely recognized as the first 3D printing process; it was certainly the first to be commercialized. *SLA* is a laser-based process that works with photopolymer resins. The photopolymer resins react with the laser and cure to form a solid in a very precise way to produce very accurate parts. It is a complex process, but simply put, the photopolymer resin is held in a container with a movable platform inside. A laser beam is directed in the X-Y axes across the surface of the resin according to the 3D data supplied to the machine. The resin hardens precisely as the laser hits the designated area. Once the current layer is completed, the platform within the container drops down by a fraction (in the Z axis) and the subsequent layer is traced out by the laser. This 2D layer tracing continues until the entire object is completed.

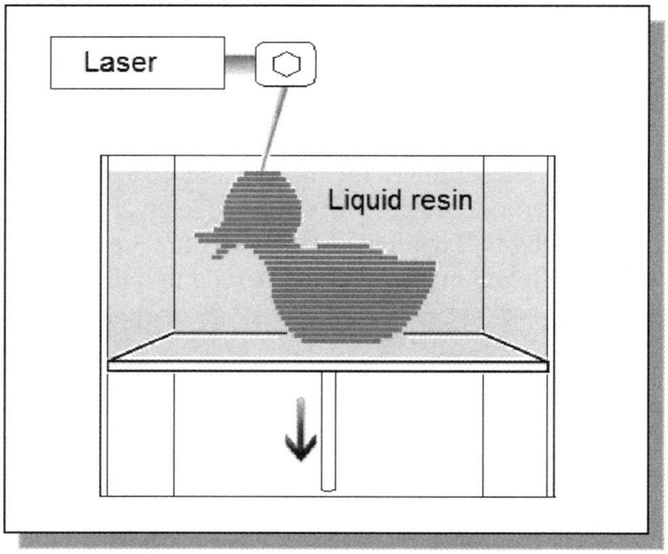

Because of the nature of the SLA process, support structures are needed for some parts, specifically those with overhangs or undercuts. These support structures need to be removed once the part is created. Many 3D printed objects using SLA need to be further cleaned and/or cured. Curing involves subjecting the part to intense light in an oven-like machine to fully harden the resin. SLA is generally accepted as being one of the most accurate 3D printing processes with excellent surface finish.

Fused Deposition Modeling (FDM) / Fused Filament Fabrication (FFF)

3D printing utilizing the extrusion of thermoplastic material is probably the most popular 3D printing process. The original name for the process is **Fused Deposition Modeling** (**FDM**), which was developed in the early 1990s and is a trade name registered by *Stratasys*. However, a similar process, **Fused Filament Fabrication** (**FFF**), has emerged since 2009. The majority of the desktop 3D printers, both open source and proprietary, utilize the FFF process, which is a more basic extrusion form of FDM.

The FDM and FFF processes work by melting plastic filament that is deposited, via a heated extruder, one layer at a time, onto a build platform according to the 3D data supplied to the 3D printer. Each layer hardens as it cools down and bonds to the previous layer.

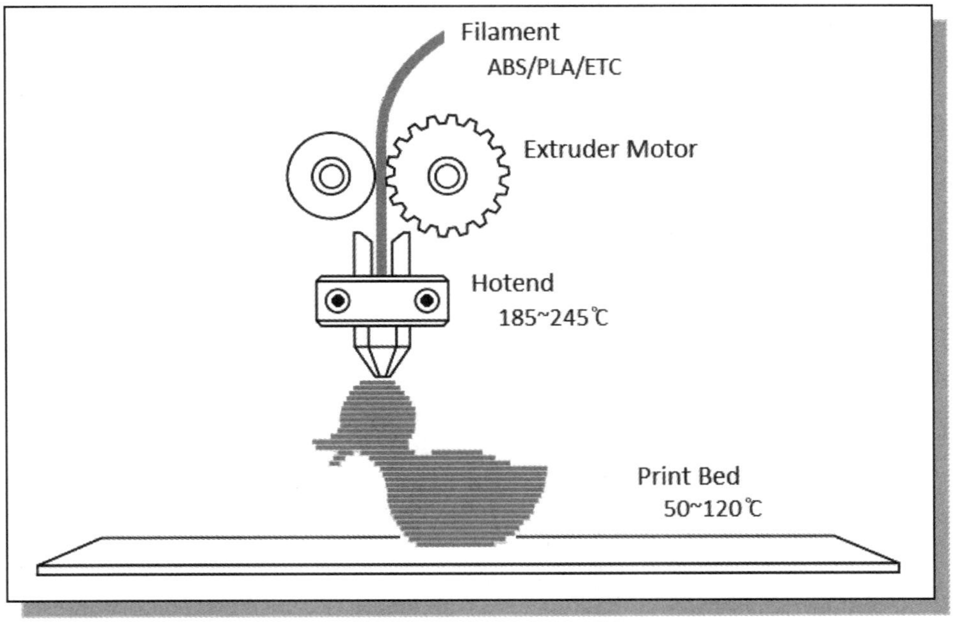

Stratasys has developed a range of proprietary industrial grade materials for its FDM process that are suitable for production applications. However, the most common materials for both FDM and FFF 3D printers are **ABS** (**Acrylonitrile Butadiene Styrene**) and **PLA** (**Polylactic Acid**). The FDM and FFF processes require support structures for any applications with overhanging geometries. This generally entails a second, typically water-soluble or breakaway material, which allows support structures to be easily removed once the print is complete.

The FDM and FFF printing processes can be slow for large parts or parts with complex geometries. The layer-to-layer adhesion can also be a problem, resulting in parts that warp or separate easily. The surface finish of FDM and FFF printed parts might appear a bit rough as the thin layers are generally visible. To improve the appearance, several options are feasible, such as using acetone, sanding and/or spray paint.

Laser Sintering / Laser Melting

Laser Sintering (**LS**) or **Selective Laser Sintering** (**SLS**) creates tough and geometrically intricate parts using a high-powered CO_2 laser to fuse/sinter/melt powdered thermoplastics. The main advantage of SLS *3D printing* is that as a part is made, it remains encased in powder; this eliminates the need for support structures and allows for very complex 3D geometries to be 3D printed. SLS can be used to produce very strong parts as exceptional materials such as Nylon and metal powders are commonly used.

Laser sintering refers to a laser-based 3D printing process that works with powdered materials. The laser is traced across a powder bed of tightly compacted powdered material, according to the 3D data provided to the machine, in the X-Y axes. As the laser interacts with the powdered material it sinters and fuses the particles to each other forming a solid. As each layer is completed the powder bed drops incrementally and a roller is used to compact the powder over the top surface of the bed prior to the next pass of the laser for the subsequent layer.

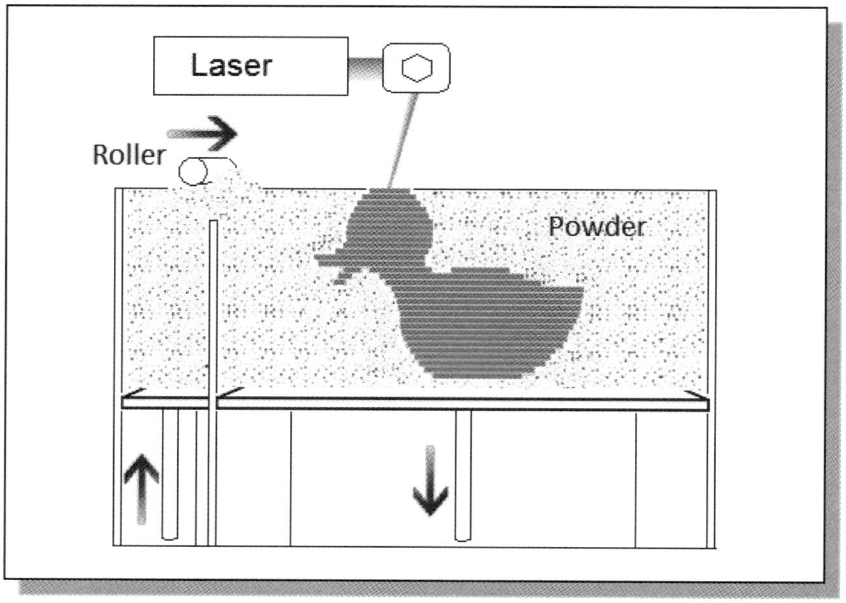

The build chamber is completely sealed as it is necessary to maintain a precise temperature during the process specific to the melting point of the powdered material of choice. One of the key advantages of this process is that the powder bed serves as an in-process support structure for overhangs and undercuts, and therefore complex shapes that could not be manufactured in any other way become possible with this process. Because of the high temperatures required for laser sintering, cooling can take a long time. Porosity is also a common issue with this process; additional metal infiltration processes may be required to improve mechanical characteristics.

Laser sintering can process plastic and metal materials, although metal sintering does require a much higher-powered laser and higher in-process temperatures. Parts produced with this process are much stronger than parts made with SLA or FDM, although generally the surface finish and accuracy is not as good.

Primary 3D Printing Materials for FDM and FFF

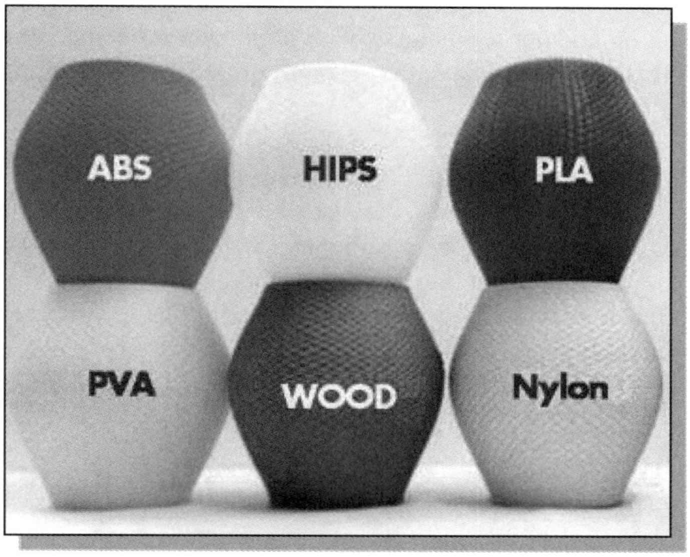

ABS (Acrylonitrile Butadiene Styrene)
ABS is a popular choice for 3D printing. It is a strong thermoplastic that is among the most widely used plastics. It is tough with mild flexibility, making it more durable to stress and has a higher heat resistance of up to 200 degrees Fahrenheit. However, this material has a tendency to shrink, which can affect the accuracy of designs. ABS has a pretty high melting point and can experience warping if cooled while printing. Because of this, ABS objects are printed typically on a heated surface. ABS also requires ventilation when in use, as the fumes can be unpleasant. The aforementioned factors make ABS printing difficult for hobbyist printers, though it's the preferred material for professional applications.

PLA (Polylactic Acid)
PLA is a staple and it is becoming one of the most popular choices for 3D printing with good reason. In addition to the fact that it is a biodegradable thermoplastic derived from renewable resources such as corn starch, tapioca roots, chips or starch, or sugarcane, *PLA* is a very rigid material that is easy to use for 3D printing, and it is able to withstand a good amount of impact and weight. It also has a glossier finish than ABS and in most scenarios, PLA is the preferred material for 3D printing large objects. The main disadvantage of PLA is it's not as heat resistance as ABS; it should not be placed in environments that exceed 140 degrees Fahrenheit.

Flexible (Thermoplastic Elastomer)
Flexible material is for applications that require incredible rubbery flex in their applications. Flexible filament goes beyond bending; it is more like rubber. When it comes to Flexible filament, it's all about finding a balance between flexibility (softness) and printability. This softness is sometimes indicated with a *Shore* value (like 85A or 60D). A higher Shore value means less flexibility. Harder filaments (less flexible) are easier to 3D print when compared to softer, more flexible filaments.

PETG (Polyethylene Terephthalate)
PETG is a material that is similar to *PLA*, with more attractive characteristics such as being generally a tougher and denser material with good heat resistance of up to 190 degrees Fahrenheit. It is reported to have the strength of *ABS*, while printing as easily as *PLA*.

HIPS (High Impact Polystyrene) and **PVA (Polyvinyl Alcohol)**
HIPS and *PVA* are relatively new materials that are growing in popularity for their dissolvable properties. They are used for creating support material. Their ability to dissolve in certain liquids means that they can be easily removed. These materials can be hard to print with, because they don't stick well to the build plates. It is also important not to print *PVA* at too high a temperature, as it can turn into tar and jam the extruder.

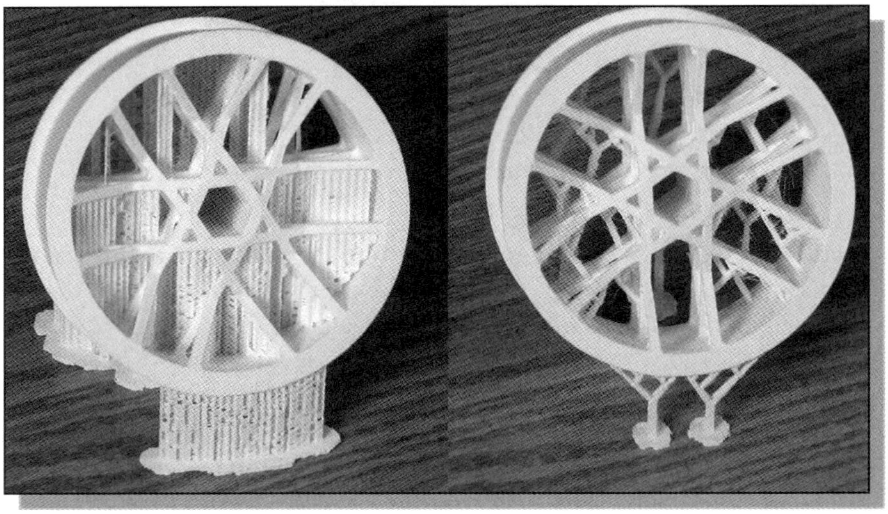

Wood Fiber
Wood Fiber filament contains a mixture of recycled wood with a binding polymer. Thus, a 3D printed object can look and smell like real wood. Due to its wooden nature, it's difficult to tell that the object is 3D printed. Using Wood filament is similar to using a thermoplastic filament like ABS or PLA. However, a 3D object having a wooden-like appearance can be created with this material.

From 3D Model to 3D Printed Part

To create a 3D printed part, it all starts with making a virtual design of the object. This virtual design may be created with a computer-aided design (CAD) package, via a 3D scanner, or by a digital camera and photogrammetry software. 3D scanning and photogrammetry software can process the collected digital data on the shape and appearance of a real object and create a digital 3D model. The 3D virtual design can generally be modified with 3D CAD packages, allowing verification of the virtual design before it is 3D printed.

Once the virtual design is verified, the 3D data will then be transferred to the 3D printing software. There is a multitude of file formats that 3D printing software supports. However, the most popular are the STL file format and the OBJ file format. The STL file format is the most commonly used file format for 3D printing. Most CAD software has the capability of exporting models in the STL format. The STL file contains only the surface geometry of the modeled object. The OBJ file format is considered to be more complex than the STL file format as it is capable of displaying texture, color and other attributes of the three-dimensional object. However, the STL file format holds the top spot for 3D printing, as this file format is simpler to use, and most CAD packages work better with STL files than OBJ files.

Once the 3D data of the virtual design is transferred into the 3D printing software, further examination and/or repair can be performed if necessary. The 3D printing software will also process the imported 3D data by the special software known as a **Slicer**, which converts the model into a series of thin layers and produces a G-code file containing instructions tailored to a specific type of 3D printer. G-code is the common name for the most widely used numerical control (NC) programming language. It is used mainly in computer-aided manufacturing to control automated machine tools. The generated G-code file can be sent to the 3D printer and create the 3D printed part.

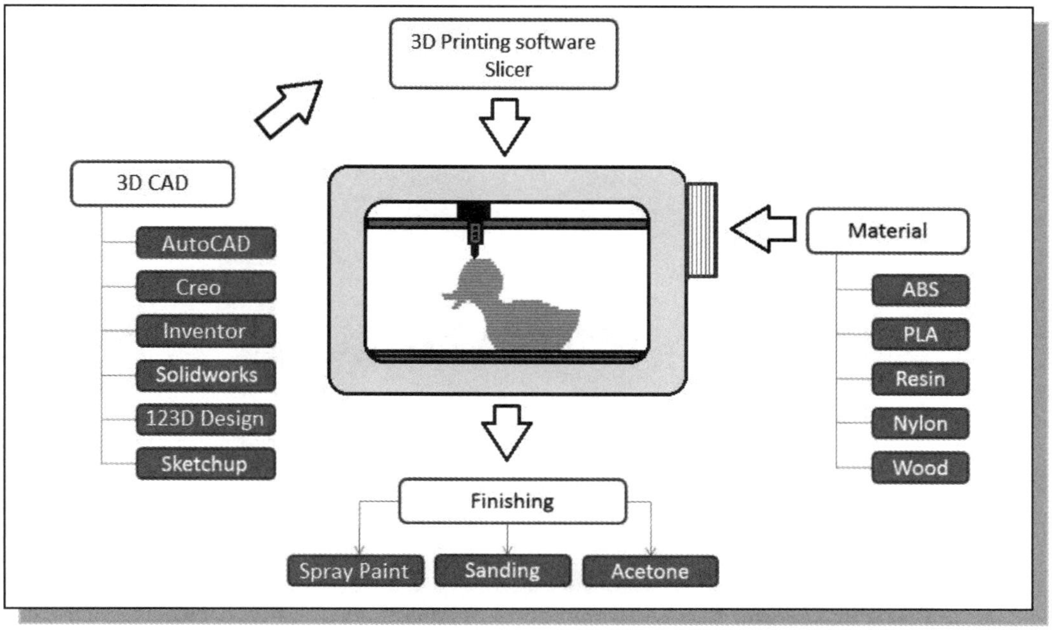

Starting Autodesk Inventor

1. Select the **Autodesk Inventor** option on the *Start* menu or select the **Autodesk Inventor** icon on the desktop to start Autodesk Inventor. The Autodesk Inventor main window will appear on the screen.

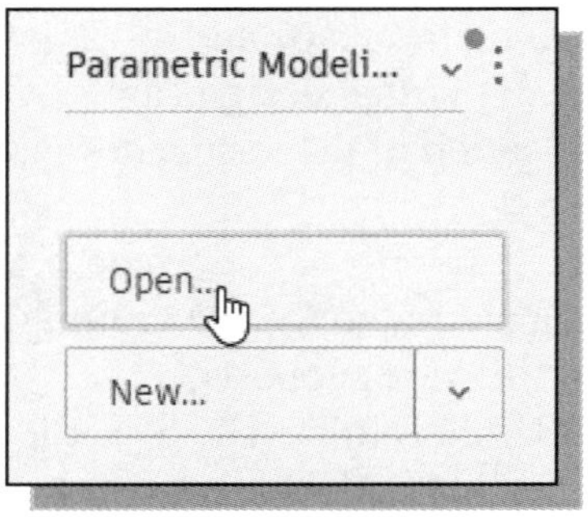

2. In the Autodesk Inventor *Startup* dialog box, select **Open** with a single click of the left-mouse-button as shown.

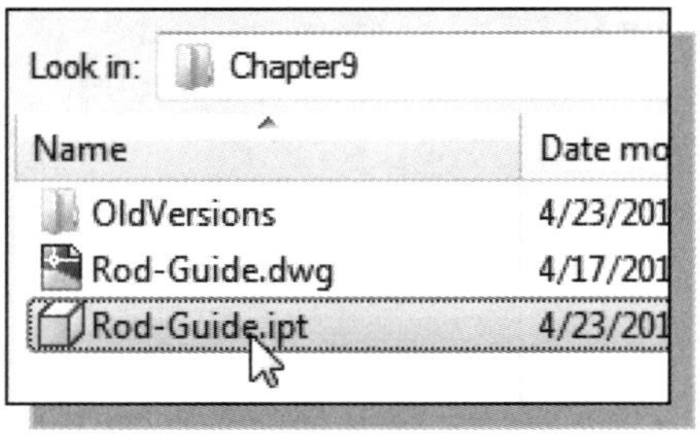

3. In the *Open* window, select the **Rod-Guide.ipt** file. Use the *browser* to locate the file if it is not displayed in the *File name* list box.

4. Click on the **Open** button in the *Open* dialog box to accept the selected selection.

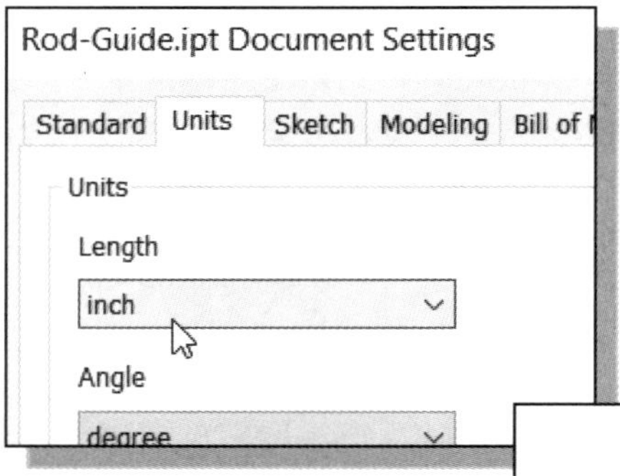

5. On your own, confirm the *Length Units* is set to **inches** under the Document Settings options as shown. Note that the majority of the 3D printer control settings are generally measured in millimeters, for example the filament diameter, layer height, and extruder size.

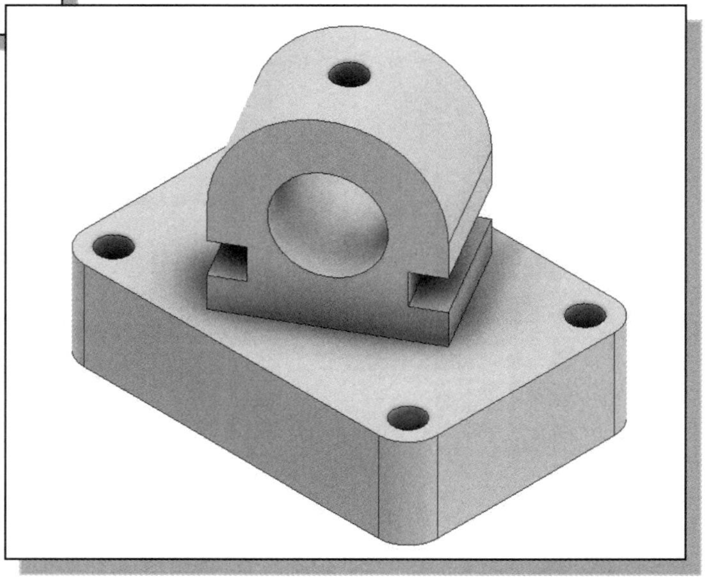

Export the Design as an STL File

3D printers will generally accept a 3D model with the STL or OBJ file formats. Autodesk Inventor offers two options for saving the 3D model in STL format: 1. Using the Autodesk Inventor's **3D Print Preview** command or 2. Using the **Save As** option. The *Save As* option can be used to very quickly export the 3D model, while the *3D Print Preview* command provides a graphics preview of the model prior to exporting the model.

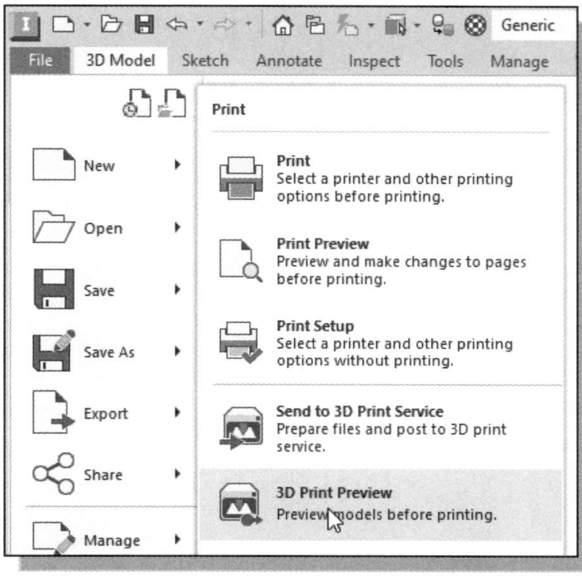

1. In the *File Toolbar*, select the **Print → 3D Print Preview** command as shown.

2. Once the *3D Print Preview* command is activated, Autodesk Inventor's graphics window now displays a faceted model. Note the number of facets of the model is 396 under the default settings, which produces a very rough looking top section of the model.

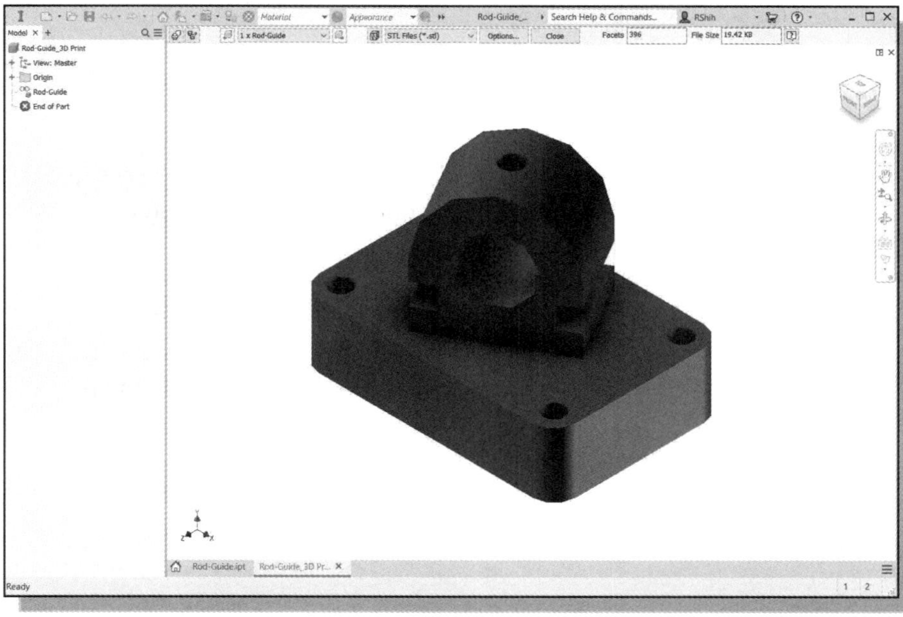

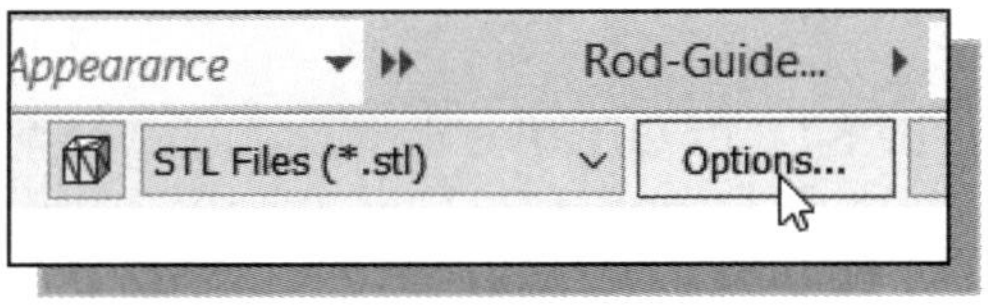

3. Click on the **Options** button to examine the STL file options as shown. Note that we can also use the **Save As** command to adjust these settings.

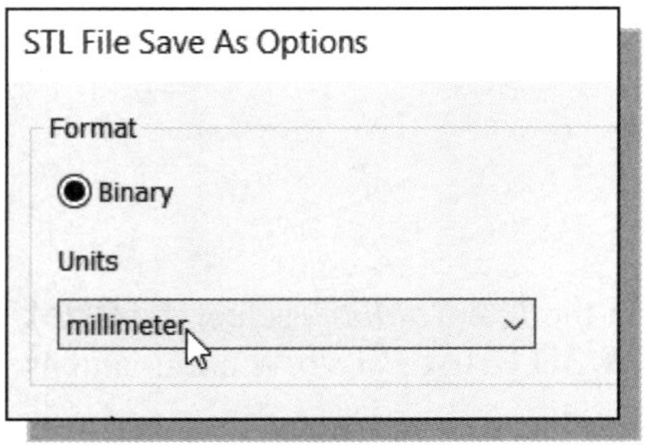

4. In the *STL File Save As Options* dialog box, set *Format* to **Binary** and *Units* to **Millimeter** as shown.

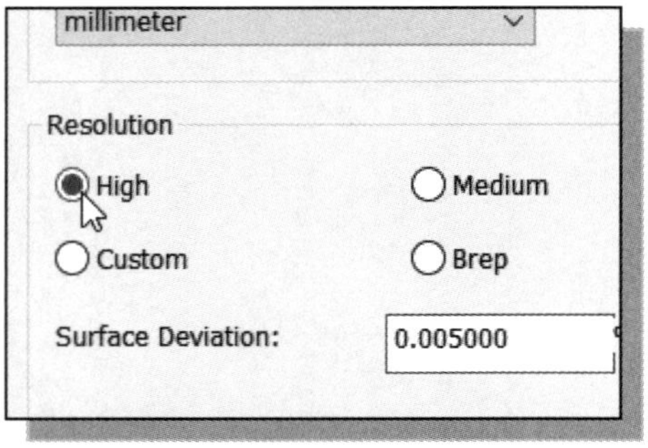

5. Set the *Resolution* to **High**; this will increase the number of facets used on curved surfaces.

6. On your own, examine the changes to the settings by selecting Medium or Low resolution.

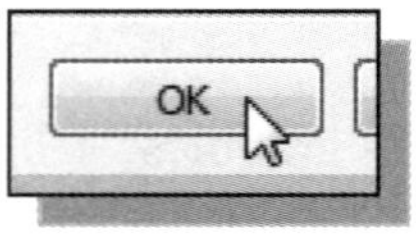

7. Click on the **OK** button to accept the changes in the settings.

- Notice the number of facets has been increased to 3,574, and the model looks much better with this adjustment.

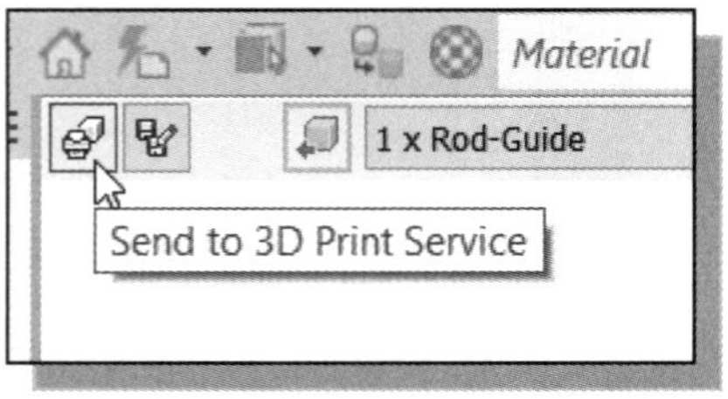

- Note that the STL file can be exported to a 3D Print Service directly by clicking on the associated icon as shown.

8. Click on the **Save Copy As** icon to export the model as an STL file.

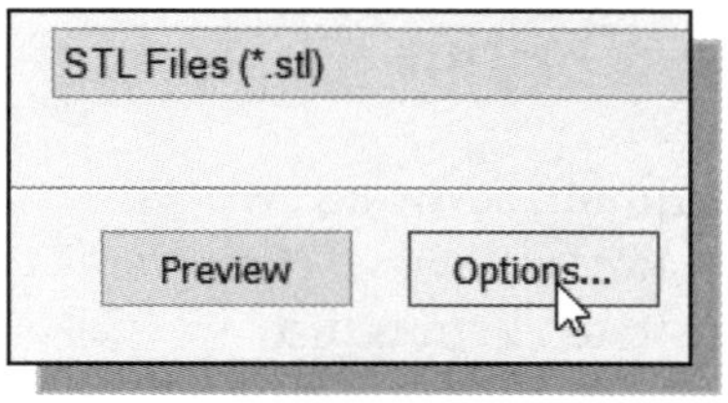

- Note that the **Options** icon is available near the bottom of the dialog box.

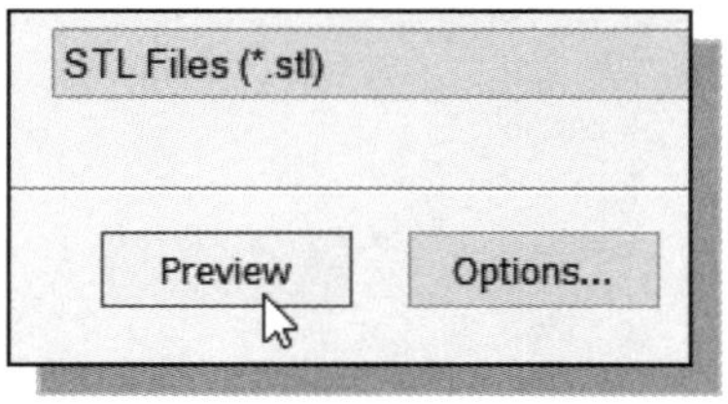

- We can also return to the Preview window by clicking on the **Preview** button as shown.

9. On your own, switch to Chapter 10 as the location for the new file. Click **Save** to export the model using the STL file format.

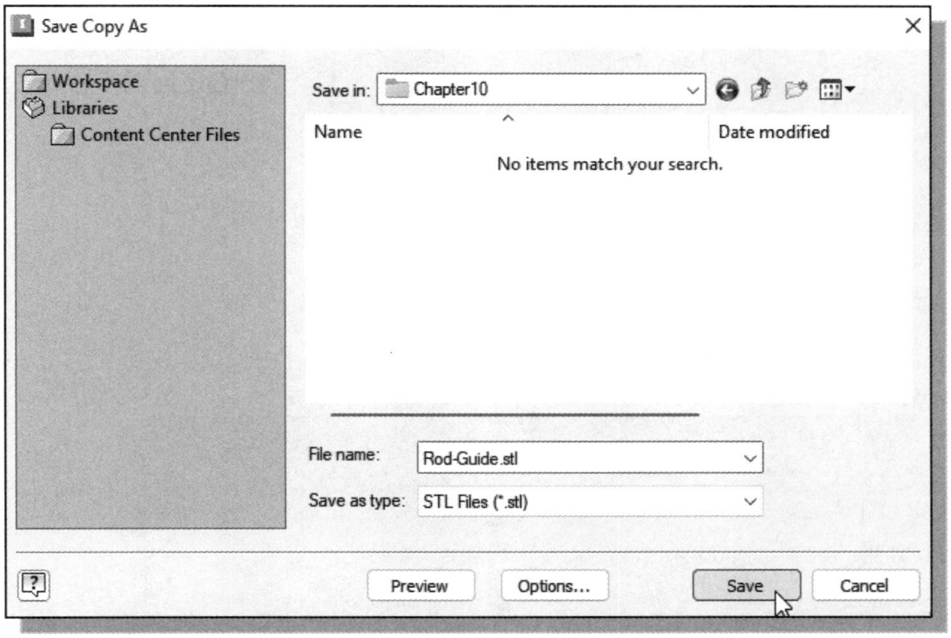

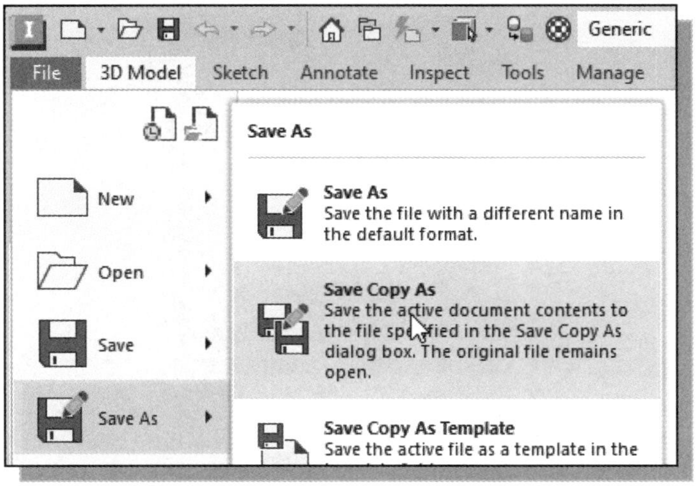

- Note that we can also use the **Save As** command to directly reach the above dialog box.

Using the 3D Printing Software to Create the 3D Print

To 3D print the model, we will open the STL file in the 3D printing software. We will use **Matter Control** to demonstrate the procedure. Note that *Matter Control* (Freeware) supports quite a few desktop 3D printers. The procedure illustrated here is also applicable to other similar software.

1. Start the **Matter Control** software.

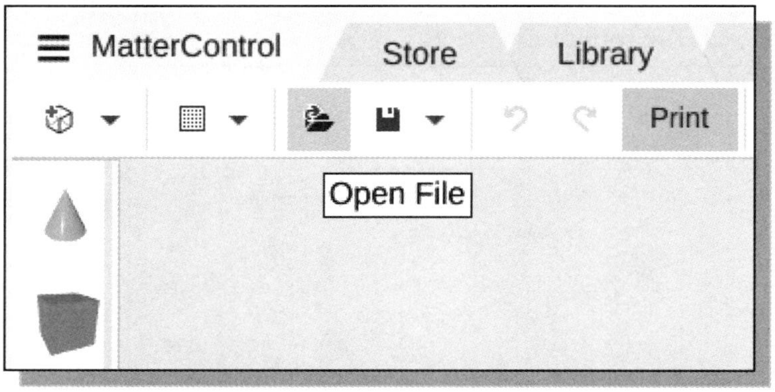

2. In the *File pull-down menu*, select **Add File to Queue** as shown.

3. On your own, switch to the saved STL file folder and select the Rod-Guide.STL file as shown.

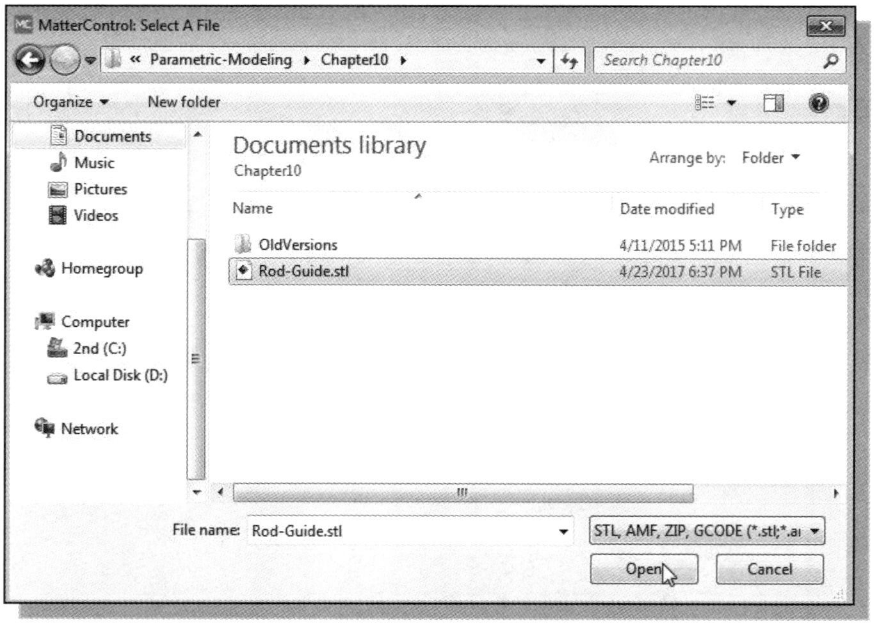

4. Click **Open** to import the STL file into Matter Control.

5. Once the STL file is imported into the program, the STL model is displayed in the *View* window. Note that the model is imported with the incorrect size; it is very small as shown in the figure below. Most 3D Printers expect the model to be in millimeters—the Rod-Guide design was created in inches.

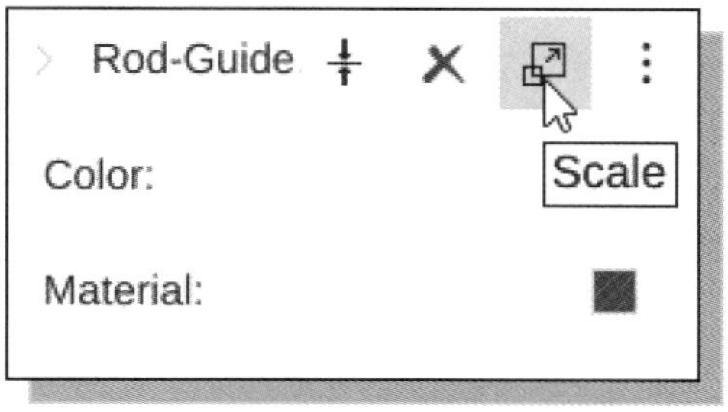

6. Click on **Scale** to expand the *Scale control panel.*

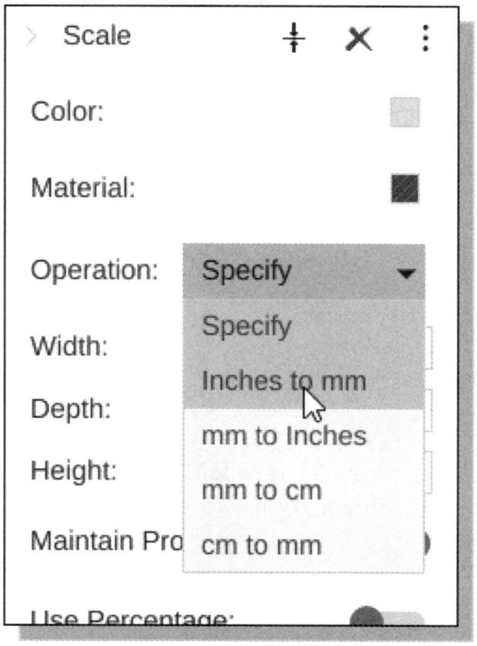

7. Set the *Scale ratio* to **in to mm (25.4)** as shown. Note that 1 inch equals 25.4 mm.

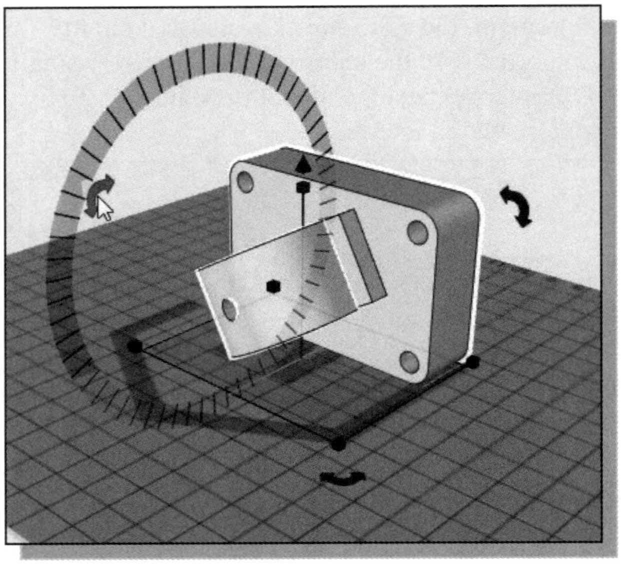

8. To **Rotate** the design, move the cursor on the curve-arrow icon and drag with the left mouse button to perform the rotate operation.

9. On your own, rotate the design 90 degrees, so that the flat face is aligned to the print-bed as shown.

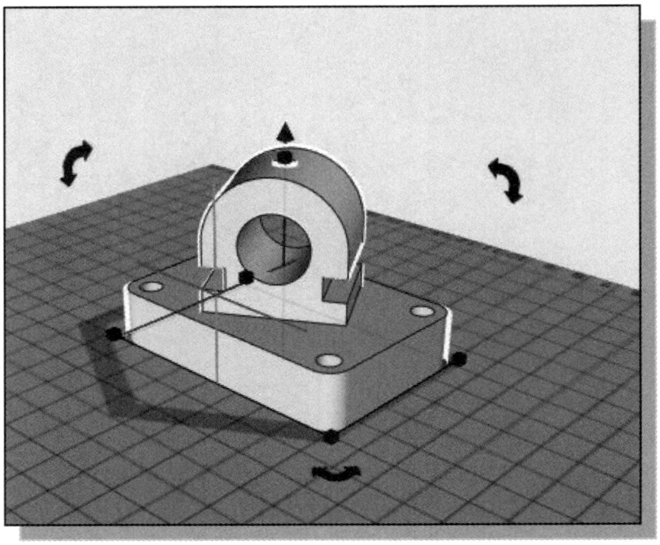

- Note the straight-arrows icons can be used to move/translate the design in X/Y/Z directions.

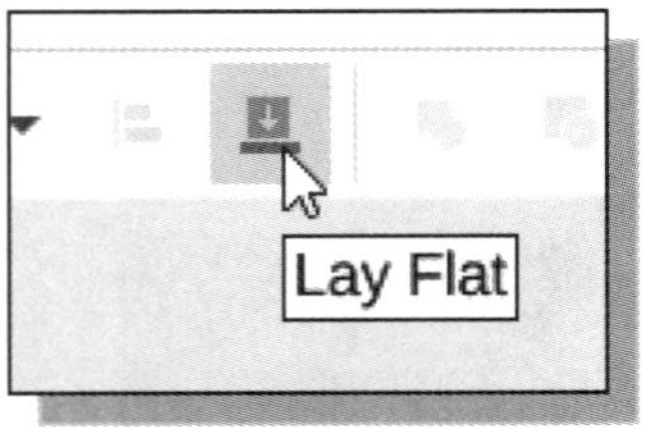

10. Click on the Lay Flat command to align the bottom face of the design to the print-bed.

11. On your own, examine the different display options, such as **Home** and **View Cube**, near the upper right corner of the main window.

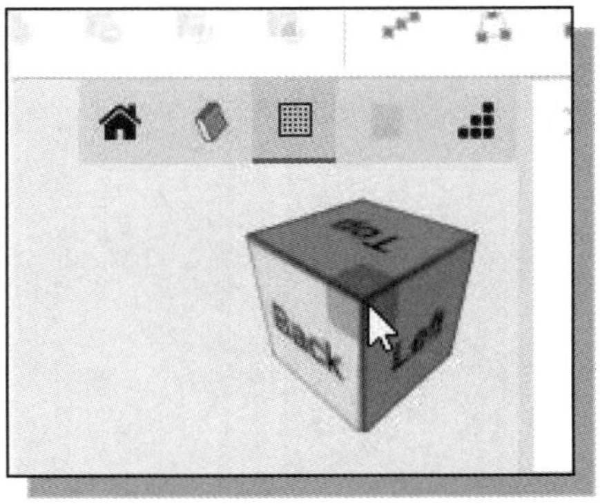

12. Use the right-mouse-button on the design to display the option list, and note the different options available to modify the design.

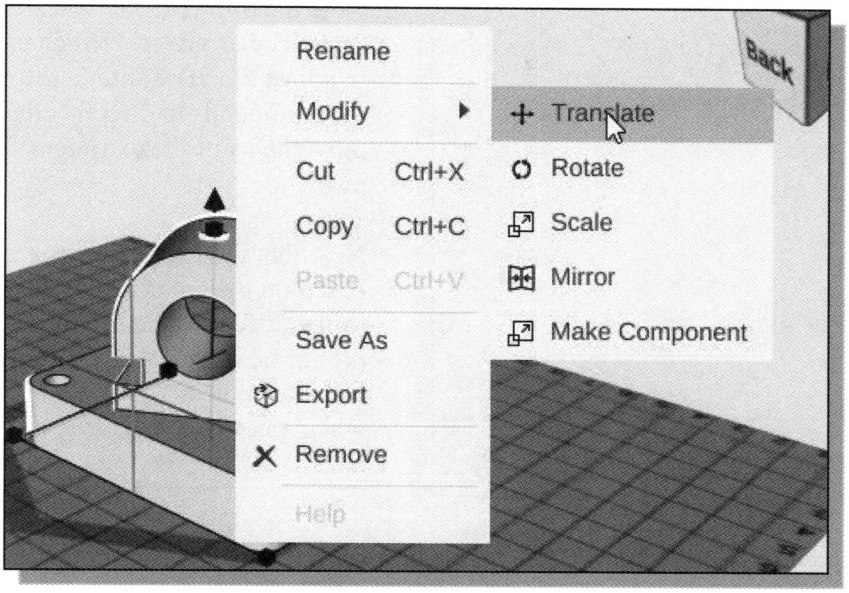

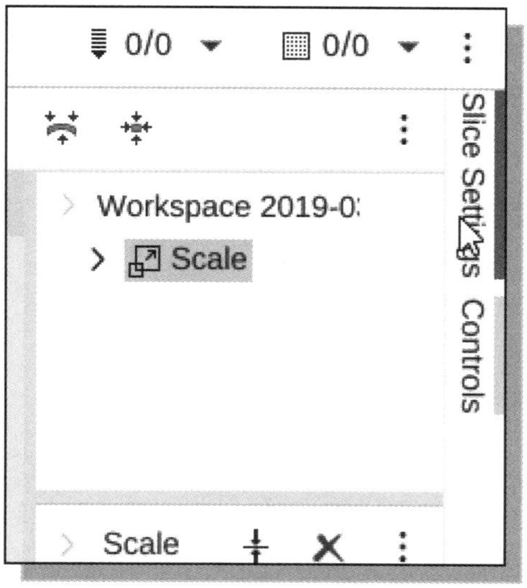

13. Near the upper right corner of the main window, click **Slice Settings** to review the Slicer program settings.

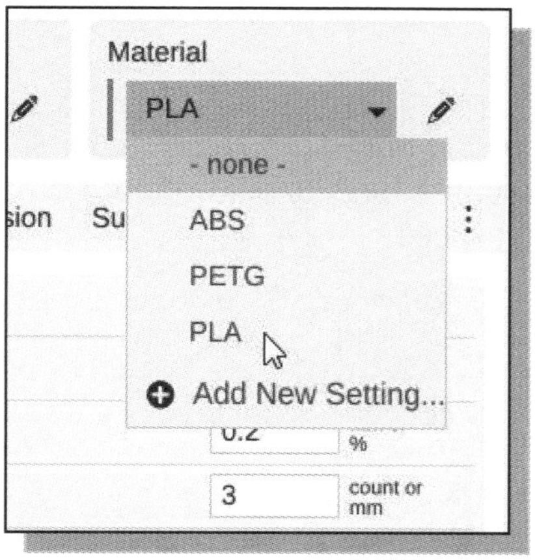

14. In the **Material list**, choose **PLA** as the type of filament properties. Note that the related properties, such as the diameter, extruder temperature, and bed temperature can be adjusted to match the actual filament being used.

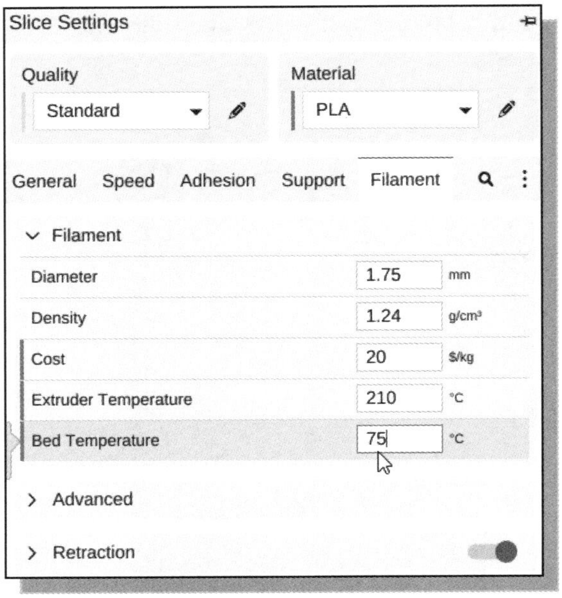

15. Note the different settings are organized as groups, which can be used to adjust the 3D printer's settings and the commands to directly control the movements of the extruder of the printer.

- Note that the temperatures are different based on the types of filaments, as well as the different brands and the type of printer bed. It may be necessary to do some testing and/or experimenting when a new roll of filament is used.

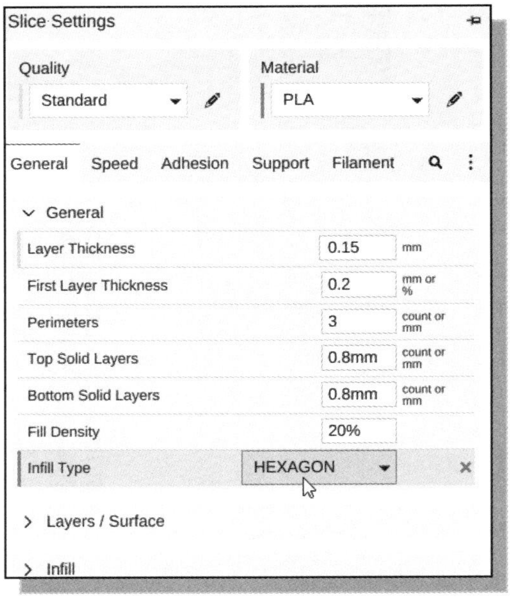

16. Switch to the **General** Tab, and examine the Layer control settings, such as the layer thickness, top layer and bottom layer thickness.

17. Also note the settings for **Infill Type**, the **Fill Density**.

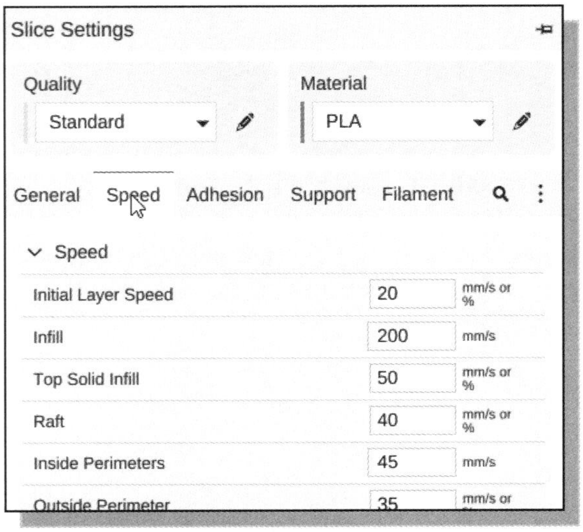

18. Click on the **Speed** tab and examine the related settings to control the 3D printing speed.

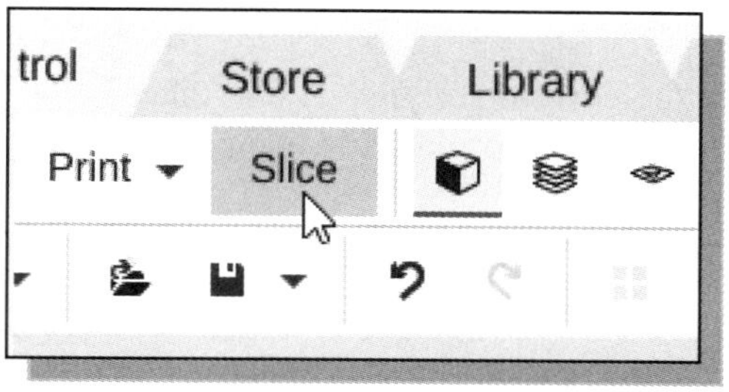

19. Click Slice to process the 3D model, which includes slicing and generating the G-code for the specific 3D printer.

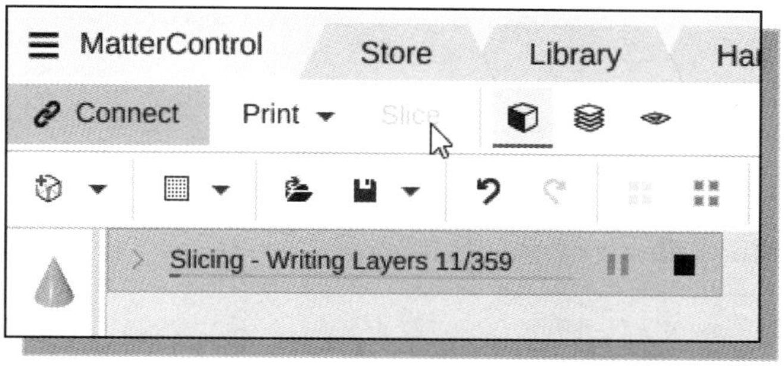

20. Matter Control will display a status bar showing the slicing progress.

- Note that Matter Control allows the use of other Slicers, such as Cura or Slic3r, which are the two extremely popular open source slicer programs.

21. On your own, drag the vertical slider to review the thin layers generated by the slicer.

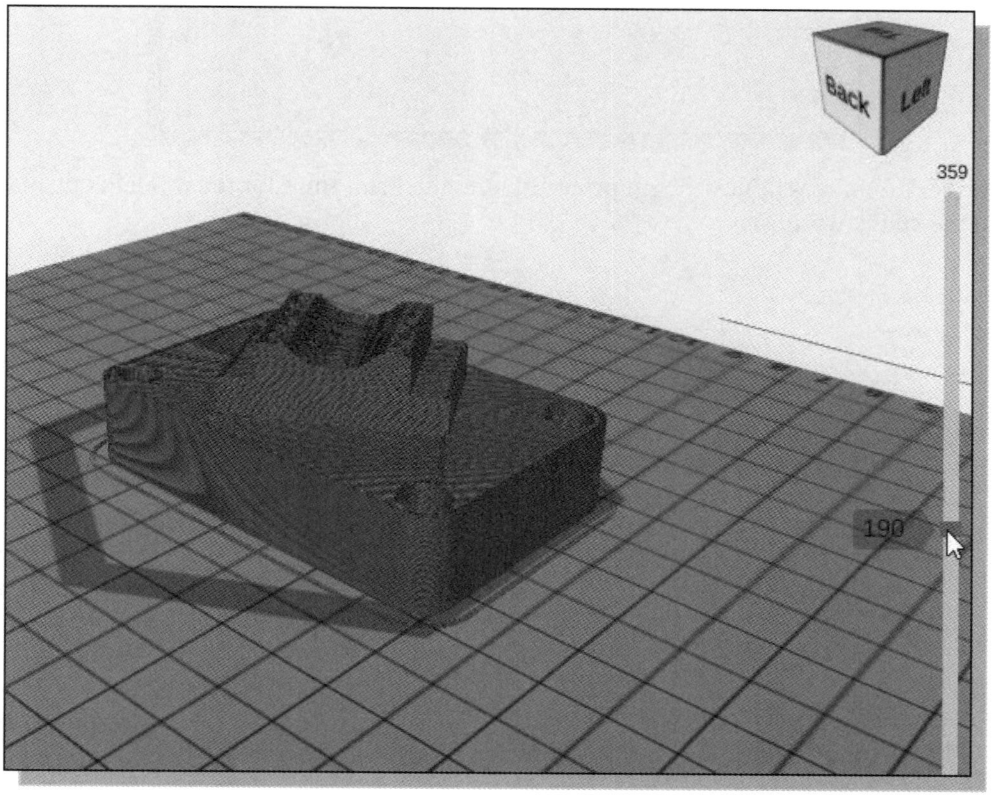

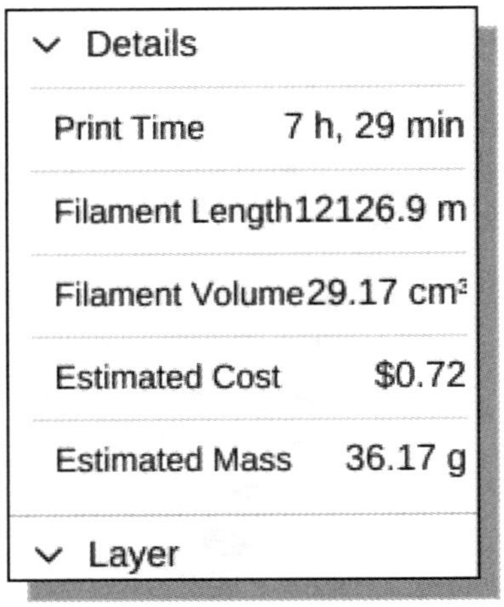

- Notice that with the current settings, it will take 7 hours and 29 minutes to complete the print using 12126.9 mm of filament.

22. Click **Connect → Print** to start the printing of the 3D model.

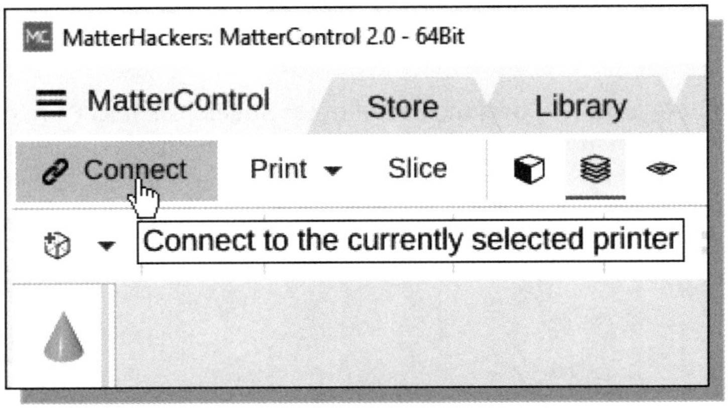

23. *Matter Control* will now begin printing; note the Print time for the model is displayed in the status window.

Questions:

1. What is the main difference between **Additive Manufacturing** and the traditional **Subtractive Manufacturing** technologies?

2. Which 3D printing process is recognized as the first 3D printing process?

3. Describe the general procedure to create a 3D printed part.

4. What are the three primary types of 3D printing processes?

5. Which 3D printing process is the most popular 3D printing process?

6. What is the main advantage of using PLA over ABS for FFF process?

7. Which are the most popular file formats for 3D printing?

8. What is the main function of a **Slicer** program?

Notes:

Chapter 11

Symmetrical Features in Designs

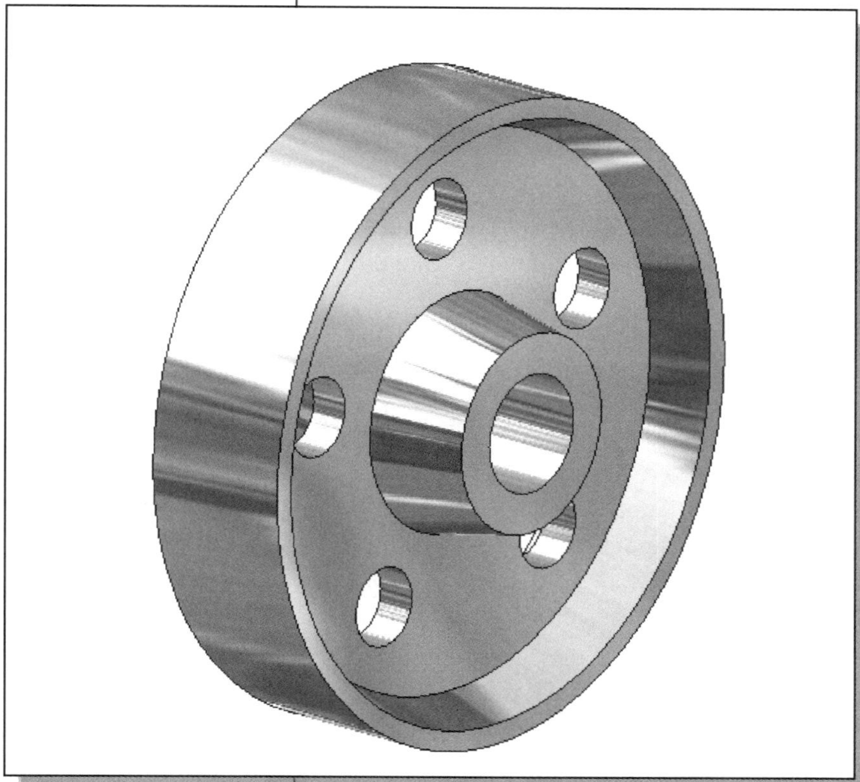

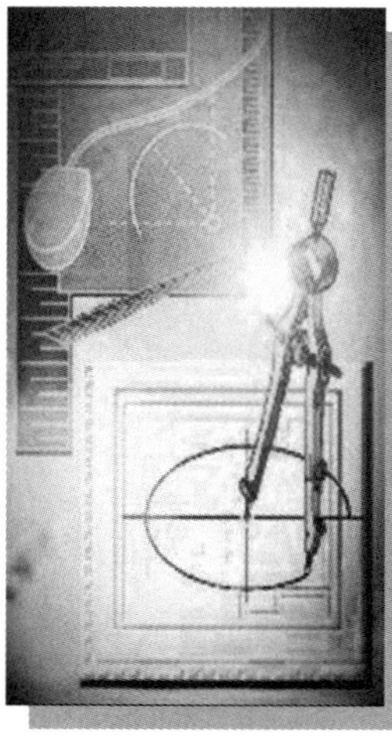

Learning Objectives

- Create Revolved Features
- Use the Mirror Feature Command
- Create New Borders and Title Blocks
- Create Circular Patterns
- Create and Modify Linear Dimensions
- Use Autodesk Inventor's Associative Functionality
- Identify Symmetrical Features in Designs

Autodesk Inventor Certified User Exam Objectives Coverage

Parametric Modeling Basics

Section 3: Sketches

Objectives: Creating 2D Sketches, Draw Tools, Sketch Constraints, Pattern Sketches, Modify Sketches, Format Sketches, Sketch Doctor, Shared Sketches, Sketch Parameters.

Section 4: Parts

Objectives: Creating parts, Work Features, Pattern Features, Part Properties.

Section 6: Drawings

Objectives: Create drawings.

Introduction

In parametric modeling, it is important to identify and determine the features that exist in the design. *Feature-based parametric modeling* enables us to build complex designs by working on smaller and simpler units. This approach simplifies the modeling process and allows us to concentrate on the characteristics of the design. Symmetry is an important characteristic that is often seen in designs. Symmetrical features can be easily accomplished by the assortment of tools that are available in feature-based modeling systems, such as Autodesk Inventor.

The modeling technique of extruding two-dimensional sketches along a straight line to form three-dimensional features, as illustrated in the previous chapters, is an effective way to construct solid models. For designs that involve cylindrical shapes, shapes that are symmetrical about an axis, revolving two-dimensional sketches about an axis can form the needed three-dimensional features. In solid modeling, this type of feature is called a ***revolved feature***.

In Autodesk Inventor, besides using the **Revolve** command to create revolved features, several options are also available to handle symmetrical features. For example, we can create multiple identical copies of symmetrical features with the **Feature Pattern** command or create mirror images of models using the **Mirror Feature** command. We can also use *construction geometry* to assist the construction of more complex features. In this lesson, the construction and modeling techniques of these more advanced options are illustrated.

A Revolved Design: Pulley

❖ Based on your knowledge of Autodesk Inventor, how many features would you use to create the design? Which feature would you choose as the **base feature** of the model? Identify the symmetrical features in the design and consider other possibilities in creating the design. You are encouraged to create the model on your own prior to following through the tutorial.

Modeling Strategy – A Revolved Design

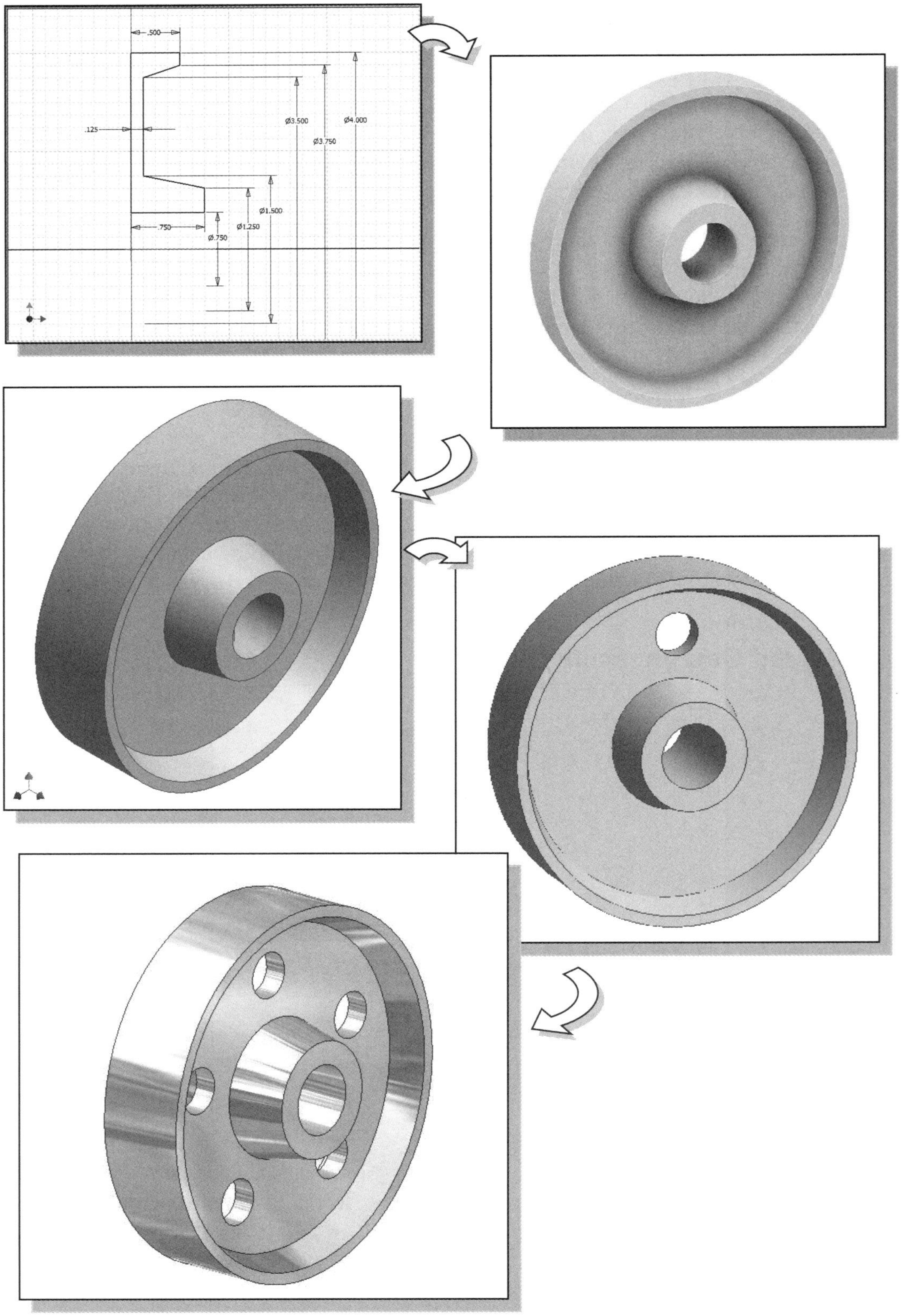

Starting Autodesk Inventor

1. Select the **Autodesk Inventor** option on the *Start* menu or select the **Autodesk Inventor** icon on the desktop to start Autodesk Inventor. The Autodesk Inventor main window will appear on the screen.

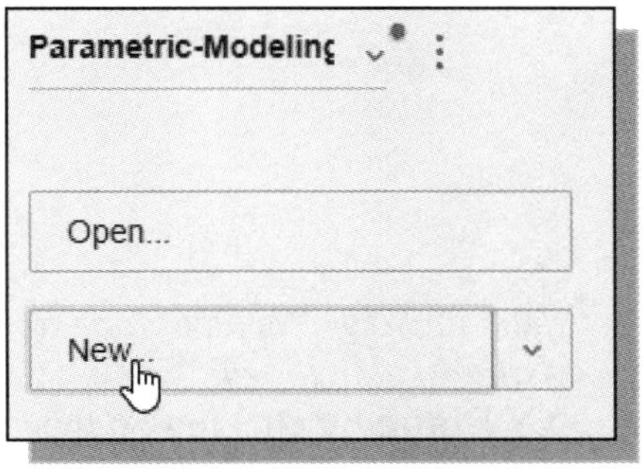

2. Select the **New File** icon with a single click of the left-mouse-button.

3. Select the **English** units set and in the *Part File* area, select **Standard(in).ipt**.

4. Click **Create** in the *New File* dialog box to start a new model.

Set Up the Display of the Sketch Plane

1. In the *Part Browser* window, click on the [**+**] symbol in front of the **Origin** feature to display more information on the feature.

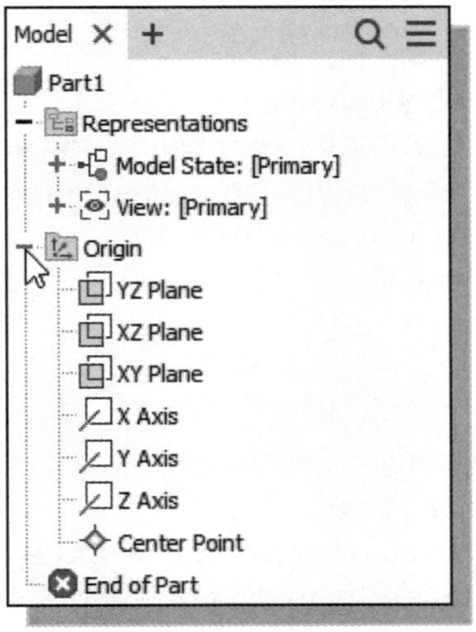

❖ In the *browser* window, notice a new part name appeared with seven work features established. The seven work features include three *work planes*, three *work axes* and a *work point*. By default, the three work planes and work axes are aligned to the **world coordinate system** and the work point is aligned to the *origin* of the **world coordinate system**.

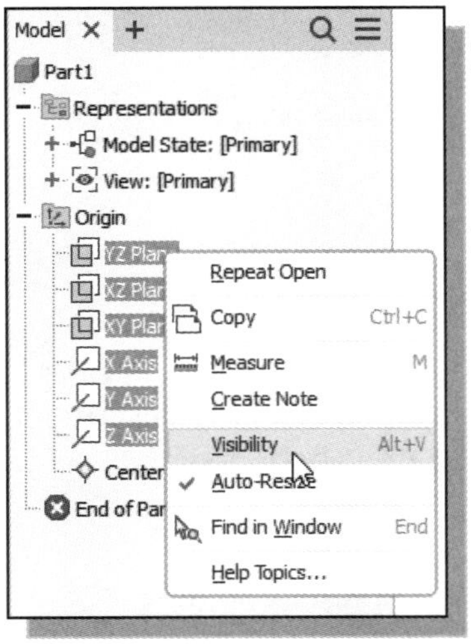

2. Inside the *browser* window, select the three work planes by holding down the **[Ctrl]** key and clicking with the left-mouse-button.

3. Click the right-mouse-button on any of the work features to display the option menu. Click on **Visibility** to toggle ***ON*** the display of the plane.

Creating the 2D Sketch for the Base feature

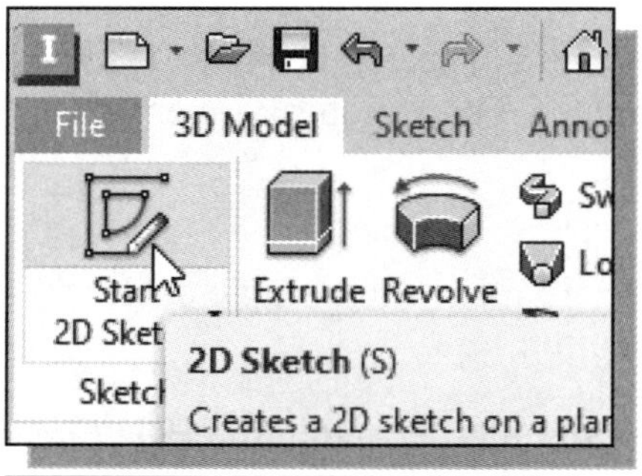

1. In the *Sketch* toolbar select the **Start 2D Sketch** command by left-clicking once on the icon.

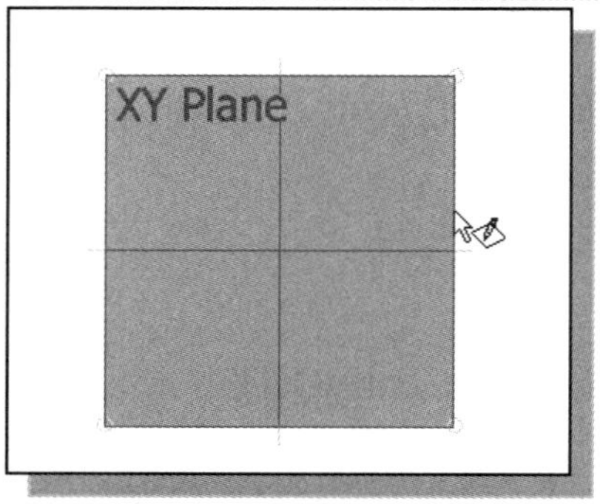

2. In the *Status Bar* area, the message "*Select plane to create sketch or an existing sketch to edit.*" is displayed. Select the **XY Plane** by clicking on the plane inside the graphics window, as shown.

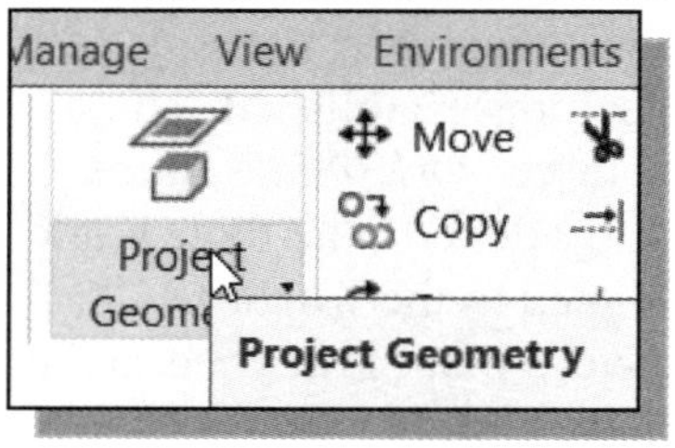

3. Select the **Project Geometry** command in the *2D Sketch* panel. The Project Geometry command allows us to project existing features onto the active sketching plane. Left-click once on the icon to activate the Project Geometry command.

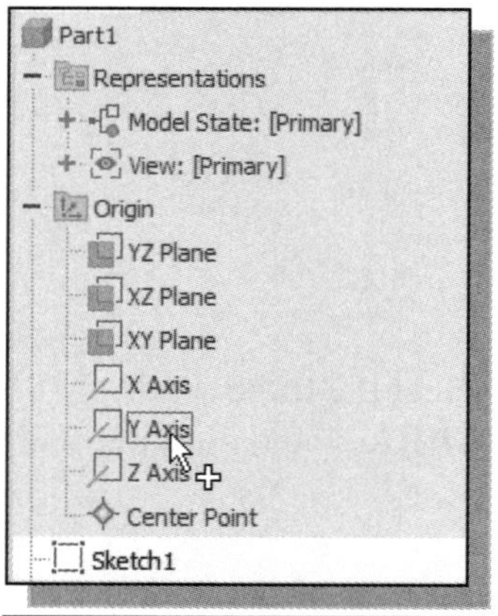

4. In the *Status Bar* area, the message "*Select edge, vertex, work geometry or sketch geometry to project.*" is displayed. Inside the *browser* window, select the **X-axis** and **Y-axis** to project these entities onto the sketching plane.

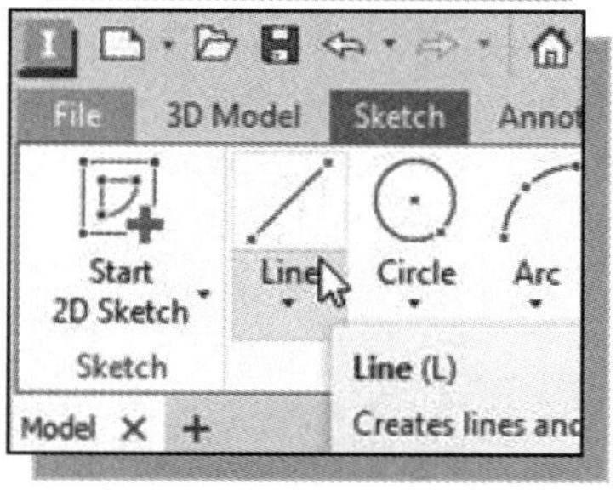

5. Select the **Line** option in the *2D Sketch* panel. A *Help-tip box* appears next to the cursor and a brief description of the command is displayed at the bottom of the drawing screen: "*Creates Straight lines and arcs.*"

6. Create a closed-region sketch with the starting point aligned to the projected **Y-axis** as shown below. (Note that the ***Pulley*** design is symmetrical about a horizontal axis as well as a vertical axis, which allows us to simplify the 2D sketch as shown below.)

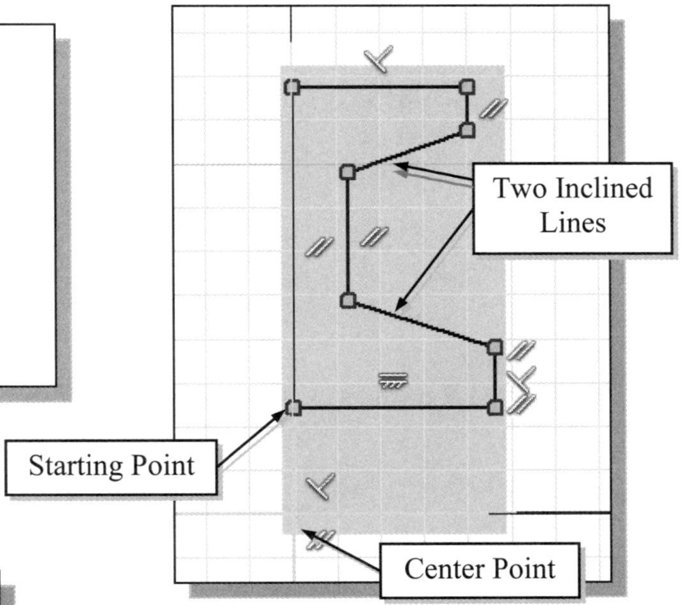

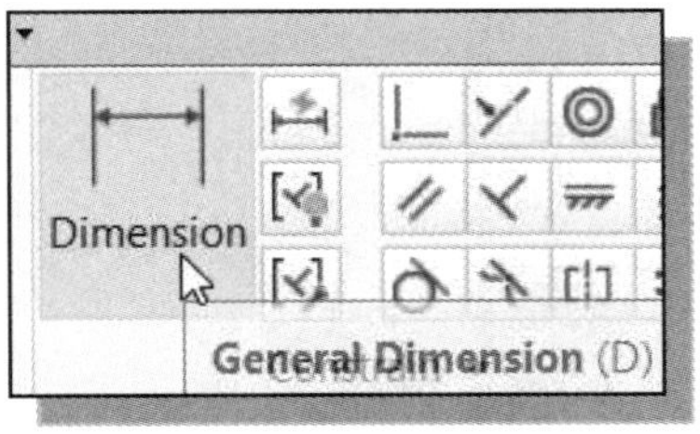

7. Select the **General Dimension** command in the *2D Sketch* panel.

8. Pick the **X-axis** as the first entity to dimension as shown in the figure below.

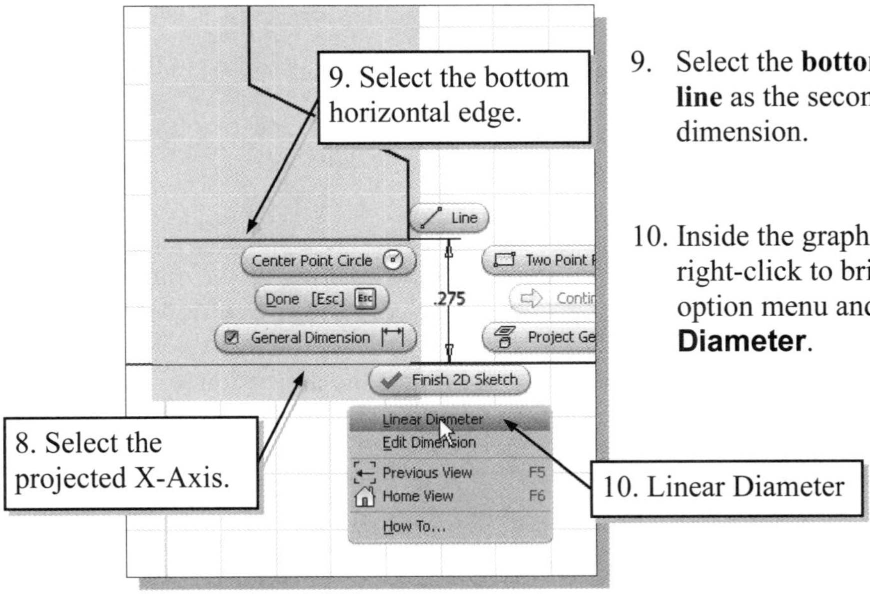

9. Select the **bottom horizontal line** as the second object to dimension.

10. Inside the graphics window, right-click to bring up the option menu and select **Linear Diameter**.

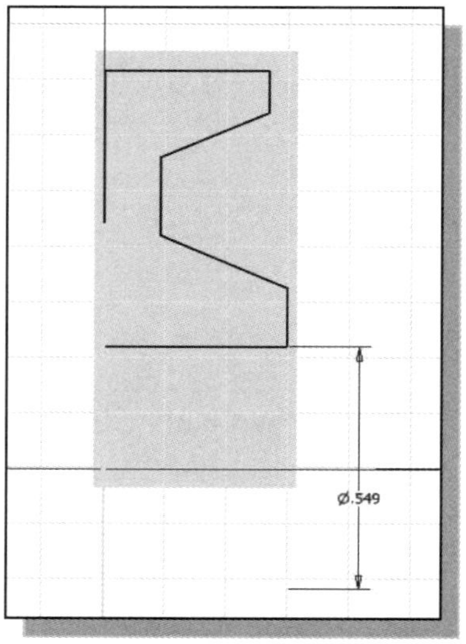

11. Place the dimension text to the right side of the sketch.

- To create a dimension that will account for the symmetrical nature of the design, pick the axis of symmetry, pick the entity, select **Linear Diameter** in the option menu, and then place the dimension.

12. Pick the projected **X-axis** as the first entity to dimension as shown in the figure below.

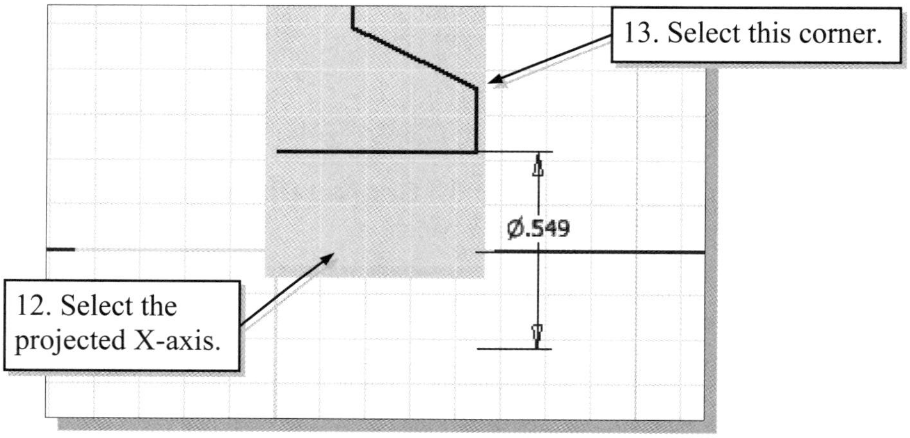

13. Select the **corner point** as the second object to dimension as shown in the above figure.

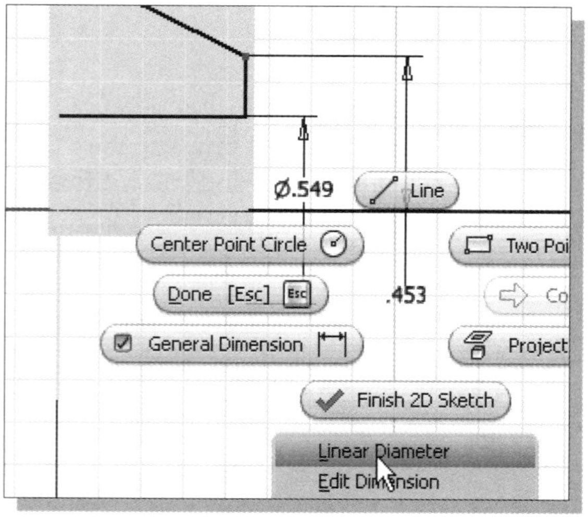

14. Inside the graphics window, right-click to bring up the option menu and select **Linear Diameter**.

15. Place the dimension text toward the right side of the sketch.

16. On your own, create and adjust the vertical size/location dimensions as shown below. (Hint: Modify the larger dimensions first.)

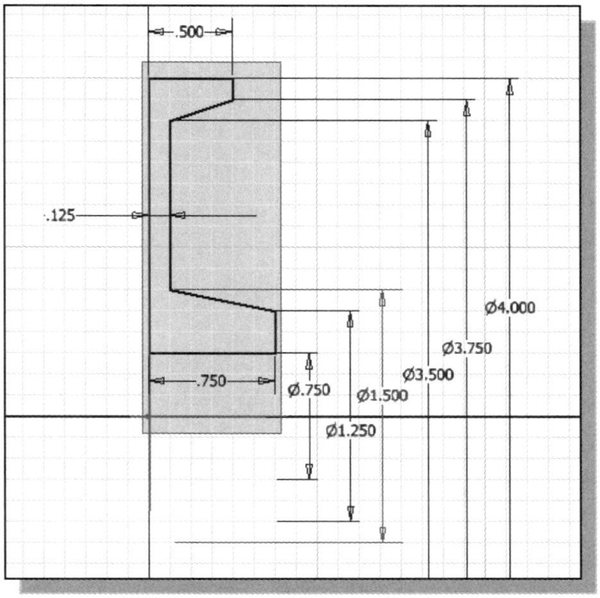

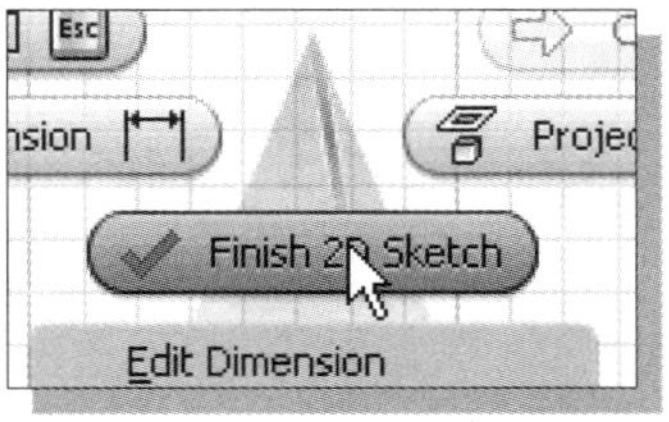

17. Inside the graphics window, click once with the right-mouse-button to display the option menu. Select **Finish 2D Sketch** in the pop-up menu to end the Sketch option.

➢ On your own, use the **3D-Rotate** command to confirm the completed sketch and dimensions are on a 2D plane.

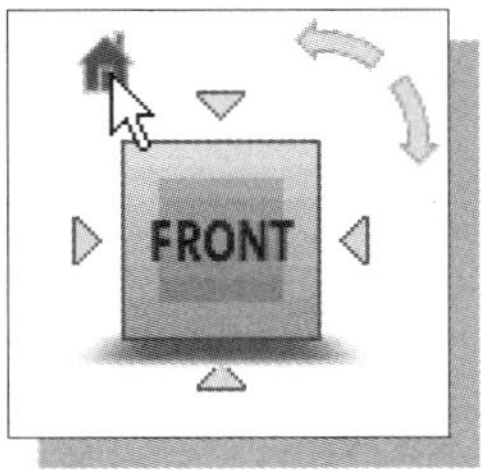

18. Select **Home View** in the ViewCube to adjust the display of the 2D sketch to the isometric view.

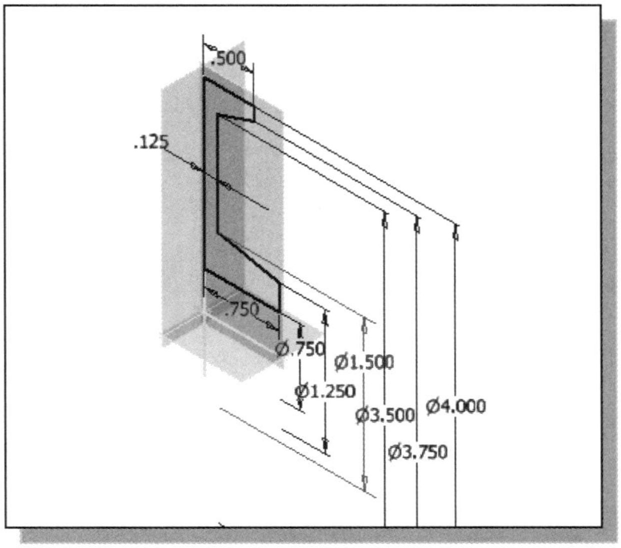

Create the Revolved Feature

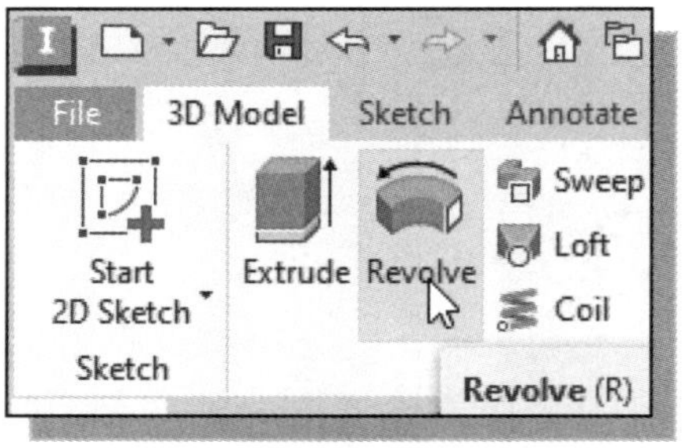

1. In the *Create Feature* panel select the **Revolve** command by clicking the left-mouse-button on the icon.

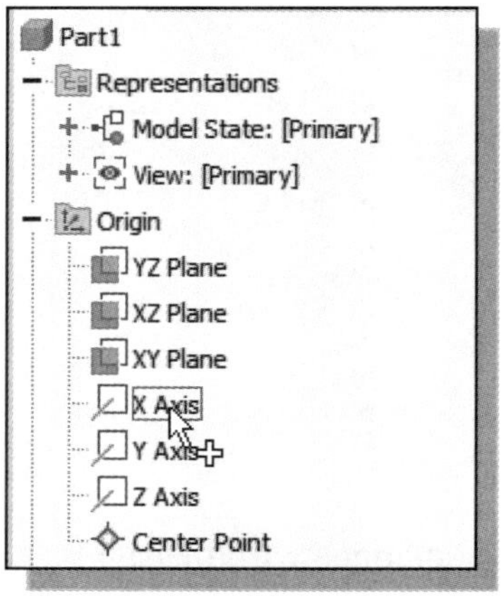

2. In the *Revolve* dialog box, the **Axis** button is activated indicating Autodesk Inventor expects us to select the revolution axis for the revolved feature. Select **X-Axis** as the axis of rotation in the *browser* window as shown.

3. In the *Revolve* dialog box, confirm the termination *Extents* option is set to **Full** as shown.

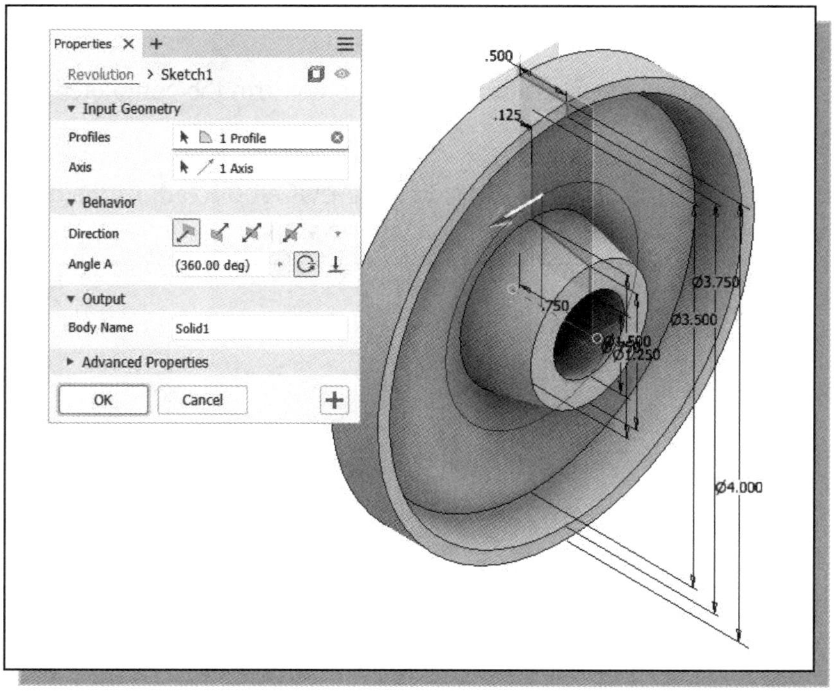

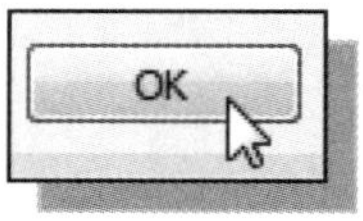

4. Click on the **OK** button to accept the settings and create the revolved feature.

Mirroring Features

In Autodesk Inventor, features can be mirrored to create and maintain complex symmetrical features. We can mirror a feature about a work plane or a specified surface. We can create a mirrored feature while maintaining the original parametric definitions, which can be quite useful in creating symmetrical features. For example, we can create one quadrant of a feature, and then mirror it twice to create a solid with four identical quadrants.

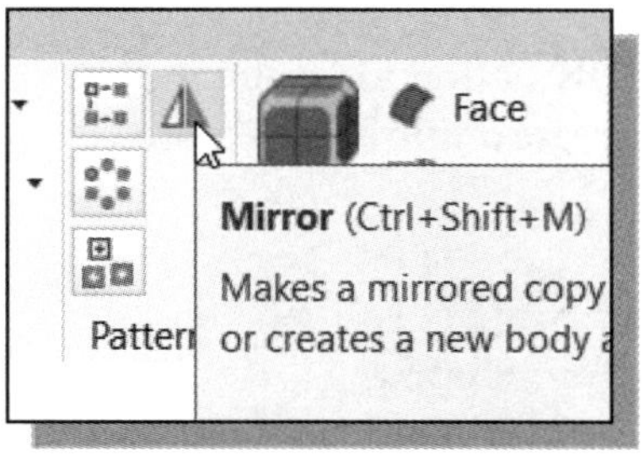

1. In the *Pattern* panel, select the **Mirror Feature** command by left-clicking on the icon.

2. In the *Mirror Pattern* dialog box, the **Features** button is activated. Autodesk Inventor expects us to select features to be mirrored. In the prompt area, the message *"Select feature to pattern"* is displayed. Select **any edge** of the 3D base feature.

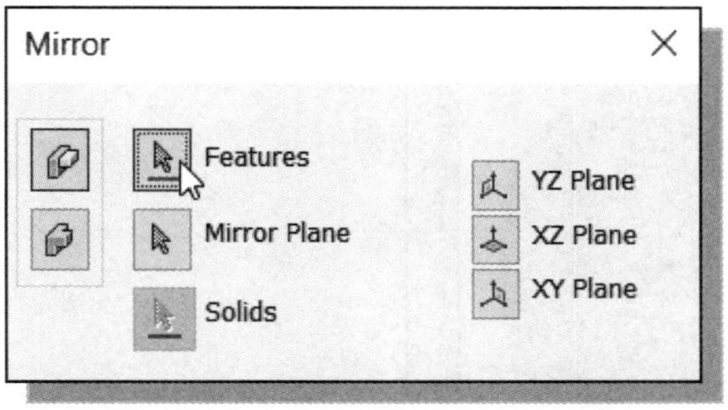

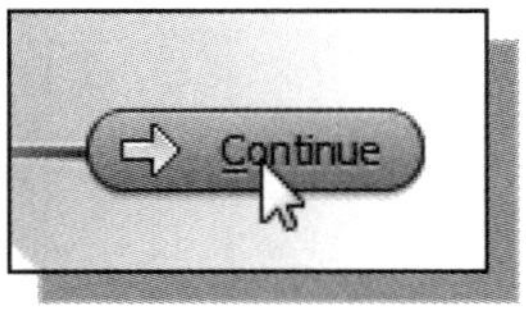

3. Inside the graphics window, **right-click** to bring up the **option menu**.

4. Select **Continue** in the option list to proceed with the Mirror Feature command.

➢ In the *Mirror Pattern* dialog box, we can also activate the **Mirror Plane** option by clicking on the icon. Autodesk Inventor expects us to select a planar surface about which to mirror.

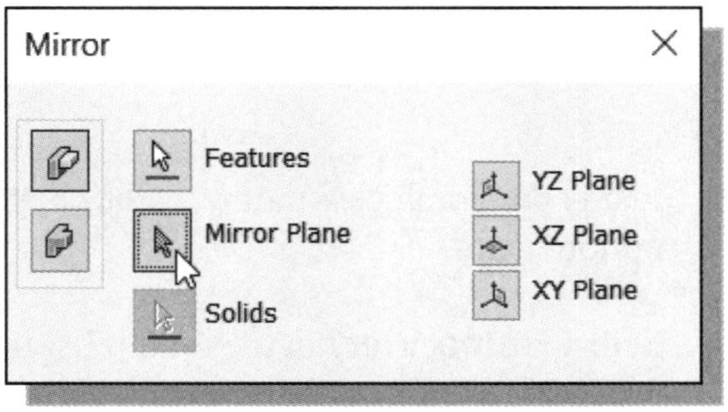

5. On your own, use the **ViewCube** or the **3D-Rotate** function key [**F4**] to dynamically rotate the solid model so that we are viewing the back surface as shown.

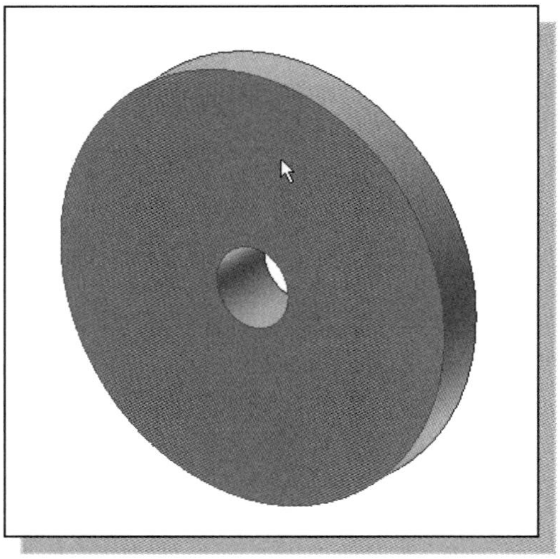

6. Select the surface as shown to the left as the planar surface about which to mirror.

7. Click on the **OK** button to accept the settings and create a mirrored feature.

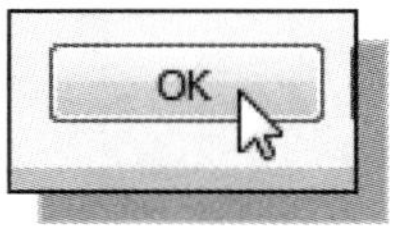

8. On your own, use the **ViewCube** or the **3D-Rotate** function key [**F4**] to dynamically rotate the solid model and view the resulting solid.

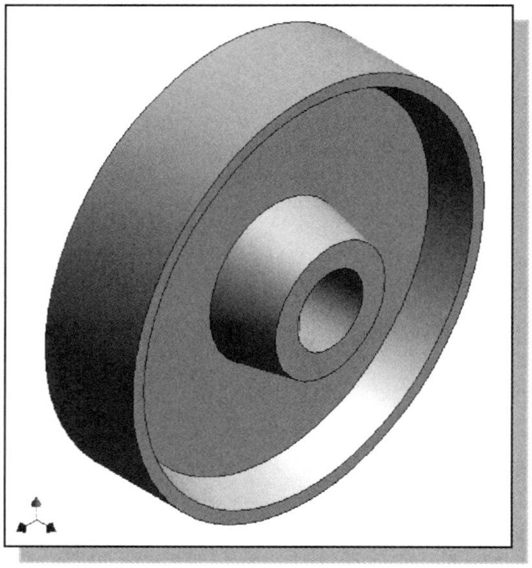

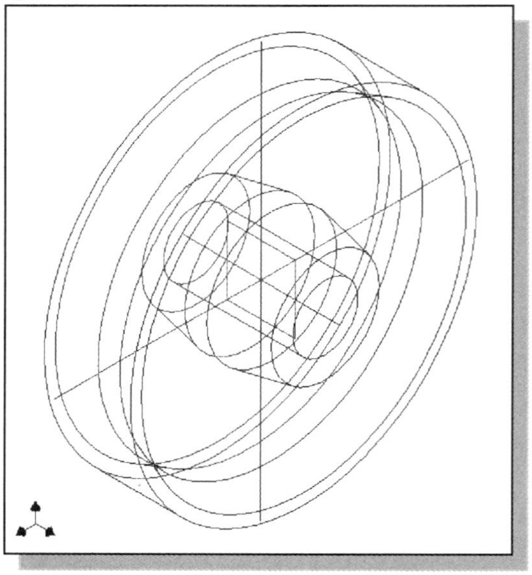

➢ Now is a good time to save the model (quick-key: [**Ctrl**] + [**S**]). It is a good habit to save your model periodically, just in case something might go wrong while you are working on it. You should also save the model after you have completed any major constructions.

9. Inside the graphics window, right-click to bring up the option menu.

10. Select **Home View** in the option list to adjust the display of the 2D sketch on the screen.

Create a Pattern Leader Using Construction Geometry

In Autodesk Inventor, we can also use **construction geometry** to help define, constrain, and dimension the required geometry. **Construction geometry** can be lines, arcs, and circles that are used to line up or define other geometry but are not themselves used as the shape geometry of the model. When profiling the rough sketch, Autodesk Inventor will separate the construction geometry from the other entities and treat them as construction entities. Construction geometry can be dimensioned and constrained just like any other profile geometry. When the profile is turned into a 3D feature, the construction geometry remains in the sketch definition but does not show in the 3D model. Using construction geometry in profiles may mean fewer constraints and dimensions are needed to control the size and shape of geometric sketches. We will illustrate the use of the construction geometry to create a cut feature.

- The *Pulley* design requires the placement of five identical holes on the base solid. Instead of creating the five holes one at a time, we can simplify the creation of these holes by using the **Pattern** command, which allows us to create duplicate features. Prior to using the **Pattern** command, we will first create a ***pattern leader***, which is a regular extruded feature.

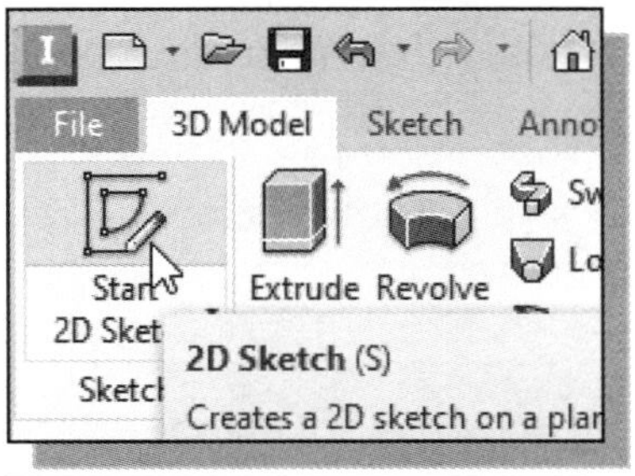

1. In the *Sketch* toolbar select the **Start 2D Sketch** command by left-clicking once on the icon.

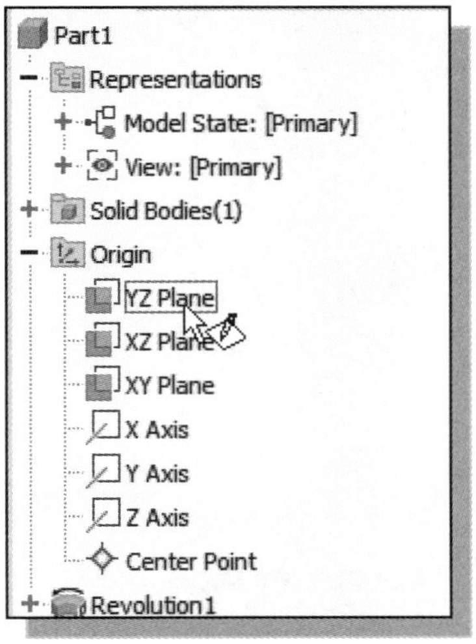

2. In the *Status Bar* area, the message "*Select face, work plane, sketch or sketch geometry.*" is displayed. Pick the **YZ Plane**, inside the *browser* window, as shown.

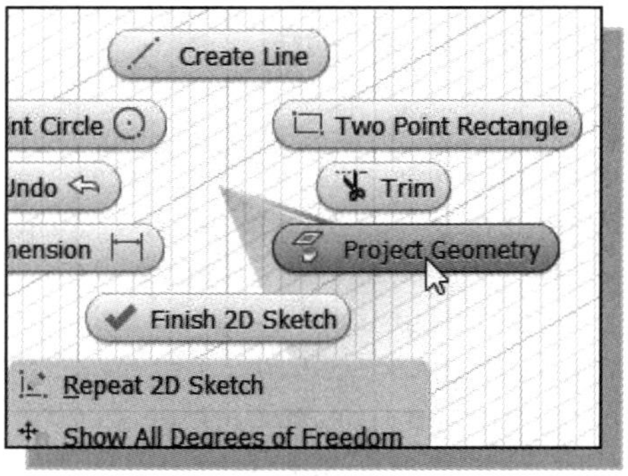

3. Inside the graphics window, **right-click** to bring up the option menu.

4. Select **Project Geometry** in the option menu. The **Project Geometry** command allows us to project existing geometry to the active sketching plane.

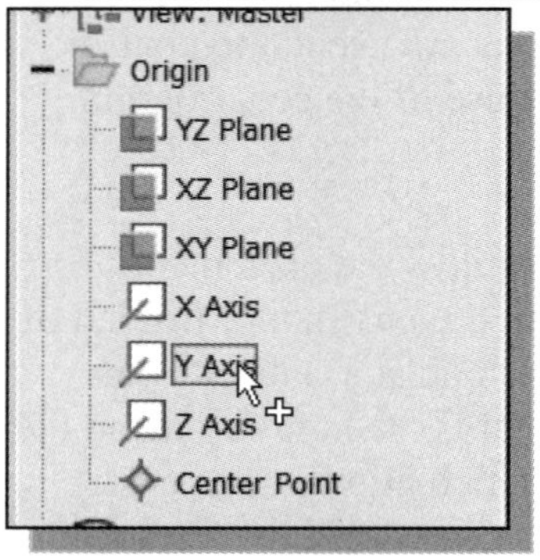

5. In the *Status Bar* area, the message "*Select edge, vertex, work geometry or sketch geometry to project.*" is displayed. Inside the *browser* window, select the **Y-axis** and **Z-axis** to project these entities onto the sketching plane.

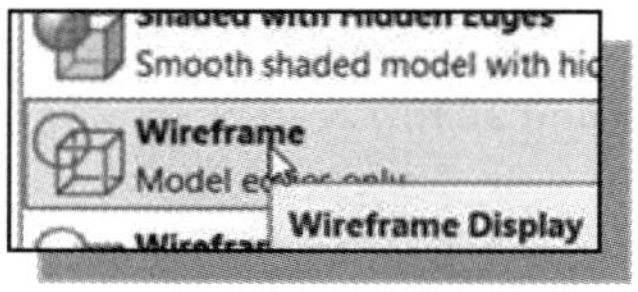

6. On your own, set the display to *wireframe* by clicking on the **Wireframe Display** icon in the *View* panel.

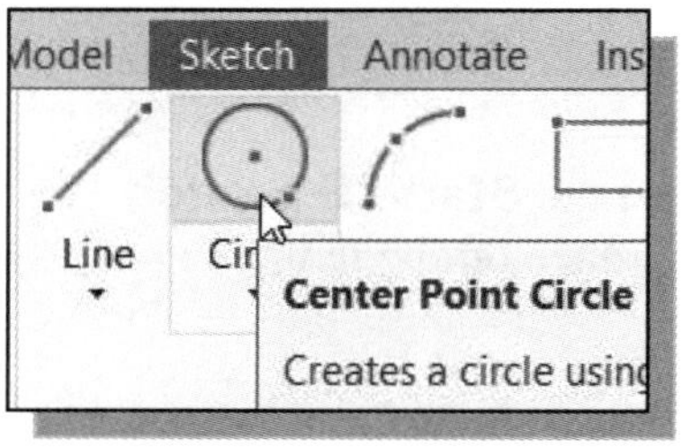

7. Select the **Center Point Circle** command by clicking once with the left-mouse-button on the icon in the *2D Sketch* panel.

8. Create a circle of arbitrary size as shown below.

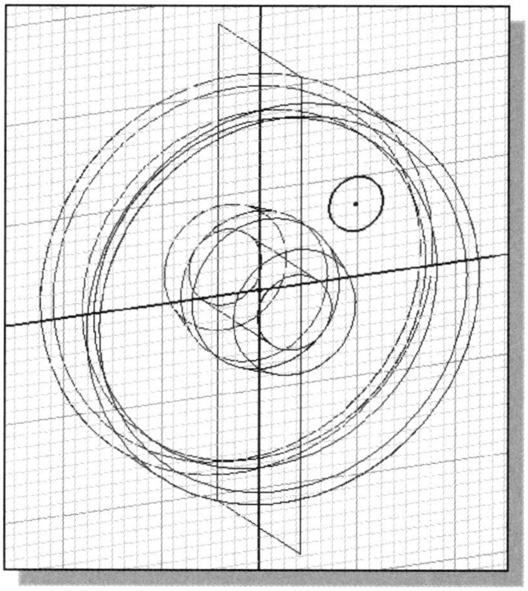

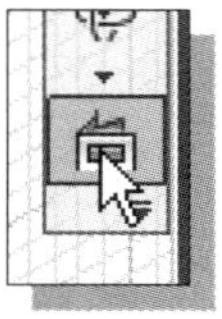

9. In the *Standard* toolbar area, click on the **Look At** button.

10. Select the circle we just created to orient the display of the sketching plane on the screen.

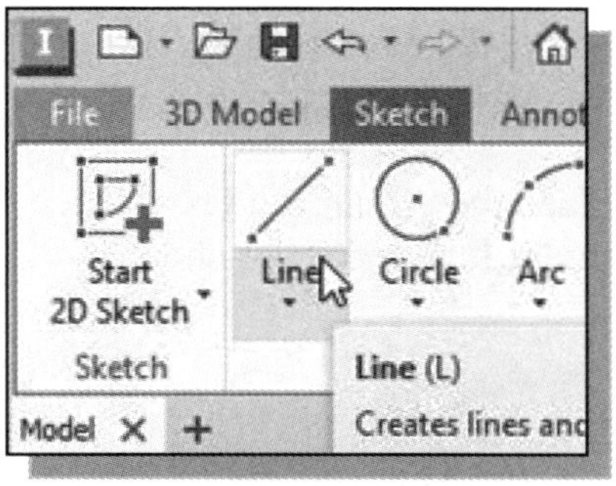

11. Select the **Line** command in the *2D Sketch* panel. A brief description of the command is displayed at the bottom of the drawing screen: "*Creates Straight lines and arcs.*"

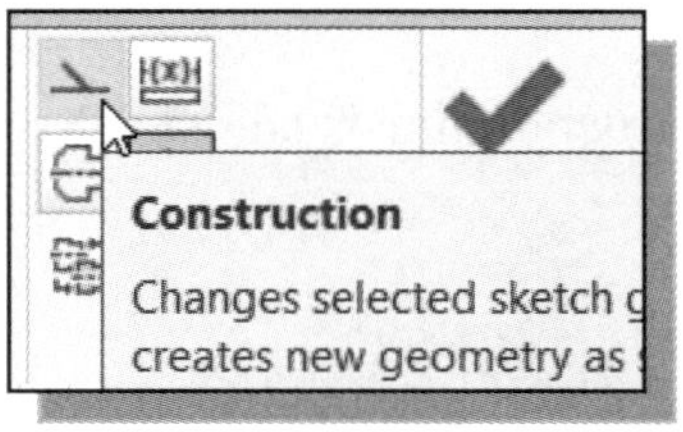

12. On the right side of the ribbon toolbars, set the *Style* option to **Construction** as shown.

13. Create a *construction line* by connecting from the center of the circle we just created to the projected center point (***origin***) at the center of the 3D model as shown below.

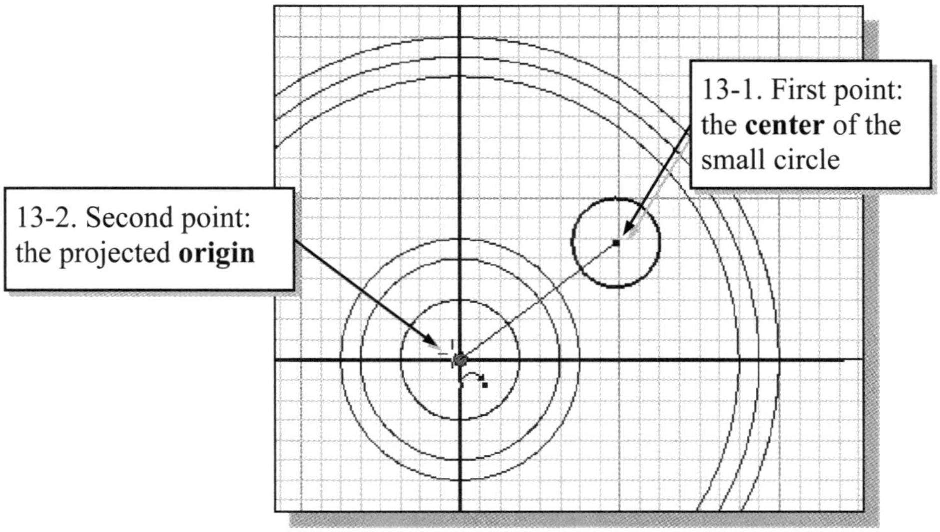

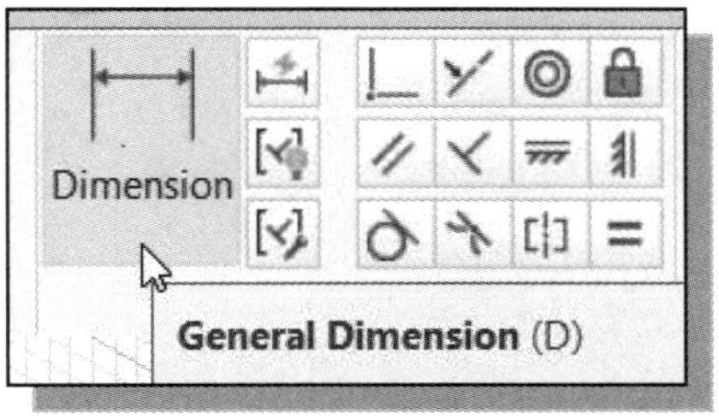

14. Select the **General Dimension** command in the *Constrain* panel.

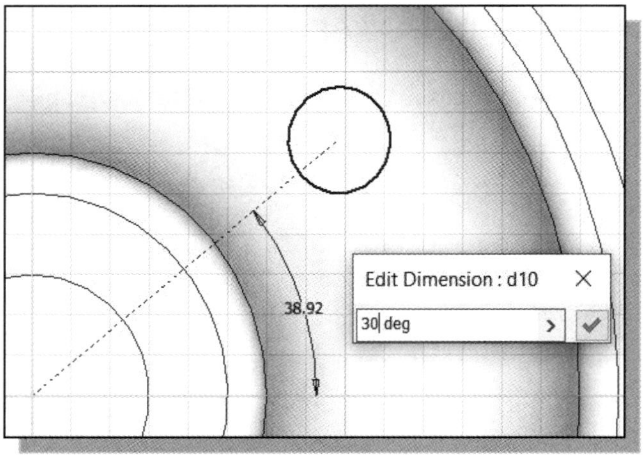

15. Pick the **horizontal axis** as the first entity to dimension as shown in the figure.

16. Select the **construction line** as the second object to dimension.

17. Place the dimension text to the right of the model as shown.

18. On your own, set the angle dimension to **30** as shown in the above figure.

➢ Note that the small circle moves as the location of the construction line is adjusted by the *angle dimension* we created.

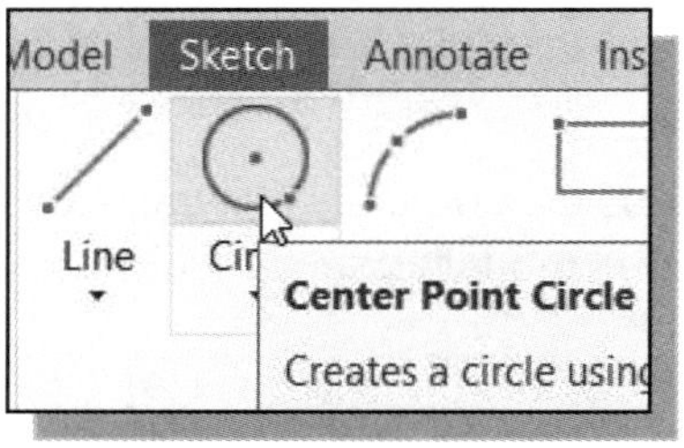

19. Select the **Center Point Circle** command by clicking once with the left-mouse-button on the icon in the *Draw* panel.

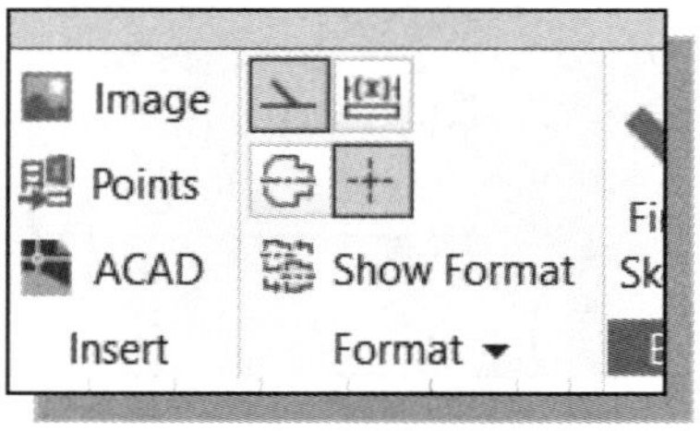

20. Confirm the *Style* option is still set to **Construction**.

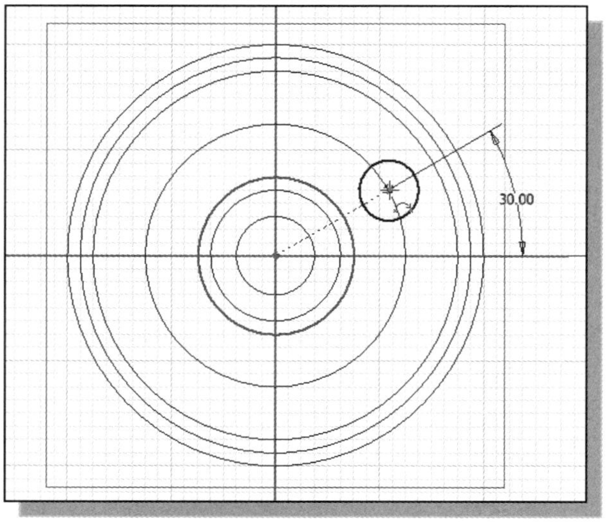

21. Create a *construction circle* by placing the center at the projected center point (***origin***).

22. Pick the **center** of the small circle to set the size of the construction circle as shown.

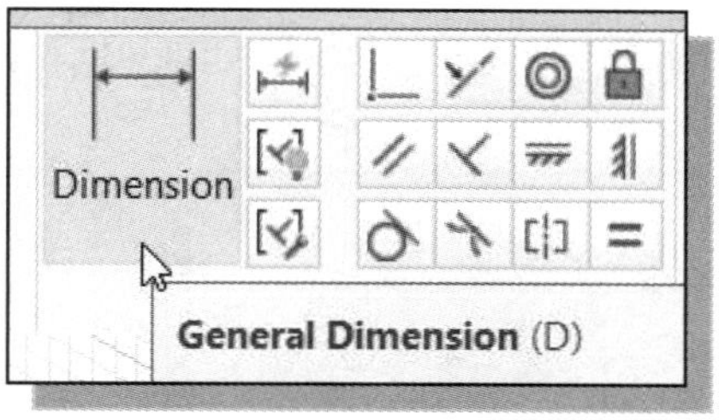

23. Select the **General Dimension** command in the *Constrain* panel.

24. On your own, create the two diameter dimensions, **.5** and **2.5**, as shown below.

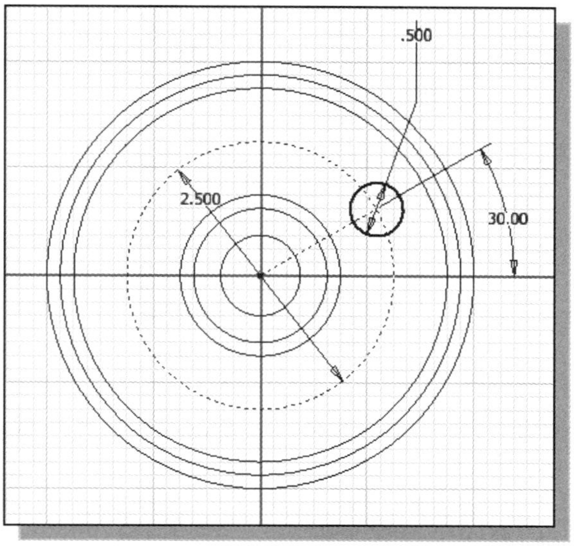

25. Inside the graphics window, click once with the right-mouse-button to display the option menu. Select **Finish Sketch** in the pop-up menu to end the Sketch option.

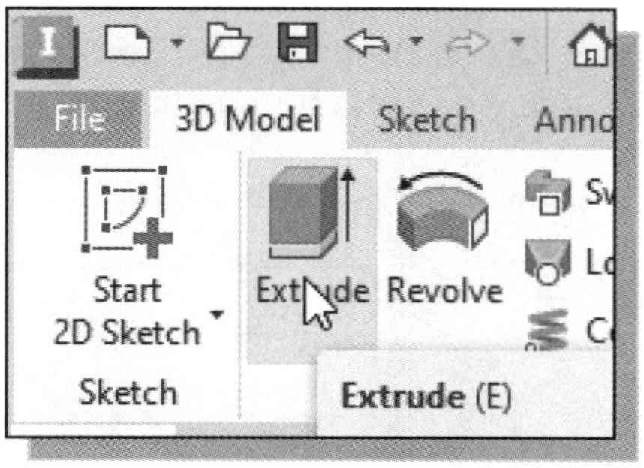

26. In the *Create Features* panel select the **Extrude** command by left-clicking on the icon.

27. Pick the inside region of the circle to set up the profile of the extrusion.

28. Inside the *Extrude* dialog box, select the **Cut** operation for **Symmetric (both directions)**, and set the *Extents* to **All** as shown.

29. Click on the **OK** button to accept the settings and create the cut feature.

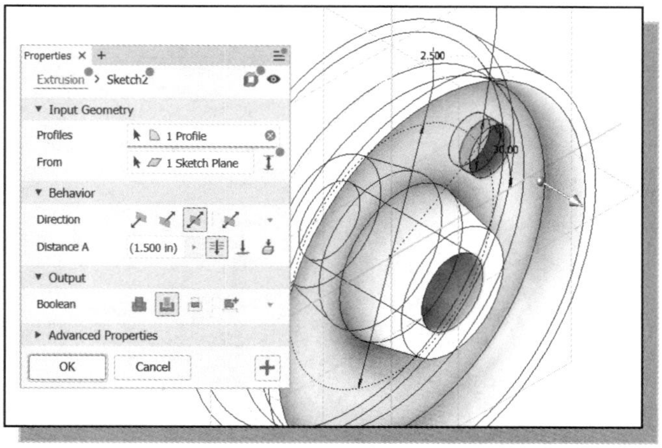

30. On your own, adjust the angle dimension applied to the construction line of the cut feature to **90** and observe the effect of the adjustment.

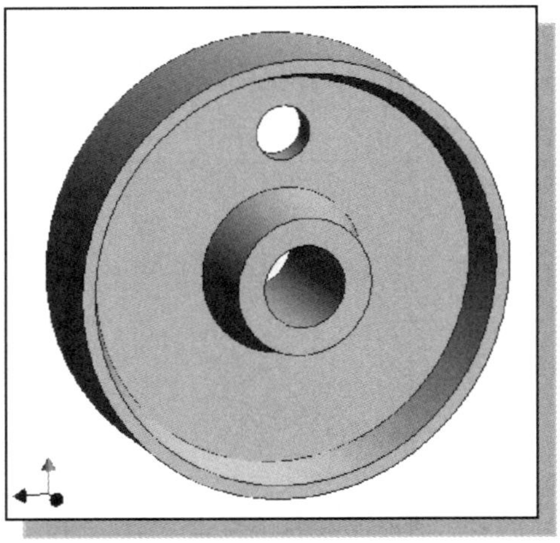

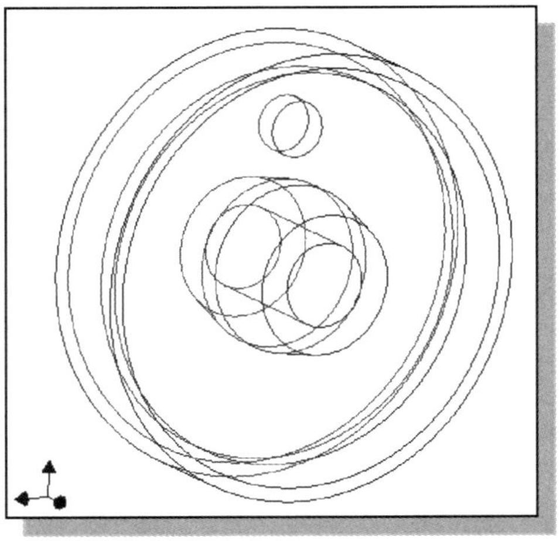

Circular Pattern

In Autodesk Inventor, existing features/solids can be easily duplicated. The **Pattern** command allows us to create both rectangular and polar arrays of features. The patterned features are parametrically linked to the original feature; any modifications to the original feature are also reflected in the arrayed features.

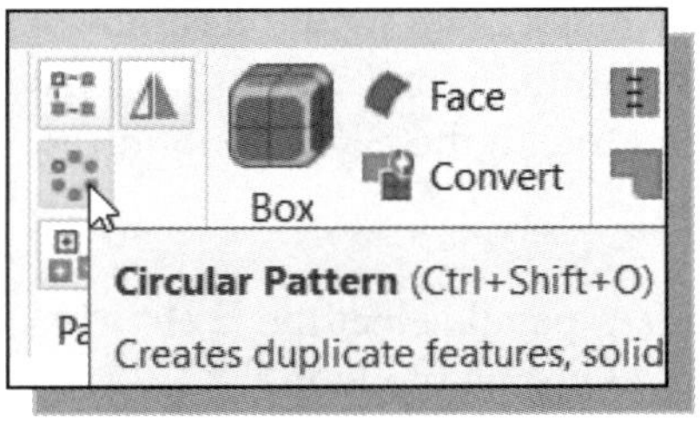

1. In the *Pattern Features* panel, select the **Circular Pattern** command by left-clicking once on the icon.

2. The message "*Select Feature to be arrayed*:" is displayed in the command prompt window. Select the **circular cut feature** when it is highlighted as shown.

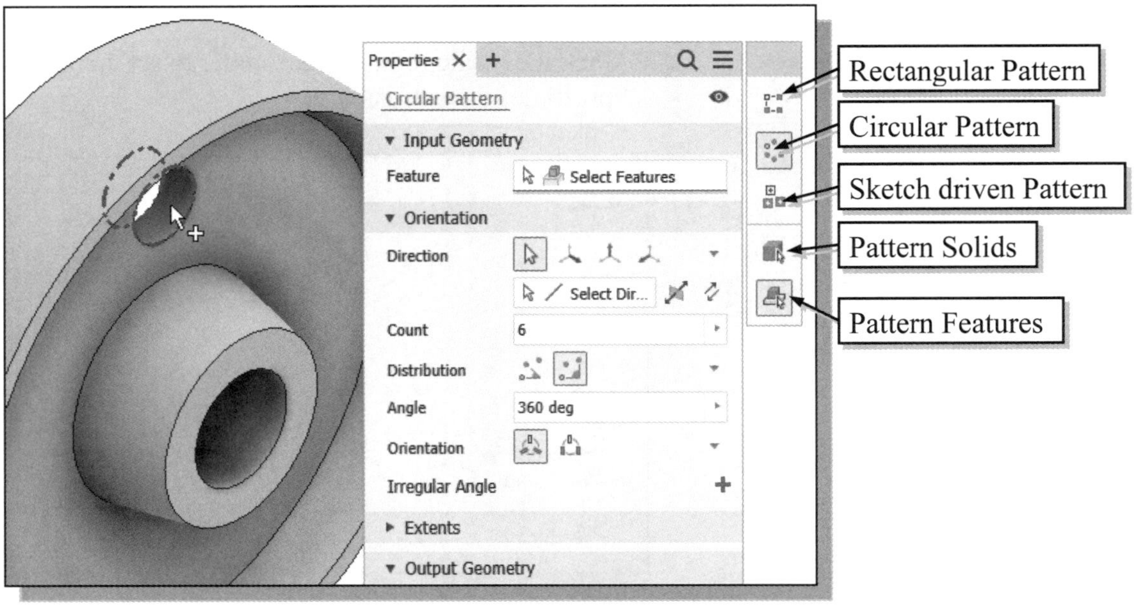

3. Inside the graphics window, **right-click** to bring up the **option menu**.

4. Select **Continue** in the option list to proceed with the Circular Pattern command.

5. Autodesk Inventor expects us to select an axis to pattern about. In the prompt area, the message *"Define the axis of revolution"* is displayed. Select the **X-Axis** in the *browser* window or the inside cylindrical surface in the graphics window.

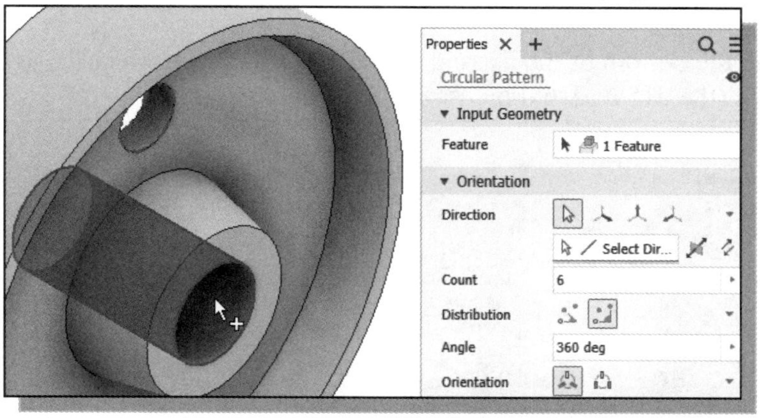

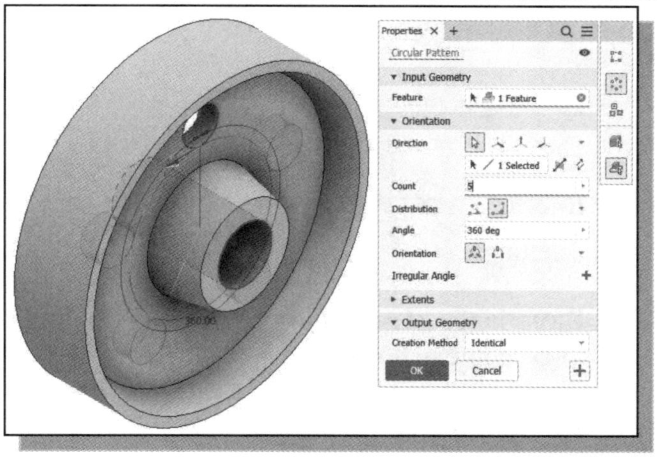

6. In the *Circular Pattern* dialog box, enter **5** in the ***Count*** box and **360** in the ***Angle*** box as shown. (Note the different options available for the Circular Pattern command.)

7. Click on the **OK** button to accept the settings and create the *circular pattern*.

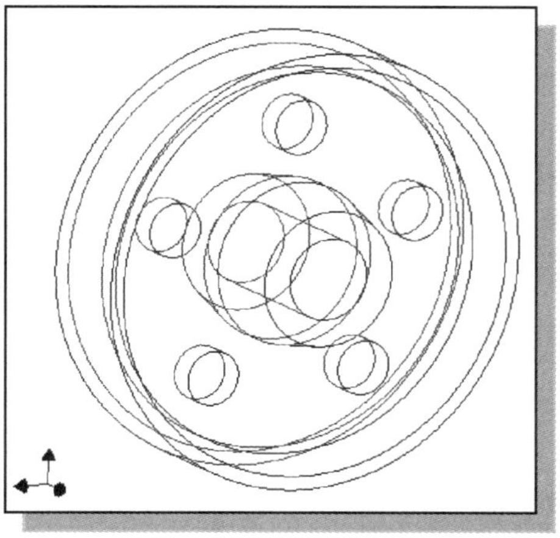

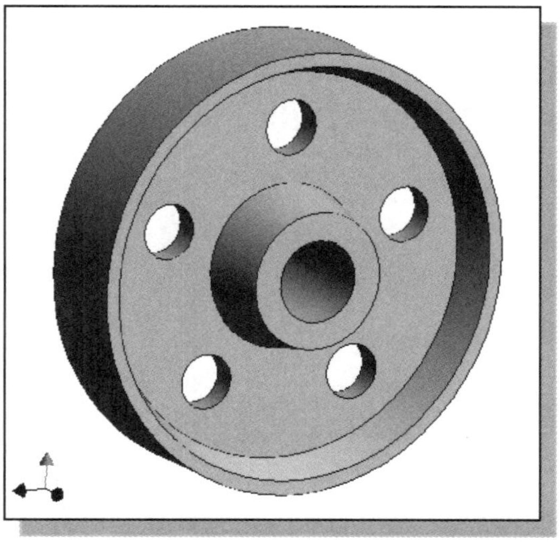

Examine the Design Parameters

1. In the *Manage* tab select the **Parameters** command by left-clicking once on the icon. The *Parameters* pop-up window appears.

2. Scroll down to the bottom of the list and locate the two dimensions used to create the circular pattern (the d15 & d16 dimensions in the below figure).

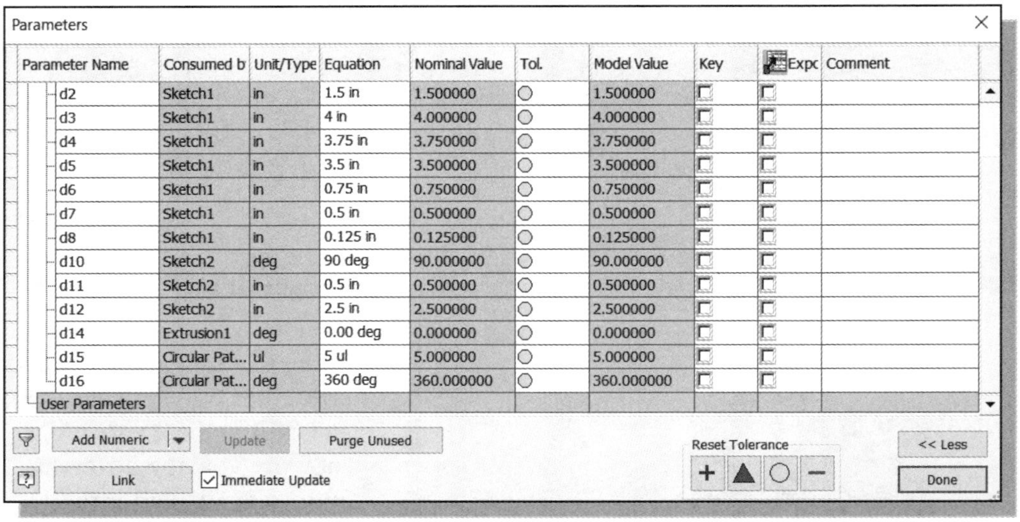

Parameters

Parameter Name	Consumed b	Unit/Type	Equation	Nominal Value	Tol.	Model Value	Key	Expc	Comment
d2	Sketch1	in	1.5 in	1.500000		1.500000			
d3	Sketch1	in	4 in	4.000000		4.000000			
d4	Sketch1	in	3.75 in	3.750000		3.750000			
d5	Sketch1	in	3.5 in	3.500000		3.500000			
d6	Sketch1	in	0.75 in	0.750000		0.750000			
d7	Sketch1	in	0.5 in	0.500000		0.500000			
d8	Sketch1	in	0.125 in	0.125000		0.125000			
d10	Sketch2	deg	90 deg	90.000000		90.000000			
d11	Sketch2	in	0.5 in	0.500000		0.500000			
d12	Sketch2	in	2.5 in	2.500000		2.500000			
d14	Extrusion1	deg	0.00 deg	0.000000		0.000000			
d15	Circular Pat...	ul	5 ul	5.000000		5.000000			
d16	Circular Pat...	deg	360 deg	360.000000		360.000000			
User Parameters									

3. Click on the **Done** button to accept the settings.

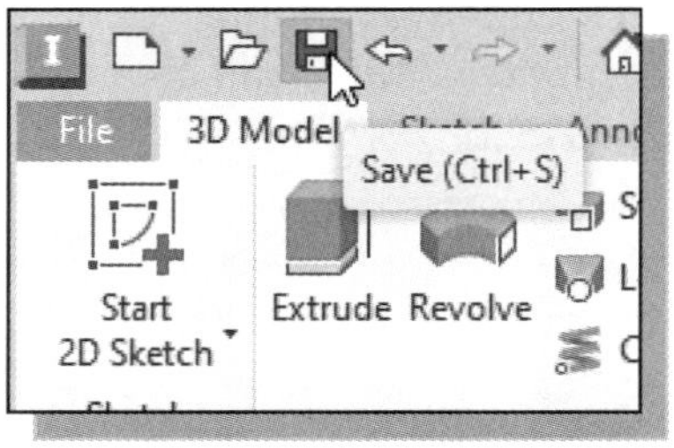

4. Select **Save** in the *Standard* toolbar; we can also use the "**Ctrl-S**" combination (press down the [Ctrl] key and hit the [S] key once) to save the part as ***Pulley*** in the **Chapter11** folder.

Drawing Mode – Defining a New Border and Title Block

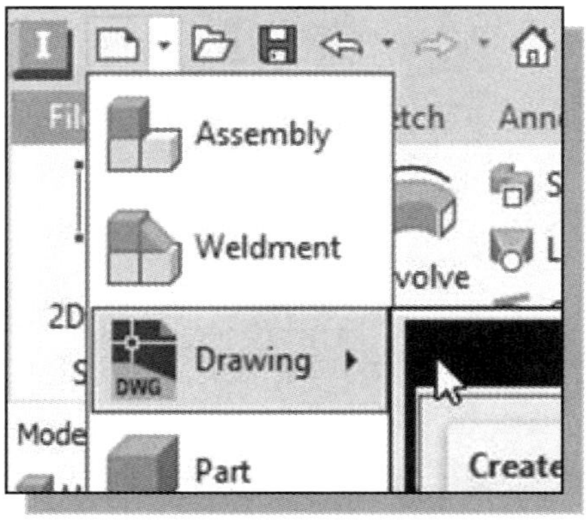

1. Click on the **drop-down arrow** next to the **New File** button in the *Quick Access* toolbar area to display the available new file options.

2. Select **Drawing** from the drop-down list.

➢ Note that a new graphics window appears on the screen. We can switch between the solid model and the drawing by clicking the corresponding graphics window.

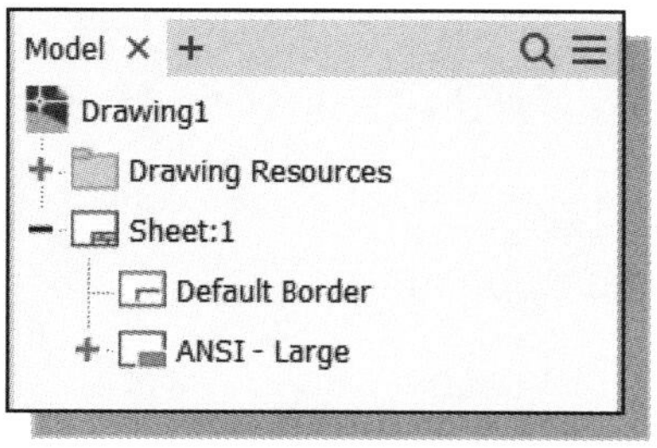

3. Inside the drawing *browser* window, click on the [**+**] symbol in front of **Sheet1** to display the settings.

4. On your own, **delete** the **Default Border** and the **ANSI-Large** ***title block***. (Hint: right-click on the names to bring up the option menu.)

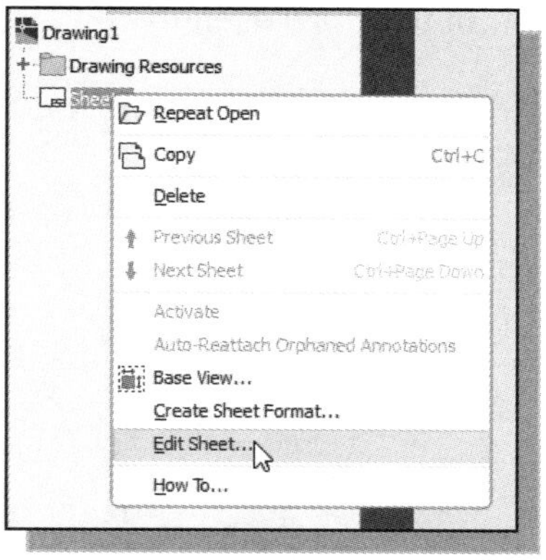

5. Inside the drawing *browser* window, **right-click** on **Sheet1** to bring up the option menu.

6. Select **Edit Sheet** in the option list.

7. In the *Edit Sheet* dialog box, set the sheet size to **A** (8.5 × 11).

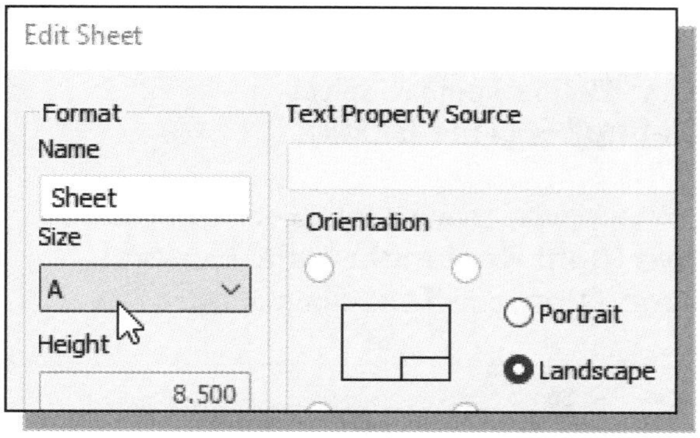

8. Confirm the page *Orientation* is set to **Landscape** and click on the **OK** button to accept the settings.

9. In the *Ribbon* toolbar panel, select **Manage → Define New Border**.

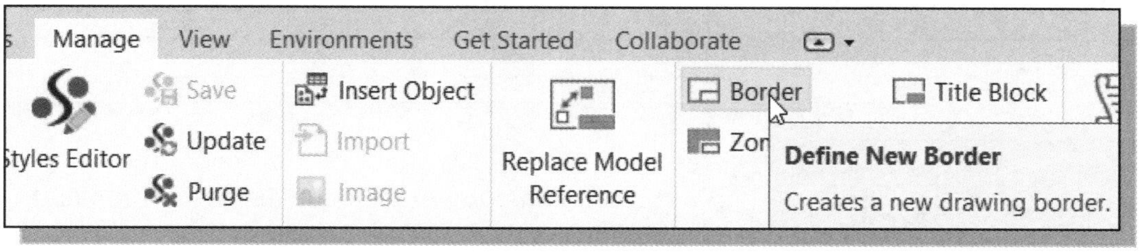

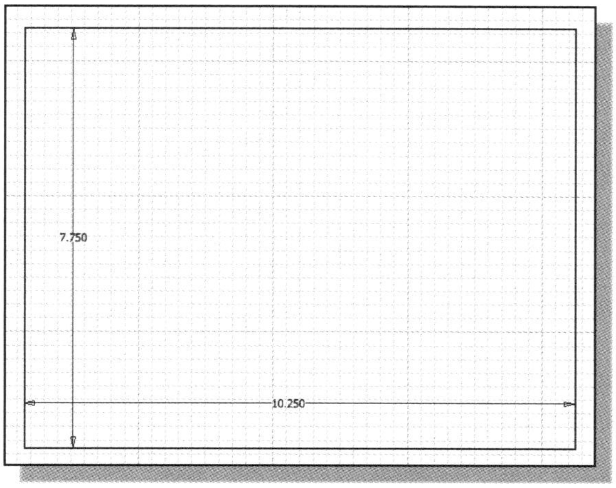

10. On your own, create a rectangle (**7.75** × **10.25**) using the **Two Point Rectangle** command and the **General Dimension** command.

11. Drag and drop one of the edges of the rectangle to position the rectangle to the center of the sheet.

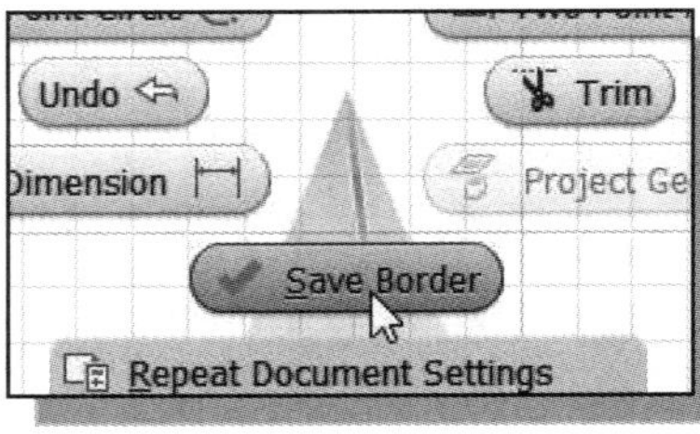

12. Inside the graphics window, **right-click** to bring up the option menu.

13. Select **Save Border** in the option list.

14. In the *Border* dialog box, enter **A-Size** as the new border name.

15. Click on the **Save** button to end the Define New Border command.

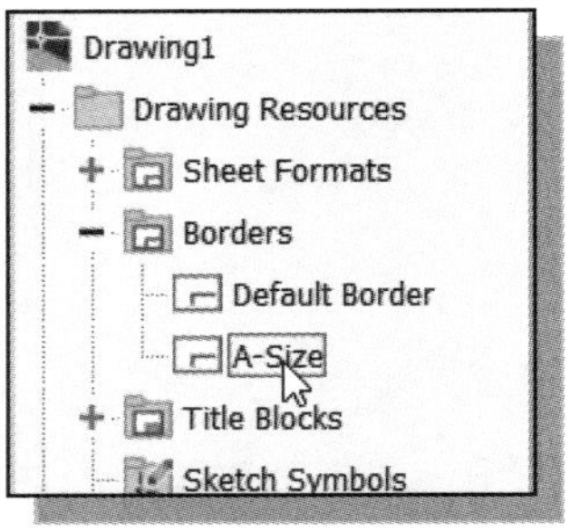

16. Inside the *browser* window, expand the **Drawing Resources** list by clicking on the [**+**] symbol.

17. **Double-click** on the **A-size** border, the border we just created, to place the border in the current drawing. Note that none of the dimensions used to construct the border are displayed.

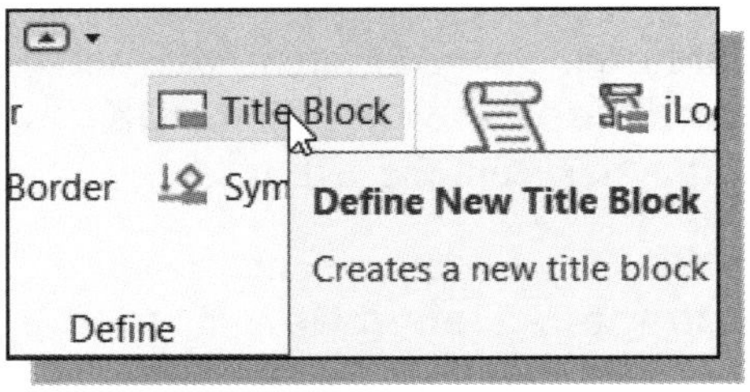

18. In the Define Toolbar, select **[Define] → [Title Block]**.

19. On your own, create a title block using the **Two Point Rectangle**, **Line**, **General Dimension**, and **Text** commands.

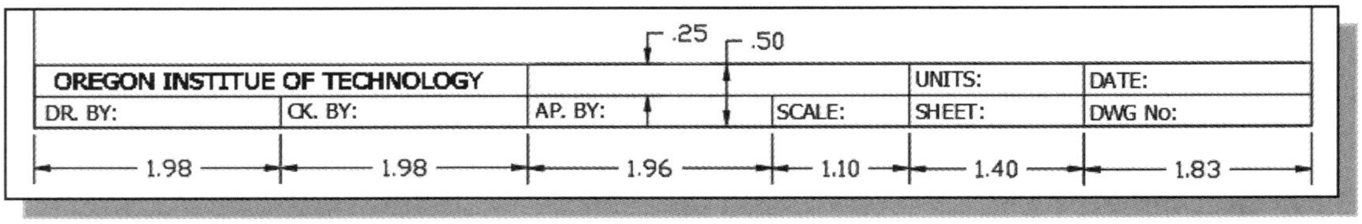

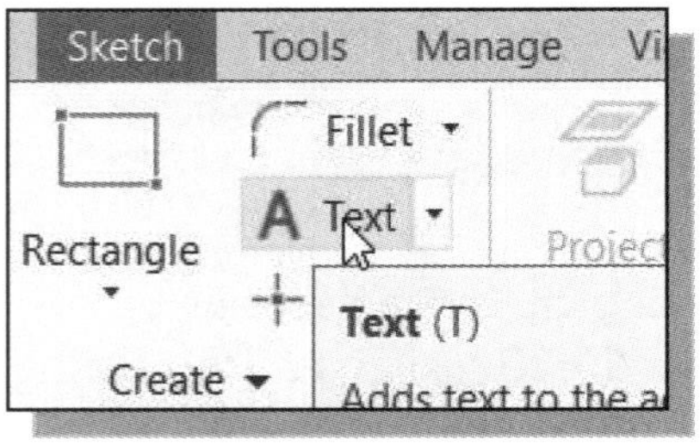

20. In the *Create toolbar*, activate the **Text** command as shown.

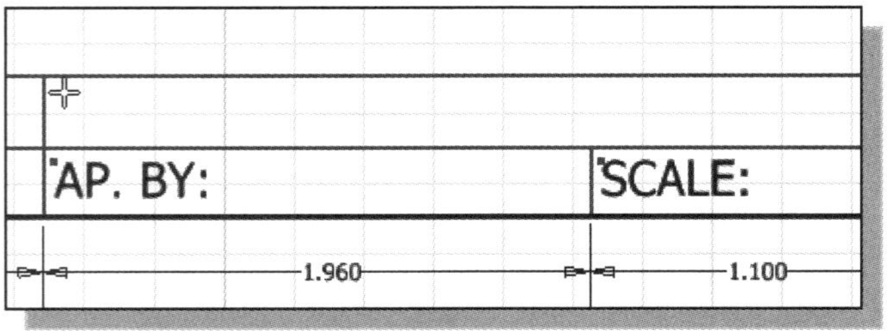

21. Click near the top left corner of the *Title Box*, the second box of the top section in the title block we just created, as shown.

22. In the *Format Text dialog* box, adjust the text size to **0.125** as shown.

23. In the *Format Text dialog* box, select **Standard iProperties** and **Title** as the *Type* and *Property* parameter as shown.

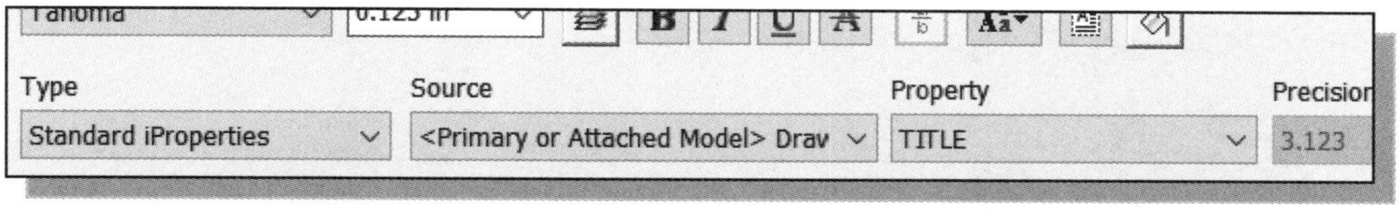

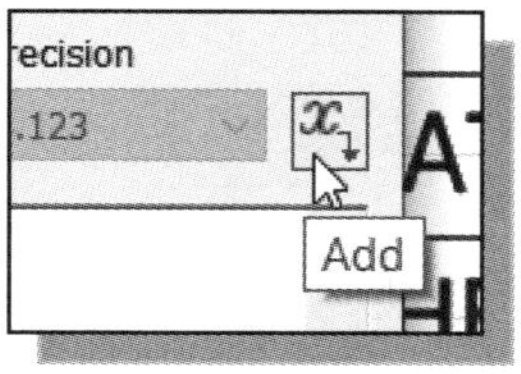

24. Click on the **Add Parameter** icon to insert the associated iProperty information to the title block.

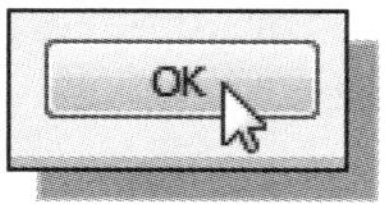

25. Click on the **OK** button to accept the settings.

26. On your own, repeat the above process to add the **Standard iProperty - Creation Date** to the *Date Box* as shown.

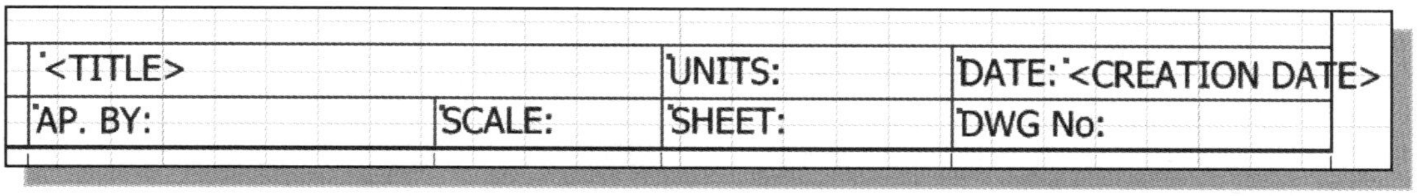

27. On your own, repeat the above steps and add in the associated Standard iProperties for the bottom line of the title block as shown.

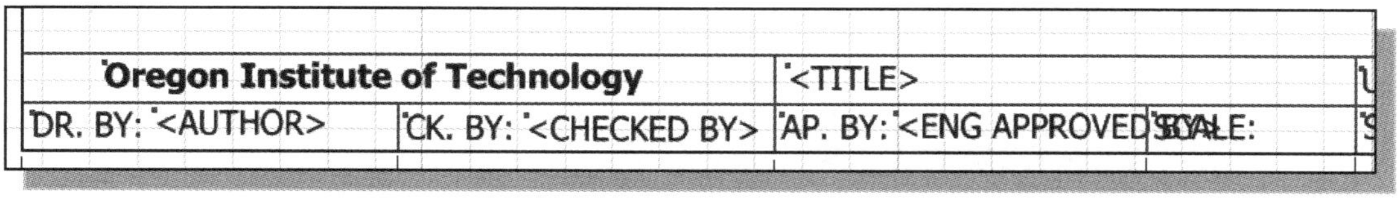

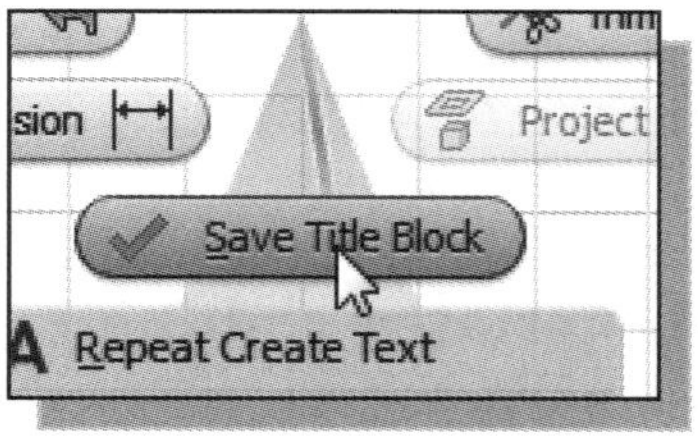

28. Inside the *graphics window*, right-click to bring up the option menu.

29. Select **Save Title Block** in the option list.

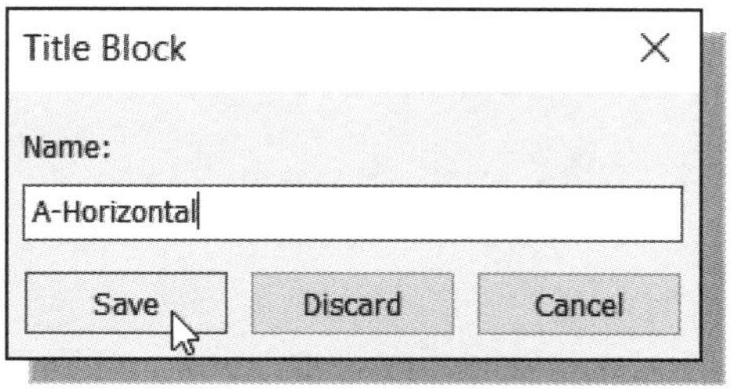

30. In the *Title Block* dialog box, enter **A-Horizontal** as the new title block name.

31. Click on the **Save** button to end the Define New Title Block command.

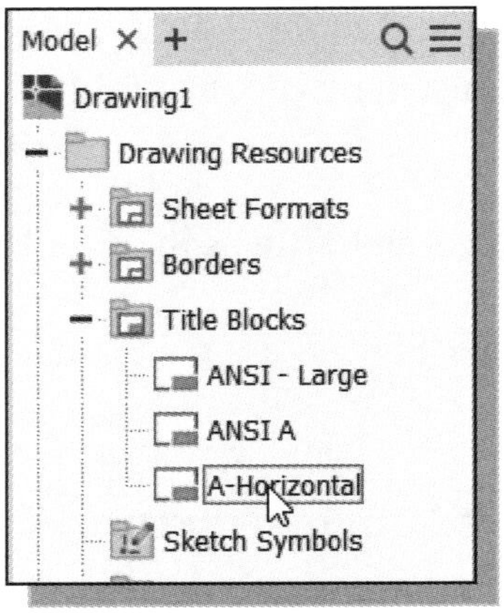

32. Inside the *browser* window, expand the **Title Blocks** list by clicking on the [+] symbol.

33. Double-click on the **A-Horizontal** title block, the title block we just created, to place it into the current drawing.

Create a Drawing Template

In Autodesk Inventor, each new drawing is created from a template. During the installation of Autodesk Inventor, a default drafting standard was selected which sets the default template used to create drawings. We can use this template or another predefined template, modify one of the predefined templates, or create our own templates to enforce drafting standards and other conventions. Any drawing file can be used as a template; a drawing file becomes a template when it is saved in the *Templates* folder. Once the template is saved, we can create a new drawing file using the new template.

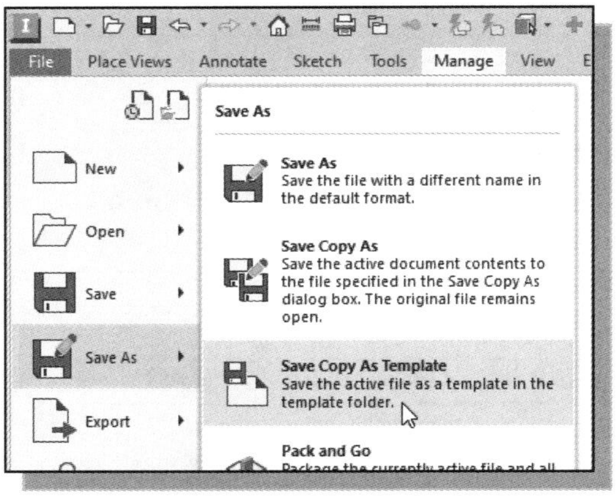

1. Select **Save Copy As Template** in the *File Menu*; we will save the current drawing as a template file.

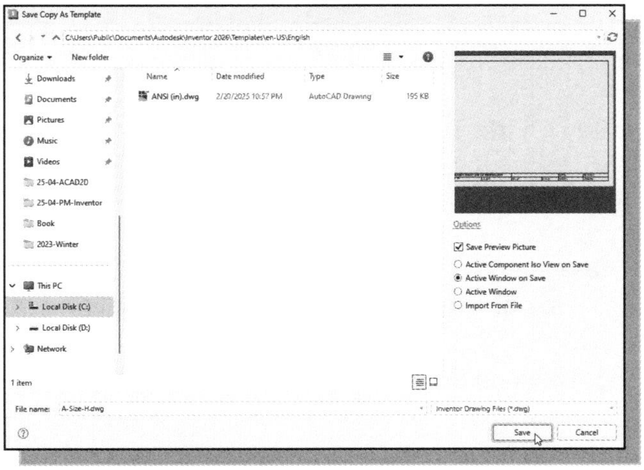

2. In the *Save As* dialog box, switch to the *Public User - Inventor* folder: **Inventor 2027 → Templates → en-US→ English** directory.

3. Enter **A-Size-H.dwg** as the template filename.

4. Click on the **Save** button to create a drawing template file.

Create the Necessary Views

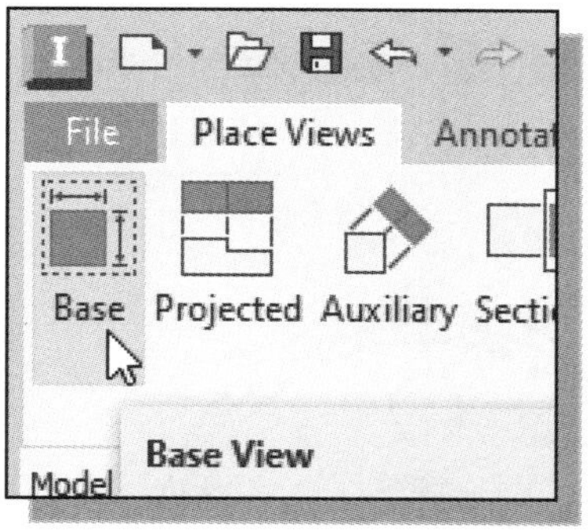

1. Click on **Base View** in the *Place Views* panel to create a base view.

2. In the *Drawing View* dialog box, set the scale to **3 : 4** and select the ***left*** view as shown in the figure below.

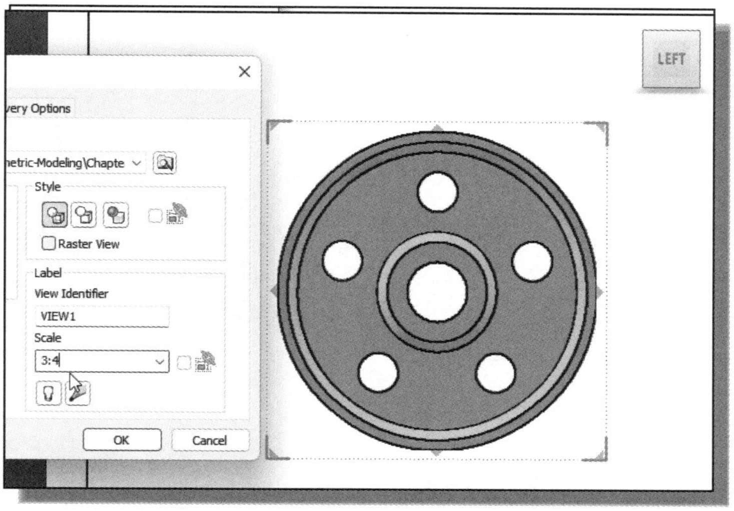

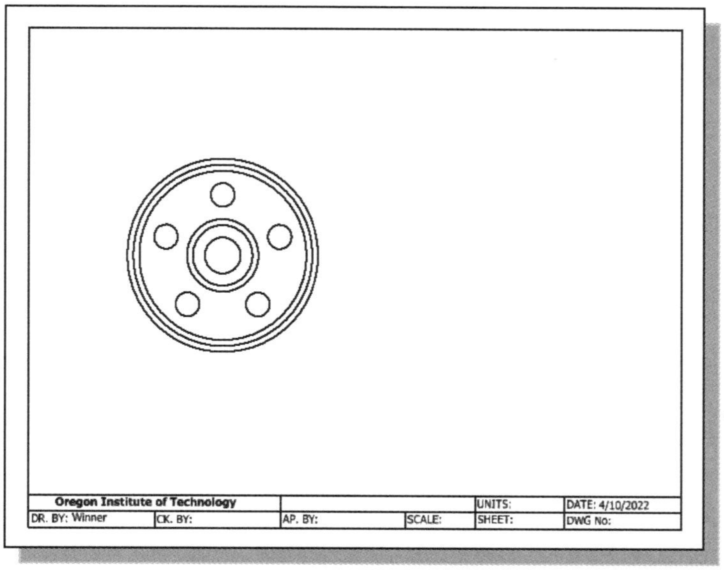

3. Move the cursor inside the graphics window and place the **base view** toward the left side of the border as shown. (If necessary, drag the *Drawing View* dialog box to another location on the screen.)

4. Click on the **OK** button to accept the setting.

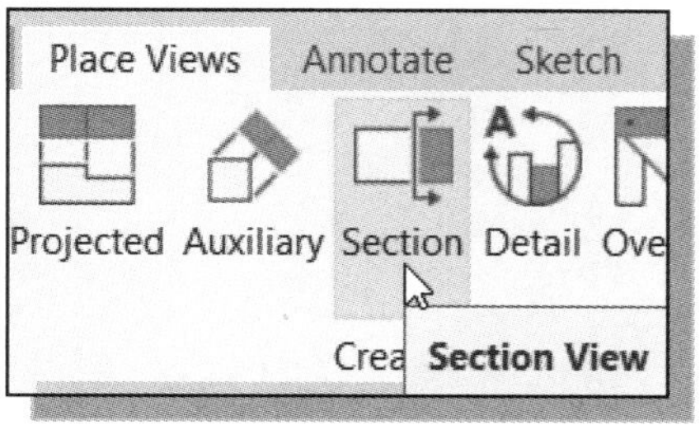

5. Click on the **Section View** button in the *Place Views* panel to create a section view.

6. Click on the **base view** to select the view where the section line is to be created.

7. Inside the graphics window, align the cursor to the center of the base view and create the vertical cutting plane line as shown.

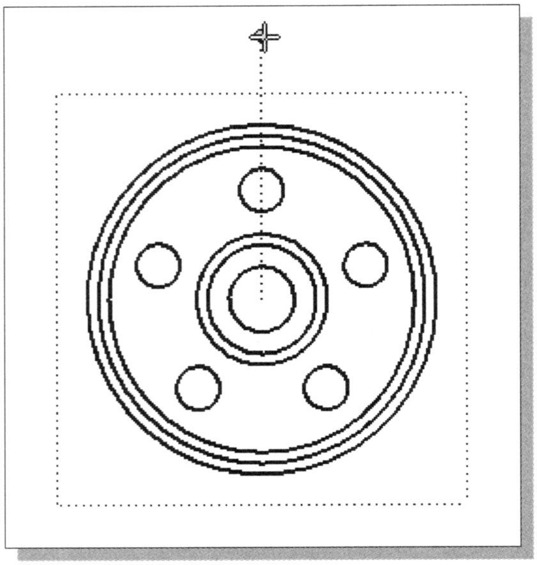

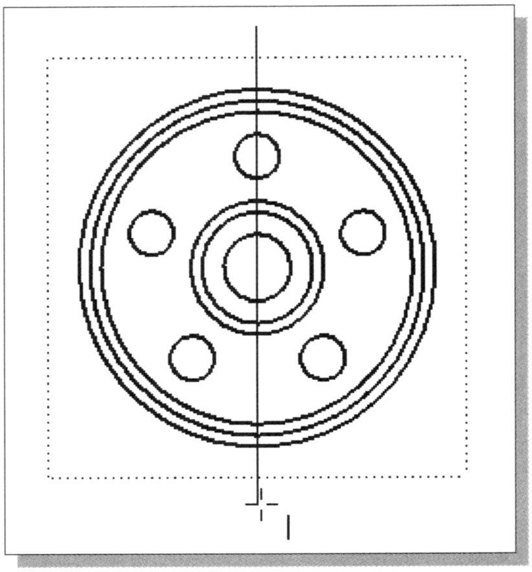

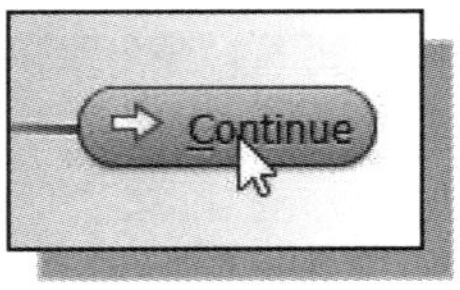

8. Inside the *graphics window*, right-click once to bring up the **option menu**.

9. Select **Continue** to proceed with the Section View command.

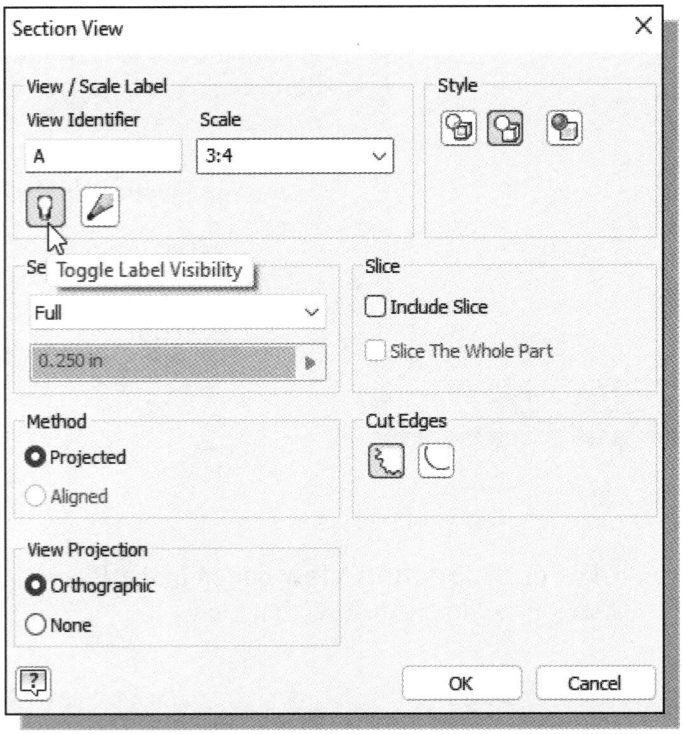

10. Inside the *Section View* dialog box, toggle ***OFF*** the **Label Visibility** option as shown.

11. Set the *Display Style* to **Hidden Line Removed**, as shown. (**DO NOT** click on the **OK** button.)

12. Next, Autodesk Inventor expects us to place the projected section. Select a location that is toward the right side of the base view as shown in the figure.

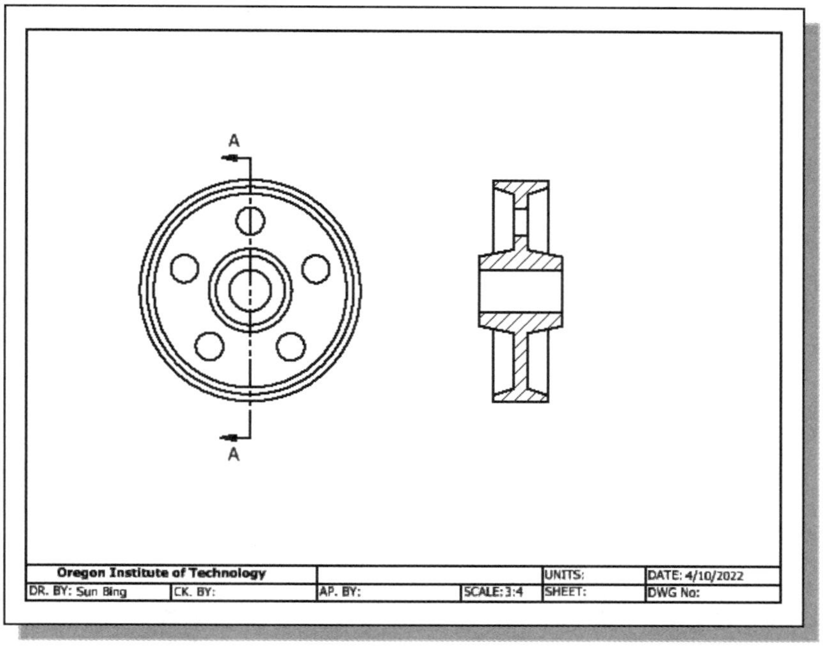

13. On your own, use the Projected View option and create an ***isometric* view** (Iso Top Left, ¾ scale) of the design and place the view toward the right side of the section view as shown below.

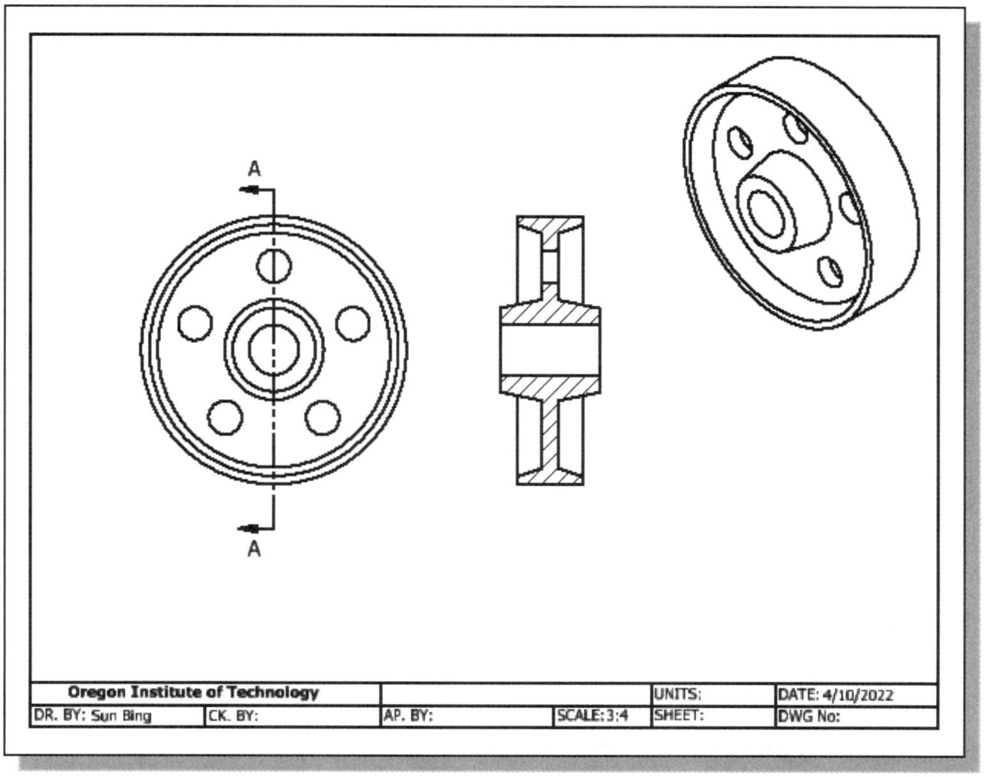

Retrieve Model Annotations – Features Option

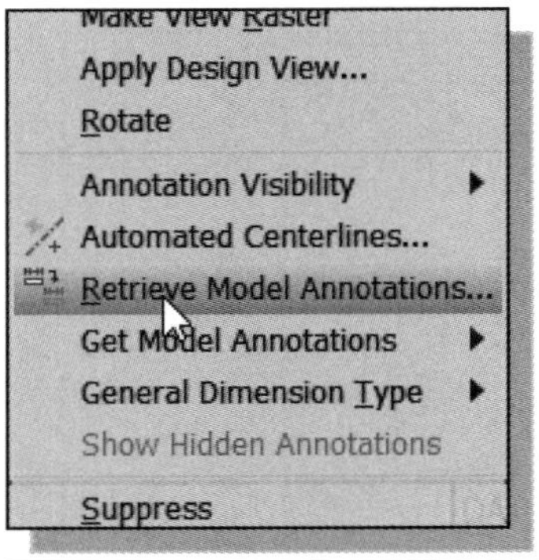

1. Move the cursor on top of the ***section* view** of the model and watch for the box around the entire view indicating the view is selectable as shown in the figure.

2. Inside the *graphics window*, **right-click** once to bring up the option menu.

3. Select **Retrieve Model Annotations** to display the parametric dimensions used to create the model.

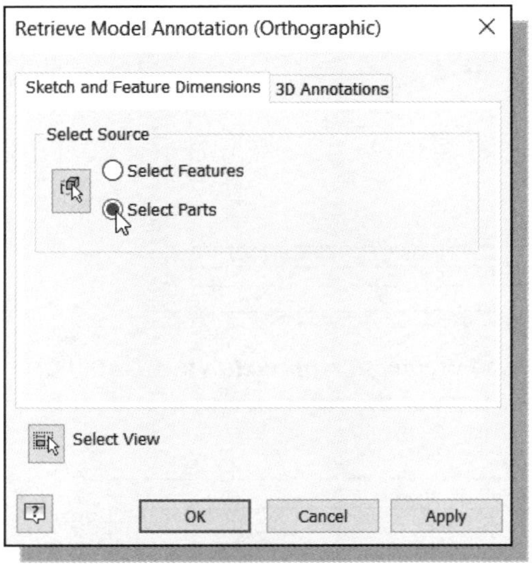

4. In the *Retrieve Model Annotations* dialog box, set the *Select Source* option to **Select Parts** as shown.

➢ The dimensions used to create the revolved feature are now displayed in the section view.

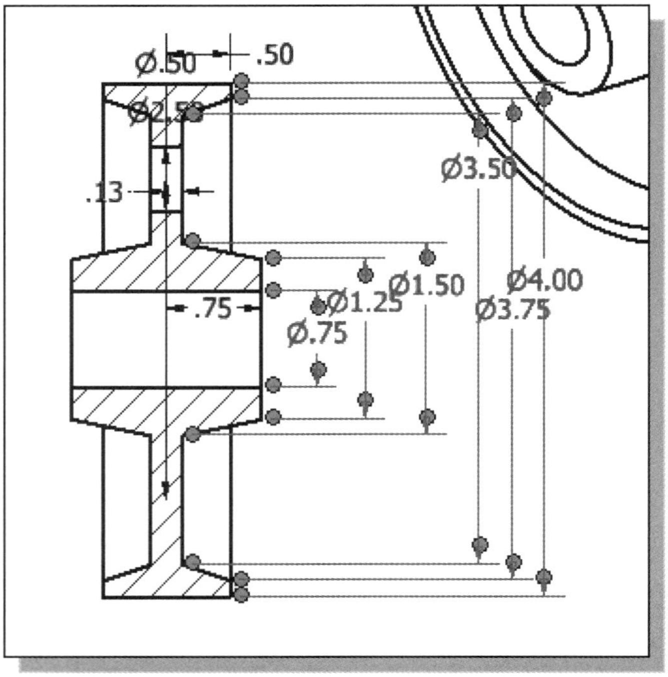

5. On your own, select the desired **model dimensions** to be displayed.

6. Click on the **Apply** button and notice the selected dimensions changed color indicating they have been retrieved.

7. Note that the **Select View** button is activated; click on the Base view to continue with the Retrieve Model Annotations command.

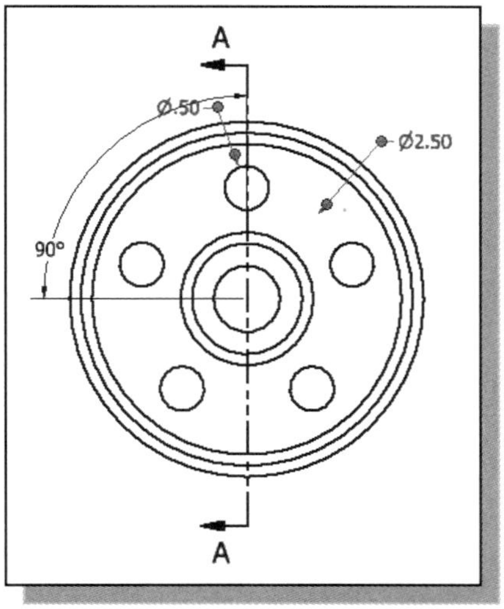

8. On your own, retrieve the model dimensions on the base view as shown.

9. On your own, reposition the views and dimensions as shown in the figure.

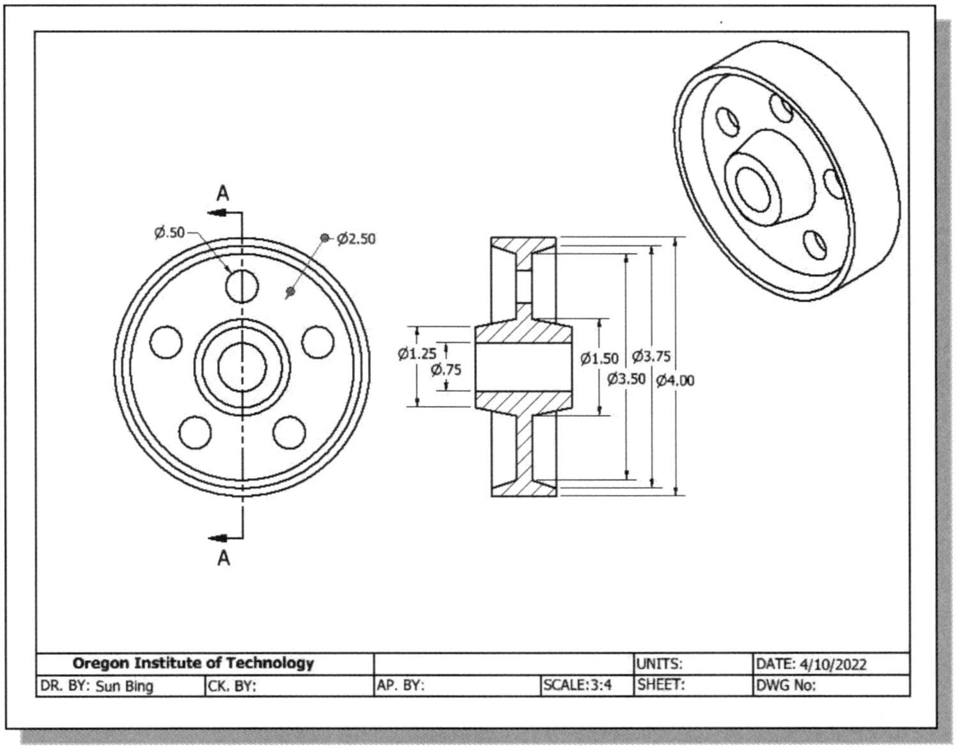

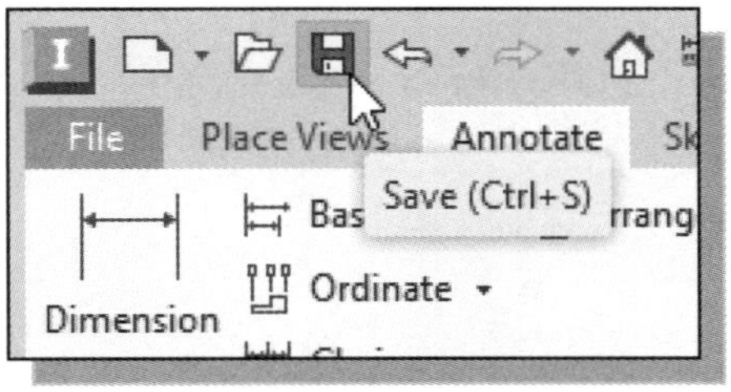

10. On your own, use the **Save** command and save the drawing as ***pulley.dwg***.

Associative Functionality – A Design Change

Autodesk Inventor's *associative functionality* allows us to change the design at any level, and the system reflects the changes at all levels automatically. We will illustrate the associative functionality by changing the circular pattern from five holes to six holes.

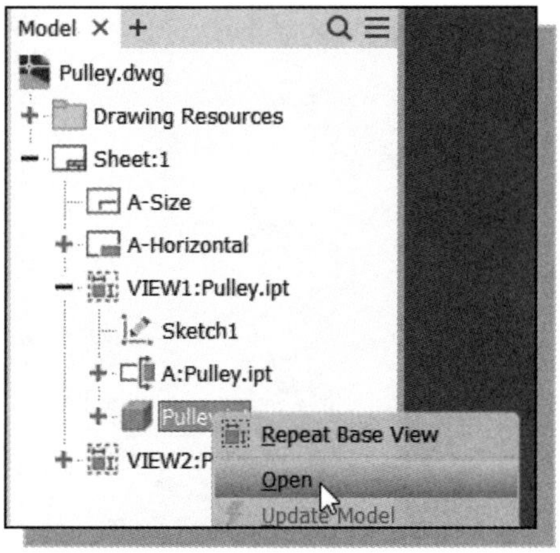

1. Inside the *Model Tree* window, below the VIEW1 list, right-click on the **Pulley.ipt** part name to bring up the option menu.

2. Select **Open** in the pop-up menu to switch to the associated solid model.

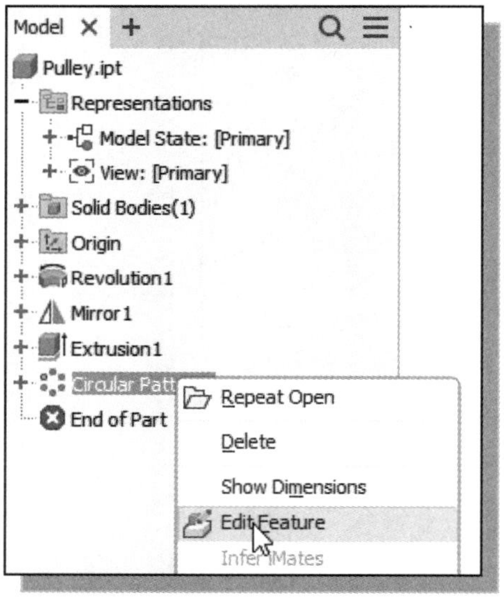

3. Inside the *Model Tree* window, right-click on the CircularPattern1 feature to bring up the option menu.

4. Select **Edit Feature** in the pop-up menu to bring up the associated feature option.

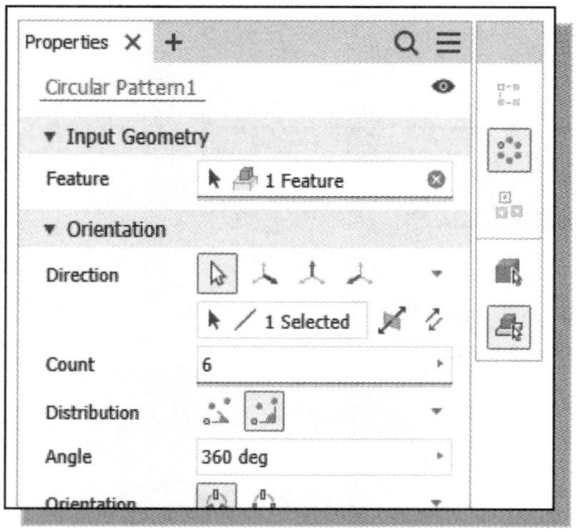

5. In the *Circular Pattern* dialog box, change the number to **6** as shown.

6. Click on the **OK** button to accept the setting.

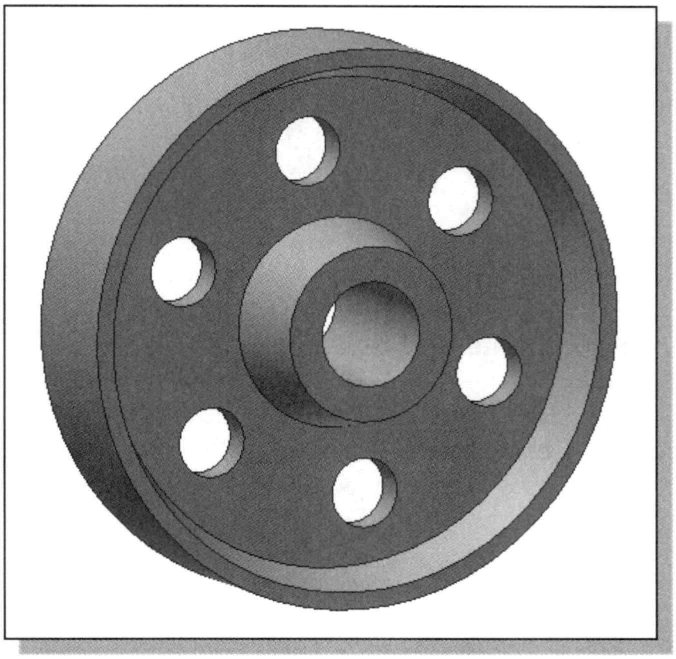

❖ The solid model is updated showing the 6 equally spaced holes as shown.

7. Switch back to the *Pulley* drawing and notice the drawing is also updated.

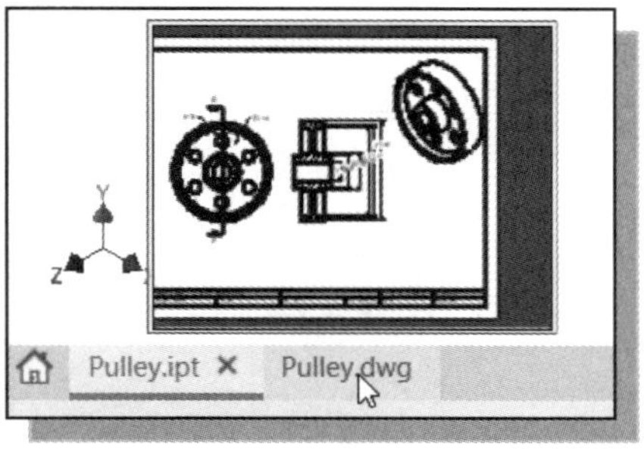

❖ Notice, in the *Pulley* drawing, the circular pattern is also updated automatically in all views.

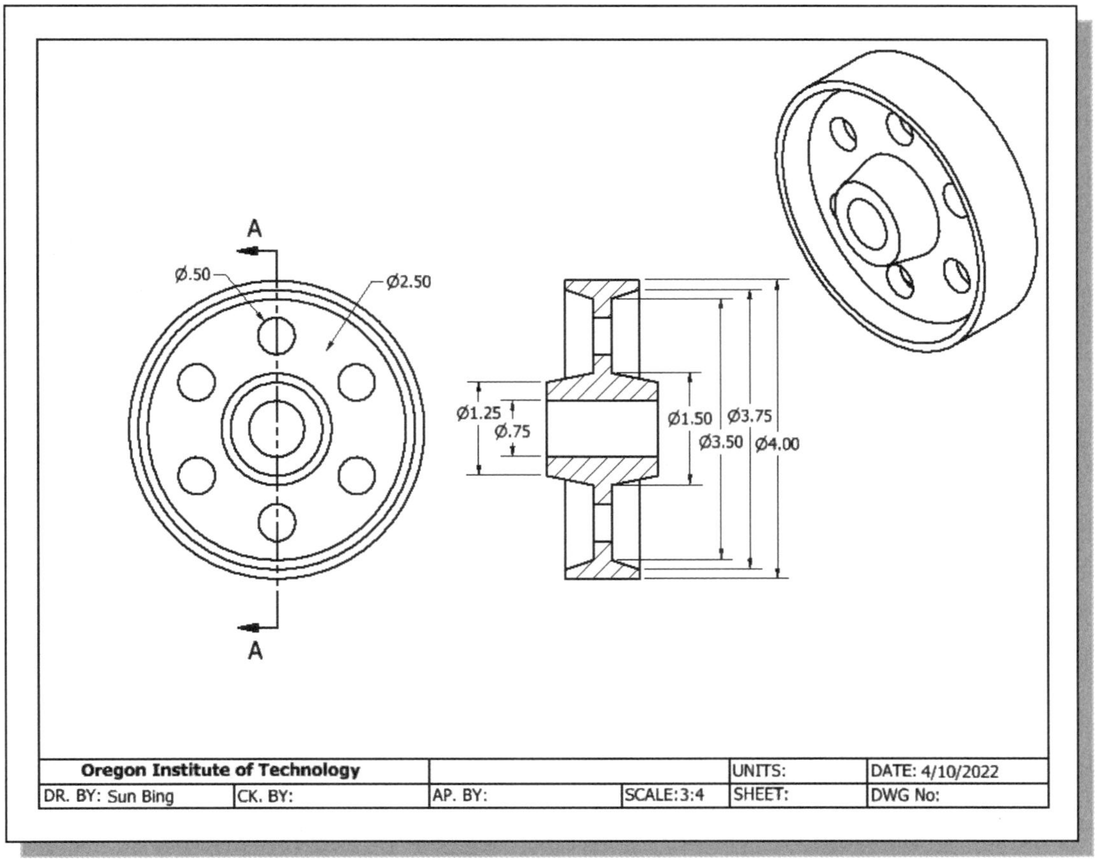

Add Centerlines to the Pattern Feature

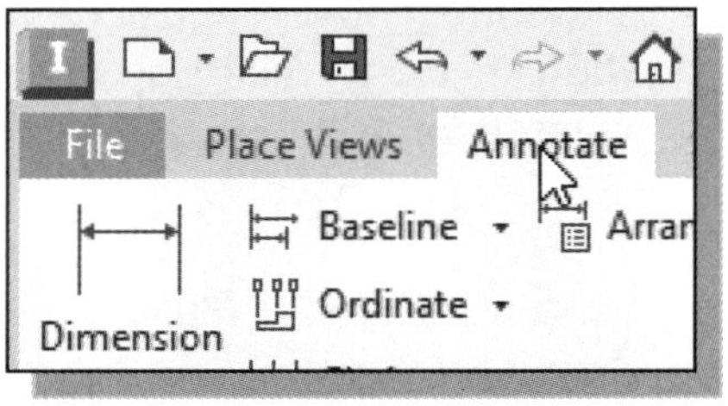

1. On your own, switch to the ***Drawing Annotation*** panel.

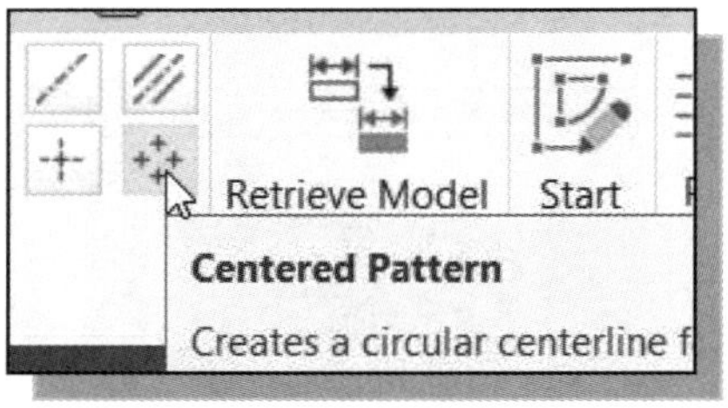

2. Select **Centered Pattern** from the symbol toolbar.

➢ The **Centered Pattern** option allows us to add centerlines to a patterned feature.

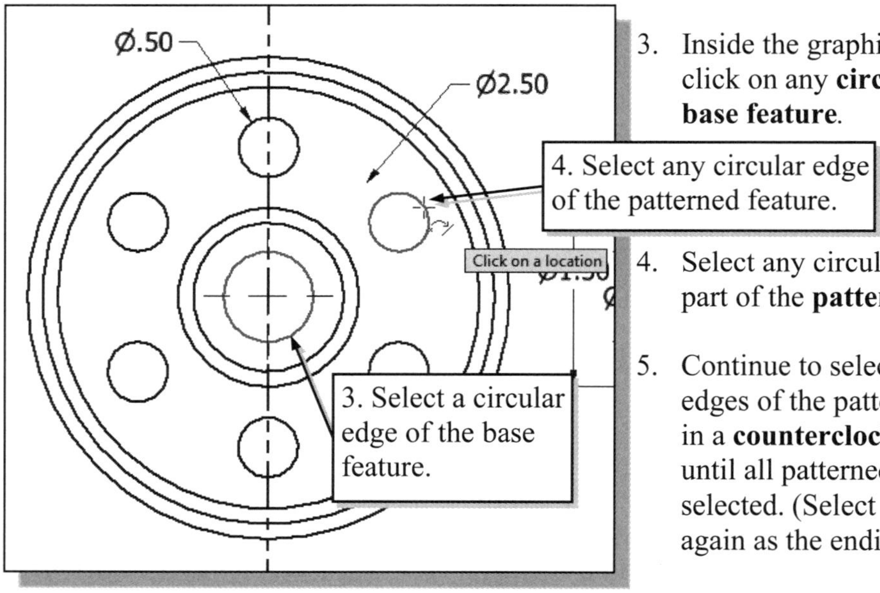

3. Inside the graphics window, click on any **circular edge** of the **base feature**.

4. Select any circular edge that is part of the **patterned feature**.

5. Continue to select the circular edges of the patterned features, in a **counterclockwise** manner, until all patterned items are selected. (Select the **first circle** again as the ending item.)

6. Inside the graphics window, **right-click** once to bring up the option menu.

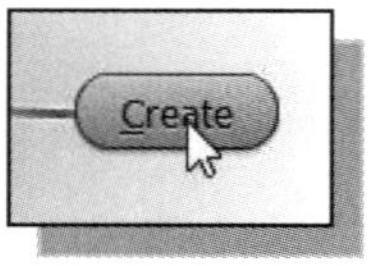

7. Select **Create** to create the centerlines around the selected items.

8. Inside the graphics window, **right-click** once to bring up the option menu.

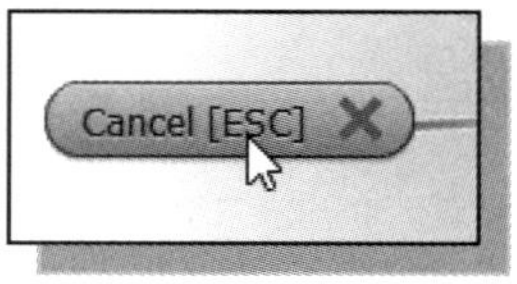

9. Select **Cancel [ESC]** to end the Centered Pattern command.

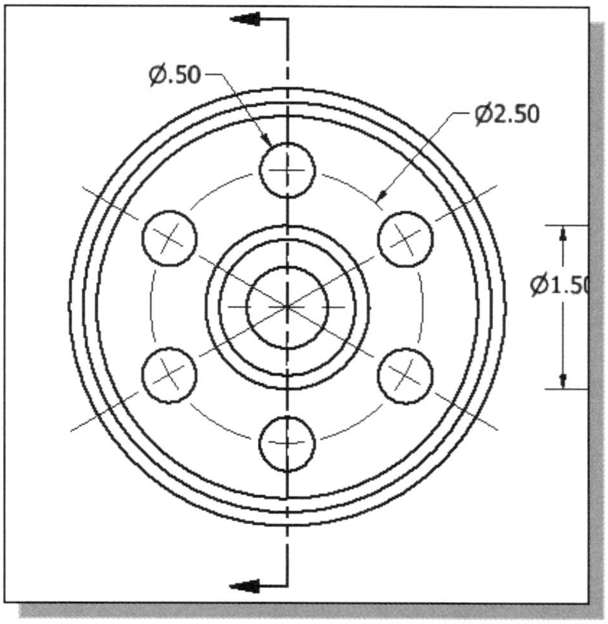

10. On your own, extend the segments of the centerlines so that they pass through the center of the base feature.

Complete the Drawing

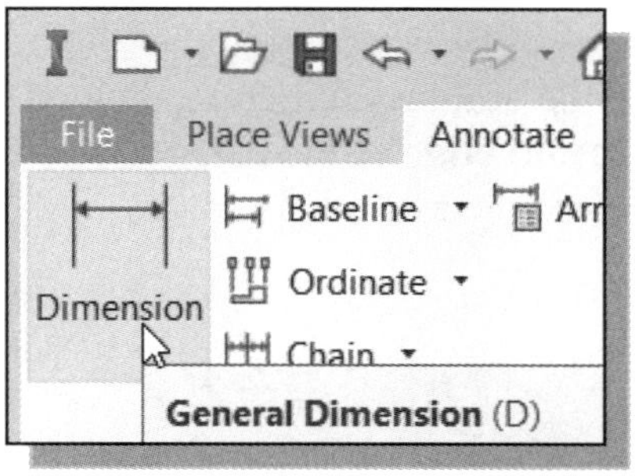

1. On your own, use the **General Dimension** command to create the angle dimension as shown in the figure below.

2. For the diameter of the circular pattern, set the options so that the **Arrowheads Inside** option is switched *ON* and the **Single Dimension Line** option is switched *OFF*.

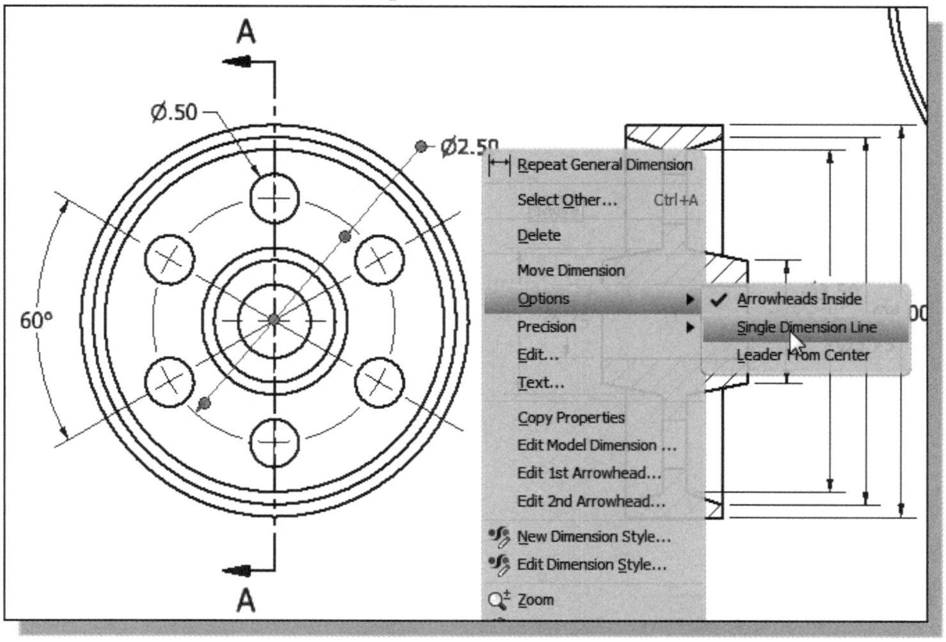

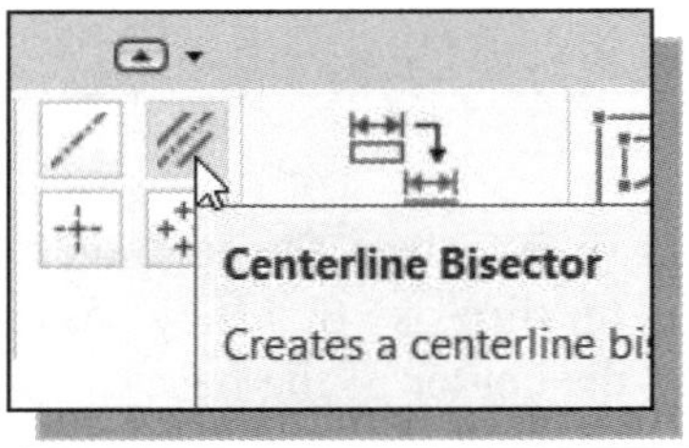

3. Select **Centerline Bisector** from the *Symbols* toolbar.

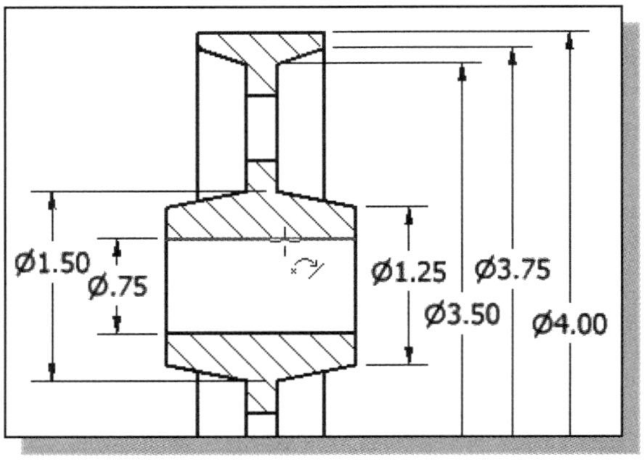

4. Inside the graphics window, click on the **top edge** of the 0.75 diameter hole as shown in the figure.

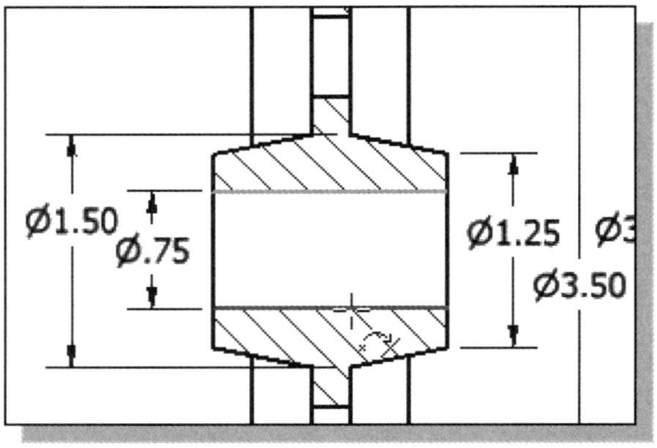

5. Inside the graphics window, click on the **bottom edge** of the 0.75 diameter hole as shown.

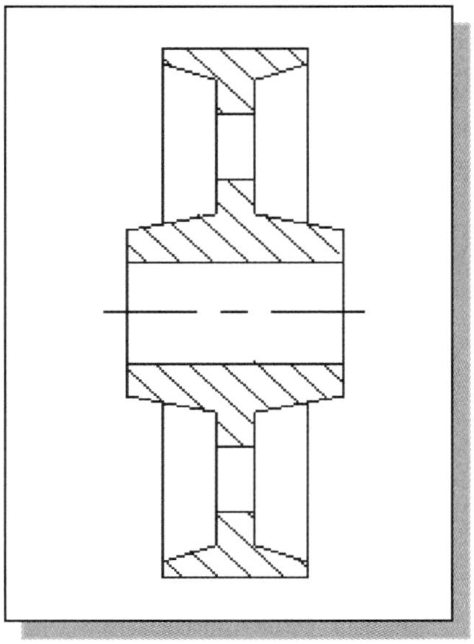

- ➢ Notice that a centerline is created through the center of the section view as shown in the figure.

6. Repeat the above step and create the centerlines through the other two holes.

7. On your own, complete the drawing and the title block so that the drawing appears as shown on the next page.

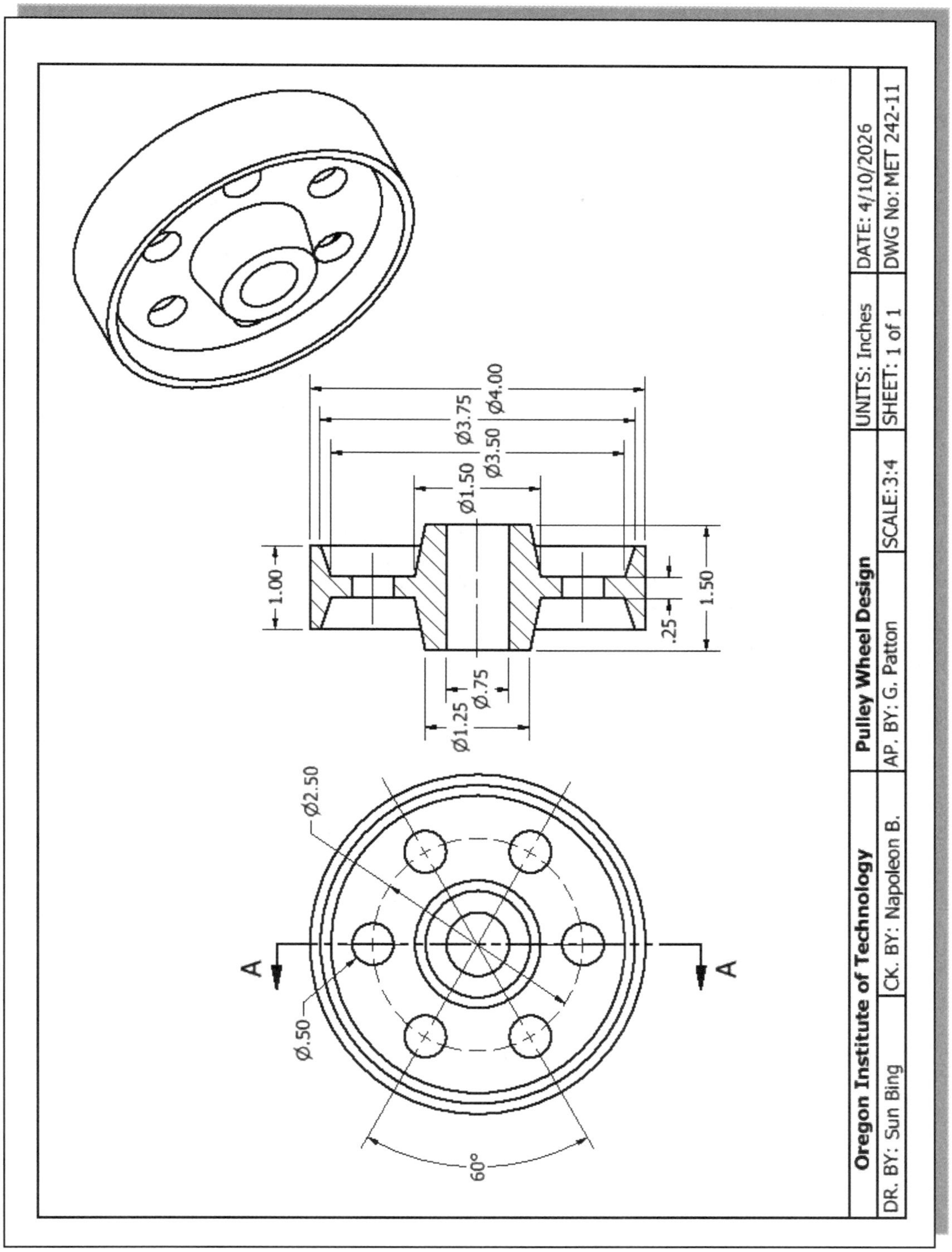

Ø2.50
Ø.50
60°
A
A
Ø1.25
Ø.75
Ø1.50
Ø3.50
Ø3.75
Ø4.00
1.00
.25
1.50
Oregon Institute of Technology
Pulley Wheel Design
UNITS: Inches
DATE: 4/10/2026
DR. BY: Sun Bing
CK. BY: Napoleon B.
AP. BY: G. Patton
SCALE:3:4
SHEET: 1 of 1
DWG No: MET 242-11

Additional Title Blocks

Drawing Paper and Border Sizes

The standard drawing paper sizes are as shown in the below tables. The edges of the title block border are generally 0.5 ~ 1 inches or 10~20 mm from the edges of the paper.

American National Standard	Suggested Border Size
A – 8.5″ X 11.0″	**A – 7.75″ X 10.25″**
B – 11.0″ X 17.0″	**B – 10.0″ X 16.0″**
C – 17.0″ X 22.0″	**C – 16.0″ X 21.0″**
D – 22.0″ X 34.0″	**D – 21.0″ X 33.0″**
E – 34.0″ X 44.0″	**E – 33.0″ X 43.0″**

International Standard	Suggested Border Size
A4 – 210 mm X 297 mm	**A4 – 190 mm X 276 mm**
A3 – 297 mm X 420 mm	**A3 – 275 mm X 400 mm**
A2 – 420 mm X 594 mm	**A2 – 400 mm X 574 mm**
A1 – 594 mm X 841 mm	**A1 – 574 mm X 820 mm**
A0 – 841 mm X 1189 mm	**A0 – 820 mm X 1168 mm**

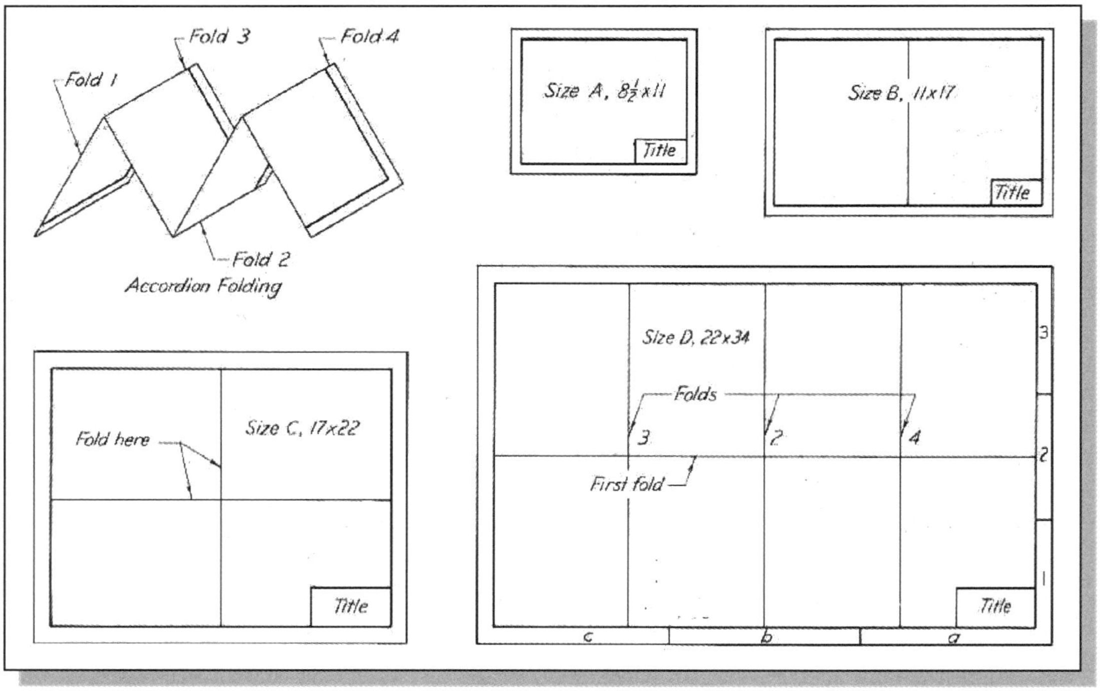

- **English Title Block** (For A size paper, main units are in inches.)

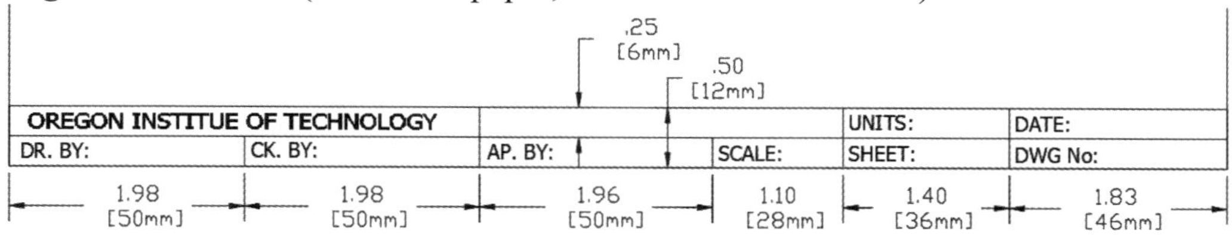

- **Metric Title Block** (For A4 size paper, main units are in mm.)

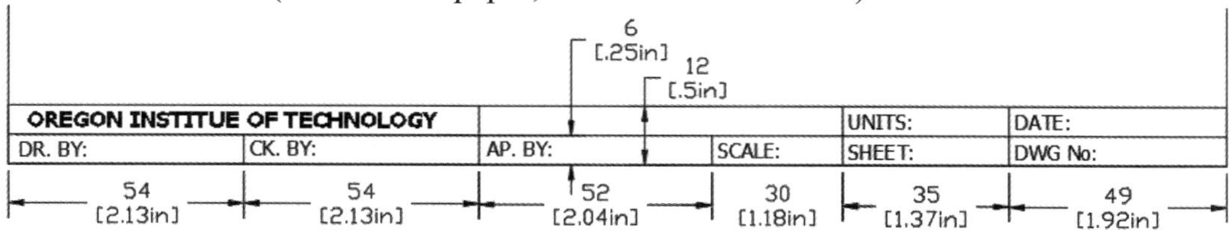

- **English Title Block** (Dimensions are in inches.)

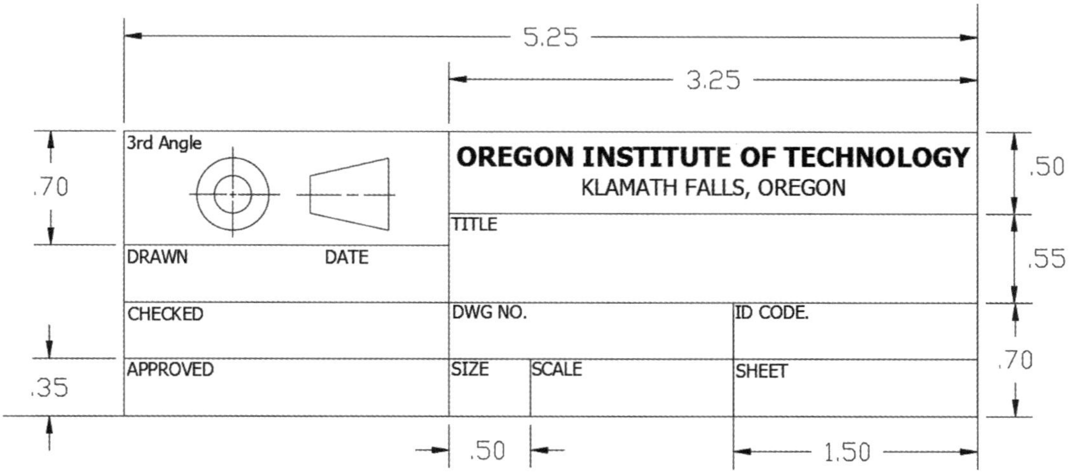

- **Metric Title Block** (Dimensions are in mm.)

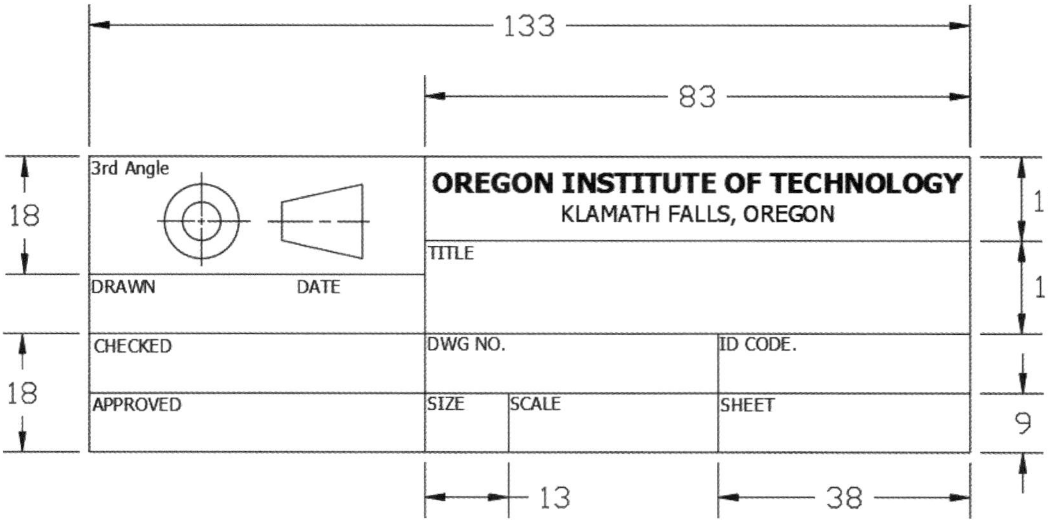

Review Questions: (Time: 25 minutes)

1. List the different symmetrical features created in the *Pulley* design.
2. What are the advantages of using a *drawing template*?
3. Describe the steps required in using the **Mirror Feature** command.
4. Why is it important to identify symmetrical features in designs?
5. When and why should we use the **Pattern** option?
6. What are the required elements in order to generate a sectional view?
7. How do we create a *Linear Diameter dimension* for a revolved feature?
8. What is the difference between *construction geometry* and *normal geometry*?
9. List and describe the different centerline options available in the *Drawing Annotation* panel.
10. What is the main difference between a sectional view and a projected view?

Exercises: Create the Solid models and the associated 2D drawings and save the exercises in the Chapter11 folder. (Time: 180 minutes.)

1. **Switch Base** (Dimensions are in inches.)

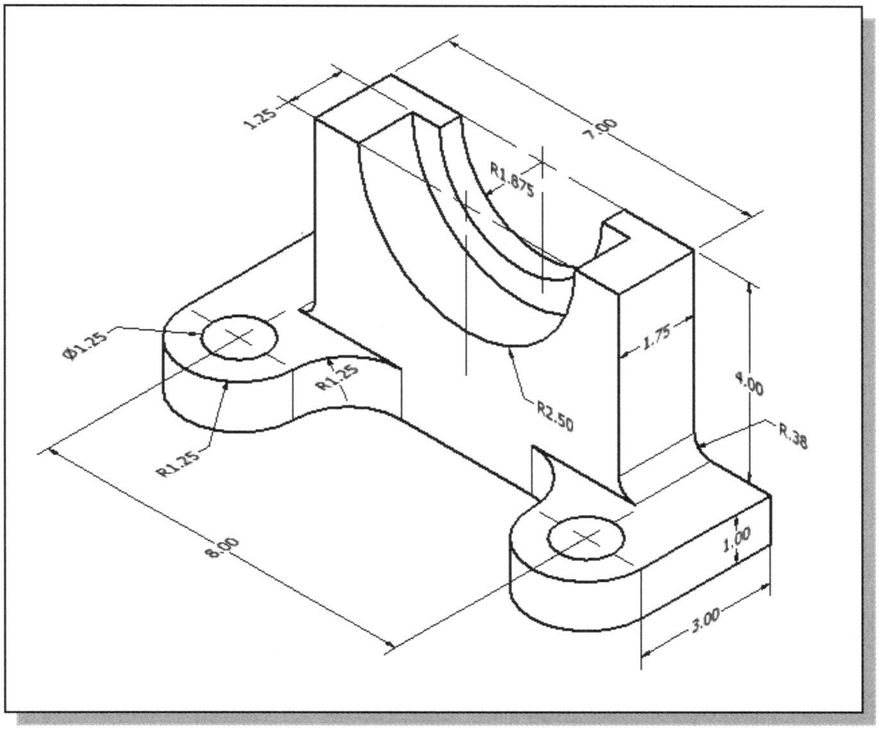

2. **Ratchet Plate** (Dimensions are in inches. Thickness: 0.125 inch.)

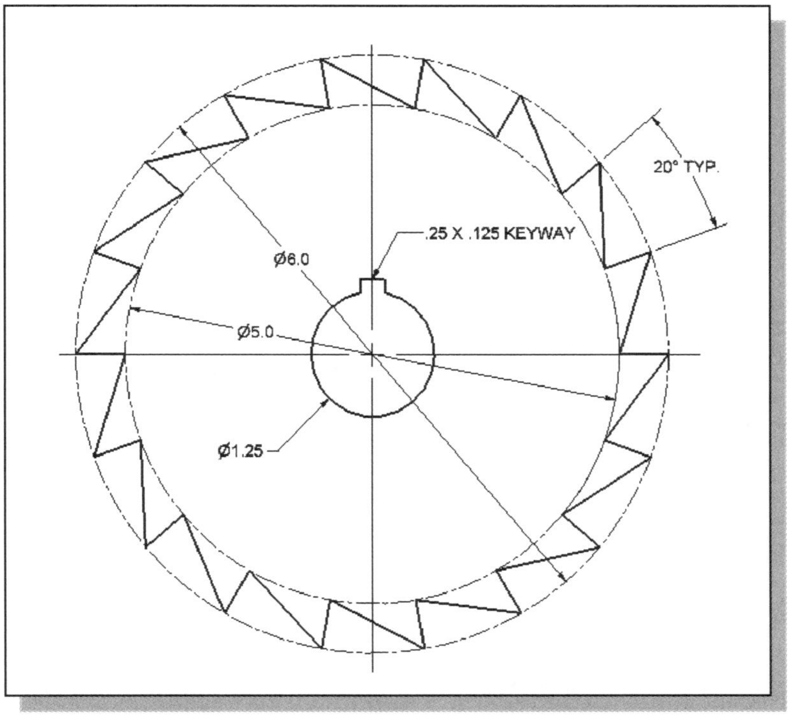

3. **Geneva Wheel** (Dimensions are in inches.)

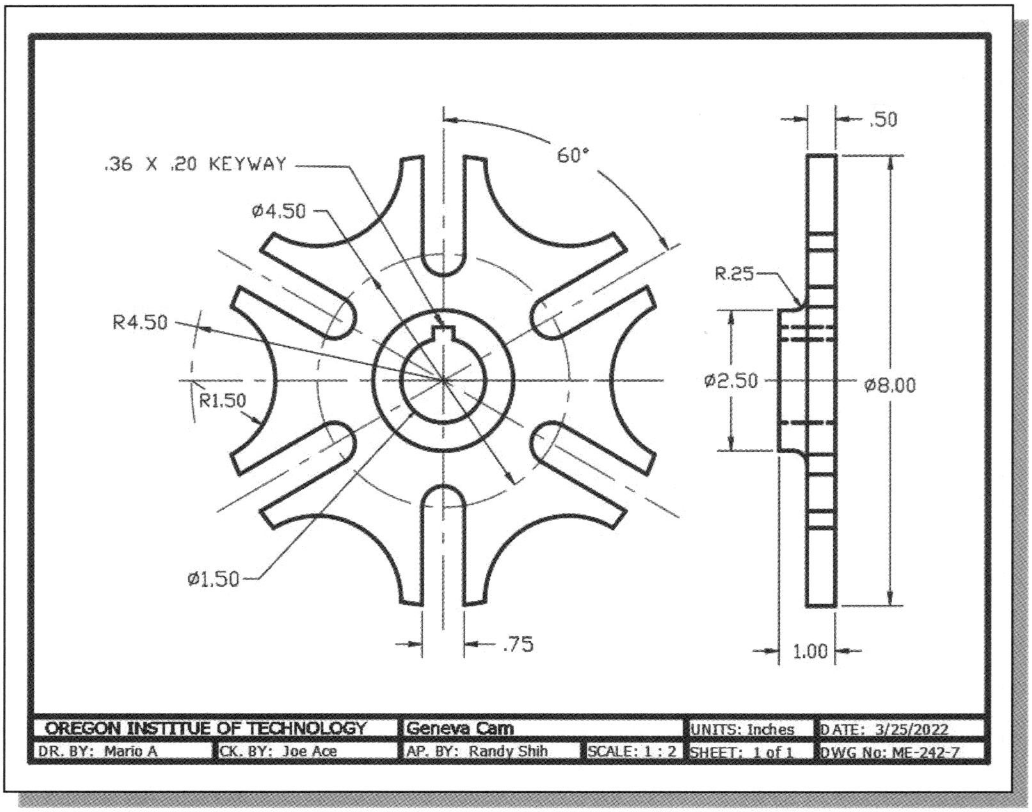

4. **Hub** (Dimensions are in inches.)

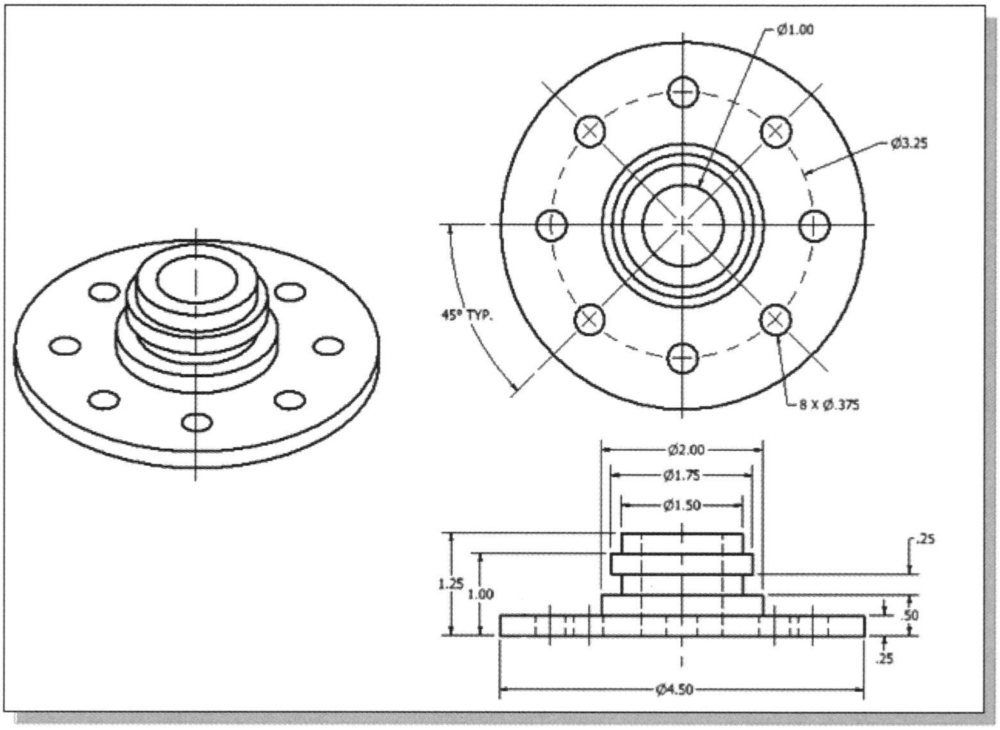

5. **Pulley Wheel** (Dimensions are in inches.)

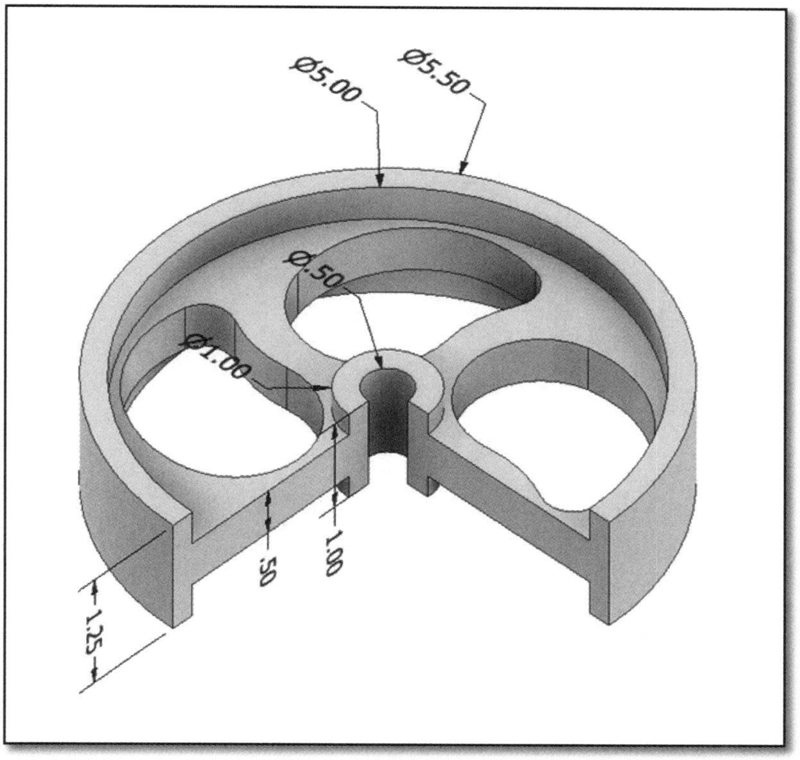

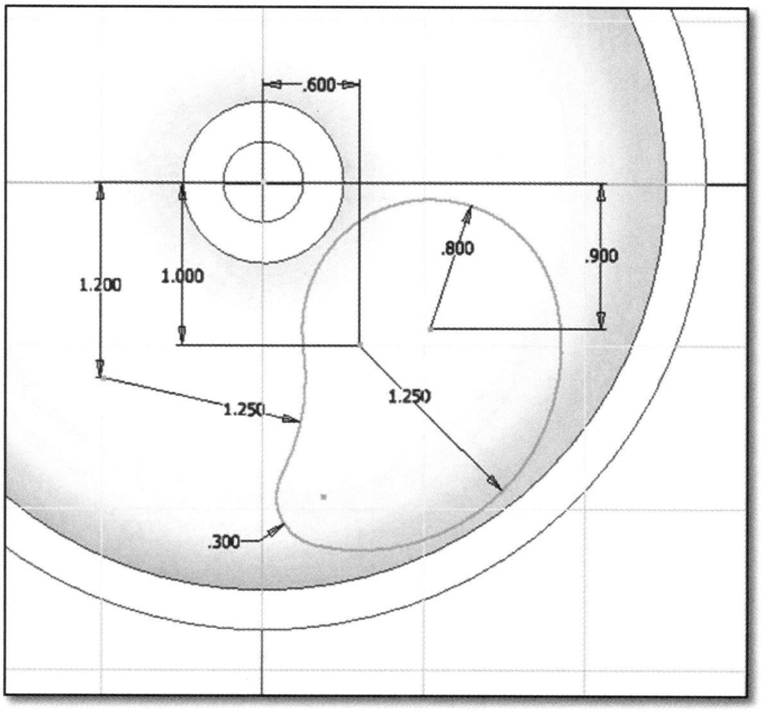

6. **Bearing Seat** (Dimensions are in inches.)

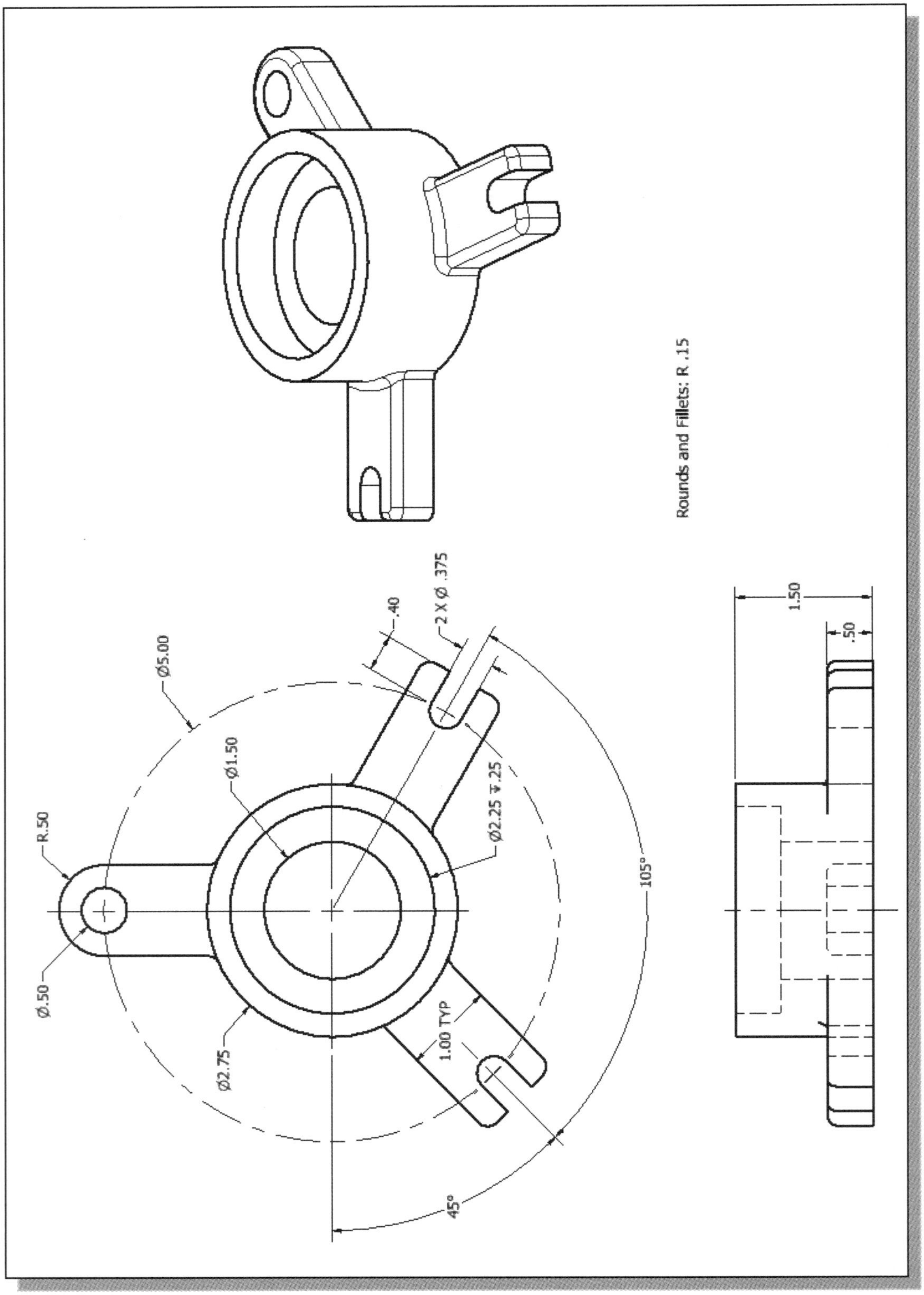

7. **Pivot Arm** (Dimensions are in inches.)

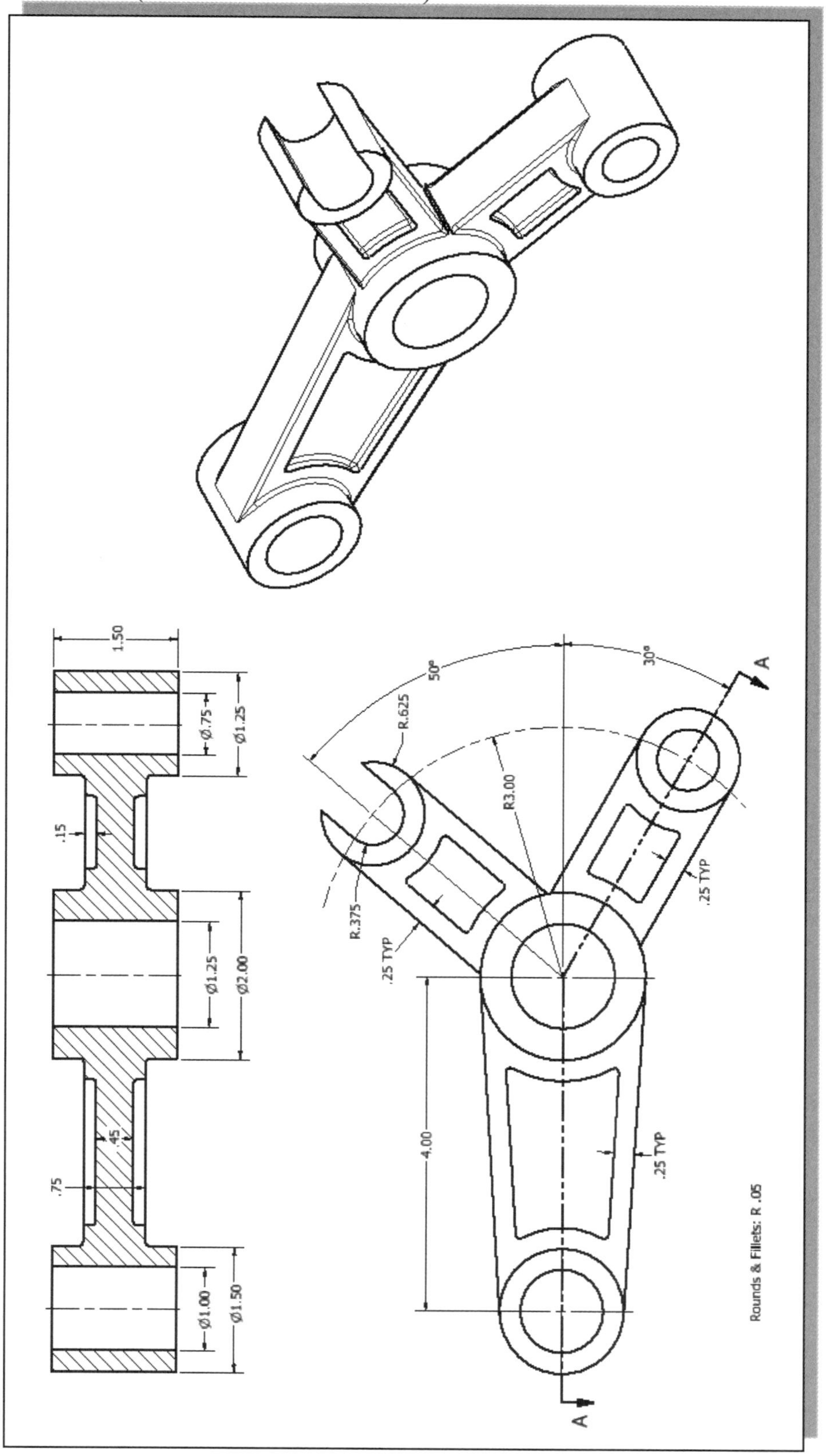

Notes:

Chapter 12
Advanced 3D Construction Tools

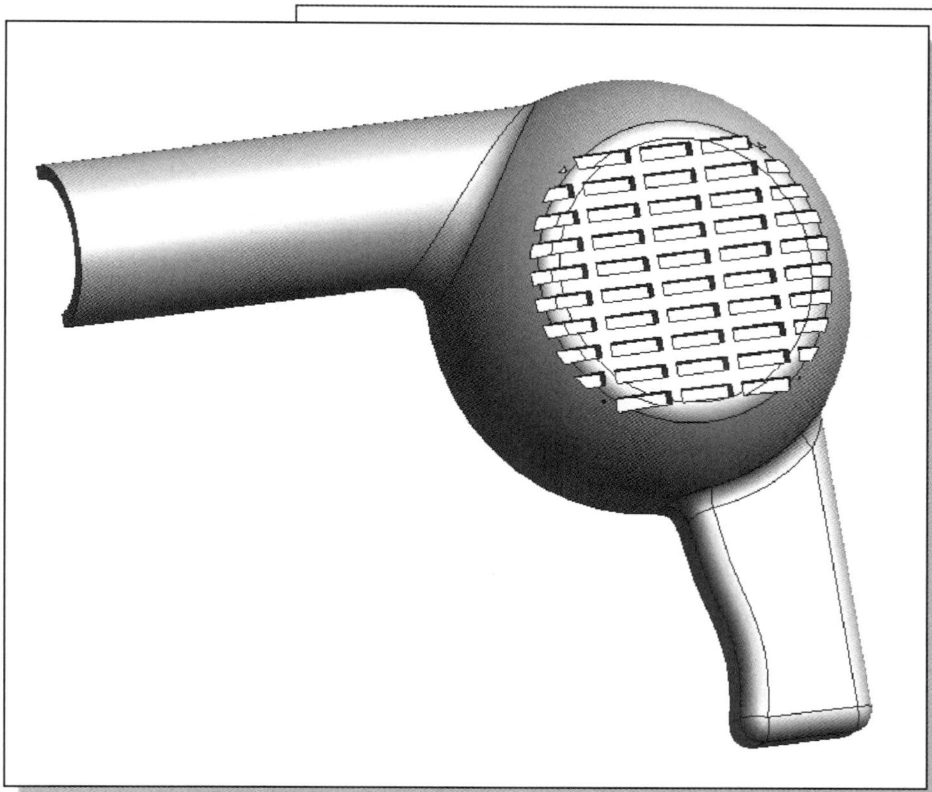

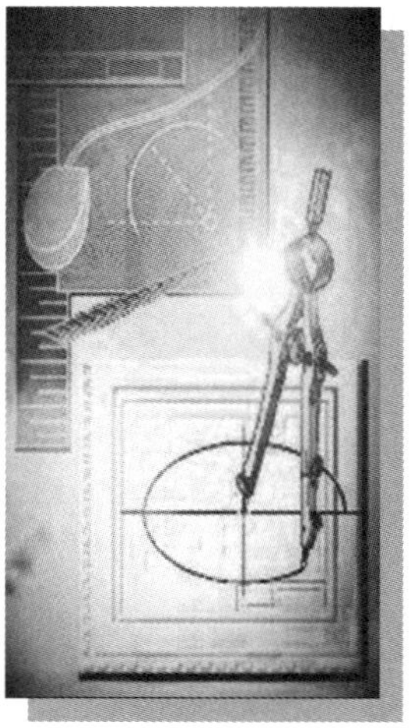

Learning Objectives

- Understand the Concepts Behind the Different 3D Construction Tools
- Set up Multiple Work Planes
- Create Swept Features
- Create Lofted Features
- Use the Shell Command
- Create 3D Rounds & Fillets

Autodesk Inventor Certified User Exam Objectives Coverage

Parametric Modeling Basics

Section 3: Sketches

Objectives: Apply *Geometric Constraints*; *Dimension* sketches.

Section 4: Parts

Objectives: Creating parts, Work Features, Pattern Features, Part Properties.

Introduction

Autodesk Inventor provides an assortment of three-dimensional construction tools to make the creation of solid models easier and more efficient. As demonstrated in the previous lessons, creating **extruded** features and **revolved** features are the two most common methods used to create 3D models. In this next example, we will examine the procedures for using the **Sweep** command, the **Loft** command, and the **Shell** command, and also for creating **3D rounds** and **fillets** along the edges of a solid model. These types of features are common characteristics of molded parts.

The **Sweep** option is defined as moving a cross-section through a path in space to form a three-dimensional object. To define a sweep in Autodesk Inventor, we define two sections: the trajectory and the cross-section.

The **Loft** command allows us to blend multiple profiles with varying shapes on separate planes to create complex shapes. Profiles are usually on parallel planes; however, non-parallel planes can also be used. We can use as many profiles as we wish but, to avoid twisting the loft shape, we should map points on each profile that align along a straight vector.

The **Shell** option is defined as hollowing out the inside of a solid, leaving a shell of specified wall thickness.

A Thin-Walled Design: Dryer Housing

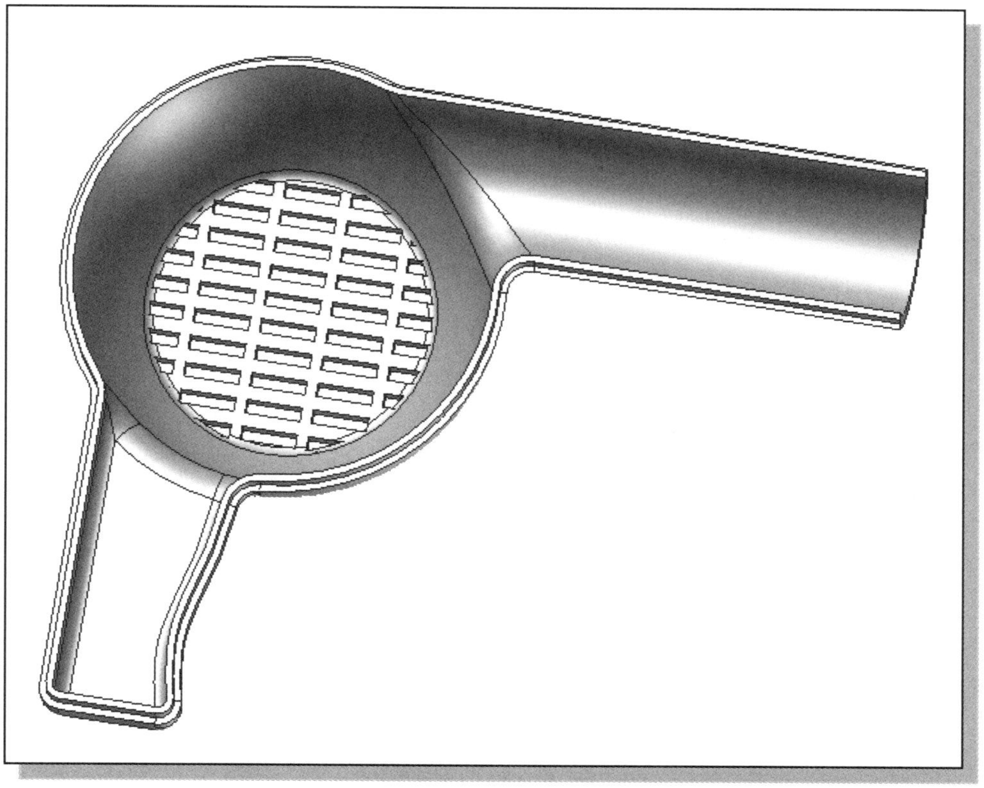

Modeling Strategy

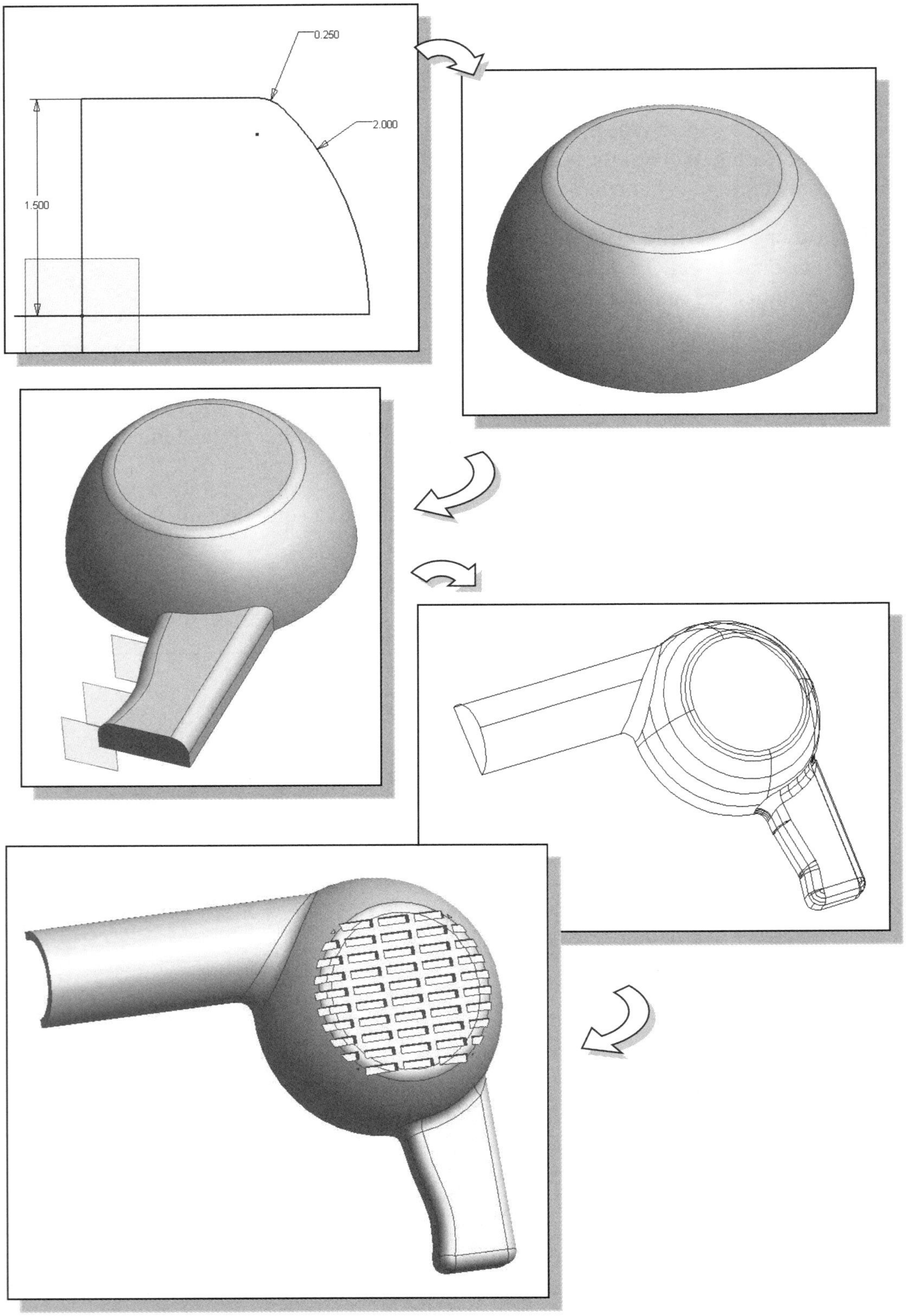

Starting Autodesk Inventor

1. Select the **Autodesk Inventor** option on the *Start* menu or select the **Autodesk Inventor** icon on the desktop to start Autodesk Inventor. The Autodesk Inventor main window will appear on the screen.

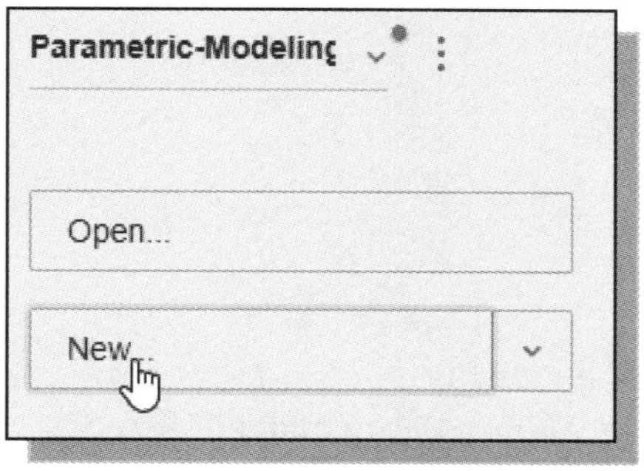

2. Select the **New File** icon with a single click of the left-mouse-button.

3. Select the **English** units set and in the *Part File* area, select **Standard(in).ipt**.

4. Click **Create** in the *New File* dialog box to start a new model.

Set Up the Display of the Sketch Plane

1. In the *Part Browser* window, click on the [**+**] symbol in front of the **Origin** feature to display more information on the feature.

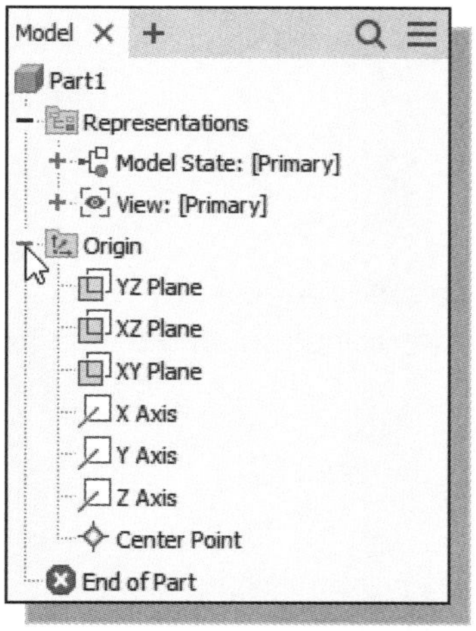

❖ In the *browser* window, notice a new part name appeared with seven work features established. The seven work features include three *work planes*, three *work axes* and a *work point*. By default, the three work planes and work axes are aligned to the **world coordinate system** and the work point is aligned to the *origin* of the **world coordinate system**.

2. Inside the *browser* window, move the cursor on top of the third work plane, **XY Plane**. Notice a rectangle, representing the work plane, appears in the graphics window.

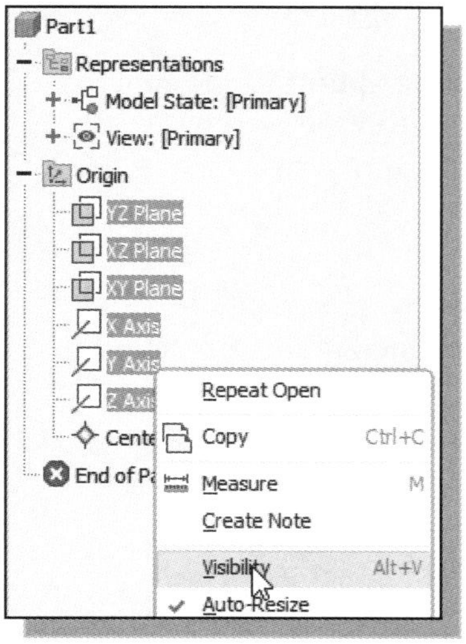

3. Inside the *browser* window, click once with the right-mouse-button on **XY Plane** to display the option menu. Click on **Visibility** to toggle *ON* the display of the plane.

4. On your own, repeat the above step and toggle ***ON*** the display of the **X** and **Y work axes** and the **Center Point** on the screen.

Create the 2D Sketch for the Base Feature

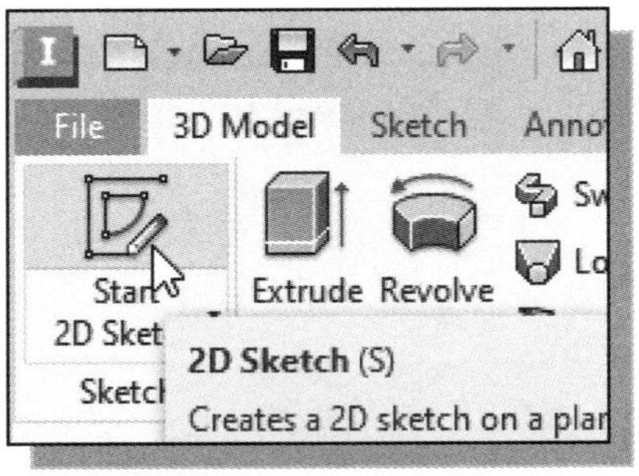

1. In the *Sketch* toolbar select the **Start 2D Sketch** command by left-clicking once on the icon.

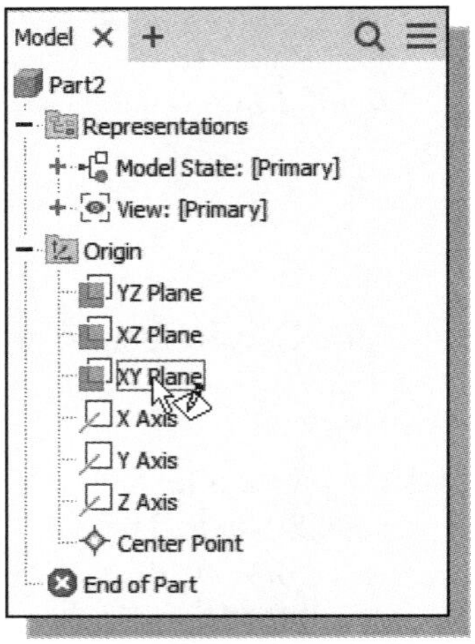

2. In the *Status Bar* area, the message "*Select face, work plane, sketch or sketch geometry.*" is displayed. Select the **XY Plane** by left-clicking once on any edge of the XY Plane in the graphics window or in the *browser* window as shown.

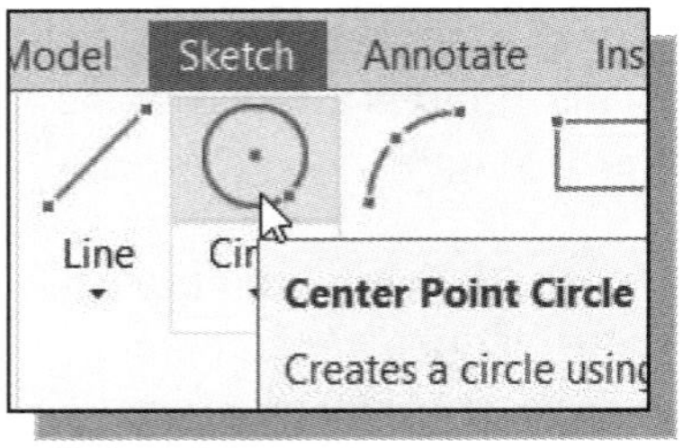

3. Select the **Center Point Circle** command by clicking once with the left-mouse-button on the icon in the *2D Sketch* panel.

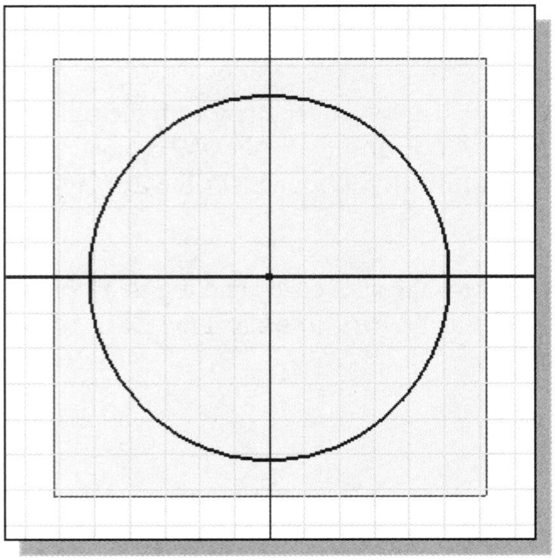

4. Pick the **center point** as the center location of the circle.

5. Create a **circle** of arbitrary size, with its center aligned to the projected center point as shown.

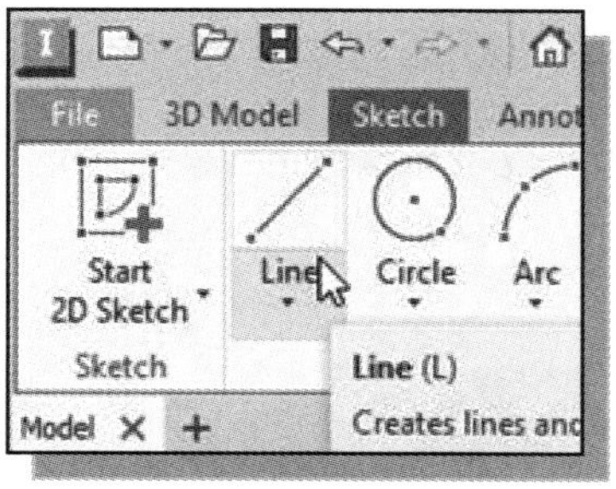

6. Click on the **Line** icon in the *2D Sketch* panel.

7. Create the three line segments as shown. The lines are either horizontal or vertical with the **lower left corner** aligned to the **Center Point**.

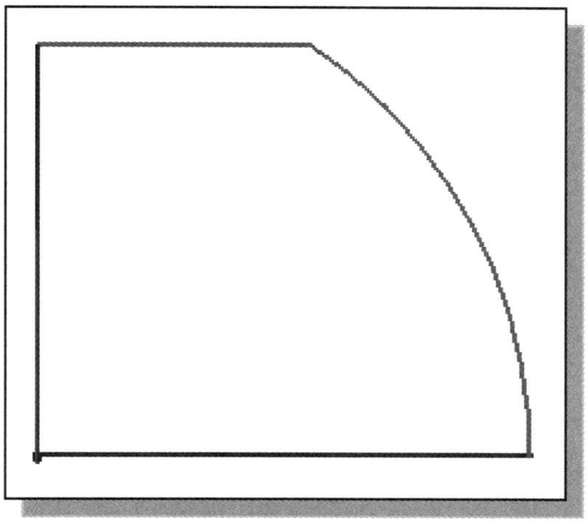

8. Modify the sketch by trimming the circle as shown.

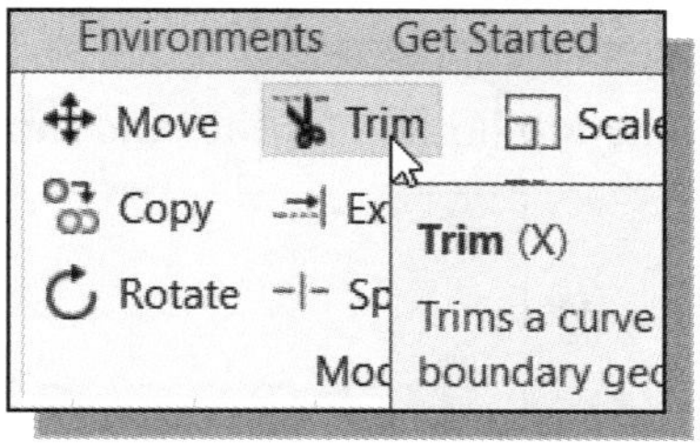

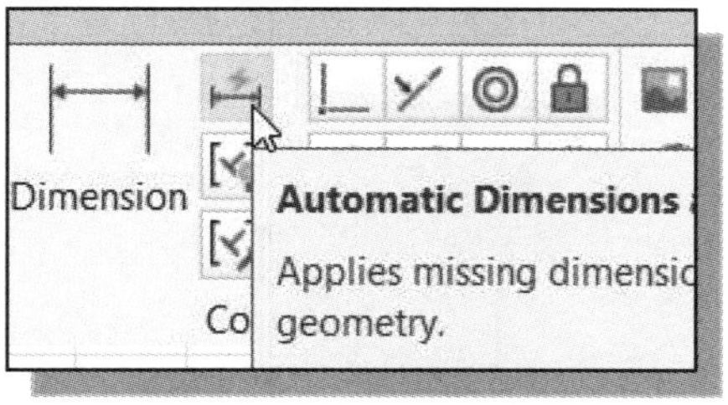

9. On your own, use the **Auto Dimension** command to create and modify the dimensions as shown in the figure below.
(Hint: Modify the radius dimension first.)

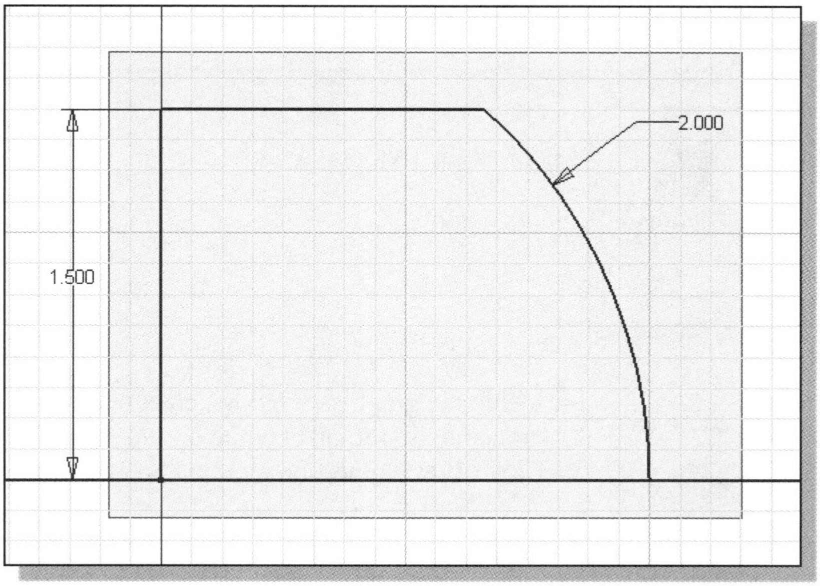

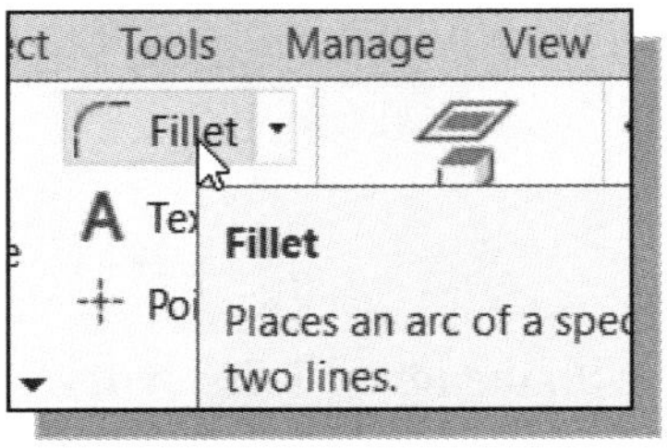

10. Click on the **Fillet** icon in the *2D Draw* panel.

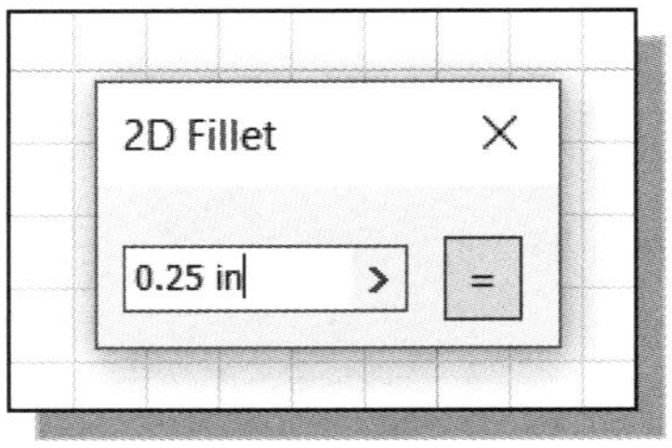

11. In the *2D Fillet* dialog box, set the ***radius*** to **0.25**.

12. Select the top horizontal line and the arc to create a rounded corner as shown.

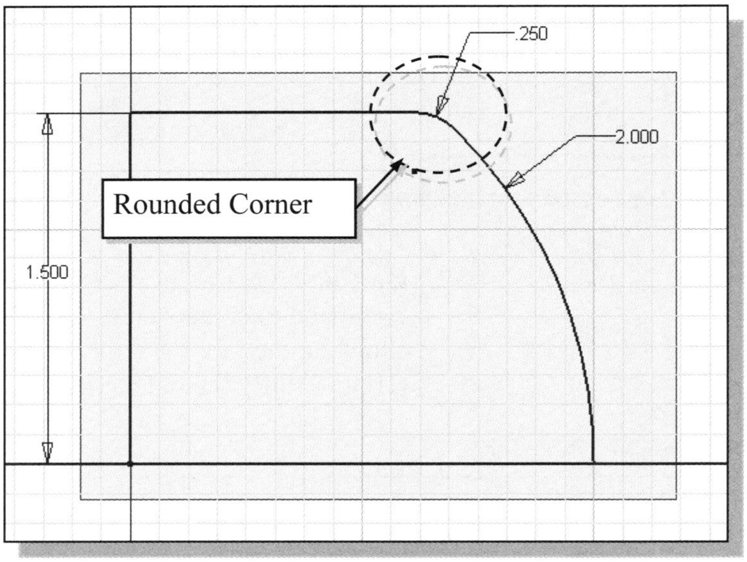

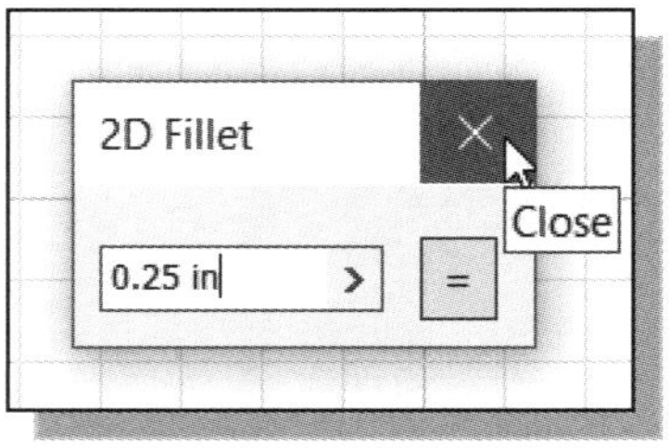

13. Click the [**X**] icon to close the *2D Fillet* dialog box.

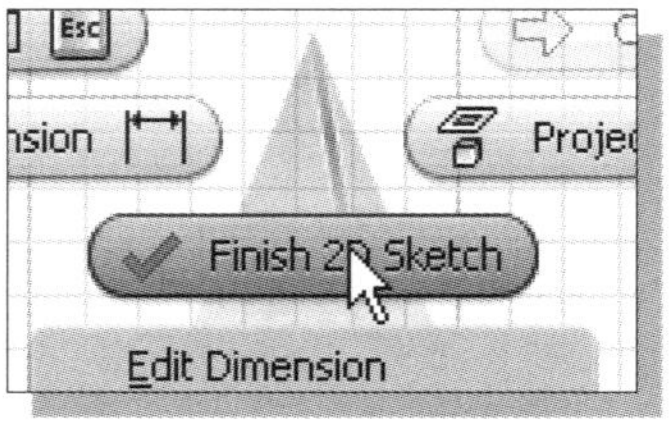

14. Inside the graphics window, click once with the right-mouse-button to display the option menu. Select **Finish Sketch** in the pop-up menu to end the Sketch option.

Create a Revolved Feature

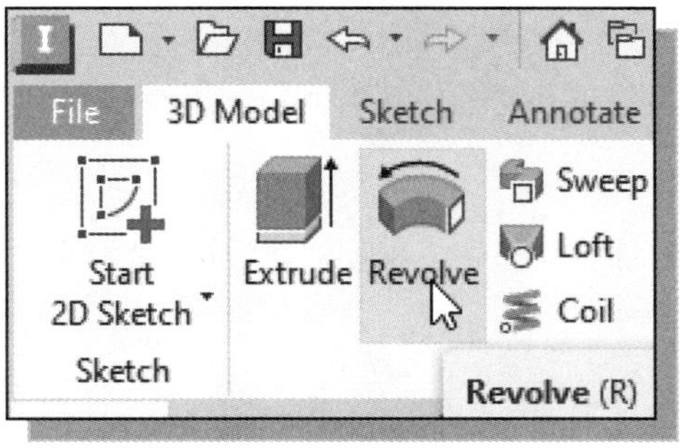

1. In the *3D Model Tab* select the **Revolve** command by clicking the left-mouse-button on the icon.

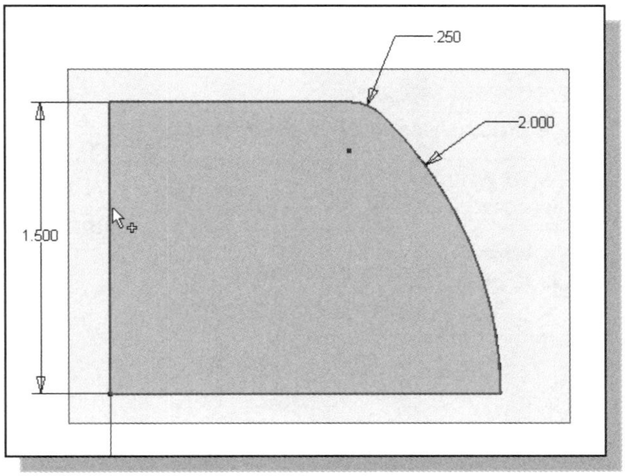

2. In the *Revolve* dialog box, the **Axis** button is activated indicating Autodesk Inventor expects us to select the revolution axis for the revolved feature. Select the **vertical edge** of the sketch as the axis of rotation as shown.

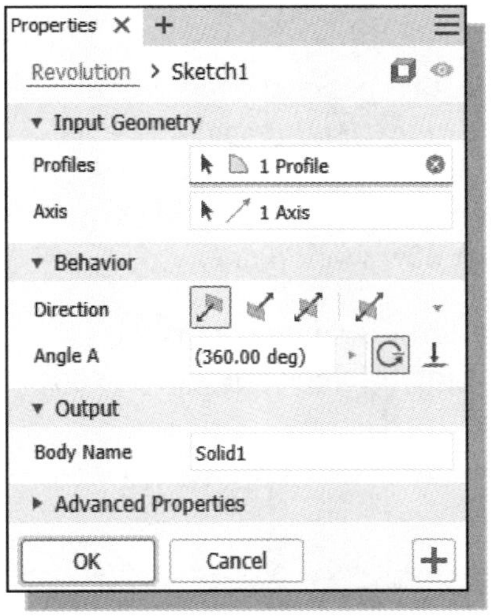

3. In the *Revolve* dialog box, set the termination *Extents* option to **Full** as shown.

4. Click on the **OK** button to accept the settings and create the revolved feature.

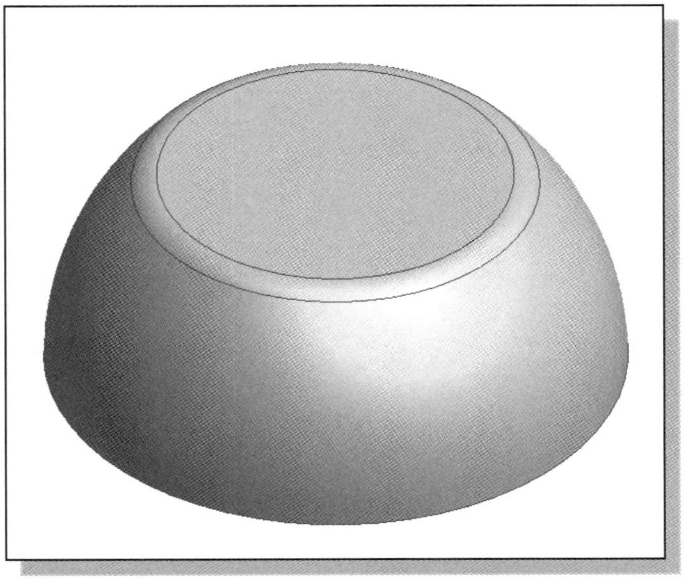

Create Offset Work Planes

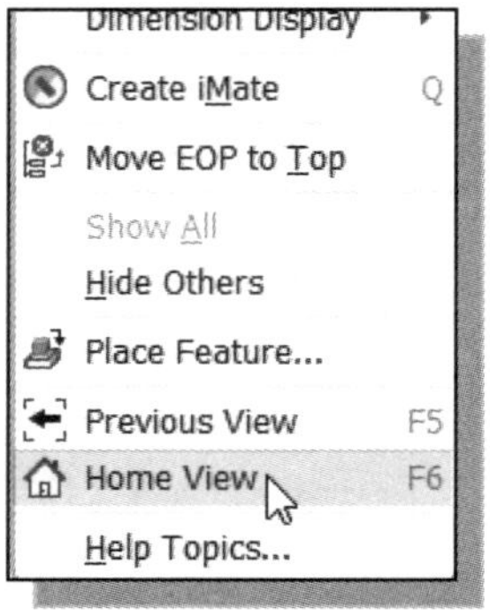

1. Inside the *graphics window*, right-click to bring up the option menu.

2. Select **Home View** in the option list to adjust the display of the 3D object on the screen.

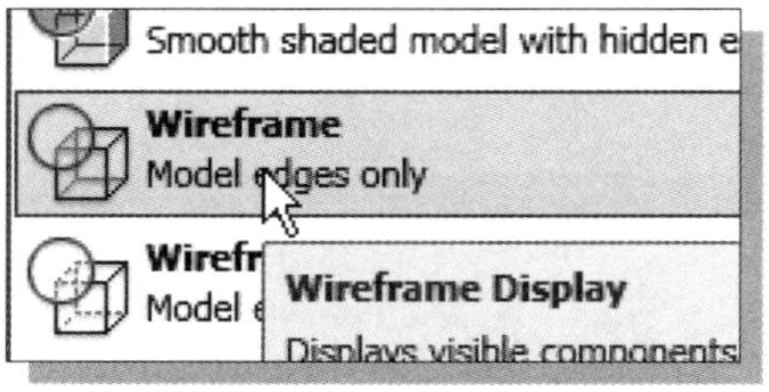

3. On your own, switch the display mode to **Wireframe** by choosing the corresponding icon under the **View** tab.

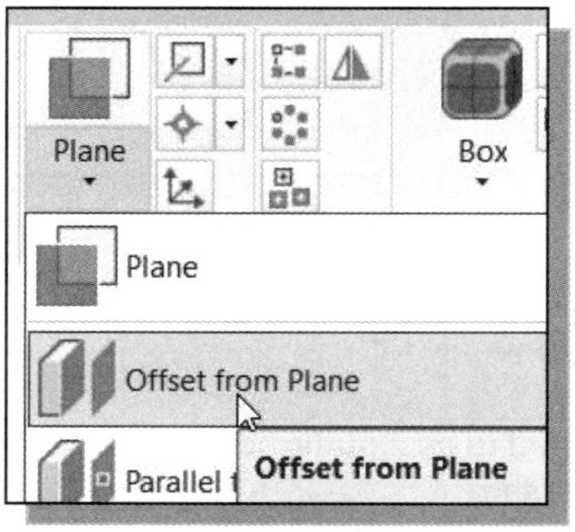

4. Switch back to the **3D Model** tab and select the **Offset from Plane** command by left-clicking on the icon.

5. In the *Status Bar* area, the message "*Define work plane by highlighting and selecting geometry*" is displayed. Autodesk Inventor expects us to select any existing geometry, which will be used as a reference to create the new work plane.

6. Select the **XY Plane** in the graphics window, or in the *browser* window, as the reference plane. Left-click once to set the XY Plane as the reference of a new work plane.

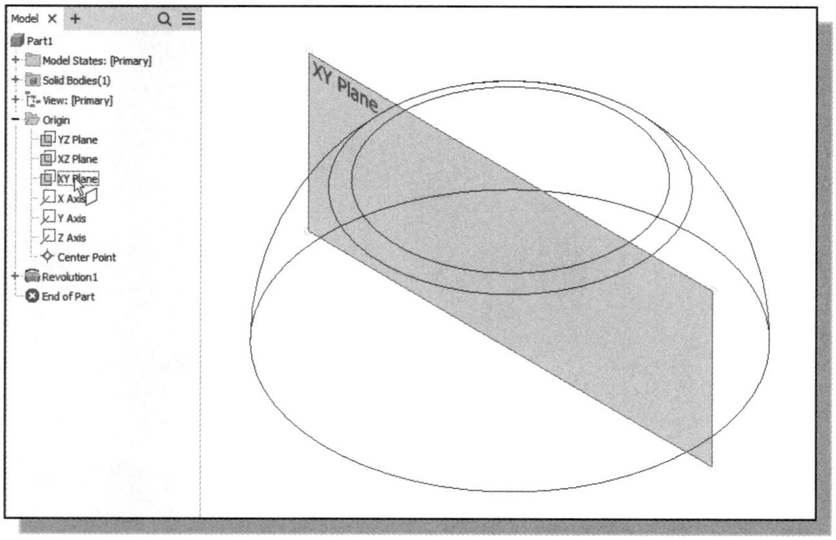

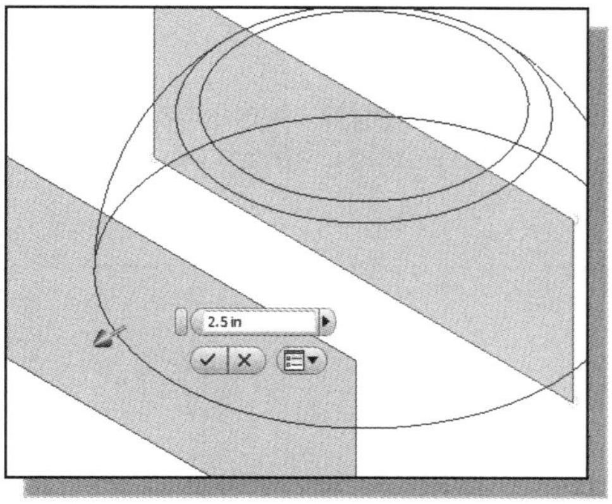

7. In the ***Offset*** pop-up window, enter **2.5** as the offset distance for the new work plane.

8. Click on the **check mark** button to accept the setting.

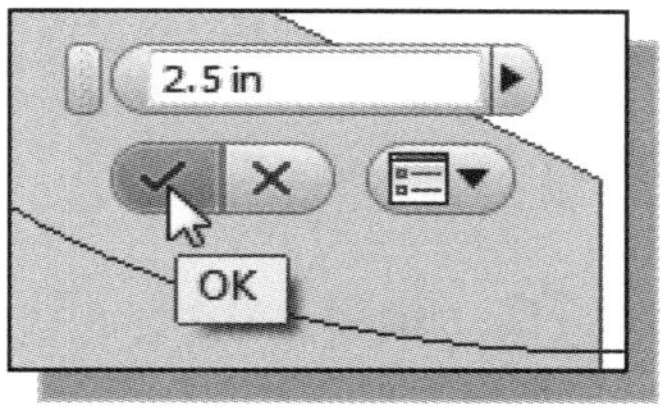

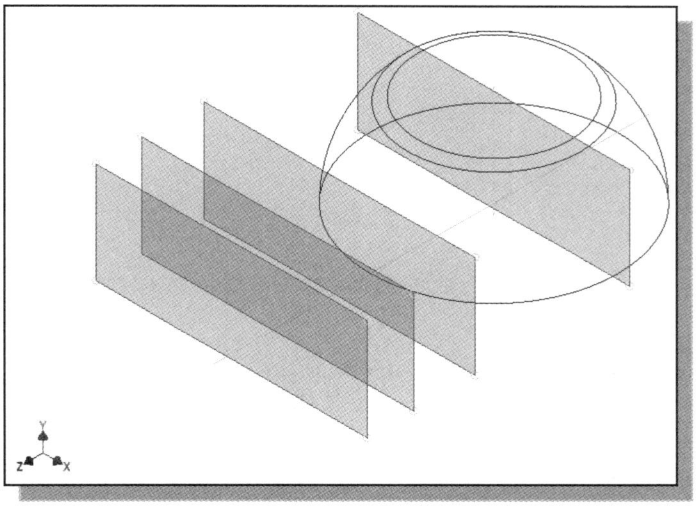

9. On your own, repeat the above steps and create two additional work planes that are **3.5** inches and **4.25** inches away from the XY Plane.

Start 2D Sketches on the Work Planes

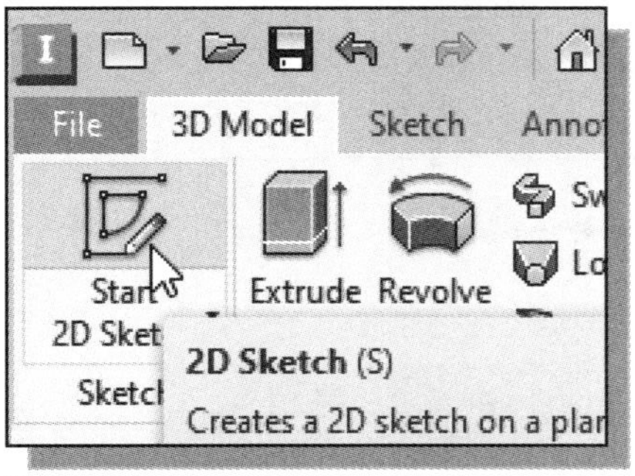

1. In the *3D Model tab* select the **Start 2D Sketch** command by left-clicking once on the icon.

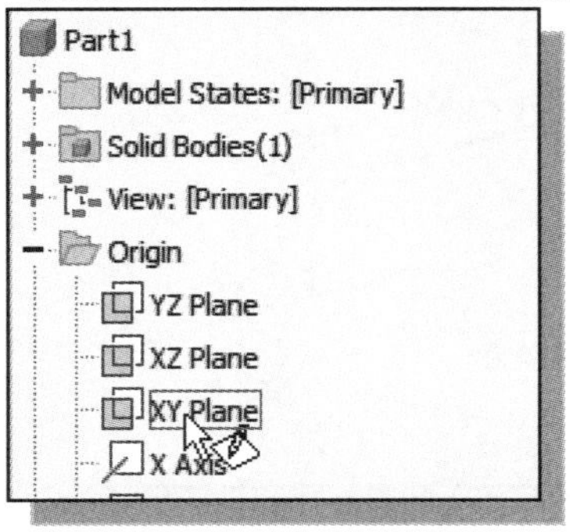

2. In the *Status Bar* area, the message "*Select face, work plane, sketch or sketch geometry*" is displayed. Select the **XY Plane** by left-clicking once on any edge of the XY Plane in the graphics window.

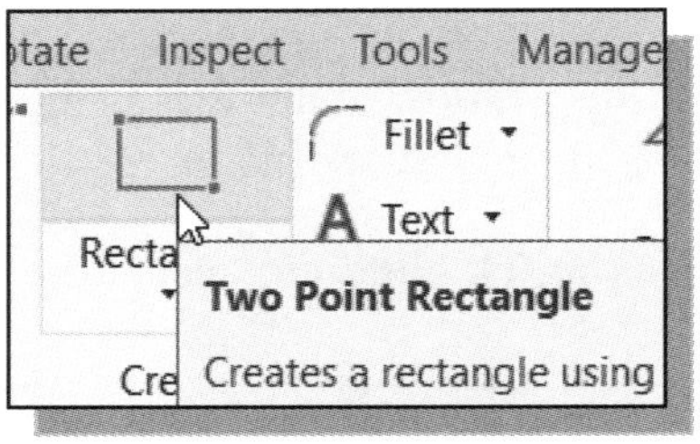

3. Select the **Two point rectangle** command by clicking once with the left-mouse-button on the icon in the *Draw* panel.

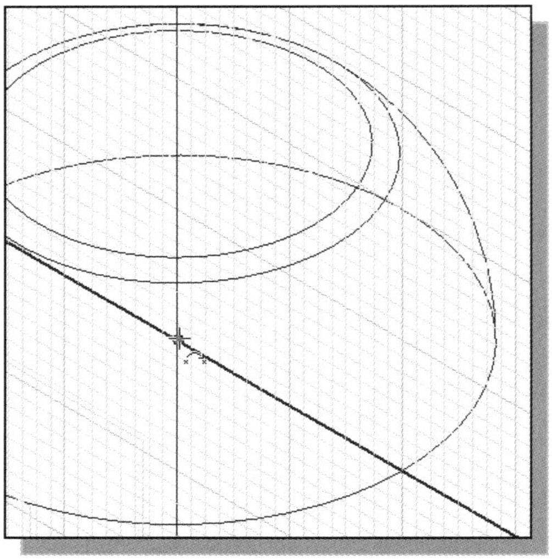

4. Click on the **Center Point** to align the first corner of the rectangle.

5. Create a rectangle of arbitrary size by selecting a location that is toward the right side of the graphics window as shown.

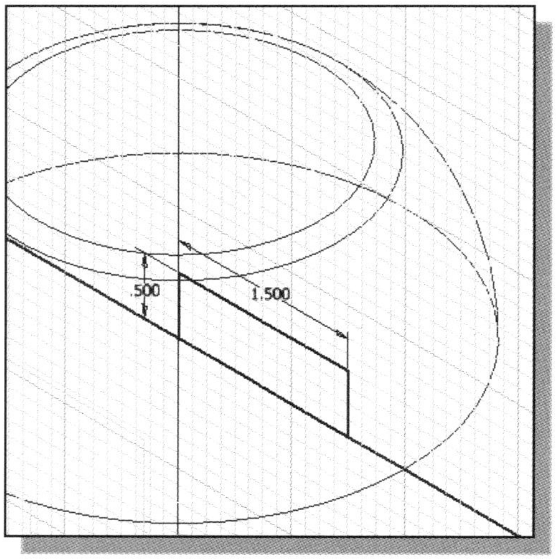

6. On your own, create and modify the two dimensions, **1.5 X 0.5**, as shown in the figure.

7. On your own, create two rounded corners (**radius 0.25**) as shown.

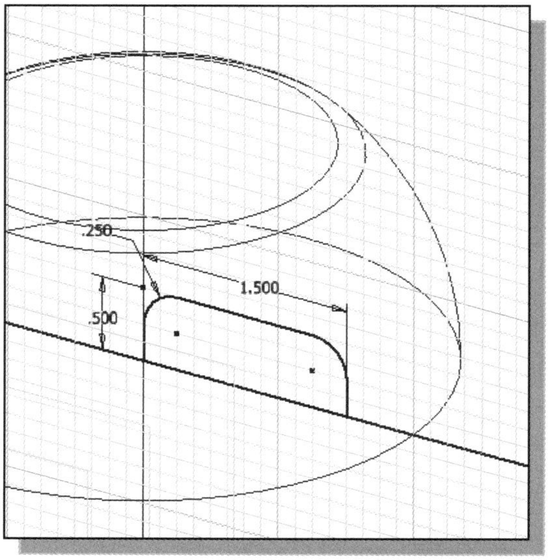

8. Inside the graphics window, click once with the right-mouse-button to display the option menu. Select **Finish 2D Sketch** in the pop-up menu to end the Sketch option.

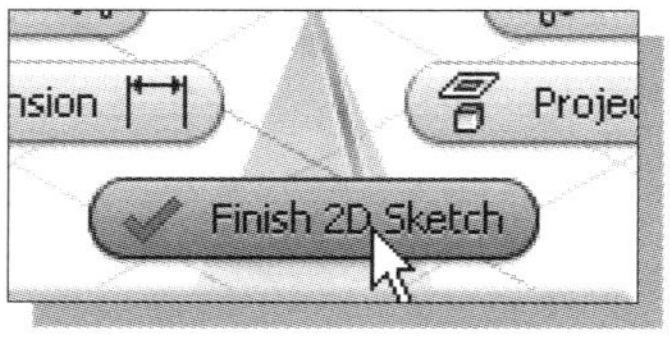

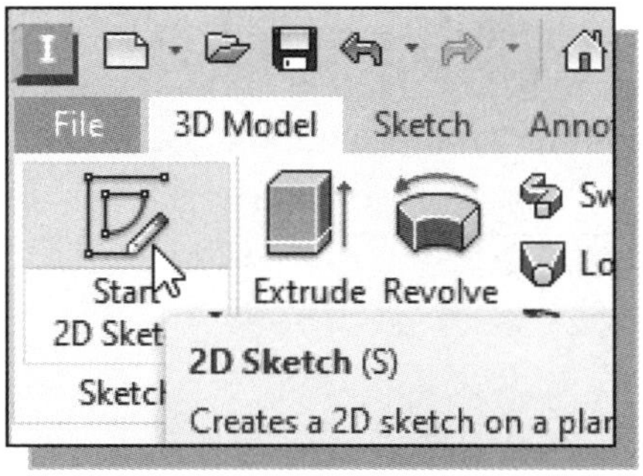

9. In the *3D Model tab* select the **Start 2D Sketch** command by left-clicking once on the icon.

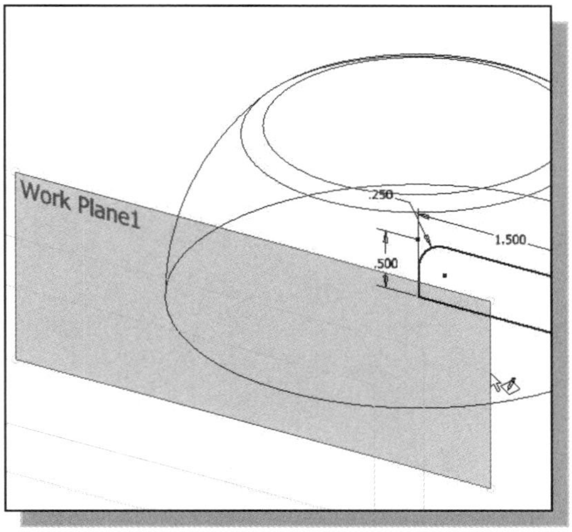

10. In the *Status Bar* area, the message "*Select face, work plane, sketch or sketch geometry*" is displayed. Select **WorkPlane1**, by left-clicking once on any edge of the first offset work plane in the graphics window as shown.

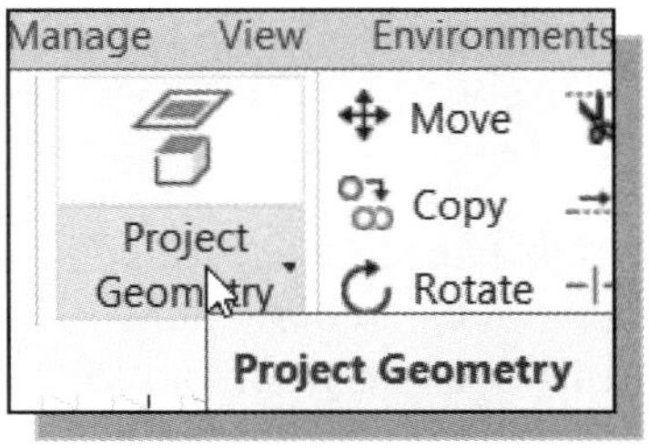

11. In the *Ribbon* toolbar area, select **Project Geometry**.

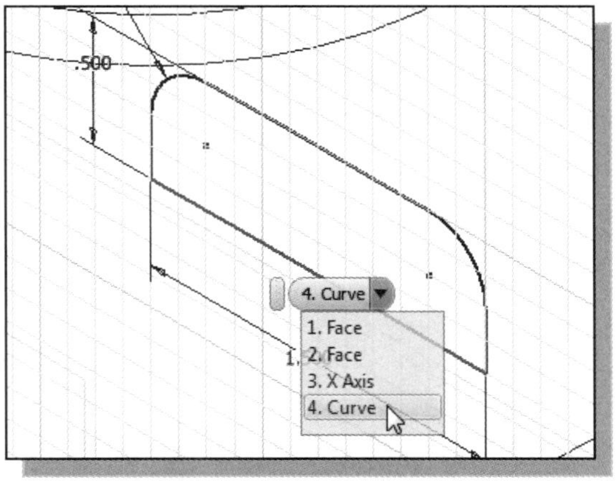

12. Select the bottom edge of the 2D sketch we just created, as shown in the figure. (Hint: Use the **Select** option if necessary.)

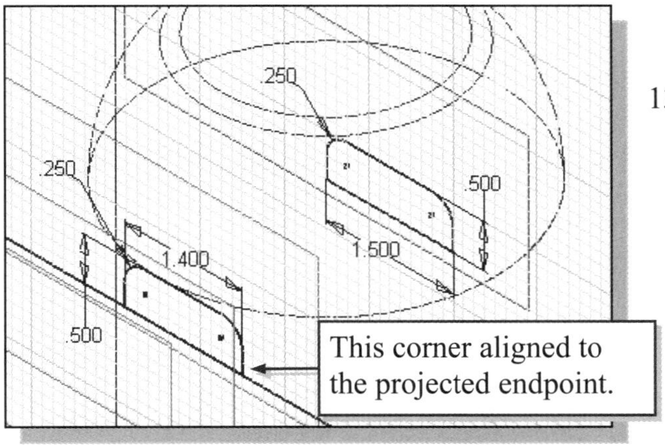

13. On your own, create and modify the 2D sketch as shown in the figure. (The lower right corner of the sketch is aligned to the lower right corner of the previous sketch.)

14. On your own, repeat the above steps and create two additional sketches on **WorkPlane2** and **WorkPlane3** as shown in the figure below. Note that the lower right corners of the four sketches are aligned in the Z-direction through the use of the projected bottom edge.

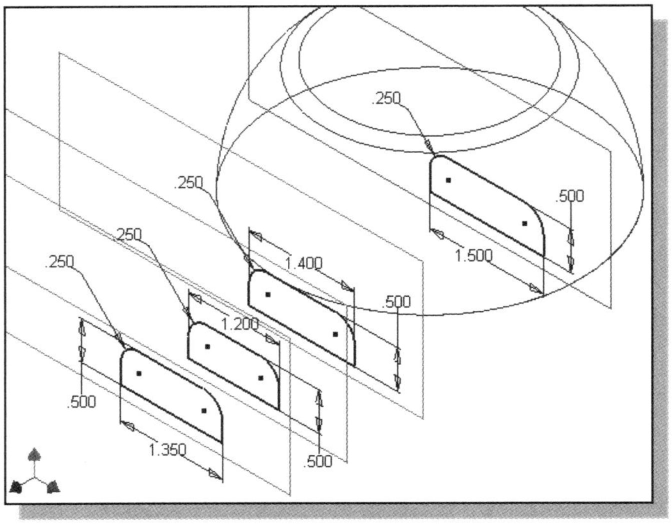

Create a Lofted Feature

The **Loft** option allows us to blend multiple profiles with varying shapes on separate planes to create complex shapes. Profiles are usually on parallel planes, but any non-perpendicular planes can also be used. We can use as many profiles as we wish, but to avoid twisting the loft shape, we should map points on each profile that align along a straight vector.

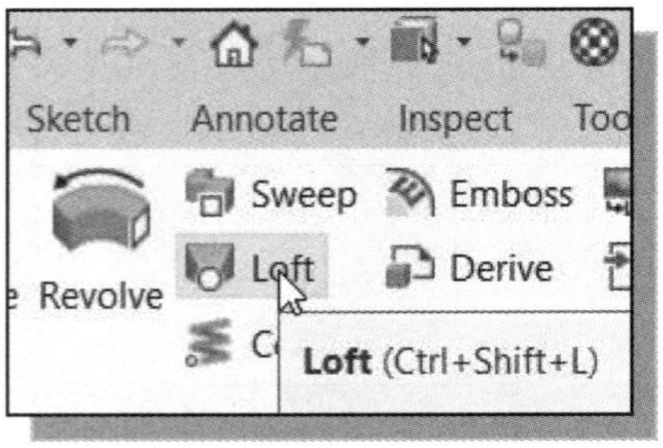

1. In the *3D Model tab*, select the **Loft** command by left-clicking on the icon.

2. In the *Loft* dialog box, the **Curves** option is activated. Autodesk Inventor expects us to select a number of existing profiles, which will be used to create the lofted feature.

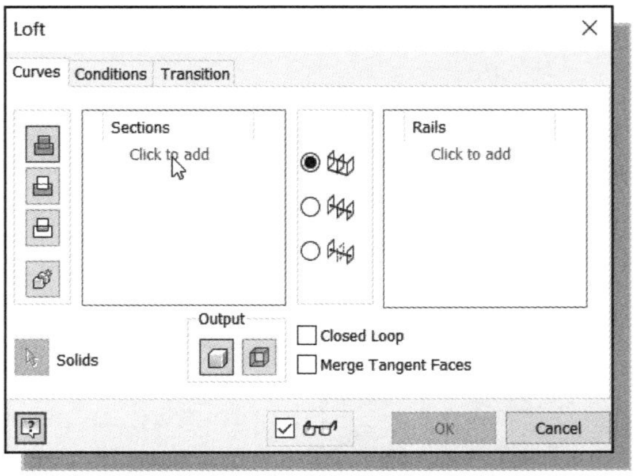

3. Click inside the **Sections** box to activate the selection of sketches.

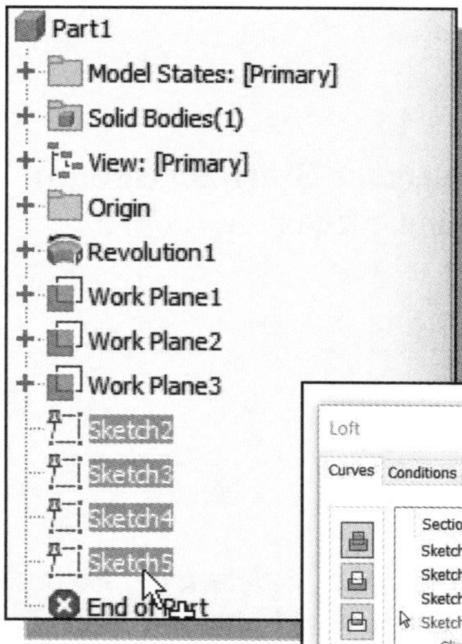

4. Pick the four sketched sections, in the order that they were created, by selecting the sketches in the ***Graphics window*** or ***Model Browser***. Three arrows are displayed showing the blending direction of the sketches.

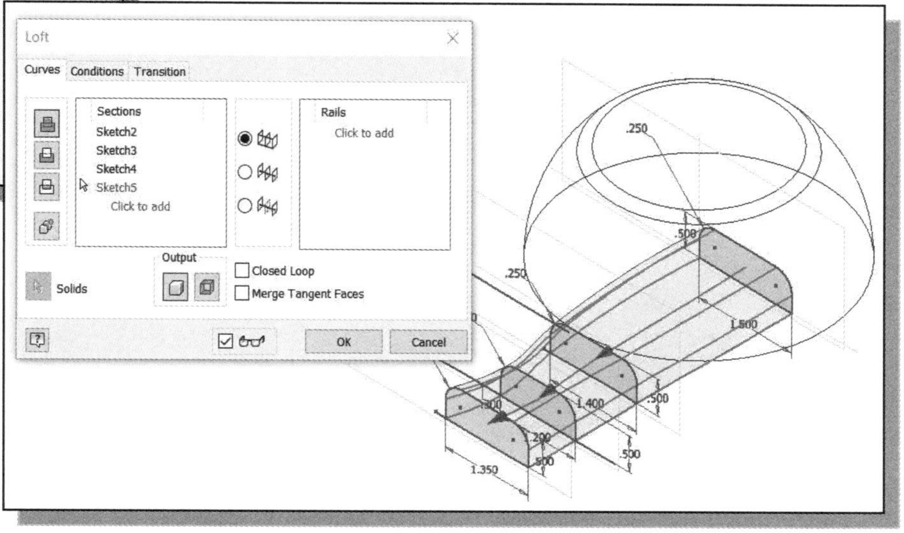

5. Click on the **OK** button to accept the settings and create the lofted feature.

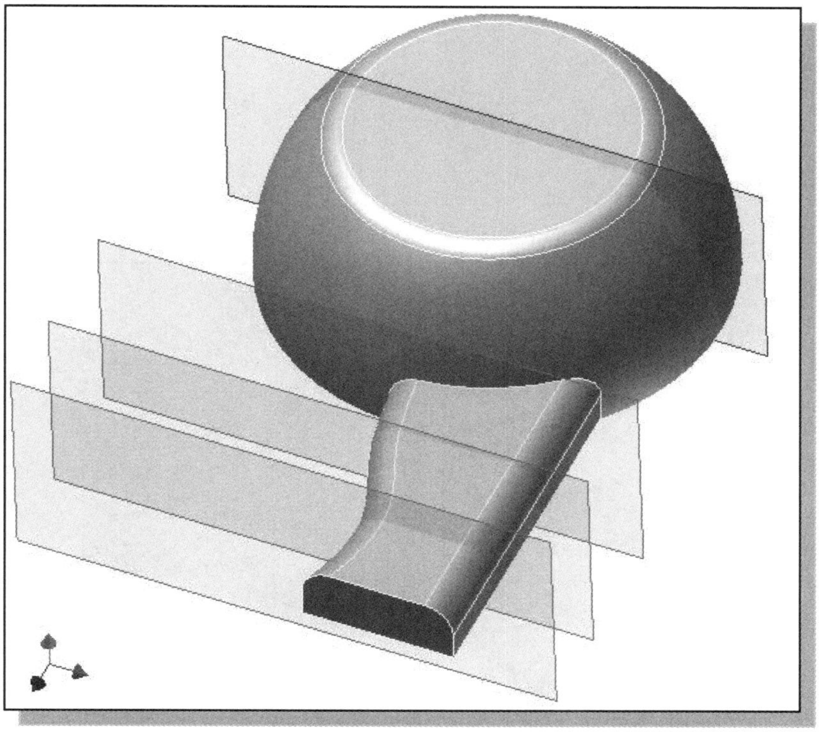

Create an Extruded Feature

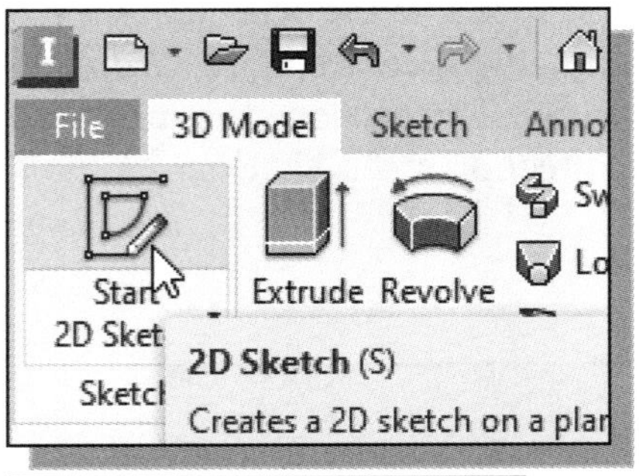

1. In the *3D Model tab* select the **Start 2D Sketch** command by left-clicking once on the icon.

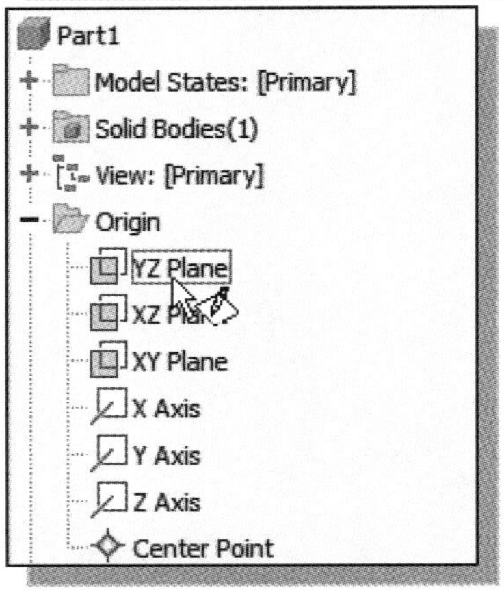

2. In the *Status Bar* area, the message "*Select face, work plane, sketch or sketch geometry.*" is displayed. Select the **YZ Plane**, by left-clicking once on the YZ Plane, in the *browser* window.

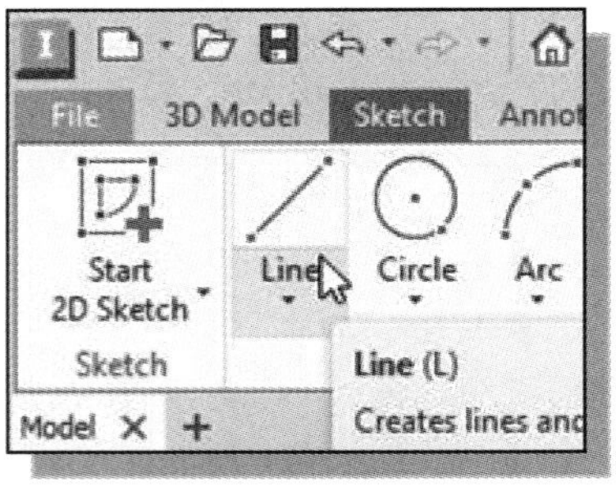

3. Click on the **Line** icon in the *2D Sketch* panel.

4. Create a line that is parallel (nearly aligned) to the horizontal axis as shown.

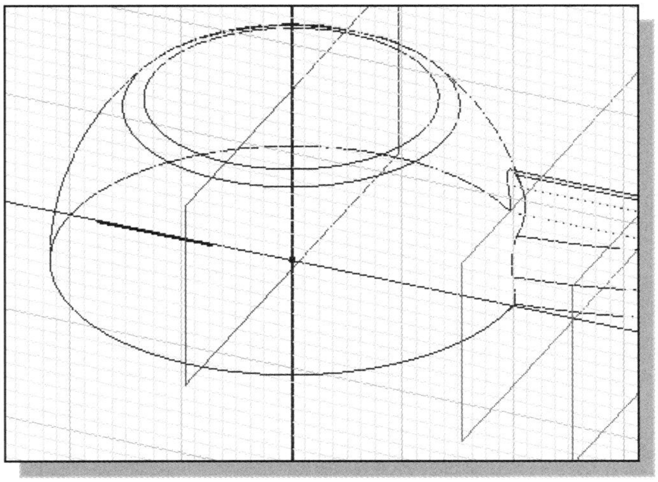

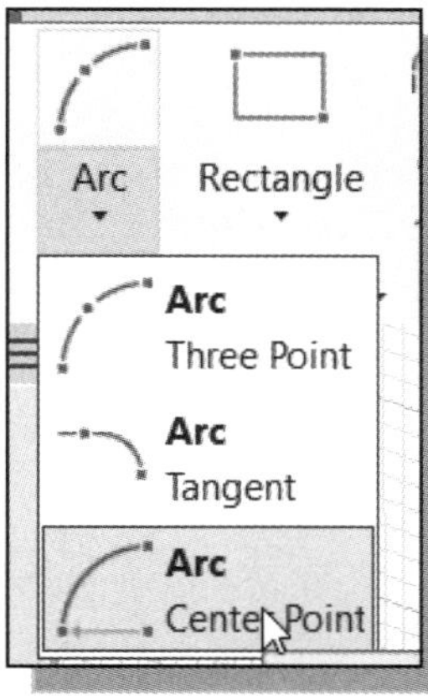

5. Select the **Center Point Arc** option in the *2D Sketch* panel as shown.

6. On your own create an arc aligned to the **mid-point** (Hint: use the Midpoint-Snap option) and **endpoints** of the previously created line as shown in the figure below.

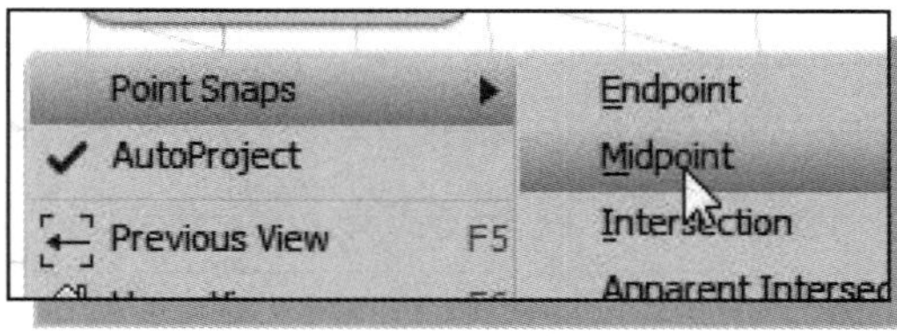

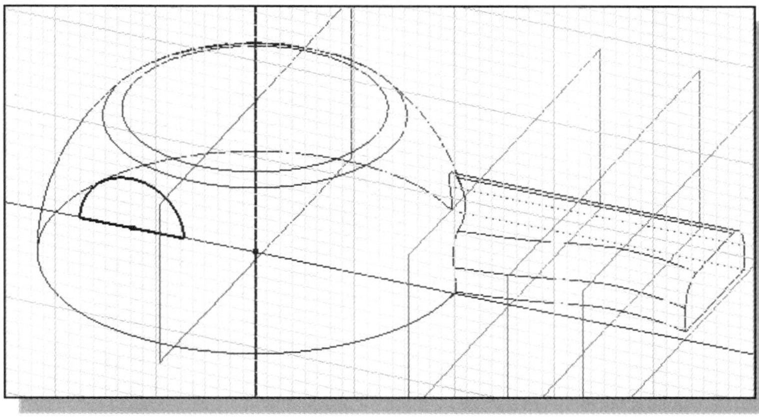

7. On your own, create and modify the three dimensions as shown in the figure below. (Hint: Use the **Auto Dimension** command after adding the radius.)

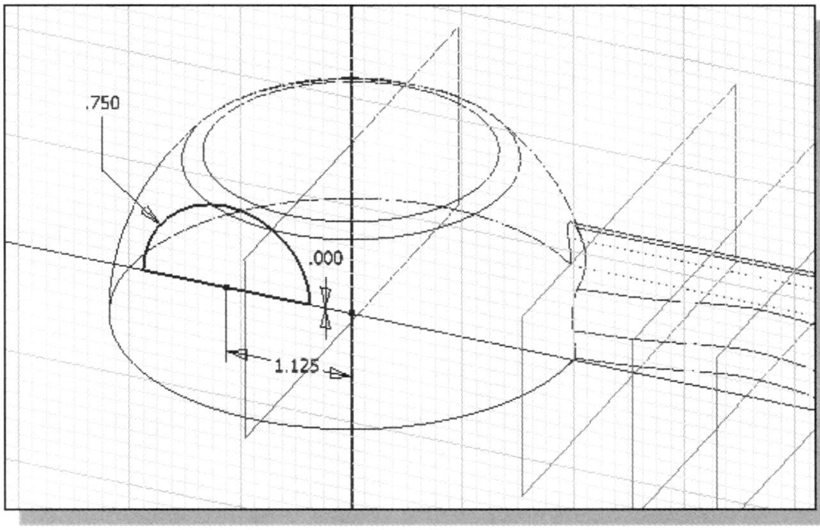

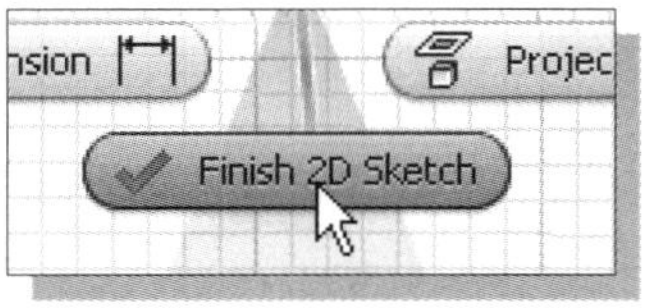

8. Inside the graphics window, click once with the right-mouse-button to display the option menu. Select **Finish 2D Sketch** in the pop-up menu to end the Sketch option.

Complete the Extruded Feature

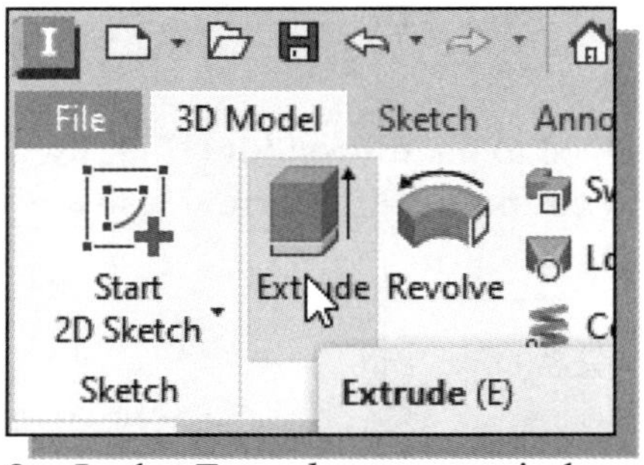

1. In the *3D Model tab*, select the **Extrude** command by left-clicking once on the icon.

2. In the *Extrude* pop-up window, enter **5.5** as the extrusion distance.

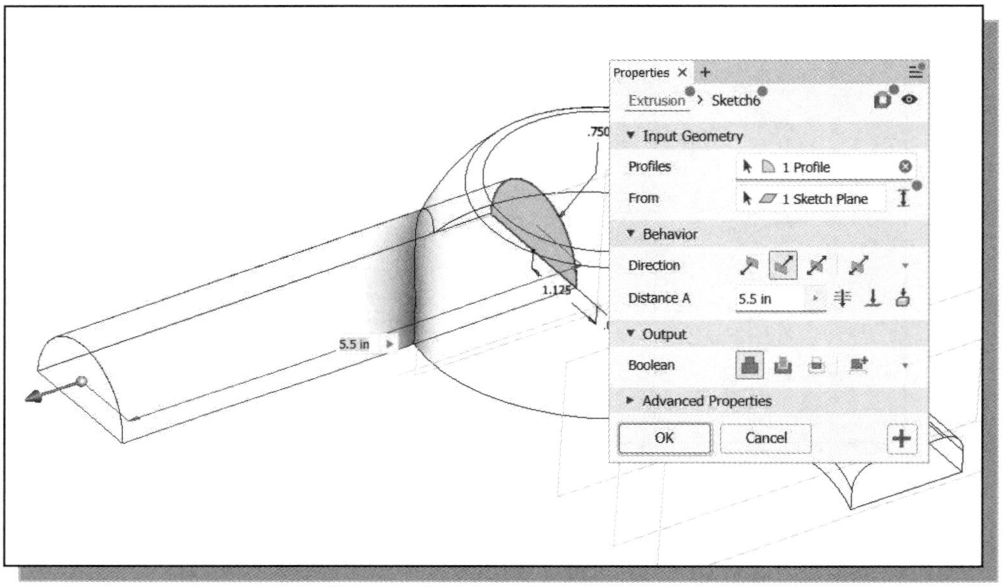

3. In the *Extrude* dialog box, set the extrusion direction as shown and confirm the operation is set to **Join**, and then click on the **OK** button to accept the settings to create the feature.

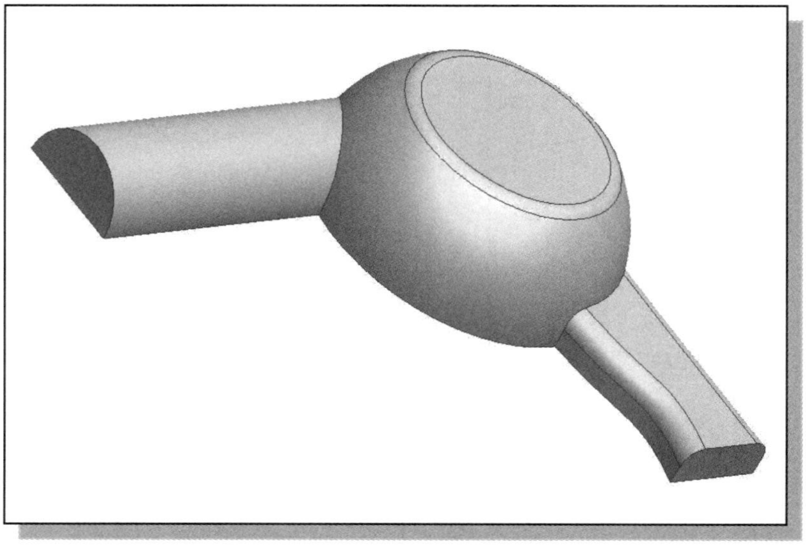

3D Rounds and Fillets

The fillet command makes it easy to apply fillets and rounds of varying radii and conditions as one feature. The property panel provides fillet types, selection modes, and selection sets to achieve a robust filleting operation. For each fillet type you can have multiple selection sets of differing sizes.

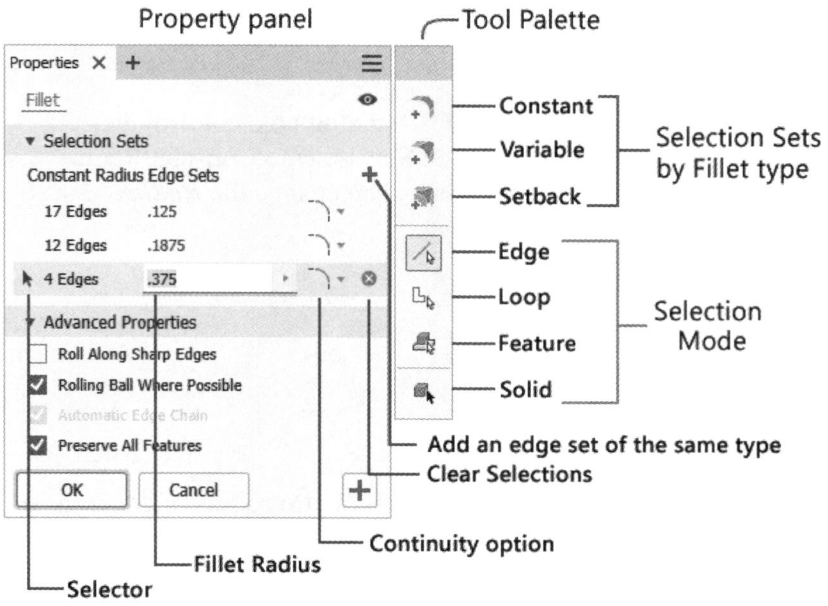

The **tool palette** controls enable adding multiple fillet sizes and types in one command session:

- Add different edge fillet selection set types - constant or variable, and create fillet setbacks. To add different selection set type, in the tool palette, click the type you want to use and select the edges for that type. To add fillets with a different radius, add another selection set to the current fillet type.
- Change selection filter priorities - edge, loop, feature, or solid body.
    - Selecting the Solid selection filter will replace an empty selection set.
    - Selecting a different fillet type with a Solid selection filter empty deletes the solid selection group and adds the selected filter type.

Selection Mode:

- Edge. Adds or removes (Ctrl+click) single edges.
- Loop. Selects or removes the edges of a closed loop on a face.
- Feature. Applies a fillet to all edges of the selected feature that do intersect with other features. If used in the assembly environment, only part features can be selected.
- Solid. Selects all edges of a solid body enabling rapid fillet creation. All body edges share the same radius value.

Create 3D Rounds and Fillets

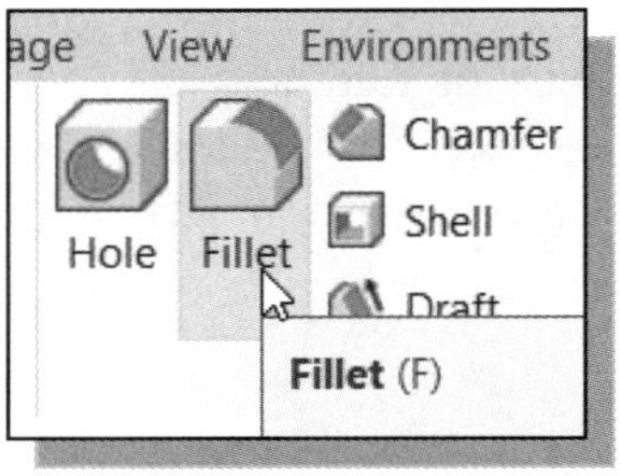

1. In the *Modify Features* toolbar, select the **Fillet** command by left-clicking once on the icon.

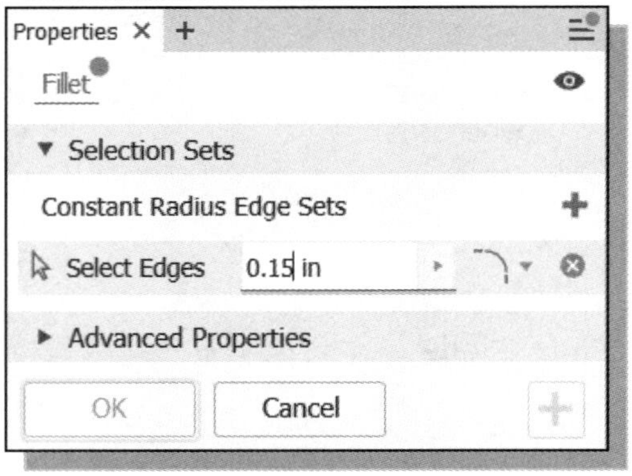

2. In the *Fillet* dialog box, confirm the Selection Sets is set to Constant Radius Edge Sets, then change the ***Radius*** to **0.15** as shown.

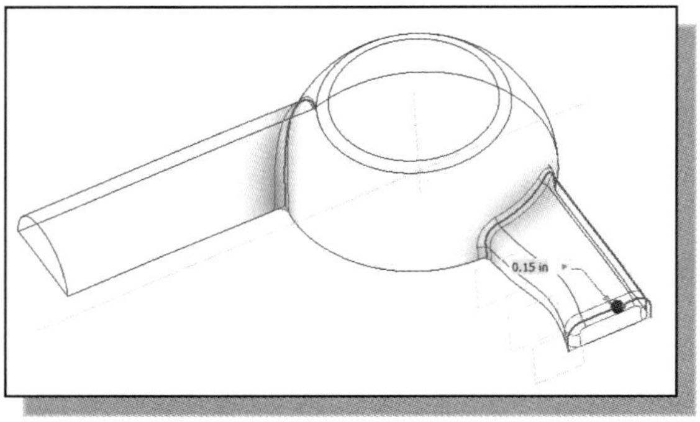

3. Click on the **three** sets of **edges** as shown.

4. Click on the **OK** button to accept the settings and create the 3D rounds and fillets.

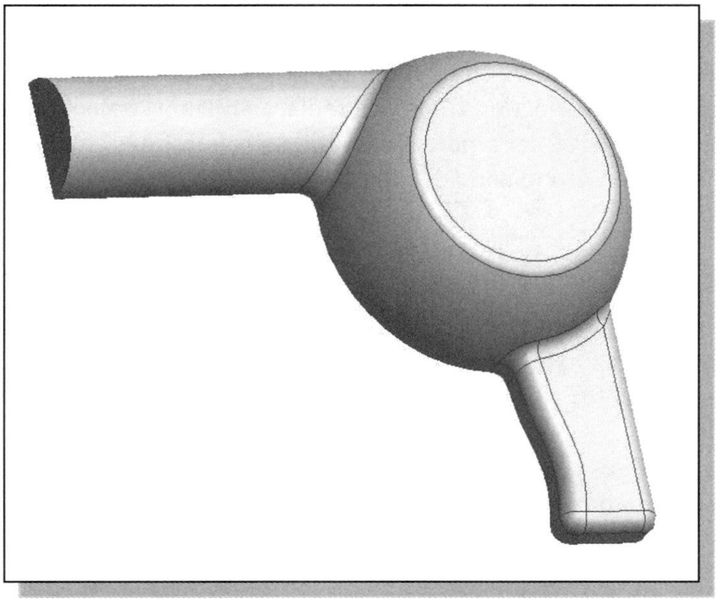

Create a Shell Feature

The **Shell** command can be used to hollow out the inside of a solid, leaving a shell of specified wall thickness.

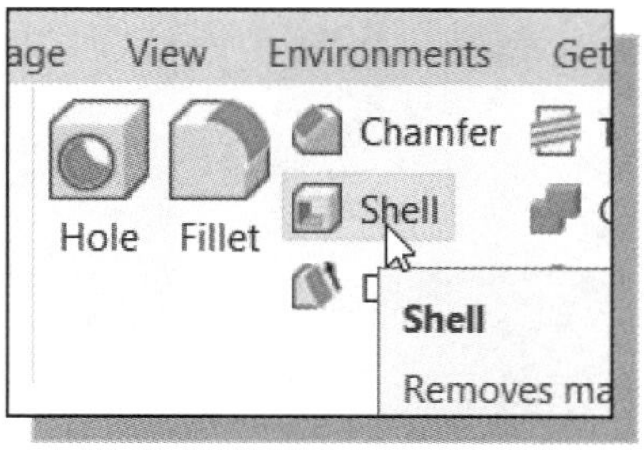

1. In the *Modify Features* toolbar, select the **Shell** command by left-clicking once on the icon.

2. On your own, use the **ViewCube** or the **3D Rotation** quick-key [**F4**] to display the back faces of the model as shown below.

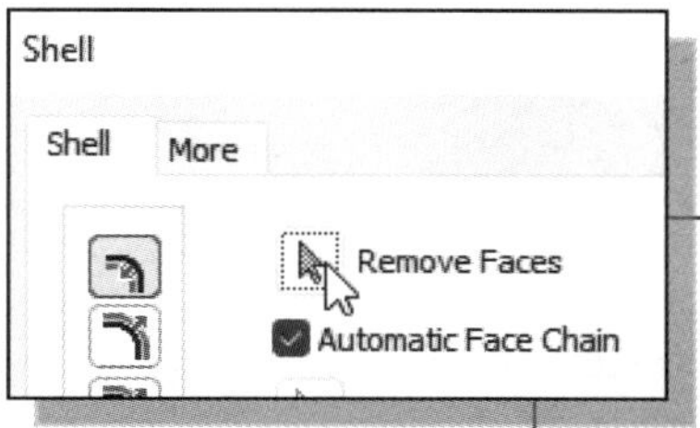

3. In the *Shell* dialog box, activate the **Remove Faces** option, then select the two faces as shown below.

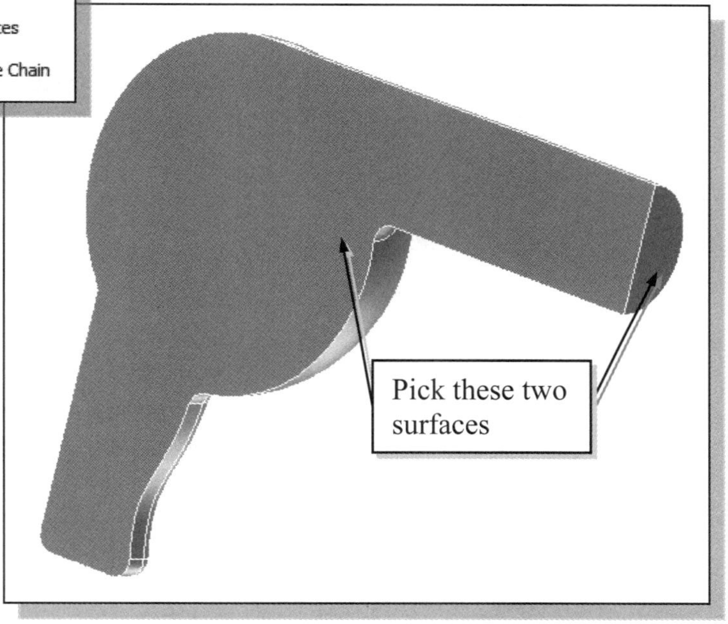

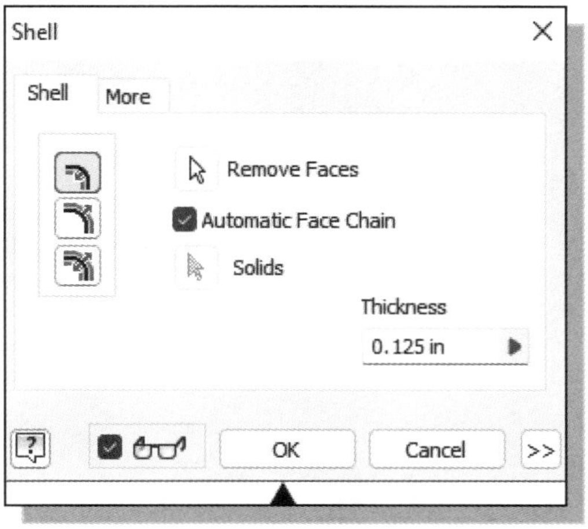

4. In the *Shell* dialog box, set the option to **Inside** with a value of **0.125** as shown.

5. In the *Shell* dialog box, click on the **OK** button to accept the settings and create the shell feature.

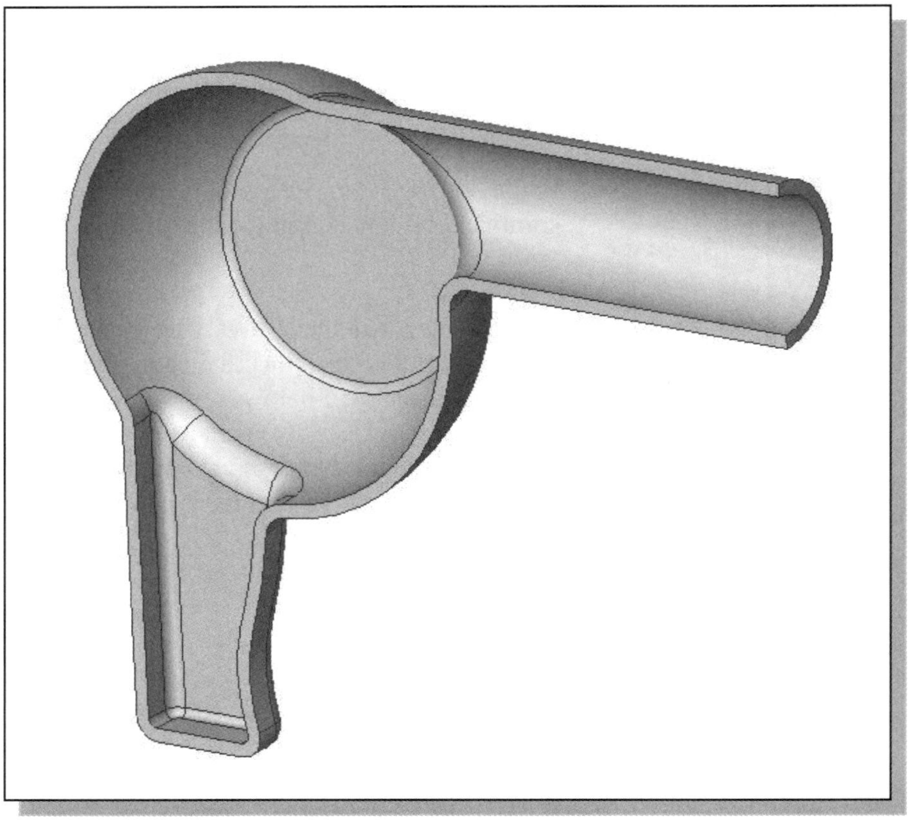

Create a Pattern Leader

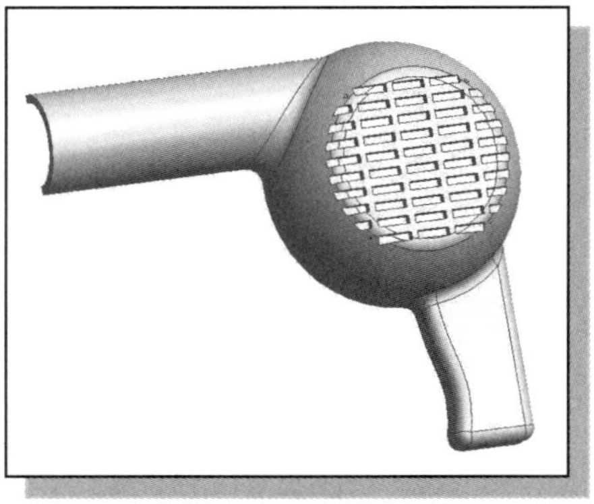

The *Dryer Housing* design requires the placement of identical holes on the top face of the solid. Instead of creating the holes one at a time, we can simplify the creation of these holes by using the **Pattern** command to create duplicate features. Prior to using the Pattern command, we will first create a ***pattern leader***, which is a regular cut feature.

1. Select **Home View** in the **ViewCube** to adjust the display of the 2D sketch to the isometric view.

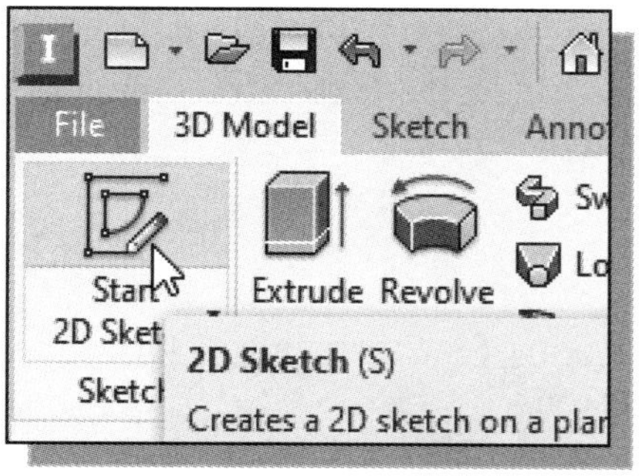

2. In the *3D Model tab* select the **Start 2D Sketch** command by left-clicking once on the icon.

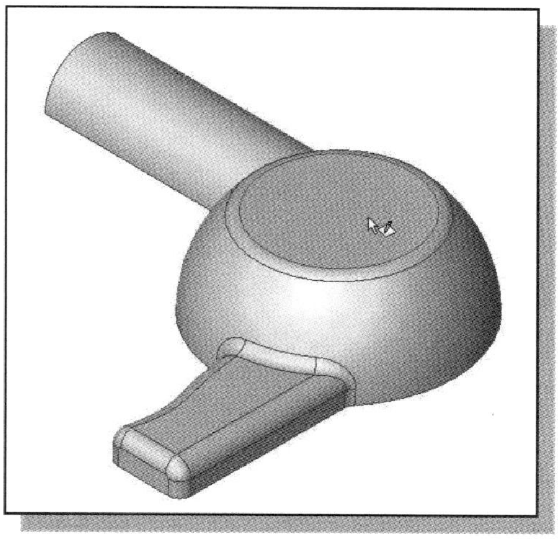

3. In the *Status Bar* area, the message "*Select face, work plane, sketch or sketch geometry.*" is displayed. Pick the top face of the base feature as shown.

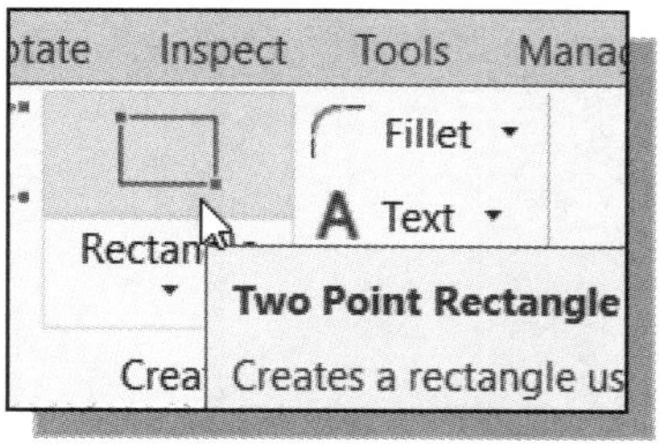

4. Select the **Two point rectangle** command by clicking once with the left-mouse-button on the icon in the *Draw* panel.

5. Create a rectangle and modify the dimensions as shown.

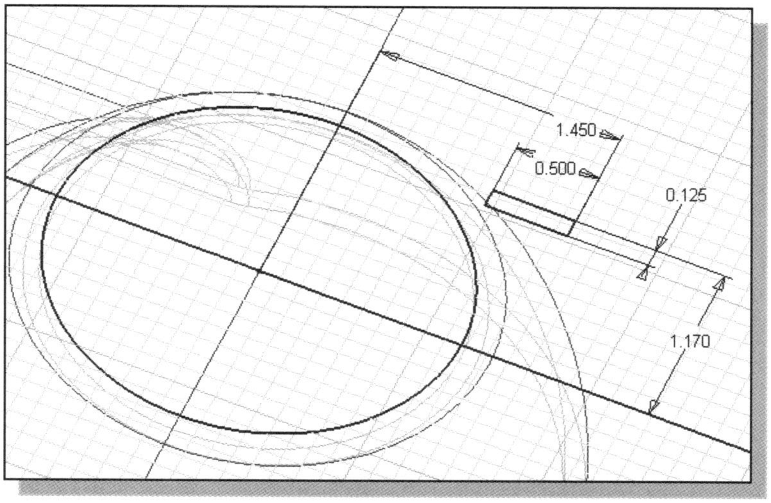

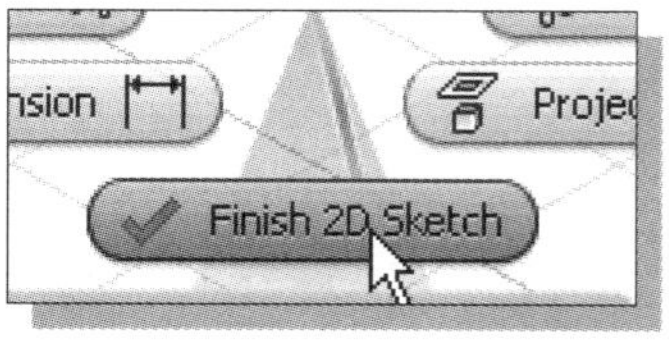

6. Inside the graphics window, click once with the right-mouse-button to display the option menu. Select **Finish 2D Sketch** in the pop-up menu to end the Sketch option.

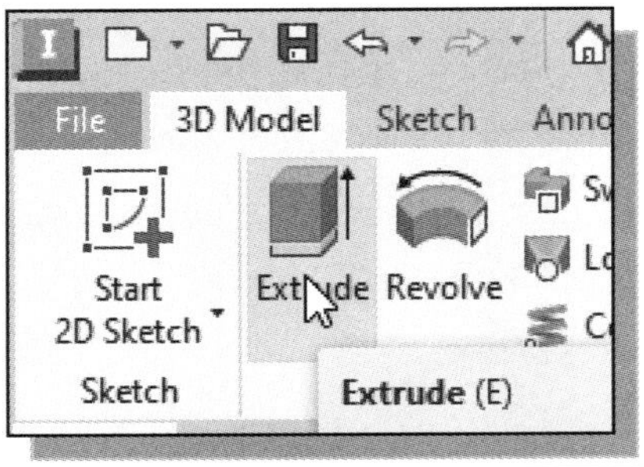

7. In the *3D Model tab*, select the **Extrude** command by left-clicking on the icon.

8. Select the inside region of the rectangle as the profile of the extrusion.

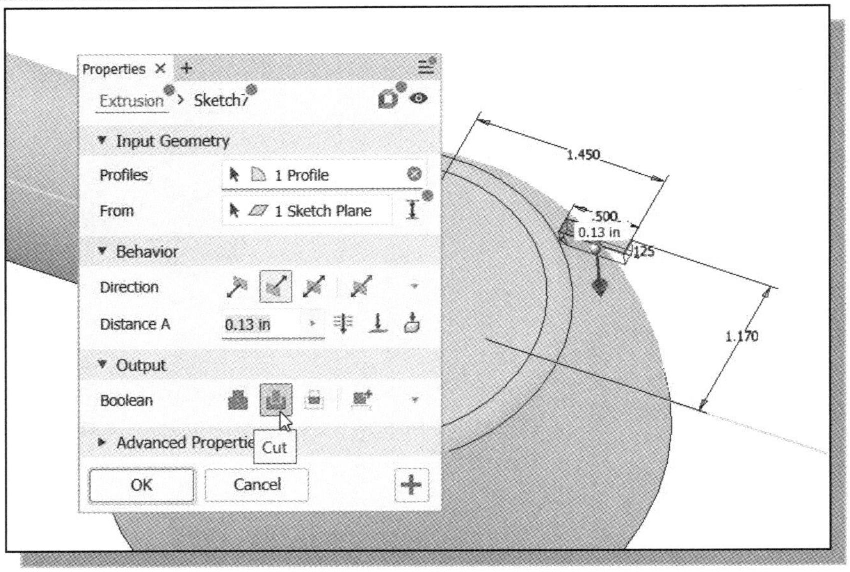

9. Inside the *Extrude* dialog box, select the **Cut** operation and set the *Extents* **Distance** to **0.13 in** as shown.

10. In the *Extrude* dialog box, click on the **OK** button to proceed with creating the cut feature.

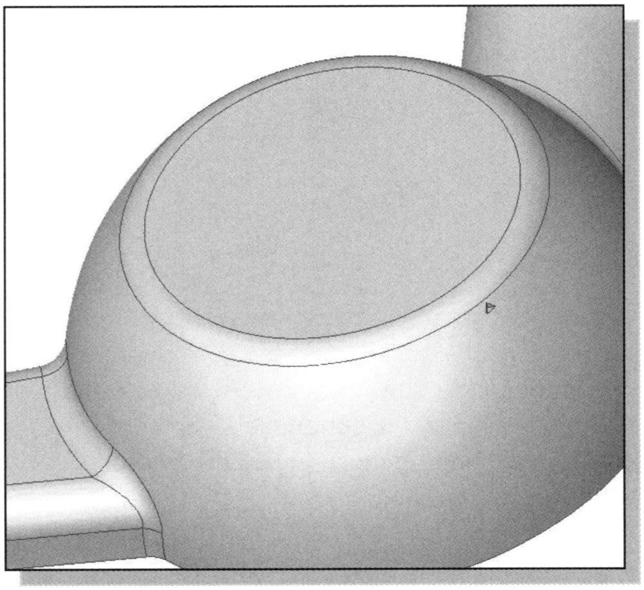

- Note that the pattern leader creates a fairly small cut on the solid model.

Create a Rectangular Pattern

In Autodesk Inventor, existing features can be easily duplicated. The **Pattern** command allows us to create both rectangular and polar arrays of features. The patterned features are parametrically linked to the original feature; any modifications to the original feature are also reflected on the arrayed features.

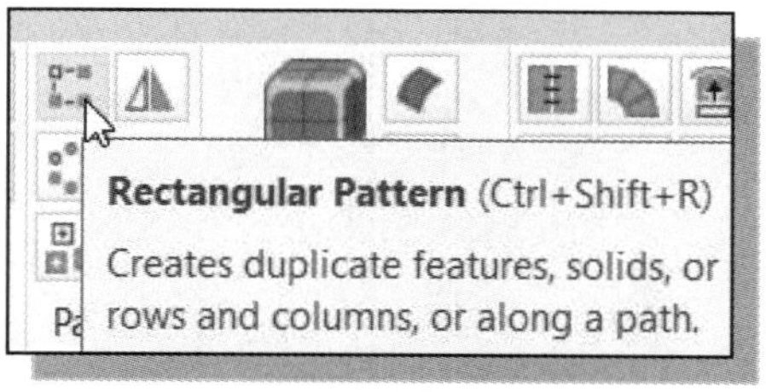

1. In the *Pattern Features* toolbar, select the **Rectangular Pattern** command by left-clicking once on the icon.

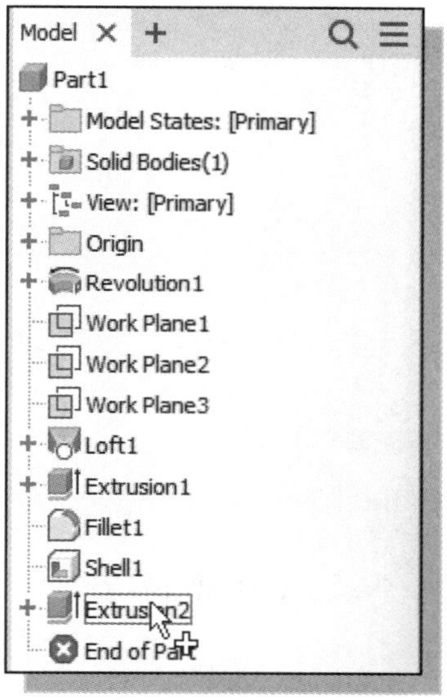

2. The message "*Select Feature to be patterned:*" is displayed in the command prompt window. Select **Extrusion2**, the *cut feature created in the last section*, in the *browser* window.

3. Inside the *graphics window*, **right-click** to bring up the option menu.

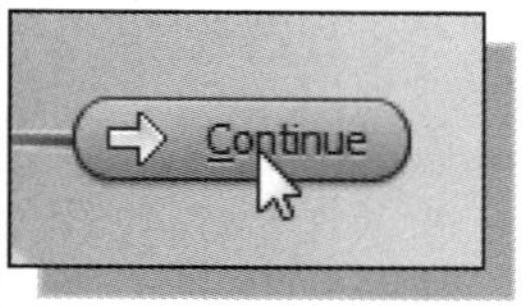

4. Select **Continue** in the option list to proceed with the Rectangular Pattern command.

5. In the *Rectangular Pattern* dialog box, notice the **selection** option for *Direction 1* is activated. Click on an edge along the swept feature as shown in the figure.

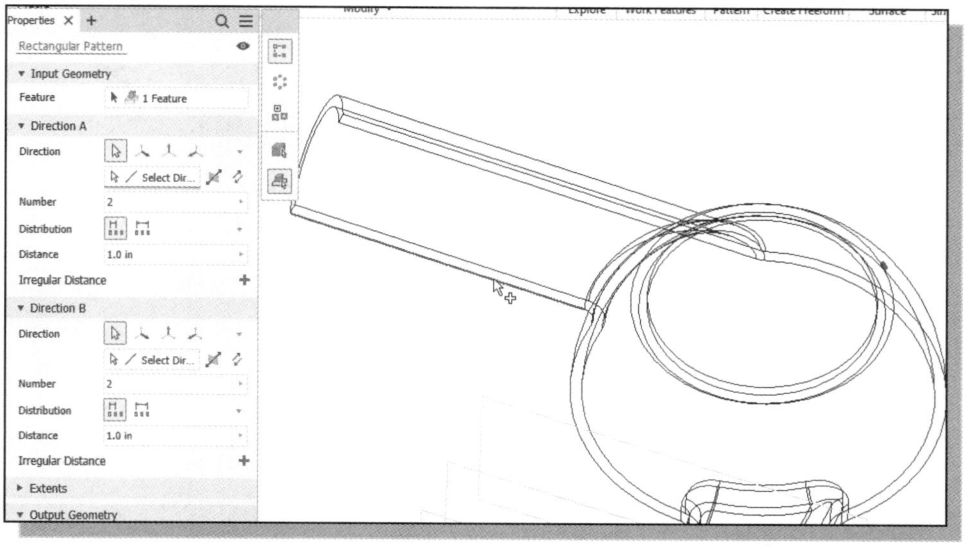

6. In the *Rectangular Pattern* dialog box, enter **5** in the *Count* box and **0.6** in the *Spacing* box as shown.

7. In the *Rectangular Pattern* dialog box, click on the **direction selection** icon for *Direction 2*.

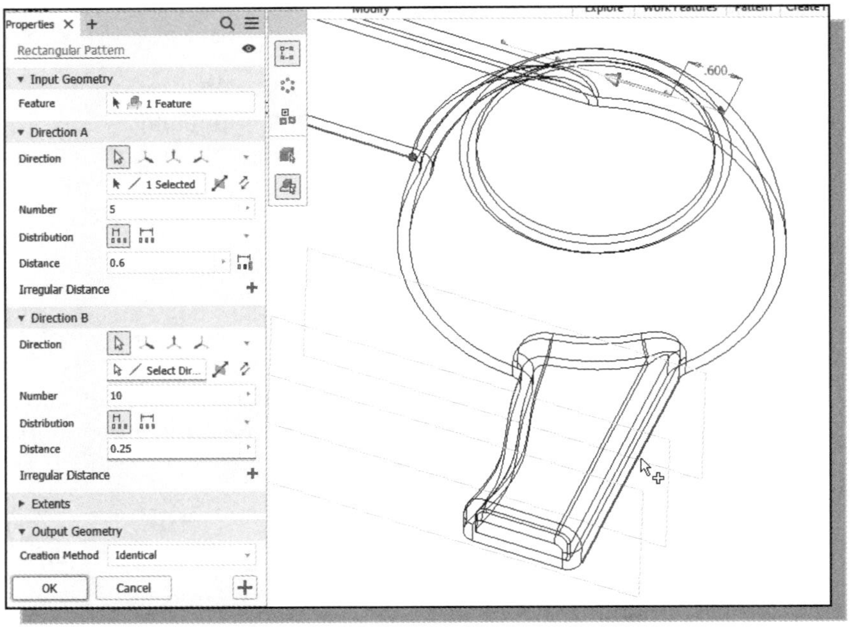

8. Click on a straight edge along the lofted feature as shown in the figure.

9. In the *Rectangular Pattern* dialog box, enter **10** in the *Count* box and **0.25** in the *Spacing* box as shown.

10. Click on the **OK** button to accept the settings and create the *rectangular pattern*.

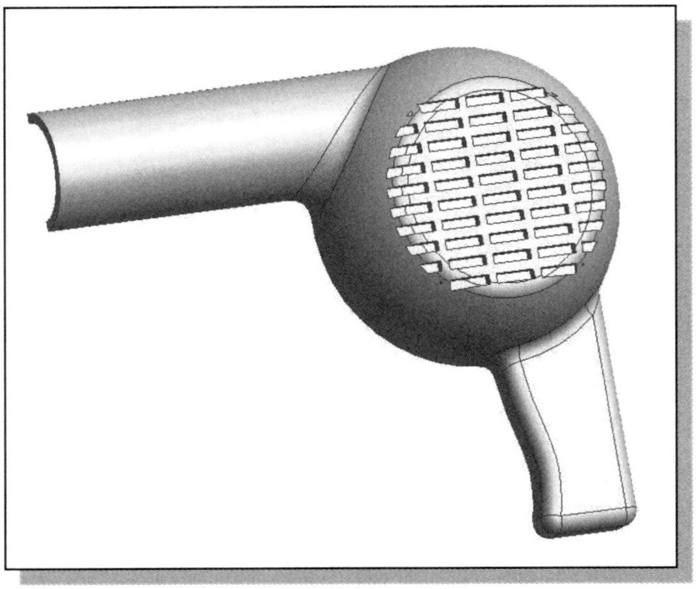

Create a Swept Feature

The **Sweep** operation is defined as moving a planar section through a planar (2D) or 3D path in space to form a three-dimensional solid object. The path can be an open curve or a closed loop but must be on an intersecting plane with the profile. The **Extrusion** operation, which we have used in the previous lessons, is a specific type of sweep. The **Extrusion** operation is also known as a ***linear sweep*** operation, in which the sweep control path is always a line perpendicular to the two-dimensional section. Linear sweeps of unchanging shape result in what are generally called ***prismatic solids*** which means solids with a constant cross-section from end to end. In Autodesk Inventor, we create a ***swept feature*** by defining a path and then a 2D sketch of a cross section. The sketched profile is then swept along the planar path. The Sweep operation is used for objects that have uniform shapes along a trajectory.

♦ Define a Sweep Path

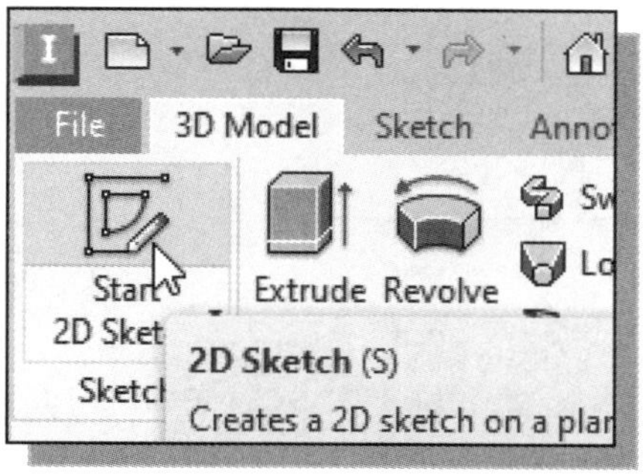

1. In the *3D Model tab* select the **Start 2D Sketch** command by left-clicking once on the icon.

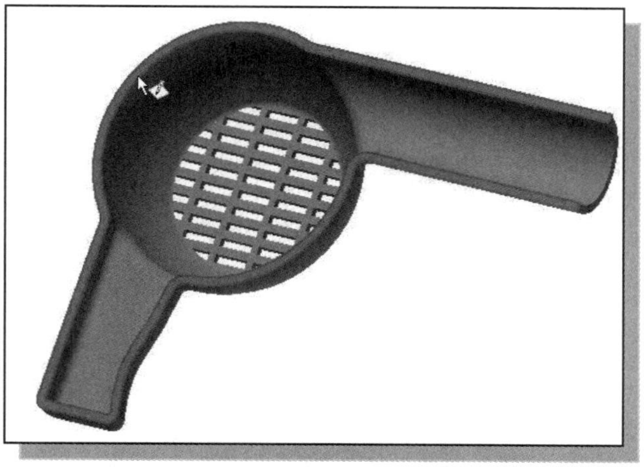

2. Select the **bottom face of the model**, by left-clicking once, as shown.

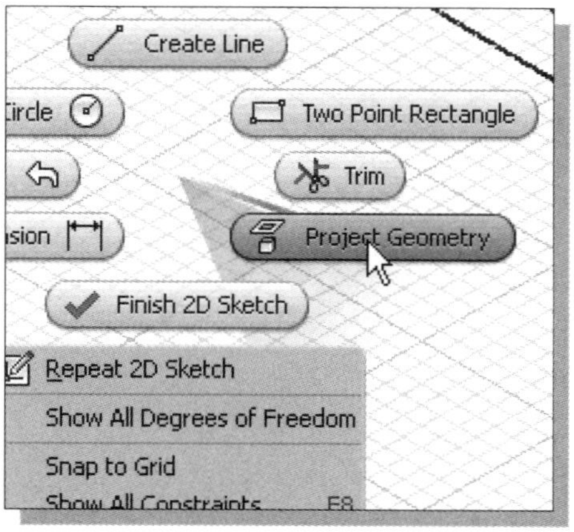

3. Inside the graphics window, right-click to bring up the option menu and select **Project Geometry**.

4. In the *Status Bar* area, the message: "*Select face, work plane, sketch or sketch geometry*" is displayed. Select the **outer edges** of the bottom surface of the model as shown. (Hint: Use the Dynamic Zoom function to assist selecting all neighboring edges.)

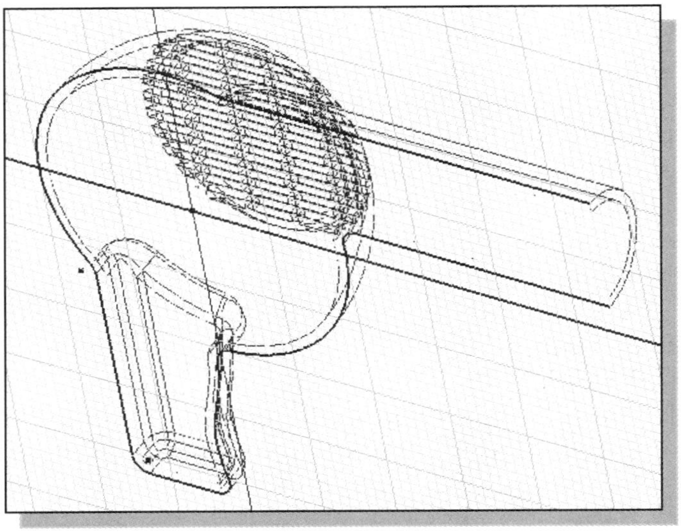

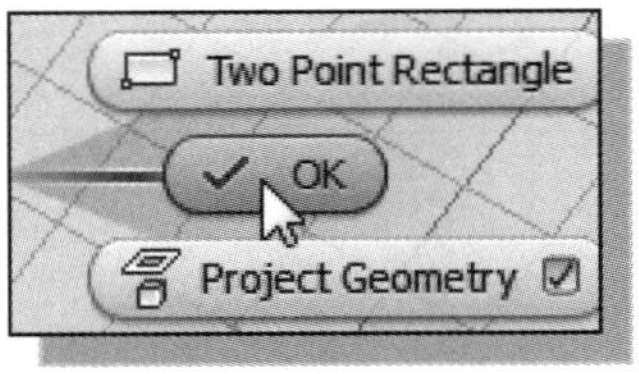

5. Inside the graphics window, right-click to bring up the option menu and select **OK** to end the Project Geometry command.

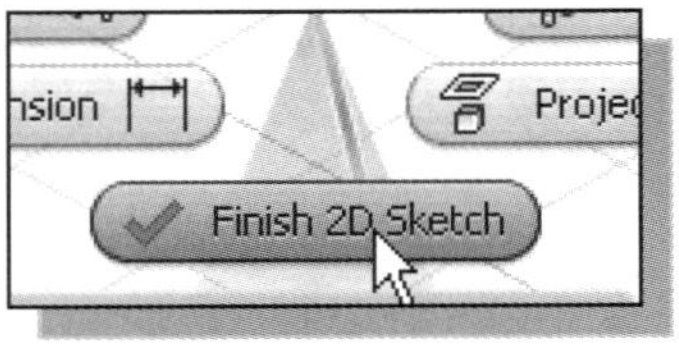

6. Inside the graphics window, click once with the right-mouse-button to display the option menu. Select **Finish 2D Sketch** in the pop-up menu to end the Sketch option.

- The projected geometry will be used as the 2D sweep path of the feature.

♦ Define the Sweep Section

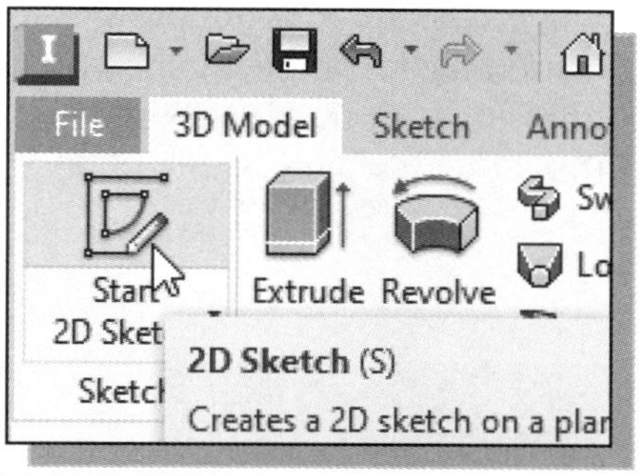

1. In the *3D Model tab* select the **Start 2D Sketch** command by left-clicking once on the icon.

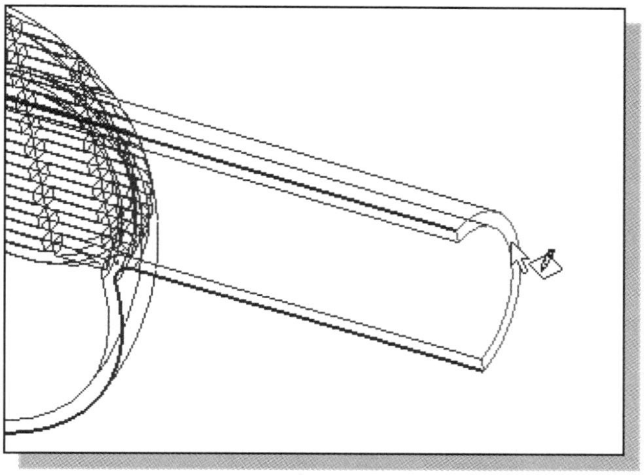

2. In the *Status Bar* area, the message: "*Select face, work plane, sketch or sketch geometry*" is displayed. Select the small circular surface of the model as shown.

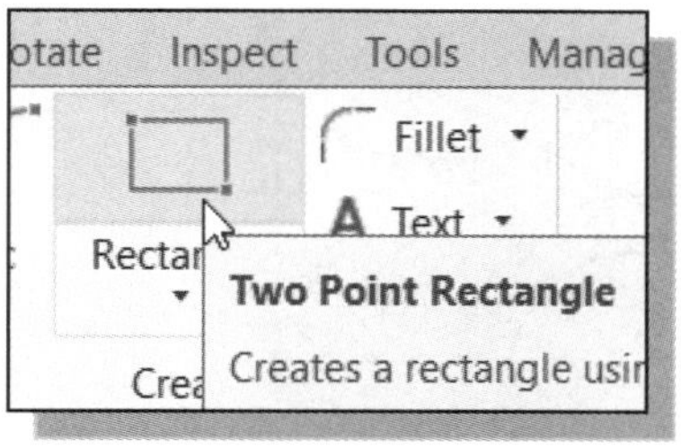

3. Select the **Two point rectangle** command by clicking once with the left-mouse-button on the icon in the *2D Sketch* panel.

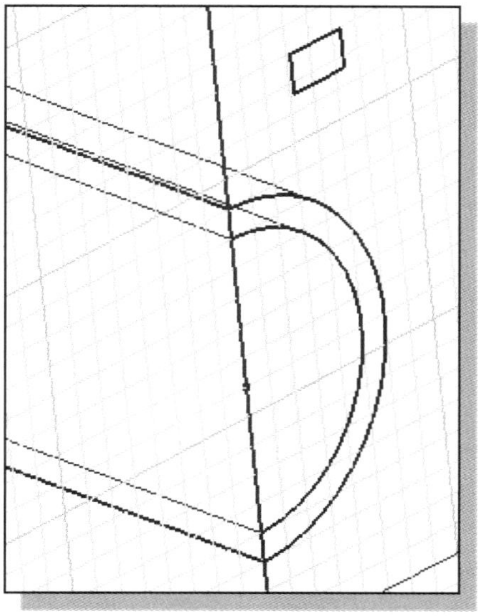

4. On your own, create a rectangle that is toward the top right side of the model as shown.

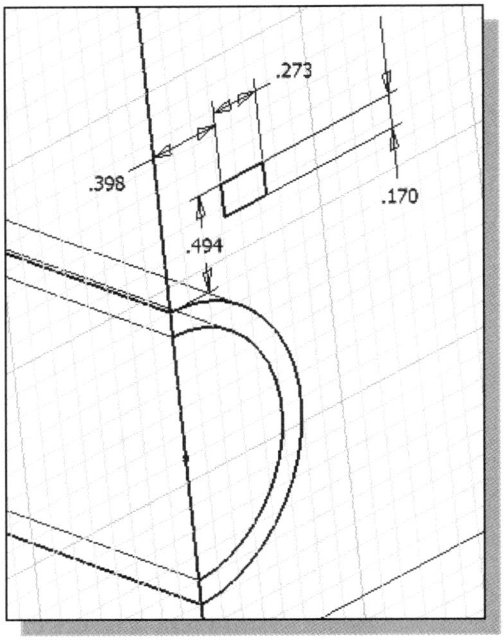

5. On your own, create the four dimensions to size and position the rectangle as shown. (Note that the values of the dimensions may be different than what is shown in the figure.)

6. Modify the dimensions as shown in the figure.

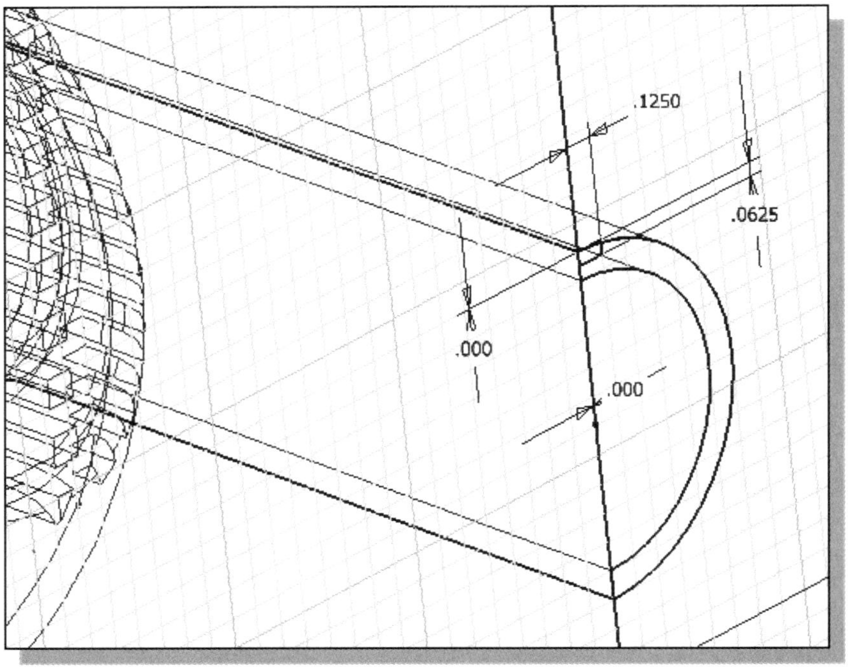

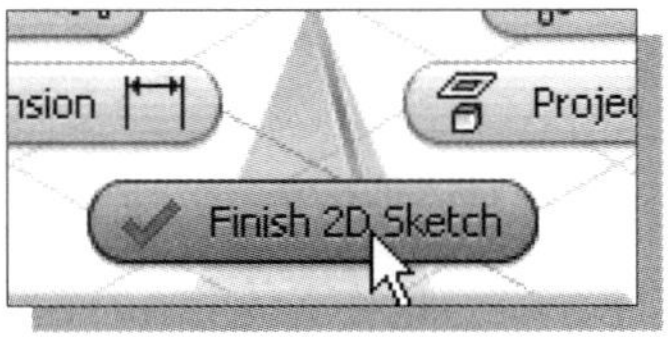

7. Inside the graphics window, click once with the right-mouse-button to display the option menu. Select **Finish 2D Sketch** in the pop-up menu to end the Sketch option.

- The created 2D sketch will be used as the sweep section of the feature.

♦ Complete the Swept Feature

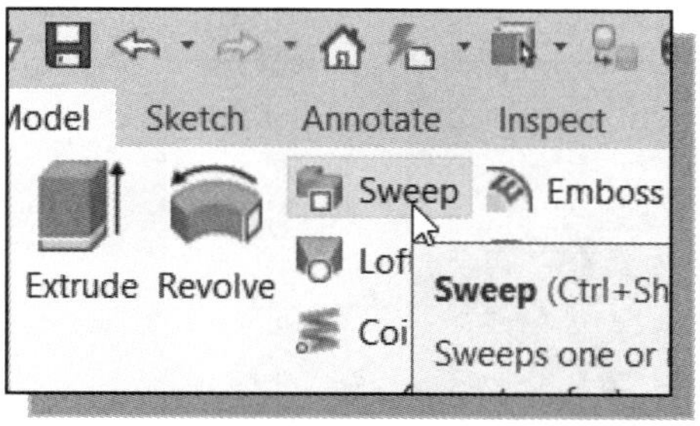

1. In the *3D Model tab*, select the **Sweep** command by left-clicking once on the icon.

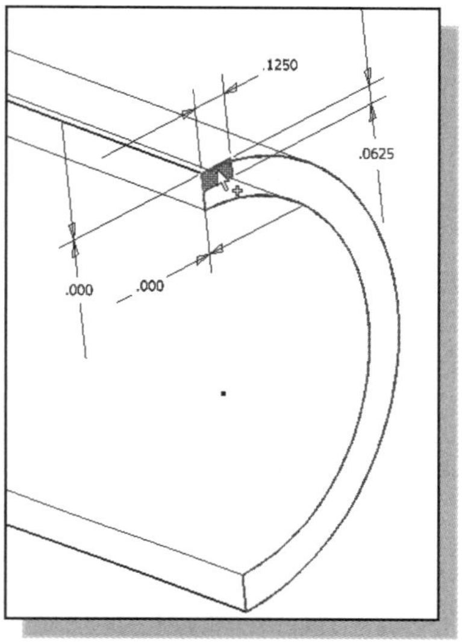

2. Confirm the rectangle region is selected as the sweep section. (Hint: if the rectangle region is not selected, use the Dynamic Viewing function to assist the selection.)

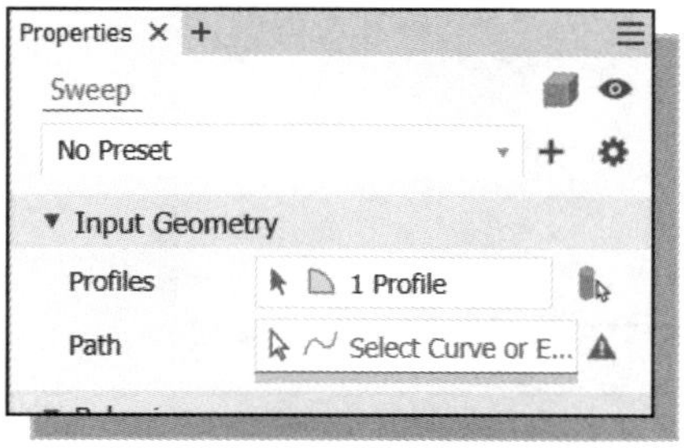

3. Notice the **Path** option is activated once the profile has been defined.

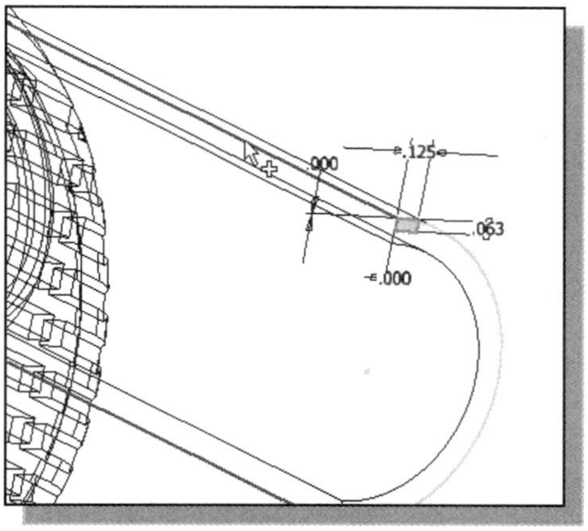

4. Click on the projected open curve we created as the sweep path as shown in the figure.

➢ On your own, confirm the selected path follows the projected outline of the *Dryer Housing* as shown in the figure.

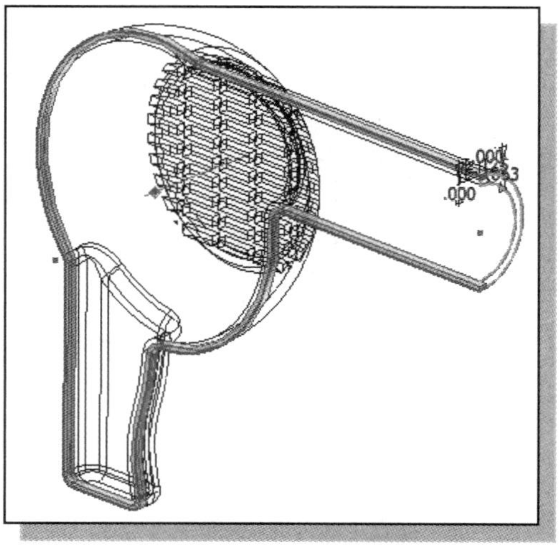

5. In the *Sweep* dialog box, confirm the operation to **Cut**, and then click on the **OK** button to accept the settings and create the *swept* feature.

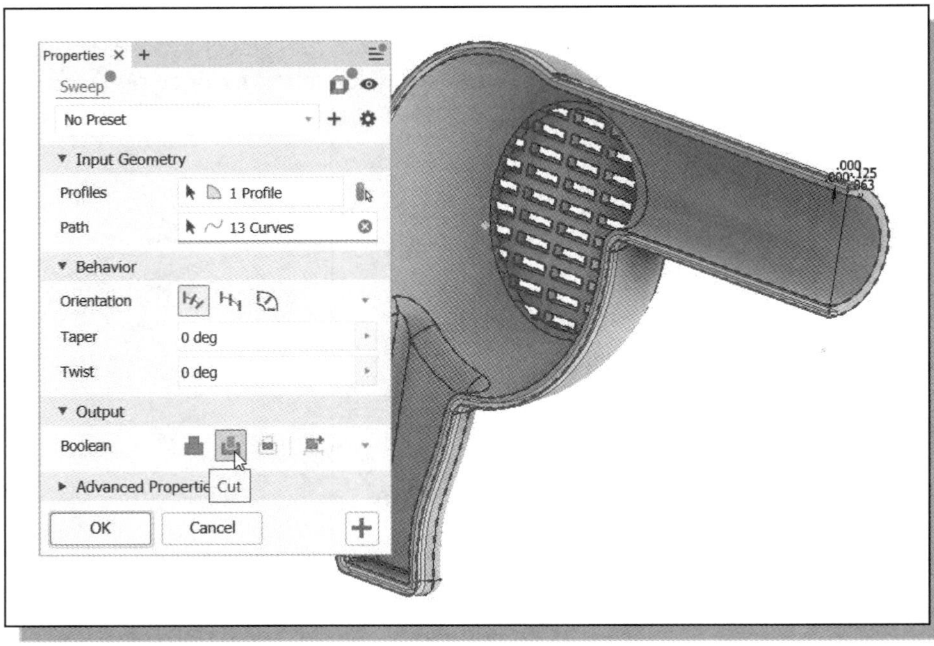

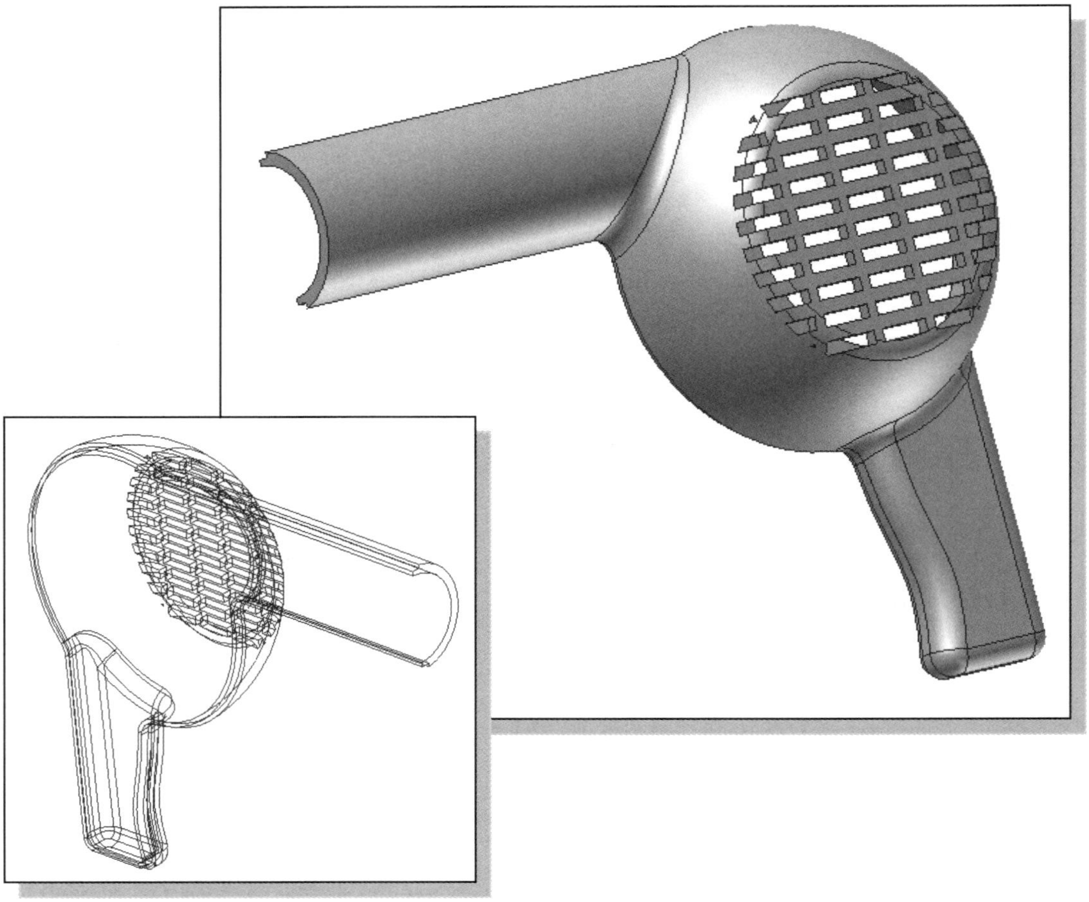

Review Questions: (Time: 25 minutes)

1. Keeping the *History Tree* in mind, what is the difference between *cut with a pattern* and *cut each one individually*?

2. What is the difference between Sweep and Extrude?

3. What are the advantages and disadvantages of creating fillets using the 3D Fillets command and creating fillets in the 2D profiles?

4. Describe the steps used to create the *Shell* feature in the lesson.

5. How do we modify the *Pattern* parameters after the model is built?

6. Describe the elements required in creating a *Swept* feature.

7. Create sketches showing the steps you plan to use to create the model shown on the next page:

Exercises: Create and save the exercises in the Chapter12 folder. (Time: 200 minutes.)

1. **Motor Housing** (Dimensions are in inches.)

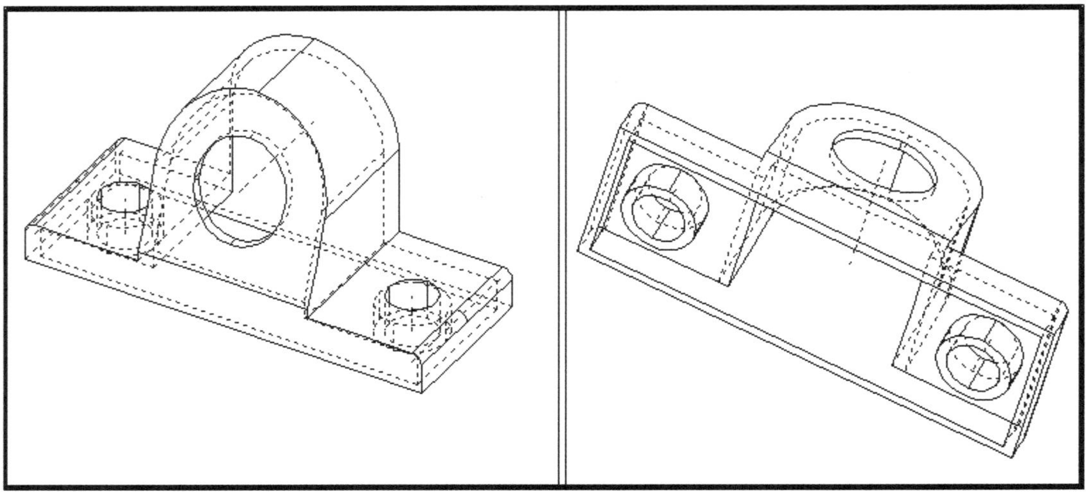

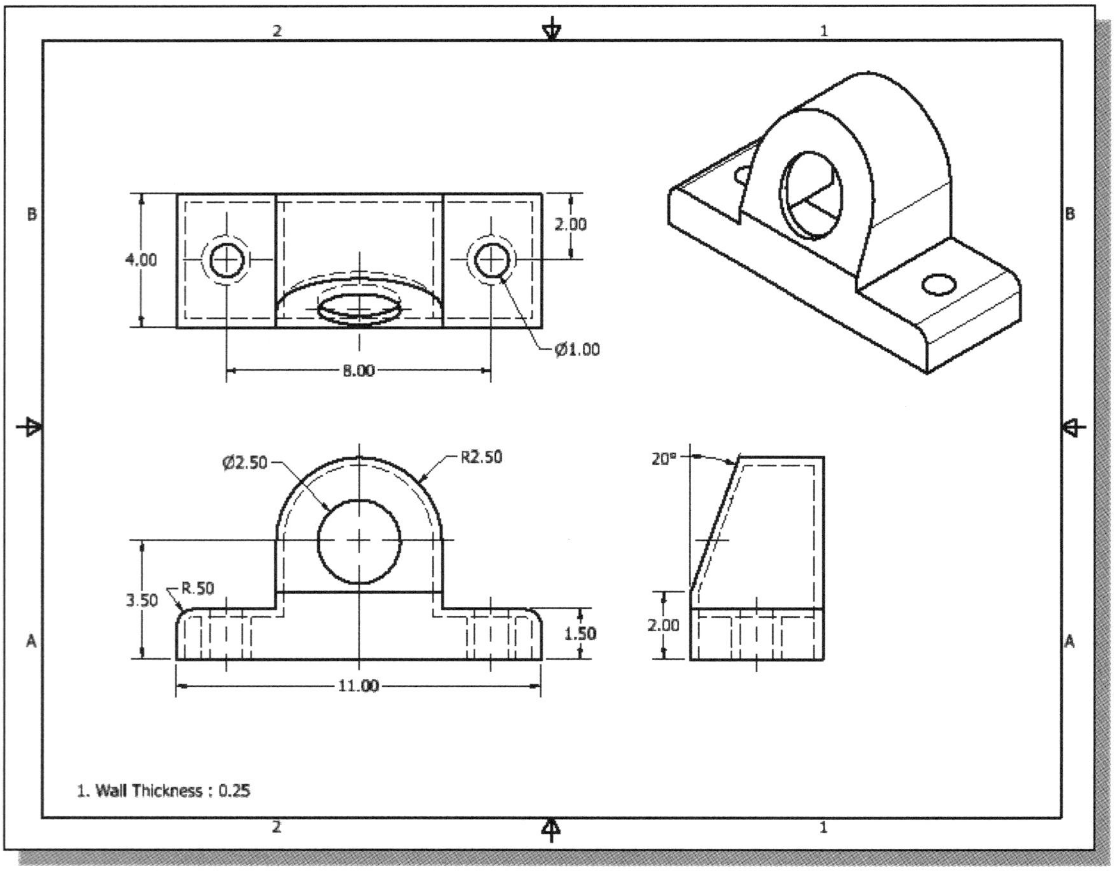

2. **Pivot Latch** (Dimensions are in inches.)

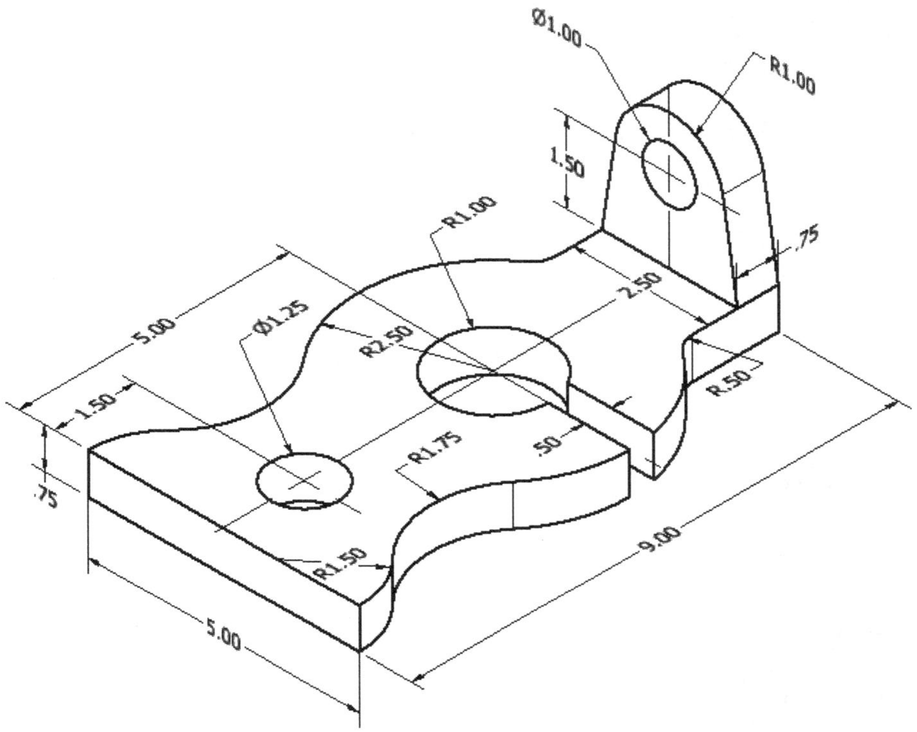

3. **Piston Cap** (Dimensions are in inches.)

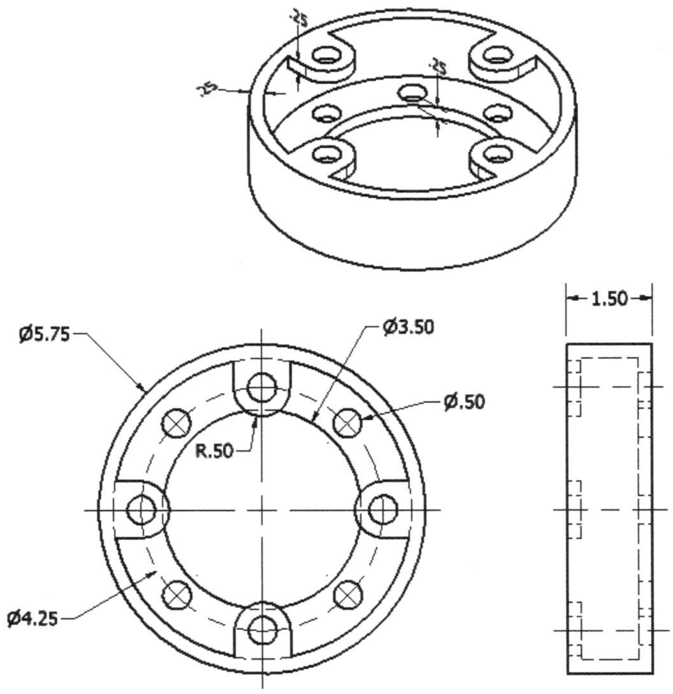

4. Using the same dimensions given in the tutorial, construct the other half of the dryer housing. Plan ahead and consider how you would create the matching half of the design. (Save both parts so that you can create an assembly model once you have completed Chapter 14.)

5. **Intake Flange** (Dimensions are in inches. Thickness: .25 inches.)

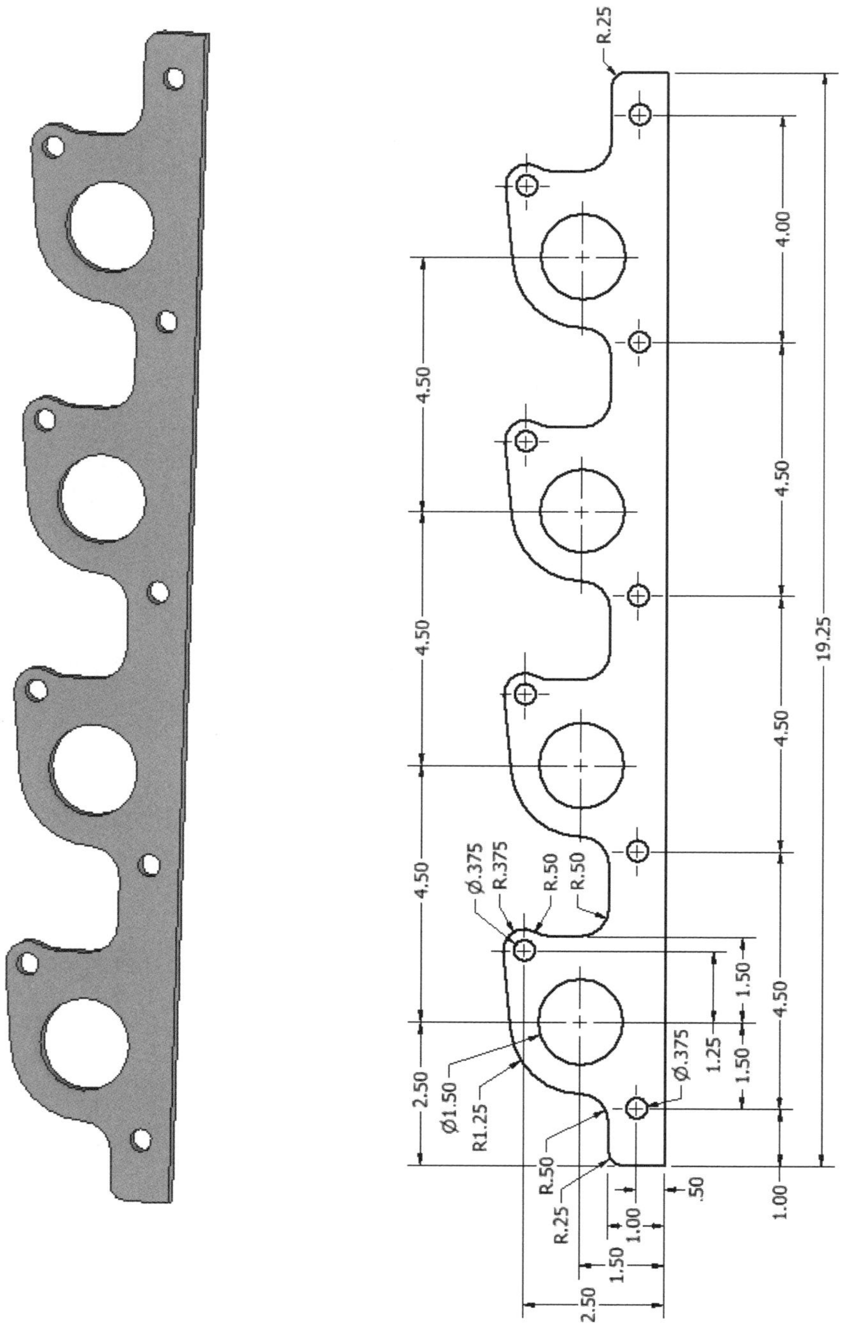

6. **Shaft Support** (Dimensions are in inches.)

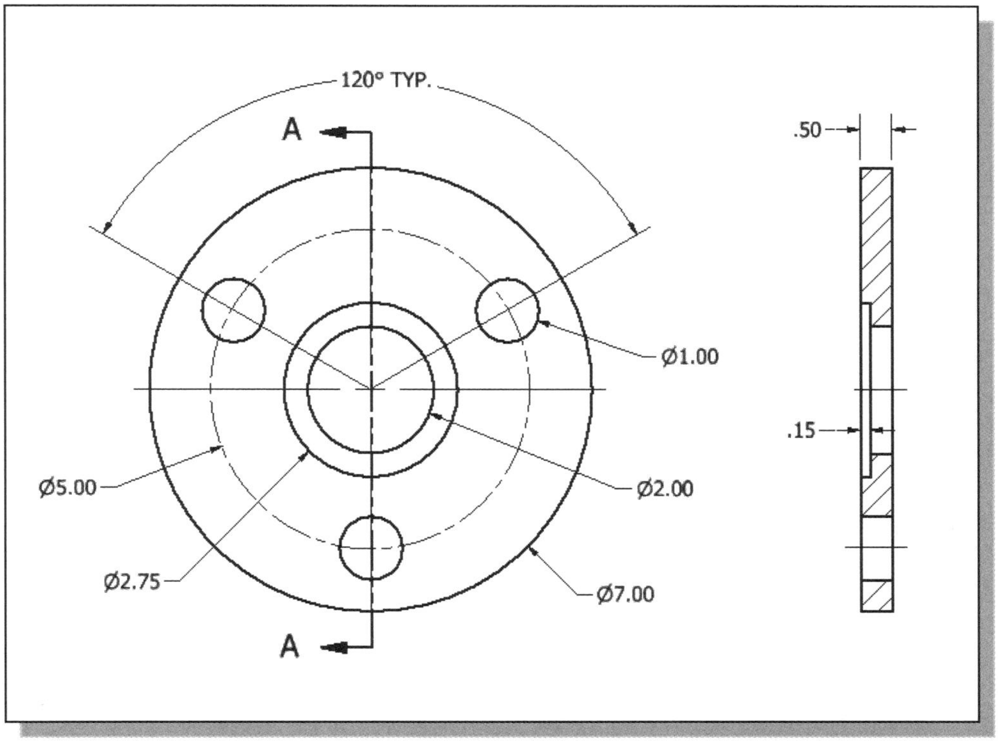

7. **Anchor Base** (Dimensions are in inches.)

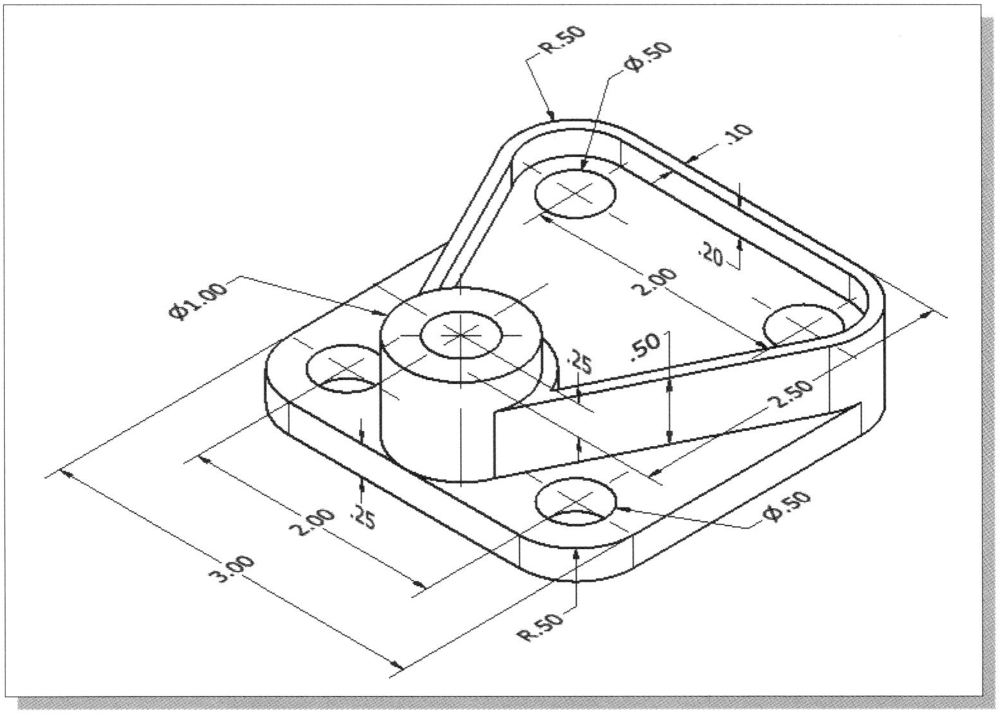

Notes:

Chapter 13
Sheet Metal Designs

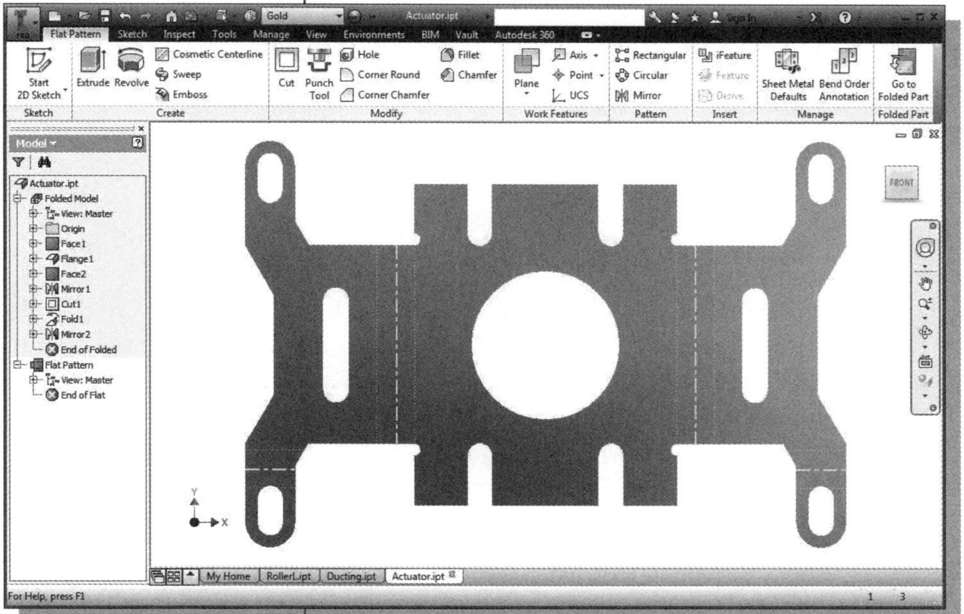

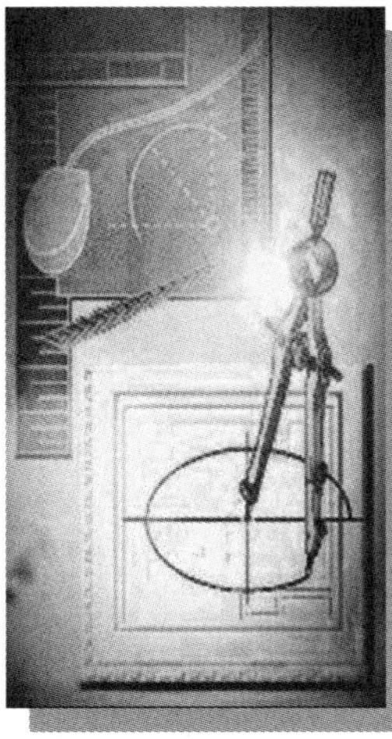

Learning Objectives

- **Understand the Sheet Metal Manufacturing Processes**
- **Understand the Autodesk Inventor Sheet Metal Modeling Methodology**
- **Create Parts in the Sheet Metal Modeler Mode**
- **Utilize the Autodesk Inventor Sheet Metal Tools to Create Bends and Flanges**
- **Create Flat Pattern Layouts**

Autodesk Inventor Certified User Exam Objectives Coverage

Parametric Modeling Basics

Section 4: Parts

Objectives: Creating parts, Work Features, Pattern Features, Part Properties.

Section 6: Drawings

Objectives: Create drawings.

Sheet Metal Processes

Sheet metal is one of the most commonly used materials in our everyday life. Sheet metal is simply a thin and flat piece of metal, which can be cut and bent into a variety of different shapes. The thicknesses of sheet metal can vary significantly, but the thickness is generally between 0.006" and 0.250".

Sheet metal is generally produced by reducing the thickness of a work piece by compressive forces applied through a set of rolls. This process is known as rolling and has been around since 1500 AD. Sheet metal is identified by the thickness, or gauge, of the metal and is generally available as flat pieces or in coils. The gauge of sheet metal (see Appendix A) ranges from 30 gauge to about 6 gauge. The higher the gauge number, the thinner the metal is. Aluminum, brass, copper, cold rolled steel, tin, nickel and titanium are some of the more commonly available sheet metal materials. Typical sheet metal applications are seen in cars, boats, airplanes, casing for electronic devices and many other things.

The main feature of sheet metal is its ability to be formed and shaped by a variety of processes, such as **bending** and **cutting**. Different processes can be used to achieve the desired shape and form. Some of the more commonly used sheet metal processes include:

Drawing
Drawing forms sheet metal into parts by using a punch, where the punch presses a sheet metal blank into a die cavity. This process is generally used to create shallow or deep parts with relatively simple shapes. **Soft punches** can also be utilized to create more arbitrary shapes. **Deep drawing** is generally done by making multiple steps; this process is known as *draw reductions*.

Stretch forming
Stretch forming is a process where the sheet metal is clamped around its edges and stretched over a die. This process is mainly used for the manufacturing of large parts with shallow contours, such as aircraft wings or automotive door and window panels.

Spinning
Spinning is the process used to make axis-symmetric parts by applying a work piece to a rotating mandrel with the help of rollers. *Spinning* is commonly used to make cylindrical shapes, such as missile nose cones and satellite dishes.

Stamping
Stamping is the general term used to describe a variety of operations, such as bending, flanging, punching, embossing, and coining. The main advantage of stamping is its speed; designs containing simple or complex shapes can be formed at relatively high production rates.

Flanging
Flanging is a process used to strengthen different sections of a sheet metal part and also to form various shapes. This process is commonly used for a variety of parts, for example, aluminum cans for soft drinks.

Bending

Bending is a process by which sheet metal can be deformed by plastically deforming the material and changing its shape. The material is stressed beyond the yield strength but below the ultimate tensile strength. With this process, the surface area of the material does not change much. *Bending* usually refers to deformation about one axis.

Bending is a flexible process by which many different shapes can be produced. Standard die sets are used to produce a wide variety of shapes. The material is placed on the die and positioned in place with *stops* and *gauges*. The material is held in place with *hold-downs*. The upper part of the press, the ram, with the appropriately shaped punch descends and forms the v-shaped bend.

Bending is usually done using ***press brakes***. The lower die of the press contains a V-shaped groove. The upper part of the press contains a punch that will press the sheet metal down into the v-shaped die, causing it to bend.

The most commonly used modern *Bending* method is the ***air bending*** method, where a sharper die angle is used; for example, an 85 degree angle is used for a 90 degree bend. *Air Bending* is done with the punch touching the work piece, and the work piece not bottoming in the lower die. By controlling the push stroke of the upper punch, the metal is pushed down to the required bend angle. The groove width of the lower die is typically 8 to 10 times the thickness of the metal to be bent. The *press brake* can also be computer controlled to allow the making of a series of bends to assure a high degree of accuracy in manufactured parts.

Cutting

Cutting sheet metal can be done in various ways, from using a variety of hand tools to very large powered shears. Today, computer-controlled cutting is also available for very precise cutting. Most modern computer-controlled sheet metal cutting operations are ***CNC laser cutting*** and ***CNC punch press***.

CNC laser cutting is done by moving the laser beam over the surface of the sheet metal. The sheet metal is heated and then burnt by the laser beam. The quality of the edge can be extremely smooth. *CNC punching* is performed by moving the sheet metal between the computer-controlled punch. The top punch mates with the bottom die, cutting a simple shape, such as a square, circle, or hexagon from the sheet. An area can be cut out by making several hundred small square cuts around the perimeter. A *CNC punch* is less flexible than a laser for cutting compound shapes, but it is faster for repetitive shapes. A typical *CNC punch* has a choice of up to 60 tools in a ***turret***. A modern *CNC punch* can run as fast as 600 blows per minute. A *CNC punch* or a *CNC laser* machine can typically cut a blank sheet into the desired shapes in less than 15 seconds with very high precision.

Sheet Metal Modeling

In reality, a sheet metal part is made from a piece of flat metal sheet of uniform thickness by cutting out a flat pattern and then folding it into the desired shape. To construct a computer sheet metal part, we can (1) simulate the actual production methods and start with a flat pattern layout to make the model; (2) use the building block approach which concentrates on the different sections of the formed 3D design; or (3) construct a solid model first, then convert it into a sheet metal model. All three methods are applicable in modern parametric modeling software such as Autodesk Inventor.

Since the actual sheet metal manufacturing process requires a flat pattern layout, the accurate generation of the flat pattern layout in the computer modeling software is critical. The conversion between the 3D formed designs and 2D flat pattern layouts requires the use of the correct ***K-Factor***, which can be used to determine the required ***Bend Allowance***.

Bend allowance is the term used to describe how much material is needed between two panels to accommodate a given bend. Determining bend allowance is commonly referred to as ***Bend Development***.

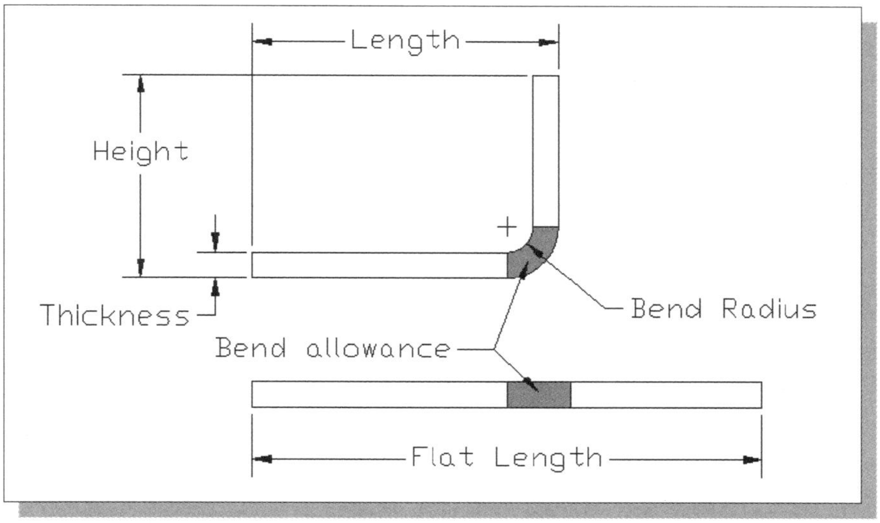

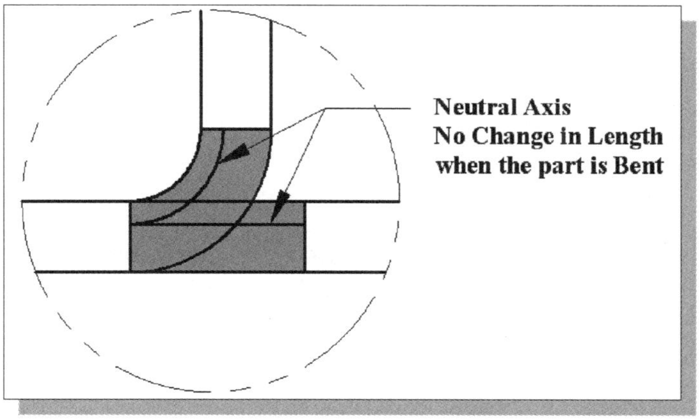

In sheet metal, the ***Neutral Axis*** is defined as the location where there is no change in length when the part is bent.

On the inside of the bend, above the neutral axis in the figure, the material is in compression, where the area below the neutral axis is in tension.

K-Factor

The location of the neutral axis in a bend is called the ***k-factor***. Since the amount of inside compression is always less than the outside tension, the k-factor can never exceed **0.50** in practical use. To the other extreme, a reasonable assumption is that the k-factor cannot be less than **0.25**.

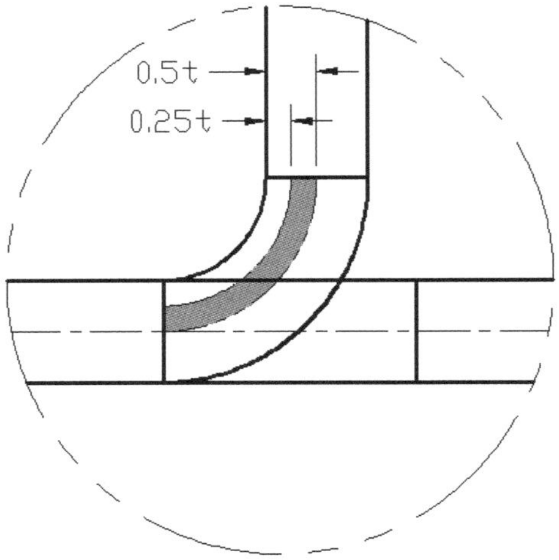

Several factors can change the k-factor, such as the type of bending (free vs. constrained), tool geometry, rate of bend, material (Mild Steel, Cold Rolled Steel, Aluminum, etc.), and even grain direction. With some grades of aluminum, the age of the material can also be a factor.

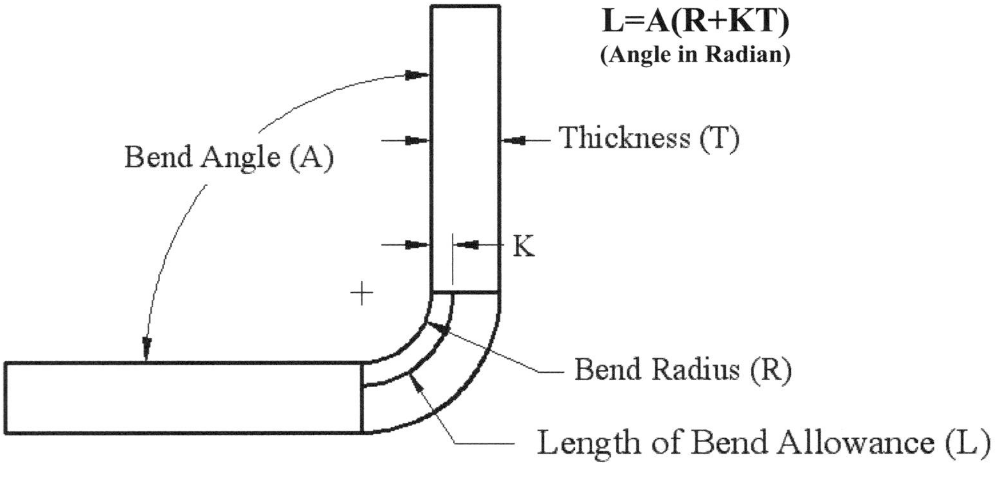

Sheet metal fabricators will typically have developed a k-factor table (usually through trial and error) to use. Autodesk Inventor is set up to allow the k-factor to be added to create material specific profiles. By using the correct data, Autodesk Inventor can be used to create fairly accurate and reliable flat patterns.

The Actuator Bracket Design

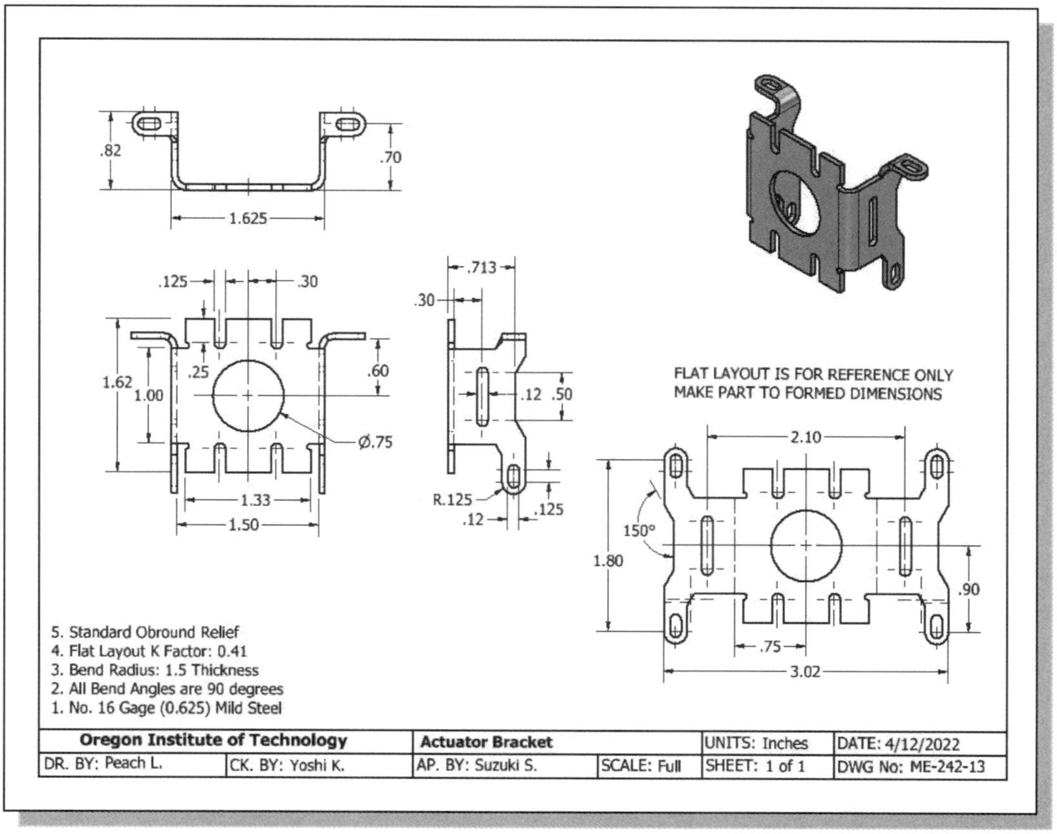

Starting Autodesk Inventor

1. Select the **Autodesk Inventor** option on the *Start* menu or select the **Autodesk Inventor** icon on the desktop to start Autodesk Inventor. The Autodesk Inventor main window will appear on the screen.

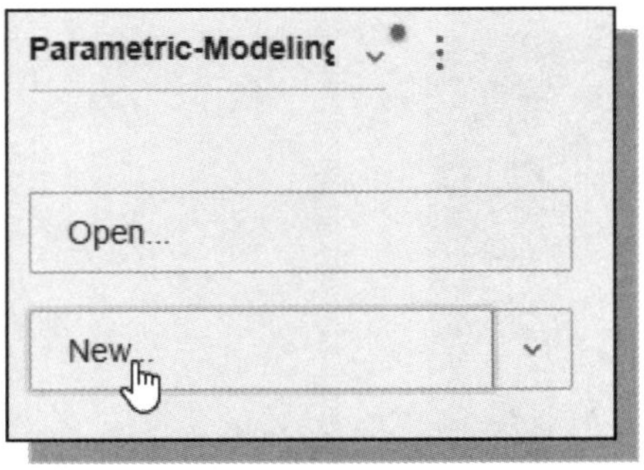

2. Select the **New File** icon with a single click of the left-mouse-button in the *Launch* toolbar.

3. Select the **English** tab, and in the *Template* list select **Sheet Metal (in).ipt** (*Standard Inventor Sheet Metal* template file).

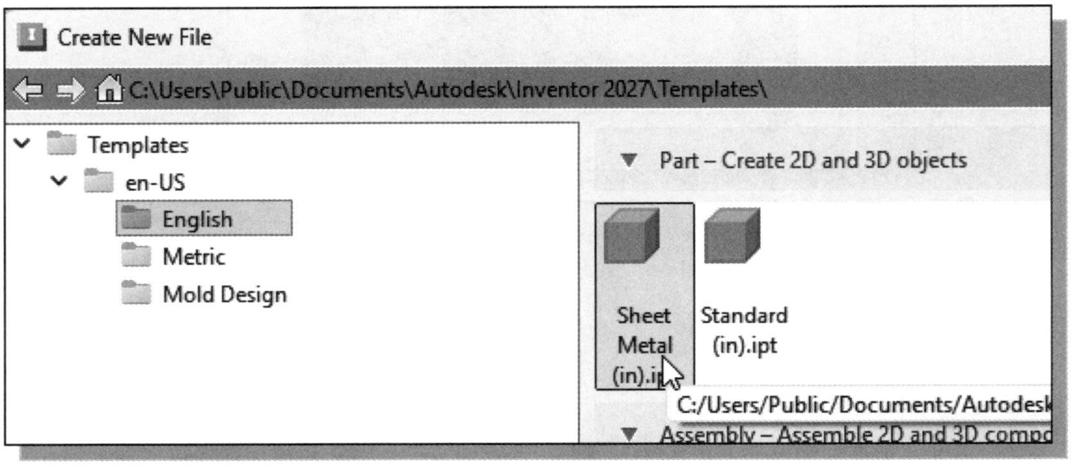

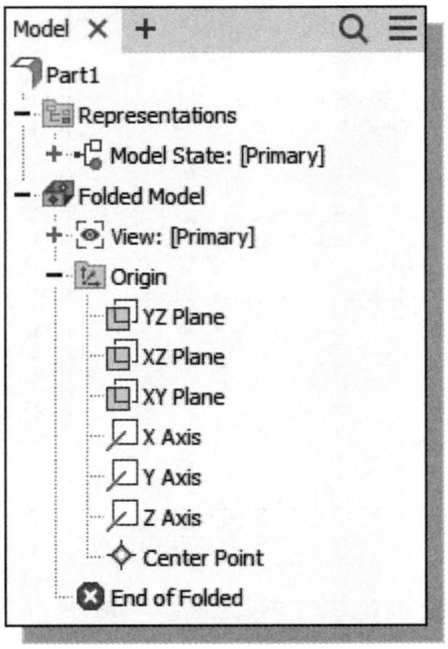

4. Click on the **Create** button in the *New File* dialog box to accept the selected settings.

- In the *browser* window, sheet metal part **Part1** is displayed with a set of work planes, work axes and a work point. In most aspects, the usage of work planes, work axes and work point is very similar to that of the *Inventor Part Modeler*.

- Notice, in the *Sheet Metal Features* panel, some of the basic sheet metal related options are available: **Sheet Metal Defaults**, **Face** and **Contour Flange**. As the names imply, these are the options to set up the sheet metal design and should be used as the starting point of the sheet metal design.

Sheet Metal Defaults

The **Sheet Metal Defaults** command allows us to specify material style, thickness, and rules for the active sheet metal part. The active sheet metal style contains the definitions of the **Unfold**, **Bend** and **Corner Relief** options, and alternative flat pattern representations for punches.

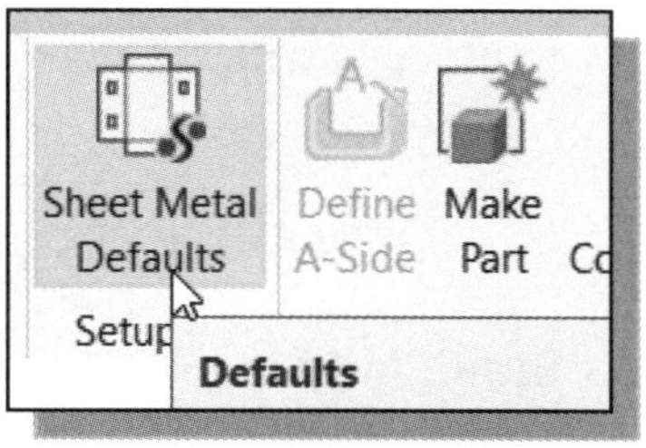

1. In the *Sheet Metal Setup* panel, select the **Sheet Metal Defaults** command by left-clicking the icon.

❖ Note the three *Sheet Metal Style* settings are listed as shown.

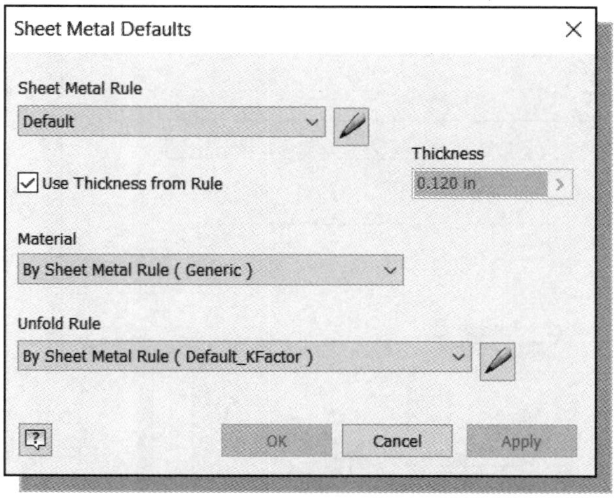

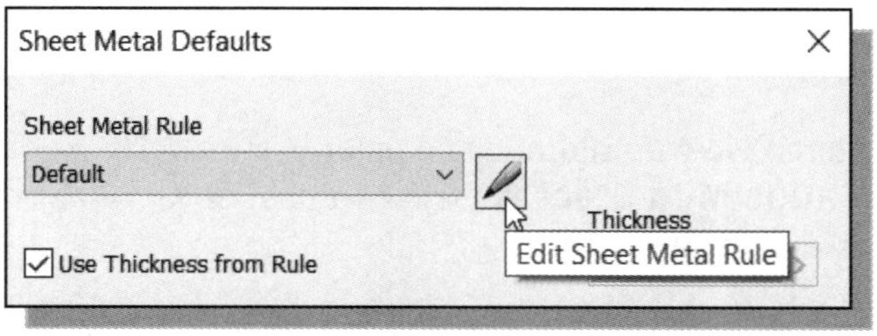

2. Click on the **Edit Sheet Metal Rule** icon as shown in the figure.

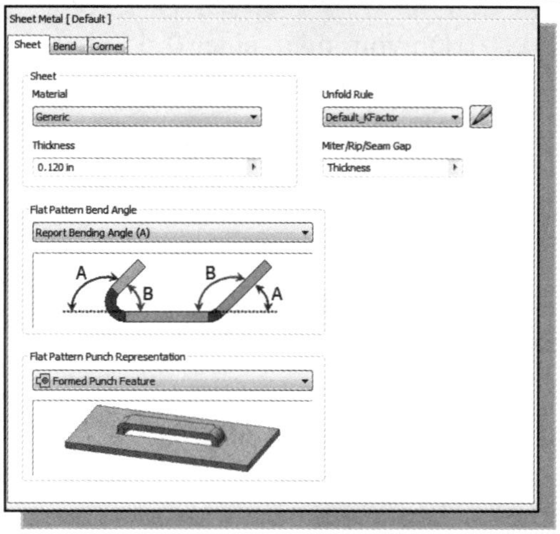

❖ Note the *Default* sheet metal style is **Read Only**.

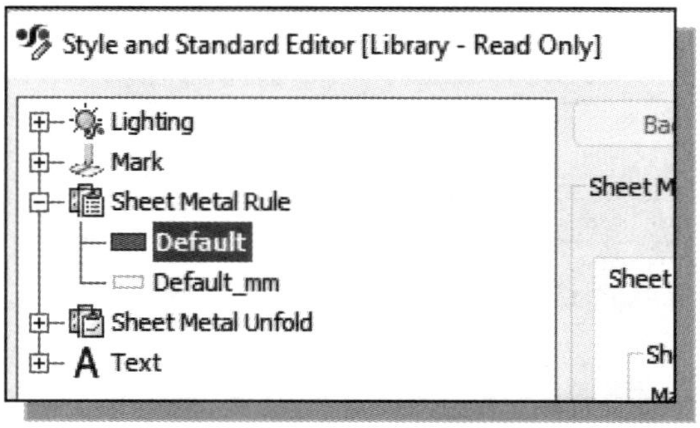

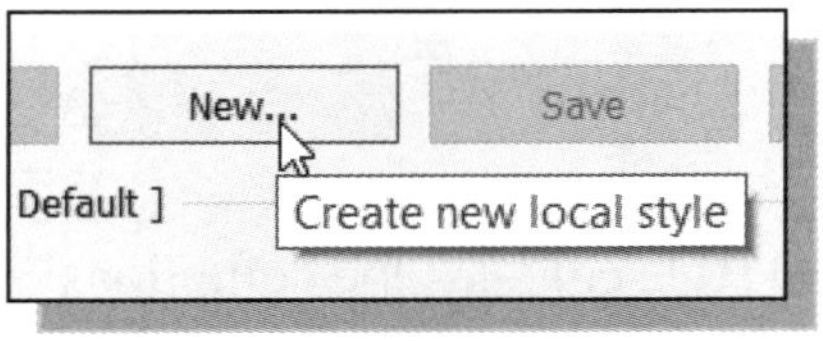

3. Select the **New** icon to create a new style in the list window.

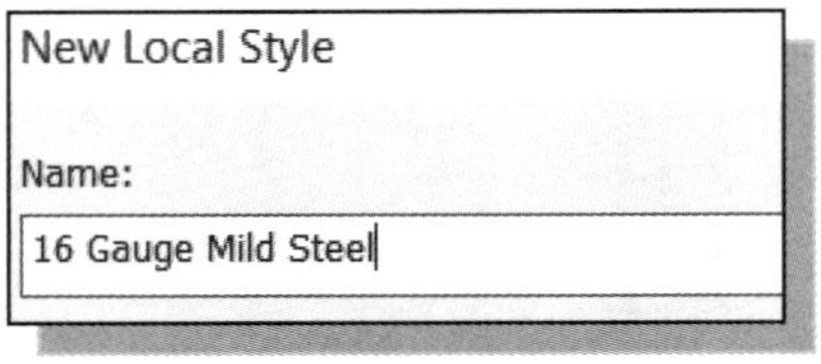

4. Enter **16 Gauge Mild Steel** as the *New Style Name* as shown. Click OK to proceed with the new style name.

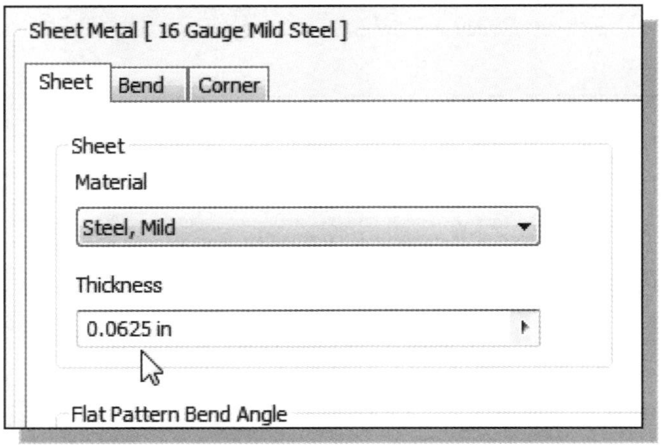

5. Select **Steel, Mild** as the sheet metal *Material* as shown.

6. Enter **0.0625** as the *Material Thickness* as shown.

❖ On your own, examine the other options in the different tabs under the *Edit Sheet Metal Rule* settings.

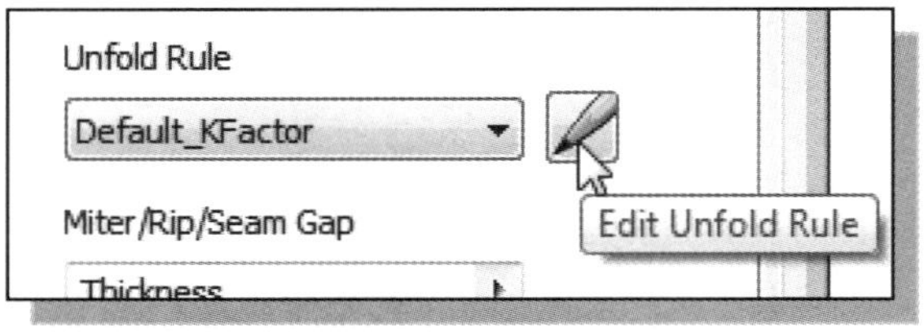

7. In the Unfold Rule section, pick the **Edit Unfolding Rule** icon as shown.

8. Click **Yes** to save the *edits*.

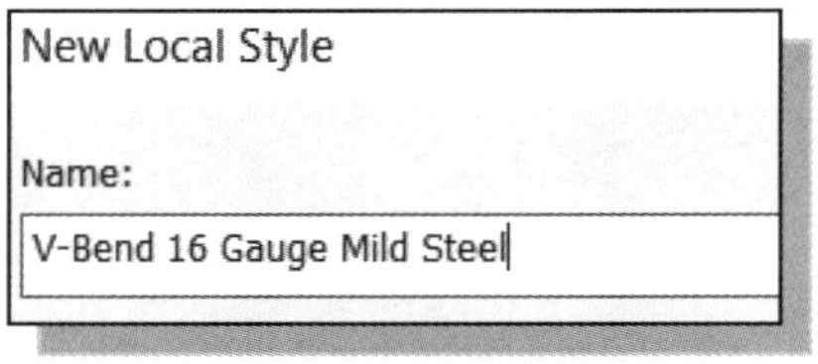

9. Start a **New** unfold rule and enter **V-Bend 16 Gauge Mild Steel** as the *New Local Style Name* as shown.

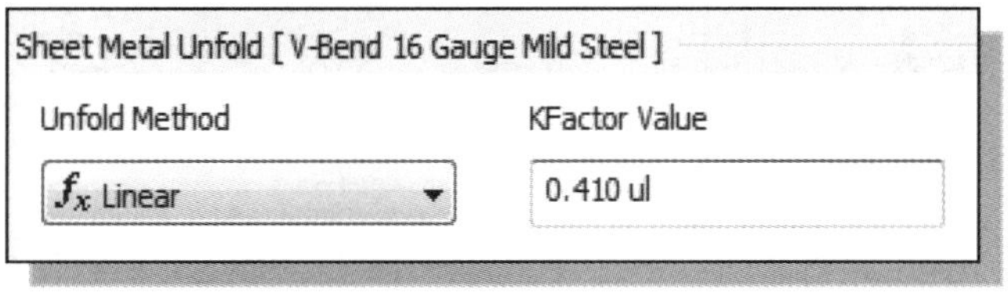

10. On your own, enter **0.41** as the *Linear KFactor Value* as shown.

11. On your own, **Save** and **Close** the *K-Factor* setting option.

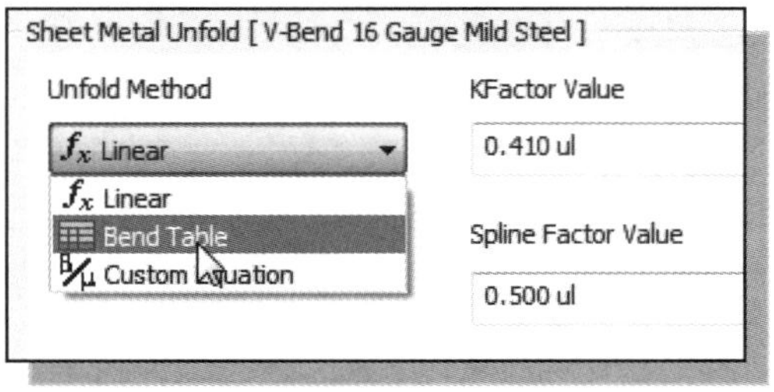

❖ Note, for more complex settings, the **Bend Table** option is also available.

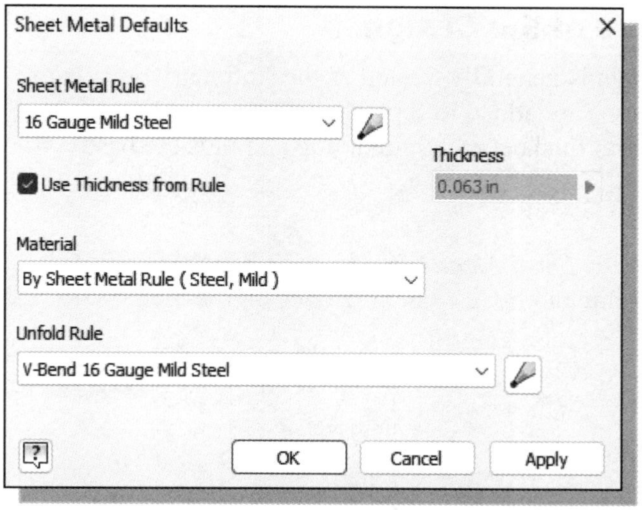

12. In the *Sheet Metal Defaults* dialog box, set the *Sheet Metal Rule* to use the newly created **16 Gauge Mild Steel** style as shown.

13. Set the *Material Style* to **By Sheet Metal Rule (Steel, Mild)** and the *Unfolding Rule* to **V-Bend 16 Gauge Mild Steel** as shown in the figure.

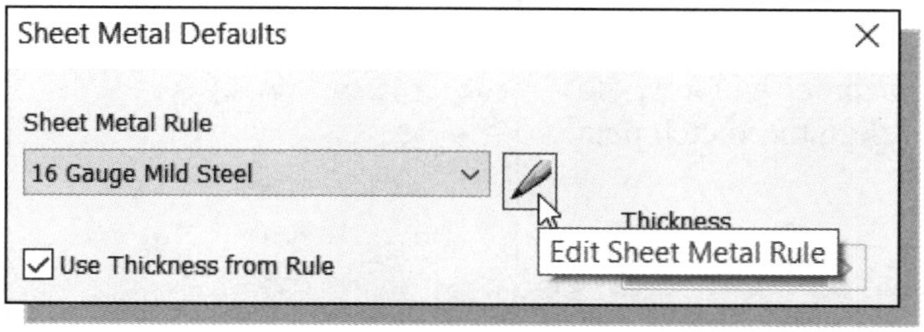

14. Click on the **Edit Sheet Metal Rule** icon as shown.

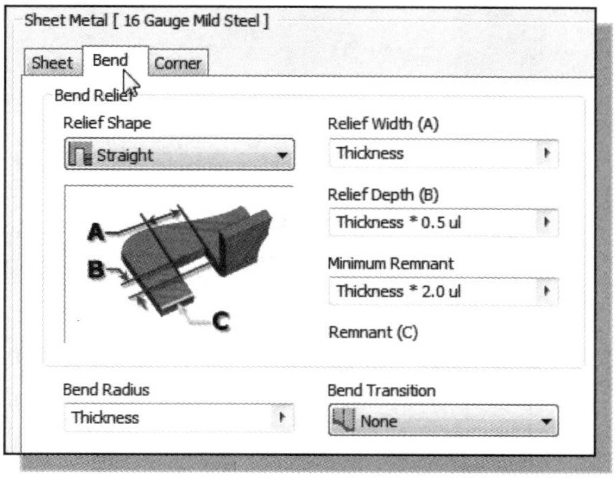

15. Click on the **Bend** tab to examine the *Bend Relief* settings.

16. Choose **Round** as the *Relief Shape*.

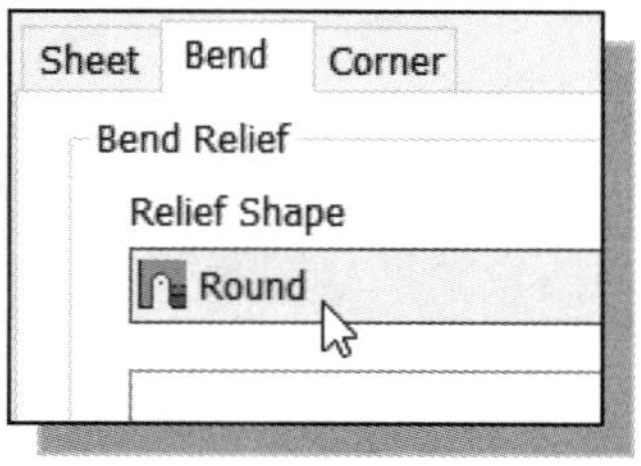

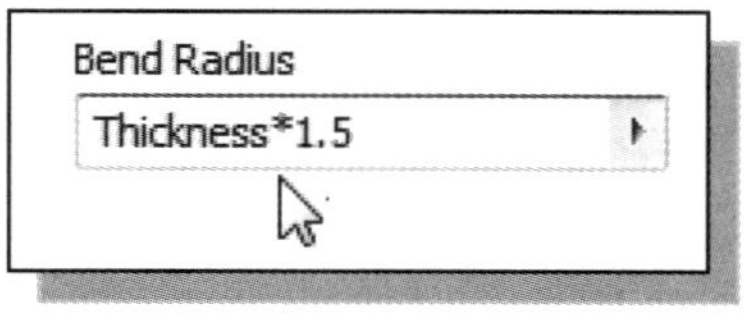

17. Enter **Thickness * 1.5** as the *Bend Radius* as shown in the figure.

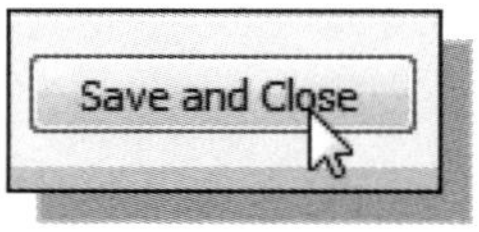

18. Click **Save and Close** to accept the settings and end the Edit Sheet Style command.

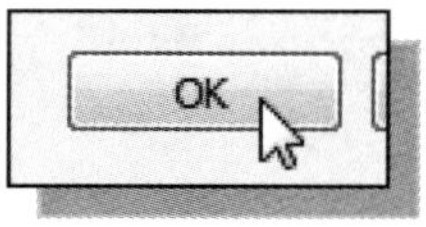

19. Click **OK** to accept the settings and end the Sheet Metal Defaults command.

Create the Base Face Feature of the Design

The main section of a sheet metal design is generally treated as the stationary portion of the design, to which all the other sections are added to form the final design. The main section is also typically the starting point of sheet metal modeling and thus the base feature in parametric modeling.

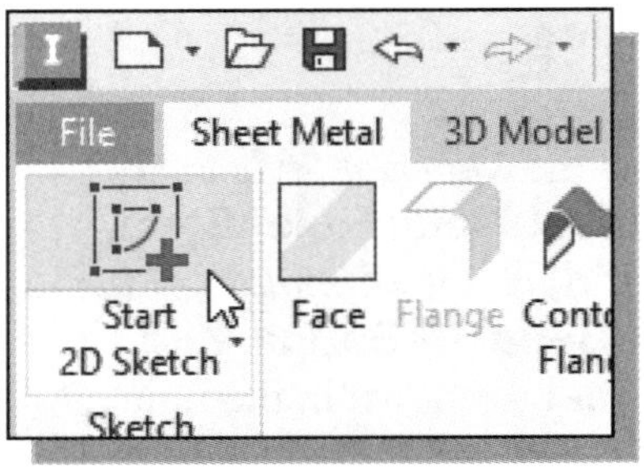

1. In the *Sheet Metal* tab select the **Start 2D Sketch** command by left-clicking once on the icon.

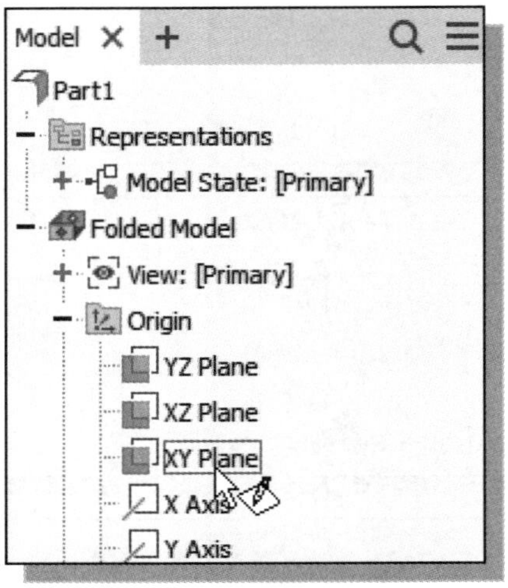

2. Select the **XY Plane**, in the *browser* window, of the sheet metal part Part1 to align the sketch plane.

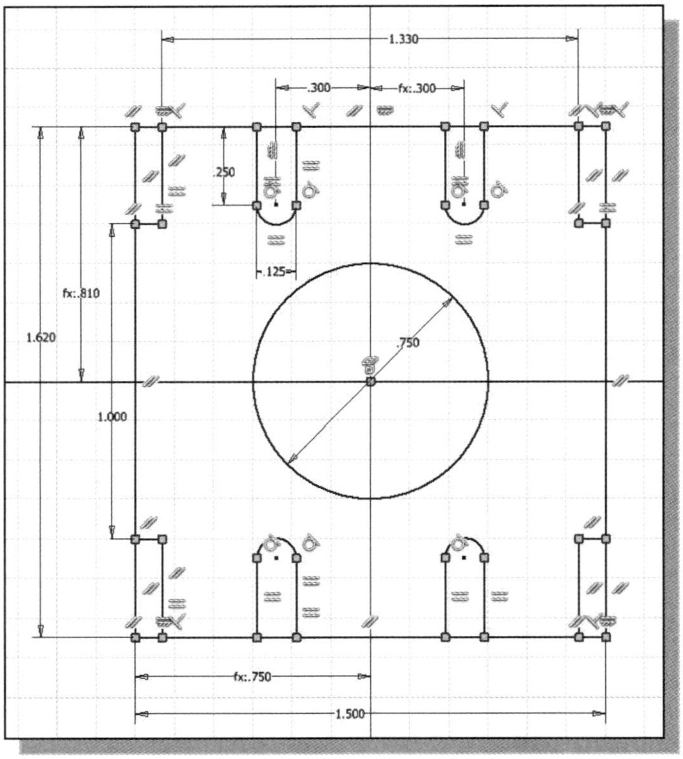

3. On your own, construct the 2D sketch and apply the proper constraints and dimensions as shown. (Hint: Use the constraint tools to align the symmetrical geometry.)

❖ Note the design is symmetrical about the horizontal and vertical axes.

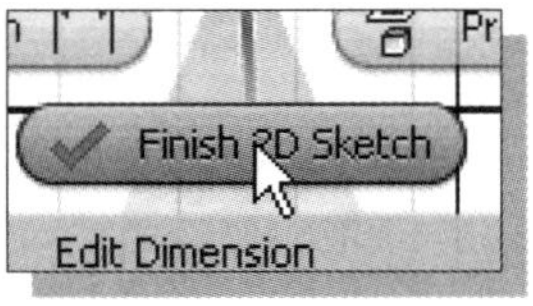

4. Use the *option menu* and select **Finish 2D Sketch** to end the *2D Sketch* mode and return to the *Sheet Metal Features* panel.

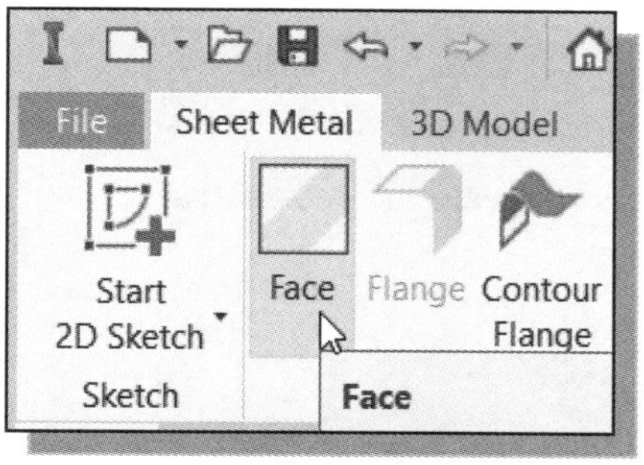

5. In the *Sheet Metal Features* panel, select the **Face** command by left-clicking the icon.

6. Select the profile as shown in the figure below.

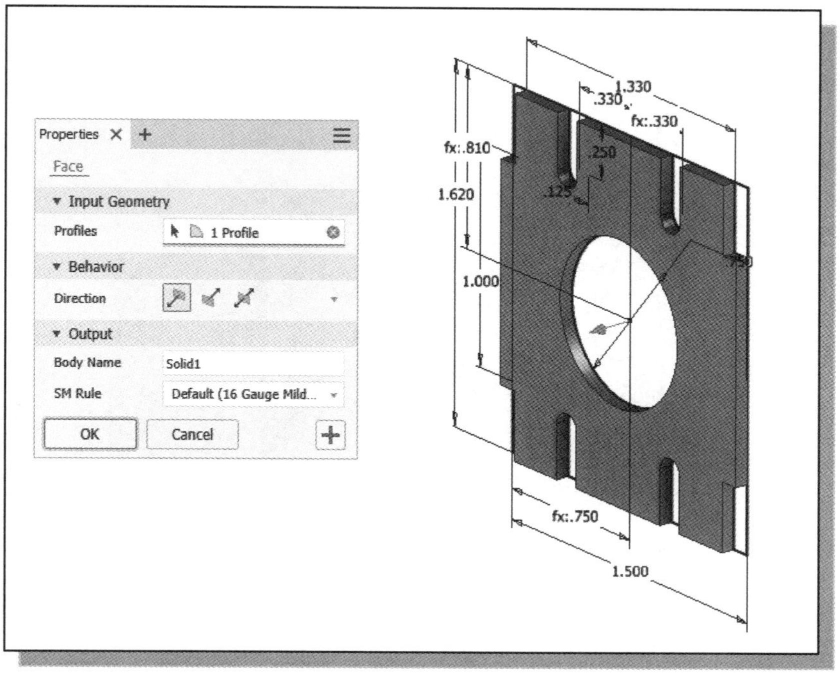

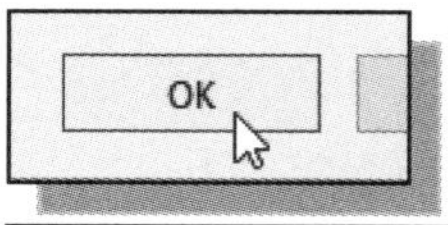

7. Click **OK** to accept the settings and proceed to creating the base feature of the design.

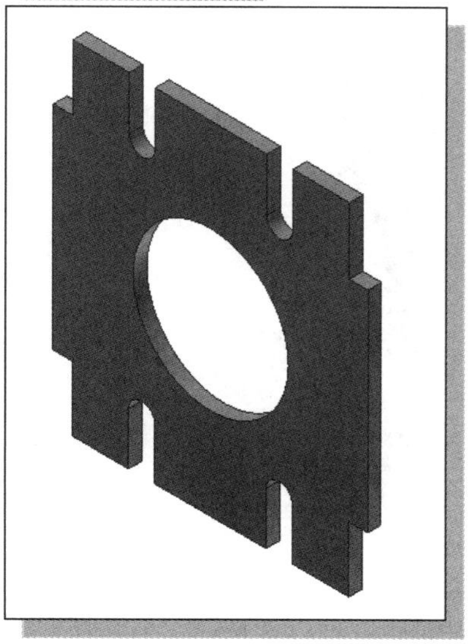

- The base feature of the design is constructed using the defined *Sheet Metal Defaults* settings.

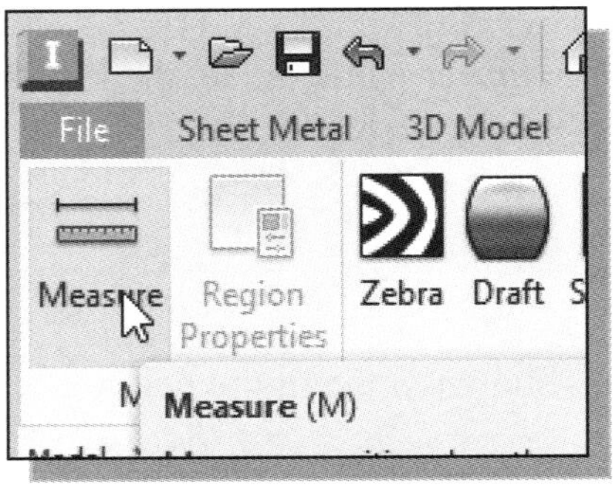

8. In the *Inspect menu*, activate the **Measure command** as shown.

9. Select any edge in the thickness direction and confirm the distance measured is **0.0625**, matching our setting.

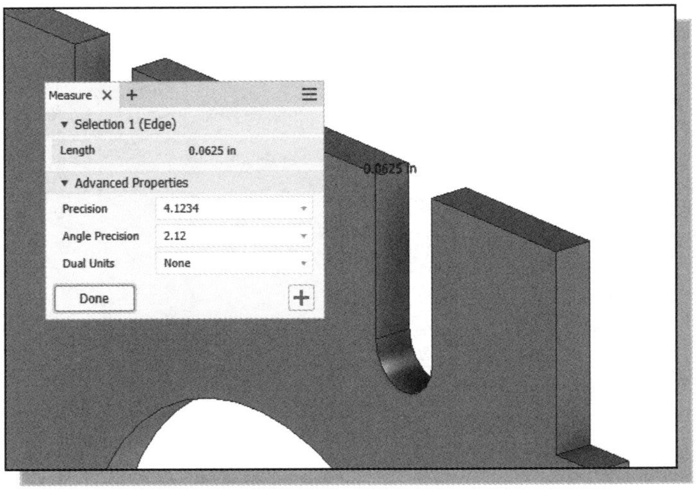

10. On your own, confirm the overall width of the design is at 1.5 inches.

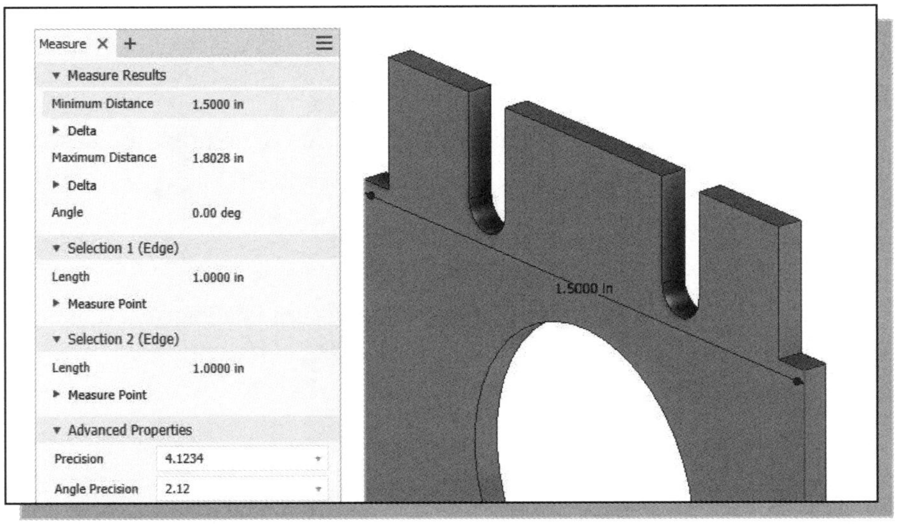

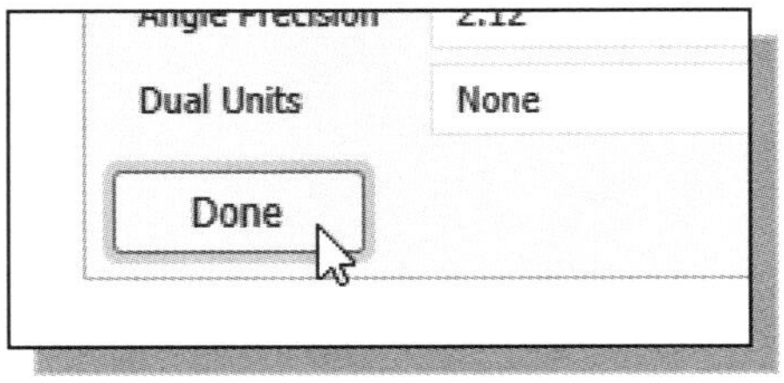

11. Click on the **Done** button to exit the Measure command.

Using the Flange Command

Sheet metal flange features consist of a flat face, which has a bend that connects the flat face to an existing straight edge of the sheet metal model. Flange features are added by selecting one or more edges and by specifying a set of options which determines the size and position of the material added. The **Flange** command can be used to build additional sections on a selected edge and with extra controls on the bend.

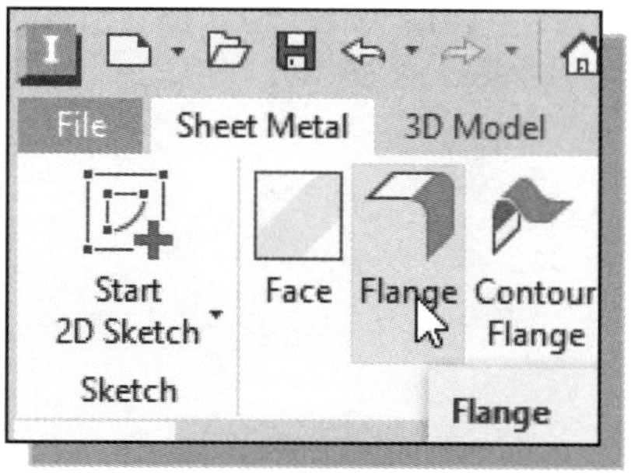

1. In the *Sheet Metal Features* panel, select **Flange** to activate the command.

❖ The **Flange** command requires the selection of an existing straight edge and creates a flat feature, while the **Contour Flange** command allows the creation of 3D surfaces that are not flat.

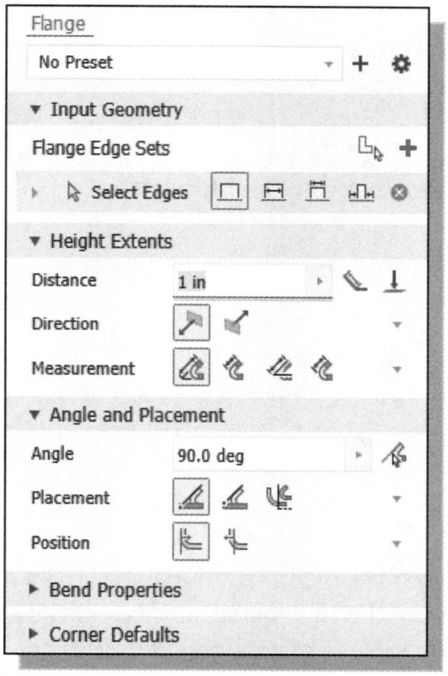

❖ The *Flange* dialog box contains many settings to create a flange feature.

2. In the *Height Extents* section, set the **Distance** to **0.5** inches as shown in the figure.

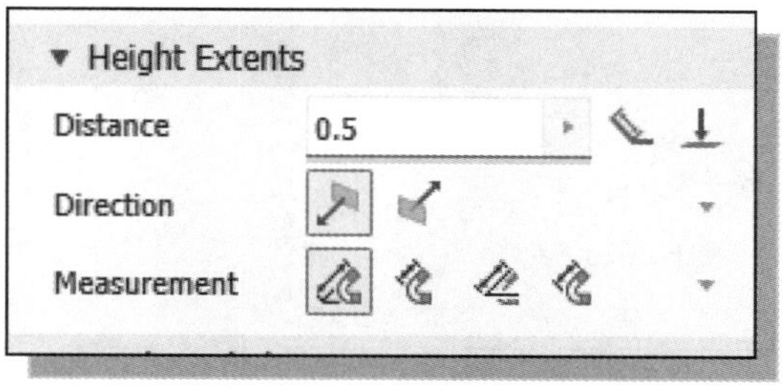

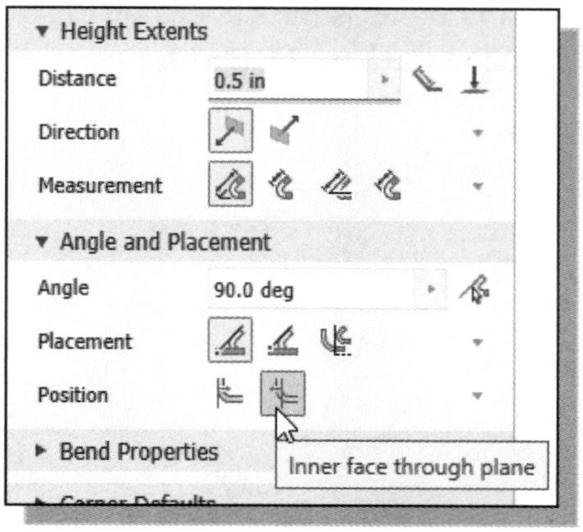

3. Set the *Bend Position* to **Inner face through plane** as shown.

❖ The ***Position*** option is used to control the position of the bend that connects to the new flange section.

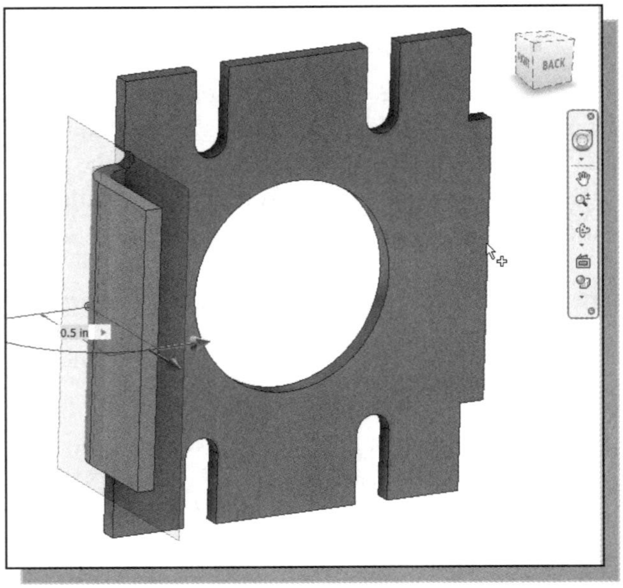

4. Rotate the display by clicking on the **ViewCube** so that you are viewing the back side of the base feature.

5. Select the **two vertical side edges** on the back face to add the two flanges as shown.

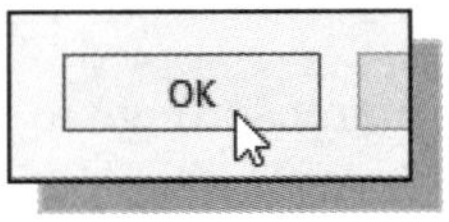

6. In the *Flange* dialog box, click **OK** to accept the selections and settings to create the features.

7. In the *Inspect menu*, activate the **Measure command** as shown.

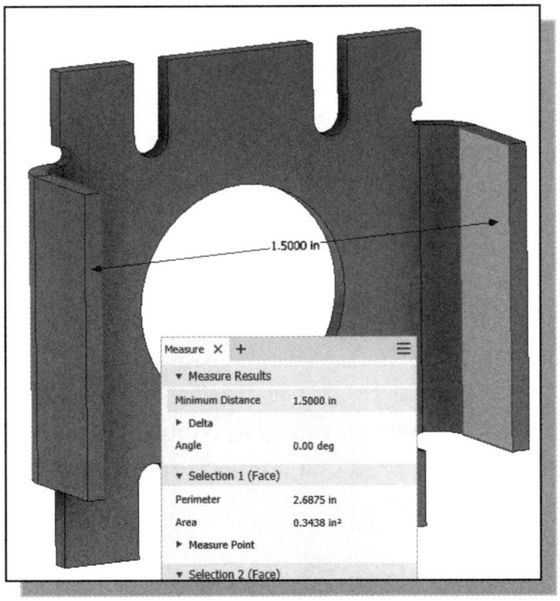

8. On your own, confirm the distance between the two inside surfaces of the two flanges is maintained as **1.500**. This is the same as the dimension specified in the base feature.

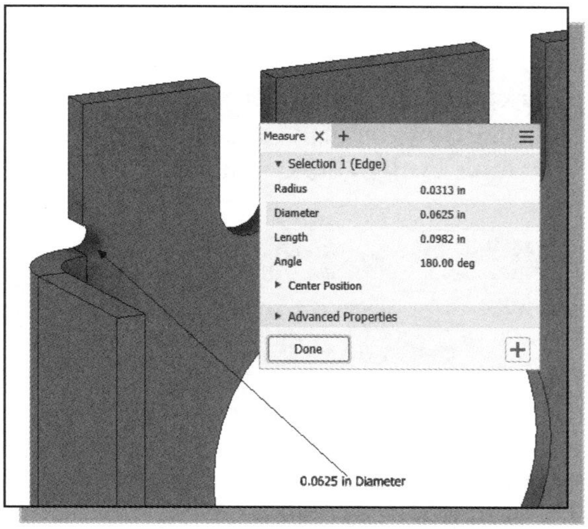

❖ Note the relief cut is also added when the two flanges are created.

9. On your own, measure the size of the relief and confirm it is equal to the thickness of the material.

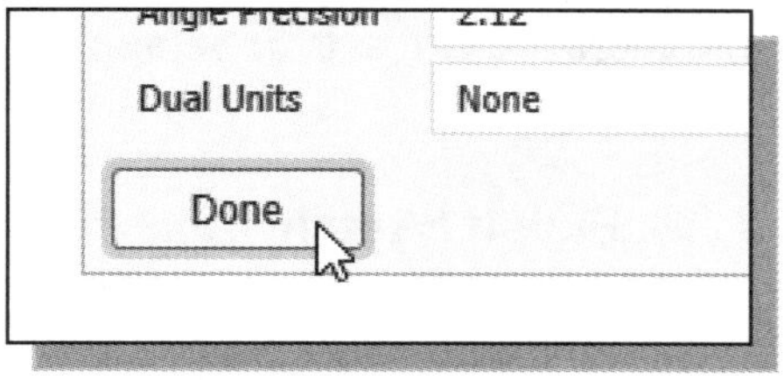

10. Click on the **Done** button to exit the Measure command.

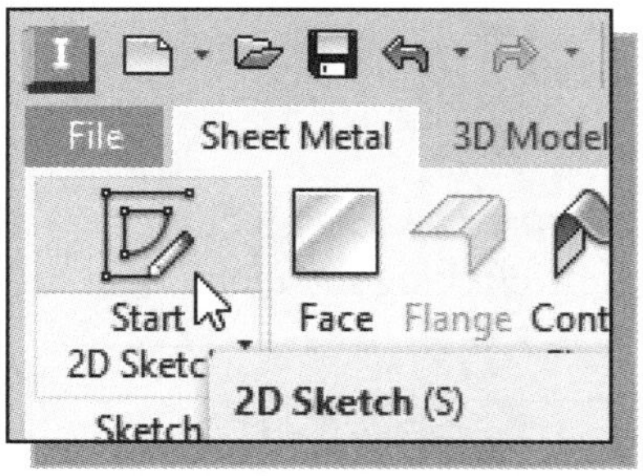

11. In the *Standard* toolbar select the **Sketch** command by left-clicking once on the icon.

12. Select one of the outside surfaces of the *Flanges* we just created.

❖ We will next create a more complex sketched section on the existing flange surface. Note that it is also feasible to create this new section directly aligned to the edges of the base feature without first creating the flange feature.

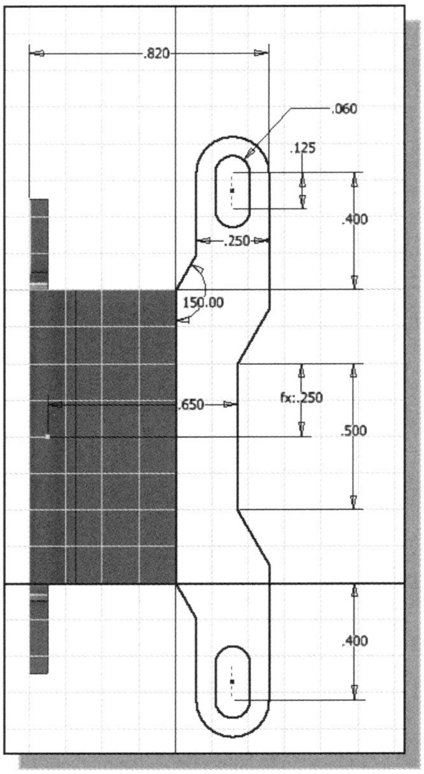

13. On your own, construct the 2D sketch and apply the proper constraints and dimensions as shown. (Hint: Use the **Slot** command and use the constraint tools to align the symmetrical geometry.)

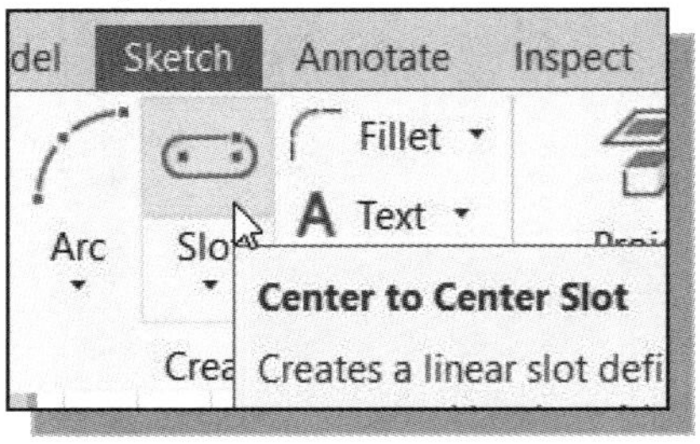

❖ Note the design is symmetrical about the horizontal axis and there are two sets of parallel inclined lines.

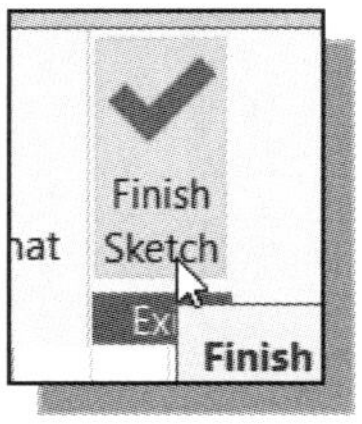

14. Click **Finish Sketch** to end the *2D Sketch* mode and return to the *Sheet Metal Features* panel.

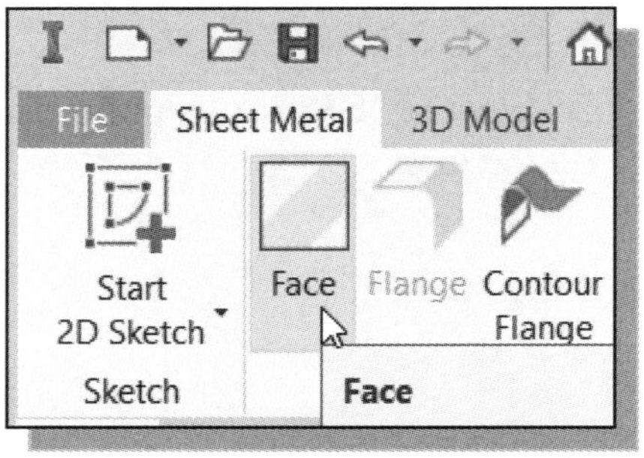

15. In the *Sheet Metal Features* panel, select the **Face** command by left-clicking the icon.

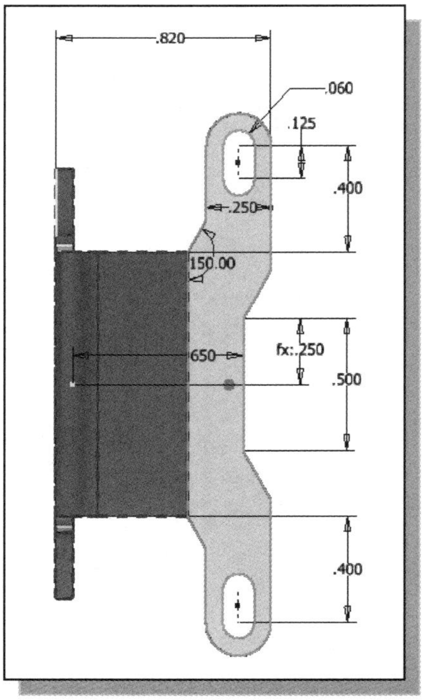

16. Select the **profile** as shown in the figure.

17. Click **OK** to accept the settings and proceed to creating the face feature.

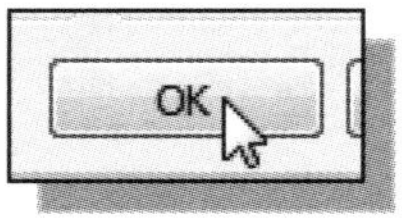

Mirroring Features

In Autodesk Inventor, a sheet metal feature can be mirrored just like a regular solid feature. The mirrored feature is parametrically linked to the original parametric definitions.

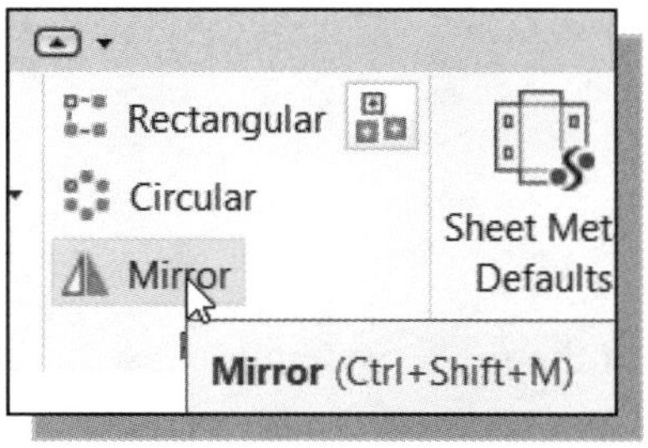

1. In the *Sheet Metal Pattern* panel, select the **Mirror Feature** command by clicking the left-mouse-button on the icon.

➢ In the *Mirror Pattern* dialog box, the **Features** button is activated. Autodesk Inventor expects us to select features to be mirrored.

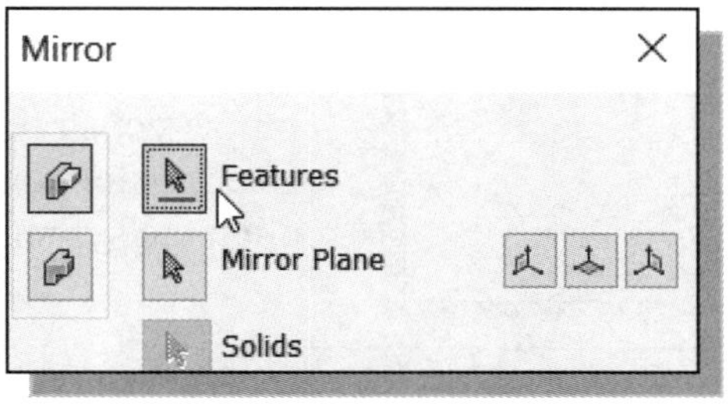

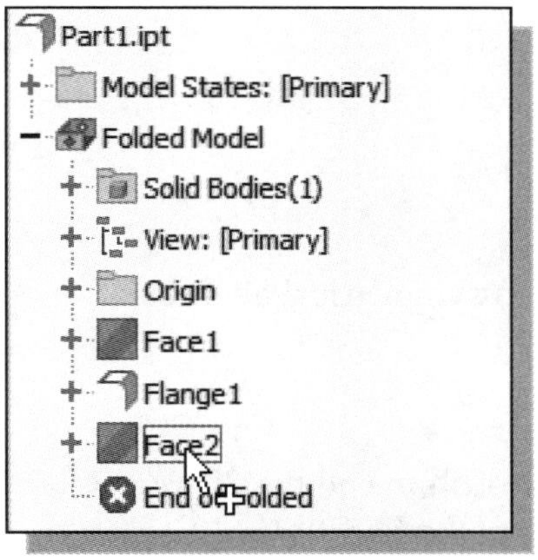

2. In the prompt area, the message "*Select feature to pattern*" is displayed. Select the last face feature in the *browser* window as shown.

3. Inside the graphics window, right-click to bring up the option menu.

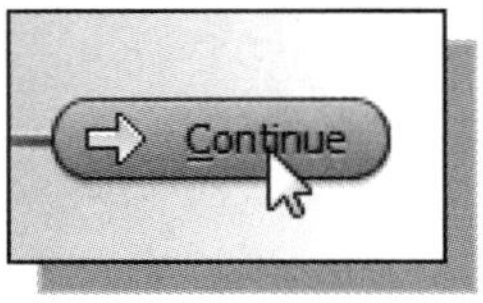

4. Select **Continue** in the option list to proceed with the Mirror Feature command.

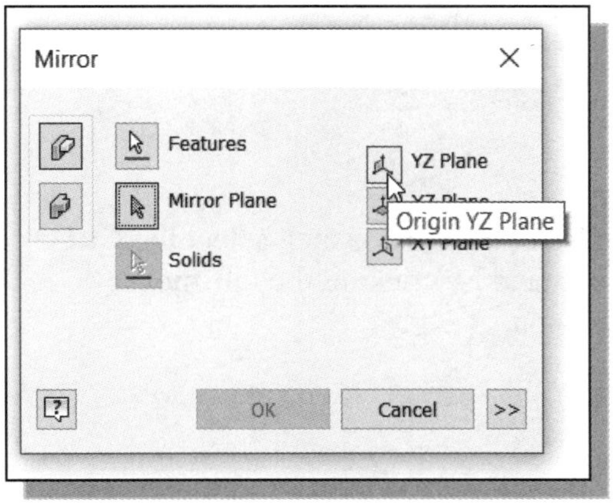

5. Select the **YZ Plane** in the *dialog box* as shown.

6. Click **OK** to create the *Mirror* feature.

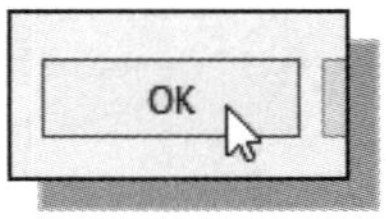

Create a Cut Feature

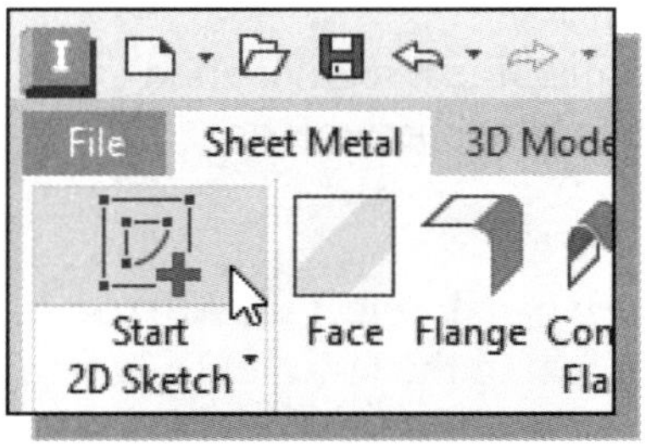

1. In the *Sheet Metal* tab select the **Start 2D Sketch** command by left-clicking once on the icon.

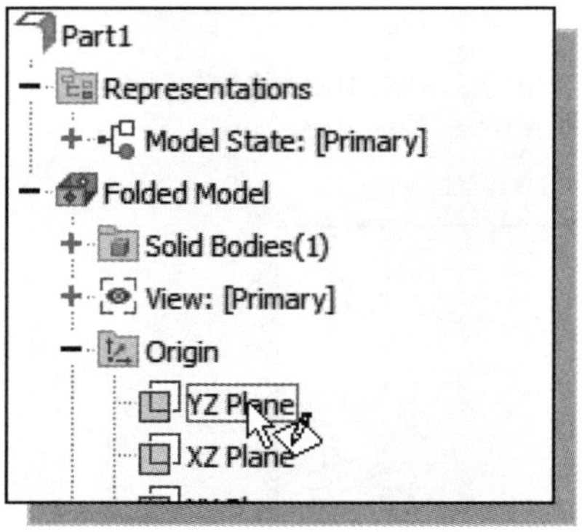

2. Select the **YZ Plane** in the *browser* window as shown.

3. On your own, construct the 2D sketch and apply the proper constraints and dimensions as shown. (Hint: Use the **slot** command and/or constraint tools to align the symmetrical geometry.)

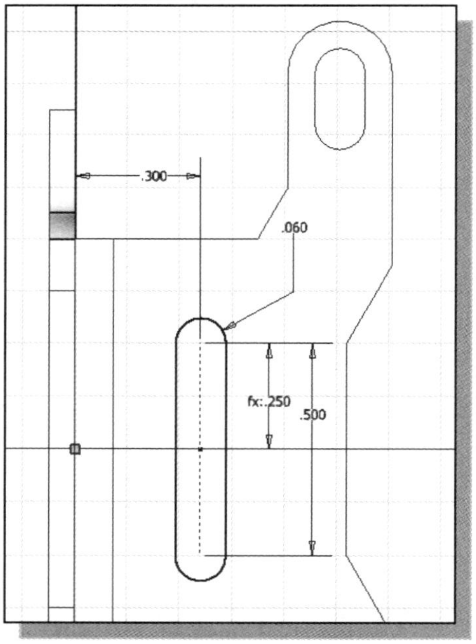

❖ Note the design is symmetrical about the horizontal axis.

4. Click **Finish sketch** to end the *2D sketch* mode and return to the *Sheet Metal Features* panel.

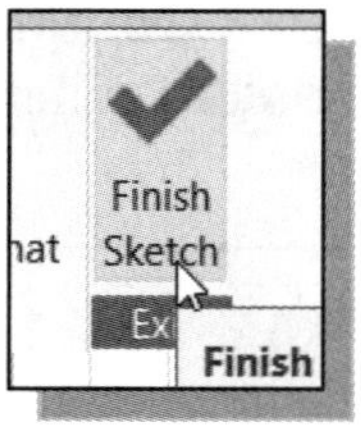

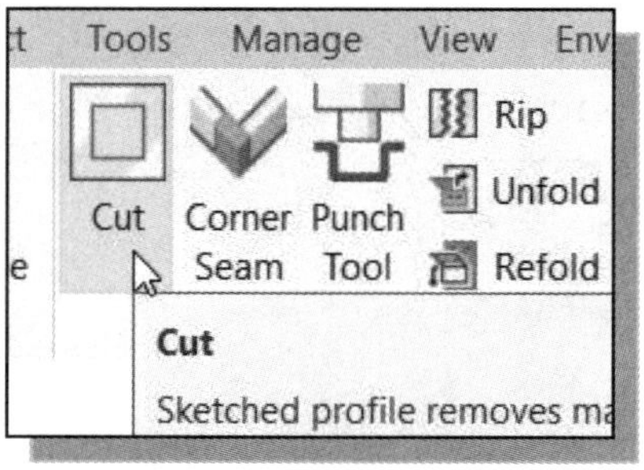

5. In the *Sheet Metal Features* panel, select the **Cut Feature** command by clicking the left-mouse-button on the icon.

6. In the *Cut* feature dialog box, set the *Cut Extents* to **All** and *Direction* to **Both Sides** as shown in the figure.

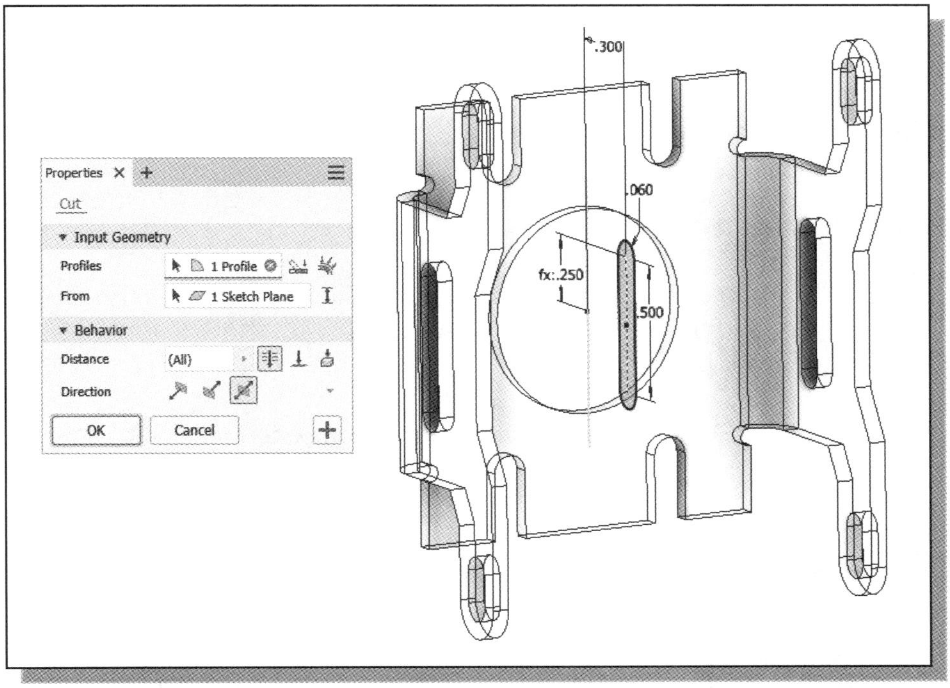

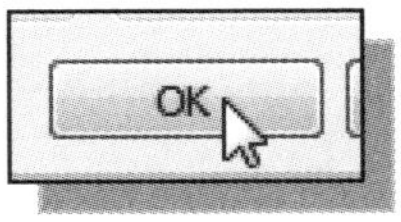

7. Click **OK** to create the ***Cut*** feature.

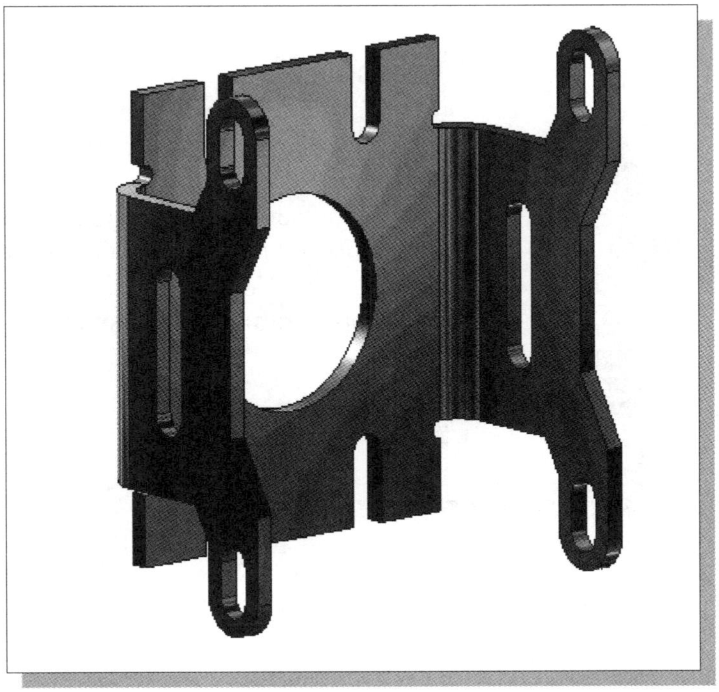

Create a Fold Feature

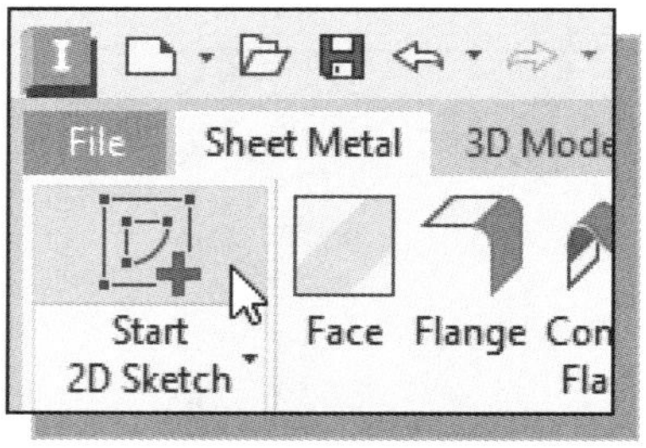

1. In the *Sketch* toolbar select the **Start 2D Sketch** command by left-clicking once on the icon.

2. Select the outside surface of the ***Face*** feature as shown.

3. In the *2D Sketch* panel, select the **Line** command as shown.

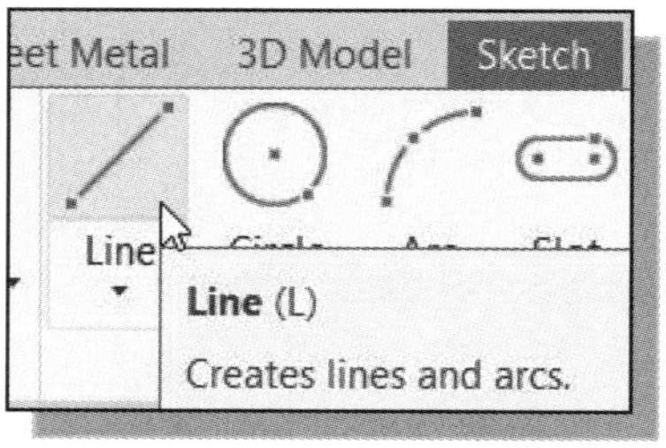

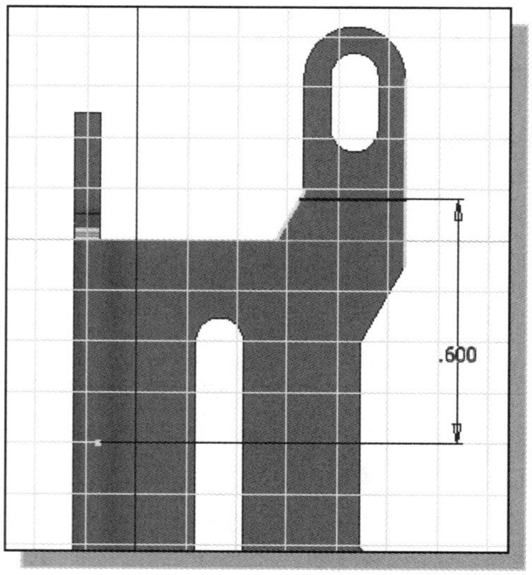

4. On your own, construct a **horizontal line** and apply a vertical dimension, measuring to the origin of the work planes, as shown.

❖ Note that the two endpoints of the *fold line* **must be on the two edges** of an existing surface. You can use the **trim** command or apply **coincident constraint** to achieve this.

5. Click **Finish sketch** to end the *2D Sketch* mode and return to the *Sheet Metal Features* panel.

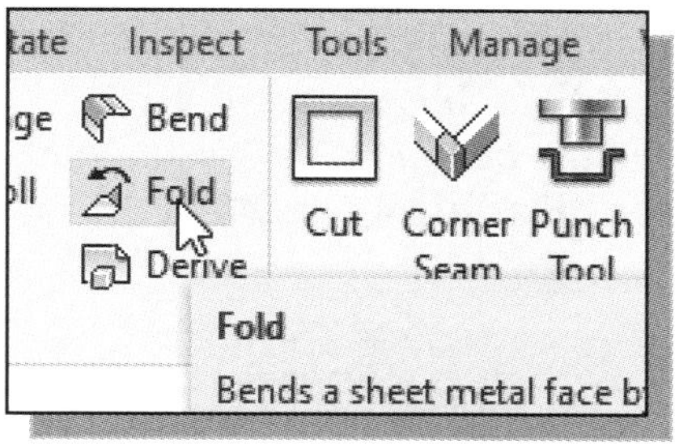

6. In the *Sheet Metal Features* panel, select the **Fold Feature** command by clicking the left-mouse-button on the icon.

7. Activate the **Bend Line selection** and pick the line we just created as shown in the figure below.

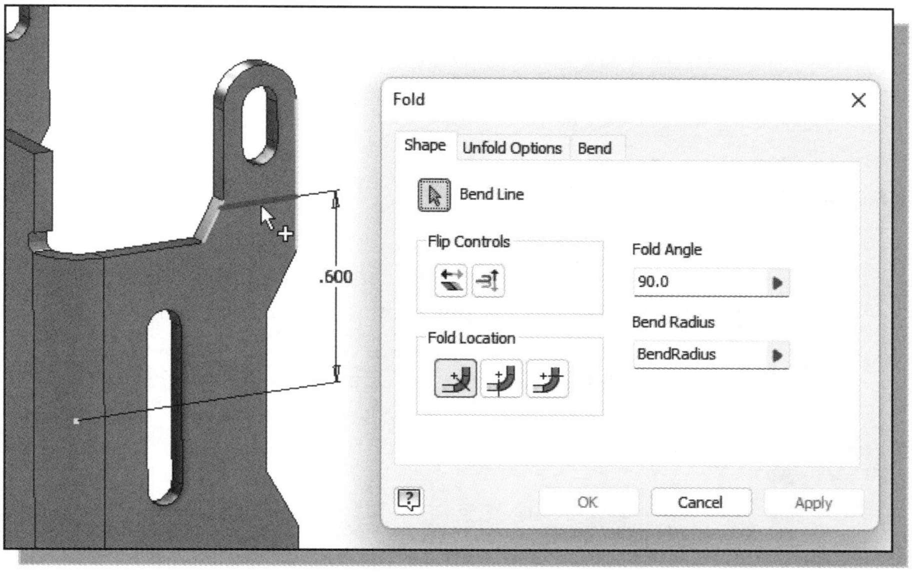

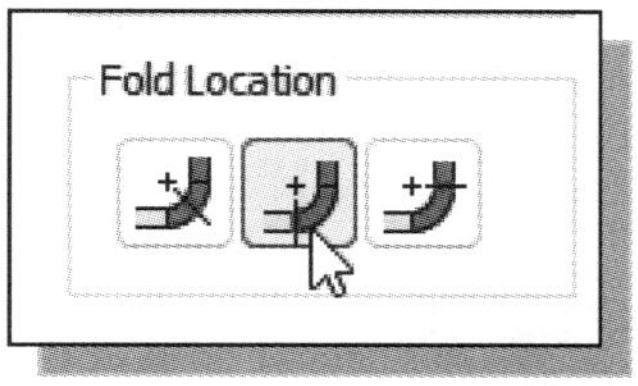

8. Set the *Fold Location* to **Start of Bend** as shown in the figure.

9. On your own, click on the two icons located in the ***Flip Controls*** area and observe the effect of the two controls. Reset the control arrows as shown in the figure before continuing to the next section.

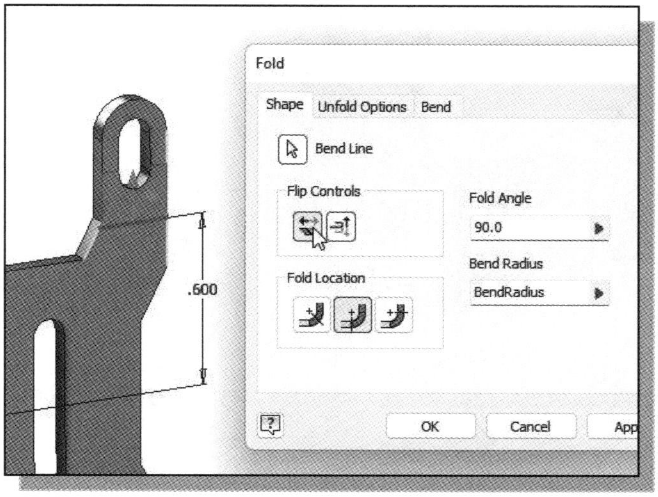

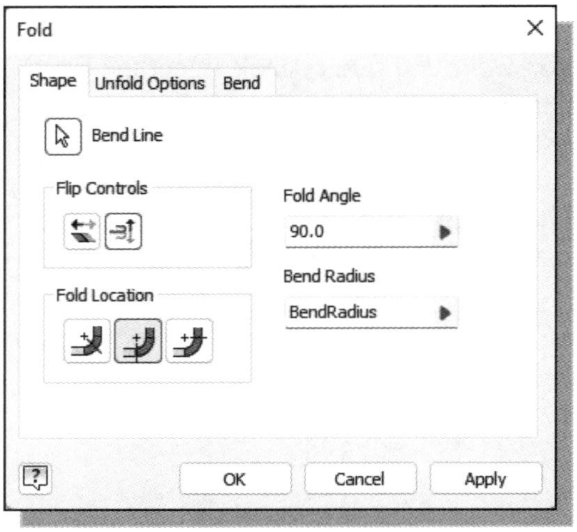

10. Confirm the *Fold Angle* is set to **90** degrees and the *Bend Radius* is set to the default **BendRadius** as shown.

11. Click **OK** to create the ***Fold*** feature.

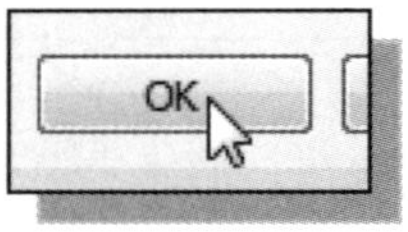

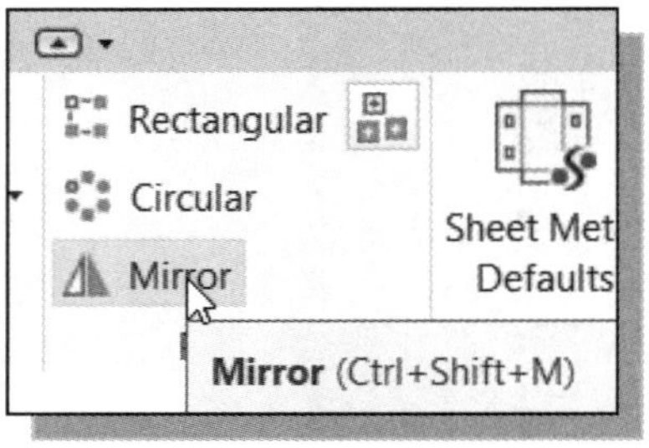

12. On your own, create a mirror image of the *Fold* feature, mirroring about the **YZ** work plane.

➢ The ***Fold*** feature requires the use of a ***bend line***, simulating the edge where the bend will occur.

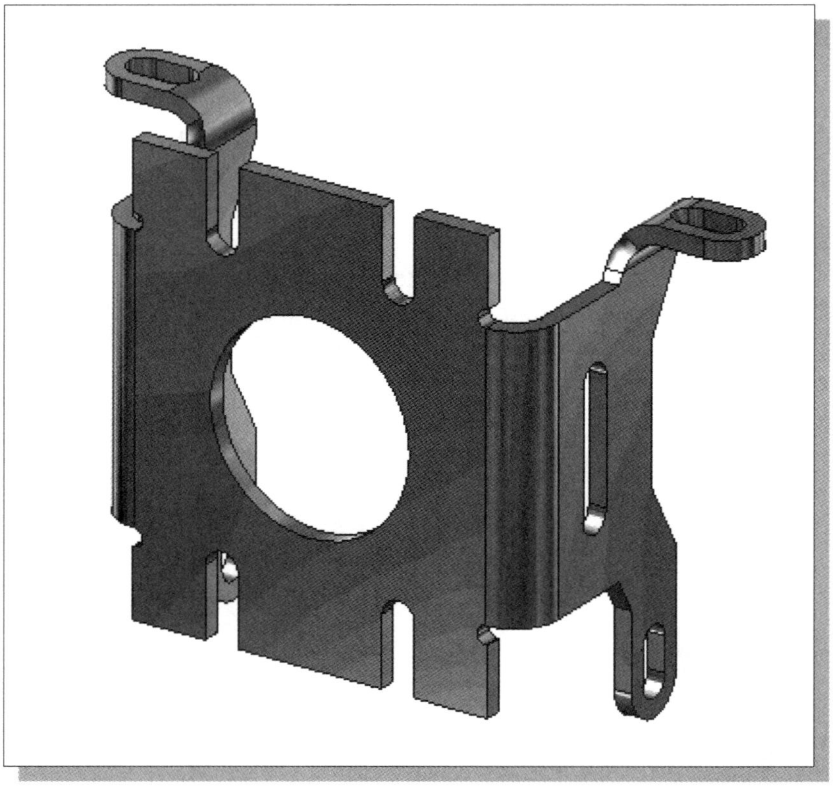

Create the Associated Flat Pattern

A sheet metal flat pattern is the shape of the sheet metal part before it is formed. A flat pattern is required to create drawings for manufacturing. The flat pattern shows the shape of the sheet metal part before it is formed showing all the bend lines, bend zones, punch locations, and the shape of the entire part with all bends flattened and bend factors considered.

The Inventor **Flat Pattern** command calculates the material and layout required to flatten a 3D sheet metal model. Once the flat pattern is created, the part *browser* window displays a *Flat Pattern* item and the flattened state of the model is displayed whenever this item is active. The flat pattern is typically created normal to the sketched base face feature, but this can be edited if necessary.

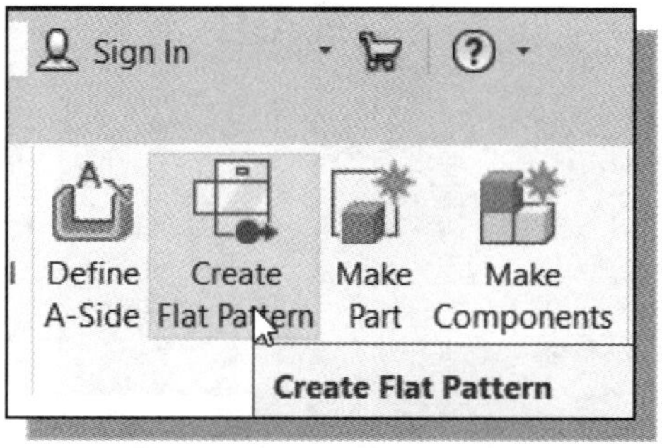

1. In the *Sheet Metal Features* panel, select the **Create Flat Pattern** command by clicking the left-mouse-button on the icon.

❖ Autodesk Inventor calculates the material and layout required to flatten the 3D sheet metal model and displays the flat pattern as shown.

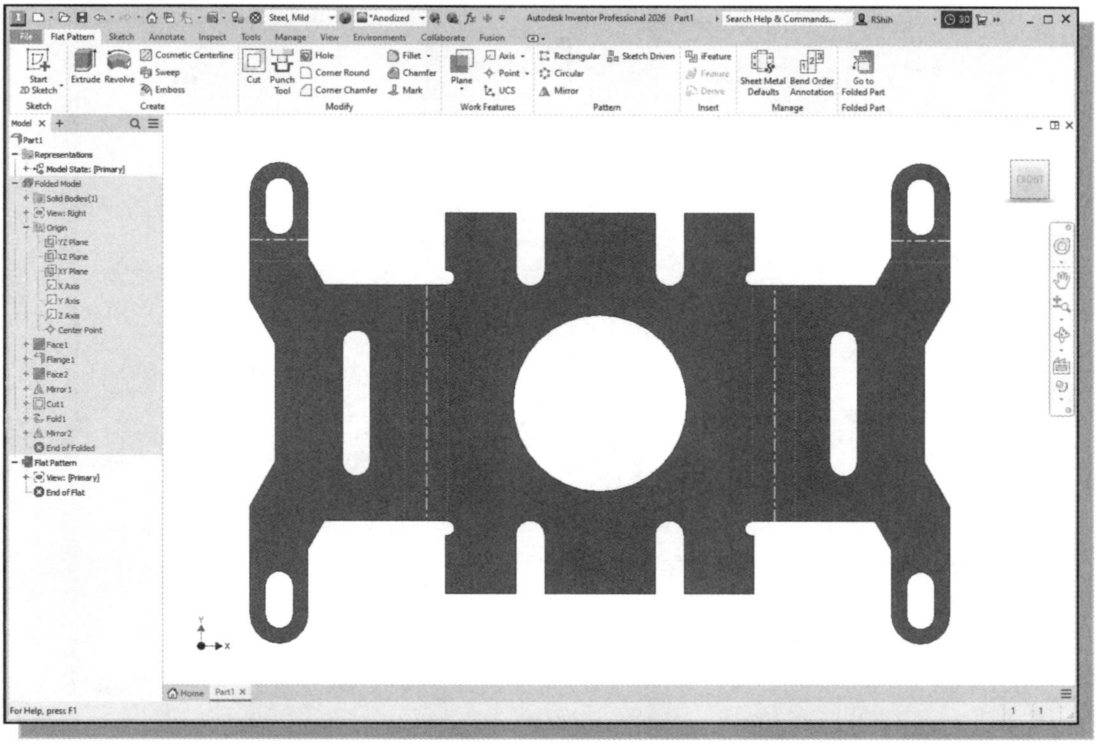

❖ Note that the displayed center lines and also dashed lines identify the locations and sizes of the bends.

Confirm the Flattened Length

1. In the *Inspect* Ribbon tab, activate the **Measure** command as shown.

2. Select the right edge, as shown in the figure below, and the large circle to measure.

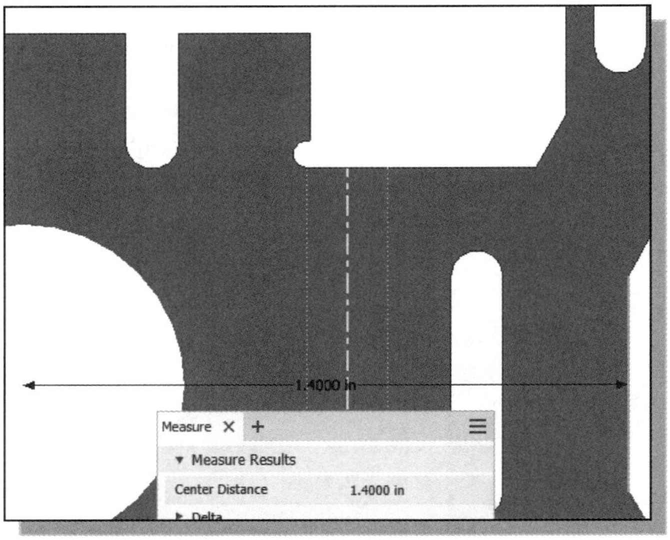

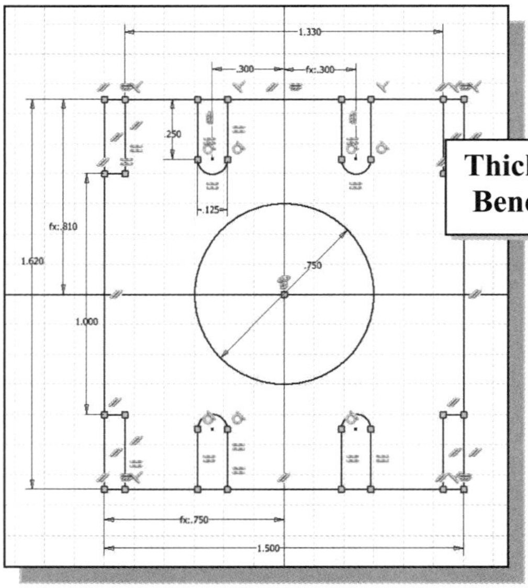

Thickness t=0.0625
Bend Radius=1.5t

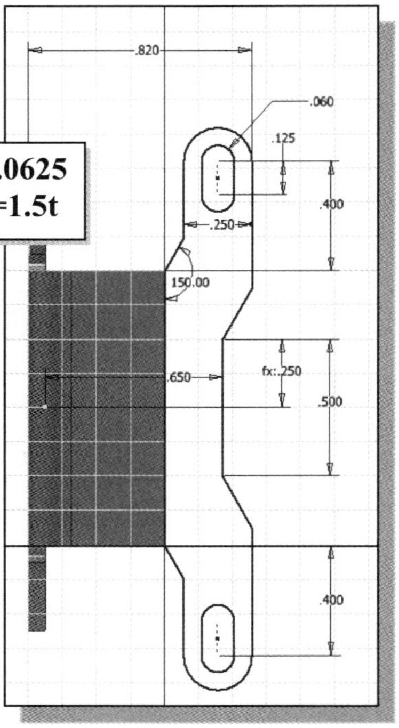

L=A(R+KT)
= π/2(1.5 t +0.41t)
= 0.1875

Flattened Length = W+H+L
=(1.500/2-1.5t)+(0.650-1.5t)+0.1875=1.400

Create a 2D Sheet Metal Drawing

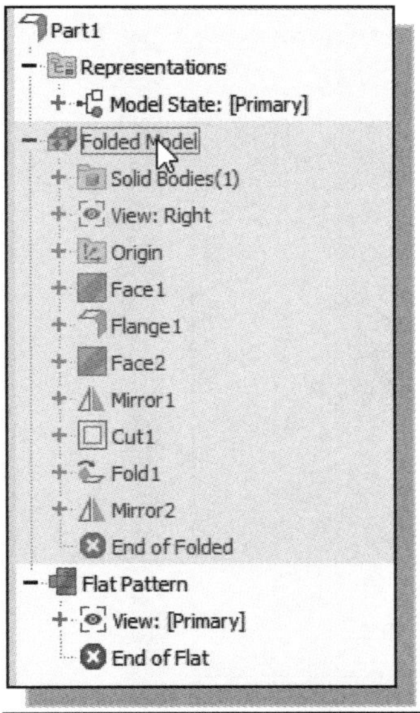

❖ Note in the part *browser* window, a **Flat Pattern** item is added once the associated flat pattern has been created.

1. In the part *browser* window, double-click on the **Folded Model** item to switch back to the 3D folded model.

2. On your own, **save** the design. Use the following filename: ***Actuator-Bracket.ipt***.

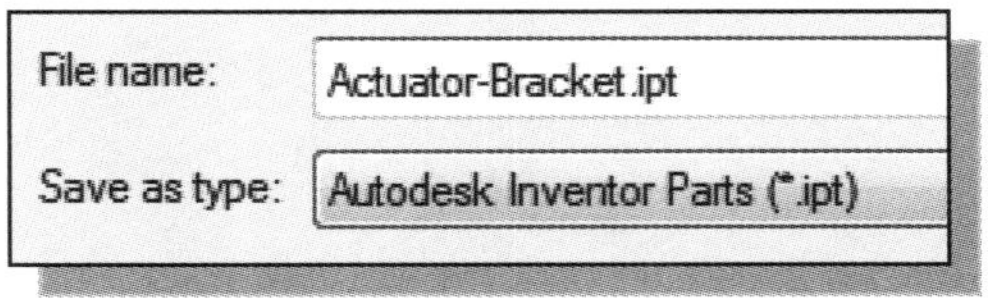

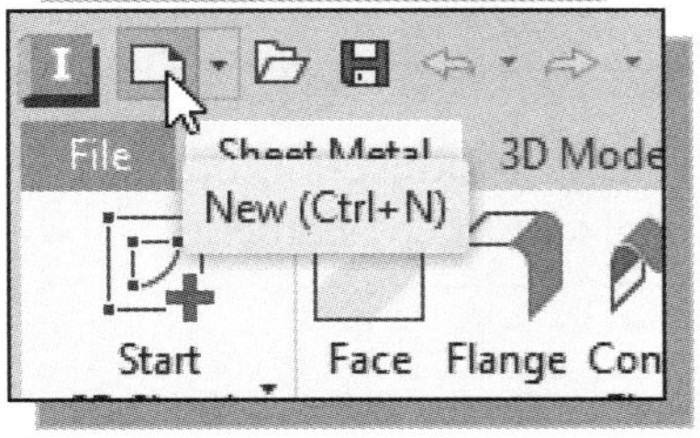

3. Click on the **New** icon in the *Quick Access* toolbar area to start a new file.

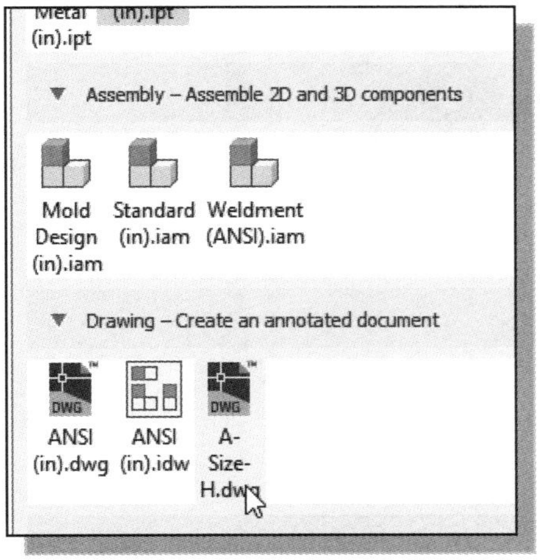

4. Select the **A-Size-H.dwg** template from the template file list.

5. Click **Create** to accept the selection.

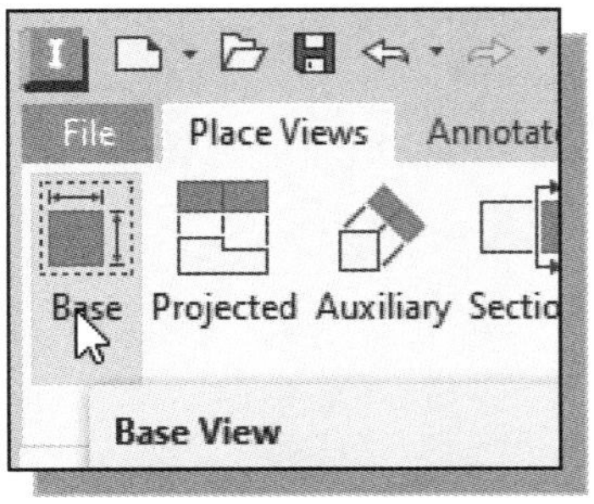

6. Click on the **Base View** in the *Drawing Views* panel to create a base view.

7. In the *Drawing View* dialog box, set *Sheet Metal View* to **Folded Model**, *Scale* to **1:1**, and *Orientation* to **Iso Top Right** and *Style* to **Shaded** as shown in the figure.

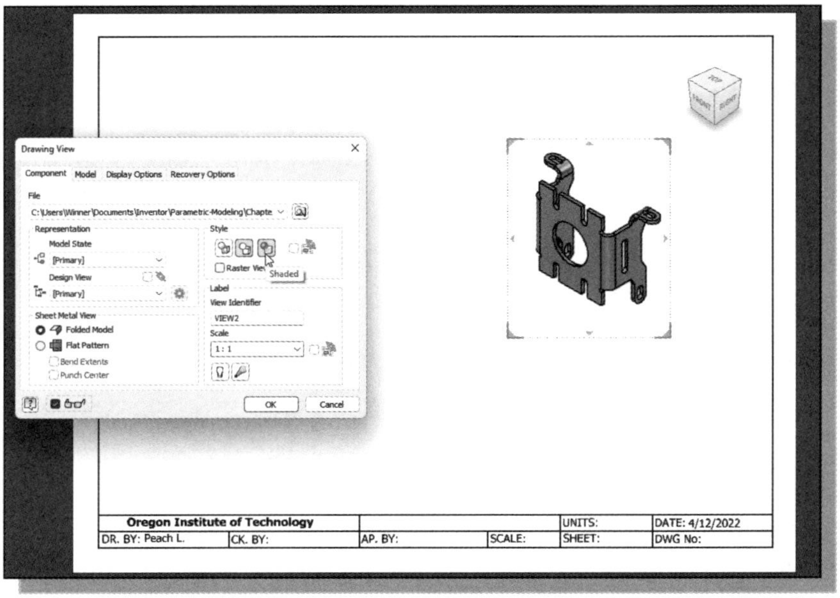

8. Place the ***isometric*** view, by clicking and dragging with the left-mouse-button, near the upper right side of the border as shown. Click **OK** to accept the placement.

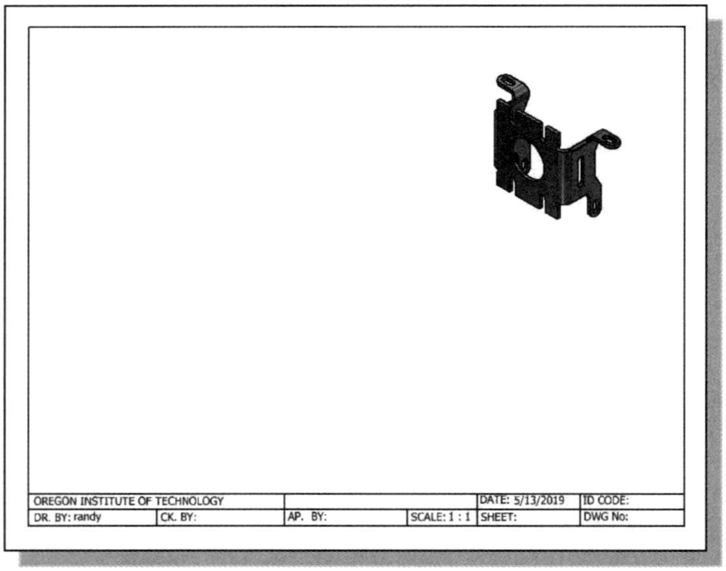

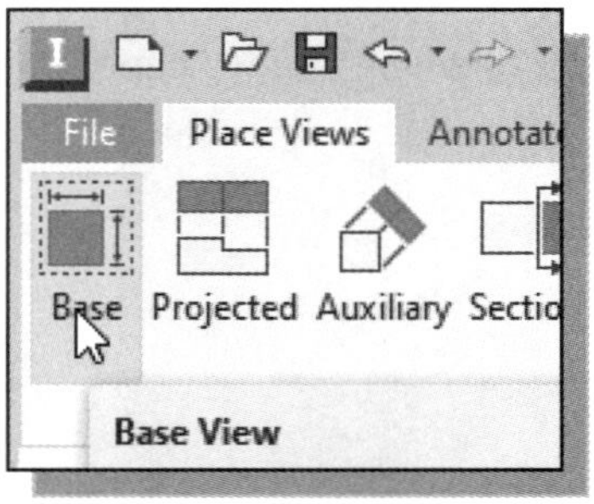

9. Click on the **Base View** in the *Drawing Views* panel to create another base view.

10. In the *Drawing View* dialog box, accept the default settings: *Sheet Metal View* set to **Folded Model**, *Scale* to **1:1**, *Orientation* to **Front** and *Style* to **Hidden Line** as shown in the figure. (**DO NOT** click on the **OK** button at this point.)

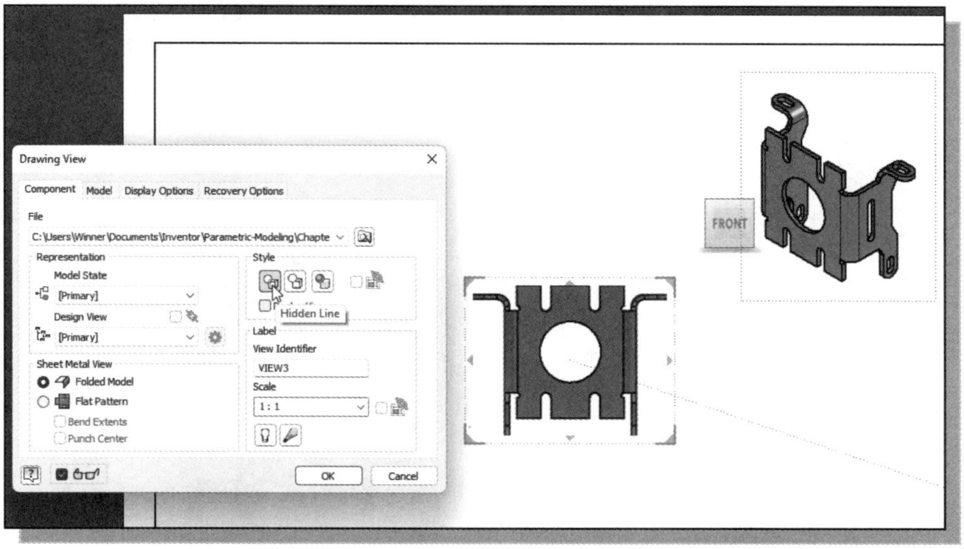

11. Place the ***front*** view, by clicking and dragging with the left-mouse-button, near the left side. (Do not click on the **OK** button at this point.)

12. Note that the associated projected views can also be added at this point; add the **top** and **side** views as shown.

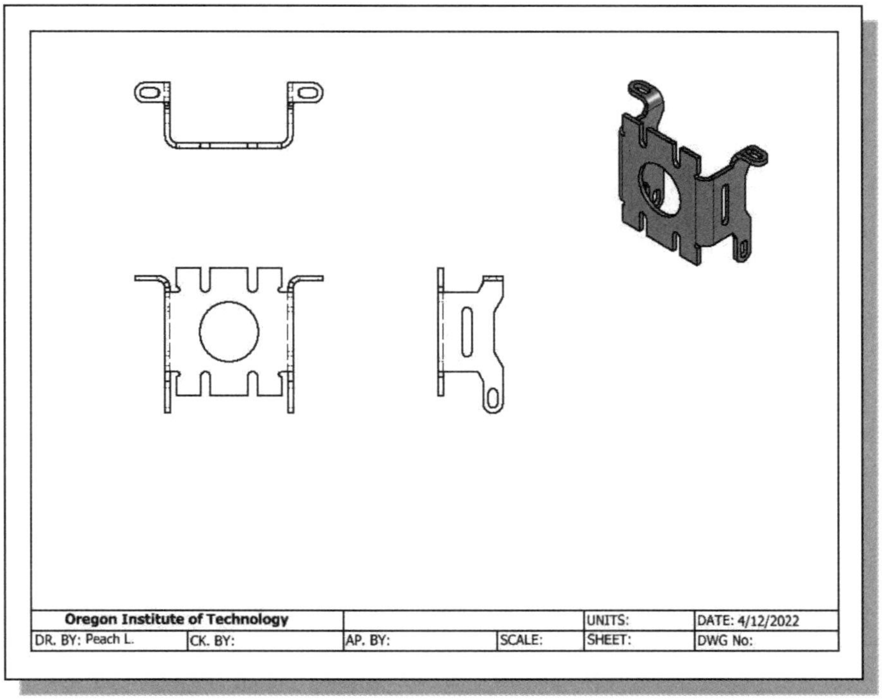

13. Click the **OK** button to accept the placed views.

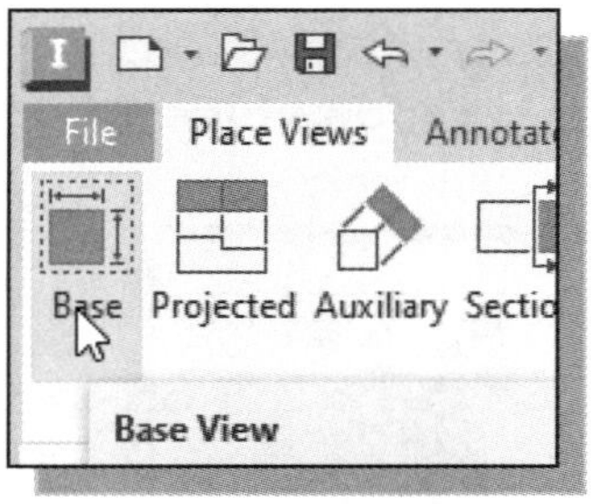

14. Click on the **Base View** in the *Drawing Views* panel to create another base view.

15. In the *Drawing View* dialog box, set *Sheet Metal View* to **Flat Pattern**, *Scale* to **1:1**, *Orientation* to **Default** and *Style* to **Hidden Line** as shown in the figure. (**DO NOT** click on the **OK** button at this point.)

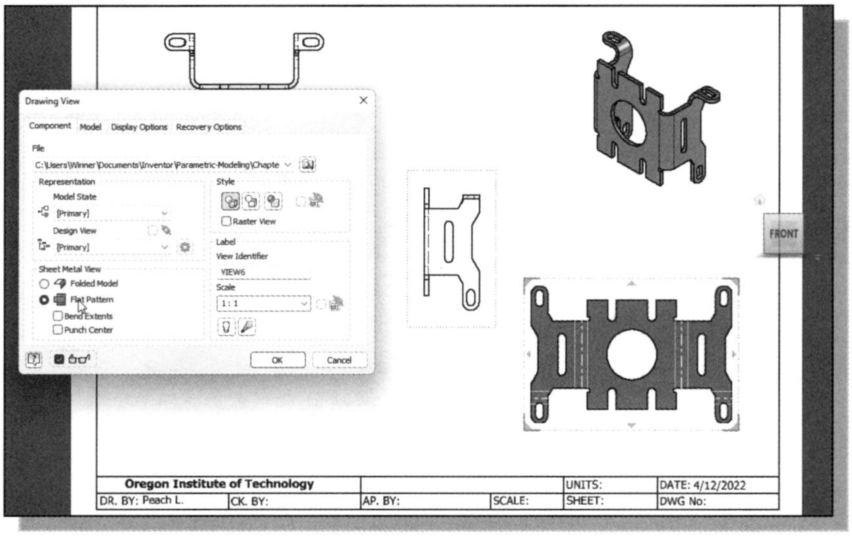

16. Place the ***flat pattern*** view, by clicking and dragging with the left-mouse-button, below the *isometric* view as shown. Click the **OK** button to accept the settings.

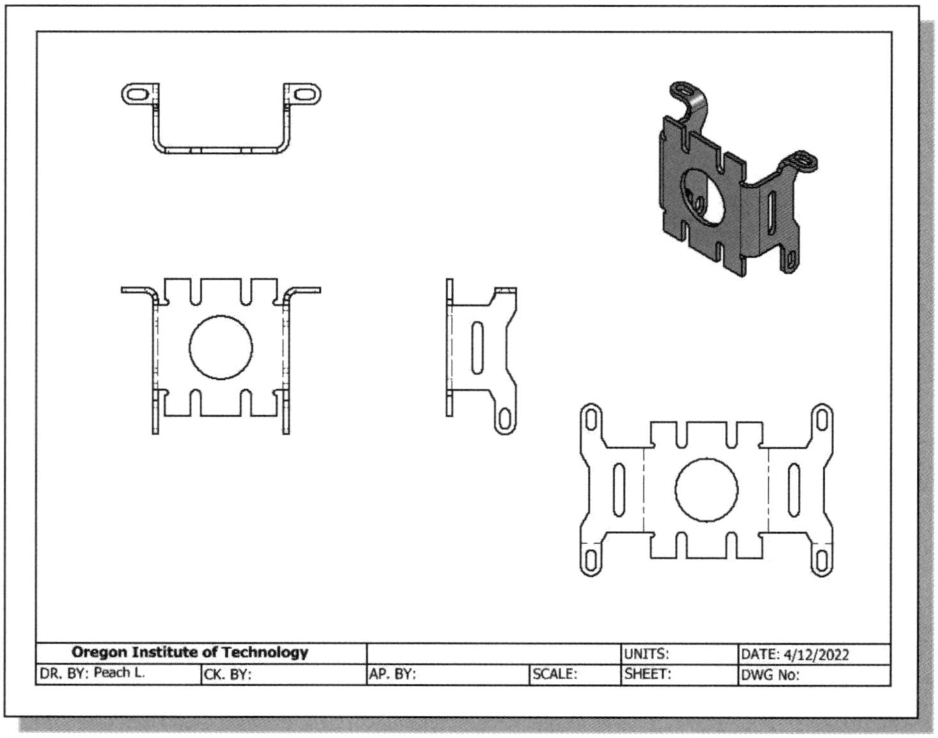

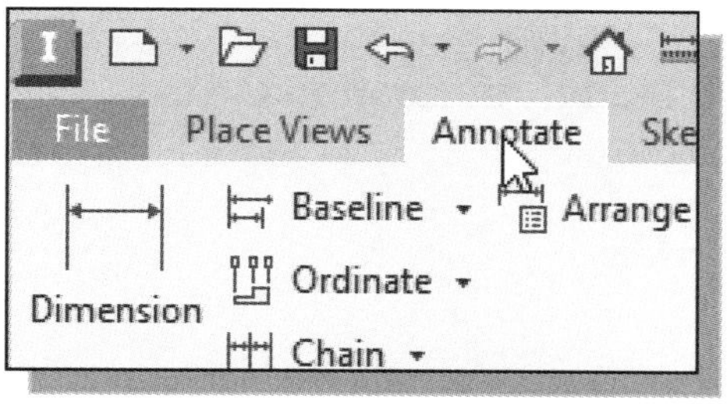

17. Select the **Drawing Annotation Tool** by clicking in the associated tab of the *Ribbon toolbar* as shown.

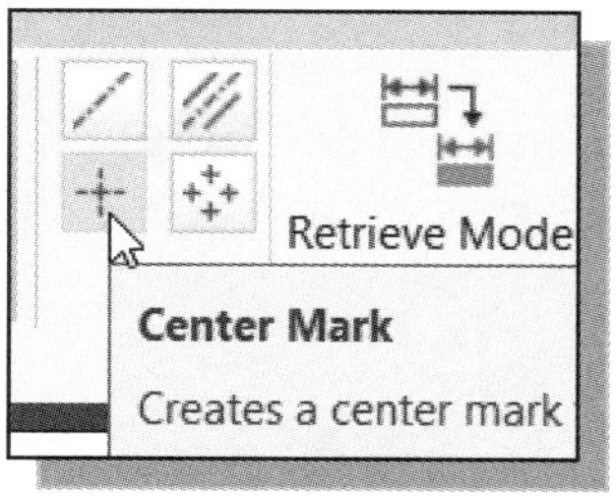

18. Select **Center Mark** in the *Drawing Annotation* panel as shown.

19. Click on the circle to place the associated center lines.

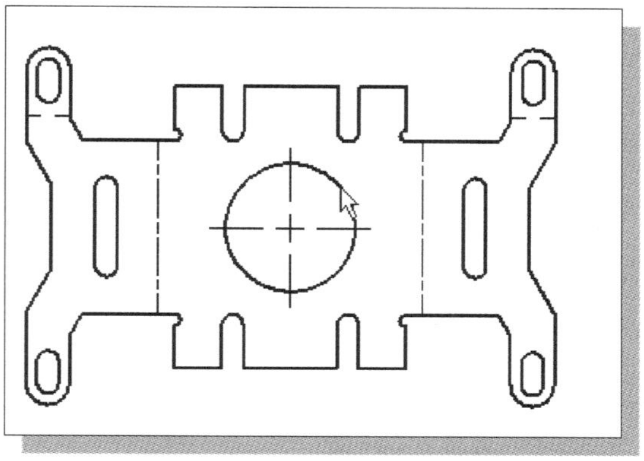

20. On your own, create and adjust the center marks for all of the circular features as shown.

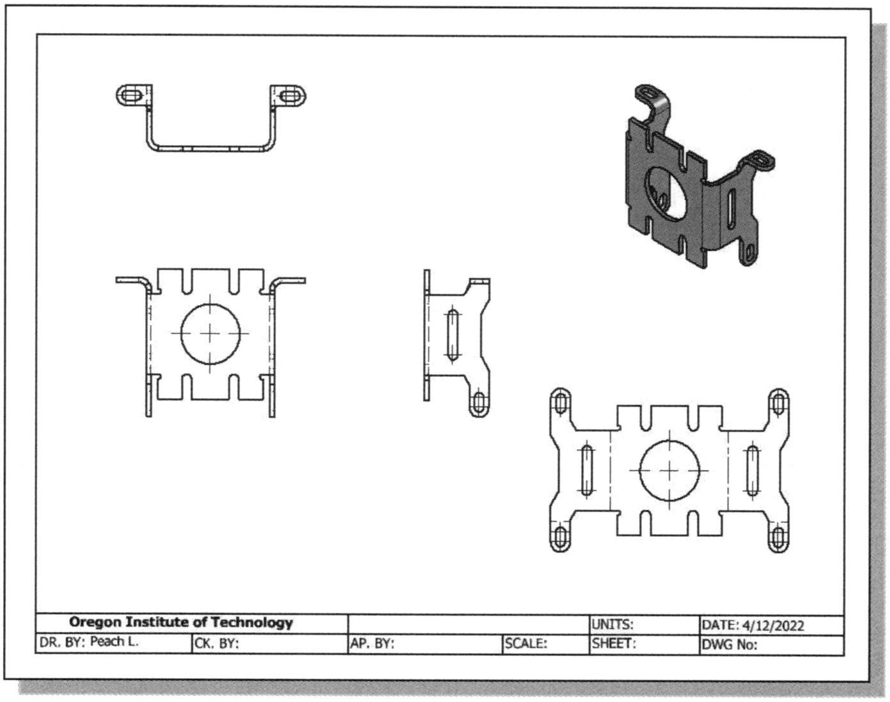

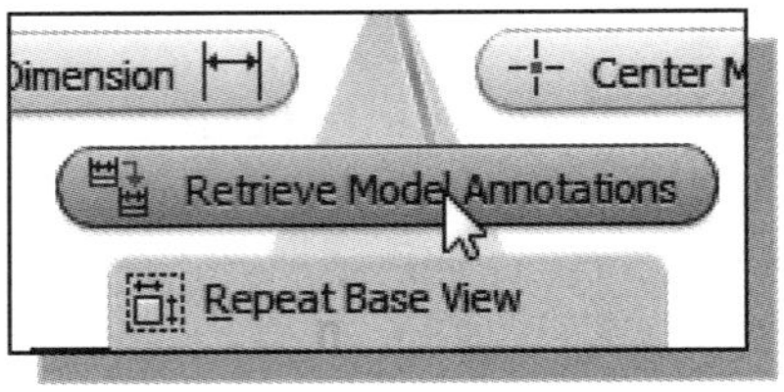

21. Select **Retrieve Model Annotations** in the *Option menu* as shown.

22. Select the **front view**.

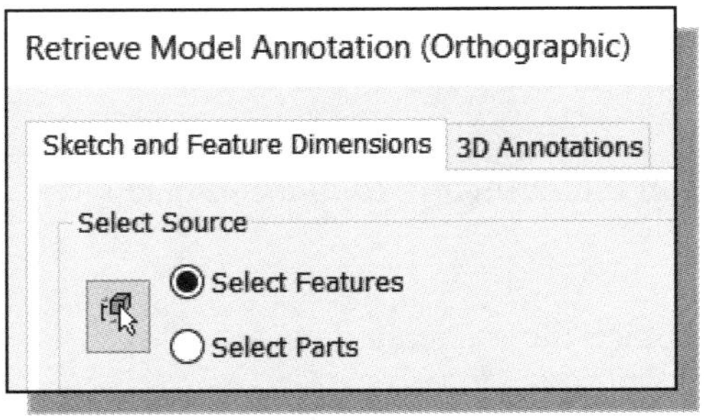

23. Confirm the *Select Source* is set to **Select Feature** as shown.

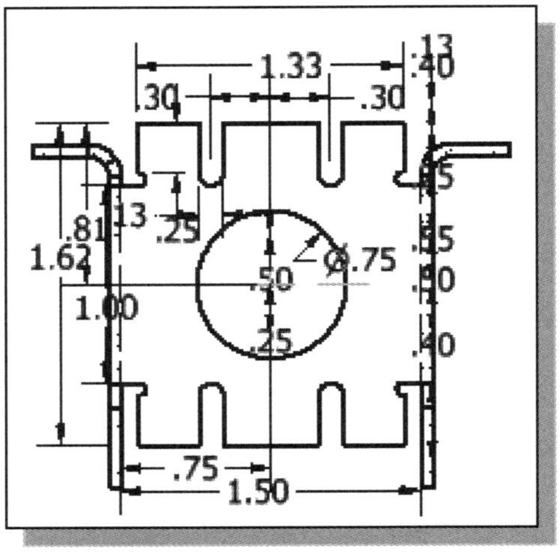

24. Click **Select Dimension Source** to activate the selection option.

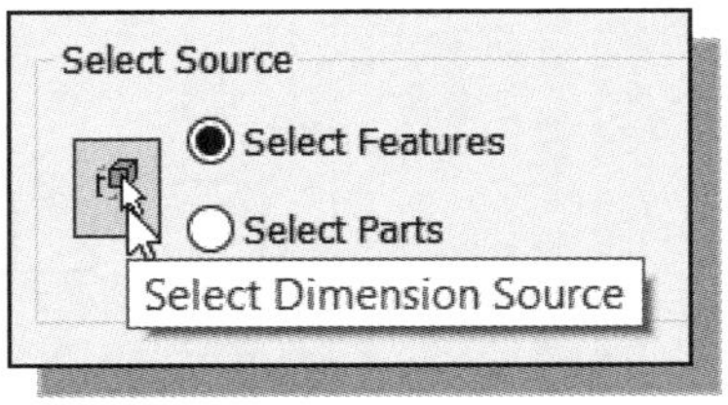

25. Click on the *base* feature to retrieve the parametric dimensions as shown.

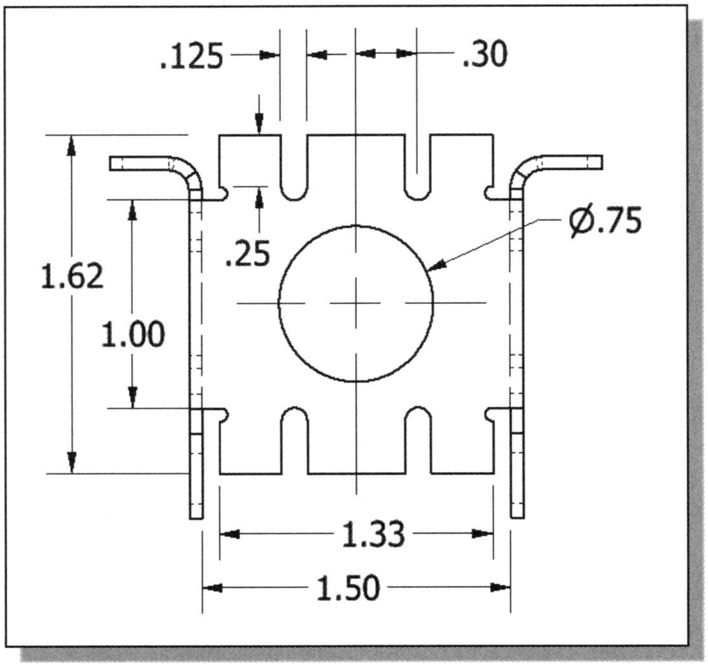

26. Select the desired dimensions to keep.

27. On your own, adjust the locations and the appearances of the dimensions as shown.

❖ On your own, complete the drawing by adding the necessary dimensions and text as shown on the next page.

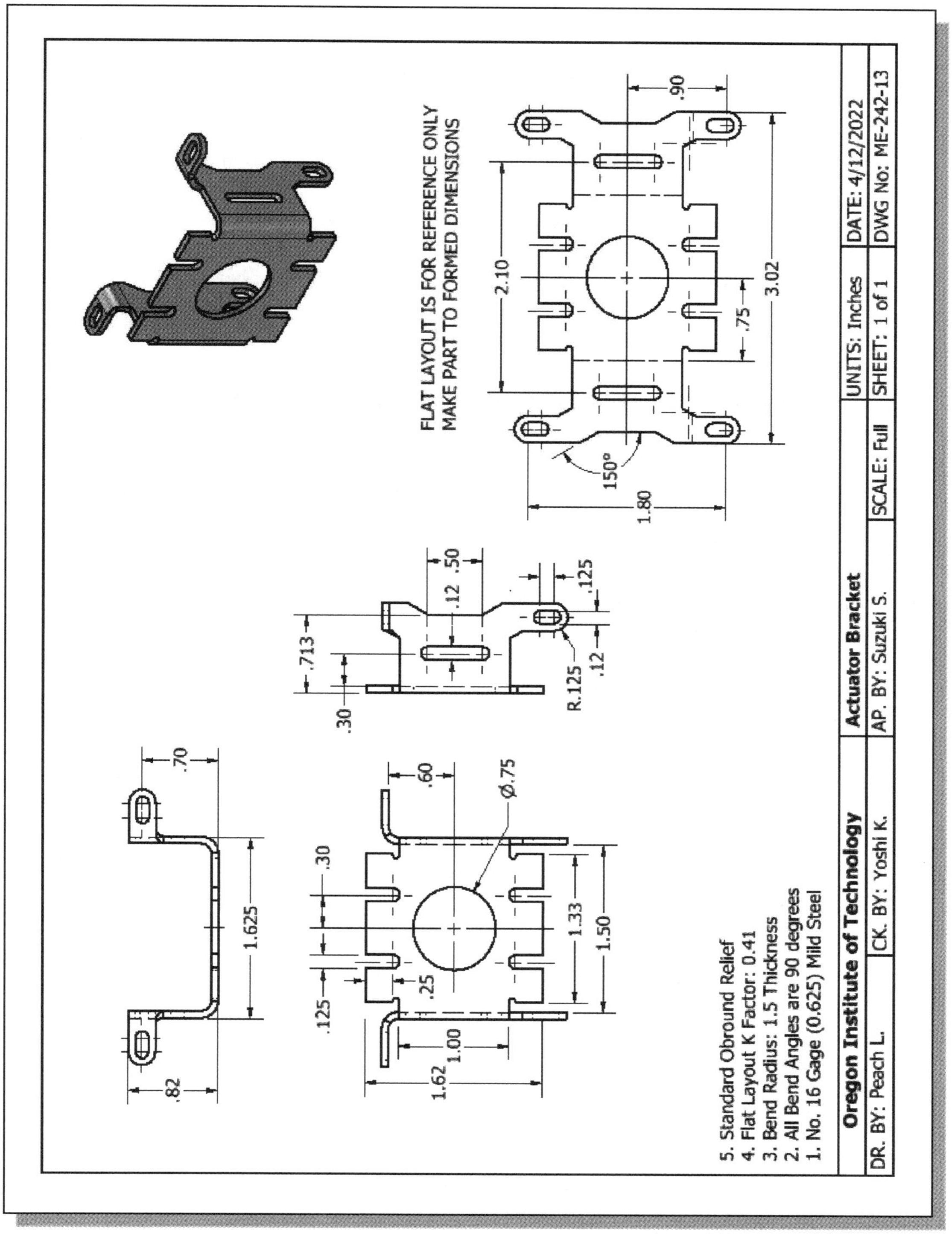

FLAT LAYOUT IS FOR REFERENCE ONLY
MAKE PART TO FORMED DIMENSIONS
.82
1.625
.70
.60
Ø.75
.30
.125
.25
1.33
1.50
1.62
1.00
.713
.30
.12
.50
.125
R.125
.12
.90
2.10
.75
3.02
150°
1.80
5. Standard Obround Relief
4. Flat Layout K Factor: 0.41
3. Bend Radius: 1.5 Thickness
2. All Bend Angles are 90 degrees
1. No. 16 Gage (0.625) Mild Steel
Oregon Institute of Technology
Actuator Bracket
UNITS: Inches
DATE: 4/12/2022
DR. BY: Peach L.
CK. BY: Yoshi K.
AP. BY: Suzuki S.
SCALE: Full
SHEET: 1 of 1
DWG No: ME-242-13

Review Questions: (Time: 30 minutes)

1. List and describe two of the more commonly used sheet metal processes.

2. Is it possible to construct a solid model first, and then convert it into a sheet metal model in Autodesk Inventor?

3. Which command do we issue to display the flat pattern of a 3D sheet metal design?

4. What is the **k-factor** used in sheet metal processes?

5. How is the **k-factor** used to calculate the flattened length in sheet metal flat patterns?

6. List and describe two of the factors that can change the k-factor value.

7. List and describe two of the settings available in the **Sheet Metal Defaults** in Autodesk Inventor.

8. List and describe the differences between the **Flange** and **Face** commands.

9. Can the **Retrieve Dimensions** command be used on a ***flat pattern*** view?

10. In the Autodesk Inventor sheet metal module, can the feature-duplicating commands, such as **Mirror** and **Pattern**, be used on sheet metal features?

11. In the *Drawing View* dialog box, which option allows us to display the *flat pattern* of the sheet metal design?

12. Is the *flat pattern* item always available in the sheet metal part *browser* window?

13. Can we create a sheet metal feature that is at a 30 degree angle to the base face feature? Which command would you use if the new feature contains fairly complex 2D geometry?

Exercises: (Time: 125 minutes.)
(Create the 3D model and the associated 2D drawings. All dimensions are in inches.)

1. **Cooling Fan Cover**

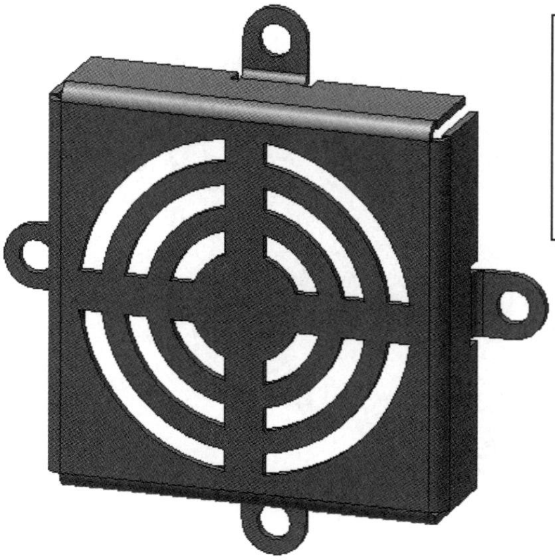

1. No. 16 Gauge (0.0625) Mild Steel
2. Standard Straight Relief
3. Flat Layout K-Factor: 0.44
4. Bend Radius: Thickness
5. All Bend Angles are 90 degrees

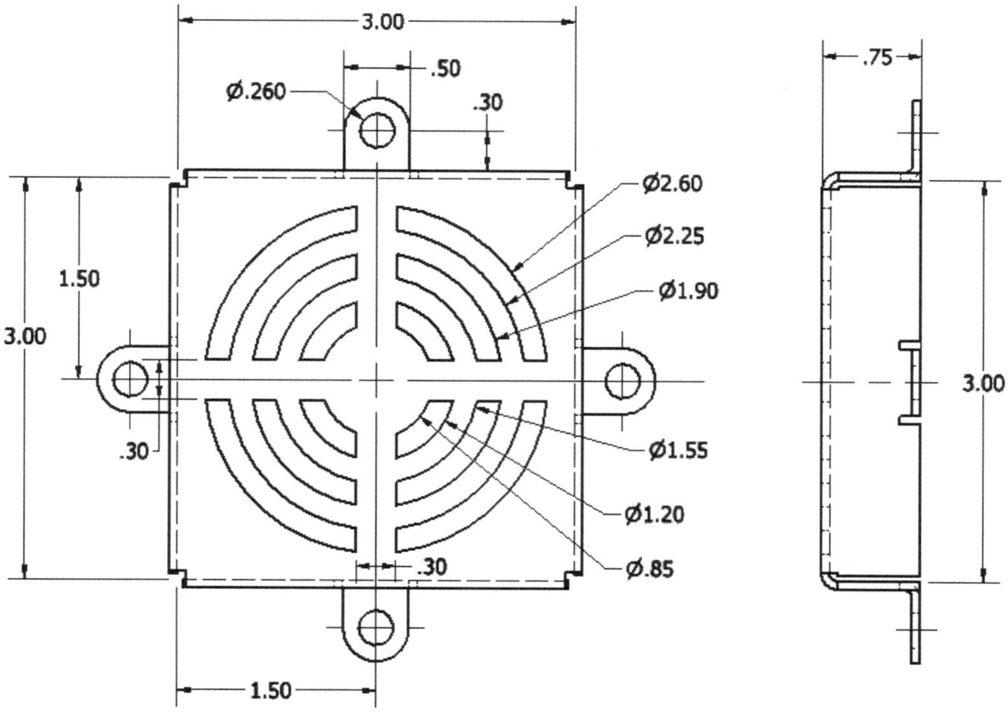

2. **Sheet Metal Rectangle to Square Transition** (Create and convert a Solid model to a Sheet Metal model.)

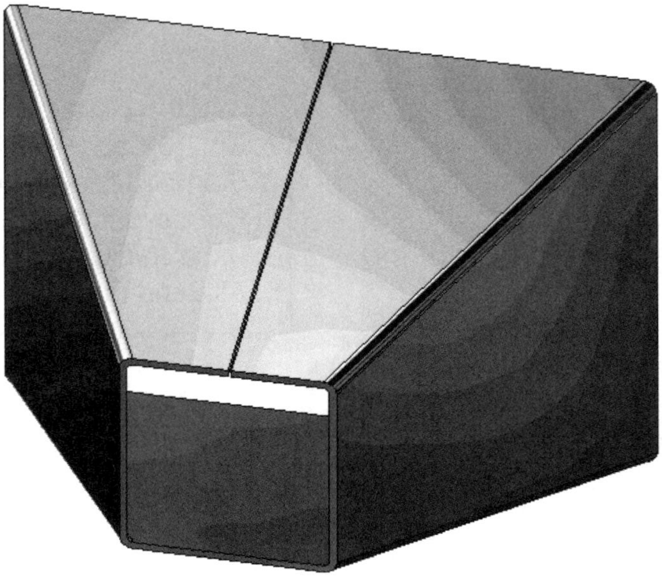

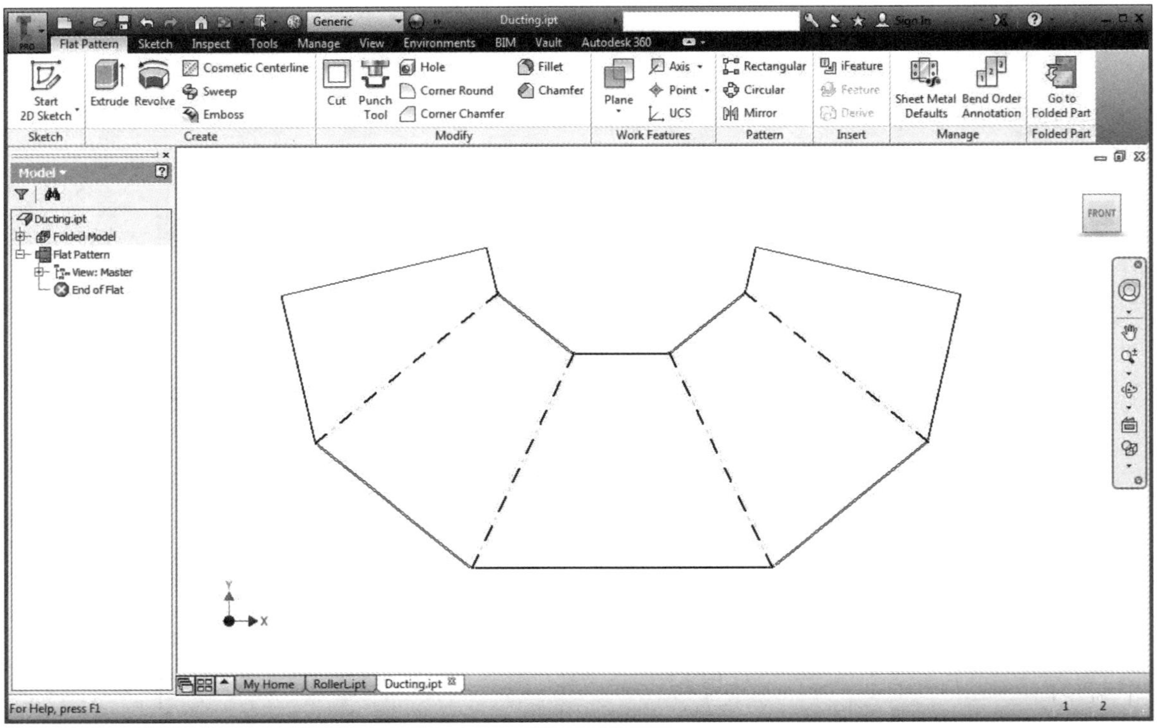

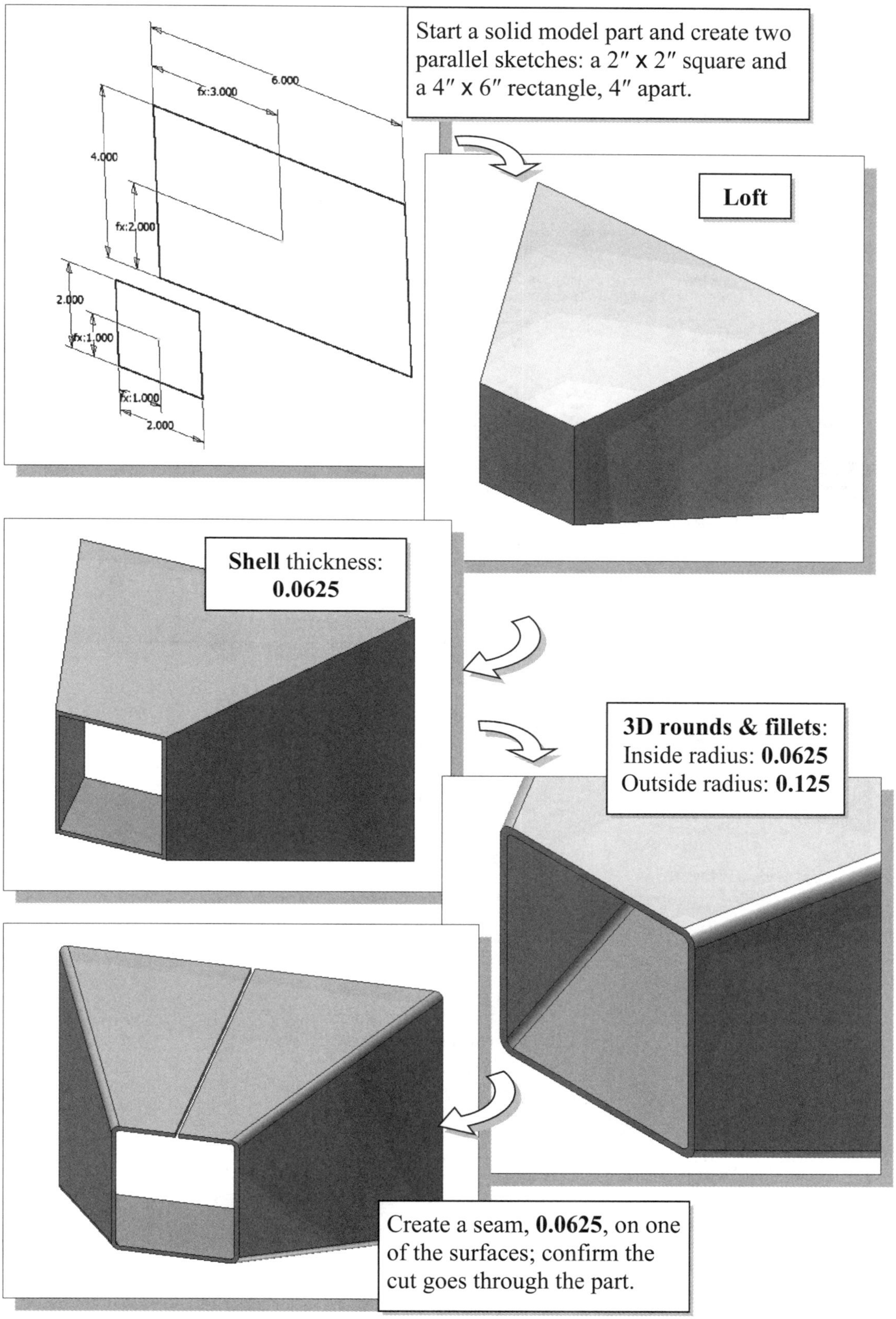
Start a solid model part and create two parallel sketches: a 2″ x 2″ square and a 4″ x 6″ rectangle, 4″ apart.
6.000
fx:3.000
4.000
fx:2.000
2.000
fx:1.000
fx:1.000
2.000
Loft
Shell thickness: 0.0625
3D rounds & fillets:
Inside radius: 0.0625
Outside radius: 0.125
Create a seam, 0.0625, on one of the surfaces; confirm the cut goes through the part.

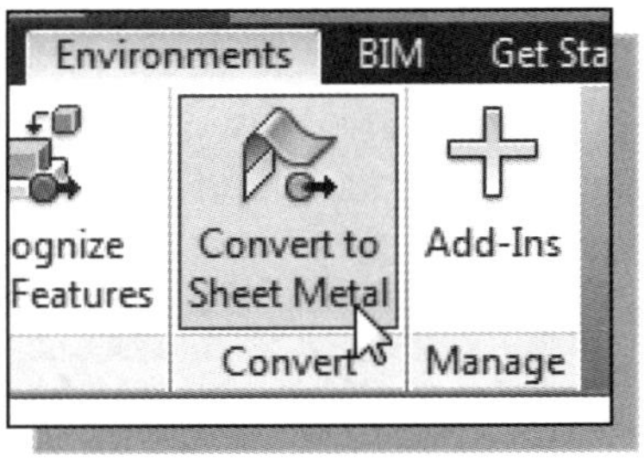

In the *Environments tab*, use the **Convert to Sheet Metal** option.

Sheet Metal **Thickness** and **Bend Radius** settings must match the model for proper unfolding.

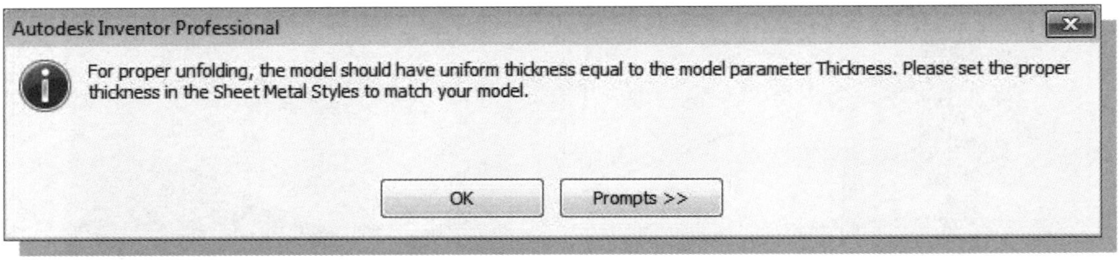

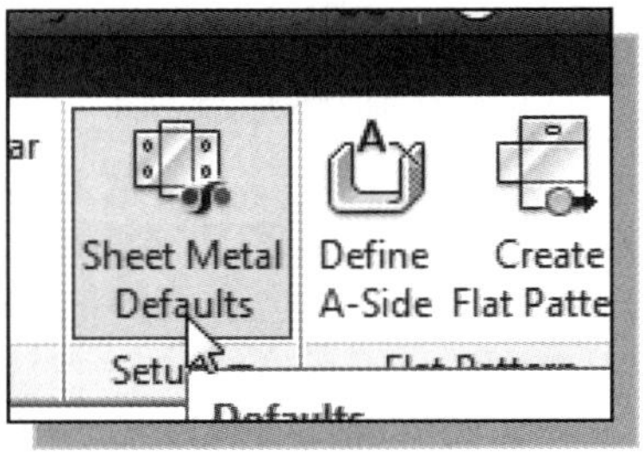

Sheet Metal **Thickness** should be set to match the shell thickness **0.0625**. The **Bend Radius** should be set to the *Thickness* value by default.

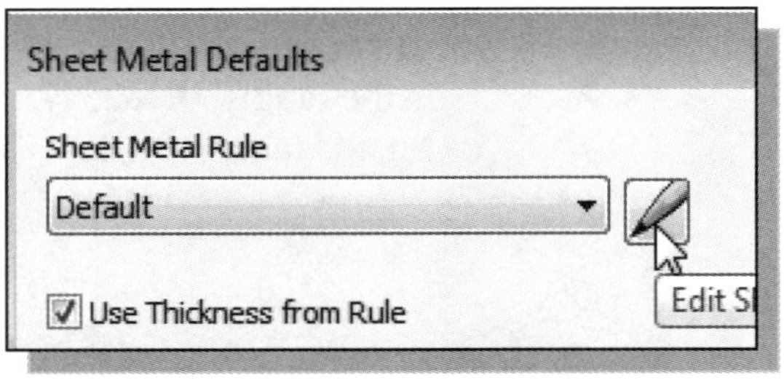

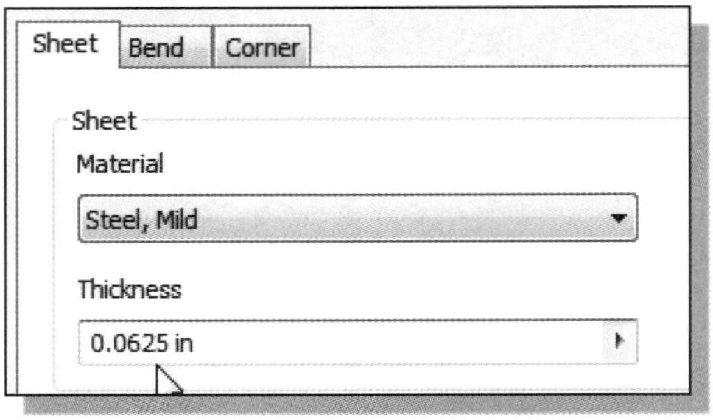

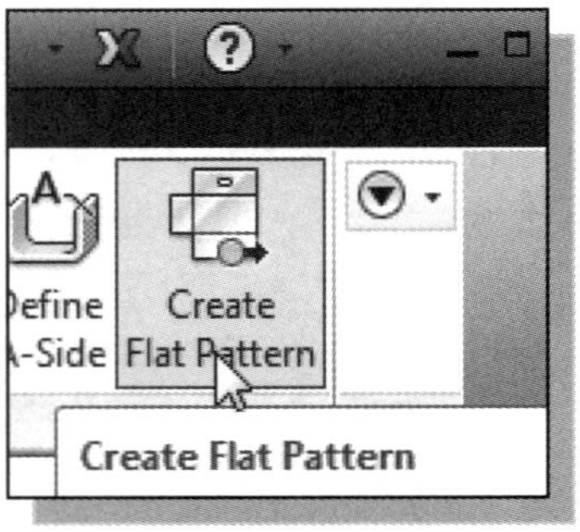

Flat Pattern can only be created if the *Sheet Metal Defaults* are set correctly.

Chapter 14
Assembly Modeling – Putting It All Together

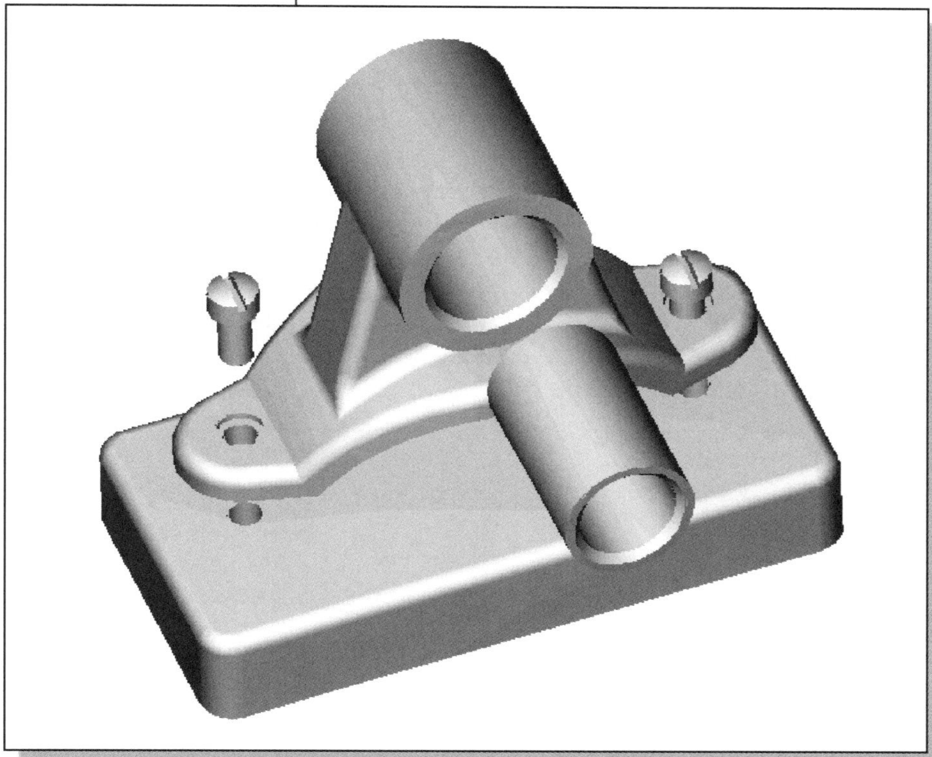

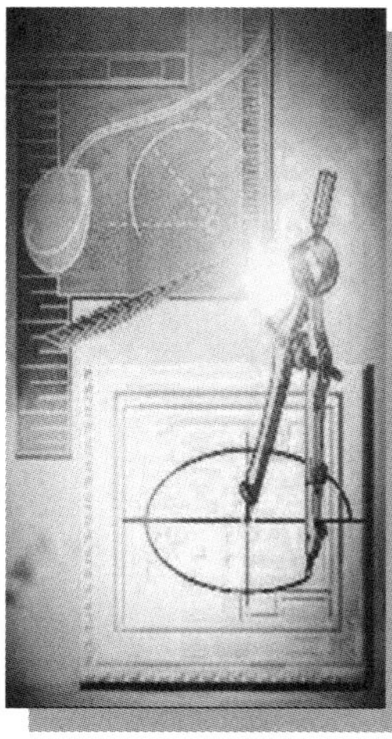

Learning Objectives

- ♦ Understand the Assembly Modeling Methodology
- ♦ Create Parts in the Assembly Modeler Mode
- ♦ Understand and Utilize Assembly Constraints
- ♦ Understand the Autodesk Inventor DOF Display
- ♦ Utilize the Autodesk Inventor Adaptive Design Approach
- ♦ Create Exploded Assemblies

Autodesk Inventor Certified User Exam Objectives Coverage

Section 5: Assemblies

Objectives: Creating Assemblies, Viewing Assemblies, Animation Assemblies, Adaptive Features, Parts, and Subassemblies.

Section 6: Drawings

Objectives: Create drawings.

Introduction

In the previous lessons, we have gone over the fundamentals of creating basic parts and drawings. In this lesson, we will examine the assembly modeling functionality of Autodesk Inventor. We will start with a demonstration on how to create and modify assembly models. The main task in creating an assembly is establishing the assembly relationships between parts. To assemble parts into an assembly, we will need to consider the assembly relationships between parts. It is good practice to assemble parts based on the way they would be assembled in the actual manufacturing process. We should also consider breaking down the assembly into smaller subassemblies, which helps the management of parts. In Autodesk Inventor, a subassembly is treated the same way as a single part during assembling. Many parallels exist between assembly modeling and part modeling in parametric modeling software such as Autodesk Inventor.

Autodesk Inventor provides full associative functionality in all design modules, including assemblies. When we change a part model, Autodesk Inventor will automatically reflect the changes in all assemblies that use the part. We can also modify a part in an assembly. **Bi-directional full associative functionality** is the main feature of parametric solid modeling software that allows us to increase productivity by reducing design cycle time.

One of the key features of Autodesk Inventor is the use of an assembly-centric paradigm, which enables users to concentrate on the design without depending on the associated parameters or constraints. Users can specify how parts fit together and the Autodesk Inventor *assembly-based fit function* automatically determines the parts' sizes and positions. This unique approach is known as the **Direct Adaptive Assembly** approach, which defines part relationships directly with no order dependency.

In this lesson, we will also illustrate the basic concept of Autodesk Inventor's **Adaptive Design** approach. The key element in doing **Adaptive Design** is to ***under-constrain*** features or parts. The applied *assembly constraints* in the assembly modeler are used to control the sizes, shapes, and positions of ***under-constrained*** sketches, features, and parts. No equations are required, and this approach is extremely flexible when performing modifications and changes to the design. We can modify adaptive assemblies at any point, in any order, regardless of how the parts were originally placed or constrained.

In Autodesk Inventor, features and parts can be made adaptive at any time during the creation or assembly. The features of a part can be defined as adaptive when they are created in the part file. When we place such a part in an assembly, the features will then resize and change shape based on the applied assembly constraints. We can make features and parts adaptive from either the part modeling or assembly modeling environments.

The *Adaptive Design approach* is a unique design methodology that can only be found in Autodesk Inventor. The goal of this methodology is to improve the design process and allows you, the designer, to ***design the way you think***.

Assembly Modeling Methodology

The Autodesk Inventor assembly modeler provides tools and functions that allow us to create 3D parametric assembly models. An assembly model is a 3D model with any combination of multiple part models. *Parametric assembly constraints* can be used to control relationships between parts in an assembly model.

Autodesk Inventor can work with any of the assembly modeling methodologies:

The Bottom Up Approach

The first step in the *bottom up* assembly modeling approach is to create the individual parts. The parts are then pulled together into an assembly. This approach is typically used for smaller projects with very few team members.

The Top Down Approach

The first step in the *top down* assembly modeling approach is to create the assembly model of the project. Initially, individual parts are represented by names or symbols. The details of the individual parts are added as the project gets further along. This approach is typically used for larger projects or during the conceptual design stage. Members of the project team can then concentrate on the particular section of the project to which he/she is assigned.

The Middle Out Approach

The *middle out* assembly modeling approach is a mixture of the bottom-up and top-down methods. This type of assembly model is usually constructed with most of the parts already created, and additional parts are designed and created using the assembly for construction information. Some requirements are known, and some standard components are used, but new designs must also be produced to meet specific objectives. This combined strategy is a very flexible approach for creating assembly models.

The different assembly modeling approaches described above can be used as guidelines to manage design projects. Keep in mind that we can start modeling our assembly using one approach and then switch to a different approach without any problems.

In this lesson, the *bottom up* assembly modeling approach is illustrated. All of the parts (components) required to form the assembly are created first. Autodesk Inventor's assembly modeling tools allow us to create complex assemblies by using components that are created in part files or are placed in assembly files. A component can be a subassembly or a single part, where features and parts can be modified at any time. The sketches and profiles used to build a 3D Model can be fully or partially constrained. Partially constrained features may be adaptive, which means the size or shape of the associated parts are adjusted in an assembly when the parts are constrained to other parts. The basic concept and procedure of using the adaptive assembly approach is demonstrated in the tutorial.

The Shaft Support Assembly

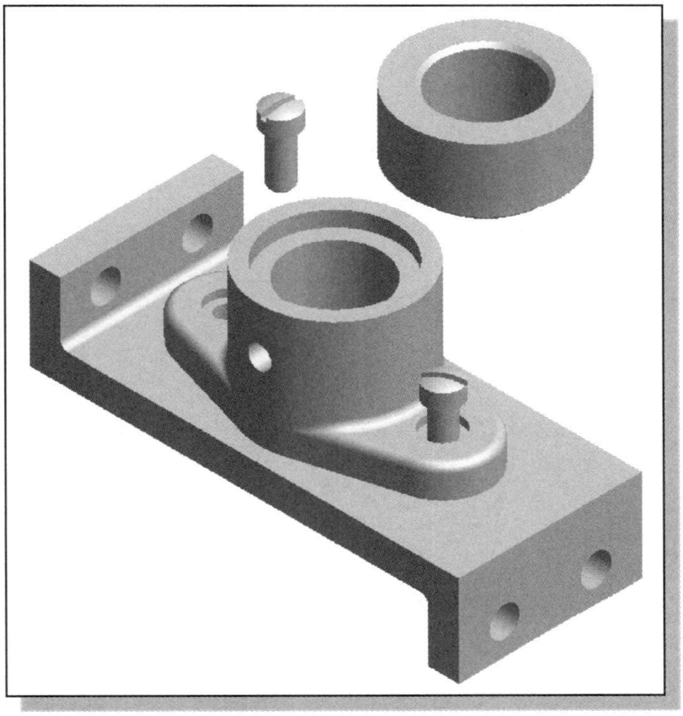

Additional Parts

Four parts are required for the assembly: (1) ***Collar***, (2) ***Bearing***, (3) ***Base-Plate*** and (4) ***Cap-Screw***. On your own, create the four parts as shown below; save the models as separate part files: *Collar, Bearing*, *Base-Plate*, and *Cap-Screw*. (Place all parts in the **Chapter14** folder and close all part files or exit *Autodesk Inventor* after you have created the parts.)

(1) ***Collar***

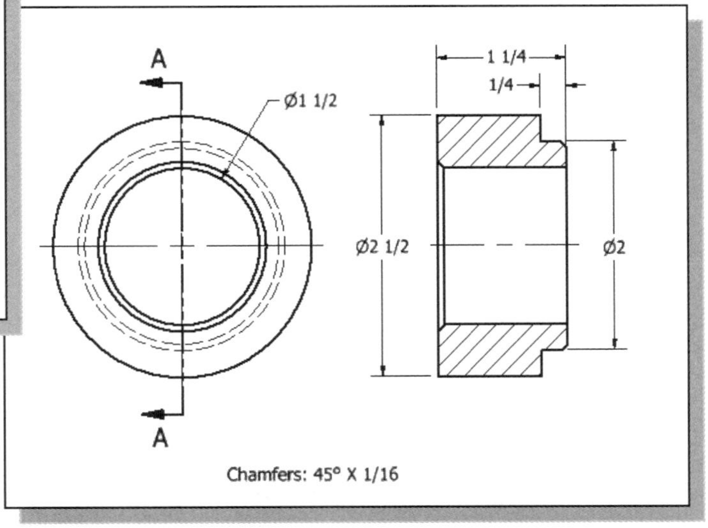

(2) ***Bearing*** (Construct the part with the datum origin aligned to the bottom center.)

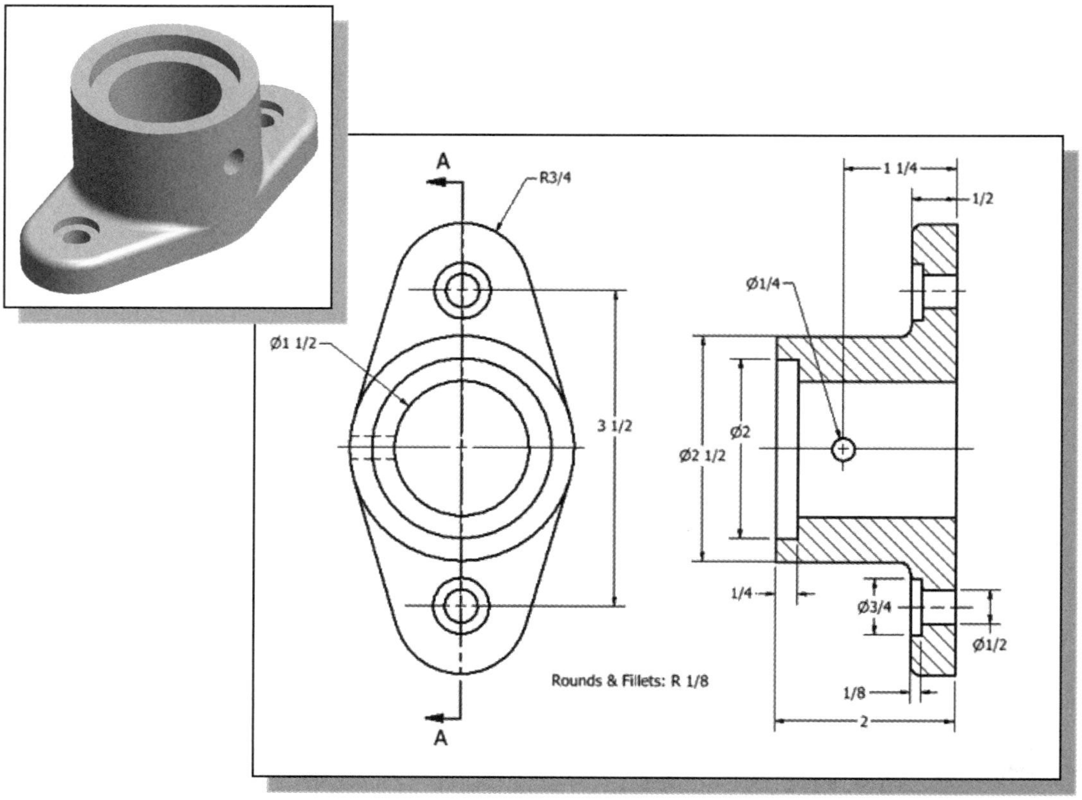

(3) ***Base-Plate*** (Construct the part with the datum origin aligned to the bottom center of the large hole.)

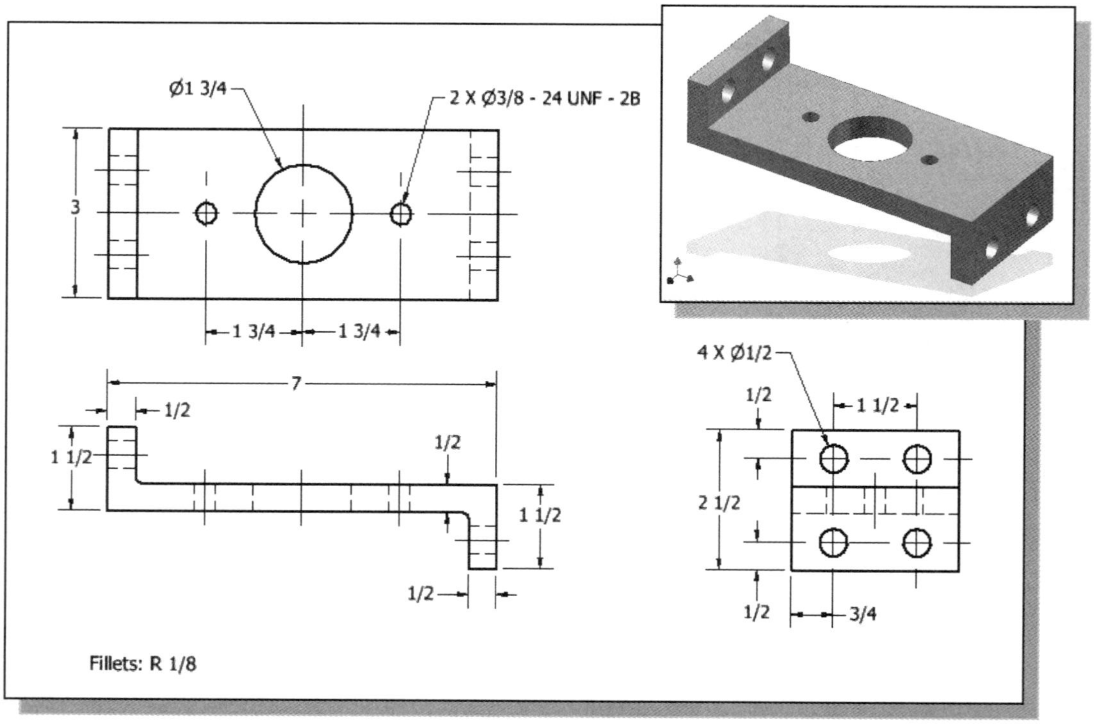

(4) ***Cap-Screw***

- Autodesk Inventor provides two options for creating threads: **Thread** and **Coil**. The **Thread** command does not create true 3D threads; a pre-defined thread image is applied on the selected surface, as shown in the figure. The **Coil** command can be used to create true threads, which contain complex three-dimensional curves and surfaces. You are encouraged to experiment with the **Coil** command and/or the **Thread** command to create threads.

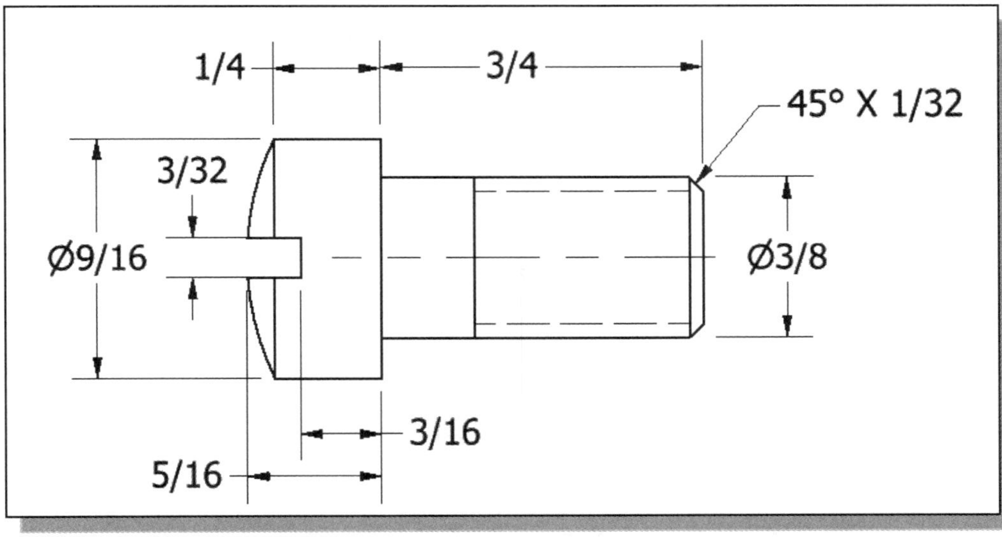

- Hint: First create a revolved feature using the profile shown below.

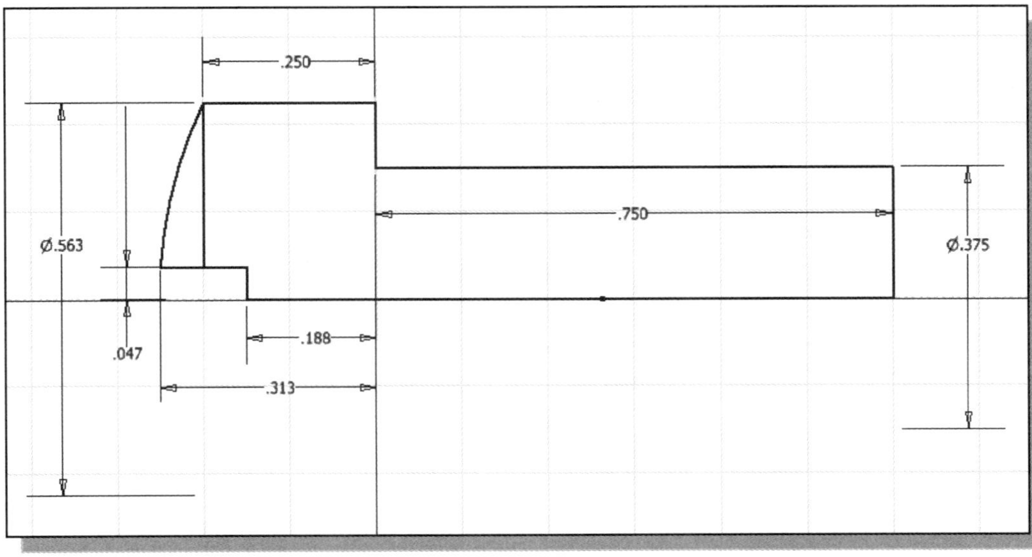

Starting Autodesk Inventor

1. Select the **Autodesk Inventor** option on the *Start* menu or select the **Autodesk Inventor** icon on the desktop to start Autodesk Inventor. The Autodesk Inventor main window will appear on the screen.

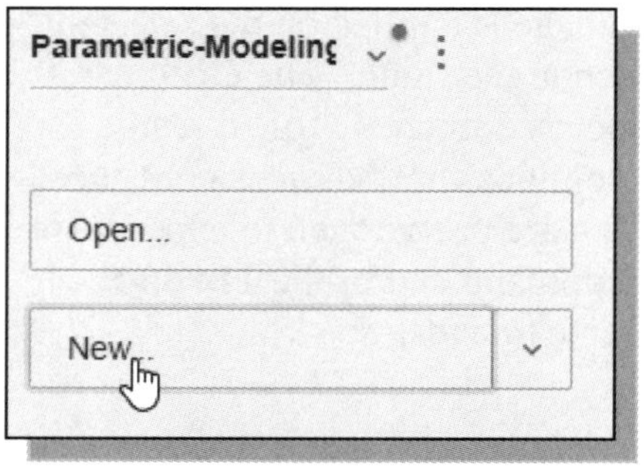

2. Select the **New File** icon with a single click of the left-mouse-button in the *Launch* toolbar as shown.

3. Confirm the ***Parametric-Modeling-Exercises*** project is activated; note the **Projects** button is available to view/modify the active project.

4. Select the **English** tab, and in the *Template* list select **Standard(in).iam** (*Standard Inventor Assembly Model* template file).

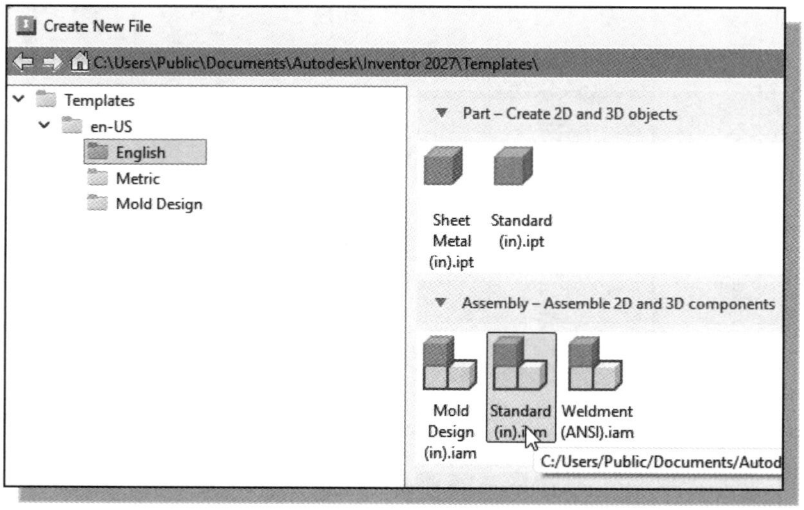

5. Click on the **Create** button in the *New File* dialog box to accept the selected settings.

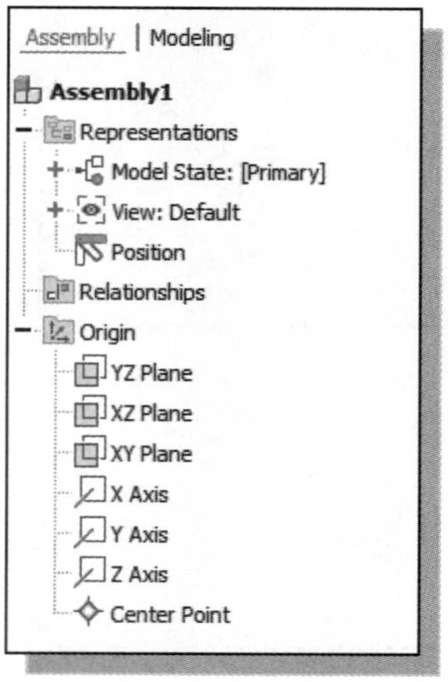

- In the *browser* window, **Assembly1** is displayed with a set of work planes, work axes and a work point. In most aspects, the usage of work planes, work axes and work point is very similar to that of the *Inventor Part Modeler*.

- Notice, in the *Ribbon* toolbar panels, several *component* options are available, such as **Place Component**, **Create Component** and **Place from Content Center**. As the names imply, we can use parts that have been created or create new parts within the *Inventor Assembly Modeler*.

Placing the First Component

The first component placed in an assembly should be a fundamental part or subassembly. The first component in an assembly file sets the orientation of all subsequent parts and subassemblies. The origin of the first component is aligned to the origin of the assembly coordinates and the part is grounded (all degrees of freedom are removed). The rest of the assembly is built on the first component, the ***base component***. In most cases, this *base component* should be one that is **not likely to be removed** and **preferably a non-moving part** in the design. Note that there is no distinction in an assembly between components; the first component we place is usually considered as the *base component* because it is usually a fundamental component to which others are constrained. We can change the base component to a different base component by placing a new base component, specifying it as grounded, and then re-constraining any components placed earlier, including the first component. For our project, we will use the ***Base-Plate*** as the base component in the assembly.

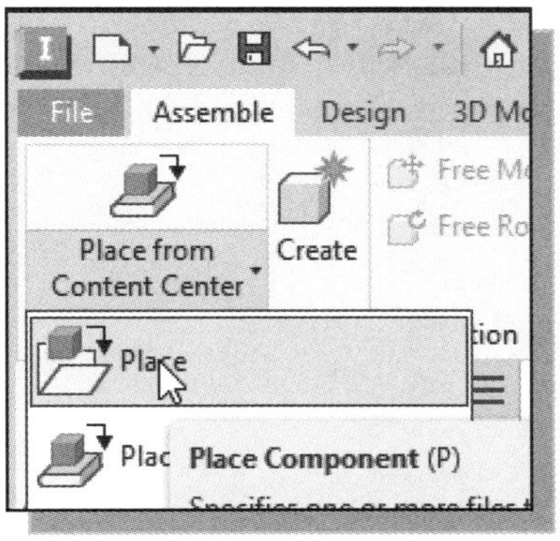

1. In the *Assemble* panel (the toolbar that is located to the left side of the graphics window) select the **Place Component** command by left-clicking the icon.

2. Select the ***Base-Plate*** (part file: ***Base-Plate.ipt***) in the list window.

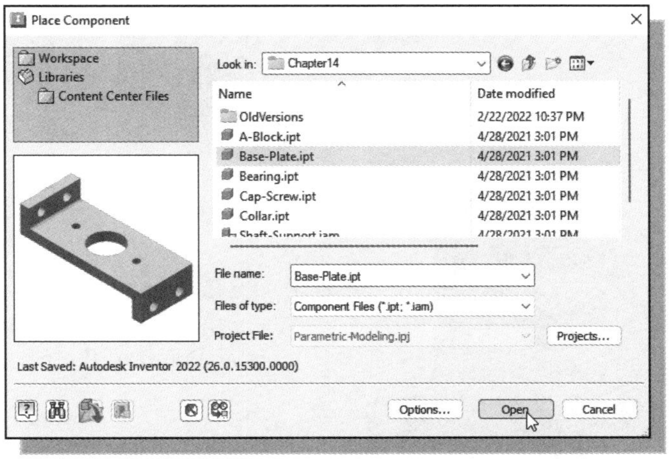

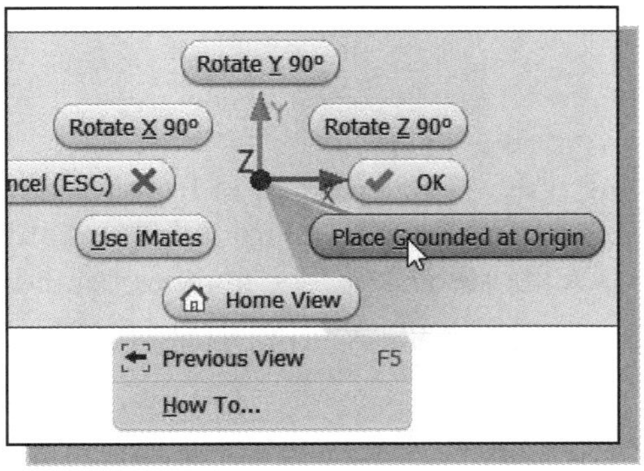

3. Click on the **Open** button to retrieve the model.

4. Right-click once to bring up the option menu and select **Place Grounded at Origin.**

5. Right-click again to bring up the option menu and select **OK** to end the placement of the *Base-Plate* part.

Placing the Second Component

We will retrieve the *Bearing* part as the second component of the assembly model.

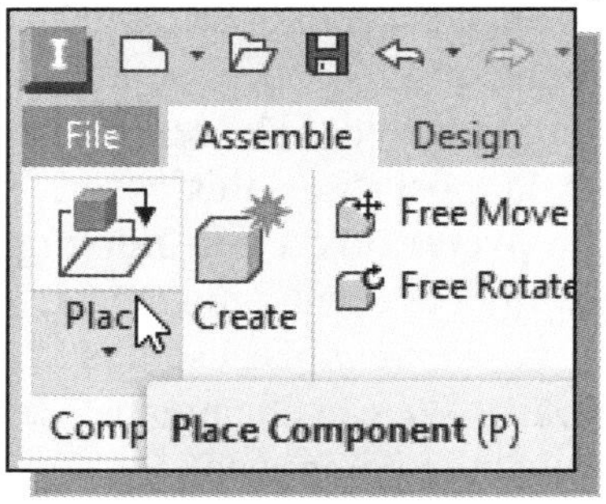

1. In the *Assemble* panel (the toolbar that is located to the left side of the graphics window) select the **Place Component** command by left-clicking the icon.

2. Select the ***Bearing*** design (part file: ***Bearing.ipt***) in the list window. And click on the **Open** button to retrieve the model.

3. Place the *Bearing* toward the upper right corner of the graphics window, as shown in the figure.

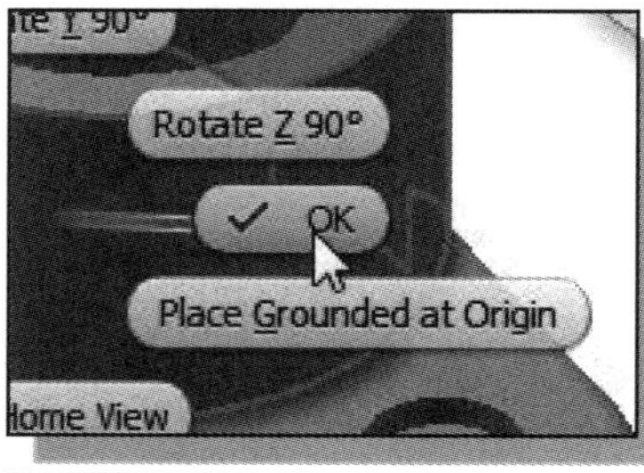

4. Inside the graphics window, right-click once to bring up the option menu and select **OK** to end the placement of the *Bearing* part.

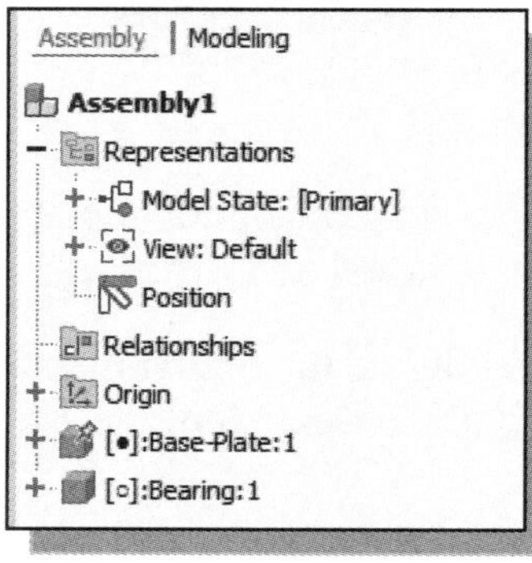

- Inside the *browser* window, the retrieved parts are listed in their corresponding order. The **Pin** icon in front of the *Base-Plate* filename signifies the part is grounded and all ***six degrees of freedom*** are restricted. The number behind the filename is used to identify the number of copies of the same component in the assembly model.

Degrees of Freedom and Constraints

Each component in an assembly has six **degrees of freedom** (**DOF**), or ways in which rigid 3D bodies can move: movement along the X, Y, and Z axes (translational freedom), plus rotation around the X, Y, and Z axes (rotational freedom). *Translational DOF*s allow the part to move in the direction of the specified vector. *Rotational DOF*s allow the part to turn about the specified axis.

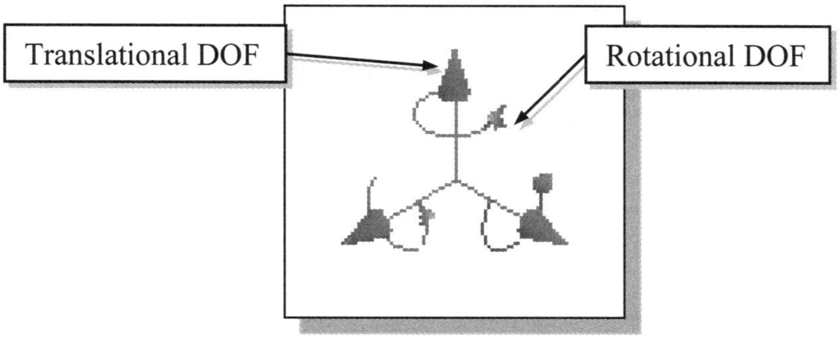

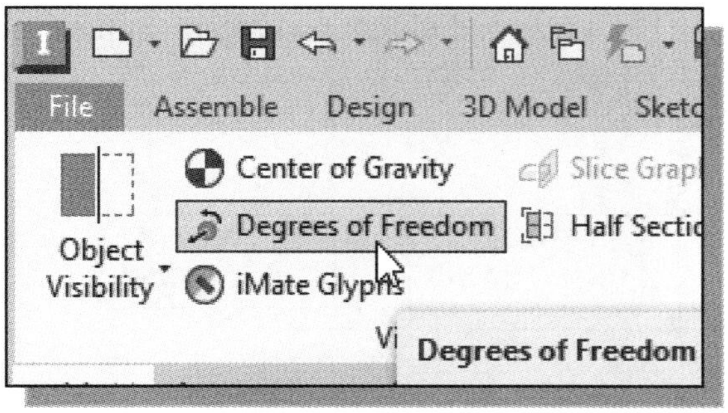

➢ Select the **Degrees of Freedom** option in the **View** tab to display the DOF of the unconstrained component.

In Autodesk Inventor, the degrees-of-freedom symbol shows the remaining degrees of freedom (both translational and rotational) for one or more components of the active assembly. When a component is fully constrained in an assembly, the component cannot move in any direction. The position of the component is fixed relative to other assembly components. All of its degrees of freedom are removed. When we place an assembly constraint between two selected components, they are positioned relative to one another. Movement is still possible in the unconstrained directions.

It is usually a good idea to fully constrain components so that their behavior is predictable as changes are made to the assembly. Leaving some degrees of freedom open can sometimes help retain design flexibility. As a general rule, we should use only enough constraints to ensure predictable assembly behavior and avoid unnecessary complexity.

Assembly Constraints

We are now ready to assemble the components together. We will start by placing assembly constraints on the ***Bearing*** and the ***Base-Plate***.

To assemble components into an assembly, we need to establish the assembly relationships between components. It is good practice to assemble components the way they would be assembled in the actual manufacturing process. **Assembly constraints** create a parent/child relationship that allows us to capture the design intent of the assembly. Because the component that we are placing actually becomes a child to the already assembled components, we must use caution when choosing constraint types and references to make sure they reflect the intent.

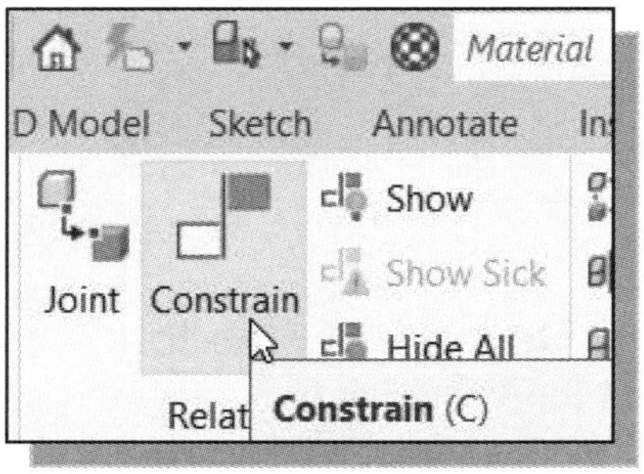

➤ Switch back to the *Assemble* panel; select the **Constrain** command by left-clicking once on the icon.

• The *Place Constraints* dialog box appears on the screen. Five types of assembly constraints are available.

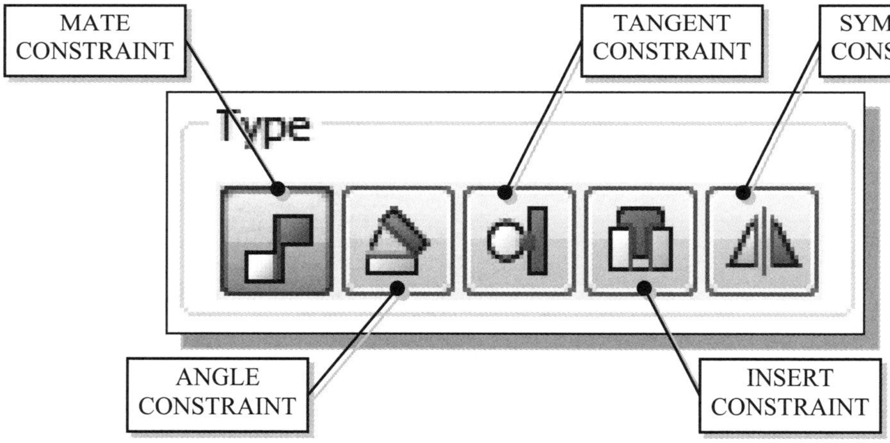

➤ Assembly models are created by applying proper *assembly constraints* to the individual components. The constraints are used to restrict the movement between parts. Constraints eliminate rigid body degrees of freedom (**DOF**). A 3D part has *six degrees of freedom* since the part can rotate and translate relative to the three coordinate axes. Each time we add a constraint between two parts, one or more DOF is eliminated. The movement of a fully constrained part is restricted in all directions. Five basic types of assembly constraints are available in Autodesk Inventor: Mate, Angle, Tangent, Insert and Symmetry. Each type of constraint removes different combinations of rigid body degrees of freedom. Note that it is possible to apply different constraints and achieve the same results.

➢ **Mate** – Constraint positions components face-to-face, or adjacent to one another, with faces flush. Removes one degree of linear translation and two degrees of angular rotation between planar surfaces. Selected surfaces point in opposite directions and can be **offset** by a specified distance. Mate constraint positions selected faces normal to one another, with faces coincident.

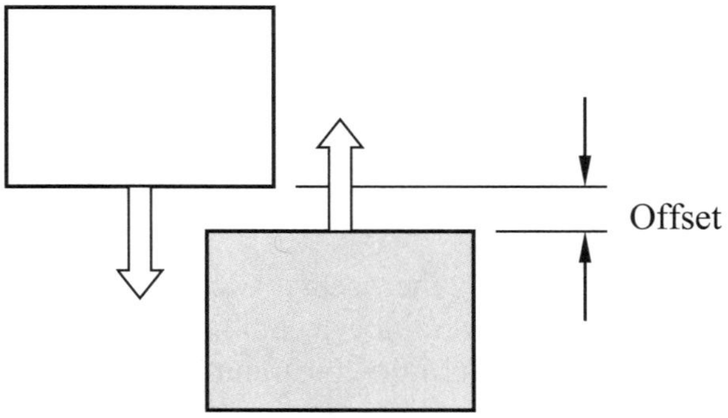

➢ **Flush** – Makes two planes coplanar with their faces aligned in the same direction. Selected surfaces point in the same direction and are offset by a specified distance. Flush constraint aligns components adjacent to one another with faces flush and positions selected faces, curves, or points so that they are aligned with surface normals pointing in the same direction. (Note that the Flush constraint is listed as a selectable option in the Mate constraint.)

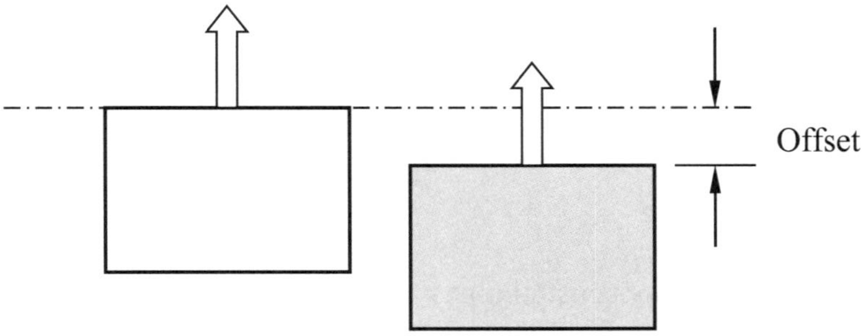

➢ **Angle** – Creates an angular assembly constraint between parts, subassemblies, or assemblies. Selected surfaces point in the direction specified by the angle.

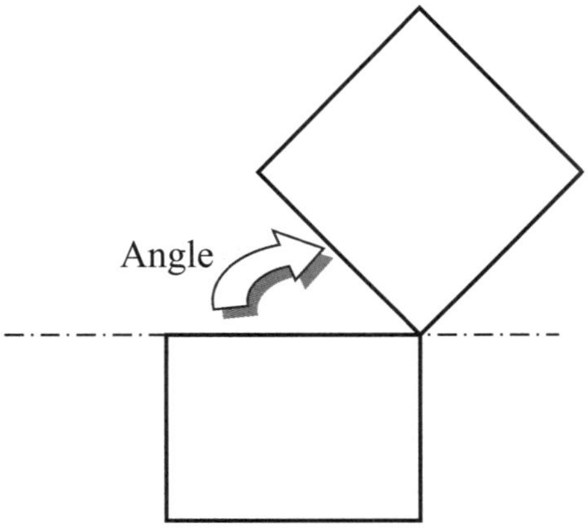

- **Tangent** – Aligns selected faces, planes, cylinders, spheres, and cones to contact at the point of tangency. Tangency may be on the inside or outside of a curve, depending on the selection of the direction of the surface normal. A Tangent constraint removes one degree of translational freedom.

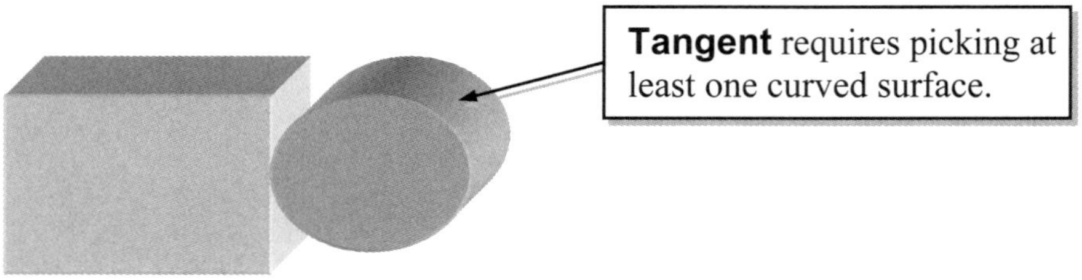
Tangent requires picking at least one curved surface.

- **Insert** – Aligns two circles, including their center axes and planes. Selected circular surfaces become co-axial. Insert constraint is a combination of a face-to-face Mate constraint between planar faces and a Mate constraint between the axes of the two components. A rotational degree of freedom remains open. The surfaces do not need to be full 360-degree circles. Selected surfaces can point in opposite directions or in the same direction and can be offset by a specified distance.

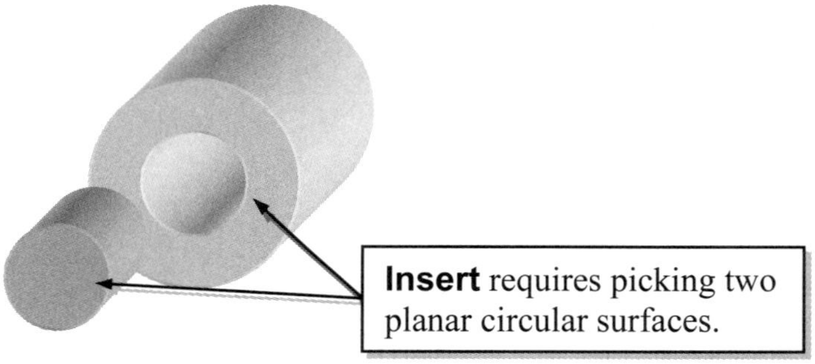
Insert requires picking two planar circular surfaces.

- **Symmetry** – The Symmetry constraint positions two objects symmetrically according to a plane or planar face. The Symmetry constraint is available in the Place Constraint dialog box.

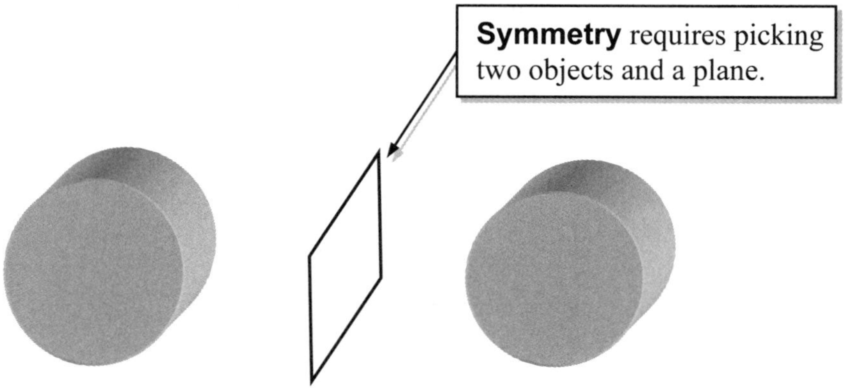
Symmetry requires picking two objects and a plane.

Apply the First Assembly Constraint

1. In the *Place Constraint* dialog box, confirm the constraint type is set to **Mate** constraint and select the **top horizontal surface** of the base part as the first part for the Mate alignment command.

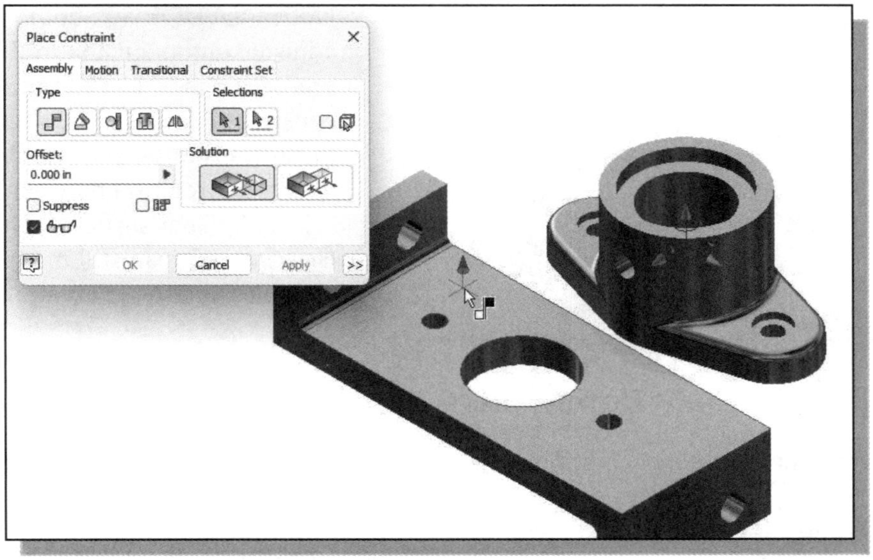

2. On your own, dynamically rotate the displayed model to view the bottom of the *Bearing* part, as shown in the figure below.

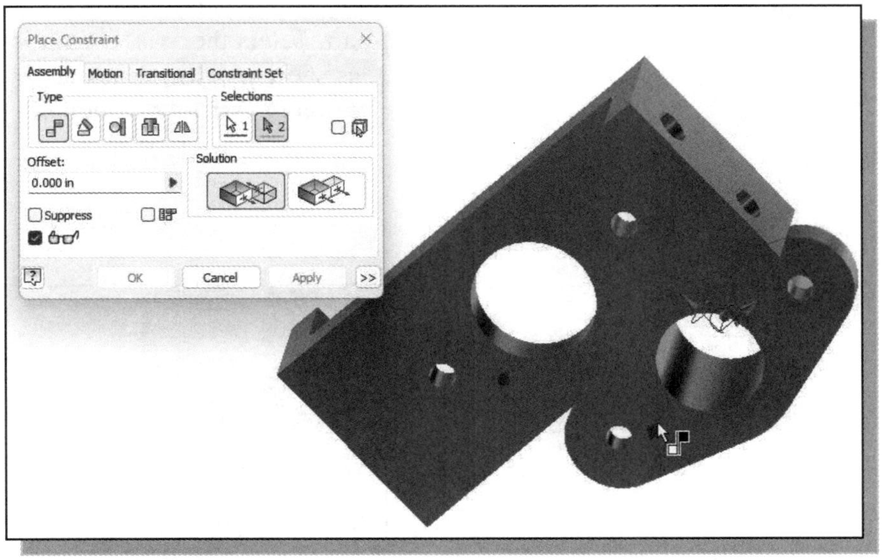

3. Click on the bottom face of the *Bearing* part as the second part selection to apply the constraint. Note the direction normals shown in the figure; the Mate constraint requires the selection of opposite direction of surface normals.

4. Click on the **Apply** button to accept the selection and apply the Mate constraint.

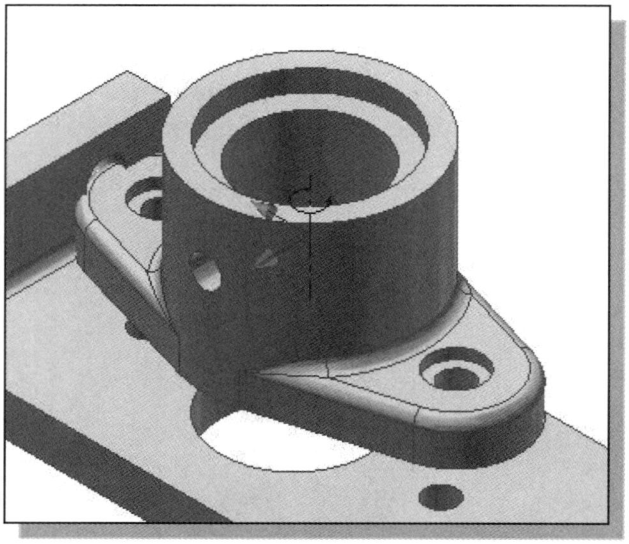

5. On your own, reset the display to **Isometric view** by clicking on the home view in the View Cube.

❖ Notice the DOF symbol is adjusted automatically in the graphics window. The Mate constraint removes one degree of linear translation and two degrees of angular rotation between the selected planar surfaces. The *Bearing* part can still move along two axes and rotate about the third axis.

Apply a Second Mate Constraint

The Mate constraint can also be used to align axes of cylindrical features.

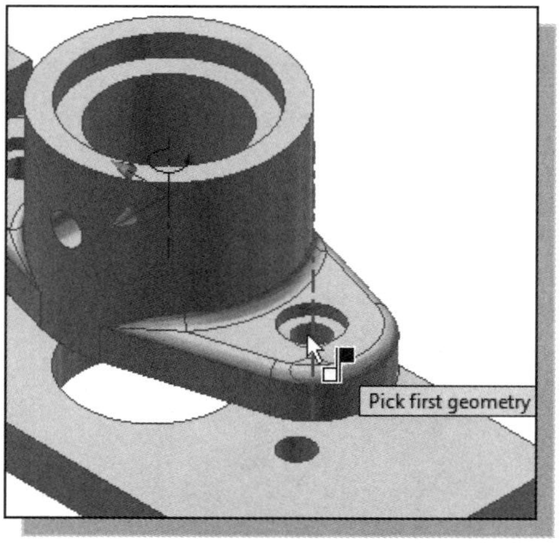

1. In the *Place Constraint* dialog box, confirm the constraint type is set to **Mate** constraint and move the cursor near the cylindrical surface of the right counter bore hole of the *Bearing* part. Select the axis when it is displayed as shown. (Hint: Use the *Dynamic Rotation* option to assist the selection.)

2. Move the cursor near the cylindrical surface of the small hole on the *Base-Plate* part. Select the axis when it is displayed as shown.

3. In the *Place Constraint* dialog box, set the *Solution* option to **Aligned** (if not aligned) and click on the **Apply** button to accept the selection and apply the Mate constraint.

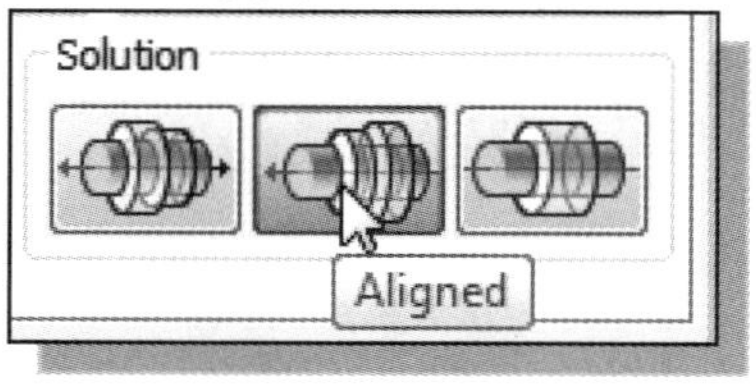

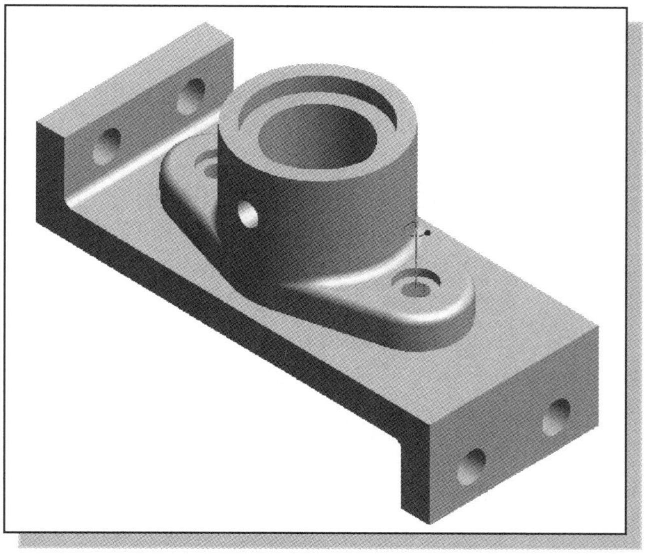

4. In the *Place Constraint* dialog box, click on the **Cancel** button to exit the Place Constraint command.

❖ The *Bearing* part appears to be placed in the correct position. But the DOF symbol indicates that this is not the case; the bearing part can still rotate about the displayed vertical axis.

Constrained Move

To see how well a component is constrained, we can perform a constrained move. A constrained move is done by dragging the component in the graphics window with the left-mouse-button. A constrained move will honor previously applied assembly constraints. That is, the selected component and parts constrained to the component move together in their constrained positions. A grounded component remains grounded during the move.

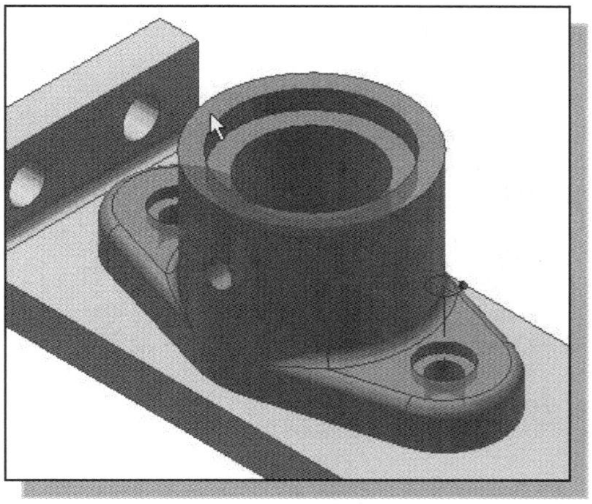

1. Inside the *graphics window*, move the cursor on top of the top surface of the *Bearing* part as shown in the figure.

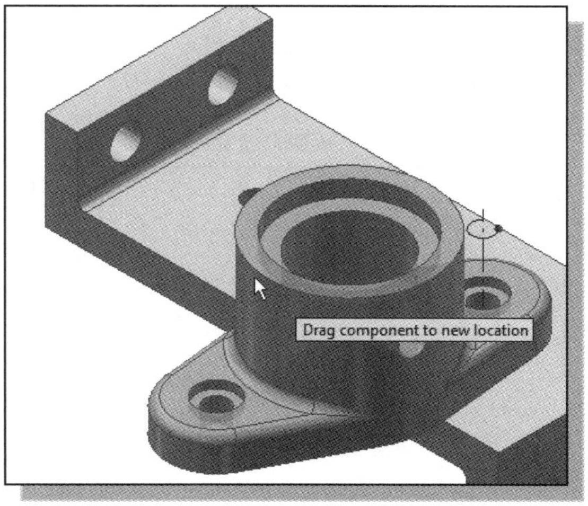

2. Press and hold down the left-mouse-button and drag the *Bearing* part downward.

❖ The *Bearing* part can freely rotate about the displayed axis.

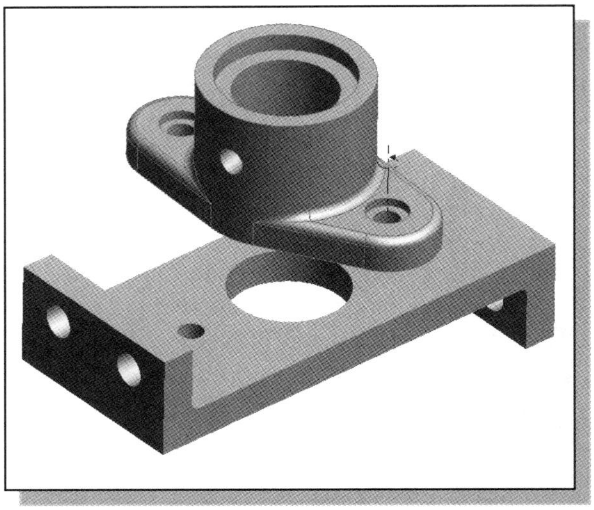

3. On your own, use the dynamic rotation command to view the alignment of the *Bearing* part.

4. Rotate the *Bearing* part and adjust the display roughly as shown in the figure.

Apply a Flush Constraint

Besides selecting the surfaces of solid models to apply constraints, we can also select the established work planes to apply the assembly constraints. This is an additional advantage of using the *BORN technique* in creating part models. For the *Bearing* part, we will apply a **Mate** constraint to two of the work planes and eliminate the last rotational DOF.

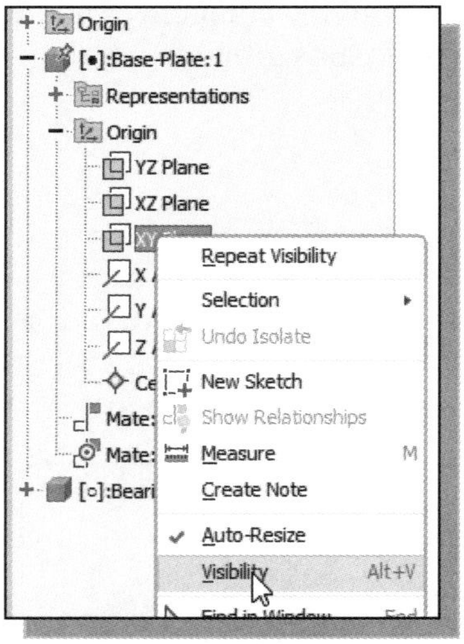

1. On your own, inside the *browser* window toggle *ON* the **Visibility** for the two corresponding work planes, shown on the following page, which will be used for alignment of the *Base-Plate* and the *Bearing* parts.

2. In the *Relationship* panel, select the **Constrain** command by left-clicking once on the icon.

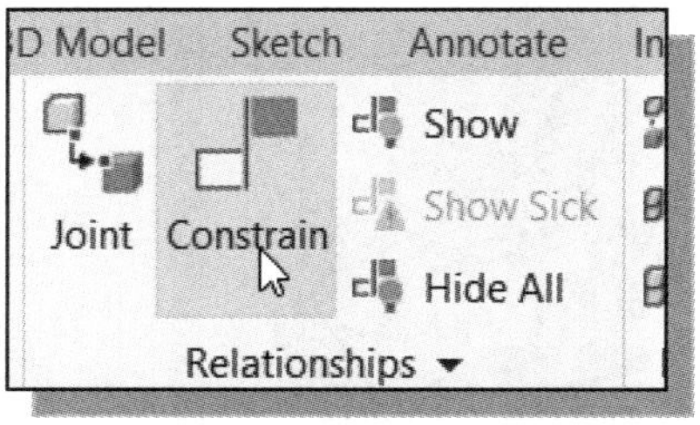

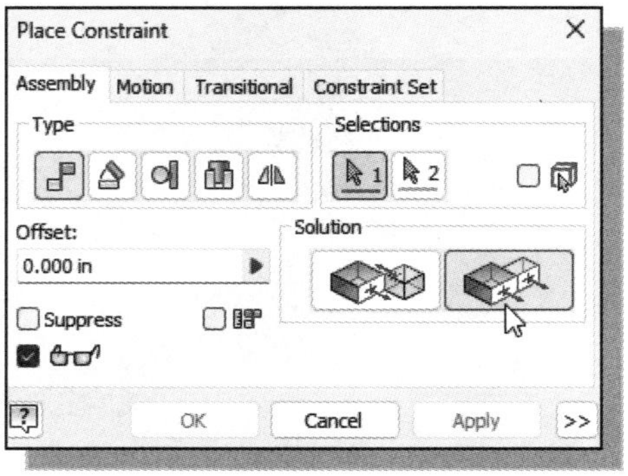

3. In the *Place Constraint* dialog box, switch the *Solution* option to **Flush** as shown.

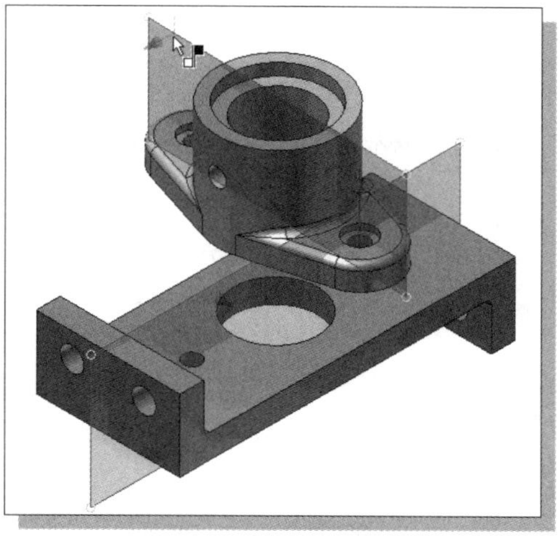

4. Select the *work plane* of the ***Base-Plate*** part as the first part for the **Flush** alignment command.

5. Select the corresponding *work plane* of the ***Bearing*** part as the second part for the **Flush** alignment command.

❖ Note that the **Flush** constraint makes two planes coplanar with their faces aligned in the same direction.

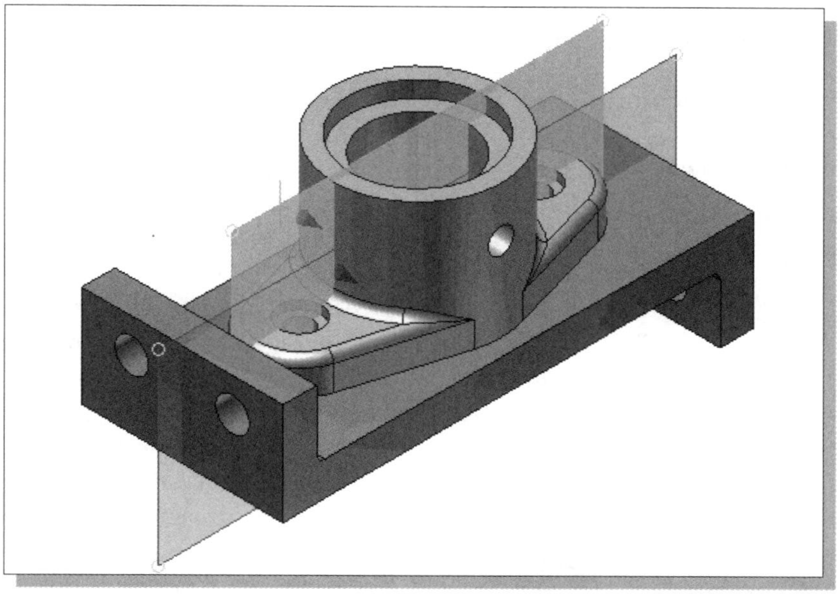

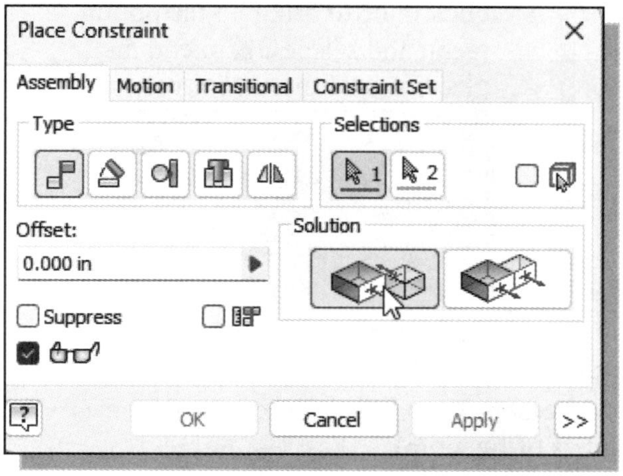

6. In the *Place Constraint* dialog box, switch the *Solution* option to **Mate** and notice the ***Bearing*** part is rotated 180 degrees of the **Flush** option.

❖ Note that the *Show Preview* option allows us to preview the result before accepting the selection.

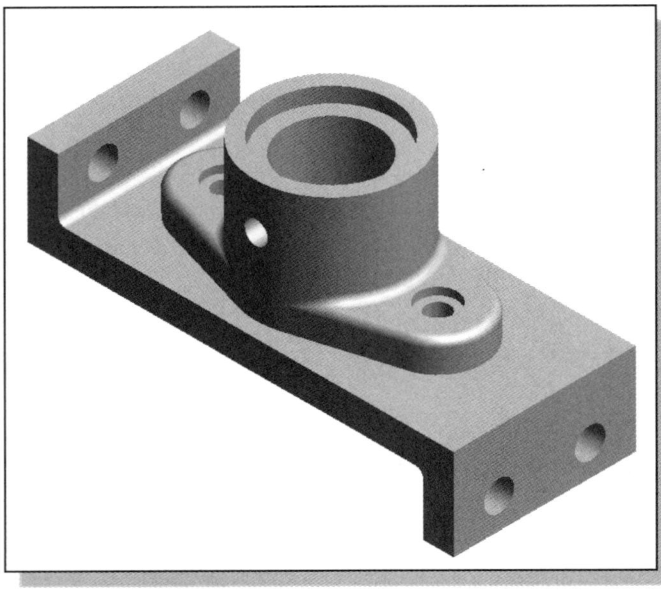

7. On your own, align the two parts as shown and click on the **Apply** button to accept the settings.

8. In the *Place Constraint* dialog box, click on the **Cancel** button to exit the Place Constraint command.

❖ Note the DOF symbol disappears, which indicates the assembly is fully constrained.

Placing the Third Component

We will retrieve the *Collar* part as the third component of the assembly model.

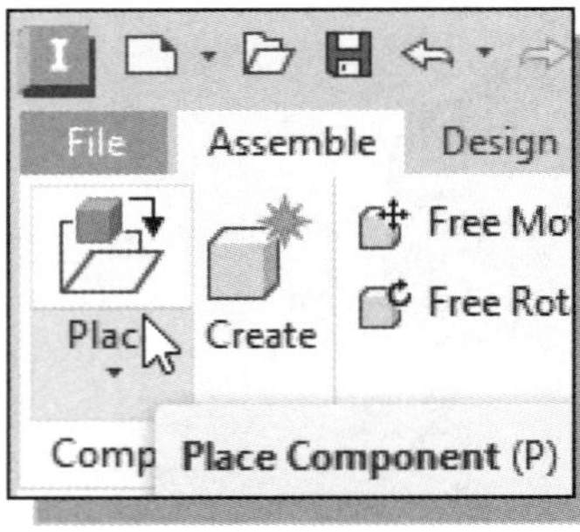

1. In the *Assemble* panel (the toolbar that is located to the left side of the graphics window) select the **Place Component** command by left-clicking the icon.

2. Select the ***Collar*** design (part file: ***Collar.ipt***) in the list window. And click on the **Open** button to retrieve the model.

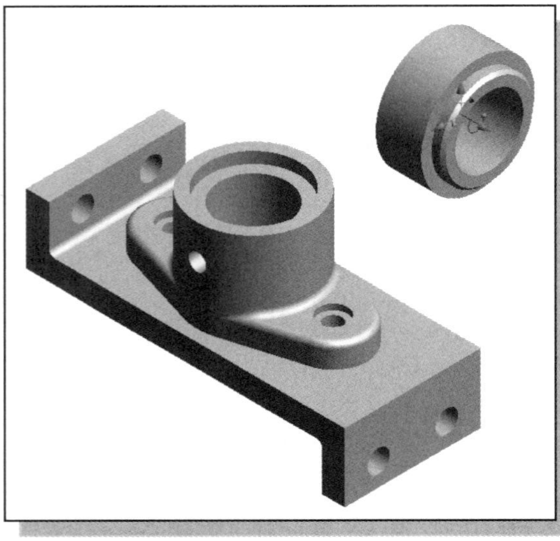

3. Place the *Collar* part toward the upper right corner of the graphics window, as shown in the figure.

4. Inside the *graphics window*, **right-click** once to bring up the option menu and select **OK** to end the placement of the *Collar* part.

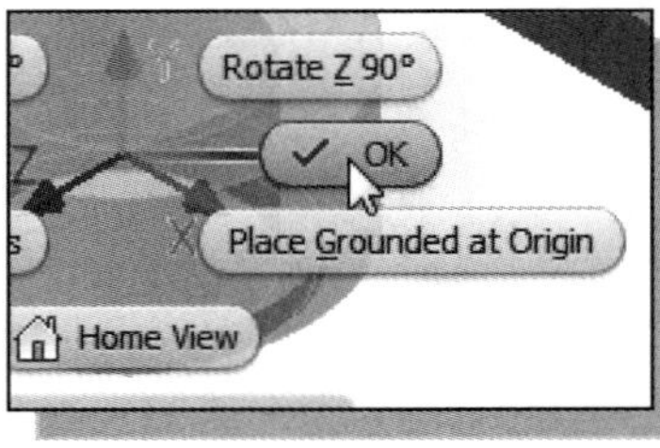

❖ Notice the DOF symbol displayed on the screen. The *Collar* part can move linearly and rotate about the three axes (six degrees of freedom).

Applying an Insert Constraint

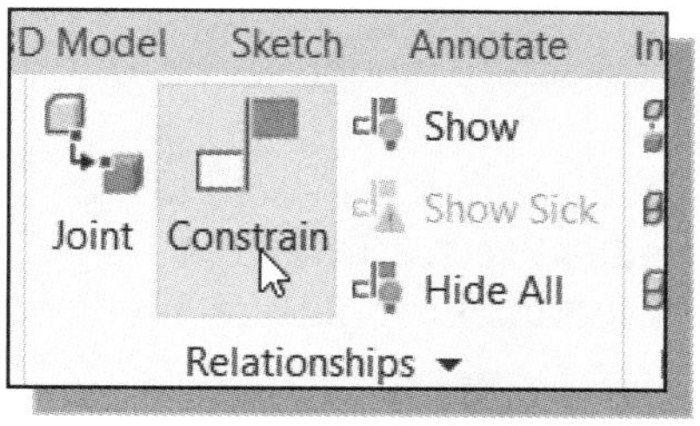

1. In the *Relationships* panel, select the **Constrain** command by left-clicking once on the icon.

2. In the *Place Constraint* dialog box, switch to the **Insert** constraint.

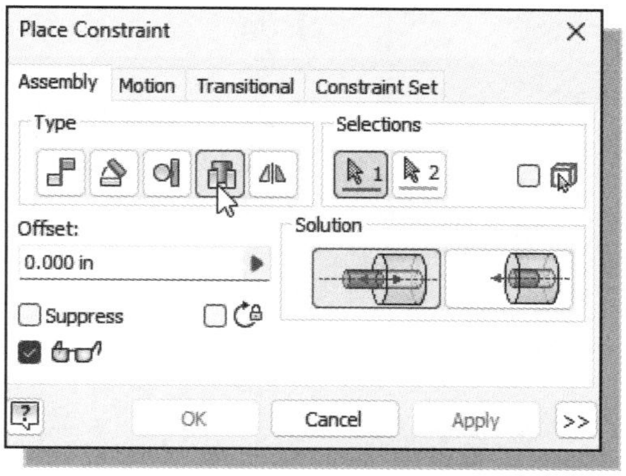

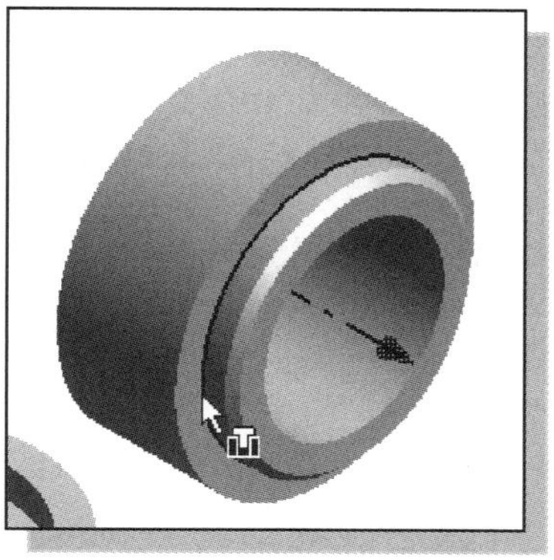

3. Select the inside corner of the *Collar* part as the first surface to apply the Insert constraint, as shown in the figure.

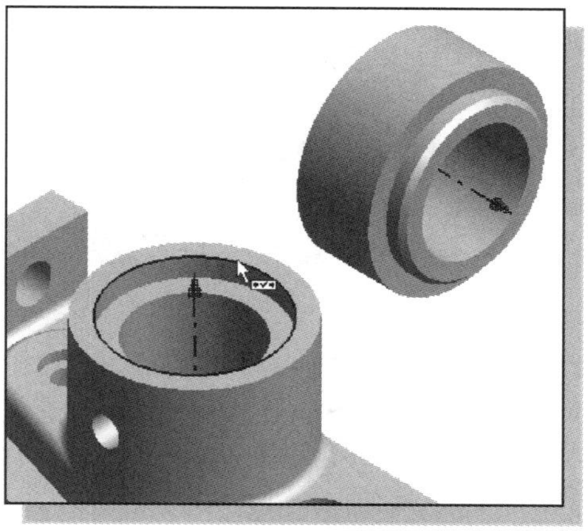

4. Select the inside circle on the top surface of the *Bearing* part as the second surface to apply the Insert constraint, as shown in the figure.

5. Click on the **Apply** button to accept the settings.

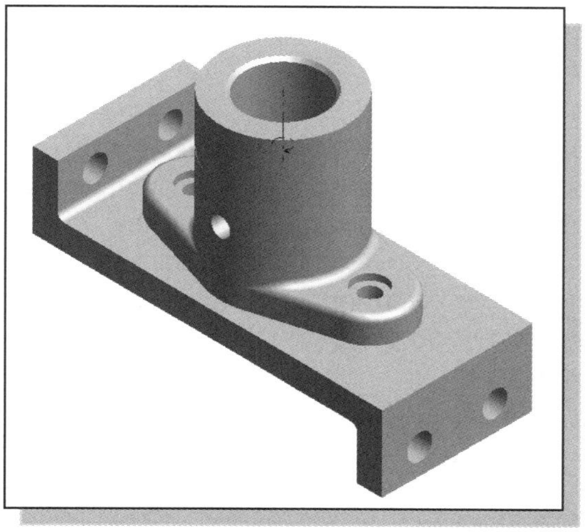

➢ Note that one rotational degree of freedom remains open; the *Collar* part can still freely rotate about the displayed DOF axis.

Assemble the Cap-Screws

We will place two of the *Cap-Screw* parts to complete the assembly model.

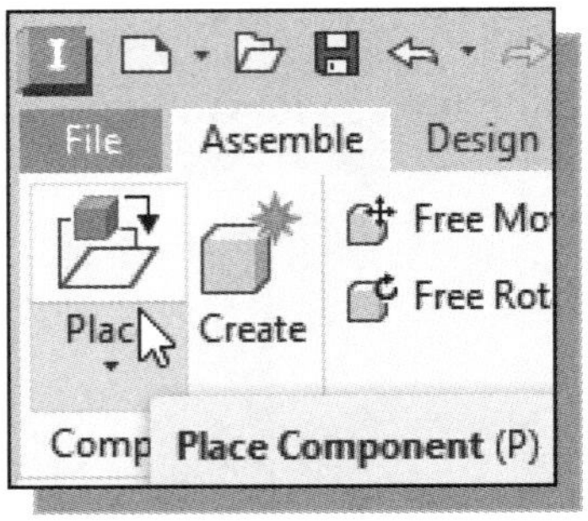

1. In the *Assemble* panel (the toolbar that is located to the left side of the graphics window) select the **Place Component** command by left-clicking once on the icon.

2. Select the ***Cap-Screw*** design (part file: ***Cap-Screw.ipt***) in the list window. And click on the **Open** button to retrieve the model.

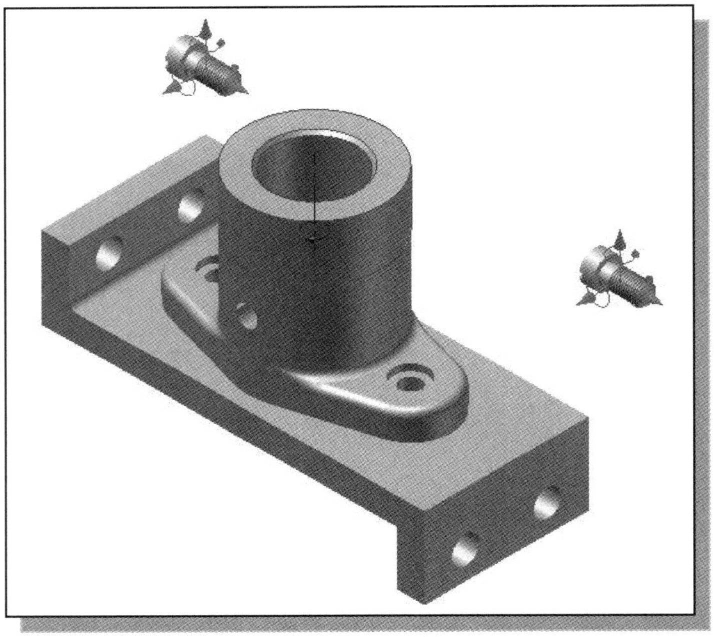

3. Place two copies of the *Cap-Screw* part on both sides of the *Collar* by clicking twice on the screen as shown in the figure.

4. Inside the *graphics window*, **right-click** once to bring up the option menu and select **OK** to end the Place Component command.

❖ Notice the DOF symbols displayed on the screen. Each *Cap-Screw* has six degrees of freedom. Both parts are referencing the same external part file, but each can be constrained independently.

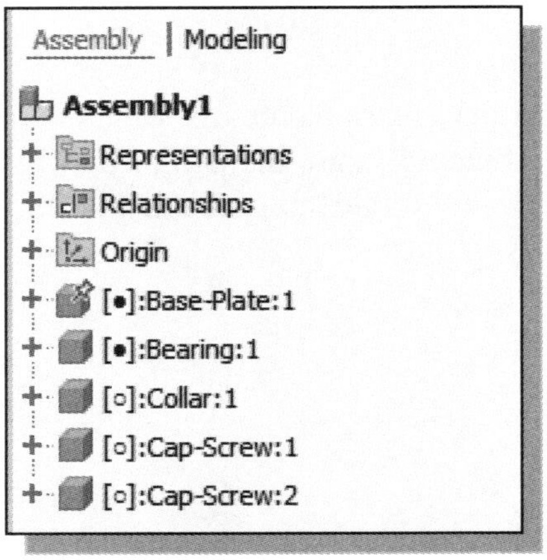

- Inside the *browser* window, the retrieved parts are listed in the order they are placed. The number behind the part name is used to identify the number of copies of the same part in the assembly model. Move the cursor to the last part name and notice the corresponding part is highlighted in the graphics window.

➢ On your own, use the Place Constraints command and assemble the *Cap-Screws* in place as shown in the figure below.

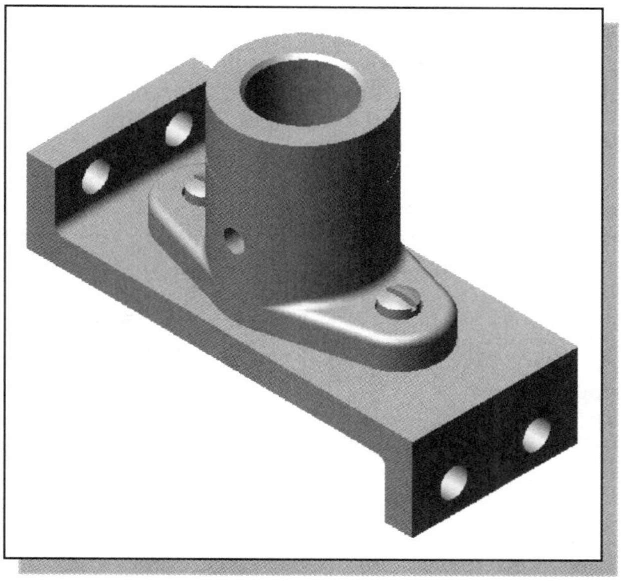

Exploded View of the Assembly

Exploded assemblies are often used in design presentations, catalogs, sales literature, and in the shop to show all of the parts of an assembly and how they fit together. In Autodesk Inventor, an exploded assembly can be created by two methods: (1) using the Move Component and Rotate Component commands in the *Assembly Modeler*, which contains only limited options for the operation but can be done very quickly; (2) transferring the assembly model into the *Presentation Modeler*. For our example, we will create an exploded assembly by using the Move Component command that is available in the *Assembly Modeler*.

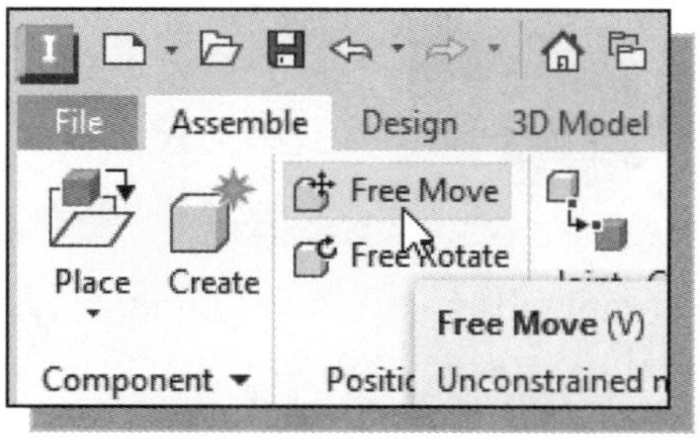

1. In the *Position* toolbar panel, select the **Free Move** command by left-clicking once on the icon.

2. Inside the graphics window, move the cursor on top of the top surface of the *Collar* part as shown in the figure.

3. Press and hold down the left-mouse-button and drag the *Collar* part toward the right side of the assembly as shown.

4. On your own, repeat the above steps and create an exploded assembly by repositioning the components as shown in the figure below.

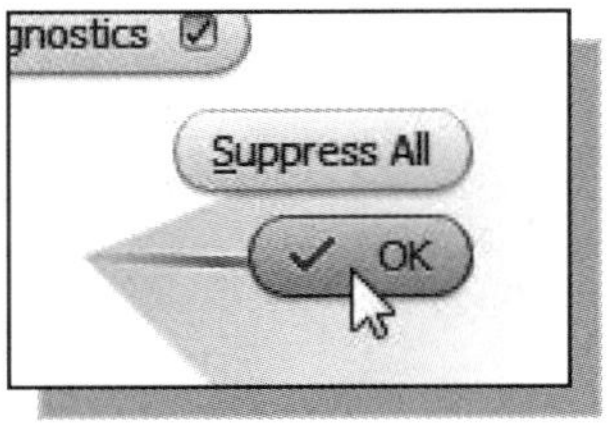

5. Inside the graphics window, right-click once to bring up the option menu and select **OK** to end the Move Component command.

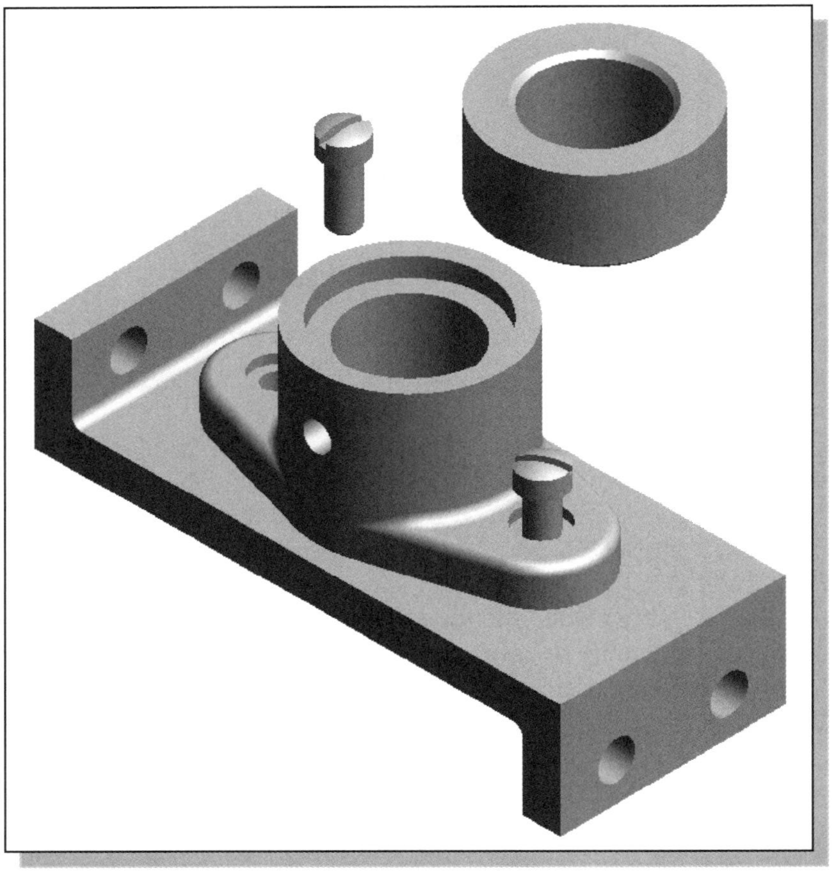

❖ The Move Component and Rotate Component commands are used to temporarily reposition the components in the graphics window. The displayed image is temporary, but it can be printed with the **Print** command through the pull-down menu.

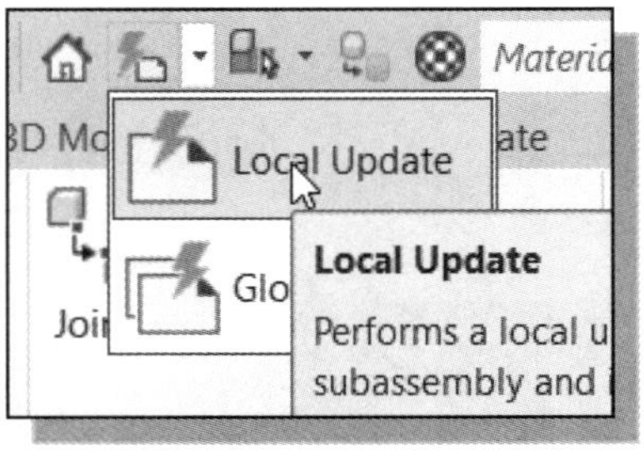

6. Click on the **Update** button in the *Standard* toolbar area.

❖ Note that the components are reset back to their assembled positions, based on the applied assembly constraints.

Editing the Components

The *associative functionality* of Autodesk Inventor allows us to change the design at any level, and the system reflects the changes at all levels automatically.

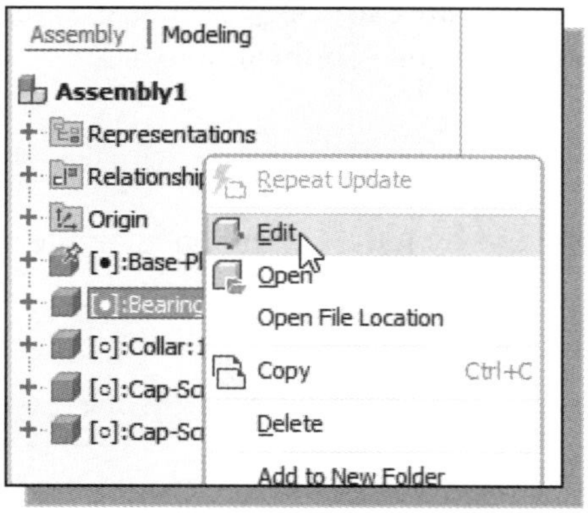

1. Inside the *Desktop Browser*, move the cursor on top of the ***Bearing*** part. Right-click once to bring up the option menu and select **Edit** in the option list.

❖ Note that we are automatically switched back to *Part Editing Mode.*

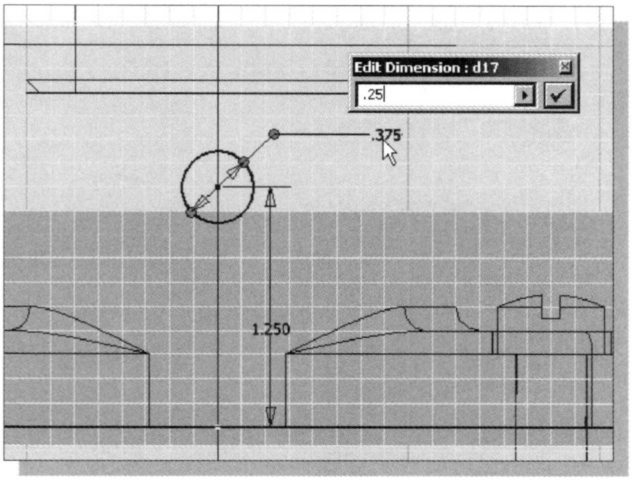

2. On your own, adjust the diameter of the small *Drill Hole* to **0.25** as shown.

3. Click on the **Update** button in the *Standard* toolbar area to proceed with updating the model.

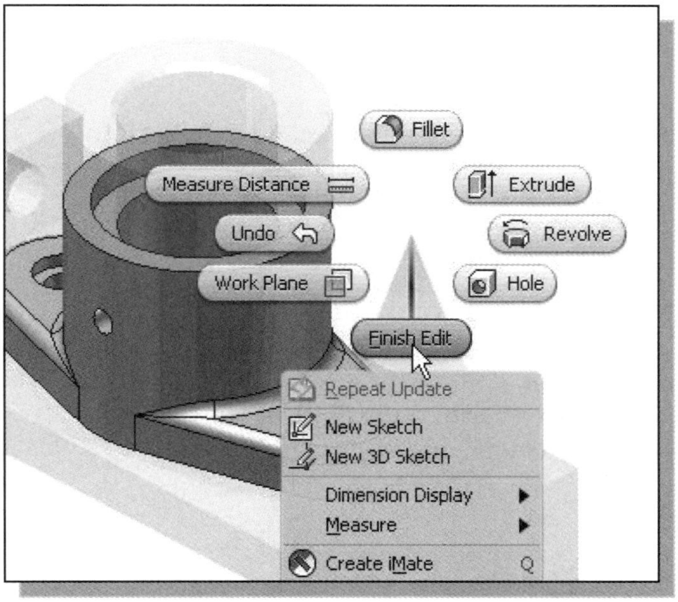

4. Inside the *graphics window*, click once with the **right-mouse-button** to display the *option menu.*

5. Select **Finish Edit** in the pop-up menu to exit *Part Editing Mode* and return to *Assembly Mode.*

➢ Autodesk Inventor has updated the part in all levels and in the current *Assembly Mode*. On your own, open the *Bearing* part file to confirm the modification is performed.

Adaptive Design Approach

Autodesk Inventor's ***Adaptive Design*** approach allows us to use the applied *assembly constraints* to control the sizes, shapes, and positions of ***underconstrained*** sketches, features, and parts. In this section, we will examine the procedure to apply the constraints directly on 3D parts; note that the adaptive design approach is also applicable to 2D sketches.

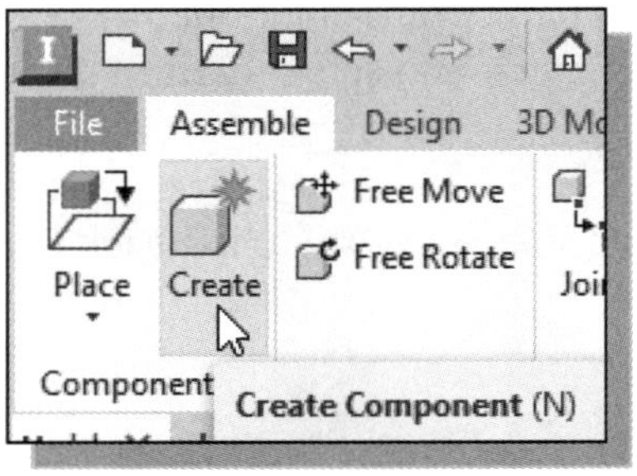

1. In the *Assemble* panel, select the **Create Component** command by left-clicking once on the icon.

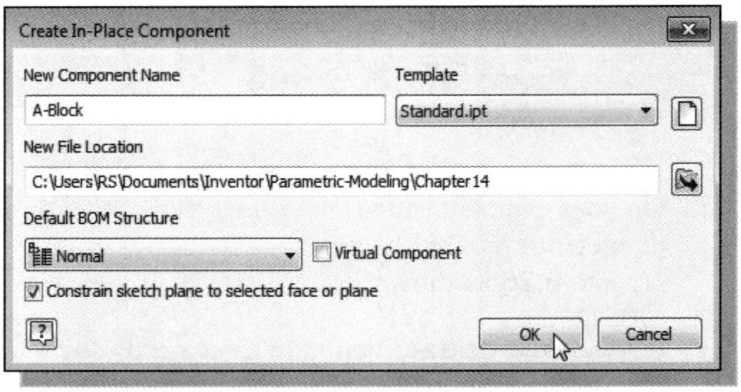

2. In the *Create In-Place Component* dialog box, enter **A-Block** as the new file name.

3. Set the *File Location* to **Chapter14** as shown.

4. Click on the **OK** button to accept the settings.

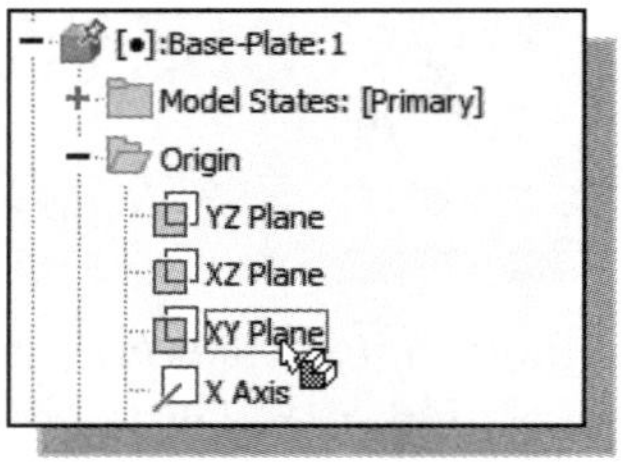

5. Click on the ***XY Plane*** of the *Base-Plate* part, inside the *browser* window, to apply a Flush constraint.

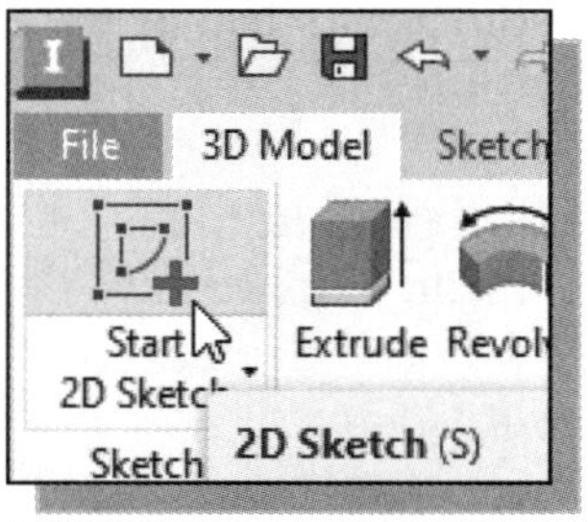

6. Activate the *Model* toolbar and select the **2D Sketch** command by left-clicking once on the icon.

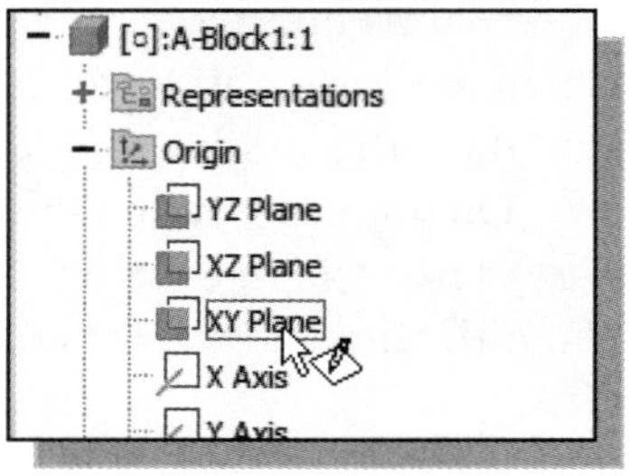

7. Select the ***XY Plane***, in the *browser* window of the *A-Block* part, to align the sketch plane.

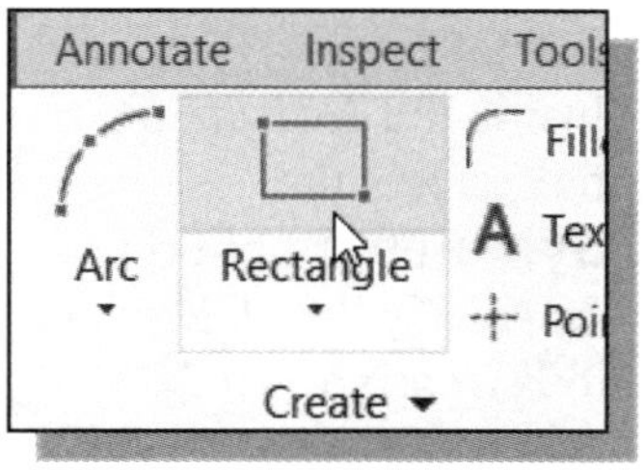

8. Select the **Two point rectangle** command by clicking once with the left-mouse-button on the icon in the *2D Sketch* panel.

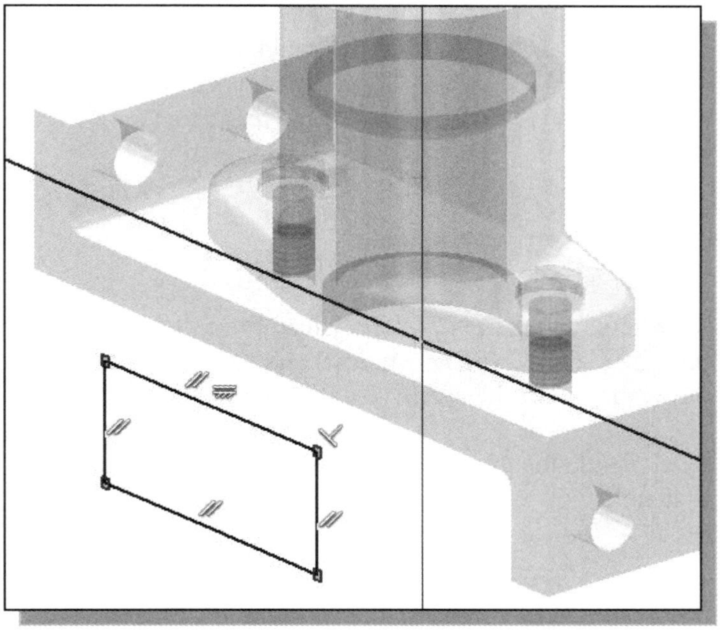

9. On your own, create a rectangle of arbitrary size below the assembly model as shown in the figure.

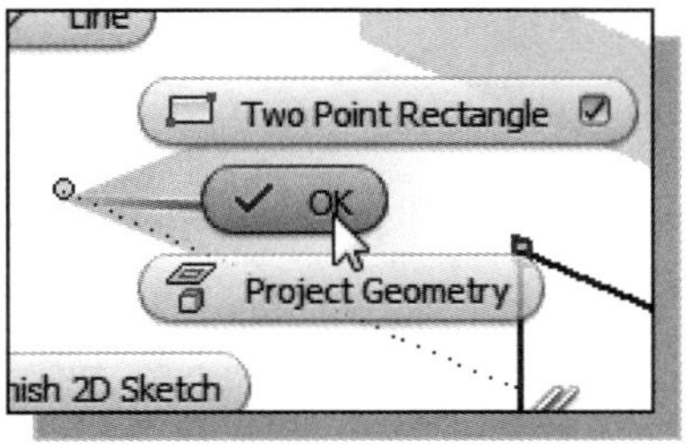

10. Inside the *graphics window*, click once with the **right-mouse-button** to display the option menu and select **OK** to end the Rectangle command.

11. Exit the sketch mode by clicking the **Finish Sketch** button.

12. In the *Ribbon* toolbar panel, select the **Extrude** command in the *Model tab* as shown in the figure below.

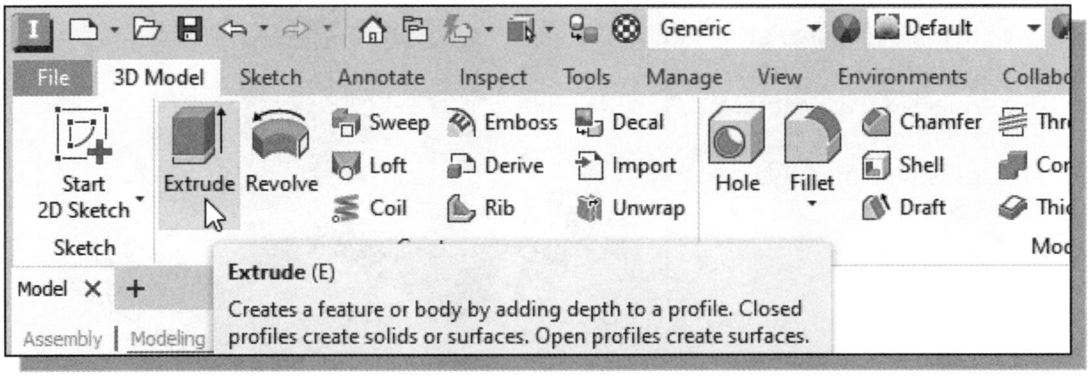

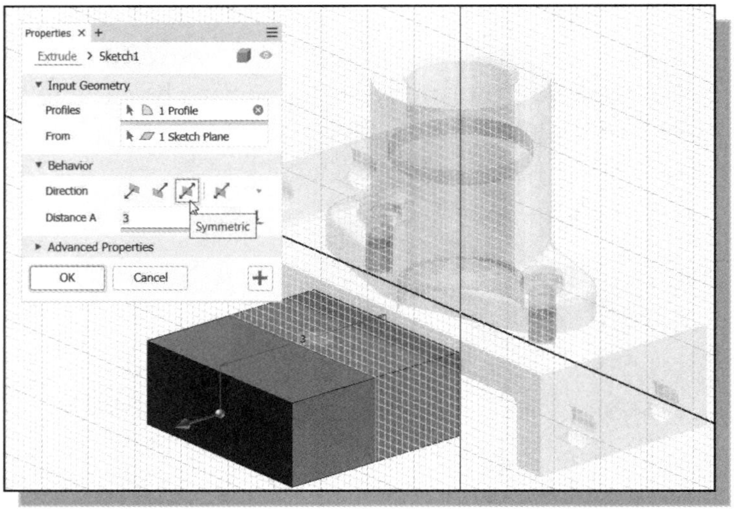

13. Inside the *Extrude* dialog box, select the **Symmetric** option and set the *Extents Distance* to **3 in** as shown.

14. Click on the **OK** button to proceed with creating the feature.

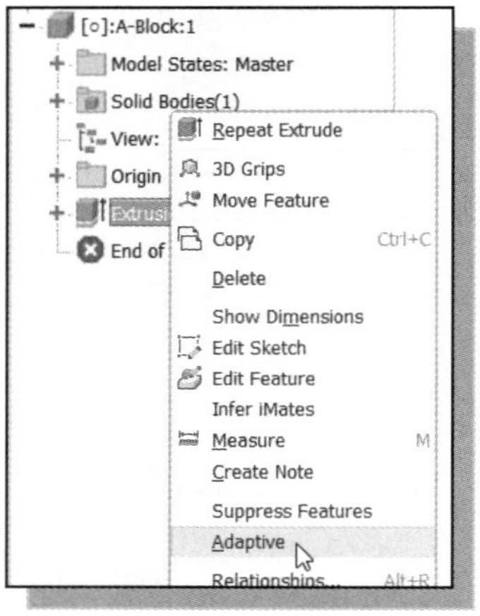

15. Right-click on the ***Extrusion1*** feature of the *A-Block* part, inside the *browser* window, and select **Adaptive** to allow the use of the adaptive design approach.

16. Click inside the graphics window to deselect any part or assembly.

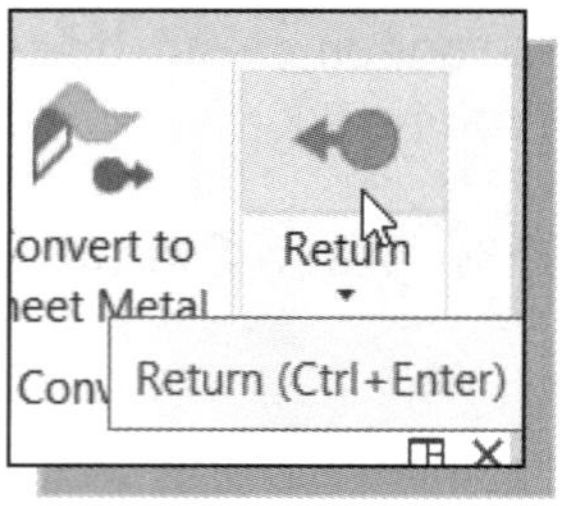

17. Select **Return** in the Ribbon toolbar to exit *Part Editing Mode* and return to *Assembly Modeling Mode.*

- Note that the rectangular block is created with only one dimension, the extrude distance. The 2D sketch of the part is intentionally under-constrained.

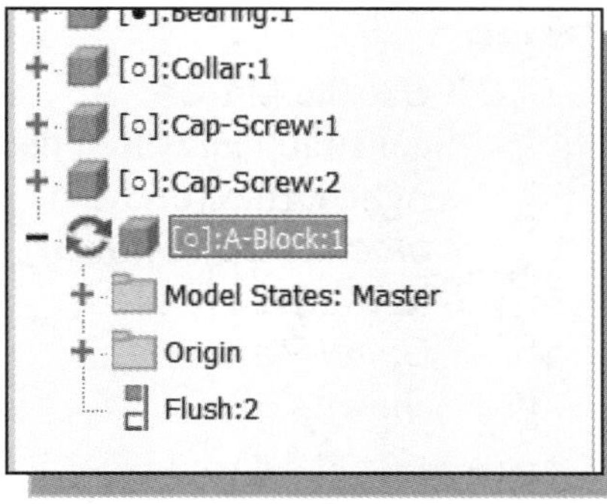

- Note the **Adaptive** icon, the two-rotate-arrows symbol, appears in front of the part name in the *browser* window.

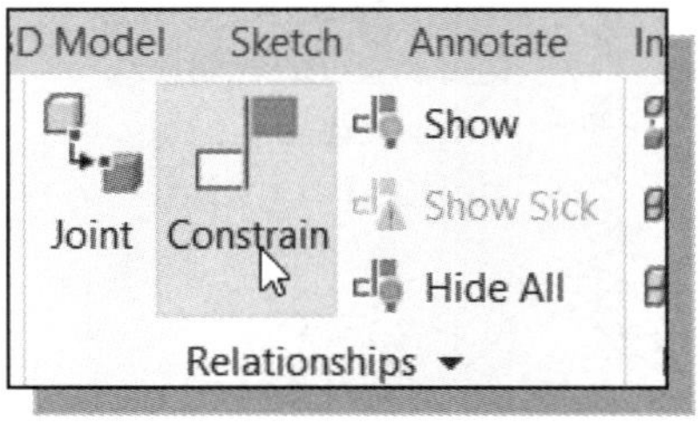

18. In the *Assemble* panel, select the **Constrain** command by left-clicking once on the icon.

19. On your own, use the **Flush** constraint to align the *A-block* to the left-edge of the *Base-Plate* as shown.

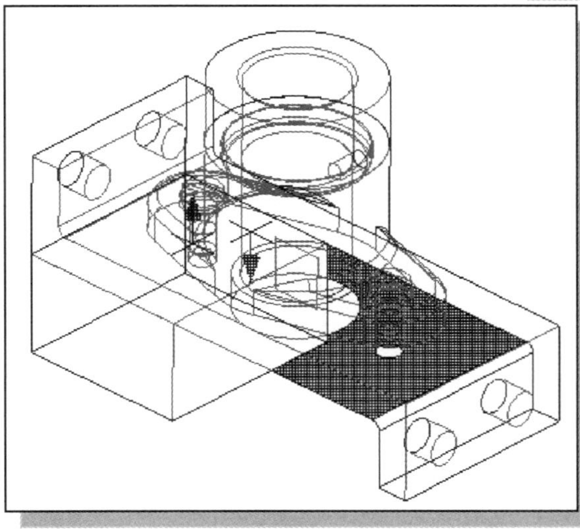

20. Create a **Mate** constraint to align the top of the *A-block* to the bottom of the *Base-Plate* part as shown.

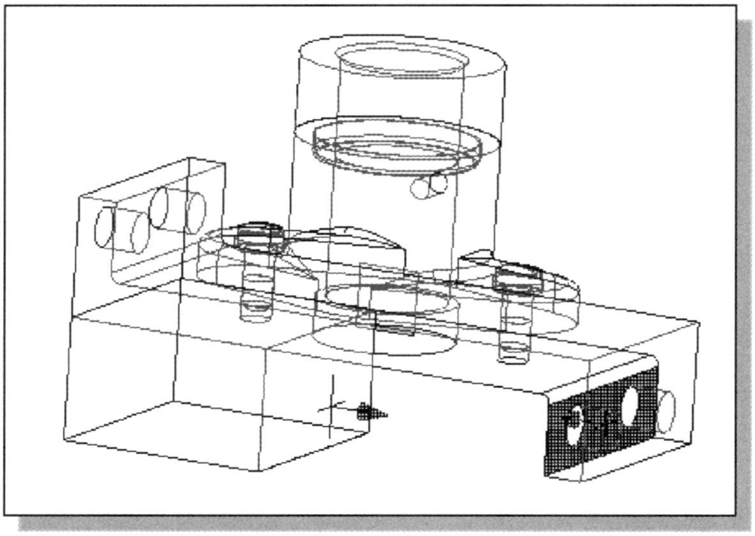

21. Use the **Mate** constraint and align the right surface of the *A-block* to the bottom left surface of the *Base-Plate* part as shown.

- Note that the length of the *A-block* part is adjusted to fit the defined constraint.

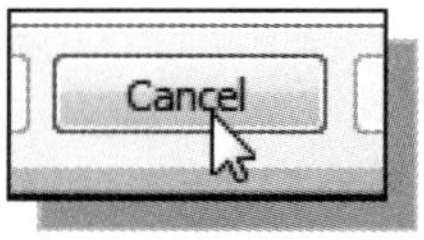

22. In the *Place Constraint* dialog box, click on the **Cancel** button to end the Place Constraint command.

Delete and Re-apply Assembly Constraints

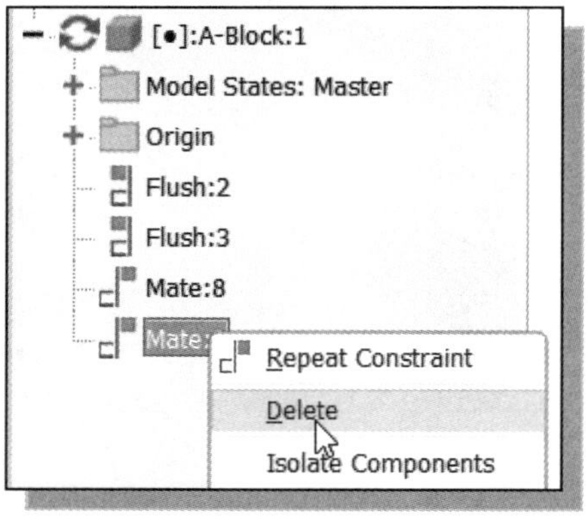

1. Inside the *browser* window, right-click on the last **Mate** constraint of the *A-Block* part to bring up the option menu.

2. Select **Delete** to remove the applied constraint.

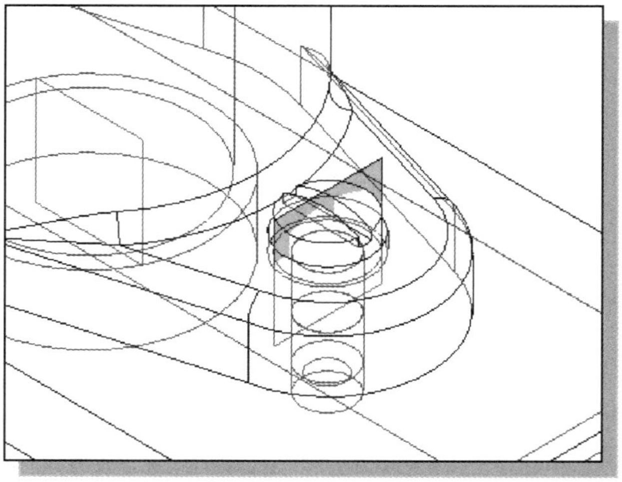

3. On your own, switch ***ON*** the **Visibility** of the vertical work plane of the *Cap-Screw* part, the plane that is perpendicular to the length of the *Base-Plate* part, as shown.

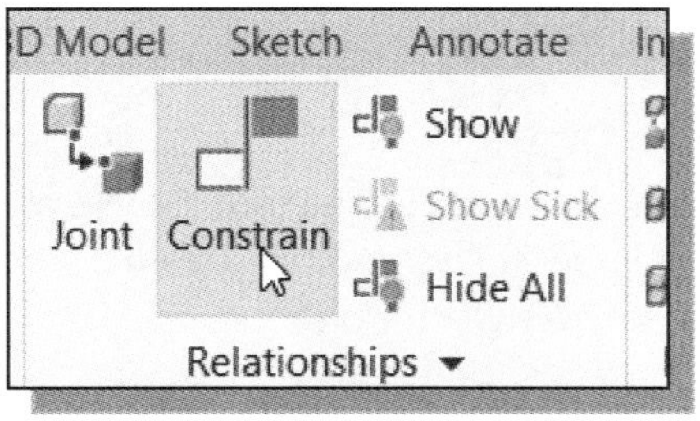

4. In the *Assembly* position panel, select the **Constrain** command by left-clicking once on the icon.

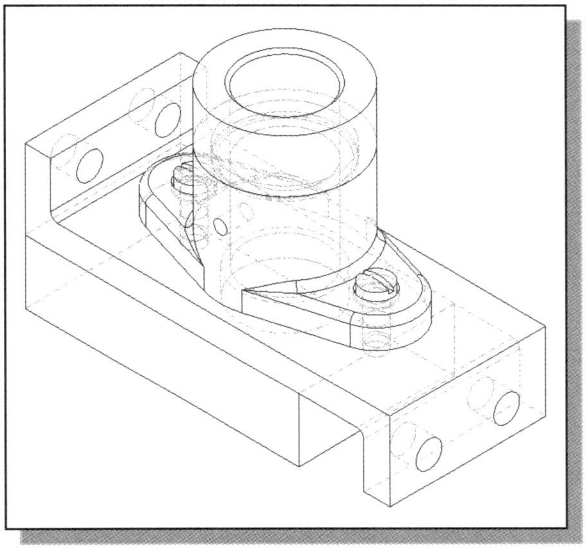

5. On your own, align the right vertical surface of the ***A-block*** part to the vertical work plane as shown in the figure.

6. On your own, experiment with aligning the right vertical surface of the ***A-block*** part to other vertical surfaces of the assembly model.

- As can be seen, the length of the ***A-block*** is adjusted to the newly applied constraint. The ***Adaptive Design*** approach allows us to have greater flexibility and simplifies the design process.

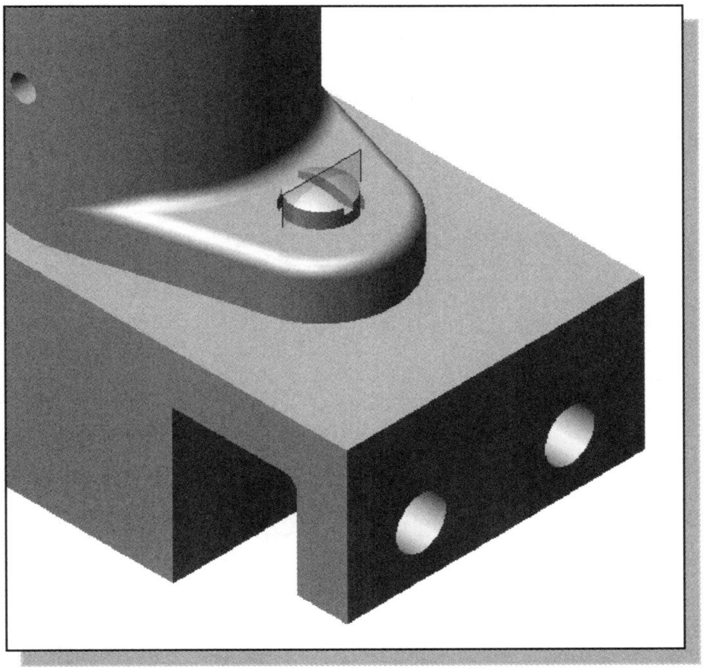

7. On your own, save the assembly model as **Shaft-Support.iam** under the Chapter14 folder.

Set up a Drawing of the Assembly Model

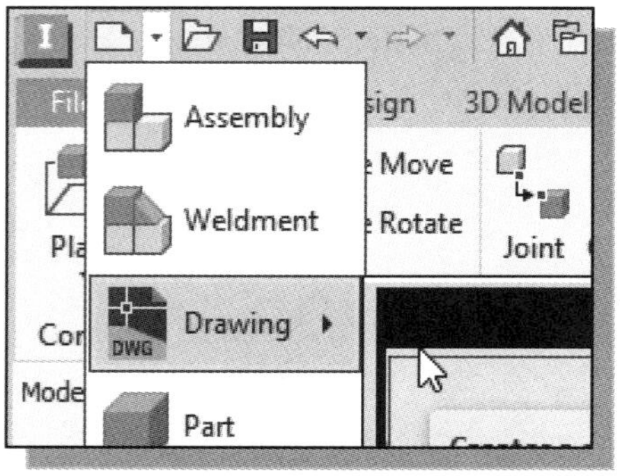

1. Click on the **drop-down arrow** next to the **New File** icon in the *Quick Access* toolbar area to display the available **New File** options.

2. Select **Drawing** from the option list.

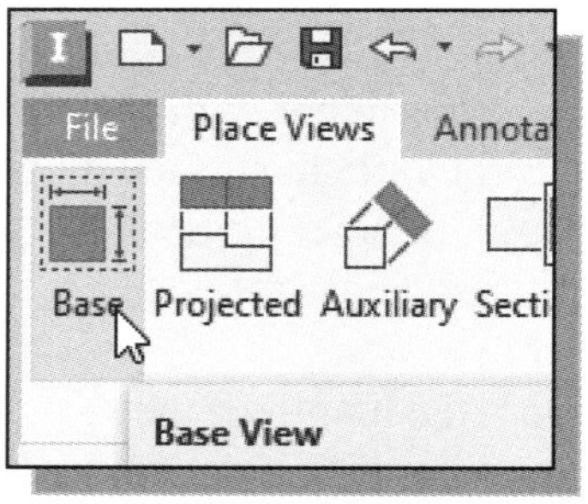

3. Click on the **Base View** in the *Drawing Views* panel to create a base view.

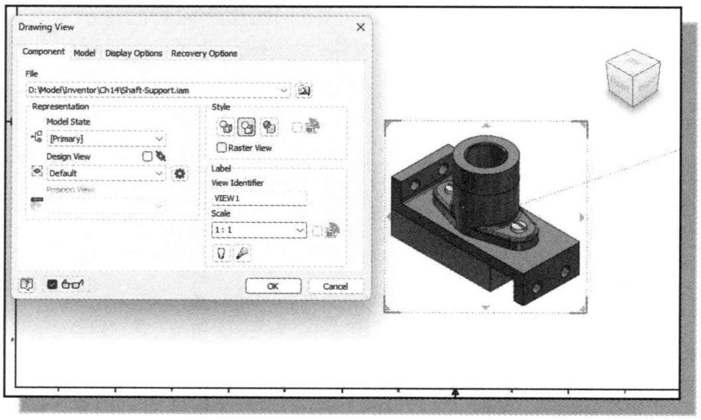

4. In the *Drawing View* dialog box, set *Orientation* to **Iso Top Right View, Scale 1:1** and **Hidden Line Removed** as shown in the figure.

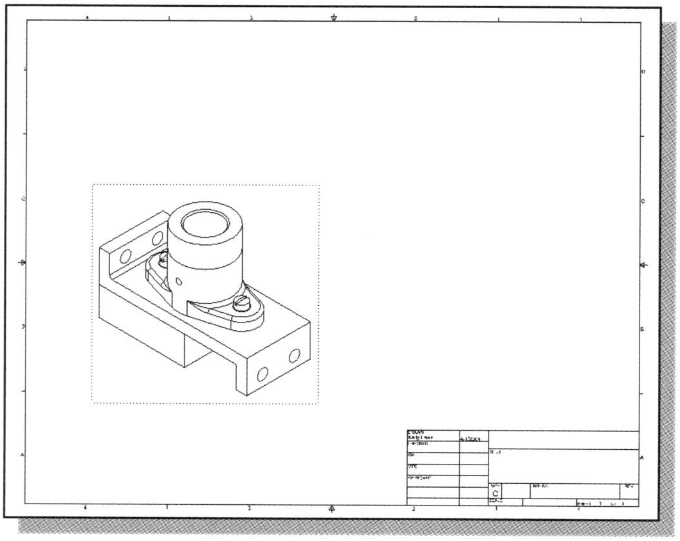

5. Move the cursor inside the graphics window and place the ***base*** view near the lower left side of the *Border* as shown.

6. Click the **OK** button to end the command.

❖ Note that the default sheet size is much bigger than the created view. *Inventor* allows us to adjust the sheet size even when views have been created.

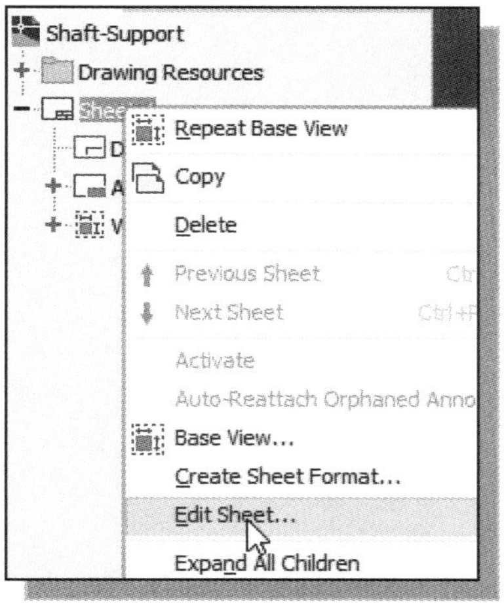

7. Inside the *Drawing Browser* window, right-click on ***Sheet:1*** to display the option menu.

8. Select **Edit Sheet** in the option menu to display the settings for the drawing.

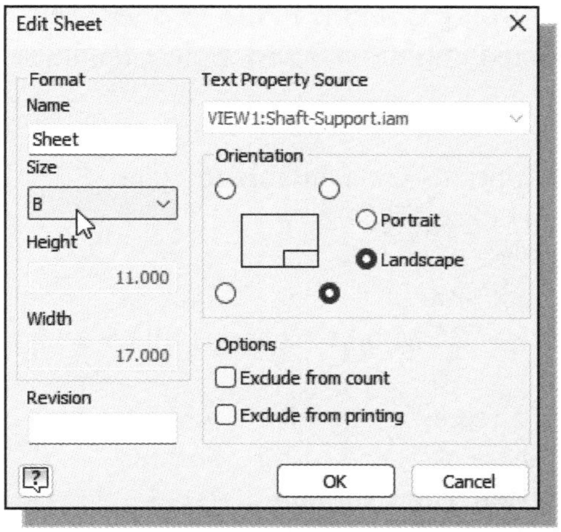

9. Set the sheet size to ***B-size*** as shown.

10. Click on the **OK** button to accept the settings and exit the Edit Sheet command.

11. On your own, reposition the Isometric view as shown in the figure.

❖ Note that the *Border* and *Title Block* are automatically replaced as the sheet size is adjusted.

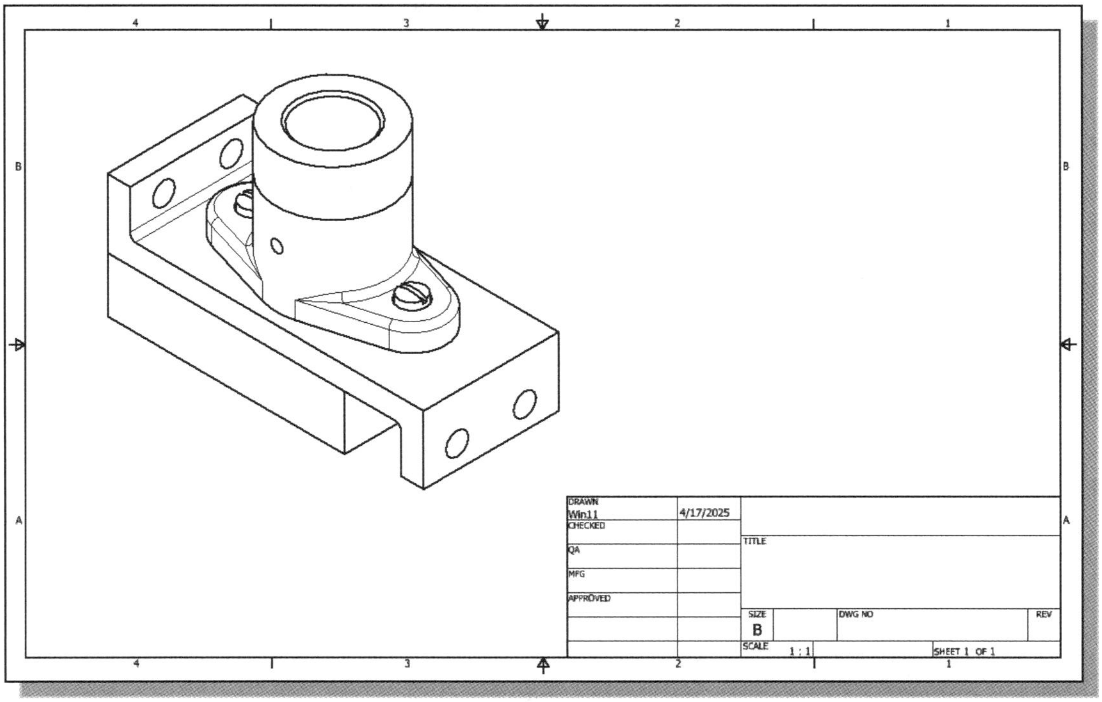

Creating a Parts List

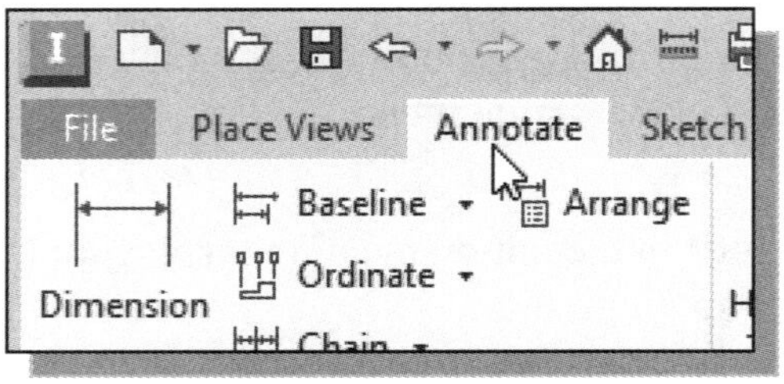

1. In the *Ribbon Toolbar*, click on the **Annotate** tab as shown.

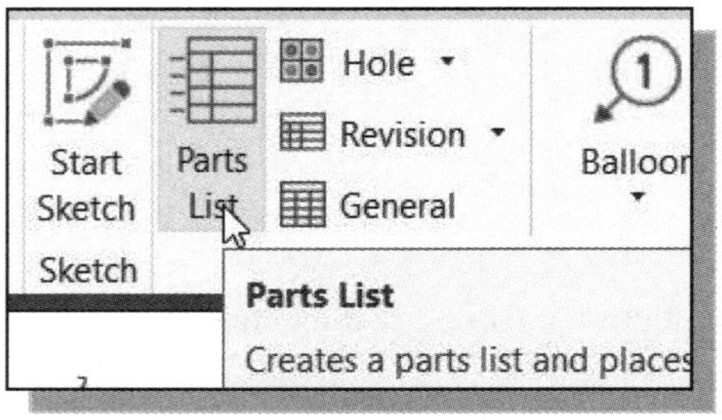

2. In the *Drawing Annotation* window, click on the **Parts List** button.
3. In the *prompt area*, the message "*Select a view*" is displayed. Items in the selected view will be listed in the *Parts List*. Select the **base view**.

❖ The *Parts List* dialog box appears on the screen; options are available to make adjustments to the numbering system and table wrap settings.

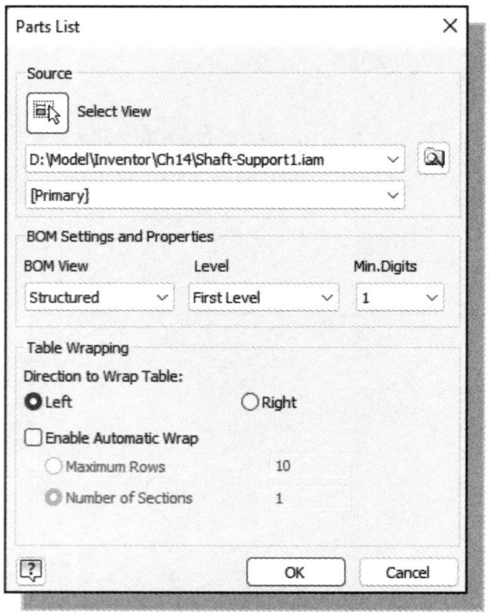

➢ The *Parts List* – BOM options are as follows:

Structured: Creates a parts list in which subassemblies are assigned using a nested numbering system (for example, 1, 1.1, 1.1.1). The nested number extends as many levels as needed for the assembly levels in the model.

Only Parts: Creates a parts list that sequentially numbers all parts in the assembly, including parts that are contained in subassemblies.

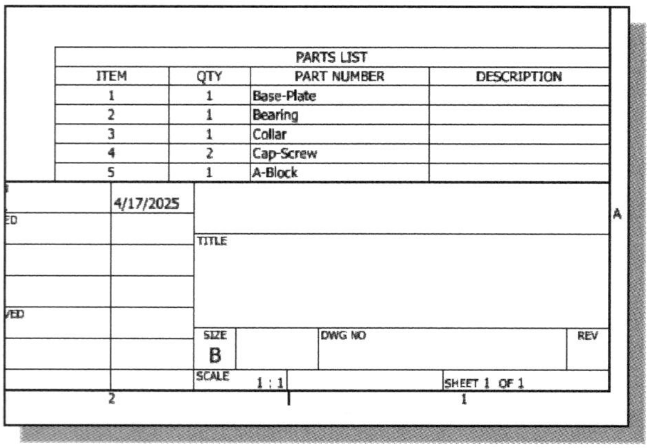

❖ Note that no subassemblies are currently used in the assembly; we will accept the default settings.

4. Click **OK** to accept the default settings and also enable the **BOM View** option.
5. Place the ***Parts List*** above the *Title Block* as shown.

Edit the Parts List

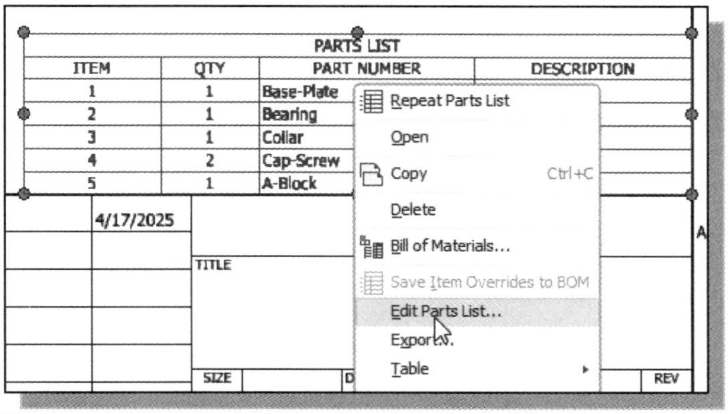

1. Move the cursor on top of the ***Parts List*** and **right-click** once to display the option menu.

2. Choose **Edit Parts List** in the option menu as shown.

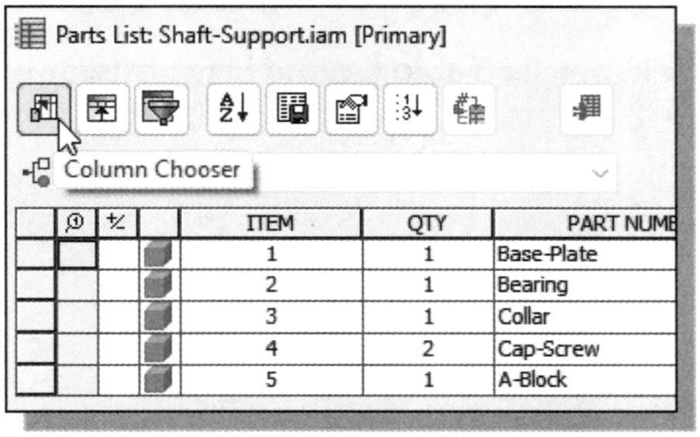

3. Click on the **Column Chooser** button as shown.

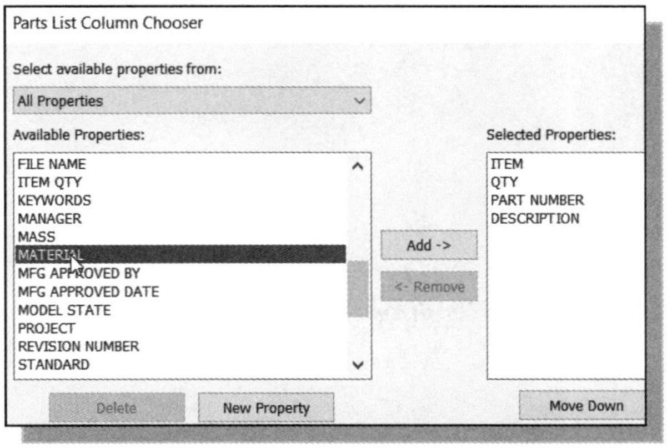

4. Select **MATERIAL** in the *Available Properties* list as shown.

5. Click **Add** to add the selected item to the *Selected Properties* list.

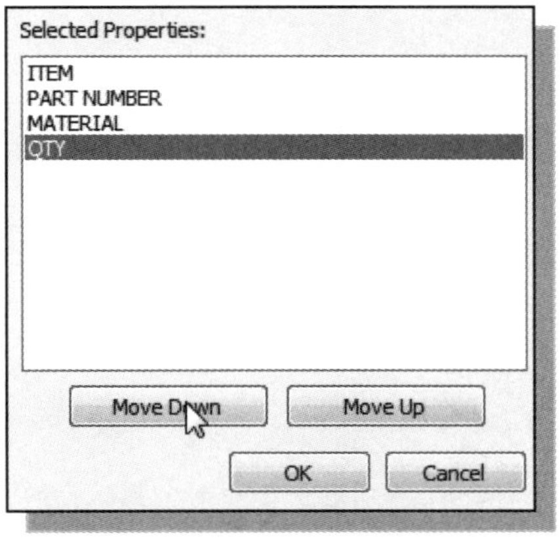

6. On your own, adjust the ***Selected Properties*** list as shown. (Hint: Use the **Move Up** and **Move Down** buttons to arrange the order of the list.)

7. Click **OK** to accept the settings.

❖ The ***Parts List*** is adjusted using the new settings. Note that, currently, all of the parts are using the **Generic** material type.

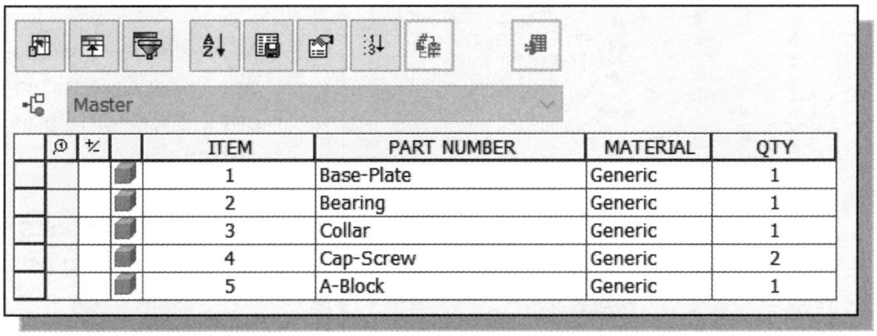

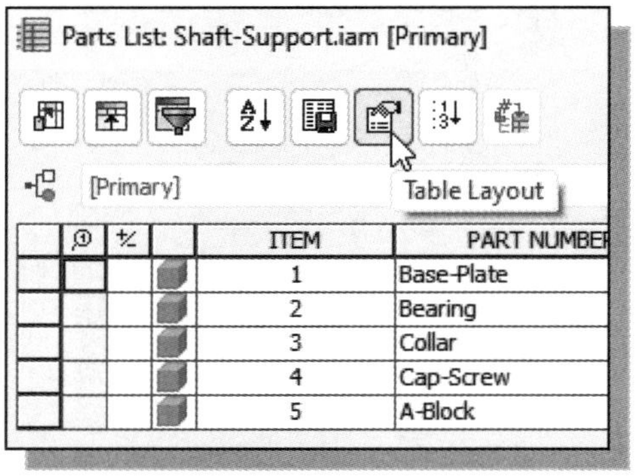

8. Click on the **Table Layout** button as shown.

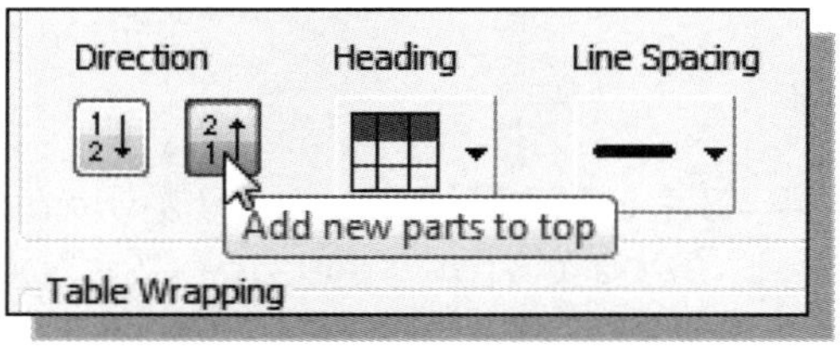

9. Set the ***Table Direction*** to **Add new parts to top** as shown.

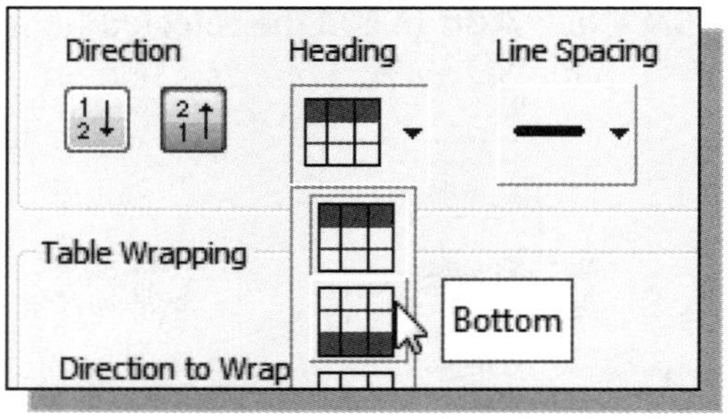

10. Set the ***Heading Placement*** to **Bottom** as shown.

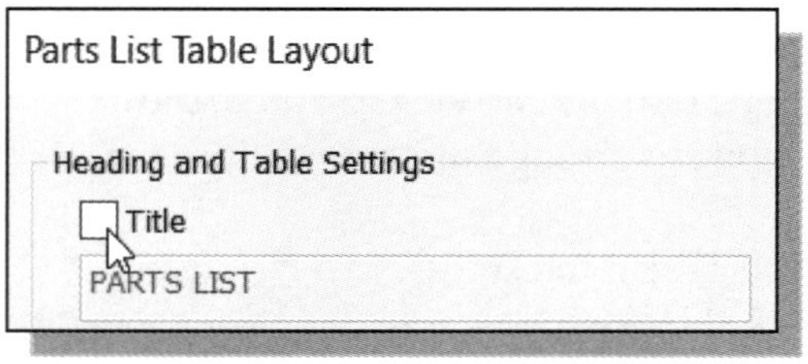

11. Turn **OFF** the ***Parts List Title***.

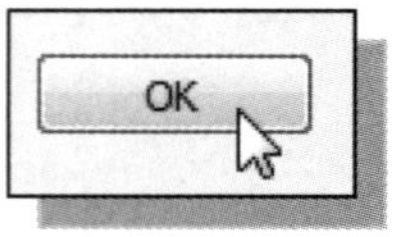

12. Click **OK** twice to accept the ***Parts List*** settings.

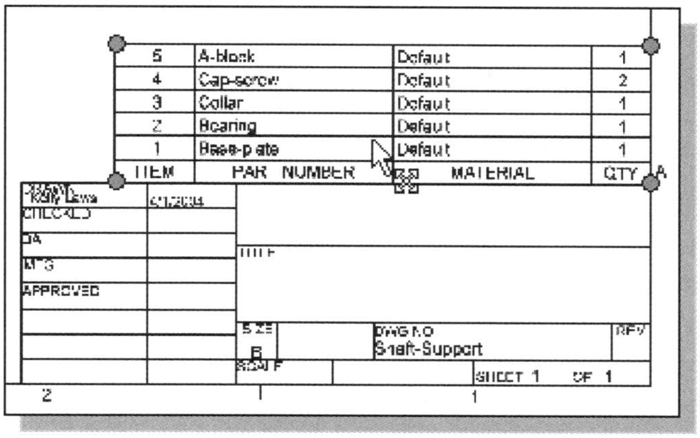

13. On your own, adjust the position of the ***Parts List*** so that it is aligned to the top edge of the *Title Block* as shown. (Hint: Look for the MOVE symbol next to the cursor.)

Change the Material Type

We will switch back to the assembly model to change the assignments of the material type.

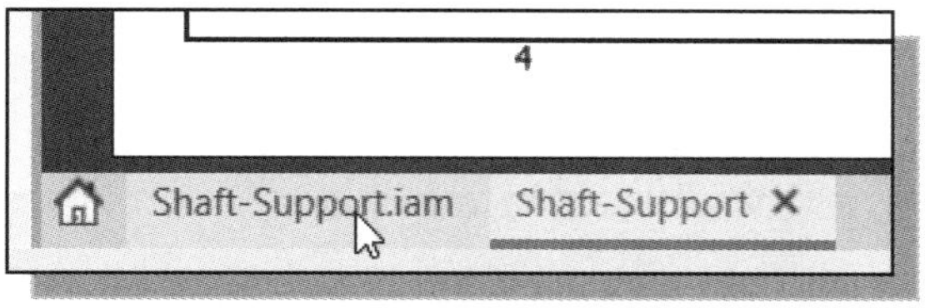

1. Click on the *Shaft-Support.iam* tab to switch back to the assembly model.

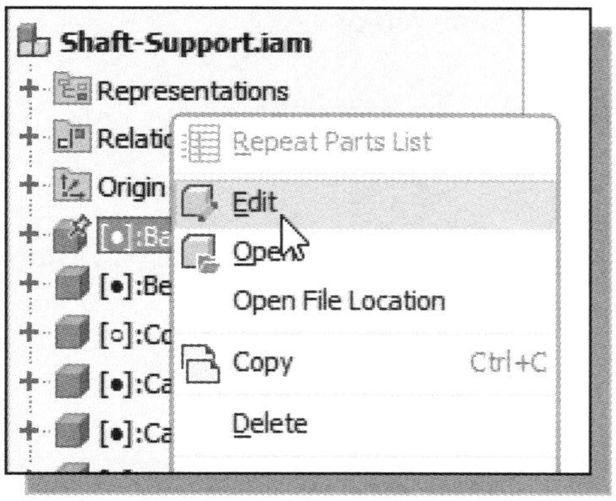

2. Inside the *Model Browser* window, right-click on **Base-Plate:1** to display the option menu.

3. Select **Edit** in the option menu to enter the *Edit Mode* for the selected part.

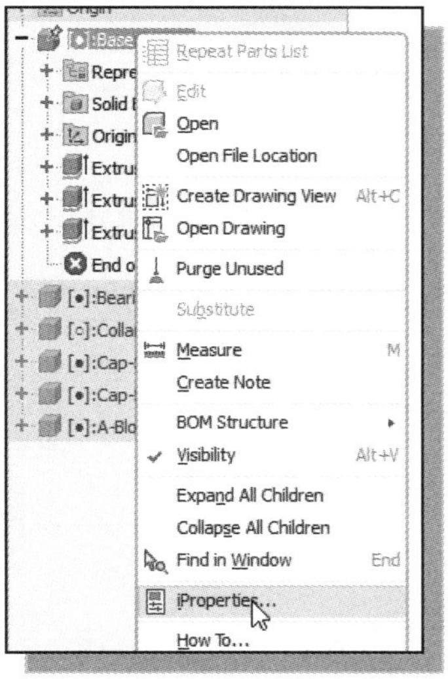

4. Inside the *Model Browser* window, select the **Base-plate** part by clicking once with the left-mouse-button.

5. Right-click on **Base-plate:1** to display the option menu and choose **iProperties** in the options list.

6. Click on the **Physical** tab in the *Base-plate.ipt Properties* window.

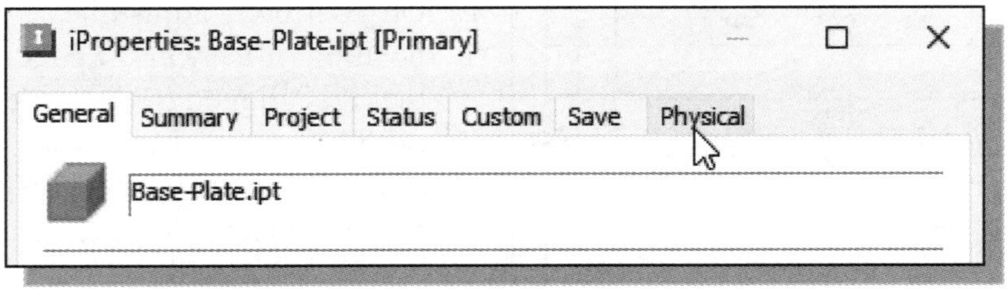

7. Choose the **Steel, Mild** in the *Material* list. Note the properties of the selected material are also displayed in the *Properties* list as shown in the figure.

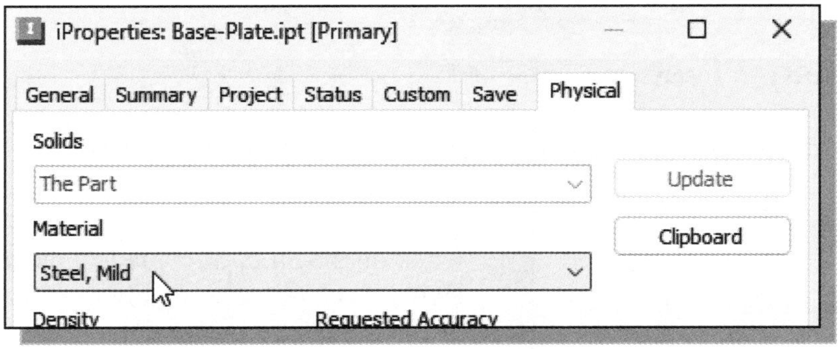

❖ Autodesk Inventor comes with many materials that have pre-entered information; additional material types/properties can also be added/changed as well.

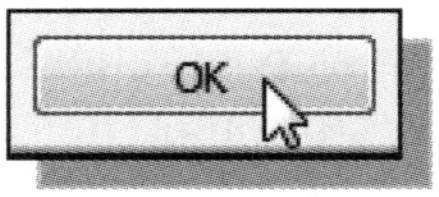

8. Click **OK** to accept the setting and exit the *Materials* dialog box.

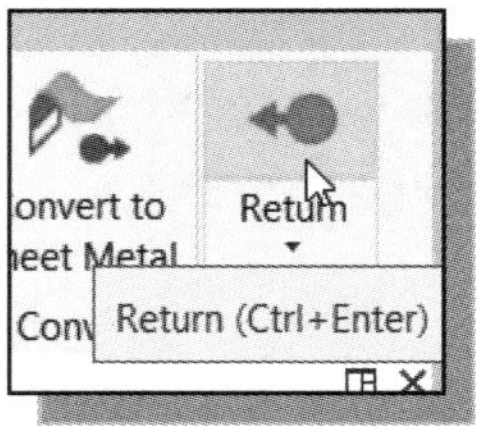

9. Select **Return** in the Standard toolbar to exit the *Part Editing Mode*.

10. On your own, switch back to the *Shaft-Support.idw* window and notice the *Material* information for the ***Base-plate*** part is now updated.

11. On your own, repeat the above steps to change the material information for the other parts as shown in the figure below.

5	A-Block	Aluminum 6061	1
4	Cap-Screw	Steel, Mild	2
3	Collar	Bronze, Soft Tin	1
2	Bearing	Iron, Cast	1
1	Base-Plate	Steel, Mild	1
ITEM	PART NUMBER	MATERIAL	QTY

Add the Balloon Callouts

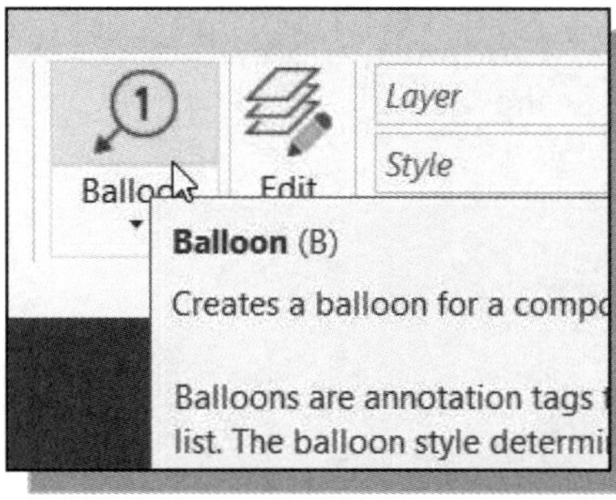

1. In the *Drawing Annotation* window, click on the **Balloon** button.

2. In the prompt area, the message "*Select a location*" is displayed. Click on the ***Collar*** part to attach an arrowhead to the part.

3. Pick another location to **place the balloon** as shown in the figure below.

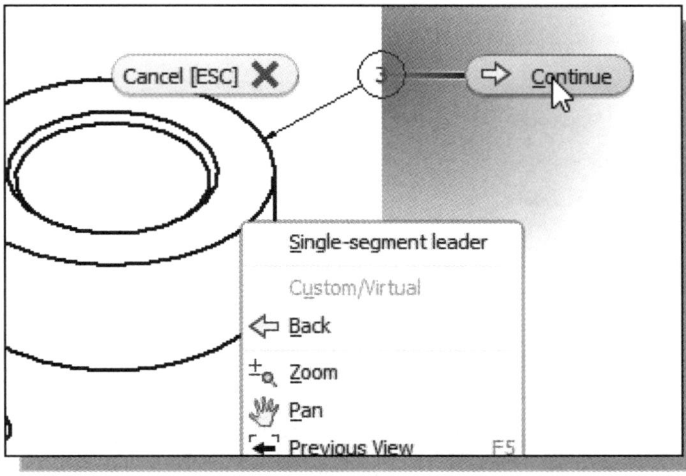

4. Inside the graphics window, click once with the right-mouse-button to display the **option menu**.

5. Select **Continue** in the pop-up menu to proceed with the creation of the balloon.

Completing the Title Block Using the iProperties option

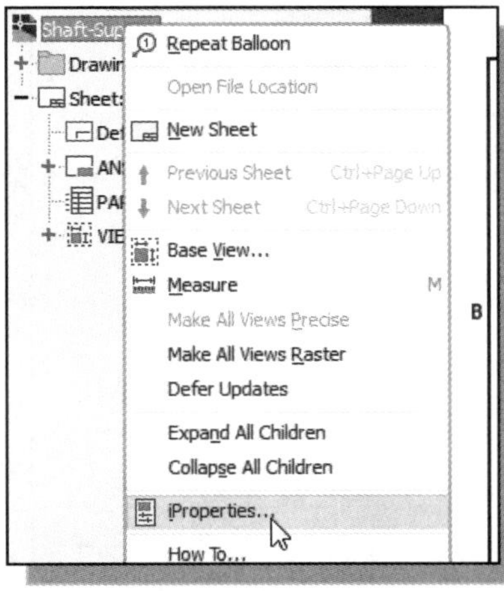

1. Inside the *Model Browser* window, select the **Shaft-Support** drawing by clicking once with the left-mouse-button.

2. Right-click on **Shaft-Support** to display the option menu and choose **iProperties** in the options list.

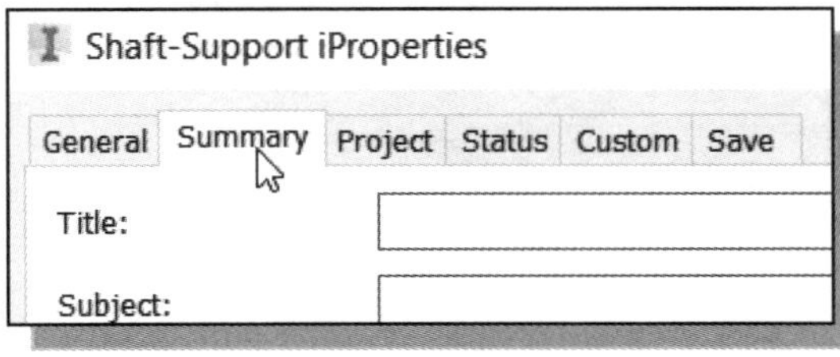

3. Click on the **Summary** tab to view the list of general information regarding the design.

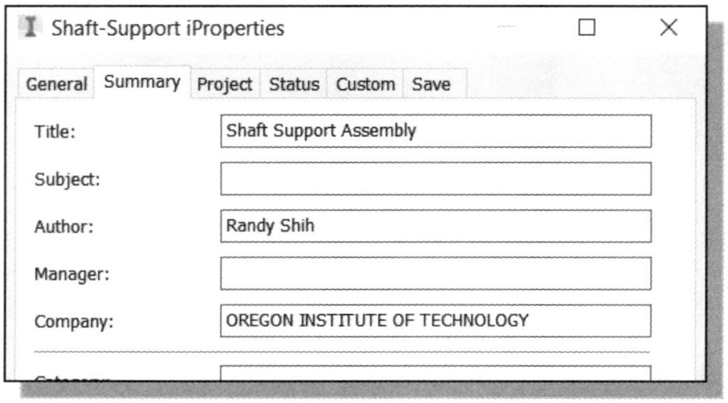

4. Enter the **Title**, **Author** and **Company** information as shown.

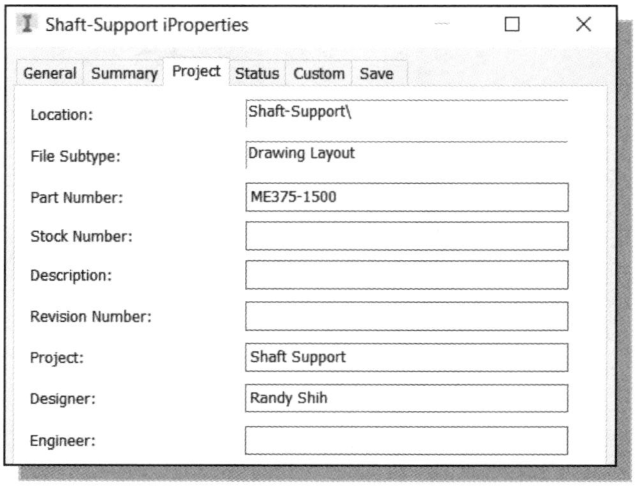

5. Click on the Project tab and enter the Part Number, Project name as shown.

6. Click **OK** to accept the setting and exit the *iProperties* dialog box.

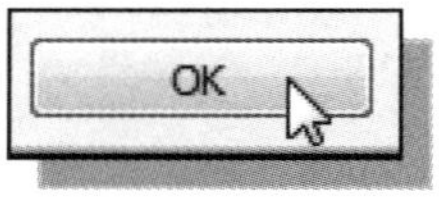

7. On your own, add the additional balloons and complete the drawing as shown.

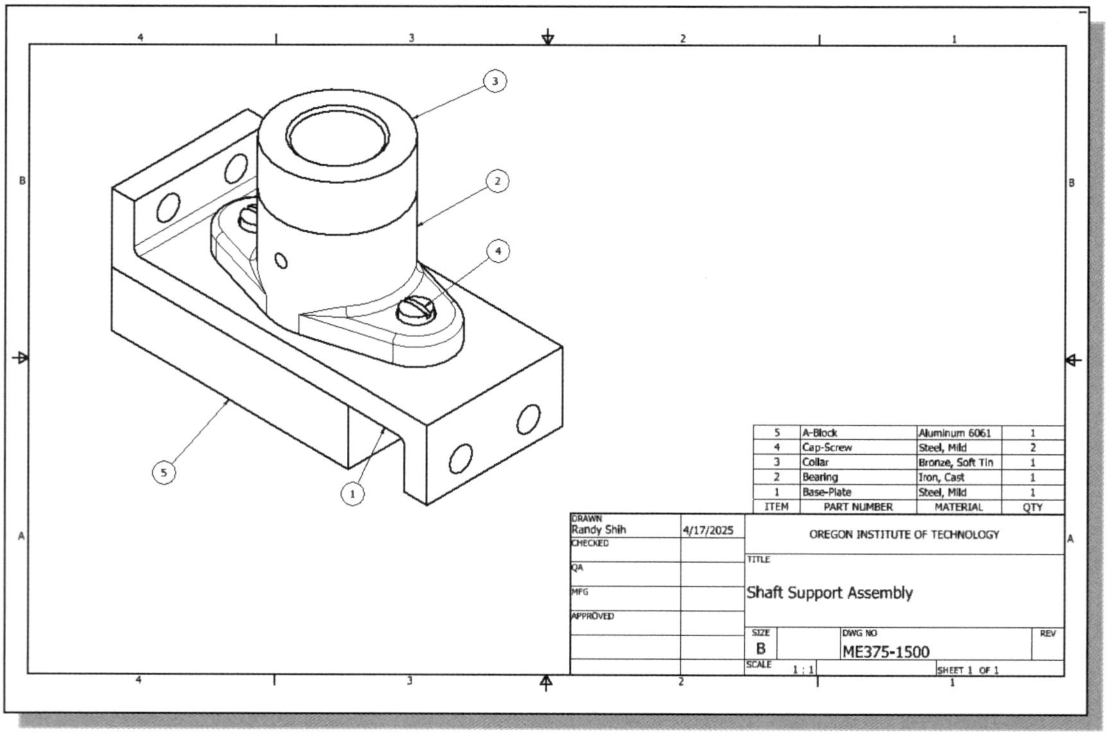

Bill of Materials

A bill of materials (BOM) is a table that contains information about the parts within an assembly. The BOM can include information such as part names, quantities, costs, vendors, and all of the other information related to building the part. The ***parts list***, which is used in an assembly drawing, is usually a partial list of the associated BOM.

In Autodesk Inventor, both the *bill of materials* and *parts list* can be derived directly from data generated by the assembly and the part properties. We can select which properties to be included in the *bill of materials* or *parts list*, in what order the information is presented, and in what format to export the information. The exported file can be used in an application such as a spreadsheet or text editor.

(a) BOM from Parts List

1. Move the cursor on top of the ***Parts List*** and right-click once to display the option menu.

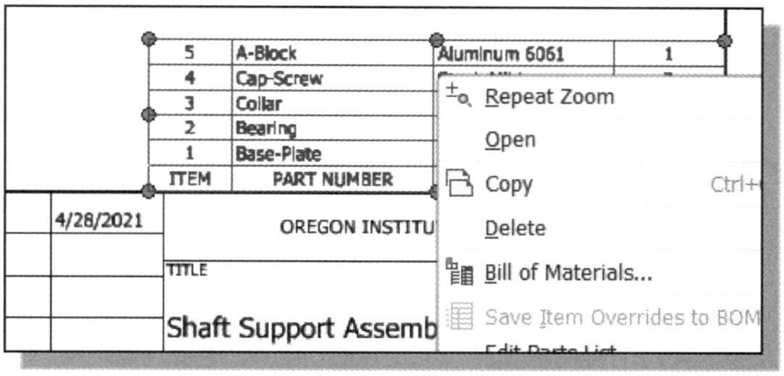

2. Choose **Export...** in the option menu as shown.

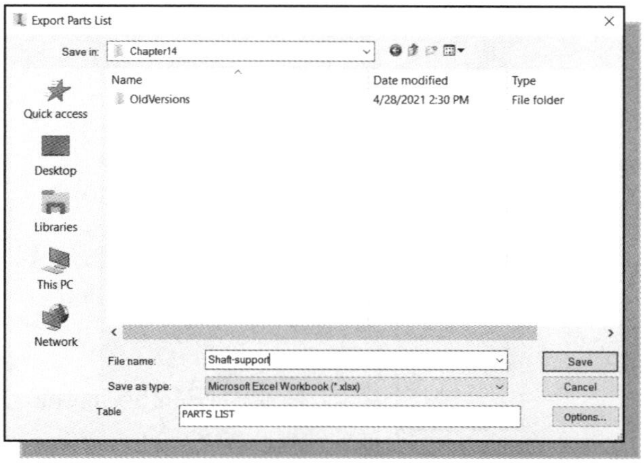

3. Confirm the ***Save as type*** list is set to **Microsoft Excel**.

4. Enter ***Shaft-support*** as the filename and click **Save** to export the BOM.

❖ On your own, examine the exported BOM by opening up the file in *Excel*.

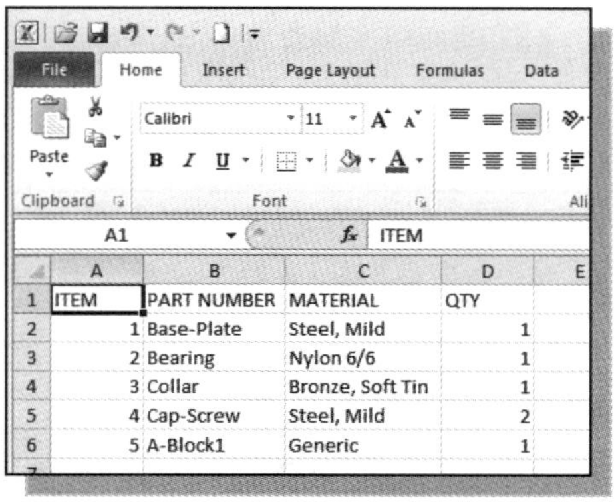

(b) BOM from Assembly Model

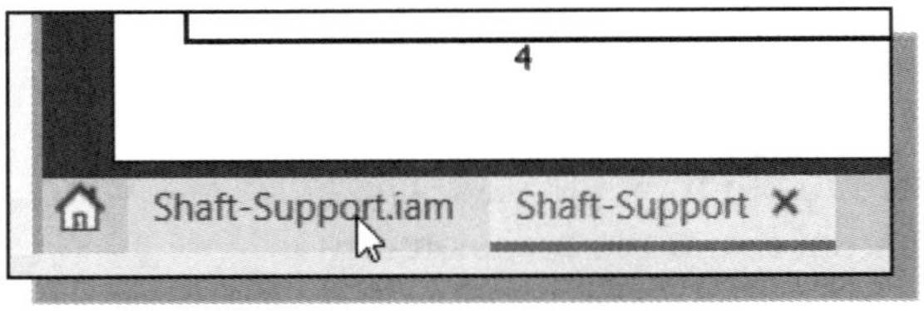

1. Click on the ***Shaft-Support.iam*** tab to switch back to the assembly model.

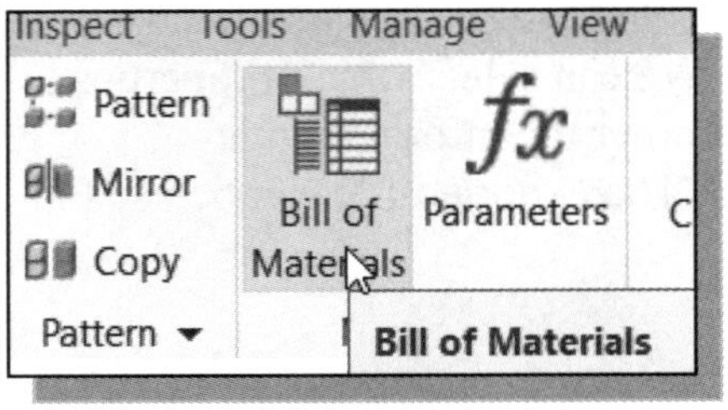

2. Select **Bill of Materials** in the *Manage* tab of the *Ribbon* toolbar panel.

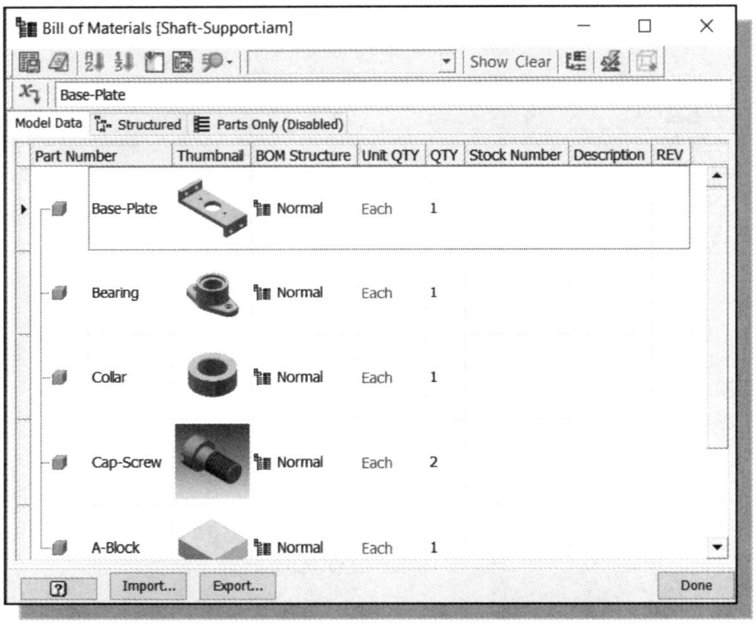

❖ Note that many of the controls and options are similar to those of the **Parts List** command in the *Drawing Mode*.

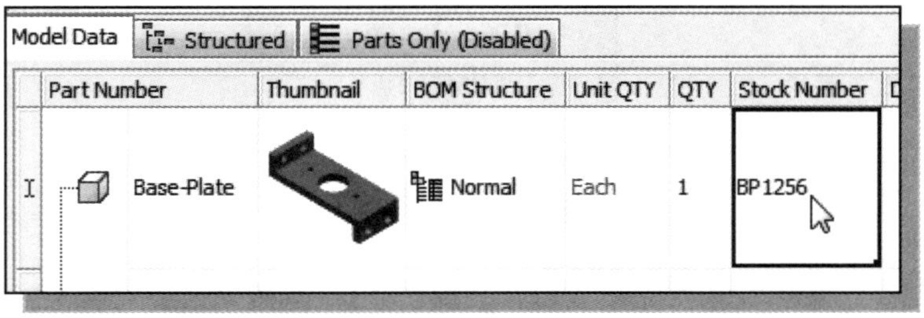

3. Click inside the ***Stock Number*** box to enter the *Edit Mode*.

4. Enter **BP1256** as the new *Stock Number*.

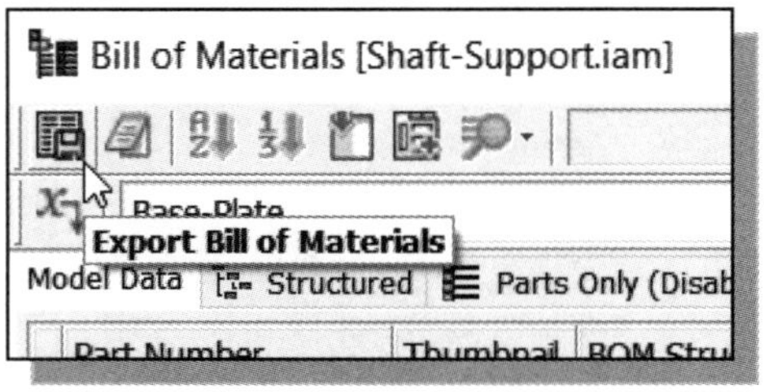

5. Click the **Export Bill of Materials** button.

6. On your own, export using the Microsoft Excel format and examine the *BOM* in Microsoft Excel.

Review Questions: (Time: 35 minutes)

1. What is the purpose of using **assembly constraints**?
2. List three of the commonly used **assembly constraints**.
3. Describe the difference between the **Mate** constraint and the **Flush** constraint.
4. In an assembly, can we place more than one copy of a part? How is it done?
5. How should we determine the assembly order of different parts in an assembly model?
6. How do we adjust the information listed in the **parts list** of an assembly drawing?
7. In Autodesk Inventor, describe the procedure to create a **bill of materials** (BOM)?
8. Create sketches showing the steps you plan to use to create the four parts required for the assembly shown on the next page:

Ex.1)

Ex.2)

Ex.3)

Ex.4)

Exercises: (Time: 180 minutes.)

1. **Wheel Assembly** (Create a set of detail and assembly drawings. All dimensions are in mm.)

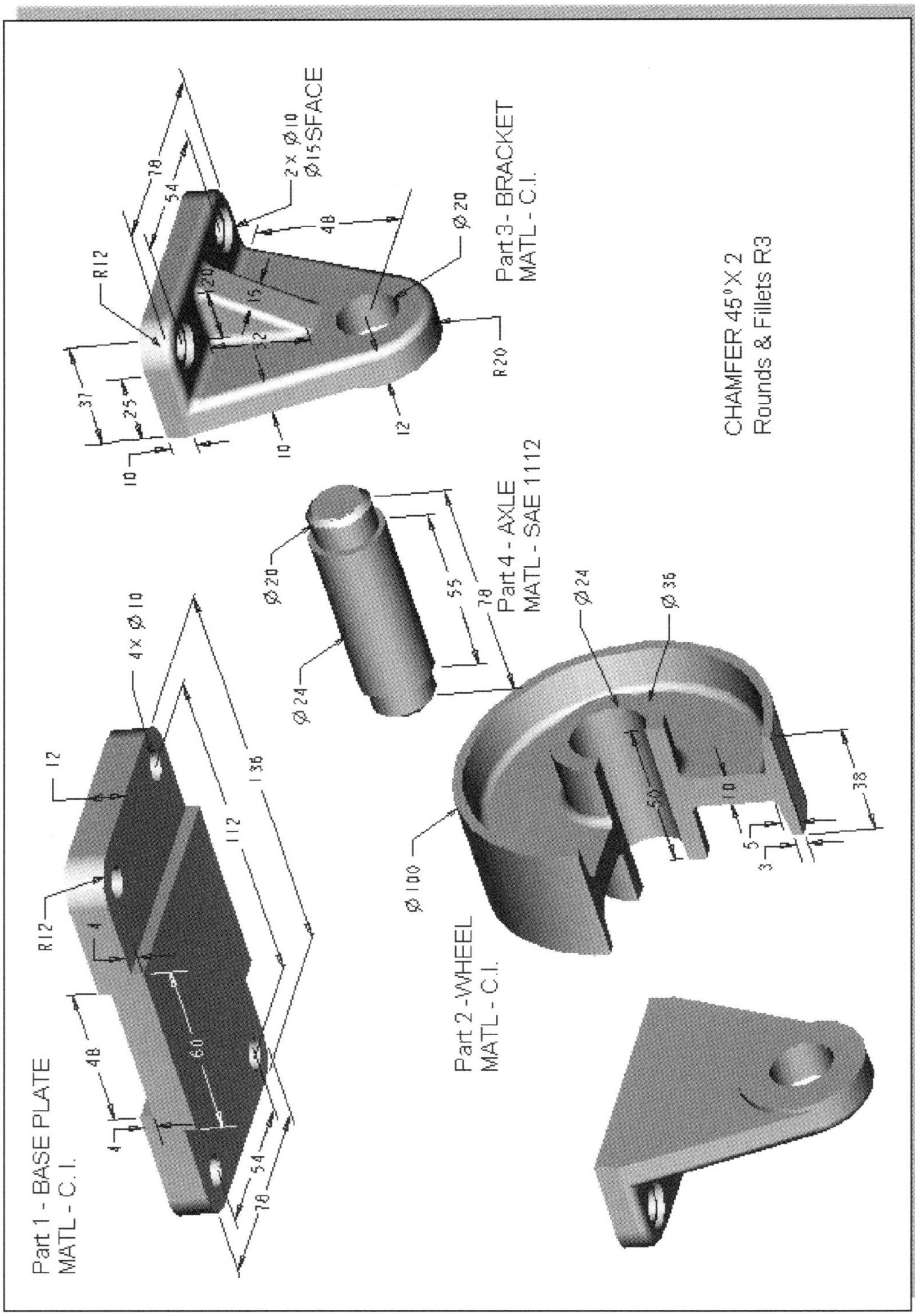

2. **Vise Assembly** (Create a set of detail and assembly drawings. All dimensions are in inches.)

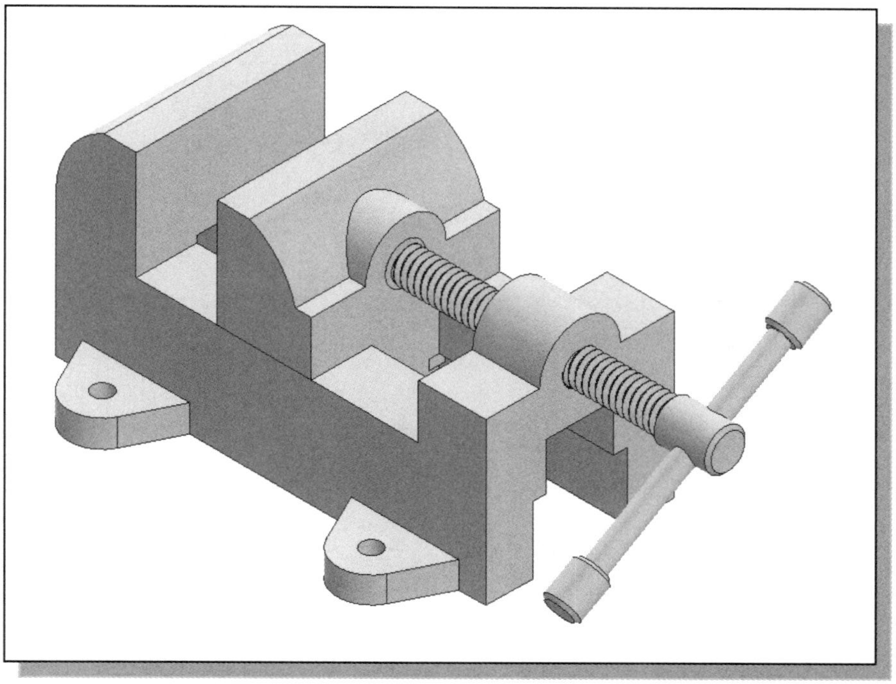

(a) **Base:** The 1.5 inch wide and 1.25 inch wide slots are cut through the entire base. Material: Gray Cast Iron. Threads: 5/8”-8-UNC-3B

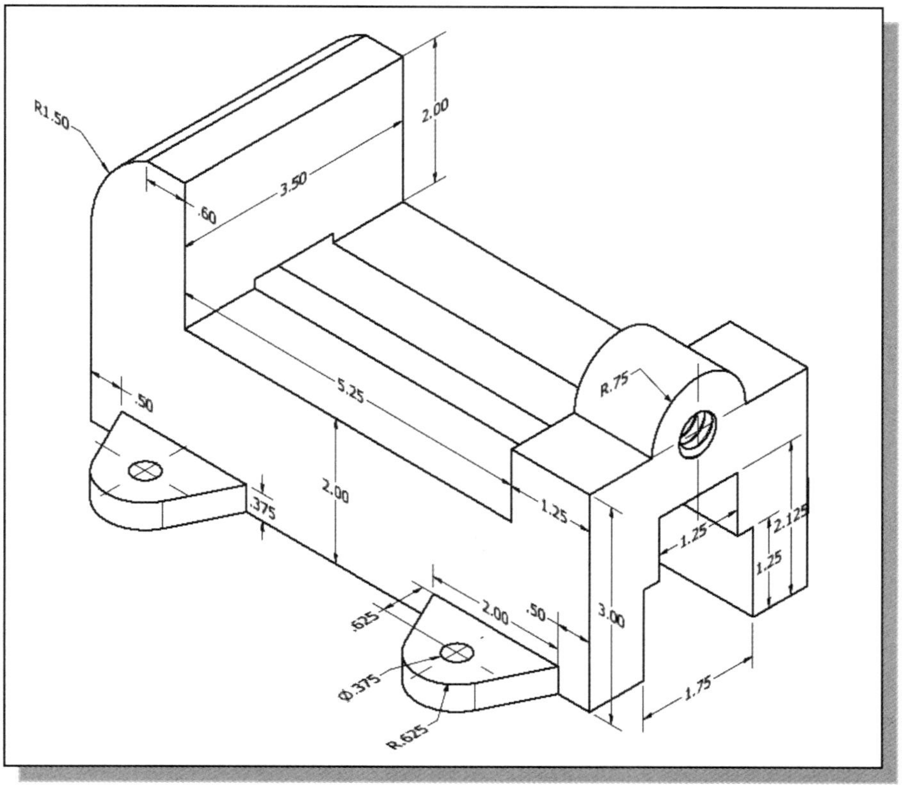

(b) **Jaw:** The shoulder of the jaw rests on the flat surface of the base and the jaw opening is set to 1.5 inches. Material: Gray Cast Iron.

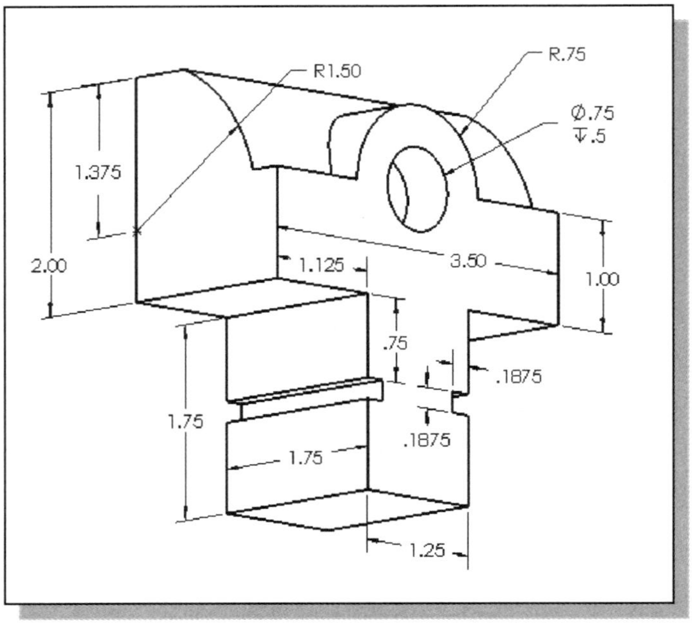

(c) **Key:** 0.1875 inch H x 0.3 inch W x 1.75 inch L. The keys fit into the slots on the jaw with the edge faces flush as shown in the sub-assembly to the right. Material: Alloy Steel.

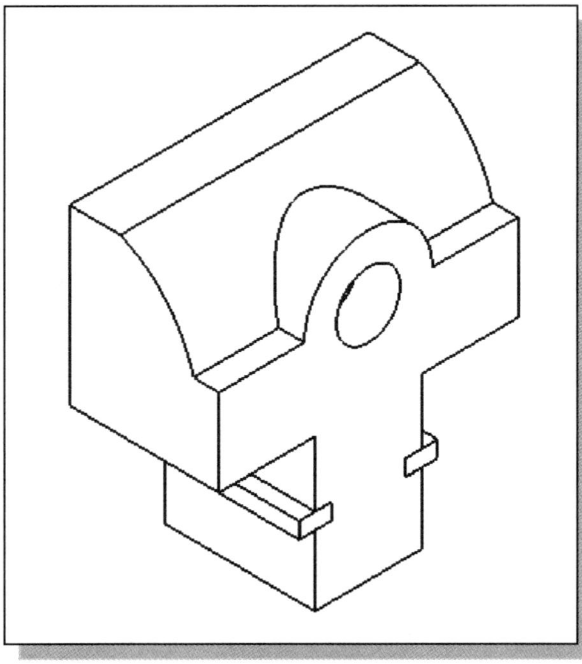

(d) **Screw:** 5/8"-8-UNC-3A There is one chamfered edge (0.0625 inch x 45°). The flat ∅ 0.75" edge of the screw is flush with the corresponding recessed ∅ 0.75 face on the jaw. Material: Alloy Steel. (Use the **Threads** or **Coil** command to create the threads.)

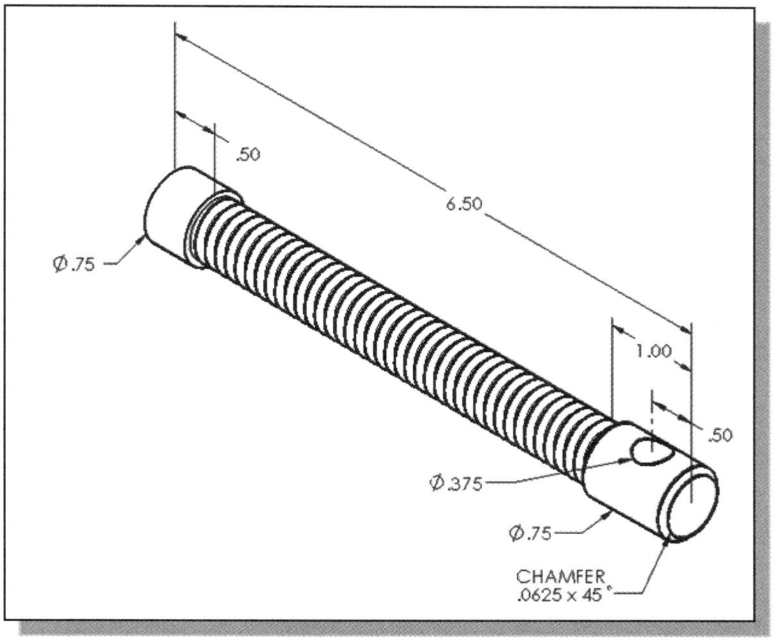

(e) **Handle Rod:** ∅ 0.375" x 5.0" L. The handle rod passes through the hole in the screw and is rotated to an angle of 30° with the horizontal as shown in the assembly view. The flat ∅ 0.375" edges of the handle rod are flush with the corresponding recessed ∅ 0.735 faces on the handle knobs. Material: Alloy Steel.

(f) **Handle Knob:** There are two chamfered edges (0.0625 inch x 45°). The handle knobs are attached to each end of the handle rod. The resulting overall length of the handle with knobs is 5.50". The handle is aligned with the screw so that the outer edge of the upper knob is 2.0" from the central axis of the screw. Material: Alloy Steel.

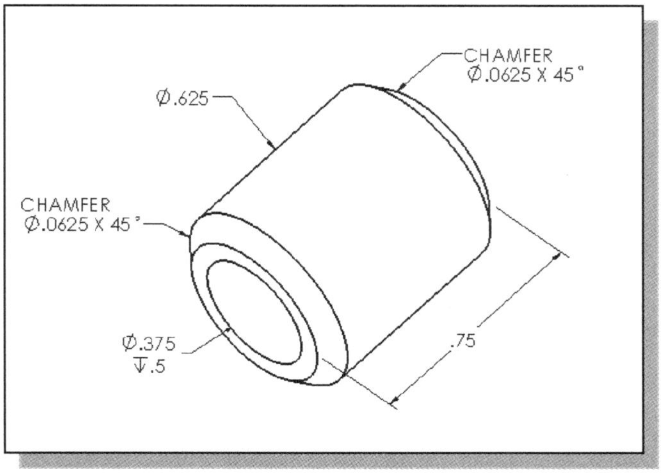

Notes:

Chapter 15
Content Center and Basic Motion Analysis

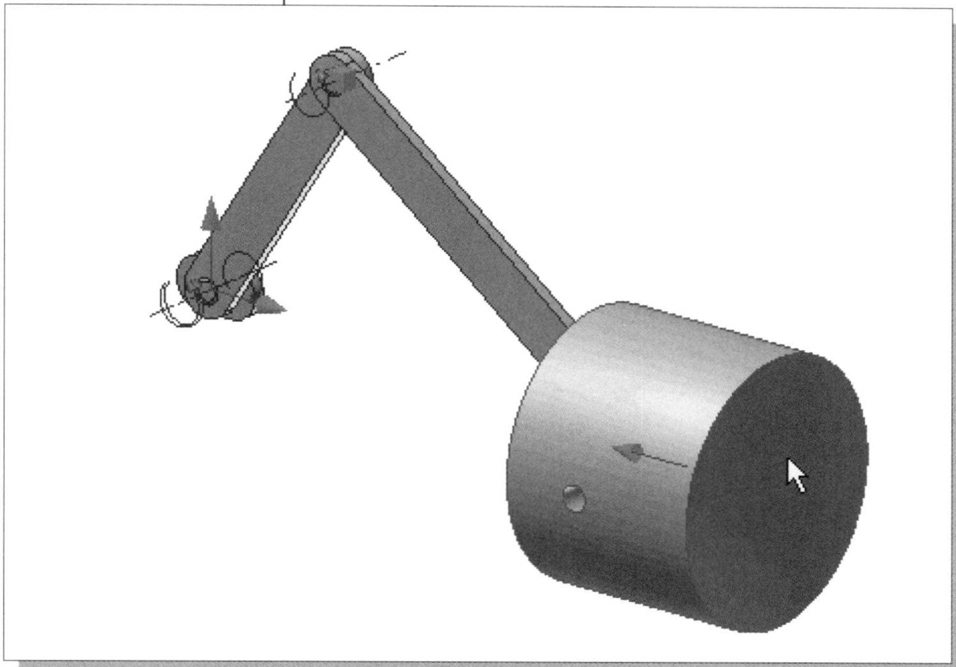

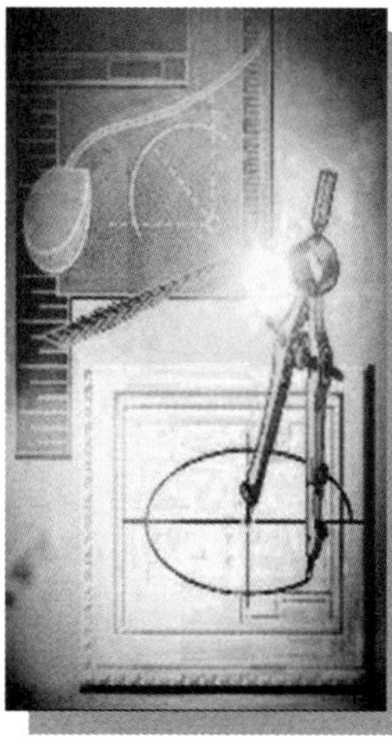

Learning Objectives

- Use the Autodesk Content Center Libraries
- Use the Assemble Command
- Use the Analyze Interference Tool
- Use the Drive Constraint Tool to Create Basic Motion Analysis
- Understand and Utilize the 3D Grips Editing Approach
- Output the Associated Simulation Video File

Autodesk Inventor Certified User Exam Objectives Coverage

Parametric Modeling Basics

Section 5: Assemblies

Objectives: Creating Assemblies, Viewing Assemblies, Animation Assemblies, Adaptive Features, Parts, and Subassemblies.

Section 7: Visualization

Objectives: Create Rendered Images, Animate an Assembly

Autodesk Inventor Certified User Reference Guide

Introduction

In this lesson, we will examine some of the procedures that are available in Autodesk Inventor to reuse existing 2D data and 3D parts. We will also build an assembly model and resolve any interference problems by performing a basic motion analysis.

In Autodesk Inventor, we also have the option of using the standard parts library through what is known as the ***Content Center***. The *Content Center* consists of multiple libraries of standard parts that have been created based on industry standards. Significant amounts of time can be saved by using these parts. Note that we can also create and publish our libraries so that others can reuse our parts. The *Content Center* has two modes for two distinct roles: **consumer** and **editor**. In the *consumer* mode, we access the *Content Center* libraries and use the parts as a consumer. In the *editor* mode, we can define the different categories and also define the iterations for the parts the consumer can select and use.

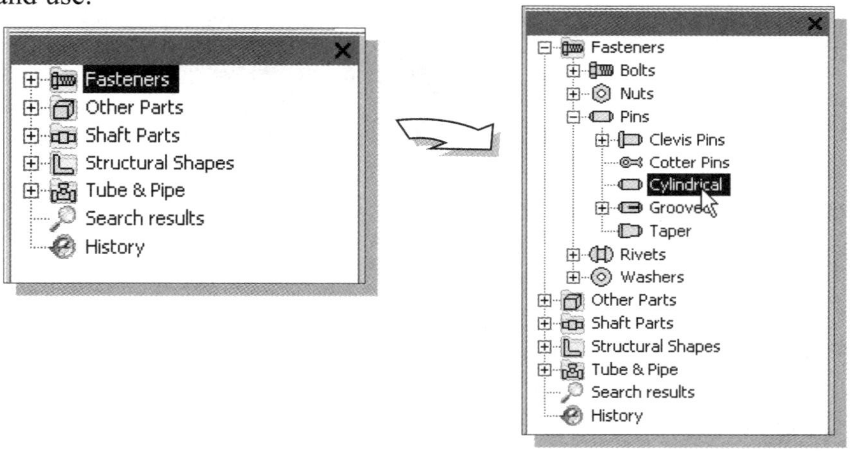

One main advantage of using parametric parts and assemblies is the ability to check for potential problems without actually creating a physical prototype. In Autodesk Inventor, ***interference analysis*** can be performed in several different ways. The **Analyze Interference** tool is a quick and easy option to examine interference between assembled components. ***Motion analysis*** can also be used for *interference analysis*.

Motion analysis can also be performed to visually confirm the proper assembly of the designs, and also to check for any interference between mating parts and any other potential problems. In Autodesk Inventor, several options are available to perform motion analysis, for example, the **Dynamic Simulation** module, the **Inventor Studio** module, the **Joint Constraints** and the **Drive Constraint** tool. The Dynamic Simulation module can be used to perform a fairly in-depth motion analysis, while the Drive Constraint tool provides a relatively simple motion analysis that can be done in a matter of seconds.

In this chapter, the concepts and procedures of creating assemblies by reusing existing 2D data to create 3D parametric models and using the *Content Center* for standard parts are illustrated. The procedure for basic motion analysis of an assembly is also illustrated using the basic Drive Constraint tool. The use of the more advanced **Joint Constraints** to perform motion analysis will be illustrated in the following chapter.

The Crank-Slider Assembly

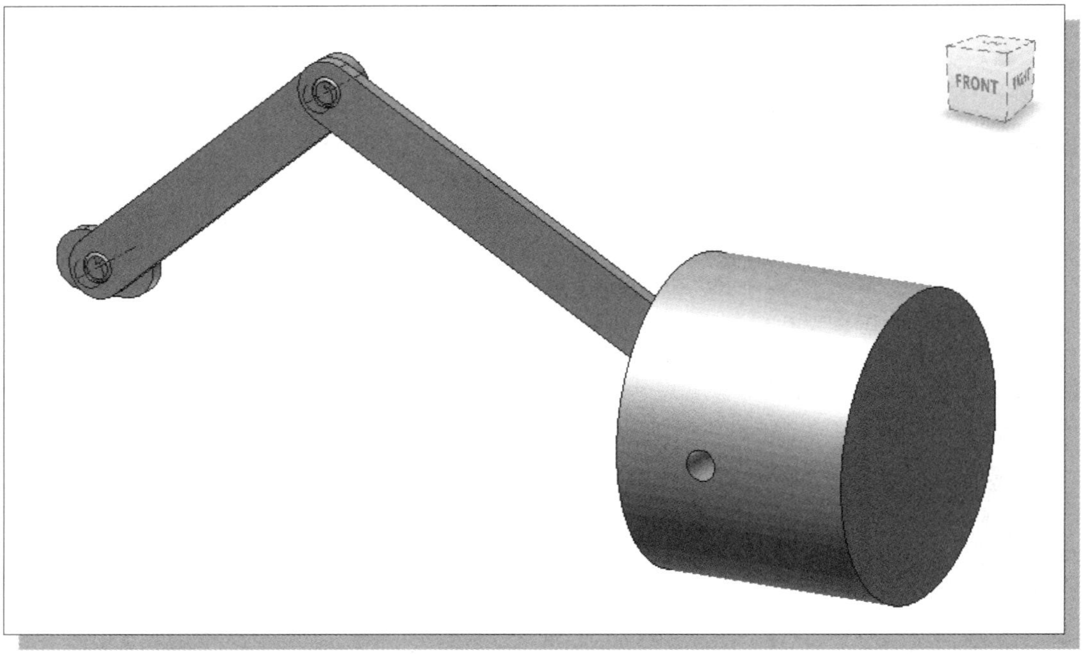

Create the Required Parts

Five parts are required for this assembly: (1) ***Base***, (2) ***Crank***, (3) ***Connecting Rod***, (4) ***Slider*** and (5) ***Pin***. On your own, create the four parts shown below. Save the models as separate part files: *CS-Base, CS-Crank*, *CS-Rod*, and *CS-Slider*. (Note the location of the parts relative to the datum planes and close all part files or exit Autodesk Inventor after you have created the parts.) We will use a standard *Pin* that is available in the *Content Center*.

(1) ***CS-Base***

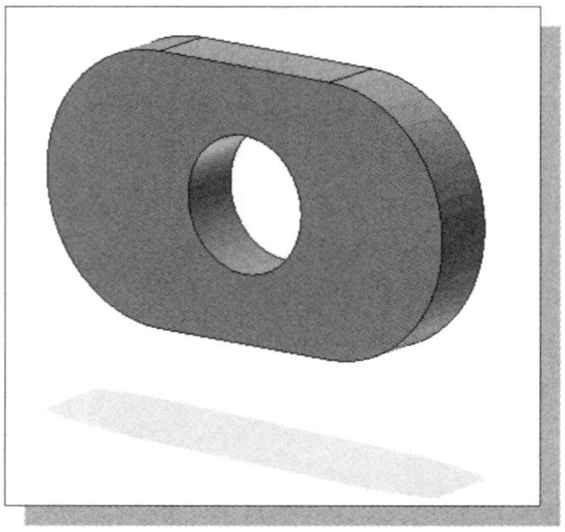

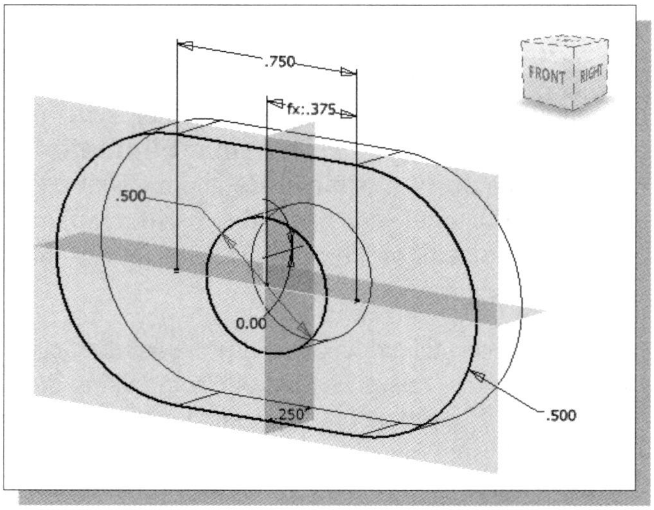

(2) ***CS-Crank*** (Construct the part with the datum origin aligned to the center of the part.)

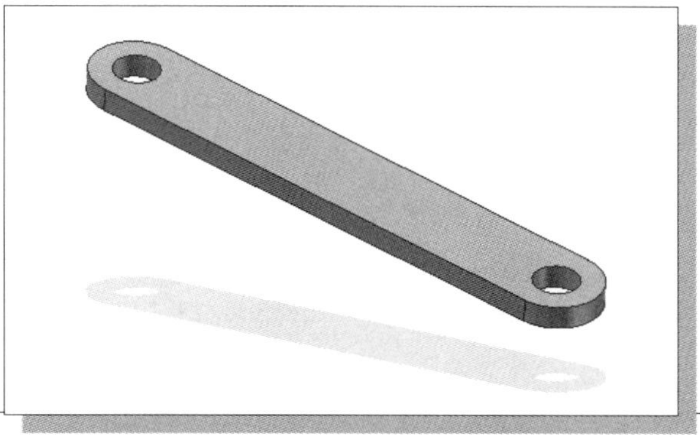

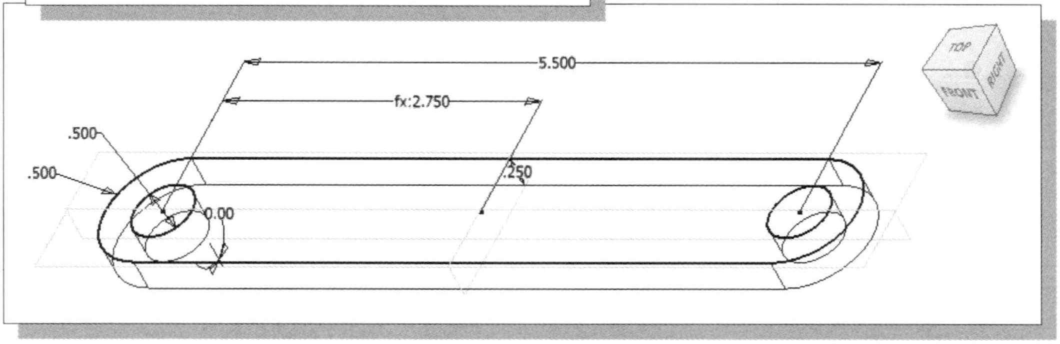

(3) ***CS-Rod*** (Same as the CS-Crank part but with a length dimension of 9.0. Hint: Use the **Save As** option to save the design as a new file.)

(4) ***CS-Slider*** (Construct the part with the vertical datum pass through the center.)

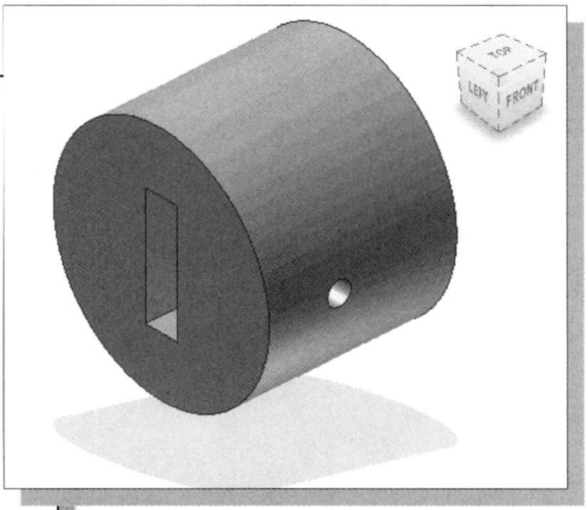

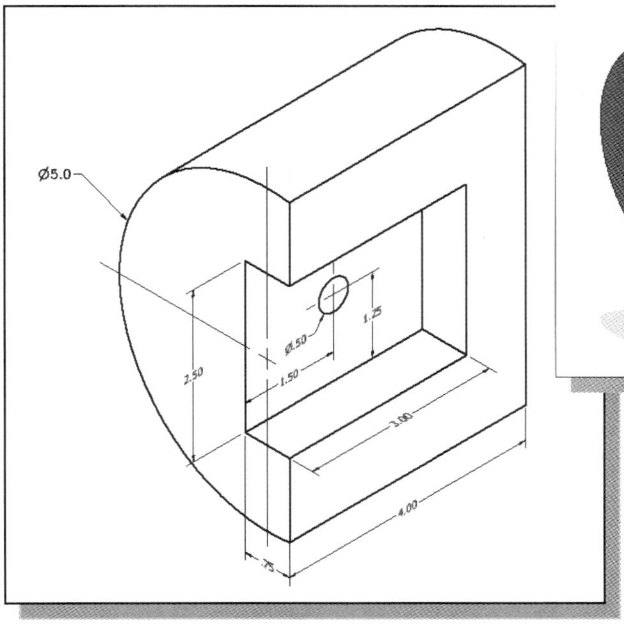

Starting Autodesk Inventor

1. Select the **Autodesk Inventor** option on the *Start* menu or select the **Autodesk Inventor** icon on the desktop to start Autodesk Inventor. The Autodesk Inventor main window will appear on the screen.

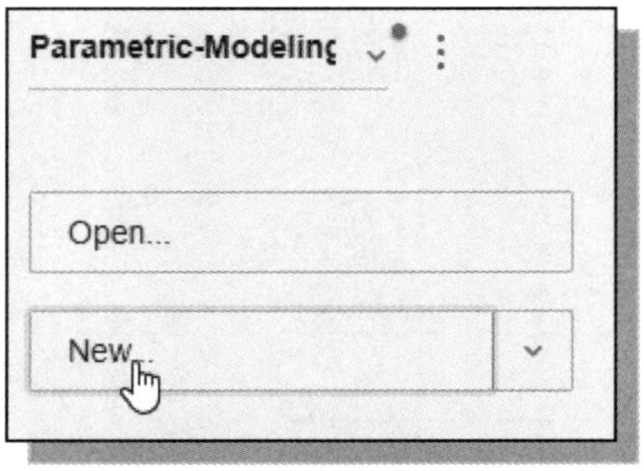

2. Select the **New File** icon with a single click of the left-mouse-button in the *Launch* toolbar as shown.

3. Select the **English** units set, and in the *Template* list select **Standard(in).iam** (*Standard Inventor Assembly Model* template file).

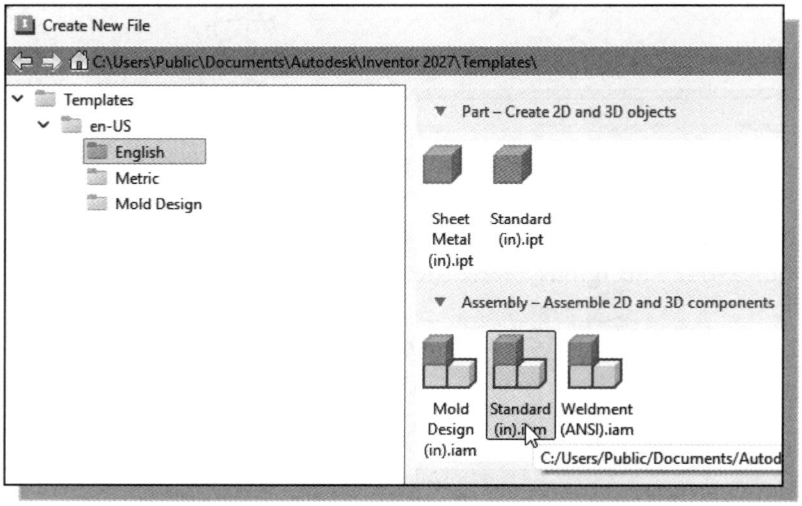

4. Click on the **Create** button in the *New File* dialog box to start a new model.

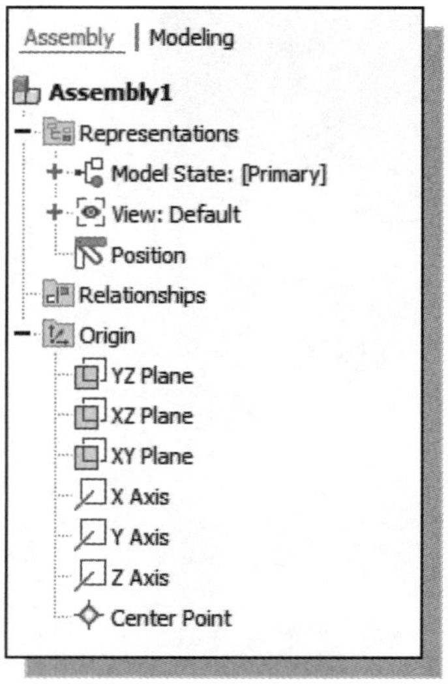

- In the *browser* window, **Assembly1** is displayed with a set of work planes, work axes and a work point. In most aspects, the usage of work planes, work axes and work point is very similar to that of the *Inventor Part Modeler*.

- Notice, in the *Ribbon* toolbar panels several *component* options are available, such as **Place Component, Create Component** and **Place from Content Center**. As the names imply, we can use parts that have been created or create new parts within the *Inventor Assembly Modeler*.

Placing the First Component

The first component in an assembly file sets the orientation of all subsequent parts and subassemblies and therefore should be one that is **not likely to be removed** and **preferably a non-moving part** in the design. For our project, we will use the ***CS-Base*** as the base component in the assembly.

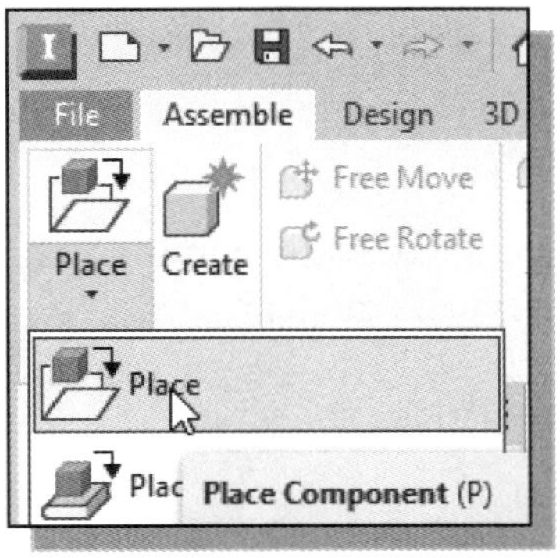

1. In the *Assemble* panel, select the **Place Component** command by left-clicking the icon.

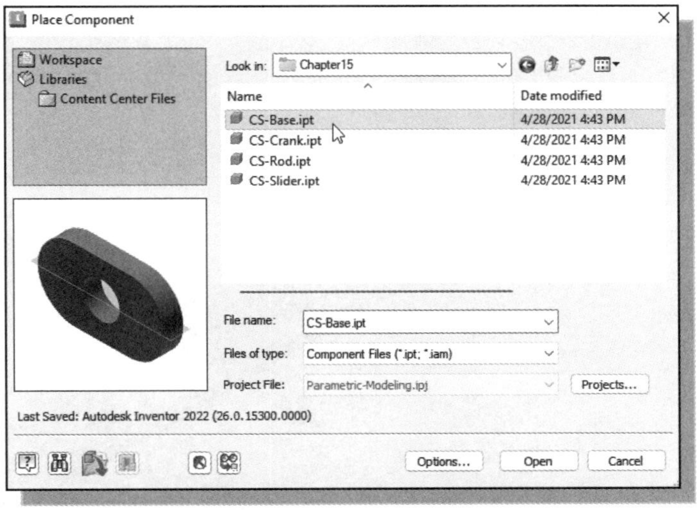

2. Select the **CS-Base** (part file: ***CS-Base.ipt***) in the list window.

3. Click on the **Open** button to retrieve the model.

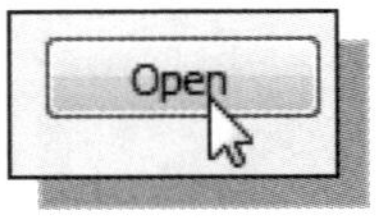

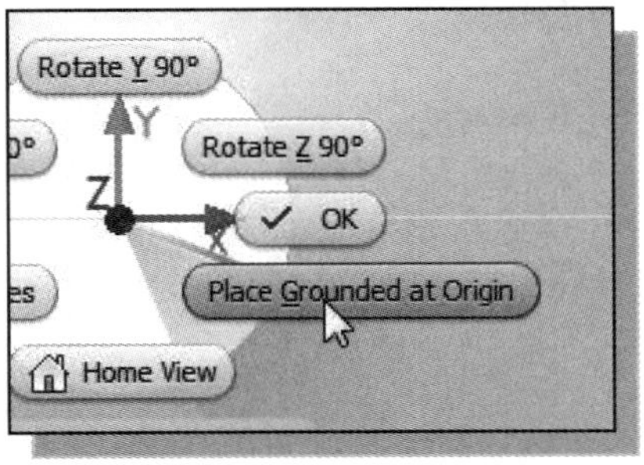

4. We will align the CS-Base part to the origin of the assembly coordinates. Right-click once to bring up the option menu and select **Place Grounded at Origin** as shown.

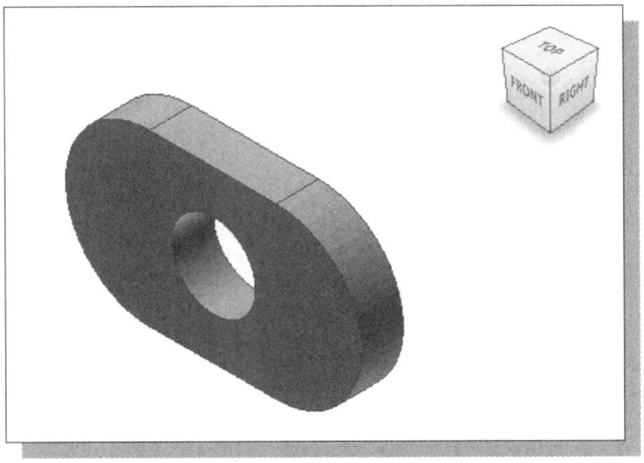

5. **Right-click** once to bring up the *option menu* and select **OK** to end the placement of the CS-Base part.

Placing the Second Component

We will retrieve the *CS-Crank* part as the second component of the assembly model.

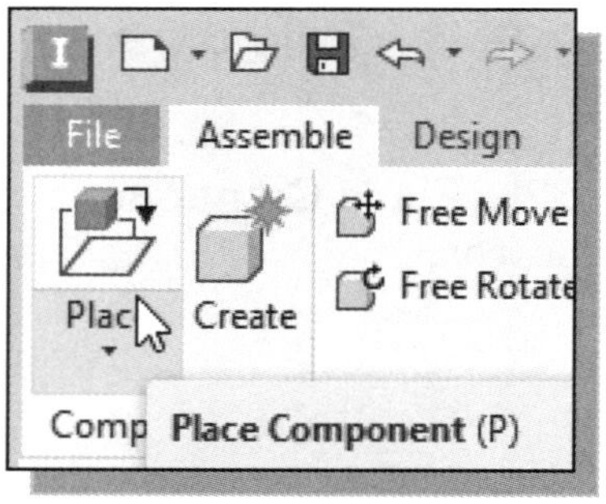

1. In the *Assemble* panel, select the **Place Component** command by left-clicking the icon.

2. Select the **CS-Crank** design in the list window. Click on the **Open** button to retrieve the model.

3. Place the *CS-Crank* toward the front side of the *CS-Base* part, as shown in the figure.

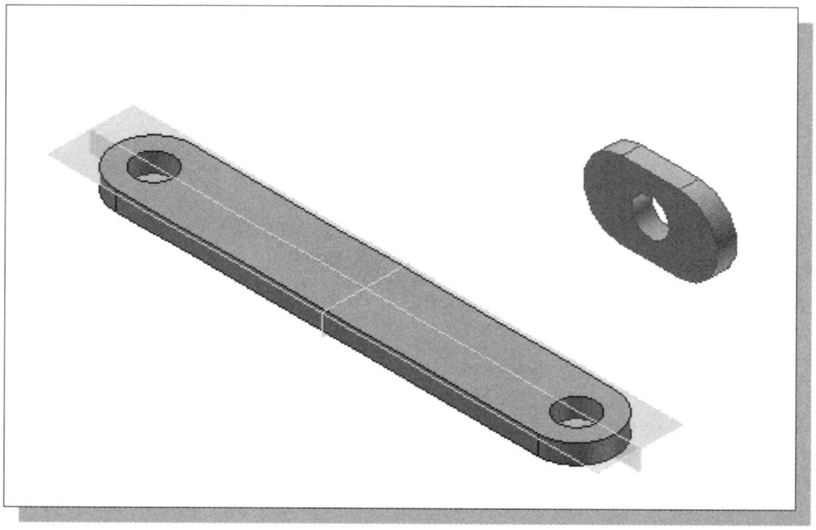

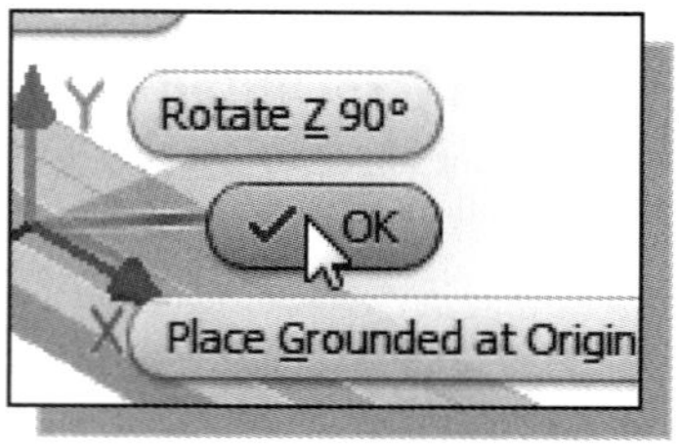

4. Inside the graphics window, right-click once to bring up the option menu and select **OK** to end the placement of the *CS-Crank* part.

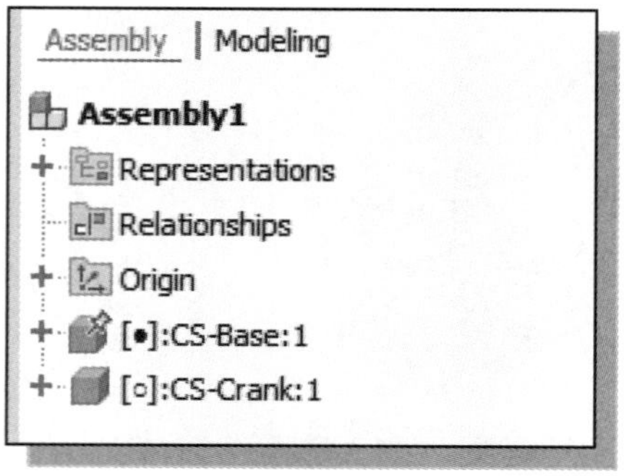

- Inside the *browser* window, the retrieved parts are listed in their corresponding order. The **Pin** icon in front of the CS-Base filename signifies the part is grounded and all ***six degrees of freedom*** are restricted. The number behind the filename is used to identify the number of copies of the same component in the assembly model.

Apply the Assembly Constraints

We are now ready to assemble the *CS-Crank* component to the *CS-Base* part.

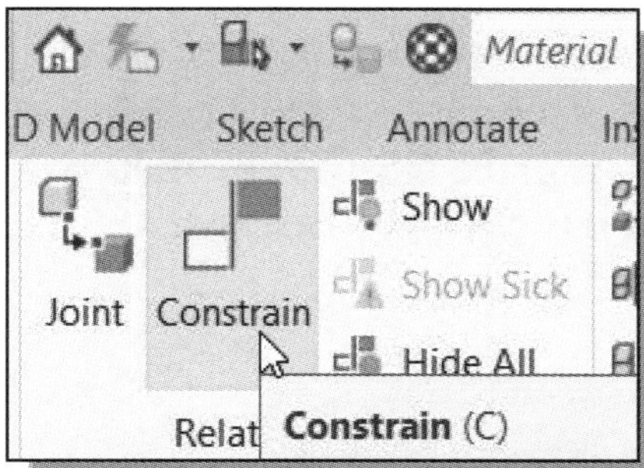

1. In the *Relationships* panel, select the **Constrain** command by left-clicking once on the icon.

2. In the *Place Constraint* dialog box, confirm the constraint type is set to the **Mate** constraint and select the top horizontal surface of the *CS-Crank* part as the first part for the Mate alignment command.

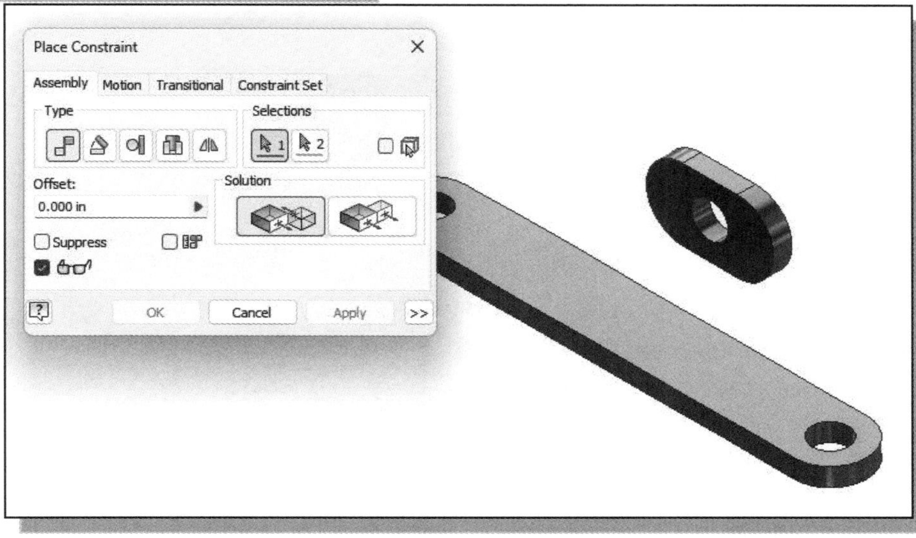

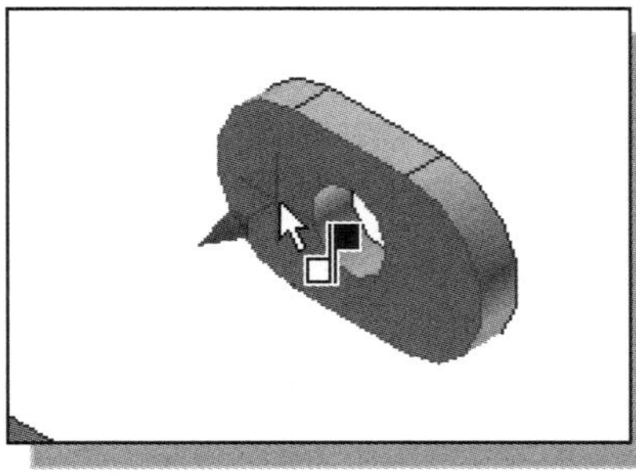

3. Click on the *front* face of the *CS-Base* part as the second part selection to apply the constraint. Note the direction normals shown in the figure; the Mate constraint requires the selection of opposite direction of surface normals.

4. Click on the **Apply** button to accept the selection and apply the Mate constraint.

5. On your own, use the dynamic viewing functions to confirm the alignment of the two parts.

Apply a Second Mate Constraint

The Mate constraint can also be used to align axes of cylindrical features.

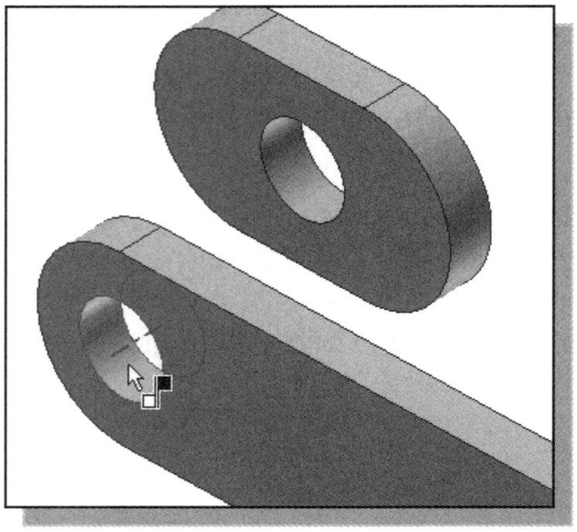

1. Move the cursor near the cylindrical surface of the left side hole of the *CS-Crank* part. Select the axis when it is displayed as shown.

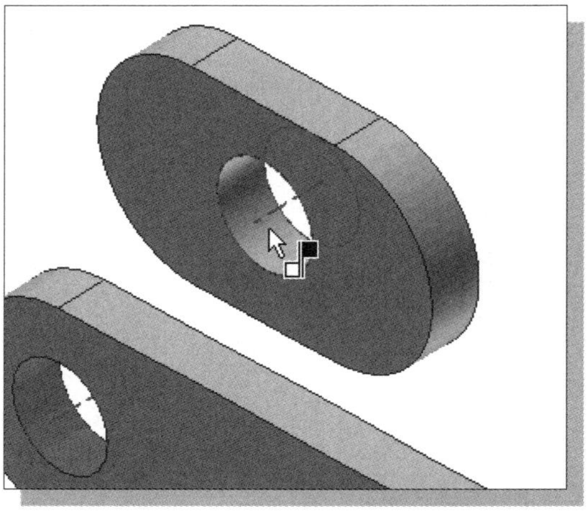

2. Move the cursor near the cylindrical surface of the small hole on the *CS-Base* part. Select the axis when it is displayed as shown.

3. In the *Place Constraint* dialog box, click on the **Apply** button to accept the selection and apply the Mate constraint.

4. In the *Place Constraint* dialog box, click on the **Cancel** button to exit the Place Constraint command.

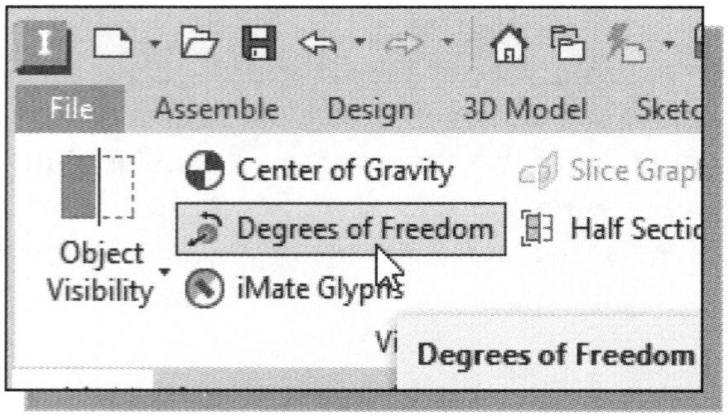

5. On your own, turn on the display of the **Degrees of Freedom**. (Hint: The icon is listed under the **View** tab.)

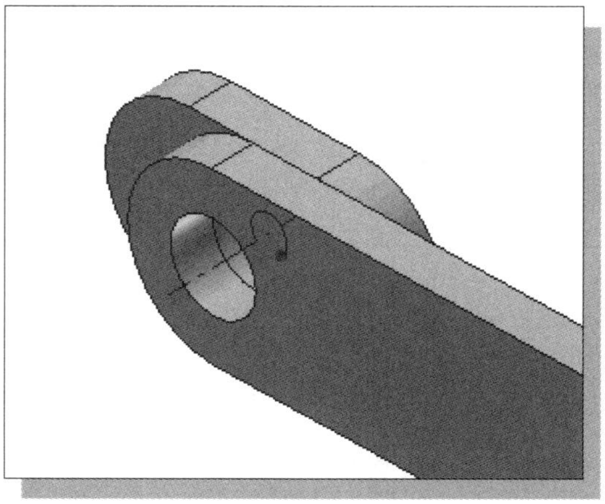

❖ The *CS-Crank* part is placed in the correct position, but the DOF symbol indicates that the *CS-Crank* part can still rotate about the displayed axis.

Constrained Move

A constrained move is done by dragging the component in the graphics window with the left-mouse-button. A constrained move will honor previously applied assembly constraints. That is, the selected component and parts constrained to the component move together in their constrained positions. A grounded component remains grounded during the move.

1. Press and hold down the left-mouse-button and drag the *CS-Crank* part and notice the *CS-Crank* part can freely rotate about the displayed axis.

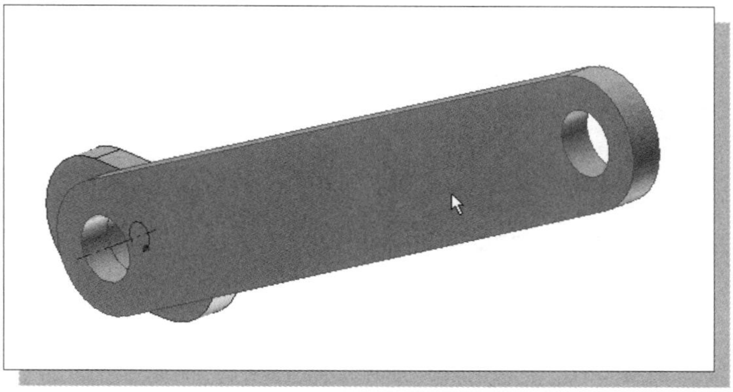

Place the Third Component

We will next use the **Content Center** and use a standard *Pin* as the third component of the assembly model.

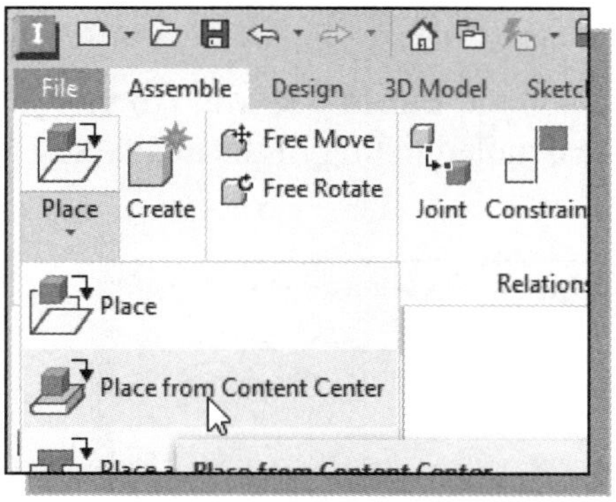

1. In the *Assemble* panel select the **Place from Content Center** command by left-clicking the icon.

2. Select the **Fastener** category and then **Pins** as shown.

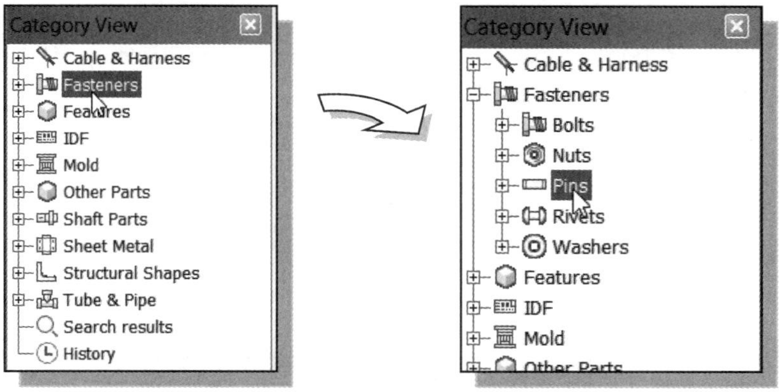

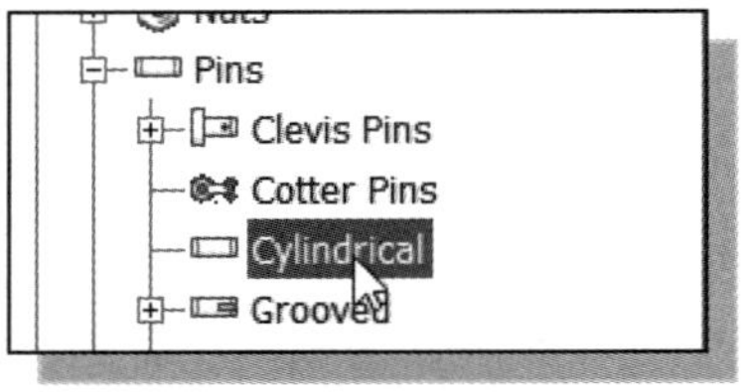

3. Choose **Cylindrical** under **Pins** as shown in the figure.

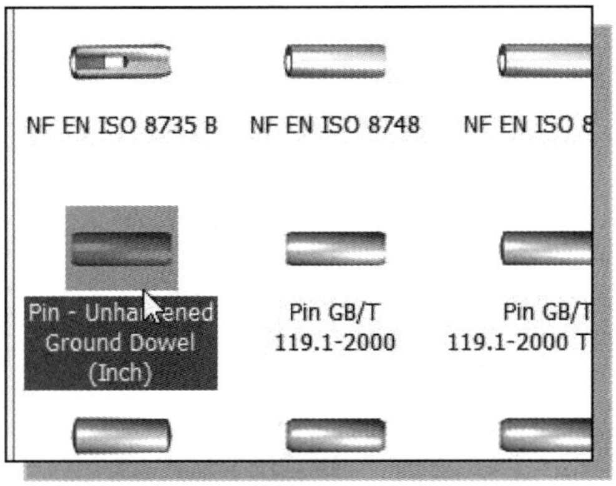

4. Select **Pin - Unhardened Ground Dowel (Inch)** as shown in the figure.

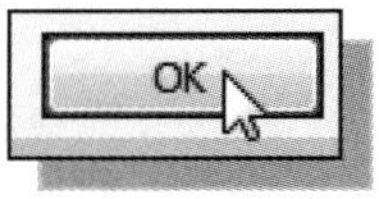

5. Click **OK** to enter the selection dialog box.

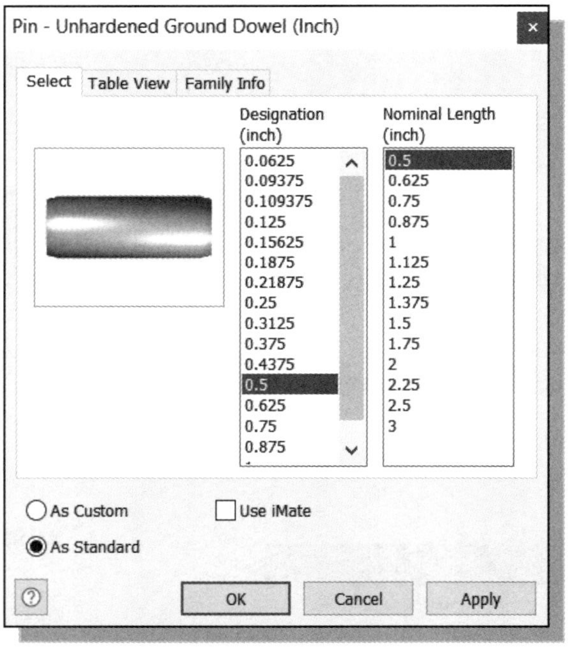

6. Set the designation diameter to **0.5** inch.

7. Set the nominal length to **0.5** inch.

8. Click on the **Table View** tab to view a more detailed listing of the available pins.

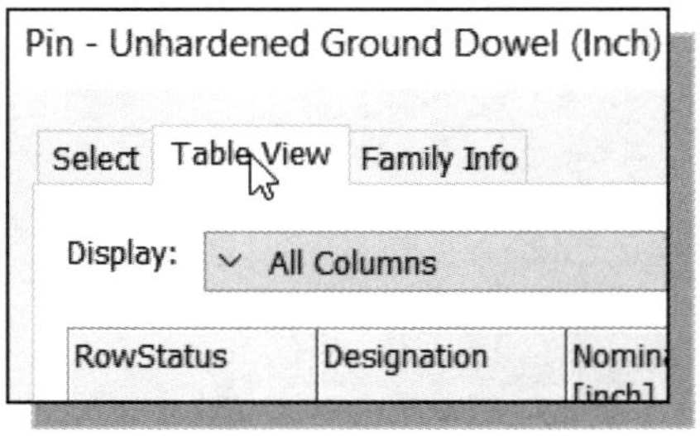

➢ Note the selected pin is highlighted in the list.

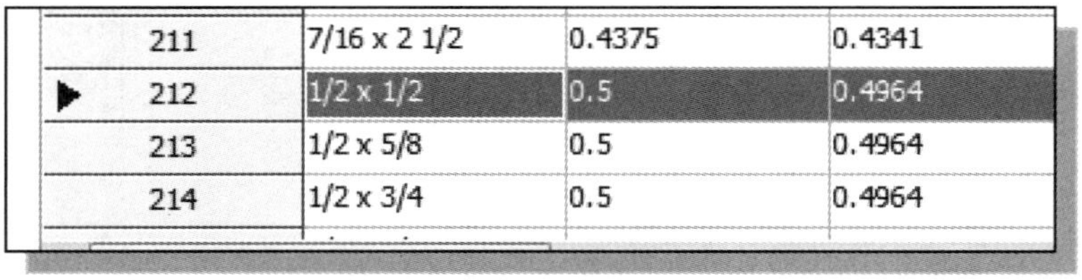

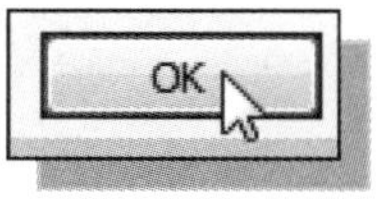

9. Click **OK** to accept the selection.

10. Place a copy of the *Pin* on the screen next to the other pieces as shown.

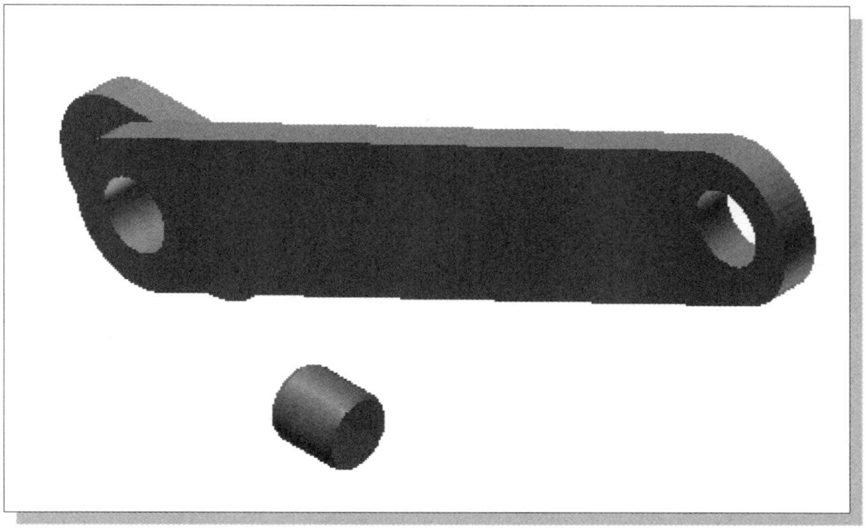

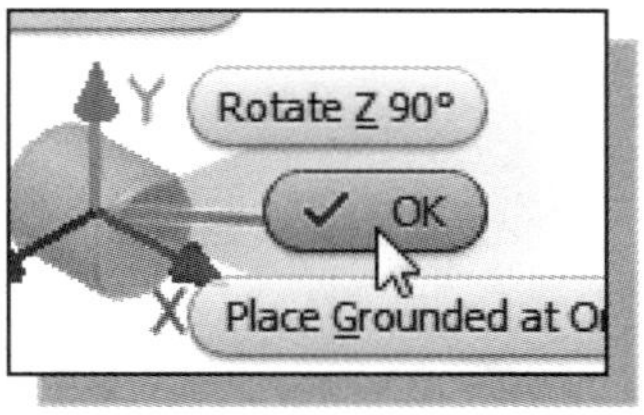

11. Inside the graphics window, right-click once to bring up the option menu and select **OK** to end the placement of the *Pin*.

12. On your own, constrain the *Pin* as shown in the figure below. (Hint: Use the **Mate** and **Flush** constraints on the parts and datum planes.)

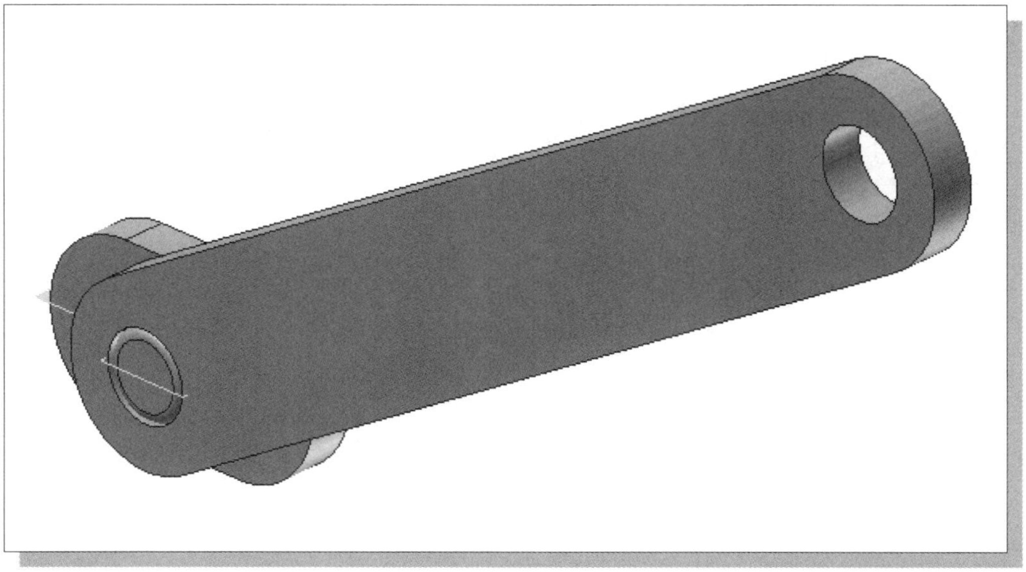

Assemble the CS-Rod Part

Next, we will place the *CS-Rod* part in the assembly model.

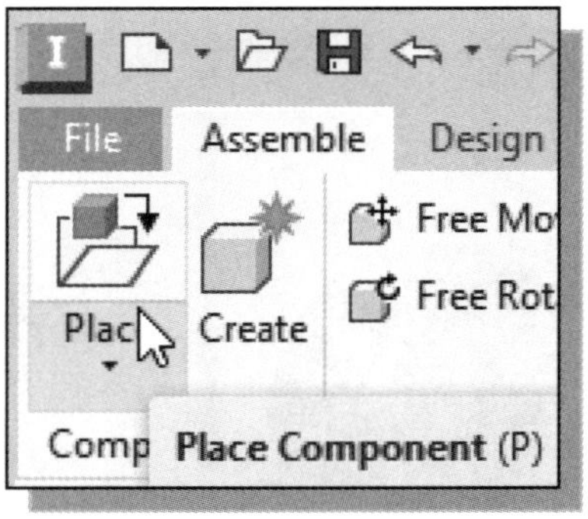

1. In the *Assemble* panel, select the **Place Component** command by left-clicking the icon.

2. Select the **CS-Rod** design in the list window. Click on the **Open** button to retrieve the model.

3. Place the part toward the right side of the assembly as shown in the figure.

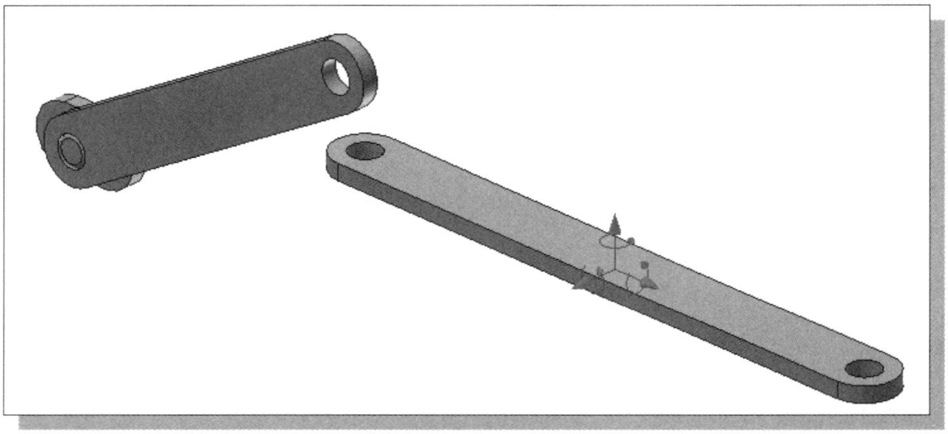

4. Inside the graphics window, right-click once to bring up the option menu and select **OK** to end the placement of the *CS-Rod* part.

5. On your own constrain the *CS-Rod*, leaving only one DOF as shown in the figure below. (Hint: Use the **Mate** and **Flush** constraints.)

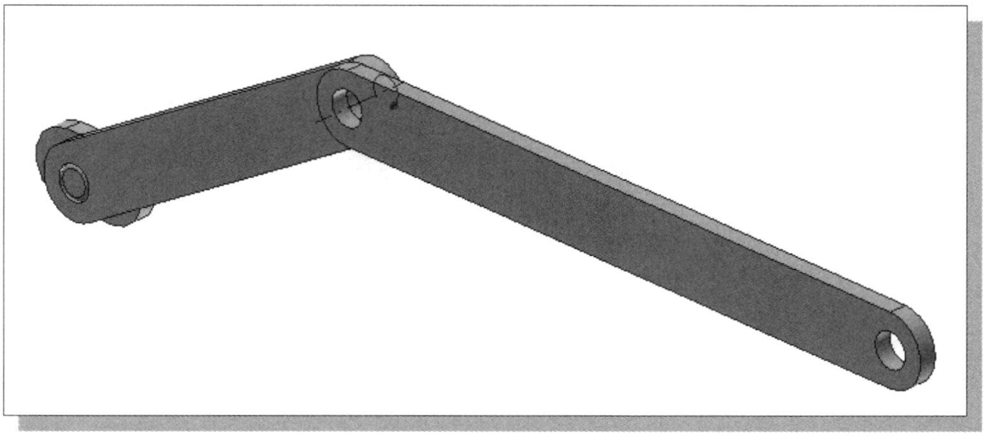

Make a Copy of the Pin

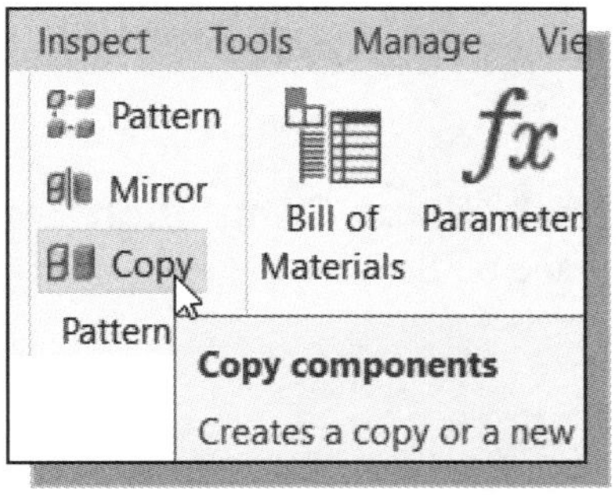

1. **De-select** any pre-selected part in the *Browser window.*

2. In the *Assemble* panel, select the **Copy Components** command by left-clicking once on the icon in the pattern toolbar.

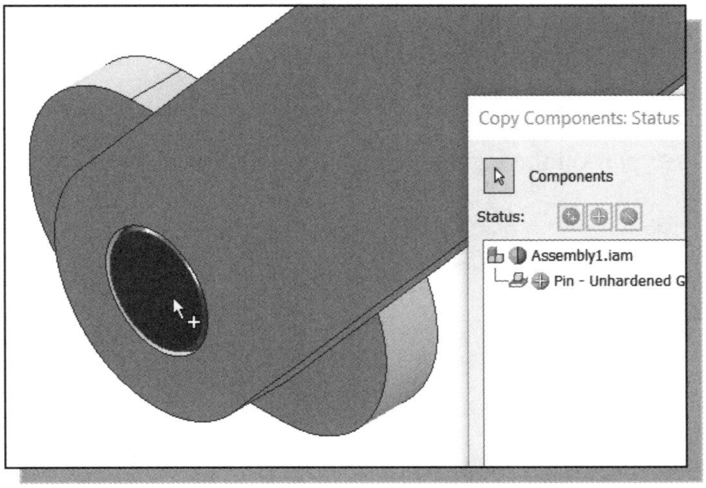

3. Select the ***Pin*** part by clicking on the part in the graphics window as shown.

4. Click **Next** to proceed with the selection.

5. Click **OK** to accept the name of the part.

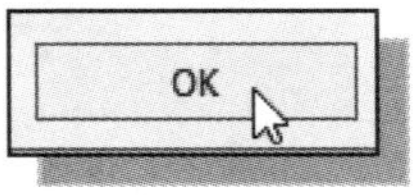

6. Place a copy of the part on the screen and exit the Copy Component command.

7. On your own, constrain the *Pin* as shown in the figure below. (Hint: Try using the **Insert** constraint.)

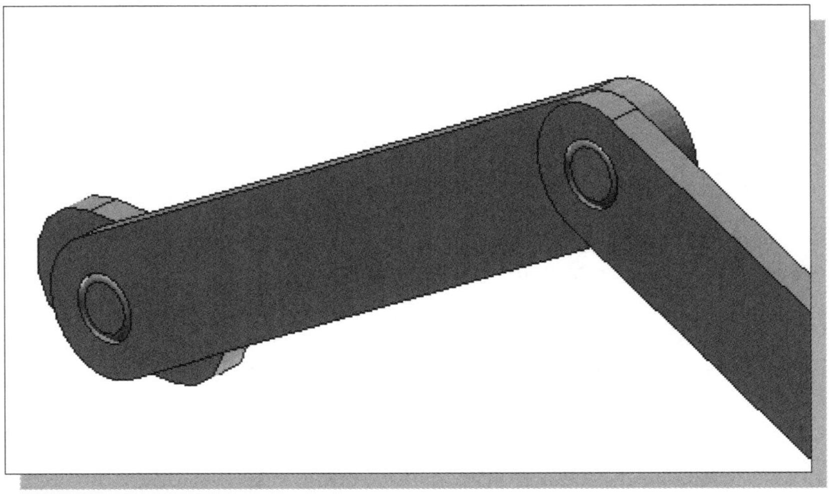

8. On your own, drag the different components in the graphics window with the left-mouse-button and confirm the parts are properly constrained.

Assemble the CS-Slider Part

We will place a copy of the *CS-Slider* part to complete the assembly model.

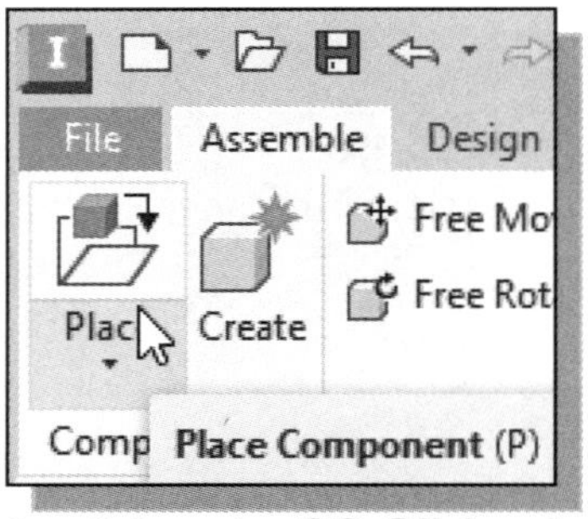

1. In the *Assemble* panel, select the **Place Component** command by left-clicking once on the icon.

2. Select the **CS-Slider** design in the list window and click **Open** to retrieve the model.

3. Place one copy of the *CS-Slider* part on the right side of the assembly as shown in the figure.

4. Inside the graphics window, right-click once to bring up the option menu and select **OK** to end the Place Component command.

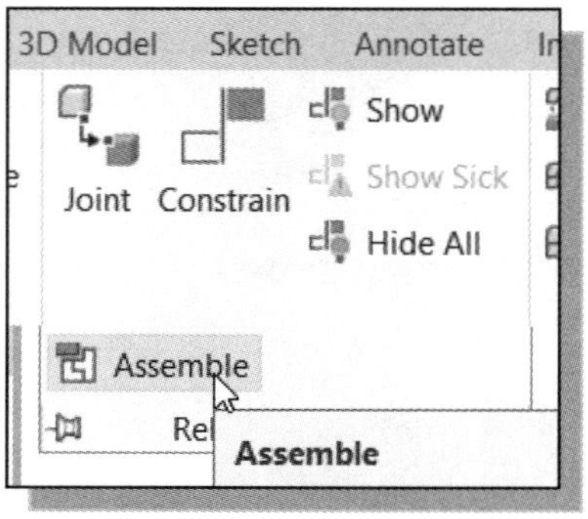

5. In the *Relationships* panel, select the **Assemble** command by left-clicking once on the icon.

➢ The *Assemble* command displays a mini constraints toolbar in the graphics window. Different constraints can be applied based on the entity selections.

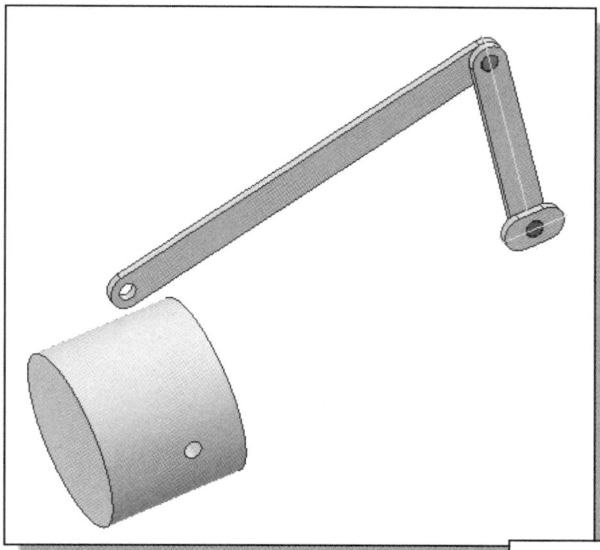

6. Using the **ViewCube**, rotate the display of the model roughly as shown.

7. Select the cylindrical hole surface of the slider part as the first part for the **Automatic constrain** command.

➢ Note the selected object now will move with the cursor.

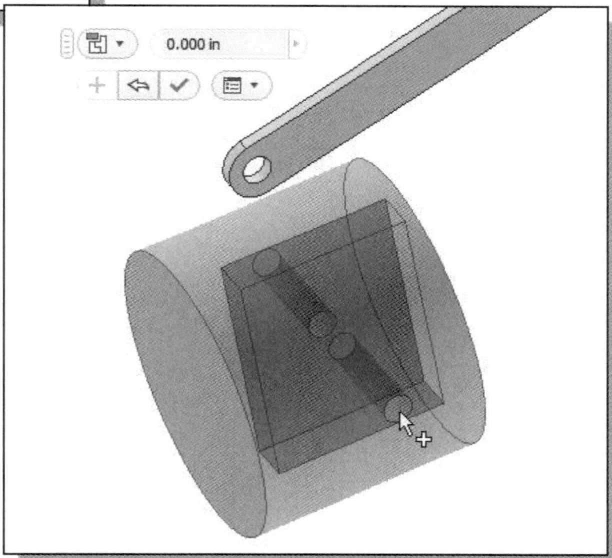

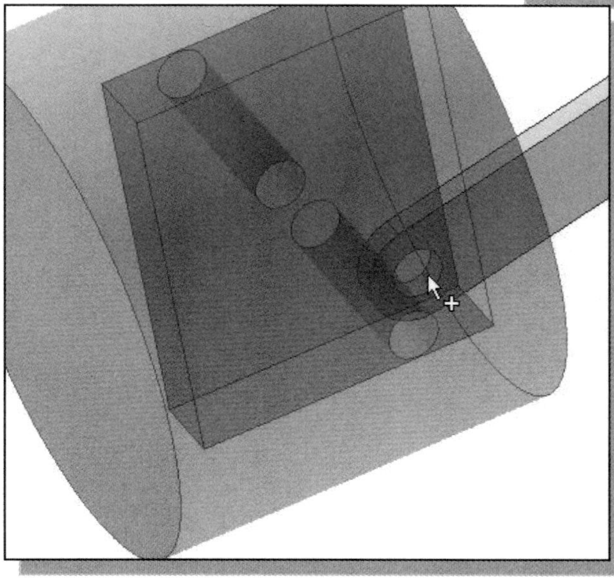

8. Select the cylindrical surface of the *CS-Rod* part as shown.

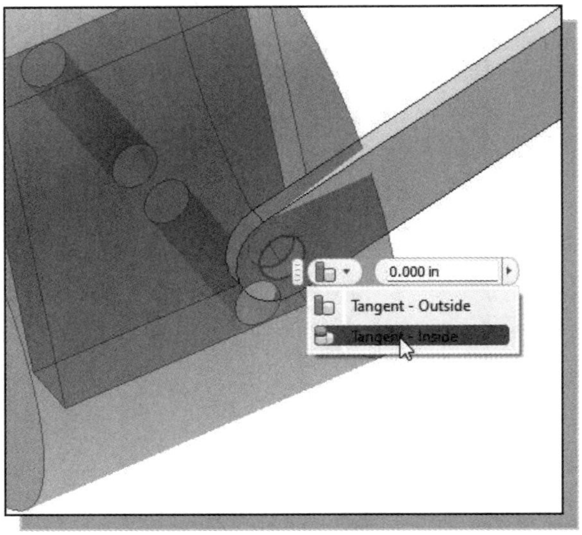

9. Select the **Tangent-Inside** from the option list as shown.

10. Click **Apply** to apply the displayed constraint.

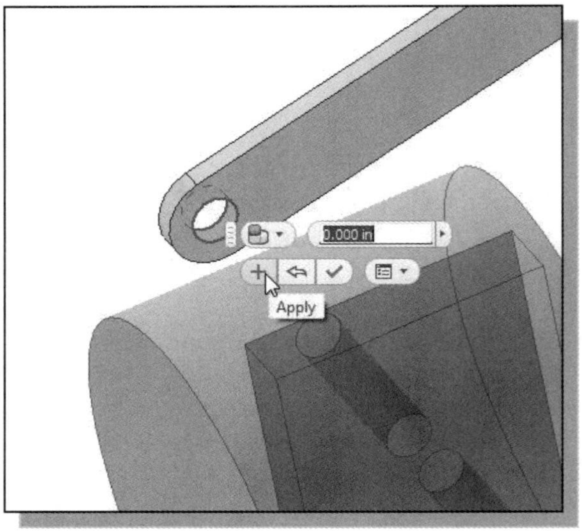

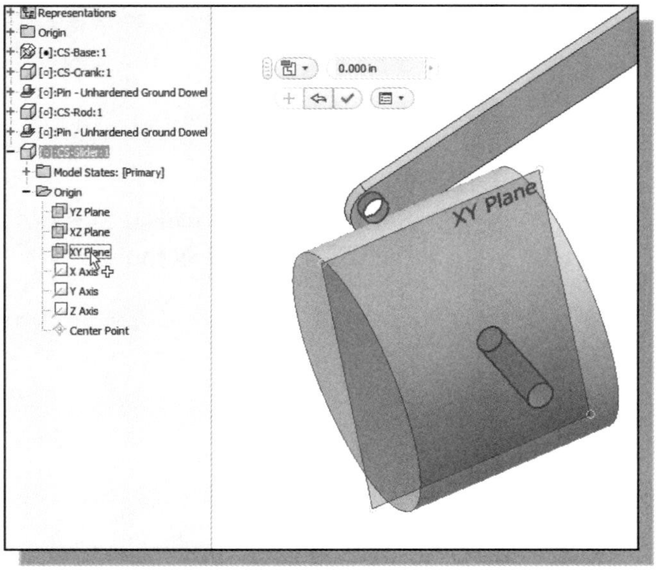

11. Select the vertical datum plane, in the model history tree, of the *CS-Slider* part as the first part for the **Auto constrain** command.

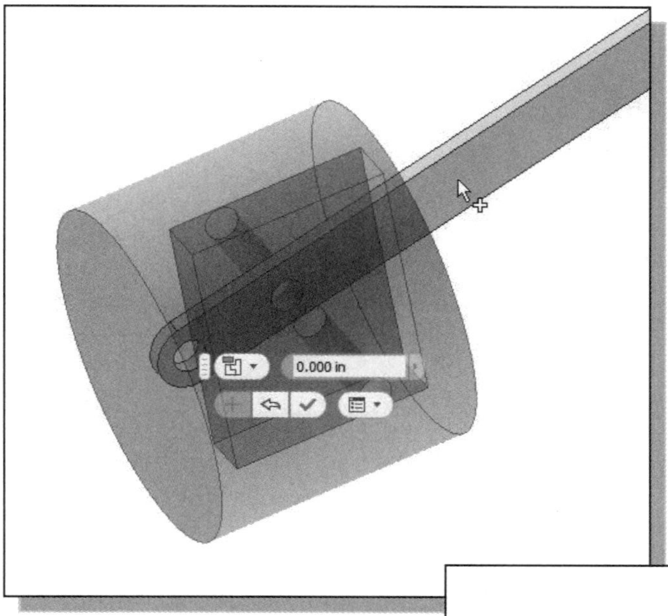

12. Select the back flat side of the *CS-Rod* part as the second surface for the Auto Constrain as shown.

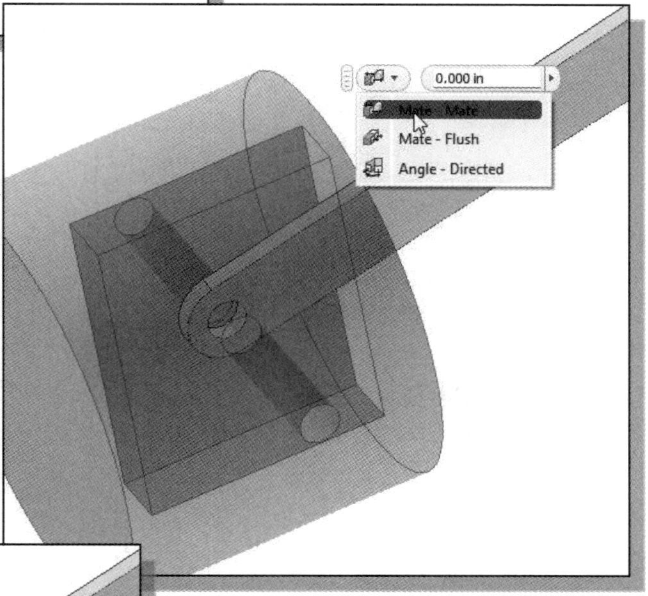

13. Based on the selections, the constraint icon is now set to the **Mate-Mate** constraint.

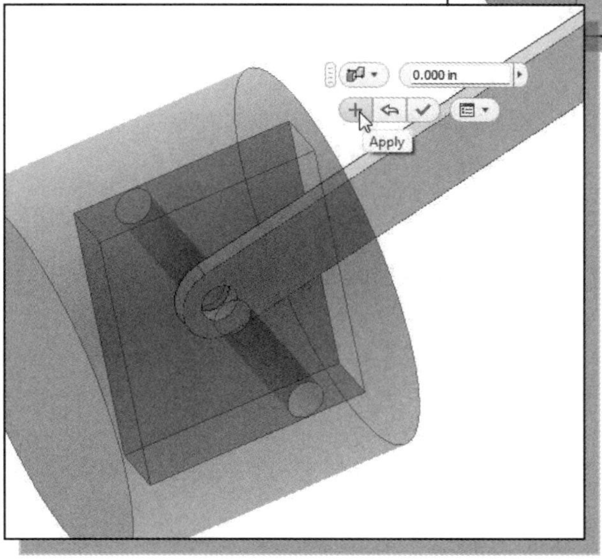

14. To apply the constraint click **Apply**.

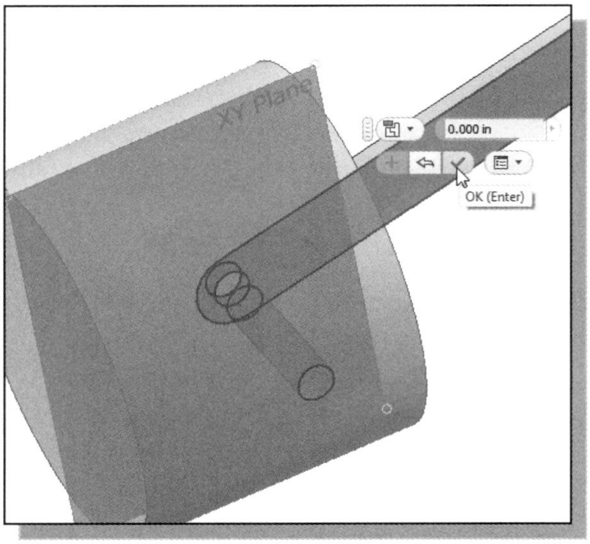

15. Click **OK** to accept the applied constraints and exit the assemble command.

16. Drag the Slider part and notice the *CS-Slider* is not fully constrained; we will next restrict the motion of the slider to the horizontal direction only.

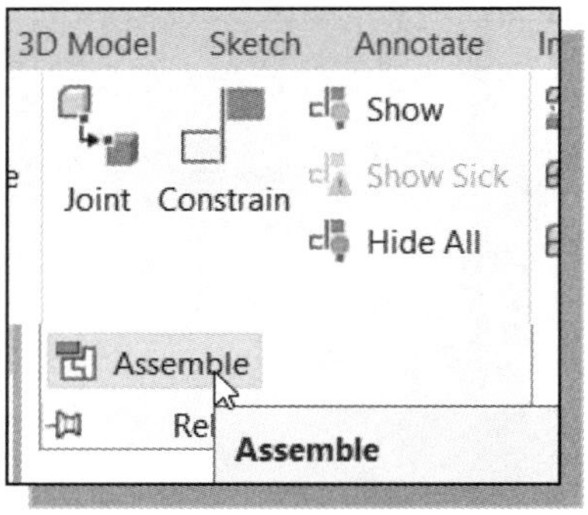

17. In the *Assembly Relationship* panel, select the **Assemble** command by left-clicking once on the icon.

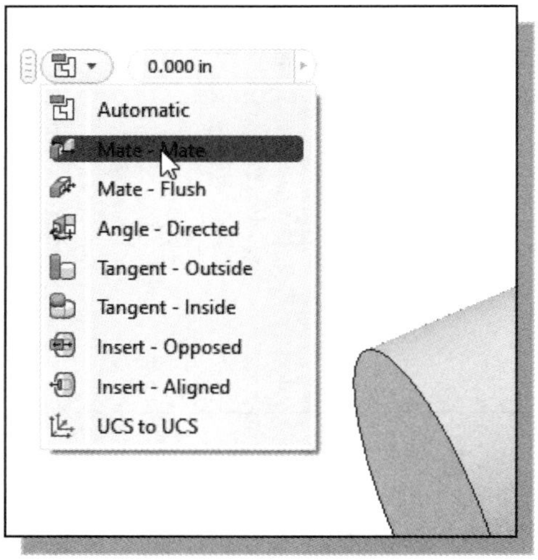

18. Choose **Mate-Mate** constraint from the mini toolbar as shown.

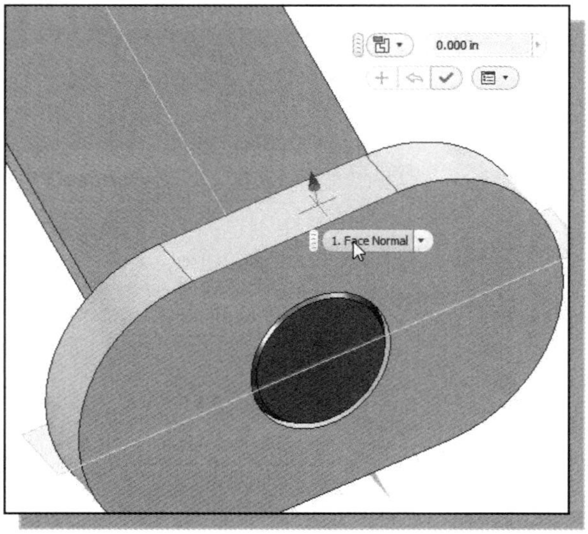

19. Select the **top surface** of the *CS-Base* part as shown.

20. A warning message appears on the screen, stating the selected object is already fully constrained. Click **NO** to redo the selection.

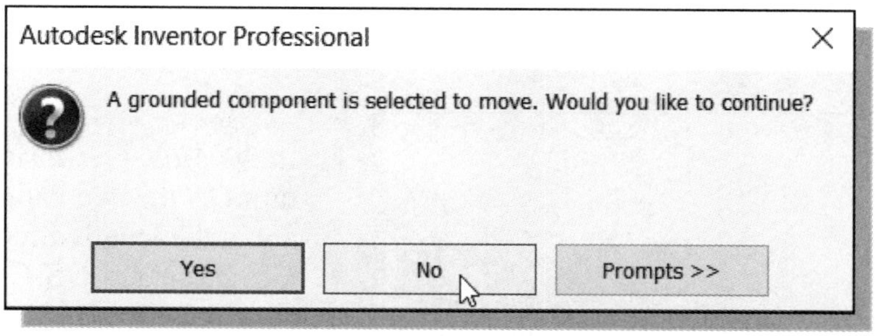

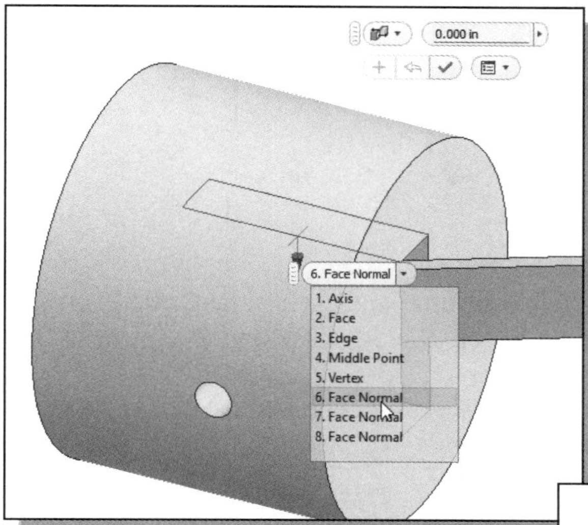

- The surface on the moving part needs to be selected first before choosing the surface on the fixed part.

21. On your own, use the selection tool and select the top surface of the inside cut feature of the *CS-Slider* part as shown.

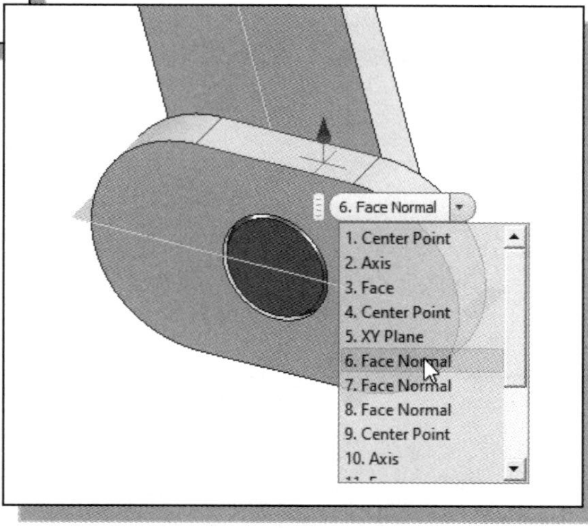

22. Select the top surface of the *CS-Base* part.

23. Click **Apply** to accept the constraint.

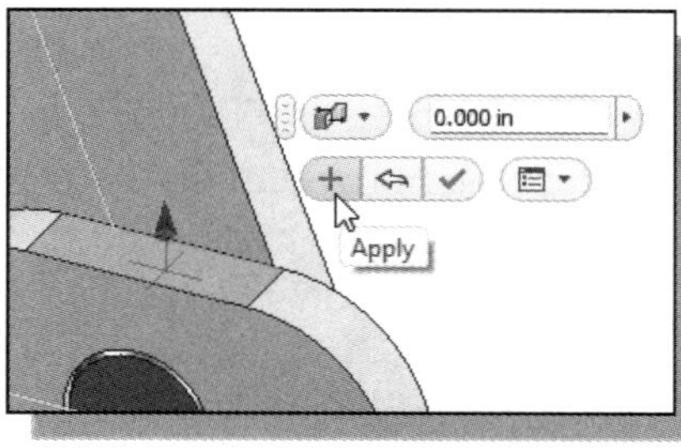

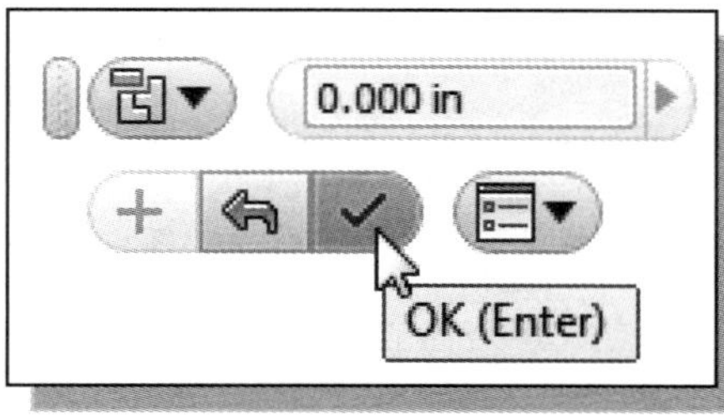

24. Click **OK** to accept the applied constraints and exit the assemble command.

25. On your own, drag the *CS-Slider* part and confirm the slider is properly constrained.

Add an Angle Constraint to Fully Constrain the Assembly

The *Crank* is the driver of the assembly, and its position can be controlled by adding an **Angle** constraint. This **Angle** constraint will also be used to create the simulation of the assembly.

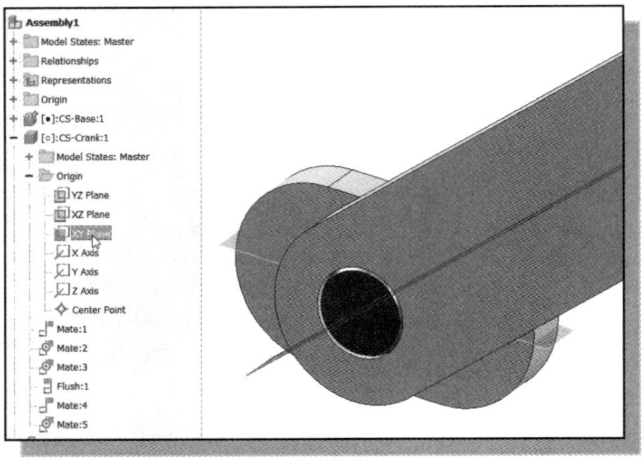

1. In the *Model History Tree* window, turn *ON* the two datum planes that are in the length directions of the *CS-Base* and *CS-Crank* parts as shown in the figure.

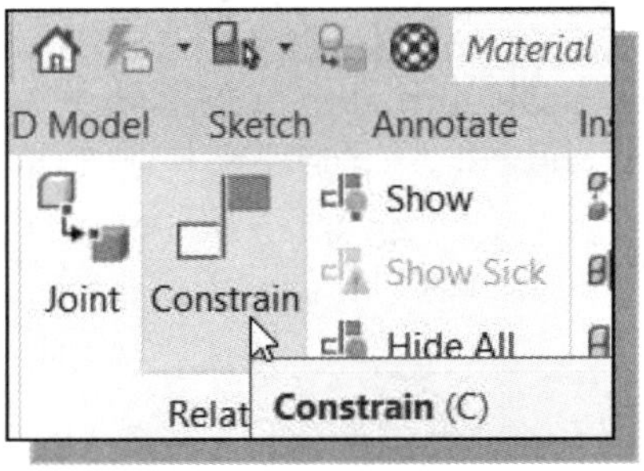

2. In the *Relationships* panel, select the **Constrain** command by left-clicking once on the icon.

3. In the *Place Constraint* dialog box, select the **Angle** constraint as shown.

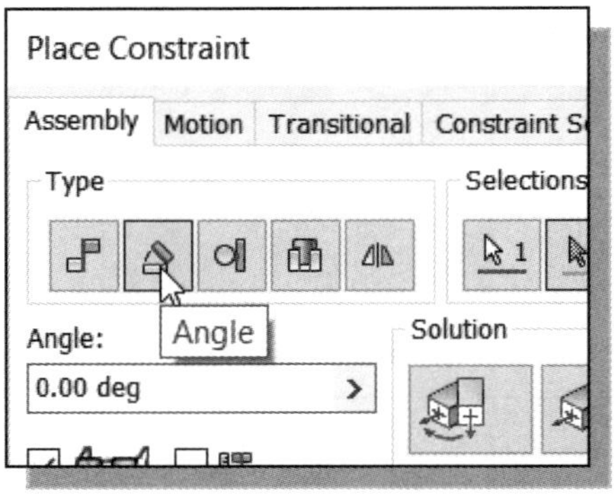

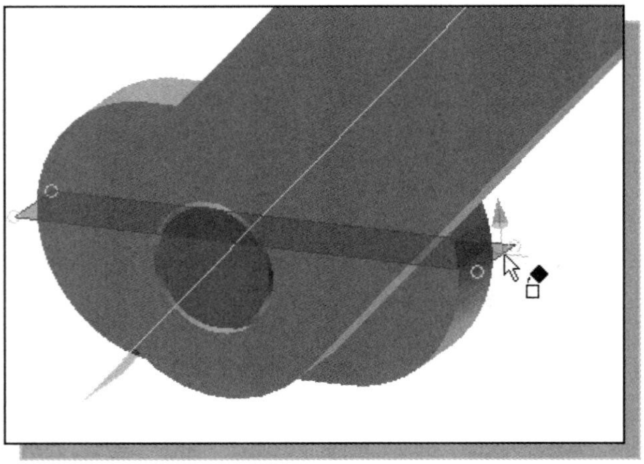

4. Select the horizontal datum plane of the *CS-Base* part as the first selection as shown.

5. Set the *Solution* option to **Directed Angle** as shown.

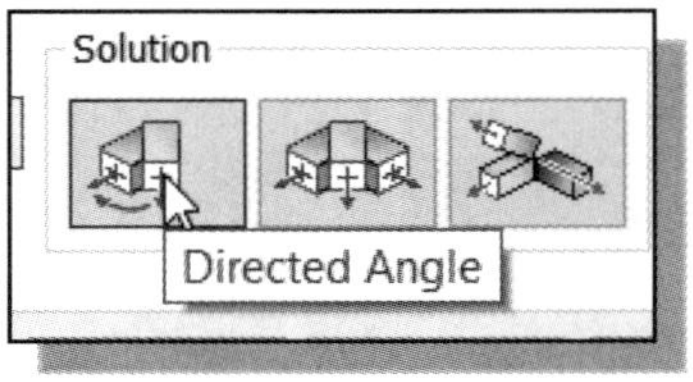

6. Select the datum plane of the *CS-Crank* part as the second selection as shown.

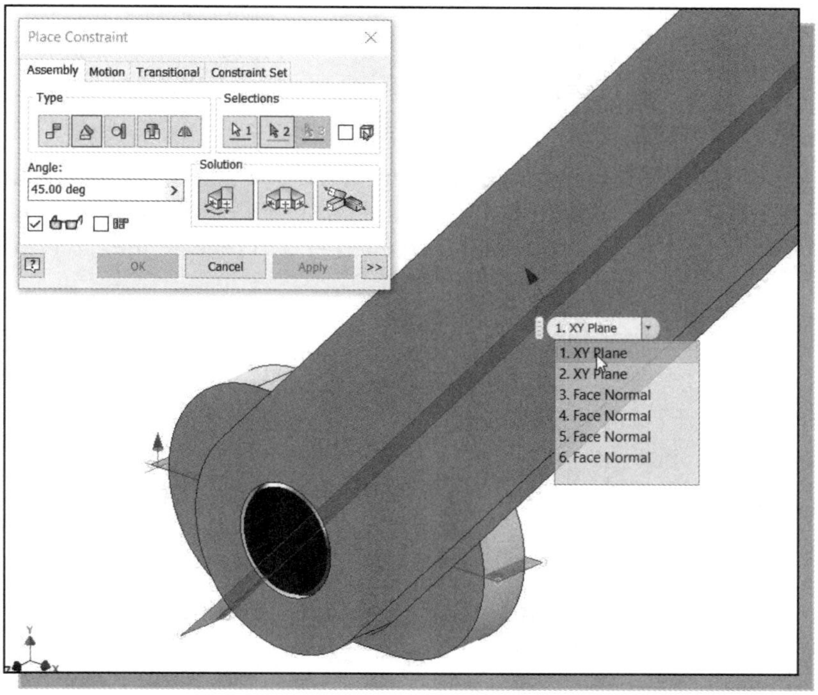

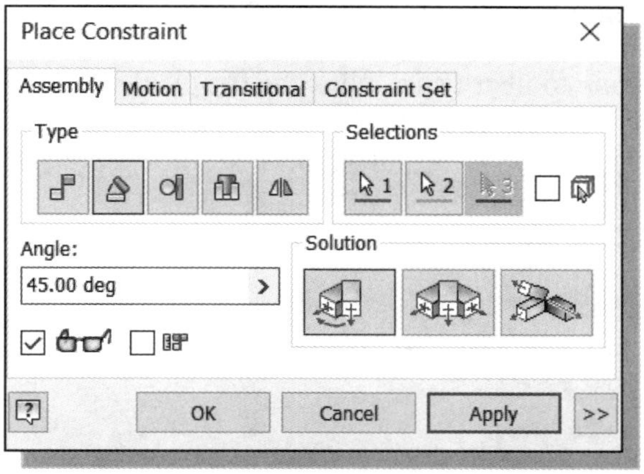

7. Enter **45** deg in the *Angle* box and notice the crank position is adjusted accordingly.

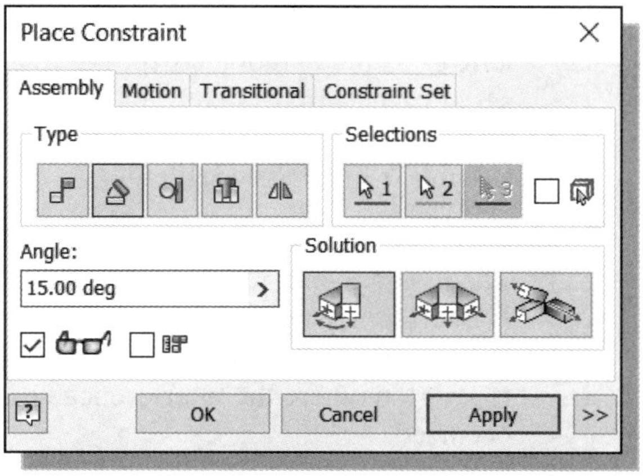

8. Enter **15** deg in the *Angle* box and confirm the *Solution* option is set to **Directed Angle** as shown.

9. On your own, apply the constraint and exit the Constraint command.

❖ Using the Directed Angle option assures the angle is set in one direction of the reference plane.

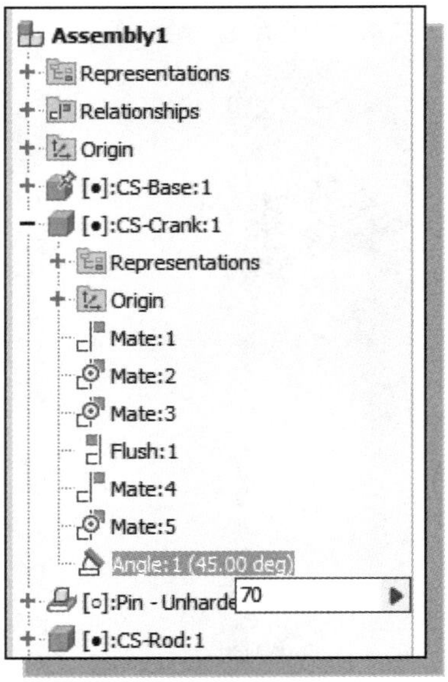

10. In the *Model History Tree* window, expand the *CS-Crank* part and click on the Angle constraint. The **Edit Angle** box appears below the *Model History Tree* window as shown.

11. Enter **70.00 deg** in the box and the crank position is adjusted accordingly. (If the crank is rotated below horizontal then enter -70 to reverse the direction.)

12. On your own, adjust the angle to other values by repeating the above steps. Also confirm the *Crank*, *Connecting Rod* and *Slider* parts are properly constrained.

❖ Reset the angle back to 70 degrees.

Interference Analysis

Autodesk Inventor allows us to perform *Interference Analysis* based on the current assembled positions of the assembled parts.

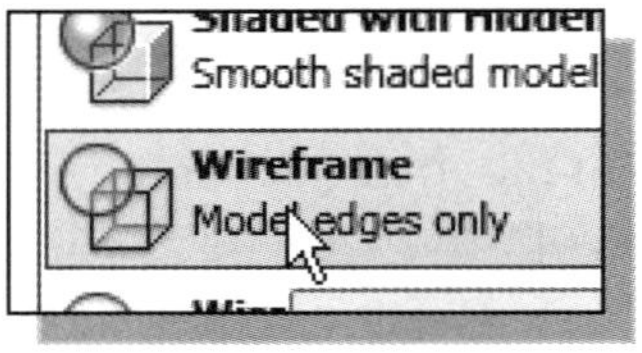

1. Switch to **Wireframe Display** through the *View* toolbar.

2. Inside the *Ribbon* toolbar panel, click the **Inspect** tab and then **Analyze Interference** as shown in the figure.

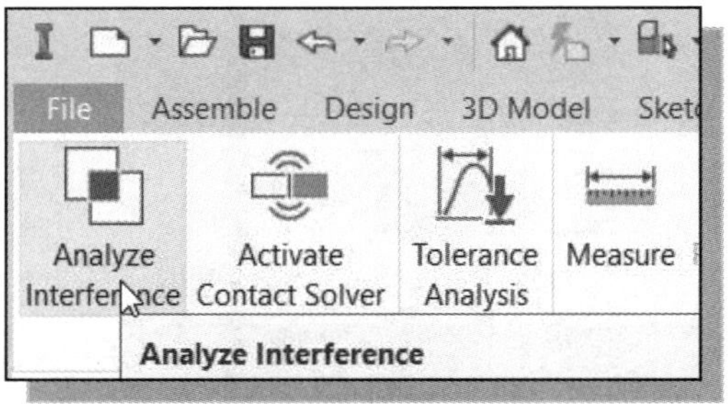

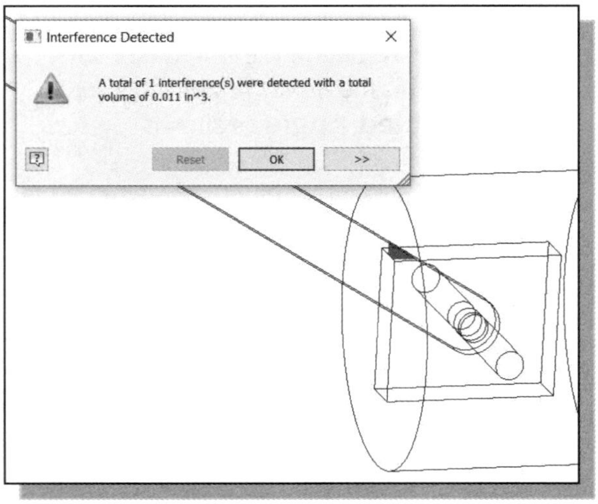

3. Select the ***CS-Rod*** and the ***CS-Slider*** as the two objects for the first set of objects to be analyzed.

4. Click **OK** to proceed with the analysis.

5. The message box identifies that there is interference between the two selected parts. Note the highlighted red region where the interference has occurred.

6. Click **OK** to exit the command.

Basic Motion Analysis

Autodesk Inventor's **Drive Constraint** tool allows us to perform basic motion analysis by creating simulations of assemblies with moving parts.

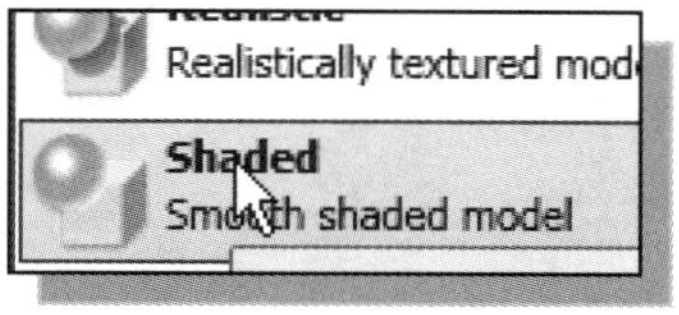

1. In the *View* toolbar, select the **Shaded Display** command by left-clicking once on the icon.

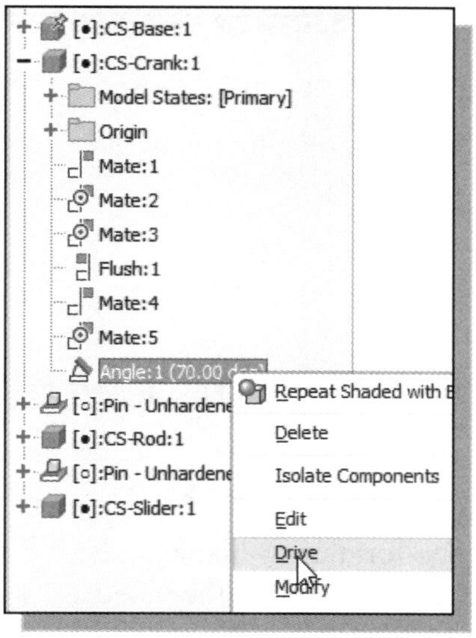

2. In the *Model History Tree* window, right-click on the *Angle* constraint to bring up the option menu and select **Drive** as shown.

3. In the *Drive Constraint* dialog box, enter **0.00** as the starting angle and **360.00** as the ending angle as shown.

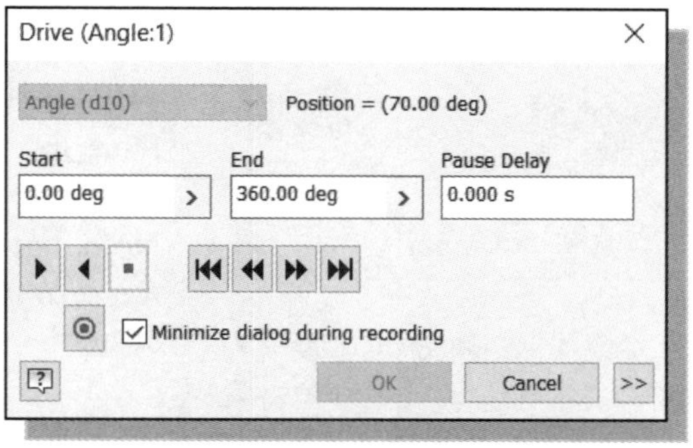

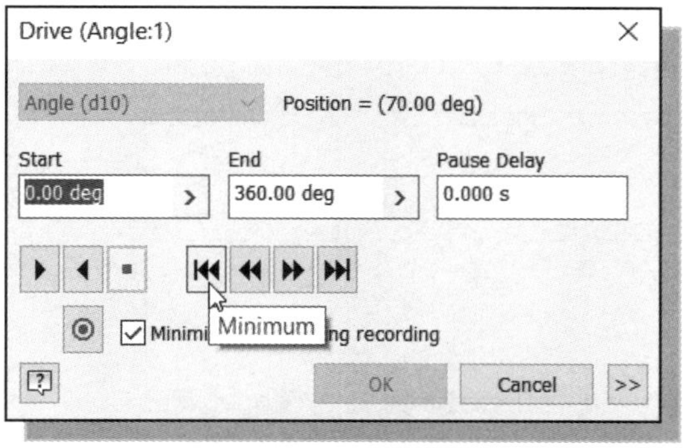

4. Click on the **Minimum** button to reset the crank, as well as all the components in the assembly, to the **Minimum** position.

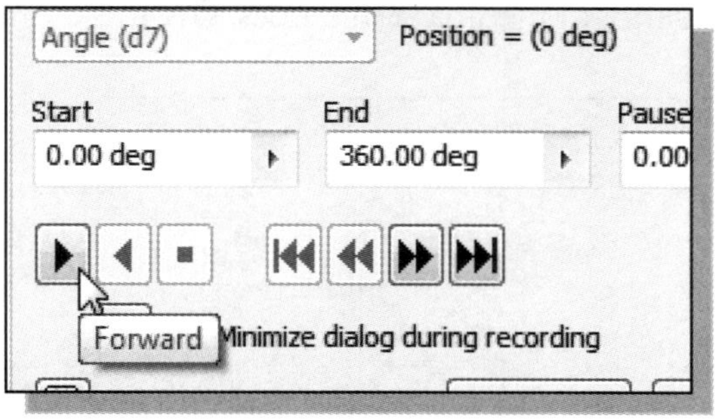

5. Click on the **Forward** button to begin the simulation. Notice the forward direction of the motion is defined by the applied angle constraint.

6. Click on the **Reverse** button to run the simulation in the backward direction.

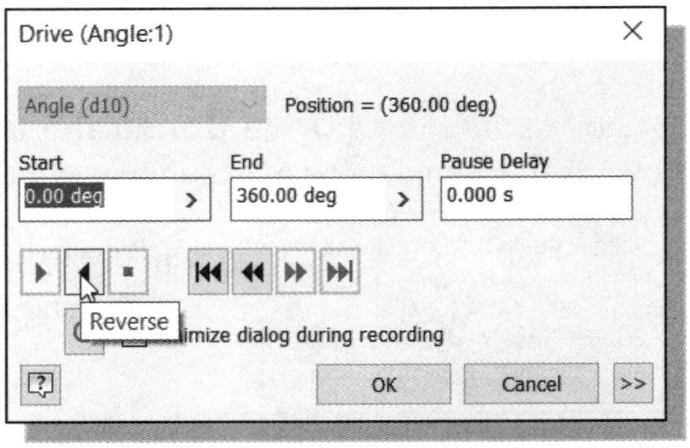

7. On your own, start the simulation and dynamically rotate the display while the simulation is running.

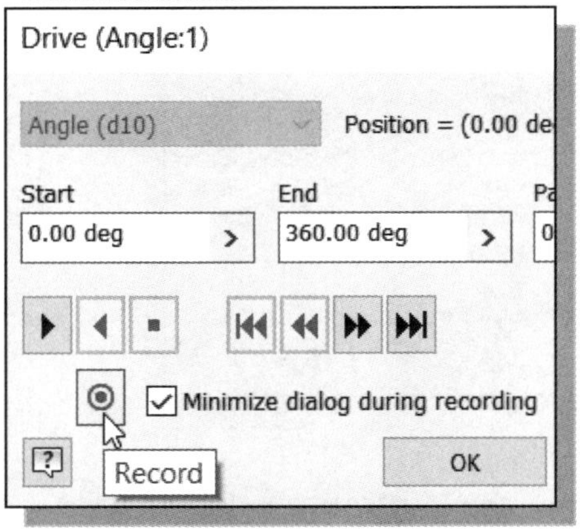

8. The simulation can also be saved as an AVI or MS WMV movie format file. Click the **Record** button as shown.

- Note that the video formats available under the *Record option* are dependent on the video systems installed under the Operating System.

9. Enter **CS-Assembly.wmv** as the *File name* and click **Save** to accept the file name.

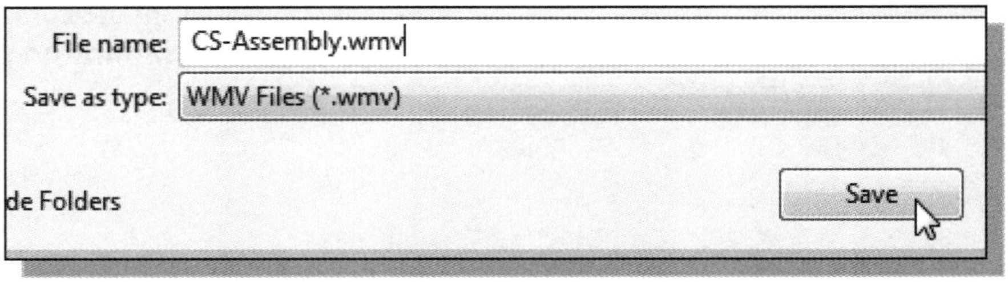

10. On your own, adjust the settings to set the quality of the image and click **OK** to proceed.

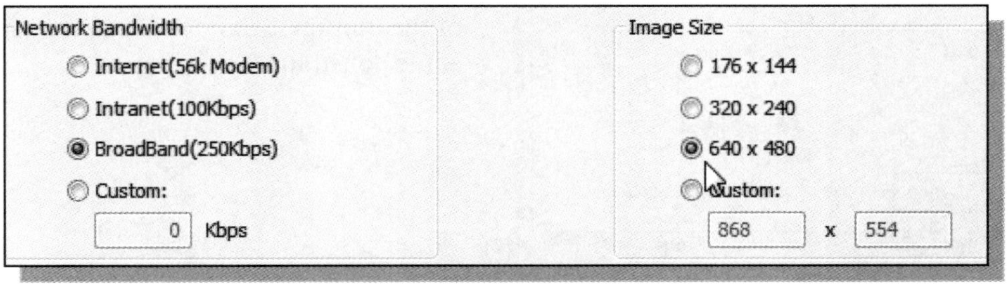

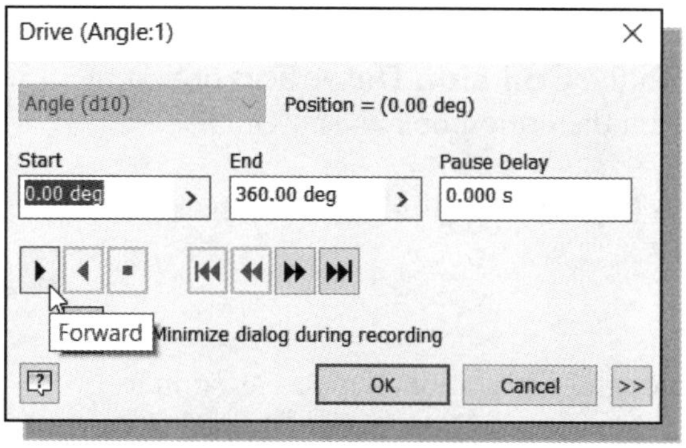

11. Click on the **Forward** button to begin the simulation, and also record the simulation.

❖ Note the *Minimize dialog during recording* option is turned *ON* by default.

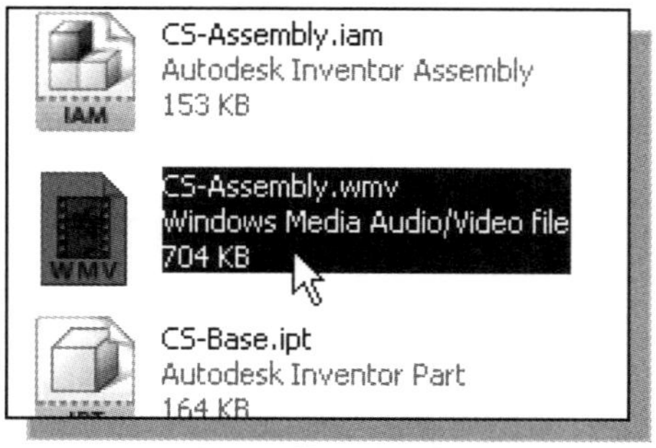

12. On your own, view the recorded video file.

13. Turn off the **Record** option and click on the **Expand** option button to reveal more control options.

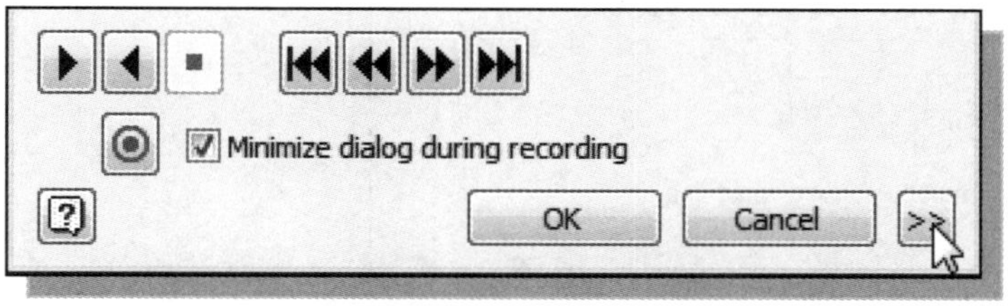

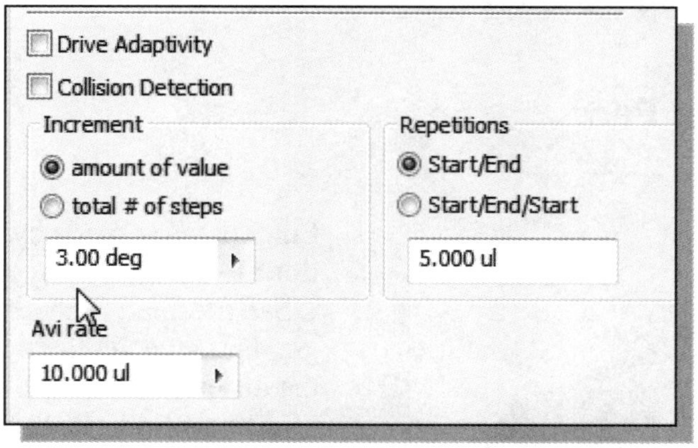

14. Set the *Increment* option to **3.00 deg** and the *Repetitions* option to **5**; these settings control the speed and number of repeated simulations.

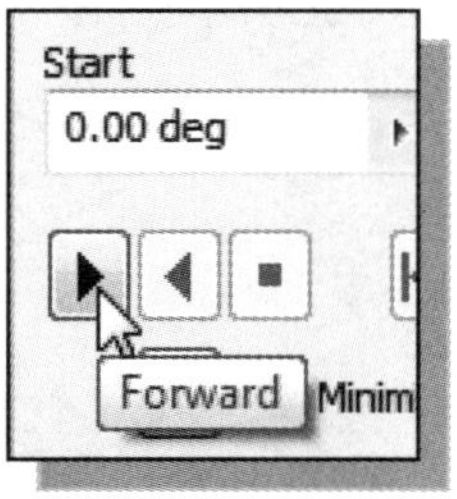

15. Click on the **Forward** button to begin the simulation. Notice the speed is 3 times faster than before, and the simulation is repeated 5 times.

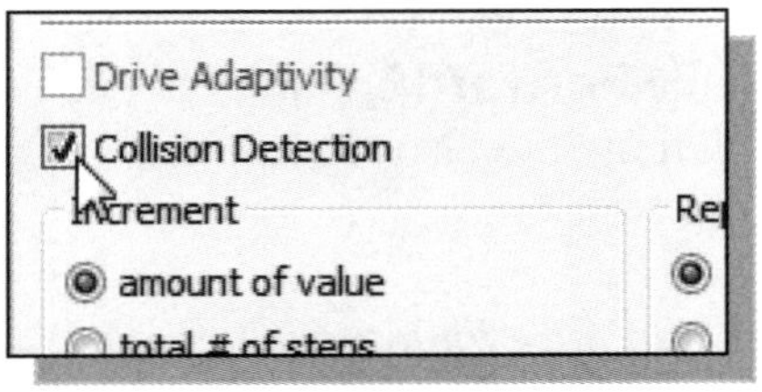

16. Switch on the **Collision Detection** option by clicking on the option box as shown.

17. Click the **backward** button to start the motion again.

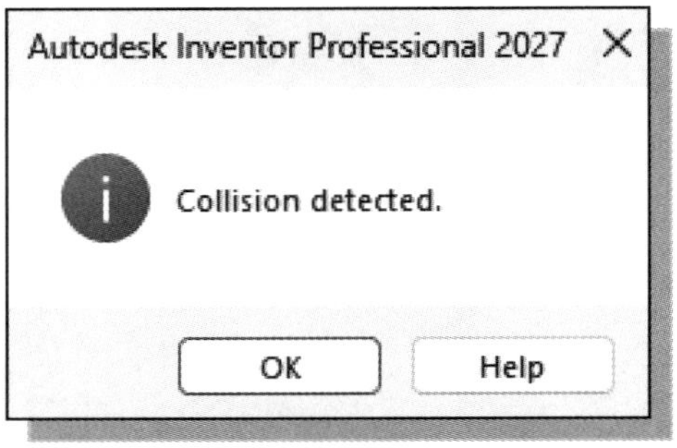

18. Notice the simulation stopped as soon as collision occurred. Notice it stopped at the angle of **129** degrees.

19. Click **OK** to close the message box.

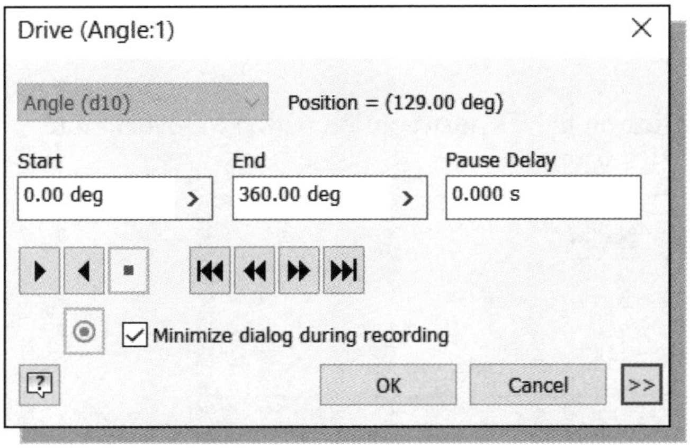

20. Click **OK** to close the *Drive Constraint* dialog box.

3D Grip Editing the CS-Slider Part

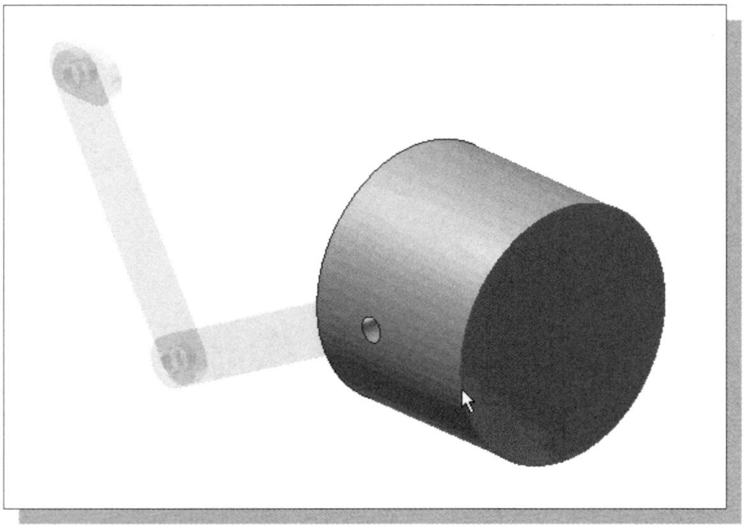

1. Enter the *Edit* mode by **double-clicking** the *CS-Slider* part in the graphics area. Notice the different shading of the parts indicates the *CS-Slider* part is activated.

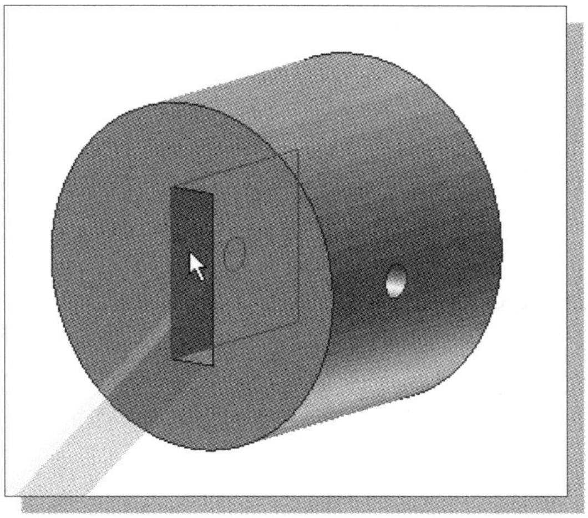

2. Rotate the 3D display so that the rectangular cut is visible and select **one of the inside vertical surfaces** as shown.

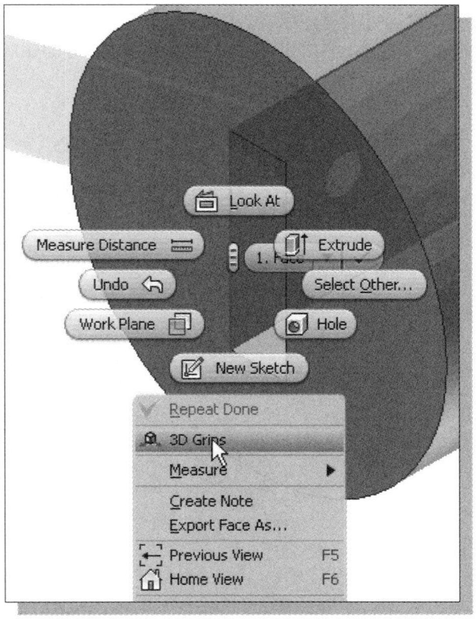

3. Inside the graphics window, **right-click** once to bring up the option menu.

4. Select **3D Grips** to display the 3D grip points associated with the selected surface.

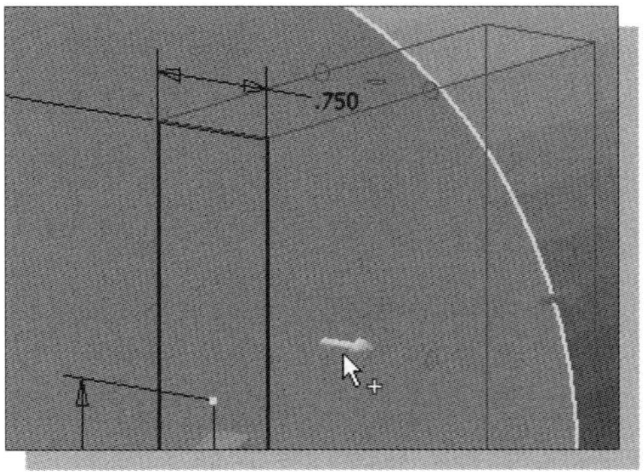

5. Move the cursor on top of the small circles, the 3D grip points, and notice the associated editing arrow is displayed. (The displayed arrow indicates the moving direction of the surface.)

6. Drag the 3D grip point on the bottom surface of the rectangular cut downward and increase the vertical length dimension to **3.75** as shown in the figure.

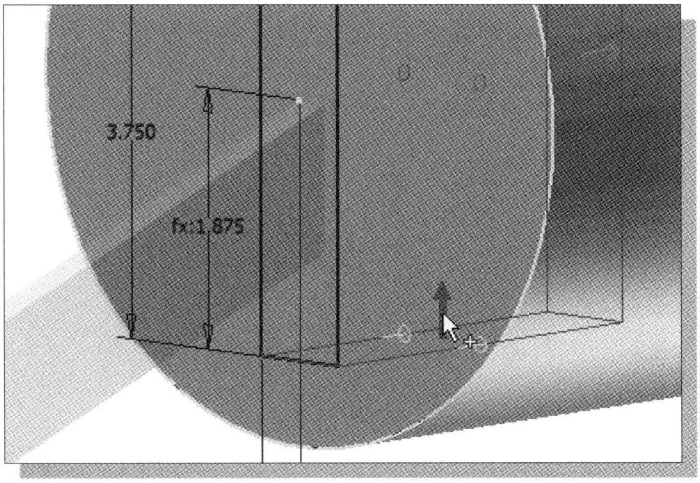

7. Right-click once and select **Done** to accept the modification.

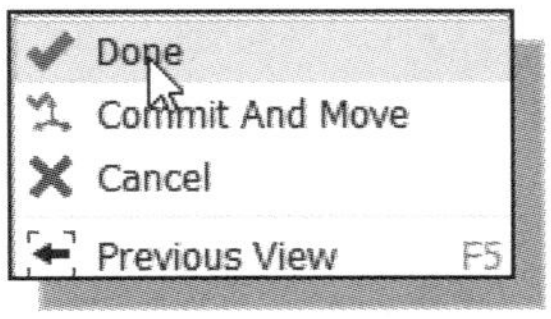

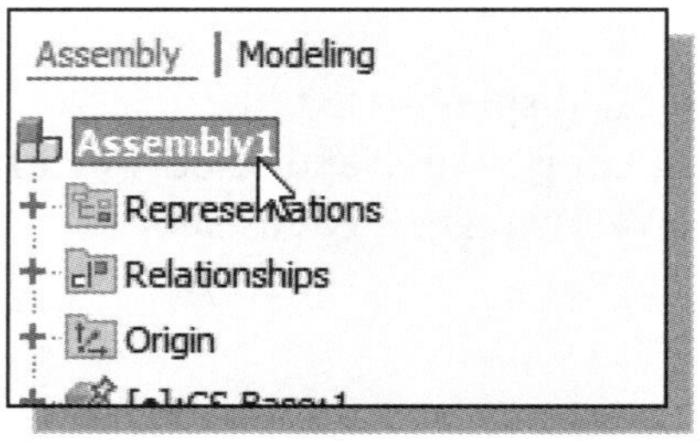

8. Reactivate the assembly model by **double-clicking** on the assembly model name in the *Model History Tree* as shown.

9. On your own, activate the *Edit* mode for the *CS-Rod* part by double-clicking on the part.

10. Select the ***front*** face of the *CS-Rod* part and **right-click** once to bring up the option menu. Select the **3D Grips** option as shown.

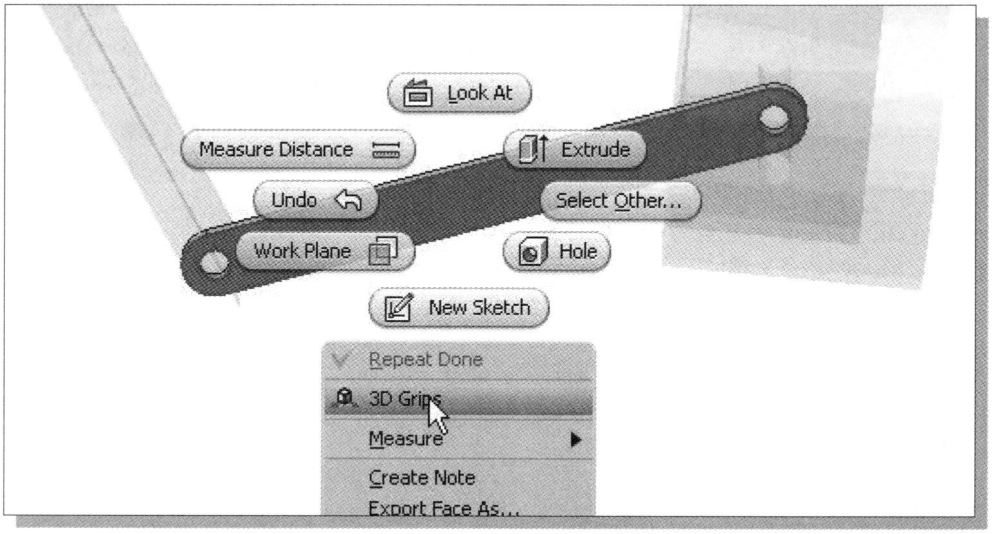

11. Adjust the length of the part to **11.5** inches by dragging the right grip point as shown in the figure.

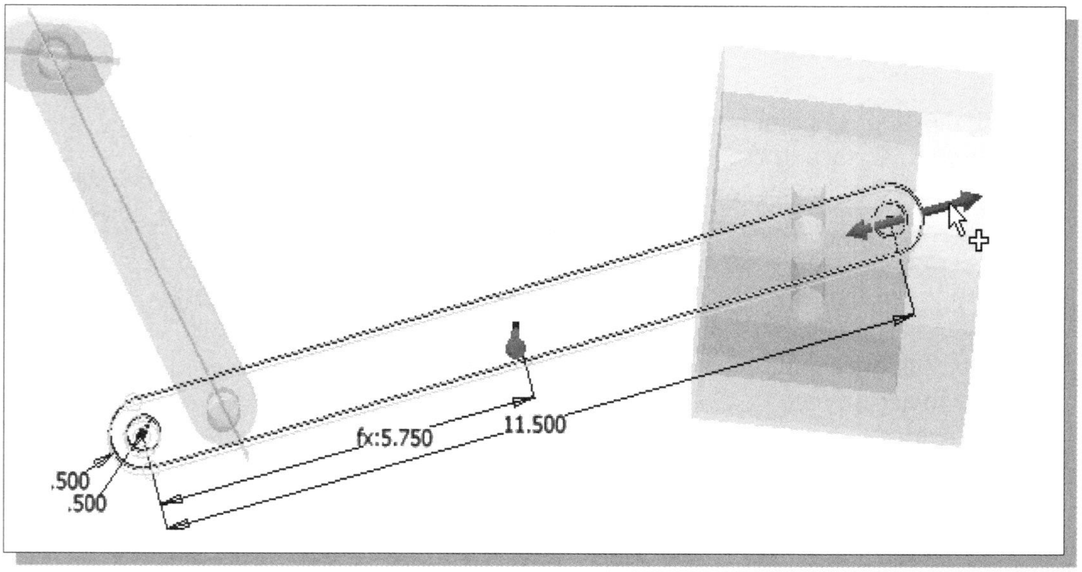

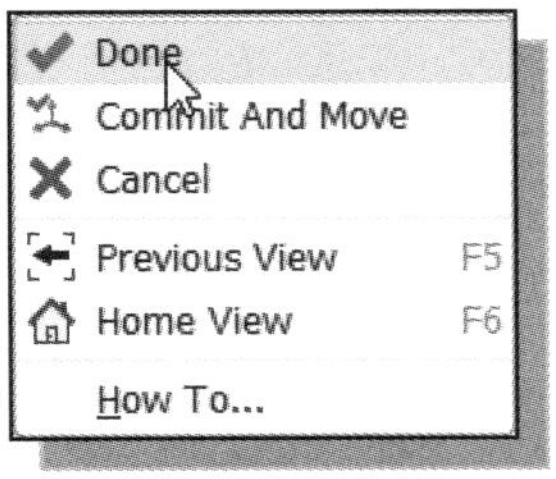

12. Click **Done** to accept the modification.

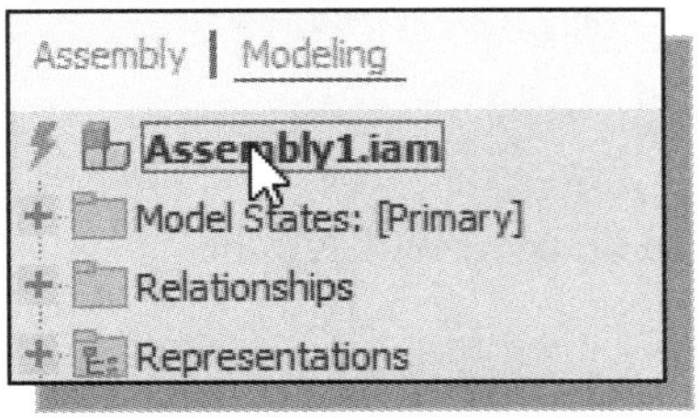

13. Re-activate the assembly model by double-clicking on the assembly model name in the *Model History Tree* as shown.

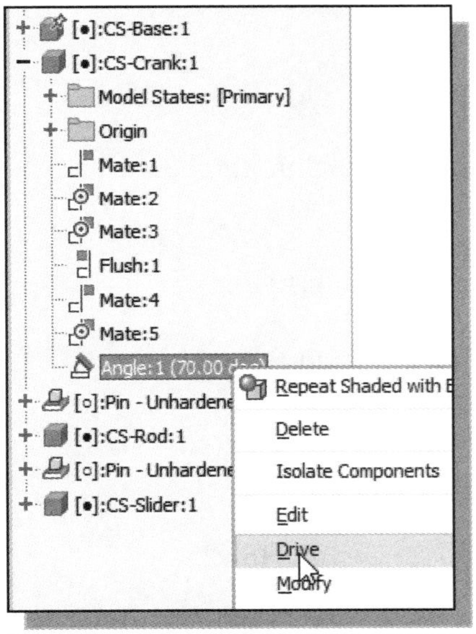

14. In the *Model History Tree* window, right-click on the *Angle* constraint to bring up the option menu and select **Drive** as shown.

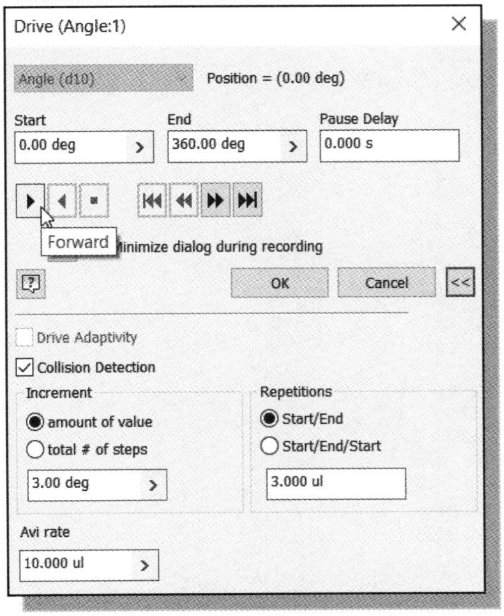

15. In the *Drive Constraint* dialog box, confirm the ***Collision Detection*** option is turned *ON* and enter **1.00** as the ***Increment angle*** and **2.00** as the ***Repetitions*** number as shown.

16. Click on the **Forward** button to begin the simulation and confirm the interference problem has been resolved.

❖ The design was modified quite quickly and effectively through the use of the very powerful analysis and editing tools.

Review Questions: (Time: 30 minutes)

1. What is included in the ***Autodesk Content Center***?
2. What is the usage of **3D grips**?
3. How do you save the Inventor simulation as an AVI or MS WMV movie?
4. How do you access the **Drive Constraint** command?
5. List and describe two of the assembly constraints that can be used to drive the simulation with the **Drive Constraint** command.
6. How do we enter the *Edit* mode in an assembly model?
7. Can we access the 2D sketch of a part in an assembly?
8. How do we end the *Edit* mode and return to the assembly model in *Autodesk Inventor*?
9. List and describe two methods to edit the angle of an **Angle** constraint.
10. Describe the procedure to perform an *Interference Analysis* in an assembly.
11. What information is available through the *Interference Analysis* in *Autodesk Inventor*?
12. Can we control the speed of the simulation in *Autodesk Inventor*? How?

Exercises: (Time: 220 minutes) (Create a set of detail and assembly drawings.)

1. **Leveling Assembly: placed at the feet of high precision machineries.** (Dimensions are in millimeters.)

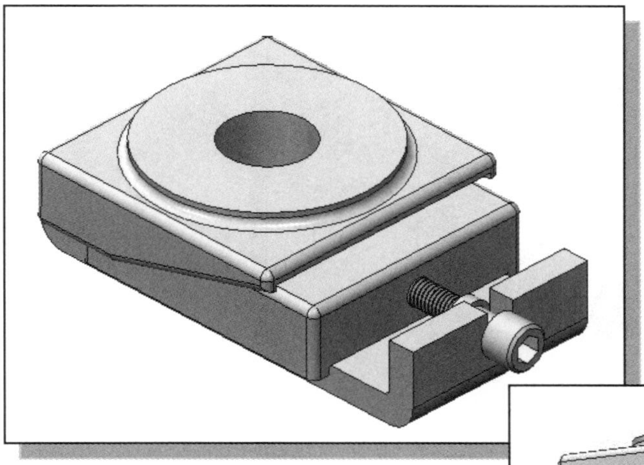

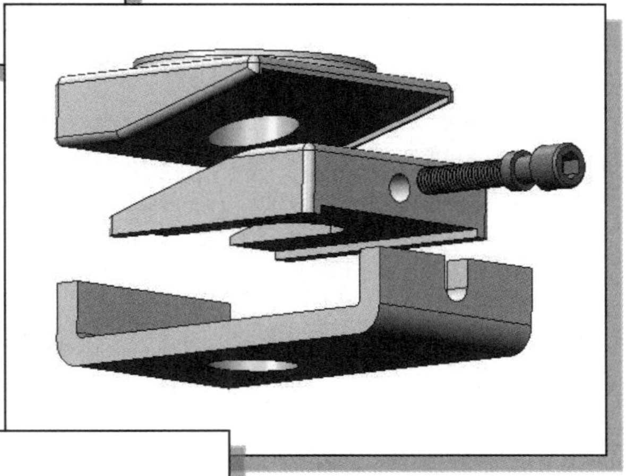

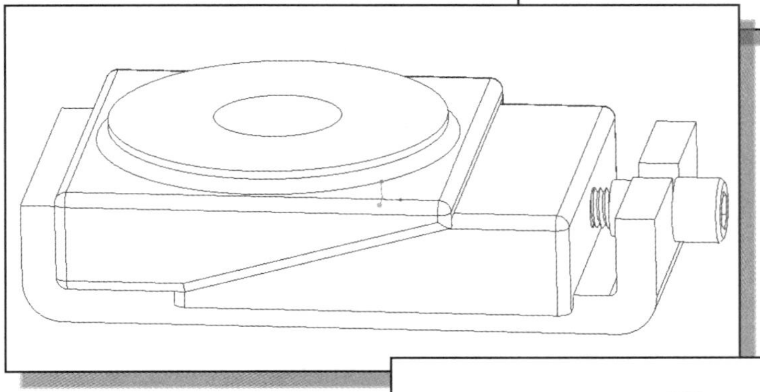

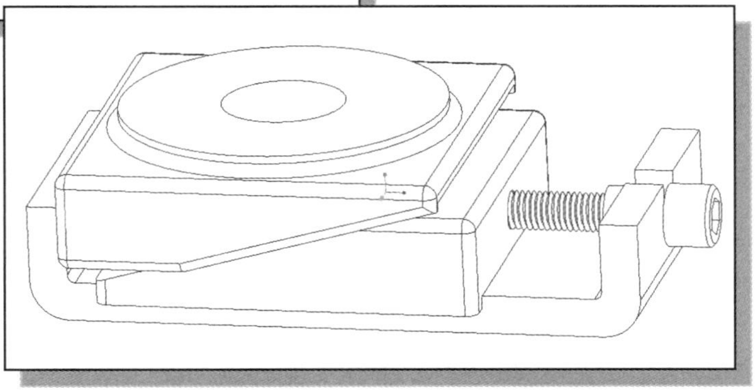

(a) **Base Plate**

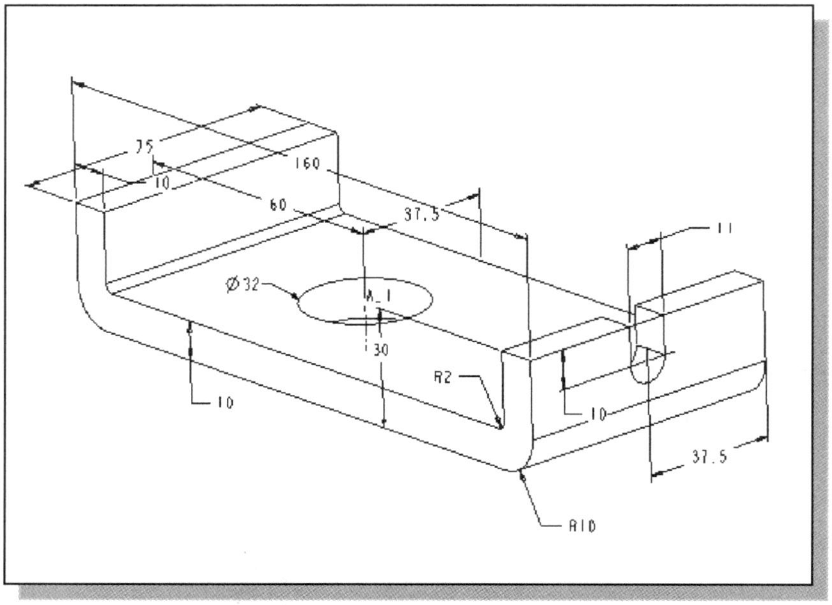

(b) **Sliding Block** (Rounds & Fillets: R3)

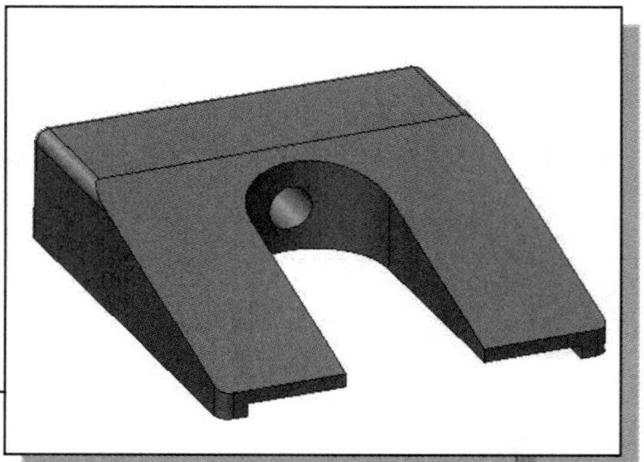

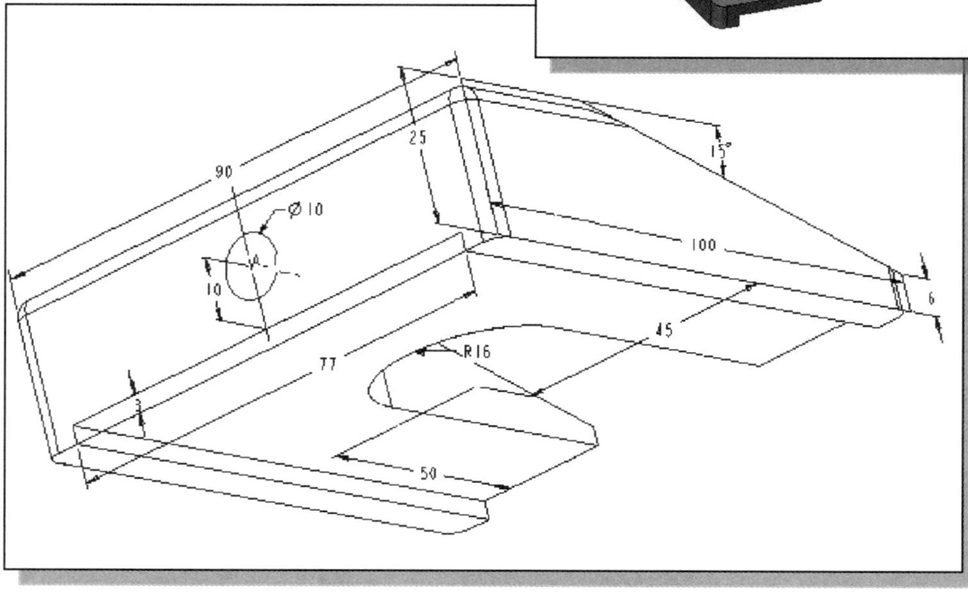

(c) **Lifting Block** (Rounds & Fillets: R3)

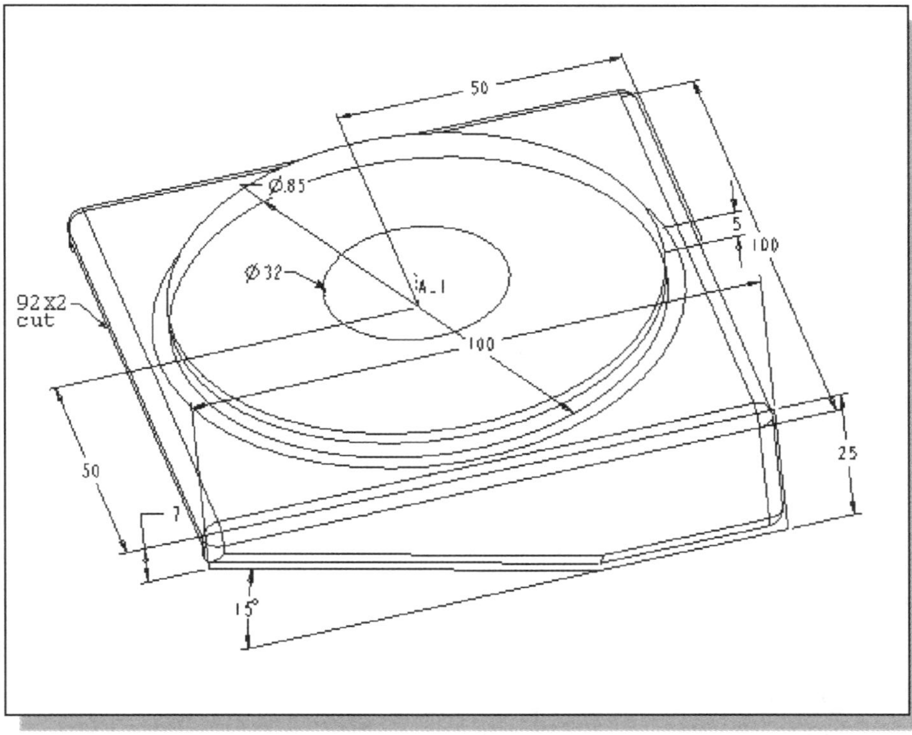

(d) **Adjusting Screw** (M10 × 1.5) (Use the **Threads** or **Coil** command to create the threads.)

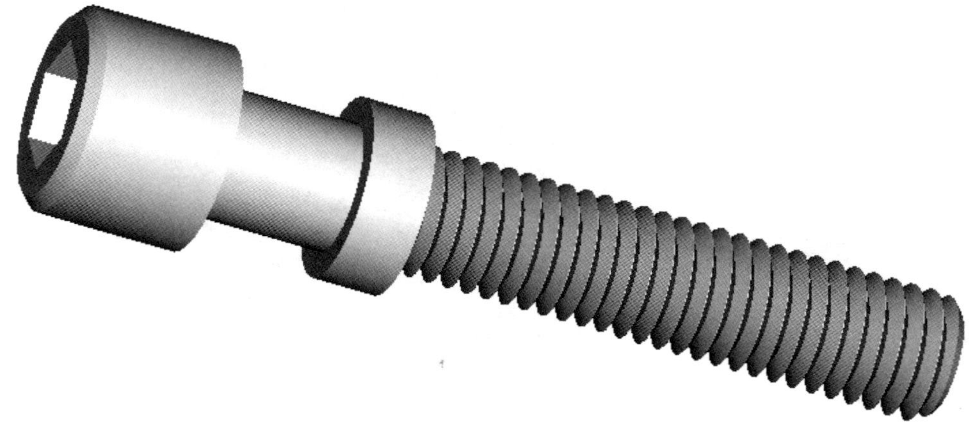

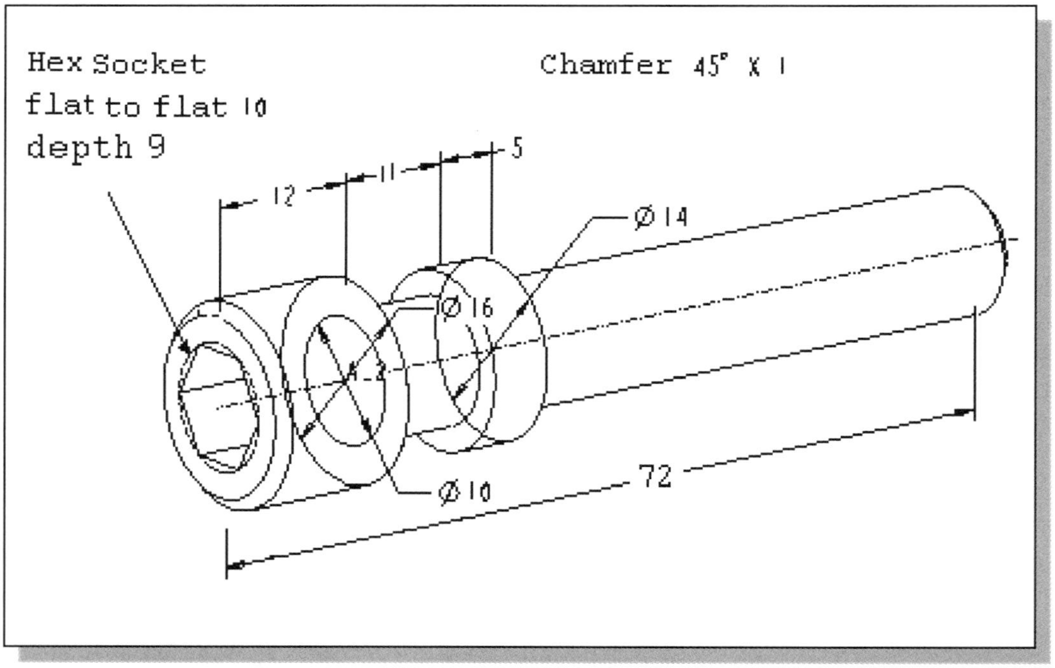

Basic Motion Analysis: The **Sliding Block** is to slide on top of the **Base Plate** as the adjusting screw is turned. The **Lifting Block** is pushed against the **Base Plate** and will slide up and down as the Sliding Block moves.

2. **Toggle-Clamp Assembly**

(Create a set of detail and assembly drawings. All dimensions are in inches.)

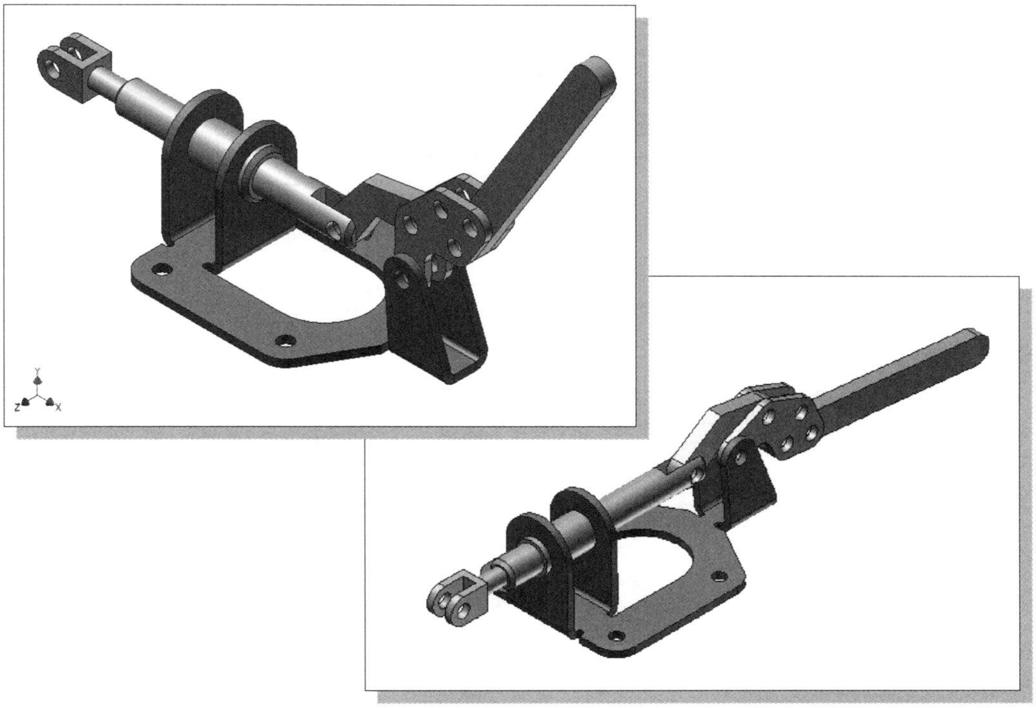

(a) **Sheet Metal Base**

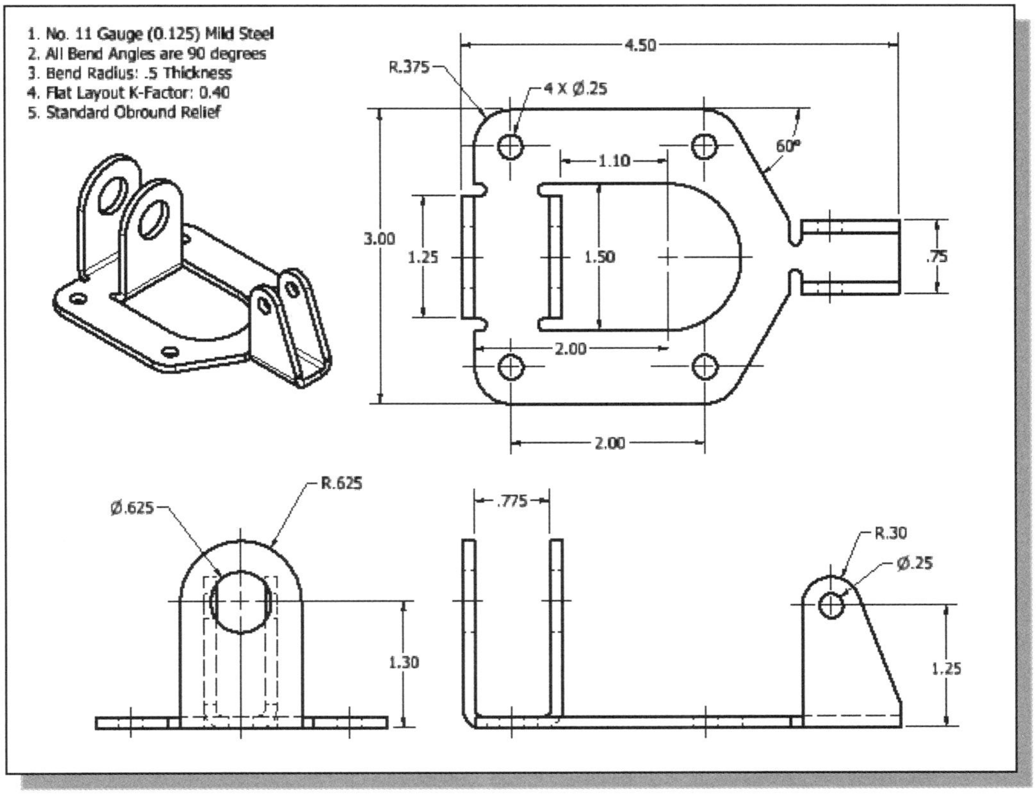

(b) **Connector**

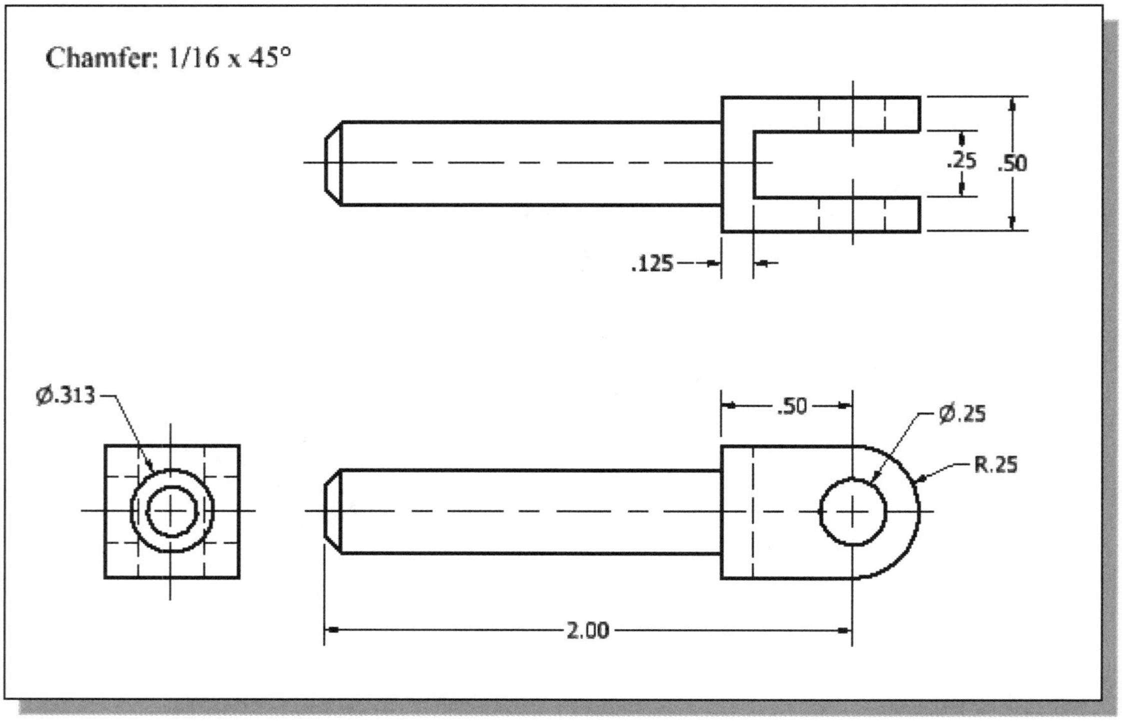

(c) **Handle**

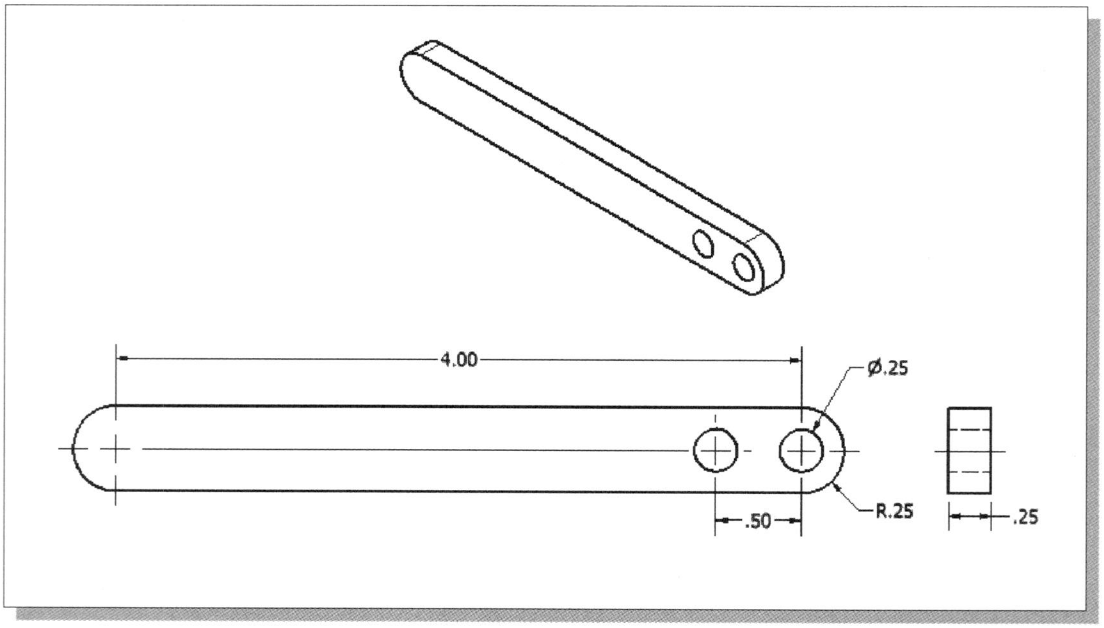

(d) **Joint Plate**

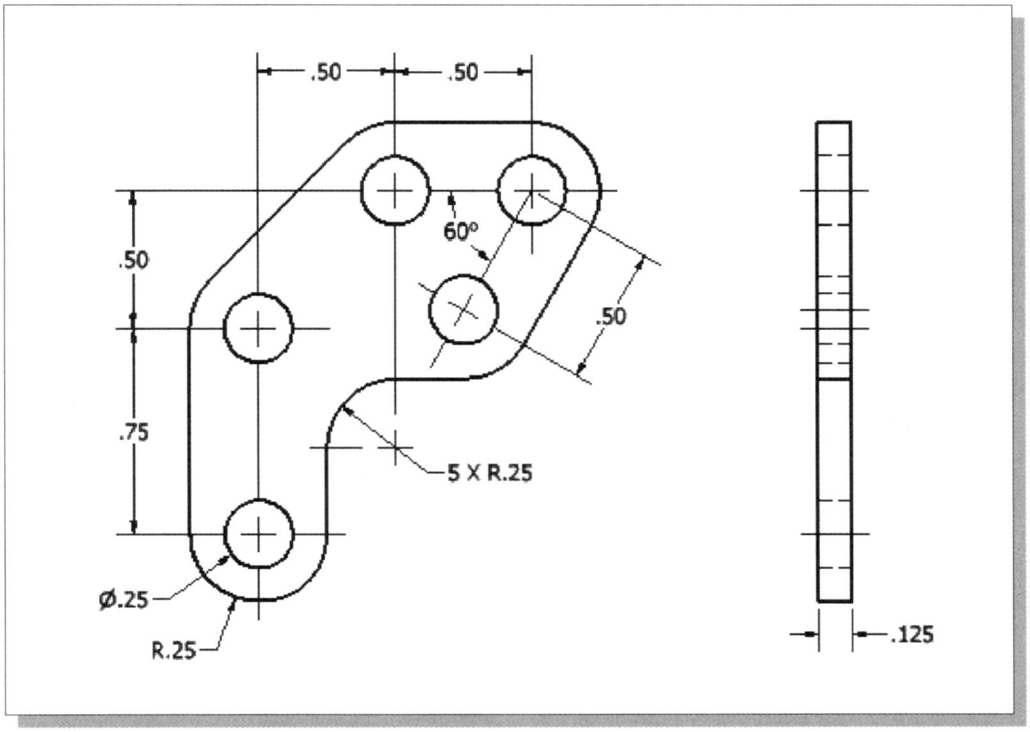

(e) **V-Link**

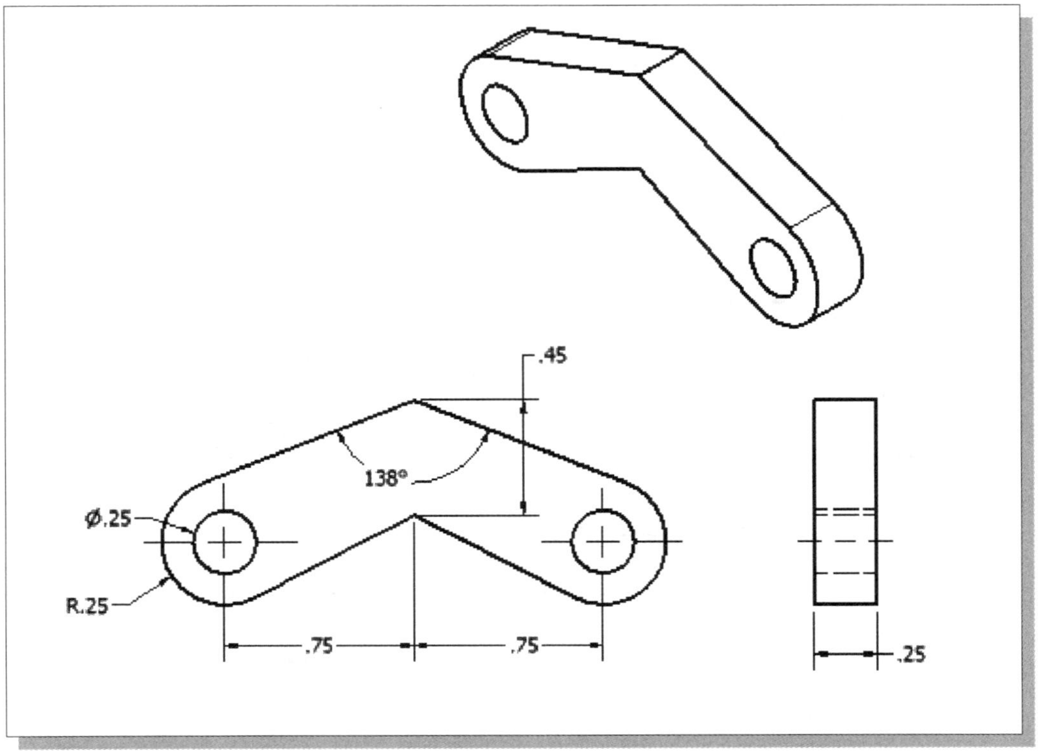

(f) **Rod**

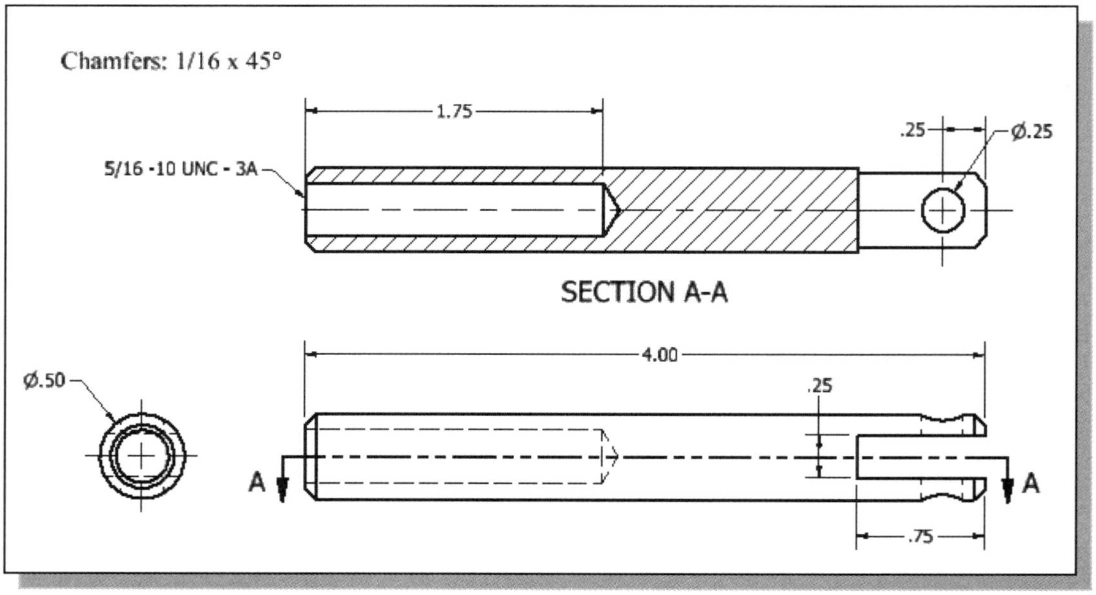

(g) **Bushing**

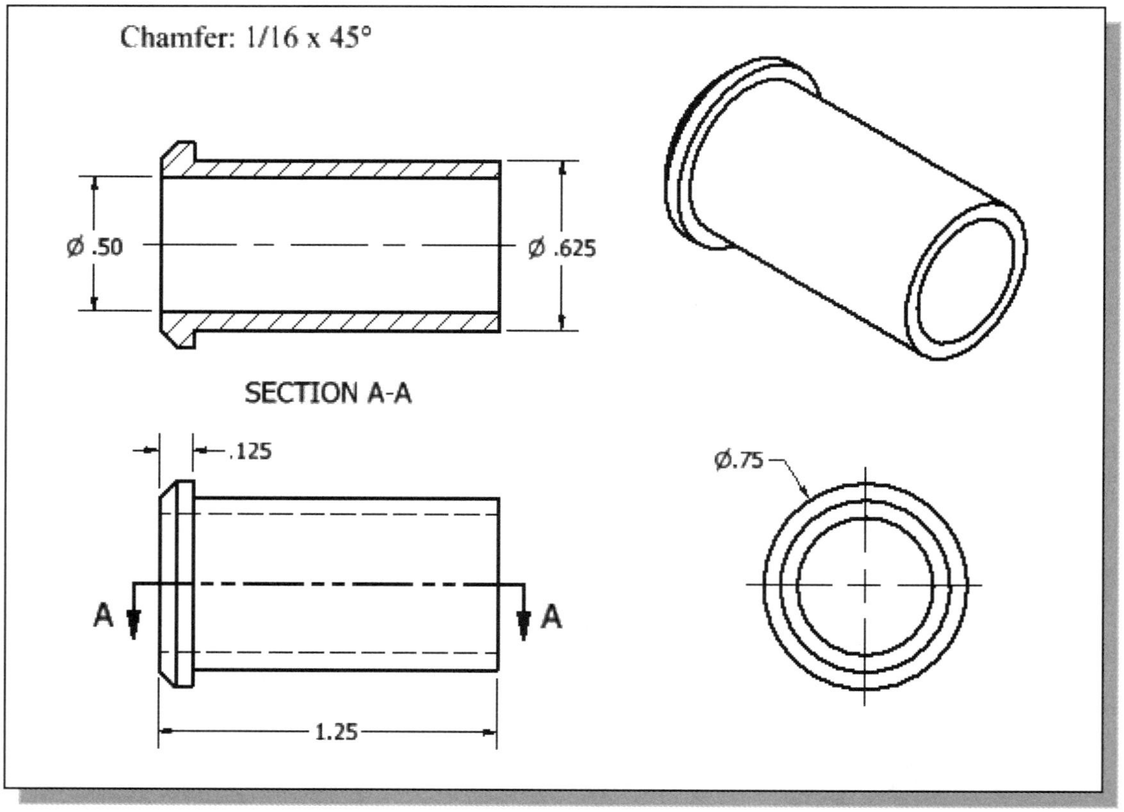

Chapter 16
2D Design Reuse, Collision and Contact

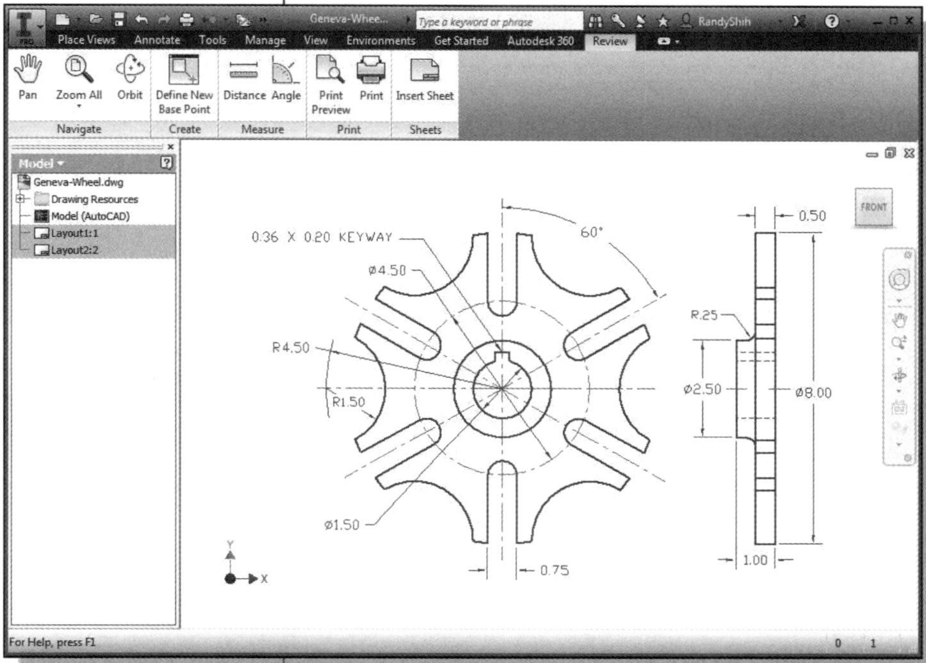

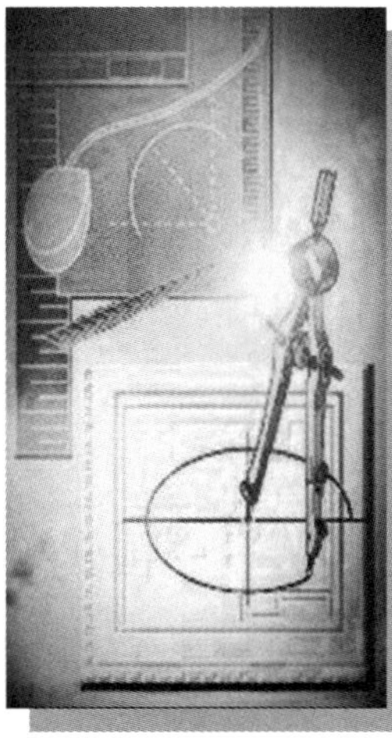

Learning Objectives

- Understand the 2D Design Reuse Concepts
- Open AutoCAD DWG Files in Autodesk Inventor
- Measure Distances and Angles
- Reuse 2D AutoCAD Files to Create a Solid Model in Autodesk Inventor
- Use the Autodesk Inventor Contact Solver
- Utilize the Autodesk Inventor Suppression Option

Autodesk Inventor Certified User Exam Objectives Coverage

Parametric Modeling Basics

Section 5: Assemblies

Objectives: Creating Assemblies, Viewing Assemblies, Animation Assemblies, Adaptive Features, Parts, and Subassemblies.

Section 7: Visualization

Objectives: Create Rendered Images, Animate an Assembly.

Introduction

In this lesson, we will examine the procedures to reuse existing 2D AutoCAD data in Autodesk Inventor. The AutoCAD DWG file format is one of the most commonly used CAD formats in industry; in 2020, it was estimated that there were over 6 million AutoCAD users worldwide.

In parametric modeling, 2D sketches are commonly used in building 3D features. In Autodesk Inventor, we can use existing 2D AutoCAD data exactly as it is, or as a reference, and thus create the same design in 3D. This process is generally known as **2D Design Reuse**. One of the easiest ways to reuse existing 2D AutoCAD data is to open the AutoCAD DWG file in Autodesk Inventor, and simply Copy and Paste the desired geometry into a new part model sketch. When copying and pasting the 2D geometry, if the associative dimensions are included in the selection, they will be converted into parametric dimensions. The pasted geometry can then be modified and edited as if it were created in Autodesk Inventor.

Here are some guidelines for opening DWG files in Autodesk Inventor:

1. The AutoCAD **Model Space** and **Layouts** are accessible through the **Inventor Browser window**.

2. The objects that are in *Model Space* are **Read-Only** in Autodesk Inventor. By default, **Read-Only Drawing Review** mode is activated when an AutoCAD DWG file is opened.

3. The layouts are displayed as drawing sheets in Autodesk Inventor and can be **viewed**, **plotted** and **deleted**.

4. The objects that are in *Layouts* can be **deleted**. New geometry can be **added** in *Layouts*, using the Inventor annotation tools and sketching tools.

5. Additional **Views** and **Annotations** can be created on an *AutoCAD Layout* with Inventor sketching tools. The modified file can be saved in Autodesk Inventor, still as an AutoCAD DWG file.

6. The data in the DWG file, either in *Model Space* or *Layouts*, are all viewable in Inventor. The Inventor **viewing** and **measuring** commands are available for determining distances and angles.

7. The data in the DWG file, either in *Model Space* or *Layouts*, can be **selected** to **Copy** and **Paste** into any type of **Inventor sketches** in part, assembly and drawing files.

In this chapter, we will also look at the use of the **Inventor Contact Solver** to detect, and resolve, *collisions* among parts in conjunction with the ***Joint*** tool. The Inventor *Contact Solver* allows us to produce very realistic animations for more complex assemblies.

The Geneva CAM Assembly

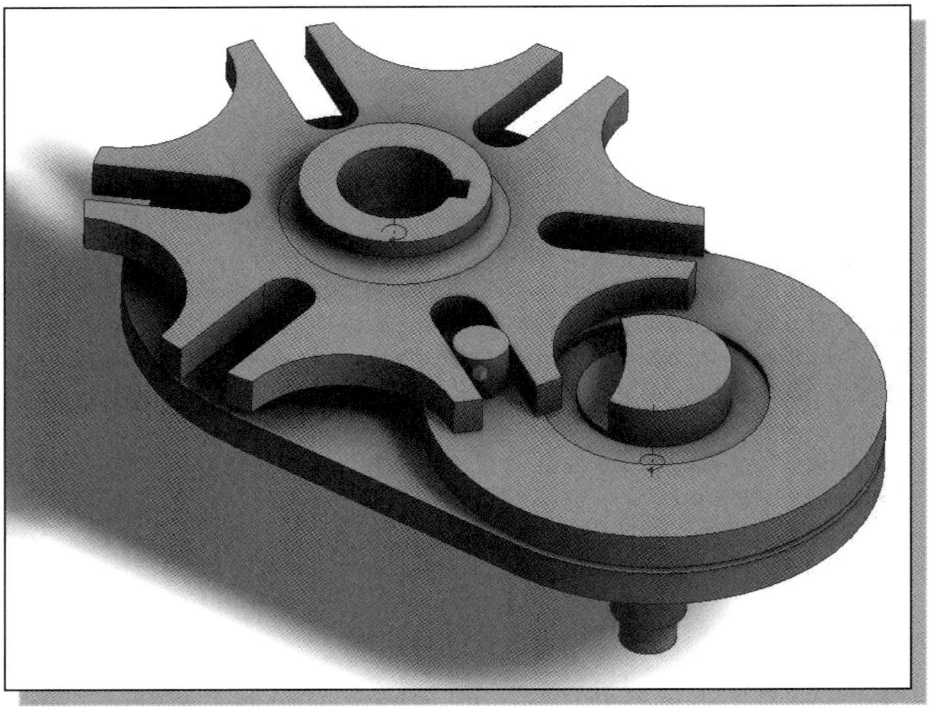

Download the Geneva-Wheel DWG File

To illustrate the concepts and procedures of reusing existing 2D AutoCAD data, we will first download the *Geneva Wheel DWG* drawing from the SDC Publications website.

1. Launch your internet browser, such as *MS Edge* or *Mozilla Firefox*.

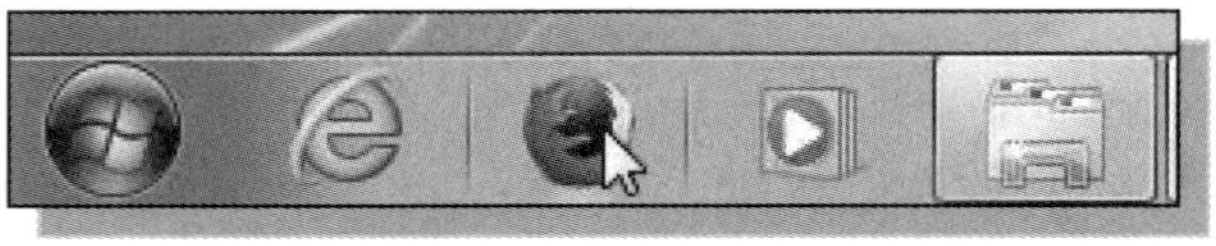

2. Visit this book's website through www.SDCPublications.com, and download the **Geneva-Wheel.dwg** file.

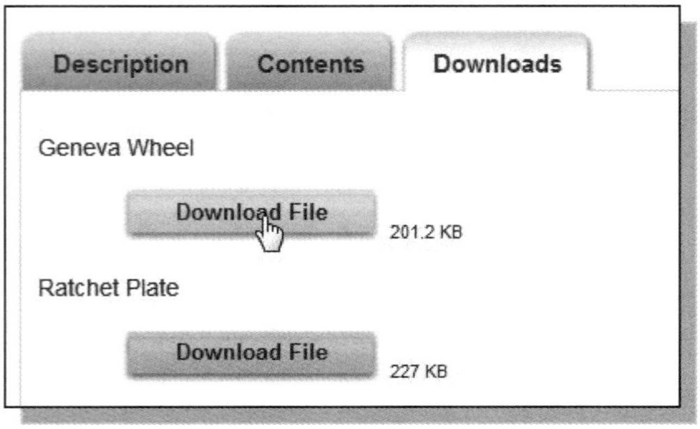

3. On your own, save the *AutoCAD* ***Geneva-Wheel.dwg*** to the *Inventor* Geneva Project folder. (Hint: depending on the internet browser, you may need to use the right-mouse-button to save the file.)

Opening AutoCAD DWG File in Inventor

1. Select the **Autodesk Inventor** option on the *Start* menu or select the **Autodesk Inventor** icon on the desktop to start Autodesk Inventor. The Autodesk Inventor main window will appear on the screen.

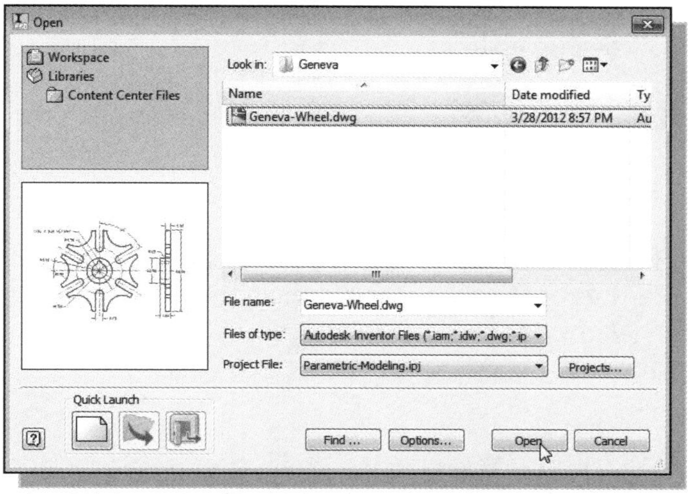

2. Once the program is loaded into memory, select the **Open File** option.

3. Select the **Geneva-Wheel.dwg** file with a single click of the left-mouse-button in the *File name* list box.

4. Click on the **Open** button in the *Open* dialog box to proceed with opening the file.

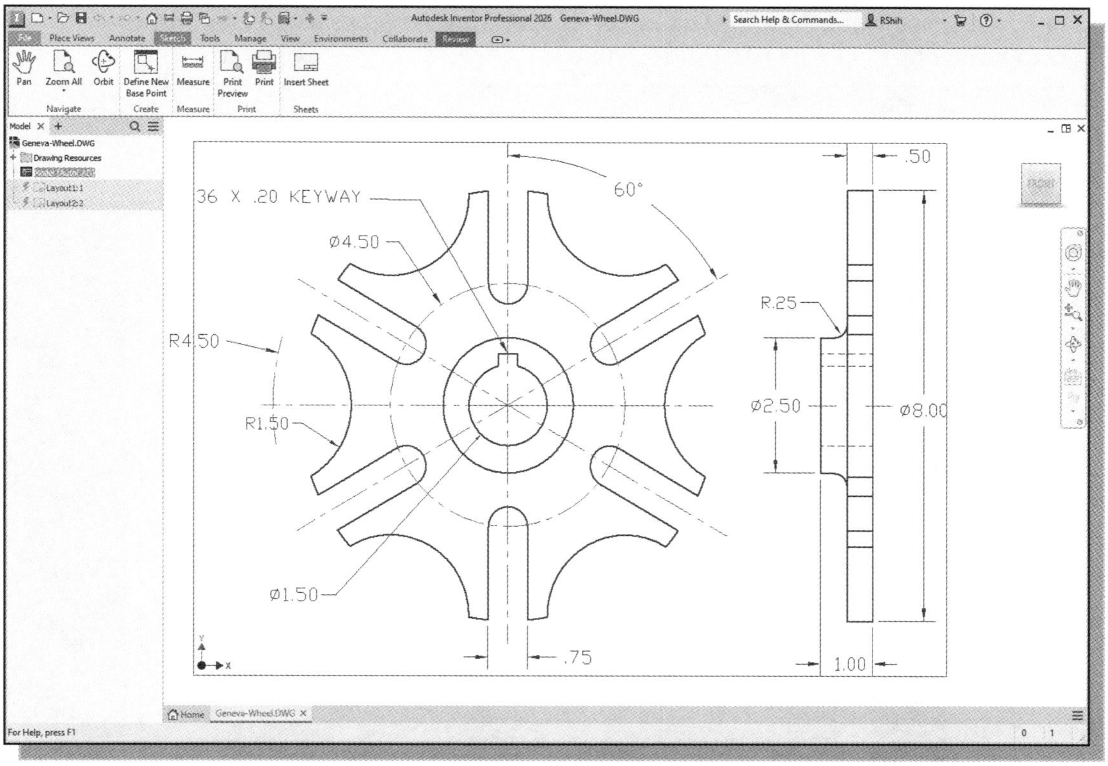

❖ By default, the **Read-Only Drawing Review** mode is activated when an *AutoCAD* DWG file is opened. Note that objects that are in *Model Space* are also **Read-Only** in Autodesk Inventor.

Switch to the AutoCAD DWG Layout

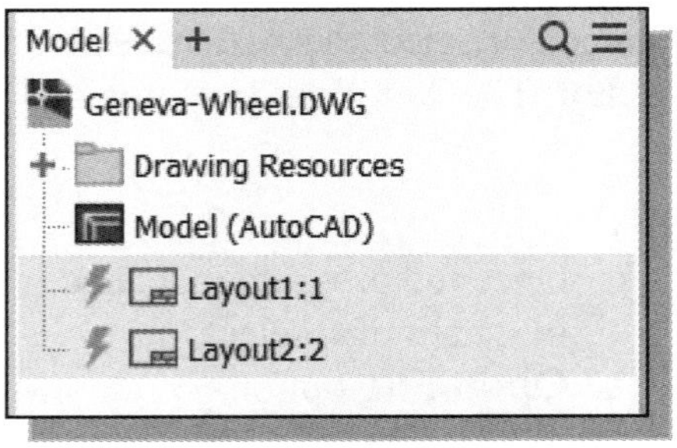

- Note that in the *browser* window, the three items listed are corresponding to the *AutoCAD Model Space* and *Layouts* associated with the ***Geneva-Wheel*** drawing.

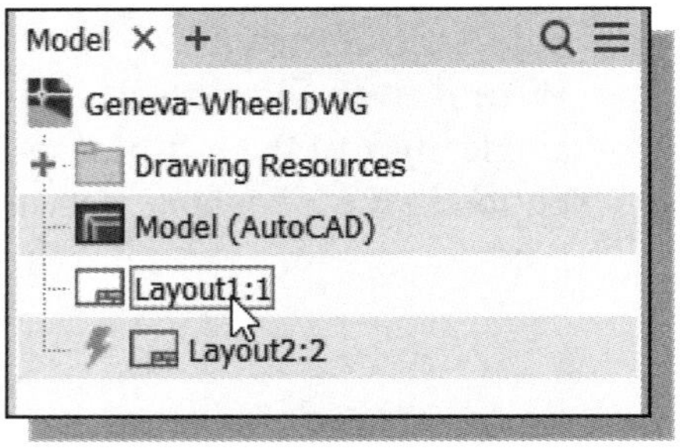

1. In the *browser* window, **double-click** on **Layout1** to open the associated *AutoCAD DWG Layout.*

➢ Inventor now opens the AutoCAD **Layout1** in the *Drawing Sheet* mode. Note that the layout contains a title block and one viewport as shown in the below figure. The Layout space is the AutoCAD's Paper Space, which represents a 2D drafting environment.

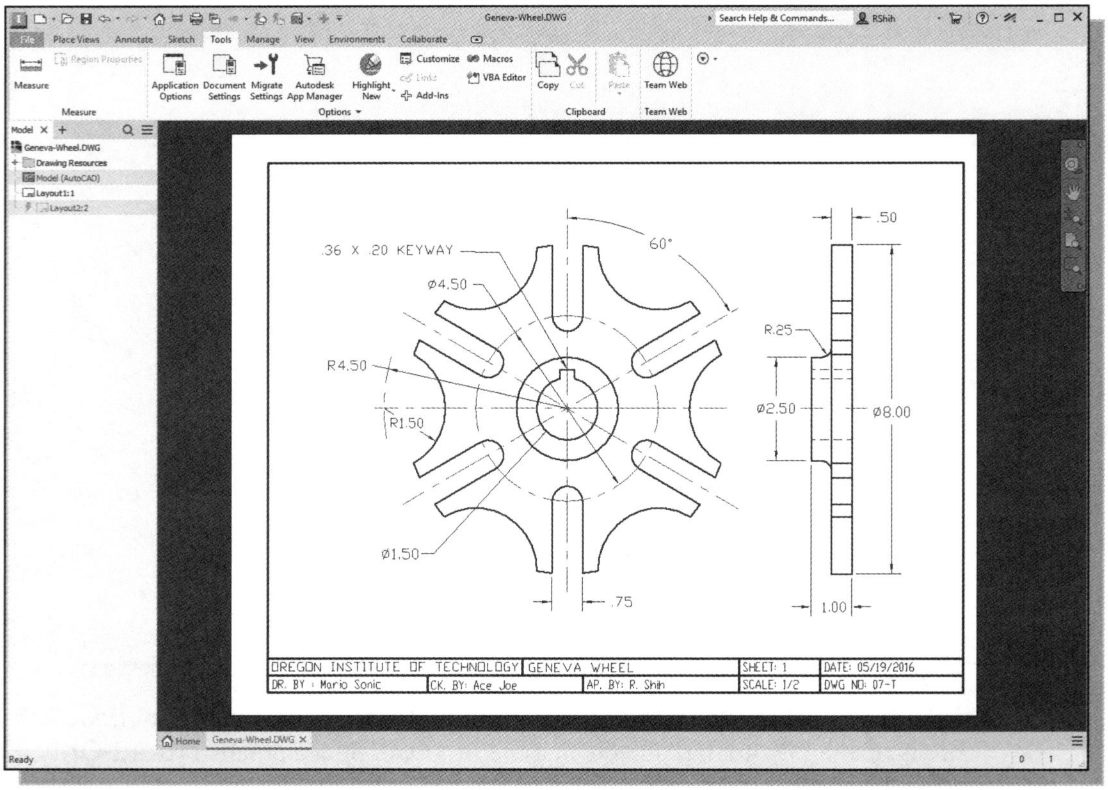

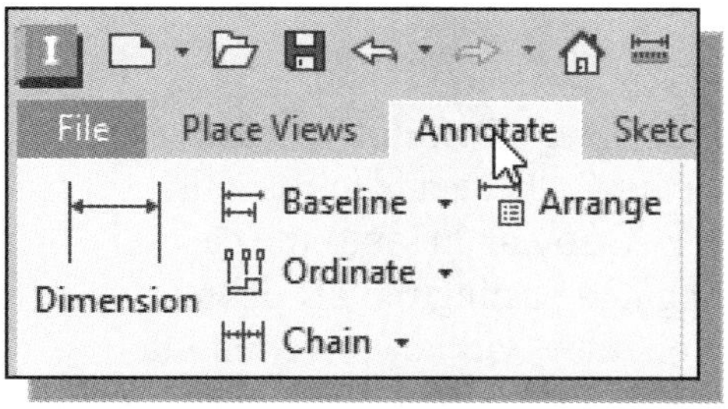

2. Click on the *Annotate* tab to switch to the annotation toolbar panel as shown.

➤ Notice the available Annotation commands, such as the **General Dimension** command and the **Center Mark** command.

➤ Note that objects in *Layouts* can be **deleted**. New geometry can also be **added** in *Layouts*, using the Inventor annotation tools and sketching tools.

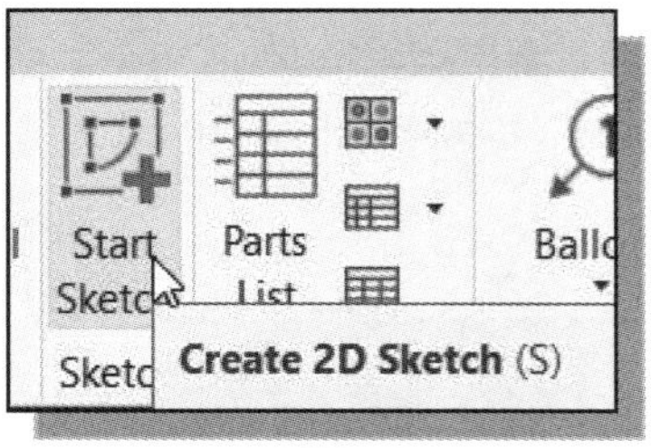

3. In the *Sketch* toolbar, click **Start Sketch** as shown in the figure.

➤ The *2D Drawing Sketch* panel appears above the *browser* window.

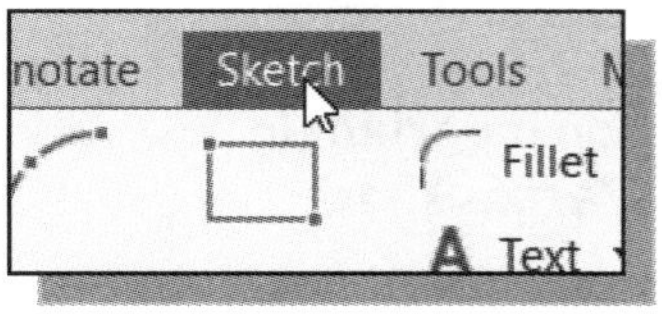

4. If necessary, click on the highlighted *Sketch* tab to switch to the 2D Sketch toolbar panel as shown

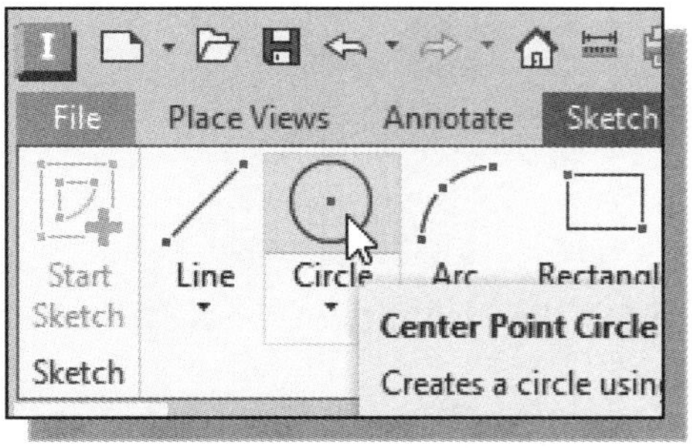

5. Select the **Center Point Circle** icon as shown.

6. Move the cursor near the center of the *front* view and notice that no *SNAP* option is available. Any new entities added will remain only in the *Layout*, not in the *Model Space*.

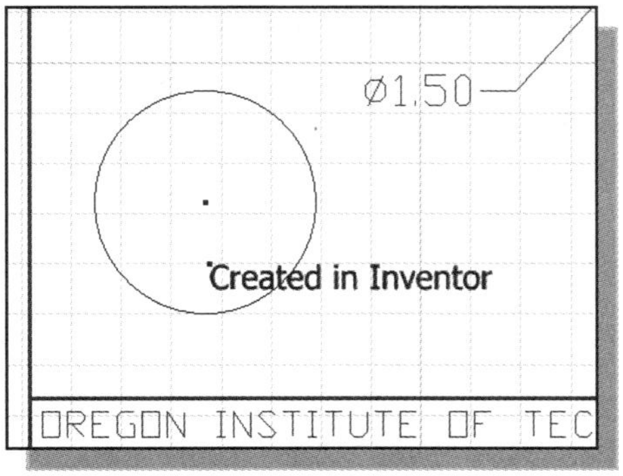

7. On your own, create a circle and some text at the lower left corner of the viewport as shown.

❖ Note that Inventor can be used to create new geometry and annotations in *AutoCAD Layouts*.

8. In the toolbar area, click **Finish Sketch** to exit the Drawing Sketch mode.

2D Design Reuse

The main concept of **2D Design Reuse** is to reuse existing 2D AutoCAD data exactly as it is. One of the easiest ways to reuse existing 2D AutoCAD data is to open the AutoCAD DWG file in Autodesk Inventor, and simply Copy and Paste the desired geometry into a new part model sketch. When copying and pasting the 2D geometry, if the associative dimensions are included in the selection, they will be converted into parametric dimensions. The pasted geometry can then be modified and edited as if it were created in Autodesk Inventor.

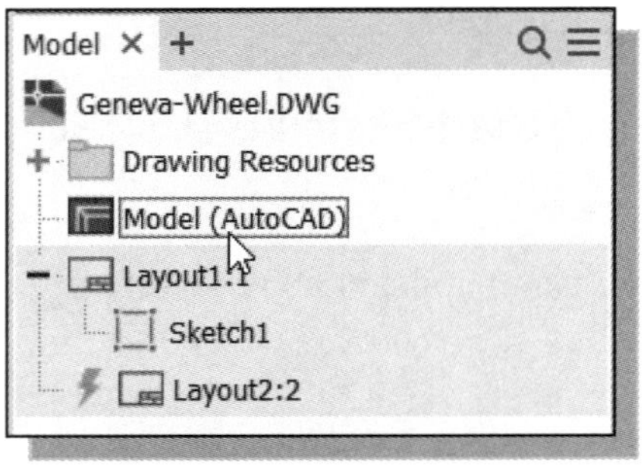

1. In the *browser* window, **double-click** on the **Model (AutoCAD)** icon to switch back to *AutoCAD Model Space.*

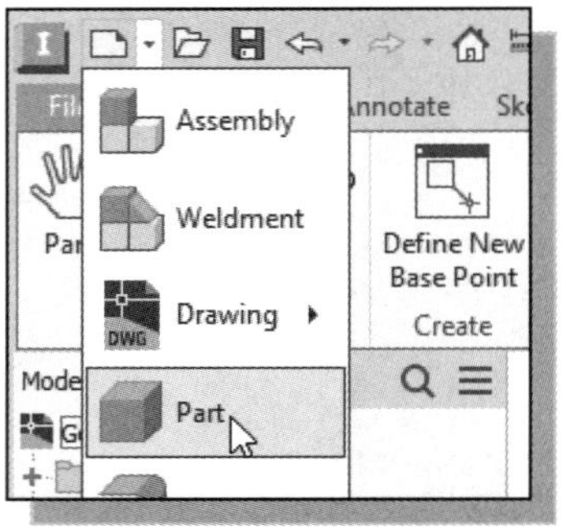

2. Start a new part file by clicking on the **triangle** next to the **New File** icon and choose **Part** as shown.

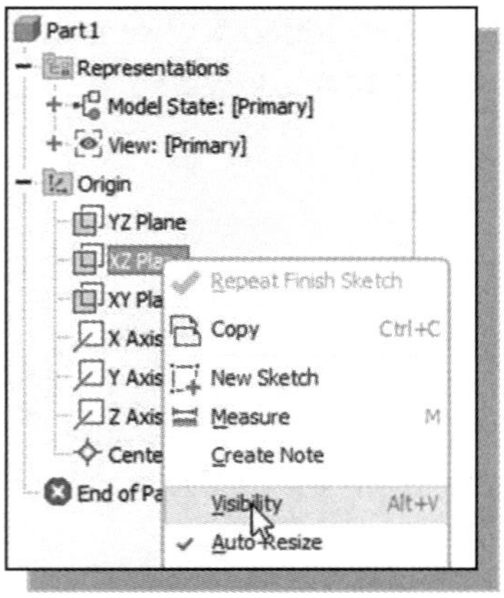

3. On your own, switch *ON* the visibility of the **XZ Plane** as shown.

4. Reset the display to *isometric* view by clicking on the Home icon of the ViewCube as shown.

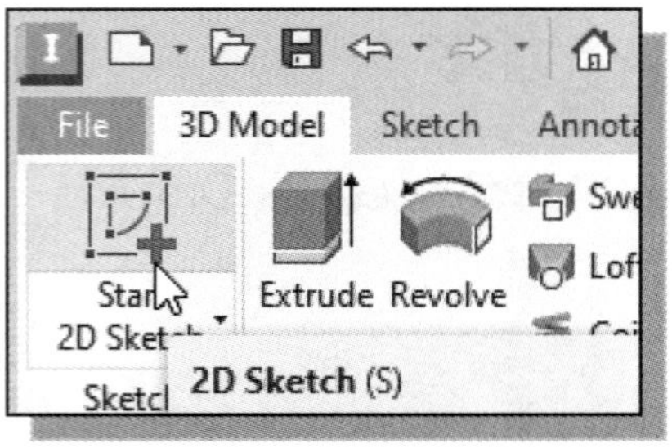

5. Click **Start 2D Sketch** to enter the *2D Sketching* mode.

6. Select the **XZ plane** to align the sketching plane.

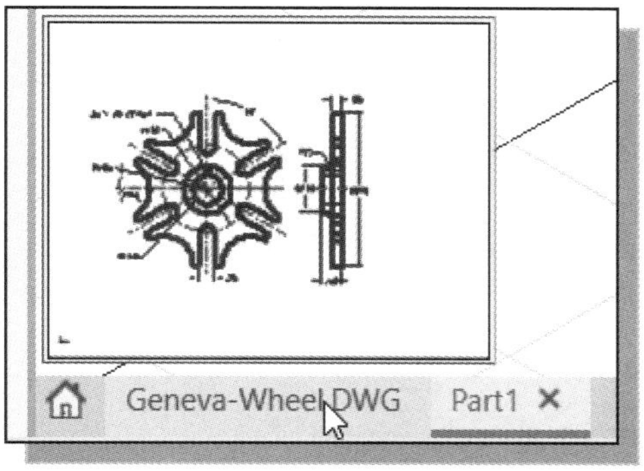

7. Switch to the ***Geneva-Wheel.DWG*** window by clicking on the associated file name tab as shown.

8. Select all the entities in the *front* view by using the Inventor selection window as shown. (Click and drag with the left-mouse-button to enclose all entities inside the window.)

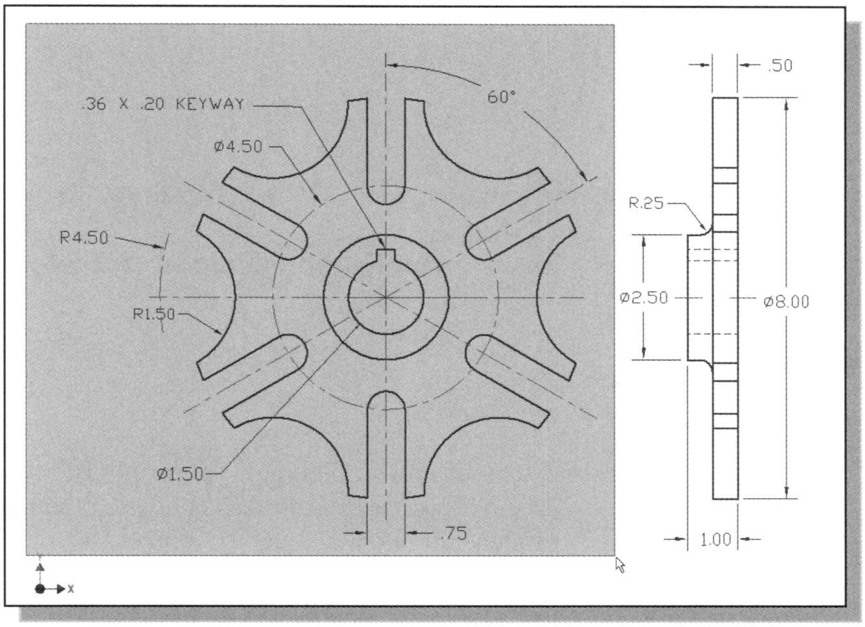

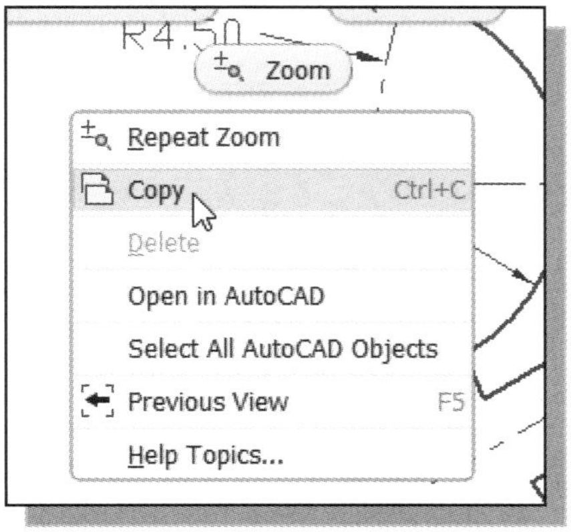

9. Right-mouse click once to bring up the option menu and select **Copy** or use the quick-key option **[Ctrl+C]**.

10. Switch to the **Part1** file window by choosing the associated name in the **tabs** as shown.

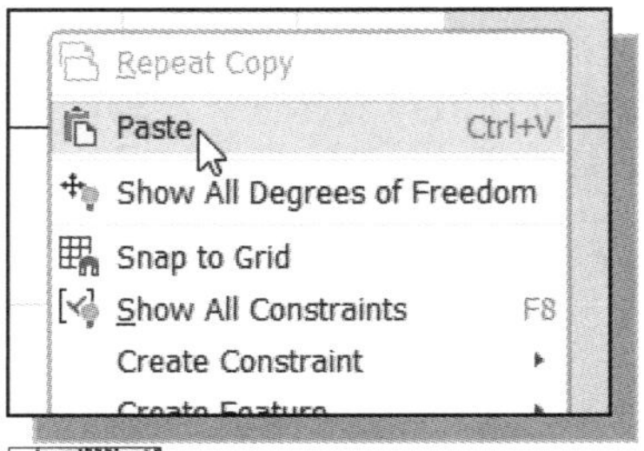

11. Select **Paste** in the **Edit** pull-down menu or use the quick-key option **[Ctrl+V]**.

12. Place the copied entities on the **XZ Plane** by clicking the left-mouse-button at an arbitrary location.

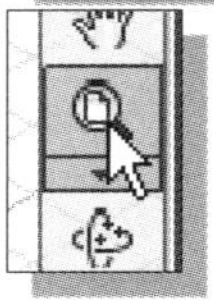

13. Click **Zoom All** in the *Display* toolbar as shown.

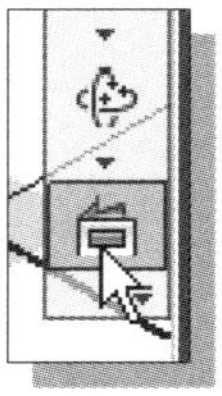

14. Select any of the imported geometry to align the view angle.

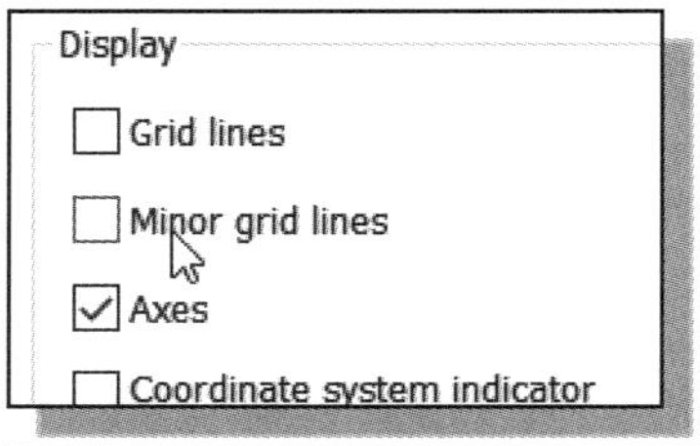

15. On your own, turn *OFF* the **Grid lines** display options as shown.
[Tools → Application Options → Sketch]

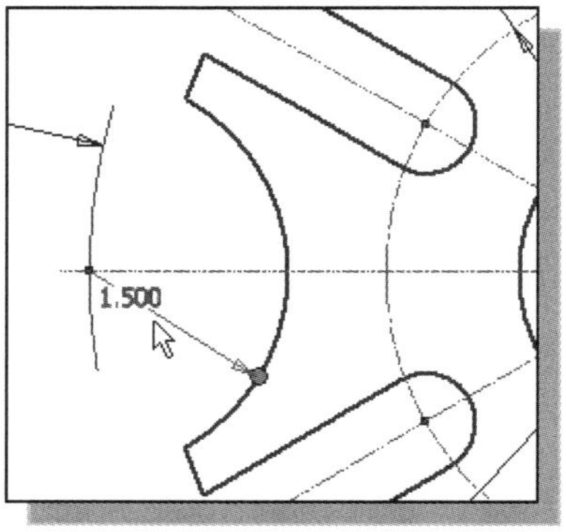

16. Move the cursor on top of the **radius 1.5** dimension and notice that Inventor has converted the radius dimension into a *parametric dimension.*

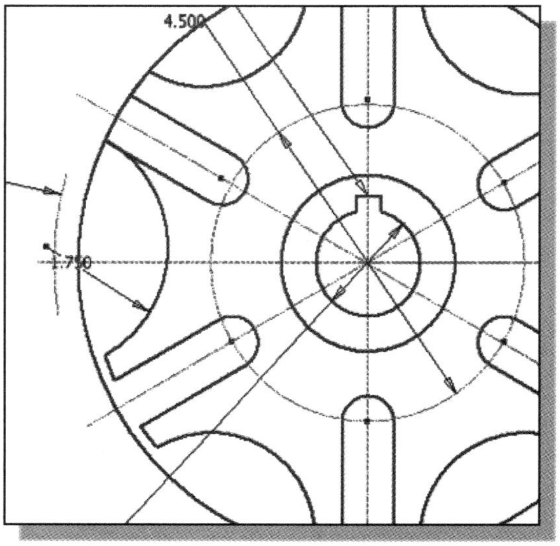

17. On your own, adjust the radius dimension to **1.75** and notice that some of the other geometric entities become distorted when the size of the related arc is updated.

❖ On the *Status Bar* of the Inventor window, Inventor indicates that over 100 additional dimensions are needed to fully define the current geometry.

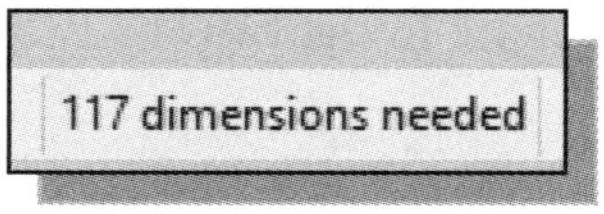

❖ By copying and pasting the entire view, we have extra entities that are not necessary to create the 3D part, such as the note for the keyway and the centerlines. As far as building the 3D model is concerned, we can take advantage of the symmetrical nature of the design and simply copy a relatively small portion of the original geometric entities.

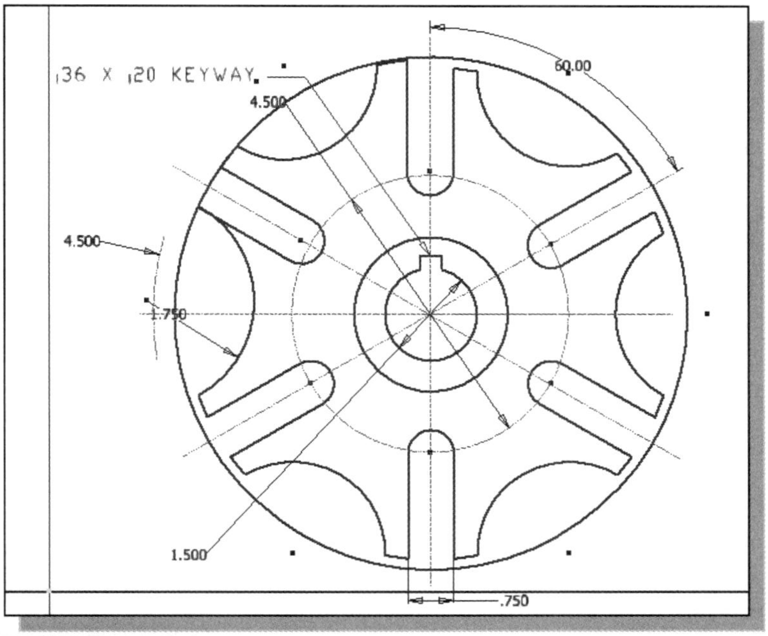

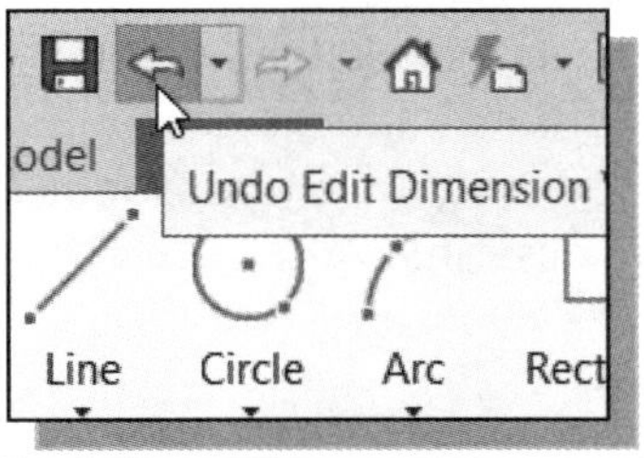

18. In the *Quick Access* toolbar, click **Undo** once to undo the last step.

19. Continue to click **Undo** until all of the pasted geometry entities are removed from the screen.

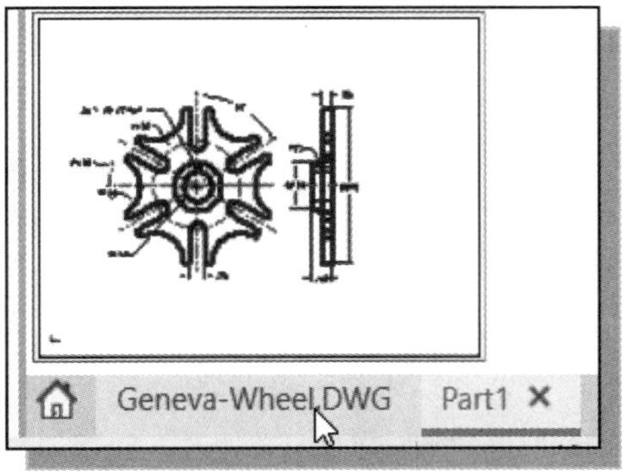

20. Switch to the **Geneva-Wheel.DWG** window by clicking on its *window*, in the toolbar area, or selecting the associated *file tab* as shown.

21. Select any set of four entities that form a 30 degree section as shown. (Hint: Hold down the **[Ctrl]** key while clicking on the entities.)

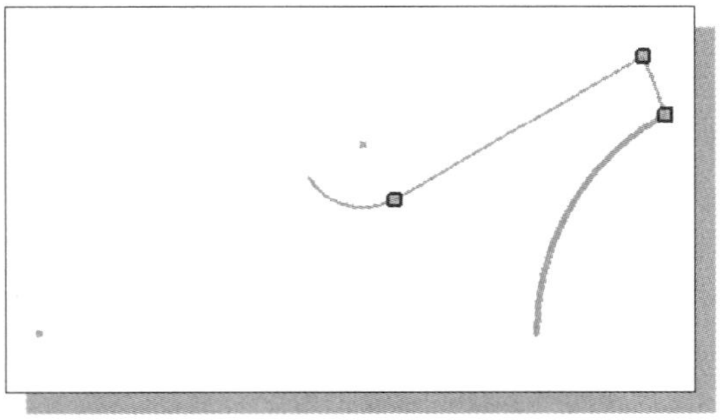

22. Right-mouse click once to bring up the option menu and select **Copy** or use the quick-key option **[Ctrl+C]**.

Complete the Imported Sketch

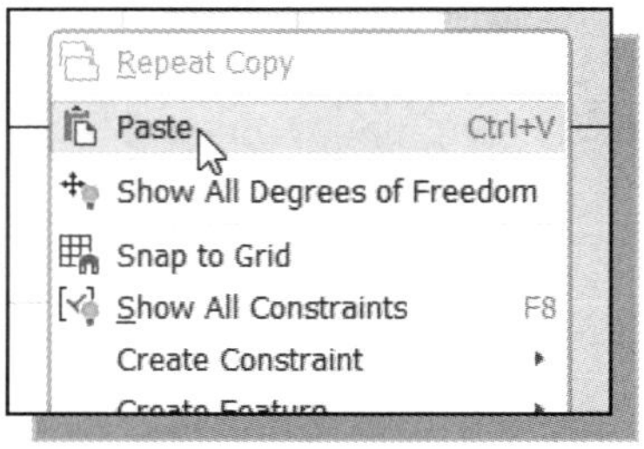

1. On your own, **Copy** and **Paste** the four entities onto the **XZ Plane** of the 2D sketch in the **Part1** file window.

❖ To create an extruded feature, two additional lines are necessary to form a closed region.

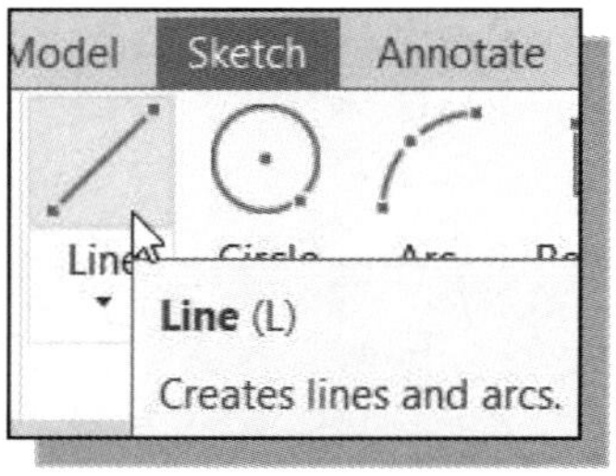

2. Select the **Line** option in the *2D Sketch* panel as shown.

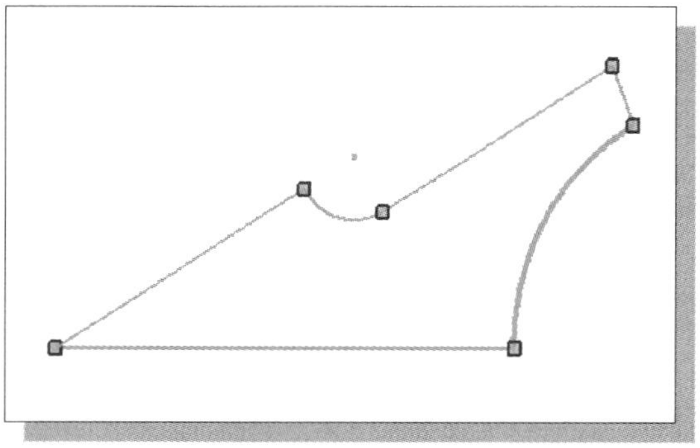

3. Add the two lines connecting to the lower left center point and apply coincident constraint to all corners.

❖ Note that we could simply create the 3D model with the current sketch, without adding any additional dimensions or constraints. However, a fully constrained sketch is more desirable.

4. On your own, create the following six dimensions as shown.

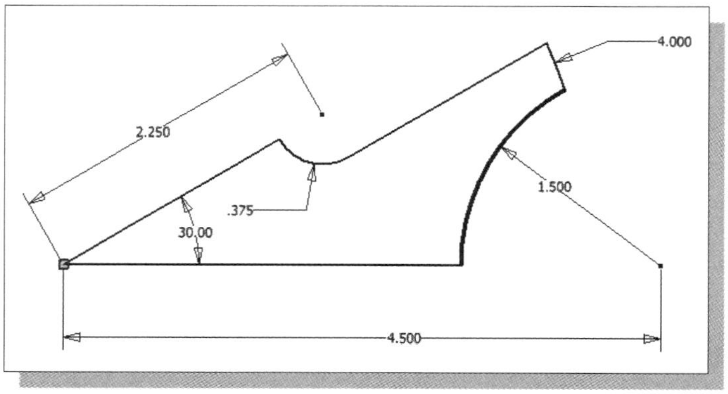

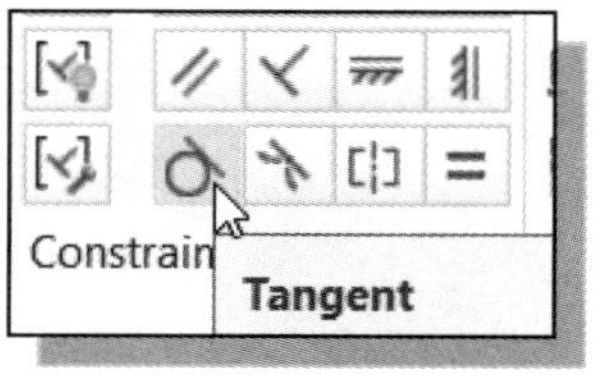

5. Select the **Tangent** constraint in the *Constrain* panel as shown.

6. On your own, apply a tangent constraint between the top-right inclined line and the .375 radius in the above figure.

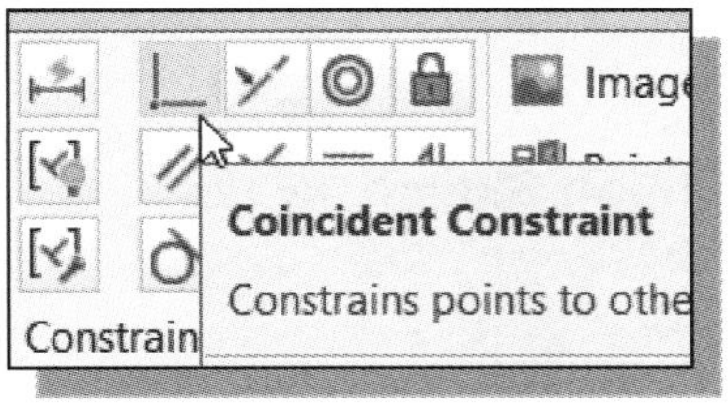

7. Select the **Coincident Constraint** in the *Constrain* panel as shown.

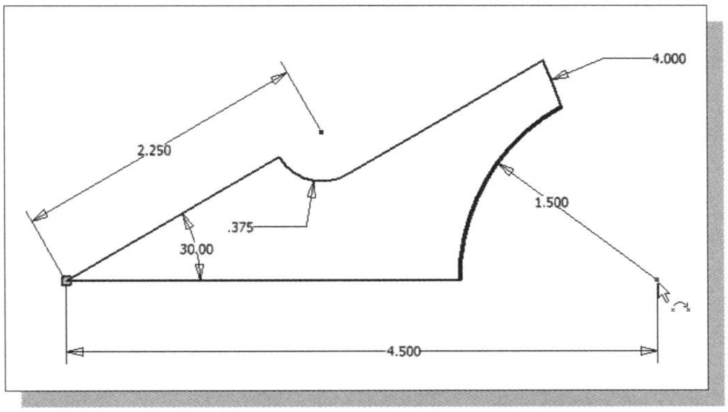

8. Select the center point on the right as shown.

9. Select the bottom horizontal line to align the center of the lower arc to the horizontal line.

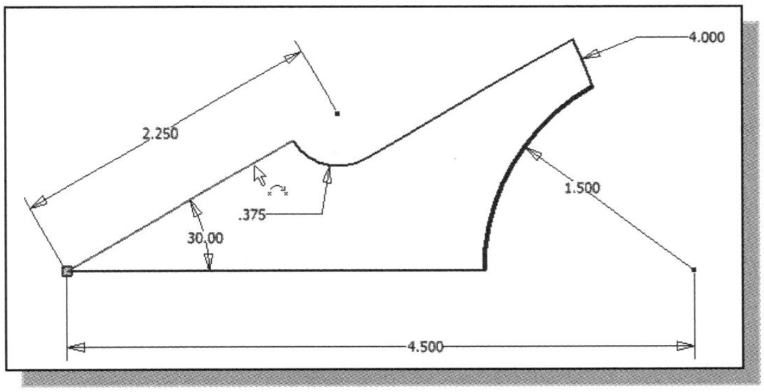

10. Repeat the above steps and align the center point of the radius 0.375 arc to the inclined line as shown.

11. On your own, add a **parallel constraint** to the two parallel inclined lines.

12. On your own, align the lower left corner of the sketch to the origin as shown.

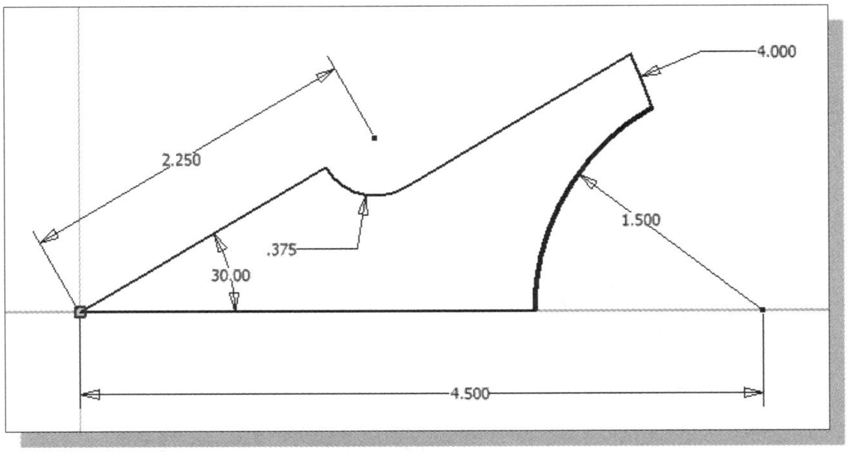

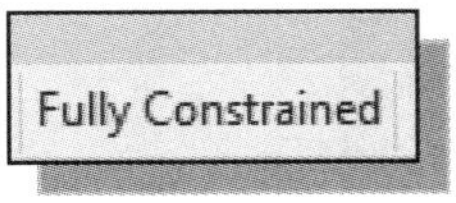

13. Note the sketch is now fully constrained, as indicated in the *Status Bar*.

Create the First Solid Feature

1. In the *Standard* toolbar, select **Finish Sketch**, by left-clicking on the icon, to exit the *2D Sketch Mode.*

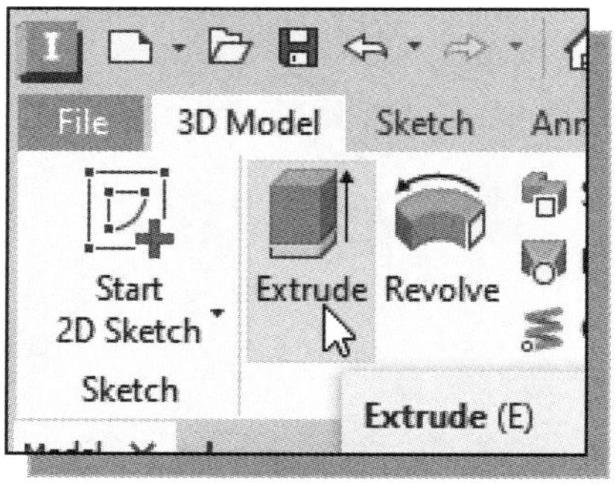

2. In the *3D Model* panel, select the **Extrude** command by left-clicking the icon.

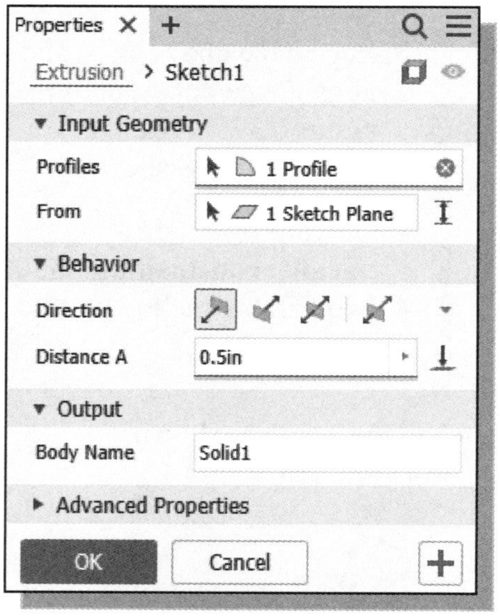

3. In the *Extrude* dialog box, enter **0.5** as the extrusion *Distance* as shown.

4. Click **OK** to accept the settings to create an extruded solid feature as shown.

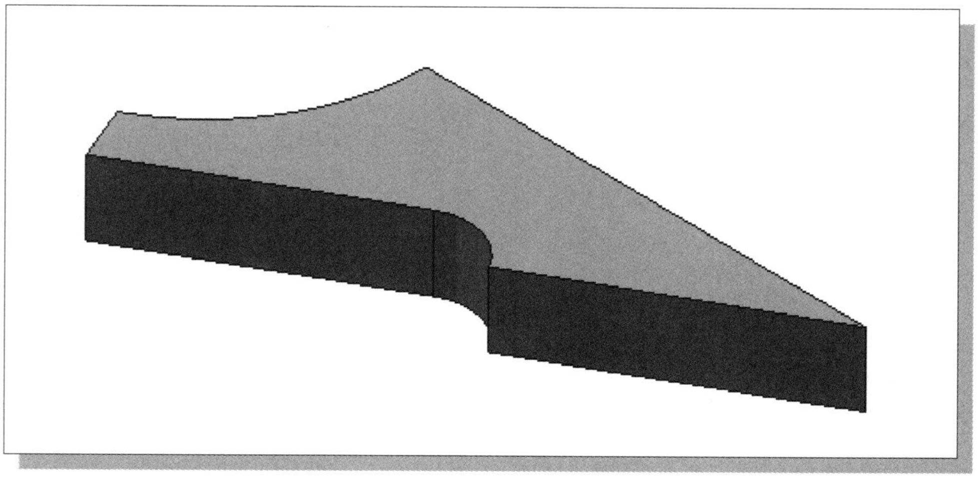

Create a Mirrored Feature

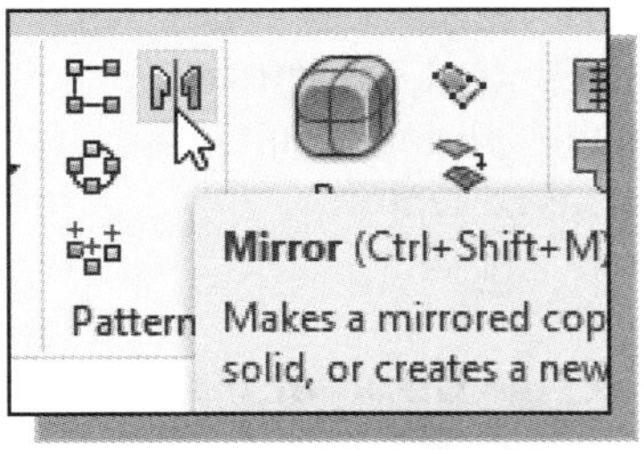

1. In the *Pattern toolbar* panel, select the **Mirror** command by left-clicking the icon.

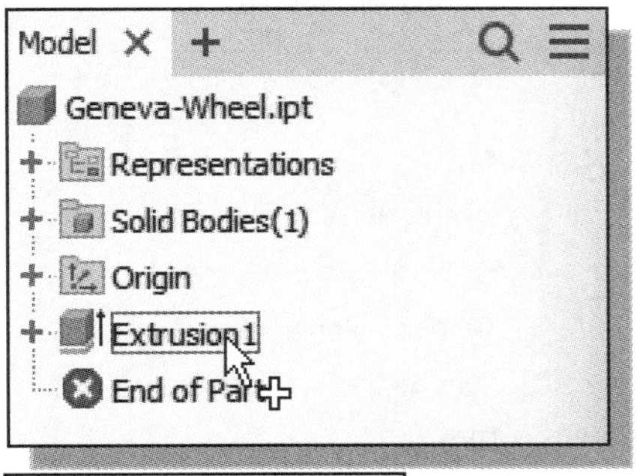

2. Select **Extrusion1** in the *browser* window as shown in the figure.

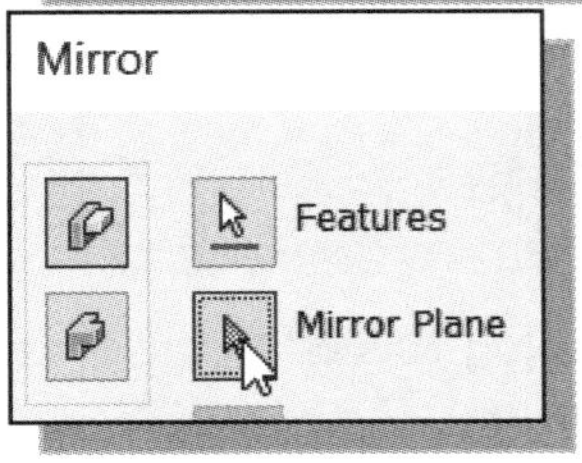

3. Click the **Mirror Plane** button to proceed to the next step, defining the *mirror image plane*.

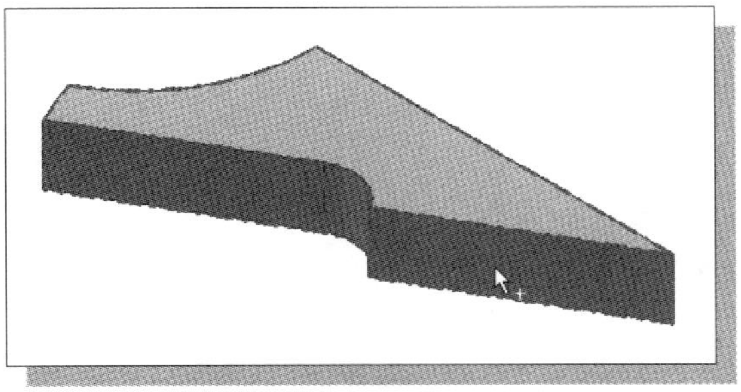

4. Select the small flat plane as shown in the figure.

5. Click **OK** to create the mirror image.

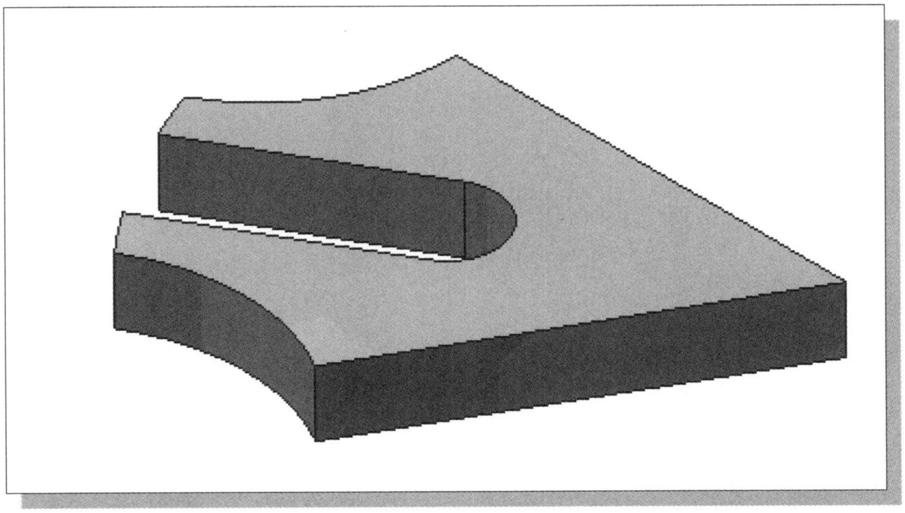

Circular Pattern

In Autodesk Inventor, existing features can be easily duplicated with the **Pattern** command. The patterned features are parametrically linked to the original feature; any modifications to the original feature are also reflected in the arrayed features.

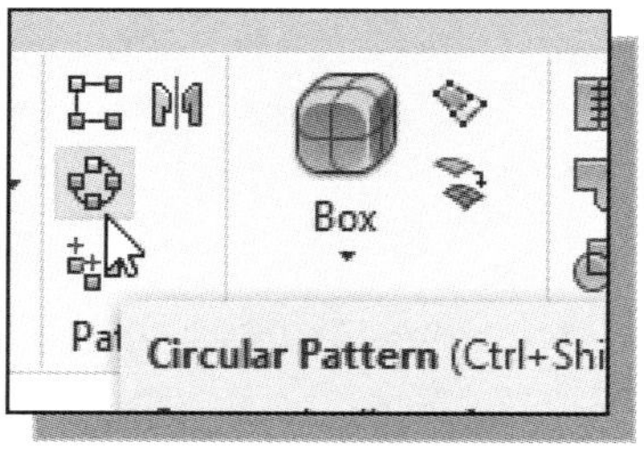

1. In the *3D Model* panel, select the **Circular Pattern** command by left-clicking once on the icon.

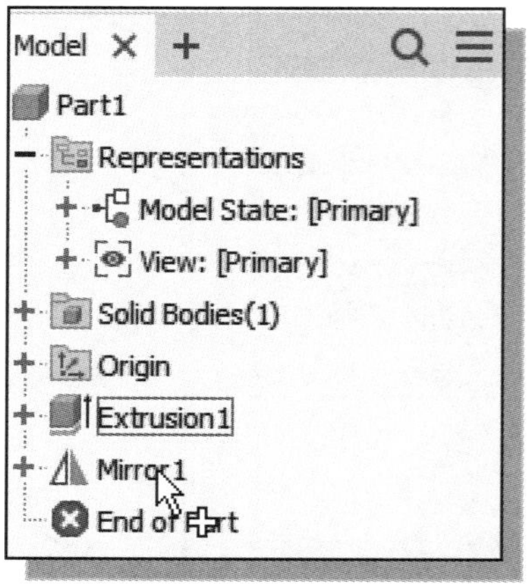

2. The message "*Select Feature to be arrayed*:" is displayed in the command prompt window. Select the **Mirror1** feature in the *browser* window as shown.

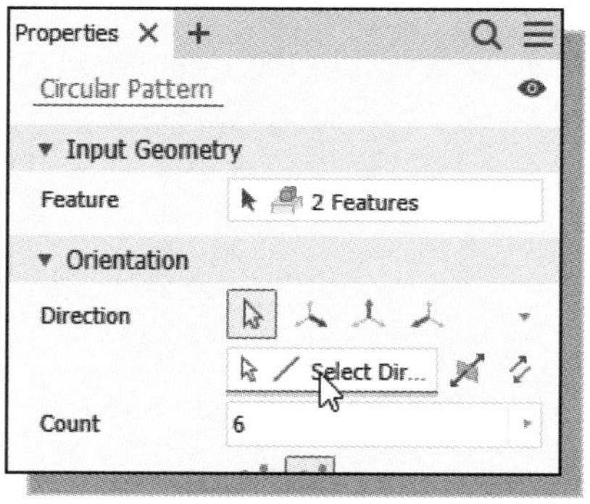

3. Click the **Select Axis** button to proceed to the next step, defining the *axis of rotation.*

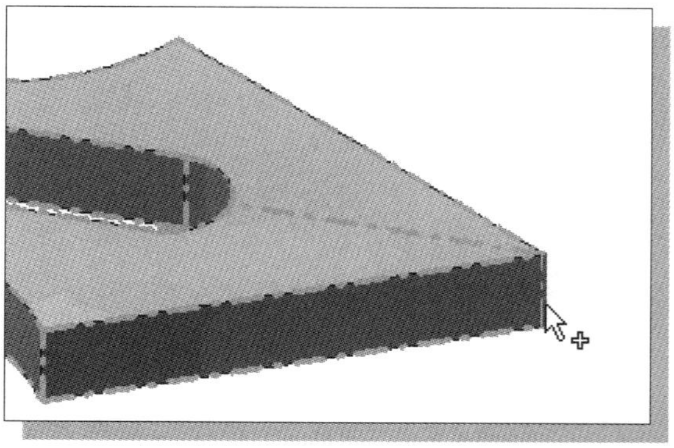

4. Select the sharp edge of the extruded feature as shown.

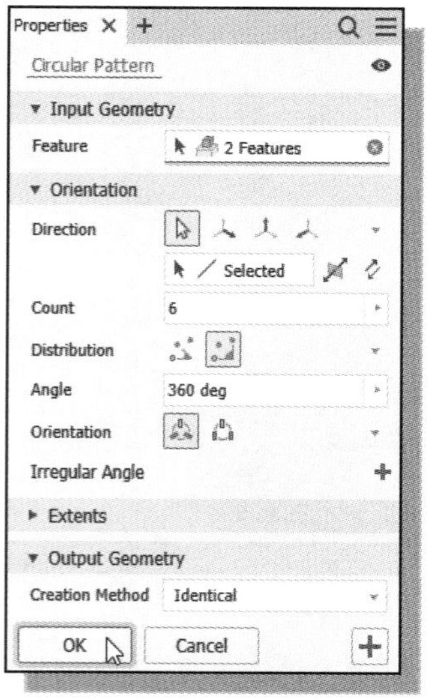

5. Confirm the placement options are set to **6** and **360 deg** as shown.

6. Click **OK** to accept the settings and create the *patterned* feature.

Complete the Geneva Wheel Design

On your own, complete the **Geneva-Wheel** part by adding another ***extruded join*** feature, a ***fillet*** feature and an ***extruded cut*** feature. The benefits of **2D Design Reuse** are more obvious for designs with complex geometry. Parametric modeling software, such as Autodesk Inventor, provides many different tools that allow us to quickly build 3D models.

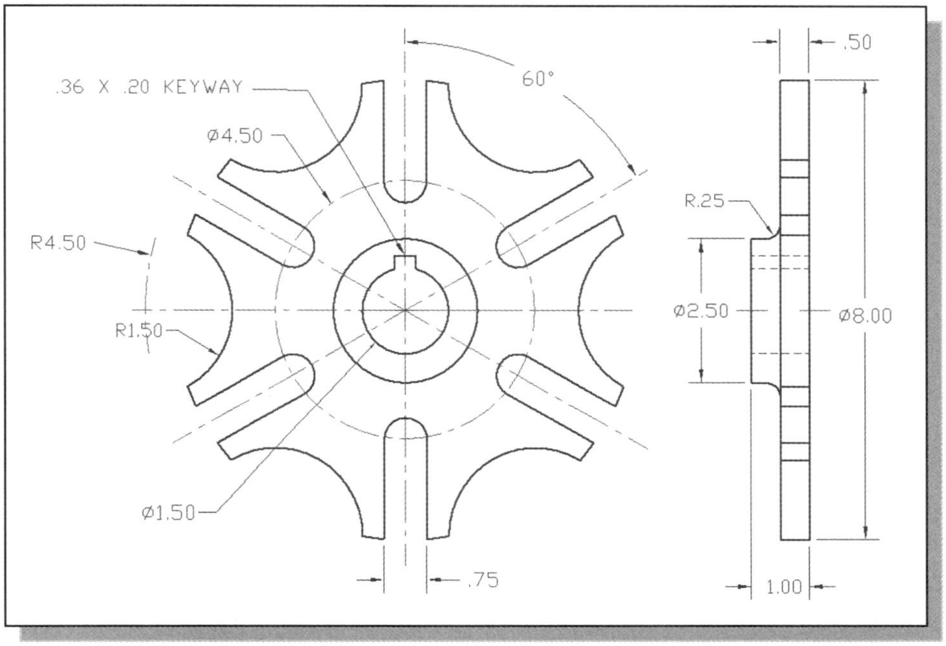

Additional Parts

Three additional parts are required for this assembly: (1) ***Geneva-Driver***, (2) ***Geneva-Base***, and (3) ***Geneva-Pin***. On your own, create the three parts as shown below. Save the models as separate part files in the Geneva folder.

(1) *Geneva-Driver*

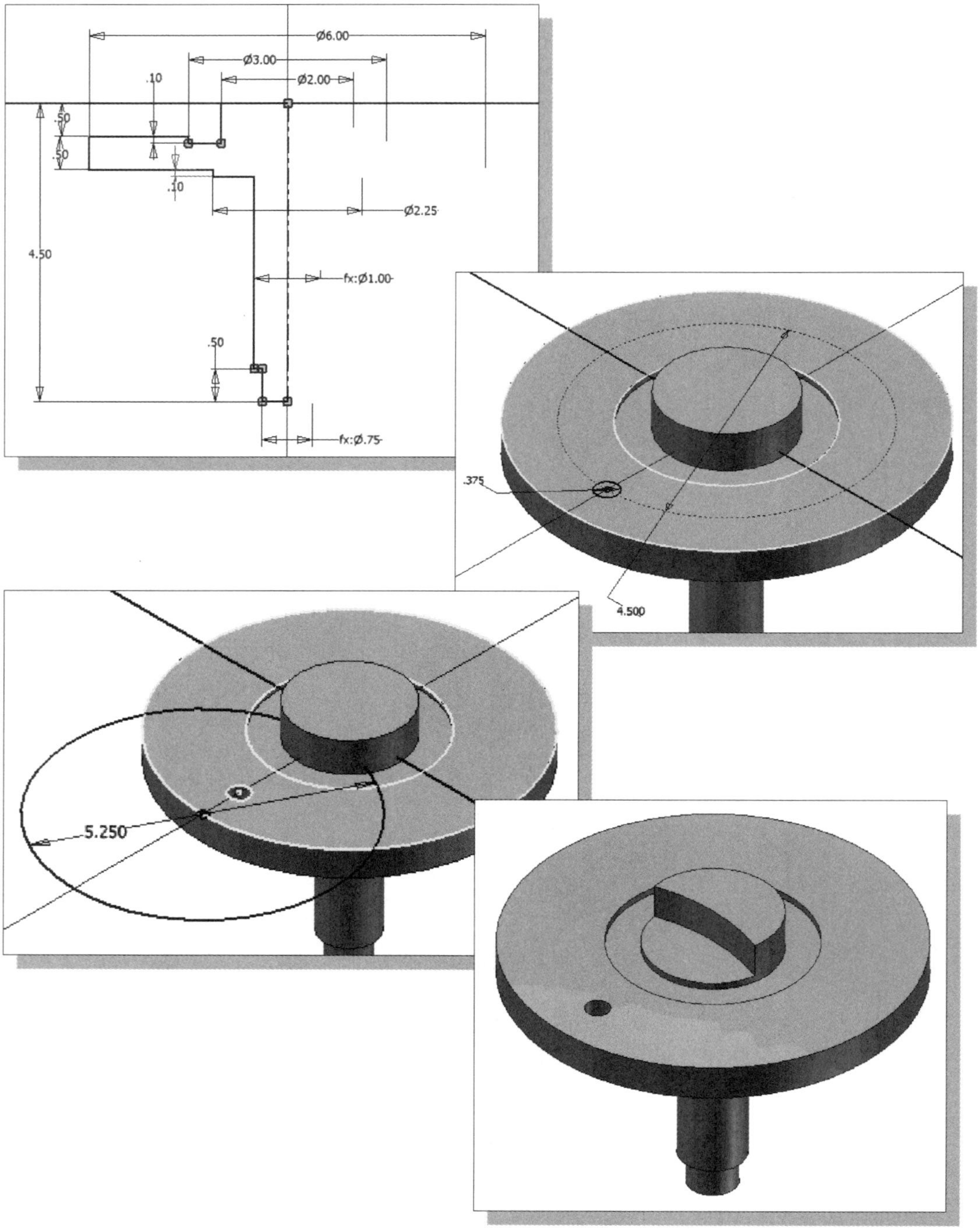

(2) ***Geneva-Base***

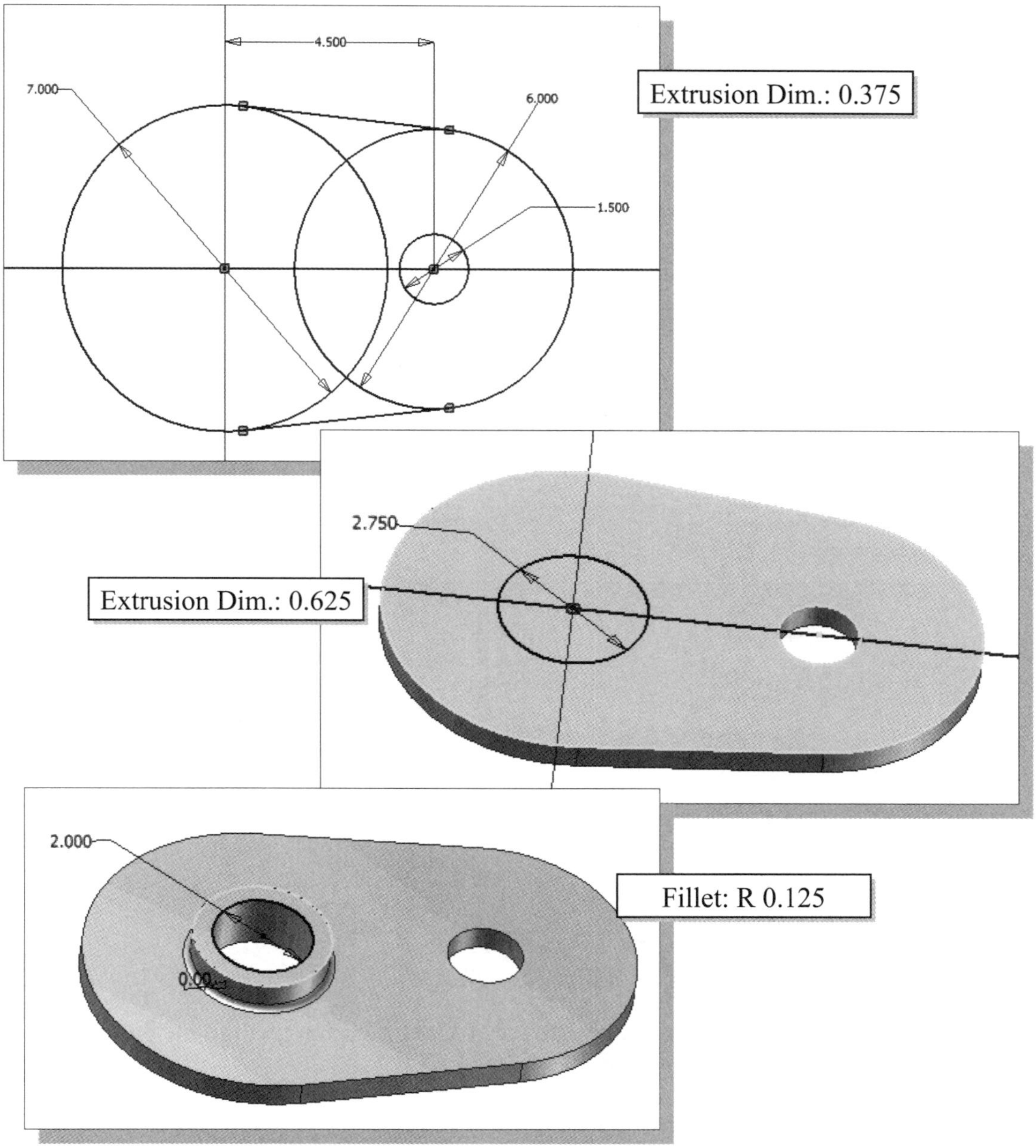

(3) ***Geneva-Pin***

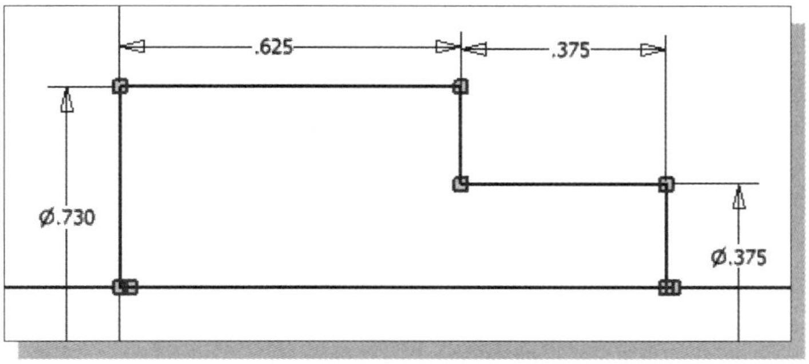

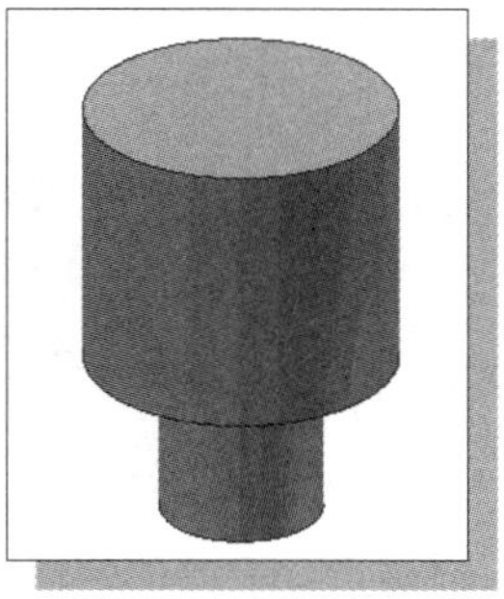

Start a New Assembly

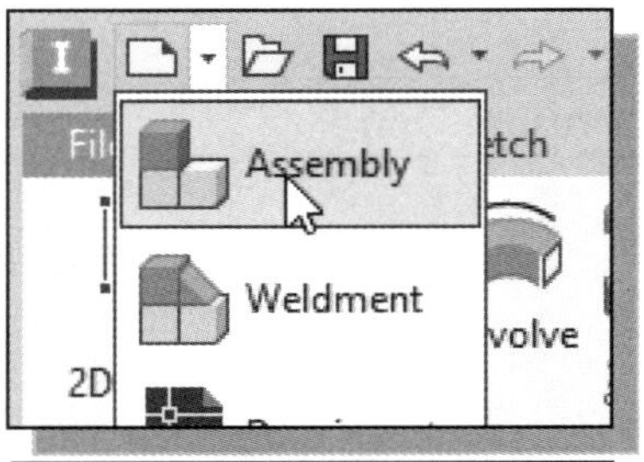

1. Select **Assembly** in the *New File* toolbar with a single click of the left-mouse-button in the *Quick Launch* box.

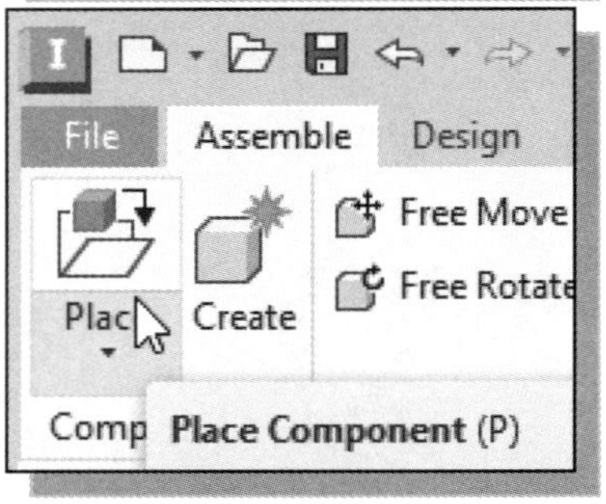

2. In the *Assemble* panel (the toolbar that is located to the left side of the graphics window) select the **Place Component** command by left-clicking the icon.

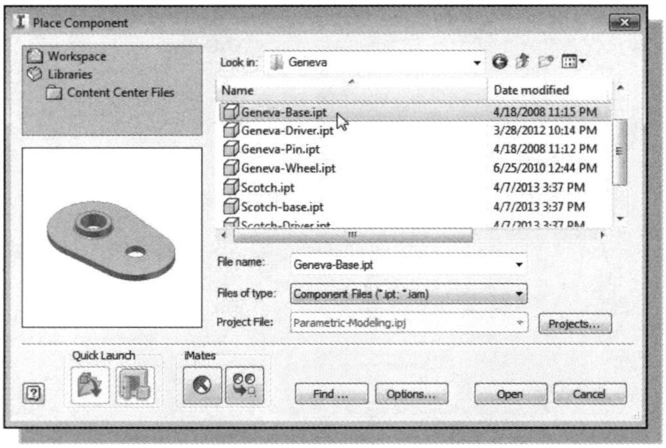

3. Switch to the Geneva folder and select the **Geneva-Base** part in the list window.
4. Click on the **Open** button to retrieve the model.
5. On your own, use the **Place Grounded at Origin** option to align the part to the origin of the assembly coordinates.

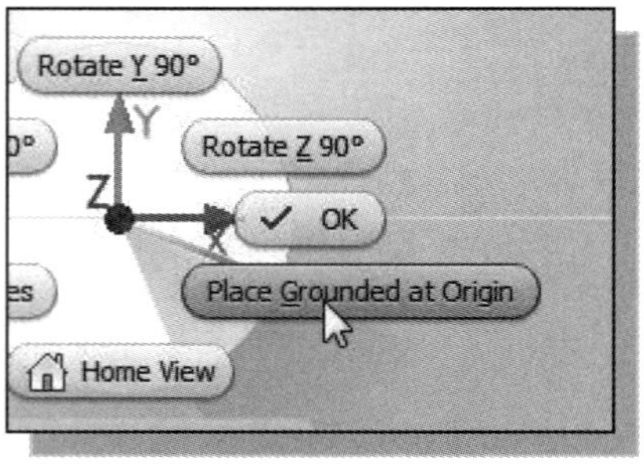

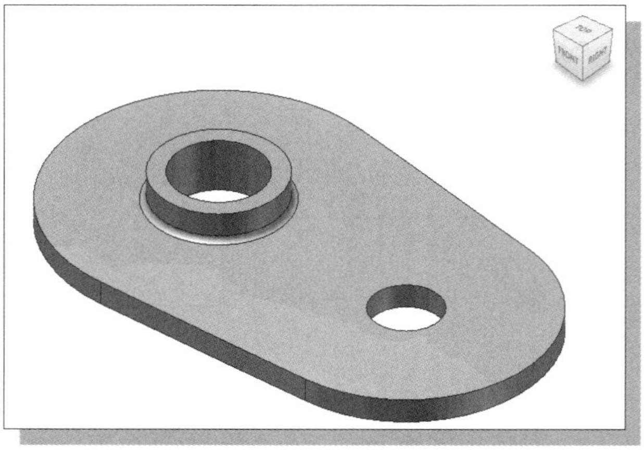

Placing the Second Component

We will retrieve the *Geneva-Wheel* part as the second component of the assembly model.

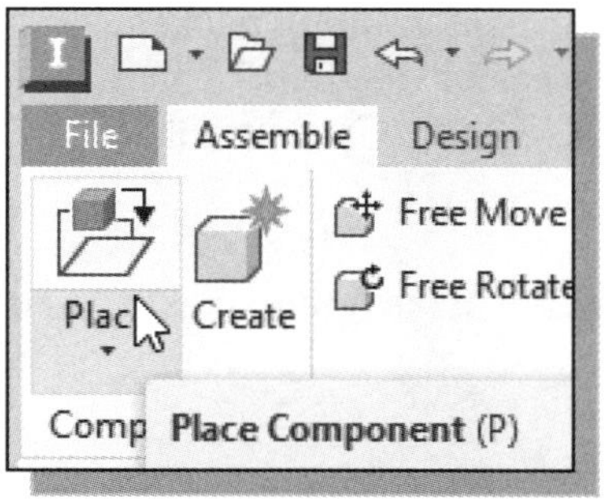

1. In the *Assemble* panel (the toolbar that is located to the left side of the graphics window) select the **Place Component** command by left-clicking the icon.

2. Select the **Geneva-Wheel** design in the list window and click on the **Open** button to retrieve the model.

3. Place the *Geneva-Wheel* toward the front side of the *Geneva-Base* part, as shown in the figure.

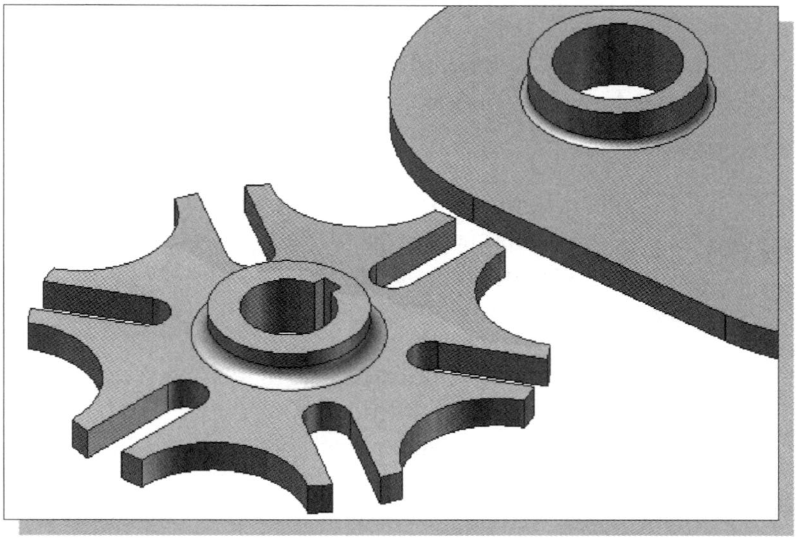

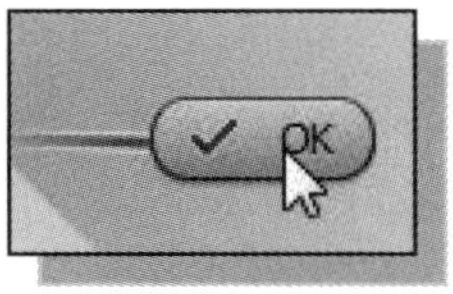

4. Inside the graphics window, right-click once to bring up the option menu and select **OK** to end the placement of the part.

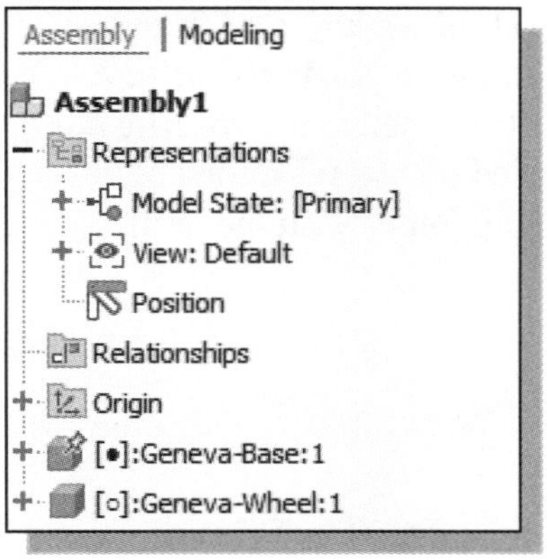

- Inside the *browser* window, the retrieved parts are listed in their corresponding order. The **Pin** icon in front of the *Base* part signifies the part is grounded and all ***six degrees of freedom*** are restricted. The number behind the filename is used to identify the number of copies of the same component in the assembly model.

The Assembly Joint Command

The Joint command is a simple way to position parts and define any motions in between parts. Creating a joint connection allows the user to very quickly define the relations of moving parts in an assembly. The Joint command can be viewed as creating packaged constraint sets that can be used to allow specific motion (degrees of freedom.) There are six options available under this command: Rigid, Rotational, Slider, Cylindrical, Planar and Ball connections.

This option will remove all degrees of freedom. This option can be used for parts that are welded or bolted together.

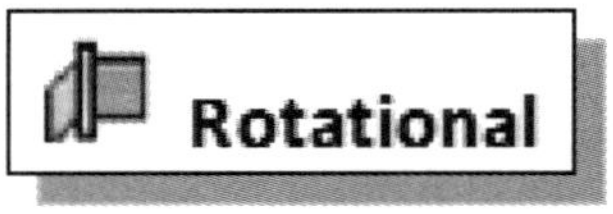

This option will simulate a part that has one rotational degree of freedom. This option can be used for rotating objects such as a level or a hinge.

This option will simulate a part that has one translational degree of freedom. This option can be used for an object to move along a specific track.

This option will simulate a part that has one translational and one rotational degree of freedom. This option can be used for objects rotating about an axis. Note the part can also move along the axis of rotation.

This option will simulate a part that has two translational and one rotational degree of freedom. This option can be used for parts that slide and/or rotate on a plane.

This option will simulate a part that has three rotational degrees of freedom. This option can be used for objects that cannot move but can rotate freely, such as a Ball and Socket joint.

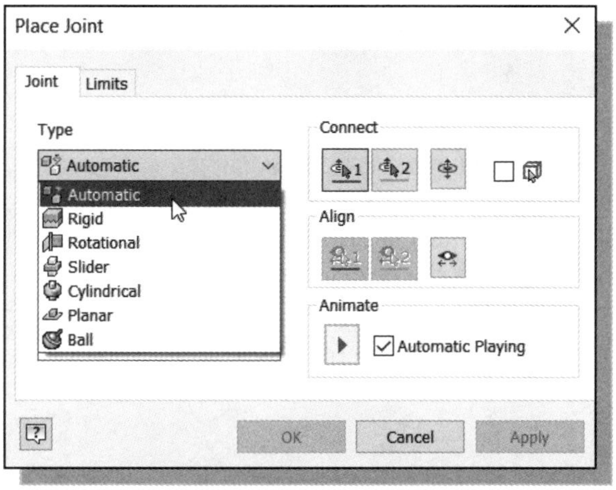

- Note that the *default type* in the *Joint* command is set to **Automatic**. Inventor will select the specific *Joint type* based on the selections of different geometry on the parts.

Create a Joint connection

We are now ready to assemble the *Geneva-Wheel* component to the *Geneva-Base* part.

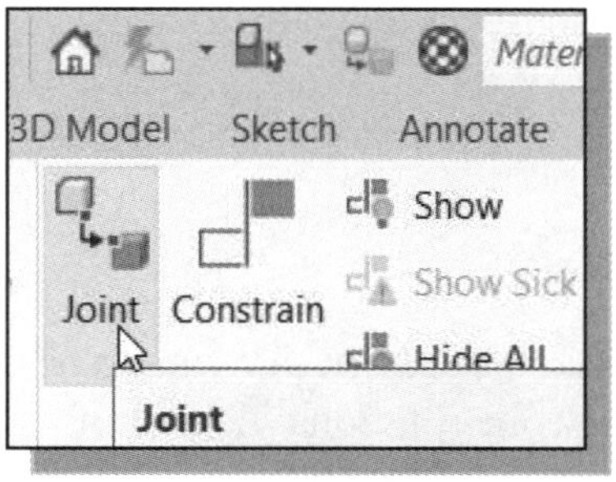

1. In the *Assemble* panel, select the **Joint** command by left-clicking once on the icon.

2. In the *Place Joint* dialog box, confirm the joint type is set to the **Automatic** and select the arc at the bottom surface of the *Geneva-Wheel* part as the first origin for the Joint alignment.

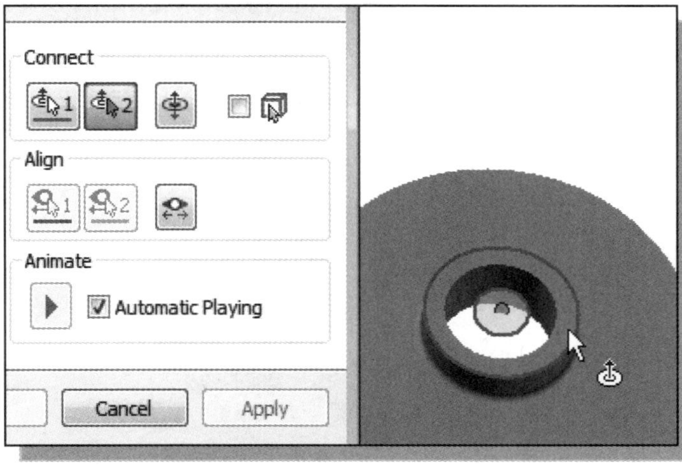

3. Click on the circle on the top face of the *Geneva-Base* part as the second origin to align the two parts as shown.

4. Confirm the *Joint type* has been set to **Rotational** after the selection.

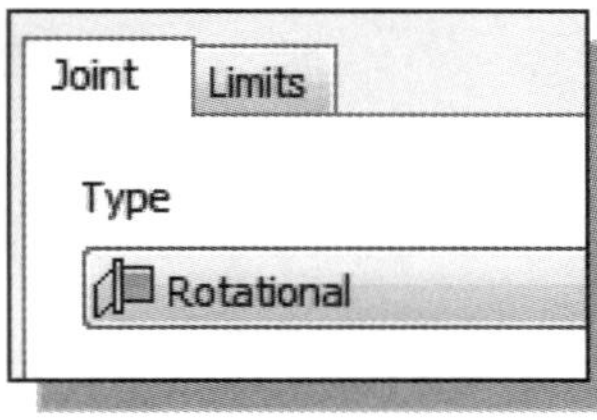

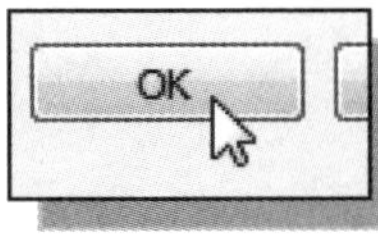

5. Click on the **OK** button to accept the selection and apply a Rotational Joint between the two parts.

Constrained Move

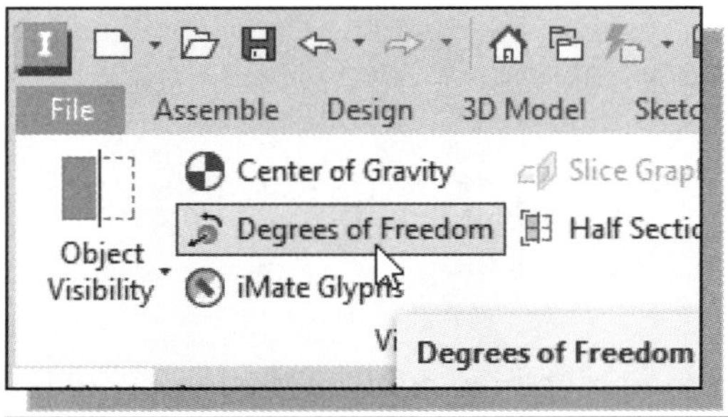

1. On your own, turn *ON* the display of the **Degrees of Freedom**. (Hint: The option is available under the **View** tab.)

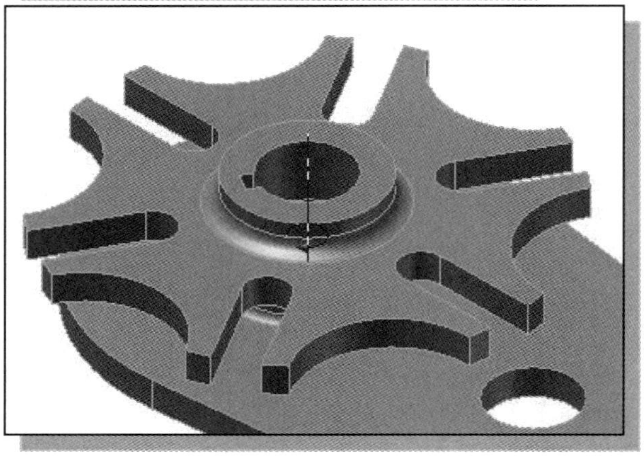

- Note that the *Geneva-Wheel* part has only one rotational degree of freedom; all other DOF have been removed.

2. On your own, perform a constrained move by dragging the *Geneva-Wheel* component in the graphics window with the left-mouse-button. Confirm the two parts are properly constrained.

Placing a Copy of the Geneva-Driver Part

Next, we will assemble the *Geneva-Driver* part.

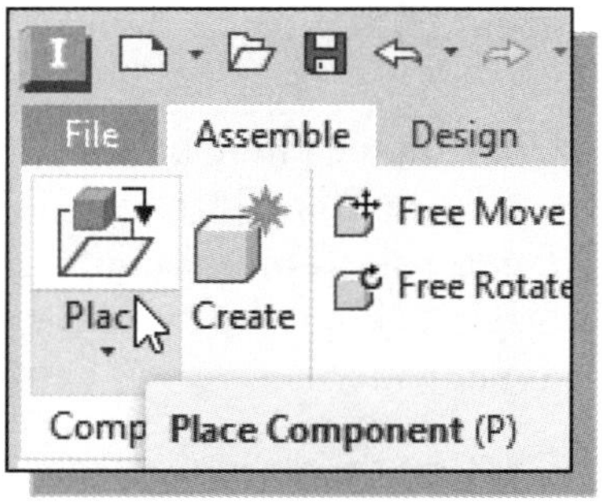

1. In the *Assemble* panel, select the **Place Component** command by left-clicking the icon.
2. On your own, open and place a copy of the *Geneva-Driver* design in the graphics window as shown.

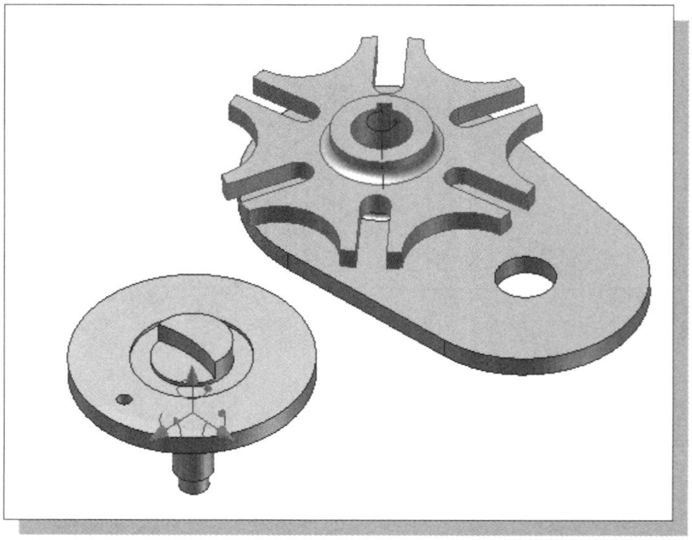

Create a Second Joint Connection

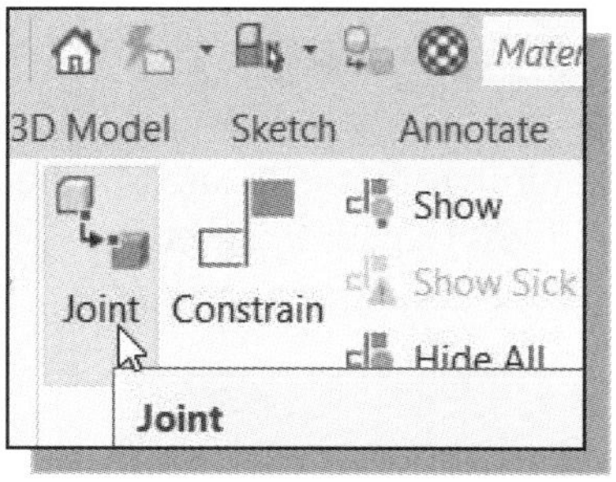

1. In the *Assemble* panel, select the **Joint** command by left-clicking once on the icon.

2. In the *Place Joint* dialog box, confirm the joint type is set to the **Automatic** and select the circle at the base of the shaft of the *Geneva-Driver* part as the first origin for the Joint alignment.

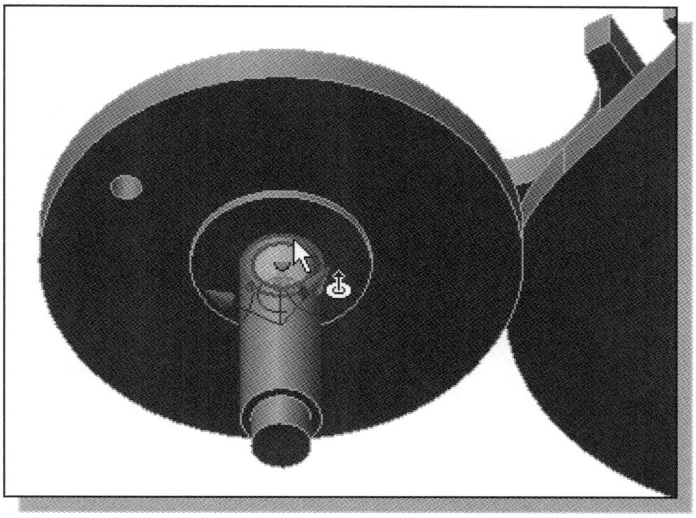

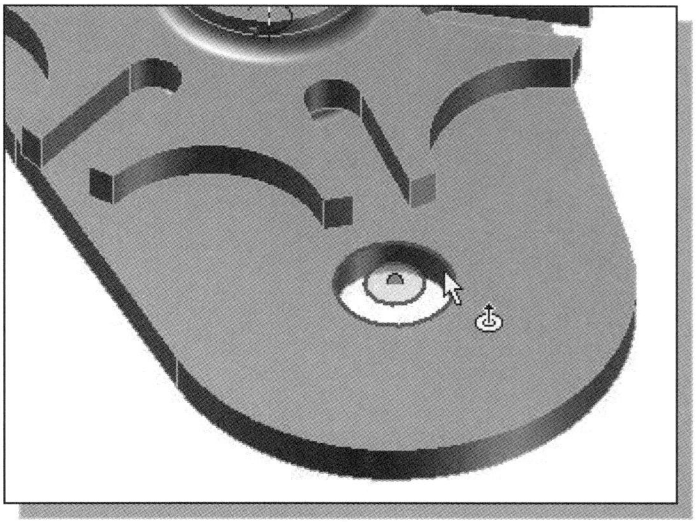

3. Click on the smaller circle on the top face of the *Geneva-Base* part as the second origin to align the two parts as shown.

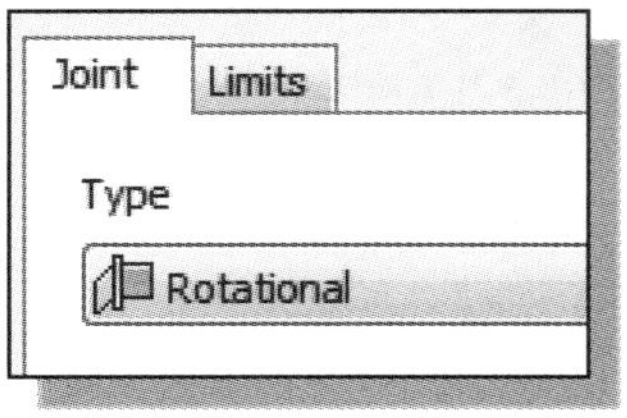

4. Set the *Joint type* to **Rotational** as shown.

5. Click on the **OK** button to accept the selection and apply a Rotational Joint between the two parts.

Assemble the Geneva-Pin Part

We will next assemble the *Geneva-Pin* part to complete the assembly model.

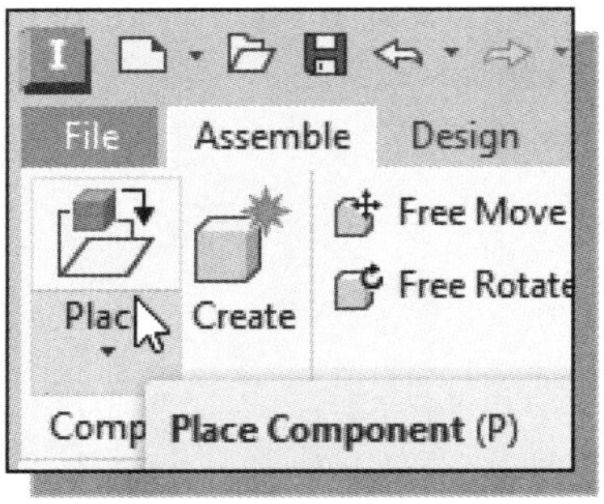

1. In the *Assemble* panel (the toolbar that is located to the left side of the graphics window) select the **Place Component** command by left-clicking once on the icon.

2. On your own, open and place a copy of the *Geneva-Pin* design in the graphics window as shown.

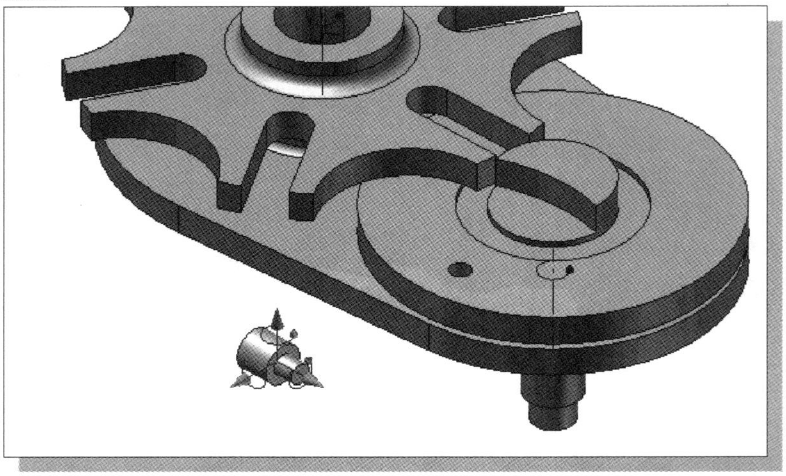

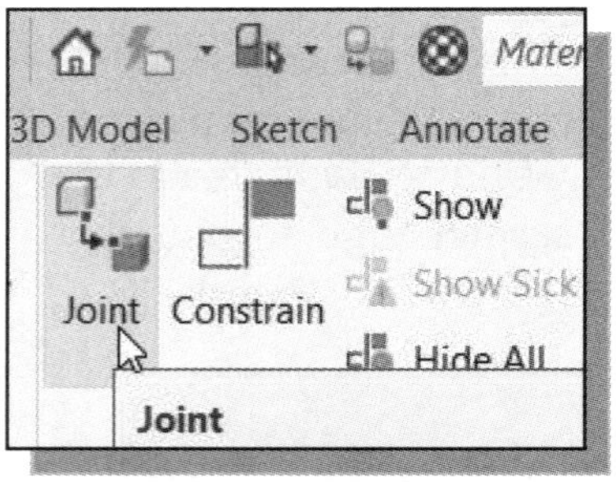

3. In the *Assemble* panel, select the **Joint** command by left-clicking once on the icon.

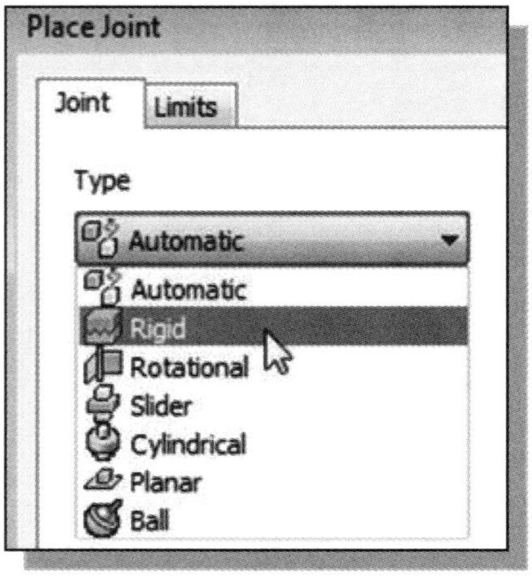

4. In the *Place Joint* user interface, click on the downward arrow and select **Rigid** as the joint type.

- Note that the *joint type* can be set manually.

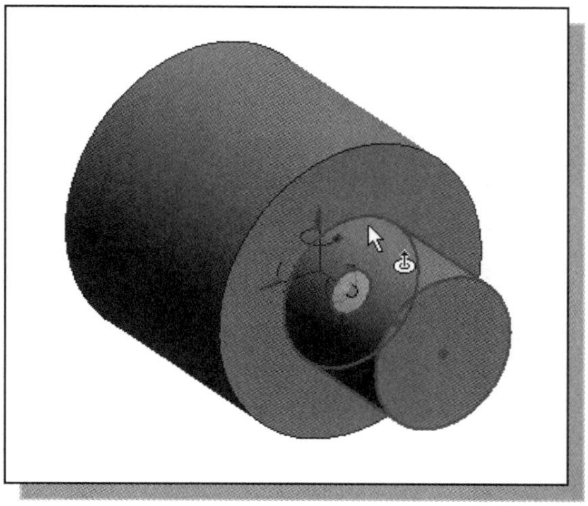

5. Select the circle at the base of the shaft of the *Geneva-Pin* part as the first origin for the **Joint** alignment.

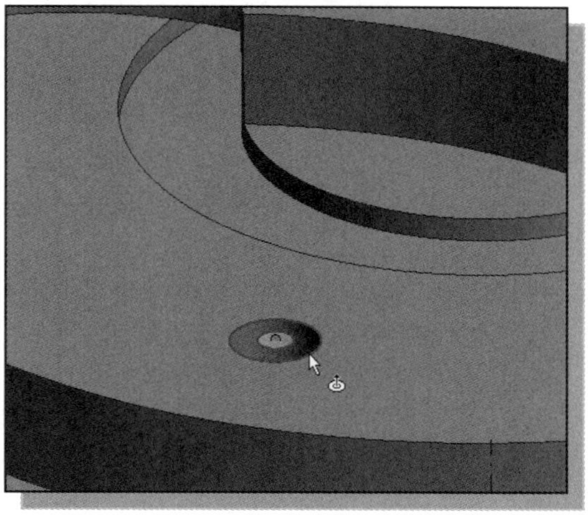

6. Select the small hole at the top surface of the *Geneva-Driver* part as the second origin for the **Joint** alignment.

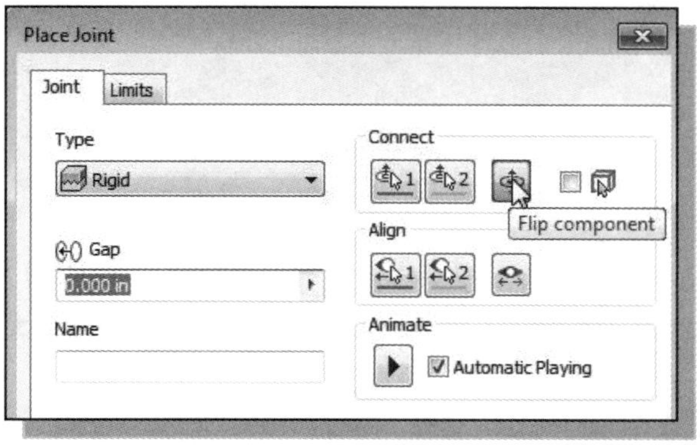

7. Click on the **Flip component** icon to toggle the orientation of the *Geneva-Pin* part.

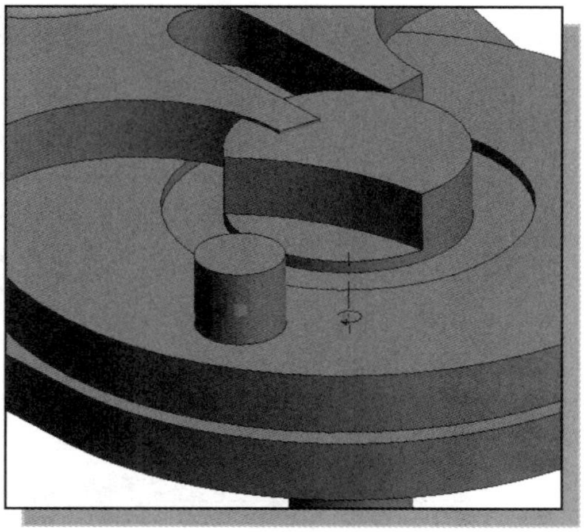

8. Click on the **OK** button to accept the selection and apply a **Rigid** connection between the two parts.

Repositioning the Pieces

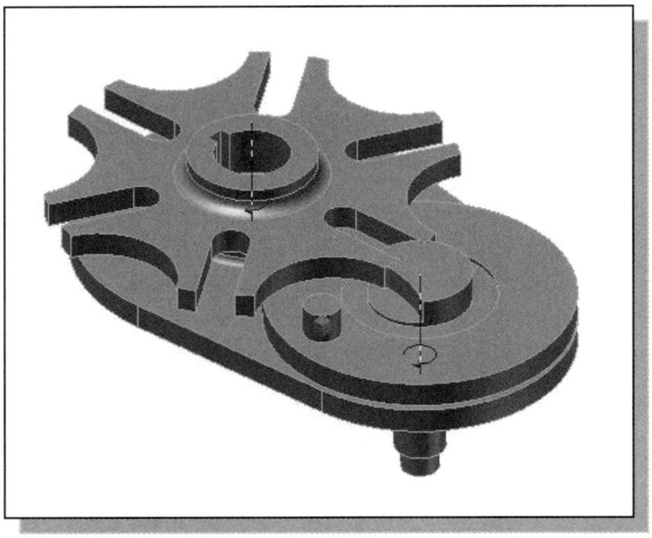

- Note the small green symbol displayed at the applied joint location indicating the two selected parts are rigidly joined together. All DOF have been removed for the *Geneva-Pin* part.

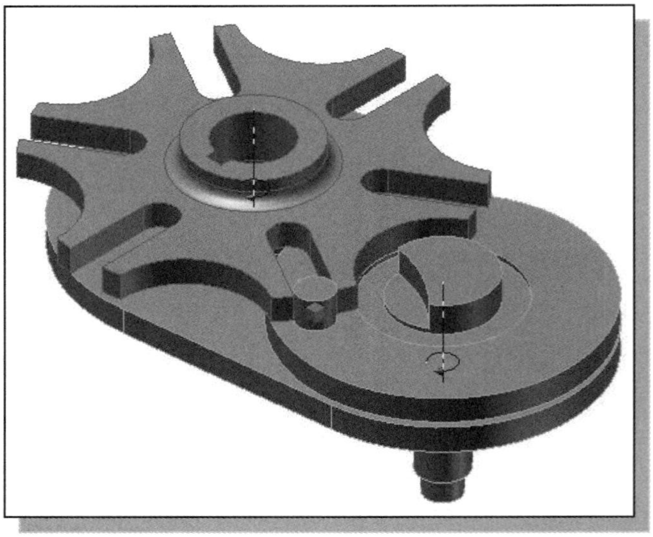

1. On your own, reposition the assembly by dragging the *Geneva-Wheel* part and the *Geneva-Driver* part in the graphics window with the left-mouse-button.

- Note that the *Geneva-Pin* part is rigidly locked to the *Geneva-Driver* part and therefore cannot be moved by dragging.

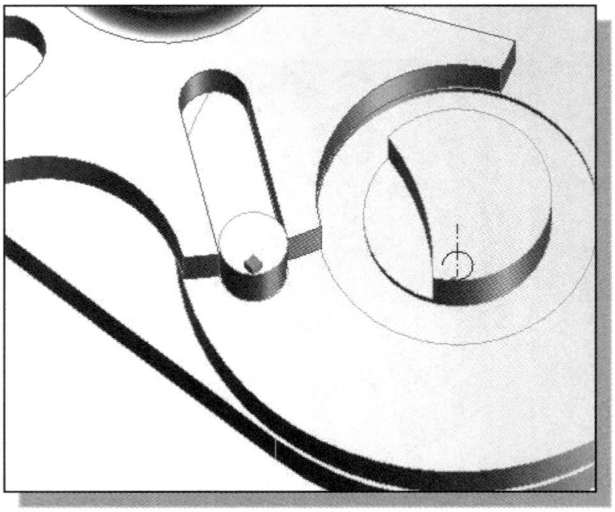

2. Reposition the pieces so that the *Geneva-Pin* part is roughly aligned to enter one of the grooves of the *Geneva-Wheel* part as shown.

Animation with Drive Tool

Autodesk Inventor's **Drive** tool allows us to perform basic motion analysis by creating animations of assemblies with moving parts.

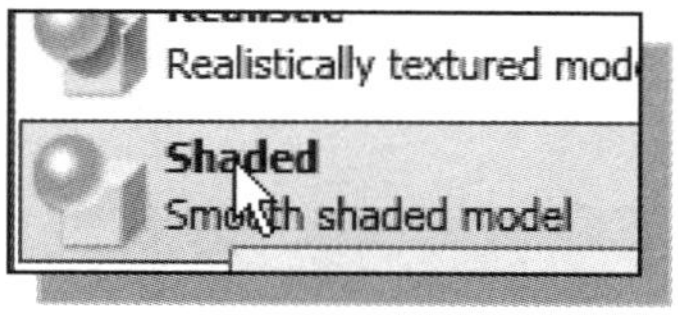

1. In the *View* toolbar, select the **Shaded Display** command by left-clicking once on the icon.

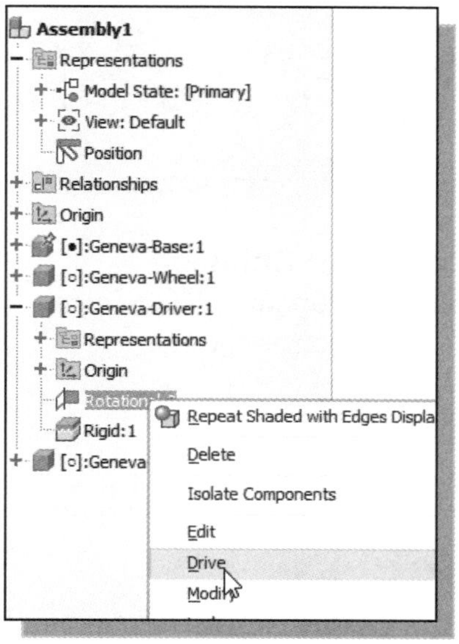

2. In the *Model History Tree* window, right-click on the Rotational Joint to bring up the option menu and select **Drive** as shown.

3. In the *Drive* dialog box, notice the current position is identified on top, modify the **End position** so that the part will make a full revolution as shown.

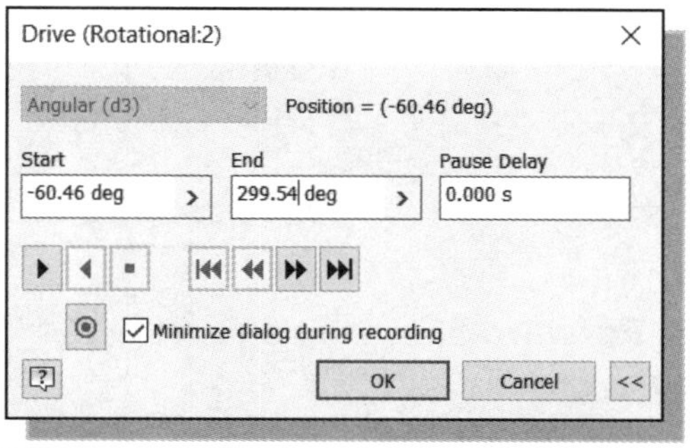

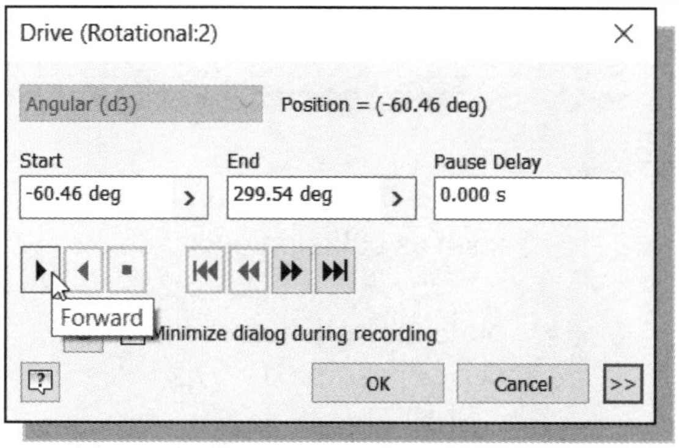

4. Click on the Minimum button to reset the driver part back to the **Start** position.

5. Click on the **Forward** button to begin the animation. Notice the forward direction of the motion is defined by the applied angle constraint.

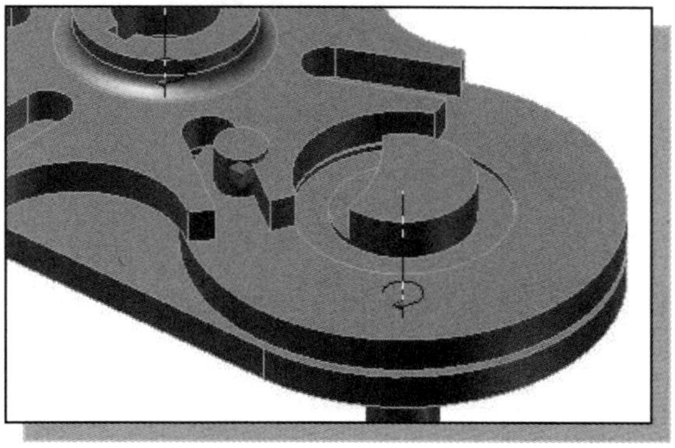

❖ Notice the *Geneva-Pin* part runs through the *Geneva-Wheel* part as if the part does not exist.

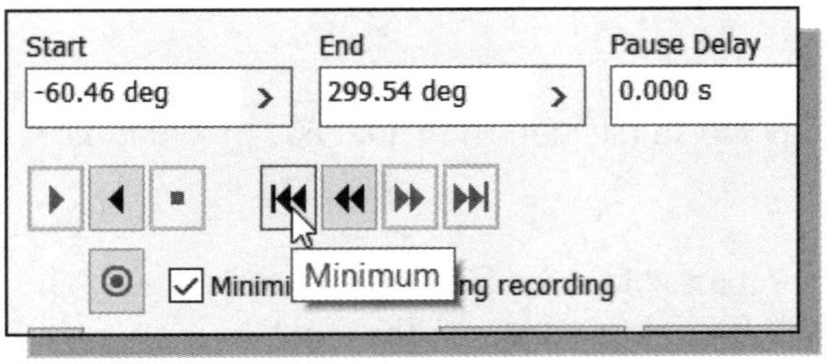

6. Click on the Minimum button to reset the driver part back to the **Start** position.

7. Click on the **Expand** option button to reveal more control options.

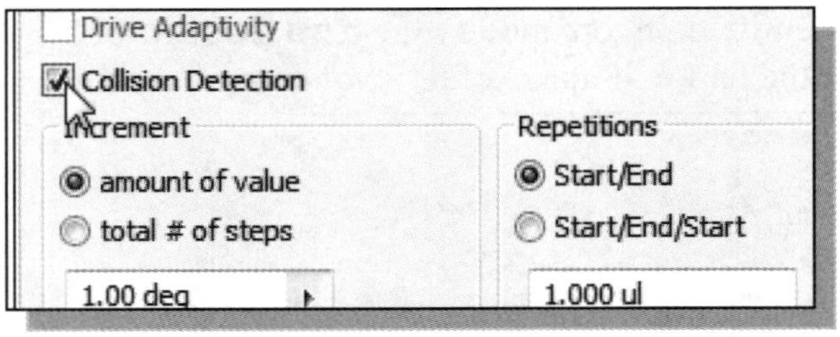

8. Switch on the **Collision Detection** option by clicking on the option box as shown.

9. Click the **Forward** button to start the motion again.

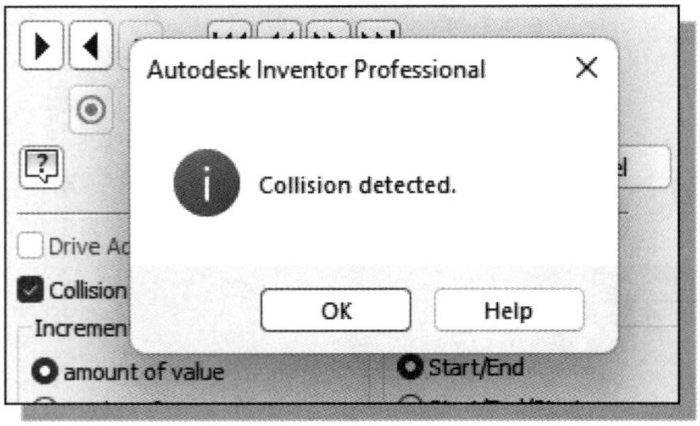

10. Notice the simulation stops as soon as collision occurs.

- Note the collision occurs at the driver position of -57.46 deg.

11. Click **OK** to close the message box.

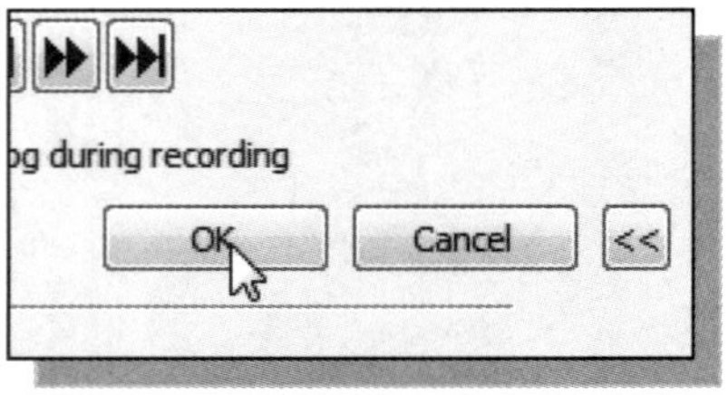

12. Click **OK** to exit the **Drive Constraint** command.

Use the Inventor Contact Solver

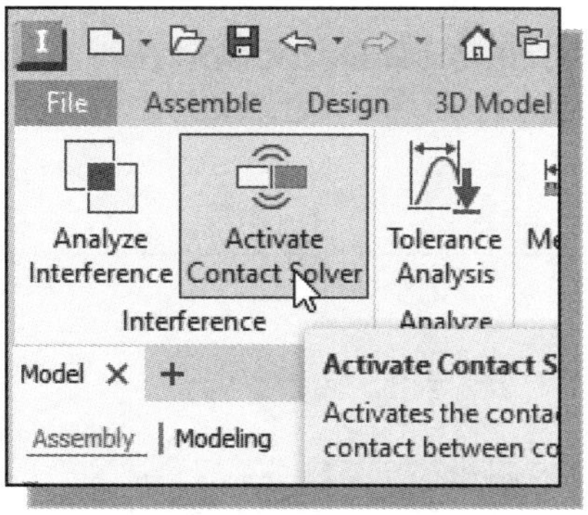

1. Select **Activate Contact Solver** through the **Inspect** tab in the *Ribbon toolbars* as shown.

❖ The **Contact Solver** can be used to detect contact between assembly components. The Contact Solver isolates selected components in a user defined contact set. Related components need to be included in a ***contact-set***; the Contact Solver will then treat the components as rigid bodies and simulate the correct interaction between parts.

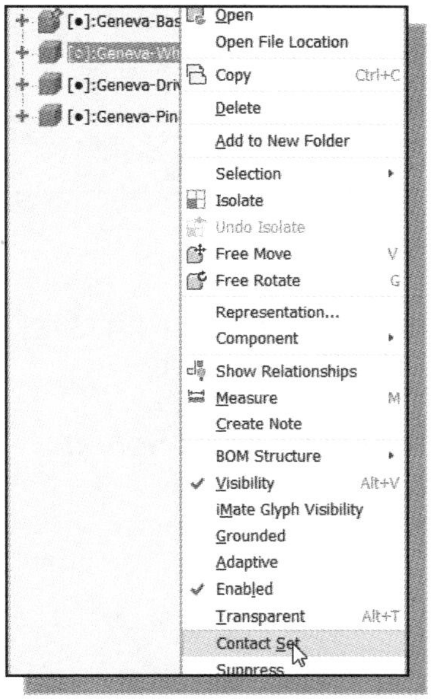

2. In the *Model History Tree* window, right-click on the **Geneva-Wheel** part to bring up the option menu and select **Contact Set** as shown.

3. Repeat the above step and right-click on the **Geneva-Pin** part to bring up the option menu and select **Contact Set** as shown.

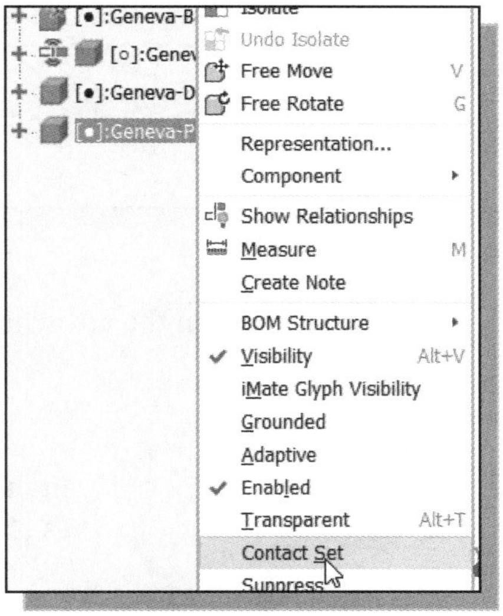

❖ In the *browser* window, the **Contact Solver** icons in front of the part names indicate the parts that are included in the analysis set.

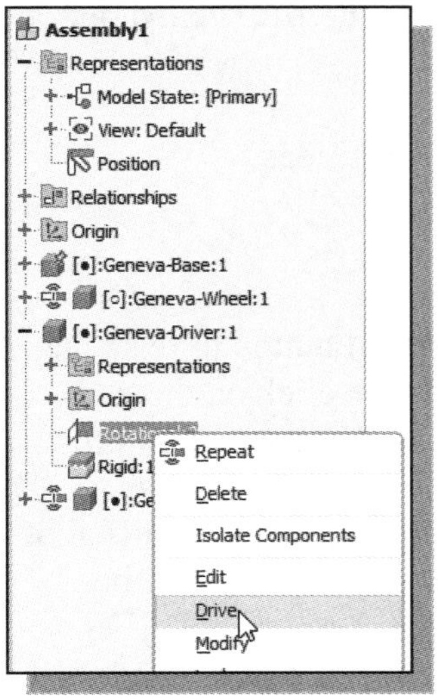

4. In the *Model History Tree* window, right-click on the **Rotational Joint** of the **Geneva-Driver** part to bring up the option menu and select **Drive Constraint** as shown.

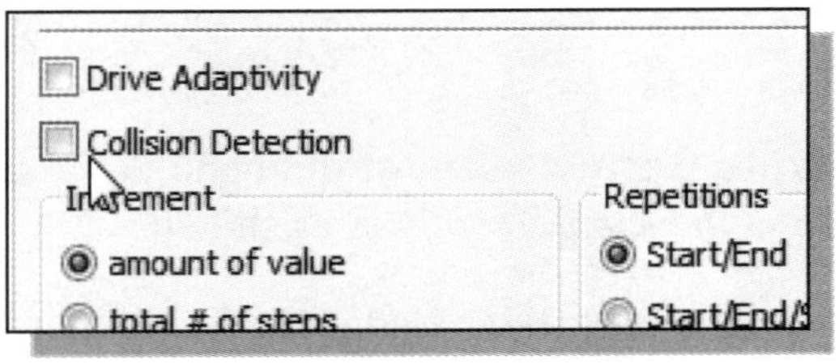

5. Switch the **Collision Detection** option ***OFF*** as shown.

6. In the *Drive* dialog box, modify the **Start** to match the collision position.

7. Also adjust the **End position** so that the part will make a full revolution as shown.

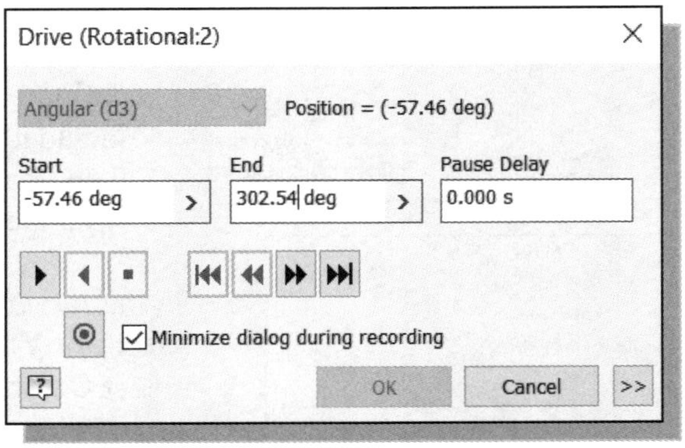

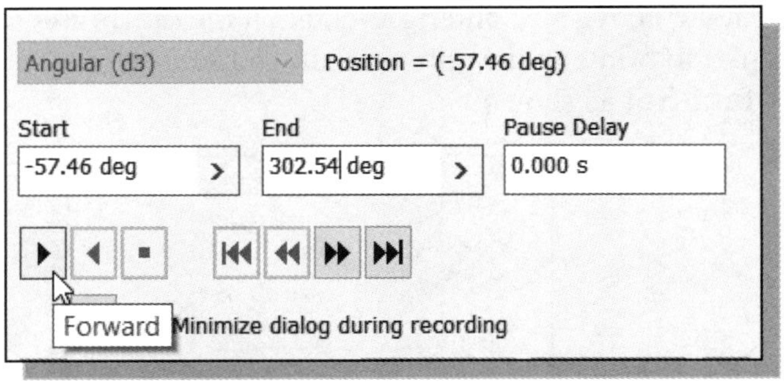

8. Click the **Forward** button to start the driver motion.

❖ Notice the **Contact Solver** correctly simulates the interaction between parts.

9. Click on the arrow icon to expand the option list.

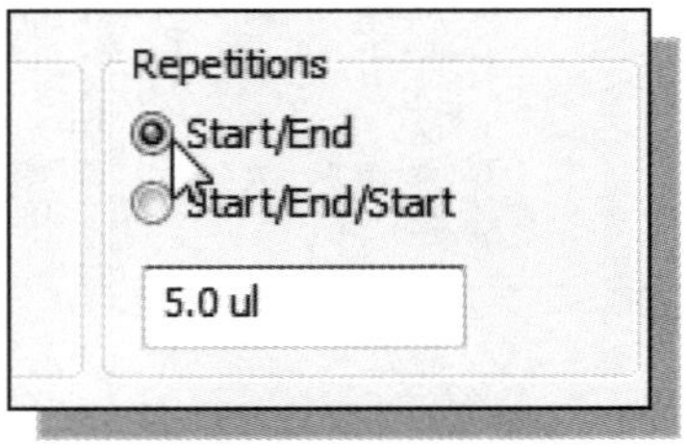

10. In the *Repetitions section*, choose the **Start/End** option as shown.

11. Enter **5.0 ul** as the number of repetitions.

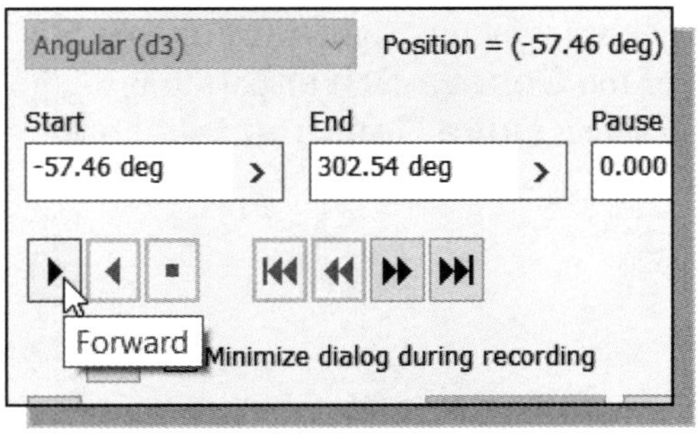

12. Click on the **Minimum** button to reset the driver position and then click on the **Forward** button to begin the animation.

13. On your own, experiment with the different options available in the Drive dialog box.

Constrained Move with Contact Solver

The Contact Solver treats the components as rigid bodies and simulates the correct interaction between parts. The contacts between components exist even when we drag and drop the components.

1. On your own, perform a constrained move by dragging the *Geneva-Driver* component in the graphics window with the left-mouse-button.

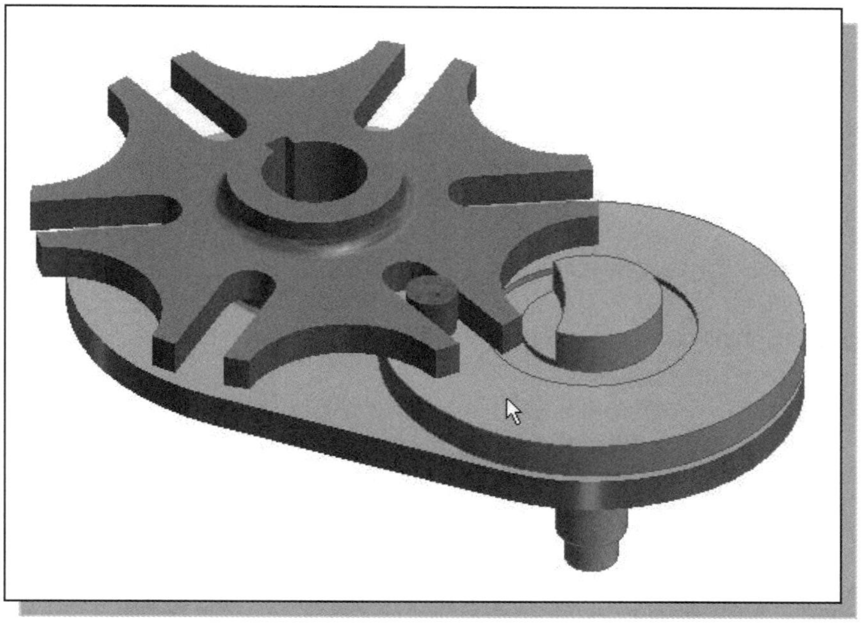

❖ Note the **Contact Solver** correctly simulates the interaction between parts by treating the components as rigid bodies.

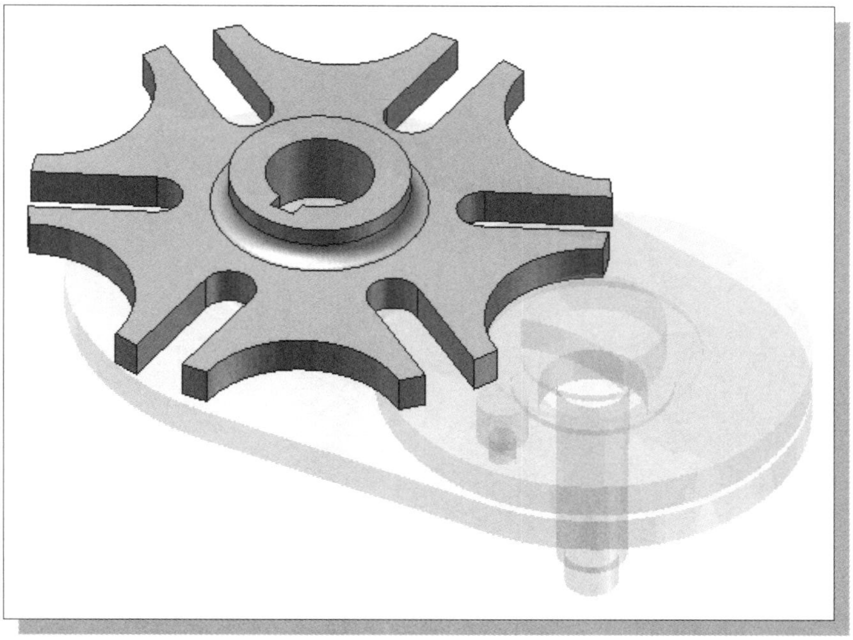

Review Questions: (Time: 25 minutes)

1. Briefly describe the concept of **2D Design Reuse**. Describe some of the advantages of **2D Design Reuse**.
2. Can AutoCAD DWG files be opened and saved with Autodesk Inventor?
3. Can we edit the AutoCAD DWG entities that are in *AutoCAD Model Space* under Autodesk Inventor?
4. Can we edit the AutoCAD DWG entities that are in *AutoCAD Layout* under Autodesk Inventor?
5. List and describe three commands available in the *Drawing Review* panel.
6. What does the **Contact Solver** allow us to do?
7. Describe the steps involved in using the Inventor **Contact Solver**.
8. Can we perform a constrained move on fully constrained components?
9. Can Autodesk Inventor calculate the center of gravity of an assembly model? How do you activate this option?
10. When and why would you use the **Collision Detection** option available in Autodesk Inventor?
11. Can Autodesk Inventor calculate the weight of an assembly model? How is this done?

Exercises: (Time: 200 minutes)

1. **Ratchet Plate** (Using the 2D Design Reuse approach, create the Ratchet Plate.)

Download the *RatchetPlate.dwg* on the book's webpage:
www.SDCpublications.com/downloads/978-1-63057-800-8

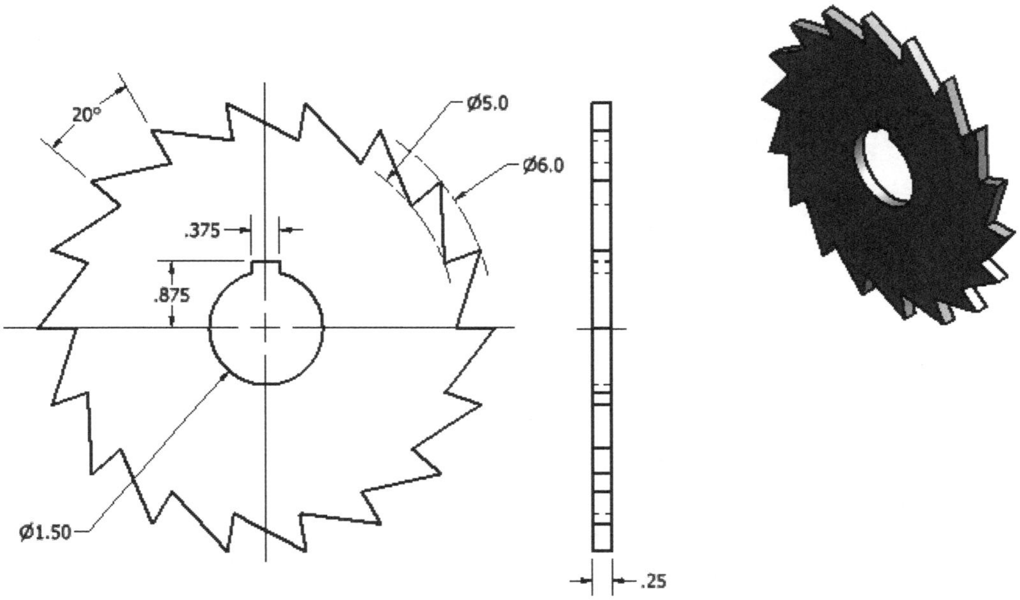

2. **Scotch Yoke mechanism** (Design and create the necessary parts for the design.) Use the Contact Solver to perform a motion analysis.

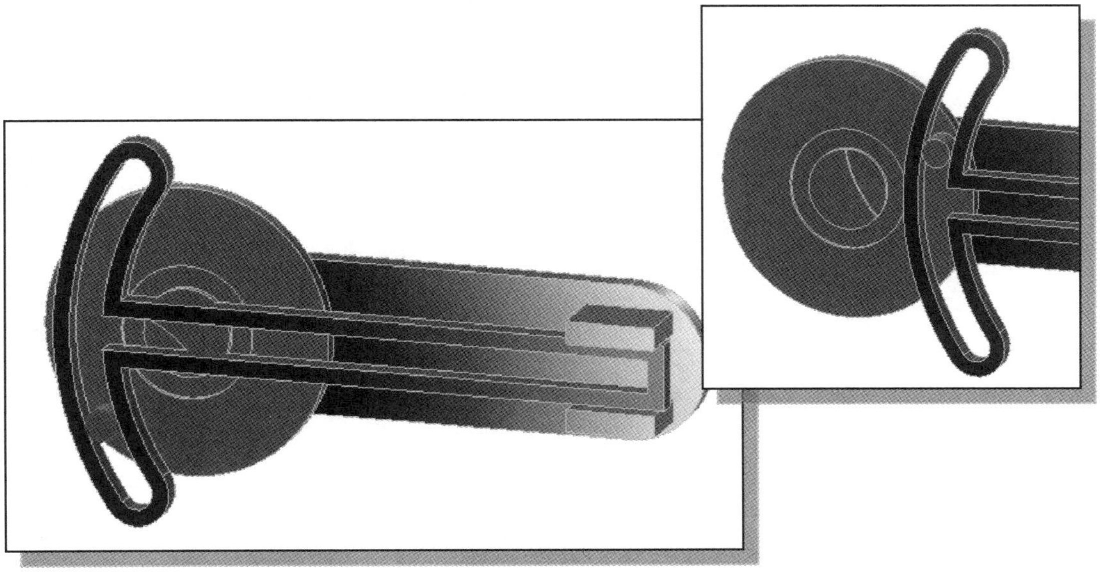

3. **Quick Return Mechanism** (Design and create the necessary parts.)

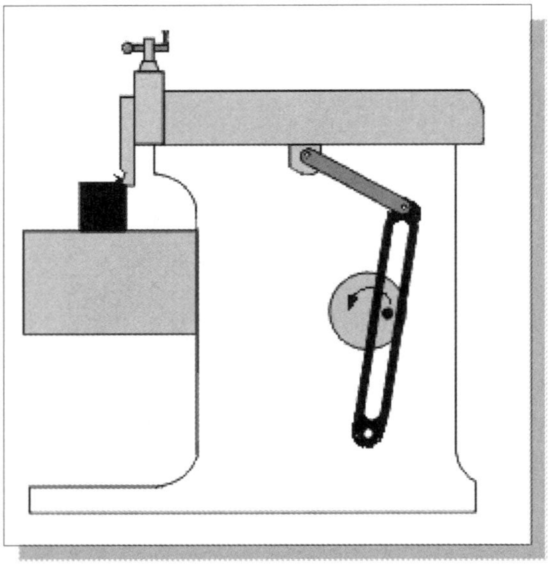

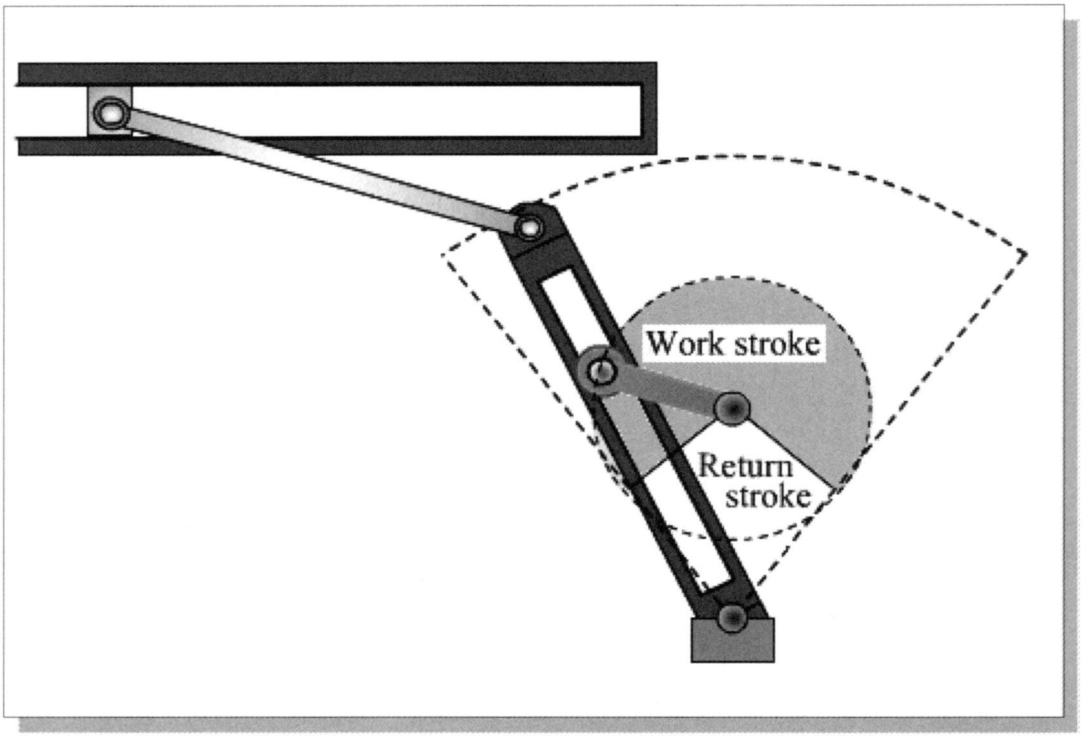

Chapter 17
Introduction to Stress Analysis

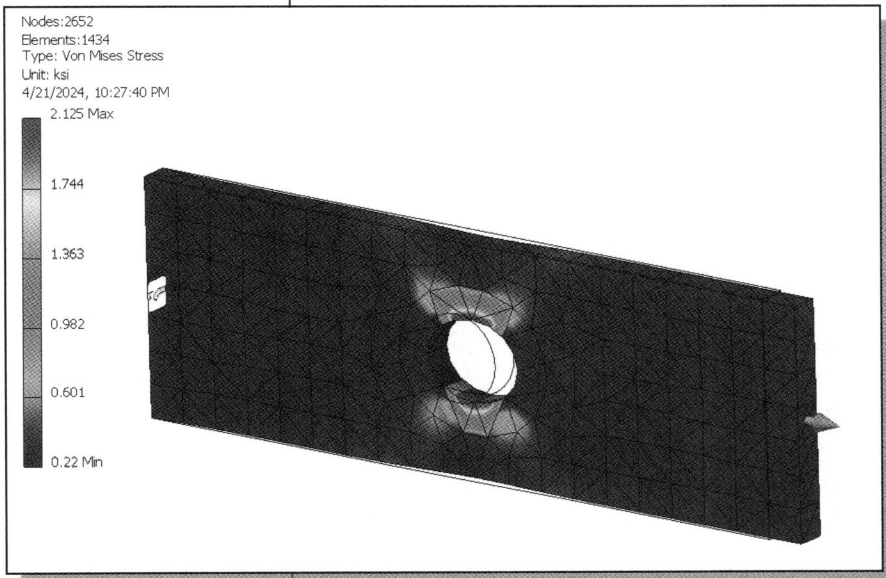

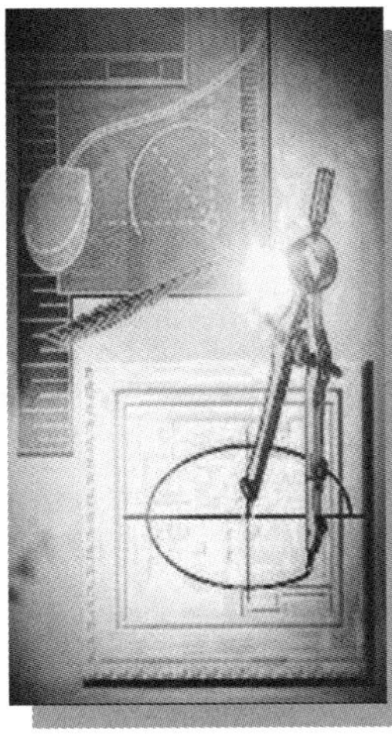

Learning Objectives

- **Create FEA Study**
- **Apply Fixtures and Loads**
- **Perform Basic Stress Analysis**
- **View Results**
- **Assess Accuracy of Results**
- **Output the Associated FEA Report File**

Introduction

In this chapter we will explore basic design analysis using the *Inventor Stress Analysis Module*. The *Stress Analysis Module* is a special module available for part, sheet metal, and assembly documents. The *Stress Analysis Module* has commands unique to its purpose. With Autodesk Inventor, *contact analysis*, *frame analysis* and *dynamic analysis* can also be performed.

Inventor Stress Analysis Module provides a tool for basic stress analysis, allowing the user to examine the effects of applied forces on a design. Displacements, strains and stresses in a part are calculated based on material properties, fixtures, and applied loads. Stress results can be compared to material properties, such as yield strength, to perform failure analysis. The results can also be used to identify critical areas, calculate safety factors at various regions, and simulate deformation. *Inventor Stress Analysis Module* provides an easy-to-use method within the Autodesk Inventor's *Stress Analysis Module* to perform an initial stress analysis. The results can be used to improve the design.

In *Inventor Stress Analysis Module*, stresses are calculated using **linear static analysis** based on the **finite element method**. *Linear static analysis* is appropriate if deflections are small and vary only slowly. *Linear static analysis* omits time as a variable. It also excludes plastic action and deflections that change the way loads are applied. The *finite element method (FEM)* is a numerical method for finding approximate solutions to complex systems. The technique is widely used for the solution of complex problems in engineering mechanics. Analysis using the method is called *finite element analysis (FEA)*.

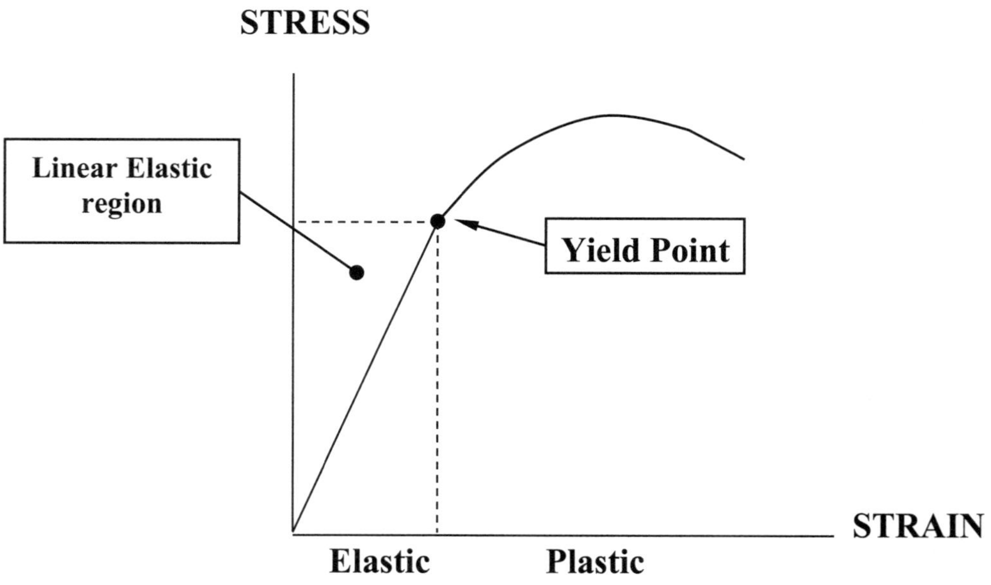

Stress-Strain diagram of typical ductile material

In the finite element method, a complex system is modeled as an equivalent system of smaller bodies of simple shapes, or *elements*, which are interconnected at common points called *nodes*. This process is called *discretization*; an example is shown in the figures below. The mathematical equations for the system are formulated first for each finite element, and the resulting system of equations is solved simultaneously to obtain an approximate solution for the entire system. In general, a better approximation is obtained by increasing the number of elements, which will require more computing time and resources.

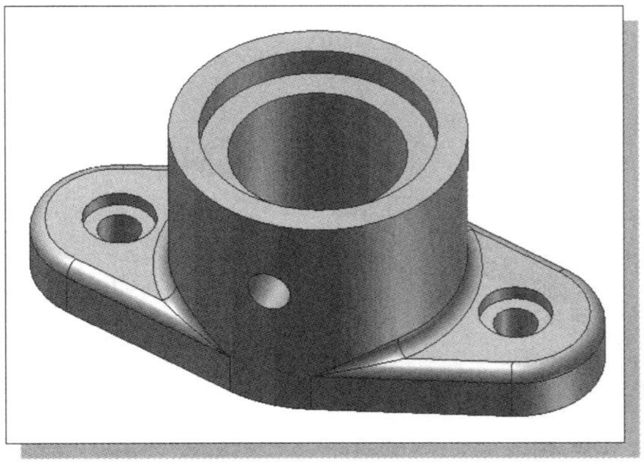

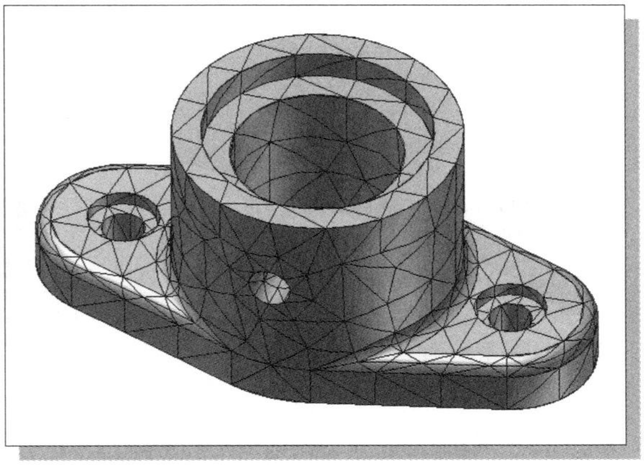

Inventor Stress Analysis Module utilizes tetrahedral elements for which the edges and faces can be curvilinear and which allow the modeling of curved surfaces, as seen in the figure above. The behavior of these elements is analyzed using linear static analysis and the appropriate material properties to relate local coordinate nodal displacements to local forces. The motion of each node is described by displacements in the X, Y, and Z directions, called *degrees of freedom* (DOFs). The equations describing the behavior of each element are assembled into a global system of equations, incorporating compatibility requirements based on connectivity among elements. Using the known material properties, supports, and loads, *Inventor Stress Analysis Module* solves the system of equations for the unknown displacements at each node. These displacements are used in the results stage to calculate strains and stresses.

While *Inventor Stress Analysis Module* is a powerful and easy-to-use tool, it is important to appreciate that it is the designer's responsibility to properly assess the accuracy of the results. A better FEA approximation is generally obtained by increasing the number of elements. An assessment must be made regarding the mesh used to discretize the model to ensure it is adequate. There are other important factors affecting the accuracy of the results. The material properties used in the analysis must accurately characterize the behavior of the material. The supports and loads must be applied in a manner which accurately reflects the actual conditions. Proper meshing and application of boundary conditions often require significant experience in FEA and may require tools and capabilities not available in *Inventor Stress Analysis Module*. *Inventor Stress Analysis Module* is an easy-to-use tool for a quick stress analysis.

Problem Statement

Determine the maximum normal stress that loading produces in the aluminum-6061 plate.

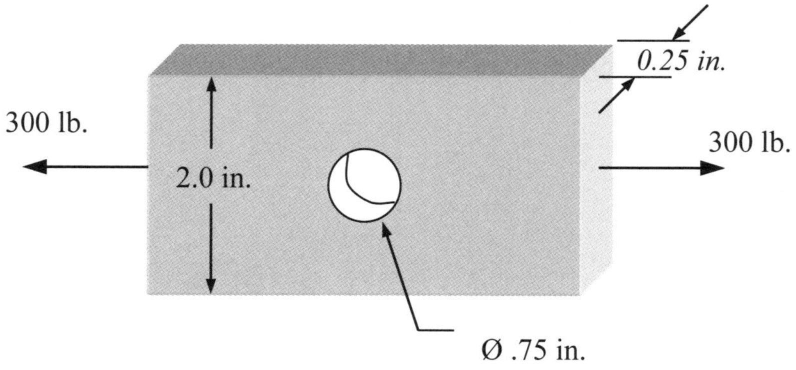

Preliminary Analysis

- **Maximum Normal Stress**
  The nominal normal stress developed at the smallest cross section (through the center of the hole) in the plate is

$$\sigma_{nominal} = \frac{P}{A} = \frac{300}{(2 - 0.75) \times .25} = 960 \text{ psi.}$$

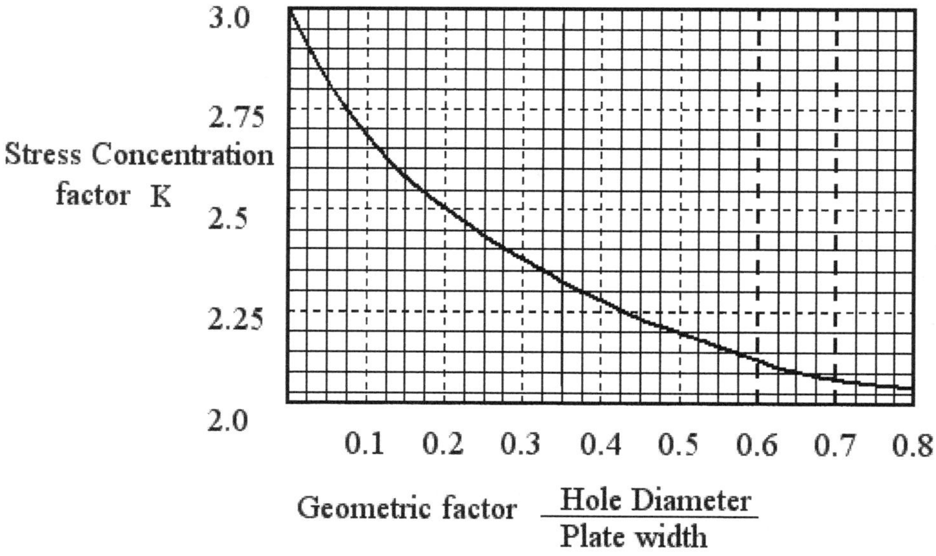

Geometric factor = .75/2 = 0.375

Stress concentration factor K is obtained from the graph, **K = 2.27**

$$\sigma_{MAX} = K\,\sigma_{nominal} = 2.27 \times 960 = 2180 \text{ psi.}$$

- **Maximum Displacement**
  We will also estimate the displacement under the loading condition. For a statically determinant system, the stress results depend mainly on the geometry. The material properties can be in error and still the FEA analysis comes up with the same stresses. However, the displacements always depend on the material properties. Thus, it is necessary to always estimate both the stress and displacement prior to a computer FEA analysis.

  The classic one-dimensional displacement can be used to estimate the displacement of the problem:

$$\delta = \frac{PL}{EA}$$

Where P=force, L=length, A=area, E= elastic modulus, and δ = deflection.

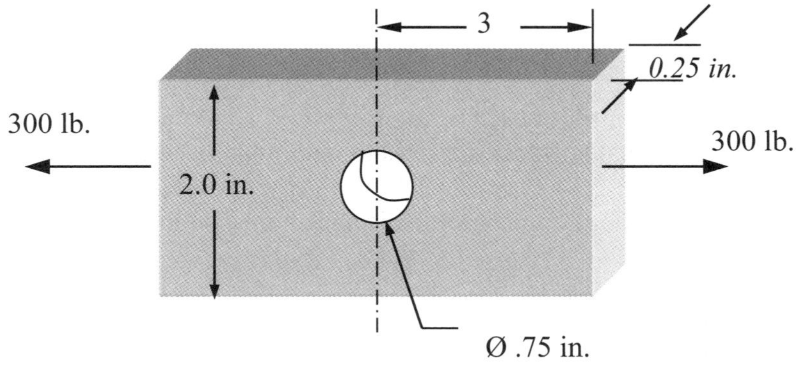

A lower bound of the displacement of the right edge, measured from the center of the plate, is obtained by using the full area:

$$\delta_{lower} = \frac{PL}{EA} = \frac{300 \times 3}{10E6 \times (2 \times 0.25)} = 1.8E\text{-}4 \text{ in.}$$

and an upper bound of the displacement would come from the reduced section:

$$\delta_{upper} = \frac{PL}{EA} = \frac{300 \times 3}{10E6X(1.25 \times 0.25)} = 2.88E\text{-}4 \text{ in.}$$

but the best estimate is a sum from the two regions:

$$\delta_{average} = \frac{PL}{EA} = \frac{300 \times 0.375}{10E6 \times (1.25 \times 0.25)} + \frac{300 \times 2.625}{10E6 \times (2.0X0.25)}$$

$$= 3.6E\text{-}5 + 1.58E\text{-}4 = 1.94E\text{-}4 \text{ in.}$$

Finite Element Analysis Procedure

In the previous section, an approximate preliminary analysis was performed prior to carrying out the finite element analysis; this will help us to gain some insights into the problem and also serves as a means of checking the finite element analysis results.

For a typical linear static analysis problem, the finite element analysis requires the following steps:

1. Preliminary Analysis.

2. Preparation of the finite element model:
   a. Model the problem into finite elements.
   b. Prescribe the geometric and material information of the system.
   c. Prescribe how the system is supported.
   d. Prescribe how the loads are applied to the system.

3. Perform calculations:
   a. Generate a stiffness matrix of each element.
   b. Assemble the individual stiffness matrices to obtain the overall, or global, stiffness matrix.
   c. Solve the global equations and compute displacements, strains, and stresses.

4. Post-processing of the results:
   a. Viewing the stress contours and the displaced shape.
   b. Checking any discrepancy between the preliminary analysis results and the FEA results.

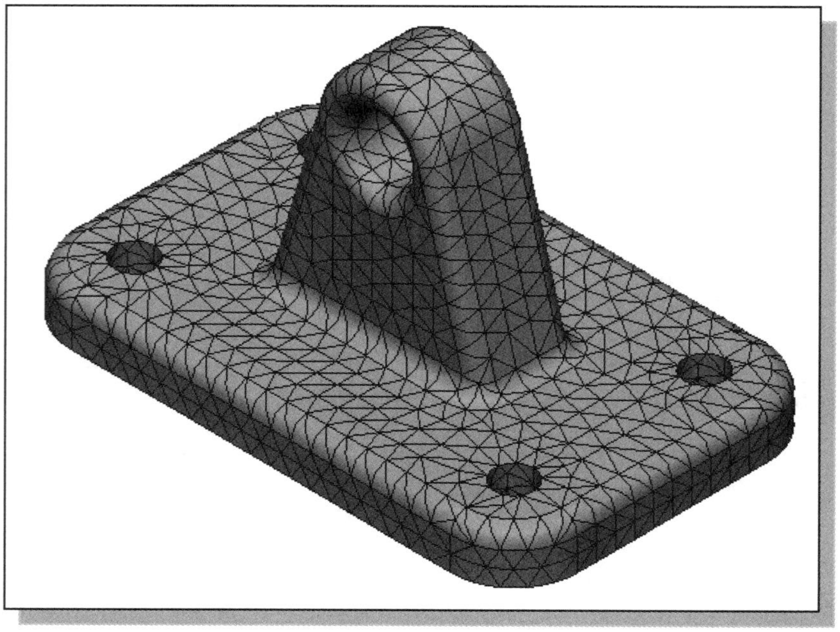

Create the Autodesk Inventor Part

1. Select the **Autodesk Inventor** option on the *Start* menu or select the **Autodesk Inventor** icon on the desktop to start Autodesk Inventor. The Autodesk Inventor main window will appear on the screen.

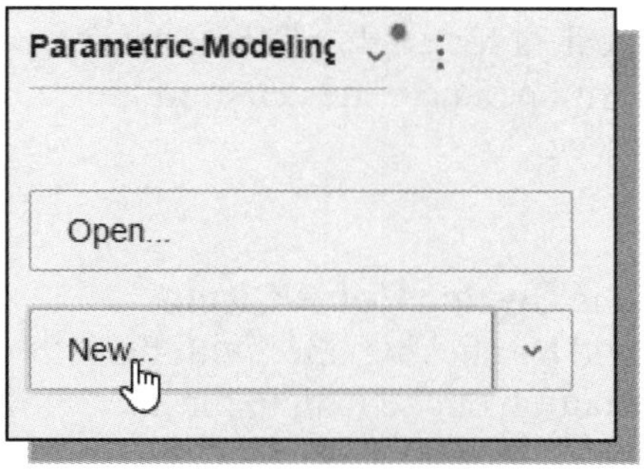

2. Select the **New File** icon with a single click of the left-mouse-button as shown.

3. Select the **English** tab, and in the *New File* area select **Standard(in).ipt**.

4. Pick **Create** in the *New File* dialog box to accept the selected settings.

Create the 2D Sketch for the Plate

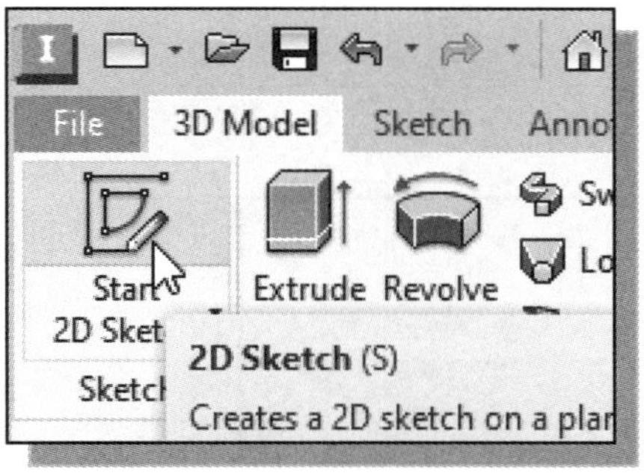

1. In the *3D Model* tab select the **Start 2D Sketch** command by left-clicking once on the icon.

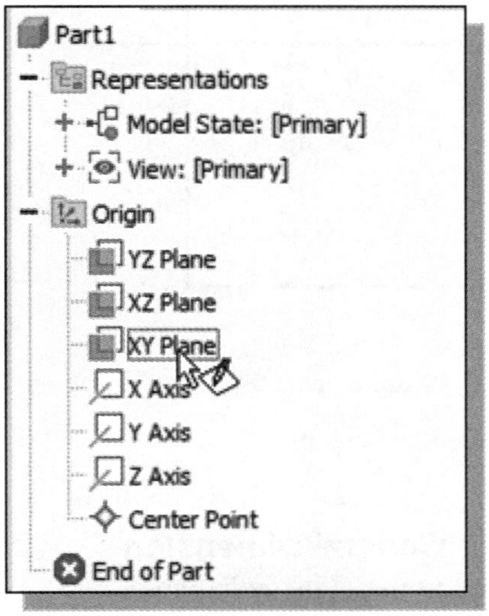

2. In the *Status Bar* area, the message "*Select plane to create sketch or an existing sketch to edit.*" is displayed. Select the **XY Plane** by clicking the associated item in the graphics area or inside the *Model* history tree window or in the graphics window.

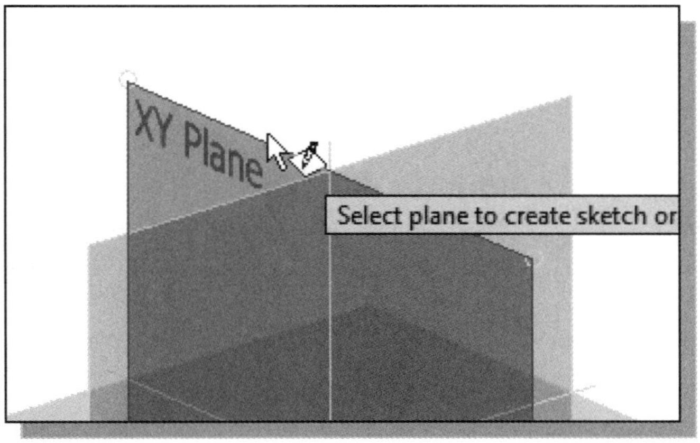

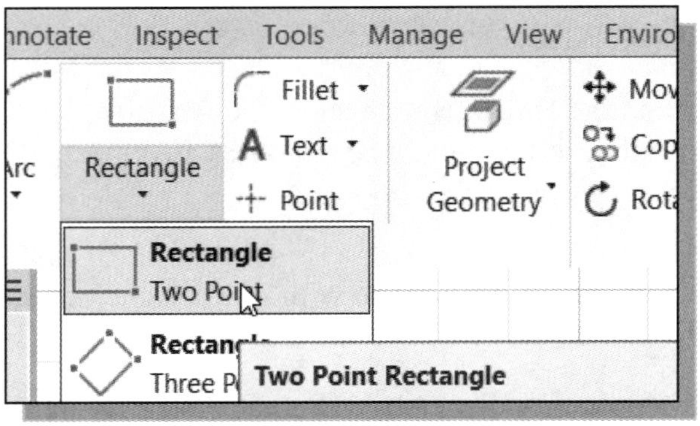

3. Select the **Two point rectangle** command by clicking once with the left-mouse-button on the icon in the *Sketch* toolbar.

4. Create a rectangle of arbitrary size positioned near the center of the screen.

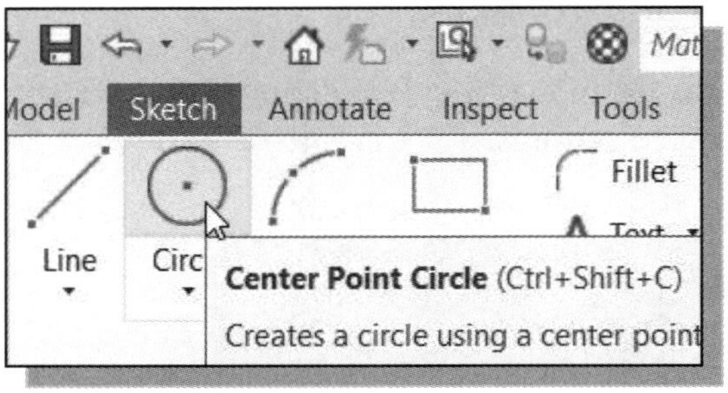

5. Select the **Center Point Circle** command by clicking once with the left-mouse-button on the icon in the *Sketch* toolbar.

6. Create a **circle** of arbitrary size and aligned to the *center point* as shown.

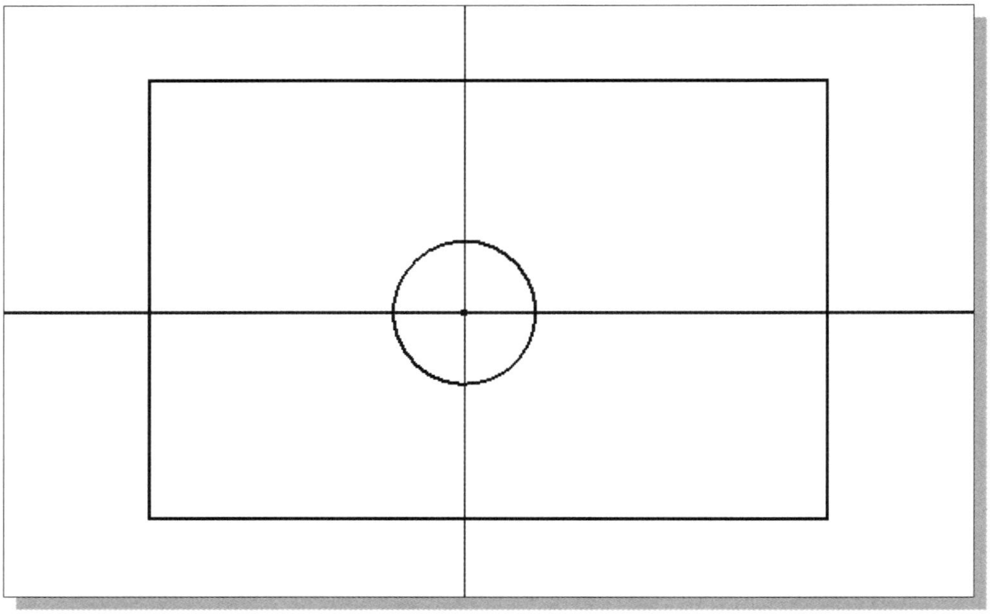

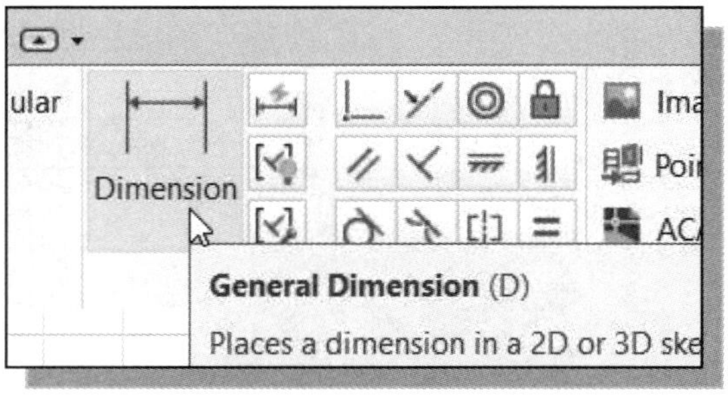

7. Select the **General Dimension** command in the *Constrain* panel.

8. On your own, create and adjust the dimensions of the rectangle and circle as shown. (Hint: Use parametric equations to position the geometry.)

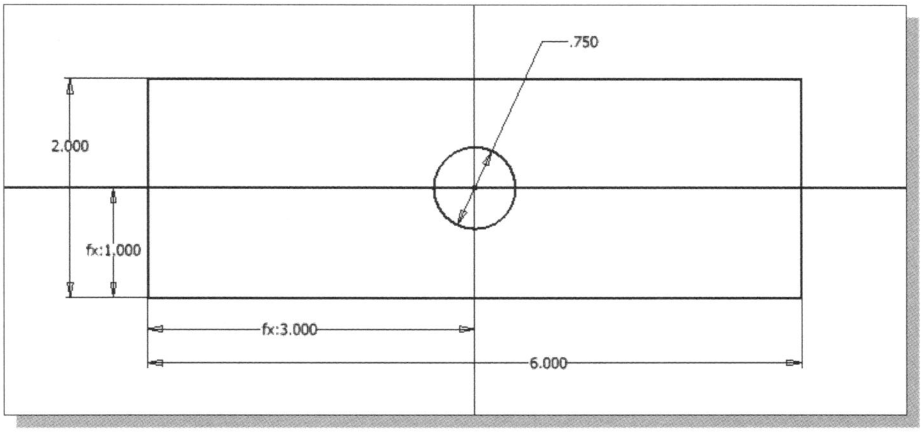

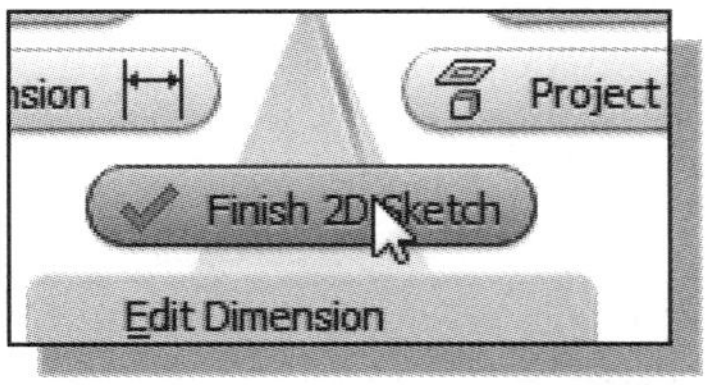

9. Inside the *graphics window*, click once with the right-mouse-button to display the option menu. Select **Finish 2D Sketch** in the pop-up menu to end the Sketch option.

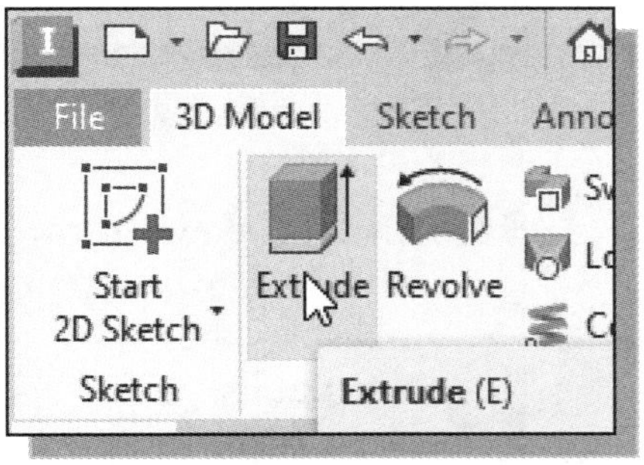

10. In the *3D Model* toolbar, select the **Extrude** command by left-clicking on the icon.

11. On your own, create an **Extruded** feature with a thickness of **0.25 in** as shown.

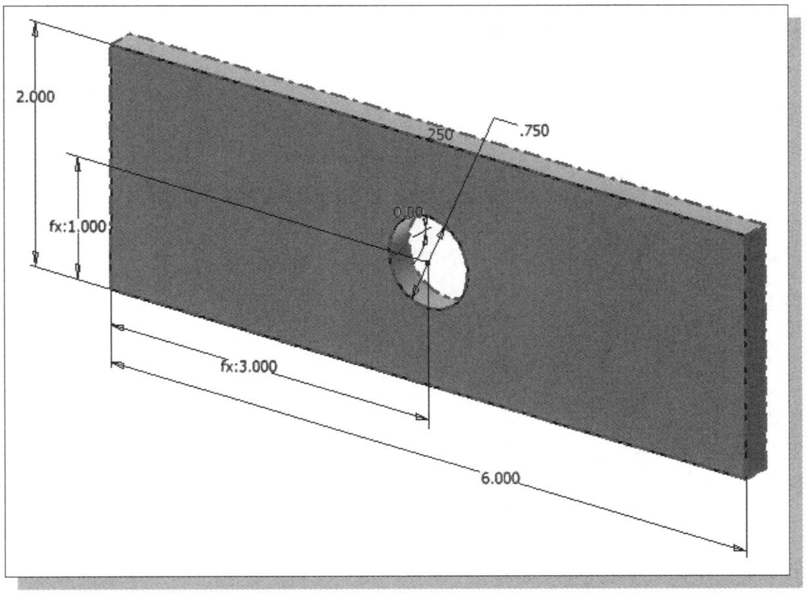

Assigning the Material Properties

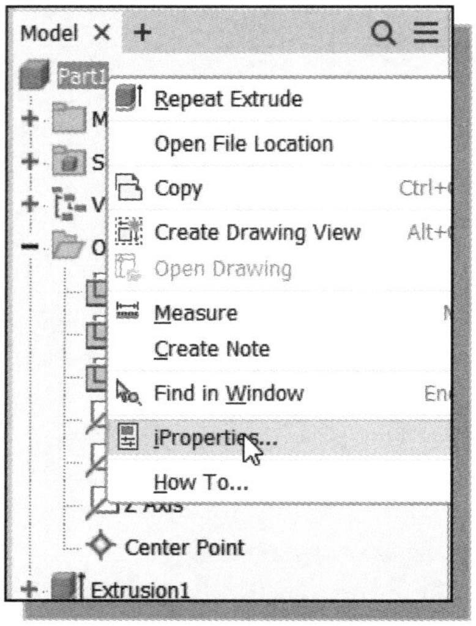

1. In the *browser*, **right-click** once on the ***part name*** to bring up the option menu, and then pick **iProperties** in the *pop-up* menu.

2. On your own, look at the different information listed in the *iProperties* dialog box.

3. Click on the **Physical** tab; this is the page that contains the physical properties of the selected model.

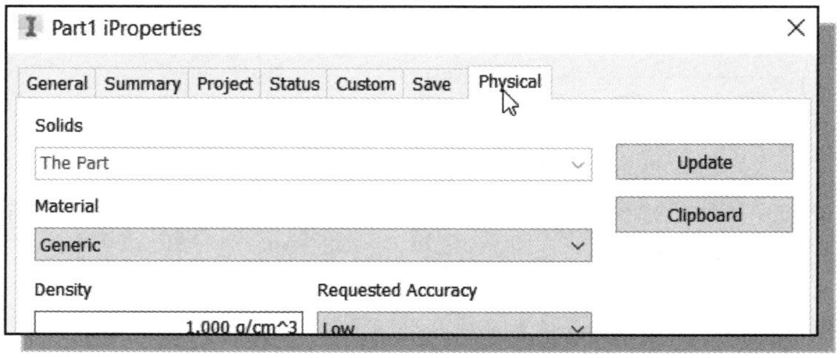

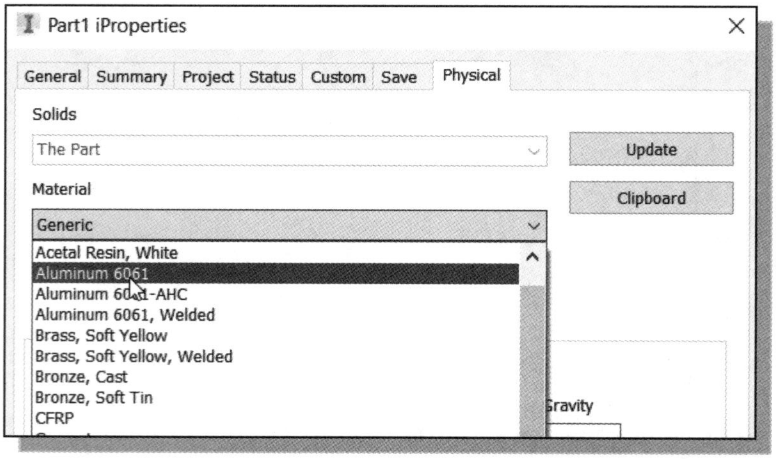

4. Click the down-arrow in the *Material* option to display the material list, and select **Aluminum-6061** as shown.

5. Click **OK** to accept the settings.

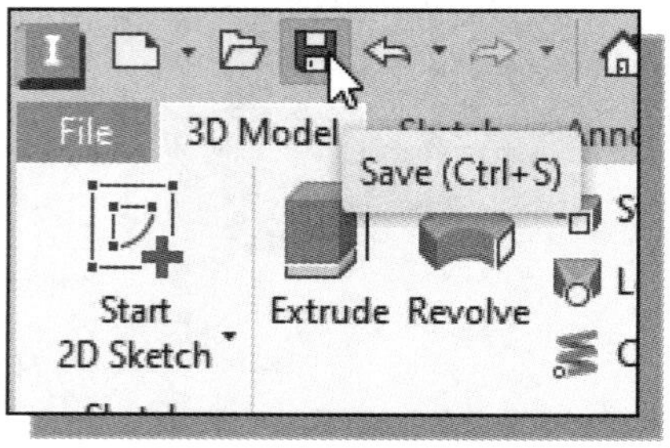

6. On your own, save the part with the filename **FEA_Al_Plate.ipt**.

Switch to the Stress Analysis Module

1. In the *Ribbon* toolbar, select the **Environments** tab as shown.

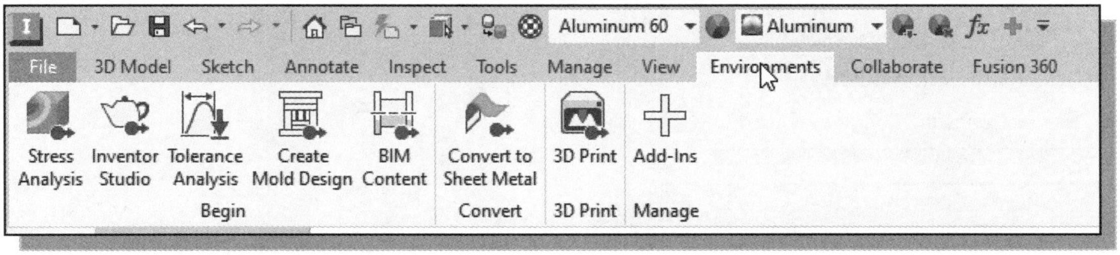

2. Click **Stress Analysis** to enter the *Inventor Stress Analysis Module.*

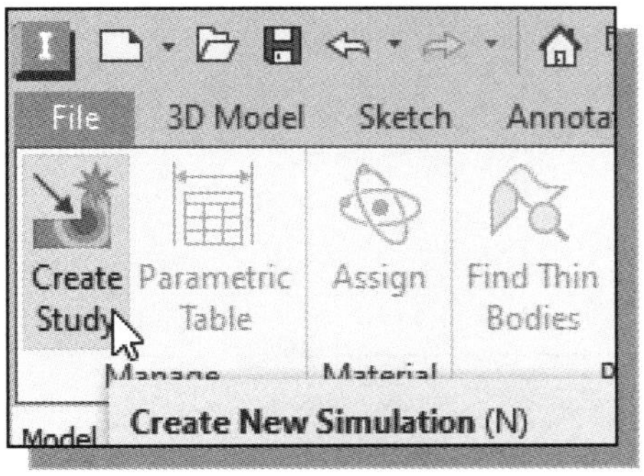

3. Click **Create New Simulation** to start a new simulation.

4. Note the default settings include (1) *Simulation Type* set to perform **Static Analysis** and (2) *Design Objective* set to **Single Point**. These settings are used for basic *linear static analysis*.

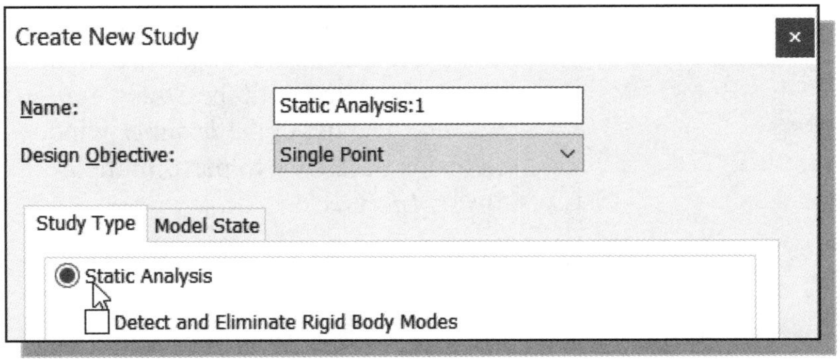

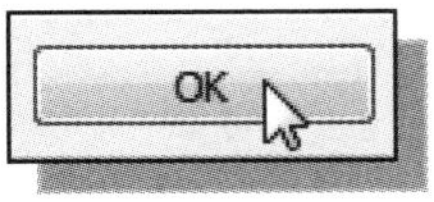

5. Click **OK** to accept the default settings.

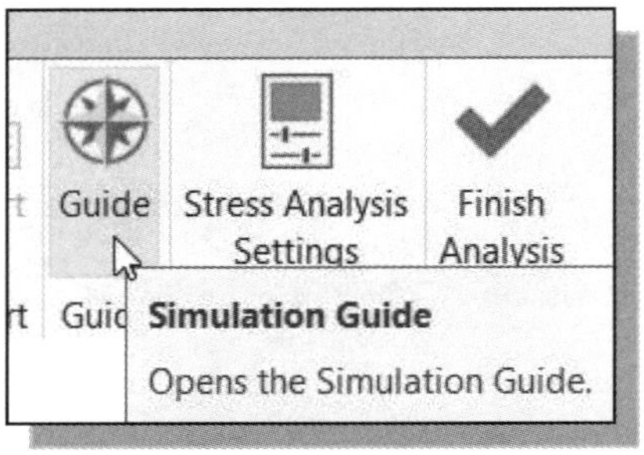

6. In the *Ribbon* toolbar, select the **Simulation Guide** icon as shown.

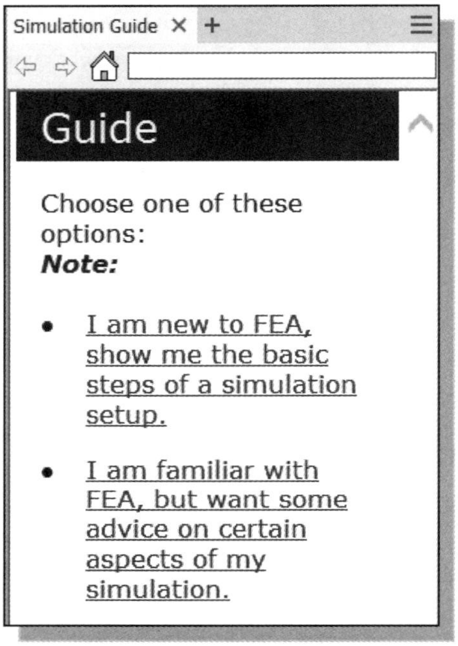

➢ Note the *Simulation Guide* provides an overall view of the Inventor FEA procedure. You are encouraged to read through the *Simulation Guide* to get familiar with the general procedure in performing the Inventor FEA simulation.

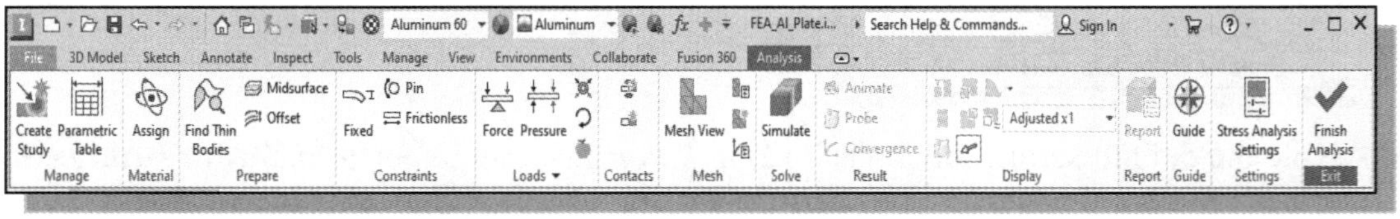

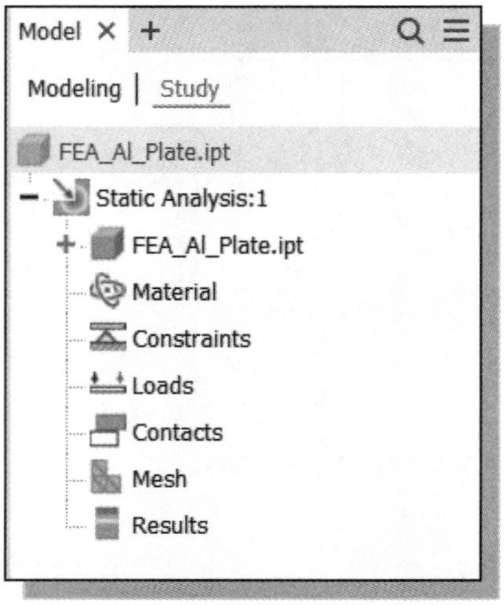

➢ Notice the *Ribbon* toolbar and the *browser* window, to the left side of the graphics area, now display items associated with the *Stress Analysis Simulation Module*. The list in the *browser* window shows the elements necessary to perform the *Finite Element Stress Analysis*.

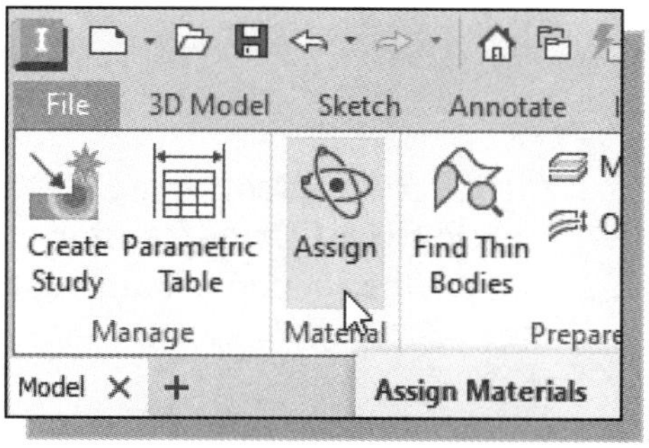

7. In the *Ribbon* toolbar area, select **Assign Materials** to examine the material assignment for the part.

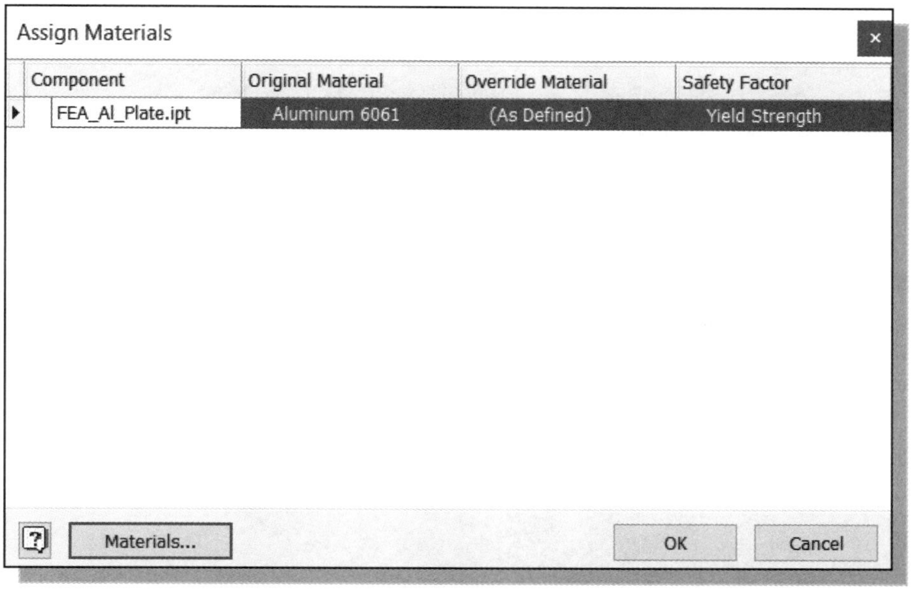

- Note that the assigned material, **Aluminum-6061**, is shown under the *Original Material*; the *Override Material* option is available for us to examine the effects of using different materials.

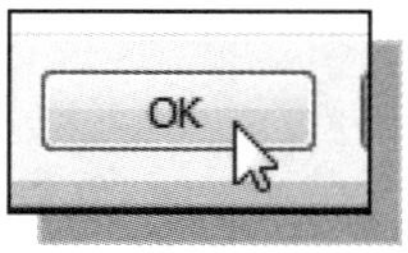

8. Click **OK** to accept the settings.

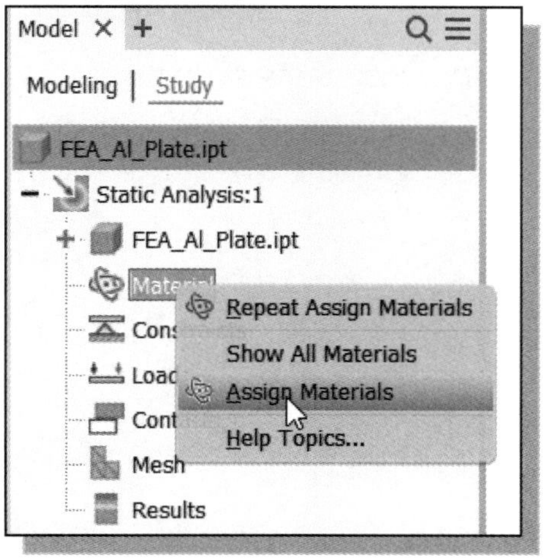

- Note that the same command can also be accessed through the *browser* window.

Apply Constraints and Loads

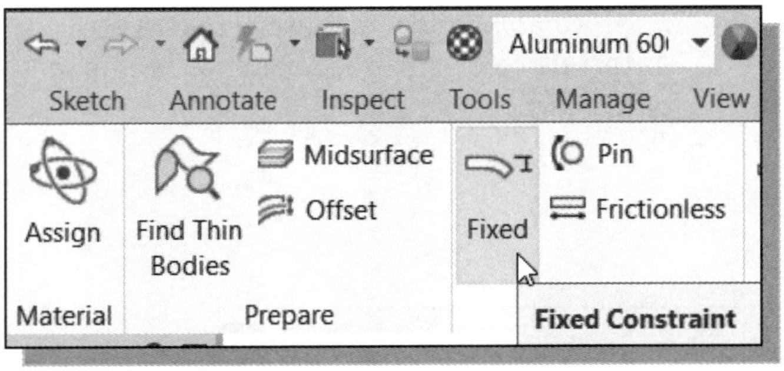

1. In the *Ribbon* toolbar area, select **Fixed Constraint** to assign the support condition of the plate.

2. Click on the **top left corner** of the *ViewCube* to rotate the display; the plate will be rotated 90 degrees showing the left vertical face of the plate.

3. Select the small **vertical surface** of the left-end of the plane as shown.

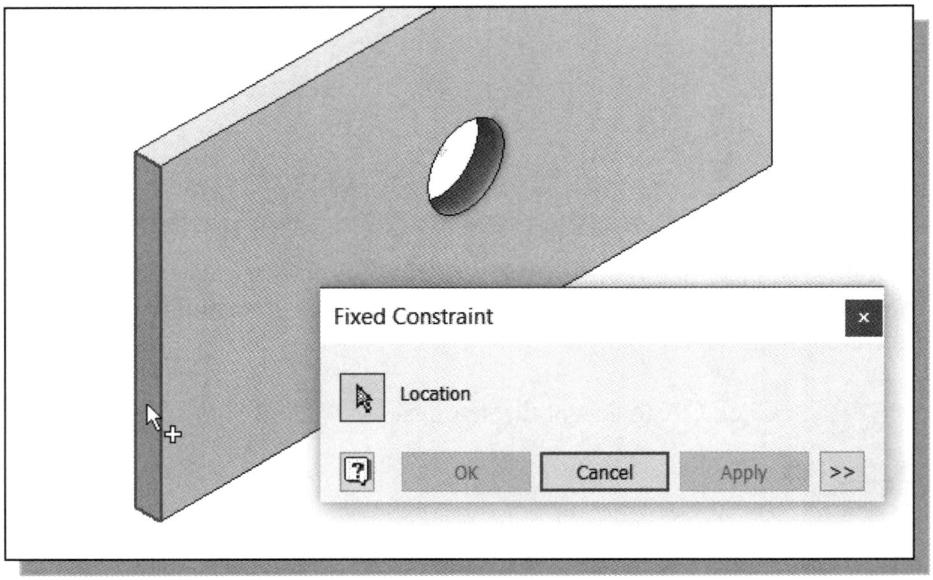

4. Note the selection of the surface is recognized as the selection label is changed to **Faces**. Click **OK** to accept the selection.

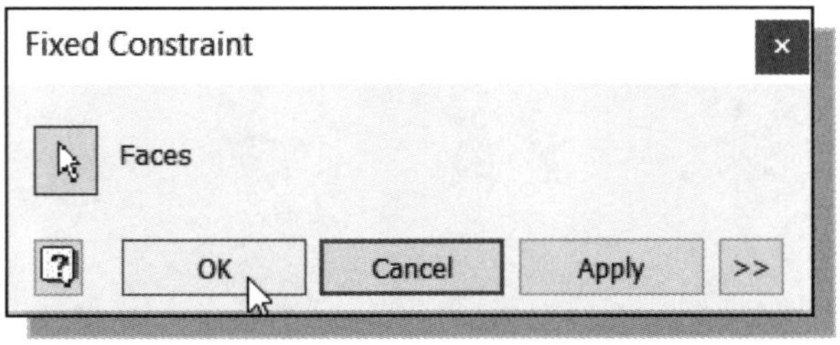

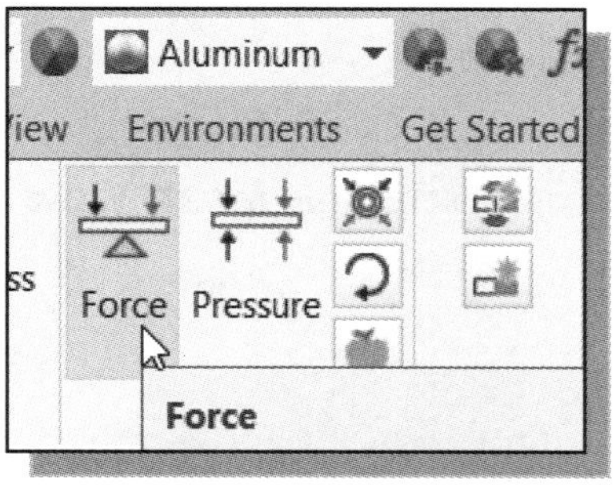

5. In the *Ribbon* toolbar area, select the **Force** command as shown.

6. Click on the **Home** icon above the *ViewCube* to reset the display back to the isometric view.

7. Click on the small vertical surface to the right end of the plate. In the *Force* dialog box, enter **300 lbf** as the force value and check the **Reverse Direction** option box. Notice the direction of the force load is now outward.

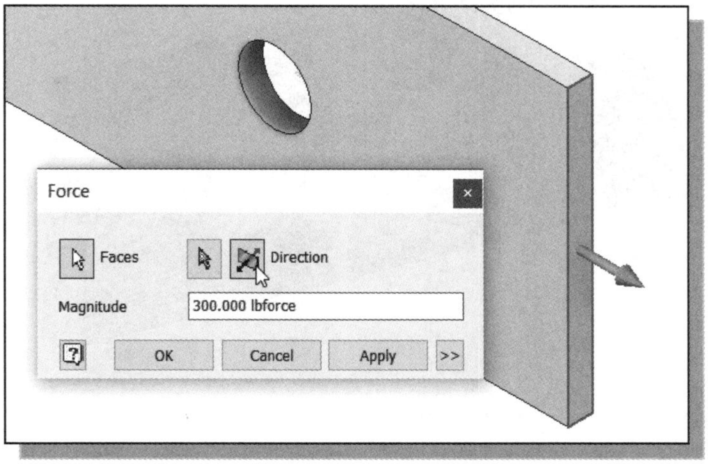

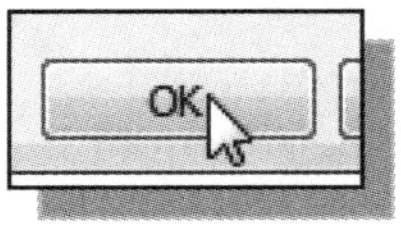

8. Click **OK** to accept the selection and exit the Force command.

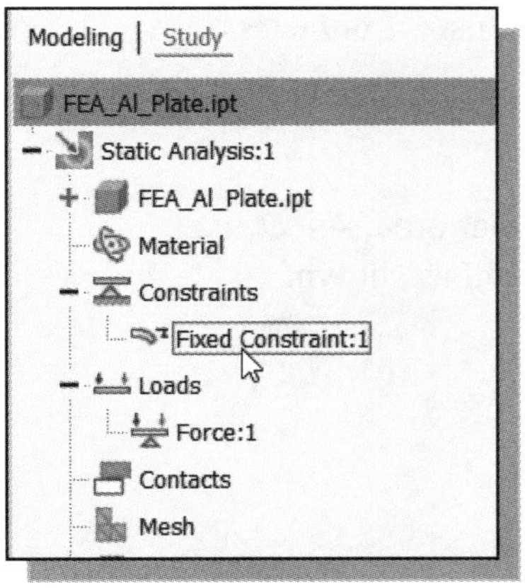

➢ Notice the Fixed constraint and Force load appear in the *browser* window to the left of the graphics area with the default names Constraint:1 and Force:1.

Create a Mesh and Run the Solver

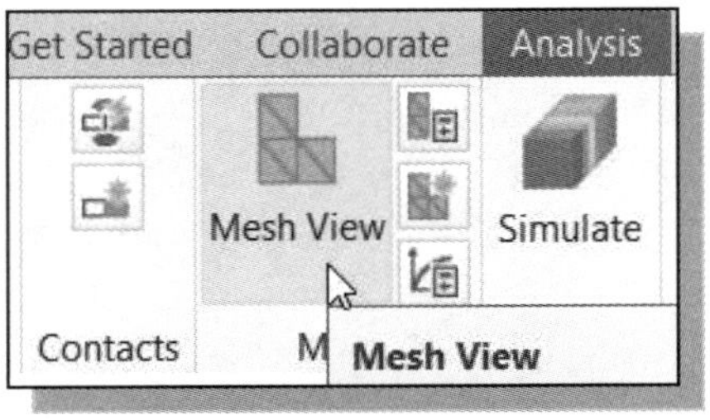

1. In the *Ribbon* toolbar area, select the **Mesh View** command as shown.

- Note that with the default settings, *Inventor Simulation* generated 653 nodes and 291 elements, which is a relatively coarse mesh.

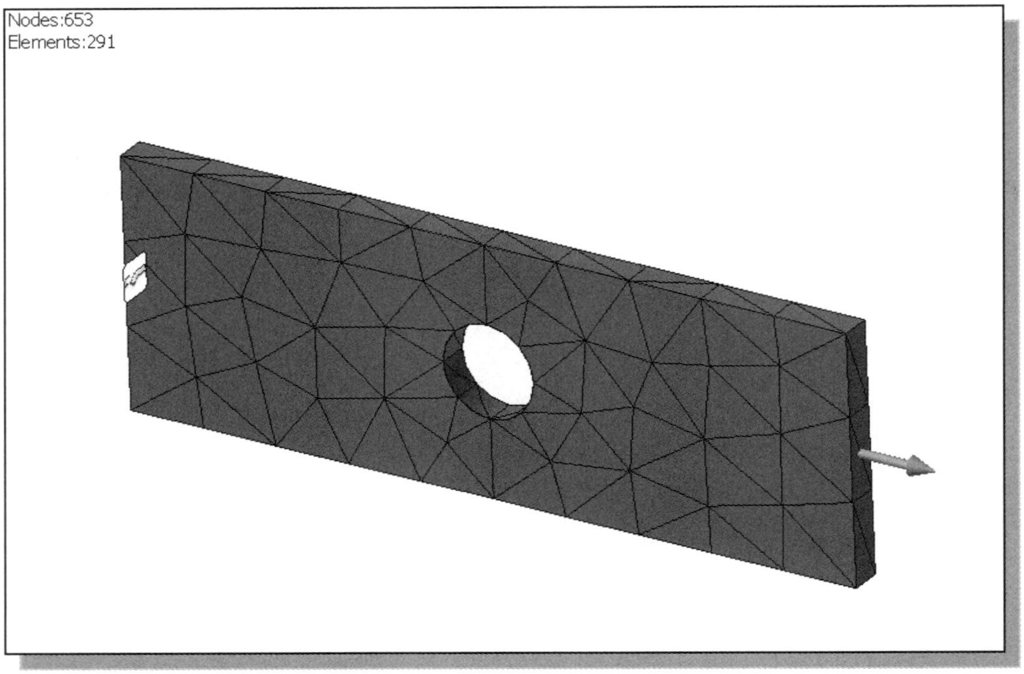

- As a rule in creating the first FEA mesh, start with a relatively small number of elements and progressively move to more refined models. The main objective of the first coarse mesh analysis is to obtain a rough idea of the overall stress distribution. In most cases, use of a complex and/or a very refined FEA model is not justifiable since it most likely provides computational accuracy at the expense of unnecessarily increased processing time.

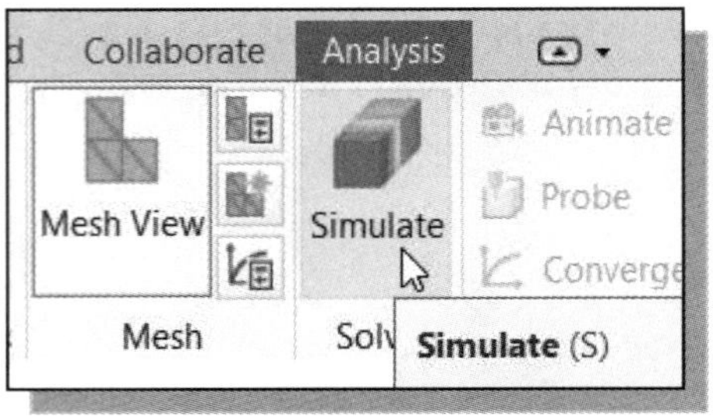

2. In the *Ribbon* toolbar area, select the **Simulate** command as shown.

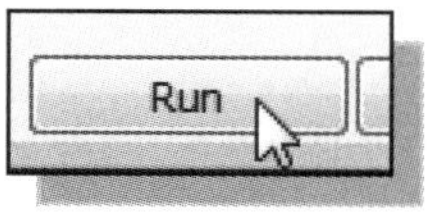

3. Click the **Run simulation** button to proceed with the FEA simulation.

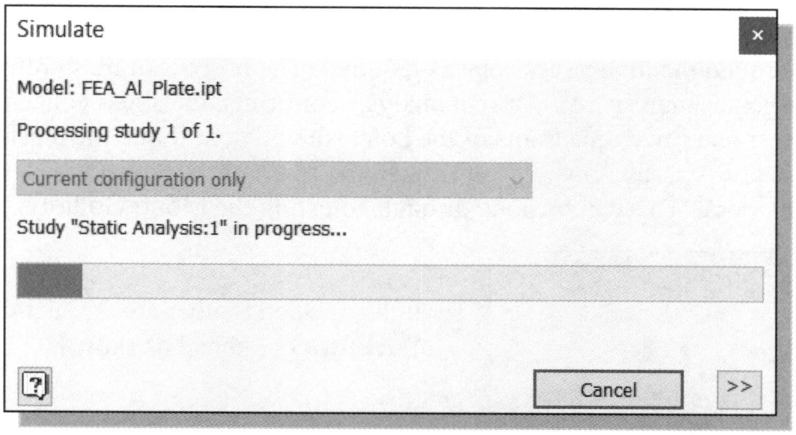

- Note that if the stress display does not look good, you might need to choose a different option of the *Visual Style* to find a better viewing of the stress result. Depending on the computer hardware, some settings might provide better viewing than others.

➢ A plot of equivalent stress (Von Mises stress) is generated and displayed in the graphics area. A color-coded scale is used with the associated scale bar displayed at the right. Red represents the regions of highest stress. The maximum stress is 2074 psi, which is well below the yield strength of the material (3999.3 psi) and is located at the stress concentration points at the upper and lower quadrants of the hole.

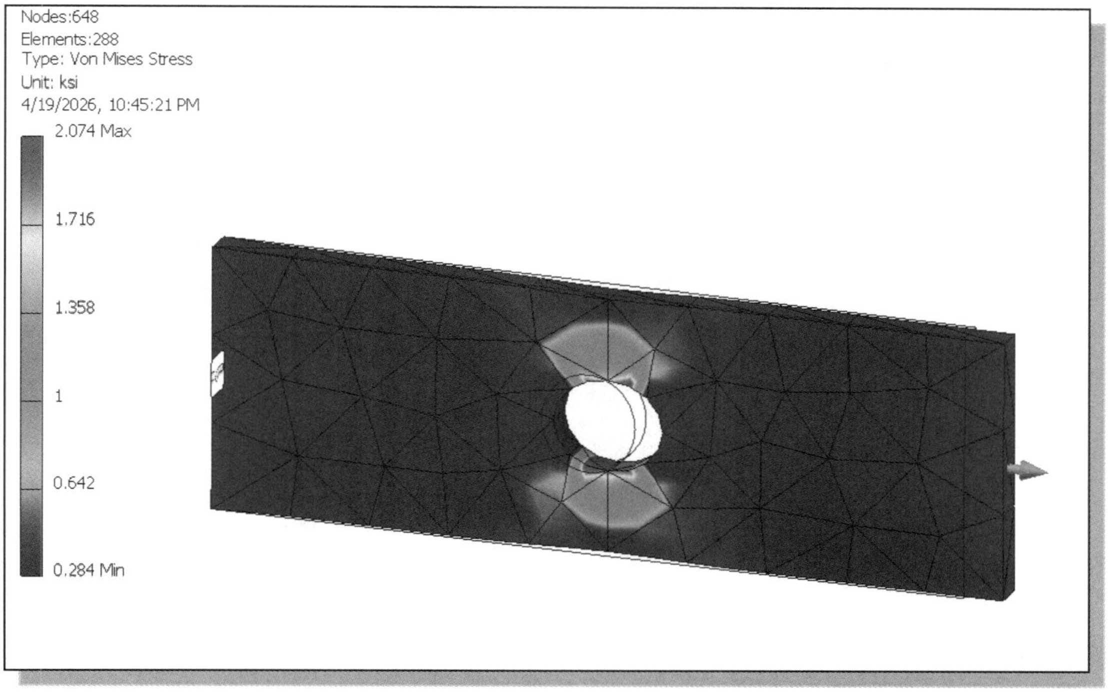

Refinement of the FEA Mesh – Global Element Size

In order to gain accuracy for complex geometries or to represent a highly varying stress distribution, finer elements must be used. The increase in the number of elements will also increase the solution time and computer disk space. This is why the refinement of mesh around the high stress areas is required. The process of mesh refinement is called convergence analysis. As our first analysis confirmed the stress concentration points at the upper and lower quadrants of the hole, we will next refine the mesh to obtain a more accurate FEA result. One way of refinement is simply to adjust the element size to a smaller value. This can be done through adjusting the Mesh Settings.

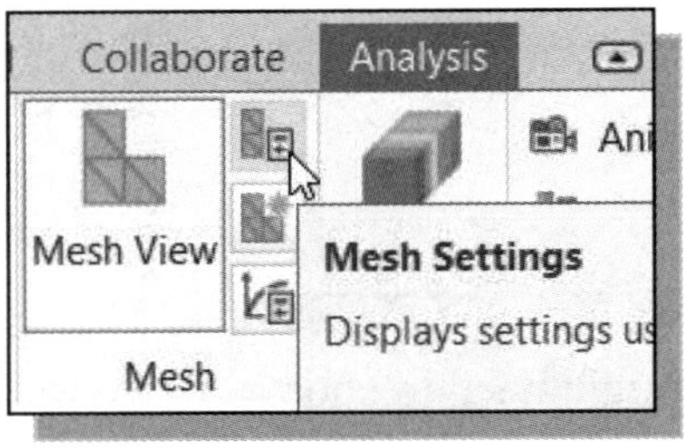

1. In the *Ribbon* toolbar area, select the **Mesh Settings** command as shown.

2. In the *Mesh Settings* dialog box, adjust the *Average Element Size* to **0.05**, which will create twice as many elements. Note the number is a fraction of the length dimension of the plate.

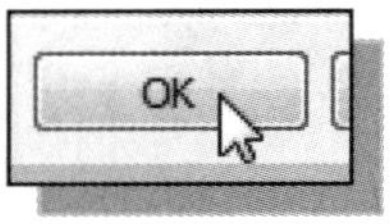

3. Click **OK** to accept the new settings and exit the Mesh Settings command.

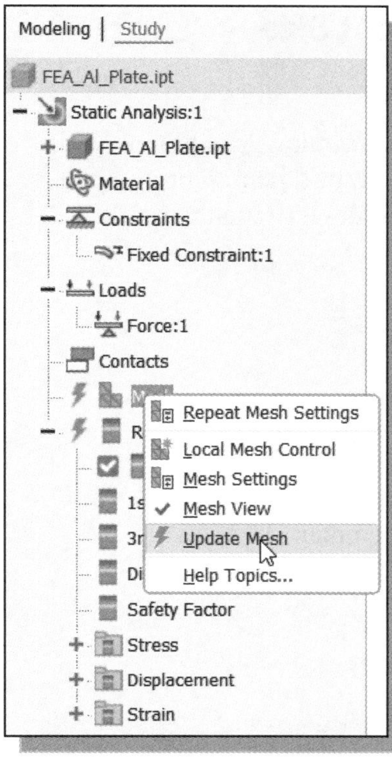

4. It is necessary to update the mesh with the new settings through the *browser* window. Right-click on the **Mesh** item to bring up the **option list**.

5. Select **Update Mesh** in the option list to activate the mesh update.

- Note that with the new settings, *Inventor Simulation* generated 1931 nodes and 907 elements, which is almost three times more than the original coarse mesh.

6. On your own, perform the FEA analysis.

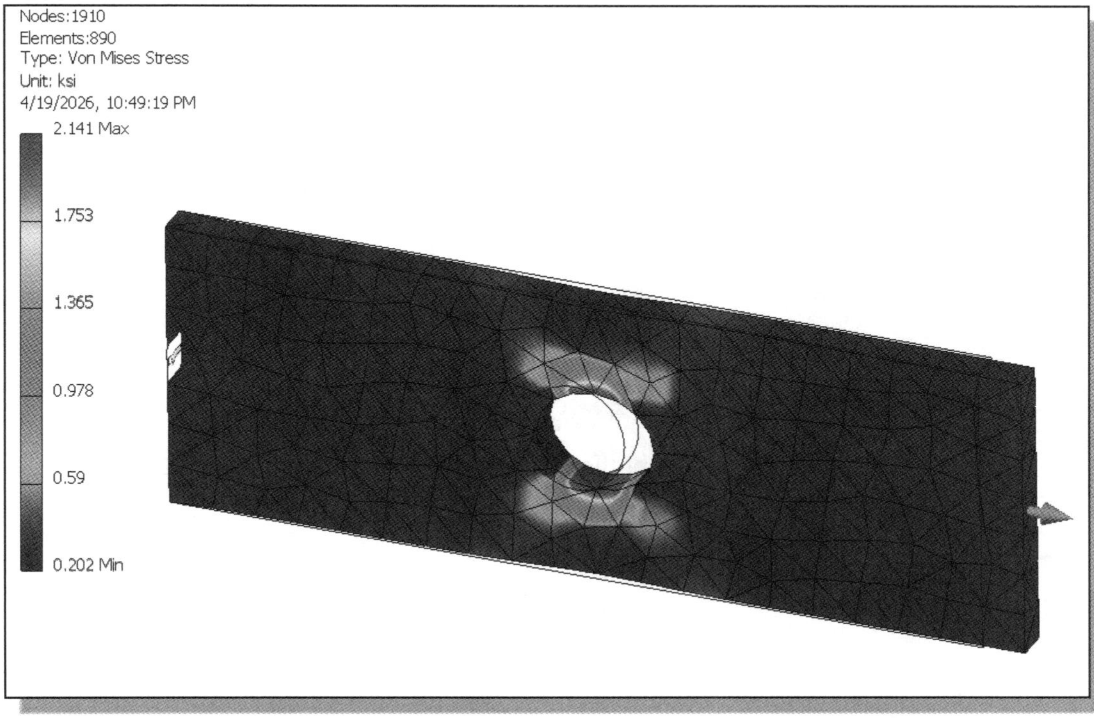

➢ The maximum stress with the refinement is now 2141psi, which is higher than the previous FEA analysis, and it is closer to the preliminary analysis. Next, we will further refine the mesh in the high stress area.

Refinement of the FEA Mesh – Local Element Size

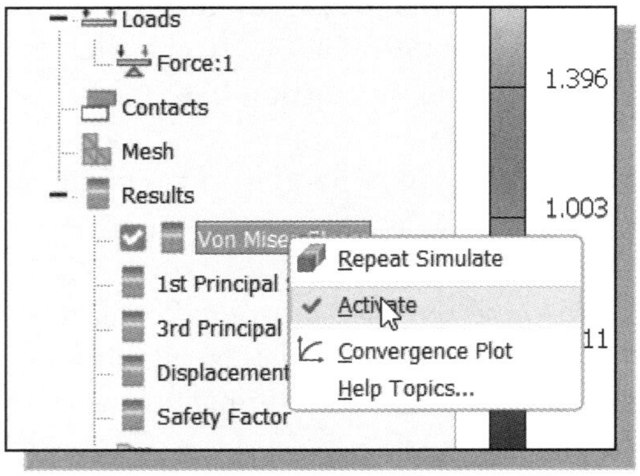

1. In the *browser* window, right-click on the **Von Mises Stress** item to bring up the *option list* and **de-activate** the result as shown.

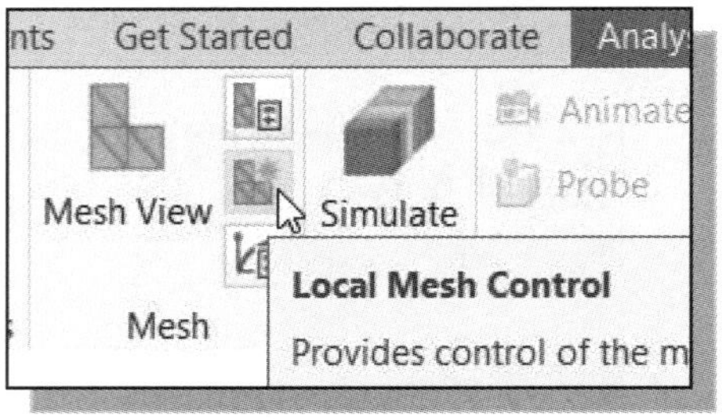

2. In the *Ribbon* toolbar area, select the **Local Mesh Control** command as shown.

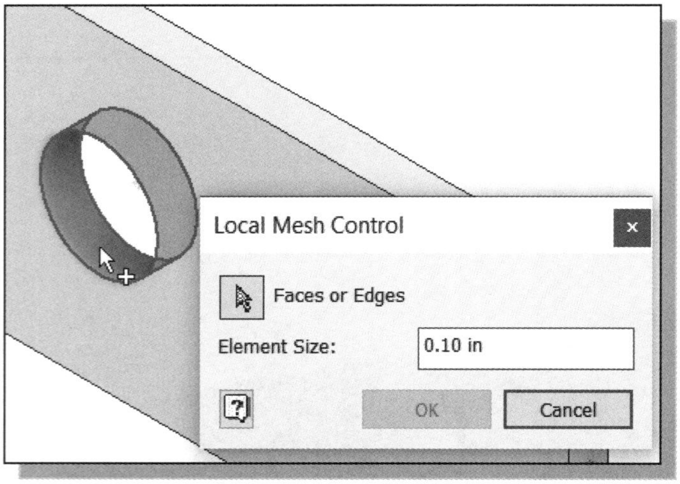

3. Select the **inside cylindrical surface** of the hole as shown.

4. Set the local *Element Size* to **0.1** as shown.

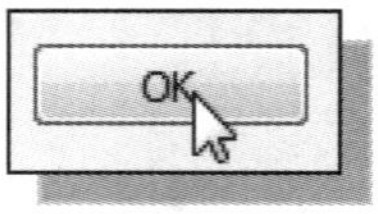

5. Click **OK** to accept the new setting and exit the Local Mesh Control command.

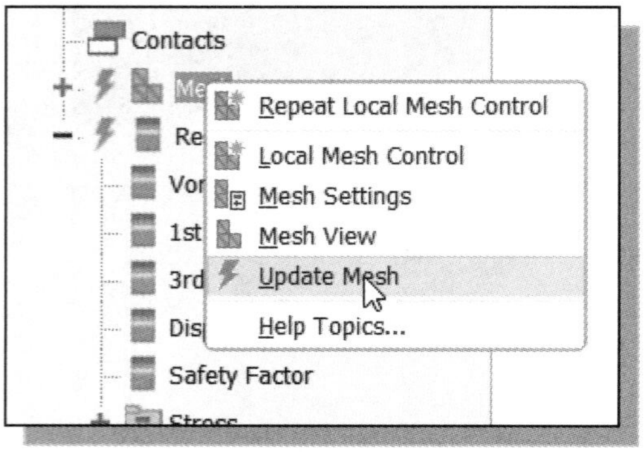

6. On your own, activate the **Update Mesh** command as shown.

- Note that with the new settings, *Inventor Simulation* generated 3363 nodes and 1663 elements, which is about seven times the original coarse mesh.

7. On your own, perform the FEA analysis.

- The maximum stress with the refinement is now 2189 psi, which is just a bit higher than the previous FEA analysis. The refinement caused only a small difference compared to the previous mesh.

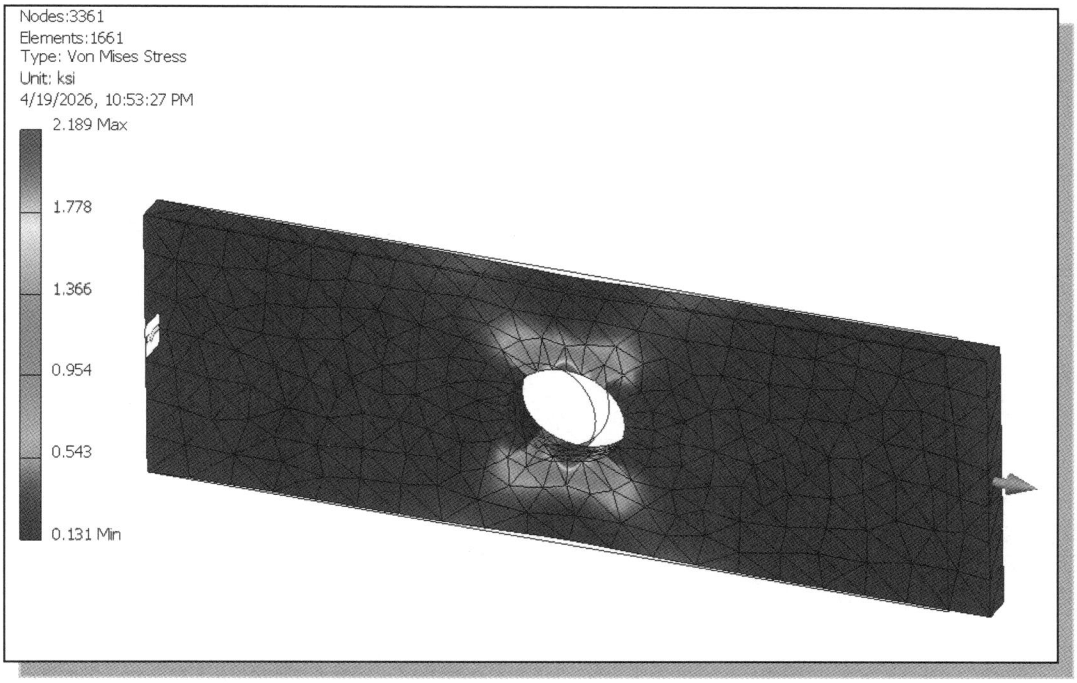

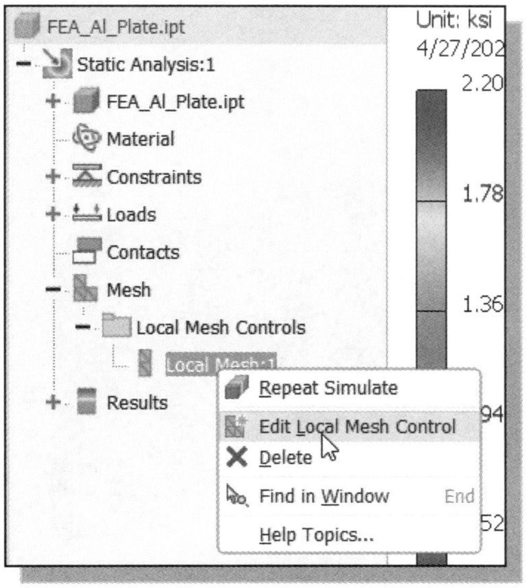

8. On your own, adjust the element size to 0.05 of the Local Mesh Control value as shown.

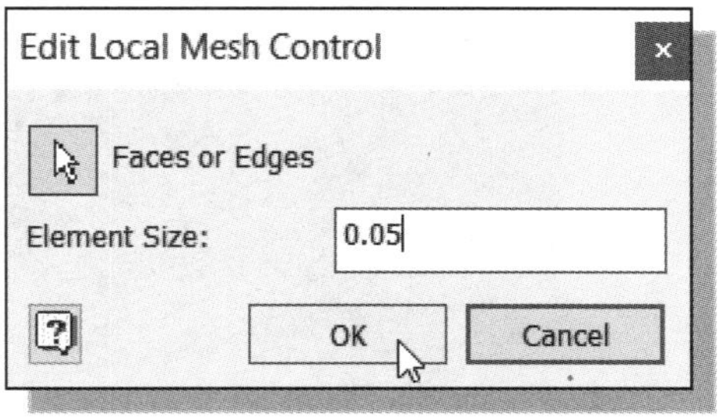

9. On your own, perform another FEA with the current settings.

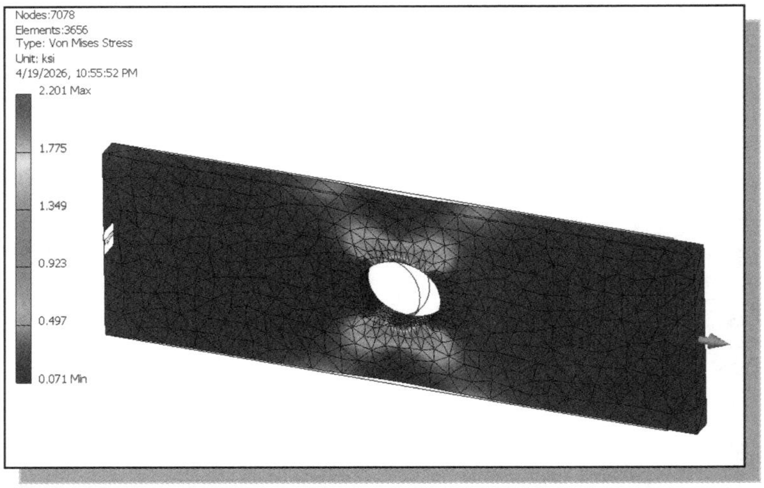

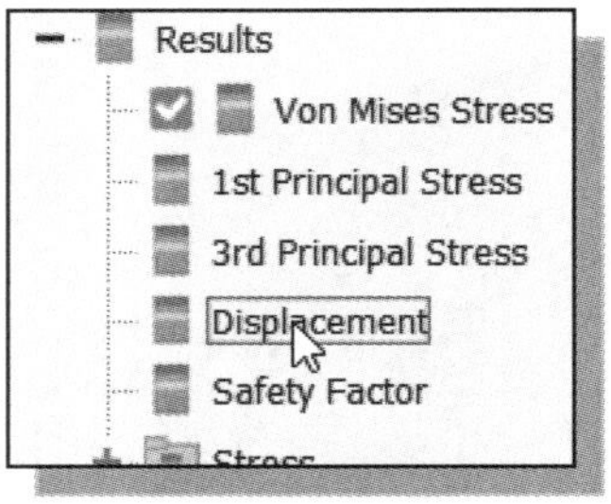

10. Double-click on the **X Displacement** item in the *browser* window to display the displacement of the plate.

- The FEA result showed the total displacement of 4.057e-4 inch, which matches quite well with the 3.88e-4 inch preliminary analysis estimate.

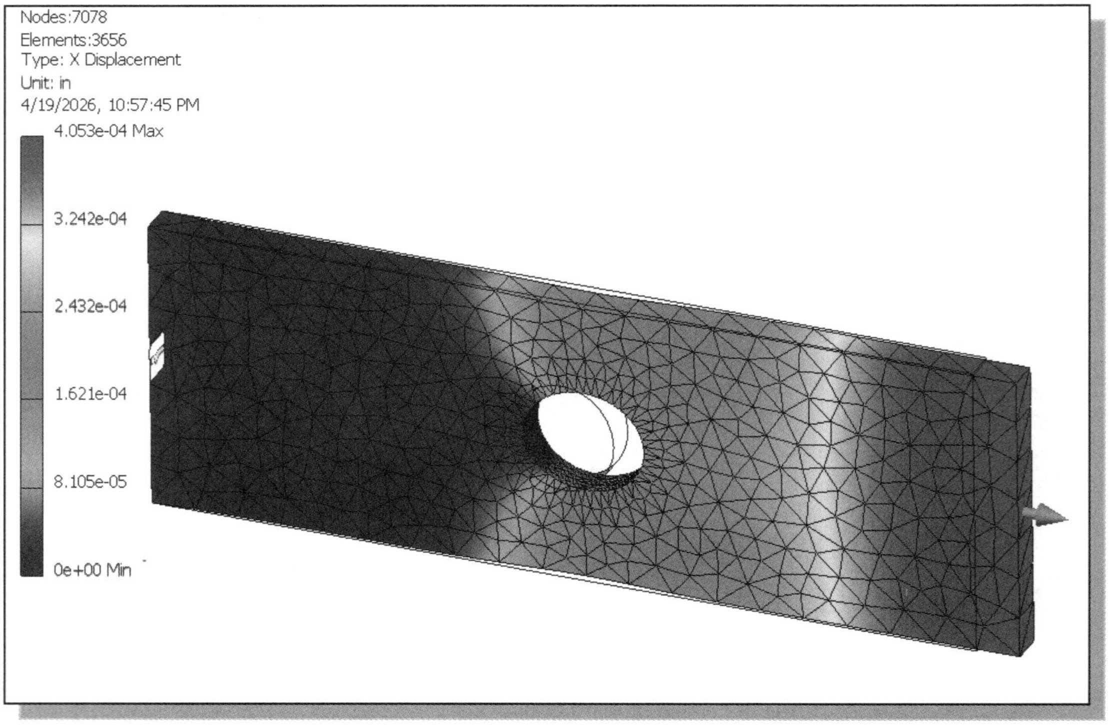

Comparison of Results

The accuracy of the *Inventor Simulation* results for this problem can be checked by comparing them to the analytical results presented earlier. In the Preliminary Analysis section, the maximum stress was calculated using a stress concentration factor and the value obtained was **2180 psi**. One should realize the analytical result is obtained through the use of charts from empirical data and therefore involves some degrees of error. The maximum stress obtained by finite element analysis using *Inventor Simulation* ranges from **2074** to **2201 psi**. In the Preliminary Analysis section, the maximum displacement was also estimated to be around **1.94E-4 inches**, measured from the center of the hole to one end of the plate. The maximum displacement obtained by finite element analysis using *Inventor Simulation* was around **2.0285E-4 inches**. The agreement between the analytical results and those from *Inventor Simulation* demonstrate the potential of *Inventor Simulation* as a very powerful design tool.

In FEA, the process of mesh refinement is called convergence analysis. For our analysis, the refinement of the mesh does show the FEA results converging near the analytical results. The refinement to the third mesh is quite adequate for our analysis. Any further refinement does not provide any additional insight and is therefore not necessary.

Number of Elements	σ_{max} (psi)	D_{max} (in)
288	2074	2.024e-4
890	2141	2.027e-4
1661	2189	2.028e-4
3656	2201	2.027e-4

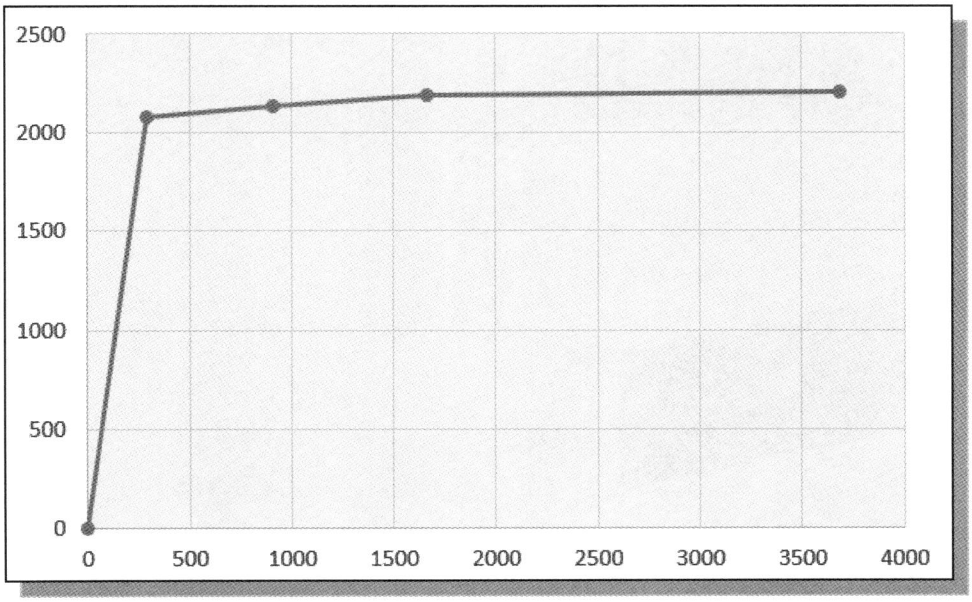

Create an HTML Report

Inventor Stress Analysis Simulation module also includes options to create a report in HTML format which contains data related to the simulation.

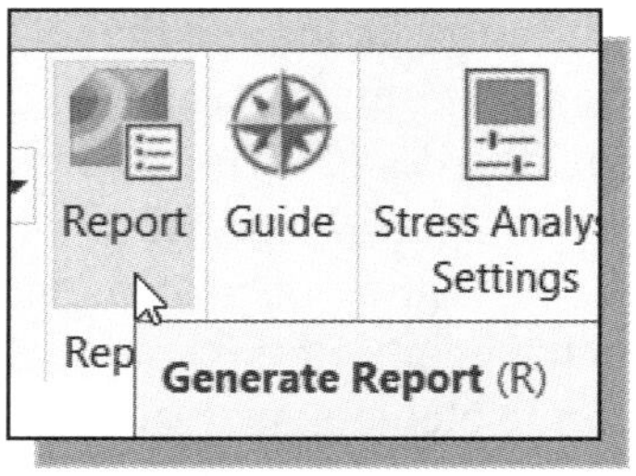

1. Select the **Report** option as shown.

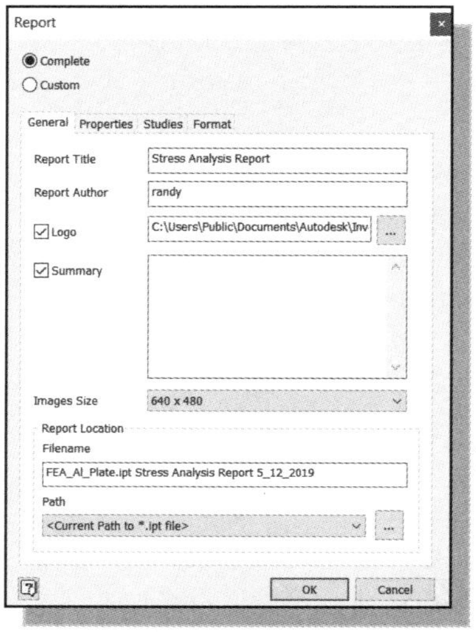

2. On your own, examine the different options available for the report. Click **OK** to generate a report.

➢ *Inventor Stress Analysis Environment* will create an HTML file and automatically open it in your default Web browser. The HTML file is saved in the location shown in the *File Information* section at the beginning of the report.

3. On your own, read through the report. Notice that all the relevant data are recorded.

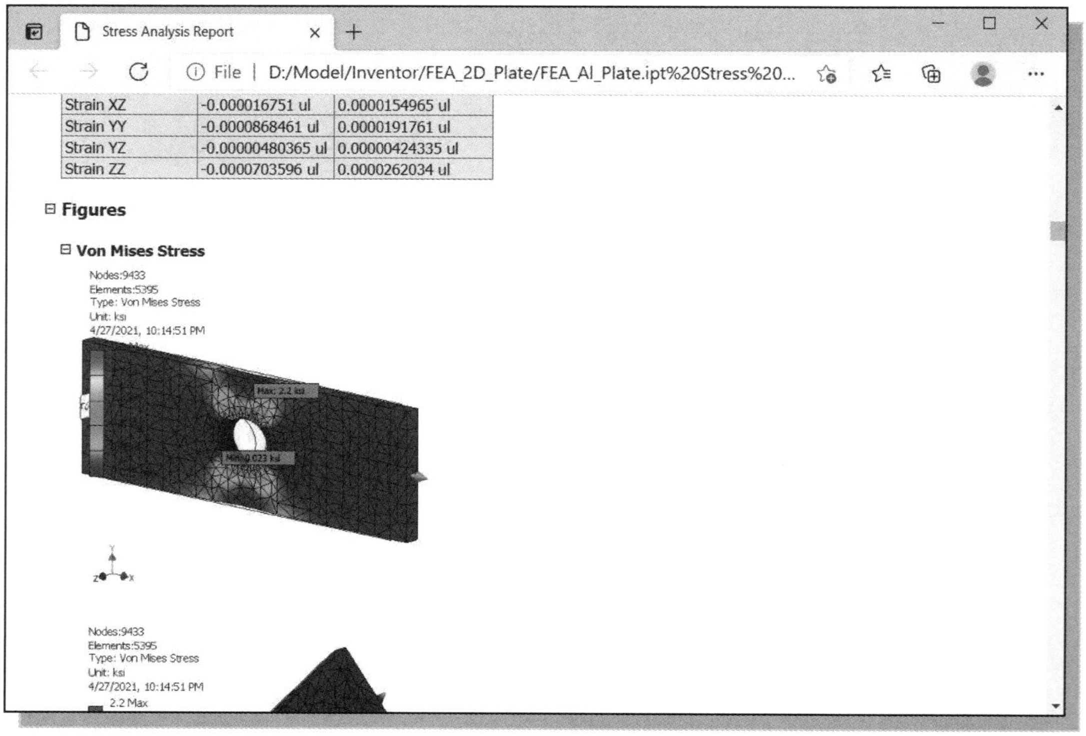

Geometric Considerations of Finite Elements

In the previous sections, the entire plate was created and analyzed, but a closer examination of the associated geometry suggests a more effective approach can be used to analyze the plate.

For *linear statics analysis*, designs with symmetrical features can often be reduced to expedite the analysis.

For our plate problem, there are two planes of symmetry. Thus, we only need to create an FE model that is one-fourth of the actual system. By taking advantage of symmetry, we can use a finer subdivision of elements that will provide more accurate and faster results.

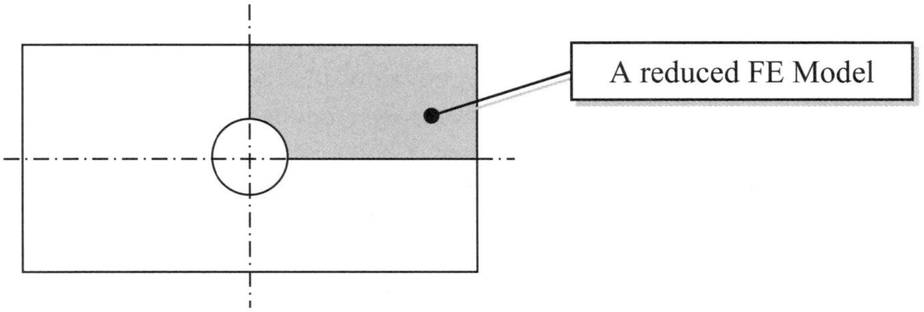

In performing a stress analysis, it is necessary to consider the constraints in all directions. For our plate model, deformations will occur along the axes of symmetry; we will therefore place roller constraints along the two center lines as shown in the figure below. You are encouraged to perform the FEA on this more simplified model and compare the results obtained in the previous sections.

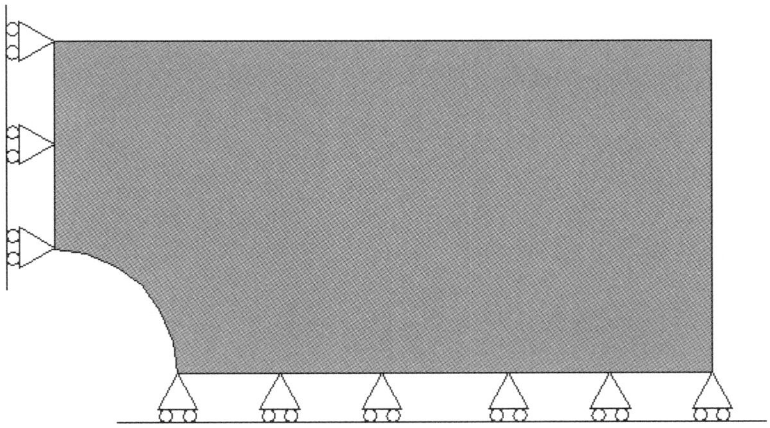

❖ One should also be cautious of using symmetrical characteristics in FEA. The symmetry characteristics of boundary conditions and loads should be considered. Also, note the symmetry characteristic that is used in *linear statics analysis* does not imply similar symmetrical results in vibration or buckling modes.

Conclusion

Design includes all activities involved from the original conception to the finished product. Design is the process by which products are created and modified. For many years designers sought ways to describe and analyze three-dimensional designs without building physical models. With advancements in computer technology, the creation of parametric models on computers offers a wide range of benefits. Parametric models are easier to interpret and can be easily altered. Parametric models can be analyzed using finite element analysis software, and simulation of real-life loads can be applied to the models and the results graphically displayed.

Throughout this text, various modeling techniques have been presented. Mastering these techniques will enable you to create intelligent and flexible solid models. The goal is to make use of the tools provided by Autodesk Inventor and to successfully capture the **DESIGN INTENT** of the product. In many instances, only one approach to the modeling tasks was presented; you are encouraged to repeat all of the lessons and develop different ways of accomplishing the same tasks. We have only scratched the surface of Autodesk Inventor's functionality. The more time you spend using the system, the easier it will be to perform parametric modeling with Autodesk Inventor.

Summary of Modeling Considerations

- **Design Intent** – determine the functionality of the design; select features that are central to the design.
- **Order of Features** – consider the parent/child relationships necessary for all features.
- **Dimensional and Geometric Constraints** – the way in which the constraints are applied determines how the components are updated.
- **Relations** – consider the orientation and parametric relationships required between features and in an assembly.

Review Questions:

1. Describe the required steps in performing a stress analysis using *Inventor Stress Analysis Environment*.

2. Describe two ways the material properties can be defined or edited.

3. What is meant by the term *Constraints* in *Inventor Stress Analysis Module*?

4. How do we control whether a load applied to a face is in the *outward* or *inward* direction?

5. Define *degrees of freedom* (DOF).

6. How do we end the *Inventor Stress Analysis Module* and return to the model in Autodesk Inventor?

Exercises:

1. The shaft shown below is fixed at the large end and a 50 kN force is applied to the small end. Find the maximum stress and maximum deflection in the shaft. The material is AISI 1020. (Dimensions in mm.)

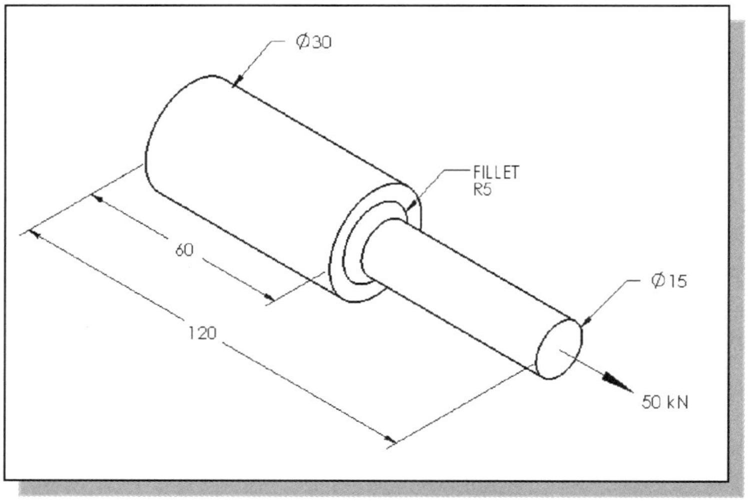

2. For the hanging bracket, the top face is fixed and a 100 psi pressure load is applied to the horizontal surface as shown. Find the maximum stress and maximum deflection in the bracket. The material is Alloy Steel. (Dimensions in inches.)

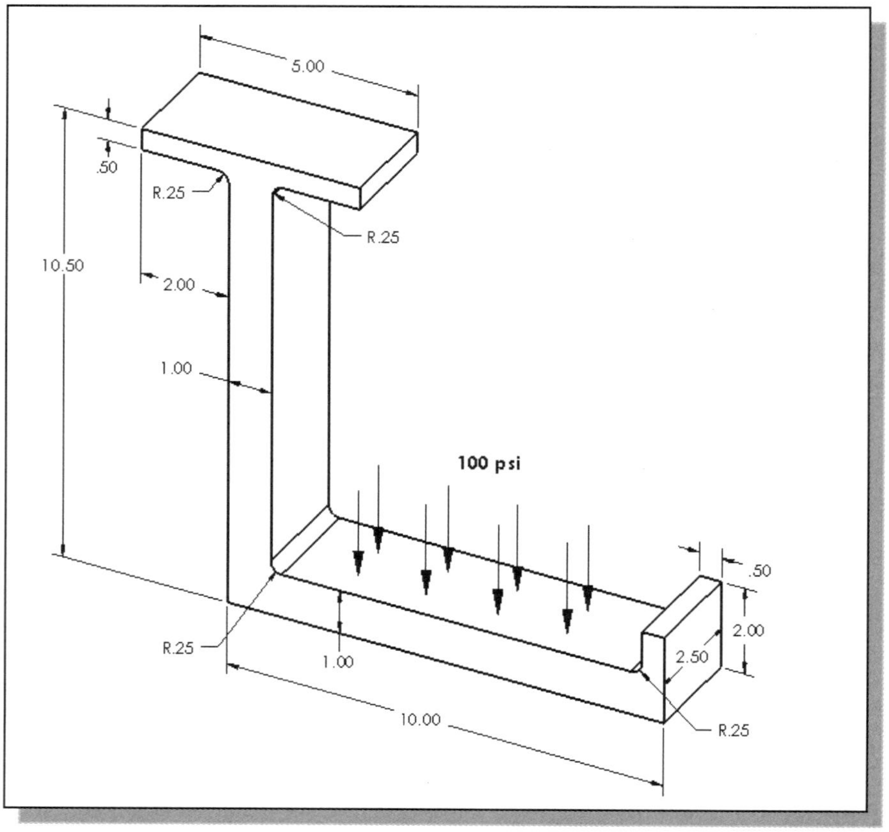

Appendix

Sec. 206. Standard gauge for sheet and plate iron and steel

For the purpose of securing uniformity, the following list is established as the only standard gauge for sheet and plate iron and steel in the United States of America, namely:

		Thickness				Weight		
Gauge	**Frac. Inch**	**Dec. Inch**	**mm**	**oz/ft²**	**lb/ft²**	**kg/ft²**	**kg/m²**	**lb/m²**
0000000	1/2	.5	12.7	320	20.00	9.072	97.65	215.28
000000	15/32	.46875	11.90625	300	18.75	8.505	91.55	201.82
00000	7/16	.4375	11.1125	280	17.50	7.983	85.44	188.37
0000	13/32	.40625	10.31875	260	16.25	7.371	79.33	174.91
000	3/8	.375	9.525	240	15	6.804	73.24	161.46
00	11/32	.34375	8.73125	220	13.75	6.237	67.13	148.00
0	5/16	.3125	7.9375	200	12.50	5.67	61.03	134.55
1	9/32	.28125	7.14375	180	11.25	5.103	54.93	121.09
2	17/64	.265625	6.746875	170	10.625	4.819	51.88	114.37
3	1/4	.25	6.35	160	10	4.536	48.82	107.64
4	15/64	.234375	5.953125	150	9.375	4.252	45.77	100.91
5	7/32	.21875	5.55625	140	8.75	3.969	42.72	94.18
6	13/64	.203125	5.159375	130	8.125	3.685	39.67	87.45
7	3/16	.1875	4.7625	120	7.5	3.402	36.62	80.72
8	11/64	.171875	4.365625	110	6.875	3.118	33.57	74.00
9	5/32	.15625	3.96875	100	6.25	2.835	30.52	67.27
10	9/64	.140625	3.571875	90	5.625	2.552	27.46	60.55
11	1/8	.125	3.175	80	5	2.268	24.41	53.82
12	7/64	.109375	2.778125	70	4.375	1.984	21.36	47.09
13	3/32	.09375	2.38125	60	3.75	1.701	18.31	40.36
14	5/64	.078125	1.984375	50	3.125	1.417	15.26	33.64
15	9/128	.0703125	1.7859375	45	2.8125	1.276	13.73	30.27
16	1/16	.0625	1.5875	40	2.5	1.134	12.21	26.91
17	9/160	.05625	1.42875	36	2.25	1.021	10.99	24.22
18	1/20	.05	1.27	32	2	.9072	9.765	21.53
19	7/160	.04375	1.11125	28	1.75	.7938	8.544	18.84
20	3/80	.0375	.9525	24	1.50	.6804	7.324	16.15
21	11/320	.034375	.873125	22	1.375	.6237	6.713	14.80
22	1/32	.03125	.793750	20	1.25	.567	6.103	13.46
23	9/320	.028125	.714375	18	1.125	.5103	5.493	12.11
24	1/40	.025	.635	16	1	.4536	4.882	10.76
25	7/320	.021875	.555625	14	.875	.3969	4.272	9.42

26	3/160	.01875	.47625	12	.75	.3402	3.662	8.07
27	11/640	.0171875	.4365625	11	.6875	.3119	3.357	7.40
28	1/64	.015625	.396875	10	.625	.2835	3.052	6.73
29	9/640	.0140625	.3571875	9	.5625	.2551	2.746	6.05
30	1/80	.0125	.3175	8	.5	.2268	2.441	5.38
31	7/640	.0109375	.2778125	7	.4375	.1984	2.136	4.71
32	13/1280	.01015625	.25796875	6 1/2	.40625	.1843	1.983	4.37
33	3/320	.009375	.238125	6	.375	.1701	1.831	4.04
34	11/1280	.00859375	.21828125	5 1/2	.34375	.1559	1.678	3.70
35	5/640	.0078125	.1984375	5	.3125	.1417	1.526	3.36
36	9/1280	.00703125	.17859375	4 1/2	.28125	.1276	1.373	3.03
37	17/2560	.006640625	.168671875	4 1/4	.265625	.1205	1.297	2.87
38	1/160	.00625	.15875	4	.25	.1134	1.221	2.69

Reformatted from: **http://www4.law.cornell.edu/uscode/15/206.html**

INDEX

G

H

I

J

T

U

V

W

X

Y

Z